Essentials of Medicinal and Aromatic Crops

Muhammad Zia-Ul-Haq
Arwa Abdulkreem AL-Huqail
Muhammad Riaz • Umar Farooq Gohar
Editors

Essentials of Medicinal and Aromatic Crops

Springer

Editors
Muhammad Zia-Ul-Haq
Office of Research, Innovation
and Commercialization
University of Engineering and Technology
Lahore, Pakistan

Muhammad Riaz
Department of Pharmacy
Shaheed Benazir Bhutto University
Sheringal Dir Upper, Pakistan

Arwa Abdulkreem AL-Huqail
Department of Biology, College of Science
Princess Nourah bint Abdulrahman University
Riyadh, Saudi Arabia

Umar Farooq Gohar
Institute of Industrial Biotechnology
Government College University, Lahore
Lahore, Pakistan

ISBN 978-3-031-35402-1 ISBN 978-3-031-35403-8 (eBook)
https://doi.org/10.1007/978-3-031-35403-8

This Springer imprint is published by the registered company Springer Nature Switzerland AG
The registered company address is: Gewerbestrasse 11, 6330 Cham, Switzerland

Paper in this product is recyclable.

I dedicate this book to my honorable parents who instilled in us the love of science, discovery, and writing since childhood. I also dedicate it to my husband, children, and family who supported and encouraged me so that this work could be completed and see the light. And I dedicate it to my university, Princess Nourah bint Abdulrahman University, who taught us that fulfilling wishes for achievements is not impossible with patience and perseverance. I also dedicate this book to my country, the Kingdom of Saudi Arabia, which seeks to meet the challenges of climate change with its pioneering environmental initiatives (the Saudi Green Initiative and the Green Middle East Initiative), which gave us all opportunities to cooperate with scientists and researchers in the world for the good of humanity and the protection of the planet. I am also pleased to dedicate this book to all official national and international bodies concerned with plants, as well as to everyone who loves plants and knows their role in the reconstruction of the planet, their solution to all environmental problems and issues of all kinds, and the achievement of environmental sustainability, God willing, in the whole world.

Dr. Arwa Abdulkreem Al-Huqail

Family and friends

Dr. Muhammad Zia-Ul-Haq

Committed to expressing my gratitude to my family, parents, and friends for their unwavering support and motivating words throughout this journey.

Dr. Muhammad Riaz
Dr. Umar Farooq Gohar

Contents

About the Editors

Muhammad Zia-Ul-Haq is working in the Office of Research, Innovation and Commercialization at University of Engineering and Technology, Lahore, Pakistan. Besides PhD, he has an LLB from Punjab University, Pakistan, and LLM-IP from Turin University, Italy. He served Office of Research, Innovation and Commercialization, Lahore College for Women University as Senior Manager. Previously he served as Patent Examiner in The Patent Office, IPO Pakistan (Ministry of Commerce) for 8 years. He received patent trainings from Japan, Korea, Malaysia, USA and France. He has published 4 books with Springer and more than 120 research and review papers with total IF of 120 and total Google Scholar Citations of 7000. He is peer reviewer of many journals published by Elsevier and Springer. He won RPA from Pakistan Council for Science and Technology (PCST), Ministry of Science and Technology (MOST) from 2010 to 2015.

Arwa Abdulkreem AL-Huqail is a Distinguished Associate Professor of Ecology (since 2021) at the Department of Biology, College of Science at Princess Nourah bint Abdulrahman University, Riyadh, Saudi Arabia. She is an environmental and community service activist having more than 65 publications in prestigious international scientific journals (web of science). She has more than 26 years of experience in university teaching and research. She has organized, attended and participated in many initiatives, seminars, workshops and scientific meetings at the national and regional level. She was the Vice Dean of the College of

Science for Community Service and Environmental Development for three consecutive years, Vice Dean of Development and Quality at the Community College for one year and Adviser to the Saudi Ministry of Environment, Water and Agriculture and the National Center for the Development of Vegetation Cover and Combating Desertification for the second year in a row.

Muhammad Riaz is an Assistant Professor of Pharmacy at Shaheed Benazir Bhutto University, Sheringal, Pakistan. He has authored over 40 peer-reviewed publications in renowned journals at the national and international level and three books: *Anthocyanins and Human Health: Biomolecular and Therapeutic Aspects* (2016) and *Carotenoids: Structure and Function in the Human Body* (2021) for Springer and *The covid-19 pandemic: A Multidisciplinary Review of Diagnosis, Prevention, and Treatment* (2022) for CRC/Apple Academic. He received his PhD in Pharmacognosy from the University of Karachi, Pakistan (2012), followed by a postdoctoral fellowship at Prof. Dou Deqiang Lab, Liaoning University of Traditional Chinese Medicine, China (2017). During his PhD, he worked at Prof. Michael Heinrich Labs, UCL School of Pharmacy, London, (2010) under International Research Support Initiative Program of the Higher Education Commission of Pakistan.

Umar Farooq Gohar has a bachelor's degree in Pharmacy (B Pharmacy) from the University of the Punjab and Master of Philosophy (MPhil) and Doctor of Philosophy (PhD) in Biotechnology from the Institute of Industrial Biotechnology, Government College University, Lahore. Dr. Gohar did some of his PhD research work at the School of Chemical and Biomolecular Engineering, University of Sydney, Australia. He has served Pharmacy Department at the University of Lahore as Lecturer; later he served Riphah International University as Assistant Professor. Currently he is working as Assistant Professor at the Institute of Industrial Biotechnology, Government College University Lahore. Dr. Gohar is currently supervising five PhD scholars. He has also supervised more than 40 MPhil projects in the field of pharmacy and biotechnology. He has completed two projects funded by ORIC Government College University and Higher Education commission of Pakistan. He has published more than 45 publications in various national and international journals. He is reviewer of many peer-reviewed journals. His area of interest is production and application of bioactive compound from microbial and plant sources. He is also guest editor of two special issues of journal *Molecules*. He also remained associated with a biannual journal *Biologia* Pakistan. He got Mevlana exchange program funding from Uşak University, Turkey.

Chapter 1
Tissue Culture of Medicinal Plants

Isha Fatima, Muhammad Akram, Hamid Mukhtar, Umar Farooq Gohar, Zahoor Ahmad Sajid, and Uzma Hameed

1.1 Introduction

Plants are a major source of medicinally important metabolites. With the increasing demand for medicines, the need to produce naturally occurring pharmaceutically important metabolites of plants has become essential. Tissue culture or micropropagation proved to be the most appropriate method for the healthy cultivation of medicinal plants. It is gaining popularity due to season-independent growth with less time and space. Tissue culture techniques are employed to obtain plants in a short time and a restricted space [1]. Plants have been used for therapeutic purposes for a very long period. A plant, whole or any of its parts, or its extracted substances that can be exploited for medicinal purposes and producing precursors for producing medicines, is called a medicinal plant [2]. A plant can be called a medicinal plant if it satisfies one or more of the following properties: its use in galenical preparations, for example, decoctions, infusions, etc. Its cultivation for the extraction of pure substances that can be used for direct therapeutic usage or its role in the semi-synthesis of a drug. For example, diosgenin, a precursor of various hormones such as sex hormones, is obtained from *Trigonella, Costus* species. Any food, spice, or fragrant plant with medicinal properties, such as *Ginger or Turmeric*. Any plant that can be used for manufacturing surgical dressings, such as cotton, jute, or flax [2].

I. Fatima · H. Mukhtar · U. F. Gohar · U. Hameed (✉)
Institute of Industrial Biotechnology, Government College University, Lahore, Pakistan
e-mail: uzmahameed@gcu.edu.pk

M. Akram
Government Shalimar Graduate College, Lahore, Pakistan

Institute of Botany, University of the Punjab, Lahore, Pakistan

Z. A. Sajid
Institute of Botany, University of the Punjab, Lahore, Pakistan

M. Zia-Ul-Haq et al. (eds.), *Essentials of Medicinal and Aromatic Crops*,
https://doi.org/10.1007/978-3-031-35403-8_1

1.2 Tissue Culture

Plant cells are totipotent and can be turned into new plantlets providing suitable medium and conditions. In tissue culture, scientists use this ability to grow plant tissues and organs on artificial media *in vitro* under a precise sterile setting. The pioneer of tissue culture is German physiologist Gottlieb Haberlandt. In 1902, he proposed the concept of *in vitro* cell culture. In his experiment, he cultured the individual palisade cells in knop's salt solution supplemented with sucrose. The cells remained alive for one month, accumulated starch, and increased size but failed to divide. This experiment laid the foundation of plant tissue culture. Thus, Haberlandt is known as the father of plant tissue culture. New techniques and methods were developed between the 1940s and the 1960s [3, 4]. Through micropropagation techniques, not only single plants but crops can also be grown that are needed to produce medicinal molecules or extracts [5]. The nutrient medium for plant tissue culture must contain all necessary nutrients (Fig. 1.1) for the growth and development of the explant [4]. Another important factor that must be considered during tissue culturing is pH. Generally, it is set in the range of 5.4 to 5.8 for both liquid and solid media [4].

The medium required for the plant *in vitro* micropropagation must contain the following components [4, 6].

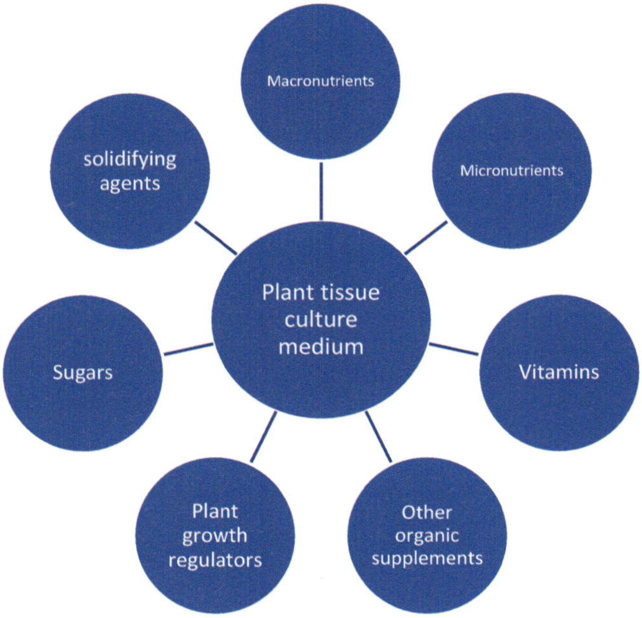

Fig. 1.1 Essential components of the tissue culture medium

1.2.1 Macroelements

The elements required in a concentration greater than 0.5 $mM.L^{-1}$ are macroelements. These are always supplemented in the medium, for example, nitrogen, potassium, sulfur, phosphorus, etc. [7].

1.2.2 Microelements

These are also known as trace elements and are required in an amount less than 0.5 $mM.L^{-1}$ concentration. These nutrients are always employed in the medium. Examples of micronutrients are iron, boron, nickel, manganese etc. [7].

1.2.3 Sugar

Sugars are the source of carbon and energy in the medium. These are always supplemented in the medium but may be skipped under particular circumstances. Examples of frequently used sugars in tissue culture medium are glucose, fructose, and sucrose [7].

1.2.4 Vitamins

Vitamins are essential for the growth and development of *in vitro* plant culture. These are needed to synthesize several compounds involved in different metabolic processes in cell growth and differentiation. Examples of vitamins used in *in vitro* micropropagation of plants are thiamin (B_1), pyridoxine (B_6) and nicotinic acid [7].

1.2.5 Solidifying Agent

A solidifying agent plays a vital role in the tissue culture medium as it solidifies the medium and provides ground for growth. Depending upon the requirement, the semisolid or solid-state of the medium can be maintained by changing the concentration of the solidifying agent. Frequently used gelling agents are agar, agarose and gellan gum [7].

1.2.6 Amino Acids and Nitrogen-Containing Compounds

Amino acids act as the source of nitrogen and raw materials for synthesizing other compounds required for optimal growth of plant culture. For this purpose, hydrolysates of casein, L-glutamine, L-asparagine, and adenine are commonly used in the plant tissue culture medium [7].

1.2.7 Undefined Supplements

Scientists also tested the effects of natural substances on the growth of *in vitro* cultures. These components contribute to the above-mentioned substances as well as regulate plant growth. Such undefined supplements include coconut milk, yeast extract, malt extract, ground banana, orange juice, tomato juice, and activated charcoal. Their effects on the culture depend on the plant species and can also have inhibitory effects on the culture. For example, activated charcoal had an inhibitory effect on soybean cultures [8].

1.2.8 Buffers

Buffers can be added to maintain the pH of the medium. Buffers are beneficial due to their ability to resist pH changes during the active growth of calli [6].

1.2.9 Plant Growth Hormones

Plant growth hormones or plant growth regulators are the most important components of plant tissue culture media as they regulate growth, induce cell proliferation, and organize culture development. Generally, "five main classes of plant growth hormones are: Auxins, abscisic acid, cytokinins, ethylene, and gibberellins" [9]. Table 1.1 shows summary of the components and their role in tissue culture medium. Preparation of plant tissue culture media is an important task and it should be done in dedicated and clean area provided with all necessary equipment that should be regularly maintained [4]. Examples of micropropagation medium for plants are Murashige and Skoog (MS) medium [10], Gamborg (B_5) medium, Linsmaier and Skoog (LS) medium, Nitsch and Nitsch (NN) medium [4].

Table 1.1 Major components of tissue culture medium and their role

Component	Example	Role	Reference
Macroelements	Nitrogen, potassium, sulfur, phosphorus	Growth, morphogenesis	[6, 11]
Microelements	Iron, boron, nickel, manganese	Tissue growth	[6, 11]
Sugar	Glucose, fructose, and sucrose	Energy source, carbon source, osmotic agent	[6]
Vitamins	Thiamin (B_1), pyridoxine (B_6) and nicotinic acid	Essential intermediates metabolic, catalysts	[6]
Solidifying agent	Agar, agarose and Gellan gum	Solidifies medium	[6]
Amino acids and other nitrogen supplements	L-glutamine, L-asparagine, and adenine	Sources of nitrogen, raw materials for synthesizing other essential compounds	[6]
Undefined supplements	Yeast extract, malt extract, ground banana, activated charcoal	Source of peptides, amino acids, fatty acids, carbohydrates, vitamins, plant growth substances	[6, 8]
Buffers	Organic acids acting as buffers, a mixture of KH_2PO_4 and K_2HPO_4	Stabilize the pH of hydroponic solutions	[6]
Plant growth hormone	Auxins, abscisic acid, cytokinins, ethylene, and gibberellins	Inducing the production of roots, shoots, play important role in development of calli	[6]

1.3 Types of Micropropagation Methods

Different methods of *In vitro* micropropagation are available for culturing of plants (Fig. 1.2). These include callus culture, organ culture, single cell culture, suspension culture, embryo culture, anther culture, pollen culture, somatic embryogenesis, protoplast culture, meristem culture [12, 13]. Various types of micropropagation techniques used are shown in Fig. 1.2. Tissue culture techniques are used for the production of artificial or synthetic seeds that are encapsulated shoot buds, cell aggregates, somatic embryos or any other meristematic tissue having the potential to regrow into new plantlets in a suitable medium after the storage conditions.

1.3.1 Callus Culture

Callus is defined as the unorganized mass of proliferative plant cells produced when an explant is grown on an artificial nutrient medium under a controlled sterile environment. Calli differ significantly in texture, appearance, and growth rate based on the explant source and nutrient medium constitution. The process of callus formation is called callogenesis. Small explants or sections from the plant organs are

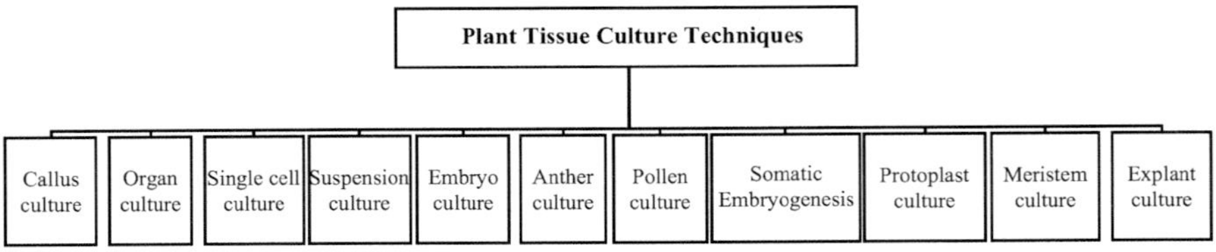

Fig. 1.2 Various types of tissue culture techniques

inoculated on solidified medium to initiate callus formation [14]. There are two steps in callus culture. First, biomass is produced with the growth and multiplication of calluses, and then biosynthesis of compounds takes place from this biomass.

Selecting an appropriate parent plant is crucial to obtain a good yield because the commencement of high producing callus and buildup of secondary metabolites is genotype-dependent [15]. Callus cultures are useful for obtaining commercial products like cosmetic, pharmaceutical, nutritional, coloring agents, flavors, fragrances and animal health-related products. Callus culture is ideal when limited space and time are available. Rather than growing the whole plant, the same specific metabolite can be obtained just by callus culturing [16]. In callus culture, the explant is first sterilized with 0.5% NaOCl for 15 min and then washed three times with sterile water. Another method of sterilization is to apply 0.1% mercuric chloride solution for 8–10 min and washing the explant six to eight times with sterile distilled water. After this, the explants are cultured into test tubes or Petri plates containing MS medium supplemented with sugar/carbon source, gelling agent, and plant growth hormones for callus induction. Sometimes callus can get dark or brown in color. To prevent this, different antioxidants such as ascorbic acid, polyvinylpyrrolidone (PVP) and activated charcoal can be added at 0.5–1.0 mg/L concentration. Petri dishes or the test tubes are sealed tightly with polyethylene film and placed in the incubator or growth culture room at a temperature between 22 and 25 °C under a photoperiod of 16 h light/8 h darkness. When calli show growth and shoot formation, they are subcultured into fresh media for root growth and acclimatization [17]. Figure 1.3 is the graphic representation of the procedure of callus culture [18, 19].

The bioactive compounds of calli are collected mainly at the stationary phase of their growth cycle because their production increases during this phase. Metabolites can also be collected at different phase depending on the plant species and callus properties. The identification and quantification of extracted compounds from calli are carried out using HPLC, LC-MS, etc. [20]. Explants for callus cultures can be obtained from different parts of plants. For example, explants in wheat can be mature or immature embryos [21], shoot apical meristems [22], coleoptile [23] an anther [24] as shown in Fig. 1.4. In other instances, other excellent materials for callus initiation and regeneration are hypocotyls and root explants.

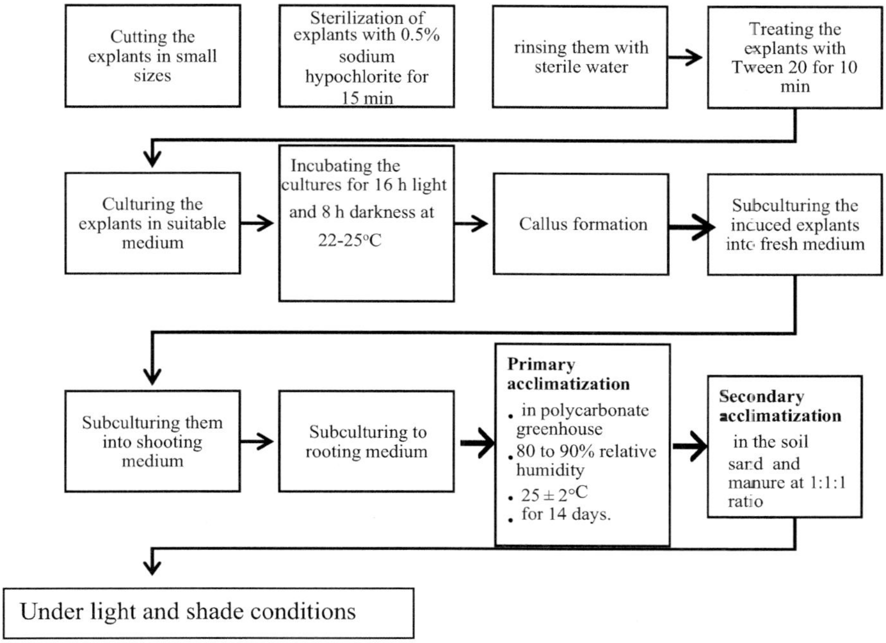

Fig. 1.3 Schematic representation of callus culture

Fig. 1.4 Different explants used for wheat tissue culturing

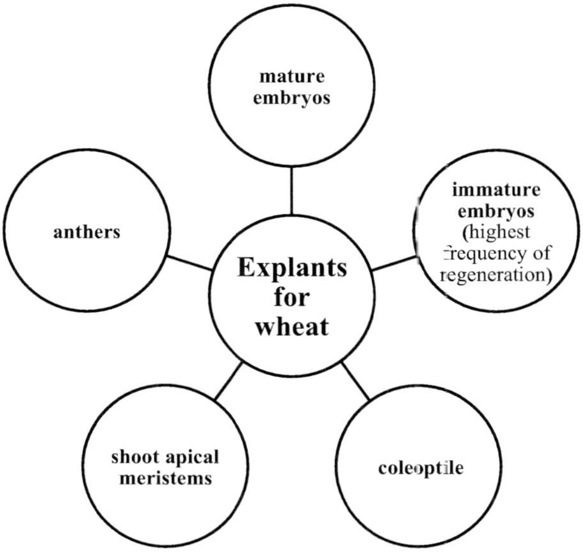

1.3.2 Organ Culture

The process of excising the whole or part of an organ or organ primordial (e.g., leaf, node, internode, shoot, axillary bud, root, or seedling) and using it for *in vitro* micropropagation is called organ culture. This technique helps to preserve the structure and function of the organs and retains their characteristics. Organ culturing provides information about growth, differentiation, and development patterns. Using appropriate phytohormones, the explant organ can differentiate into complex structures (e.g., cotyledon, shoot tips, leaf disks, hypocotyls, anther internodes, stem, roots, or thin cell layers) and then ultimately complete plant [12]. Organ culture can be used for diverse applications, including rapid clonal propagation, preservation of endangered plants, production of valuable metabolites and enzymes (Fig. 1.5) [12, 25, 26].

The first step in organ culture is the aseptic isolation of tissue. The organ/tissue is treated with 7X-detergent for 5 min and then immersed in a freshly prepared, saturated filtered solution of chlorinated lime for 20 min. After this, the explant is washed with sterile distilled water many times and cut into small pieces of tissue. These small explants are incubated in the culture medium. This method can be modified slightly depending on the nature of the explant [27]. Organ culture benefits in maintaining high-class mother plant varieties that amass greater concentration of target compounds [12]. The schematic diagram of the method is shown in the Fig. 1.6.

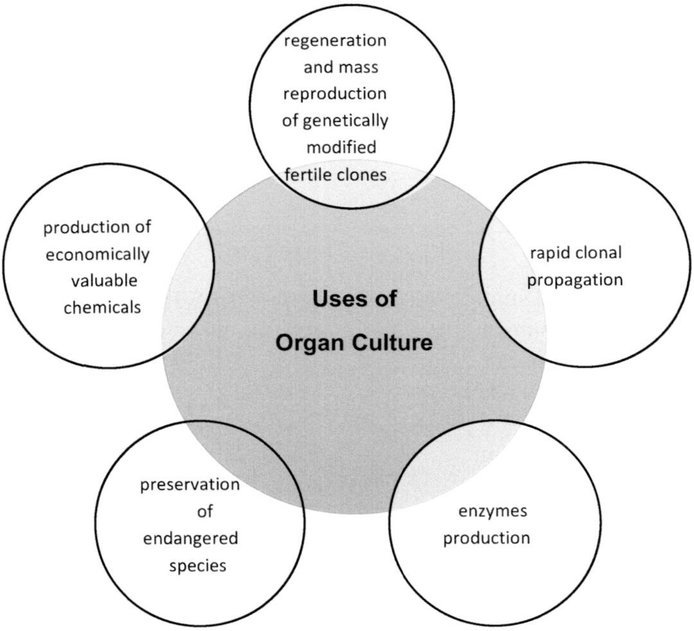

Fig. 1.5 Applications of organ culture

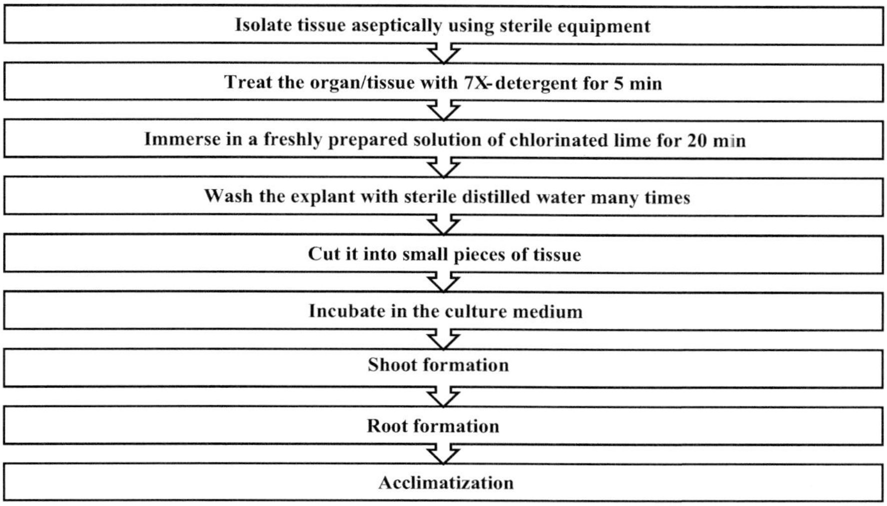

Fig. 1.6 Schematic representation of organ culture method

Large-scale plant cell, tissue and organ cultures provide an even and controlled supply of phytochemical products irrespective of the plant availability. In the past, organ culture did not prove very successful as organ tissue culture occurs in a continuous batch. Any change in conditions would disturb the equilibrium and the cultures would not grow. But the advancement of technology and the development of specific bioreactors for organ culture has dramatically improved the situation. Researchers are even more motivated to develop bioreactors to improve the procurement of secondary metabolites from organ micropropagation [25, 26].

1.3.3 Single Cell Culture

In this technique, isolated single cells are grown on a suitable nutrient medium in a controlled sterile environment. Single cell culture has numerous applications. Sources of the single cell can be a callus, cell suspension, variety of tissues and organs. Single cells can be obtained by treating intact plant tissue (leaf, stem, etc.) with enzymes or mechanical separation. The single isolated cell can be cultured in a suitable liquid or solid medium. In the culture medium, the isolated cells multiply to form a callus tissue. This callus tissue can also grow into a plantlet through organogenesis and embryogenesis.

Diffuse liquid cultures can also be used for single cells cultures. Single cells can grow effectively on filter paper over established cultures. When the cultures are moved to agar, they grow very well. Some cultures can grow well even after being transferred 25 times or more within 5–6 weeks without lessening their growth. This technique has proven successful for the clone establishment of grapes of crown gall origin and marigold. This method can also be applied to *Nicotiana tabacum* (strain WTI), *N. glutinosa* × *N. tabacum* (strain WTH). Researchers have observed alterations among clones in growth rate, texture, and color resulting from the same parent plant. Different media can be used to grow these single cells to establish shoots or roots in the culture [28].

The first callus is produced and maintained at 24 °C on a suitable medium in the dark to produce single cell cultures. One-third of the calli are refreshed and re-cultured into the fresh medium plate. The cell suspension can be initiated after 14 days of sub-culturing the calli. To initiate single-cell suspension, take 2 g of the callus and transfer it to 50 ml medium in a 250 ml conical flask. The flask is gently shaken to break open the callus at 100 rpm. The flask is kept in a shaking incubator at 23 °C for 5 days in dark. To refresh the medium, top it with 50 ml of the freshly prepared medium. After some days, a mixture of micro calli and single cells is formed. To get the desired product, the flask is allowed to stand by for some time to separate the micro callus, leaving a compound mixture of dead and viable cells in the suspension. Decant the two-thirds of the supernatant gently and only keep the one third of the supernatant. This residual mixture contains both single cells and micro calli. Refresh the inoculum by adding 100 ml of freshly prepared medium. Incubate this flask in the dark at 23 °C for 12 days. Repeat this procedure 3–4 times. This procedure allows an average of 6×10^{-5} cell/mL with 65% viable single cells in the supernatant. This suspension can be maintained and utilized for 6 months. A flowsheet diagram is shown in the Fig. 1.7 [29].

Single cell culture is a very valuable technique that can be used not only to produce plant clones but also to produce resistant cell lines and as a research tool (Fig. 1.7). It can be used to study photosynthesis, ion transport, secondary metabolite production, cell growth, differentiation, and apoptosis. For example, the cell suspension cultures of *Arabidopsis* and *Zinnia* cells were used to detect multiple genes and a wide range of gene expression studies [29]. Single cell cultures are very helpful in producing resistant cell lines. Single cells are grown on a medium with mutagenic compounds and the proliferating cell lines are isolated. The cultures which survive these mutagens are selected and grown into whole plants to confirm and compare their phenotypes with normal plants. This simple method has assisted in selecting cell lines and plants resistant to antibiotics, herbicides, fungal toxins, etc. By using this technique rather than growing the whole plant, a greater amount of commercially and medically important compounds can be produced [29–32] (Fig. 1.8).

1.3.4 Suspension Culture

In this micropropagation technique, a single cell or minute masses of cells (explants) are cultured in a liquid medium on a shaking incubator [33]. Preferably, cells should be suspended singly in cell suspension cultures through continuous agitation. But

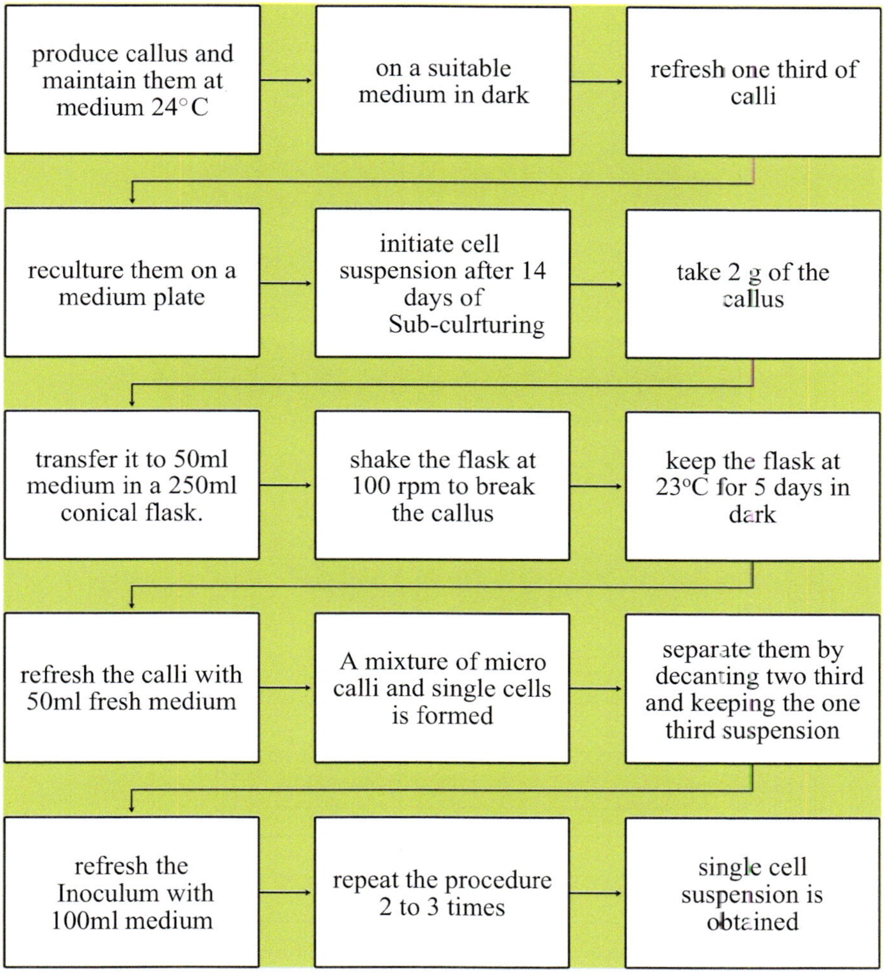

Fig. 1.7 Schematic representation of single cell culture

such culture exists only in rare cases. Researchers strive to obtain a fine culture by increased cell dissociation, which can lead to increased culture uniformity. But even after many measures are taken, some cell aggregates are formed. The cell aggregates can be divided into two groups. One is 'fine' suspension cultures which consist of micro- to sub-macroscopic colonies consisting of approximately 5–200 cells. Other group contains aggregates of cells about 0.5–1.0 mm in diameter (Fig. 1.9). The second group is often readily achievable in which cells grow perfectly well and meet all the investigation requirements depending on the purpose of the research. It is preferable to have some degree of cell aggregates in the suspension as cells in this

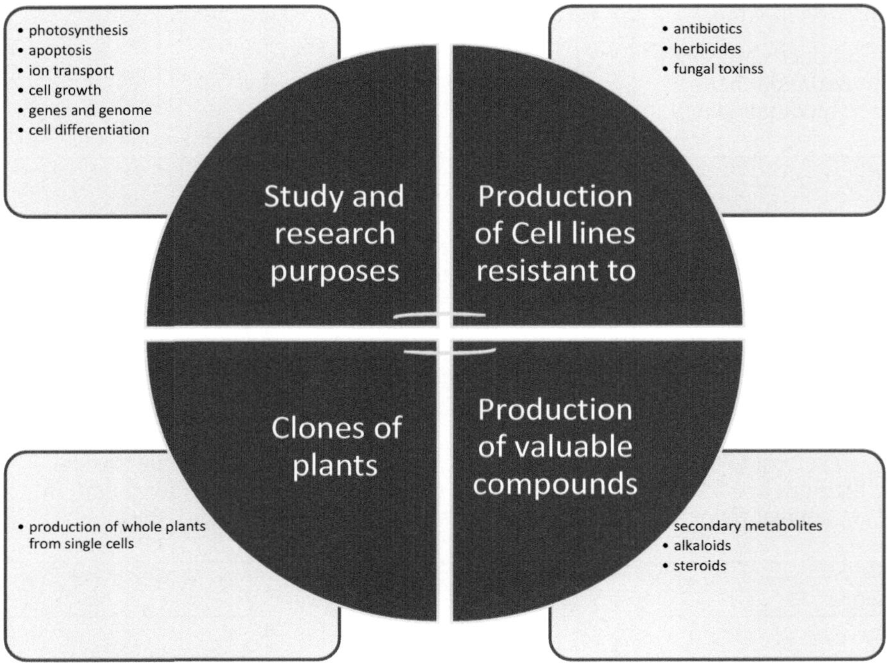

Fig. 1.8 Applications of single cell culture

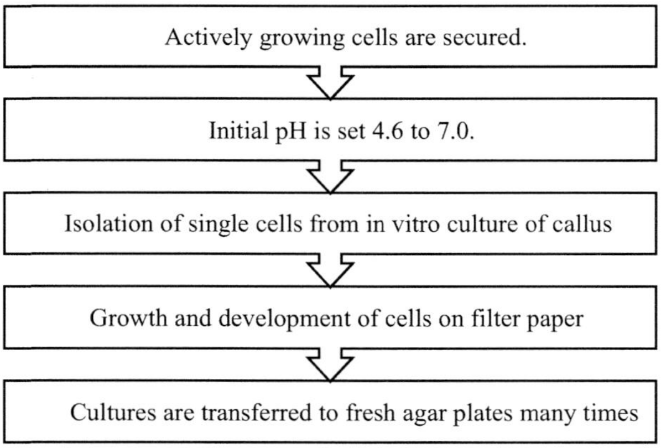

Fig. 1.9 Schematic representation of suspension culture

state can retain the totipotent character and enhance the production of desired metabolites [34].

Plant cell suspension has wide applications in research because it bypasses the structural complexity of the whole plant. Suspension cultures provide homogenous cell populations with high growth rates and reproducibility of conditions which help scientists to study complex physiological processes at molecular and cellular processes. It can also be used to produce high-value secondary metabolites of commercial interest [35]. Cultures growing homogeneously in the liquid medium are beneficial for metabolic engineering and targeting metabolite production. Metabolite production can be commercialized using a bioreactor system. Other advantages of suspension culture are the study of *in vitro* mutagenesis and genetic transformation [33, 35].

1.3.5 Embryo Culture

In this process, an embryo (in different developmental stages) is aseptically isolated from mature seeds and cultured under aseptic *in vitro* controlled conditions on a suitable medium. In most situations, sterilization of embryos is not required after collection because they are protected by the aseptic environment of the ovule. Therefore, entire ovules or ovaries are sterilized exteriorly, and embryos are removed aseptically from the adjacent tissues. Researchers also use harsh methods to disinfect the surface due to hard protecting tissues. Direct decontamination is only necessary if the seed coat is cracked or there is any sign of endophytic pathogens. Endophytic infections can occur in the seeds of fescue (*Festuca* spp. L.), corn (*Zea mays* L.), and dogwood (*Cornus* spp. L.) [36].

Embryos of larger size are not hard to excise, but small embryos pose dissection problems because researchers need microdissection tools and a dissecting microscope for excision without injury. It is necessary to ensure that the embryo is not desiccated or injured during culturing. The process may vary from specie to specie. Still, it is generally performed by making an incision at the micropylar end of the young ovule and applying pressure at opposite ends to force the embryo out through the opening. If a liquid endosperm surrounds the embryo, the pressure must be applied carefully because it may damage the fragile embryonic tissue[36]. Figure 1.10 shows the process of embryo culture.

Similarly, in somatic embryogenesis, a single or group of cells can regenerate from non-zygotic embryos and germinate to form complete plants. Immature zygotic embryos and seedlings are appropriate explants for somatic embryogenesis. Somatic embryogenesis can be either direct or indirect. In direct embryogenesis, embryos are developed from the exterior of explants, whereas during indirect embryogenesis, embryo formation involves the intermediate callus phase. This method is beneficial for metabolic engineering. Mature organs (leaves or roots) have a low capacity for somatic embryogenesis [12]. Embryo culture is useful in studying the factors involved in the dormancy of seeds and crosses between plants of different ploidy within the same species, e.g., Iris and Zea. This technique also helps to procure inter-specific hybrids, e.g., *Gossypium, Datura,* and *Lycopersicum.* This tissue culturing technique has shortened breeding cycles. This method has served as a tool for studying the effect of specific substances on the morphology of embryos [37].

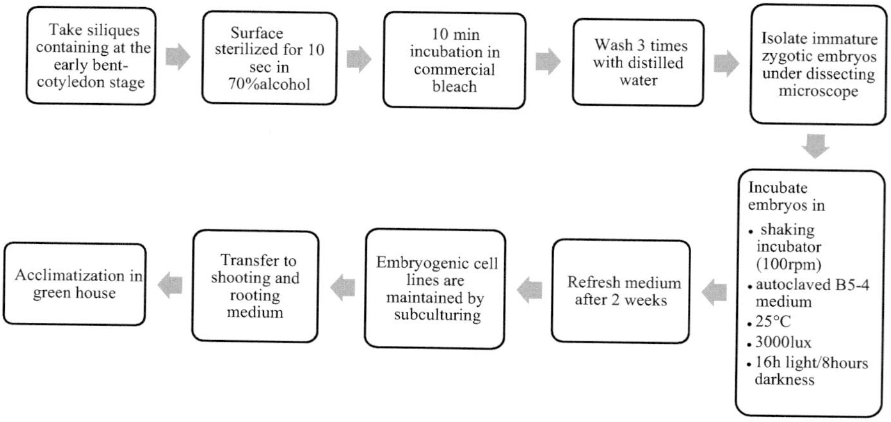

Fig. 1.10 Schematic representation of embryo culture

1.3.6 Anther Culture

Anther culture is also known as "Androgenesis." In this method, haploid plants are developed *in vitro* from effective pollen grains through a series of cell divisions and differentiation. Androgenesis can be performed by both *in vivo* and *in vitro* methods. *In vivo* androgenesis is induced by irradiation or temperature-shock treatments, as reported in *Crepis tectorum* and *Antirrhinum majus* [38].

In vitro androgenesis involves the separation of anthers from the bud. The bud from a young flower is surface sterilized and rinsed with water when the pollens reach a suitable development phase. The bud is dissected after removing the sepals and petals. Each anther is gently detached from the filament and placed horizontally on the nutrient medium (either liquid or agar medium). Only intact and uninjured anthers are selected for culturing. Figure 1.11 represents the schematic diagrams for anther culture. Androgenesis has many advantages. It creates haploid and spontaneous diploid plants *in vitro*. These plants have genetic potential that is rapidly expressed phenotypically. This technique lessens the time needed to grow self-pollinated crops, reduces the breeding cycle, and requires less space and labour [39, 40].

1.3.7 Protoplast Culture

A cell without a cell wall is known as a protoplast. A protoplast can regenerate its cell wall, divide, and grow into a whole plant if a suitable medium is provided under aseptic conditions. The technique which uses protoplast as an explant is known as protoplast culture. Protoplasts can be obtained by removing the cell wall by

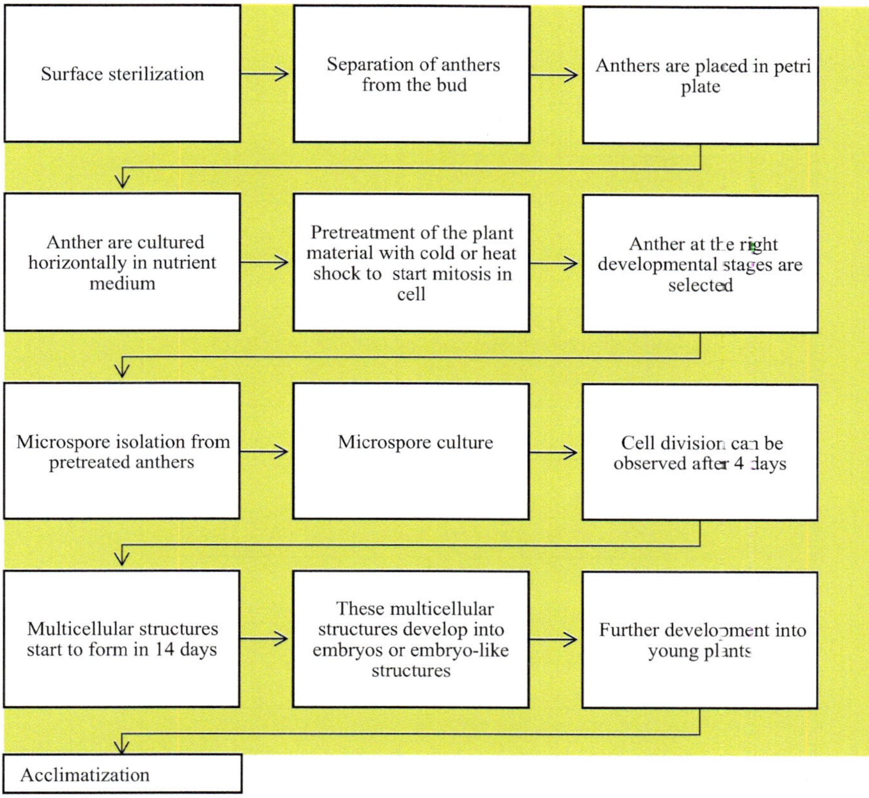

Fig. 1.11 Schematic representation of anther culture

mechanical disruption or treating the plant tissue or cultured cells with enzymes [41]. In order to get protoplasts, fresh-weight tissues such as leaves or embryos are chopped and incubated in 0.5 M mannitol for an hour. After this, mannitol is replaced with an enzymatic mixture containing cellulase, macerase and driselase. The suspension is then incubated in the dark for 16 h with gentle shaking (50 rpm). The digest is then passed through a 50 mm nylon sieve. After repeated centrifugation, the pellet is resuspended in 0.5 M mannitol agar. These protoplasts can be cultured after protoplast fusion or without fusion. For culturing, both liquid and solid media can be used. Figure 1.12 shows the flow chart of the process [42].

Protoplasts are used when selection or hybridization at the cellular level is required. Commonly protoplasts are fused by either chemical or electrical means. The mostly employed chemical is polyethylene glycol (PEG). Specific equipment that can induce the electrical fusion of cells is needed for electrical fusion. It is equipped with an AC field or a DC source to apply the AC field or DC pulses and a suitable fusion chamber [41].

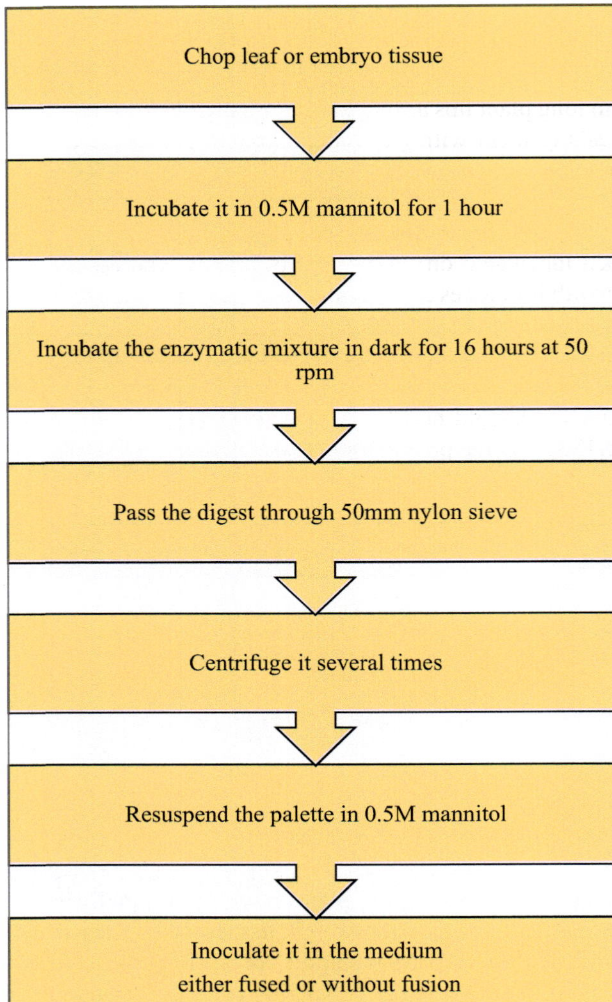

Fig. 1.12 Schematic diagram of Protoplast culture

1.3.8 Meristem Culture

Meristem refers to the tips of shoots that contain the shoot apical meristem. Meristem tip comprises the meristematic dome and often a pair of leaf primordia which can be used as an explant. This explant is cultured into a suitable nutrient medium under aseptic conditions and can develop into whole plants. Occasionally, a liquid medium is used. This method is used for contaminant-free axillary shoot multiplication and sustaining diverse superior mother plants that amass a greater number of metabolites of interest. Meristem cultures are an excellent method for virus-free cultivation [12].

About 10 mm long plant tips are cut off and used as explants. These are sterilized carefully by washing them with running tap water and then detergent for 5 min. After this, the explant is washed with water to eliminate the detergent, then treated with 1% sodium hypochlorite solution for 15 minutes and rinsed three times with sterile distilled water to clean all sodium hypochlorite. The explants should be 0.5–1.0 cm when inoculated on the medium. The medium must be supplemented with suitable growth hormones along with sugar. The cultures are then incubated at 25 °C at 3000 lux and provided with cool white fluorescent lamps. When cultures show growth, they should be re-cultured into fresh medium. After 2–4 weeks of further growth, plantlets are transferred for rooting and acclimatization. To acclimatize plantlets, the soil should be disinfected and saturated with water in a greenhouse. Figure 1.13 shows the procedure in a schematic diagram [43, 44].

Fig. 1.13 Schematic representation of meristem culture

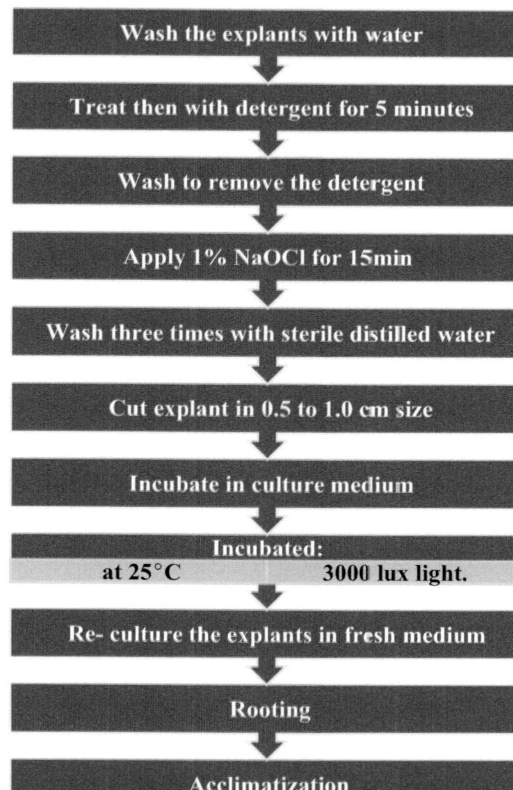

1.3.9 *Pollen Culture*

In pollen culture, pollen grains are separated from intact anther at microscopic stages and transferred to a suitable medium to produce microspores without producing male gametes. This technique enables scientists to culture haploid plants for propagating haploid plants and for genetic studies [45]. The details of the procedure are given in Fig. 1.14. First, the pollen grains are isolated from anthers by crushing. The pollen grains are filtered by passing them through a sieve of a pore size of 100 pm. Isolated pollen grains are washed with distilled water twice and centrifuged at 100× for 2 min. Pollen suspension of known concentration is prepared and inoculated into the nutrient medium. The medium used by Matsushima and colleagues was supplemented with 1% sucrose and glucose, 0.1% yeast extract and casein hydrolysate, and 10–5 M 2, 4-dichlorophenoxyacetic acid (pH 5.8). They got the best results with the medium diluted multiple times. A concentrated medium does not give good results with pollen culture [45].

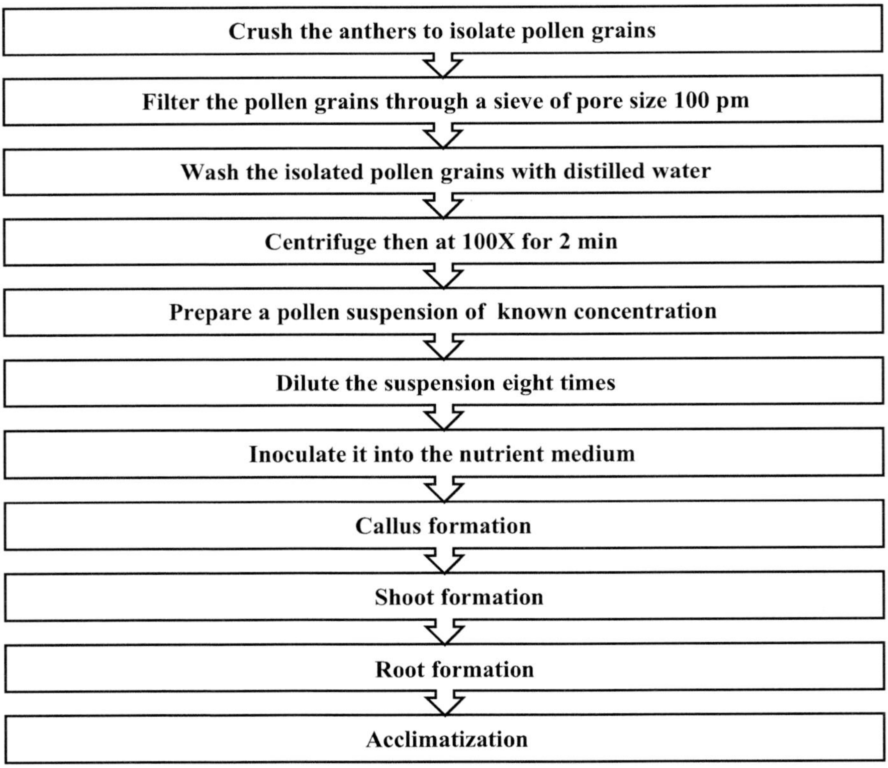

Fig. 1.14 Schematic representation of pollen culture

1.4 Tissue Culturing of Various Medicinal Plants

Different plants have different requirements when they are cultured *in vitro*. Concentration of plant growth hormones and essential additional components vary from species to species.

1.4.1 Micropropagation of Neem (**Azadirachta indica** *L.*)

A. indica is one of the major therapeutic plants that contains a plethora of compounds with therapeutic uses. Neem also has anticancerous, anti-tumor, and anti-inflammatory effects. Many researchers have investigated the antioxidant and anti-cancerous properties of neem. Cervical cancer patients, when treated with neem, showed apoptosis of cancerous cells due to the anti-oxidant properties of neem. The enhanced activities of caspase 3, 8, and 9, IFN-γ levels and a reduction in TNF-α in monocytes were observed in these patients [46]. Azadirachtin is a secondary metabolite with a complex structure found in neem seeds. It is a valuable pesticide and anticancerous compound [47]. Table 1.2 shows compounds present in neem [48]. Meliatetraolenone is a new tetranortriterpenoid compound separated

Table 1.2 Medicinal properties of neem plant

Active component	Activity	Effects	Reference
Azadirachtin	Anti-oxidant	Activity in cervical cancer patients	[46]
Azadirachtin Salannin deacetylgedunin	Insecticide	Can kill the larvae of *A. stephensi*	[61]
Azadirachtin	Antibacterial	Against *Staphylococcus aureus* and MRSA	[62]
Nimbolide	Cytotoxic effect	On fibroblasts	[63]
Nimbolide	Anti-tumor effects	Against human choriocarcinoma	[64]
Gedunin	Anti-breast-cancer activity	Inhibition of a protein Hsp90	[65]
Azadirachtin and nimbolide	Cytotoxic activity	Decreased HeLa cell viability	[66]
Nimbolide azadirachtin	Anticancerous effect	Induction of cell apoptosis	[67]
Gedunin	Anticancer activity against ovarian, colon, and prostate cancer cells	Intervention with principal signaling pathways	[50, 68]
Quercetin	Anti-inflammatory and anti-cancerous	Represses the viability of HeLa cells in dose dependent manner	[69]
Quercetin	Antimetastatic agent	Effective in cervical cancer	[70]
Quercetin	Diminished multiplication of ovarian cancer cells	Increased action of caspase-3 and caspase-9	[115, 116]

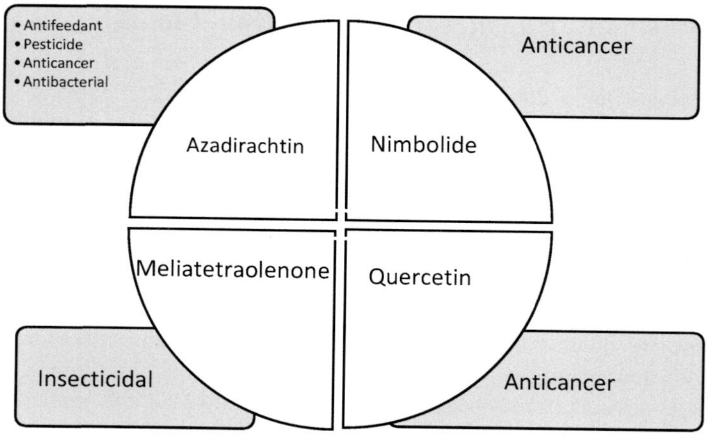

Fig. 1.15 Pharmaceutically important compounds of neem with their activities

from the fresh neem leaves. It shows insecticidal action [49]. Gedunin, an extract of the neem plant, shows antiproliferative activity that has been verified by many researchers [50–52]. The neem leaf extract is also advantageous in the antiproliferation of breast cancer cell lines [47, 53–56]. Figure 1.15 diagram shows the effects of different components of neem extract.

The micropropagation of neem shoot tips is more likely to develop into a callus. The presence of cotyledons affects the developmental stage [57]. The highest concentration for shoot elongation was observed with 0.5μM BAP for young calli [58]. Quraishi and colleagues used the crown and basal-sprout explants. They controlled the leaching of phenol growth inhibitors by supplementing the nutrient medium with 12.5 μM PVP-40. They found that Driver Kuniyuki Walnut (DWK) medium supplemented with 0.22 μM benzyl adenine was better than the MS medium for shoot propagation. 100% of basal-sprout and seedling explants grew in half-strength DWK medium containing 4.9 μM IBA and root formation. Plantlets developed from both explant types showed a 90% survival rate after acclimatization [59]. When the immature flower was used as an explant for neem micropropagation, the best callus formation response was observed in the M9 medium (MS medium containing 3.0% sucrose, 1.0 mg/L 2,4-D, 1.0 mg/L BAP and 0.2 mg/L NAA). Researchers reported that the extent of callus formation was improved when the same explants were first cultured on MS medium containing 2,4dichlorophenoxyacetic acid (2,4-D) 1.0 mg/L, BAP1.0 mg/L and 1Naphthaleneacetic acid (NAA) (0.2 mg/L) along with 10% sucrose for 15 days and then sub-cultured on a same medium supplemented with 3.0% sucrose [60].

1.4.2 *Tissue Culturing of* Pinus roxburghii *Sarg*

Pinus roxburghii Sarg. (common name Chir Pine) is also a pharmaceutically important plant found in the Himalayan region in Kashmir, Tibet, Bhutan, Sikkim, Nepal, and North India. Extract of *P. roxburghii* Sarg. shows high analgesic activities and is reported to inhibit carrageenan-induced edema in mice by inhibiting cyclooxygenase synthesis. Thus it can act as a nonsteroidal anti-inflammatory drug like indomethacin because polyphenolic compounds, bioflavonoids, quercetin, and rutin are present in the extract [71] (Fig. 1.16). Various chemical constituents that are present in turpentine oil of *P. roxburghii* Sarg. (Fig. 1.17) include α-pinene, β-pinene, car-3-ene, turpine longifolene hydrocarbons (d- and l-pinene), resin acids, camphene, fenchene, dipentene, and polymeric terpenes [72].

The maximum number and length of shoots was shown by the medium supplemented with 10 μM BAP at 5.8 pH. BAP (10 μM) showed the best axillary bud induction and proliferation as compared to Kn and Bisphenol A (BPA) with the addition of 0.5 and 0.25 μM α-NAA, respectively. Best rooting hormone for this plant was α-NAA at 2.5 μM [73].

Fig. 1.16 Medicinal properties of *Pinus roxburghii* Sarg

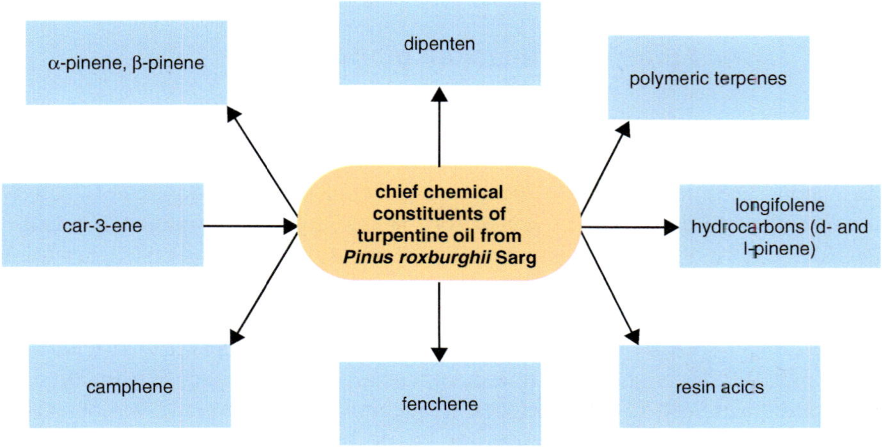

Fig. 1.17 Showing chief chemical components *of Pinus roxburghii* Sarg

1.4.3 *Tissue Culture of* Ziziphora tenuior

Ziziphora tenuior L. belongs to family Lamiaceae is an annual scented herb whose extract shows antifungal and antibacterial properties. Pulegone is the main component of its volatile oil. In addition, *Z. tenuior* has been used to cure fever, dysentery, diarrhea, gut inflammation and cough [74–77]. For micropropagation of *Z. tenuior*, the best concentration for callus initiation from leaf explants was 0.5 mg/L NAA. The maximum shoot number was obtained at 2 mg/L BAP. Moreover, 1 mg/L Kin and 0.1 mg/L α-NAA provided the best shoot multiplication and length [78].

1.4.4 *Micropropagation of* Ajuga bracteosa

It is a healing herb for the treatment of gout, rheumatism, palsy, and amenorrhoea. *A. bracteosa* is used to treat headaches, pimples, measles, stomach acidity, burns, boils, jaundice, hypertension, cough and sore throat by the local population [79]. Researchers reported anticancerous and anti-inflammatory properties of this herb. The leaves of *A. bracteosa* are reported to have anti-malarial properties. Thus, can also be used as a substitute for quinine [80, 81]. Kuria and colleagues reported that two components of this herb (ajugarin-1 and ergosterol-5,8-endoperoxide) have antibiotic properties in contradiction of chloroquine-sensitive *Plasmodium falciparum* [80].

Fast callusing of *A. bracteosa* was obtained on MS medium supplemented with IAA (2 mg/L) and BA (5 mg/L) after 10 days of culture. Medium supplemented with IAA 2 mg/L and BA 5 mg/L encouraged the maximum number of shoots. Best well-differentiated roots were obtained on the medium having 0.5 mg/L IBA [20].

1.4.5 *Tissue Culture of* Pongamia pinnata

P. pinnata plant contains various derivatives of flavonoids such as flavans, flavones, chalcones, and other compounds, including steroids, terpenes, and fatty acids. This plant shows antimicrobial, antioxidant, anti-diabetic, and anti-inflammatory activities. Extracts of this plant are reported for low toxicity towards mammalian cells, therefore, can be used as a potential medicinal agent [82]. Tan and colleagues used Woody Plant Medium [83] as a basal medium solidified with 0.28% (w/v) Gelrite at 5.7 pH. Maximum shoot formation was shown in the medium having thidiazuron (TDZ). Longer shoots of *P. pinnata* were obtained when supplemented with 2IP, zeatin and BAP. The highest multiple shoot bud initiation was reached using Woody Plant Medium at 5 μM TDZ. The highest frequency of rooting in plants was observed with Woody Plant Medium encompassing 20 μM Indole-3-Butyric Acid (IBA) and 200 μM silver thiosulphate (STS) [84].

1.4.6 *Tissue Culture* Linum usitatissimum

Table 1.3 shows the therapeutic effects of *L. usitatissimum.*

The highest root induction from hypocotyl explants of *L. usitatissimum* is reported when incubated in MS medium covering a combination of 0.5 mg/L TDZ + 0.5 mg/L kinetin. TDZ (0.5 mg/L) alone is effective in producing maximum shoot induction. Success in shoot induction (69%) from nodal explants on medium contained 2 mg/L BAP along with 0.5 mg/L NAA is recorded [85]. Another study showed that TDZ was better for shoot regeneration than other plant growth regulators [86].

Stem explants of *L. usitatissimum* cultured in a medium holding 2.0 mg/L TDZ combined with 0.1 mg/L NAA resulted in considerable quantity of shoot regeneration [87]. Medium encompassing 2,4-D (2 mg/L) and BAP 1 mg/L exhibited a considerable shoot regeneration [88]. For rooting, IBA (0.1 mg/mL) with half strength MS medium showed the best results [89]. Anjum and colleagues observed that the callus cultures established from the leaf explant grown on TDZ (2.0 mg/L) showed the highest antioxidant activity [90]. A higher amount of pharmacologically active lignan secoisolariciresinol was observed in cultures grown in a medium containing TDZ + Kin at 0.5 mg/L for each [89].

1.4.7 *Micropropagation of Mountain Mulberry*

Mulberry is a valuable tree with many applications in the pharmaceutical, food, and construction industries [97]. Mulberry silkworms (*Bombyx mori*) feed on these plants. In traditional medicine, products of the mulberry plant are used to treat

Table 1.3 Therapeutic effects of *Linum usitatissimum*

Class of compound	Example	Therapeutic effects	References
Lignans	Secoisolariciresinol Diglucoside (SDG) Matairesinol (MAT)	Treatement of breast, colon, and prostate cancer	[91, 92]
	Secoisolariciresinol (SECO)		
	Lariciresinol diglucoside (LDG)		
Other activity	Phytoestrogens	Regulate estrogen level in the human body	[93, 89]
Polyunsaturated fatty acids	Omega-6-fatty acid Omega-3-fatty acid linoleic acid	Treatment of eart diseases, stroke, high blood pressure, type-2 diabetes, Alzheimer's disease, and rheumatoid arthritis	[94]
Water and Fat-soluble vitamins	Vitamin E	Treatment of cardiovascular diseases, Alzheimer's disease	[95, 96]

ailments and diseases like throat inflammations, dysentery, helminthiasis, constipation, and diabetes [97, 98].

Dubey and colleagues found that the rate of bud break and longest shoot development was maximum in MS medium having BAP (1.0 mg/L) and NAA (0.5 mg/L). Maximum shoot multiplication was reported in the medium supplemented with BAP (1.0 mg/L), NAA (0.5 mg/L) and 10% coconut water. For root formation, better results were obtained when 20% activated charcoal was added to half-strength MS medium with the highest mean number of roots [99].

1.4.8 Micropropagation of **Hoslundia opposita** *Vahl*

This plant is effective against parasites like ticks, a common problem in Ghana [100]. The active component extracted from the leaves of this plant is ursolic acid, a triterpene. The crude methanolic extract exhibits high anti-larval action, followed by the ethyl acetate segment. For *Hoslundia oppsita*, best callus induction was observed in MS medium containing 30 g/L sucrose, 4.4 μM BA solidified with 0.2% gelrite [101].

1.4.9 Micropropagation of **Aloe** *species*

Aloe peglerae, a endangered *Aloe* specie [102] is used in the treatment of infections, blisters, wound curing and laxative [103, 104]. Figure 1.18 shows different uses of *Aloe* sp. [105–108].

For *Aloe vera,* the highest shooting formation was observed with 0.5 mg/L BAP along with 0.5 mg/L NAA [109]. To obtain more shoots, a mixture of 4.0 mg/L BAP and 0.2 mg/L NAA showed the best results in *A. vera*. Different genotypes respond differently to *in vitro* culture [110, 111]. For *Aloe trichosantha*, best callus induction was reported when 0.5 mg/L BAP and 0.5 mg/L NAA were used. For *A. percrassa* Todaro, 0.20 mg/L BAP and 0.20 mg/L NAA were effective [112]. Other researchers found 0.10 BAP and 0.50 mg/L NAA for the best callus induction in *A. vera* explants [113, 114]. Maximum shoot numbers were observed in medium combined with 2.0 mg/L of BAP. Rooting with the better number was only observed in MS medium at 0.25 mg/L to 1.5 mg/L NAA (Hailu et al. 2020). For *Aloe peglerae,* the highest number of shoots were observed in MS medium with 2.5 μM *m*TR. Hlatshwayo and colleagues ranked the effectiveness of cytokinins in shoot proliferation as follows *m*TR > *m*T > Kinetin > BA [117]. For *Aloe species*, the best shooting response was reported in the medium enriched with 1.0 mg/L BAP + 0.5 mg/L NAA. Enhanced rooting was observed with IBA at greater concentrations (1.0 and 1.5 mg/L), but similar results were also obtained with lesser concentrations (0.5 and 1.0 mg/L) of NAA [118].

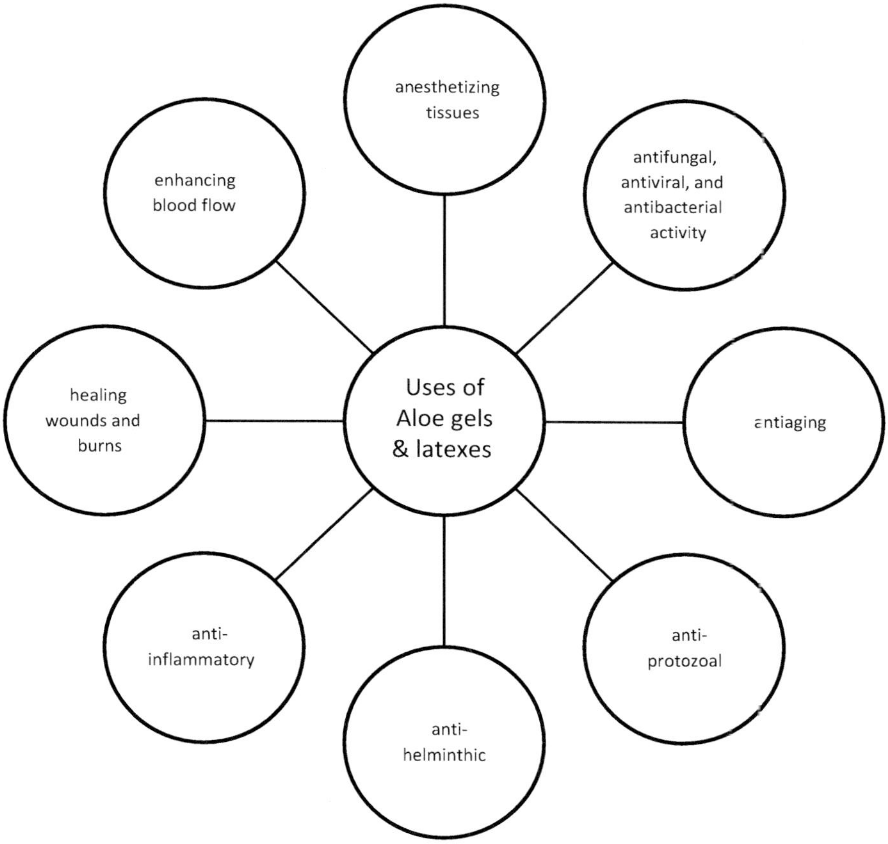

Fig. 1.18 Therapeutic properties of *Aloe vera*

1.5 Conclusion

The demand for therapeutic compounds derived from plants has increased with an escalating human population. This demand cannot be fulfilled by growing whole plants owing to the challenges associated with the growth, space, and time requirements. Micropropagation is a reasonable solution to this problem. Micropropagation techniques not only help to overcome space and time issues related to plant growth but also offer bulk production of required metabolites quickly.

References

1. Sidhu, Y. (2011). In vitro micropropagation of medicinal plants by tissue culture. *The Plymouth Student Scientist, 4*(1), 432–449.
2. Sofowora, A., Ogunbodede, E., & Onayade, A. (2013). The role and place of medicinal plants in the strategies for disease prevention. *African Journal of Traditional, Complementary, and Alternative Medicines, 10*(5), 210–229.
3. Thorpe, T. A. (2007). History of plant tissue culture. *Molecular Biotechnology, 37*(2), 169–180.
4. Hussain, A., Qarshi, I. A., Nazir, H., & Ullah, I. (2012). Plant tissue culture: Current status and opportunities. In *Recent advances in plant in vitro culture*. IntechOpen. https://doi.org/10.5772/50568
5. Moraes, R. M., Cerdeira, A. L., & Lourenço, M. V. (2021). Using micropropagation to develop medicinal plants into crops. *Molecules, 26*(6), 1752.
6. George, E. F., Hall, M. A., & Klerk, G.-J. D. (2008). The components of plant tissue culture media I: Macro- and micro-nutrients. In E. F. George, M. A. Hall, & G.-J. D. Klerk (Eds.), *Plant propagation by tissue culture* (Vol. 1, pp. 65–113). Springer.
7. Maucieri, C., Nicoletto, C., van Os, E., Anseeuw, D., Havermaet, R. V., & Junge, R. (2019). Hydroponic technologies. In S. Goddek, A. Joyce, B. Kotzen, & G. M. Burnell (Eds.), *Aquaponics food production systems: Combined aquaculture and hydroponic production technologies for the future* (pp. 77–110). Springer.
8. Pan, M. J., & van Staden, J. (1998). The use of charcoal in in vitro culture – A review. *Plant Growth Regulation, 26*(3), 155–163.
9. Gaspar, T., Kevers, C., Penel, C., Greppin, H., Reid, D. M., & Thorpe, T. A. (1996). Plant hormones and plant growth regulators in plant tissue culture. *In Vitro Cellular & Developmental Biology – Plant, 32*(4), 272–289.
10. Murashige, T., & Skoog, F. (1962). A revised medium for rapid growth and bio assays with tobacco tissue cultures. *Physiologia Plantarum, 15*(3), 473–497.
11. Saad, A. I. M., & Elshahed, A. M. (2012). Plant tissue culture media. In *Recent advances in plant in vitro culture*. IntechOpen. https://doi.org/10.5772/50569
12. Ogita, S. (2015). Plant cell, tissue and organ culture: The most flexible foundations for plant metabolic engineering applications. *Natural Product Communications, 10*(5).
13. Gupta, N., Jain, V., Joseph, M. R., & Devi, S. (2020). A review on micropropagation culture method. *Asian Journal of Pharmaceutical Research and Development, 8*(1), 86–93.
14. Davies, K. M., & Deroles, S. C. (2014). Prospects for the use of plant cell cultures in food biotechnology. *Current Opinion in Biotechnology, 26*, 133–140.
15. Grunennvaldt, R. L., Degenhardt-Goldbach, J., Brooks, P., Tomasi, J., Cá, D., Hansel, F. A., Tran, T., Gomes, E. N., & Deschamps, C. (2020). Callus culture as a new approach for the production of high added value compounds in Ilex paraguariensis: Genotype influence, medium optimization and compounds identification. *Anais Da Academia Brasileira De Ciencias, 92*(3), e20181251.
16. Espinosa-Leal, C. A., Puente-Garza, C. A., & García-Lara, S. (2018). In vitro plant tissue culture: Means for production of biological active compounds. *Planta, 248*(1), 1–18.
17. Roy, S. C., & Sarkar, A. (1991). In vitro regeneration and micropropagation of Aloe vera L. *Scientia Horticulturae, 47*(1), 107–113.
18. Benderradji, L., Brini, F., Kellou, K., Ykhlef, N., Djekoun, A., Masmoudi, K., & Bouzerzour, H. (2011). Callus induction, proliferation, and plantlets regeneration of two bread wheat (Triticum aestivum L.) genotypes under saline and heat stress conditions. *ISRN Agronomy*, e367851.
19. Molsaghi, M., Moieni, A., & Kahrizi, D. (2014). Efficient protocol for rapid Aloe vera micropropagation. *Pharmaceutical Biology, 52*(6), 735–739.
20. Kaul, S., Das, S., & Srivastava, P. S. (2013). Micropropagation of Ajuga bracteosa, a medicinal herb. *Physiology and Molecular Biology of Plants, 19*(2), 289–296.

21. Ozgen, M., Turet, M., Altinok, S., & Sancak, C. (1998). Efficient callus induction and plant regeneration from mature embryo culture of winter wheat (Triticum aestivum L.) genotypes. *Plant Cell Reports, 18*, 3–4.
22. Ahmad, A., Zhong, H., Wang, W., & Sticklen, M. B. (2002). Shoot apical meristem: In vitro regeneration and morphogenesis in wheat (Triticum aestivum L.). *In Vitro Cellular & Developmental Biology – Plant, 38*(2), 163–167.
23. Benkirane, H., Sabounji, K., Chlyah, A., & Chlyah, H. (2000). Somatic embryogenesis and plant regeneration from fragments of immature inflorescences and coleoptiles of durum wheat. *Plant Cell, Tissue and Organ Culture, 61*(2), 107–113.
24. Armstrong, T. A., Metz, S. G., & Mascia, P. N. (1987). Two regeneration systems for the production of haploid plants from wheat anther culture. *Plant Science, 51*(2), 231–237.
25. Verpoorte, R., van der Heijden, R., Schripsema, J., Hoge, J. H. C., & Ten Hoopen, H. J. G. (1993). Plant cell biotechnology for the production of alkaloids: Present status and prospects. *Journal of Natural Products, 56*(2), 186–207.
26. El Meskaoui, A. (2013). Plant cell tissue and organ culture biotechnology and its application in medicinal and aromatic plants. *Medicinal & Aromatic Plants*. https://doi.org/10.4172/2167-0412.1000e147
27. Lee, E. (1974). Tissue and organ culture of eucalyptus. *New Zealand Journal of Forestry Science, 4*(2), 267–278.
28. Muir, W. H., Hildebrandt, A. C., & Riker, A. J. (1958). The preparation, isolation, and growth in culture of single cells from higher plants. *American Journal of Botany, 45*(8), 589–597.
29. Dong, J., Bowra, S., & Vincze, E. (2010). The development and evaluation of single cell suspension from wheat and barley as a model system; a first step towards functional genomics application. *BMC Plant Biology, 10*(1), 239.
30. Sheen, J. (2001). Signal transduction in maize and Arabidopsis mesophyll protoplasts. *Plant Physiology, 127*(4), 1466–1475.
31. Rao, S. R., & Ravishankar, G. A. (2002). Plant cell cultures: Chemical factories of secondary metabolites. *Biotechnology Advances, 20*(2), 101–153.
32. Menges, M., Hennig, L., Gruissem, W., & Murray, J. A. H. (2003). Genome-wide gene expression in an Arabidopsis cell suspension. *Plant Molecular Biology, 53*(4), 423–442.
33. Xu, J., Ge, X., & Dolan, M. C. (2011). Towards high-yield production of pharmaceutical proteins with plant cell suspension cultures. *Biotechnology Advances, 29*(3), 278–299.
34. Hall, R. D. (1997). The initiation and maintenance of plant cell suspension cultures. In K. Lindsey (Ed.), *Plant tissue culture manual: Supplement 7* (pp. 45–65). Springer.
35. Moscatiello, R., Baldan, B., & Navazio, L. (2013). Plant cell suspension cultures. *Methods in Molecular Biology (Clifton, N.J.), 953*, 77–93.
36. Fathi, H., & Jahani, U. (2012). Review of embryo culture in fruit trees. *Scholars Research Library Annals of Biological Research, 3*(9), 4276–4281.
37. West, M. A. L., & Harada, J. J. (1993). Embryogenesis in higher plants: An overview. *The Plant Cell, 5*(10), 1361–1369.
38. Schwander, T., & Oldroyd, B. P. (2016). Androgenesis: Where males hijack eggs to clone themselves. *Philosophical Transactions of the Royal Society B: Biological Sciences, 371*(1706), 20150534.
39. Pockovska, M., Trajkova, F., & Koleva Gudeva, L. (2018). *Current application of anther culture as a tool for improvement of horticultural crops* (p. 20). Goce Delcev University.
40. Wang, M., van Bergen, S., & Van Duijn, B. (2000). Insights into a key developmental switch and its importance for efficient plant breeding. *Plant Physiology, 124*(2), 523–530.
41. Aoyagi, H. (2011). Application of plant protoplasts for the production of useful metabolites. *Biochemical Engineering Journal, 56*(1), 1–8.
42. Duquenne, B., Eeckhaut, T., Werbrouck, S., & Huylenbroeck, J. (2007). Effect of enzyme concentrations on protoplast isolation and protoplast culture of Spathiphyllum and Anthurium. *Plant Cell Tissue and Organ Culture, 91*, 165–173.

43. Ali, A., Sajid, A., Naveed, N. H., Majid, A., Saleem, A., Khan, U. A., Jafery, F. I., & Naz, S. (2011). Initiation, proliferation and development of micro-propagation system for mass scale production of banana through meristem culture. *African Journal of Biotechnology, 10*(70), 15731–15738.
44. Mori, K. (1971). Production of virus-free plants by means of meristem culture. *Japan Agricultural Research, 6*, 1–7.
45. Matsushima, T., Kikuchi, S., Takaiwa, F., & Oono, K. (1988). Regeneration of plants by pollen culture in rice (Oryza sativa L.). *Plant Tissue Culture, 5*, 78–81.
46. Vasenwala, S. M., Seth, R., Haider, N., Islam, N., Khan, T., Maheshwari, V., & Ur Rehman, S. (2012). A study on antioxidant and apoptotic effect of Azadirachta Indica (neem) in cases of cervical cancer. *Archives of Gynecology and Obstetrics, 286*(5), 1255–1259.
47. Moga, M. A., Bălan, A., Anastasiu, C. V., Dimienescu, O. G., Neculoiu, C. D., & Gavriş, C. (2018). An overview on the anticancer activity of Azadirachta indica (neem) in Gynecological cancers. *International Journal of Molecular Sciences, 19*(12), 3898.
48. Alzohairy, M. A. (2016). Therapeutics role of Azadirachta indica (neem) and their active constituents in diseases prevention and treatment. *Evidence-Based Complementary and Alternative Medicine*, 7382506.
49. Siddiqui, B. S., Afshan, F., Gulzar, T., Sultana, R., Naqvi, S. N.-H., & Tariq, R. M. (2003). Tetracyclic triterpenoids from the leaves of Azadirachta indica and their insecticidal activities. *Chemical & Pharmaceutical Bulletin, 51*(4), 415–417.
50. Kamath, S. G., Chen, N., Xiong, Y., Wenham, R., Apte, S., Humphrey, M., Cragun, J., & Lancaster, J. M. (2009). Gedunin, a novel natural substance, inhibits ovarian cancer cell proliferation. *International Journal of Gynecological Cancer: Official Journal of the International Gynecological Cancer Society, 19*(9), 1564–1569.
51. Patwardhan, C. A., Fauq, A., Peterson, L. B., Miller, C., Blagg, B. S. J., & Chadli, A. (2013). Gedunin inactivates the co-chaperone p23 protein causing cancer cell death by apoptosis. *The Journal of Biological Chemistry, 288*(10), 7313–7325.
52. Tharmarajah, L., Samarakoon, S. R., Ediriweera, M. K., Piyathilaka, P., Tennekoon, K. H., Senathilake, K. S., Rajagopalan, U., Galhena, P. B., & Thabrew, I. (2017). In vitro anticancer effect of gedunin on human teratocarcinomal (NTERA-2) cancer stem-like cells. *BioMed Research International*, 2413197.
53. Subapriya, R., Kumaraguruparan, R., Abraham, S. K., & Nagini, S. (2004). Protective effects of ethanolic neem leaf extract on N-methyl-N'-nitro-N-nitrosoguanidine-induced genotoxicity and oxidative stress in mice. *Drug and Chemical Toxicology, 27*(1), 15–26.
54. Othman, F., Motalleb, G., Lam Tsuey Peng, S., Rahmat, A., Basri, R., & Pei Pei, C. (2012). Effect of neem leaf extract (Azadirachta indica) on c-Myc oncogene expression in 4T1 breast cancer cells of BALB/c mice. *Cell Journal, 14*(1), 53–60.
55. Elumalai, P., Gunadharini, D. N., Senthilkumar, K., Banudevi, S., Arunkumar, R., Benson, C. S., Sharmila, G., & Arunakaran, J. (2012). Induction of apoptosis in human breast cancer cells by nimbolide through extrinsic and intrinsic pathway. *Toxicology Letters, 215*(2), 131–142.
56. Arumugam, A., Agullo, P., Boopalan, T., Nandy, S., Lopez, R., Gutierrez, C., Narayan, M., & Rajkumar, L. (2014). Neem leaf extract inhibits mammary carcinogenesis by altering cell proliferation, apoptosis, and angiogenesis. *Cancer Biology & Therapy, 15*(1), 26–34.
57. Houllou, L. M., de Souza, R. A., dos Santos, E. C. P., da Silva, J. J. P., Barbosa, M. R., Sauvé, J. P. G., & Harand, W. (2015). Clonal propagation of neem (Azadirachta indica A. Juss.) via direct and indirect in vitro regeneration. *Revista Árvore, 39*, 439–445.
58. Chaturvedi, R., Razdan, M. K. & Bhojwani, S. S. (2003). Production of haploids of neem (Azadirachta indica A. Juss.) by anther culture. *Plant Cell Reports, 21*, 531–537.
59. Quraishi, A., Koche, V., Sharma, P., & Mishra, S. K. (2004). In vitro clonal propagation of neem (Azadirachta indica). *Plant Cell, Tissue and Organ Culture, 78*(3), 281–284.
60. Rafiq, M., & Dahot, M. U. (2010). Callus and azadirachtin related limonoids production through in vitro culture of neem (Azadirachta indica A. Juss). *African Journal of Biotechnology, 9*(4), 4.

61. Mulla, M. S., & Su, T. (1999). Activity and biological effects of neem products against arthropods of medical and veterinary importance. *Journal of the American Mosquito Control Association, 15*(2), 133–152.
62. Sarmiento, W. C., Maramba, C. C., & Gonzales, M. L. M. (2011). An in-vitro study on the antibacterial effect of neem (azadirachta indica) leaf extract on methicillin-sensitive and methicillin-resistant staphylococcus aureus. *PIDSP Journal, 12*, 6.
63. Kashif, M., Hwang, Y., Hong, G., & Kim, G. (2017). In vitro comparative cytotoxic effect of Nimbolide: A limonoid from Azadirachta indica (neem tree) on cancer cell lines and normal cell lines through MTT assay. *Pakistan Journal of Pharmaceutical Sciences, 30*(3(Suppl.)), 967–973.
64. Harish Kumar, G., Chandra Mohan, K. V. P., Jagannadha Rao, A., & Nagini, S. (2009). Nimbolide a limonoid from Azadirachta indica inhibits proliferation and induces apoptosis of human choriocarcinoma (BeWo) cells. *Investigational New Drugs, 27*(3), 246–252.
65. Brandt, G. E. L., Schmidt, M. D., Prisinzano, T. E., & Blagg, B. S. J. (2008). Gedunin: A novel hsp90 inhibitor: Semisynthesis of derivatives and preliminary structure-activity relationships. *Journal of Medicinal Chemistry, 51*(20), 6495–6502.
66. Priyadarsini, R. V., Murugan, R. S., Sripriya, P., Karunagaran, D., & Nagini, S. (2010). The neem limonoids azadirachtin and nimbolide induce cell cycle arrest and mitochondria-mediated apoptosis in human cervical cancer (HeLa) cells. *Free Radical Research, 44*(6), 624–634.
67. Sharma, C., Vas, A. J., Goala, P., Gheewala, T. M., Rizvi, T. A., & Hussain, A. (2014). Ethanolic neem (Azadirachta indica) leaf extract prevents growth of MCF-7 and HeLa cells and potentiates the therapeutic index of cisplatin. *Journal of Oncology*, 321754.
68. Uddin, S. J., Nahar, L., Shilpi, J. A., Shoeb, M., Borkowski, T., Gibbons, S., Middleton, M., Byres, M., & Sarker, S. D. (2007). Gedunin, a limonoid from Xylocarpus granatum, inhibits the growth of CaCo-2 colon cancer cell line in vitro. *Phytotherapy Research, 21*(8), 757–761.
69. Vidya Priyadarsini, R., Senthil Murugan, R., Maitreyi, S., Ramalingam, K., Karunagaran, D., & Nagini, S. (2010). The flavonoid quercetin induces cell cycle arrest and mitochondria-mediated apoptosis in human cervical cancer (HeLa) cells through p53 induction and NF-κB inhibition. *European Journal of Pharmacology, 649*(1–3), 84–91.
70. Zhang, W., & Zhang, F. (2009). Effects of quercetin on proliferation, apoptosis, adhesion and migration, and invasion of HeLa cells. *European Journal of Gynaecological Oncology, 30*(1), 60–64.
71. Kaushik, D., Kumar, A., Kaushik, P., & Rana, A. C. (2012). Analgesic and anti-inflammatory activity of Pinus roxburghii Sarg. *Advances in Pharmacological Sciences*, 245431.
72. Smaleh, M., Sharma, O. P., & Dobhal, N. P. (1976). Chemical composition of turpentine oil from pleoresin (Pinus roxburghii Sargent) Indian oerfumer. *Chemistry of Forest Products Branch, 20*, 15–19.
73. Arya, A., Kumar, S., & Kasana, M. S. (2012). Effect of plant growth regulators and pH of medium on in vitro regeneration of Pinus roxburghii. *Sarg, 2*, 10.
74. Naeini, A., Khosravi, A., Tadjbakhsh, H., Ghazanfari, T., Yaraee, R., & Shokri, H. (2009). Evaluation of the immunostimulatory activity of Ziziphora tenuior extracts. *Comparative Clinical Pathology, 19*, 459–463.
75. Pirbalouti, A. G., Malekpoor, F., & Hamedi, B. (2012). Ethnobotany and antimicrobial activity of medicinal plants of Bakhtiari Zagross mountains, Iran. *Journal of Medicinal Plants Research, 6*(5), 675–679.
76. Mahboubi, M., Bokaee, S., Dehdashti, H., & Mehdi, F. (2012). Antimicrobial activity of Mentha piperitae, Zhumeria majdae, Ziziphora tenuior oils on ESBLs producing isolates of Klebsiella pneumoniae. *Biharean Biologist, 6*, 5–9.
77. Safa, O., Soltanipoor, M. A., Rastegar, S., Kazemi, M., Nourbakhsh Dehkordi, K., & Ghannadi, A. (2013). An ethnobotanical survey on Hormozgan province, Iran. *Avicenna Journal of Phytomedicine, 3*(1), 64–81.

78. Dakah, A., Zaid, S., Suleiman, M., Abbas, S., & Wink, M. (2014). In vitro propagation of the medicinal plant Ziziphora tenuior L. and evaluation of its antioxidant activity. *Saudi Journal of Biological Sciences, 21*(4), 317–323.
79. Hamayun, M., Afzal, S., & Khan, M. A. (2006). Ethnopharmacology, indigenous collection and preservation techniques of some frequently used medicinal plants of Utror and Gabral, district Swat, Pakistan. *African Journal of Traditional, Complementary and Alternative Medicines, 3*(2), 2.
80. Kuria, K. A., De Coster, S., Muriuki, G., Masengo, W., Kibwage, I., Hoogmartens, J., & Laekeman, G. M. (2001). Antimalarial activity of Ajuga remota Benth (Labiatae) and Caesalpinia volkensii Harms (Caesalpiniaceae): In vitro confirmation of ethnopharmacological use. *Journal of Ethnopharmacology, 74*(2), 141–148.
81. Njoroge, G. N., & Bussmann, R. W. (2006). Diversity and utilization of antimalarial ethnophytotherapeutic remedies among the Kikuyus (Central Kenya). *Journal of Ethnobiology and Ethnomedicine, 2*(1), 8.
82. Al Muqarrabun, L. M. R., Ahmat, N., Ruzaina, S., Ismail, N. H., & Sahidin, I. (2013). Medicinal uses, phytochemistry and pharmacology of Pongamia pinnata (L.) Pierre: A review. *Journal of Ethnopharmacology, 150*(2), 395–420.
83. Lloyd, G., & McCown, B. (1980). Commercially-feasible micropropagation of mountain laurel, Kalmia latifolia, by use of shoot-tip culture. *Commercially-Feasible Micropropagation of Mountain Laurel, Kalmia Latifolia, by Use of Shoot-Tip Culture, 30*, 421–427.
84. Tan, S. N., Tee, C. S., & Wong, H. L. (2018). Multiple shoot bud induction and plant regeneration studies of Pongamia pinnata. *Plant Biotechnology (Tokyo, Japan), 35*(4), 325–334.
85. Khan, I., Khan, M. A., Shehzad, M. A., Ali, A., Mohammad, S., Ali, H., Alyemeni, M. N., & Ahmad, P. (2020). Micropropagation and Production of Health Promoting Lignans in Linum usitatissium. *Plants, 9(3)*, 495–499.
86. Janowicz, J., Niemann, J., & Wojciechowski, A. (2012). The effect of growth regulators on the regeneration ability of flax (Linum usitatissimum L.) hypocotyl explants in in vitro culture. *Biotechnologia, 93*(2), 135–138.
87. Burbulis, N., & Blinstrubien, A. (2009). Regeneration of adventitious shoots of linseed (Linum usitatissimum L.) from hypocotyl explants. *Zemdirbyste-Agriculture, 96*(3), 168–175.
88. Chen, J., Wang, L., & Thompson, L. U. (2006). Flaxseed and its components reduce metastasis after surgical excision of solid human breast tumor in nude mice. *Cancer Letters, 234*(2), 168–175.
89. Khan, I., Khan, M. A., Shehzad, M. A., Ali, A., Mohammad, S., Ali, H., Alyemeni, M. N., & Ahmad, P. (2020). Micropropagation and production of health promoting lignans in Linum usitatissimum. *Plants, 9*(6), 728.
90. Anjum, S., Abbasi, B. H., & Hano, C. (2016). Trends in accumulation of pharmacologically important antioxidant-secondary metabolites in callus cultures of Linum usitatissimum L. *Plant Cell, Tissue and Organ Culture (PCTOC), 1*(129), 73–87.
91. Hano, C., Addi, M., Bensaddek, L., Crônier, D., Baltora-Rosset, S., Doussot, J., Maury, S., Mesnard, F., Chabbert, B., Hawkins, S., Lainé, E., & Lamblin, F. (2006). Differential accumulation of monolignol-derived compounds in elicited flax (Linum usitatissimum) cell suspension cultures. *Planta, 223*(5), 975–989.
92. Szewczyk, M., Abarzua, S., Andr, S., Nebe, B., Piechulla, B., Volker, B., & Dagmar-Ulrike, R. (2014). Effects of extracts from Linum usitatissimum on cell vitality, proliferation and cytotoxicity in human breast cancer cell lines. *Journal of Medicinal Plants Research, 8*, 237–245.
93. Desmawati, D. & Sulastri, D. (2019). Phytoestrogens and Their health effect. *Open Access Macedonian Journal of Medical Sciences, 7(3)*, 237–245.
94. Katare, C., Saxena, S., Agrawal, S., & Prasad, G. (2012). Flax seed: A potential medicinal food. *Journal of Nutrition & Food Sciences, 02*, 01.
95. Sen, C. K., Khanna, S., & Roy, S. (2006). Tocotrienols: Vitamin E beyond tocopherols. *Life Sciences, 78*(18), 2088–2098.

96. Morris, M. C., Evans, D. A., Tangney, C. C., Bienias, J. L., Wilson, R. S., Aggarwal, N. T., & Scherr, P. A. (2005). Relation of the tocopherol forms to incident Alzheimer disease and to cognitive change. *The American Journal of Clinical Nutrition, 81*(2), 508–514.
97. Lochynska, M. (2015). Energy and nutritional properties of the white mulberry (Morus alba L.). *Journal of Agricultural Science and Technology A, 5*.
98. Vijayan, K. (2014). Biotechnology of mulberry (Morus L.) – A review. *Emeritus Journal of Food and Agriculture, 26*, 472–496.
99. Dubey, V., Khan, D. S., Shah, D. K. W., & Raghuwanshi, D. R. K. (2020). Standardization of protocol for in vitro micropropagation of Morus alba L. *An Important Economical and Medicinal Plant. Pharmaceutical and Biosciences Journal*, 46–51.
100. Annan, K., Jackson, N., Dickson, R. A., Sam, G. H., & Komlaga, G. (2011). Acaricidal effect of an isolate from Hoslundia opposita vahl against Amblyomma variegatum (Acari: Ixodidae). *Pharmacognosy Research, 3*(3), 185–188.
101. Prakash, S., & Staden, J. (2007). Micropropagation of Hoslundia opposita Vahl – A valuable medicinal plant. *South African Journal of Botany, 73*, 60–63.
102. Pfab, M. F., & Victor, J. E. (2002). Threatened plants of Gauteng, South Africa. *South African Journal of Botany, 68*(3), 370–375.
103. George, J., Laing, M. D., & Drewes, S. E. (2001). Phytochemical research in South Africa: Review article. *South African Journal of Science, 97*(3), 93–105.
104. Cock, I. E. (2015). The genus aloe: Phytochemistry and therapeutic uses including treatments for gastrointestinal conditions and chronic inflammation. Progress in Drug Research. Fortschritte Der Arzneimittelforschung. *Progres Des Recherches Pharmaceutiques, 70*, 179–235.
105. Surjushe, A., Vasani, R., & Saple, D. G. (2008). Aloe vera: a short review. *Indian Journal of Dermatology, 53*(4), 163–166.
106. Yao, H., Chen, Y., Li, S., Huang, L., Chen, W., & Lin, X. (2009). Promotion proliferation effect of a polysaccharide from Aloe barbadensis Miller on human fibroblasts in vitro. *International Journal of Biological Macromolecules, 45*(2), 152–156.
107. Grace, O. M. (2011). Current perspectives on the economic botany of the genus Aloe L. (Xanthorrhoeaceae). *South African Journal of Botany, 77*(4), 980–987.
108. Dwivedi, N. K., Indiradevi, A., Asha, K. I., Nair, R. A., & Suma, A. (2014). A protocol for micropropagation of Aloe vera L. (Indian Aloe) a miracle plant. *Research in Biotechnology*. https://updatepublishing.com/journal/index.php/rib/article/view/2447
109. Zakia, S., Zahid, N. Y., Yaseen, M., Abbasi, N. A., Hafiz, A. A., & Mahmood, N. (2013). Standardization of micropropgation techniques for Aloe vera: A pharmaceutically important plant. *Pakistan Journal of Pharmaceutical Sciences, 26*(6), 1083–1087.
110. Nayanakantha, N. M. C., Singh, B. R., & Kumar, A. (2011). Improved culture medium for micropropagation of Aloe vera L. *Tropical Agricultural Research and Extension, 13*(4), 87–93.
111. Lobine, D., Soulange, J. G., Sanmukhiya, M. R., & Lavergne, C. (2015). A tissue culture strategy towards the rescue of endangered mascarene aloes. *ARPN Journal of Agricultural and Biological Science, 10*(1), 28.
112. Hailu, A., Sbhatu, D. B., & Abraha, H. B. (2020). In vitro micropropagation of industrially and medicinally useful plant aloe trichosantha Berger using offshoot cuttings. *The Scientific World Journal*, e3947162.
113. Gupta, S., Sahu, P. K., Sen, D. L., & Pandey, P. (2014). In-vitro propagation of Aloe vera (L.) Burm. F. *Biotechnology Journal International*, 806–816.
114. Khanam, N., & Sharma, G. K. (2014). Rapid in vitro propagation of Aloe vera L. with some growth regulators using lateral shoots as explants. *World Journal of Pharmacy and Pharmaceutical Sciences, 3*(3), 2005–2018.
115. Scambia, G., Panici, P. B., Ranelletti, F. O., Ferrandina, G., De Vincenzo, R., Piantelli, M., Masciullo, V., Bonanno, G., Isola, G., & Mancuso, S. (1994). Quercetin enhances transforming growth factor beta 1 secretion by human ovarian cancer cells. *International Journal of Cancer, 57*(2), 211–215.

116. Gao, X., Wang, B., Wei, X., Men, K., Zheng, F., Zhou, Y., Zheng, Y., Gou, M., Huang, M., Guo, G., Huang, N., Qian, Z., & Wei, Y. (2012). Anticancer effect and mechanism of polymer micelle-encapsulated quercetin on ovarian cancer. *Nanoscale, 4*(22), 7021–7030.
117. Hlatshwayo, N. A., Amoo, S. O., Olowoyo, J. O., & Doležal, K. (2020). Efficient micropropagation protocol for the conservation of the endangered Aloe peglerae, an ornamental and medicinal species. *Plants, 9*(4), 506.
118. Niguse, M., Sbhatu, D. B., & Abraha, H. B. (2020). In vitro micropropagation of aloe adigratana Reynolds using offshoot cuttings. *The Scientific World Journal*, e9645316.

Chapter 2
Mentha

Muhammad Akram, Muhammad Tayyab Akhtar, Fatima Akram, and Umar Farooq Gohar

2.1 Introduction

The family Lamiaceae is a large group of plants with annual and perennial basils, thymes and sages. These plants have been used by man for pharmaceutically important compounds for many years in life saving medicines and to create aroma and taste in food. Amongst the group, the genus *Mentha* also known as mint comprised of well-known plants for essential oil production in varying composition. Essential oils in pure forms have a worldwide demand in trade. They are cultivated for highly valuable monoterpene present in their essential oils [1]. The most cultivated forms of the genus are *Mentha arvensis*, *M. aquatica*, *M. saveolense*, *M. piperita* and *Oscimum bacilicum*, *O. sanctum*, etc. are the other species of this family.

The extract of mint leaves and inflorescence may be known as mint oil or essential oil. The term essential oil is derived from aroma chemicals and is used as a flavoring, fragrance and preservative agent in various food items obtained from large number of plant species [2]. The chemical composition of mint oil is determined through chromatographic techniques, where menthol is the chief component along with carvone, menthone, pulegone are the other components in varying composition. Peppermint has been reported to be in use in China and Japan at least 2000 years ago [3] and value of menthol is also very old (200 years ago) in Japan where they kept menthol with them in silver boxes for regular use hanging from their belts. Mint oil is used for indigestion, nausea, sore throat, diarrhea, cold, headache and used in toothpastes for cooling sensation and fresh breath, confectionaries, cosmetics and pesticides also prefer for their products synthesis. Menthol is widely used

M. Akram (✉)
Department of Biology, Government Shalimar Graduate College, University of the Punjab, Lahore, Pakistan

M. T. Akhtar · F. Akram · U. F. Gohar
Institute of Industrial Biotechnology, Government College University, Lahore, Pakistan

M. Zia-Ul-Haq et al. (eds.), *Essentials of Medicinal and Aromatic Crops*,
https://doi.org/10.1007/978-3-031-35403-8_2

for the treatment of burns, sunburns, poison ivy rash and athlete's foot [4]. Peppermint essential oil is used to kill pathogens of 21 human and plants [5]. Later on, Salehi et al. [6] described the antioxidant and antimicrobial activities of the genus *Mentha*. Zhao et al. [7] elaborated and explained the applications of biological activities of peppermint essential oil as an anti-inflammatory, antibacterial, antiviral, scolicidal and for many more problems.

Mints are generally cultivated in moist sub-tropical and tropical areas. The damp conditions are not tolerated by mints due to plant damage by root rot. It is generally propagated by vegetative methods by creeping stolons, suckers or shoot cuttings. Stolons are obtained from previous year planting. However, these vegetative methods of propagation require considerable time and therefore limit the speed of plant propagation. Moreover, the method of propagation by cutting is not a reliable method for production of good planting material because these are prone to attack by variety of pathogens that cause many fungal (rust, powdery mildew and stolon rot) and viral (spot wilt) diseases, which reduce the yield and also cause the gradual degeneration of cultivar. Quality of plant is very important to get high yielding plant material and for maximum oil extraction. In this scenario, plant tissue culture offers for quality oil biosynthesis in micropropagated plants, breeding and disease eradication has been recognized as potentially valuable tools in crop improvement programs. Since last decades this technique has gained momentum on commercial application in the field of plant propagation. Improvement in terpenoids biosynthesis and production especially menthol is delimited in in vitro differentiation of mint plants [1]. Once in vitro plants get maturation, the origin and biosynthesis of terpenoids started in leucoplasts of secretory cells and stored in specialized non-photosynthetic cells of glandular trichomes [8, 9].

Due to highly valuable monoterpenes, mint is subjected to steam distillation and solvent extraction. For callus and in vitro plants having little raw material, solvent extraction cum distillation method is used to minimize the evaporation of volatile substances in the oil. For whole plant material having hard texture steam distillation method is used or depending upon the nature of the plant material for extraction of essential oils. Mint oil has been obtained by steam distillation for the extraction of essential oil and the techniques Gas Chromatography Mass Spectrometry (GC-MS) [10] and High-Performance Liquid Chromatography (HPLC) [11] are used for the detection of constituents present in mint oil. Later on, super critical fluid extraction (SCFE) has been used along with previously described methods [12, 13].

Thus, in this chapter a comparative study of different mint species and the available propagation methods have been described. Moreover, biotechnological interventions have also been given to elucidate secondary metabolites enhancement, extraction and analysis of mint oil by different chromatographic techniques.

2.2 Habit and Habitat

Mint is a common herb that can easily be distinguished by its aromatic smell. It grows vertically as well as has creeping habit. The size of plants varies depending upon the environmental conditions of habitat. The trailing habit of mint saves itself from herbivory and other adverse environmental conditions. Commonly it is cultivated in small ponds, damp places, ditches, pots and on the agricultural land atmosphere. It grows under mesophytic and in a small drought condition can also withstand plenty of water and damp places [14].

2.3 Morphological Characters

The genus *Mentha* has variable forms on the basis of its species and environmental conditions (Table 2.1). The plant is erect and sometimes likes to run on the ground, tender rhizomatic growth. The size of the plant ranges from few inches to one foot in height. Stems are herbaceous, green and purple due to lack of acquisition of water or deficiency of phosphorus. The leaves of mint are simple, petiolate, hairy in few species, elliptic, dentate and serrate. Leaves have opposite phyllotaxy with dark green color and highly aromatic. The size of the mint leaves is different according to the species and variety [15].

2.4 Plant Propagation and Multiplication

Plant multiplication is an important step in plant germplasm conservation. Due to high economic value of *M. arvensis*, its multiplication has been carried by various methods.

2.4.1 Conventional Propagation

The propagation methods of plants adopted by localized people of the specific area may be termed as conventionally multiplication strategies.

Stem Cutting Stem cutting is used for direct initiation of adventitious roots [19] for clonal multiplication in a suitable potting media that may be a soil enriched with organic matter. For this purpose, mature stems with single, double or multiple leafy nodes may be used and lay down or stick erect on the medium. Plant growth regulators (PGRs), for example, auxins (IAA or IBA) are commonly applied at the cut side for root induction and elongation and for healthy growth [20]. Due to the flexible

Table 2.1 Morphological and reproductive characters of mints

Mint species	Common name	Stem	Leaves	Inflorescence	Reference
Mentha arvensis	Wild mint	Simple straight stems or sometimes may be branched, often purplish below	The leaves are ovate, the size of petiole is 10 mm and lamina is 12–20 mm, margins are serrate, acute	Verticillasters distantly present, group of flowers, 15 mm in diameter	[16]
M. spicata	Spear mint	300–600 mm long straight stems, bear leaves and sometime branched and glabrous	Green leaves are ovate lanceolate, 20 to 60 mm × 5 to 15 mm, serrate margins, glabrous	Petals are white and pink, verticillaster spike	[16]
M. piperita	Peppermint	Erect, branched, leafy stems 800 mm long, glabrous, sparse eglandular indumentum	Petiole are ranged from 10 to 15 mm, 30 to 80 mm long & width 15 to 40 mm, lamina is ovate	Rectangle spikes of 50 to 70 × 15 mm, verticillaster	[16]
M. longifolia	Silver mint	Branching habit having leaves, strait, vertical, length 300–1200 mm, profuse branched, tender hairs present	Different sized and color, profuse hairs, size 20–80 × 5–30 mm, ovate, lanceolate, coarsely dentate	Verticillasters, spikes terminal	[16]
M. royleana	Royle's mint	Just like above but slenderer in habit	Usually narrowly oblong-elliptic, frequently discolorous, clearly or shortly petiolate	Spikes slenderer with verticillasters generally separated and calyces usually 1.5–2 mm	[16]
M × rotundifolia	Apple mint	1–1.5 foot long generally short-hairy	Size 26–60 mm, 2 mm long petiole, ovate, indent, base pointed	Forms a spike at 3–5 nodes, delimiting bracts	[17]
M. aquatica	Water mint	Perennial 3 feet long	Ovate-lanceolate 20–60 mm long, 10–40 mm width, green opposite, toothed	Small, dense group, purple some time pink to lilac in color	[18]

nature of mint, it may also be grown in all types of soil provided with high moisture contents. Mint cannot grow under drought conditions.

Grafting Method This is useful for the acquisition of hybrid varieties. For example, seedling production *M. piperita* and *M. spicata* may be an alternative system for high yielding seedlings [21].

2.4.2 In Vitro Propagation

For disease free and true to type plants production, modern techniques are used to get desired characteristics and for enhanced metabolite production in *Mentha* species.

Callus Induction Callus is an amorphous irregular mass of parenchymatous tissue (Table 2.2) having meristematic regions [22]. Shoot regeneration from callus cultures is a useful method for healthy in vitro plant propagation, conservation and for isolation of secondary metabolites in medicinal plants [23–27]. Vegetative or somatic tissues like mature and immature leaf, stem or flower explants are generally used for culture on the basal growth medium supplemented with PGRs. Rigorous work has been carried out on in vitro callus induction from different explants of *Mentha* [28–35]. Akram et al. [29] obtained callus from leaf, internode and nodal explants on MS medium supplemented with 2,4-D + BAP of *M. arvensis* cv. Japanese mint. Similarly, callus from two cv of *M. haplocalyx* has been achieved by Yu et al. [30] from leaf explants.

Except traditional methods, cell culture technology is quite a novel way for direct plant formation via organogenesis to boost up the internal metabolism for enhanced metabolite production in mint.

Shoot Organogenesis Organogenesis is a process by which shoot, root or somatic embryos are formed under the influence of chemically controlled environment in tissue culture medium. The organogenic development from the callus tissues (Fig. 2.1, Table 2.3), a piece of any tissue is cultured on shoot regeneration medium with or without plant growth regulators to get such response. The later response of somatic embryogenesis leads to the phenomenon of tissue habituation [36]. This doesn't mean tissue lose insensivity to PGRs [37]. Somatic cells sensitive to PGRs may only be achieved by adjusting the level of both cytokinins and auxins in tissue culture medium which subsequently governs organogenesis [38–40]. Organogenesis is a chemically controlled phenomenon [41] conceptualized in tobacco organogenesis by adjusting the relative levels of kinetin and indole acetic acid and providence of organogenesis in culture is determined by two hormones is controlled by multistep process [22]. At first stage, cells acquire ability or competence to get hormonal signals. Secondly, the competent cells become able to develop the organ.

Table 2.2 Formation of callus from different explants of mint

Mint species	Optimum culture medium	Explant	Callus response	Purpose	Reference
Mentha spicata	0.2% (w/v) Casein hydrolysate + 2 ppm 2,4-D	Stem Leaf	+ +++	Determination of lipid constituents	[28]
M. spicata	MS+4 mg/L TDZ+25% coconut water	Leaf	Morphogenic	Shoot regeneration	[42]
M. arvensis Japanese mint	MS+13.5 μM 2,4-D+0.8 μM BAP	Leaf	+++	Secondary metabolites determination	[29]
M. haplocalyx	B_5 +0.5 mg/L BAP+1.5 mg/L NAA	Leaf	Morphogenic	Biomass production	[30]
M. arvensis	MS+1.5 mg/L 2,4-D	Leaf	93.8%	Bio-efficacy against bacteria	[31]
M. pulegium	MS+1 mg/L 2,4-D	Leaf	12.35 mm diameter	Production of chemical drugs	[32]
M. pulegium	MS+ 0.5 mg/L BA and 1 mg/L NAA	Hypocotyl	92%	Shoot regeneration	[33]
M. spicata	MS + 2 ng/μL BAP + 1 ng/μL NAA	Cotyledon node	Morphogenic	Shoot regeneration	[43]
M. piperita	MS+ 1.5 mg/L NAA+ 0.2 mg/L BAP	Leaf disc	Greenish, organogenic and nodular	Shoot regeneration	[34]
M ×piperita M. pulgium	MS+100 μL NAA + 600 μL IBA	Node	Morphogenic	Shoot regeneration	[44]
M. longifolia	Modified MS + BAP + 2,4-D	Young leaves	Friable	Secondary metabolite production	[35]

This tactic has extensively been practiced to get organogenesis in *M. arvensis* [45], *M. piperata* [34, 46], *M. spicata* [47] and large number of other plant species [48–50].

Micropropagation Axenic shoots either derived from in vitro cultures or surface sterilized shoot buds are usually used for culture to establish axillary shoots on the culture medium. Such phenomenon for clonal propagation is termed as micropropagation (Fig. 2.2, Table 2.4). This is very useful method to achieve true to type, disease free woody and non-woody herbaceous crops including medicinal and non-medicinal plants. Most of the plants of family lamiaceae are medicinal in nature which have been cultivated by various means to get essential oil production.

Fig. 2.1 Shoot organogenesis from callus tissues

Table 2.3 Different species of mint showing in vitro formation of shoots (organogenesis)

Mint species	Optimum culture medium	Explant	Culture's response	Reference
M. spicata	MS+4 mg/L TDZ+25% coconut water	Leaf	100%	[42]
M. arvensis	MS+5 mg/L BAP+0.5 mg/L NAA	Leaf disc	Direct shoot buds	[45]
M. arvensis	MS+8.9 μM BAP+4.4 μM IAA	Node	Direct multiple shoots	[51]
M. viridis	MS+3 mg/L BAP	Node	Multiple shoots Regeneration	[52]
M. piperita	½ MS+3 mg/L Zeatin	Internode	Multiple shoot Regeneration	[53]
M. piperita	MS+1.5 mg/L kinetin	Node	Direct organogenesis	[46]
M. piperita	MS, B_5 + 11.35 μM TDZ + Zeatin 4.54 μM + 10% coconut water + 20g sucrose	2nd internode	100% shoot regeneration	[54]
M. piperita	MS+2.0 mg/L BAP+0.5 mg/L NAA	Callus	Shoot organogenesis	[34]

The cultivation of *Mentha* through micropropagation is very old [55] using different explants, culture vessels and variable temperature conditions [56], manipulation of PGRs [57] such as BAP, NAA [43], IAA and TDZ [58] for enhanced metabolites production [59]. Ali et al. [60] used surface sterilized nodal explants of *M. arvensis* var. Explants of Japanese mint collected from open environment and cultured on MS agar-solidified medium supplemented with 1 μM BAP and 0.5 μM NAA for in vitro shoot proliferation, acclimatization and secondary metabolites production by GC-MS in the same species and cv. [29].

Raja & Arockiasamy, [52] cultured nodal explants on MS medium supplemented with BAP and Kinetin for micropropagation of *M. viridis*. Later on, Rahman et al. [61] also clonally multiplied the same species (*M. viridis*) on solid MS medium added with BA and IAA. Inspite of PGRs, endophytic fungi play significant role in

Fig. 2.2 Micropropagation of *M. arvensis* [29]

morphological and biochemical attributes and mutualistic effects on *M. viridis* [62]. Similarly, mycorrhizal fungi also play significant role in micropropagation to get long shoots and roots in *M. piperita* [63]. A multifactorial experimental design for micropropagation of *M. piperita* in which Vaidya et al. [64] demonstrated a semi-solid and liquid culture systems on MS or Chee and Pool (C_2D) medium supplemented with different cytokinins and obtained highest number of shoots with liquid medium for essential oil production.

Bird eye view the above-mentioned knowledge demonstrated micropropagation is an essential protocol for clonal multiplication for healthy plant production ultimately synthesis of metabolites production. Several species of *Mentha* have been researched but *M. piperita* has extensively been micropropagated for industrial crops to produce natural products.

Cell Suspension Culture Cell suspension culture (CSC) is used for isolation of individual cells in liquid medium or cells are suspended and cultured under the sterile in vitro conditions using divers type culture vessels (Fig. 2.3). This is an alternative method to produce menthol production [76]. For this purpose, friable calluses are usually used and transfer on to the liquid medium fortified with previously used PGRs or addition/substitution of new PGRs are tested and agitated on orbital shaker. The CSC of several medicinal plants has been reported for the production of secondary metabolites, somatic embryogenesis or regeneration purposes. Several successful stories are there to achieve desired components of *Mentha* species and in its cultivars using liquid medium supplemented with PGRs or on the simple basic media devoid of PGRs.

Table 2.4 Micropropagation from different explants of mint

Mint species	Optimum culture medium	Explant	Response/purpose	Reference
Mentha sp.	Various	Shoot tip, node	Multiple shoot formation	[55]
M. arvensis	MS + 40μM BAP + 0.5 μM NAA under red light	Node	200 multiple shoots	[65]
M. piperita	Microgravity + 1μM BAP	Node	Shoot curvature	[66]
M. arvensis	MS + 1 μM BAP + 0.5 μM NAA		35 multiple shoots	[60]
M. piperita	MS + 4.4 μM BAP + 2.32 μM Kin	Node	4.1 shoots	[67]
M. arvensis	MS liquid + 4.4 μM BAP + 3.48 μM kin	Node	38 multiple shoots	[29]
M. viridis	MS + 3mg/L BAP	Node	In vitro shoots	[52]
M. spicata	MS + 1 μM BAP + 0.5 μM NAA	Node	Multiple shoots	[68]
M. piperita	MS + 4.5 BAP mg/L + 0.009 mg/L IBA	Shoot tip	4-fold	[69]
M. gracilis Sole	MS + 2 μM TDZ	Node	Micro-shoots	[70]
M. arvensis	MS + 15μM TDZ	Pretreated nodal explants with TDZ	23.7 multiple shoots	[58]
M. pulegium	MS + 0.5 mg/L BAP	Shoot tip	100% with 14 shoots	[71]
M. piperita	MS+ 2 mg/L BAP + 0.5 mg/L NAA + 1 mg/L GA_3	Node from in vitro plants	Shoot multiplication	[72]
M. piperita	IAA producing rhizobacteria	In vitro plant	Improved growth parameters	[73]
M. piperita	MS + 3 mg/L BAP	Node	42 number of shoots	[74]
M. piperita	C_2D + 4 μM BAP	Shoot tips from field-grown plants	40.7 number of shoots	[64]
M. piperita	0.5 mg/L IBA	Shoot tip	Lateral shoots formation, rooting and odor-active compounds	[59]
M. piperita	MS + 1 mg/L BAP	Node	7.12 shoots	[75]

The precursor feeding in the CSC is favorable to get improved concentration of the desired compound. Addition of precursors in the form of chemical compounds participate in different reactions and enhance its production. Addition of menthone of 35 μM improves the yield of menthol in CSC of *M. Piperita* [76]. When they used

Fig. 2.3 Cell suspension culture derived from calluses

γ-cyclodextrin (60 μM) alone or in combination with 35 μM menthone yielded up to 92 and 110 mg/L menthol, respectively as compared to control (77 mg/L).

The elicitors have also been reported such as Jasmonic acid and methyl jasmonate have profound effect triggered Rosmarinic acid production in CSC of *M. piperita* [77]. They used 200 μM jasmonic acid in CSC allowed for 24 h produced highest rosmarinic acid (117.95 mg/g dry weight) as compared to control however, it decreased the total biomass. Due to chemical nature and variation in compound physical structure does not allow to isolate under the conditions of CSC. Such compounds may be extracted out with simple solvents system. Agar, liquid, glass gravel are the physical factors as well as replacement incidence of media, capacity of culture vessels greatly affected on the growth and carvone metabolite production in *M. spicata* [78]. Their results demonstrated that with low carvone treatment (mg carvone g/FW) produced higher vegetative growth however total carvone ((mg carvone g-FW−1) × g culture FW) increased due to greater vegetative biomass accumulation per vessel.

2.4.3 Essential Oil and Terpenoid Production

Mint cultivation is to get essential oil rich in terpenoids used in pharmaceutical and in confectionaries. A schematic representation of mint oil extraction and analysis has been described in Fig. 2.4. Mint oil rich in menthol is used for the treatment of indigestion, nausea, sore throat, diarrhea, colds, headache and in the treatment of burns, sunburns, poison ivy rash and athlete's foot [4, 79]. Plant cell culture consisted of callus and suspension culture is a competent method for production biologically important secondary metabolites that may provide continuous and reliable source for large-scale application in plant pharmaceuticals [80].

Plant growth regulators like NAA increased triterpene oleanolic acid whereas kinetin increased urosolic acid in callus cultures of *M. arvensis* var. piperascene. A wide range of variation of oil contents (0.32–1.10 %) and oil yield (0.66–5.22 ml/plant) has been observed by Kukreja et al. [81] in *M. arvensis*. They observed four major essential oil contents such as menthol ranges from 65.2% to 94.77%, menthone (1.40% to 20.89%), isomenthone (0.96% to 5.14%) and methyl acetate (0.75% to 8.52%). Secondary metabolism in tissue culture was investigated [82] in

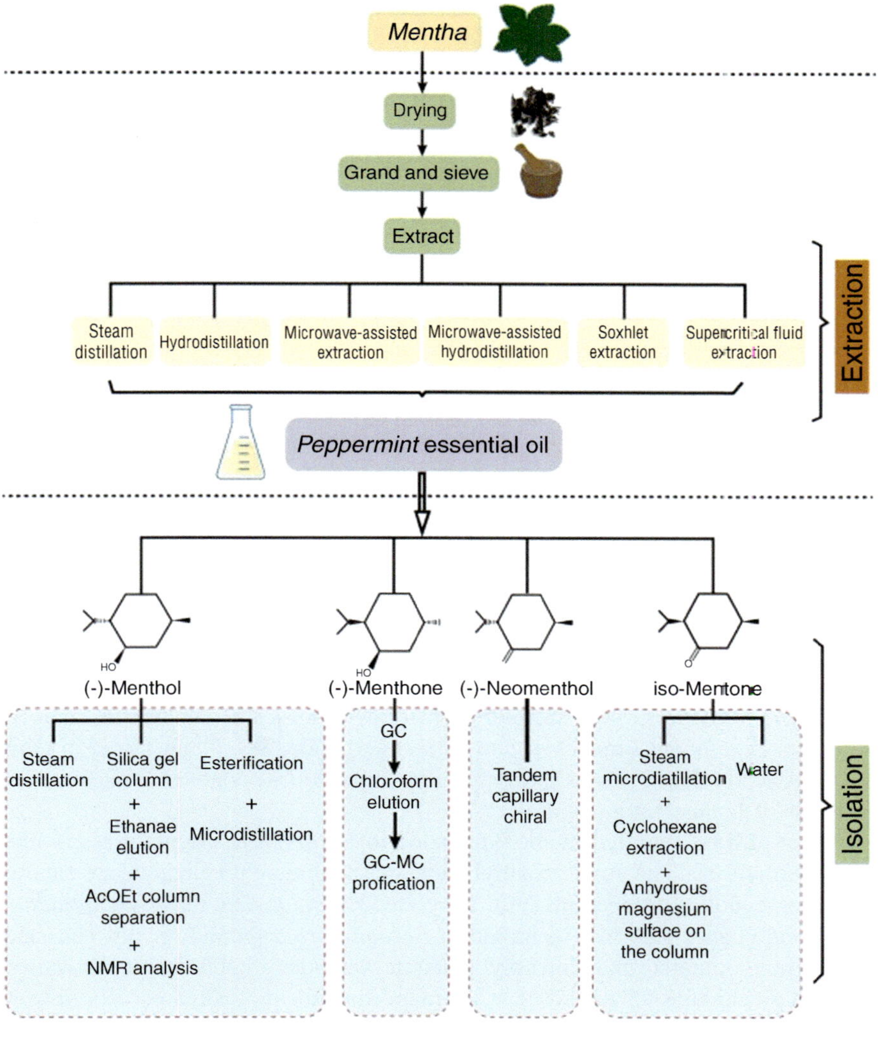

Fig. 2.4 A schematic representation of mint oil extraction and analysis. Source with permission [7]

M. spicata and *M. longifolia* failed to accumulate essential oils however produced two un-usual pigments believed to be derived from rosmarinic acid. Asai et al. [83] obtained pulegone as a chief constituent from 4 weeks old in vitro as well as acclimatized plantlets of *M. arvensis* on both liquid and agar-solidified medium under the dark conditions. However, concentration of both menthone and menthol was highest under 16-h photoperiod. Pulegone was the dominant content at the earlier stages of acclimatized pot plants whereas menthol detected as the main compound during the later stages of its cultivation. Similarly, menthol (51.68%) was also a major component amongst 21 compounds in the essential oil of *M. arvensis* var. piperascene characterized by GC-MS analysis [84] followed by menthone (26.08%) and methyl acetate (10.55%). Rosmarinic acid, lithospermic acid A and lithospermic acid B (LAB) were the main constituents of *M. spicata* as determined by High Performance Liquid Chromatography (HPLC). In the same species of *M. spicata*, the main component transpiperitone oxide was observed from mint oil collected from different localities in island of Egypt by GC-MS.

Axenic shoots of *M. piperata* were investigated for essential oil and phenolic compound production in MS basal solid medium. The major compound was reported as menthol; carvone was found when shoots were transferred on MS medium devoid of PGRs in photoperiod for 42 days whereas menthol was not observed by GC-MS in shoot culture [10]. As detected in various studies NAA increased monoterpene hydrocarbons e.g., limonene from 0.5% to 13% whereas cis- and trans-dihydrocarbon and other sesquiterpenes were reduced. The essential oil extract was analyzed with GC-MS as well as by supercritical fluid extraction (SFE), SFE provides more fractions due to least hydrolysis possibility [85].

Essential oils stored in specialized trichomes of different *Mentha* species demonstrated that natures of chemical compounds in mint oil were reported differently as determined by the HPLC and GC-MS. Flow of menthyl acetate and neomenthol observed towards the older plant parts and in young parts menthone and isomenthone were stored in *M. piperita* as determined by Solid-phase Microextraction (SPME) and GC-MS [86]. Turner et al. [9] reported higher level of menthone and menthol and level of these compounds increase with leaf development [8, 87] until 12–20 days old leaves. Such secretory structures in leaves of peppermint may be used as an experimental model system against herbicide [88]. Plants raised in vitro as well as acclimatized tissues showed similar monoterpene contents at early stages of growth of Japanese mint [45].

Hall et al. [89] reported methoxyfenozide in *M. arvensis,* Menthyl esters like monomenthyl succinate, monomenthyl glutarate and dimenthyl glutarate are natural sources as a cooling compounds [90]. They used HPLC tandem mass spectrometry, monomenthyl succinate was identified in *Lycium barbarum* and *M. piperita*, and monomenthyl glutarate and dimenthyl glutarate were identified in *Litchi chinenesis*. Essential oil contains 95% carvone as the main constituent of aerial parts of *M. spicata* determined by GC-MS [91]. Pandey et al. [92] identified menthol 71.40% followed by p-menthone (8.04%), iso-menthone (5.42%) and neo-menthol (3.18%) by GC-MS. Menthol may be a useful marker for breeding programs which is a main constituent 19 cultivars of peppermint [93]. The later studies showed that the indigenous level of menthol may be based on the experimental conditions or analysis techniques but remains round about 38–69% in *M. piperita*.

The oil composition in fresh and dried plant material is different. Fresh leaves of *M. longifolia* has pulegone as a major component, however, menthone becomes dominant in sun and air-dried leaves whereas oil from oven-dried leaves had limonene as the chief component [94]. Air-drying of leaves in shade followed by GC-MS analysis has been suggested by [95] in walnut to maintain the original chemical composition of the plant material. Chauhan et al. [96] reported that flowering stage of *M. spicata* accession is full of secondary metabolites especially carvone, limonene and cineole after analysis by GC-MS. Monthly seasonal and plant maturity periods during ontogenic growth greatly effect on the indigenously produced essential oil in *M. piperita* [97]. Such compounds extracted from air-dried leaves and analyzed by GC-MS of *M. piperita* have strong antifungal such as *Candida albicans*, antibiofilm, antioxidant, anticholinesterase, natural pesticides, insecticidal activities of wheat and stomach distension [98–104]. Such evaluated activities may be due to the major compound of mint oil associated with the medicinal properties [99]. There is greatly variation of essential oil production in plant material of *Mentha* may be classified into different chemotypes. All chemotypes are responsible for use as an antioxidants and other commercial importance based on genetic diversity of *Mentha* [7].

2.5 Ethnobotany and Ethnopharmacology

2.5.1 Health Benefits

Nature has hidden the cure for numerous diseases in herbs, fruits and vegetables. This is the reason why we see more emphasis these days on the benefits of various herbs, vegetables or fruits, with the aim of enlightening the readers on their hidden health secrets. Generally, mint is used as a decoration in food, to enhance the taste of chutney or to enjoy coffee, but apparently, this simple vegetable has many health treasures in it. If we know the benefits of this cheapest vegetable, forget about running to the doctor for every sneeze or recalling it. There is no doubt that a meal is incomplete without mint, but do you know how essential these leaves are for a healthy human body? Let's find out how.

2.5.1.1 Rich in Nutrition

We do not use mint in large quantities in our daily routine. A small amount of it is enough to make food delicious or to enhance the taste of food. Even this small amount of mint is packed with nutrients, for example, three-quarters of a cup of mint contains 6 calories, 1 g of fiber, 12% vitamin A, 9% iron, 4% folic acid, and 8% magnesium. This is why only a small amount of nutrient-dense mint is used, compared to other herbs and spices, mint is an excellent source of antioxidants. The fiber in mint lowers cholesterol levels while magnesium strengthens bones.

2.5.1.2 Useful in Dieting

Adding mint to salads, smoothies, or even water can help with weight loss. Best results can be achieved by including mint leaf tea in the diet routine. Mint is excellent for the human digestive system. It activates digestive enzymes that help in better absorption of nutrients in foods. Metabolism improves when the body absorbs these ingredients properly. Regular consumption of mint is believed to be useful in melting excess belly fat, which helps in weight loss.

2.5.1.3 The Best Cleanser, Relieves Skin Diseases

Peppermint extract is called menthol, which is used in making cosmetic products and especially various creams. Peppermint extract is recognized as the best cleanser, best for human skin. By using it, the face becomes fresh and all the skin diseases like scabies and acne on the face are eliminated. If your skin is also prone to nail acne, mix peppermint leaves with a small amount of honey and apply it on the skin for 20 min, then wash it off with warm water.

2.5.1.4 Excellent for Respiratory System

Peppermint process relieves all respiratory disorders like chest congestion, throat and lung infections. Not only this, daily use of mint is also very beneficial for asthma patients. Mint is also used in medicine. Mint can be used in case of breathing problems.

2.5.1.5 Stomach Problems

Peppermint is an excellent pain reliever. Its use is beneficial in case of nausea and abdominal pain. In case of stomach ache or nausea, it is recommended to eat a few mint leaves, while drinking a tea of its leaves is extremely beneficial in stomach ache. It is excellent for relieving indigestion, excessive belching, gas or bad breath.

2.5.1.6 Useful for Headache and Mental Health

Headaches are considered a common ailment due to the hectic life and busy schedules. In case of headache, massage a few drops of peppermint oil on the forehead, within 15 min the intensity of the pain will decrease. Not only this, consumption of this aromatic herb can help improve alertness and brain function. According to a study, the use of mint also improves memory.

2.5.1.7 Blood Pressure Control

Blood pressure can also be controlled by mint. Eating peppermint and garlic chutney is beneficial for blood pressure patients, while drinking peppermint decoction can also control blood pressure.

2.5.1.8 Restful Sleep

Mint is given great importance for restful sleep. If you also suffer from lack of sleep, its use can be beneficial.

2.5.1.9 Get Rid of Bad Breath

Mint is helpful in eliminating bad breath. Peppermint plays an important role in eliminating bad breath caused by eating garlic, onion and other such items. Peppermint oil can also be used as a good mouthwash. It naturally deodorizes breath while reducing cavities, put a drop of this oil on the tongue and get rid of bad breath.

2.6 Conclusion

It is to find out that the genus *Mentha* has various forms cultivated by means of conventional and in vitro methods. Cell culture technology is a useful method for disease free plants to get high yielding cell lines for improved essential oils production. Peppermint is the most demanded species for peppermint oil rich in menthol for therapeutic use of various ailments. Due to higher flavoring nature, the *Mentha* essential oils may be subjected to synthesize various drugs and biopesticides in future.

References

1. Shasany, A. K., Khanuja, S. P., Dhawan, S., & Kumar, S. (2000). Positive correlation between menthol content and in vitro menthol tolerance in *Mentha arvensis* L. cultivars. *Journal of Biosciences, 25*(3), 263–266.
2. Preedy, V. R. (2016). *Essential oils in food preservation, flavor and safety*. Academic.
3. Said, H. M. (1996). *Medicinal herbal* (Vol. 1, pp. 140–141). Hamdard Foundation.
4. Tyler, V. E. (1992). *The honest herbal: a sensible guide to the use of herbs and related remedies*. Pharmaceutical Products Press. xviii, 375.
5. Işcan, G., Kirimer, N., Kürkcüoğlu, M., Başer, H. C., & Demirci, F. (2002). Antimicrobial screening of *Mentha piperita* essential oils. *Journal of Agricultural and Food Chemistry, 50*(14), 3943–3946.
6. Salehi, B., Stojanović-Radić, Z., Matejić, J., Sharopov, F., Antolak, H., Kręgiel, D., et al. (2018). Plants of genus *Mentha*: From farm to food factory. *Plants, 7*(3), 70.
7. Zhao, H., Ren, S., Yang, H., Tang, S., Guo, C., Liu, M., et al. (2022). Peppermint essential oil: Its phytochemistry, biological activity, pharmacological effect and application. *Biomedicine & Pharmacotherapy, 154*, 113559.
8. Gershenzon, J., Conkey, M. E., & Croteau, B. R. (2000). Regulation of monoterpene accumulation in leaves of peppermint. *Plant Physiology, 122*(1), 205–214.
9. Turner, G. W., Gershenzon, J., & Croteau, R. B. (2000). Distribution of peltate glandular trichomes on developing leaves of peppermint. *Plant Physiology, 124*, 655–663.

10. Toshio, O., Iwao, A., Murakami, Y., & Ishimaru, K. (1998). Essential oil and phenolic production in *Mentha piperita* shoot culture. *Japanese Journal of Food Chemistry, 5*(2).
11. Li, B., Zhang, C., Peng, L., Liang, Z., Yan, X., Zhu, Y., & Liu, Y. (2015). Comparison of essential oil composition and phenolic acid content of selected *Salvia* species measured by GC–MS and HPLC methods. *Industrial Crops and Products, 69*, 329–334.
12. Ruiz del Castillo, M. L., Santa-María, G., Herraiz, M., & Blanch, G. P. (2003). A comparative study of the ability of different techniques to extract menthol from *Mentha piperita*. *Journal of Chromatographic Science, 41*(7), 385–389.
13. Lopez-Hortas, L., Rodriguez, P., Diaz-Reinoso, B., Gaspar, M. C., de Sousa, H. C., Braga, M. E., & Dominguez, H. (2022). Supercritical fluid extraction as a suitable technology to recover bioactive compounds from flowers. *The Journal of Supercritical Fluids, 105652*.
14. Abbasi, A. M., Khan, M. A., Ahmad, M., & Zafar, M. (2012). *Medicinal plant biodiversity of lesser Himalayas-Pakistan*. Springer.
15. Lawrence, B. M. (2006). *Mint: The genus Mentha*. CRC Press.
16. https://eflora.org
17. https://ucjeps.berkeley.edu/
18. https://en.wikipedia.org/wiki/Mentha_aquatica
19. Díaz-Sala, C. (2021). Adventitious root formation in tree species. *Plants, 10*(3), 486.
20. Guan, L., Tayengwa, R., Cheng, Z., et al. (2019). Auxin regulates adventitious root formation in tomato cuttings. *BMC Plant Biology, 19*, 435.
21. Akoumianaki-Ioannidou, A., Rasouli, M., Podaropoulou, L., & Bilalis, D. (2012). Seedlings production of *Mentha* × *piperita* (Peppermint) and *Mentha spicata* (Spearmint). In float system with organic and inorganic fertilization. *Acta Horticulturae, 937*, 1307–1311.
22. Bhojwani, S. S., & Dantu, P. K. (2013). Cellular totipotency. In *Plant Tissue Culture: An Introductory Text* (pp. 63–74). Springer.
23. Constabel, F. (1990). Medicinal plant biotechnology1. *Planta Medica, 56*(05), 421–425.
24. Nalawade, S. M., & Tsay, H. S. (2004). In vitro propagation of some important Chinese medicinal plants and their sustainable usage. *In Vitro Cellular & Developmental Biology-Plant, 40*(2), 143–154.
25. Murch, S. J., Peiris, S. E., Liu, C. Z., & Saxena, P. K. (2004). In vitro conservation and propagation of medicinal plants. *Biodiversity, 5*(2), 19–24.
26. Thomas, D. T., & Shankar, S. (2009). Multiple shoot induction and callus regeneration in *Sarcostemma brevistigma* Wight & Arnott, a rare medicinal plant. *Plant Biotechnology Reports, 3*(1), 67–74.
27. Shaik, S., Kumar, A., Mallikarjuna, G., Kola, G. L., Reddy, P. C. O., & Sekhar, A. C. (2023). Molecular breeding and application of molecular markers for improvement of antidiabetic medicinal plants. In *Antidiabetic potential of plants in the era of omics* (pp. 263–277). Apple Academic Press.
28. Suga, T., Hirata, T., & Yamamoto, Y. (1980). Lipid constituents of callus tissues of *Mentha spicata*. *Agricultural and Biological Chemistry, 44*(8), 1817–1820.
29. Akram, M., Afrasiab, H., Mahmood, S., & Aftab, F. (2007). Monoterpene contents in in vitro and field-grown plants of Japanese mint (*Mentha arvensis* L.). *Pakistan Journal of Biochemistry and Molecular Biology, 40*(2), 74–79.
30. Yu, Y., Lin, L. W., Yuan, L. C., & Yan, L. (2009). Induction and multiplication of leaf callus from two cultivars of *Mentha haplocalyx*. *Journal of Plant Resources and Environment., 18*(2), 84–88.
31. Johnson, M., Wesely, E. G., Kavitha, M. S., & Uma, V. (2011). Antibacterial activity of leaves and inter-nodal callus extracts of *Mentha arvensis* L. *Asian Pacific journal of tropical medicine, 4*(3), 196–200.
32. Darvishi, E., Kazemi, E., Kahrizi, D., Bahraminejad, S., Mansouri, M., Chaghakaboudi, S. R., & Khani, Y. (2014). Optimization of callus induction in Pennyroyal (*Mentha pulegium*). *Journal of Applied Biotechnology Reports, 1*(3), 97–100.

33. Jafari, A., Kahrizi, D., & Mansouri, M. (2016). Effects of plant growth regulators and explant on callus induction in pennyroyal (*Mentha pulegium* L.). *Biharean Biologist, 10*(2), 134–136.
34. Islam, A. T. M. R., & Alam, M. F. (2018). In vitro callus induction and indirect organogenesis of *Mentha piperita* (L.) – An aromatic medicinal plant. *GSC Biological and Pharmaceutical Sciences, 4*(3), 049–060.
35. Bouguemra, S., Ouafi, S., Bouguedoura, N., & Chabane, D. (2022). Enhancing bioactive potential by growth regulators in callus of *Mentha longifolia* L. leaves for anti-inflammatory and analgesic activities. *Indian Journal of Experimental Biology (IJEB), 58*(02), 122–130.
36. Bhatia, S., & Sharma, K. (2015). Technical glitches in micropropagation. In *Modern applications of plant biotechnology in pharmaceutical sciences* (pp. 393–404). Academic.
37. Kevers, C., Filali, M., Petit-Paly, G., Hagège, D., Rideau, M., & Gaspar, T. H. (1996). Habituation of plant cells does not mean insensitivity to plant growth regulators. *In Vitro Cellular & Developmental Biology-Plant*, 204–209.
38. Ali, A., Naz, S., Siddiqui, F. A., & Iqbal, J. (2008). Rapid clonal multiplication of sugarcane (*Saccharum officinarum*) through callogenesis and organogenesis. *Pakistan Journal of Botany, 40*(1), 123.
39. Cheng, Z. J., Zhu, S. S., Gao, X. Q., & Zhang, X. S. (2010). Cytokinin and auxin regulates WUS induction and inflorescence regeneration in vitro in *Arabidopsis*. *Plant Cell Reports, 29*(8), 927–933.
40. Hnatuszko-Konka, K., Gerszberg, A., Weremczuk-Jeżyna, I., & Grzegorczyk-Karolak, I. (2021). Cytokinin signaling and de novo shoot organogenesis. *Genes, 12*(2), 265.
41. Sugiyama, M. (1999). Organogenesis in vitro. *Current Opinion in Plant Biology, 2*(1), 61–64.
42. Li, X., Niu, X., Bressan, R. A., Weller, S. C., & Hasegawa, P. M. (1999). Efficient plant regeneration of native spearmint (*Mentha spicata* L.). *In Vitro Cellular & Developmental Biology-Plant, 35*(4), 333–338.
43. Ozdemir, F. A. (2017). Effects of 6-benzylaminopurine and α-naphthalene acetic acid on micropropagation from ten days old cotyledon nodes of *Mentha spicata* subsp. *Romanian Biotechnological Letters, 22*(3), 12554–12559.
44. Ayaz, E., & Memon, A. (2021). Development of the aromatic medicinal plants, *Mentha × piperita* L. and *Mentha pulegium* L. through in vitro callus induction and micropropagation. *Turkish Journal of Agriculture-Food Science and Technology, 9*(1), 159–165.
45. Phatak, S. V., & Heble, M. R. (2002). Organogenesis and terpenoid synthesis in *Mentha arvensis*. *Fitoterapia, 73*(1), 32–39.
46. Manik, S. R., Yatoo, G. M., Ahmad, Z., & Nathar, V. N. (2012). Direct organogenesis of *Mentha piperata* L. from shoot tip, nodal and sucker explants. *Journal of Agricultural Technology, 8*(2), 663–669.
47. Fadel, D., Kintzios, S., Economou, A. S., Moschopoulou, G., & Constantinidou, H. I. A. (2010). Effect of different strength of medium on organogenesis, phenolic accumulation and antioxidant activity of spearmint (*Mentha spicata* L.). *The Open Horticulture Journal, 3*(1).
48. Gopitha, K., Bhavani, A. L., & Senthilmanickam, J. (2010). Effect of the different auxins and cytokinins in callus induction, shoot, root regeneration in sugarcane. *International Journal of Pharma and Bio Sciences, 1*(3), 1–7.
49. Nawaz, M., Ullah, I., Iqbal, N., & IQBAL, M. Z., & Javed, M. A. (2013). Improving in vitro leaf disk regeneration system of sugarcane (*Saccharum officinarum* L.) with concurrent shoot/root induction from somatic embryos. *Turkish Journal of Biology, 37*(6), 726–732.
50. Ikeuchi, M., Sugimoto, K., & Iwase, A. (2013). Plant callus: mechanisms of induction and repression. *The Plant Cell, 25*(9), 3159–3173.
51. Chishti, N., Shawl, A. S., Kaloo, Z. A., Bhat, M. A., & Sultan, P. (2006). Clonal propagation of *Mentha arvensis* L. through nodal explant. *Journal of Biological Scienes, 8*(9), 1416–1419.
52. Raja, H. D., & Arockiasamy, D. I. (2008). In vitro Propagation of *Mentha viridis* L. from nodal and shoot tip explants. *Plant Tissue Culture and Biotechnology, 18*(1), 1–6.

53. Thul, S. T., & Kukreja, A. K. (2010). An efficient protocol for high-frequency direct multiple shoot regeneration from internodes of peppermint (*Mentha × piperita*). *Natural Product Communications, 5*(12), 1934578X1000501223.
54. Wang, X., Gao, Z., Wang, Y., Bressan, R. A., Weller, S. C., & Li, X. (2009). Highly efficient in vitro adventitious shoot regeneration of peppermint (*Mentha × piperita* L.) using internodal explants. *In Vitro Cellular & Developmental Biology-Plant, 45*, 435–440.
55. Čellárová, E. (1992). Micropropagation of *Mentha* L. *High-Tech and Micropropagation, III*, 262–276.
56. Islam, M. T., Dembele, D. P., & Keller, E. J. (2005). Influence of explant, temperature and different culture vessels on in vitro culture for germplasm maintenance of four mint accessions. *Plant Cell, Tissue and Organ Culture, 81*, 123–130.
57. Kumar, N. (2014). *Studies on influence of different hormonal concentrations on in vitro micropropagation of Mentha piperita* (Doctoral dissertation, Shoolini University of Biotechnology and Management Sciences).
58. Faisal, M., Alatar, A. A., Hegazy, A. K., Alharbi, S. A., El-Sheikh, M., & Okla, M. K. (2014). Thidiazuron induced in vitro multiplication of *Mentha arvensis* and evaluation of genetic stability by flow cytometry and molecular markers. *Industrial Crops and Products, 62*, 100–106.
59. Łyczko, J., Piotrowski, K., Kolasa, K., Galek, R., & Szumny, A. (2020). *Mentha piperita* L. micropropagation and the potential influence of plant growth regulators on volatile organic compound composition. *Molecules, 25*(11), 2652.
60. Ali, A., Afrasiab, H., Saeed, M., & Iqbal, J. (2004). An in vitro study of regeneration and micropropagation of *Mentha arvensis*. *International Journal of Biology and Biotechnology (Pakistan)*.
61. Rahman, M. M., Ankhi, U. R., & Biswas, A. (2013). Micropropagation of *Mentha viridis* L.: An aromatic medicinal plant. *International Journal of Pharmacy & Life Sciences, 4*(9).
62. Mucciarelli, M., Scannerini, S., Bertea, C., & Maffei, M. (2003). In vitro and in vivo peppermint (*Mentha piperita*) growth promotion by nonmycorrhizal fungal colonization. *New Phytologist, 158*(3), 579–591.
63. Fusconi, A., Trotta, A., Dho, S., Camusso, W., & Mucciarelli, M. (2010). In vitro interactions between *Mentha piperita* L. and a non-mycorrhizal endophyte: root morphogenesis, fungus development and nutritional relationships. *Journal of Plant Interactions, 5*(3), 179–188.
64. Vaidya, B. N., Asanakunov, B., Shahin, L., Jernigan, H. L., Joshee, N., & Dhekney, S. A. (2019). Improving micropropagation of *Mentha×piperita* L. using a liquid culture system. *In Vitro Cellular & Developmental Biology-Plant, 55*(1), 71–80.
65. Bhat, S., Gupta, S. K., Tuli, R., Khanuja, S. P. S., Sharma, S., Bagchi, G. D., et al. (2001). Photoregulation of adventitious and axillary shoot proliferation in menthol mint, *Mentha arvensis*. *Current Science*, 878–881.
66. Paolicchi, F., Mensuali-Sodi, A., & Tognoni, F. (2002). Effect of clinorotation on in vitro cultured explants of *Mentha piperita* L. *Scientia Horticulturae, 92*(3-4), 305–315.
67. Sunandakumari, C., Martin, K. P., Chithra, M., Sini, S., & Madhusoodanan, P. V. (2004). Rapid axillary bud proliferation and ex vitro rooting of herbal spice, *Mentha piperita* L. *Indian Journal of Biotechnology, 3*(1), 308–213.
68. Paudel, B. R., & Pant, B. (2008). Micropropagation of *Mentha spicata* L. In *Medicinal plants in Nepal: An anthology of contemporary research* (pp. 101–106).
69. Minas, G. J. (2009, March). Peppermint (*Mentha piperita*) sanitation and mass micropropagation in vitro. In *International symposium on medicinal and aromatic plants-SIPAM2009 853* (pp. 77–82).
70. Garlet, T. M. B., Flores, R., & Messchmidt, A. A. (2011). Influência de citocininas na micropropagação de *Mentha × gracilis* Sole. *Revista Brasileira de Plantas Medicinais, 13*, 30–34.
71. Zarki, K. B. L., & Elmtil, N. (2012). Micropropagation of *Mentha pulegium* L. through high-frequency shoots-tip and nodal explants culture. *Moroccan Journal of Biology, 12*, 39–50.
72. Morais, T. P., Asmar, S. A., & Luz, J. M. Q. (2014). Plant growth regulators on in vitro culture of *Mentha×piperita* L. *Revista Brasileira de Plantas Medicinais, 16*, 350–355.

73. Santoro, M. V., Cappellari, L. D. R., Giordano, W., & Banchio, E. (2015). Plant growth-promoting effects of native *Pseudomonas* strains on *Mentha piperita* (peppermint): an in vitro study. *Plant Biology, 17*(6), 1218–1226.
74. Islam, A. T. M. R., Islam, M. M., & Alam, M. F. (2017). Rapid in vitro clonal propagation of herbal spice, *Mentha piperita* L. using shoot tip and nodal explants. *Research in Plant Sciences, 5*(1), 43–50.
75. Radomir, A. M., Stan, R., Pandelea, M. L., & Vizitiu, D. E. (2022). In vitro multiplication of *Mentha piperita* L. and comparative evaluation of some biochemical compounds in plants regenerated by micropropagation and conventional method. *Acta Scientiarum Polonorum Hortorum Cultus, 21*(4), 45–52.
76. Chakraborty, A., & Chattopadhyay, S. (2008). Stimulation of menthol production in *Mentha piperita* cell culture. *In Vitro Cellular & Developmental Biology-Plant, 44*, 518–524.
77. Krzyzanowska, J., Czubacka, A., Pecio, L., Przybys, M., Doroszewska, T., Stochmal, A., & Oleszek, W. (2012). The effects of jasmonic acid and methyl jasmonate on rosmarinic acid production in *Mentha×piperita* cell suspension cultures. *Plant Cell, Tissue and Organ Culture, 108*, 73–81.
78. Tisserat, B., & Vaughn, S. F. (2008). Growth, morphogenesis, and essential oil production in *Mentha spicata* L. plantlets in vitro. *In Vitro Cellular & Developmental Biology-Plant, 44*, 40–50.
79. Peixoto, I. T. A., Furlanetti, V. F., Anibal, P. C., Duarte, M. C. T., & Höfling, J. F. (2009). Potential pharmacological and toxicological basis of the essential oil from *Mentha* spp. *Revista de Ciências Farmacêuticas Básica e Aplicada, 30*(3).
80. Mulabagal, V., & Tsay, H. S. (2004). Plant cell cultures – An alternative and efficient source for the production of biologically important secondary metabolites. *International Journal of Applied Science and Engineering, 2*(1), 29–48.
81. Kukreja, A. K., Dhawan, O. P., Mathur, A. K., Ahuja, P. S., & Mandal, S. (1991). Screening and evaluation of agronomically useful somaclonal variations in Japanese mint (*Mentha arvensis* L.). *Euphytica, 53*, 183–191.
82. Brown, G. D., & Banthorpe. (1992). Secondary metabolism in tissue culture of the Labiatae. *Lamiales Newsletters, 1*(40), 25–26.
83. Asai, I., Yoshihira, K., Omoto, T., Sakui, S., & Koichiro, S. (1994). Growth and monoterpene production in shoot culturse and regenerates of *Mentha arvensis* L. *Plant Tissue Culture Letter, 11*(3), 218–225.
84. Pino, J. A., Aristides, R., & Fuentes, V. (1996). Chemical composition of essential oil of *Mentha arvensis* L. var. Piperascens. *Journal of Essential Oil Research, 8*(5), 658–686.
85. Lemberkovics, E., Kéry, A., Marczal, G., Simándi, B., & Szöke, E. (1998). Phytochemical evaluation of essential oils, medicinal plants and their preparations. *Acta Pharmaceutica Hungarica, 68*(3), 141–149.
86. Rohloff, J. (1999). Monoterpene composition of essential oil from peppermint (Mentha× piperita L.) with regard to leaf position using solid-phase microextraction and gas chromatography/mass spectrometry analysis. *Journal of Agricultural and Food Chemistry, 47*(9), 3782–3786.
87. Lawrence, B. M. (1998). Monoterpene interrelationships in the *Mentha* genus: A biosynthetic discussion. In B. D. Mookherjee & C. J. Mussinan (Eds.), *Essential oils* (pp. 1–81). Allured Pub.
88. Lange, B. M., Wildung, M. R., Stauber, E. J., Sanchez, C., Pouchnik, D., & Croteau, R. B. (2000). Probing essential oil biosynthesis and secretion by functional evaluation of expressed sequence stages from mint glandular trichomes. *Plant Biology, 97*(5), 2934–2939.
89. Hall, G. L., Engebretson, J., Hengel, M. J., & Shibamoto, T. (2004). Analysis of methoxyfenozide residues in fruits, vegetables, and mint by liquid chromatography-tandem mass spectrometry (LS-MS/MS). *Journal of Agricultural and Food Chemistry, 52*(4), 672–276.

90. Hiserdt, R. D., Adedeji, J., John, T. V., & Dewis, M. L. (2004). Identification of monomenthyl succinate, monomenthyl glutarate and dimenthyl glutaratein nature by (HPLC). *Journal of Agricultural and Food Chemistry, 52*(11), 3536–3541.
91. Benyoussef, E., Yahiaoui, N., Nacer, N., Khelfaoui, A., & Belhadj, M. (2004). Essential oil of *Mentha spicata* L. from Algeria. *Revista Italiana, 37*, 31–35.
92. Pandey, A. K., Rai, M. K., & Acharya, D. (2004). Chemical composition and antimycotic activity of the essential oils of corn mint and lemon grass (*Cymbopogon flexuosus*) against human pathogenic fungi. *Pharmaceutical Biology, 41*(6), 421–425.
93. Stanev, S., & Zheljazkov, V. D. (2004). Study on essential oil and free menthol oil accumulation in 19 cultivars, and clones of peppermint. *Acta Horticulturae, 629*, 149–152.
94. Asekun, O. T., Grierson, D. S., & Afolayan, A. J. (2007). Effects of drying methods on the quality and quantity of the essential oil of *Mentha longifolia* L. subsp. Capensis. *Food Chemistry, 101*(3), 995–998.
95. Tewari, L. M., Rana, L., Arya, S. K., Tewari, G., Chopra, N., Pandey, N. C., et al. (2019). Effect of drying on the essential oil traits and antioxidant potential J. regia L. leaves from Kumaun Himalaya. *SN Applied Sciences, 1*, 1–9.
96. Chauhan, R. S., Kaul, M. K., Shahi, A. K., Kumar, A., Ram, G., & Tawa, A. (2009). Chemical composition of essential oils in *Mentha spicata* L. accession [IIIM (J) 26] from North-West Himalayan region, India. *Industrial Crops and Products, 29*(2–3), 654–656.
97. Grulova, D., De Martino, L., Mancini, E., Salamon, I., & De Feo, V. (2015). Seasonal variability of the main components in essential oil of *Mentha*× *piperita* L. *Journal of the Science of Food and Agriculture, 95*(3), 621–627.
98. Saharkhiz, M. J., Motamedi, M., Zomorodian, K., Pakshir, K., Miri, R., & Hemyari, K. (2012). Chemical composition, antifungal and antibiofilm activities of the essential oil of *Mentha piperita* L. *International Scholarly Research Notices, 2012*.
99. de Sousa Barros, A., de Morais, S. M., Ferreira, P. A. T., Vieira, Í. G. P., Craveiro, A. A., dos Santos Fontenelle, R. O., et al. (2015). Chemical composition and functional properties of essential oils from *Mentha* species. *Industrial Crops and Products, 76*, 557–564.
100. Brahmi, F., Adjaoud, A., Marongiu, B., Falconieri, D., Yalaoui-Guellal, D., Madani, K., & Chibane, M. (2016). Chemical and biological profiles of essential oils from *Mentha spicata* L. leaf from Bejaia in Algeria. *Journal of Essential Oil Research, 28*(3), 211–220.
101. Singh, P., & Pandey, A. K. (2018). Prospective of essential oils of the genus *Mentha* as biopesticides: A review. *Frontiers in Plant Science, 9*, 1295.
102. Chagas, E. C., Majolo, C., Monteiro, P. C., Oliveira, M. R. D., Gama, P. E., Bizzo, H. R., & Chaves, F. C. M. (2020). Composition of essential oils of *Mentha* species and their antimicrobial activity against Aeromonas spp. *Journal of Essential Oil Research, 32*(3), 209–215.
103. Piras, A., Porcedda, S., Falconieri, D., Maxia, A., Gonçalves, M. J., Cavaleiro, C., & Salgueiro, L. (2021). Antifungal activity of essential oil from Mentha spicata L. and *Mentha pulegium* L. growing wild in Sardinia island (Italy). *Natural Product Research, 35*(6), 993–999.
104. Mahboubi, M. (2021). *Mentha spicata* L. essential oil, phytochemistry and its effectiveness in flatulence. *Journal of Traditional and Complementary Medicine, 11*(2), 75–81.

Chapter 3
Amla

Sadia Javed, Tooba Nasim, and Muhammad Zia-Ul-Haq

3.1 Introduction

Amla or Indian gooseberry belongs to the family Euphorbiaceae and is a short-lived plant. The regions of China, India, Southeast Asia, Sri Lanka, Malaysia Pakistan, and Iran have the majority of Amla trees. It is a natural gift to humans with great alleviative qualities, helping to facilitate digestion, improve liver and cardiac health and reduce anxiety. It also helps to treat anemic conditions and is good for the health of the reproductive system. *P. emblica* tree has antioxidant properties and its fruits are commonly used in medicinal and cosmetic industries. Due to its great medicinal benefits, it can be used as a nutritional supplement. Emblic fruit is a natural plant source of medicine, antioxidants, and, nutraceuticals due to the existence of phenolic compounds. According to the studies, amla has anti-inflammatory, anti-hypoglycemic, anti-hyperglycemic, and anti-hyperlipidemic effects. Gallic acid, ascorbic acid and, phenolic compounds are some antioxidants present in amla which help to improve digestion and the immune system of the body [1].

S. Javed (✉) · T. Nasim
Department of Biochemistry, Government College University, Faisalabad, Pakistan
e-mail: sadiajaved@gcuf.edu.pk

M. Zia-Ul-Haq
Office of Research, Innovation and Commercialization, University of Engineering and Technology, Lahore, Pakistan

M. Zia-Ul-Haq et al. (eds.), *Essentials of Medicinal and Aromatic Crops*,
https://doi.org/10.1007/978-3-031-35403-8_3

3.2 Scientific Classification

Kingdom: Plantae
Phylum: Tracheophyta
Class: Magnoliopsida
Order: Malphigiales
Family: Phyllanthaceae
Genus: *Phyllanthus*
Specie: *P. emblica*
Botanical name: *Phyllanthus emblica* or *Emblica officinalis*
Common name: Amla

Phyllanthus emblica Linn. is a significant therapeutic herb in Ayurveda and Unani medication [2]. The amla fruit is a key dietary component and is used to make sweets, pickles, fresh juice, curries, and cookery in India [3]. It is tremendously used as a tonic for the body. It grows around 8-18 m tall with delicate light unclear bark, basic leaves, and sub-sessile, firmly set along the branchlets that appear as pinnate leaves. Amla is also reported to have anti-mutagenic, anti-radioactive, anti-chemopreventive, antimicrobial, and immunomodulatory properties. It also has anti-free radicals and cell-reinforcing properties. These qualities are useful in the prevention and treatment of several illnesses like diabetes, cancer, atherosclerosis, liver, gastric ulcers, heart infections, and other problems. The main challenge facing pharmaceutical companies is choosing the appropriate genotype of medicinal plant material. Modern quality control methods for botanicals are now necessary due to limitations in morphological and substance approaches for confirmation. Almost all components have healing qualities, especially natural products, which are used in conventional medicine and Ayurveda to treat a variety of infections such as loose bowels, jaundice, irritability, acidity, and ulcers. In the Indian system of medicine, amla organic product is frequently used as a diuretic, purgative, refrigerant, stomachic, liver tonic, remedial, hostile to pyretic, and hair tonic to treat common colds and fevers as well as to prevent ulcers and dyspepsia (Fig. 3.1) [1, 4]. Figure 3.1 presents Phyllanthus emblica Linn, commonly known as amla, which holds significant therapeutic importance in Ayurveda and Unani medicine.

3.2.1 Nutritive Value

Amla is familiar with its nutritive value. It is rich in minerals, polyphenols, and a rich source of vitamin C. Amla fruit is rich in vitamin C and pectin. Normally, amla fruit has 81.2% water, 3.4% fiber, 0.5% protein, 0.1% fat, 14% carbohydrates, 600 mg vitamin C/100 g of pulp and 0.05%cacium, 0.02% phosphorus and 1.2 mg/100 g iron. The Vitamin C content of amla juice is greater than the fruit. Dried fruit offers approximately 3470 mg of vitamin C per 100 g (dried fruit). It

Fig. 3.1 *Phyllanthus emblica Linn.* (Amla) [5]

keeps as much as 1780–2660 mg of vitamin C even if it is dried up in shade and then turned into powder. Polyphenol is present in fruit, which impedes Vitamin C oxidation. It is also a good source of vitamin B (30 mg/100 g) and nicotinic acid (0.2 mg/100 g). The fatty value of amla fruit is 59/100 g of fruit. It has great industrial, medicinal, and other applications due to its antiscorbutic, laxative, diuretic, antibiotic and other properties [6, 7].

3.3 Chemical Constituents

Different parts of plants such as fruits, leaves, and roots of amla have rich phytochemistry. The chief group of secondary metabolites is polyphenols and compounds related to phenolic acids, phenolic, flavonoids, and derivatives compounds. In the fresh fruits, the presence of hydroxybenzoic acids i.e.; 4-hydroxybenzoic acid, gallic acid, coumaric acid, protocatechuic acid, vanillic acid, and syringic acid were recognized. In leaves and branches, the reported hydroxybenzoic acid is gallic acid. In amla fruits, the occurrence of hydroxycinnamic acids (chlorogenic acid and caffeic acid) was indicated. Amla plants also comprise compounds like flavonoids (particularly flavonols, flavones, flavanones, and flavan-3-ols). From these compounds, flavonols are commonly present in the amla plant. In the fruits, leaves,branches of amla plants kampferol and their products (dihydrokaempferol, kaempferol 3-o-rhamnoside,kaempferol 3-b-dglucopyranoside, kaempferol-3-o-α-l-(6″-ethyl)- rhamnopyranoside, and kaempferol-3-o-α-l-(6″-methyl)-rhamnopyranoside) are extensively present. Quercetin and its derivatives (quercetin

3-b-D-glucopyranoside, quercetin 3-O-rhamnoside, quercetin 3-O-glucoside, and rutin) are also found in all parts of amla plants.

In the fresh fruits of amla apigenin, luteolin, and myricetin were indicated. Interestingly, in the branches and leaves of the amla tree myricetin 3-O-rhamnoside, flavanones, and flavan-3-ols were reported. Eriodictyol, naringenin, and their derivatives (S)-eriodictyol 7-O-(6″-O-trans-p-coumaroyl)-β-Dglucopyranoside, (S)-eriodictyol 7-O- (6″-O-galloyl)-β-D-glucopyranoside, naringenin 7-O-(6″-O-trans-pcoumaroyl)-glucoside, naringenin 7-O-(6″-O-galloyl)-glucoside, etc. are well-known flavanones. Epicagallocatechin and epigallocatechin 3-O-gallate were identified compounds concerning flavan-3-ols.

Tannins in amla fruits are included in the group of phenolic compounds. The fruits, leaves, and branches of amla contain ellagitannins, which contain chebulagic acid, chebulinic acid, emblicanin A and B, corilagin, pedunculagin, isocorilagin, phyllanemblinins A–F, and punigluconin. Amla plant also has Ellagic acid and its derivatives (de-carboxy ellagic acid and 30 -O-methylellagic acid 4-O-α-L-rhamnopyranoside). The leaves and branches of amla are rich in phlorotannins and hydrolyzable tannins. The exception which was addressed in amla fruit is tannic acid. Furthermore, phenolics (2,4-di-tert-butyl phenol and Phenol, 3,5-bis (1,1-dimethyl ethyl)) and alkaloids i.e.; phyllantine and phyllantidine were also identified in amla fruit (Fig. 3.2) [1, 8]. Figure. 3.2 illustrates the structure of phytochemicals in amla.

3.4 Cultivation

3.4.1 Climate and Conditions of Soil

E. officinalis is a fruitful tree that grows in the sub-tropical region. It is spread all through a variety of climatic conditions. Though, it is relatively effective in dry sub-tropical and tropical climates for the large-scale cultivation of trees. Annual rainfall of about 630–800 mm is most advantageous for the growth of the *E. officinalis* tree while a warm climate up to 46 °C is favorable for a mature tree. Being a specie of deep-rooted, deciduous tree species, it can be grown up in a wide variety of soil types. Additionally, it has been demonstrated that *E. officinalis* has enormous potential for commercial-scale cultivation in the area afflicted by salinity. A wide range of trees is present in the saline wasteland and ravine land of India [9].

3.5 Cultivars

The majority of the plantations were grownup from seeds and are very diverse. Up until the middle of the 1970s, there had been no standardization of cultivars and they were mostly identified by their size, color, and names of areas, such as Bansi Red, Green tinged, Pink tinged, Red tinged, and White-streaked. Certain identified cultivars are known to be grown in Uttar Pradesh, including Francis, Banarasi, and

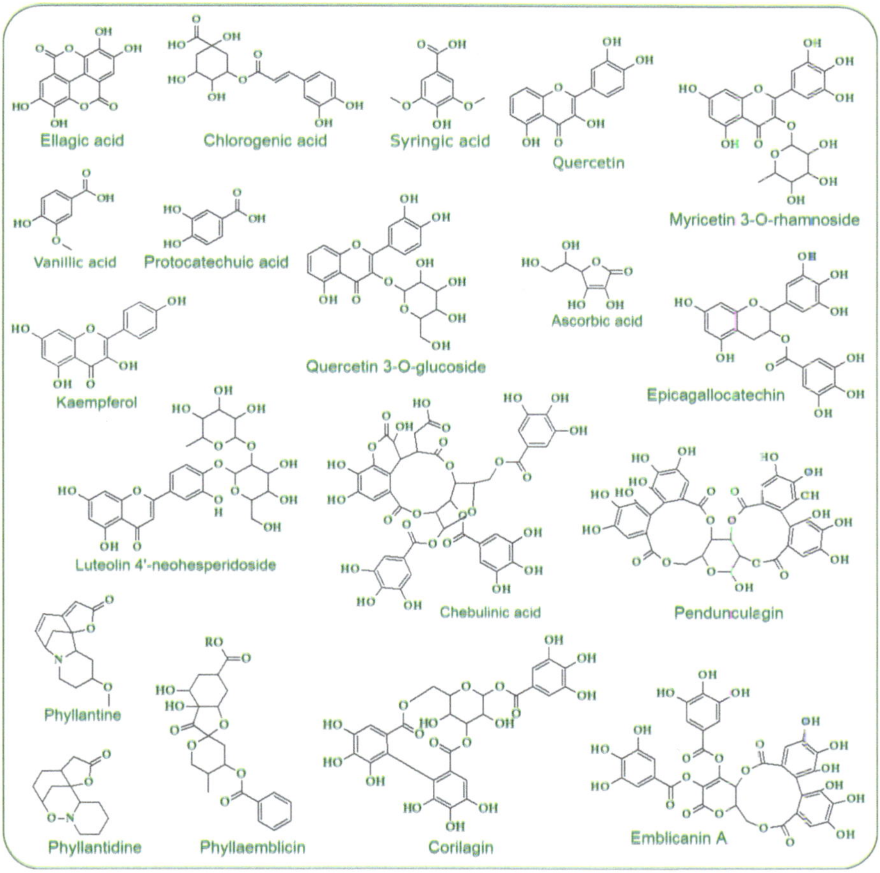

Fig. 3.2 Structure of phytochemicals in amla. (Modified from [1])

Chakaiya. They have also been made to appropriate India's northern climate conditions. A few significant types that can be grown in various regions of the nation are described as follows:

Banarasi This seedling selection comes from the Uttar Pradesh district of Varanasi. This common variation grows erect and has three branchlets per node. Long and rather deep pink inflorescence. Fruits are large in size, oblong, flattened, and have smooth, yellowish skin. The segment is raised in three sections. Soft, semi-transparent, and fairly fibrous describe the flesh. In the fourth week of March, flowers begin to bloom. Since it possesses a large number of male flowers and is incompatible with itself, it is a reserved bearer. Ascorbic acid is present in large amounts. Making preserves is done with it. Early bearing with excellent keeping qualities. A little bit prone to necrosis.

NA-4 (Krishna) An errant Banarasi seedling from Partapgarh. This tree is somewhat tall has a scattering growth practice and bears little fruit. The fruit is flattened, conical, angular, and medium to large in size. Very smooth skin has a yellowish color and a reddish blush on the surface. The fruit flesh is extremely astringent, firm, and fibreless. It is also semitransparent. Early development, reserved, and without fruit necrosis.

NA-5 (Kanchan) This seedling was likewise chosen from Chakaiya. It is a rich and regular-bearing cultivar with fruits that are medium in size and high in fiber. Industries use it more frequently to extract the pulp and manufacture a variety of goods. Fruit is tiny to medium-sized (30–40 g), with smooth, yellowish skin that is perfect for pickling. The variety grows 150–200 kg of fruit per tree and has been exceptionally well adapted to semiarid climates. Fruit that has reached late maturity is free from necrosis and has an excellent keeping quality.

NA-6 This seedling selection from Chakaiya takes on the optimal tree shape and bears copiously and abundantly. It is the perfect cultivar for sweets and preserves due to its low fiber content. This is the amla variety that has so far shown the most promise for plantations.

NA-7 This Francis clone is a prodigious, prolific, and consistent bearer. Necrosis does not occur on the fruit. In the states of Bihar, Rajasthan, Jharkhand, Andhra Pradesh, Madhya Pradesh, Jammu & Kashmir, and Tamil Nadu, this type has been well-accepted. The main limitation of NA-7 is the brittleness of the branches, which frequently snap under the weight of their fruit. This cultivar is excellent for processing and has a lot of potentials.

NA-10 Agra Bold cultivar Banarasi seedling that was accidentally created. A semi-spreading, semi-tall growth style with a long, deep pink inflorescence. Round and brown in color, the fruits begin to grow early in the season and first seem pinkish yellow before progressively disappearing later. Fruit is spherical, flattened, and of average size. Fruit skin has a pink tint and is tough and yellowish green. The flesh is delicate, juicy, faintly fibrous, pale green, and quite astringent. It has the best keeping quality and is the earliest maturing variety.

Francis It comes from Pratapgarh and is also called "Hathijhool" (U.P.). The branches frequently sag. It is a frequent carrier. Large, flattened, oval fruits have smooth skin that is a greenish-yellow color. The flesh is supple and almost fiberless. The keeping is of very low quality. It is unsuitable for preparing preserves because it is particularly vulnerable to fruit necrosis.

Chakaiya This seedling variety has an erect, tall growth habit. It bears frequently and profusely. Fruits range in size from small to medium and have smooth, greenish skin. The flesh is tough and fibrous. Maintaining quality is excellent. This type has fruits that mature slowly and have good storing qualities, making it ideal for pickles and other goods. Void of necrosis.

Laxmi-52 Also known as Francis' superior chance seedling, Laxmi-52 is being grown in a private orchard. Like its parent Francis, the tree it produces has semi-erect growth and branches that do not droop. Large, 40–60 g, 4.0–4.5 cm in diameter, with 6 ridges on the fruit. Fruit color is pale pink throughout the early stages of growth before disappearing at full development. Compared to other amla cultivars, the leaves are a wider, longer, and darker green. It is devoid of necrosis, mid-season maturing (mid-November to mid-December), and has a production potential of 2–2.5 q/tree (after 10 years). It commands a greater price on the market since the fruit is larger and more appealing in color. The fruit has been discovered to be appropriate for segment preparation in syrup, candy, and preserves [7, 8].

3.5.1 Seed Propagation

Amla is propagated by seeds and the trees grown by these seeds have inferior-quality fruits. These trees have long growing stages and show a different configuration of vegetative growth and different size, shape, and yield of fruit. These trees need a long time to reach the first reproductive stage than vegetative propagative ones. Under favorable conditions, fresh seeds typically do not propagate due to thick, hard testa and need particular treatments subsequently like soaking of seed in water, growth managing treatment of plant, scarification and, stratification. These treatments help to control inactivity. In seed-spreading practice, ripe berries from plants are picked in November and December [9]. Then the fruits that are collected are dried in sun and pulled out with a gentle press. About 1 kg of seeds is produced by one quintal of fruits. The typical time is between April and June for the sowing of *E. officinalis* seeds. The small polybags are used for sowing at the depth of 5 cm. Immersion of seeds in GA3 200 ppm solution for 24 h, considerably enhanced seed propagation and seedling strength [7]. Multiple metabolic activities are stimulated by hormonal and enzymatic mechanisms that result in the elongation of root and shoot and increased dry weight of seedlings [9].

3.5.2 Budding and Grafting

Budding is the supreme experimental method out of all ways of vegetative propagation in *E. officinalis*. Patch and shield budding is practiced for marketable propagation. During early July, 1-year-old seedlings with 1 cm width are shield budded with fit and fleshy buds. Shield budding is more favorable than patch budding. Besides budding, softwood grafting also has a 70% rate of success, particularly in dry areas. Cleft and veneer grafting are additional successful techniques.

3.5.3 Nursery Preparation

A seedling bed is essential for nurturing seedlings. Nursery beds are normally 10–15 cm raised using farmyard manure, in partial shade. The seeds are soaked for 2 days before sowing. Then these seeds are propagated in 2–3 cm of soil with 15 cm row-to-row spacing in the rainy or spring season. Healthy plants after germination are used for planting and can be restocked for budding [9].

Rootstock Preparation Six to one-year-old seedlings are usually used for restocking. From November–December mature amla fruits are taken, dried and their seeds are extracted. These seeds are sown in a raised nursery bed after April and these seedlings can also be transplanted in individual beds for consequent budding. Polybags, polytubes and root trainers are commercially used for the propagation of amla [7].

3.5.4 Orchard Establishment

An *E. officinalis* tree begins to yield fruit 3–4 years after planting, and after 10–12 years, they reach a specific level where they are mature enough to continue producing fruit for commercial purposes for another 60–70 years. The layout is carried out on land that has been extensively tilled, leveled, and cleared of any vegetation. In a subtropical environment, planting typically begins around the middle of August and is finished by the completion of the month. Early seeding makes sure that the plants receive a lengthy period of rain that is essential for the initial growth and maturation of saplings. However, the seedlings planted around mid-July had the most success in blooming in both March and July [10]. Mature *E. officinalis* trees have a height of between 8 and 18 m. Therefore, square planting is done with 8–10 m gaps between and within the rows to allow for enough amount of light, easy culturing processes (like trimming), and sufficient fruiting. Today, establishing hedgerows with spacing of 8 m between lines and 4–5 m between plants is also being studied. *E. officinalis* is normally spread directly by saplings grown in appropriate containers in unfavorable soil conditions, which are then transplanted to a permanent location where in situ budding is performed. To overcome self-incompatibility, two cultivars must be planted in alternating rows. Before planting, the fields are set out and designated. Then, throughout the months of April and May, 1 × 1 × 1 -sized pits are dug at the designated locations and left for at minimum 2 weeks to get rid of bug pests. Before the rainy season begins, pits are then filled with FYM (around 15–20 kg), neem cake (1 kg), muriate of potash (MOP) (200–300 g), single superphosphate (500 g) and Hep tachlor® and Furadan 3G®. After that, soil is left in its current state during the first few raindrops to fall and settle properly. Before providing regular irrigation after planting, immediate watering is done to aid in the appropriate establishment of the plants [7, 9].

3.5.5 Orchard Management

Taking proper care of the fertilizer and water supply, canopy architectural maintenance, field cleaning, and timely implementation of plant protection measures are all part of managing an *E. officinalis* orchard. Young plants often emerge with a particular level of yearly vegetative growth to provide an initial canopy. The plants develop a proper canopy to yield fruits after 2 years. Flowers and fruits need to be cut back in the initial 2 years, though, to maintain healthier development, and regular spraying, hoeing, tidying, plant fortification, etc. should be done as needed. Trees in young orchards of fruits (2–7 years) require additional nutrients to maintain healthy development and ripening. Early trees need adequate and judicious trimming, ideally between March and April, to allow the chief branches to develop to an altitude of 0.75 to 1 m from the soil since their excessive vegetative growth inhibits regular fruiting. Eventually, only 4–6 carefully chosen branches are permitted to continue growing [7, 9].

3.5.6 Nutrient Management

Fruit production and quality are increased when both organic and inorganic nutrients are used; nevertheless, biofertilizer utilization significantly enhances fruit quality. These nutrient sources affect the physical, chemical and biological characteristics of the soil. Depending on the soil quality, age of the plant, and regularity of fruiting, different manure and fertilizer dosages are used. Typically, a one-year-old plant receives 10 kg FYM, 100 g of potassium, 100 g nitrogen and 50 g phosphorus. Up to 10 years of annual dosage increases should be guaranteed, after which a steady dose is administered in the following years. During the months of December and January, the full dose of FYM, 50% of the nitrogen, phosphorus, and MOP are administered round the tree sinks. In August, the remaining half is implemented. Any problematical soil is supplemented with 100–500 g of boron, copper sulphate, and zinc sulphate in addition to regular fertilizers [9]. Between mid-May and mid-July, the use of synthetic auxins (α-naphthalene acetic acid (NAA) and gibberellic acid) in combination with thiourea may lead to successful resolution to reduce the rate of yield brought on by significant fruit drop. This approach can be supported in sodic soils having manufacture limits such as restricted obtainability of various inorganic nutrients for the best tree development and output. The growth, development, and quality of fruit are all greatly enhanced by plant growth regulators and a variety of other nutrients. When the pH of the soil is high and a variety of macro and micro constituents are inaccessible, nutrient sprays are comparably more effective for quick consumption in plants. In well-established orchards, two foliar treatments of NAA (30 ppm) from May to July enhance fruit quality and fruit storage [11]. According to the experimental observations, spraying a mixture of 0.4%

$CuSO_4$ + 0.5% $ZnSO_4$+ 10 ppm NAA was effective in enhancing plant growth and reducing fruit drop. To improve the physicochemical characteristics of *E. officinalis* fruits, foliar feeding of 0.4%$CuSO_4$ + 0.4%$ZnSO_4$ + 0.5%$MnSO_4$ twice between mid-May and mid-July is the most effective; however, foliar application of GA3 (150 ppm) is most active to boost vegetative growth and fruit production. Similar results were shown when zinc and boron were applied as a foliar spray, improving fruit output. The 0.2% borax+0.5% $ZnSO_4$ foliar sprays resulted in the maximum fruit quality and output (with improved vitamin-C content) per tree [12].

3.5.7 Water Management

In established orchards with typical soils, *E. officinalis* is grown as a moisture tree, thus irrigation is not necessary, especially during the winter and rainy months. The first watering should be applied to the fruit-bearing plant only after fertilizer has been applied between January and February. But, it is advised to avoid applying water throughout the flowering phase (mid-March to mid-April). For *E. officinalis*, a basin irrigation system works well. Pitcher irrigation is typically advised for the growth of orchards in places with water shortages, while drip irrigation is also a viable approach. Substantial stock width and plant height can be observed during a normal dry season with the use total of 9 irrigations together with the mulch, which finally can save up to 20 cm of irrigation water (four irrigations) on the basis of net area. The water saved in this case might be used to plant an extra area of the orchard [13]. Sometimes, using fertigation instead of irrigation or manure administration results in a substantial improvement in the frequency of flowering. By applying a 125% recommended dose of fertilizer in the form of water-soluble fertilizer by fertigation, the largest plant height, trunk diameter, and plant dispersion may be registered [9].

3.5.8 Cropping System

Because the E. officinalis tree canopy has a finite number of leaves, intercropping offers a unique potential to utilize existing orchard interspaces during the first 3–4 years following planting. Even under fully grown trees, it gives space for plenty of incoming light and encourages intercropping in open areas [9]. *E. officinalis* showed encouraging results when intercropping with turmeric, arbi, and ginger in terms of yield, accessible nitrogen, carbon, and phosphorus, as well as agricultural economics [14]. Amorphophallus is a dark-loving plant that may be produced commercially in an *E. officinalis* orchard. The growth, yield, and quality parameters, soil richness, gross and net income, the cost-benefit ratio, and other factors, growing elephant foot yam as an intercrop in an *E. officinalis* plantation was found to be the most advantageous. In the desert region, *E. officinalis* is intercropped with winter crops including fenugreek, chickpea, cumin, and mustard as well as rainy-season crops like moth bean. Many plants, including gladiolus and marigold, as well as vegetables including bottle gourd, okra, coriander, cauliflower, peas, and turmeric, have been discovered to be suitable for intercropping. Spiny sesbania can be interplanted for a few years on fragile or salt-affected soils to improve the physicochemical characteristics of the soil. Tuber crops may also be successfully produced even in orchards with a lot of shade. Cropping system types like *E. officinalis* with guava, phalsa, spiny sesbania, wheat or barley, onion or brinjal, German chamomile, etc. have all been shown to be quite profitable [9].

3.5.9 Fruit Maturity, Harvesting, and Yield

The preferred yield and processing quality determine when *E. officinalis* fruit reaches maturity. When computing the maturity index of any cultivar of *E. officinalis*, marketable features including days from sowing to maturity, total soluble sugar, fruit skin color, heat units, acid ratio, etc. are taken into consideration. The fruits first turn light green as they mature and ripen, and then they turn greenish-yellow or, sometimes brick red. Mature fruits have the highest ascorbic acid content, whereas immature fruits have lower mineral content and ascorbic acid. Fruits can be handpicked during the months of November and December. To prevent fruit falling, especially on the "Banarasi" and "Francis" cultivars, fully matured fruits are picked (either in the morning or the evening). A budded or grafted tree begins producing fruit after 3 years of implanting but a sapling tree takes 6–8 years to produce fruit. However, the latter tree may continue producing fruit for another 60–75 years after planting. *E. officinalis* tree may produce 100–300 kg of fruits per tree with a yield of 15–20 tons/ha. If greater fruit preservation and other yield-attributing traits are guaranteed, a higher yield of E. ofcinalis can be achieved. The cultivar 'Kanchan' had the highest fruit output (99.79 kg/tree), with 'Krishna' coming in second (76.55 kg/tree). However, if adequate agro-technology is used, fruit yields of up to 220–280 kg/tree may often be observed. The ripe fruits are often quite firm, which makes it easier to harvest them in large quantities, transport them, and market them even in far-off places [9].

3.6 Pests and Diseases

3.6.1 Pests

3.6.1.1 *Inderbela tetrosis*

Inderbela tetrosis (bark-eating caterpillar) feed up the tissues and destroy the bark of the trunk and branches of the trees. Injecting kerosene oil, Monocrotophos (0.03%), or Dichlorvas help to reduce its attack.

3.6.1.2 *Betousa stylophora*

Short galls are caused by *Betousa stylophora*. Spraying with 0.05% Monocrotophos or sniping the gall twigs, shoot galls can be controlled.

3.6.1.3 *Virachola isocrates* and *Cerciaphis emblica*

Anar butterfly (*Virachola Isocrates*) and, scales and aphids (*Cerciaphis emblica*) are the minor insects that cause damage to Amla. Insecticides can be used to control these insects.

3.6.2 Diseases

Aonla Rust It is also called ring rust of amla, caused by *Ravenelia emblicae* var. *fructoidae*. The visible black spots appear on fruits and leaflets which later form rings. These rings join together and cover a large area. These spots uncover black spore mass after rupture. Isolated or grouped pinkish spots develop on leaves. The affected fruits early drop off.

Control 3–4 sprays of wettable sulfur @ 5 g/l of water at an intermission of 1 month or mancozeb @ 2.5 g/l of water at an interval of 15 days.

3.6.3 Fruit Rot

Nigrospora sphearica, Phomopsis phyllanthi, Pestalotia creenta, Cladosporium tenuissium, Alternaria alternata, and *Cytospora sp.* are causes of fruit rot. Small necrotic spots of pinkish brown color appear in this disease. Complete decay of fruit takes place after the appearance of black spots and soft areas.

Control Give post-harvest treatment to the fruit with difolaton @ 1.5 g/l of water or mancozeb @ 1 g/l of water or carbendazim @ 0.5 g/l of water.

3.6.4 Anthracnose

Anthracnose is caused by *Colletotrichum sp.* Leaflets have tiny, round, grayish spots with yellow margins. The central area has dot-like fruit bodies and remains grayish. The fruit becomes dark in the center and the fruit lesions become depressed. Spore mass seems on fruit bodies at high moisture and infected fruits become wrinkled and decay.

Control Spray copper oxychloride @ 3 g/l of water or mancozeb @ 2 g/l of water or carbendazim @ 0.5 g/l of water at 15 days intervals. Repeat sprays as per severity.

3.6.5 Blue Mold Rot

Penicillium islandicum is the main cause of blue mold rot. Brown spots or patches appear on fruit and different colors i.e., bluish green, bright yellow and purplish-brown develop with the progress of the disease. Yellowish droplets of the liquid project from the patches and smell emit from the fruit.

Control Handle fruit carefully avoiding wounds. Manage good sanitary conditions in storage by gas treatment with NCl3and ozone. Treat fruit with borax @ 0.5 g/l of water [7].

3.6.6 Physiological Disorders

Major physiological disorders include chilling injury, necrosis, white spots and pink spots that distress the quality of amla fruits. Piercing of covering and irregular maturing of fruits starts after a chilling injury that finally leads to decay. The storing temperature must be adjusted to about 12 °C to avoid this injury. Pink spots seemed randomly on amla fruits due to the insufficiency of boron that ultimately deteriorate the quality of the fruit. To manage these conditions, borax spray (0.5%) is applied three times at an interval of 2 weeks from September to October which is useful to control these disorders. At the pickling and preserving stage, white spots also cause a poor appearance and a squishy surface of the fruit [9].

3.7 Medicinal Uses

3.7.1 Antioxidant Action

Amla berry is a potent free radical scavenger and broad-spectrum antioxidant. It slows down the aging process and reduces the occurrence of diseases. Superoxide dismutase, a free-radical scavenger, is present in high concentrations in amla preparations [6]. Amla shows free-radical reducing ability and also reduces erythema caused by UV. It shows the chelating ability of copper and iron and also acts as a matrix metalloproteinase inhibitor. Amla contains ellagic acid, a potent antioxidant that repairs chromosomal abnormalities and inhibits mutations in genes [5].

3.7.2 Source of Vitamin C

Amla contains Vitamin C in its most potent form. Vitamin C is fused with tannins in the amla fruit that shield it from being destroyed by light or heat [6]. As Vitamin C is abundant in amla, it is considered the best treatment for scurvy. Equal amounts of dried herb and sugar taken with milk, thrice a day are useful in the treatment of scurvy. The outbreak of scurvy and jaundice can be prevented by drinking amla juice first thing in the morning on an empty stomach [5].

3.7.3 Cardioprotective Activity

Amla helps the heart and might occasionally function as a cardiac stimulant. According to the studies, amla lowers cholesterol and guards against heart disease [6]. One of the main contributors to cardiovascular problems is hyperlipidemia but amla contains several bioactive substances that may contribute to controlling this disorder. Amla juice lessens LDL cholesterol oxidation by 90% and restricts macrophage LDL oxidation uptake. Another study revealed that hydroalcoholic amla extract lowers serum sodium levels and arterial mean blood pressure. *P. emblica* L. controls the increased expressions of MDA, COX-2, and Bax in the liver as well as serum nitric oxide (NO) activation, the endogenous antioxidant system, and electrolyte levels in serum. It also controls Bcl-2 expression. The metabolic alterations brought on by high fructose ingestion are reduced by the phenol extract of *P. emblica* L. in an animal model. Moreover, *P. emblica* L. fruit extracts lower levels of VLDL, LDL, and cholesterol while raising HDL levels and preventing atherosclerosis [1].

3.7.4 Antidiabetic Effect and Diuretic

Due to its high vitamin C concentration, Amla is extremely beneficial in the treatment of diabetes. Together the juices of bitter gourd and amla support the pancreas and permit it to release insulin. Insulin helps to reduce blood sugar levels. Amla prevents eye problems in diabetes. It increases insulin production and secretion through the regeneration of beta cells. Tannins are present in amla which enhances glucose uptake and inhibits adipogenesis. Amla extract scavenges free radicals resulting in rapid protection against lipid peroxidation. Fruits and decoctions of seeds and leaves are used in the treatment of diabetes.

Fresh fruit of amla acts as a diuretic. The retention of urine and irritation of the bladder can be treated with a fruit paste mixed with Indian saffron and rose water. It is used as an anti-diuretic and anti-inflammatory. Burning sensation in urination can be treated by Amla-berry. It shows natural diuretic action by eliminating waste from the body without overstimulating the urinary system. Regular use of a mixture of

E. officinalis pulp with water and Gur helps to cure the urinary problem. The combination of radish and amla powder has the power to dissolve bladder stones and flush them out through urination. The morning or the evening are the optimum times to eat them [5].

3.7.5 Anticancer Activity

Amla is useful in the treatment and prevention of different types of cancers. The growth of some human cancer cell lines can be inhibited by *P. emblica* extract [6]. Radiotherapy and chemotherapy are frequently used to treat cancer. Amla extract also lessens the adverse effect of these therapies [5].

Plant-derived polyphenols inhibit oxidative stress, prevent DNA damage and produce pro-inflammatory chemicals. It also increases apoptosis through various mechanisms. It plays a protective role in radio- and chemotherapy. The protective bioactive components of amla act as a preventative against the growth of cancer [1]. *P. emblica* demonstrates its anticancer actions through the suppression of AP-1 and targets the transcription of viral oncogenes important for the onset and development of cervical malignancy, suggesting that it may be useful for treating cervical malignancies brought on by HPV [15].

3.7.6 Brain-Protective and Anti-Brain Aging

The traditional Indian approach has relatively few medications accessible for the treatment of brain multifactorial illnesses like Alzheimer's disease, Parkinson's disease, and Huntington's chorea. Several semi-synthetic drugs and plant-derived phytochemicals have been used in the management of neurological disorders. The phytoconstituents of amla are known to have complement-inhibitory potential as it inhibits alternative complementary pathway. Variations in complementing alternative pathways are critical in the emergence of neuro-inflammatory diseases such as mad cow disease and dementia associated with HIV. The phytoconstituents inhibit neuroinflammation associated with CNS maladies [8].

Treatment with ayurvedic preparation of *E. officinalis*.is used to study how memory scores improve in mice and rats. Due to its many positive benefits, including memory enhancement and the reversal of memory losses, *E. officinalis* may prove to be a viable treatment for the control of Alzheimer's disease [16]. Golechha et al. studied the effectiveness of *E. officinalis* against scopolamine-induced memory loss. Results from many research suggest that E. officinalis might be used as a tonic to treat dementia. It also has antioxidant activity, the ability to enhance and cure memory loss, and anticholinesterase action. It also lowers cholesterol.

Epilepsy-related cognitive deficits can result from antiepileptic medication usage. This outlines the requirement for the right pharmacological therapy, which can stop the advancement of epilepsy and enhance rather than diminish cognitive performance. Kainic acid and pentylenetetrazole-induced seizures were eliminated by the hydroalcoholic extract of E. officinalis, which significantly improved cognitive performance. *E. officinalis* has antioxidant and anti-inflammatory potential and show a dose-dependent reduction of kainic acid-induced increased TNF-α in the brain. One of the most crucial targets for the therapy of Alzheimer's disease is acetylcholinesterase. Methanolic extract of amla fruits showed significant inhibition against acetylcholinesterase. Human neuroblastoma cells were treated with both aqueous and methanol extracts of *E. officinalis*, and the results showed significant protection against DNA damage induced by H_2O_2. Human neuroblastoma cells were treated with aqueous and methanolic extracts of *E. officinalis,* and comet test results displayed substantial protection against H_2O_2-induced DNA damage and increased cell survival. Despite its effects on amyloid protein, therapy with *E. officinalis* restored AlC13-impaired learning memory and movement. In the development of different disorders i.e.; the aging process, the primary factor is stress-related oxidative stress. E. officinalis treatment caused a decline in oxidative stress in the anterior cortex of the brain. Studies demonstrated E. officinalis preventive neuroprotective viewpoint against neuroleptic mediator-induced tardive dyskinesia caused by haloperidol. Its propensity to regulate the oxidative load brought on by stress may be the cause of this effect [8].

Dhingra et al. investigated the antidepressant effect of *E. officinalis* in mice, using the tail suspension test and the forced swim test. In both these tests, the extract dramatically reduced the immobility period, demonstrating considerable antidepressant-like effects. The extract performed better in an antidepressant-like manner at a low dose of 200 mg/kg. It was discovered that the extract's effectiveness was found to be equivalent to that of imipramine (15 mg/kg), 20 mg/kg of phenelzine and fluoxetine (20 mg/kg). The mice's locomotor activity was not significantly affected by the extract. Additionally, the extract markedly reduced brain monoamine oxidase A levels. The interaction of the aqueous extract with dopamine D2-receptors, α1-adrenoceptors, $GABA_B$, and serotonergic receptors may result in an antidepressant-like effect. This investigation discovered that the aqueous extract of amla has 2.94% ascorbic acid. Ascorbic acid, along with flavonoids, polyphenolic compounds and tannoid principles, may be the cause of the antidepressant-like effects of the aqueous *E. officinalis* extract. *E. officinalis* aqueous extract demonstrated antidepressant-like action likely by decreasing MAO-A and GABA, as well as because of its antioxidant activity [17].

Stress has been identified as a fundamental contributor to changes in behavior and mental state. Improved mental health and dose-dependent decrease against stress-induced high corticosterone levels in mice are noticed with the usage of hydroalcoholic extract of *E. officinalis.* Noise can be known as one of the key stressors which directly influence mental health. Studies on albino rats subjected to 100 dB noise for 4 h every day for 15 days revealed significant immobility, nursing, and other stress-related behavioral alterations. *E. officinalis* showed anti-stressor potential and healing was noticed from the stress induced by noise after treatment with *E. officinalis* [18]. It is estimated that treatment with *E. officinalis* and the other traditional herbal preparations having amla as the main ingredient, has overturned the diazepam and scopolamine-induced memory deficits [8]. Amla berry benefits the brain and improves the coordination of learning, retaining information, and memory. It enhances mental capacity and intellect. It improves the senses and helps the neurological system [6].

3.7.7 Enhances Food Absorption

The usage of amla-berries aids in meal digestion and absorption. While it supports all 13 digestive fires, its effects are less immediate than those of ginger and other digestive-supporting herbs. People can thus consume it because it does not produce an overabundance of stomach acid. Additionally, it facilitates the absorption of iron for healthy blood [6]. Asia has long utilized the fruits of the emblic leaf flower as food and traditional medicine. Modern research has revealed a wide range of biological activities that point to the fruits' potential as a nutritious diet and a source of bioactive food components. One of the main bioactive components of fruits is hydrolyzable tannin. A promising source material for natural food preservatives is leaf flowers [19].

3.7.8 Strengthens the Eyes

Amla helps to support the health of the eye and acts as a good tonic for the eyes. Infused leaves of amla can be applied to sore eyes. In ophthalmia, medical lotion formed by dried amla fruit immersed in water for the whole night is used as an eyewash. It may be applied warm or cold. The seeds can be infused to treat eye conditions including conjunctivitis and other inflammations of the eye. Cataract medicine also contains amla [6].

Glaucoma and red eyes (conjunctivitis) can be treated by using fresh juice of amla. It also improves eyesight. Eye problems are treated by using a mixture of amla and honey. Amla is used to lessen eye weakness and tension within the eyeball. Herbal eye drops (Ophthacare) comprising of different herbs, are used for the treatment of different ophthalmic disorders. Treatment with natural eye drop led to improvements. No side effects of eye drops were observed during this study and patients tolerated this herbal eye drop well [5].

3.7.9 Hepato-Protective

Amla helps to purify the blood and support liver functions. It also strengthens the liver and helps the liver to eliminate toxins from the body [6]. The use of natural liver disease treatments has a long history beginning with Ayurvedic medicine and expanding to the Chinese, European, and other traditional medical systems. A paradigm change toward the therapeutic assessment of herbal items in liver illnesses has occurred in the twenty-first century. It is well-recognized that plant-based medications are important in the treatment of liver conditions. It has been demonstrated that numerous plants and their extracts have hepatoprotective properties. Because of natural and genetic changes, seasonal changes, variations in soil and climate, and changes in the nutritional quality of the medicinal plant, there can be batch-to-batch variations in their efficacies. Numerous research on *Phyllanthus amarus* has demonstrated the hepatoprotective, immunomodulating, and anti-inflammatory activities of this plant as well as an antiviral against the hepatitis B and C viruses [20, 21]. Hepatic damage is mostly caused by inflammation and oxidative stress. The phytoconstituents such as gallic acid, ascorbic acid, tannoid principles and flavonoids in *E. officinalis* protect the liver cells from hazardous molecules that are produced by inflammation or oxidative stress.

Carbon tetrachloride (CCl4) usage over an extended period results in structural and functional hepatic abnormalities and fatty liver disease. Many studies have shown the hepato-protective effect of *E. officinalis* on CCl4-induced acute liver damage. *E. officinalis* also decreased liver penetration and necrosis. This treatment also decreased high levels of alkaline phosphatase, serum glutamate oxaloacetate transferase, lactate dehydrogenase, acid phosphatase, lipid peroxidation, and other oxidative stress factors and irrefutably reversed the pre-fibrogenic effects of CCl4. Acute and chronic hepatic infections, hepatic adenomas, and hepatocarcinoma are all brought on by the medication thioacetamide, which upon biodegradation produces sulfine and sulfene. *E. officinalis* has demonstrated hepatoprotective effects against oxidative stress caused by thioacetamide and hepatocarcinogenesis.

Diethylnitrosamine disrupts the nuclear enzymes involved in DNA replication and repair and causes hepatocarcinoma due to the metabolic generation of promutagens that may serve as hepatic carcinoma initiators. Cure with *E. officinalis* controlled diethylnitrosamine-induced enzymes and also well-maintained hepatic autophagy and apoptosis. The activity was accredited to its antioxidant, anti-apoptotic, anti-inflammatory, and antiautophagy properties [8].

Due to ROS generation and oxidative burden alcohol induced hepatic damage occurs which contributes to mitochondrial defect. Both hepatotoxicity and the protective effect of the E. officinalis extracts were studied in alcohol-intoxicated rat primary hepatocyte cultures. Alcohol intoxication increased the release of transaminase i.e. serum glutamate oxaloacetate transferase and serum glutamate pyruvate transferase in hepatocytes, and also serum transaminases, hepatic triglycerides levels. According to reports, treating *E. officinalis* caused the high levels of alcohol-induced serum glutamate oxaloacetate transferase, serum glutamate pyruvate transferase, Tumor Necrosis Factor and Interleukin 1β to be controlled in a dose-dependent manner. Because flavonoids and tannins can stabilize membranes and serve as antioxidants, many studies have found that different *E. officinalis* extracts are hepato-protective against hepatic mitochondrial dysfunction brought on by alcohol and changes in antioxidant levels [22].

Many studies have also revealed the hepato- and nephron-protective effect of *E. officinalis* extract against oxidative stress caused by ochratoxinA in particular organs. OchratoxinA is a contaminant found in food. Groundwater naturally contains arsenic which can damage numerous antioxidants and mutate DNA, resulting in toxicity of organs including the liver. There was a significant reduction in lipid peroxidation, elevated levels of SOD, catalase, and GSH, and lower hepatic serum transaminase levels after treatment with *E. officinalis* [23]. *E. officinalis* also offered protection from arsenic-induced liver damage by reducing the destructive fragmentation of the necrotic cell's nucleus and cytoplasmic vacuolation.

The body needs iron to function but its adverse effect includes iron toxicity. Iron toxicity cause liver damage due to the unnecessary storing of iron. The main hepatic consequences of iron overload are liver failure, hepatocellular carcinoma and liver cirrhosis. *E. officinalis* have hepatoprotective action of emblicanin-A and emblicanin-B against iron-intoxicated liver damage in rats. Pyrogallol is an important phytochemical component of *E. officinalis* that has a preventative measure against liver cancer [8].

3.7.10 Antimutagenecity and Antigenotoxicity

There is protective action of fruit extract towards lead nitrate-induce chromosomal dysplasia against the appearance of sperm head defects in mice germ cells. The findings unequivocally demonstrate that extract significantly reduced the occurrence of defective sperm heads. That is evident from the findings of above study that *P. emblica* is essential for preventing heavy metal mutagenesis in animals. The protection is related to its antioxidant ability and regulator effect on detoxifying enzymes. Fruit extract from *Emblica officinalis* shielded mice from the chromosome-damaging actions of the carcinogenic 3,4-benzopyrene. *P. emblica* inhibits the activation and mutagenicity of 2-Acetamidofluorene, the cytochrome P-450 and aniline hydroxylase (Fig. 3.3) [6]. Figure 3.3 provides an overview of the pharmacological actions and therapeutic applications of amla, highlighting its various medicinal uses and potential health benefits.

3.7.11 Diarrhea

Amla is used medically for the cure of diarrhea. In dysentery, a mixture of amla with sour milk is given to the groups. A decoction of root solution yields an astringent extract the same as catechu. In case of chronic diarrhea mixture of amla leaves and fenugreek seed is given to patients [6].

Fig. 3.3 Pharmacological actions and therapeutic applications of amla [2]

3.7.12 Promotes Healthier Hair

Amla makes teeth, bones, nails, and hair healthy by increasing the absorption of calcium. It inhibits premature graying of hair and keeps hair follicles strong so they do not thin out with age. The minced fruits promote hair development and stop hair from greying [6]. The usage of dried fruit solids that have been cooked in coconut oil stops hair from going grey. The dried amla bits that are soaked in water overnight are also beneficial to hair [5]. *Phyllanthus emblica* fruit extract has antidiarrheal activity, is used as an anti-spasmodic agent, and cause blockade of Ca^{2+} channels and muscarinic receptors, thus elucidating its medicinal use in diarrhea [24].

3.7.13 Heals Wounds and Ulcers

Both ascorbic acid and tannins (emblicanin A and emblicanin B) have potent antioxidant properties that aid in cell repair. Bactericidal mouthwash is made by a decoction of the leaves. Honey and amla root bark mixture is useful for inflammations of the mouth. The mouthwash made from the decoction of amla leaves is used to treat aphthae. For aphthous stomatitis, root extract applied with honey is recommended [5]. The fruit extract can stimulate neo-angiogenesis, aiding in the healing of stomach lesions. The anti-inflammatory properties enhance ulcer healing even further. Due to its anti-secretory and chemopreventive abilities, *Phyllanthus emblica*, whether used alone or in combination, is a viable natural method to treat numerous chronic illnesses, including ulcers [25].

3.7.14 Antitussive Effect

The airway's laryngopharyngeal and tracheobronchial mucosal regions were mechanically stimulated in conscious cats to assess *E. officinalis*-antitussive efficacy. The ethanol extract of the *E. officinalis* fruits is able to prevent mechanically-induced cough, but only at large doses (200 mg per kg body weight). This suggests that *E. officinalis* possesses antitussive efficacy in conscious cats that are dose-dependent but larger than the antitussive activity of dropropizine, a popular non-narcotic antitussive drug. The dry *E. officinalis* extract is thought to have antitussive properties due to its antispasmolytic, antiphlogistic, and antioxidant effectiveness effects, as well as it affects the airways' ability to secrete mucus [26].

3.7.15 Strengthens the Lungs

Amla is a wonderful tonic for nourishing and strengthening the lungs and the whole respiratory tract. It controls the lungs' moisture balance. Amla fruit with seeds is used for the treatment of asthma and bronchitis [6].

3.7.16 Good for the Skin

Amla supports healthy digestion, aids in liver detoxification, and is a good source of nutrients including vitamin C. Additionally, it helps improve skin tone. Amla nourishes the skin and supports the immunity of the skin against bacterial infection. It increases glow and shines [6]. Timudom et al. studied that amla has a variety of advantages for the skin, including anti-collagenase, anti-elastase, and skin-whitening properties. The goal of this study was to examine the anti-sebum effectiveness of Emblica cleanser on facial skin. The cleanser base was created, enhanced stability was tested, and 10 individuals' preferences were assessed. A single patch test on volunteers was used to determine whether the cleanser base with the greatest preference score was combined with emblica extract. The toners did not irritate the skin and were steady. The toner from Emblica is safe and appears to have anti-sebum efficacy on face skin, and the methodology of assessment of anti-sebum effectiveness utilized in this study is feasible [27].

3.7.17 Helps the Urinary System

As it supports digestion and absorption, amla is exclusively helpful to the urinary system. It is useful for the treatment of mild burning sensations while urinating. It maintains natural diuretic function but does not function like diuretic pills. In simple words, it does not overstimulate the urinary system and helps to eliminate waste from the body [6]. To treat bladder discomforts and retain urine the fruit paste is used alone or in combination with Nelumbium speciosum [28].

3.7.18 Flushes out Toxins

Amla helps the liver in flushing out toxic chemicals in those who have stored preservers and chemicals in their liver from consuming junk food over an extended period.

3.7.19 Improves Immunity

Amla is a potent immunity booster due to all of the aforementioned advantages. Studies have shown that amla fruit has antiviral antibacterial and anti-fungal properties.

3.7.20 Improves Muscle Tone

Amla is good for constructing lean muscle mass and strengthening muscles because it enhances protein synthesis. Its exclusive action offers a natural way for sportsmen and bodybuilders to tone their muscles and add lean mass [6].

3.7.21 Anti-Hyperthyroidism

To assess its ameliorative effects on the L-thyroxine (L-T4)-induced hyperthyroidism and hepatic lipid peroxidation in mice, the ethanolic extract from the fruits of Emblica officinalis was studied. According to the research, *E. officinalis* caused a greater reduction in T3 and T4 levels than propyl thiouracil. In hyperthyroid animals, the plant extract preserved a normal level of glu-6-pase activity. In hyperthyroid rats, the plant extract also showed its hepatoprotective properties by reducing hepatic lipid peroxidation (LPO) and increasing catalase (CAT) and superoxide dismutase (SOD) activity [29].

3.7.22 Stop Nausea, Vomiting and Bleeding of the Nose

To stop emesis, honey is administered along with a powder made from red sandalwood and amla seeds. Seeds have been fried in ghee and crushed, then applied to the forehead to stop nose bleeding [5].

3.7.23 In Water Purification

Natives in certain areas add young branches of amla to the wells to give the water a nice flavor, especially if the water is dirty due to the buildup of plant materials or other factors [5].

3.7.24 Pharmacological Action of Amla and Mechanisms

The increased production of free radicals and their poor scavenging cause pro-oxidant conditions in the body. Disproportion in pro-oxidant and antioxidant homeostasis cause diseases in the body. Amla is a great antioxidant due to high content of Vitamin C present in it. Amla inhibit the growth of cancer due to its anti-proliferative and antioxidant activity. It employs pharmacological activities by different mechanisms. Some mechanisms are listed in Table 3.1.

3.7.25 Uses of Amla As a Home Remedy

Amla fruit has demonstrated a variety of health advantages and medical worth. It has long been used as a common household cure for various illnesses. Table 3.2 shows the conditions and how to utilize amla as a home cure.

Table 3.1 Pharmaceutical activities of Amla and Mechanisms

Pharmaceutical actions of Amla	Description	Mechanism
Antimicrobial	Useful in treatment of asthma, bronchitis and tuberculosis	Show antibacterial and antiviral activity by inhibiting *Candida albicans* adherence to epithelial cells of buccal cavity [30] Pentagalloylglucose, a polyphenolic compound of amla show antiviral activity, inhibiting replication of influenza A virus [31] Substantial decrease in the colony of *E. coli*, *S. aureus*, *Pasteurella multocida* by tube distillation technique [32]
Cardio protective	Function as cardiac stimulant Lowers VLDL, LDL and cholesterol levels Raise HDL levels Prevent atherosclerosis	Prevents oxidative stress caused by ischemia-reperfusion Emblicanin A, B (tannoid principles in fruit) have antioxidant property in vitro and in vivo and have a cardio protective effect [33]
Diabetes	Tannins present in amla, enhances glucose uptake and inhibits adipogenesis Amla and a concentrated portion of its tannoids can prevent rats from developing diabetic cataracts	*E. officinalis* inhibits oxidative stress and TGF-β in diabetic rats and show nephro-protective effect [34] Ellagic acid in *E. officinalis* fruit show antioxidant and antidiabetic activity by inhibiting α-amylase and α-glucosidase [35] Aldose reductase activity and sorbitol production in the lens are both prevented by preventing the oxidative stress brought on by the polyol pathway [36]
Anti-oxidant	Enhance the body's natural antioxidant defense enzymes Acts as a matrix metalloproteinase inhibitor Decreases raised levels of; Urea nitrogen, TBARS Serum creatinine, Decreased COX 2, iNOS expressions	Scavenging of free radicals (2,2-diphenylpicrylhydrazyl) [37, 38] Increase of oxidative stress by inhibition of nuclear factor (NF-$_{K}$B) activation [39]
Cancer	The protective bioactive components of amla act as a preventative against the growth of cancer Lessen the adverse effects of different cancer therapies	Polyphenol fraction of *E. officinalis* inhibit lipid peroxidation, DNA topoisomerase I in mutant cells of *Saccharomyces cerevisiae* and cdc25 tyrosine phosphate [40] Show anti proliferative action and inhibiting cell proliferation in tumor cell lines of humans [41]

(continued)

Table 3.1 (continued)

Pharmaceutical actions of Amla	Description	Mechanism
Hepato-protective	Helps to purify blood Strengthens liver and helps the liver to eliminate toxins from the body	Emblicanin A, B, pedunculagin, puniculagin exhibit antioxidant action, prevent hepatic lipid peroxidation caused by iron and show hepatoprotection [42]
Wound healing	Ascorbic acid and tannins have potent antioxidant properties that aid in cell repair Bactericidal mouthwash is made by a decoction of the leaves. Honey and amla root bark mixture is useful for inflammations of the mouth Laxative to relieve constipation	Due to anti-inflammatory properties, *E. officinalis* is used as potent drug against chronic and acute inflammation as compared to diclofenac [43]
Ulcer protective	Heals and prevent ulcers and infections Enhance ulcer healing due to anti-inflammatory properties	Used as gastro protective agent in NSAIDs-induced gastropathy Amla extract upregulates anti-inflammatory (concentration of cytokine) and capably decrease pro inflammatory TNF-α and 1 L-1β (cytokine) levels [44]
Immunity booster	Acts as immunity booster due to antiviral antibacterial and anti-fungal properties Remarkable source of Vitamin C	Increase the action of natural killer cells and cellular toxicity caused by antibodies [45]
Nephro protective	Protective in age-related renal disorder	Reduce the expression levels of inducible nitric oxide synthase (iNOS) and cyclooxygenase (COX-2) by inhibition of renal nuclear factor ($NF\text{-}_{K}B$) activation [39]
Memory enhancer	Convenient remedy for the management of Alzheimer's disease Improves memory Increase cholinesterase activity of brain	Increased acetylcholinesterase activity increases memory, useful in a reversal of memory deficits [16, 46]
Hair tonic	Increases the absorption of calcium and makes teeth, bones, nails, and hair healthier Inhibits premature graying of hair	Amla is the second potent inhibitor of 5α-reductase inhibitor, also involve in hair growth [47]
Skin aging	Guard the skin against harmful effects of free radicals Amla extract encourages the spread of fibroblasts and help in the production of procollagen	Extract inhibits the matrix metalloproteinase-1 in fibroblast of human skin and promotes the production of procollagen [48]
Diarrhea	Amla extracts are used in the treatment of diarrhea	Have antidiarrheal activity, causing blockage of calcium channels and muscarinic receptors Inhibit diarrhea induced by castor oil [24]

Table 3.2 Home remedies of amla [2, 4]

Home remedy	Use
Anti-aging	Due to presence of a variety of nutrients, aids in maintaining stamina in older individuals, Fresh amla fruit has a refreshing impact on the body
Eye tonic	Healthy dietary supplement to enhance nearsightedness Helpful in the treatment of glaucoma and conjunctivitis Used in eyewash and eye drops
Cough remedy	Mixture of amla with warm milk or honey is a natural remedy for cough
Hair growth	Amla oil is made by grinding dried amla fruits that have been cooked in coconut oil Amla oil acts as a powerful conditioner, stops hair greying and thinning
Treat white spots on the nail	Useful treatment for vitamin deficiency Addition of amla juice or powder to the diet alleviates this problem
Natural treatment for anemia	Vitamin C, a crucial component of amla that aids in the absorption of iron
Blood sugar stabilizer	Regular consumption of bitter gourd juice and amla seeds or dried amla powder in capsule form
Natural cholesterol remedy	It causes a decrease in a reduction of triglycerides, total cholesterol, LDL and VLDL cholesterol A daily routine diet might include dried amla powder
Reduce hypertension	High vitamin C intake helps regulate blood pressure Amla powder or Triphala (a recipe of amla and other herbs) is a potent treatment for hypertension
Menstrual disorders treatment	Amla may be consumed combined with mashed banana or honey and fennel
Treatment of jaundice	A fermented fluid of amla root is used in treatment of jaundice
For itch	A mixture of itch oil with powder of burnt amla seed relieves in itch

3.8 Conclusion

One can understand the abundant phytochemistry of amla as a substantial source of chemicals with potential health advantages. The undeviating dominance of oxidative processes and the promotion of an antioxidant defense system are two key effects of the antioxidant. In addition to its antioxidant properties, amla constituents have a role in promoting health and enhancing resistance to illness. The greatest source of vitamin C is present in it. The equivalent amount of fresh amla is the same as 16 bananas or 3 oranges. Numerous phytochemicals, including flavonoids, tannins, terpenoids, and alkaloids in amla, have been found to have pharmacological actions that include antioxidant, anti-cancer, antigenotoxic, anticancer, and anticarcinogenic activities. It is supposedly a safe herbal medicine with no side effects. Thus, it can be said that amla is a fruit that has been used for centuries and has a long history of therapeutic success.

References

1. Gul, M., Liu, Z.-W., Rabail, R., Faheem, F., Walayat, N., Nawaz, A., et al. (2022). Functional and nutraceutical significance of amla (phyllanthus emblica l.): A review. *Antioxidants, 11*(5), 816.
2. Dasaroju, S., & Gottumukkala, K. M. (2014). Current trends in the research of emblica officinalis (amla): A pharmacological perspective. *International Journal of Pharmaceutical Sciences Review and Research, 24*(2), 150–159.
3. Mathai, R. T., Tonse, R., Kalekhan, F., Colin, M. D., Prabhu, H. S., Rao, S., et al (2015). Amla in the prevention of aging: Scientific validation of the ethnomedicinal claims. In *Foods and dietary supplements in the prevention and treatment of disease in older adults* (pp. 29–35). Elsevier.
4. Kumar, K. S., Bhowmik, D., Dutta, A., Yadav, A. P., Paswan, S., Srivastava S., et al. (2012). Recent trends in potential traditional indian herbs emblica officinalis and its medicinal importance. *Journal of Pharmacognosy and Phytochemistry, 1*(1), 24–32.
5. Priya, F., & Islam, M. S. (2019). Phyllanthus emblica Linn.(amla)—A natural gift to humans: An overview. *Journal of Diseases and Medicinal Plants, 5*, 1–9.
6. Singh, E., Sharma, S., Pareek, A., Dwivedi, J., Yadav, S., & Sharma, S. (2012). Phytochemistry, traditional uses and cancer chemopreventive activity of amla (phyllanthus emblica): The sustainer. *Journal of Applied Pharmaceutical Science., 2*(1), 176–183.
7. Wali, V., Bakshi, P., Jasrotia, A., Bhushan, B., & Bakshi, M. (2015). In Directorate of Extension (Ed.), *Aonla* (pp. 1–30). SKUAST-Jammu.
8. Variya, B. C., Bakrania, A. K., & Patel, S. S. (2016). Emblica officinalis (amla): A review for its phytochemistry, ethnomedicinal uses and medicinal potentials with respect to molecular mechanisms. *Pharmacological Research, 111*, 180–200.
9. Gantait, S., Mahanta, M., Bera, S., & Verma, S. K. (2021). Advances in biotechnology of emblica officinalis gaertn. Syn. Phyllanthus emblica l.: A nutraceuticals-rich fruit tree with multifaceted ethnomedicinal uses. *3 Biotech, 11*(2), 1–25.
10. Banyal, S. K., & Banyal, A. K. (2019). Refinement of propagation techniques of aonla (emblica officinalis gaertn.) in north western himalayan region. *Int J Bio-resource Stress Manag., 10*, 87–91.
11. Singh, A., & Singh, H. (2015). Application of plant growth regulators to improve fruit yield and quality in indian gooseberry (emblica officinalis gaertn.). *Journal of AgriSearch, 2*(1), 20–23.
12. Mishra, S. M., Ram, D., Pandey, A., & Meena, A. K. (2017). Effect of foliar feeding of micronutrients yield of aonla fruits (emblica officinalis gaertn). *Progressive Research An International Journal, 12*, 932–934.
13. Vashisht, B., Singh, C., & Biwalkar, N. (2018). Establishment and growth of aonla (emblica officinalis) as affected by irrigation and mulching in the shivaliks of Punjab. *Journal of Soil and Water Conservation, 17*, 98–101.
14. Das, D., Chaturvedi, O., Jha, R., & Kumar, R. (2011). *Yield, soil health and economics of aonla (emblica officinalis gaertn.)-based Agri-horticultural systems in eastern India* (Vol. 101, pp. 786–790). Current Science.
15. Mahata, S., Pandey, A., Shukla, S., Tyagi, A., Husain, S. A., Das, B. C., et al. (2013). Anticancer activity of phyllanthus emblica Linn.(indian gooseberry): Inhibition of transcription factor ap-1 and hpv gene expression in cervical cancer cells. *Nutrition and Cancer, 65*(sup1), 88–97.
16. Vasudevan, M., & Parle, M. (2007). Effect of anwala churna (emblica officinalis g aertn.): An ayurvedic preparation on memory deficit rats. *Yakugaku Zasshi, 127*(10), 1701–1707.
17. Dhingra, D., Joshi, P., Gupta, A., & Chhillar, R. (2012). Possible involvement of monoaminergic neurotransmission in antidepressant-like activity of emblica officinalis fruits in mice. *CNS neuroscience & therapeutics., 18*(5), 419–425.
18. Wankhar, D., Devi, R. S., & Ashok, I. (2014). Emblica officinalis outcome on noise stress induced behavioural changes in wistar albino rats. *Biomedicine & Preventive Nutrition, 4*(2), 219–224.

19. Yang, B., & Liu, P. (2014). Composition and biological activities of hydrolyzable tannins of fruits of phyllanthus emblica. *Journal of Agricultural and Food Chemistry, 62*(3), 529–541.
20. Girish, C., & Pradhan, S. C. (2012). Indian herbal medicines in the treatment of liver diseases: Problems and promises. *Fundamental & Clinical Pharmacology, 26*(2), 180–189.
21. Thyagarajan, S., Jayaram, S., Gopalakrishnan, V., Hari, R., Jeyakumar, P., & Sripathi, M. (2002). Herbal medicines for liver diseases in India. *Journal of Gastroenterology and Hepatology, 17*, S370–S376.
22. Pramyothin, P., Samosorn, P., Poungshompoo, S., & Chaichantipyuth, C. (2006). The protective effects of phyllanthus emblica Linn. Extract on ethanol induced rat hepatic injury. *Journal of Ethnopharmacology, 107*(3), 361–364.
23. Kamp, H. G., Eisenbrand, G., Janzowski, C., Kiossev, J., Latendresse, J. R., Schlatter, J., et al. (2005). Ochratoxin a induces oxidative DNA damage in liver and kidney after oral dosing to rats. *Molecular Nutrition & Food Research, 49*(12), 1160–1167.
24. Mehmood, M. H., Siddiqi, H. S., & Gilani, A. H. (2011). The antidiarrheal and spasmolytic activities of phyllanthus emblica are mediated through dual blockade of muscarinic receptors and ca2+ channels. *Journal of Ethnopharmacology, 133*(2), 856–865.
25. Pal, A. D. (2018). Phyllanthus emblica: The superfood with anti-ulcer potential. *International Journal of Food Science & Nutrition, 3*(1), 84–87.
26. Gaire, B. P., & Subedi, L. (2014). Phytochemistry, pharmacology and medicinal properties of phyllanthus emblica Linn. *Chinese Journal of Integrative Medicine, 9*, 1–8.
27. Timudom, T., Chaiyasut, C., Sivamaruthi, B. S., Tiampasook, P., & Nacapunchai, D. (2020). Anti-sebum efficacy of phyllanthus emblica l.(emblica) toner on facial skin. *Applied Sciences, 10*(22), 8193.
28. Grover, M. A comprehensive review on pharmacological and ayurvedic aspect of phyllanthus emblica (amalki). *Advance Pharmaceutical Journal, 6*(3), 87–94.
29. Panda, S., & Kar, A. (2003). Fruit extract of emblica officinalis ameliorates hyperthyroidism and hepatic lipid peroxidation in mice. *Die Pharmazie-An International Journal of Pharmaceutical Sciences, 58*(10), 753–755.
30. Thaweboon, B., & Thaweboon, S. (2011). Effect of phyllanthus emblica Linn. On candida adhesion to oral epithelium and denture acrylic. *Asian Pacific Journal of Tropical Medicine, 4*(1), 41–45.
31. Liu, G., Xiong, S., Xiang, Y.-F., Guo, C.-W., Ge, F., Yang, C.-R., et al. (2011). Antiviral activity and possible mechanisms of action of pentagalloylglucose (pgg) against influenza a virus. *Archives of Virology, 156*(8), 1359–1369.
32. Patil, S. G., Deshmukh, A., Padol, A. R., & Kale, D. B. (2012). In vitro antibacterial activity of emblica officinalis fruit extract by tube dilution method. *International Journal of Toxicology and Applied Pharmacology, 2*(4), 49–51.
33. Bhattacharya, S. K., Bhattacharya, D., Sairam, K., & Ghosal, S. (2002). Effect of bioactive tannoid principles of emblica officinalis on ischemia-reperfusion-induced oxidative stress in rat heart. *Phytomedicine, 9*(2), 171–174.
34. Suryavanshi, S. V., Garud, M. S., Barve, K., Addepalli, V., Utpat, S. V., & Kulkarni, Y. A. (2020). Triphala ameliorates nephropathy via inhibition of tgf-β1 and oxidative stress in diabetic rats. *Pharmacology, 105*(11–12), 681–691.
35. Nampoothiri, S. V., Prathapan, A., Cherian, O. L., Raghu, K., Venugopalan, V., & Sundaresan, A. (2011). In vitro antioxidant and inhibitory potential of terminalia bellerica and emblica officinalis fruits against ldl oxidation and key enzymes linked to type 2 diabetes. *Food and Chemical Toxicology, 49*(1), 125–131.
36. Suryanarayana, P., Saraswat, M., Petrash, J. M., & Reddy, G. B. (2007). Emblica officinalis and its enriched tannoids delay streptozotocin-induced diabetic cataract in rats. *Molecular Vision, 13*, 1291–1297.
37. Prakash, D., Upadhyay, G., Gupta, C., Pushpangadan, P., & Singh, K. (2012). Antioxidant and free radical scavenging activities of some promising wild edible fruits. *International Food Research Journal, 19*(3), 1109–1116.

38. Prakash, D., Upadhyay, G., Pushpangadan, P., & Gupta, C. (2011). Antioxidant and free radical scavenging activities of some fruits. *Journal of Complementary and Integrative Medicine, 8*(1), 1–16.
39. Yokozawa, T., Kim, H. Y., Kim, H. J., Tanaka, T., Sugino, H., Okubo, T., et al. (2007). Amla (emblica officinalis gaertn.) attenuates age-related renal dysfunction by oxidative stress. *Journal of Agricultural and Food Chemistry, 55*(19), 7744–7752.
40. Rajeshkumar, N., Pillai, M. R., & Kuttan, R. (2003). Induction of apoptosis in mouse and human carcinoma cell lines by emblica officinalis polyphenols and its effect on chemical carcinogenesis. *Journal of experimental & clinical cancer research: CR, 22*(2), 201–212.
41. Luo, W., Zhao, M., Yang, B., Ren, J., Shen, G., & Rao, G. (2011). Antioxidant and antiproliferative capacities of phenolics purified from phyllanthus emblica l. fruit. *Food Chemistry, 126*(1), 277–282.
42. Bhattacharya, A., Kumar, M., Ghosal, S., & Bhattacharya, S. (2000). Effect of bioactive tannoid principles of emblica officinalis on iron-induced hepatic toxicity in rats. *Phytomedicine, 7*(2), 173–175.
43. Santoshkumar, J., Devarmani, M. S., Sajjanar, M., Pranavakumar, M., & Dass, P. (2013). A study of anti-inflammatory activity of fruit of emblica officinalis (amla) in albino rats. *Medica Innovatica, 2*(1), 17–26.
44. Chatterjee, A., Chattopadhyay, S., & Bandyopadhyay, S. K. (2010). Biphasic effect of phyllanthus emblica l. extract on nsaid-induced ulcer: An antioxidative trail weaved with immunomodulatory effect. *Evidence-based Complementary and Alternative Medicine, 2011*, 146808.
45. Suresh, K., & Vasudevan, D. (1994). Augmentation of murine natural killer cell and antibody dependent cellular cytotoxicity activities by phyllanthus emblica, a new immunomodulator. *Journal of Ethnopharmacology, 44*(1), 55–60.
46. Vasudevan, M., & Parle, M. (2007). Memory enhancing activity of anwala churna (emblica officinalis gaertn.): An ayurvedic preparation. *Physiology & Behavior, 91*(1), 46–54.
47. Kumar, N., Rungseevijitprapa, W., Narkkhong, N.-A., Suttajit, M., & Chaiyasut, C. (2012). 5α-reductase inhibition and hair growth promotion of some thai plants traditionally used for hair treatment. *Journal of Ethnopharmacology, 139*(3), 765–771.
48. Fujii, T., Wakaizumi, M., Ikami, T., & Saito, M. (2008). Amla (emblica officinalis gaertn.) extract promotes procollagen production and inhibits matrix metalloproteinase-1 in human skin fibroblasts. *Journal of Ethnopharmacology, 119*(1), 53–57.

Chapter 4
Belladonna

Sadia Javed, Asmara Ahmad, Muhammad Sajid Hamid Akash, Kanwal Rehman, and Arwa A. Al-Huqail

4.1 Introduction

Belladonna is a perennial herbaceous plant that produces an unpleasant odor [1]. The scientific name of belladonna is *Atropa belladonna* belonging to he Solanaceae, which is a widely distributed and famous botanical family [2]. The genus name Atropa is derived from the Greek word "Atropos", meaning inexorable, and the specie name belladonna is an italic word that means beautiful lady. The genus Atropa includes four species, each with five-lobed flowers with an alternative leaf pattern [3, 4]. The plant is given this name because of its use as a beautifying agent by women in Renaissance, Italy [5]. The common names of this plant are belladonna, deadly nightshade, gray morel, devil's herb, dwale, poison black cherry, naughty man's cherries, devil's cherries, divale, and dwayberry [6].

During the Middle Ages, the *Atropa belladonna* plant extract was employed by women as eye drops that made their facial features attractive by dilating the pupils of their eyes [7]. However, belladonna was no longer employed as a cosmetic product due to its harmful effects on humans, including enhanced heart rate, blurred vision, and ultimately results in blindness [4]. The Romans military people

S. Javed (✉) · A. Ahmad
Department of Biochemistry, Government College University, Faisalabad, Pakistan
e-mail: sadiajaved@gcuf.edu.pk

M. S. H. Akash
Department of Pharmaceutical Chemistry, Government College University Faisalabad, Faisalabad, Pakistan

K. Rehman
Department of Pharmacy, The Women University Multan, Multan, Pakistan

A. A. Al-Huqail
Department of Biology, College of Science, Princess Nourah bint Abdulrahman University, Riyadh, Saudi Arabia

M. Zia-Ul-Haq et al. (eds.), *Essentials of Medicinal and Aromatic Crops*,
https://doi.org/10.1007/978-3-031-35403-8_4

commonly employed the belladonna plant extract as a biological agent for intoxicating the food supply of their enemies. Likewise, the poison-tipped arrows were made by the archers using the toxic paste of belladonna extract [8]. It has been stated that recently more than 15,000 lethal cases related to exposure to greater doses of belladonna alkaloids have occurred [9].

The Solanaceae are well known for their applications in the food industries, but unfortunately, these are absurdly lethal due to the higher contents of alkaloids in them. However, a subgroup of these plants is hallucinogenic and has medicinal properties [2]. One of these plants is *Atropa belladonna* has been reported to have been employed as a narcotic, pharmaceutical, haunting, and warfare agent for many years in Europe [10–12].

The different parts of *Atropa belladonna* contain different amounts of tropane alkaloids, which exhibit both therapeutic and toxic properties when applied as herbal medicine [5]. This chapter will highlight the morphological characteristics, agronomy, and diseases caused by pests' attacks to belladonna. In addition, it will also briefly discuss the therapeutic applications of belladonna in the field of medicine.

4.2 Morphology

Atropa Belladonna has a height ranging from 1.5 to 2 m, a branched pubescent stem, and large acute, ovate-lanceolate, and petiolate leaves of 8–15 cm in length [13]. As depicted in Fig. 4.1, it has a bell-shaped, purple-colored flowers that are solitary, located in the axil between stem and leaves, and grow laterally [4]. At the

Fig. 4.1 *Atropa belladonna*

time of fruiting, the calyx part of the flower is campanulate and pubescent, having a length of 3 cm and being separated into five ovate-lanceolate, acute sepals. The corolla part, which helps in plants' reproduction, is campanulate, tubular, and has five thin lobes at its upper terminal, with greenish-yellow shade at the bottom end and blue-violet shade at the top end. There are five stamens containing unequal filaments with rounded or ovate anthers that are placed at the base of the corolla [1]. Moreover, the fruits belladonna produces are spherical-shaped berries whose size range is 13–18 mm in diameter. These berries comprise numerous kidney-shaped seeds with a sweet taste and purple juice [4].

4.3 Agronomy

Atropa belladonna starts flourishing in May, keeps healthy until summer, and blossoms again in the autumn [1]. The plant species are cultivated around the globe but are particularly native to Western Asia, North Africa, Europe, and occasionally as ornamental plants in the United States. The Eastern Hemisphere has been known for practicing the hallucinogenic effect of *Atropa belladonna* [5]. It requires moderate climatic conditions, such as clear and sunny weather especially during the crop harvesting season. It is unable to survive water-logged situations as the increased level of humidity and constant dampness cause root deterioration [14].

The cultivation of belladonna can be carried out in several places, like wooded hills, hedgerows, shady areas, quarries, grasslands, and slopes [15]. But the favorable soil conditions include shady, humid, well-drained, slightly acidic, nitrified, silty-loam to clayey-loam, and an abundant amount of humus-containing area with an altitude of 400–2000 m [1]. Studies have demonstrated that it can grow 10–1800 m in height in shaded trees, in shear-cut open areas, specifically in beech and spruce forests [13].

Regarding the crop plantation, it is done either by the vegetative method via cutting the roots and shoots or by propagating the seeds. The seed propagation method is the easiest and most cost-effective approach. In this method, seeds are carefully chosen from a plant with a high content of alkaloids. The germination of seeds is enhanced by treating them either with ethanol for 3 min or with petroleum for 6 min, and then the seeds are disseminated in the nursery beds at the start of the spring season. 15 to 40% of the germination occurred within 10–21 days. Then, during the month of August, the transplantation of germinating seedlings carrying 1–3 leaves is carried out by planting them in the field at a distance of 45–60 cm. Moreover, before plantation, farmyard manure, potassium oxide, and diammonium phosphate are also applied to the cultivated land, whereas at the time of branching and picking, nitrogen fertilizer is applied. However, during summer, irrigation of crops is done every 10–15 days [14].

4.4 Pests and Viruses Infecting Belladonna

The pests that mainly attack *Atropa belladonna* are flea beetles, Epitrix atropae, Foudras, and Epitrix pubescens, Koch. The genus Epitrix is widely distributed and comprises almost 180 species [16]. These small beetles mostly attack the plants of Solanaceae and riddle the leaves of plants [17]. During the summer season, the flea beetles oviposit in the soil, and within 7–8 days, eggs hatch into larvae. Larvae remain in the soil and are converted to pupa after about 1 week. The other pests that attack the belladonna are Psylliodes hyoscyami, L., and caterpillars of Barathra (Mamestra) brassicae, L., and Heliothis peltigera, Schiff [18].

Studies also demonstrated that the members of the Solanaceae family are being infected by several pathogenic viruses. Employing high-throughput sequencing technologies, the International Committee on Taxonomy of Viruses (ICTV) has identified six orders, 32 families, and 141 genera, containing 1901 species of plant viruses. The mode of transmission of plant viruses may be vertical or horizontal, depending on the viral species. The viruses attacking the Solanaceae plants use insect vectors, especially aphids, as their mode of transmission. The viruses identified to infect belladonna are belladonna mottle virus (BeMV) and henbane mosaic virus (HMV). BeMV is transmitted through Epitrix atropae, whereas HMV is transmitted via aphids [19].

4.5 Chemical Constituents

The principle chemical constituents of *Atropa belladonna* are 20 different types of tropane alkaloids, including atropine, atropamine, apoatropine, belladonnine, cuscohygrine, 1-hyoscyamine, 6-β-hyoscyamine, norhyoscyamine, N-methylpyrroline, N-methylpyrrolidine, scopolamine, and tropine [4, 15, 20]. l-hyoscyamine and atropine are present in the highest amounts in leaves, stems, roots and fruits. They both contribute 99% of the total alkaloid content present in the leaves [1, 3]. Out of the total alkaloid complex, hycosamine was observed to contribute 87.6% in the leaves and 68.7% in the roots [15]. The concentration of atropine in the fruit is 0.1%, whereas in the roots it ranges between 0.4% and 0.6%. It has been observed that the concentration of alkaloids varies from one plant to other due to varying environmental factors like geographical area, harvesting season, soil type, and climatic conditions [1]. The plan's chemical defense system involves several alkaloids like hyoscyamine, scopolamine, and atropine which help protect the plant against various abiotic and biotic stresses [15].

In order to identify the alkaloids present in the plant extract of belladonna, several separating techniques are being employed. These include capillary electrophoresis coupled to UV detection, high-performance liquid chromatography (HPLC) coupled to UV detection, thin layer chromatography (TLC) coupled to UV detection, high-performance liquid chromatography-mass spectrometry (HPLC-MS) and gas chromatography-mass spectrometry (GC-MS), quadrupole coupled to time-of-flight analyzers (Q-TOF), and high-resolution mass analyzers like Orbitrap [21]. The analysis carried out by GLC and GLC-MS showed that the roots of belladonna plant contain 13 alkaloids while 7 alkaloids are distributed in other parts of the plant [4].

4.6 Biosynthesis of Belladonna Alkaloids

The biosynthesis of tropane alkaloids is mainly carried out in the roots of belladonna, from where they are transported to the stem, leaves, and fruits. The key enzyme involved in the formation of tropane alkaloids is S-adenosylmethionine dependent Putrescine N-Methyltransferase (PMT) causes the methylation of a diamine, putrescine, into N-methyl putrescine, which is a primary metabolite in these biosynthetic reactions [22, 23]. The N-methyl putrescine, then converted to tropinone through a series of intermediate compounds like 4-methylaminobutanal, N-methyl-pyrrolium cation, and hygrine. Then tropinone reductase I catalyzes the conversion of tropinone into tropine [24]. Afterwards, the condensation reaction of phenyl-lactate with tropine results in the synthesis of littorine, which undergoes oxidation and enzymatic rearrangement to form hyoscyamine. Furthermore, hyoscyamine-6-hydroxylase catalyzes the conversion of hyoscyamine to scopolamine [4] (Fig. 4.2).

4.7 Mechanism of Action of Belladonna Alkaloids

Belladonna tropane alkaloids work as competitive inhibitors of all the subtypes (M1–M5) of muscarinic acetylcholine receptors (mAChRs) [25, 26]. These are G-protein-coupled receptors, found in several parts of human body. M1 receptors are excitatory in nature, found both in the central and peripheral nervous system, and involve the excitation of the CNS, regulation of homeostasis, and gastric parietal cells for secretion of gastric acid. The M2 receptor has inhibitory function and is located in cardiac cells. The M3 subtype receptors are found in exocrine glands, endothelium of the vascular system, and smooth muscles of the bronchi, gastrointestinal tract, and urinary system. These receptors seem to be involved in vasodilation, glandular secretions, and the contraction of smooth muscles. Whereas M4 and M5 receptors are mostly present in the CNS, controlling attention, numbness, arousal, and memory [4].

The mechanism of action of tropane alkaloids is depicted in Fig. 4.3. The affinity of atropine molecules is stronger for bronchial, heart, and gastrointestinal muscles. While scopolamine acts strongly on the iris and ciliary body and greatly influences the reduction of the bronchial, salivary, and sweat gland secretions [25, 26].

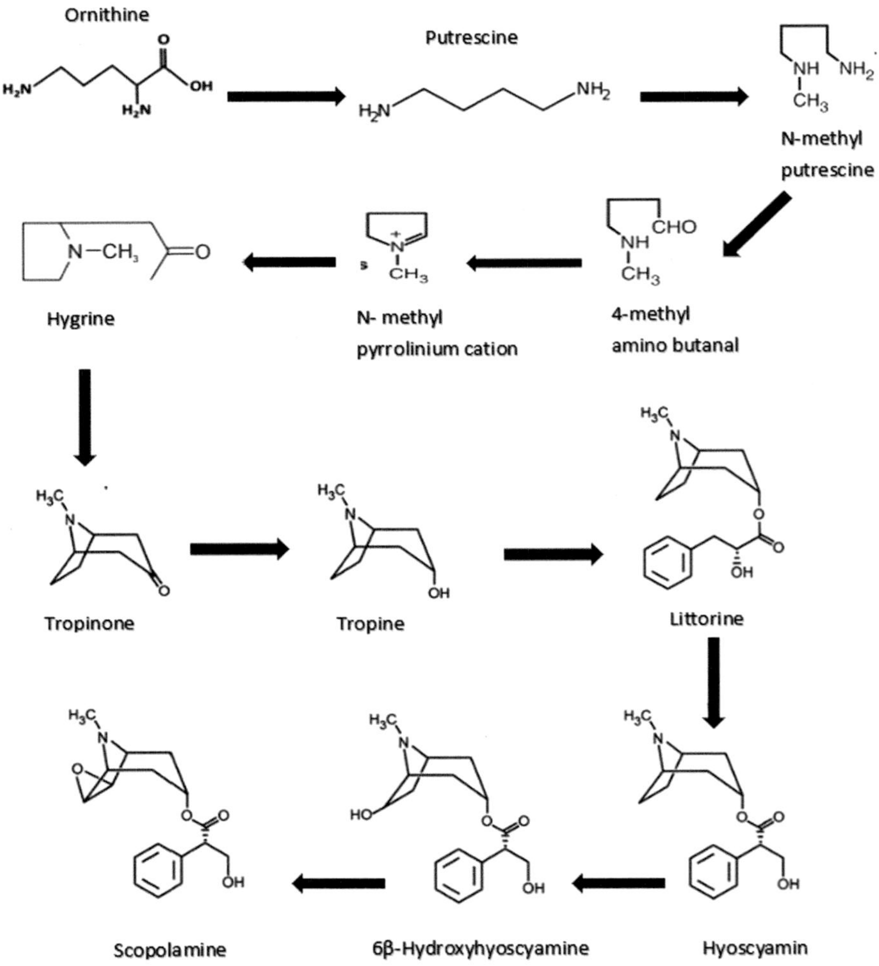

Fig. 4.2 Biosynthetic pathway of belladonna alkaloids

4.8 Pharmacodynamics of Belladonna

Belladonna alkaloids stimulate the CNS by functioning as antagonist of muscarinic receptors present in the brain [26]. The lower concentration of tropine shows no significant effect, whereas hyoscyamine is effective even at low dosages. Atropine also involves the stimulation of respiratory, vagal, and vasomotor centers located in the medulla. Belladonna has a significant effect on the cardiovascular system. It is responsible for causing tachycardia, especially in adults. However, no noticeable or consistent impact of belladonna alkaloid on blood pressure was observed [27]. The

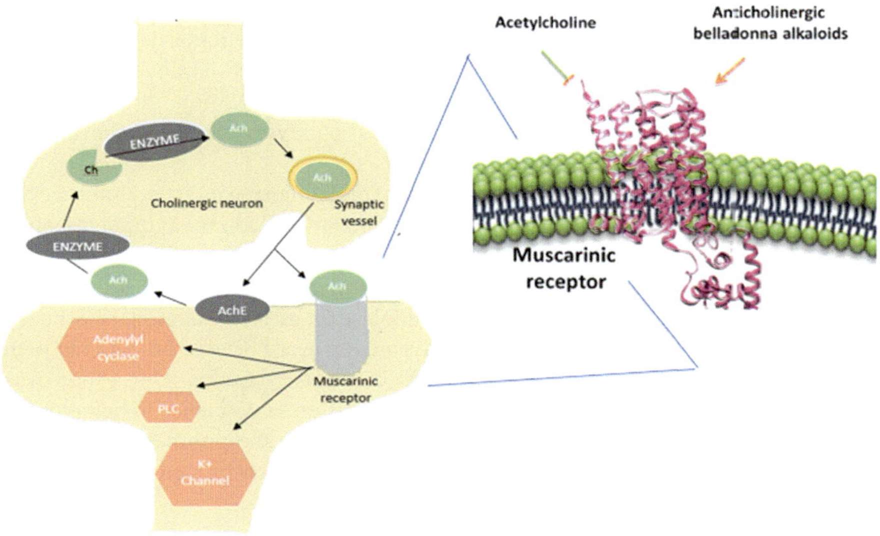

Fig. 4.3 Mechanism of action of belladonna alkaloids. (Modified from Ref. [4])

mydriatic property of belladonna is being employed in ophthalmology. Additionally, it helps to examine the retina and other deep eye structures [1].

Atropine also seems to be involved in the relaxation of all the visceral smooth muscles receiving signals from parasympathetic motor neurons. It functions to relax the stomach, intestine, and urinary system [28]. Belladonna Alkaloids are also able to paralyze the nerves of the pharynx by controlling the reflex action of swallowing. The berries of belladonna are utilized in Moroccan culture as aphrodisiacs and euphoria, and in students for improving memory [1].

4.9 Pharmacotherapeutic Role of Belladonna

Atropa belladonna, being rich in tropane alkaloids, has a number of medicinal applications. It possesses anticholinergic, antimicrobial, anti-inflammatory, anticonvulsant, antispasmodic, analgesic, anesthetic, and mydriatic characteristics [1, 4]. Plant extracts are employed in several homeopathic and herbal drugs to treat local inflammation, headache, and scarlet fever [21]. It is also utilized as an anticholinergic agent in cough suppressants, as a sedative and bronchodilator in asthma, whopping cough, and chronic obstructive pulmonary diseases for prevention against bronchospasm, and as an analgesic agent in rheumatoid arthritis, Parkinson's disorder, colic, and neuralgia. Moreover, it is also the fundamental part of plasters used in the treatment of psychiatric disorders [15, 27, 29].

The topical instillation of atropine helps to abolish cycloplegia and light reflexes as well as cause the enhancement of intraocular pressure, particularly in the condition of narrow-angle glaucoma [27]. Furthermore, it is widely employed as an antidote drug to treat organophosphate poisoning. The application of atropine as an antidote had been carried out in the past to treat paralysis due to lethal nerve gas used in the Second World War [4].

A study was conducted to observe the effect of belladonna extract on viral encephalitis. This study demonstrated that the ultra-dilution of belladonna is quite effective in reducing the encephalitis virus infection in the chorioallantoic membrane of chicks due to calystegines that act as inhibitors of glycosidase [23].

Another clinical study was conducted on adolescent girls to examine the efficacy of belladonna in the treatment of primary dysmenorrhea. For this, 30 patients with primary dysmenorrhea were taken and divided into two groups. One group was treated as a placebo and the other was treated with different potencies of belladonna. The results demonstrated that, as compared to placebo, the number of improved patients was greater in belladonna treatment group, which proved belladonna is quite efficient in curing primary dysmenorrhea [30].

4.9.1 Role in the Treatment of Depression

The cholinergic system of the CNS might be the reason for depression when it works hyperactively. So, the problems associated with depression can be cured by utilizing the anticholinergic compounds. Belladonna alkaloids, particularly scopolamine, prove to be effective therapeutic substances that are employed to treat depression [31]. Scopolamine has been observed to show resistance against conventional antidepressants such as serotonin-nor-adrenaline reuptake inhibitors (SNRIs) or selective serotonin reuptake inhibitors (SSRIs). It was observed that scopolamine quickly helps in relieving the symptoms of depression for a longer period of time by working as a non-selective muscarinic acetylcholine receptors (mAChRs) inhibitor [32].

4.9.2 Role in the Treatment of Parkinsonism

The disproportion of neurotransmitters like acetylcholine or dopamine in the basal ganglia causes parkinsonism. The signs and symptoms of this disorder are tremor, inflexibility, bradykinesia, and postural instability. Belladonna plant extracts seem to be quite effective in the management of parkinsonism. Scopolamine works as an inhibitor of the acetylcholine receptor and thus helps to control the extreme salivation in Parkinson's patients [4, 33].

4.9.3 Role in the Treatment of Motion Sickness

Motion sickness is related to travelling, tempting bewilderment, queasiness, and exhaustion due to head motion [34]. The treatment employed presently is not very efficient in controlling motion sickness. However, scopolamine, being anticholinergic in nature, seems to be an effective therapeutic approach to deal with motion sickness. Studies have shown that, in comparison to conventional therapeutic approaches, transdermal application of scopolamine gave good results in alleviating the secretions and motility of the gastrointestinal system, sweat, and salivary glands [4].

4.10 Toxicology of Belladonna Alkaloids

The dosage belladonna higher than the recommended amount (10 mg of atropine) leads to acute toxicity. This is due to anticholinergic activity, which causes anticholinergic syndrome [7, 35]. The symptoms of this syndrome include ataxia, bewilderment, hallucinations, amnesia, seizures, coma, agitated delirium, psychosis, cortical excitation, respiratory failure, and cardiovascular collapse [27].

The other lethal effects of belladonna affecting the peripheral system are acute hypertension, cycloplegia, mydriasis, hyperreflexia, urinary retention, dry mucous membranes, flushed skin, and reduced bowel sounds [4, 20]. The 0.3 mg/L and 0.005 mg/L concentrations of atropine and scopolamine, respectively, in the peripheral blood seemed to be lethal [36].

Studies demonstrated that the chronic toxicity of belladonna is very rare and occurrs by adverse effects of acute overdose. For instance, the condition of induced psychosis is linked with chronic toxicity due to an overdose of alkaloids for an extended time. In addition, no immunotoxicity has been observed related to alkaloids of belladonna. However, IgE antibody reactions develop in those patients that are allergic to atropine or scopolamine. Furthermore, its genotoxic or carcinogenic studies has not been seen in the literature yet [26].

4.11 Conclusion

We have found that *Atropa belladonna* is quite an attractive and interesting plant. Due to limited research and insufficient information, it is mostly regarded as a poisonous plant. But it needs to be explored more for its medicinally important alkaloids, and then, if it is treated carefully, it might have far more therapeutic applications than unwanted toxic side effects. The study of literature also tells us that its safe handling and proper regulation of its dosage is extremely helpful in minimizing its toxicology. In order to get greater therapeutic benefit from *Atropa Belladonna,* there is a need to synthesize dosage dependent customized drugs.

References

1. Cano Ortiz, A., Piñar Fuentes, J. C., & Cano, E. (2022). Some medicinal plants of interest for their content in alkaloids I. *Biomedical Journal of Scientific & Technical Research, 42*(3), 33702–33705.
2. Fatur, K., & Kreft, S. (2021). Nixing the nightshades: Traditional knowledge of intoxicating members of the solanaceae among hallucinogenic plant and mushroom users in Slovenia. *PLoS One, 16*(2), e0247688.
3. Passos, I. D., & Mironidou-Tzouveleki, M. (2016). *Hallucinogenic plants in the mediterranean countries. Neuropathology of drug addictions and substance misuse* (pp. 761–772). Elsevier.
4. Maurya, V. K., Kumar, S., Kabir, R., Shrivastava, G., Shanker, K., Nayak, D., et al. (2020). Dark classics in chemical neuroscience: An evidence-based systematic review of belladonna. *ACS Chemical Neuroscience, 11*(23), 3937–3954.
5. Kwakye, G. F., Jiménez, J., Jiménez, J. A., & Aschner, M. (2018). Atropa belladonna neurotoxicity: Implications to neurological disorders. *Food and Chemical Toxicology, 116*, 346–353.
6. Lacković, Z. (2017). "Bunanje": XX century abuse of atropa belladonna halucinogenic berries in continental Croatia. *Psychiatria Danubina, 29*(3), 379–382.
7. Berdai, M. A., Labib, S., Chetouani, K., & Harandou, M. (2012). Case report-atropa belladonna intoxication: A case report. *Pan African Medical Journal, 11*(1).
8. Smulyan, H. (2018). *The beat goes on: The story of five ageless cardiac drugs* (pp. 441–450). Elsevier.
9. Zhang, X. C., Farrell, N., Haronian, T., & Hack, J. (2017). Postoperative anticholinergic poisoning: Concealed complications of a commonly used medication. *The Journal of Emergency Medicine, 53*(4), 520–523.
10. Carruthers, D. M. (2015). Lines of flight of the deadly nightshade: An enquiry into the properties of the magical plant, its literature and history. *Mosaic: A Journal for the Interdisciplinary Study of Literature, 48*(2), 119–132.
11. Ruck, C. A. (2014). Entheogens in ancient times. In *History of toxicology and environmental health: Toxicology in antiquity* (Vol. 2, pp. 116–125). Elsevier.
12. Dajić Stevanović, Z., Petrović, M., & Aćić, S. (2014). Ethnobotanical knowledge and traditional use of plants in Serbia in relation to sustainable rural development. In *Ethnobotany and biocultural diversities in the balkans* (pp. 229–252). Springer.
13. Öz, M., Fidan, M. S., Baltaci, C., Ücüncü, O., & Karatas, S. M. (2021). Determination of antimicrobial and antioxidant activities and chemical components of volatile oils of atropa belladonna L. growing in Turkey. *Journal of Essential Oil Bearing Plants, 24*(5), 1072–1086.
14. Alamgir, A. (2017). Cultivation of herbal drugs, biotechnology, and in vitro production of secondary metabolites, high-value medicinal plants, herbal wealth, and herbal trade. In *Therapeutic use of medicinal plants and their extracts: Volume 1* (pp. 379–452). Springer.
15. Tian, H., Ghorbanpour, M., & Kariman, K. (2018). Manganese oxide nanoparticle-induced changes in growth, redox reactions and elicitation of antioxidant metabolites in deadly nightshade (atropa belladonna L.). *Industrial Crops and Products, 126*, 403–414.
16. Bieńkowski, A. O., & Orlova-Bienkowskaja, M. J. (2016). Key to holarctic species of epitrix flea beetles (coleoptera: Chrysomelidae: Galerucinae: Alticini) with review of their distribution, host plants and history of invasions. *Zootaxa, 4175*(5), 401–435.
17. Deczynski, A. M. (2016). Morphological systematics of the nightshade flea beetles epitrix foudras and acallepitrix bechyné (coleoptera: Chrysomelidae: Galerucinae: Alticini) in America North of Mexico. *All theses*. 2479.
18. Parfentjev, I. (1921). Insect pests of medicinal plants in the crimea. *Bulletin de la Société de Pathologie Exotique, 14*(3), 164–167.
19. Hančinský, R., Mihálik, D., Mrkvová, M., Candresse, T., & Glasa, M. (2020). Plant viruses infecting solanaceae family members in the cultivated and wild environments: A review. *Plants, 9*(5), 667.

20. Wendt, S., Lübbert, C., Begemann, K., Prasa, D., & Franke, H. (2022). Poisoning by plants. *Deutsches Ärzteblatt International, 119*(18), 317.
21. Marín-Sáez, J., Romero-González, R., Garrido Frenich, A., & Egea-González, F. J. (2018). Screening of drugs and homeopathic products from atropa belladonna seed extracts: Tropane alkaloids determination and untargeted analysis. *Drug Testing Analysis, 10*(10), 1579–1589.
22. Biastoff, S., Brandt, W., & Dräger, B. (2009). Putrescine n-methyltransferase–the start for alkaloids. *Phytochemistry, 70*(15–16), 1708–1718.
23. Bandyopadhyay, B., Das, S., Sengupta, M., Saha, C., Das, K. C., Sarkar, D., et al. (2010). Decreased intensity of Japanese encephalitis virus infection in chick chorioallantoic membrane under influence of ultradiluted belladonna extract. *American Journal of Infectious Diseases, 6*(2), 24–28.
24. Ziegler, J., & Facchini, P. J. (2008). Alkaloid biosynthesis: Metabolism and trafficking. *Annual Review of Plant Biology, 59*, 735.
25. Riad, M., & Hithe, C. C. (2021). *Scopolamine. Statpearls [internet]*. StatPearls Publishing.
26. Prusakov, P. (2014). Belladonna alkaloids. E*ncyclopedia of Toxicology* (Vol. 1, pp. 244–245). Elsevier.
27. Agrawal, S. (2018). Atropine: The sagacious molecule. *Journal of Traditional Medicine and Clinical Naturopathy, 7*(264), 2.
28. Fatur, K., & Kreft, S. (2020). Common anticholinergic solanaceaous plants of temperate Europe-a review of intoxications from the literature (1966–2018). *Toxicon, 177*, 52–88.
29. Vengamma, R., Ramani, U., & Swapna, P. (2019). Bioactive alkaloid markers-an overview. *International Journal of Research and Analytical Reviews, 6*(1), 2349–5138.
30. Reddy, T. A., & Sreevidhya, J. (2022). A clinical study to assess the effectiveness of belladonna in various potencies in primary dysmenorrhoea of adolescent girls. *International Journal of Research in AYUSH Pharmaceutical Sciences, 6*, 610–614.
31. Dulawa, S. C., & Janowsky, D. S. (2019). Cholinergic regulation of mood: From basic and clinical studies to emerging therapeutics. *Molecular Psychiatry, 24*(5), 694–709.
32. Anacker, C. (2018). New insight into the mechanisms of fast-acting antidepressants: What we learn from scopolamine. *Biological Psychiatry, 83*(1), e5–e7.
33. Kwakye, G. F., Jiménez, J., Jiménez, J. A., & Aschner, M. (2018). Atropa belladonna neurotoxicity: Implications to neurological disorders. *Food and Chemical Toxicology* (Vol. 116, pp. 346–353). Elesevier.
34. Keshavarz, B., & Golding, J. F. (2022). Motion sickness: Current concepts and management. *Current Opinion in Neurology, 35*(1), 107–112.
35. Demirhan, A., Tekelioğlu, Ü. Y., Yıldız, İ., Korkmaz, T., Bilgi, M., Akkaya, A., et al. (2013). Anticholinergic toxic syndrome caused by atropa belladonna fruit (deadly nightshade): A case report. *Turkish Journal of Anaesthesiology and Reanimation, 41*(6), 226.
36. Glatstein, M., Alabdulrazzaq, F., & Scolnik, D. (2016). Belladonna alkaloid intoxication: The 10-year experience of a large tertiary care pediatric hospital. *American Journal of Therapeutics, 23*(1), e74–e77.

Chapter 5
Babchi

Muhammad Azeem, Sadia Javed, and Arwa A. AL-Huqail

5.1 Introduction

In the traditional medical approach or in indigenous medicinal practices, native plants are employed as treatments for a variety of ailments. Due to the enormous chemical diversity, they provide, chemicals from natural sources have been increasing prominence over the past few decades. In the past 20 years, the demand for herbal treatment has dramatically increased as a result of this. They are widely accessible, reasonably safe, and reasonably priced. These medications have provided crucial information for drug development, leading to the identification of novel compounds [1]. From the invention of medicine, natural substances, particularly those made from plants, were used to support human health. Throughout the beginning of time, humans have recognized and administered traditional medicine, and this has continued throughout history. Plants have been a great source of medicines since the dawn of mankind [2]. Scientists from all over the world have been interested in plant-derived medicines for a long time because of their few side effects and beneficial impact on human health [3, 4].

The natural products, especially those derived from plants, have been used to help mankind sustain human health since the dawn of medicine. The traditional medicine has been used since time immemorial and has been well accepted and utilized by the people throughout history. Since ancient times, plants have been an excellent source of medicines [1]. Plant-derived medicinal products have attracted the attention of scientists around the world for many years due to their minimum

M. Azeem · S. Javed (✉)
Department of Biochemistry, Government College University, Faisalabad, Pakistan
e-mail: sadiajaved@gcuf.edu.pk

A. A. AL-Huqail
Department of Biology, College of Science, Princess Nourah bint Abdulrahman University, Riyadh, Saudi Arabia

M. Zia-Ul-Haq et al. (eds.), *Essentials of Medicinal and Aromatic Crops*,
https://doi.org/10.1007/978-3-031-35403-8_5

side effects and positive effects on human health [2, 3]. The *Psoralea corylifolia* L. family is Fabaceae (Leguminosae) and is an endangered herbaceous and medicinally useful plant. It is mostly found in tropical and subtropical region of the world, and it grows in the plains of Central Asia.

One of the most well-known Traditional Chinese Medicines is the dry fruit of the green manure herb *Psoralea corylifolia* Linn. (syn: Cullen corylifolium Linn.), which is officially classified in the Chinese Pharmacopoeia. It has been used for millennia to treat a number of skin disorders, including psoriasis, leukoderma, and leprosy, and has significant biological relevance [5]. *Psoralea corylifolia* L. belongs to the Fabaceae (Leguminosae) family and is herbaceous, medicinally valuable plant that is in risk of extinction. It grows on the plains of Central Asia and is primarily found in tropical and subtropical regions of the world [4, 6].

Leguminosae, one of the largest families of plant species, has more than 12,000 species and 500 genera, some of which contain biologically active chemicals. Psoralea, Pongamia, Alhegi, and Indigofera species have been found to treat a wide range of illnesses. A native Asian Psoralea species that is often found in Pakistan, China as well as India has garnered a lot of interest as a medicine. The word "genus" comes from the Greek word "psoraleos," which translates to "afflicted with the itch or with leprosy."

Psoralea corylifolia L. is a species of significant plant (Leguminosae). *P. corylifolia* is an annual herb that stands upright. This plant can grow anywhere between 30 and 180 cm tall, prefers a warm environment, and doesn't perform well in shadow. This plant requires loam, clay, and sand-type soils. The plant can survive in basic, acidic, and neutral conditions. March to April is the ideal period to sow this shrub. In November, seeds reach maturity. If given the correct care, the plant could grow for up to 5–7 years. Psoralea produces perennial fruit. Fruit cannot endure frigid temperatures. Although the fruit is generally odorless, chewing it releases a pungent flavor. The fruit has a terrible, bitter, and caustic flavor. The flowers are tiny and have a crimson clover shape [5, 7].

Racemes make up the leaf arrangement. Simple leaves have broad, dented borders and are elliptic in shape. About five main nerves came from the base of the leaf in total. On both their upper and below surfaces, the leaves are pubescent and covered in white hairs. The flowers are purple with blue undertones when they bloom during rain. The axils are more congested. The peduncles have lengthy heads, and each raceme might have between 10 and 30 blooms. The pods resemble black chocolate in hue. The pods measure only 3.5–4.5 by 2.0–3.0 mm in size. Pods can be flat or ovoid or oblong in form. Psoralea only has one seed. The seed has an elongated form and a smooth exterior. The seed is compact, hairless, and densely pitted. The seed is a dark brown hue. The seeds lack endosperm, have an oily feel, and no starch. In 7–8 months, the crop reaches its full maturity. Since it takes time for the seeds to mature, collection can be done four to five times between December and March [7].

5.2 Classification

Kingdom: Plantae.
Division: Angiospermae.
Class: Dicotyledoneae.
Order: Fabales
Family: Fabaceae.
Subfamily: Papilionaceae.
Genus: Psoralea
Species: P. corylifolia Linn [5, 8].

5.3 Binomial Name

Psoralea corylifolia L.

5.4 Common Names

- Urdu: Babchi
- English: Babachi, Babchi seeds
- Arabic: Babchi
- Telegu: Bhavanji, Bavanchalu, Baaranchal, Karubogi, Bapurlen, Baranchalu
- Tamil: Karpurarishi, Karpokarishi, Karporgam
- Sanskrit: Bakuchi, Somaraji, Sugandha Kantak
- Persian: Babchi
- Marathi: Babchi, Babachi, Bavachi
- Kannada: Baukuchi, Baranchigida, Bauchige, Bhavanchigid, Bhavanti buja, Karbekhiya
- Gujarati: Babchi, Bawachi, Bavacha, Bhava, Bakchi
- Bengali: Bavachi, Hakuchi, Lata Kasturi [5, 8, 9]

5.5 Distribution

Despite being an Asian native (Pakistan, China, India, Ceylon, Yunnan, Burma, Arabia, Szechuan, Socotra, Somali Republic), *Psoralea corylifolia* L. is grown in Australia, North America, and Africa, among other places [5, 9]. It is widely cultivated on the world's plains, as well as in tropical and subtropical regions. Its wide geographic range in Asia includes Pakistan, China, India and Japan [1].

5.6 Description

Psoralea corylifolia L. commonly called Babchi is a medicinal plant belong to the subfamily Papilionaceae of the Leguminaseae family. In the indigenous medical system, leukoderma, leprosy, and psoriasis are treated with babchi seeds. A sticky, oily pericarp that surrounds the seed contains coumarins, of which Psoralen and Isopsoralen are crucial for therapeutic purposes. Psoralen is being researched as a treatment for a number of disorders, including AIDS, in addition to psoriasis. Additionally, it aids in wound and ulcer healing as well as the treatment of intestinal amoebiasis. Numerous studies on the anti-microbial, anti-feedant, and insecticidal properties of babchi have been published in the literature, pointing to additional potential uses [5, 9].

Fruits are ovoid-oblong or bean-shaped, 3–4.5 mm long, 2–3 mm wide, slightly compressed, ovate rounded or mucronate, and heavily pitted. The seeds are campylotropous, non-endo spermous, fatty, and starch-free. Fruits come in a variety of colors, from dark chocolate to almost black. The fruit has a sour, caustic, and bitter taste. The flowers are tiny and have a crimson clover-like shape [6, 9].

Fruit on Psoralea is perennial. Fruit that freezes will lose its freshness. Racemes of leaves are present. The simple leaves have dented margins and are broad and oval in shape. The flowers are purple with blue undertones when they bloom during rain. In 7–8 months, the crop is fully grown. The seeds can be collected four to five times between December and March because they require some time to mature. On the abaxial side, there are cotyledons of polyhedral parenchyma, testa, an outer layer of the palisade epidermis, a layer of bearer cells with noticeably thicker inner tangential and basal radial walls, and 2–3 layers of parenchyma. There are also three layers of palisade cells. The pericarp is made up of collapsed parenchyma and massive secretory glands, and it features noticeable ridges and depressions [5, 6, 9].

5.7 Agronomy

5.7.1 Climate

For successful growth and yield, the crop prefers a dry tropical climate with a comparatively warmer climate. With the arrival of the monsoons, it is commonly grown as a rainy season crop. It's a hardy plant that does best in summertime climates with little to moderate rainfall. Under Pakistani conditions, the crop sown in June has produced the best results [9].

5.7.2 Cultivation

Babchi is a type of plant that is propagated by seeds. Because of the hard seed coat, the seeds have a problem with dormancy, and germination is always poor (6–8%). It has been discovered that sulphuric acid treatment or mechanically puncturing seed coverings to break dormancy can increase the germination rate of seeds. Before sowing the seed, the sulphuric acid should be washed away with several washes in water [9].

5.7.3 Fertilizers

Fertilizers, whether organic or inorganic, are advantageous to the crop. A 20 t/ha base application of farmyard manure (FYM) has been found to promote strong early growth and greatly raise seed production. Additionally recommended is a nourishment application of 60 kg P, 50 kg K and 100 kg N/ha. After 45 days of sowing, the leftover 50 kg Nitrogen/ha is used in one dosage as a top-covering, with a basal dose of full doses of P and K and half of N [9].

5.7.4 Irrigation

Babchi is predominantly farmed as a rainy season crop, therefore it requires little irrigation. The crop can be irrigated every 2 weeks once the rainy season is finished. The crop may receive between six and eight irrigations before being harvested [9].

5.7.5 Land

Before the monsoon season begins, light ploughing followed by two harrowing should be done to prepare the field for good seedbeds.

5.7.6 Soil

In nature, it grows on a wide range of soils, from fine sand medium loam to alluvial soil. However, it grows and yields best in sandy loam soil with plenty of organic matter. It can also tolerate a wide range of soil pH, but soils that are too acidic or alkaline should be avoided [5, 9].

5.7.7 Sowing

Depending on the fertility of the land, the seeds are planted in rows 45–60 cm apart, with plant to plant spacing of 30–45 cm. A seed rate of 7 kg will cover an area of one hector [9].

5.8 Medicinal Uses

It is believed that seeds have medicinal value. Both oral ingestion of babchi seeds as well as topical external application in the way of ointment (Marham) and a paste (Zamad) are recommended. The plant contains volatile oil, chalcones, coumarins, flavones, lipids, monoterpenoids, resins, phenols and stigma steroids. The nature and quantity of phytochemicals in *P. corylifolia* varies depending on the climatic conditions, according to a phytochemical study. Researchers discovered that compared to other plant parts, seeds had the highest concentration of active ingredients. Below is a description of each part's phytochemistry.

5.8.1 *Entire Plant*

The petroleum ether and chloroform extract of *P. corylifolia's* entire plant yielded the isomers psoralen, isopsoralen, corylifolin, corylin, and psoralidin. Neo-psoralen, a novel chemical, was identified from *P. corylifolia* in 1996 by Peng et al. The structures have been determined using spectroscopic techniques and chemical evidence.

5.8.2 *Leaves*

The leaves of *P. corylifolia* are traditionally used to relieve diarrhea [5].

5.8.3 *Fruits*

Fruit is bitter, which aids in the prevention of vomiting, the treatment of micturition difficulties, the treatment of piles and anemia, and the improvement of complexion. Babchi plant fruit has herbal supplement characteristics and is used as a natural remedy for the genital areas. The fruits are used to treat fevers, incontinence, premature ejaculation, bedwetting, frequent urination, impotence, and lower backaches [10]. Using a high-performance liquid chromatographic technique, the concentration of three important isoflavonoids, biochanin A, genistein and daidzein, mainly in the fruit of *P. corylifolia* was improved. The dried fruits of *P. corylifolia* contain seven known and one novel isoflavone, (7,4′-dihydroxy-3′-[(E)-3,7-dimethyl2,6-octadienyl] isoflavone) known as corylinin, as well as psoralen, isopsoralen, neobavaisoflavone, sophoracoumestan A, and uracil. A phenolic monoterpene called bakuchiol was discovered in hexane extract of plant fruits. Yao et al. measure the bakuchiol content of *P. corylifolia* fruits harvested from numerous places [11].

5.8.4 Oil

P. corylifolia oil is used to treat skin diseases such as ringworm, tinea versicularis, psoriasis and scabies. Babchi oil is used to treat vitiligo [6].

5.8.5 Roots

The plant's root is beneficial in the treatment of dental caries. *P. corylifolia* promotes bone calcification, making it useful in the treatment of osteoporosis and bone fractures. The roots of the *P. corylifolia* plant are used to treat dental problems [6, 12].

5.8.6 Seeds

The seeds and isolate powder are employed as diuretics, antihelminthics, laxatives, and to heal wounds. The seeds have antipyretic and alexiteric properties. Seeds have stomachic, stimulant, aphrodisiac, and diaphoretic properties. It works well for impotence, menstrual disorders, and uterine hemorrhage. It demonstrates coronary vasodilatory activity. It is used to treat gynecologic bleeding. It is also employed in the treatment of spermatorrhea and premature ejaculation. Seeds are also used in the production of perfumed oil [6, 13, 14].

The biosynthesis of the highly significant pharmacological compound bakuchiol occurred in 1983, and it was determined that it is a product of the phenylpropane pathway. The dimeric monoterpenoid skeleton was discovered to contain two monoterpenes linked together by a dioxane bridge, comparable towards how Bisbakuchiols A and B were assessed. The presence of a number of chemicals containing ketones and aldehydes was discovered in the low polarity ether extract of *P. corylifolia* seeds, which would include C-formylated chalcone, Isoneobayachalcone, Corylinal. The seeds also included Psorlenal, a brand-new isoflavone compound [12, 15]. Other chromatographical methods, for instance high-speed countercurrent chromatography, were used to isolate the same compounds, psoralen and isopsoralen [7].

The focus of the exploration for biological active compounds has been the seed sample; as a consequence of an isolation strategy comprising spectroscopic techniques and crystal X-ray diffraction, five new compounds were isolated. The names of these substances are psoracorylifols A–E, chalcone, and bavachromanol [5, 16]. The bavachalcone and all three types of Corylifols (A, B, & C), three novel flavonoid compounds, were separated from *P. corylifolia* plant seeds. From the seeds, a different substance called bakuchicin was also discovered. Additional flavonoids, including bavachin, bavachinin, isobavachalcone and isobavachin, are reportedly present in the seeds (IBC). Psoralenoside and isopsoralenoside, benzofuran-type glycosides, were found in the seeds of *P. corylifolia* [5, 12, 17].

Additionally, polar chemicals such as neobavachalcone, 7-methylbavachin, and bavachromene have been found in seeds of *P. corylifolia*. These substances were extracted and identified from the ethanolic extract's insoluble section. These novel molecules, known as Cyclobakuchiol C, were extracted from the less polar portion of *P. corylifolia* seeds [5, 16, 18]. Psoralester and psorachromene, two novel metabolites, were discovered while studying several esters in *P. corylifolia*. The latter is an isomer of the previously identified compound bayachromene and a 10 membered compound lactone which is lactone in nature [5, 12].

5.9 Pharmacological Actions

1. Anthelmintic (Qatil-e-Deedan-e-Amaa) [5]
2. Anti-asthmatic (Dafe-e-Damah) [5, 16]
3. Antibacterial (Maney-e-Jarasim) [5, 12, 17]
4. Anti-inflammatory (Muhallil-e-Waram) [5, 12]
5. Anti-Leprosy (Daf-e-Juzam) [5, 18]
6. Anti-Leucoderma (Daf-e-Bars) [5]
7. Anti-phlegmetic Fever (Daf-e-Tap-e-Balghamiya) [5, 15]
8. Anti-psoriatic (Daf-e-Daussadaf) [5, 15]
9. Anti-stomacache (Daf-e-Waja-e-Meda) [12]
10. Appetizer (Mushtahi) [5, 12]
11. Blood Purifier (Musaffi-e-Khoon) [5, 16]
12. Cardiac Tonic (Muqavvi-e-Qalb) [5, 12, 19]
13. Carminative (Kasir-e-Riyah) [5]
14. Anti-pruritic (Daf-e-Kharish) [5]
15. Detergent (Jali) [5]
16. Diaphoretic and Diuretic (Mu'arriq wa Mudirr-e-Baul) [5, 12]
17. Gastric tonic (Muqavvi-e-Medah) [5]
18. Laxative (Mulayyen-e-Am'aa) [5, 16, 18]
19. Purgative (Mushil) [5, 12]
20. Sedative (Musakkin) [12]
21. Stimulant and Aphrodiasiac (Muharrik wa Muqavvi-e-Bah) [12, 15]

5.10 Pharmacological Active Compounds

Among the chemical compounds found in the plant are astragalin, areneobavaisoflavone, bavachalcone, isobavachalcone, bavachin, bavachinin, corylin, corylifolin corylifol, dadzin, dadzein, psoralen, isopsoralen, psoralidin, angelicin, bakuchiol, 3-hydroxybakuchiol, 6-prenylnaringenin and many others (Table 5.1). *P. corylifolia*

Table 5.1 The bioactive compounds from babchi (*P. corylifolia*) and their potential pharmaceutical properties

Compound	Nature	Plant Part	Activity	Reference
Angelicin	Furanocoumarin	Seeds	Antibacterial	[17]
Aryl coumarin	Coumarin	Seeds	Anticancer	[23, 24]
Astragalin	Flavonoid	Seeds	Antioxidant	[25]
Bakuchiol	Meroterpene	Seeds/ fruits	Anti-acne, antibacterial, antifungal, anti-aging	[12, 17]
Bavachinin	Flavone	Seeds	Antibacterial, anti-Alzheimer	[25, 26]
Bakuisoflavone	Flavone	Fruits	Antibacterial	[17]
Bakuflavanone	Flavone	Fruits	Antibacterial	[27]
Bavachin	Flavnonoid	Seeds/ fruits	Osteoblast	[28]
Bakuchicin	Coumarin	Seeds	Topoisomerase inhibitor	[29, 30]
Bavachalcone	Chalcone	Seeds	Anticancer	[24]
Bavachinone A	Flavonoid	Fruits	Antibacterial, antifungal	[31]
Bavachinone B	Flavonoid	Fruits	Antibacterial	[32]
Bavacoumestan C	Flavonoid	Fruits	Antibacterial	[17]
Corylifolinin	Chalcone	Seeds	Antibacterial, carboxylesterase inhibitors	[32, 33]
Corylifols	Prenyl flavonoid	Seeds	Antibacterial	[32]
Corylifol A	Flavonoid	Seed/fruit	Antibacterial, carboxylesterase inhibitors	[17, 32]
Corylifol B	Flavonoid	Seeds	Carboxylesterase inhibitors	[33]
Corylifol C	Flavonoid	Seeds	Protein kinase inhibition, anticancer	[34]
Corylifol D	Flavonoid	Seeds	Anticancer	[24]
Corylifol E	Flavonoid	Seeds	Anticancer	
Coryfolin	Flavonoid	Whole plant	Antioxidant, anti-diabetic	[21, 35]
Corylin	Flavonoid	Whole plant	Anticancer, carboxylesterase inhibitors, osteoblast	[29]
Coryaurone A	Flavonoid	Fruit	Antibacterial	[17]
Dadzin	Isoflavnoid	Fruit	Antioxidant	
Dadzein	Isoflavnoid	Fruit	Antioxidant, antidiabetic, topoisomerase inhibitor	[29, 30, 35]
Dihydroxy coumestan	Essential oil component	Seeds	Estrogenic, insecticidal, genotoxic	[36]
Genistein	Isoflavone	Fruit	Antidiabetic, anti-obesity	[21]
Hydroxy bukuchio	Meroterpene	Seeds	Lymph angiogenesis inhibition	[37]
Hydroxypsoralenol A	Flavonoid	Fruit	Antibacterial	[32]
Hydroxypsoralenol B	Flavonoid	Fruit	Antibacterial	[32]

(continued)

Table 5.1 (continued)

Compound	Nature	Plant Part	Activity	Reference
Isobavachalcone	Chalcone	Seeds	Estrogen receptor agonist, neuroprotective, anti-Alzheimer, carboxylesterase inhibitors	[26, 38]
Isobavachin	Flavonoid	Seed/fruit	Osteoblast	[28]
Isopsoralen	Furanocoumarin	Whole plant	Antiprotozoal	[39]
Psoralen	Furanocoumarin	Whole plant	Leucoderma, psoriasis, anticancer, antioxidant, anti-Alzheimer, Collagengenesis	[24, 38, 40]
Psoralidin	Coumarin	Whole plant	Estrogen receptor modulator, antioxidant, antibacterial, anti-diabetic, antiprotozoal	[24, 35, 39]
Psoracorylifol D	Flavonoid	Seeds	Lymphangiogenesis inhibition, antidepressant	[41]
Psoracoumestan	Coumestans	Seeds essential oil	Anticancer	[34]
Xanthoangelol	Chalcone	Seeds	Anticancer	[24]

leaves contain extremely high levels of genistein [20]. Numerous research studies have shown that plants and foods high in polyphenolic material are effective free radical scavengers, aiding in disease prevention through antioxidant capacity [21]. In diabetic individuals, antioxidants from food, medicinal herbs, and plants aid to avoid vascular disorders. Secondary plant metabolites like flavonoids and tannins are thought to be a natural supply of antioxidants that prevent beta cell degeneration and ROS production brought on by diabetes. *P. corylifolia* plant exhibit strong enzyme inhibitory and antioxidant activity are utilized to treat diabetes overall [22].

5.11 Pharmacological Properties

5.11.1 Anthelmintic Activity

The antiworm property of *P. corylifolia* seeds has been clinically proven on roundworms and flatworms [42]. *Psoralea corylifolia* essential oil from the leaves and ether and subsequent alcoholic extracts of the seeds were tested on two enzymatic reactions using rat brain as a model in order to elucidate the method of anthelmintic activities [43].

5.11.2 Anti-acne Activity

It is a reliable acne treatment. Bakuchiol, a phenolic chemical, is present, and it is utilized in anti-acne preparations. Due to its safety and lack of irritating properties, it can be used throughout the day for longer lengths of time without becoming sensitization [44].

5.11.3 Anti-Alzheimer

IBC and BCN, two substances derived from the *P. corylifolia* plant that is frequently used in healthcare situations, alter amyloid (A) peptides, especially those with approximately fourty residues, considered crucial for the development of amyloid plaques in Alzheimer's illness [26].

5.11.4 Antibacterial Activity

The plant has been assessed for antibacterial activity against *S. aureus* and other bacterial strains [17]. Due to the presence of bakuchiol, a substance that has been previously identified as an antibacterial agent, during in vitro tests, the ethanolic seed extract of *P. corylifolia* shown remarkable growth inhibitory effect against *S. epidermidis* ATCC strain 12228 and *S aureus* ATCC strain 25923 and [27]. Corylifols A–C (1–3), three novel prenylflavonoids that were extracted from the seeds of *P. corylifolia* and demonstrated significant antibacterial activity against *S. aureus* and *S. epidermidis*, when analyzed for their structure activity relations [32].

The fruits and seeds of *P. corylifolia* also demonstrated excellent antibacterial activity in extract and phytoconstituent form. The dried fruits of *P. corylifolia* included the novel flavonoid bavachinone B, together with bavacoumestans B and C, which had a mildly inhibitive effect on Staphylococcus mutans [45]. Furthermore, an ethanol extract of *P. corylifolia* seeds was found to be active against methicillin resistant strains of *Staphylococcus aureus* (MRSA) and *Listeria monocytogenes*, with MIC values of 100 and 50 g/mL, respectively [46]. Following extract treatment, these bacteria showed alterations in cellular membrane permeability and damage to the integrity of their cell membranes, suggesting that *P. corylifolia* metabolites may have antibacterial capabilities that need further research. The primary component of *P. corylifolia* seeds, bakuchiol, was extracted in methanol and demonstrated potency as a quorum sensing inhibitor [47]. Quorum sensing is a signaling system that bacteria use to adapt to population density by changing the expression of certain genes [48]. At sub-lethal concentrations, *P. corylifolia* seeds extract and bakuchiol both exhibited quorum sensing inhibitory activity as well as prevention of the development of biofilms from *Chromabacterium violaceum* CV12472, *Listeria monocytogenes*, *Serratia marcescens*, and *Pseudomonas aeruginosa* PAO1 [47]. This study demonstrated the antibacterial potential of *P. corylifolia* isolate and its phytoconstituent bakuchiol, with the goal of reducing biofilm formation and QS-associated virulence.

5.11.5 Anticancer Activity

Arylcoumarin and psoracoumestan, two phytochemical components of *P. corylifolia*, shown strong anticancer action by blocking the phosphorylation of MAPK/ERK kinase enzymes. The fundamental method was programmed cell death. Other compounds, such as xanthoangelol and corylifol C, have been shown to be effective kinase inhibitors [23].

5.11.6 Anti-coagulant Effect

Compared to the antidote used as a control, the plant extract of *P. corylifolia* prevented the coagulation caused by Naja karachiensis snakebite. The effects of the snake poison on activated thrombin time, prothrombin time and partial thromboplastin time and were examined in human plasma that had been citrated [49]. The snake venom was tested on humanplasma to see how it affected the activated partial prothrombin time (PT), thromboplastic time (aPTT), and thrombin time (TT). Snake venom (200 g/ml) was discovered to delay PT (13 ± 0.57 to 23 ± 0.57 s), aPTT (35 ± 1.52 to 48 ± 2.0 s), and TT (13 ± 0.57 to 33 ± 0.57 s). The fact that PT and TT were prolonged suggested the presence of thrombin like or plasminogen activating enzymes [50].

5.11.7 Antidepressant Activity

Additionally, it was shown that *P. corylifoia* had antidepressant qualities. The mechanism of action of antidepressant plants as well as the chemical constituents identified from them were discussed by Marzieh Sarbandi Farahani and colleagues [51]. Several mouse studies have indicated a potential anti-stress and anti-depressant impact, however not to a notable extent. The most effective and helpful medication for treating geriatric depression is *P. corylifolia*. Furthermore, demonstrated the antidepressant properties of psoralen utilizing a forceful swimming test model of depression in male mice [41]. Psoralea and conventional stimulants may also interact.

5.11.8 Anti-diabetic Activity

The hexokinase, glucose-6-phosphatase, and G-6-P dehydrogenase activities were significantly increased by aqueous extract of *P. corylifolia* seed after a thorough biochemical investigation. The levels of serum transaminase, lipid peroxidation in liver tissue, and antioxidant enzymes such catalase, peroxidase, and superoxide dismutase as well as lowered fasting blood glucose levels in streptozotocin-induced diabetic rats [52]. Five flavonoids and a meroterpene bakuchiol were found in the ethyl acetate fraction of *P. corylifolia* fruits, and their potential to treat diabetes was tested. Bavachin dramatically boosted fat accumulation and PPAR (proliferator-activated receptor) transcriptional activity, among other things, in a dose-dependent manner. According to a subsequent mechanistic investigation, bavachin has the potential to treat type 2 diabetes mellitus by facilitating insulin-induced glucose absorption by activating the insulin signalling pathway in differentiated adipocytes [53].

In a test involving streptozotocin-induced diabetic mice, it was discovered that the water-soluble extract of *P. corylifolia* seeds had a protective effect against diabetic nephropathy. The extract's antifibrotic and antiapoptotic properties were demonstrated after oral treatment when many genes linked to renal fibrosis and apoptosis had their expression downregulated. The extract was also noted for inhibiting mesangial cell death, similar to what was shown when its primary metabolites, bakuchiol, psoralen, and isopsoralen, were administered [54]. In addition, the study found that the seeds of *P. corylifolia* contained three novel metabolites with substantial IC_{50} values, including a novel bavacoumestan E, as well as 11 recognized metabolites. All of the newly discovered flavonoids demonstrated considerable inhibition of diacylglycerol acyltransferase 1 when tested for their antidiabetic potential (DGAT1) [55].

The triacylglycerol (TG) production pathway uses two isoforms of DGAT, known as DGAT1 and DGAT2. It is thought that DGAT1 targeting has potential for reducing diabetes and obesity [56]. In contrast to other isolated natural products, the known coumestans, bavacoumestans B and C, demonstrated much better inhibition against protein tyrosine phosphatase 1B (PTP1B) as well as stronger inhibition against -glucosidase [55]. A standardized flavonoid-rich fraction of *P. corylifolia* seeds was reported for its promising effect in obese mice caused by a high-fat supplemented diet, in keeping with prior findings on the antidiabetic potency of flavonoids from *P. corylifolia* [53, 55]. By administering a flavonoid-rich fraction, body weight, fat mass, and insulin sensitivity were all significantly reduced. Its effects come from the encouragement of various thermogenic gene expressions that help avoid obesity. By the stimulation of insulin signalling and sugar transport in adipose tissue, this flavonoid fraction additionally enhanced glucose homeostasis [57]. Moreover, bavacoumestan D, a brand-new coumestan discovered in *P. corylifolia* seed ethyl acetate extract, showed only little inhibition of DGAT1. Moreover, bavacoumestan D somewhat inhibited glucosidase [58], similar to the activity shown with its counterparts bavacoumestans B and C [55].

5.11.9 Anti-eczema Activity

In one investigation, hexane and oil in water were used to extract *P. corylifolia* seeds, and stearic acid was used as the base to make cream. After then, 30 people with eczema participated in an open clinical trial that lasted 30 days. According to this study's findings, this plant can be used to successfully cure eczema [42].

5.11.10 Antifungal Activity

Numerous pathogenic fungus strains, including *Epidermophyton floccosum, Microsporum gypseum, Trichophyton mentagrophytes*, and *Trichophyton rubrum* were resistant to the antifungal effects of the phenolic compound bakuchiol that was isolated from *P. corylifolia* (seeds) [59].

5.11.11 Anti-inflammatory Activity

Bavachini was extracted from the fruits of *P. corylifolia* and showed distinct anti-inflammatory, antipyretic, and mild analgesic effects. The maximum fatal dose was larger than 1000 mg/kg in mice, and it showed better antipyretic action than paracetamol while having no influence on the central nervous system. By preventing IL-6-induced STAT3 activation and phosphorylation, a number of flavonoids from *P. corylifolia* may be effective treatments for inflammatory disorders. Moreover, it reduced the edoema that carrageenan-induced in rats [36]. Overall, bakuchiol was discovered to be the most active metabolite among the tested compounds for its neuroprotective and neuroinflammation inhibitory effects [24].

Furthermore, according to a recent study three new isoflavone derivatives, 7-Oisoprenylcorylifol A, 7-Oisoprenylneobavaisoflavone, and 7-O-methylcorylifol A were reported from *P. corylifolia* fruits, along with bakuchiol, epoxybakuchiol, psoralidin, and other known compounds [40]. Bakuchiol had the strongest inhibitory effect in RAW264.7 cells when evaluated for its anti-inflammatory potential against LPS-induced NO production, which is consistent with its activity previously described [20]. Furthermore, bakuchiol was found to have a remarkable protective effect against sepsis-induced acute kidney inflammation in kidneys by inhibiting NF-kB and p38 MAPK signalling [40]. These findings highlight the anti-inflammatory properties of *P. corylifolia* and its bioactive metabolites, which protect organs.

5.11.12 Anti-leucoderma Activity

It was discovered that *P. corylifolia* works well as an anti-leucoderma treatment. Leucoderma is treated using "soralen," one of the bioactive isolated chemicals, because it has been shown to be able to increase the production of melanin [60].

5.11.13 Anti-obesity

Various animal studies have shown that genistein can help people lose weight by reducing their food intake. It also decreased the weight of fat pads and increased apoptosis in adipose tissues. One such experiment was carried out on ovariectomized mice [61]. Several research conducted on animals shown that genistein can lower body weight by lowering food intake. Moreover, it decreased the weight of the theft pad and improved adipose tissue apoptosis. For instance, ovariectomized mice were used in one such investigation. Genistein, a well-known trihydroxy flavone that was also extracted from *P. corylifolia*, demonstrated potential anti-obesity and low-grade inflammation-related actions through a variety of pathways and cell signal transduction [61, 62]. The action of *P. corylifolia* extract on the adipocyte life cycle, obesity-related low-grade inflammatory, and oxidative stress results in anti-obesity and anti-diabetic activities.

5.11.14 Antioxidant Activity

That can be obtained from food sources, herbs, and plants help diabetics avoid vascular problems. Secondary plant metabolites called tannins and flavonoids are hypothesized to be a natural supply of antioxidants that stop cell death and reactive oxygen species generation brought on by diabetes [22]. As antioxidative components, the seed contained a meroterpene and four flavonoids: bakuchiol, corylin, corylifolin, and psoralicin. In rat liver microsomes and mitochondria, barachin, isobarachalcane, isobavachin, and bakuchiol demonstrated broad antioxidative activities. In microsomes, they inhibited NADH-dependent ascorbate and CCl_4-induced lipid peroxidation. The most powerful antioxidant in microsomes was bakuchiol, and lipid peroxidation's ability to limit oxygen consumption was time dependent. Moreover, bakuchiol guarded against oxidative hemolysis in human red blood cells [28]. It has been demonstrated that the phenolic compounds in *P. corylifolia* are efficient at shielding cellular membranes from a variety of oxidative stressors.

5.11.15 Antiprotozoal Activity

An external protozoan parasite called *Ichthyophthirius multifiliis*, or "ich," has been identified in freshwater fish. *P. corylifolia* extracted with methanol showed excellent action against *I. multifiliis* theronts when treated for 4 h at doses of 1.25 mg/L or higher. The *P. corylifolia* extract killed all pro tomonts and 88.9% of encysted tomonts at a dosage of 5.00 mg/L [39]. When exposed for 4 h, the *P. coryfolia* extract with methanol shown excellent action against *I. multifiliis* theronts at concentrations of 1.25 mg/L or greater. At a concentration of 5.00 mg/L, the *Psoralea corylifolia* extract results in 100% protomont mortality. It has been discovered that *Psoralea corylifolia* can be used to suppress the external protozoan parasite *I. multifiliis* instead of malachite green. According to the screening, *Psoralea corylifolia* extract had higher activity against *I. multifiliis* pathogens. In vivo experiments with 1.25 mg/L or higher concentrations of *P. corylifolia* methanol extract resulted in 0% death rates of theronts after 4 h of exposure. This study demonstrated the damaging effect of *P. corylifolia* on the trophont of *I. multifiliis* [39, 63]. The same study led to the assessment of the action of antiprotozoal compounds extracted from *P. corylifolia* against *I. multifiliis*.

5.11.16 Anti-psoriatic Activity

The skin ailment psoriasis is also treated with an extract from the *P. corylifolia* plant [26]. At a dose of 6 grammes administered twice daily on an empty stomach for 45 days, *P. corylifolia* seeds powder (Safoof) was found to be successful in the treatment of (psoriasis) in more than 40 patients [6]. A different study revealed that the seed extract from *Psoralia corylifolia* may have anti-psoriatic potential. Another study found that the administration of psoralen and its chemical compounds, including trioxalen, combined with sunshine exposure is a more successful treatment for psoriasis [4].

5.11.17 Antiviral Activity

The papain-like protease (PLpro) of the severe acute respiratory syndrome corona virus (SARS-CoV) was reported to have strong activity against the crude ethanol isolate of *P. corylifolia* seeds, with an IC50 value of 15 g/ml. A crucial enzyme in the SARS virus's reproduction is SARS-CoV (PLpro) [64].

5.11.18 DNA Polymerase and Topoisomerase II Inhibitors

In an activity directed isolation assay that resulted in the separation of novel compounds bakuchiol, corylifolin, resveratrol and neobavaisoflavone, it was found that the crude extract of *P. corylifolia* with solvent ethanol was a powerful DNA polymerase inhibitor of DNA replication enzyme. Daidzein and bakuchicin, two topoisomerase II inhibitors, were also discovered in a related enzyme assay [30].

5.11.19 Estrogenic Effects

Phytoestrogens are naturally occurring substances generated from plants that resemble endogenous estrogens structurally and have estrogenic action. Moreover, phytoestrogens have been linked to a number of biological processes, such as osteoporosis prevention, menopause syndrome relief, and the reduction of proliferative and cardiovascular illnesses. It has been consistently observed that the phenolic chemicals stilbenes, isoflavones, lignans and coumestans have estrogenic effects [65]. The estrogenic effects of many flavonoids from *P. corylifolia* fruits is investigated utilizing isoflavones, flavanones, and chalcones [66, 67]. The flavonoids that were put to the test were chalcones (bavachalcone, isobavachalcone, and 4′-Omethylbavachalcone), flavanones (bavachin, isobavachin, and bavachinin), and isoflavones (corylin and neobavaisoflavone). With the exception of corylin, which lacked any binding potential, all of the examined flavonoids in a fluorescence polarisation assay demonstrated the ability to bind to the protein human oestrogen receptor ligand-binding domain (hERLBD) in a dose-dependent manner. Neobavaisoflavone demonstrated the highest binding capacity to hER-LBD among the active flavonoids [65]. Because corylin lacks estrogenic activity, a quantitative structure-activity relationship (QSAR) analysis concluded that the presence of hydroxyl and prenyl groups is necessary for the estrogenic activities of flavonoid compounds [67].

5.11.20 Osteoporosis and Estrogenic Activity

A metabolic bone disease called osteoporosis is characterized by decreasing bone mass and deteriorating bone microstructure, which causes bone fragility [68]. Previous investigations have established the potential therapeutic use of *P. corylifolia* in the treatment of osteoporosis. The osteogenic activity of neobavaisoflavone, which was isolated from *P. corylifolia*, was described. The activation of p38 and subsequent upregulation of the transcription factors Runx2 and Osx are how it stimulates osteogenesis [69]. Furthermore, in glucocorticoid-induced osteoporosis rats, ethanolic extract of *P. corylifolia* seeds significantly controlled 18 possible biomarkers connected to the etiology of osteoporosis. These biomarkers were discovered to be involved in the pathways for the metabolism of arginine, nicotinamide, and tryptophan [66]. Evidence for its possible therapeutic application in the treatment of osteoporosis was provided by the influence of *P. corylifolia* extract on the regulation of several metabolic pathways and the related biomarkers. Corylin from *P. corylifolia* fruits has been demonstrated to have osteogenic activity in addition to neobavaisoflavone. Runx2, Osterix, Co11, and ALP found to be key osteogenesis biomarkers that were induced by corylin [70]. Its osteogenic activity involved two pathways through oestrogen and Wnt/catenin signalling, indicating its therapeutic promise for osteoblasts-mediated osteoporosis, according to a thorough analysis of its mode of action.

Using various in vitro assays, the estrogenic effects of the alcoholic extract and its active ingredients from *P. corylifolia* L. were investigated. According to analysis of the ethanol extract, bakuchiol, bavachinin, isobavachalcone, isobavachromene, and psoralen were present. Hexane and chloroform fractions from a fractionation process demonstrated estrogenic activity in a yeast transactivation experiment. At a concentration of 1.0 ng/ml, the ethanolic extracts demonstrated noticeably higher activities, while bakuchiol demonstrated the highest activity, that was greater than genistein at the same concentration in the yeast transactivation assay. Bakuchiol demonstrated the maximum ER-binding capacity for ER α in the ER binding experiment, and it demonstrated a fivefold stronger affinity for ER α than for ER β [36].

5.11.21 Immunomodulatory Activity

When tested in mice, the extract of *P. corylifolia* seeds was found to have stimulant activity against natural killer cells. According to this study, the extract also regulates antibody-dependent cellular toxicity [24].

5.11.22 Insecticidal and Genotoxic Activity

The southern house mosquito, *Culex quinquefasciatus* (earlier known as Culex fatigans), larvae and adults were highly susceptible to the extracted oil from the seeds of *P.* corylifolia [71].

5.11.23 Inhibition of Lymph Angiogenesis

The bioactivity-guided fractioned compounds from the Psoraleae extract are angelicin, isobavachalone, p-hydroxybenzaldehyde, psoralen, bakuchiol hydroxybakuchiol and psoracorylifol, significantly slowed down the growth of temperature-sensitive rat lymphatic endothelial (TRLE) cells in vitro [6]. These bioactive substances prevented TR-LE cell proliferation and the development of capillary-like tubes. A concentration of 10 μm of bakuchiol was used in the tube emergence assay to analyze the cell cycle of TR-LE cells, and it was incubated for 6–48 h [72]. Propodeum iodide was employed to stain after reaping. Some substances examined shown selectivity. The substances investigated may make promising candidates for the creation of lymph angiogenesis-blocking chemotherapy and anti-metastatic medicines.

5.11.24 Neuroprotective Properties

P. corylifolia, which is used to treat a number of central nervous system problems, incorporating neurotropic action and as an agent that maintains the central nervous system, is a common ingredient in ayurvedic formulations [38]. The *P. corylifolia* L. seed extract was the subject of a study based on this knowledge. Against the cytotoxicity of 3 nitropropionic acid, the findings demonstrated a potent protective effect. The therapeutic potential of *P. corylifolia* seed extracts for neurological diseases was identified in this study [40]. As a result, *P. corylifolia* Linn seed extracts might be used therapeutically to treat neurological diseases.

5.11.25 Osteoblastic Activity

In vitro cultures of the UMR106 cell line revealed osteoblastic proliferation-stimulating activity from *P. corylifolia* L. fruit extracts. According to a study, the extract contained two flavonoids, bavachin and corylin, which were found to be effective against osteoporosis [28]. The bakuchiol as hepatoprotective molecule, as well as two moderately active compounds, psoralen and bakuchicin on tacrine-induced cytotoxicity in human liver-derived Hep G-2 cells are analyzed from the aqueous extract of *P. corylifolia* seed.

5.11.26 Skin Diseases Control

P. corylifolia seeds have been used for ages to treat a variety of skin conditions. In keeping with its conventional uses, a number of clinical trials have also revealed encouraging results from formulations made from the powdered seeds or natural substances isolated from the seeds or fruits of *P. corylifolia* to treat dermatological conditions like vitiligo and skin irritation [73]. When compared to retinal damage caused by oxidative stress, the compounds from *P. corylifolia* were found to have a protective effect [60]. In 2016, researchers ran a clinical trial to examine if a hydrophilic ointment containing 10% w/w powdered *P. corylifolia* seeds might affect the depigmentation of vitiligo patients. A skin ailment called vitiligo causes white, pale patches to appear on the skin as a result of damaged cutaneous melanocytes, which affects the skin's colour [74]. Many treatment plans for vitiligo have been created, however none of them are truly effective cures.

Additionally, a cream formulation with 0.5% meroterpene phenol bakuchiol extracted from *P. corylifolia* seeds was reported to have similar effects to retinol in a double-blind, randomized clinical experiment in improving facial photoaging, minimizing wrinkles, and lowering hyperpigmentation [75]. A total of 44 healthy volunteers were included in the 12-week trial. Each set of participants got either a face cream containing 0.5% bakuchiol or 0.5% retinol after being randomly assigned to one of two groups. It has previously been reported that bakuchiol can activate genes involved in the cellular uptake and stimulation of retinol as well as govern the creation of extracellular matrix proteins, similar to what has previously been described for retinol [76]. In addition to its capacity as a free radical scavenger, bakuchiol increased cellular resilience to oxidative stress through activation of Nrf2 known as nuclear factor erythroid 2-related factor 2 [77]. These activities taken together help bakuchiol's antiaging properties.

The potential for bakuchiol and four other natural compounds isolated from *P. corylifolia* seeds including psoralidin, psoralen, isopsoralen, and 8-methoxypsoralen to treat psoriasis-like lesions in an in vivo experiment was investigated further [78]. It was discovered that isopsoralen and 8-methoxypsoralen improved psoriasis-like lesions by reducing epidermal thickness, cytokine release, and skin barrier deficiencies brought on by UVA radiation. Ultimately, this result suggested that isopsoralen, along with the well-known 8-methoxypsoralen, is a good photosensitizing option for photochemotherapy against psoriasis-like lesions.

5.12 Industrial Applications

Psoralea corylifolia has a number of isolated chemicals that are commercially available and have numerous industrial applications. The following are a few examples: Methoxsalen (CAS 298-81-7), also known as 8-Methoxypsoralen, is a powerful cytochrome P450 suicide inhibitor. There is an antifungal substance called angelicin (CAS 523-50-2). Available as a mild antioxidant that promotes bone growth is bavachin (CAS 19879-32-4) [79]. PTP1B and DNA polymerase inhibitors are both accessible in the form of bakuchiol (CAS 10309-37-2). Genistein (CAS 44672-0), a substance, is a highly effective protein tyrosine kinase inhibitor [61]. The substance Daidzin (CAS 55266), which has been isolated from many plant species exhibits anti-chemotherapeutic properties and effective inhibitor of human mitochondrial metabolic enzymes including aldehyde dehydrogenase [6, 80]. These demonstrations plainly highlighted the significance of bioactive substances extracted from *P. corylifolia*, an indispensable traditional plant with pharmaceutical applications.

5.13 Clinical Studies

For a very long period, Psoralea species have been utilized in traditional medicine and folklore. Many Psoralea products have been successfully manufactured and are on the market. You can eat the parts of *P. argophylla*, also known as silvery leaf Indian breadroot, fresh or cooked. Because the plant possesses analgesic qualities, a poultice made from the roots is applied to sore body parts [1, 81]. Colds, coughs, headaches, and sore throats can all be treated with a combination of the plants stem and roots. In Chinese traditional medicine, *Psoralea corylifolia* L. (Bu Gu Zhi) is renowned as a tonic that boosts general vitality. The plant has demonstrated efficacy in the Unani system against skin conditions, high fever, and internal ulcers [1].

Due to the plant's blood-purifying characteristics, it is used to cure ringworm infection, boils, eruptions or red papules, itchy, severe eczema, rough and discolored dermatosis with scabies and fissures [82]. Streptococcal skin infection is said to be strongly affected by the essential oil from the plant [83]. It also has a reputation for improving the hue of nails, hair, and skin. Seeds are described as being sour, bitter, caustic, and astringent. Japan has used the ethanol extract of the seeds to preserve pickles and other processed goods [1, 84].

Additionally, *P. corylifolia* fruits can be used medicinally. Since the seeds have potent aphrodisiac properties, they are used as a tonic to strengthen the sexual organs [1]. The fruits are known to improve complexion, have a bitter flavor, and help treat piles, bronchitis, anemia, and difficulties urinating in addition to preventing vomiting [85]. Fruit extract also inhibit *Mycobacterium tuberculosis* growth [86]. *P. corylifolia* seeds and roots are used to stop tooth decay. They are advantageous for treating bone fractures and osteoporosis because they also induce bone calcification [1].

The traditional remedy with *P. corylifolia* as its primary component was applied locally to 30 vitiligo patients as part of a clinical experiment, in addition to taking Gandhaka rasayana orally. Early vitiligo cases improved significantly between 1 to 10 months, however chronic cases of vitiligo on the lips responded poorly. Very positive outcomes were obtained when 8-methoxypsoralen was taken orally and exposed to sunshine for 5–30 minutes each day for 1–7 weeks. In a different study, it was discovered that psoralen and its chemical compounds, particularly trioxalen, work better as psoriasis treatments when combined with sun exposure. 49 patients received *P. corylifolia* treatment for 6 months in one research. A further 19% of these patients repaired at least two-thirds of the damaged skin's pigmentation., while 14% of them were cured [87].

About 76 patients with grade II and III acne vulgaris between the ages of 16 and 24 participated in a clinical study. Patients were told to use Clarina cream as a topical treatment, as well as herb medicines Purim pills with *P. corylifolia* listed among the ingredients. According to the findings, people with the acne disease of grade II had an incredible response in 56 percent of situations and a positive outcome in 43 percent of cases [8, 88].

5.14 Negative Impacts and Cytotoxicity

In everyday life, the prospective hepatotoxicity of medicinal herbs is usually overlooked. *Psoralea corylifolia* seemed to be interconnected to the development of acute cholestatic hepatotoxicity. Some alternative therapy research indicates that *P. corylifolia* can treat osteoporosis. They discovered a specific instance of intense cholestatic hepatitis in a post - menopausal woman who took more than ten times the usual dose of *P. corylifolia* seeds [34]. A biopsy revealed region multiple tissue damage, degenerating cells, cholestasis, and inflammatory cell infiltrations. This case emphasizes the importance of warning about the prospective hepatotoxicity of *P. corylifolia* seed particularly in large doses [6]. Psoralea has great bioactivities but can have negative consequences if used excessively. It has been claimed that Psoralea genus is harmful to horses and cattle, hence it is not advised for use as fodder. There have been reports of cutaneous allergic responses following the administration of oral and injectable Psoralea formulations [89]. Psoralea overdose symptoms include lightheadedness, generalized weakness, blurred vision, fast breathing, and vomiting. Blood has been spat out, people have lost consciousness, and others have gone into a coma in severe overdose situations [90]. In a related investigation, rats received various amounts of *P. corylifolia* solutions daily for 90 days. The medication reduced body weight and reproductive organ weight (testes and ovaries), showing reproductive toxicity caused by psoralen [91]. There are numerous studies on the effects of long-term psoralen or isopsoralen use on the livers of mice and rats. The potential toxicity of *P. corylifolia* and its natural products (neobavaisoflavone, bavachin, and corylifol) was assessed in a study. According to the findings, it exhibited a significant inhibitory effect on human UDP-glucuronosyltransferase 1A1 (UGT1A1), a known stimulator of P. corylifolia-related toxicity including liver damage and elevated bilirubin levels [92]. Three occurrences of liver damage in Chinese people have been linked to eating dried *P. corylifolia* seedlings. The tablets manufactured from *P. corylifolia* leaves were determined to be the hepatotoxicity's most likely culprit. The underlying hepatotoxic processes of *P. corylifolia* and its primary constituents, however, are not fully understood [93, 94]. According to a book on traditional Chinese medicine, Psoralea doesn't have any negative side effects at doses within the typical range [89]. One thing was discovered to be common to all case studies and research on psoralea-induced toxicity, and that is the deliberate and extended consumption of a high dose. We may infer from the literature that is now available that there are differing views regarding the toxicity caused by Psoralea, and as a result, there is still a study gap regarding its method of action and its pharmacological standardization. The strongest inhibition against all examined cancer cell lines was seen with bakuchiol among the tested meroterpenoids, and its novel cyclic counterparts did not exhibit any cytotoxic effect [95]. This result shows that the meroterpenoid metabolites of *P. corylifolia* have less cytotoxic effect due to side-chain cyclization. Moreover, another subsequent chemical analysis of *P. corylifolia* fruits produced 17 meroterpene phenols, which were given the unimportant names psocorylins A–Q as the new meroterpene analogues [96].

5.15 Conclusion

The most significant disorders that the belonging to the family Psoralea has the most potential to treat are vitiligo, leprosy, and psoriasis. Pre-clinical research has shown encouraging findings, but further research is needed to fully understand the molecular processes behind the aforementioned pharmacological activities as well as to assess the efficacy of isolated drugs in tests that are appropriately constructed. Before the start of clinical trials, the estimated risk of extracts must also be established by long-term toxicity studies and information on drug interactions. The effectiveness and efficiency of the bioactive substances found in the Psoralea genus have occasionally been assessed. Since there is already knowledge of the medical benefits supported by scientific evidence. Further investigation should concentrate on bulk cultivation and the extraction of bioactive compounds. Millions of patients with psoriasis, leprosy, and vitiligo will experience natural alleviation with the development of pharmaceutical medications comprising psoralea as a sole or partial constituent. It is abundantly obvious from reviewing the literature on the genus Psoralea that *P. corylifolia* is a crucial component of a traditional biological, medicinal plant, phytoconstituents and ethnopharmacological point of view. This includes figuring out how many bioactive elements are present in each species. To sum up, the Psoralea species have enormous potential to be a cure-all for many illnesses, so protecting them before overusing them should be a precondition. Since olden days, the plant has been used as a medicinal agent. The use of traditional herbal remedies or their derivatives for preventative and curative purposes is becoming more popular around the world. A thorough examination of the literature disclosed that *P. corylifolia* is a significant indigenous herbal asset with a wide range of pharmacological activities. Clinical studies have confirmed that it plays a significant part in the prevention and control of a variety of diseases.

References

1. Koul, B., Taak, P., Kumar, A., Kumar, A., & Sanyal, I. (2019). Genus Psoralea: A review of the traditional and modern uses, phytochemistry and pharmacology. *Journal of Ethnopharmacology, 232*, 201–226.
2. Farnsworth, N. R., Akerele, O., Bingel, A. S., Soejarto, D. D., & Guo, Z. J. (1985). Medicinal plants in therapy. *Bulletin of the World Health Organization, 63*(6), 965.
3. Mehta, G., Nayak, U., & Dev, S. J. T. (1973). Meroterpenoids—I: Psoralea corylifolia Linn.—1. bakuchiol, a novel monoterpene phenol. *Tetrahedron, 29*(8), 1119–1125.
4. Prasad, N. R., Anandi, C., Balasubramanian, S., & Pugalendi, K. (2004). Antidermatophytic activity of extracts from Psoralea corylifolia (Fabaceae) correlated with the presence of a flavonoid compound. *Journal of Ethnopharmacology, 91*(1), 21–24.
5. Masihuzzaman, A. M., Iftikhar, M. Y., Alia, B., & Farheen, K. (2022). Babchi (Psoralea corylifolia Linn.): an effective Unani medicine for dermatological disorders: A review. *International Journal of Unani and Integrative Medicine, 6*(1), 17–21.
6. Shaikh, H. S., & Shaikh, S. S. (2021). Babchi (Psoralea corylifolia): From a variety of traditional medicinal application to its novel roles in various diseases: A review. *Asian Journal of Pharmacy and Technology, 11*(3), 238–244.

7. Alam, F., Khan, G. N., & Asad, M. H. H. B. (2018). Psoralea corylifolia L: Ethnobotanical, biological, and chemical aspects: a review. *Phytotherapy Research, 32*(4), 597–615.
8. Khushboo, P., Jadhav, V., Kadam, V., & Sathe, N. (2010). Psoralea corylifolia Linn.—"Kushtanashini". *Pharmacognosy Reviews, 4*(7), 69.
9. Farooqi, A. A., & Sreeramu, B. (2004). *Cultivation of medicinal and aromatic crops*. Universities Press.
10. Joshi, A. R., & Joshi, K. (2000). Indigenous knowledge and uses of medicinal plants by local communities of the Kali Gandaki Watershed Area, Nepal. *Journal of Ethnopharmacology, 73*(1–2), 175–183.
11. Santao, Y., Bin, Y., & Zhiling, X. (1995). Determination of bakuchiol in the fruit of Psoralea corylifolia L. *Zhongguo Zhong Yao Za Zhi, 20*(11), 681–683.
12. Pandey, P., Mehta, R., & Upadhyay, R. (2013). Physico-chemical and preliminary phytochemical screening of Psoralea corylifolia. *Archives of Applied Science Research, 5*(2), 261–265.
13. Agharkar S. Medicinal plants of Bombay presidency. 1953.
14. Nadkarni, A. (1954). *Nadkarni's Indian Materia Medica*.
15. Nadkarni, K., & Nadkarni, A. J. L. (1976). *Bombay. Indian Materia Medica* (Vol. 1, p. 799). Popular Prakashan Pvt.
16. Azhar, M., Naushin, S., & Alam, M. (2020). Zahar Mohra (Bezoar) an Alexipharmic Unani mineral drug. *Journal of Drug Delivery and Therapeutics, 10*(6), 236–238.
17. Wang, T.-X., Yin, Z.-H., Zhang, W., Peng, T., & Kang, W.-Y. (2013). Chemical constituents from Psoralea corylifolia and their antioxidant alpha-glucosidase inhibitory and antimicrobial activities. *China Journal of Chinese Materia Medica, 38*(14), 2328–2333.
18. Bi, S., Akhtar, J., Bashir, F., & Alvi, R. (2020). Pharmacological investigations of Babchi (Psoralea corylifolia Linn) – An important drug of Unani system of medicine. *International Journal of Unani and Integrative Medicine, 4*(1), 32–36.
19. Kirtikar, K., & Basu, B. (1935). *Indian medicinal plants*. Lalit Mohan Publication.
20. Zhao, L., Huang, C., Shan, Z., Xiang, B., & Mei, L. (2005). Fingerprint analysis of Psoralea corylifolia L. by HPLC and LC–MS. *Journal of Chromatography B, 821*(1), 67–74.
21. Büyükbalci, A., & El, S. N. (2008). Determination of in vitro antidiabetic effects, antioxidant activities and phenol contents of some herbal teas. *Plant Foods for Human Nutrition, 63*, 27–33.
22. Aslan, M., Orhan, N., Orhan, D. D., & Ergun, F. (2010). Hypoglycemic activity and antioxidant potential of some medicinal plants traditionally used in Turkey for diabetes. *Journal of Ethnopharmacology, 128*(2), 384–389.
23. Limper, C., Wang, Y., Ruhl, S., Wang, Z., Lou, Y., Totzke, F., et al. (2013). Compounds isolated from Psoralea corylifolia seeds inhibit protein kinase activity and induce apoptotic cell death in mammalian cells. *Journal of Pharmacy and Pharmacology, 65*(9), 1393–1408.
24. Latha, P., Evans, D., Panikkar, K., & Jayavardhanan, K. (2000). Immunomodulatory and antitumour properties of Psoralea corylifolia seeds. *Fitoterapia, 71*(3), 223–231.
25. Tang, S. Y., Whiteman, M., Peng, Z. F., Jenner, A., Yong, E. L., Halliwell, B., et al. (2004). Characterization of antioxidant and antiglycation properties and isolation of active ingredients from traditional Chinese medicines. *Free Radical Biology and Medicine, 36*(12), 1575–1587.
26. Chen, X., Yang, Y., & Zhang, Y. (2013). Isobavachalcone and bavachinin from Psoraleae Fructus modulate Aβ42 aggregation process through different mechanisms in vitro. *FEBS Letters, 587*(18), 2930–2935.
27. Chopra, B., Dhingra, A. K., & Dhar, K. L. (2013). Psoralea corylifolia L. (Buguchi) – Folklore to modern evidence. *Fitoterapia, 90*, 44–56.
28. Wang, D., Li, F., & Jiang, Z. (2001). Osteoblastic proliferation stimulating activity of Psoralea corylifolia extracts and two of its flavonoids. *Planta Medica, 67*(08), 748–749.
29. Bhawya, D., & Anilakumar, K. (2011). Antioxidant, DNA damage protection and antibacterial effect of Psoralea corylifolia. *Asian Journal of Pharmaceutical and Clinical Research, 4*(2), 149–155.

30. Sun, N. J., Woo, S. H., Cassady, J. M., & Snapka, R. M. (1998). DNA polymerase and topoisomerase II inhibitors from Psoralea corylifolia. *Journal of Natural Products, 61*(3), 362–366.
31. Borate, A., Udgire, M., & Khambhapati, A. (2014). Antifungal activity associated with Psoralea corylifolia Linn. (Bakuchi) seeds and chemical profile crude methanol seed extract. *Mintage Journal of Pharmaceutical and Medical Sciences, 3*(3), 4–6.
32. Yin, S., Fan, C.-Q., Wang, Y., Dong, L., & Yue, J.-M. (2004). Antibacterial prenylflavone derivatives from Psoralea corylifolia, and their structure–activity relationship study. *Bioorganic & Medicinal Chemistry, 12*(16), 4387–4392.
33. Kaufman, P. B., Duke, J. A., Brielmann, H., Boik, J., & Hoyt, J. E. (1997). A comparative survey of leguminous plants as sources of the isoflavones, genistein and daidzein: Implications for human nutrition and health. *Journal of Alternative and Complementary Medicine, 3*(1), 7–12.
34. Toyooka, T., & Ibuki, Y. (2009). Histone deacetylase inhibitor sodium butyrate enhances the cell killing effect of psoralen plus UVA by attenuating nucleotide excision repair. *Cancer Research, 69*(8), 3492–3500.
35. Kamboj, J., Sharma, S., & Kumar, S. (2011). In vivo Anti-diabetic and Anti-oxidant potential of Psoralea corylifolia seeds in Streptozotocin induced type-2 diabetic rats. *Journal of Health Science, 57*(3), 225–235.
36. Lim, S.-H., Ha, T.-Y., Ahn, J., & Kim, S. (2011). Estrogenic activities of Psoralea corylifolia L. seed extracts and main constituents. *Phytomedicine, 18*(5), 425–430.
37. Jeong, D., Watari, K., Shirouzu, T., Ono, M., Koizumi, K., Saiki, I., et al (2013). Studies on lymphangiogenesis inhibitors from Korean and Japanese crude drugs. *Biological and Pharmaceutical Bulletin, 36*(1), 152–157.
38. Goel, S., & Ojha, N. K. (2015). Ashtang Ghrita: A noble Ayurveda drug for central nervous system. *Journal of Ayurveda and Holistic Medicine, 3*(2), 18–24.
39. Ling, F., Lu, C., Tu, X., Yi, Y., Huang, A., Zhang, Q., et al. (2013). Antiprotozoal screening of traditional medicinal plants: Evaluation of crude extract of Psoralea corylifolia against Ichthyophthirius multifiliis in goldfish. *Parasitology Research, 112*, 2331–2340.
40. Im, A., Chae, S.-W., & Lee, M.-Y. (2014). Neuroprotective effects of Psoralea corylifolia Linn seed extracts on mitochondrial dysfunction induced by 3-nitropropionic acid. *BMC Complementary and Alternative Medicine, 14*(1), 1–8.
41. Xu, Q., Pan, Y., Yi, L.-T., Li, Y.-C., Mo, S.-F., Jiang, F.-X., et al. (2008). Antidepressant-like effects of psoralen isolated from the seeds of Psoralea corylifolia in the mouse forced swimming test. *Biological & Pharmaceutical Bulletin, 31*(6), 1109–1114.
42. Gidwani, B., Alaspure, R., Duragkar, N., Singh, V., Rao, S. P., & Shukla, S. (2010). Evaluation of a novel herbal formulation in the treatment of eczema with Psoralea corylifolia. *Iranian Journal of Dermatology, 13*, 122–127.
43. Shilaskar, D., & Parasar, G. (1985). Studies on effect of Psoralea corylifolia and Piper betle on cholinesterase and succinic dehydrogenase enzymes as possible targets of their anthelmintic action. *The Indian Veterinary Journal, 62*, 557–562.
44. Iwamura, J., Dohi, T., Tanaka, H., Odani, T., & Kubo, M. (1989). Cytotoxicity of corylifoliae fructus. II. Cytotoxicity of bakuchiol and the analogues. *Pharmaceutical Science Journal, 109*(12), 962–965.
45. Won, T. H., Song, I.-H., Kim, K.-H., Yang, W.-Y., Lee, S. K., Oh, D.-C., et al (2015). Bioactive metabolites from the fruits of Psoralea corylifolia. *Journal of Natural Products, 78*(4), 666–673.
46. Li, H.-N., Wang, C.-Y., Wang, C.-L., Chou, C.-H., Leu, Y.-L., Chen, B.-Y., et al. (2019). Antimicrobial effects and mechanisms of ethanol extracts of Psoralea corylifolia seeds against Listeria monocytogenes and methicillin-resistant Staphylococcus aureus. *Foodborne Pathogens and Disease, 16*(8), 573–580.
47. Husain, F. M., Ahmad, I., Khan, F. I., Al-Shabib, N. A., Baig, M. H., Hussain, A., et al. (2018). Seed extract of Psoralea corylifolia and its constituent bakuchiol impairs AHL-based quorum sensing and biofilm formation in food-and human-related pathogens. *Frontiers in Cellular and Infection Microbiology, 8*, 351.

48. Rutherford, S., & Bassler, B. (2012). Quorum sensing: Its role in virulence and possibilities for its control. *Cold Spring Harbor Perspectives in Medicine, 2*(11), 1.
49. Asad, M. H. H. B., Durr-e-Sabih, Choudary, B. A., Asad, A. F., Muratza, G., & Hussain, I. (2014). Compensatory effects of medicinal plants of Pakistan upon prolongation of coagulation assays induced by Naja naja karachiensis bite. *Current Science, 106*, 870–873.
50. Asad, M. H. H. B., Razi, M. T., Durr-e-Sabih, Najamus-Saqib, Q., Nasim, S. J., Murtaza, G., et al. (2013). Anti-venom potential of Pakistani medicinal plants: Inhibition of anticoagulation activity of Naja naja karachiensis toxin. *Current Science*, 1419–1424.
51. Farahani, M. S., Bahramsoltani, R., Farzaei, M. H., Abdollahi, M., & Rahimi, R. (2015). Plant-derived natural medicines for the management of depression: An overview of mechanisms of action. *Reviews in the Neurosciences, 26*(3), 305–321.
52. Ghosh, D., Bera, T., Chatterjee, K., Ali, K., & Debasis, D. (2009). Antidiabetic and antioxidative effects of aqueous extract of seed of Psoralea corylifolia (somraji) and seed of Trigonella foenum-graecum L., (methi) in Separate and composite manner in streptozotocin-induced diabetic male Albino rat. *International Journal of Pharmaceutical Research and Development, 1*(7), 1–10.
53. Lee, H., Li, H., Noh, M., & Ryu, J.-H. (2016). Bavachin from Psoralea corylifolia improves insulin-dependent glucose uptake through insulin signaling and AMPK activation in 3T3-L1 adipocytes. *International Journal of Molecular Sciences, 17*(4), 527.
54. Seo, E., Oh, Y. S., & Jun, H.-S. (2016). Psoralea corylifolia L. seed extract attenuates nonalcoholic fatty liver disease in high-fat diet-induced obese mice. *Nutrients, 8*(2), 83.
55. Zhu, G., Luo, Y., Xu, X., Zhang, H., & Zhu, M. (2019). Anti-diabetic compounds from the seeds of Psoralea corylifolia. *Fitoterapia, 139*, 104373.
56. He, S., Hong, Q., Lai, Z., Yang, D. X., Ting, P. C., Kuethe, J. T., et al. (2014). Discovery of a potent and selective DGAT1 inhibitor with a piperidinyl-oxy-cyclohexanecarboxylic acid moiety. *ACS Medicinal Chemistry Letters, 5*(10), 1082–1087.
57. Liu, J., Zhao, Y., Huang, C., Li, Y., & Guo, F. (2019). Prenylated flavonoid-standardized extract from seeds of Psoralea corylifolia L. activated fat browning in high-fat diet–induced obese mice. *Phytotherapy Research, 33*(7), 1851–1864.
58. Chai, M.-Y. (2019). A new bioactive coumestan from the seeds of Psoralea corylifolia. *Journal of Asian Natural Products Research, 22*(3), 295–301.
59. Hosamani, A., Lakshman, H. C., & Sandeepkumar, K. (2012). Antimicrobial activity of leaf extract of Psoralea corylifolia L. *Life Sciences Leaflets, 30*, 35–39.
60. Kim, K.-A., Shim, S. H., Ahn, H. R., & Jung, S. H. (2013). Protective effects of the compounds isolated from the seed of Psoralea corylifolia on oxidative stress-induced retinal damage. *Toxicology and Applied Pharmacology, 269*(2), 109–120.
61. Behloul, N., & Wu, G. (2013). Genistein: A promising therapeutic agent for obesity and diabetes treatment. *European Journal of Pharmacology, 698*(1–3), 31–38.
62. Anusha, S., Haja Sherief, S., Sindhura, S., Jaya Preethi, P., & Siva Kumar, T. (2013). Synergistic effet of indigenous medicinal plant extracts on Psoriasis. *International Journal of Phytopharmacy, 3*(1), 23–29.
63. Rajendra Prasad, N., Anandi, C., Balasubramanian, S., & Pugalendi, K. V. (2004). Antidermatophytic activity of extracts from Psoralea corylifolia (Fabaceae) correlated with the presence of a flavonoid compound. *Journal of Ethnopharmacology, 91*(1), 21–24. https://doi.org/10.1016/j.jep.2003.11.010
64. Kim, D. W., Seo, K. H., Curtis-Long, M. J., Oh, K. Y., Oh, J.-W., Cho, J. K., et al. (2014). Phenolic phytochemical displaying SARS-CoV papain-like protease inhibition from the seeds of Psoralea corylifolia. *Journal of Enzyme Inhibition and Medicinal Chemistry, 29*(1), 59–63.
65. Desmawati, D., & Sulastri, D. (2019). Phytoestrogens and their health effect. *Open Access Macedonian Journal of Medical Sciences, 7*(3), 495.
66. Zhao, F.-J., Zhang, Z.-B., Ma, N., Teng, X., Cai, Z.-C., & Liu, M.-X. (2019). Untargeted metabolomics using liquid chromatography coupled with mass spectrometry for rapid discovery of metabolite biomarkers to reveal therapeutic effects of Psoralea corylifolia seeds against osteoporosis. *RSC Advances, 9*(61), 35429–35442.

67. Zhang, T., Zhong, S., Meng, Y., Deng, W., Hou, L., Wang, Y., et al. (2018). Quantitative structure-activity relationship for estrogenic flavonoids from Psoralea corylifolia. *Journal of Pharmaceutical and Biomedical Analysis, 161*, 129–135.
68. Sözen, T., Özışık, L., & Başaran, N. Ç. (2017). An overview and management of osteoporosis. *European Journal of Rheumatology, 4*, 46–56.
69. Fu, P.-K., Yang, C.-Y., Tsai, T.-H., & Hsieh, C.-L. (2012). Moutan cortex radicis improves lipopolysaccharide-induced acute lung injury in rats through anti-inflammation. *Phytomedicine, 19*(13), 1206–1215.
70. Yu, A. X. D., Xu, M. L., Yao, P., Kwan, K. K. L., Liu, Y. X., Duan, R., et al. (2020). Corylin, a flavonoid derived from Psoralea Fructus, induces osteoblastic differentiation via estrogen and Wnt/β-catenin signaling pathways. *The FASEB Journal, 34*(3), 4311–4328.
71. Dua, V. K., Kumar, A., Pandey, A. C., & Kumar, S. (2013). Insecticidal and genotoxic activity of Psoralea corylifolia Linn.(Fabaceae) against Culex quinquefasciatus Say, 1823. *Parasites & Vectors, 6*(1), 1–8.
72. Nam, S. W., Baek, J. T., Lee, D. S., Kang, S. B., Ahn, B. M., & Chung, K. W. (2005). A case of acute cholestatic hepatitis associated with the seeds of Psoralea corylifolia (Boh-Gol-Zhee). *Clinical Toxicology (Philadelphia, Pa.), 43*(6), 589–591.
73. Zhang, X., Zhao, W., Wang, Y., Lu, J., & Chen, X. (2016). The chemical constituents and bioactivities of Psoralea corylifolia Linn.: A review. *The American Journal of Chinese Medicine, 44*(01), 35–60.
74. Hussain, I., Abro, H. A., & Mubarak, N. (2019). Skin pigmentation effects of Psoralea corylifolia: A case study of Vitiligo. *Journal of Islamic International Medical College, 14*(1), 48–50.
75. Dhaliwal, S., Rybak, I., Ellis, S., Notay, M., Trivedi, M., Burney, W., et al. (2019). Prospective, randomized, double-blind assessment of topical bakuchiol and retinol for facial photoageing. *The British Journal of Dermatology, 180*(2), 289–296.
76. Chaudhuri, R. K., & Bojanowski, K. (2014). Bakuchiol: A retinol-like functional compound revealed by gene expression profiling and clinically proven to have anti-aging effects. *International Journal of Cosmetic Science, 36*(3), 221–230.
77. Shoji, M., Arakaki, Y., Esumi, T., Kohnomi, S., Yamamoto, C., Suzuki, Y., et al. (2015). Bakuchiol is a phenolic isoprenoid with novel enantiomer-selective anti-influenza A virus activity involving Nrf2 activation. *The Journal of Biological Chemistry, 290*(46), 28001–28017.
78. Alalaiwe, A., Hung, C.-F., Leu, Y.-L., Tahara, K., Chen, H.-H., Hu, K.-Y., et al. (2018). The active compounds derived from Psoralea corylifolia for photochemotherapy against psoriasis-like lesions: The relationship between structure and percutaneous absorption. *European Journal of Pharmaceutical Sciences, 124*, 114–126.
79. Jing, H., Wang, S., Wang, M., Fu, W., Zhang, C., & Xu, D. (2017). Isobavachalcone attenuates MPTP-induced Parkinson's disease in mice by inhibition of microglial activation through NF-κB pathway. *PLoS One, 12*(1), e0169560.
80. Panda, H. (1999). *Herbs cultivation and medicinal uses*. National Institute of Industrial Research.
81. Lau, K.-M., Wong, J. H., Wu, Y.-O., Cheng, L., Wong, C.-W., To, M.-H., et al. (2014). Anti-dermatophytic activity of bakuchiol: In vitro mechanistic studies and in vivo tinea pedis-inhibiting activity in a guinea pig model. *Phytomedicine, 21*(7), 942–945.
82. Khare, C. (2004). *Encyclopedia of indian medicinal plants: Rational western therapy, ayurvedic and other traditional usage, botany*. Springer.
83. Bankoti, K., Rana, M., & Bharadwaj, M. (2012). Accelerated stability study of herbal capsules. *IOSR Journal of Pharmacy, 2*(5), 1–6.
84. Qiao, C. F., Han, Q. B., Song, J. Z., Mo, S. F., Kong, L. D., Kung, H. F., et al. (2007). Chemical fingerprint and quantitative analysis of Fructus Psoraleae by high-performance liquid chromatography. *Journal of Separation Science, 30*(6), 813–818.
85. Kunwar, R. M., Shrestha, K. P., & Bussmann, R. W. (2010). Traditional herbal medicine in Far-west Nepal: A pharmacological appraisal. *Journal of Ethnobiology and Ethnomedicine, 6*(1), 1–18.

86. Flaws, B., & Sionneau, P. (2001). *The treatment of modern Western medical diseases with Chinese medicine: A textbook & clinical manual*. Blue Poppy Enterprises.
87. Sharma, P., Yelne, M., Dennis, T., Joshi, A., & Billore, K. (2000). *Database on medicinal plants used in Ayurveda*. Central Council for Research in Ayurveda & Siddha.
88. Gopal, M., Farahana, B., & Pramesh, R. (2001). Effectiveness of herbal medications in the treatment of acne vulgaris – A pilot study. *The Indian Practitioner, 54*(10), 723.
89. Bensky, D., Clavey, S., & Stõger, E. (2004). *Materia medica* (pp. 3–6).
90. Chen, J. K., Chen, T. T., & Crampton, L. (2004). *Chinese medical herbology and pharmacology*. Art of Medicine Press City of Industry.
91. Takizawa, T., Mitsumori, K., Takagi, H., Nasu, M., Yasuhara, K., Onodera, H., et al. (2004). Sequential analysis of testicular lesions and serum hormone levels in rats treated with a Psoralea corylifolia extract. *Food and Chemical Toxicology, 42*(1), 1–7.
92. Wang, X.-X., Lv, X., Li, S.-Y., Hou, J., Ning, J., Wang, J.-Y., et al. (2015). Identification and characterization of naturally occurring inhibitors against UDP-glucuronosyltransferase 1A1 in Fructus Psoraleae (Bu-gu-zhi). *Toxicology and Applied Pharmacology, 289*(1), 70–78.
93. Teschke, R., & Bahre, R. (2009). Severe hepatotoxicity by Indian Ayurvedic herbal products: A structured causality assessment. *Annals of Hepatology, 8*(3), 258–266.
94. Teschke, R., Wolff, A., Frenzel, C., & Schulze, J. (2014). Herbal hepatotoxicity – An update on traditional Chinese medicine preparations. *Alimentary Pharmacology & Therapeutics, 40*(1), 32–50.
95. Xu, Q.-X., Xu, W., & Yang, X.-W. (2020). Meroterpenoids from the fruits of Psoralea corylifolia. *Tetrahedron, 76*(31–32), 131343.
96. Xu, Q.-X., Zhang, Y.-B., Liu, X.-Y., Xu, W., & Yang, X.-W. (2020). Cytotoxic heterodimers of meroterpene phenol from the fruits of Psoralea corylifolia. *Phytochemistry, 176*, 112394.

Chapter 6
Ashwagandha

Sadia Javed, Ayesha Nazir, Ameer Fawad Zahoor, and Arwa A. AL-Huqail

6.1 Introduction

Ashwagandha, or *Withania somnifera* (WS), is an Indian herb. Winter cherry and ginseng has both been significant herbs in the indigenous and Ayurvedic medical traditions for more than 3000 years. The classification of plant's roots is named as Rasayanas. They are known for fostering wellness and longevity increasing disease defense and slowing down ageing procedure, restoring life to the body in sickly situations, strengthening a person's ability to withstand negative environmental elements and by fostering a sense of mental well-being wellbeing [1]. The plant has been used to treat senile dementia, ulcers, bacterial infections, liver ailments, venom toxins, and more recently, aphrodisiacs, adaptogens, liver tonics, and antioxidants. The use of WS for anxiety, neurological diseases, inflammation, hyperlipidemia, and Parkinson's disease is supported by clinical trials and animal studies. Because of its chemo-preventive qualities, WS may be a helpful adjuvant for patients receiving radiation and chemotherapy. Recently, WS has also been utilized to prevent continuous use of certain psychotropic medicines from leading to the development of tolerance and dependency [2].

S. Javed (✉) · A. Nazir
Department of Biochemistry, Government College University, Faisalabad, Pakistan
e-mail: sadiajaved@gcuf.edu.pk

A. F. Zahoor
Department of Chemistry, Government College University, Faisalabad, Pakistan

A. A. AL-Huqail
Department of Biology, College of Science, Princess Nourah bint Abdulrahman University, Riyadh, Saudi Arabia

M. Zia-Ul-Haq et al. (eds.), *Essentials of Medicinal and Aromatic Crops*,
https://doi.org/10.1007/978-3-031-35403-8_6

6.2 Taxonomical Classification

Ashwagandha scientifically called *Withania somnifera* is taxonomically classified as; it belongs to Kingdom Plantae, Plants and subkingdom Tracheobionta, Vascular plants. The super division is Spermatophyta, Seeds plants. Division Angiosperma, class Dicotyledons, order Tubiflorae, family Solanaceae, genus Withania and Species somnifera Dunal (Table 6.1) [2].

6.3 Botanical Structure

This plant is a small in size. Its branches extend from central stem in the manner of stars (stellate), and it is covered with a dense matte woolly hair called tomentose. The fruit is mature when it is orange-red and contains milk-coagulating characteristics, while the blooms are tiny and green. Long, brown, tuberous roots of the plant are used as a medicine.

6.4 Occurrence

The semi-arid area of India and other Southeast Asian nations is home to the plant known as Ashwagandha. Additionally, the plant may be found in parts of Africa including the Congo, South Africa, Egypt, Morocco, and the Middle East [3]. This plant is also grown in Pakistan and Sri Lanka. It is also called Indian ginseng and it has been used by Ayurvedic practitioners for thousands of years [4]. Madhya Pradesh, Haryana, Gujarat, Punjab, Maharashtra, Uttar Pradesh, and Rajasthan are the Indian regions where most of WS is grown. More than 5000 acres of Madhya Pradesh are dedicated to its cultivation. With an annual demand of over 7000 tons and predictable output from its Indian sources of more than 15,000 tons, it is necessary to enhance agricultural production and improve production. Ayurvedic practitioners have used WS for thousands of years a cure for various diseases [4].

Table 6.1 Taxonomical classification of *Withania sominefera*

Kingdom	Plantae, plants
Subkingdom	Tracheobionta; vascular plants
Super division	Spermatophyta, seed plants
Division	Angiosperms
Class	Dicotyledons
Order	Tubiflorae
Family	Solanacae
Genus	Withania
Species	sominefera Dunal

6.5 Pests and Diseases

No significant pests have been identified in the crop. Two or three Neem Astra sprays as the foliar spray was found to be extremely helpful against aphids, mites, and insect attacks at 10 days per time the crop is harmed by insects. Some diseases resemble seedling rot and blight, according to reports. High humidity and temperature levels drastically increase seedling mortality. Usage of seeds which are disease free and proper seed preparation before the process of sowing can lower the likelihood of illness. Neem cake can also be utilized. It will protect nematodes and insects from root damage. Additionally, by using crop rotation, planting at the right time, and ensuring appropriate soil drainage, the crop will be protected [4].

In an experimental study, Ashwagandha is affected by a number of insect pests, such as the Epilachna vigintioctopunctata which is a leaf beetle having spots on it. By applying Azophos (2 kg/ha), decomposed mixture of dung or manure (FYM) (12.5 t/ha) and cake of neem (1000 kg/ha), it was discovered that the damage caused by spotted leaf beetles could be reduced by 69.79%. With a minimal feeding area of 6.75 cm^2, Azophos and neem cake combination was not able to reduce oviposition (laying of eggs), which was noted 62.00 eggs per plants Ravikumar, Rajedran, Chinniah, & Irulandi). The lowest Epilachna beetle infestation and maximum production were seen with dimethoate 30 EC (1 ml/l). Endosulfan NSKE fiver percent, Fenvalerate and chlorpyriphos, were potentially effective treatments that were comparable to dimethoate [5, 6]. On W. somnifera, gonim and B. thuringensis were shown to be the most efficient pesticides against red bugs and aphids also decrease the larval population of defoliators [7]. 250 LE (63.61) of HaNPV, which was comparable to nimbecideine at 3 ml/L (56.66) and then NSKE at 5% (47.42) and oil extracted from neem at 5 ml/l (45.73), recorded the highest mean percent decrease of H. armigera infestation on Ashwagandha above control [8]. The gram caterpillars, H. armigera, eat the fragile leaves of Ashwagandha plants and bore into the fruits. As part of an integrated strategy, field release of T. chilonis (100,000 eggs/ha), application of HaNPV (250 LE/ha) or B. thuringiensis (0.5 kg/ha) and deployment of pheromone traps (12 traps/ha) were advised. Deilephila nerri, a minor defoliator that feeds on hawk moth caterpillars, is also managed by exposing pupae in the soil, removing and destroying larvae by hand, setting up 1 trap each ha of light traps, and also setting *Nerium oleander* (L.) as a trap plant surrounding the field [9].

6.6 Historical Background

The Indian traditional medical system known as Ayurveda traces back to the 6000 BC [10]. Historicaly, Ashwagandhaa was used as therapy called Rasayana. Its root is valued for its thermogenic, narcotic, tonic, anthelmuntic, diuretic, anthelmintic, stimulant and astringent properties.The name "Ashwagandha"refers to the fact that the root has a horse-like odor. In ancient times it was believed to give horse lie powers when consumed. Emaciation in children (when consumed with milk, it becomes the excellent source of nutrition for children), leukoderma, old age, debility, constipation, rheumatism sleeplessness, neurological disorders, goiter, etc. are all ailments for which it is frequently used [11]. Application of WS root paste which is formed by crushing its roots with water helps to decrease inflammation in joints [12]. Additionally, it is used locally to painful swellings such carbuncles and ulcers [13]. For both scorpion stings and snake venom, the root is administered in association with other medications. Additionally, it aids in the treatment of piles, worms, boils, and leucorrhoea [14]. Out all of its varieties available the Nagori Ashwagandha is the best one. When using fresh Ashwagandha powder, the benefits are the maximum [15]. The bitter leaves are suggested for fever and throbbing swellings. The flowers have aphrodisiac, diuretic, astringent, and depurative properties. The seeds are anthelmintic and erase white spots from the cornea when mixed with an astringent and rock salt. It is used to make Ashwagandharishta, which is used for The Nagori Ashwagandha is the best among hysteria, syncope, memory loss, anxiety, etc. Additionally, it stimulates the body and enhances the count of sperm in males [15].

6.7 Characteristics

Ashwagandha is a perennial to annual, branching shrub with tiny stellate and tomentose branches that grows from 30 to 120 cm. The roots are brownish-white, meaty, and tapered. Oval leaves and greenish blooms are present. Fruits that are orange and red are mature.

6.7.1 Variety

A variety called "Jawahar" from Madhya Pradesh, is a small in height and best suited for planting in dense populations. Within 180 days, the variety produces dry roots with a cumulative withanolide concentration of 0.30%.

6.8 Conditions for Growth

It is grown as a crop during the Kharif (late rainy season). The best places to cultivate crops that are nourished by rainfall are semi-tropical regions with 500–750 mm of annual precipitation. Natural rain enhances the roots increases the growth. This plant usually requires dry season to grow. Temperatures between 20 and 38 °C, as well as low temperatures as low as 10, can be tolerated. It grows to 1500 m of height above the sea level. Ground and soil.

Ashwagandha thrives on loam soil or redish soil with 7.5–8.0 of the pH range and adequate drainage.

6.8.1 *Land Preparation*

Before it rains, two to three plowing, discing, or harrowing operations should be carried out to get the soil to a very good tilth. For application, mixing, and leveling of the land, FYM 25 tons per hectare.

6.9 Conservatory

Ashwagandha is grown from seeds. Then freshly taken seeds are taken and sown in the prepared beds of nursery. While the broadcast method may be used to seed it in the field, the transplanting approach is preferable for increased efficiency and export purposes. A well-kept nursery is essential for export. The nursery bed, which is normally elevated off the ground, is prepared with sand and the compost. Approximately five kilograms of seed is needed to plant on 1 ha of the field. The nursery is built up throughout June and July. The seeds are spread and gently covered with sand just before the monsoon. The seeds begin to sprout after 5–7 days. Older seedlings that are around 35 days old are transferred into the main sector.

6.10 Irrigation

Water or excessive rains harms this crop. After transplanting, light rain helps plants establish more quickly. If necessary, irrigation that can save lives may be offered. For improved outcomes and a greater root output in an irrigated condition, the crop might be watered once every 10 days.

6.11 Transplanting

After the well incorporation of manure into the soil, ridges are placed 60 cm apart. For seedlings of good quality, the distance is kept of 30 cm. In various places, spacing of 60 cm × 60 cm or 45 cm × 30 cm is frequently used. The spacing of 60 cm into 30 cm is thought to be the most advantageous with the plant density of roughly per hectare of 55,000 seedlings.

6.12 Sowing and Seeding Rate

For broadcasting techniques the seed rate of 10–12 kg each hectare is acceptable. Alternatively, rows may be planted. The method of line-to-line is recommended because it promotes the development of roots and aids in the efficient execution of cross-cultural activities. Typically, the seeds are spread between 1 and 3 cm deep. In all processes, light soil should be applied to the seeds. Maintaining a line to line distance is kept 20–25 cm and a plant to plant distance of 8–10 cm is crucial. The spacing may be changed based on the soil fertility. In marginal soils, the population is often maintained at higher levels.

6.13 Treatment of Seed with *Trichoderma viride*

To protect seedlings against illnesses transmitted by seed, treatment of *Trichoderma viride* can give at a rate of 3 g/Kg of seed before planting. For protection of seedlings from illnesses due to bad seeds, Dithane M45 (Indofil M45) or Thiram at a rate of 3 g/Kg of seed for handling.

6.14 Intercultural Practices

Mature seedlings are cultivated sown by hand 25–30 days or by the broadcasting method or after sowing to keep plant density of around 30–60 plants per square meter. Kind and fertility of the soil can affect the number of plants that can be maintained at prior levels. The population needs to ideally be kept at a lower level if fertilizer is used. In general, weeding is crucial to keep the field weeds free for about initial 25 days after planting and subsequent 20–25 days.

6.15 Manures and Fertilizers

Growth of WS doesn't require any significant amounts of manure or fertilizer. Inorganic fertilizers are rare to be used. The crop responds favorably to compost, vermin compost, and organic manures. For each ha, it is advised to apply 10 t of FYM or 1 t of vermin-compost. The application of 15 kg of nitrogen and 15 kg of phosphorus per acre is useful for more output in typical rich ground. For enhanced root output in rich stumpy soils, 40 kg of N and P per hectare is sufficient.

6.16 Reaping

In well-managed crops, Ashwagandha yields dry root of 3–5 q and 50–75 kg of seeds per hectare in 180 days. Under scientific crop management, the dry root production increases to 6.5–7.0 q/ha. Farmers have occasionally gotten root yields as high as 1 t/ha. Commercially, roots with a diameter about 15 mm and a length of 7–10 cm are chosen. Roots contain an alkaloid content ranging from 0.13% to 0.31%.

6.17 Demand in Market

Every year, makers of Ayurvedic and Siddha medicines, as well as importers, global buyers, processors, and herbalists, visit these marketplaces to purchase Ashwagandha roots. Recognition of these plants therapeutic and financial benefits is growing in both developing and developed nations. As was already mentioned, there is a 7000-ton yearly domestic demand for Ashwagandha roots. The domestic market in India has a large potential because the output there is substantially lower than the average of 1500 tons [4].

6.18 Chemical Constituents

The Ashwagandha root and leaf extracts in methanol, hexane, and diethyl ether were discovered. Roots contain an alkneuroaloid content ranging from 0.13% to 0.31%. The alterative, narcotic, deobstruent, diuretic, restorative, aphrodisiac, sedative and properties of *Withania somnifera's* roots are well known. Ateroidal lactones and Alkaloids are known to be responsible for the root's therapeutic effect. Although substantially higher yields (up to 4.3%) have been observed, the alkaloid content of roots of Indian varieties varies between 0.13 and 0.3. There are a variety of heterogeneous biochemical alkaloids, such as isopelletierine, tropanol, cuscokygrene, choline, 3-tigioyloxytropana, pseudotopanol, and a number of many other lactons steroidal in nature. From the plant's roots, researchers have identified 12 alkaloids, 35 withanolides, and many sitoindosides. Withanolide, a physiologically active component having a C27 glucose molecule at, is known as sitoindoside. Withaferin D and withanolide A are the primary withanolides thought to be responsible for Indian ginseng's pharmacological effects. According to reports, leaves contain withaferin-A, a withanolide that is therapeutically effective. WS roots also contain starch, dulcitol, glycosides reducing sugars, withancil, also alkaloids, neutral and substances. Tyrosine, glycine, glutamic acid, aspartic acid, alanine, and cysteine are among the amino acids identified from the roots [16]. Additionally, this plant contains a high amount of iron as well as chemical constituents like withaniol, starch, acylsteryl glucosides, ducitol, reducing sugar, hantreacotane and reducing sugar (Fig. 6.1) [17].

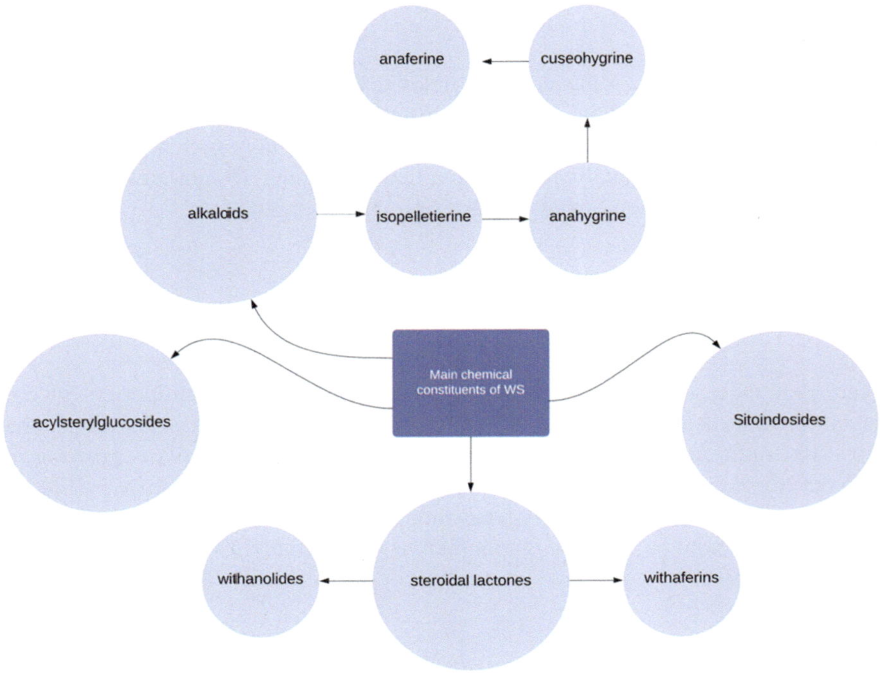

Fig. 6.1 Chemical constituents of *Withania somnifera* (WS)

6.19 Biosynthesis of Withanolides

WS produces withanolides, which are ergostane skeleton-based C28-steroidal lactones produced from triterpenoids. De novo biogenesis and withanolide accumulation are most active in young leaves and begin to wane as leaves mature. In plants, isoprenoid biosynthesis occurs via two distinct pathways: the cytosolic mevalonate (MVA) system and the plastid-localized 2-C-methyl-d-erythritol-4-phosphate (MEP) pathway. Isoprenoids produced by these pathways are channelled into a variety of metabolic pathways, where they create a variety of specialised metabolites engaged in a variety of cellular and regulatory functions. Condensation of acetyl CoA molecules with acetoacetyl CoA results in 3-hydroxy-3-methylglutaryl-CoA (HMG-CoA) and mevalonic acid. HMG-CoAs are irreversibly converted into mevalonic acids by HMGR. 3-isopentenyl pyrophosphate is formed by mevalonate-5-pyrophosphate decarboxylase (IPP). FPPS catalyses the condensation process of IPP with another IPP molecule. The MEP pathway is an important part of withanolides production, involving the conversion of pyruvate into d-glyceraldehyde-3-phosphate (DXP), DXP synthase (DXS), MEP (MEP) and 4-diphospho-cytidyl-2-methyl-d-erythritol (CDP-ME) from a CTP-dependent reaction catalyzed by CMS and CMK respectively (Fig. 6.2) [18].

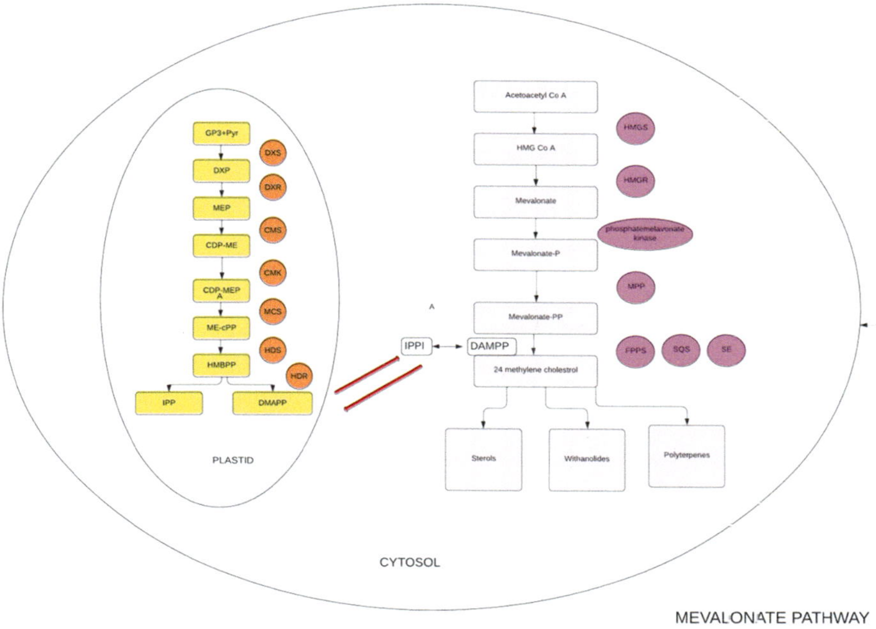

Fig. 6.2 Biosynthesis of withnolides via mevalonate pathway

6.20 Therapeutic Potential of WS in Clinical Field

WS has been used as power therapeutic source in past and modern studies and research also proved its strong potential to develop cure against various diseases (Fig. 6.3).

6.20.1 *Attention Deficit Hyperactivity Disorder (ADHD)*

A combined herbal supplement containing Ashwagandha, according to some clinical studies, can enhance attention and impulsive control in kids with ADHD. The effects of Ashwagandha alone are unknown. ADHD is a diverse brain condition characterized by distractibility or inattention, which may or may not be accompanied by hyperactivity. It is more frequent in younger age and and may persist into maturity, with males being more affected than females

6.20.2 *Cerebellar Ataxia*

According to preliminary studies, Ashwagandha may enhance balance in people with cerebellar ataxia when used combined with Ayurvedic treatment, another form of medicine.

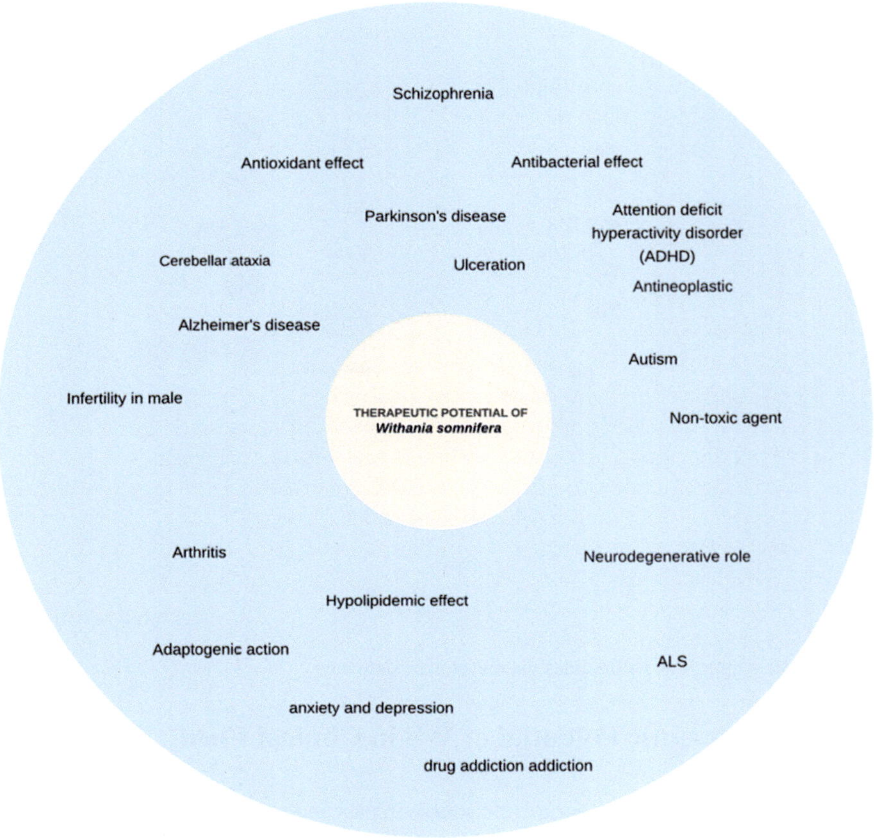

Fig. 6.3 Therapeutic potential of *Withania somnifera* (WS)

6.20.3 Infertility in Male

According to some clinicaltrials, WS enhances the quality of sperm but it does not affect the count of sperm in infertile males. It is unknown if consuming WS can enhance fertility.

6.20.4 Arthritis

The analgesic Ashwagandha calms the neurological system's pain response, helpin reducing pain in bones [19]. Ashwagandha may help reduce the symptoms of arthritis when combined with other substances in a supplement called Articulin-F. The effectiveness of Ashwagandha alone in osteoarthritis is unknown.

6.20.5 Ulceration

Using Ashwagandha for a prolonged period, reduced the tendency for uterine bleeding and caused the fibroids to decrease in patients with uterine fibroid tumors [16].

6.20.6 Antioxidant Effect

Due to their abundance in lipids and iron, two substances known to play significant roles in the generation of reactive oxygen species nervous system and brain are considerably vulnerable to damage of free radicals than the other tissues. Neurodegenerative illnesses and the process of normal aging such as Parkinson's, Alzheimer's, Schizophrenia, epilepsy, and other conditions, may be accompanied by free radical damage to neural tissue. In a study catalase (CAT). superoxide dismutase (SOD) and glutathione peroxidase (GPX) and levels in the rat brain frontal striatum and cortex were used to investigate the activity of antioxidant components of WS, withaferin-A also called glycowithanolides and sitoindosides VII-X. These enzymes' decreased activity causes harmful oxidative free radicals to build up and cause degenerative consequences. Increase in protective impact on neural tissue and antioxidant activity would be represented by an increase in these enzymes. Once daily for 21 days, active glycowithanolides of WS were administered; increases in all enzymes related to dose were noticed; the rises were equivalent to these reported with the dosage of a well-known antioxidant, deprenyl. This clearly depicts the impact of WS on brain due to its effective antioxidant properties [20].

6.20.7 Antineoplastic

In research employing mice administered WS during and prior to the cancer causing agents 7,12-dimethylbenz anthracene. And the chemo-preventive effect of WS root extract was shown. When compared to the control group, there was reduction in the incidence and average rate of lesions on skin. In addition, when the extract was administered, the levels of reduced glutathione, GPX, SOD, and CAT in the exposed tissue almost reached regular levels. It is believed that the extract's antioxidant and free radical-scavenging properties contribute to the chemopreventive effect Withanolides from WS reduced the proliferation of colon, central nervous system and lung cell lines similarly to doxorubicin, according to in vitro research. Withaferin Doxorubicin was less effective at preventing the development of colon and breast cancer cell lines. These findings point to the possibility of the creation of novel chemotherapeutic drugs and imply that WS extracts can also reduce or prevent the tumor growth in cancer patients [21]. In a different research, adult male albino mice with urethane-induced lung adenomas were given WS to see if it had any anticancer effects. The incidence of tumors was dramatically decreased when WS, given orally 200 mg/kg each day for the time of 7 months. And urethane given 125 mg/kg every 2 weeks for the time period of 7 months were administered concurrently. Animals treated by WS had lungs that shared the same histological characteristics as those seen in control animals. The negative effects of urethane on body weight, morality. Leukocyte count and lymphocyte count were similarly overturned by WS therapy [22]. The Ayurvedic medical system frequently used WS to treat inflammation, tumors, arthritis, hypertension and asthma. Bioactive withanolides have been discovered chemically in this plant's roots and leaves. Withanolides prevent the growth of tumor cells, lipid peroxidation, and cyclooxygenase enzymes. Nuclear factor kappa B (NF-κB) is a protein transcription factor activity controls numerous genes that control inflammation, metastasis, cellular proliferation, and cancer. TNF, interleukin-1 beta, doxorubicin, and cigarette smoke condensate are just a few of the inflammatory and carcinogenic substances that withanolides inhibited from making NF-kappaB active. Withanolides prevented both inducible and constitutive NF-kappaB activation, demonstrating that the suppression was not cell-type specific. The inhibition of IkappaB alpha phosphorylation, IkappaB alpha kinase inhibitory subunit activation, IkappaB alpha degradation, subsequent p65 nuclear translocation and p65 phosphorylation, caused the suppression. Additionally, TNF receptor (TNFR) 1, TNFR-associated factor 2, TNF, TNFR linked to the death causing domain, and IkappaB alpha kinase inhibited the expression of NF-kappaB-dependent reporter genes. Since withanolide reduced the expression of TNF-induced NF-kappaB-regulated antiapoptotic (inhibitor of apoptosis protein 1, Bfl-1/A1, and FADD-like interleukin-1beta-converting enzyme-inhibitory proTNFtein) and metastatic (cyclooxygenase-2 and intercellular adhesion molecule-1) gene products, it increased the apoptosis caused by TNF and chemotherapy. Overall, withanolides are thought to decrease NF-kappaB activation and NF-genetic regulation by kappaB, which can account for their capacity to increase apoptosis and prevent invasion and osteoclastogenesis [23].

6.20.8 Non-toxic Agent

By disrupting the outer membrane of the extracellular matrix of the effected tissues, venom hyaluronidases aid in rapid dissemination of the poisons. Hyaluronidase inhibitors (WSG) can be purified from WS Naja naja (Cobra) and Daboia russelii (viper) venoms' hyaluronidase activity was decreased by the glycoprotein, as shown by the zymogram test and differential activity by staining of tissues of skin. By keeping concentration of 1:1 w/w of venom to WSG, the enzyme's activity was entirely inhibited by the WSG. Some studies prove usage of a plant extract as an antidote for snakebite sufferers externally in rural India [24]. An antitoxin-PLA2 glycoprotein derived from WS was shown to inhibit the PLA2 action of Cobra venom. The effects of new PLA2 toxin inhibitors on snake biology and the creation of cutting-edge therapeutic medicines for the treatment of snake venoms [25].

6.20.9 Effect on Lipid Peroxidation or Hypolipidemic Effect

In hypercholesteremic mice, powder of WS root reduced total triglycerides cholesterol, and lipid. Alternatively, there was a major rise in the levels of HMG-CoA reductase activity, cholesterol, high density lipoprotein in plasma, and in liver the concentration of bile acid. Animals which are hypercholesteremic when given WS treatment, a parallel trend was shown in the excretion of bile acid, cholesterol, and neutral sterol. In addition, WS-treated hypercholesteremic mice showed much less lipid peroxidation than their control counterparts. But normal participants also saw a reduction in lipid profiles while using WS root powder. In another research, rats given a high-fat diet to produce hyperlipidemia had their raised blood levels of cholesterol, triglycerides, and lipoprotein levels dramatically lowered after being given extract aqueous in nature of the fruits of WS coagulans for 7 weeks [26]. In hypercholesteremic mice, powder of roots WS reduced total cholesterol level and fats in liver. In the hypercholesteremic animals with WS treatment, a similar trend was also noted in the cholesterol, neutral sterol and excretion of bile acid. In addition, WS-treated hypercholesteremic mice showed much less lipid peroxidation than their control counterparts. But normal participants also saw a reduction in lipid profiles while using WS root powder. In another research, rats given a high-fat diet to produce hyperlipidemia had their raised blood levels of cholesterol, triglycerides, and lipoprotein levels dramatically lowered after being given fruits of Withania fruit extract coagulans for 7 weeks [27]. In a different study, human volunteers were used to evaluating the hypoglycemic, diuretic, and hypocholesterolemic effects of WS roots. The powdered roots of WS were administered to six mild NIDDM participants and six mild hypercholesterolemic subjects for 30 days. The individuals' blood and urine samples, as well as their eating habits both before and after the therapy period, were examined for the right criteria. Blood glucose reduction was equivalent to that of oral hypoglycemic medication. Significant increases in urine sodium and volume as well as significant decreases in serum triglycerides, LDL (low-density lipoproteins), VLDL (very low-density lipoproteins) and cholesterol were found, suggesting that WS root may be a source of, and hypocholesterolemic, hypoglycemic and diuretic agents [28].

6.20.10 Antibacterial Effect

Agar Well Diffusion Method in vitro, it was shown that the plant's alcoholic and aqueous extracts of leaves and roots have effective activity against many bacteria. The butanolic sub-fraction of the extracts using methanol, which was subsequently sub-fractionated using different solvents, had highest inhibitory efficacy against a variety of bacteria, including Salmonella typhimurium. Furthermore, these extracts did not cause lysis when incubated with human erythrocytes, in contrast to the synthetic antibiotic chloramphenicol, demonstrating their safety to live cells. Mice with nfection caused by Salmonella in Balb/C was efficiently eradicated by oral management of the aqueous extracts, as shown by the animals' improved survival rate and decreased bacterial load in numerous important organs [29]. In a different research, the antibacterial/synergistic activity of diethyl ether, methanol and hexane extracts from the roots and leaves of WS was assessed using disc-diffusion assay on agar plate against S.typhimurium and E.coli. The minimum inhibitory concentration was evaluated at various doses of the drug combination Tibrim, which contains the antibiotics rifampicin and isoniazid for E.coli and for S. typhimurium determined to be 0.1 mg/ml. Only the hexane and methanolic extracts of the roots and leaves showed significant antibacterial activity out of the six extracts examined. When these extracts were added to the MIC of Tibrim, the antibacterial action of Tibrim was seen to enhance synergistically [30].

6.20.11 Adaptogenic Action

WS is frequently used to relieve patients' tension, serving as a form of antistress treatment. In a study conducted in our lab, in model of mouse having chronic fatigue syndrome showed positive effects due to WS. Chronic fatigue syndrome is a condition marked by recurrent bouts of weariness. The mice in this study were made exhausted by making them swim for 6 min every day for 15 days. Daily doses of antioxidants and WS were given to the animals before stress was applied. Every day, the mean immobility duration was computed and compared to control animals. When stressed mice were compared to control animals, WS resulted in a considerable increase in mobility time, demonstrating the extract's anti-stress properties.[31]

6.20.12 Neurodegenerative Role

Acetylcholinesterase and butylcholinesterase are known to be inhibited by withanoloids extracted from the WS in a dose-dependent manner. WS is a strong therapeutic agent for the cure of Alzheimer's disease and related issues due to its potential as a cholinesterase inhibitor and calcium antagonist [32]. Sitoindosides VII-X and Withaferin-A, which were discovered in methanolic aqueous extracts from the roots of WS types, are used in Indian medication to decrease forgetfulness and other brain functioning deficiencies in elderly people [33]. The impact of these WS active ingredients was also examined for potential nootropic effects in an Alzheimer's disease model that has undergone experimental validation. Lesioning of the nucleus magnocellularis in rats due to ibotenic acid caused the condition. After 2 weeks of therapy, WS effectively corrected the cognitive loss caused by ibotenic acid as well as the drop in cholinergic indicators. These results confirmed WS promoters of learning and memory (Medharasayan) impact [34]. Mice lacking in memory exhibited neuronal shrinkage and brain synaptic reduction, these effects were reversed due to withanolide therapy. It may enhance memory function since it significantly increased both axonal and dendritic regeneration in the neurons as well as the restoration of pre and postsynapses [35]. In cultured rat cortical neurons, it was discovered that withanoside IV, a component of W. somnifera roots, stimulated neurite outgrowth. Withanoside IV used orally may improve neuronal dysfunction in Alzheimer's disease since it contains the aglycone sominone [36]. Similar to this, withanolide-A (1 μM) significantly promotes axonal and dendritic regeneration as well as the repairing or reformation of neurological pre and postsynapses, making it a key contender for the treatment of neurodegenerative illnesses [35].

6.20.13 Effective on Urethane Induced Lung-Adenoma

In its raw form, Ashwagandha has proven to be quite beneficial in experimental carcinogenesis. 'It protected mice against lung adenomas brought on by urethane. Leucopoenia and other urethane side effects were also avoided. Withania avoided all of the negative consequences that the chemical stressor urethane induces. The medication can be used in conjunction with radiation or chemotherapy for cancer. It will lessen the adverse effects of medicines use for treatment of cancer, which cause many biological problems and also lower immunity. In addition of having anticancer impact, it functions in situations where reduced immune statuses of the patient are a concern, it functions as an immunomodulator and can thereby lengthen the life duration of cancer patients [19].

6.20.14 To Relief anxiety and Depression

According to the definition, anxiety is an emotion marked by feelings of tension, worry-filled thoughts, and also physical changes. Many individuals overeat in anxiety. Overeating appetizing foods high in calories leads to chronic positive energy balance, when energy intake exceeds energy expenditure, resulting in body fat storage, weight increase, and obesity [37]. Studies on WS leaf powder extract reveal that it contains anti-inflammatory, antianxiety and anti-apoptotic qualities that may be advised to prevent/slow the negative consequences of obesity and its related disorder. Alcoholic extract of WS seed and root was administered to mice (100 mg/kg intra peritoneal as a single dose) and assessed swimming performance. The swimming endurance of Ashwagandha-treated mice was found to be twice that of normal control mice. It appears that Ashwagandha generated a stage of nonspecific enhanced resistance during stress.WS has a considerable anti-stress adaptogenic effect. They looked at how WS affected chronic stress in rodents. In rats, a 21-day electric foot shock caused male sexual dysfunction, hyperglycemia, glucose intolerance, cognitive impairments, stomach ulcerations, mental distress and immunosuppression. One hour before the shock, an extract of Ashwagandha dramatically lowered stress levels. Ashwagandha reduced neuron activity and prevented nerve cells from firing excessively. Ashwagandha generates GABA-like action, which may imply that it has anti-anxiety properties [38].

Depression is a diverse condition characterized by mood swings and thoughts, misbehaviour, disappointments, melancholy, hopelessness, and a lack of physical activity and self-worth. Moreover, depression is associated with changes in diet, sleep patterns, and other everyday activities, as well as anxiety symptoms [39]. Experimental investigations have shown that Ashwagandha can help with depression. The researchers extracted the bioactive component glycowithanolides from Ashwagandha roots and studied its potential as an antidepressant at doses of 20 and 50 mg/kg in an animal investigation. Glycowithanolides were shown to have an antidepressant effect equivalent to imipramine in forced swim tests-induced behavioural despair and learned helplessness. As a result, the use of Ashwagandha as a mood stabiliser is supported [40]. At a dosage of 40 mg/kg, Ayurveda formulations including Ashwagandha, a fat extract of WS, and significantly reduced immobility time in the forced swim test generated behavioural despair, tail suspension test, and reserpine antagonism in the anti-reserpine test [41].

6.20.15 Decrease Chances of Amyotrophic Lateral Sclerosis (ALS)

Amyotrophic lateral sclerosis, commonly known as frontotemporal lobar degeneration, is a neurodegenerative condition that affects neurons involved in motor activity and voluntary muscles. It is characterized by changes in upper and lower motor neurons in the cerebral cortex, as well as in the medulla and anterior horn of the spinal cord. Failure of higher motor neurons causes muscular rigidity and spasticity, whereas failure of lower motor neurons causes muscle twitching, which leads to degeneration and loss of connection in the synapse, culminating in atrophy [42]. It is discovered to cause localised weakness, which progresses to muscle degeneration, including respiratory muscles. Ashwagandha was reported to decrease disease development, enhance motor function, and increase the number of motor neurons in the lumbar spinal cord in SOD1G93A mice through inducing autophagy activity [43].

6.20.16 Alzheimer's Disease Treatment

Alzheimer's disease is a neurodegenerative illness characterized mostly by gradual memory loss and permanent impairment in cognitive functions. Many in vitro and in vivo investigations have shown that Ashwagandha and its phytoconstituents can help with Alzheimer's disease. Recently, a study was undertaken to analyze the anti-ingredient Alzheimer's found in Ashwagandha root extract and revealed that

Withanone has substantial effectiveness, specifically by inhibiting amyloid-42. Withanone was also discovered to increase the activity of acetyl choline, glutathione, and the secretase enzyme to enhance the elevation of pro-inflammatory cytokines levels [44]. Docking modelling studies predicted that withanolide-A inhibited human acetyl cholinesterase with a high binding affinity [45]. Semi-purified root extract of Ashwagandha containing withanolides to reverse Alzheimer's disease by producing neuroprotective effects against H_2O_2- and – Amyloid cytotoxicity in APP/PS1 transgenic mice and APPSwlnd mice (line J20) of Alzheimer's disease by up-regulation of lipoprotein receptor-related protein in liver [46].

6.20.17 Effectiveness on Parkinson's Disease

Parkinson's disease is an age-related neurodegenerative condition defined predominantly by dopaminergic substantia nigra neurodegeneration. It can be due to hereditary and environmental factors. It is linked to oxidative stress, mitochondrial dysfunction, and protein aggregation abnormalities. Several research have been conducted to investigate the efficacy of ashwagandha in the treatment of Parkinson's disease. The ethanolic root extract of Ashwagandha has been shown to treat Parkinson-like symptoms in MPTP-induced Parkinson in Balb/c mice [47]. Through suppressing oxidative stress and mitochondrial dysfunctions, ashwagandha conferred reduced cholinergic function and dopamine depletion in a rotenone model of Drosophila melanogaster [48].

When mice were given 1-methyl-4-phenyl-1,2,3,6-tetrahydropyridine, their catecholamine, antioxidant, and lipid peroxidation marker levels changed. Dopamine, 3,4-dihydroxy-phenylacetic acid, homovanillic acid, antioxidants (glutathione and glutathione peroxidase), and thiobarbituric acid reactive substance levels in the striatum are normalized with 100 mg/kg Ashwagandha treatment, and motor skills improve [49].

6.20.18 Reduce Symptoms of Schizophrenia

Schizophrenia is a chronic mental condition characterized by disruptions in thinking, perception, behaviour, and other cognitive features. It also has an impact on language and induces hallucinations, delusions, and other psychotic symptoms [50]. A number of preclinical and clinical investigations found that Ashwagandha can be used to treat schizophrenia. Neuroleptics are commonly utilised in schizophrenia treatment. In haloperidol-induced orofacial dyskinesia, Ashwagandha root extract at doses ranging from 100 to 300 mg/kg improves vacuous chewing movements and tongue protrusions superiorly by lowering lipid peroxidation and increasing forebrain SOD and catalase levels while having no effect on glutathione levels. As a result, Ashwagandha may be effective in reducing extrapyramidal neuroleptic-induced symptoms [51].

6.20.19 Effect on Autism

Autism is a brain inflammation illness characterized by difficulties with learning and social interaction. It is characterized by oxidative stress, activation of astrocytes and microglia, changes in pro-inflammatory cytokines, 8-oxo-guanosine, and neuronal depression. It also involves high food intolerance, anxiety, increased anti-brain protein autoantibodies, decreased levels of reduced glutathione, sulfation, and methylation [52]. A recent study found that Ashwagandha in sodium valproate might produce autism in rodents by improving altered behavioural and oxidative stress. In histo-architecture investigations of the cerebellum, restoration of the number of purkinje fibres, neuronal degeneration, and chromatolysis shows its ameliorative impact in autism [53].

6.20.20 Effectiveness in Drug Addiction Addiction

Addiction is described as a complicated, chronic brain illness characterized by physical and psychological reliance on a chemical, substance, or activity. It is distinguished by changes in behavior, thinking, physical functioning, learning, memory, and decision making [54]. It mostly refers to the use of alcohol, cocaine, nicotine, marijuana, opioids, caffeine, inhalants, or gambling. In molecular research, transduction and transcription factors have been linked to the development and maintenance of addiction [55]. According to a recent study, Ashwagandha inhibits neuron circulation and dopamine transmission specifically in the ventral tegmental region of dopaminergic neurons and the nucleus accumbens shell, preventing behavioral and biochemical changes caused by electrochemical and neurochemical modifications induced by morphine and ethanol [56]. Another study looked into the effectiveness of Ashwagandha extract in reducing nicotine addiction. It was determined that Ashwagandha reduces nicotine-induced location preference in mice, showing anti-addictive potential due to nicotine cholinergic receptor regulation [57].

6.21 Conclusion

Ashwagandha has been used since ancient times to cure various diseases and modern studies fully support its strong therapeutic potential. It has antitumor, immunomodulatory, anti-inflammatory, antioxidant, antistress, hemopoietic, and rejuvenating properties and positively influences the endocrine, cardiopulmonary, and central nervous systems. Its roots and leaves are rich with components that have high medicinal properties. Studies reveal it is a safe compound, and it has no associated toxicity.

References

1. Bhatnagar, M., Sisodia, S. S., & Bhatnagar, R. (2005). Antiulcer and antioxidant activity of Asparagus racemosus Willd and Withania somnifera Dunal in rats. *Annals of the New York Academy of Sciences, 1056*(1), 261–278.
2. Gupta, G. L., & Rana, A. (2007). PHCOG MAG.: Plant review Withania somnifera (Ashwagandha): A review. *Pharmacognosy Reviews, 1*(1), 129–136.
3. Kulkarni, S., & Dhir, A. (2008). Withania somnifera: An Indian ginseng. *Progress in Neuro-Psychopharmacology and Biological Psychiatry, 32*(5), 1093–1105.
4. Moharana, D., Bahadur, V., Rout, S., Prusty, A. K., & Kumar, R. (2020). Ashwagandha: The miracle ginseng.
5. Chandranath, H., & Pramod, K. (2010). Management of epilachna beetle on ashwagandha. *Karnataka Journal of Agricultural Sciences, 23*(1).
6. MH, S. K., Srinivas, M., Hanumatharaya, L., & Revannavar, R. (2018). A review on integrated pest management in medicinal and aromatic plants in India. *Journal of Pharmacognosy and Phytochemistry, 7*(3S), 220–224.
7. Meshram, P. B., Mawai, N. S., & Malviya, R. (2015). Biological control of insect pests of medicinal plants, Abelmoschus moschatus, Gloriosa superba and Withania somnifera in forest nursery and plantation in Madhya Pradesh, India. *American Journal of Agriculture and Forestry, 3*(2), 47–51.
8. Das, U., Pal, S., & Kumar, N. (2021). Insect-pest of common medicinal and aromatic plants and their sustainable management. In *Biointensive integrated pest management for horticultural crops* (pp. 285–294). CRC Press.
9. Krishnaveni, T. S., & Arunachalam, R. (2018). Constraints perceived by agricultural scientists in teaching undergraduate students in Tamil Nadu Agricultural University, India. *Asian Journal of Agricultural Extension, Economics & Sociology, 23*(3), 1–10.
10. Samhita, C. (1949). Shree Gulab Kunverba Ayurvedic Society. *Jamnagar, India, 1*.
11. Meena, M. K. (2017). A review article on muscular dystrophy and its management through ayurveda.
12. Bhandari, C. (1970). Ashwagandha (Withania somnifera) "Vanaushadhi Chandroday" (an encyclopedia of Indian herbs). *Publisher: CS Series of Varanasi Vidyavilas Press, Varanasi, India, 1,* 96–97.
13. Kritikar, K., & Basu, B. (1935). *Withania somnifera, Indian medicinal plants: IIIrd.* Lalit Mohan Basu.
14. Misra, B. (2004). Ashwagandha-Bhavprakash Nigantu (Indian Materia Medica) Varanasi. *Chaukhambha Bharti Academy*, 393–394.
15. Sharma, P. (1999). Ashwagandha. *Dravyaguna Vijana, Chaukhambha Viashwabharti, Varanasi*, 763–765.
16. Abbas, S., Bhalla, M., & Singh, N. (2005). *A clinical study of Organic Ashwagandha in some cases of uterine tumors (fibroids) and dermatofibrosarcoma.* Paper presented at the Proc. workshop on essential medicines, adverse drug reactions and therapeutic drug monitoring.
17. Abraham, A., Kirson, I., Glotter, E., & Lavie, D. (1968). A chemotaxonomic study of Withania somnifera (L.) dun. *Phytochemistry, 7*(6), 957–962.
18. Singh, P., Guleri, R., Singh, V., Kaur, G., Kataria, H., Singh, B., et al. (2015). Biotechnological interventions in Withania somnifera (L.) Dunal. *Biotechnology and Genetic Engineering Reviews, 31*(1–2), 1–20.
19. Singh, N., & Gilca, M. (2010). *Herbal medicine: Science embraces tradition: A new insight into ancient.* Lambert Academic Pub.
20. Bhattacharya, S. K., Satyan, K. S., & Ghosal, S. (1997). Antioxidant activity of glycowithanolides from Withania somnifera. *Indian Journal of Experimental Biology, 35*(3), 236–239.
21. Jayaprakasam, B., Zhang, Y., Seeram, N. P., & Nair, M. G. (2003). Growth inhibition of human tumor cell lines by withanolides from Withania somnifera leaves. *Life Sciences, 74*(1), 125–132.

22. Singh, N., Singh, S., Nath, R., Singh, D., Gupta, M., Kohli, R., et al. (1986). Prevention of urethane-induced lung adenomas by Withania somnifera (L.) Dunal in albino mice. *International Journal of Crude Drug Research, 24*(2), 90–100.
23. Ichikawa, H., Takada, Y., Shishodia, S., Jayaprakasam, B., Nair, M. G., & Aggarwal, B. B. (2006). Withanolides potentiate apoptosis, inhibit invasion, and abolish osteoclastogenesis through suppression of nuclear factor-κB (NF-κB) activation and NF-κB–regulated gene expression. *Molecular Cancer Therapeutics, 5*(6), 1434–1445.
24. Girish, K., Machiah, K., Ushanandini, S., Harish Kumar, K., Nagaraju, S., Govindappa, M., et al. (2006). Antimicrobial properties of a non-toxic glycoprotein (WSG) from Withania somnifera (Ashwagandha). *Journal of Basic Microbiology, 46*(5), 365–374.
25. Lizano, S., Domont, G., & Perales, J. (2003). Natural phospholipase A2 myotoxin inhibitor proteins from snakes, mammals and plants. *Toxicon, 42*(8), 963–977.
26. Bhattacharya, A., Ramanathan, M., Ghosal, S., & Bhattacharya, S. (2000). Effect of Withania somnifera glycowithanolides on iron-induced hepatotoxicity in rats. *Phytotherapy Research: An International Journal Devoted to Pharmacological and Toxicological Evaluation of Natural Product Derivatives, 14*(7), 568–570.
27. Hemalatha, S., Wahi, A., Singh, P., & Chansouria, J. (2006). Hypolipidemic activity of aqueous extract of Withania coagulans Dunal in albino rats. *Phytotherapy Research: An International Journal Devoted to Pharmacological and Toxicological Evaluation of Natural Product Derivatives, 20*(7), 614–617.
28. Andallu, B., & Radhika, B. (2000). Hypoglycemic, diuretic and hypocholesterolemic effect of winter cherry (Withania somnifera, Dunal) root. *Indian Journal of Experimental Biology, 38*(6), 607–609.
29. Owais, M., Sharad, K., Shehbaz, A., & Saleemuddin, M. (2005). Antibacterial efficacy of Withania somnifera (ashwagandha) an indigenous medicinal plant against experimental murine salmonellosis. *Phytomedicine, 12*(3), 229–235.
30. Arora, S., Dhillon, S., Rani, G., & Nagpal, A. (2004). The in vitro antibacterial/synergistic activities of Withania somnifera extracts. *Fitoterapia, 75*(3–4), 385–388.
31. Singh, A., Naidu, P. S., Gupta, S., & Kulkarni, S. K. (2002). Effect of natural and synthetic antioxidants in a mouse model of chronic fatigue syndrome. *Journal of Medicinal Food, 5*(4), 211–220.
32. Choudhary, M. I., Nawaz, S. A., Lodhi, M. A., Ghayur, M. N., Jalil, S., Riaz, N., et al. (2005). Withanolides, a new class of natural cholinesterase inhibitors with calcium antagonistic properties. *Biochemical and Biophysical Research Communications, 334*(1), 276–287.
33. Schliebs, R., Liebmann, A., Bhattacharya, S. K., Kumar, A., Ghosal, S., & Bigl, V. (1997). Systemic administration of defined extracts from Withania somnifera (Indian Ginseng) and Shilajit differentially affects cholinergic but not glutamatergic and GABAergic markers in rat brain. *Neurochemistry International, 30*(2), 181–190.
34. Glotter, E. (1991). Withanolides and related ergostane-type steroids. *Natural Product Reports, 8*(4), 415–440.
35. Kuboyama, T., Tohda, C., & Komatsu, K. (2005). Neuritic regeneration and synaptic reconstruction induced by withanolide a. *British Journal of Pharmacology, 144*(7), 961–971.
36. Kuboyama, T., Tohda, C., & Komatsu, K. (2006). Withanoside IV and its active metabolite, sominone, attenuate Aβ (25–35)-induced neurodegeneration. *European Journal of Neuroscience, 23*(6), 1417–1426.
37. Dinh, C. H., Szabo, A., Camer, D., Yu, Y., Wang, H., & Huang, X.-F. (2015). Bardoxolone methyl prevents fat deposition and inflammation in the visceral fat of mice fed a high-fat diet. *Chemico-Biological Interactions, 229*, 1–8.
38. Zahiruddin, S., Basist, P., Parveen, A., Parveen, R., Khan, W., & Ahmad, S. (2020). Ashwagandha in brain disorders: A review of recent developments. *Journal of Ethnopharmacology, 257*, 112876.
39. Ahmed, R., Khan, N. A., Waseem, M., & Khan, Z. J. (2017). Holistic approach in the management of depression: A review. *Journal of Integrated Community Health, 6*, 10–14.

40. Bhattacharya, S., Bhattacharya, A., Sairam, K., & Ghosal, S. (2000). Anxiolytic-antidepressant activity of Withania somnifera glycowithanolides: An experimental study. *Phytomedicine, 7*(6), 463–469.
41. MK, M. J., Prathima, C., Huralikuppi, J., Suresha, R., & Murali, D. (2012). Anti-depressant effects of Withania somnifera fat (Ashwagandha ghrutha) extract in experimental mice. *International Journal of Pharma and Bio Sciences, 3*(1), 33–42.
42. Brown, R. H., & Al-Chalabi, A. (2017). Amyotrophic lateral sclerosis. *New England Journal of Medicine, 377*(2), 162–172.
43. Dutta, K., Patel, P., & Julien, J.-P. (2018). Protective effects of Withania somnifera extract in SOD1G93A mouse model of amyotrophic lateral sclerosis. *Experimental Neurology, 309*, 193–204.
44. Pandey, A., Bani, S., Dutt, P., Satti, N. K., Suri, K. A., & Qazi, G. N. (2018). Multifunctional neuroprotective effect of Withanone, a compound from Withania somnifera roots in alleviating cognitive dysfunction. *Cytokine, 102*, 211–221.
45. Grover, A., Shandilya, A., Agrawal, V., Bisaria, V. S., & Sundar, D. (2012). Computational evidence to inhibition of human acetyl cholinesterase by withanolide a for Alzheimer treatment. *Journal of Biomolecular Structure and Dynamics, 29*(4), 651–662.
46. Sehgal, N., Gupta, A., Valli, R. K., Joshi, S. D., Mills, J. T., Hamel, E., et al. (2012). Withania somnifera reverses Alzheimer's disease pathology by enhancing low-density lipoprotein receptor-related protein in liver. *Proceedings of the National Academy of Sciences, 109*(9), 3510–3515.
47. Bhatnagar, M., Goel, I., Roy, T., Shukla, S. D., & Khurana, S. (2017). Complete Comparison Display (CCD) evaluation of ethanol extracts of Centella asiatica and Withania somnifera shows that they can non-synergistically ameliorate biochemical and behavioural damages in MPTP induced Parkinson's model of mice. *PLoS One, 12*(5), e0177254.
48. Manjunath, M. (2015). Standardized extract of Withania somnifera (Ashwagandha) markedly offsets rotenone-induced locomotor deficits, oxidative impairments and neurotoxicity in Drosophila melanogaster. *Journal of Food Science and Technology, 52*, 1971–1981.
49. RajaSankar, S., Manivasagam, T., & Surendran, S. (2009). Ashwagandha leaf extract: A potential agent in treating oxidative damage and physiological abnormalities seen in a mouse model of Parkinson's disease. *Neuroscience Letters, 454*(1), 11–15.
50. Wei, Y.-Y., Lin, W.-F., Zhang, T.-H., Tang, Y.-X., Wang, J.-J., & Zhong, M.-F. (2018). Effectiveness of traditional Chinese medicineas as an adjunct therapy for refractory schizophrenia: A systematic review and meta analysis. *Scientific Reports, 8*(1), 6230.
51. Naidu, P. S., Singh, A., & Kulkarni, S. K. (2003). Effect of Withania somnifera root extract on haloperidol-induced orofacial dyskinesia: Possible mechanisms of action. *Journal of Medicinal Food, 6*(2), 107–114.
52. Rasool, M., Malik, A., Qureshi, M. S., Manan, A., Pushparaj, P. N., Asif, M., et al. (2014). Recent updates in the treatment of neurodegenerative disorders using natural compounds. *Evidence-Based Complementary and Alternative Medicine, 2014*.
53. Veeresh, B., Pratyusha, G., Mallika, S., & Sudarshini, K. (2016). Research Article Withania somnifera ameliorates sodium valproate induced austism in BALB/c mice: Behavioral and biochemical evidences.
54. Everitt, B. J., & Robbins, T. W. (2016). Drug addiction: Updating actions to habits to compulsions ten years on. *Annual Review of Psychology, 67*, 23–50.
55. Koob, G. F., & Volkow, N. D. (2016). Neurobiology of addiction: A neurocircuitry analysis. *The Lancet Psychiatry, 3*(8), 760–773.
56. Bassareo, V., Talani, G., Frau, R., Porru, S., Rosas, M., Kasture, S. B., et al. (2019). Inhibition of morphine-and ethanol-mediated stimulation of mesolimbic dopamine neurons by Withania somnifera. *Frontiers in Neuroscience, 13*, 545.
57. Dumore, N. G., Umekar, M. J., Taksande, B. G., Aglawe, M. M., & Kotagale, N. R. (2019). Effects of Withania somnifera nicotine induced conditioned place preference in mice. *Pharmacognosy Journal, 11*(1).

Chapter 7
Cowhage

Sana Aslam, Ayesha Rafiq, Matloob Ahmad, Syed Ali Raza Naqvi, and Arwa A. AL-Huqail

7.1 Introduction

Family	Fabaceae
Subfamily	Faboideae
Scientific name	*Mucuna pruriens*
English/Common Name	Velvet Beans

7.2 Plant Description

There are roughly one hundred and fifty types of annual and perennial leguminous plants in this genus of Mucuna, which belongs to the Fabaceae family, subfamily Faboideae. *Mucuna pruriens* is ubiquitous in tropical and sub-tropical areas and is one of the most commonly used wild legumes. Mucuna beans (also called as Velvet beans), like other typical pulses, have been proven to be high in protein and carbohydrates, as well as a good supply of macro- and microelements. The ripe beans and

S. Aslam
Department of Chemistry, Government College Women University, Faisalabad, Pakistan

A. Rafiq · M. Ahmad (✉) · S. A. R. Naqvi
Department of Chemistry, Government College University, Faisalabad, Pakistan
e-mail: matloob.ahmad@gcuf.edu.pk

A. A. AL-Huqail
Department of Biology, College of Science, Princess Nourah bint Abdulrahman University, Riyadh, Saudi Arabia

M. Zia-Ul-Haq et al. (eds.), *Essentials of Medicinal and Aromatic Crops*,
https://doi.org/10.1007/978-3-031-35403-8_7

the green pods are both cooked and eaten [1]. It is thought to be an effective source of dietary proteins when compared to other pulses such as rice bean, lima bean, rice and soybean [2–4]. As a result, it is recognized as an excellent dietary source.

It's a twining, climbing, annual herbaceous leguminous plant with long, slender twigs as well as opposing trifoliolate, lanceolate leaflets, 15-30 cm in length. Generally, ovate, asymmetric or rhomboid oval-shaped leaflets are present, which are uneven at ground level. It bears white to dark purple blooms that grow in two or three racemes and swing in bunches. The pod, which is thick and leathery, is the plant's fruit. It is covered in long, stiff, reddish-orange hairs that can be easily removed but can also cause irritation in harvesters. The effect of touching the seed-pod hairs is referred to as "pruriens" (from Latin, "itching feeling"). Mucuna bean seeds (4–6 in a pod) are black curved, nocturnally uneven, bloated, heavily covered with irritating bristles, oval (ranging from 6 to 12 mm long), and have a funicular hilum [5, 6].

When the plant is immature, it is totally enclosed by soft, fluffy hairs; however, as it matures, it becomes virtually hairless [7]. Trifoliate with alternating or spiralling leaves and a gray-silky underside; stalks are silky and long.

7.3 Agronomy of Plant

7.3.1 Soil Condition

The plant may flourish in a wide range of soils, although it gives preference to gritty silt soils with adequate flow and a pH of 5.50–7.50. With winter temperatures as low as 15 °C and summer temperatures as high as 38 °C, it flourishes in a sub-tropical to tropical climate.

7.3.2 Climatic Conditions

Velvet beans thrive in locations with extended growing seasons and favour hot, humid climates with 650–2500 mm of annual rainfall. It can withstand prolonged periods of drought, especially if planted during the growing season. As a result, velvet bean growth is becoming more popular in areas receiving 10–20 inches of rain per year (e.g., the natural regions of Zimbabwe). It will continue to rise until frozen or deep soil moisture runs out in the dry season. Its seeds reach maturity in the months of May and June [8].

7.4 Land Preparation

The crop does not need extensive land preparation because of its big seeds. In CA systems, whether utilizing human or mechanical operations, little soil disturbance is desired [8].

7.5 Planting

Seed is spread at an amount of 35–40 kms per hectare in one crop at the start of the period of rainfall, with inter-row spacing of 0.9–1 m and within row spacing of 30–40 cm. In semi-arid circumstances, a reduced seed rate (wider spacing) is recommended to avoid moisture competition. Mucuna seeds are huge; thus, they should be planted 3–7 cm deep [8]. Despite the lack of a designated cultivar of Mucuna, locally accessible seeds have acceptable viability and germination rates [9]. For growth, the plant needs assistance. By providing assistance, it results in a 25% rise in production and a 25% decline in insect infestation. Flowering usually starts 45–50 days after seeding [10].

7.6 Manuring

The treatment of 50 kg P_2O_5 ha^{-1}greatly boosted velvet bean growth, production of components, and seed [11]. *M. pruriens* produced the most organic matter (approximately 7.3 t ha^{-1}) with phosphorous pentaoxide treatment [12]. It was shown in a previous study that when nitrogen and phosphorus were not present in the entire treatment of fertilizer in Mucuna, organic matter output was reduced by 69% (nitrogen) and 33% (phosphorus) on average [13]. Instead, 250–300 kg of compound fertilizer can be used (preferably in a ratio of 7:14:7 of nitrogen, phosphorus, and potassium, respectively).

7.7 Pest and Diseases

Farmers should weed-free the land as soon as weeds develop to maintain the crop. This will also help keep pests at bay. The velvet bean is widely renowned for its resistance to pests and diseases. Leaf-eating caterpillars, on the other hand, have been reported to be completely destructive. Farmers should seek advice on reducing infection outbursts or insect damage while applying herbicides and follow crop chemical compatibility advice [8].

Velvet bean is among the best crops for recovering land soiled with wild flowers such as *Saccharum spontaneum*, *Cynodondactylon*, *Imperata cylindrica*, and *Cyperus species*, among others [14]. To prevent *Fusarium oxysporum* infection, it is advised to be used in rotation with cotton in Brazil. It can also efficiently reduce nematode infestations caused by Meloidogyne incognita and other species [14, 15].

7.8 Origin and Distribution of COWHEDGE Plant/Velvet Beans

It is a tropical plant native to southern regions of Asia and Malaysia. It is now widely distributed throughout tropical areas. It was first introduced to the southern United States in the late 1800s, and then to the tropical regions at the beginning of 1900 [16].Velvet beans can be found from sea level to 2100 meters in elevation. It demands a hot, humid environment with annual rainfall (from 650 to 2500 mm in height) and a long growth season free of frost throughout the season of rainfall. It may cultivate in a variety of soils, from clays to sands, although it prefers drainage at a steady rate and light-textured soils with a high acidity level [17]. Bangladesh, India, Sri Lanka, Southeast Asia, and Malaysia are among the countries where their plantations are extensively reported [18]. It is one of India's most widely used herbal medicines. It is also grown in Uttar Pradesh, Madhya Pradesh, and in the Andaman and Nicobar Islands. It may be found in the form of shrubs, hedges, and dry-deciduous low woodland types all over the Indian's grasslands. It nurtures itself naturally from the lower Himalayan range to Indian's vast humid grasslands [19].

7.9 Important Phytochemical Constituents of the Plant

Mucuna prureins seeds contain a variety of valuable phytochemicals (primary metabolites), including proteins, lipids, dietary fibres, carbohydrates and minerals such as in mg/100 g of *M. prureins* seed flour contains sodium (Na) 43.1–150.1 mg, potassium (K) 778.1–1846.0 mg, calcium (Ca) 393.4–717.7 mg, magnesium (Mg) 174.9–387.6 mg, iron (Fe) 10.8–15.0 mg, zinc (Zn) 5.0–10.9 mg, copper (Cu) 0.9–2.2 mg, manganese (Mn) 3.9–4.3 mg, and phosphorus (P) 98.4–592.1 mg [20].

Phytochemical analysis (secondary metabolites analysis) of *Mucuna pruriens* seed extract showed the presence of flavonoids, alkaloids, glycosides, steroids,saponins, terpenoids and tannins. The presence of functional groups of amides, amines, phosphine, fluorides, iodides, bromides, and nitro-substituted aliphatic and aromatic compounds was revealed by IR spectral data. Alkaloids, carboxylic acids, terpenoids, polyphenols and other secondary metabolites have been detected through GC-MS analysis [21]. L-Dopa is claimed to be a substantial ingredient of the plant, mostly present in the seeds [22–24]. Mucunadine, mucunine, prurienidine, and prurienine [25] are alkaloidal compounds testified from seeds [26, 27]. Number of amino acids also reported as a nutritious source [28, 29]. Epoxy fatty acids such as *cis*-12, 13-epoxyoctadec-*trans*-9-*cis*-acid and *cis*-12, 13-epoxyoctadec-*trans*-9-enoic acid were also described in the seeds of cowhedge plant [30]. Seeds have been shown to contain lecithin [31].

7.10 L-Dopa

The existence of the levorotatory form of Dopa (L-Dopa), a starting material of dopamine, present in the seeds of the *Mucuna prureins* rendered the plant useful in curing Parkinson's disease [32]. In Ayurvedic medicine, the specie of this plant is utilized to treat disorders of the CNS and geriatric conditions. L-Dopa is being produced in the leaves as well as the roots of *M. prurien* at a concentration of around 1% by fresh weight. The content of L-Dopa in *Mucuna pruriens* did not alter significantly when cultivated in the shade or in an open area. Fully grown seeds, pod-pericarp, leaves, stems, and roots possess 3.6 to 4.2%, 0.14 to 0.22%, 0.17 to 0.35%, 0.19 to 0.31%, and 0.12 to 0.16%, respectively, and the highest concentration of L-Dopa was detected in immature seeds. The percentage of L-Dopa in the seeds of various accessions ranged from 7.62 to 8.37%. L-Dopa is isolated from the seeds of the cowhedge plant using various extraction techniques. The extraction and quantitative measurement of L-Dopa, present in the seeds of the cowhedge plant, were done using a high-performance liquid chromatographic technique. L-Dopa concentrations in Mucuna seeds varied from 3.9 to 6.2%, according to an HPLC analysis report [33]. Velvet bean is thought to emit 100–450 kg of L-DOPA per hectare into the soil. Furthermore, its ability to manage wildflowers and nematodes reduces the need for crops to be treated with artificial pesticides (Fig. 7.1) [34].

These major biological functions have been directed to chemical studies of *M. pruriens* seeds, as a result of which various fatty acids and amino acids, in addition to L-Dopa, have been isolated [35]. Alkaloids like prurienidine, prurienine, prurieninine,etc. have also been found [36, 37]. Linoleic, palmitic, stearic, oleic, decanoic, lauric, behenic, arachidic, and vernolic acids are among the oils found in the seeds (Fig. 7.2).

The existence of derivatives of tetrahydroisoquinoline alkaloids shown below, whose structures have been identified using spectroscopic techniques, is described in this work. Because their bicyclic structure imposes conformational limitations and considerably decreased flexibility, tetrahydroisoquinoline-3-carboxylic acids

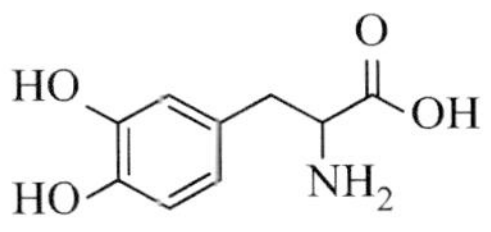

Fig. 7.1 Structure of L-dopa

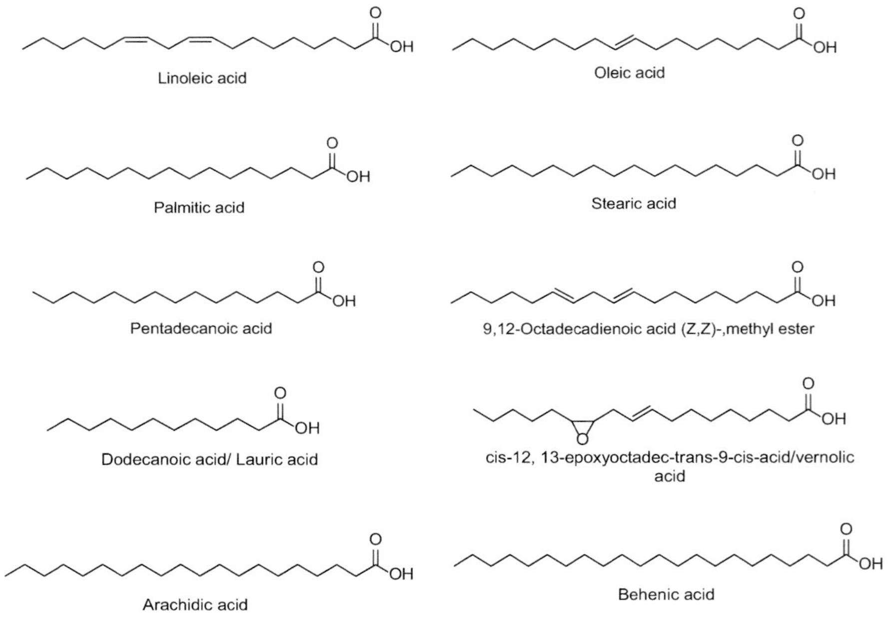

Fig. 7.2 Phytochemical constituents present in *M. Pruriens*

Fig. 7.3 Structure of tetrahydroisoquinoline alkaloids isolated from *Mucuna purines* seeds

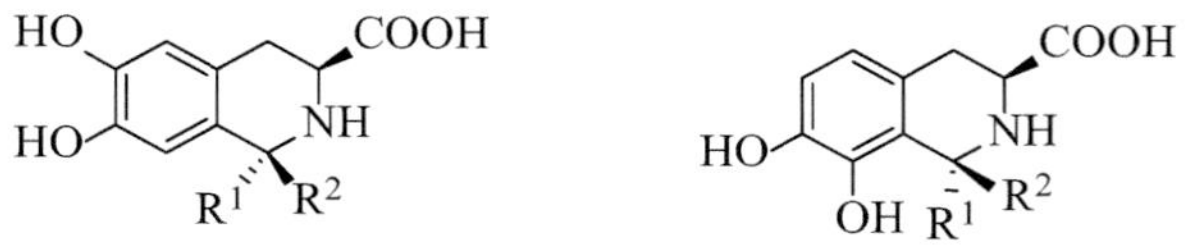

1, R^1=R^2= H
2, R^1=H,**R^2=** CH_3
3, R^1=R_2=CH_3
4, R^1= R^2= CH_3

(Tic) have been demonstrated to be particularly effective as well as selective to opioid receptors for neurotransmitters other than peptide hormones (Fig. 7.3) [38, 39].

Some of the phytochemicals have been discussed here;

B-Sitosterol: Structurally related to cholesterol [40]. It is well suited for breast cancer [41], colon cancer [42], and hypercholesterolemia [43].

Gallic acid: A study reported the antioxidant [44] and neuroprotective [45] effects of gallic acids in rats.

Bufotenine: It is found in *Mucuna prureins* [46], inhibits lipid metabolism pathway to exhibit anti-inflammatory and analgesic actions (Fig. 7.4) [47].

Genistein: An isoflavonoid from Mucuna seeds acts as an anticancer [48] and anti-inflammatory agent [49].

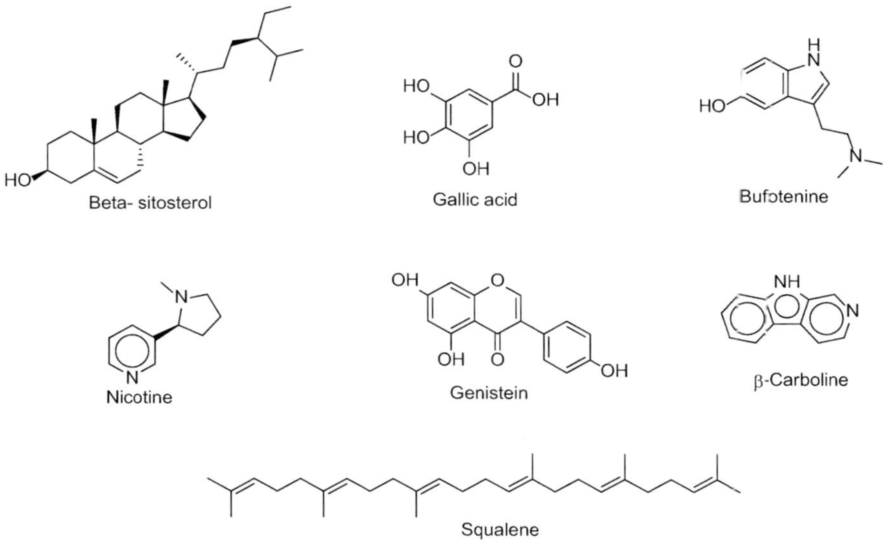

Fig. 7.4 Other phytochemical constituents present in *M. Pruriens*

β-Carboline: *Mucuna prureins*contains β-carboline which is Neuroprotective [50], an antioxidant [51], a MAO inhibitor, promotes dimethyl tryptamine activity [52].

Nicotine: Reduces levodopa-induced dyskinesia in Parkinsons' disease rat model [53, 54].

Squalene: Found in *Mununa prureins* [55], it is a dietary lipid and has potential uses in cosmetic dermatology [56], and shows antitumor activities [57].

Glutathione: It is found in *M. prureins* [58], exhibits a comprehensive role in Parkinson's disease [59].

Serotonin: Present in the pods of *M. prureins* [58], can be used as a prognostic marker of urological tumors [60].

Harmine: It is found in *Mucuna prureins* [58] and acts as a glutamate receptor antagonist (Fig. 7.5) [61].

Stizolamine: It is found in *Mucuna prureins* [62].

Flavone: It is present in *M. prureins* and can be useful for drug development [63].

Melanin: It is found to be present in the seeds of *Mucuna prureins*. The seeds get darker as a result of L-dopa being converted to melanin [58].

DMT & 5-Methoxy-DMT: reduces dyskinesia [64].

Phytic acid: suppresses 1-methyl-4-phenyl-1,2,3,6- tetrahydropyridine- (MPTP) induced hydroxyl radical generation [65].

Saponins & Tannins: *M. prureins* contains Saponins [66] and tannins (Fig. 7.6) [58].

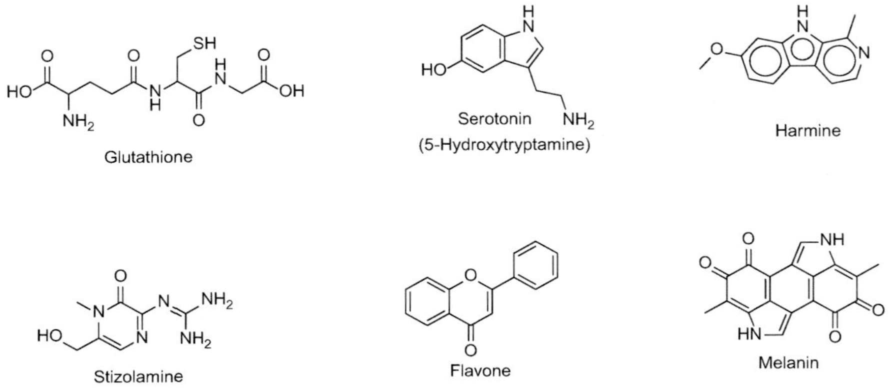

Fig. 7.5 Phytochemical constituents of *Mucuna prureins*

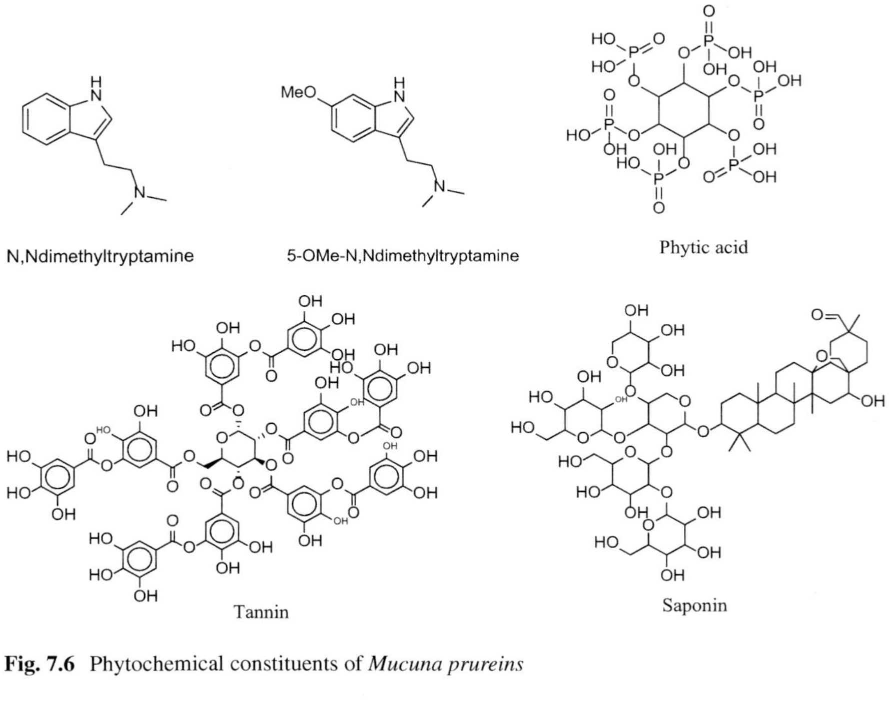

Fig. 7.6 Phytochemical constituents of *Mucuna prureins*

7.11 Medicinal Importance

Some pharmacological roles of *Mucuna prureins* are elaborated below (Fig. 7.7).

Mucuna has therapeutic properties in all of its components [67]. The presence of L-dopa is most likely the reason why *M. pruriens* extracts have been shown to include substances with a wide variety of pharmacological effects, including

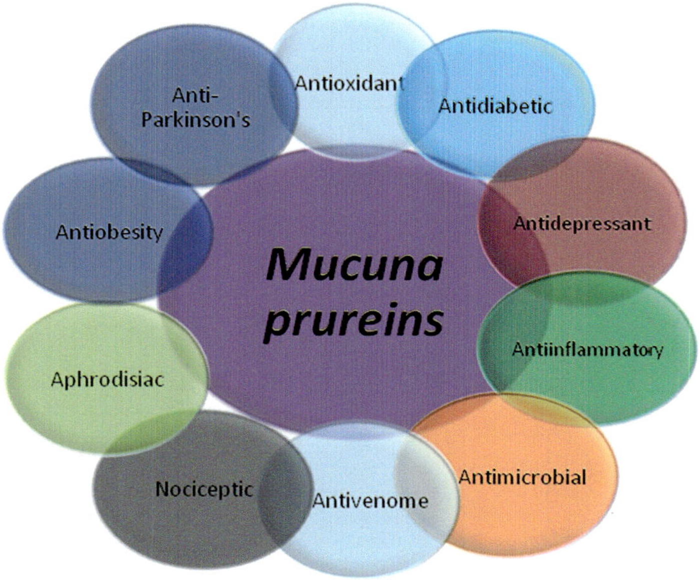

Fig. 7.7 Medicinal activities of *Mucuna prureins*

neuroprotective, anti-diabetic, anti-inflammatory, and anti-oxidant properties. [68, 69]. The primary phenolic ingredient in Mucuna seeds is L-dopa (about 5% phenolic substances) [70].

Mucuna is now being extensively researched since L-dopa is utilized as a first-line therapy for Parkinson's disease (PD). *Mucuna pruriens* has been employed as a carminative, hypertensive, and hypoglycemic agent. Diarrhea, cough, asthma, cancer, oedema, craziness, inflammation of the membranes that surround the lungs, cholera, ringworm, sores, snakebite, tumors, and bacterial infection (arising due to sexual contact)are all treated with it (Fig. 7.8).

The rasa, or flavour, of Mucuna is mostly sweet and bitter. It has a warming virya (body action) and a pleasant vipaka (post-digestive impact). This herb is great for creating ojas, the body's reservoir of necessary energy, immunity, and vigour, due to its feeding and building capabilities.

Mucuna pruriens is an excellent source of dietary protein [71]. In African nations, *M. pruriens* consumption is crucial in preventing malnutrition, for instance Nigeria and Benin [72], as well as in Central American countries for example Honduras [73]. In the eighteenth and nineteenth centuries, *M. pruriens* was also used as a food crop in Ghana and Mozambique. In the mid-1980s [74], the ladies of the World Neighbors' advancement programme held in El Rosario, Honduras, used *M. pruriens* as a substitute for espresso, cocoa, and wheat flour, and produced 22 formulations that were simple to design, deeply nutritious, and made with locally available materials. The programme highlighted the benefits of *Mucuna pruriens* construct nutria-chocolate on nursing mothers' milk production and how their

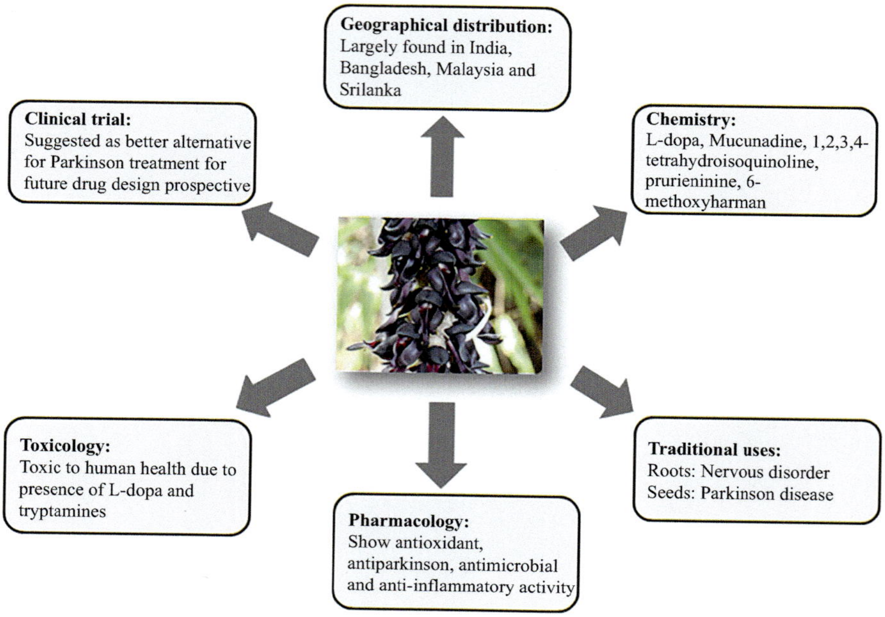

Fig. 7.8 Overall graphical representation of *Mucuna prureins*

breastfed children went from having second-degree malnutrition to having none at all in just 2 months [75]. *M. pruriens* is traditionally boiled and pulverized in Mexico to produce Nescafé, the most popular espresso in Central America.

7.12 Anti-oxidant Activity

Polyphenols are well known for their antioxidant activities, and are found in the seeds, leaves and peptide fractions of *Mucuna prureins*. Studies have been reported with different plant extracts.

Seed: Aqueous extract of *Mucuna prureins* seeds was observed, which was found to be remarkably rich in phenol (3730.1 ± 15.52 mg) and gallic acid equivalent (GAE)/g). The phenolic contents have a direct relationship to antioxidant effects. Gallic acid (a Standard drug) demonstrated the greatest activity at all concentrations, but the extract demonstrated greater activity at lower concentrations (0.02, 0.01, 0.005 and 0.0025 mg/ml) and lower activity at higher concentration 0.4 mg/ml when compared to another standard drug rutin [76].

Among phenolic components, appreciable amount of L.dopa is present in the seed of *M. prureins*. Rima et al. recently reported the antioxidant activity of L.dopa (isolated from dilute methanol extract) in terms of IC_{50}, value which was found to be 8.92 ± 0.03 ppm [77] whereas Biswas & and his coworkers previously reported the same work in aqueous media was with a 9.22 mcg/ml IC_{50} value. The whole analysis was done by using DPPH (2,2-diphenyl-1-picrylhydrazyl) assay [78]. Dhanani et al. also worked on the seed extract of *M. prureins* by employing three different methods; conventional refluxing method, ultrasound assisted solvent extraction (UASE) method and microwave assisted solvent extraction (MASE) method. It was reported that MASE method produced the best results in shorter time. The IC_{50} value recorded in this study was 5.0 μg/ml in 5 minute duration, L.dopa and total phenolic content recorded in the same duration were 5.4 ± 0.05 and 9.2 ± 0.11 respectively [79].

Leaf: A study was carried out by using an aqueous leaf extract of *M. prureins,* which showed antioxidant activity, ABTS (2,2′-Azino-bis(3-ethylbenzothiazoline-6-sulfonic acid) diammonium salt) analysis indicated better antioxidant activity at lower concentrations [80]. The phytochemicals in the leaf extract, including the flavonoids, saponins, and tannins, have antioxidant properties [81].

Peptide fraction: In a separate investigation, the in-vitro antioxidant impact of peptide fractions obtained from *M. prureins* on the HeLa cell line was investigated. It was elucidated that the peptide fractions with a molecular weight of <3 kDa showed the highest antioxidant activity [82].

7.13 Anti-diabetic Activity

Plants could also be the potent anti-diabetic agents. This plant has been found to be advantageous for the treatment of diabetes in various countries such as India, Brazil, Spain, and Germany [83]. It has been observed that *M. pruriens* seed alcohol extract revealed to have a hypoglycaemic effect in normal [84], streptozotocin [83] and alloxan [84] mice suffering from diabetes. The anti-hyperglycaemic effect was only detected after 3 weeks at high doses, and most of the effect was noticed after 15 weeks of regular intake of 200 mg/kg/day, which was 47%. In alloxan-induced diabetic rats (plasma glucose >450 mg/dL), the intake of 5, 10, 20, 30, 40, 50, and 100 mg/kg of M. edulis coarse extract of ethanol indicated the reduction in blood glucose levels in 18.6, 24.9, 30.8, 41.4, 49.7, 53.1, and 55.4% by the administration of pruriens seeds after 8 hours of treatment. Whereas, 59.7% reduction resulted by the ingestion of glibenclamide (5 mg/kg/day) [85].

Similar research has been conducted by injecting streptozotocin (65 mg/kg body weight) into normal healthy rats. Four groups were made; Group A (normal rats), Group B (diabetic control), Group C (*Mucuna prureins*; experimental drug, 200 mg/kg), Group D (glibenclamide; standard drug, 1 mg/kg body weight). *Mucuna Prureins* extract was found to be effective in showing hypoglycemic activity when compared to the diabetic control group. Weight loss is a common symptom in diabetics, but when administered with *M. prureins* and Glibenclamide drugs, weight gain was observed, more in the case of *M. prureins* than Glibenclamide [86]. In similar study as above, it was revealed that *M. prureins* alleviated blood glucose levels from 242.4 ± 9.2 mg/dl to 91.0 ± 5.2 mg/dl, hence making it a green tool to be consumed as therapy in hyperglycemic conditions [87].

7.14 Anti-depressant Activity

Depression is a life-threatening disease; this psychiatric condition affects 20% of the world's population [88]. *Mucuna prureins* seeds have antidepressant properties. Hydroalcoholic extract of *M. prureins* seeds was investigated for serotonin and noradrenaline mediated antidepressant action. Results revealed that the seed extract interacted with serotonergic and noradrenergic systems to show antidepressant behavior [89]. To clarify the antidepressant properties of *Mucuna pruriens* seeds, a similar investigation was carried out using the hydroalcoholic extract of the seeds and dealt with the dopaminergic modulating action. Results explored that there was a reduction in stress levels, which were investigated by using Forced swimming test (FST) and Tail suspension test (TST) [90].

7.15 Anti-inflammatory Activity

Mucuna prureins also played its role in reducing inflammation as it is linked to the pathophysiology of various diseases like arthritis, cancer, and gout. *Mucuna prureins* essential oil (MPEO) obtained from the leaf of a plant expressed an anti-iflammatory response. A study was carried out with five different groups to assess MPEO anti-inflammatory activity via a formalin induced assay. Group I: Control (Saline solution), Group II: Standard (peroxicam), Group III: MPEO (100 mg/kg), Group IV: MPEO (200 mg/kg), Group V: MPEO (400 mg/kg). The standard drug peroxicam 10 mg/kg showed 100% inhibition with 16.5 ± 2.8 licks, while groups IV and V showed 100% inhibition with different numbers of licks (57.2 ± 5.1) and 33.3 ± 7.1, respectively [91]. A similar study has been conducted to see the anti-inflammatory action of *Mucuna prureins* seed in five groups of mice. Group I (control; tragacanth solution), Group II (standard; Aspirin, 100 mg/kg), Group III, IV, V contained 1,2 and 3 g/kg of *M. prureins* seed respectively. Inflammation was induced by carageenan which is attributed to the release of histamine. It was observed that by increasing the dose of *M. prureins* seed, there was a subsequent increment in its anti-inflammatory response [92].

M. prureins nanoparticles (NPs) were synthesized and analyzed for their anti-inflammatory action. Copper nanoparticles and titanium dioxide nanoparticles have been synthesized, characterized and recently reported for their anti-inflammatory activity. TiO_2 nanoparticles at 50 μL concentration exhibited 90% inhibition [93] while with Cu-nanoparticles the percentage inhibition was 71% at 50 μL [94].

7.16 Antimicrobial Activity

M. prureins seed extract functionalized with zinc oxide nanoparticles (NPs) showed an antibacterial response against *Bacillus subtilis*. The activity was assessed with growth kinetics evaluation and broth dilution assay. The ZnO nanoparticles indicated MIC value of 20 µg/ml and IC_{50} value of 70 µg/ml. Inorganic metals are well known for their antibacterial effects. Nanoparticles using *M. prureins* extract act as biocidal agent in a cheap and nontoxic manner [95].

The phytochemicals, phenols, and tannins in *M. prureins* leaves demonstrated antibacterial activity in a crude methanolic extract. Activity was evaluated against *E.coli*, *Bacillus subtilis*, *Salmonella typhi* and *Shigella disenteriae*. A higher value of the zone of inhibition for *E.coli* (2.8 cm) shows higher antimicrobial potency than *B.subtilis* (2.1 cm) [96]. After that another study was made to see the antimicrobial action of *M. prureins* seed using disc fusion method, for which alcoholic extract was used. Zone of inhibition values for different microbes were observed; *Vibrio cholera* (9.4 mm inhibition zone), *Vibrio harveyi* (16.8 mm inhibition zone), *S. aureus* (3.4 mm inhibition zone) and *E. coli* (4.1 mm inhibition zone) [97].

7.17 Antivenom Activity

The seeds of *M. pruriens* are often used in traditional medicine to prevent the negative consequences of a snake bite brought on by cytotoxins, neurotoxins, the PLA2 enzyme, cardiotoxins, and proteases. It has been demonstrated that *M. pruriens* is more effective than snake poison. Traditional practitioners in Plateau State, Nigeria, recommend the seed as an oral medicine for snake bite, claiming that the ingestion of seeds resulted in the person's being protected from the symptoms of any snake bite for the whole year [98].

A study was investigated on the anti-venomous function of *M. prureins* leaf extract against cobra snake venom by dividing mice into six different groups; Group A (normal control), Group B (test control; snake venom was induced i.e., 0.075 mg/kg), Group C (standard control; antivenin) and Groups D, E & F were given ethanolic extract of *M. prureins* in 40 mg/kg, 60 mg/kg, 80 mg/kg respectively. The extract at 80 mg/kg was found to be more effective than the standard drug antivenin [99]. In vitro analysis was done to see anti-venial potential of *M. prureins* seeds against various snake venoms. Experiments were carried out on rats which were pretreated with *M. prureins* seed extract and then injected with venom. Anti-MPE antibodies prevented fatalities caused by various venoms [100].

7.18 Nociceptic Activity

Pain is a very unpleasant event. *M. prureins* seeds are widely used as analgesics for mellenia. The investigation was done in four separate groups: Group A: (Control), Group B: (Standard; Aspirin), Group C: (flavonoids), Group D: (alkaloids) via formalin paw licking test. It was observed that flavonoids (mean number of lickings: 24.25 ± 8.97) showed significantly more analgesic activity than alkaloids (mean number of lickings: 39 ± 6.78) [101].

In-vivo analysis of the nociceptic activity of *Mucuna prurein* was done by using Eddy's hot plate method. Four groups were categorized to carry out to the reaction; Group A; (standard; tramadol, 25 mg/kg), Group B, C, D of 100 mg/kg, 200 mg/kg and 400 mg/kg were used as trials. By using oral feeding, analgesic activity was noted at different time intervals, such as 0, 30, 60, 90, 120 and 150 minutes. The results obtained were most effective at 400 mg/kg *M. pruriens* extract [102].

7.19 Anti-obesity

Obesity is becoming a global epidemy [103] which causes many pathophysiological complications such as diabetes [104], cardiovascular diseases [105], immunological changes [106], psychiatric changes [107] etc. A recent study revealed that the milk of *Mucuna prureins* seeds has the ability to influence weight gain and blood lipid levels. The two varieties of *Mucuna prureins* seeds (*var. Cochinchinensis* and *var. Veracruz* mottle) were studied. For this purpose, 7 groups of healthy male Wistar rats were made for the hyperlipidaemic study. Group I; (Fed with normal diet; control group), Group II; (Fed with high fat; Hyperlipidaemic control group), Group III; (Fed with high fat with standard drug; Atorvastatin) Group IV; (Fed with high fat +20 ml dehulled *Cochinchinensis* milk/ day), Group V; (Fed with high fat +20 ml of whole *Cochinchinensis* milk/ day), Group VI; (Fed with high fat +20 ml dehulled *Veracruz* milk/ day), Group VII; (Fed with high fat +20 ml whole *Veracruz* milk/ day). The results found were astonishing in terms of triglycerides, total cholesterol levels, low density lipoprotein cholesterol, high density lipoprotein cholesterol and very low-density lipoprotein cholesterol. All the test groups had the most effective outcomes in comparison to the standard drug, Atorvastatin [108].

Similar study was held to check *Mucuna prureins'* anti-obesity functions on obese and healthy rats. Four groups were made to observe changes by administering *Mucuna prureins* extract. Group I; (Healthy group; HG), Group II; (Healthy group administered with *Mucuna prureins* extract; HGMP) Group III; (Obese group; OG); Group IV; (Obese group administered with *Mucuna prureins* extract; OGMP). *Mucuna prureins* were found to lower the cholesterol levels in all groups that were given *Mucuna prureins* extract as an antihyperlipidemic [109].

7.20 Aphrodisiac Activity

Sexual dysfunction is a problem that can be treated through medications and surgeries. Herbal remedies, on the other hand, were thought to be a solution to improve sexual life [110]. For this, *Mucuna prureins* plant seeds were used for therapeutic purposes in Tibb-e-Unani [111]. Mucuna helps to promote fertility, healthy sperm and ova, correct reproductive organ functions, and healthy vaginal secretions. Aphrodisiac *Mucuna pruriens* encourages healthy sexual energy and desire.

To determine the impact of including *Mucuna prureins* seeds in the diet, a study on rabbits was conducted. *Mucuna prureins* seed meal (MSM) was administered in different concentrations for 3 months trial. Results collected indicated positive effects on rabbits' sexual behavior, reproductive organ weight, and semen characteristics [112]. Diabetes patients frequently experience sexual dysfunction. Suresh et al. studied the efficacy of *M. prureins* on sexual behaviors in induced diabetic male rats. Diabetes was induced by using streptozotocin. Groups were classified into 6 categories; Group I (control group), Group II (diabetic rats; diabetes was induced by streptozotocin), Group III (diabetic rats were given 200 mg/kg b.w. of *M. prureins* extract), Group IV (diabetic rats were given 5 mg/kg b.w. of sildenafil citrate), and Group V (normal rats were given *Mucuna prureins* extract). The results showed that group II had considerably lower levels of daily sperm production (DSP), luteinizing hormone, and testosterone, but group III had improved sexual behavior, sperm parameters, DSP, and hormonal levels compared to group II [113].

7.21 Anti-Parkinson's Activity

Various clinical trials were conducted in Parkinsons' disease patients, and the results showed that *M. pruriens* seeds are useful for the diagnosis of Parkinson's disease [114, 115]. The bioavailability experiments were conducted with the HP-200 Mucuna seed formulation, and it was revealed that the pharmacokinetic profile was comparable to that of L-dopa [116]. It was reported in previous studies that Mucuna seed powder had a quick beginning of action and lasted longer without causing dyskinesias, suggesting that Mucuna contains elements that boost L-dopa activity (Fig. 7.9).

L-Dopa correlates to the derivatives of methylated and non-methylated tetrahydroisoquinoline in modest concentrations (about 0.25%) [117, 118]. Mucuna seeds, leaves, stem, and roots contain these compounds. *N,N*-dimethyl tryptamine and certain indole compounds are among the other chemical constituents found in various regions of the plant [119].L-dopa and its metabolites were discovered to be pro-oxidant and anti-oxidant compounds that affect oxidative DNA damage and may protect tissues from damage caused by neurodegenerative illnesses such as parkinsonism [120].

Mucuna prureins seed extract was thought to contain 12.55% L.dopa content, used to treat Parkinson's disease [121], in addition to other phytochemicals [122]. The medicinal interventions of *M. prureins* have been discussed in a number of studies.

L.dopa pharmacokinetics was observed in three different *M. prureins* samples: roasted, dried and boiled. Roasted *M. Prureins* seeds contain 5.3% and dried *M. Prureins* seeds contain 5.29% L.Dopa while boiling reduced *M. Prureins*' L.dopa content up to 70%. *M. prureins*' L-dopa dose will be effective if administered at a dose greater than 3.5X that of standard levodopa + dopa decarboxylase inhibitor [123]. Ethanolic extract of *M. prureins* seed was assessed in a

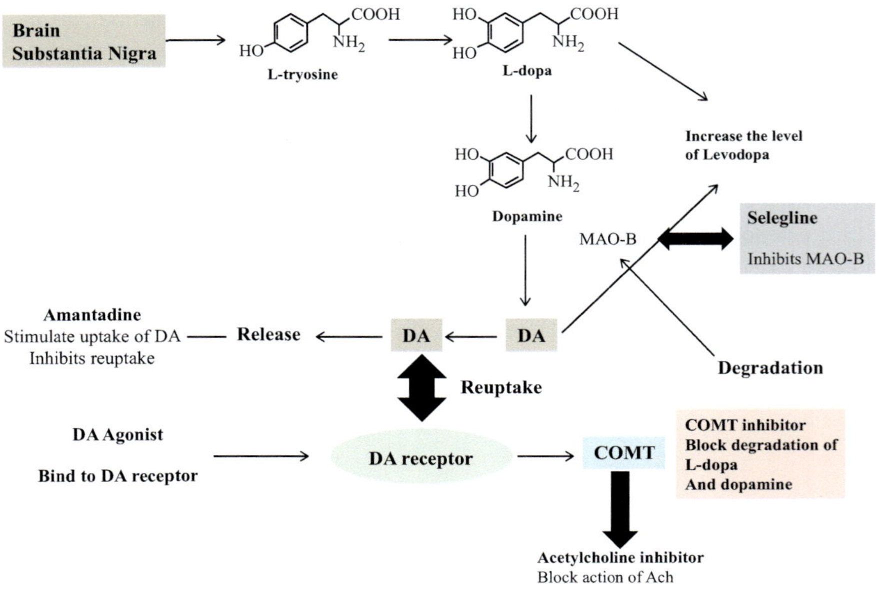

Fig. 7.9 Metabolic pathway of L-dopa agent against Parkinson disease

1-methyl-4-phenyl-1,2,3,6-tetrahydropyridine (MPTP) model. Standard was estrogen, a well-known therapeutic agent for Parkinson's disease. The whole experiment was conducted on mice which were divided into four different groups; Control, MPTP, MPTP & *M. prureins*, MPTP & estrogen. It was revealed that MPTP + *M. prureins* group was found to be more effective than MPTP + estrogen group in recovering all of the deficiencies induced by MPTP including dopaminergic neuron reduction resulting in low dopamine level and accumulation of abnormal protein synuclein resulting in lewy body formation [124].

7.22 Synthesis of L-Dopa

Due to the catalytic amount of osmium tetraoxide and chiral ligands such as hydroquinine-1,4-phthalazinediyl diether [(DHQ)2-PHAL], beta-methyl styrene and,-unsaturated ester (1) were subjected to an AD reaction to produce chiral diol (2) in perfect enantiomeric excess. The cyclic sulfite was obtained as an intermediate by treating vicinal diol **(2)** with $SOCl_2$ in the presence of NEt_2 in CH_2Cl_2 at zero degrees celcius. When the cyclic sulfite intermediate was reacted with sodium azide in a solvent (DMF) at 80 °C, high yields of the derivative azido alcohol **(3)** were produced [125, 126]. The enantiomeric excess of azido alcohol **(3)** was identified using ^{1}H and ^{19}F NMR of its Mosher ester, and the compound **(3)** was then

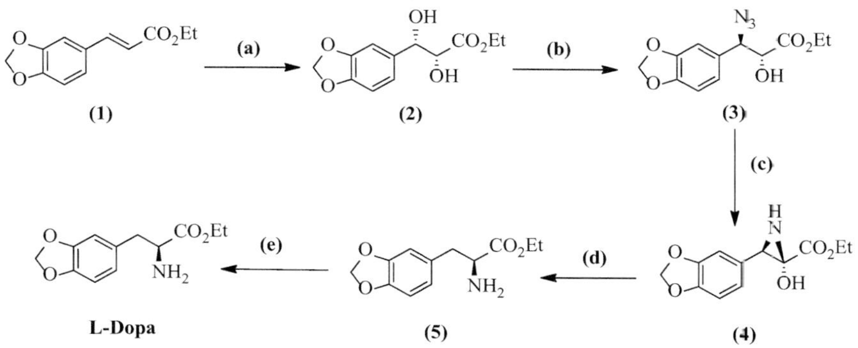

Reagents and Conditions;

(a) cat. OsO_4' $(DHQ)_2$-PHAL, $K_3Fe(CN)_6$' K_2CO_3' t-BuOH-H_2O, 0°C, 85%; (b)$SOCl_2$' NEt_3' CH_2Cl_2' 0°C, 88%; NaN_3' acetone-H_2O, 80°C, 82% yield, 95% ee; (c) PPh_3' CH_3CN, 90%; (d) 10% Pd/C, HCO_2NH_4· MeOH, reflux, 92%; (e) 6 M HCl, phenol, acetic acid, 130-135°C, 24 h, 70% yield, 85% ee.

Scheme 7.1 Synthesis of L-dopa

subsequently reacted with PPh_3 in CH_3CN to produce chiral aziridines **(4)** in high yield. When aziridine **(4)** was exposed to Pd-catalyzed reductive ring opening using NH_4HCO_2 (ammonium formate) as a source of hydrogen, at the benzylic position, it underwent the stereospecific and regioselective ring opening, yielding amines **(5)**.

Finally, after reacting with 6 molar of HCl in phenol and acetic acid at 130–135 °C, aminoester (5) yielded 85% L-dopa (Scheme 7.1) [127].

7.23 Metabolic Pathway of L-dopa

L-Dopa is a catecholamine that is formed when tyrosine is hydroxylated, and it is a precursor to neurotransmitters including adrenaline, noradrenaline, and dopamine. In particular, L-dopa is considered as a significant precursor in the production of melanin, which is present in a number of tissues in both plants and mammals. The Cu-consisting enzyme tyrosine hydroxylase used to hydroxylate the residues of tyrosine, in the existence of molecular O_2, resulting in the production of L-Dopa. Plants have an L-Dopa production pathway that is similar to that of mammals. Tyrosine hydroxylase is a Cu-containing enzyme that hydroxylates tyrosine residues in the presence of molecular O_2, resulting in the production of L-Dopa (Scheme 7.2) [128–130].

7.24 Future Perspectives of Plant

Mucuna beans have been shown to be effective in the treatment of ageing, rheumatoid arthritis, diabetes, infertility in men, and neurological disorders. As a result, more study is needed to understand the specific mechanism by which Mucuna beans

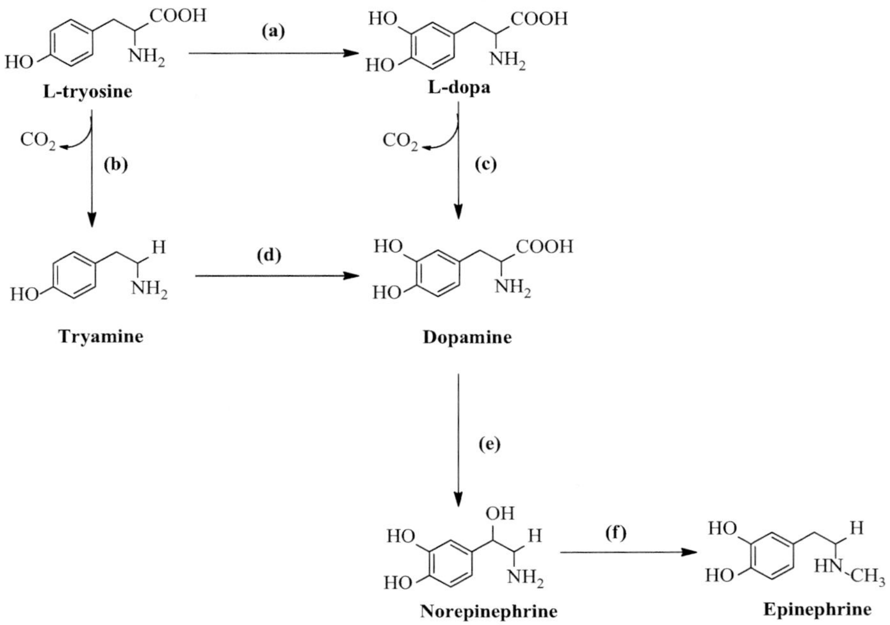

Scheme 7.2 Metabolic pathway of L-dopa. Reaction conditions; (**a**) tryosinehydrolyase, (**b**) tryosine decarboxylase, (**c**) DOPA-decarboxylase, (**e**) monophenol hydroxylase, (**e**) dopamine β-hydroxylase, (**f**) phenylethanolamine*N*-methyltransferase

are effective against various diseases so that they may be used to design and create effective therapies for these disorders. Another research study suggests that this plant has shown numerous healing properties. As a result of its multidirectional properties, we can say that its phytochemicals and their derivatives has potential to be proven as potent drugs. The existence of bioactive substances like tannins and polyphenols demonstrates that this herb's constituents are diverse. These differences are caused by differences in topography, production rate, preparation methods, climatic conditions, and different environmental components. More investigation into the phytochemical properties of Mucuna beans is therefore required to see whether they may be used as natural pesticides to benefit the environment, human and animal health and nutrition, and both. As a result, further clinical studies should be conducted to determine its full potential.

References

1. Siddhuraju, P., Becker, K., & Makkar, H. P. S. (2000). Studies on the nutritional composition and antinutritional factors of three different germplasm seed materials of an under-utilized tropical Legume, Mucuna p ruriens Var. Utilis. *Journal of Agricultural and Food Chemistry, 48*(12), 6048–6060.

2. Janardhanan, K., Gurumoorthi, P., & Pugalenthi, M. (2003). Nutritional potential of five accessions of a South Indian tribal pulse, Mucuna pruriens var utilis I. The effect of processing methods on the content of l-dopa, phytic acid, and oligosaccharides. *Tropical and subtropical agroecosystems, 1*(2–3), 141–152.
3. Pugalenthi, M., Vadivel, V., & Siddhuraju, P. (2005). Alternative food/feed perspectives of an underutilized legume Mucuna pruriens var. utilis-a review. *Plant Foods for Human Nutrition, 60*(4), 201–218.
4. Gurumoorthi, P., Pugalenthi, M., & Janardhanan, K. (2003). Nutritional potential of five accessions of a South indian tribal pulse Mucuna pruriens var utilis: ii. Investigations on total free phenolics, tannins, trypsin and chymotrypsin inhibitors, phytohaemagglutinins, and in vitro protein digestibility. *Tropical and Subtropical Agroecosystems, 1*(2–3), 153–158.
5. Rastogi, R. P., Mehrotra, B. N., Sinha, S., Pant, P., & Seth, R. (1990). *Compendium of Indian medicinal plants: 1985–1989* (Vol. 4). Central Drug Research Institute and Publications & Information Directorate.
6. Okoli, B. J. (2015). In vitro anthelmintic activity of Mucuna pruriens (DC) and Canarium schweinfurthii (Engl) on Ascaris suum. *Open Access Library Journal, 2*(02), 1–8.
7. Sahaji, P. (2011). Acute oral toxicity of Mucuna pruriens in albino mice. *International Research Journal of Pharmacy, 2*(5), 162–163.
8. Chakoma, I., Manyawu, G. J., Gwiriri, L., Moyo, S., & Dube, S. (2016). *The agronomy and use of Mucuna pruriens in smallholder farming systems in southern Africa*. ILRI Extension Brief.
9. Oudhia, P. (2001, January). My experiences with world's top ten Indian medicinal plants: Glimpses of research at farmer's field in Chhattisgarh (India). In *Abstract. Workshop cum Seminar on Sustainable Agriculture for 21st Century* (pp. 20–21). *IGAU*.
10. Oudhia, P., & Tripathi, R. S. (2001, April). The possibilities of commercial cultivation of rare medicinal plants in Chhattisgarh (India). In *Abstract. VII national science conference, Bhartiya Krishi Anusandhan Samittee* (pp. 12–14). Directorate of Cropping System Research.
11. Thomas, L., & Palaniappan, S. (1998). Seed production of velvet beans, sunnhemp and pillipesara as influenced by plant density and phosphorus application. *Madras Agricultural Journal, 85*, 35–37.
12. Kumwenda, J. D., Gilbert, R., Waddington, S., Murwira, H., Hikwa, D., & Tagwira, F. (1998). Biomass production by legume green manures on exhausted soils in Malawi: A soil fertility network trial. In *Soil fertility research for maize-based farming systems in Malawi and Zimbabwe. Harare* (pp. 85–86). SFNET and CIMMYT.
13. Houngnandan, P., Sanginga, N., Okogun, A., Vanlauwe, B., & Merckx, R. (2001). Van Cleemput, O., Assessment of soil factors limiting growth and establishment of Mucuna in farmers' fields in the derived savanna of the Benin Republic. *Biology and Fertility of Soils, 33*(5), 416–422.
14. Wulijarni-Soetjipto, N., & Maligalig, R. (1997). Mucuna pruriens (L.) DC. cv. group Utilis. *Plant resources of South-East Asia (PROSEA), 11*, 199–203.
15. Wolf, B., & Snyder, G. (2003). *Sustainable soils: The place of organic matter in sustaining soils and their productivity*. CRC Press.
16. Eilittä, M., & Carsky, R. (2003). Efforts to improve the potential of Mucuna as a food and feed crop: background to the workshop. *Tropical and Subtropical Agroecosystems, 1*(2–3), 47–55.
17. Pengelly, B., Whitbread, A., Mazaiwana, P., & Mukombe, N. (2003). Tropical forage research for the future-better use of research resources to deliver adoption and benefits to farmers. *Tropical grasslands, 37*(4), 207–216.
18. Farooqi, A. A., & Sreeramu, B. (2004). *Cultivation of medicinal and aromatic crops*. Universities Press.
19. Muralia, S., & Pathak, A.(2003). Database of medicinal plant used in ayurveda. In *Medicinal and aromatic plants cultivation and uses* (pp. 185–187).
20. Janardhanan, V. V. K. (2000). Nutritional and anti-nutritional composition of velvet bean: an under-utilized food legume in South India. *International Journal of Food Sciences and Nutrition, 51*(4), 279–287.

21. Shanmugavel, G., & Krishnamoorthy, G. (2018). Nutraceutical and phytochemical investigation of Mucuna pruriens seed. *Journal of Pharmaceutical Innovation, 7*, 273–278.
22. Bell, E. A., & Janzen, D. H. (1971). Medical and ecological considerations of L-dopa and 5-HTP in seeds. *Nature, 229*(5280), 136–137.
23. Damodaran, M., & Ramaswamy, R. (1937). Isolation of l-3: 4-dihydroxyphenylalanine from the seeds of Mucuna pruriens. *Biochemical Journal, 31*(12), 2149–2152.
24. Daxenbichler, M. E., VanEtten, C. H., Hallinan, E. A., Earle, F. R., & Barclay, A. S. (1971). Seeds as sources of L-DOPA. *Journal of Medicinal Chemistry, 14*(5), 463–465.
25. Majumdar, D. N., & Zalani, C. D. (1953). Mucuna pruriens DC, Alkaloidal constituents III, isolation of water soluble alkaloids and a study of their chemical and physiological characterization. *The Indian Journal of Pharmacy, 5*, 62–65.
26. Mehta, J. C., & Majumdar, D. N. (1994). Indian medicinal plants-V. *Indian Journal of Pharmacy and Pharmacology, 6*, 92–94.
27. Majumdar, D. N., & Santra, D. K. (1953). The Mucuna pruriens DC. Part II. Isolation of water insoluble alkaloids. *Indian Journal of Pharmacy and Pharmacology, 15*, 60–61.
28. Pant, R., Nair, C. R., Singh, K. S., & Koshti, G. S. (1974). Amino acid composition of some wild legumes. *Current Science, 43*, 235–239.
29. Niranjan, G. S., & Katiyar, S. K. (1979). Chemical-composition of some legumes. *Journal of the Indian Chemical Society, 56*(8), 822–823.
30. Hasan, S. Q., MRK, S., & SM, O. (1980). *Epoxy acids of Mucuna prurita seed oil.*
31. Panikkar, K. R., Majella, V. L., & Pillai, P. M. (1987). Lecithin from Mucuna pruriens. *Planta Medica, 53*(05), 503–503.
32. Mackenbach, J. P., Stirbu, I., Roskam, A. J. R., Schaap, M. M., Menvielle, G., Leinsalu, M., & Kunst, A. E. (2008). Socioeconomic inequalities in health in 22 European countries. *New England Journal of Medicine, 358*(23), 2468–2481.
33. Deokar, G., Kakulte, H., & Kshirsagar, S. (2016). Phytochemistry and pharmacological activity of Mucuna pruriens: A review. *Pharmaceutical and Biological Evaluations, 3*(1), 50–59.
34. Vargas-Ayala, R., Rodríguez-Kábana, R., Morgan-Jones, G., McInroy, J. A., & Kloepper, J. W. (2000). Shifts in soil microflora induced by velvetbean (Mucuna deeringiana) in cropping systems to control root-knot nematodes. *Biological Control, 17*(1), 11–22.
35. Siddhuraju, P., Vijayakumari, K., & Janardhanan, K. (1996). Chemical composition and protein quality of the little-known legume, velvet bean (Mucuna pruriens (L.) DC.). *Journal of Agricultural and Food Chemistry, 44*(9), 2636–2641.
36. Rakshit, S., & Majumdar, D. N. (1956). Mucuna pruriens DC. Part V. Alkaloidal constituents and their characterization. *The Indian Journal of Pharmacy, 18*, 285–287.
37. Ghosal, S., Singh, S., & Bhattacharya, S. K. (1971). Alkaloids of Mucuna pruriens chemistry and pharmacology. *Planta Medica, 19*(01), 279–284.
38. Wang, C., & Mosberg, H. I. (1995). Synthesis of a novel series of topographically constrained amino acids: Benzo-1, 2, 3, 4-tetrahydroisoquinoline-3-carboxylic acids. *Tetrahedron Letters, 36*(21), 3623–3626.
39. Kazmierski, W., & Hruby, V. J. (1988). A new approach to receptor ligand design: synthesis and conformation of a new class of potent and highly selective μ opioid antagonists utilizing tetrahydroisoouinoline carroxylic acid. *Tetrahedron, 44*(3), 697–710.
40. Murthy, K., & Mishra, S. (2009). Quantification of β-Sitosterol from Mucuna pruriens by TLC. *Chromatographia, 69*(1), 183–186.
41. Awad, A. B., & Fink, C. S. (2000). Phytosterols as anticancer dietary components: evidence and mechanism of action. *The Journal of Nutrition, 130*(9), 2127–2130.
42. Awad, A. B., Gan, Y., & Fink, C. S. (2000). Effect of β-sitosterol, a plant sterol, on growth, protein phosphatase 2A, and phospholipase D in LNCaP cells. *Nutrition and Cancer, 36*(1), 74–78.
43. Law, M. (2000). Plant sterol and stanol margarines and health. *BMJ, 320*(7238), 861–864.
44. Kasture, V. S., Katti, S. A., Mahajan, D., Wagh, R., Mohan, M., & Kasture, S. B. (2009). Antioxidant and antiparkinson activity of gallic acid derivatives. *Pharmacology, 1*, 385–395.

45. Lu, Z., Nie, G., Belton, P. S., Tang, H., & Zhao, B. (2006). Structure–activity relationship analysis of antioxidant ability and neuroprotective effect of gallic acid derivatives. *Neurochemistry International, 48*(4), 263–274.
46. Kavitha, C., & Thangamani, C. (2014). Amazing bean â€ œMucuna pruriensâ€: A comprehensive review. *Journal of Medicinal Plants Research, 8*(2), 138–143.
47. Wang, J., Xu, D., Shen, L., Zhou, J., Lv, X., Ma, H., et al. (2021). Anti-inflammatory and analgesic actions of bufotenine through inhibiting lipid metabolism pathway. *Biomedicine & Pharmacotherapy, 140*, 111749.
48. Banerjee, S., Li, Y., Wang, Z., & Sarkar, F. H. (2008). Multi-targeted therapy of cancer by genistein. *Cancer Letters, 269*(2), 226–242.
49. Verdrengh, M., Jonsson, I. M., Holmdahl, R., & Tarkowski, A. (2003). Genistein as an anti-inflammatory agent. *Inflammation Research, 52*(8), 341–346.
50. Kelly, A. C., Uddin, L. Q., Biswal, B. B., Castellanos, F. X., & Milham, M. P. (2008). Competition between functional brain networks mediates behavioral variability. *NeuroImage, 39*(1), 527–537.
51. Moura, D. J., Richter, M. F., Boeira, J. M., Pêgas Henriques, J. A., & Saffi, J. (2007). Antioxidant properties of β-carboline alkaloids are related to their antimutagenic and antigenotoxic activities. *Mutagenesis, 22*(4), 293–302.
52. Kasture, S., Mohan, M., & Kasture, V. (2013). Mucuna pruriens seeds in treatment of Parkinson's disease: Pharmacological review. *Oriental Pharmacy and Experimental Medicine, 13*(3), 165–174.
53. Huang, L. Z., Campos, C., Ly, J., Carroll, F. I., & Quik, M. (2011). Nicotinic receptor agonists decrease L-dopa-induced dyskinesias most effectively in partially lesioned parkinsonian rats. *Neuropharmacology, 60*(6), 861–868.
54. Quik, M., Huang, L. Z., Parameswaran, N., Bordia, T., Campos, C., & Perez, X. A. (2009). Multiple roles for nicotine in Parkinson's disease. *Biochemical Pharmacology, 78*(7), 677–685.
55. Bhaskar, A., & Nithya, V. (2021). Phytochemical evaluation by GC-MS and antihyperglycemic activity of Mucuna pruriens on streptozotocin induced diabetes in rats. *Journal of Chemical and Pharmaceutical Research, 3*(5), 689–696.
56. Huang, Z. R., Lin, Y. K., & Fang, J. Y. (2009). Biological and pharmacological activities of squalene and related compounds: potential uses in cosmetic dermatology. *Molecules, 14*(1), 540–554.
57. Senthilkumar, S., Devaki, T., Manohar, B. M., & Babu, M. S. (2006). Effect of squalene on cyclophosphamide-induced toxicity. *Clinica Chimica Acta, 364*(1–2), 335–342.
58. Sridhar, K. R., & Bhat, R. (2007). Agrobotanical, nutritional and bioactive potential of unconventional legume–Mucuna. *Livestock Research for Rural Development, 19*(9), 126–130.
59. Zeevalk, G. D., Razmpour, R., & Bernard, L. P. (2008). Glutathione and Parkinson's disease: is this the elephant in the room? *Biomedicine & Pharmacotherapy, 62*(4), 236–249.
60. Jungwirth, N., Haeberle, L., Schrott, K. M., Wullich, B., & Krause, F. S. (2008). Serotonin used as prognostic marker of urological tumors. *World Journal of Urology, 26*(5), 499–504.
61. Johnson, K. A., Conn, P. J., & Niswender, C. M. (2009). Glutamate receptors as therapeutic targets for Parkinson's disease. *CNS & Neurological Disorders-Drug Targets (Formerly Current Drug Targets-CNS & Neurological Disorders), 8*(6), 475–491.
62. Hernández-Orihuela, A. L., Castro-Cerritos, K. V., López, M. G., & Martínez-Antonio, A. (2022). Compounds characterization of a Mucuna Seed Extract: L-Dopa, Arginine, Stizolamine, and Some Fructooligosaccharides. *Compounds, 3*(1), 1–16.
63. Ushie, O. A., Iyen, S. I., Abeng, F. E., Azuaga, T. I., Okpaegbe, U. C., & Aikhoje, E. F. (2019). *Quantification of alkaloids, flavonoids and saponins in Physalis angulata and Mucuna pruriens.*
64. Riahi, G., Morissette, M., Parent, M., & Di Paolo, T. (2011). Brain 5-HT2A receptors in MPTP monkeys and levodopa-induced dyskinesias. *European Journal of Neuroscience, 33*(10), 1823–1831.

65. Obata, T. (2003). Phytic acid suppresses 1-methyl-4-phenylpyridinium ion-induced hydroxyl radical generation in rat striatum. *Brain Research, 978*(1–2), 241–244.
66. Siddhuraju, P., & Becker, K. (2005). Nutritional and antinutritional composition, in vitro amino acid availability, starch digestibility and predicted glycemic index of differentially processed mucuna beans (Mucuna pruriens var. utilis): an under-utilised legume. *Food Chemistry, 91*(2), 275–286.
67. Caius, J. F. (1989). *Medicinal and poisonous legumes of India*. Scientific Publishers.
68. Mandal, P., Babu, S. S., & Mandal, N. C. (2005). Antimicrobial activity of saponins from Acacia auriculiformis. *Fitoterapia, 76*(5), 462–465.
69. Misra, L., & Wagner, H. (2007). Extraction of bioactive principles from Mucuna pruriens seeds. *Indian Journal of Biochemistry & Biophysics, 44*(1), 56–60.
70. Vadivel, V., & Pugalenthi, M. (2008). Removal of antinutritional/toxic substances and improvement in the protein digestibility of velvet bean (Mucuna pruriens) seeds during processing. *Journal of Food Science and Technology-Mysore, 45*(3), 242–246.
71. Mohan, V. R., & Janardhanan, K. (1995). Chemical analysis and nutritional assessment of lesser known pulses of the genus, Mucuna. *Food Chemistry, 52*(3), 275–280.
72. Flores, M., Eilitta, M. C., & Myhrman, R. (2002). *Food and feed from Mucuna: current uses and the way forward (No. 633.33 F663f Ej. 1 019782)*. CIDICCO.
73. Diallo, O. K., Kante, S., Myhrman, R., Soumah, M., Cisse, N. Y., & Berhe, T. (2000, April). Increasing farmer adoption of Mucuna pruriens as human food and animal feed in the Republic of Guinea. In *International workshop on food and feed from Mucuna, proceedings*. Tegucigalpa (pp. 60–72).
74. Lim, T. K. (2012). Mucuna pruriens. In *Edible medicinal and non-medicinal plants* (pp. 779–797). Springer.
75. Carew, L. B., & Gernat, A. G. (2006). Use of velvet beans, Mucuna spp., as a feed ingredient for poultry: a review. *World's Poultry Science Journal, 62*(1), 131–144.
76. Jimoh, M. A., Idris, O. A., & Jimoh, M. O. (2020). Cytotoxicity, phytochemical, antiparasitic screening, and antioxidant activities of Mucuna pruriens (Fabaceae). *Plants, 9*(9), 1249.
77. Ishmayana, S., Malini, D. M., & Soedjanaatmadja, U. M. (2022). Nutritional Content and The Activities of L-Dopa (L-3, 4-Dihydoxyphenyalanine) from Mucuna pruriens L. DC Seeds of Central Java Accession. *Arabian Journal of Chemistry, 16*, 104390.
78. Biswas, S., Mukherjee, A., Mallick, U. K., Ghosh, G., & De, B. (2010). Cultured callus and L-DOPA. *Indian Drugs, 47*, 12.
79. Dhanani, T., Singh, R., Shah, S., Kumari, P., & Kumar, S. (2015). Comparison of green extraction methods with conventional extraction method for extract yield, L-DOPA concentration and antioxidant activity of Mucuna pruriens seed. *Green Chemistry Letters and Reviews, 8*(2), 43–48.
80. Chester, K., Zahiruddin, S., Ahmad, A., Khan, W., Paliwal, S., & Ahmad, S. (2019). Bioautography-based identification of antioxidant metabolites of Solanum nigrum L. and exploration its hepatoprotective potential against D-galactosamine-induced hepatic fibrosis in rats. *Pharmacognosy Magazine, 15*(62), 104.
81. Agbafor, K. N., & Nwachukwu, N. (2011). Phytochemical analysis and antioxidant property of leaf extracts of. *Vitex Doniana. Biochemistry Research International, 2011*, 459839.
82. Martínez-Leo, E. E., Martín-Ortega, A. M., Acevedo-Fernández, J. J., Moo-Puc, R., & Segura-Campos, M. R. (2019). Peptides from Mucuna pruriens L., with protection and antioxidant in vitro effect on HeLa cell line. *Journal of the Science of Food and Agriculture, 99*(8), 4167–4173.
83. Rathi, S. S., Grover, J. K., & Vats, V. (2002). The effect of Momordica charantia and Mucuna pruriens in experimental diabetes and their effect on key metabolic enzymes involved in carbohydrate metabolism. *Phytotherapy Research, 16*(3), 236–243.
84. Silva, F. R. M. B., Szpoganicz, B., Pizzolatti, M. G., Willrich, M. A. V., & de Sousa, E. (2002). Acute effect of Bauhinia forficata on serum glucose levels in normal and alloxan-induced diabetic rats. *Journal of Ethnopharmacology, 83*(1–2), 33–37.

85. Majekodunmi, S. O., Oyagbemi, A. A., Umukoro, S., & Odeku, O. A. (2011). Evaluation of the anti–diabetic properties of Mucuna pruriens seed extract. *Asian Pacific Journal of Tropical Medicine, 4*(8), 632–636.
86. Rajesh, R., Singh, S. A., Vaithy, K. A., Manimekalai, K., Kotasthane, D., & Rajasekar, S. S. (2016). The effect of Mucuna pruriens seed extract on pancreas and liver of diabetic wistar rats. *International Journal of Current Research and Review, 8*(4), 61.
87. Bhaskar, A., & Nithya, V. (2011). Phytochemical evaluation by GC-MS and antihyperglycemic activity of Mucuna pruriens on streptozotocin induced diabetes in rats. *Journal of Chemical and Pharmaceutical Research, 3*(5), 675–684.
88. Ravikumar, P., & Jeyam, M. (2019). Antidepressant activity and HPTLC fingerprinting of stearic acid in different days of wheat seedlings. *Grain & Oil Science and Technology, 2*(1), 6–10.
89. Patel, J. S., & Galani, V. J. (2013). Investigation of noradrenaline and serotonin mediated antidepressant action of Mucuna pruriens (L) DC seeds using various experimental models. *Oriental Pharmacy and Experimental Medicine, 13*(2), 143–148.
90. Rana, D. G., & Galani, V. J. (2014). Dopamine mediated antidepressant effect of Mucuna pruriens seeds in various experimental models of depression. *Ayu, 35*(1), 90.
91. Avoseh, O. N., Ogunwande, I. A., Ojenike, G. O., & Mtunzi, F. M. (2020). Volatile composition, toxicity, analgesic, and anti-inflammatory activities of Mucuna pruriens. *Natural Product Communications, 15*(7), 1934578X20932326.
92. Javed, N., Alam, S. S., Subhani, H., Akhtar, M. S., & Khan, A. H. (2010). Evaluation of anti-inflammatory activity of Mucuna pruriens Linn. seeds. *Proceeding SZPGMI, 24*, 97–102.
93. Thangavelu, L., Rajeshkumar, S., Arivarasu, L., & Aditya, B. S. (2021). Antioxidant and Antiinflammatory Activity of Titanium Dioxide Nanoparticles Synthesised Using Mucuna pruriens. *Journal of Pharmaceutical Research International, 33*, 414–422.
94. Anushya, P., Geetha, R. V., & Rajesh Kumar, S. (2021). Evaluation of Anti Inflammatory and Cytotoxic Effect of Copper Nanoparticles Synthesised Using Seed Extract of Mucuna pruriens. *Journal of Pharmaceutical Research International, 33*, 816–824.
95. Agarwal, H., Menon, S., & Shanmugam, V. K. (2020). Functionalization of zinc oxide nanoparticles using Mucuna pruriens and its antibacterial activity. *Surfaces and Interfaces, 19*, 100521.
96. Gupta, S., & Saxena, U. (2022). Ethnopharmacology and its phytochemical investigation of Mucuna Pruriens seed: A comprehensive review. *International Research Journal of Modernization in Engineering Technology and Science, 4*(07), 3366–3375.
97. Shanmugavel, G., & Krishnamoorthy, G. (2018). Nutraceutical and phytochemical investigation of Mucuna pruriens seed. *Pharma Innov, 7*, 273–278.
98. Guerranti, R., Aguiyi, J. C., Errico, E., Pagani, R., & Marinello, E. (2001). Effects of Mucuna pruriens extract on activation of prothrombin by Echis carinatus venom. *Journal of Ethnopharmacology, 75*(2–3), 175–180.
99. Shekins, O. O., Anyanwu, G. O., Nmadu, P. M., & Olowoniyi, O. D. (2014). Anti-venom activity of Mucuna pruriens leaves extract against cobra snake (Naja hannah) venom. *International Journal of Biochemistry Research & Review, 4*(6), 470–480.
100. Guerranti, R., Ogueli, I. G., Bertocci, E., Muzzi, C., Aguiyi, J. C., Cianti, R., et al. (2008). Proteomic analysis of the pathophysiological process involved in the antisnake venom effect of Mucuna pruriens extract. *Proteomics, 8*(2), 402–412.
101. Javed, N., Alam, S. S., Subhani, H., Akhtar, M. S., & Khan, A. H. (2011) Evaluation of Analgesic Activity of Isolated Flavonoids from Mucuna pruriens Seeds. *Proceeding SZPGMI, 25*(2), 91–94.
102. Singh, S., Sachan, A., Singh, H., Shankar, P., Kumar, D., Sachan, A. K., et al. (2015). Study Of Analgesic Activity Of Mucuna pruriens Extract On Swiss Albino Mice. *World Journal of Pharmaceutical and Medical Research, 4*(5), 1124–1132.
103. Chooi, Y. C., Ding, C., & Magkos, F. (2019). The epidemiology of obesity. *Metabolism, 92*, 6–10.

104. Sah, S. P., Singh, B., Choudhary, S., & Kumar, A. (2016). Animal models of insulin resistance: A review. *Pharmacological Reports, 68*(6), 1165–1177.
105. Vekic, J., Zeljkovic, A., Stefanovic, A., Jelic-Ivanovic, Z., & Spasojevic-Kalimanovska, V. (2019). Obesity and dyslipidemia. *Metabolism, 92*, 71–81.
106. Rojas-Osornio, S. A., Cruz-Hernández, T. R., Drago-Serrano, M. E., & Campos-Rodríguez, R. (2019). Immunity to influenza: impact of obesity. *Obesity Research & Clinical Practice, 13*(5), 419–429.
107. Delgado, I., Huet, L., Dexpert, S., Beau, C., Forestier, D., Ledaguenel, P., et al. (2018). Depressive symptoms in obesity: relative contribution of low-grade inflammation and metabolic health. *Psychoneuroendocrinology, 91*, 55–61.
108. Nicolas, N. Y., Armand, A. B., Edith, D. M. J., Dimitry, M. Y., & Thérèse, B. A. M. (2022). Effects of Mucuna Milk (Mucuna pruriens L.) on Body Weight and Serum Biochemistry in Rats Fed Hyperlipidaemic Diet. *European Journal of Nutrition & Food Safety, 14*, 43–57.
109. Tavares, R. L., de Araújo Vasconcelos, M. H., Dorand, V. A. M., Junior, E. U. T., Toscano, L. D. L. T., de Queiroz, R. T., et al. (2021). Mucuna pruriens treatment shows anti-obesity and intestinal health effects in obese rats. *Food & Function, 12*(14), 6479–6489.
110. Suresh, S., Prithiviraj, E., & Prakash, S. (2009). Dose-and time-dependent effects of ethanolic extract of Mucuna pruriens Linn. seed on sexual behaviour of normal male rats. *Journal of Ethnopharmacology, 122*(3), 497–501.
111. Muthu, K., & Krishnamoorthy, P. (2011). Evaluation of androgenic activity of Mucuna pruriens in male rats. *African Journal of Biotechnology, 10*(66), 15017–15019.
112. Mutwedu, V. B., Ayagirwe, R. B. B., Bacigale, S. B., Mwema, L. M., Butseme, S., Kashosi, T., et al. (2019). Effect of dietary inclusion of small quantities of Mucuna pruriens seed meal on sexual behavior, semen characteristics, and biochemical parameters in rabbit bucks (Oryctolagus cuniculus). *Tropical Animal Health and Production, 51*(5), 1195–1202.
113. Suresh, S., & Prakash, S. (2012). Effect of Mucuna pruriens (Linn.) on sexual behavior and sperm parameters in streptozotocin-induced diabetic male rat. *The Journal of Sexual Medicine, 9*(12), 3066–3078.
114. Vaidya, A. B., Rajgopalan, T. G., Mankodi, N. A., Antarkar, D. S., Tathed, P. S., Purohit, A. V., & Wadia, N. H. (1978). Treatment of Parkinson's disease with the cowhage plant-Mucuna pruriens Bak. *Neurology India, 26*(4), 171–176.
115. Nagashayana, N., Sankarankutty, P., Nampoothiri, M. R. V., Mohan, P. K., & Mohanakumar, K. P. (2000). Association of L-DOPA with recovery following Ayurveda medication in Parkinson's disease. *Journal of the Neurological Sciences, 176*(2), 124–127.
116. Mahajani, S. S., Doshi, V. J., Parikh, K. M., & Manyam, B. V. (1996). Bioavailability of l-DOPA from HP-200—a Formulation of Seed Powder of Mucuna pruriens (Bak): a Pharmacokinetic and Pharmacodynamic Study. *Phytotherapy Research, 10*(3), 254–256.
117. Siddhuraju, P., & Becker, K. (2001). Rapid reversed-phase high performance liquid chromatographic method for the quantification of L-Dopa (L-3, 4-dihydroxyphenylalanine), non-methylated and methylated tetrahydroisoquinoline compounds from Mucuna beans. *Food Chemistry, 72*(3), 389–394.
118. Misra, L., & Wagner, H. (2004). Alkaloidal constituents of Mucuna pruriens seeds. *Phytochemistry, 65*(18), 2565–2567.
119. Tripathi, Y. B., & Upadhyay, A. K. (2001). Antioxidant property of Mucuna pruriens Linn. *Current Science, 80*(11), 1377–1378.
120. Spencer, J. P., Jenner, A., Butler, J., Aruoma, O. I., Dexter, D. T., Jenner, P., & Halliwell, B. (1996). Evaluation of the pro-oxidant and antioxidant actions of L-DOPA and dopamine in vitro: implications for Parkinson's disease. *Free Radical Research, 24*(2), 95–105.
121. Caruana, M., Högen, T., Levin, J., Hillmer, A., Giese, A., & Vassallo, N. (2011). Inhibition and disaggregation of α-synuclein oligomers by natural polyphenolic compounds. *FEBS Letters, 585*(8), 1113–1120.
122. Pathania, R., Chawla, P., Khan, H., Kaushik, R., & Khan, M. A. (2020). An assessment of potential nutritive and medicinal properties of Mucuna pruriens: a natural food legume. *3. Biotech, 10*(6), 1–15.

123. Cassani, E., Cilia, R., Laguna, J., Barichella, M., Contin, M., Cereda, E., Isaias, I. U., Sparvoli, F., Akpalu, A., & Budu, K. O. (2016). Mucuna Pruriens for Parkinson's Disease: Low-Cost Preparation Method, Laboratory Measures and Pharmacokinetics Profile. *Journal of the Neurological Science, 365*, 175–180.
124. Yadav, S. K., Prakash, J., Chouhan, S., Westfall, S., Verma, M., Singh, T. D., & Singh, S. P. (2014). Comparison of the Neuroprotective Potential of Mucuna Pruriens Seed Extract with Estrogen in 1-Methyl-4-Phenyl-1,2,3,6-Tetrahydropyridine (MPTP)-Induced PD Mice Model. *Neurochemistry International, 65*, 1–13.
125. Lohray, B. B. (1992). Cyclic sulfites and cyclic sulfates: epoxide like synthons. *Synthesis, 1992*(11), 1035–1052.
126. Byun, H.-S., He, L., & Bittman, R. (2000). Cyclic sulfites and cyclic sulfates in organic synthesis. *Tetrahedron, 37*(56), 7051–7091.
127. Sayyed, I. A., & Sudalai, A. (2004). Asymmetric synthesis of l-DOPA and (R)-selegiline via, OsO4-catalyzed asymmetric dihydroxylation. *Tetrahedron: Asymmetry, 15*(19), 3111–3116.
128. Kulma, A., & Szopa, J. (2007). Catecholamines are active compounds in plants. *Plant Science, 172*(3), 433–440.
129. Kong, K. H., Lee, J. L., Park, H. J., & Cho, S. H. (1998). Purification and characterization of the tyrosinase isozymes of pine needles. *IUBMB Life, 45*(4), 717–724.
130. Steiner, U., Schliemann, W., & Strack, D. (1996). Assay for tyrosine hydroxylation activity of tyrosinase from betalain-forming plants and cell cultures. *Analytical Biochemistry, 238*(1), 72–75.

Chapter 8
Costus

Sana Aslam, Matloob Ahmad, Salma Shahid, Ameer Fawad Zahoor, and Arwa A. AL-Huqail

8.1 Introduction

Family	Costaceae
Subfamily	Coastaceae
Scientific name	*Costus speciosus*
English/Common Name	Crepe Ginger

8.2 Plant Description

In 1753, a Swedish Botanist, Linnaeus first designed the Costaceae family which includes herbaceous perennial plants such as Costus (previously acknowledged as Hellenia, after a Finnish botanist, Carl Niclas von Hellens). The Costus plant species are found in Africa, Asia, and America's subtropical and tropical environments

S. Aslam
Department of Chemistry, Government College Women University, Faisalabad, Pakistan

M. Ahmad (✉) · A. F. Zahoor
Department of Chemistry, Government College University, Faisalabad, Pakistan
e-mail: matloob.ahmad@gcuf.edu.pk

S. Shahid
Department of Chemistry, Government College Women University, Faisalabad, Pakistan

Department of Biochemistry, Government College Women University, Faisalabad, Pakistan

A. A. AL-Huqail
Department of Biology, College of Science, Princess Nourah bint Abdulrahman University, Riyadh, Saudi Arabia

M. Zia-Ul-Haq et al. (eds.), *Essentials of Medicinal and Aromatic Crops*,
https://doi.org/10.1007/978-3-031-35403-8_8

[1]. *Costus* plants are linked to ginger and were once included in the same Zingiberaceae family as ginger. Costaceae is the new family name for them. The Zingiberaceae family contains approximately 52 genera and over 1300 species that are originated in the America, tropical Asia, and Africa [2]. Costaceae is simply identified as well as different from other Zingiberales families by branched or completely developed aerial shots with a distinctive single layered spiral phyllotaxy [3]. Due to wide-ranging correspondences in floral properties as well as inflorescence, Costaceae was previously classified as a subfamily of Zingiberaceae family [1]. In the *Costus* genus 175 species are found [4]. Costus is one of the most researched genera owing to its medicinal and pharmacological qualities. There are many species in this family, including Costus chartaceus, Costus asae, Costus barbatus, Costus igneus, Costus giganteus, Costus spectabilis, and Costus cuspidatus, with variations in bloom or colour [5]. Some types have bracts and flowers that bloom with compact cones, whereas others have soft crepe or pineapple like flowers that sprout from the green cones [6]. The Costus Plants also have variety of leaves, with some being pubescent on the axial surface and others being smooth and purple [5].

Costus speciosus is a traditional plant used for treating diabetes and referred to as the "insulin plant." Crepe ginger, kemak, kebu, kostam, keyu, keyaful, keomol, kebuka, kembu, kostam, kondige, pushkarmula, as well as penva are all names used for the perennial herbaceous ornamental plant *C. speciosus* [7]. Regardless of its visuals, its traditional, therapeutic, and pharmacological implications have generated the curiosity of academics all over the world.

In damp soil, the plant thrives. The perennating part of C. *speciosus* is the underlying rhizome, that is divided into internodes or nodes and the vegetative phase of upper part of ground remains for 7 to 8 months. These plants grow from a rhizome that produces one bloom on a spike and are subtropical to tropical. Costus plants may reach heights of 6–10 feet (2–3 m) in the landscape, making them ideal for adding height to the landscape. They can withstand temperatures ranging from 7 to 12 °C. Costus is known for its spiraling stems, which separate it from cousins like Zingiber (real ginger). Spiral gingers refer to the genus as a whole, but it can also apply to *C. barbatus* specifically.

An unbranched tropical plant i.e., Costus afer Ker-Gawl having creeping rhizomes. This plant has small monocot shrub which grows in gigantic and humid woods and beside rivers [8]. It is a perennial plant with yellow and white blooms that may grow up to 4 m tall [8, 9]. It has spirally organised simple leaves. The sheath is green but spotted with purple closed, and lobular. The glabrous and leathery ligule is 4–8 mm long. The leaf blade is oval to obovate and is 15–35 cm × 3.5–9.5 cm [10]. The flower is bisexual and zygomorphic, and the border is sparsely hairy.

8.3 Agronomy of Plant

8.3.1 Soil Condition

Soil texture or structure has a significant impact in developing the plant and supplying the nutrients in the soil. Soil aggregation and soil nutrient have a significant link, with excellent soil accumulation being associated with the quantity of usable organic contents as well as biological activities in the soil [11]. Because of mechanical demolition as well as inadequate drying, poor soil compaction inhibits plant development and diminishes nutrient absorption, and also alters the growth and activity of roots [12].

All costus cultivars thrive in partial shade and early morning light. These plants require more water as the sun increases. They should always be well hydrated, regardless of where they are. The soil should be light and well-draining. The plant can thrive in a vast range of soils, from coastal alluvium to thick brown forest soil. It thrives on alluvial soils with a grimy to clay loam nature and a pH range of 5.7–7.5.

8.3.2 Climatic Condition

It may be cultivated everywhere from sea level to around 1500 metres above sea level. Areas between 400 and 600 metres above mean sea level, on the other hand. A subtropical environment with rainfall ranging from 1000 to 1500 mm produces high-quality materials. The optimal conditions for growing are high humid and a minimum temperature of 13 °C. In naturally existing plant populations, a specific dry interval between monsoon rains leads to greater diosgenin levels. In terms of diosgenin content, coastal locations and places with high yearly moisture and rain level would results in the low-quality plant population.

8.3.3 Propagation

Costus speciosus may be reproduced in a variety of ways, including vegetative propagation utilizing rhizome pieces [13], culm division, stem cuttings [14], and seed dispersal by birds [15].

Only rhizome cuttings are used to grow it commercially. It's crucial to choose the right rhizomes for the plant. There are a lot of nipple-shaped buds on the rhizomes. During April, bud production on the rhizomes is low. At least two healthy buds should be present in rhizome piece cuttings for propagation. Pieces of rhizome weighing roughly 40 kg should be chosen.

Seeds can also be used to cultivate it. Birds disseminate Costus seeds as they dine on the fruits. The germination rate of seeds is around 62%.

8.3.4 Land Preparation

The land should be cultivated 2–3 times and the topsoil should be fine tilthed. FYM is applied at a rate of 15 tonnes per acre and well mixed with the soil. Furrows are opened at 50 cm apart.

8.3.5 Planting

The portions of rhizome should be planted at a depth of 8–10 cm and spaced 50 cm apart in horizontal rows before being covered with dirt. The crop is watered as soon as it is planted. Thick-sized portions emerge slowly during the first 40–45 days after sowing. This is because the eye buds present on these rhizome pieces are inactive, taking long time period to grow, exclusively if the crop was planted in the month of April. After 70–75 days, 90–95% of the seeds have sprouted. The best time to plant is between the third week of April to the third week of May.

Rhizomes sown in sixth and seventh month, germinate in 60 days, but the production is poor since the crop has a shorter growth phase before dormancy sets in the first year. The crop's active vegetative growth cycle is from July to September, with maximal tuberization occurring from September to November before dormancy sets in (Fig. 8.1).

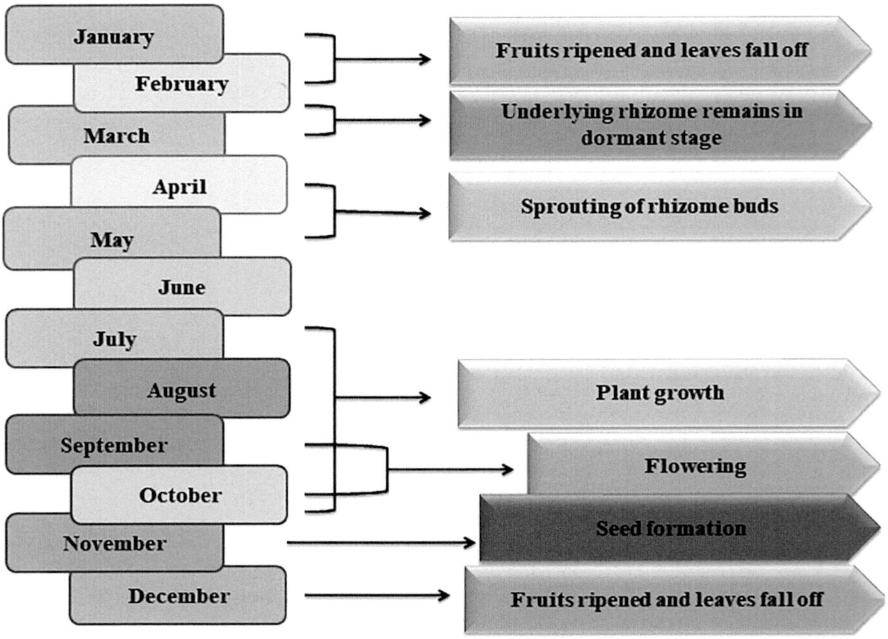

Fig. 8.1 Graphical representation of growth of *Costus* plant

8.3.6 Manuring

It is a rhizomatous crop, and considerable manuring is necessary to compensate for the biomass output. According to reports, 45 kg, 30 kg, and 30 kg of nitrogen, P_2O_5 and K_2O fertilizers, respectively, as well as 15 tonnes of FYM, are required per acre.

It was reported in an experiment that for stem height with inflorescence, the contact plant spacing as well as phosphorus dosages revealed that the 1.20 × 0.8 m spacing produced maximum outcomes regardless of the amount of phosphorus used, While for clump height treatment, the spacing was 1.20 × 0.80 m and 35 g of phosphorus produced the best results. For all other factors studied, phosphorus fertilizers had no influence on *C. speciosus* flower output, and a spacing of 1.2 × 0.8 m is suggested for development of flowers of *C. speciosus*.

8.3.7 Irrigation

For optimal growth, the crops require a large amount of water. The crop sown in April and May has to be watered at least twice a month until the monsoon arrives. If winter rainfall is scarce during the dormancy period, irrigation is required. Fourteen to seventeen irrigations are necessary for a crop to be harvested after 17–18 months.

8.3.8 Ecology

This medicative plant is native to the Indian Himalaya, but due to its limited distribution and heavy cultivation, it has been listed as Endangered and Critically Endangered by the Red Data Book (RDB) of Indian Plants and the International Union for Conservation of Nature and Natural Resources (IUCN), respectively [16]. In the Union Territory of Jammu and Kashmir, the perennial costus plant grows wild along the Indo-Pakistan border. These herbs were first grown in the Himalayan area of Himachal Pradesh (HP) and Uttarakhand (UK) in 1920 and 1929, respectively [17]. Costus has over 70 species, out of which 40 species that are found in tropical America, 25 in West tropical Africa, and 5 in South-East Asia [18]. The plant is found in Africa's woodland belt from Senegal to Ethiopia and east to Tanzania. It is found in tropical West Africa's rain forests and riverbanks in nations such as Cameroon, Senegal, Ghana, Guinea, Togo, Sierra Leone, and Nigeria [8, 19].

The Malay Peninsula of Southeast Asia is home to *Costus speciosus* (Koening) sm. In India, the plant grows wild in the sections of central India, Sub-Himalayan region and the Kerala Western Ghats, Karnataka and Maharashtra [20]. The Costus has about 100 different species. The blossom colour of the several costus species varies. Flowers and bracts on some types resemble compact cones, while others resemble pineapples or soft crepes emerging from green cones. On the abaxial surface, some leaves are hairy, while others are smooth and purple [21]. India is home to about seven species of the *Costus Linn.* genus.

8.3.9 Pest and Diseases

Pests and diseases do not appear to be a major issue for *Costus* because of the long-term health. They may be nibbled by caterpillars in the garden, and they may be harmed by red spider mite indoors, but that's about it. Particularly in light, sandy soil, mites and nematodes might be the reasons of damage to the plant crop of *Costus*.

8.4 Origin and Distribution of *Costus* Plant

C. speciosus has been found on several continents, in which North America, Africa, Asia, as well as Oceania are included, but it is mostly a Southeast Asian and Malaysian native. China, India, Bangladesh, Bhutan, Myanmar, Nepal, Philippines, Singapore, Hong Kong, Malaysia, Thailand, Vietnam, Sri Lanka, Taiwan, and Indonesia are all home to this species. However, this specie is an introduced species in Oceania as well as in North America. This plant is invasive in North American nations such as Cuba, the United States, Hawaii, and many regions of Oceania [22]. It may be found in mountain ranges such as the Western Ghats and the Himalayas in India. Andhra Pradesh, Himachal, Assam, Karnataka, Kerala, Pradesh, West Bengal and Tamil Nadu are among the other states with distribution facts [20].

8.5 Important Phytochemical Constituents of *Costus* Plant

C. speciosus is commonly utilised in different indigenous medical systems to treat a variety of maladies, including diabetes and its accompanying disorders. The leaves extracts have potential of lowering the level of blood glucose as studied in C57BLKS/J db/db mice and also exhibited hypoglycemic qualities and insulin potentiating activity [23]. Diabetics have traditionally eaten one leaf of C. *speciosus* each day to manage blood glucose levels [24]. C. *speciosus* contains large number of phytochemicals with potent therapeutical potential such as anti-helmintic, astringency, expectorant, purgative, depurative, febrifuge, aphrodisiac, and bitterness [25]. This plant's major secondary metabolites include alkaloids, glycosides, flavanoids, sterols, sesquiterpenes and phenols [26]. HPLC-MS analysis revealed that the herb contains considerable levels of saponins, with the principal saponins extracted from their seeds include dioscin, diosgenin and gracillin [27–29].

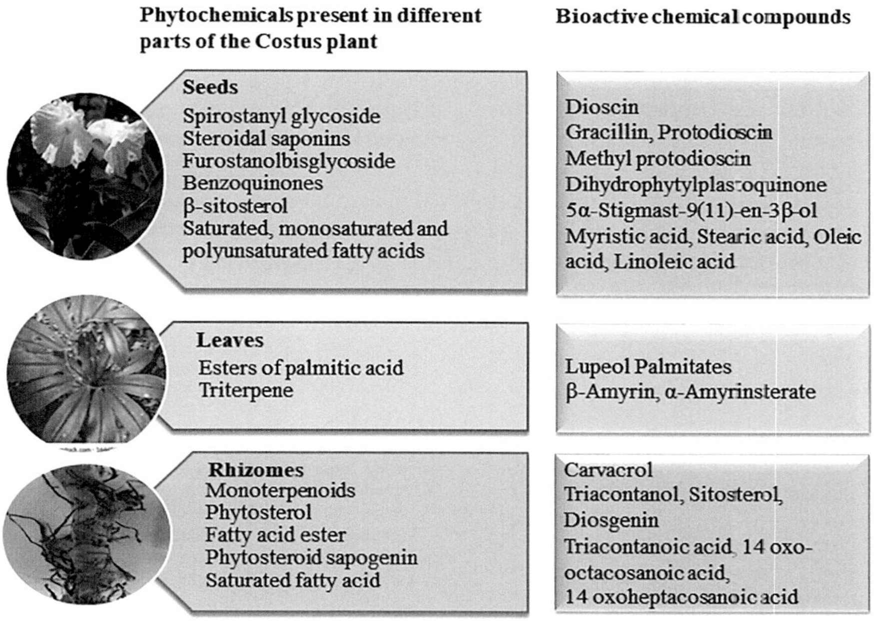

Fig. 8.2 Representation of potential phytochemical constituents present in different parts of *Costus specious*

Costus speciosus rhizomes include saponins such as alkaloids, steroids, sapogenin, tigogenin, as well as diosgenin [30, 31]. The rhizomes and stems also comprise starch mucilage, aliphatic hydroxyl ketones, triterpenes, abscisic acid, corticosteroids, fatty acids, oxa-acids, and tigogenin [32]. Lupeol, amyrin, and α-amyrin stearate have also been extracted from rhizomes while from leaves, palmitates have been extracted. Stearic acid, linoleic acid, palmitic acid, gadoleic acid and behenic acid are all found in the seed fat. Glucose, galactose, rhamnose, and diosgenin were found in defatted seeds [33]. Seeds yield two novel dihyrophytilplastoquinone and its methyl derivatives, including α-tocopherolquinone and quinones (Fig. 8.2).

Costus speciosus and *C. tonkinensis* (Zingiberaceae) from Yunnan province were studied for their chemical contents. Diosgenin, diosgenone, 25-en-cycloartenol, prosapogenin B of dioscin, and cycloartanol were extracted from the *C. speciosus* rhizomes. The rhizome of *Costus tonkinensis* possess succinic acid, tetracosanoic acid, and beta-sitosterol [34].

The Diosgenin is a key raw ingredient present in *C. afer* that is used as a precursor in the biosynthesis of many steroids such as sex hormones, oral contraceptives and corticosteroids. The saponins aferosides A-C, along with the parphyllin C, diosein and the flavonoid glycoside kaempterol, 3-O-Lrhamnopyranoside, are all found in the rhizomes [10]. In preclinical trials, diosgenin demonstrated a wide spectrum of pharmacological actions. It has neuroprotective, anticancer, immunomodulatory, cardiovascular protective, estrogenic, anti-diabetic and skin protective properties, primarily by inducing prevention of inflammatory events, apoptosis, regulation of T-cell immune response, decreasing oxidative stress, suppressing malignant transformation and promotion of cellular differentiation/proliferation, among other things. It causes apoptosis by interfering with cell death pathways as well as by their regulators i.e. the actin cytoskeleton to alter cellular motility and by altering the epithelial-mesenchymal transition, limiting the angiogenesis and decreasing matrix barrier breakdown. Diosgenin also plays a role in the inhibition of tumour spread. Diosgenin also increases antioxidant capacity and reduces lipid peroxidation [35]. It was studied that the potent phytochemical i.e. eremanthin extracted from *C. speciosus* lowered the plasma glucose levels in streptozotocin-induced diabetic Wistar rats [36]. This plant's rhizomes are also rich in sitosterol, dioscin, D-glucoside and gracillin, as evidenced by HPLC-MS and NMR analysis [7, 25] (Fig. 8.3).

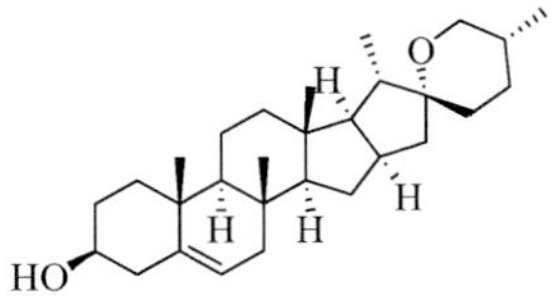

Diosgenin

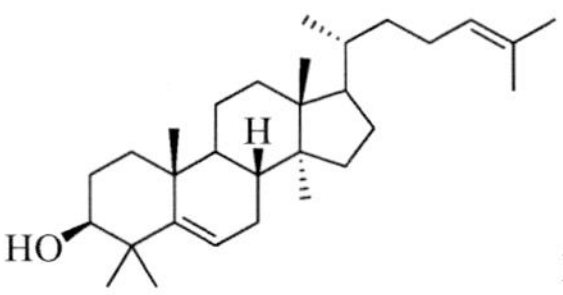

25-en-cycloartenol

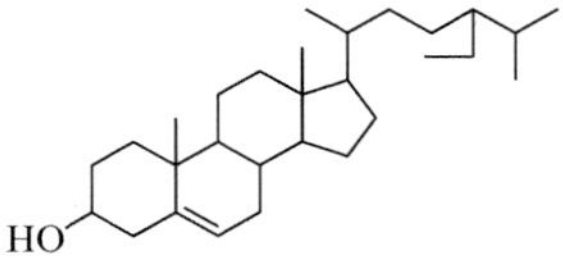

β-sitosterol

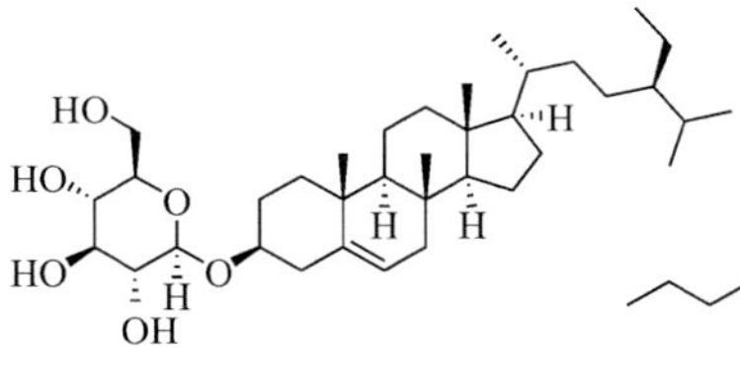

Daucosterin

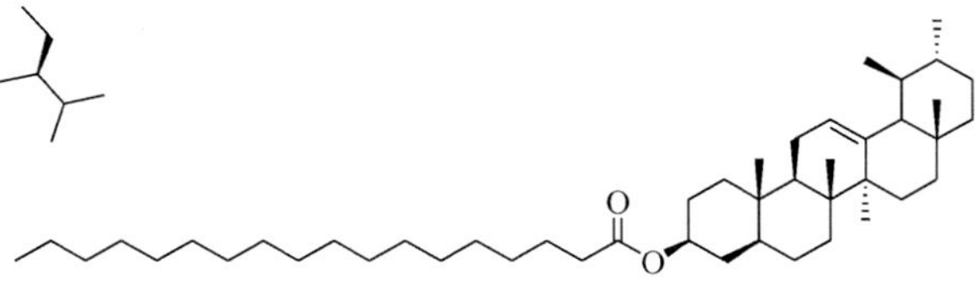

α-amyrin stearate

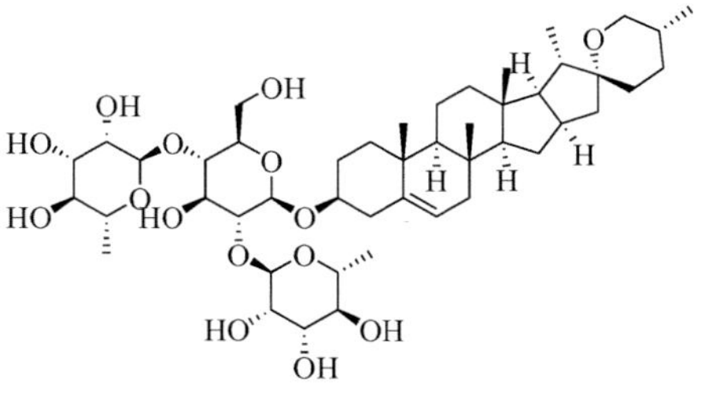

Dioscin (Saponin)

Fig. 8.3 Chemical Structures of few phytochemical constituents present in *Costus* plant

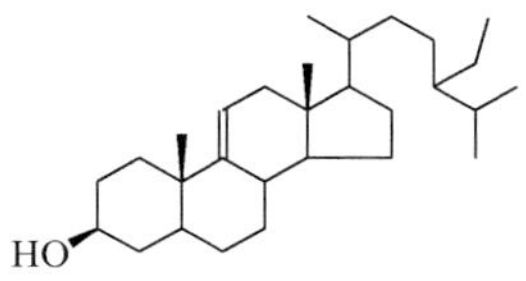

5α-stigmast9(11)-en-3ß-ol

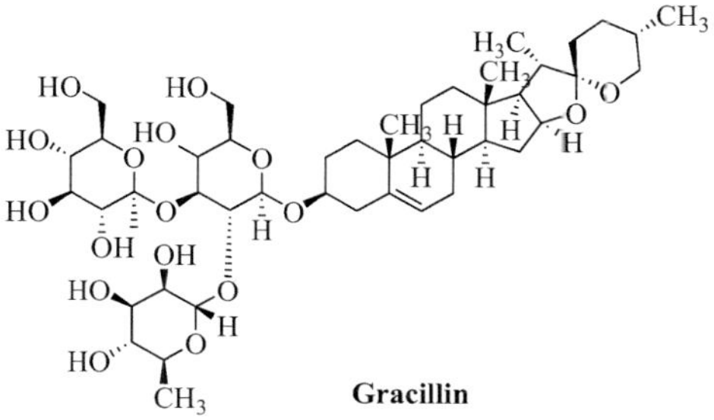

Gracillin

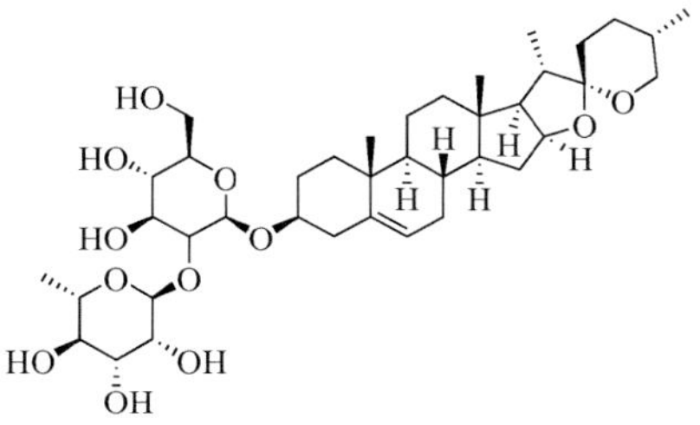

Prosapogenins A

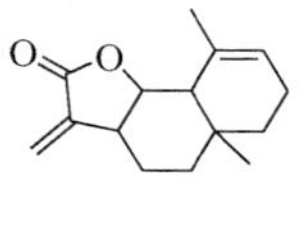

α-cyclocostunolide

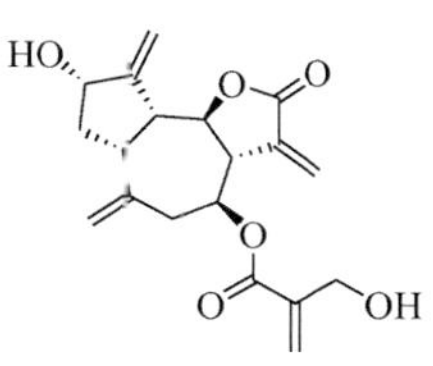

Cynaropicrin

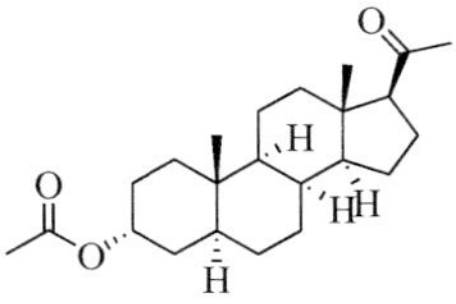

3-ß-acetoxy-9(11)-baccharene

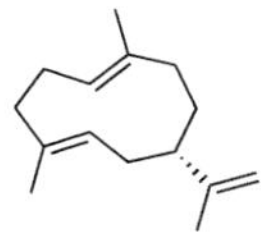

(+)-germacrene A

Fig. 8.3 (continued)

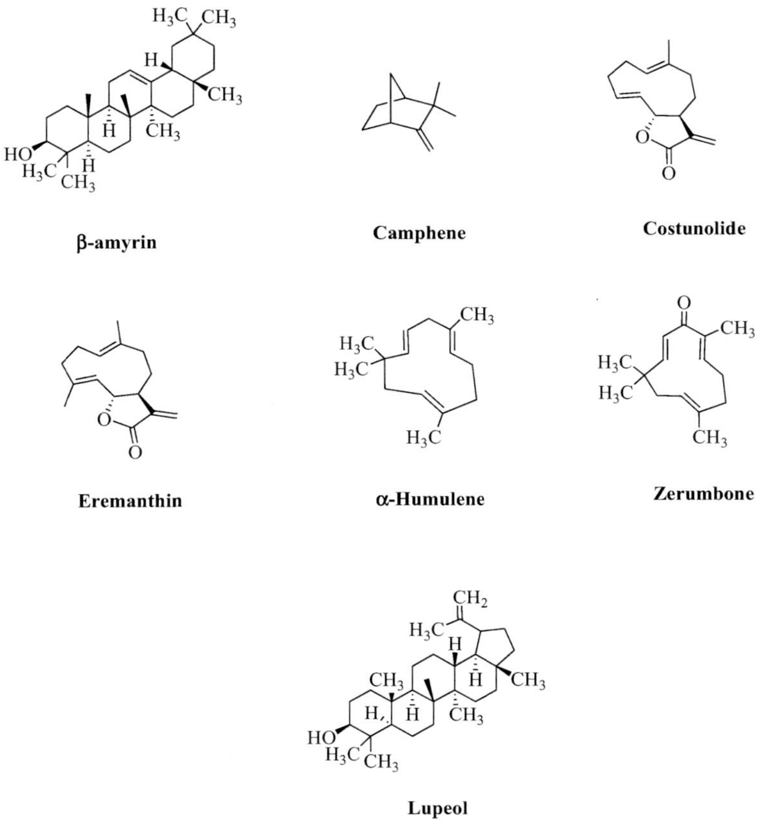

Fig. 8.3 (continued)

8.6 Medicinal Uses

Around 70–80% of the global's population is reliant on old-fashioned medicine based on medicinal plants. According to research data around 20,000 plant species are used by African native peoples for different medicinal and other uses [37]. Some Nigerian medicinal herbs are as effective as modern medication.

Antibacterial, antioxidant, antifungal, antidiuretic, anti-inflammatory, analgesic, anti-hyperglycemic, antipyretic, estrogenic, and anti-stress actions have all been discovered in *C. speciosus* [38]. The rhizomes of *C. speciosus* plant are eaten as a vegetable as well as it has also been utilized to make pickles, while flowers and leaves are utilized to make beverages in some places [26].

The bitter rhizomes have anthelmintic, astringent, and expectorant effects [39–41]. The rhizome extract is a tonic that helps with burning sensations, constipation, leprosy, anemia, asthma, and other skin conditions [42]. *Costus* rhizomes are used as a herbal treatment for fever. It was found out that the rhizome of *C. speciosus* exhibits hepatoprotective properties [43]. Boils are treated with rhizome paste, which is also used to create sexual hormones and contraception. Scabies and gastrointestinal problems are treated with the leaves. Blisters are treated with a paste made from stems. Snake bites are treated with rhizome extract [44–46]. The stimulant, carminative, diuretic, digestive, and antibacterial characteristics of *C. speciosus* have made it a popular therapeutic plant for centuries. Internally, the rhizome is used to treat stomach discomfort, liver issues, jaundice, gall bladder pain, and other ailments. *C. speciosus* leaves and rhizomes have been shown to contain the steroid diosgenin, which has anti-diabetic properties. Other properties of the rhizomes include being strongly laxative, expectorant, astringent, depurative, anthelmintic, febrifuge, tonic as well as alleged to help with ingestion [47, 48]. *C. speciosus* rhizome juice has been utilized to treat abortion besides leprosy, and it is famous for having a cooling effect, so it is used to treat severe headaches [48]. Furthermore, because of their medicinal properties, the rhizomes, stem and leaves have been utilized for traditional purposes; e.g. the infusion of stem and leaves decoction has a sudorific effect, therefore it's extract used to cure patients suffering from high fever; occasionally [49]. Additionally, the sap of stems and young leaves are used for curing infections of eye and ear, catarrhal fever, colds, and coughs, as well as snake bites [13, 47].

C. afer is a beneficial remedial plant, widely used in South-West and South-East Nigeria for its anti-arthritic, anti-diabetic, and anti-inflammatory characteristics [50]. In Ogoni country, Rivers State, the extracted juice of *C. afer* is used as drops for curing the ocular irritation and deformities. Chewing the young, fragile leaves is thought to offer vigour to the weak and dehydrated sufferer. Stomach troubles are treated with an infusion of the inflorescence [51].

The rhizome of the *Costus igenus* also known as insulin plant is bitter, purgative, astringent, cooling, anthelmintic, febrifuge, expectorant, aphrodisiac, depurative and useful in the treatment of burnt skin, worm infection, leprosy, skin diseases, constipation, breathing issues, swellings, and anaemia [25]. Research was done to investigate the effectiveness of aqueous and methanolic extracts of *C igneus* in rats with diabetes-induced hyperlipidemia. The researchers discovered that the diabetes-induced hyperlipidemia cured by the 200 mg/kg body weight of its aqueous and methanolic extracts [52]. In Triton-induced hyperlipidemic rats, an alcoholic extract of *C. igneus* at a dosage of 400 mg/kg (P.O) dramatically reduced blood cholesterol, triglycerides, and LDL [53].

The roots have a pleasant, aromatic odour and are used to treat vagotonic bronchial asthma as well as an antiseptic. Rheumatism, cholera, jaundice, dysentery, bronchitis, skin diseases, stomach illnesses, ulcers, edoema, cold, cough, and toothache temperature are all said to be treated with the herb. Costus oil, collected from the plant's roots, is used in high-quality perfume formulations and hair oils; it is also beneficial in curing leprosy. Costus roots are used as a pesticide to protect woolen clothing from insects and as incense in the Indian Himalayan area. The upper plant section is used as fuel and fodder in the Himachal Pradesh area of Lahaul and Spiti (L&S), while the dried leaves are used as tobacco [54].

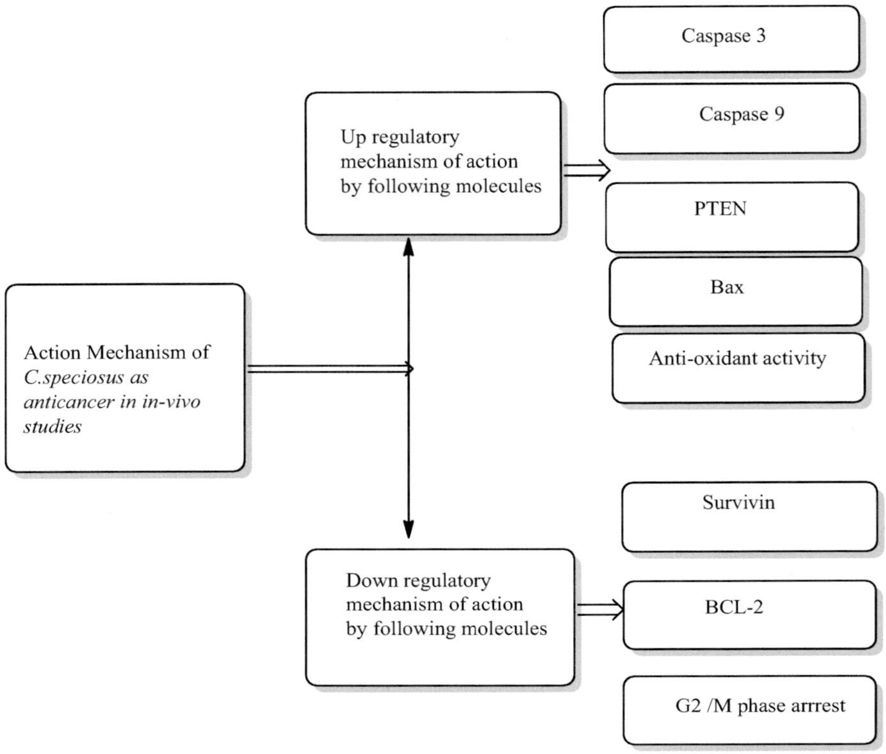

Fig. 8.4 Action mechanism of *C.speciosus* as anticancer potential in *in-vivo* studies [55]

The active phytochemical compounds of *C. speciosus* like Camphene, Lupeol, Beta-amyrin, Costunolide, Diosgenin Zerumbone, and Alpha-humulene, exhibited potent anticancer potential. In addition several *in vitro and in vivo* studies have shown that these phytochemicals followed many down-regulatory and up-regulatory mechanisms (Figs. 8.4 and 8.5). The transcription factor p53 plays an important inhibitory role in the proliferation of cancerous cells, regulation of apoptotic pathways and induction of ROS through the up-regulatory mechanisms [55].

8.7 Diosgenin

Diosgenin is one of the tenth most vital sources of steroids, as well as the most commonly authorized plant-based pharmaceutical. It is responsible for biochemical and morphological changes in megakaryocyte cells due to differences in lipoxygenase activity in human erythryoleukaemia cells and also involved in breakdown of cholesterol [56].

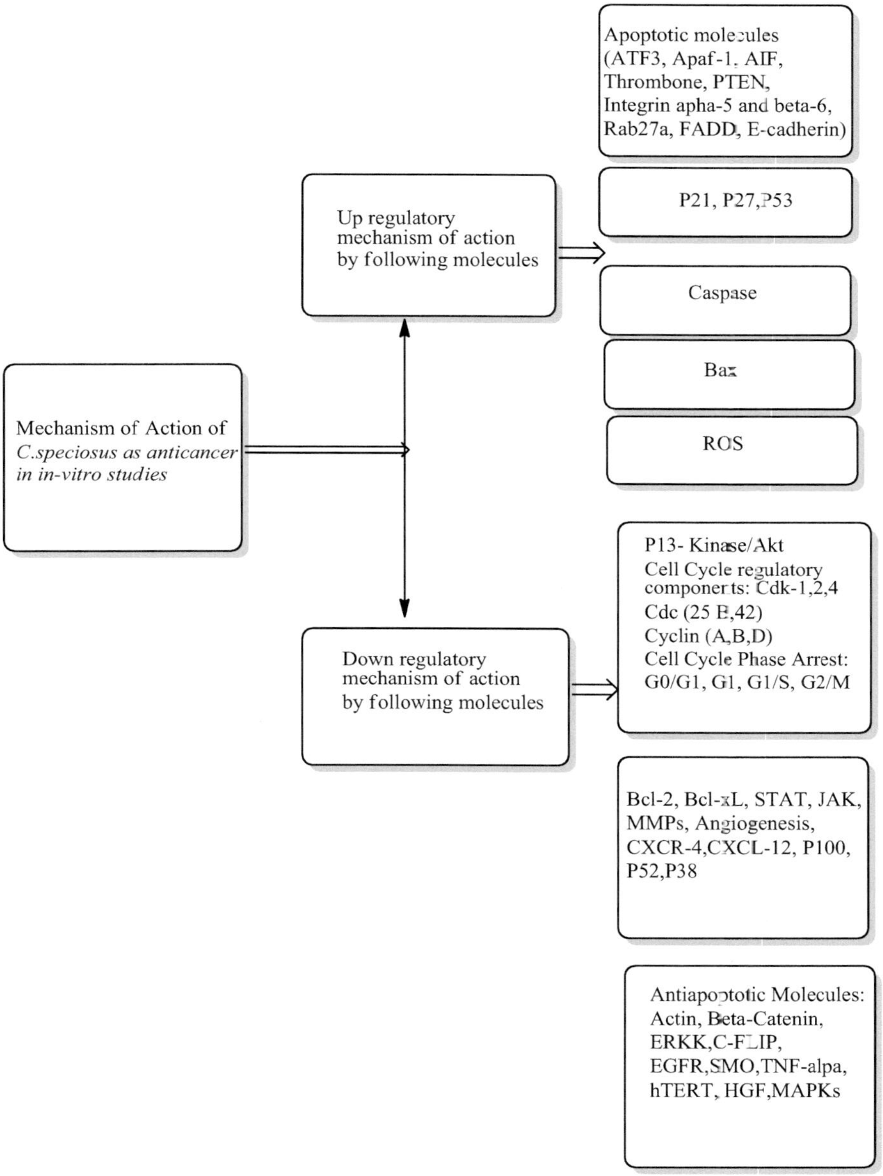

Fig. 8.5 Action mechanism of *C. speciosus* as anticancer potential in *in-vitro* studies [55]

8.7.1 Medicinal Uses of Diosgenin

Diosgenin has a number of health advantages, including the inhibition of colon cancer, climacteric syndrome, and heart disease [57]. Diosgenin was exhibited the potential to reduce postmenopausal symptoms [58]. Diosgenin is an antispasmodic that can be used to treat contractions, coughing, and muscle twinges [59]. Diosgenin is a potent phytochemical and is utilized for the purpose of inducing apoptosis in cancer cells and lower the blood pressure [60]. Diosgenin has been utilized in traditional medicine as an antihypertriacylglycerolimia, antihyperglycemic, antidiabetic, antihypercholesterolemia, and antileukemia drug [61], according to recent investigations. Diosgenin has been used to maintain the healthy blood cholesterol levels as well as generation of dehydroepiandrosterone. The continual release of diosgenin has been reported to inhibit bone loss in the same way as estrogen does [62]. Consumption of diosgenin also exhibited anti-stress and anti-inflammatory properties (Fig. 8.6) [59].

8.7.2 Role of Disogenin in Skin Aging

Skin ageing is the result of both natural ageing and ageing caused by environmental factors like UV light exposure. It is related with impaired epidermal cell degenerative and turnover alterations in dermal elastic fibres, resulting in dry skin, epidermal

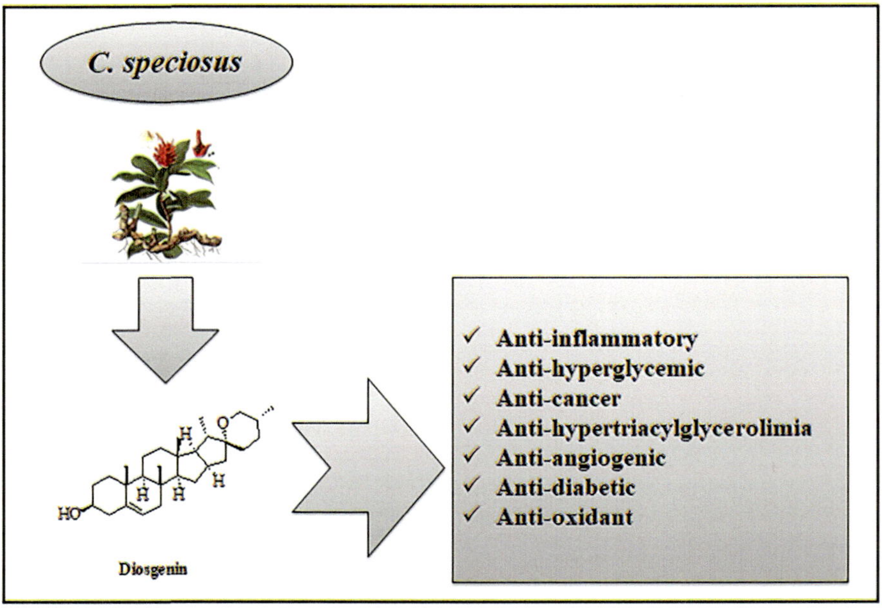

Fig. 8.6 Medicinal properties of Diosgenin

thinning, wrinkles or lentigines, [63]. A research aiming to evaluate the efficacy of diosgenin against skin ageing, it was discovered that diosgenin may increase DNA synthesis as well as keratinocyte proliferation by activating cAMP signals without involving oestrogen receptors. Diosgenin raised bromodeoxyuridine absorption and intracellular cAMP level in adult human keratinocytes in vitro and increased DNA synthesis in a human 3D skin comparable model. An adenylate cyclase inhibitor blocked the increase in bromodeoxyuridine accumulation by diosgenin but not antisense oligonucleotides against an orphan G-protein-coupled receptors GPR30, or oestrogen receptor, indicating the involvement of cAMP but not oestrogen receptor, or GPR30. *In vivo*, diosgenin treatment enhanced epidermal thickness in ovariectomized mice, a climacteric model, without affecting fat formation. To test the safety of diosgenin, breast cancer-ridden mice were given diosgenin and 17-estradiol. In a climacteric mouse model, diosgenin (0.01%, 0.02%, and 0.04% mixed in baseline diet) enhances epidermal thickness [64]. Furthermore, diosgenin (1–50 mol/L) suppresses melanogenesis in B16 melanoma cells via the PI3K pathway activation, suggesting that diosgenin may be an effective hyperpigmentation inhibitor in the therapeutical treatment of skin illnesses such as acquired hyperpigmentation situations [65]. The findings demonstrated that, whereas 17-estradiol increased tumour development, diosgenin did not. Moreover, discovery of keratinocyte proliferation restoration in aged skin shows that diosgenin may have promise as a safe healthy food for climacterics [64].

8.7.3 Role of Disogenin in Diabetes

Diabetes mellitus is a severe as well as widespread metabolic disorder that poses a major threat to human health. Diabetes' hyperglycemia and long-term metabolic abnormalities will harm the tissues and organs of entire body, culminating in catastrophic consequences. Diabetes is classified into two types: type I (T1DM) and type II (T2DM) (T2DM). T1DM is an insulin-dependent illness caused by inadequate insulin secretion [66]. T2DM is caused by insulin release abnormalities or is the result of insulin usage disorders [67, 68]. According to the International Federation of Diabetes, the global diabetes population has reached 425 million, with T1DM accounting for 5–10% of the diabetic population and T2DM accounting for 90–95% [69–71]. Diabetes affects a large number of individuals worldwide, and diabetes treatment remains a challenging issue. Because the medications on the market for treating diabetes have severe adverse effects, new therapies are urgently needed. Natural therapy using phytochemicals is a potential way towards the safe and effective treatment of diabetes. Diosgenin is a potent bioactive natural steroidal sapogenin [72]. Diosgenin has been shown in studies to help treat diabetes by reducing oxidative stress and malfunctioning lipid metabolism [73]. Based on the pharmacological actions of Diosgenin including hypoglycemic activity, hypolipidemic, anti-proliferative, anti-inflammatory, and as a potent anti-oxidant, it has a positive impact on diabetes and its consequences via many targets and pathways. In recent years, an increasing number of anti-diabetic studies have revealed that diosgenin exhibited excellent anti-diabetic potential both in *in vitro* and *in vivo,* consequently it is becoming increasingly popular as a natural treatment. However, no comprehensive investigation has been conducted to assess the preventive and possible mechanisms of diosgenin for diabetes and its consequences [74–76].

8.7.4 Medicinal Uses of Dioscin

Dioscin is a biologically active molecule with antioxidative, antiobesity [77], hepatoprotective [78], and antitumor [79] capabilities, anti-inflammatory, among other ethnopharmacological and physiological properties. This phytochemical protects against cancer, gastrointestinal illnesses, cardiovascular and cerebrovascular diseases, organ toxicity by adjusting various cellular targets and influencing different signaling pathways in the human body. Dioscin's advantages have been widely recognised in traditional Chinese medicine for decades [78].

Millions of individuals throughout the world are suffering from diabetes mellitus. By altering illness-associated signalling pathways, dioscin controls disease development in diabetics. The effects of dioscin on glycolipid metabolism in insulin-induced HepG2 cells, and spontaneous T2DM KK-Ay mice were reported. Dioscin effectively decreased the lipid accumulation, hyperglycemia, and hyperlipidaemia, as well as enhanced insulin resistance, and improved the concentration of hepatic glycogen. It also caused miR-125a-5p to have an inhibitory impact on STAT3 (signal transducer and activator of transcription 3) signalling, which is generally stimulated throughout the sickness [79]. The pharmacological activities of dioscin are widely recognized [80, 81].

Obesity is regarded as a severe health issue, since it is linked to hypertension, heart disease, and an increased risk of death. To combat the side effects of the high cost of bariatric surgery and antiobesity medicines, phytochemicals like dioscin are becoming a new target of study for reducing disease severity [82].

8.8 Synthesis of Potent Phytochemicals

In this section synthesis of different potent phytochemicals of Costus plant are briefly discussed.

8.8.1 Biosynthesis of Diosgenin

Many steroidal medications, including anti-inflammatory, antioxidants, steroids, cortisone, sex hormones, fertility control chemicals, contraceptives, as well as anabolic agents, are made from diosgenin [83–85]. C-16,22-Dihydroxylase and C-26

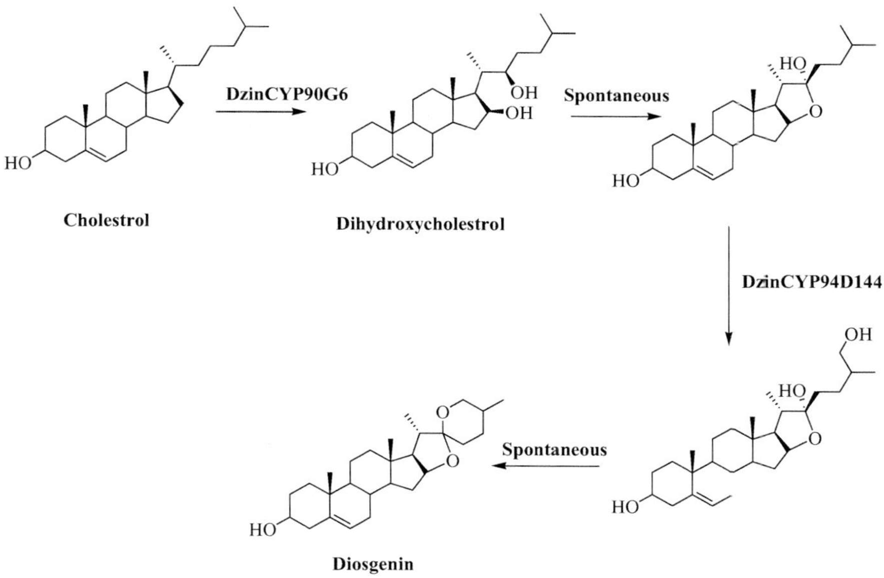

Scheme 8.1 Biosynthetic pathway of Diosgenin

hydroxylase [86] are two P450 enzymes that are used to catalyze the production of diosgenin from the starting material such as cholesterol in *Dioscorea* [87] (Scheme 8.1).

8.8.2 Chemical Synthesis of Dioscin

The most frequent steroid saponins in plants are spirostan-type saponins. They have a hexacyclic aglycone, like tigogenin or diosgenin or with an oligosaccharide chain attached to the 3-OH group. Dioscin is among the most extensively distributed steroid saponins in plants, has been discovered in a variety of traditional Chinese herbal remedies and possesses anti-inflammatory, antifungal, anticancer, immune-stimulatory, and antiviral properties. To synthesize dioscin, a number of methods were explored, the most easy and efficient of which was produced in five stages from diosgenin. The TMSOTf-catalyzed glycosylation reaction between diosgenin and perbenzoylated glucopyranosyl *N*-phenyl trifluoroacetimidate donor **1** was the first step in the synthesis. In 92% of cases, this reaction yielded the equivalent 3-O-glycoside intermediate **2**. The 2′,4′-diol acceptor was then obtained in 60% yield after successive elimination of benzoyl groups and selective 3′,6′-OPiv protection. In a 66 percent yield, intermediate **2** was combined with peracetylated l-rhamnopyranosyl *N*-phenyl trifluoroacetimidate to generate the desired monodesmosidic trisaccharide derivative, which was then deprotected globally to obtain dioscin [88] (Schemes 8.2 and 8.3).

Scheme 8.2 Synthesis of Dioscin. Reaction conditions: (a) TMSOtf, CH_2Cl_2, rt; NaOMe, MeOH, rt, 92%; PivCl, pyr., 0 °C, 60%; (b) TMSOtf, CH_2Cl_2, rt, 66%, NaOH, MEOH, H_2O, THF, rt, 90%

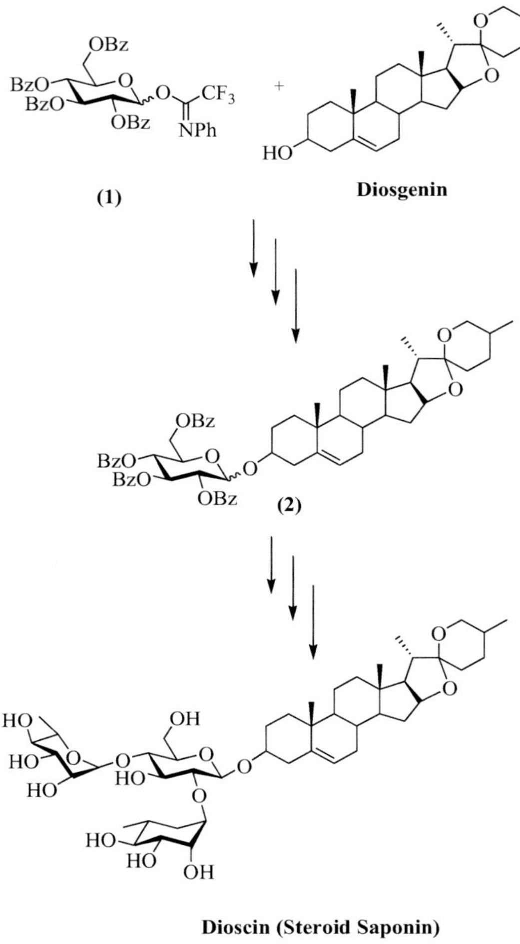

8.9 Future Prospective of Costus Plant

C. speciosus is a rich source of bioactive metabolites of economic value, such as diosgenin and dioscin, as well as with strong therapeutic potential. In addition, different investigations are underway to uncover the long-term medicinal and healing qualities of this plant for designing the novel drugs. The mode of action of the bioactive phytochemicals found in *C. speciosus*, as well as the *in vivo* toxicity of various extracts, should be further investigated. This might lead to the identification of new biological active lead compounds and treatment approaches. To maintain the conservation and long-term use of this vital medicinal plant across the world, researchers must look at molecular diversity, *in vitro* growth approaches, and other biotechnological factors such as metabolic pathway engineering to produce optimal

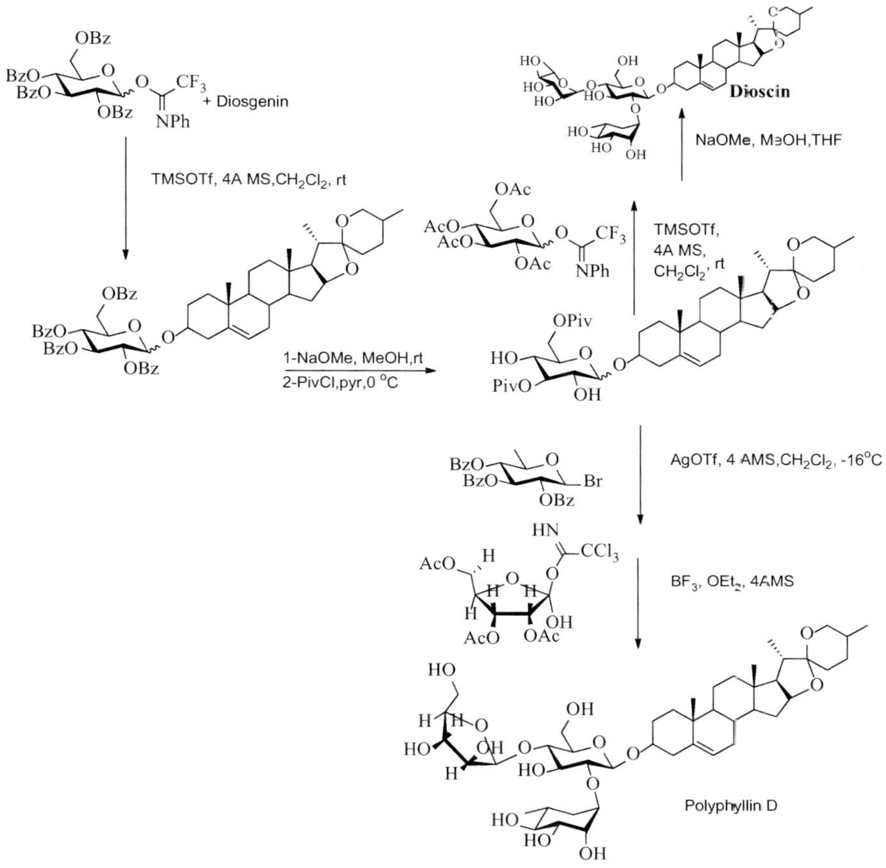

Scheme 8.3 Chemical synthesis of Dioscin and Polyphyllin D

levels of bioactive metabolites. Finally, the next drive area of study for economic viability would be elite chemotype discovery and manufacturing of therapeutically relevant lead compounds from *C. speciosus*.

References

1. Specht, C. D., & Stevenson, D. W. (2006). A new phylogeny-based generic classification of Costaceae (Zingiberales). *Taxon, 55*(1), 153–163.
2. El-Far, A., & Abou-Ghanema, I. (2013). Biochemical and hematological evaluation of Costus speciosus as a dietary supplement to Egyptian buffaloes. *African Journal of Pharmacy and Pharmacology, 7*(42), 2774–2779.
3. Kirchoff, B. K., & Rutishauser, R. (1990). The phyllotaxy of costus (Costaceae). *Botanical Gazette, 151*(1), 88–105.

4. Ariharan, V., Devi, V. M., Rajakokhila, M., & Prasad, P. N. (2012). Antibacterial activity of Costus speciosus rhizome extract on some pathogenic bacteria. *International Journal of Advanced Life Sciences, 4*, 24–27.
5. Arunvanan, M., Sasi, S., Mubarak, H., & Kanagarajan, A. (2013). An overview on anti diabetic activity of Siddha medicinal plants. *Asian Journal of Pharmaceutical and Clinical Research, 6*, 46–50.
6. Trease, G. E. (1945). *A text-book of pharmacology*. Bailliere Tindall & Cassell.
7. Pawar, V., & Pawar, P. (2014). Costus speciosus: An important medicinal plant. *International Journal of Science and Research, 3*(7), 28–33.
8. Ekpo, B. A., Bala, D. N., Essien, E. E., & Adesanya, S. A. (2008). Ethnobotanical survey of Akwa Ibom state of Nigeria. *Journal of Ethnopharmacology, 115*(3), 387–408.
9. Edeoga, H., & Okoli, B. (1996). Apomictic behaviour in Costus afer–C. lucanusianus (Costaceae) complex in Nigeria. *Feddes Repertorium, 107*(1–2), 75–82.
10. Aweke, G. (2007). Costus afer (Ker Grawl). In G. H. Schmelzer & A. Guribraukin (Eds.), *Plant Resources of Tropical Africa (PROTA)*.
11. Agbede, O. O. (2009). *Understanding soil and plant nutrition*. Petra Digital Press.
12. Smith, R., & Smith, T. (2001). *Ecology and field biology* (6th ed.). Benjamin Cummings. An imprint of Addison Wesley Longman.
13. Srivastava, S., Singh, P., Jha, K., Mishra, G., Srivastava, S., & Khosa, R. (2011). Anthelmintic activity of aerial parts of Costus speciosus. *International Journal of Green Pharmacy (IJGP), 5*.
14. Rani, A. S., Sulakshana, G., & Patnaik, S. (2012). Costus speciosus, an antidiabetic plant-review. *FS Journal of Pharmacy Research, 1*(3), 51–53.
15. Najma, C., Chandra, K., & Ansarul, H. (2012). Effect of Costus speciosus Koen on reproductive organs of female albino mice. *International Research Journal of Pharmacy, 3*(4), 200–202.
16. Kuniyal, C. P., Heinen, J. T., Negi, B. S., & Kaim, J. C. (2019). Is cultivation of Saussurea costus (Asterales: Asteraceae) sustaining its conservation? *Journal of Threatened Taxa, 11*(13), 14745–14752.
17. Kuniyal, C. P., Rawat, Y. S., Oinam, S. S., Kuniyal, J. C., & Vishvakarma, S. C. (2005). Kuth (Saussurea lappa) cultivation in the cold desert environment of the Lahaul valley, northwestern Himalaya, India: Arising threats and need to revive socio-economic values. *Biodiversity and Conservation, 14*(5), 1035–1045.
18. Atere, T., Akinloye, O., Ugbaja, R., Ojo, D., & Dealtry, G. (2018). In vitro antioxidant capacity and free radical scavenging evaluation of standardized extract of Costus afer leaf. *Food Science and Human Wellness, 7*(4), 266–272.
19. Omokhua, G. (2011). Medicinal and socio-cultural importance of Costus afer (Ker Grawl) in Nigeria. *African Research Review, 5*(5), 282–287.
20. Sarin, Y., Bedi, K., & Atal, C. (1974). Costus speciosus rhizome as source of diosgenin. *Current Science*, 569–570.
21. Whistler, W. A. (2000). Tropical ornamentals: A guide.
22. Sohrab, S., Mishra, P., & Mishra, S. K. (2021). Phytochemical competence and pharmacological perspectives of an endangered boon—Costus speciosus (Koen.) Sm.: A comprehensive review. *Bulletin of the National Research Centre, 45*(1), 1–27.
23. Malviya, N., Jain, S., & Malviya, S. (2010). Antidiabetic potential of medicinal plants. *Acta Poloniae Pharmaceutica, 67*(2), 113–118.
24. Bever, B. O. (1980). Oral hypoglycaemic plants in West Africa. *Journal of Ethnopharmacology, 2*(2), 119–127.
25. Urooj, A. (2010). Nutrient profile and antioxidant components of Costus speciosus Sm. and Costus igneus Nak. Indian Journal of Natural Products and Resources 11:116-118
26. Van Wyk, B.-E., & Wink, M. (2018). *Medicinal plants of the world*. Centre for Agriculture and Bioscience International.
27. Jayaweera, D. (1982). *Medicinal plants* (p. 151). National Science Council of Sri Lanka.
28. Bavarva, J. H., & Narasimhacharya, A. (2008). Antihyperglycemic and hypolipidemic effects of Costus speciosus in alloxan induced diabetic rats. *Phytotherapy Research: An International Journal Devoted to Pharmacological and Toxicological Evaluation of Natural Product Derivatives, 22*(5), 620–626.

29. Waisundara, V., Watawana, M., & Jayawardena, N. (2015). Costus speciosus and Coccinia grandis: Traditional medicinal remedies for diabetes. *South African Journal of Botany, 98*, 1–5.
30. Muniyandi, S., Nandanan, A., Veeti, S., Narayanan, A., & Ganesan, B. (2013). Studies on Costus speciosus Koen alcoholic extract for larvicidal activity. *International Journal of Pharmacognosy and Phytochemical Research, 5*(4), 328–329.
31. Dubey, S., Vijendra, K., Amit, K., Amit, K., & Tiwari, A. (2010). Evaluation of diuretic activity of aqueous and alcoholic rhizomes extracts of Costus speciosus Linn in wister Albino rats. *International Journal of Research in Ayurveda and Pharmacy (IJRAP), 1*(2), 648–652.
32. Rajesh, M., Harish, M., Sathyaprakash, R., Shetty, A. R., & Shivananda, T. (2009). Antihyperglycemic activity of the various extracts of Costus speciosus rhizomes. *Journal of natural remedies, 9*(2), 235–241.
33. Rastogi, R., & Mehrotra, B. (2004). *Compendium of medicinal plants* (p. 406). Central Drug Research Institute (CDRI)/National Institute of Communication and Information Resources.
34. Qiao, C.-f., Li, Q.-W., Dong, H., Xu, L.-S., & Wang, Z.-t. (2002). Studies on chemical constituents of two plants from Costus. *Zhongguo Zhong yao za zhi = Zhongguo zhongyao zazhi = China Journal of Chinese Materia Medica, 27*(2), 123–125.
35. Yan, C., You-Mei, T., Su-Lan, Y., Yu-Wei, H., Jun-Ping, K., Bao-Lin, L., et al. (2015). Advances in the pharmacological activities and mechanisms of diosgenin. *Chinese Journal of Natural Medicines, 13*(8), 578–587.
36. Eliza, J., Daisy, P., Ignacimuthu, S., & Duraipandiyan, V. (2009). Antidiabetic and antilipidemic effect of eremanthin from Costus speciosus (Koen.) Sm., in STZ-induced diabetic rats. *Chemico-Biological Interactions, 182*(1), 67–72.
37. Melchias, G. (2001). *Biodiversity and conservation*. Science.
38. Shobana, S., & Naidu, K. A. (2000). Antioxidant activity of selected Indian spices. *Prostaglandins, Leukotrienes and Essential Fatty Acids (PLEFA), 62*(2), 107–110.
39. Gupta, A. K. (2003). Quality standards of Indian medicinal plants.
40. Bown, D. (1995). *Encyclopaedia of herbs and their uses* (International Standard Book Number, 7513) (pp. 20–31). Dorling Kindersley.
41. Chopra, R., Nayar, S., & Chopra, I. (2006). *Glossary of Indian medicinal plants, National Institute of Science Communication and Information Resources* (p. 203). Council of Scientific & Industrial Research.
42. Sivarajan, V., & Balachandran, I. (1994). *Ayurvedic drugs and their plant sources* (pp. 374–376). Oxford and IBH Publishing Co. Pvt. Ltd.
43. Bhuyan, B., & Zaman, K. (2008). Evaluation of hepatoprotactive activity of rhizomes of Costus speciosus (J Konig) Smith. *Pharmacology, 3*, 119–126.
44. Khanna, P., Sharma, G., Rathore, A., & Manot, S. (1977). Effect of cholesterol on in vitro suspension tissue cultures of Costus speciosus (Koen) Sm., Dioscorea floribunda Mart. & Gal., Solanum aviculare Forst. & Solanum xanthocarpum Schard & Wendl. *Indian Journal of Experimental Biology*.
45. Rathore, A. K., & Khanna, P. (1978). Production of Diosgenin from Costus speciosus (Koen) Sm., and Solanum Nigrum L., suspension cultures. *Current Science, 47*(22), 870–871.
46. Rastogi, R. P., & Mehrotra, B. (1990). *Compendium of Indian medicinal plants*. Central Drug Research Institute.
47. Nadkarni, K. (2009). Indian materia medica, reprinted. *Bombay Popular Prakashan, 1*, 21.
48. Gupta, R. (2010). *Medicinal and aromatic plants: Traditional and commercial uses agrotechniques biodiversity conservation*. CBS Publishers & Distributors.
49. Malabadi, R. B. (2005). Antibacterial activity in the rhizome extract of Costus speciosus (Koen.). *Journal of Phytological Research, 18*(1), 83–85.
50. Soladoye, M., & Oyesika, O. (2008). *A textbook of medicinal plants from Nigeria*. University of Lagos Press.
51. Omukhua, G., & Godwin-Egein, M. (2011). Root rot disease of five fruit tree seedlings in the nursery. *Journal of Agriculture and Social Research (JASR), 11*(1).
52. Panagal, M., Kumar, R. A., Bastin, T. J., Jenifer, S., & Muthuvel, A. (2010). Comparative evaluation of extracts of C. igneus (or C. pictus) for hypoglycemic and hypolipidemic activity in alloxan diabetic rats. *International Journal of Pharmacy and Technology, 2*(1), 183–195.

53. Jayasri, M., MATHEWS, L. & Radha, A. (2009). A report on the antioxidant activity of leaves and rhizomes of Costus pictms.
54. Butola, J. S., & Samant, S. S. (2010). Saussurea species in Indian Himalayan region: Diversity, distribution and indigenous uses. *International Journal of Plant Biology, 1*(1), e9.
55. El-Far, A. H., Shaheen, H. M., Alsenosy, A. W., El-Sayed, Y. S., Jaouni, S. K., & Mousa, S. A. (2018). Costus speciosus: Traditional uses, phytochemistry, and therapeutic potentials. *Pharmacognosy Reviews, 12*, 120–127.
56. Rezaeian, S. (2011). Assessment of diosgenin production by Trigonella foenum-graecum L. in vitro condition. *American Journal of Plant Physiology, 6*(5), 261–268.
57. Lepage, C., Liagre, B., Cook-Moreau, J., Pinon, A., & Beneytout, J.-L. (2010). Cyclooxygenase-2 and 5-lipoxygenase pathways in diosgenin-induced apoptosis in HT-29 and HCT-116 colon cancer cells. *International Journal of Oncology, 36*(5), 1183–1191.
58. Attele, A. S., Wu, J. A., & Yuan, C.-S. (1999). Ginseng pharmacology: Multiple constituents and multiple actions. *Biochemical Pharmacology, 58*(11), 1685–1693.
59. Sautour, M., Mitaine-Offer, A.-C., Miyamoto, T., Dongmo, A., & Lacaille-Dubois, M.-A. (2004). Antifungal steroid saponins from Dioscorea cayenensis. *Planta Medica, 70*(01), 90–92.
60. Higdon, K., Scott, A., Tucci, M., Benghuzzi, H., Tsao, A., Puckett, A., et al. (2001). The use of estrogen, DHEA, and diosgenin in a sustained delivery setting as a novel treatment approach for osteoporosis in the ovariectomized adult rat model. *Biomedical Sciences Instrumentation, 37*, 281–286.
61. Chen, P.-S., Shih, Y.-W., Huang, H.-C., & Cheng, H.-W. (2011). Diosgenin, a steroidal saponin, inhibits migration and invasion of human prostate cancer PC-3 cells by reducing matrix metalloproteinases expression. *PLoS One, 6*(5), e20164.
62. Behera, K., Sahoo, S., & Prusti, A. (2010). Biochemical quantification of diosgenin and ascorbic acid from the tubers of different Dioscorea species found in Orissa. *Libyan Agriculture Research Center Journal International, 1*(2), 123–127.
63. McCullough, J. L., & Kelly, K. M. (2006). Prevention and treatment of skin aging. *Annals of the New York Academy of Sciences, 1067*(1), 323–331.
64. Tada, Y., Kanda, N., Haratake, A., Tobiishi, M., Uchiwa, H., & Watanabe, S. (2009). Novel effects of diosgenin on skin aging. *Steroids, 74*(6), 504–511.
65. Lee, J., Jung, K., Kim, Y. S., & Park, D. (2007). Diosgenin inhibits melanogenesis through the activation of phosphatidylinositol-3-kinase pathway (PI3K) signaling. *Life Sciences, 81*(3), 249–254.
66. Misso, M. L., Egberts, K. J., Page, M., O'Connor, D., & Shaw, J. (2010). Continuous subcutaneous insulin infusion (CSII) versus multiple insulin injections for type 1 diabetes mellitus. *Cochrane Database of Systematic Reviews, 1*.
67. Fealy, C. E., Nieuwoudt, S., Foucher, J. A., Scelsi, A. R., Malin, S. K., Pagadala, M., et al. (2018). Functional high-intensity exercise training ameliorates insulin resistance and cardiometabolic risk factors in type 2 diabetes. *Experimental Physiology, 103*(7), 985–994.
68. Castellanos, I. S., Jeremic, A., Cohen, J., & Zderic, V. (2017). Ultrasound stimulation of insulin release from pancreatic beta cells as a potential novel treatment for type 2 diabetes. *Ultrasound in Medicine & Biology, 43*(6), 1210–1222.
69. Huang, Q., Wang, L., Yu, H., & Ur-Rahman, K. (2019). Advances in phenylboronic acid-based closed-loop smart drug delivery system for diabetic therapy. *Journal of Controlled Release, 305*, 50–64.
70. Neuenschwander, M., Ballon, A., Weber, K. S., Norat, T., Aune, D., Schwingshackl, L., et al. (2019). Role of diet in type 2 diabetes incidence: Umbrella review of meta-analyses of prospective observational studies. *British Medical Journal, 366*.
71. Teng, H., & Chen, L. (2017). α-Glucosidase and α-amylase inhibitors from seed oil: A review of liposoluble substance to treat diabetes. *Critical Reviews in Food Science and Nutrition, 57*(16), 3438–3448.

72. Gan, Q., Wang, J., Hu, J., Lou, G., Xiong, H., Peng, C., et al. (2020). The role of diosgenin in diabetes and diabetic complications. *The Journal of Steroid Biochemistry and Molecular Biology, 198*, 105575.
73. Uemura, T., Goto, T., Kang, M. S., Mizoguchi, N., Hirai, S., Lee, J. Y., et al. (2011). Diosgenin, the main aglycon of fenugreek, inhibits LXRα activity in HepG2 cells and decreases plasma and hepatic triglycerides in obese diabetic mice. *The Journal of Nutrition, 141*(1), 17–23.
74. Sethi, G., Shanmugam, M. K., Warrier, S., Merarchi, M., Arfuso, F., Kumar, A. P., et al. (2018). Pro-apoptotic and anti-cancer properties of diosgenin: A comprehensive and critical review. *Nutrients, 10*(5), 645.
75. Lv, Y.-c., Yang, J., Yao, F., Xie, W., Tang, Y.-y., Ouyang, X.-p., et al. (2015). Diosgenin inhibits atherosclerosis via suppressing the MiR-19b-induced downregulation of ATP-binding cassette transporter A1. *Atherosclerosis, 24*(1), 80–89.
76. Tikhonova, M. A., Yu, C.-H., Kolosova, N. G., Gerlinskaya, L. A., Maslennikova, S. O., Yudina, A. V., et al. (2014). Comparison of behavioral and biochemical deficits in rats with hereditary defined or d-galactose-induced accelerated senescence: Evaluating the protective effects of diosgenin. *Pharmacology Biochemistry and Behavior, 120*, 7–16.
77. Kwon, C.-S., Sohn, H. Y., Kim, S. H., Kim, J. H., Son, K. H., Lee, J. S., et al. (2003). Anti-obesity effect of Dioscorea nipponica Makino with lipase-inhibitory activity in rodents. *Bioscience, Biotechnology, and Biochemistry, 67*(7), 1451–1456.
78. Siddiqui, M. A., Ali, Z., Chittiboyina, A. G., & Khan, I. A. (2018). Hepatoprotective effect of steroidal glycosides from Dioscorea villosa on hydrogen peroxide-induced hepatotoxicity in HepG2 cells. *Frontiers in Pharmacology, 9*, 797.
79. Lee, C. Y., Chou, Y. E., Hsin, M. C., Lin, C. W., Wang, P. H., Yang, S. F., et al. (2020). Dioscorea nipponica Makino suppresses TPA-induced migration and invasion through inhibition of matrix metalloproteinase-9 in human cervical cancer cells. *Environmental Toxicology, 35*(11), 1194–1201.
80. Xu, L.-N., Wei, Y.-L., & Peng, J.-Y. (2015). Advances in study of dioscin – A natural product. *Zhongguo Zhong yao za zhi = Zhongguo zhongyao zazhi = China Journal of Chinese Materia Medica, 40*(1), 36–41.
81. Bandopadhyay, S., Anand, U., Gadekar, V. S., Jha, N. K., Gupta, P. K., Behl, T., et al. (2021). Dioscin: A review on pharmacological properties and therapeutic values. *BioFactors*.
82. Balaji, M., Ganjayi, M. S., Kumar, G. E. H., Parim, B. N., Mopuri, R., & Dasari, S. (2016). A review on possible therapeutic targets to contain obesity: The role of phytochemicals. *Obesity Research & Clinical Practice, 10*(4), 363–380.
83. Fernandes, P., Cruz, A., Angelova, B., Pinheiro, H., & Cabral, J. (2003). Microbial conversion of steroid compounds: Recent developments. *Enzyme and Microbial Technology, 32*(6), 688–705.
84. Wang, Y., Zhang, Y., Zhu, Z., Zhu, S., Li, Y., Li, M., et al. (2007). Exploration of the correlation between the structure, hemolytic activity, and cytotoxicity of steroid saponins. *Bioorganic & Medicinal Chemistry, 15*(7), 2528–2532.
85. He, Z., Tian, Y., Zhang, X., Bing, B., Zhang, L., Wang, H., et al. (2012). Anti-tumour and immunomodulating activities of diosgenin, a naturally occurring steroidal saponin. *Natural Product Research, 26*(23), 2243–2246.
86. Christ, B., Xu, C., Xu, M., Li, F.-S., Wada, N., Mitchell, A. J., Han, X.-L., et al. (2019). Repeated evolution of cytochrome P450-mediated spiroketal steroid biosynthesis in plants. *Nature Communications, 10*(1), 1–11.
87. Sonawane, P. D., Pollier, J., Panda, S., Szymanski, J., Massalha, H., Yona, M., et al. (2016). Plant cholesterol biosynthetic pathway overlaps with phytosterol metabolism. *Nature Plants, 3*(1), 1–13.
88. Yang, Y., Laval, S., & Yu, B. (2014). Chemical synthesis of saponins. *Advances in Carbohydrate Chemistry and Biochemistry, 71*, 137–226.

Chapter 9
Coleus

Sana Aslam, Marriam Shahid, Matloob Ahmad, Syed Ali Raza Naqvi, and Arwa A. AL-Huqail

9.1 Introduction

Family	Lamiaceae
Scientific name	*Coleus amboinicus*
English/Common name	Coleus

9.2 Plant Description

De Loureiro named the genus 'Coleus,' which means 'sheath enclosed around style,' in 1970, and named the species 'forskohlii,' after prominent Swedish botanist Forsskhl [1]. The Sanskrit names 'Mayani' and 'Makandi' were mentioned in the Ayurvedic books [2]. Coleus is a fragrant perennial with upright stems and tuber-like roots that may grow up to 60 centimeter tall [3]. Coleus from the mint family that thrives in India's subtropical temperate regions, including Thailand, Sri Lanka, and Nepal. Coleus is a 1 to 2 feet tall plant with teardrop-shaped leaves that are

S. Aslam (✉) · M. Shahid
Department of Chemistry, Government College Women University, Faisalabad, Pakistan
e-mail: dr.sana@gcwuf.edu.pk

M. Ahmad · S. A. R. Naqvi
Department of Chemistry, Government College University, Faisalabad, Pakistan

A. A. AL-Huqail
Department of Biology, College of Science, Princess Nourah bint Abdulrahman University, Riyadh, Saudi Arabia

M. Zia-Ul-Haq et al. (eds.), *Essentials of Medicinal and Aromatic Crops*,
https://doi.org/10.1007/978-3-031-35403-8_9

dazzling green with a vibrant purple core; leaf colour varies depending on the quantity of shade. The flowers are light purple or blue in hue [4].

The leaves are thick, simple, wide, egg- or oval-shaped, have a tapering tip and measure 5–7 cm (2.0–2.8 in.) by 4–6 cm (1.6–2.4 in.) (ovate). Except at the base, the borders are coarsely crenate to dentate-crenate. They are pubescent (hairy), with the glandular hairs on the bottom surface being the most abundant, giving them an icy look. The petiole is 2 to 4.5 cm (0.79–1.77 in.) in length. The leaves have strong fragrant substances, that is a mixture of turpentine, oregano, as well as thyme. The leaves have a taste that is comparable to oregano, but with a strong minty flavor [5].

The pseudo-spicate inflorescence, which does not fit into coleus or the closely related Plectranthus, is of particular interest to taxonomists. Nonetheless, the plant is a coleus species and must be identified as *C. amboinicus Lour* [6]. It was reported that "the scarcely declinate fruiting calyx, a long, broad, uppermost lobe with 4 short, very narrow lower lobes, stamens with filaments united to beyond the middle into a long tube split on the upper-side, and compact cymes, makes *C. amboinicus* totally different from all species of the genus Coleus and from all species of Pleciranthus" [7].

9.3 Agronomy of Plant

9.3.1 Soil Conditions

Coleus requires wet, rich, loose soil that is frequently damp. Compost or another organic material should be added to the soil before planting. Any better-quality peat-based potting mix will suit for potted plants. The loose texture of potting soil is ideal for container-grown coleus, and it's always a great idea to start with a good mix with a pH range of 6.0 to 7.0 [4]. Make sure the container has drainage so the soil doesn't stay moist all the time, which can lead to root rot. It is commonly grown in several places of India, up to an elevation of 2400 m. Because it does not require extremely fertile soils, it may be cultivated at a lower cost in soils with borderline fertility [8].

9.3.2 Climatic Conditions

Temperature is directly related to photosynthetic rate, it play a significant part in determining the level of photosynthesis as well as respiration in in vitro cells [9]. As a tropical perennial plant, coleus can withstand a variety of temperatures. Temperatures between 65°–75 °F (24°–27 °C) are ideal for coleus plants. All in vitro studies on C. forskohlii were conducted at temperatures between 22 and 25 °C [10, 11]. The easiest way to care for them is to minimize unexpected temperature fluctuations and protect them from frost. Make sure that the tropical plants aren't exposed to the air. Near an open window, an air conditioning vent, or a draughty door are all bad choices. Direct heat is also harmful to plant. As a consequence, keep indoor plants away from radiators, furnaces, and other heat sources. If the plant is sheltered from direct sunshine, a sunny windowsill can give sufficient light. In the winter, though, the cold from the window at night may stress the plant.

Moisture content is also important, and its atmospheric range must be regulated since excessive amounts promote impurity and hyperhydricity in cultures [9].

9.4 Propagation of Coleus Plant

There are two methods of propagation of Coleus plant that is either by seed or by cutting roots. Propagation via seed is simple since seeds are widely available and even low-cost. Seeds are scattered on the ground and pushed down for germination; avoiding the covering of seeds since they require light to sprout. Seeds are maintained wet by growing them in a bowl and covering them with plastic, or by misting them on a regular basis. Sprouts can begin to exhibit colour after roughly 2 weeks [12].

Taking stem cuttings and rooting them is a simple way to reproduce coleus plants. Use a sharp shearing scissor to cut the stem tip to a length of 4–6 in., and then remove all leaves from the lower half of the cutting. Plant the stem by dipping the tips in a substance (such as rooting hormone compound), then place it in a wet potting mix with the exposed leaf nodes covered in soil. Place the vessel in a plastic bag, ensuring that the plastic does not come into contact with the cutting. Stem trimmings have been proven to be suitable for large proliferation of the plant. In most cases, 10–12 cm long stem clippings with 3–4 pairs of leaves are potted in well-prepared seedbed. The sprouts should be adequately cared for and watered on a regular basis [8]. Arrange the covered cutting under the sun for 2–3 weeks, or until new roots appear. Then expose the young plant in a bright light under the sun for further growth. Numerous odd varieties may be hesitant to root, so take a lot of cuttings to guarantee that to produce enough healthy plant.

9.5 Sexual and Asexual Propagation

Coleus seeds must be exposed to light in order to germinate. They must be seeded at shallow depths (0.25–0.5 cm) or immediately on the surface of a wet, sterilized medium (pH = 6.5) such as peat-lite or fine sphagnum moss to guarantee optimal penetration of light. The air temperature should be kept between 21 and 27° C for the first week after sowing, then reduced to 18° C. The soil temperature must be kept above 20° C throughout the germination phase (10–20 days) [13].

Coleus is propagated asexually through stems and shoots cuttings. Maximum roots will occur with 6 cm shoot cuttings with two pairs of leaves obtained from quickly expanding shoot tips. Hormone therapies are seldom required. Cuttings should be rooted in a wet, warm (20° C) medium [14]. After 3–4 weeks, cuttings may normally be replanted. Transplants may be cultivated at low temperatures (15–18° C), although they mature quicker above 21° C.

9.6 Planting

Because coleus seed is so minute, it has to be planted in well-drained soil, in boxes or other containers that are protected from direct sunlight and slicing rain. Use a planting media similar to the one that is utilized for growing and caring for African violets as a seed starting mix or planting medium for growing coleus seed, or use light fertile soil, with subsoil filtered through a loop of screen wire so it is reasonably small. Plant in 3 or 4-inch pots or use any vessel/container with adequate effluents. Seedlings should have plenty of room in their containers so that their stems don't seek towards the strong light. After the soil has been fully wetted and smoothed, scatter the seed delicately on top of the damp topsoil and clean fine dirt over them to obscure the seed. Mix the seed with a little fine or clean peatmoss, dry sand and scatter it on the top of seedbed for a simple planting approach. Gently firm the seedbed with the palm of the hand, instead of covering the seed. Arrange the jar in lukewarm water to allow humidity to drain from the end without disturbing the seed. When the water reaches the top of the soil, remove the pot from it and cover it with the glass piece or cellophane paper to keep the moisture in. To ensure germination, put the vessel in partially shaded area under the sun. Keep the soil wet at all times. Coleus seeds will not germinate if they are allowed to dry out or if soil temperatures fall below 60 °F [15].

9.7 Manuring

Coleus plants are commonly grown for their pink, red, yellow, or orange foliage. Use a diluted fertilizer that is high in nitrogen and low in phosphorus to produce vibrant colors. Balanced fertilizers should be avoided since they induce blooming, which depletes the plant's nutrition. Organic and inorganic fertilizers work nicely with the crop. On the 30th and 45th day after planting, 140 kg of organic manure is necessary. For the best fresh (120 t/ha) and dried (3.982 t/ha) tuber yield, a combination of 40 kg, 60 kg, and 50 kg of nitrogen, P_2O_5 and K_2O, respectively, per ha is recommended. As a base dosage, half the nitrogen, entire phosphorus, and whole potassium should be administered, followed by the addition of remaining half of N fertilizer 30 days later as a top dressing [16].

9.8 Irrigation

Coleus is a thirsty plant that has to be watered on a regular basis to maintain the soil moisture but not wet. Water anytime to the top 1″ (2.5 cm) level of the potting mix is dry throughout the growth season (April to September). The frequency of planting water is determined by air temperature, container type, and plant development.

Watering coleus is still necessary throughout the winter, although less regularly. Water only after the top half of the soil has dried out, as a general rule. Use room temperature water to water coleus. Fill the pot with enough water to allow it to drain out the bottom. Return the container to its saucer only when the water has stopped leaking [15].

9.9 Ecology

Coleus is known as rapidly-growing plant with a trailing or spreading habit and the capacity to embed where its branches come into contact with the base as they extend. This enables them to rapidly acquire favourable conditions and relocate to more desirable sites. Because the broken-off branches may root and grow into self-sustaining plants, they can tolerate being trampled and/or grazed by numerous animals. They grow 30–1400 m tall and live in Acacia and Commiphora woodlands, forested grassland, and dry rocky areas [15].

9.10 Pests and Diseases

Spider mites, whiteflies, aphids, as well as mealybugs are some of the pests that can harm coleus. In addition, root and stem rot, along with the illness carried by parasitic worms, can affect coleus plants. One of the biggest hindrances to coleus productivity is a root-knot nematode (RKN), Meloidogyne incognita [17]. Coleus spp. has been linked to 15 insect species and three crustaceans forms throughout the world [18]. Another significant pest of coleus is the mealybug. The mealybug nymphs along with the adults may eat the fragile leaves as well as shoots. In heavy infestations, mealybugs formed a thick mat of waxy excretions on the underside of leaves.. Plants that have been affected have a sickly look. Honeydew secretions from mealybugs are a common source of sooty mould formation.

Coleus (*Plectranthus scutellarioides*), a herbaceous bedding plant favoured by gardeners for its vibrant multi-coloured leaf since Victorian times, had a resurgence in popularity in recent years. Coleus is susceptible to a variety of diseases, including root rot, leaf blight, leaf spots, wilt, and RKN (root-knot nematodes); however, root rot, wilt, and nematodes are the most damaging, causing producers to lose up to 50% of their crop. It was reported that leaf spots were produced by Cercospora cassicola and Botryodiplodia theobromae [19]. Wilt-affected plants exhibit yellow stains on the leaves resulting in the withering of leaves, eventually leading to plant death. Discoloration and degradation of tap and lateral roots are seen on the damaged plants [20].

Downy mildew is the most dangerous disease for coleus. In 2005, downy mildew on coleus was first reported in the United States, it was determined to be a Peronospora species based on morphological characteristics. Depending on the pest, pesticides including azadiractin (neem), imidacloprid, endosulfan, dicofol, abamectin, horticultural oil, malathion or insecticidal soap are indicated for controlling infestations. Methods of biological controls, such as introducing ladybugs, wasps and predatory to the garden, may also be useful [21].

9.11 Distribution of Coleus Plant

Coleus amboinicus grows in woods and yellow bush, on rocky gradients and loamy or sandy areas at low heights, and spread in the areas ranging from South Africa (KwaZulu-Natal) and Eswatini through Angola and Mozambique, and north to Kenya and Tanzania [22, 23]. It would have been transported from Southern Africa to India, Arabia, and Southeast Asia via the Indian Ocean maritime trade routes by Arabs and other traders. Currently, the plant may be found growing on the Indian mainland. *Coleus amboinicus* plant was eventually transported to Europe or ultimately to the Americas, earning it the name Spanish thyme [22, 24].

C. forskohlii is widely distributed throughout the world's tropical and subtropical zones. This plant's origins may be traced back to India [25]. China, India, Nepal, Sri Lanka and Myanmar are among the nations in the Indian subcontinent where this medicinal plant is widely dispersed, as are Brazil, Thailand, Egypt and Ethiopia [4]. Its cultivation is concentrated in Garhwal as well as Himachal Pradesh's Kumaon regions, the Western Ghats and Bihar's Parasnath highlands [26]**.** It is said to be dispersed in arid hills of Western Uttarpradesh, sections of Orissa, Western ghats, Gujarat, Tamil Nadu [27] and kitchen gardens in Northern Karnataka (Belgaumand Dharward districts) for its carrot-like tubers, which are used as spices in pickle preparation [28].

9.12 Important Phytochemical Constituents of Coleus

Volatile oils and diterpenes are major components of coleus, but forskolin is the most important. Coleus'major ingredient is the diterpene forskolin, which is extracted from the plant's root. The active ingredient "forskolin"is crystalline in form, has a molecular weight of 410.5 g/mole, a melting point of 228–230° C, and exhibits maximum absorption wavelengths (λmax) of 210 and 305 nanometers. It is soluble in ethanol, dichloromethane, and methanol. Additionally, it can be dissolved

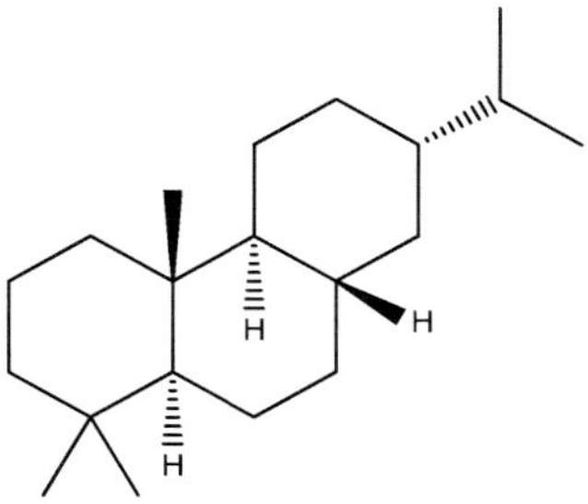

Abietanes

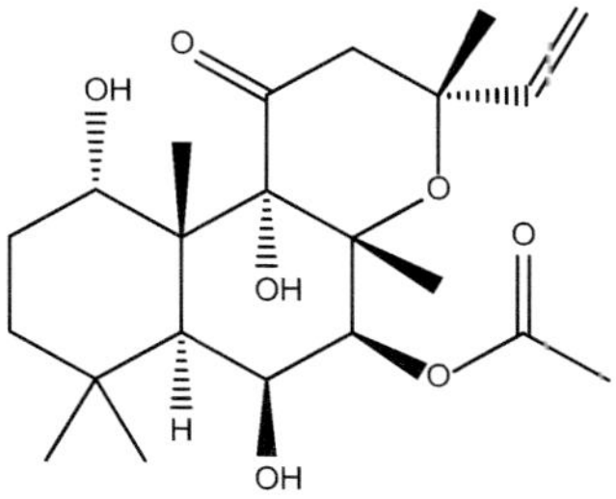

7b-Acetoxy-8,13-epoxy-1a,6b,9a-trihydroxy- labd-14-ene-11-one

(Coleonol/Colforsin)

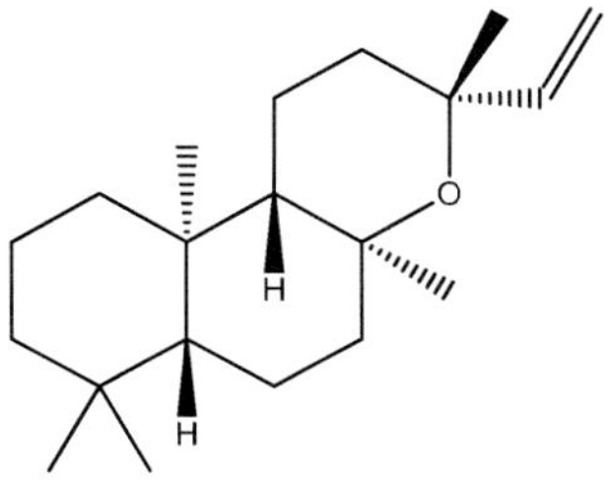

(3R,4aS,6aR,10aR,10bS)-3-ethenyl-3,4a,7,7,10a-pentamethyl-2,5,6,6a,8,9,10,10b-octahydro-1H-benzo[f]chromene

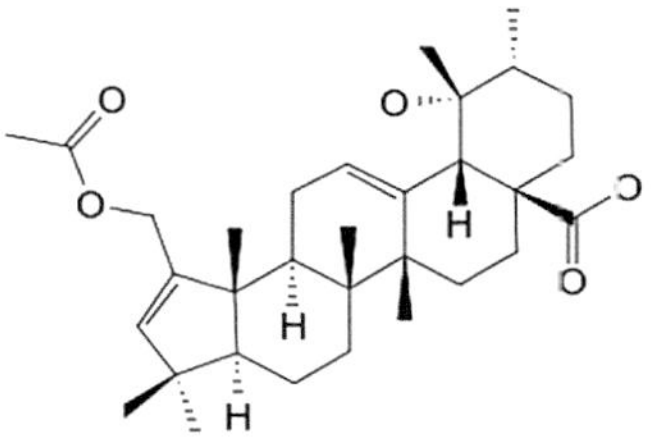

Coleonic Acid

Fig. 9.1 Structure of few potent phytochemicals of Coleus

in water with 2% volume/volume ethyl alcohol [29]. Different diterpenoids isolated from C. forskohlii are classified as follows: (1) abietane diterpenoids; (2) 8, 13-epoxy-labd-14-en-11-one-diterpenoids; (3) 8, 13-epoxy-labd-diterpenoids (Colenol, 3-hydroxy forskolin, Coleonone); (4) miscellaneous labdane diterpenoids (Forskoditerpene A, Coleonic acid, Colenolic acid), (5) 8, 13-epoxy-labd-14-en-11-onediterpene glycosides (Fig. 9.1). Moreover, the researchers identify several phytochemicals that are present in low amounts and extracted from various regions of the plant [15]. However, the other component 1, 9 dideoxyforskolin is present in comparable amounts or even greater quantities in the roots than forskolin [30]. Diterpenoids, coleonols, and volatile oils are other plant components. Coleus has roughly 20 components in several areas of the plant, although the root contains coleonols and forskolin [4].

Coleus leaf extract contains a considerable quantity of polyphenols, flavonols, and flavones, as well as a high level of antioxidant potential. High-performance liquid chromatography (HPLC) analysis of leaf and stem tissues reveals the existence of both conventional as well as extra strong antioxidative polyphenols, indicating that the coleus can be exploited as a basis of phenolic compounds with high antioxidant activity. Tannins may also be found in the coleus plant's leaves and stem

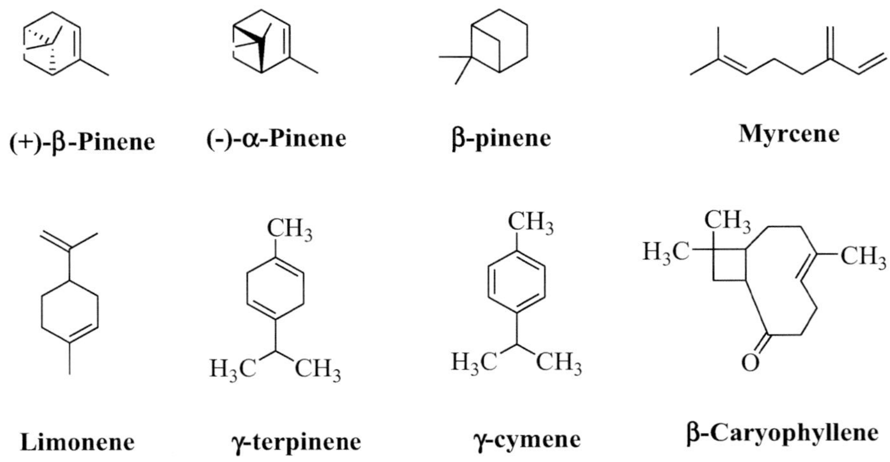

Fig. 9.2 Essential oil components found in Coleus plant

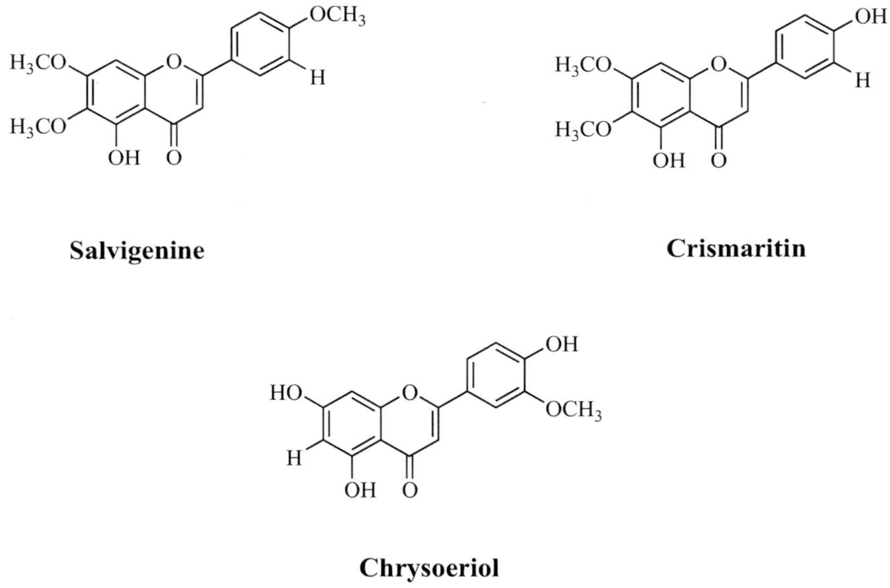

Fig. 9.3 Flavones isolated from *Coleus amboinicus*

[31]. The following essential oil compounds were identified in an earlier investigation on coleus aromaticus: α-pinene, beta-pinene, myrcene, limonene, γ-terpinene, γ-cymene, and beta-caryophylene (Fig. 9.2).

Ethyl salicylate, β-selinene phenolic fractions of thymol, carvacrol, eugenol, were also reported from an unidentified phenolic fraction [32]. Saligenine, cirstmaritin, and chrysoeriol were extracted from *Coleus amboinicus* leaves and their structures were also elucidated (Fig. 9.3) [33].

9.13 Origin and Uses of Coleus

Coleus plant is endemic to tropical Southeast Asia, India, Africa, and Australia, with Indonesia and Sri Lanka having the highest populations. By means of traders and botanists, coleus made its way into Europe and, subsequently, America. While some sources locate coleus in Europe in the seventeenth century, it is typically ascribed to Dutch botanist Karl Ludwig Blume, who named and introduced coleus to Europe about 200 years later as part of an enormous collection of plants, he researched while residing on the Indonesian island of Java.

Many coleus species are utilized as remedies to cure a variety of diseases, including skin, stomach, and chest issues, and some are also used as food, flavouring agents, or stock animal fodder. *C. amboinicus* and *C. barbatus* are the most commonly utilized for medicinal and flavouring purposes, whereas *C. esculentus*, also known as African potato/veld potato, is a well-known and extensively farmed food plant whose tuberous roots are rich in starch and are cooked and eaten like potatoes. Dietary *C. amboinicus L.* also used to improve the quality of meat and composition of fatty acid in *longissimus thoracis* muscles of lamb (Fig. 9.4) [15]

9.14 Mechanism of Action of *Coleus forskolin*

Forskolin is the main chemical present in the tuber, and its herbal formulations have an array of therapeutic effects. By raising the level of captivity, forskolin in turn prevents mast cell and basophil degranulation and histamine release. Blood pressure is lowered [35], and intraocular pressure prevents platelet aggregation, increases thyroid hormone production, bronchodilatation, vasodilatation, and lipid metabolism in fat cells [36]. Nearly all hormone-sensitive adenylate cyclases are

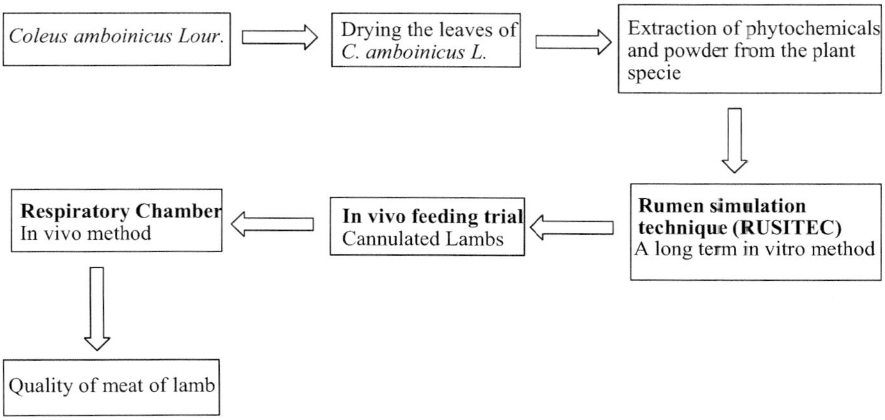

Fig. 9.4 Use of Coleus plant species (C. amboinicus) in enhancing the quality of meat [34]

immediately activated by forskolin in intact cells and tissues, as well as in a solubilized adenylate cyclase preparation. This impact of *C. forskohlii* root infusions on blood pressure has been documented. All human adenylate cyclase types, with the exception of type IX found in spermatozoa, may be activated by forskolin [37]. Nearly all adenylatecase-sensitive hormones are directly activated by forskolin in intact cells, tissues, and even a soluble form of adenylatecyclase. The novelty of this activation is that the catalytic subunit of the enzyme or a protein that is closely related to it is the active site for forskolin, which in turn raises the level of cAMP and prevents basophil and mast cell degeneration [37, 38].

9.15 Medicinal Uses of Coleus

Since the start of civilization, medicinal herbs have been used to identify numerous human ailments . Natural compounds extracted from medicinal plants, whether as pure substances or standardized extracts, present a limitless prospect for current medication discovery because to the unrivalled availability of biodynamic ingredients. [39].

The tuberous *coleus forskohlii* is the most important medicinal coleus species in India. Other species include *C. amboinicus, C. scutellaroides, C. blumei, and C. malabaricus,* which are applied for the diagnoses diarrhoea as well as digestive diseases [40]. *P. forskohlii* is commonly used for a variety of disorders in many nations. The leaf of this plant is effective as an emmenagogue, diuretic, and expectorant in Egypt and Africa. It is used as a herbal medicine for stomach problems and to treat intestinal diseases in Brazil [25].

Because of diverse bioactivity and therapeutic characteristics of *C. aromaticus,* is frequently used as an herbal medicine. The leaves as well as its juice are used for the treatment of common cold, breathing issues, bladder stones, bilious infections, kidney indigestion, bronchitis, cholera, colic, fits, cough, diarrhoea, dyspepsia, dysentery, epilepsy, renal and vesicle calculi, headache, fever, poisonous bites, and vitiated conditions of Kapha and Vata [41]. *C. aromaticus* is known as Paashanbhedi in traditional medicinal dictionary because it is used to evacuate kidneys [42].

Infections, rheumatism, and flatulence as well as, coughs, sore throats, and nasal congestion, hepatopathy, malarial infection, hiccoughs, bronchitis, helminthiasis, and epilepsy, have also been historically treated using *Coleus amboinicus* leaves [43–45].

9.16 Medicinal Uses of Forskolin

Recent scientific researchers have revealed that coleus includes a powerful chemical called forskolin, which has muscle relaxant qualities and the potential to widen blood vessels, lowering blood pressure (anti-hypertensive property). When blood

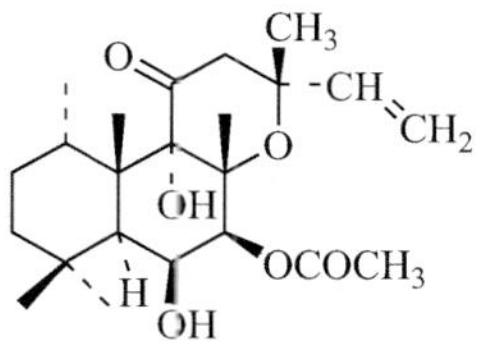

Fig. 9.5 Structure of forskolin

vessels are dilated, the heart needs to work less to pump blood around the body. Coleus has been shown to be beneficial to people with cardiomyopathy (Fig. 9.5).

C. forskohlii's morphology, phytochemical analysis, and pharmacology have all been studied. Forskolin, a labdane diterpene derived from *Coleus forskohlii* [46, 47], has been shown to stimulate adenyl cyclases and raise cyclic adenosine monophosphate (cAMP) [48].

Besides increasing stimulating activity of cAMP, forskolin prevents platelet-activating factor (PAF) binding independently of cAMP production [36]. Forskolin to decrease glucose transfer in erythrocytes, adipocytes, platelets, and other cells through affecting multiple membrane transport proteins [49]. Forskolin also has cyclic AMP independent actions via nicotinic acetyl choline receptor channel modulation, desensitization, voltage dependent potassium channel modulation, and drug resistance reversal [50]. Forskolin and Coleus forskohlii's safety have not been sufficiently studied. People with ulcers should stay away from it because it might raise their stomach acid [51].

Forskolin has been shown to be beneficial in a variety of disorders, including hypertension [52], glaucoma [53], asthma [54], cancer [55], diabetes [56, 57] and increased the rate of sensory nerve regeneration in sciatic nerves that had been frozen [58]. Additionally, its leaves are used as a flavoring agent and to treat digestive problems. Furthermore, the antioxidant properties of distinct sections of the plant has not been recorded previously, prompting the investigation of the antioxidant capacity of C. forskohlii's various portions [59].

9.17 Role of *Coleus froskolin* in Hypertension

The most prevalent psychosomatic condition, hypertension, affects 972 million individuals globally. Researchers discovered the diterpene, Forskolin, which has blood pressure lowering and antispasmodic properties, from the roots in 1974, making it the only plant source known to have this chemical thus far.The therapeutic properties of this alkaloid known as "forskolin"are well known due to its exceptional capacity to activate the enzyme Adenylate cyclase in the absence of a functional guanine nucleotide regulatory protein. It is known to possess hypotensive, positive inotrophic, anti-inflammatory, antispasmodic, smooth muscle relaxant, with anti-platelet aggregation, vasodilator, anti-metastatic properties (Fig. 9.6) [60].

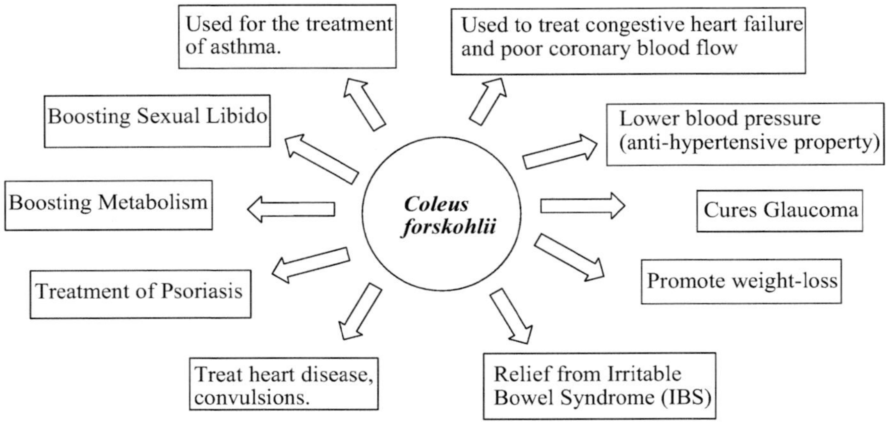

Fig. 9.6 Medicinal uses of *Coleus forskohlii*

9.18 Role of *Coleus froskolin* in Psoriasis

A common skin ailment known as Psoriasis, in which the skin thickens and becomes covered with silvery scales. Psoriasis is a widespread condition that affects millions of individuals worldwide. Coleus, an Ayurvedic herb, extracted from *Coleus Forskolin* has long been used as one of numerous natural psoriasis treatments. Coleus has been demonstrated to be helpful in the treatment of skin disorders like psoriasis because of its ability to promote regular cell formation. Herbalists utilize it to cure not just psoriasis, but also eczema and cancer [61]. Psoriasis causes cells to divide 1000 times quicker than normal. Coleus alleviates psoriasis by restoring the cAMP/cGMP ratio to baseline. Patients with psoriasis have been found to benefit from forskolin's ability to alter cAMP levels in skin cells [62].

9.19 Role of *Coleus froskolin* in Anti-obesity

To check the anti-obesity potential, *Coleus forskohlii* were examined in the overreacted betrayer and *Coleus forskohlii* extract to reduced body weight, food absorption, and growth in those rats, suggesting that *Coleus Forskohlii* may be beneficial in determining appropriateness [63]. *Coleus* Forskohlii does not appear to increase weight reduction, although it may assist obese females lose weight without any clinically relevant side effects [64, 65]. Forskolin activates adenylate cyclase, which is responsible for the synthesis of cAMP (a crucial biochemical compound in metabolic processes). cAMP causes biochemical reactions that accelerate fat loss, enhance lean body mass, activate metabolic processes and diet-induced thermogenesis [66].

9.20 Role of *Coleus froskolin* in Cancer Metastasis

Forskolin is being researched as a possible therapy for concrete.Both in vitro and in vivo platelet aggregation is a common side effect of many metastasizing cancer cell lines. When platelets clump together, they release substances that encourage the growth of tumours. Researchers discovered that forskolin can prevent platelet aggregation by enhancing intracellular Camp and activating adenylate cyclase.. Forskolin inhibited lung tumour growth by 70% [67].

The roots of *C. forskohlii* are high in diterpenes [68] such as coleonol, forskolin, and barbatusin [69–72]. These diterpenes have been demonstrated to be successful in treating a number of ailments, including coronary heart disease, hypertension [73, 74], asthma, glaucoma [75], and Alzheimer's disease [76, 77]. Its other applications in boosting lean body mass, treating mental disorders [78], and as anticancer agents [55, 79].

9.21 Role of *Coleus froskolin* in Treating Glaucoma

Glaucoma is a disorder where the pressure inside the eye is abnormally high (>22 mm/Hg) due to an imbalance between the creation and drainage of aqueous humour from the eye. If left untreated, an increase in IOP causes irreparable nerve damage and reduced vision, eventually leading to blindness. There are no professionally validated alternative remedies for glaucoma, however there are a number of useful treatments, one of which is coleus. Forskolin lowers IOP by decreasing aqueous humour input while maintaining outflow capacity, indicating forskolin's potential as a glaucoma therapy medication [80]. Topical administration of forskolin decreased intraocular pressure in rabbits, monkeys, and healthy human volunteers. Additionally, it was connected to a reduction in aqueous inflow and no alteration to the outflow facility [81].

9.22 Role of *Coleus froskolin* in Treating Asthma

Because of the intricacy of its origin, bronchial asthma is challenging to treat [82]. Early asthma is characterised by an influx of inflammatory cells as a result of an imbalance in the ratio of Th1 and Th2 cells [83, 84]. The next stage involves reshaping of the airways as the inflammation worsens. Extracellular matrix (ECM) degradation can result in airway inflammation, whereas ECM buildup results in airway remodeling [85]. Tissue inhibitors of MMPs (TIMPs) are MMP inhibitors, whereas matrix metallopeptidases (MMPs) may break down the ECM [86].

Forskolin was investigated as a potential bronchodilator for the asthma treatment [87]. Blocked bronchospasm, the main feature of bronchitis and asthma is produced in guinea pigs by histamine and leukotriene C-4 [35]. Forskolin powder formulations for inhalation were shown to be capable of producing brochodilation in asthma sufferers, according to a human investigation [88]. If administered in the right dosage, forskolin appears to be a potential medication for treating people with asthma, as well as congestive heart failure [89].

9.23 Antithrombotic Effect of Coleus

By activating adenylate cyclase, forskolin reduces platelet aggregation while enhancing prostaglandin effects [90, 91]. It was noted in rabbits that its antithrombotic activities may be strengthened by cerebral vasodilation. Adenosine did not enhance this vasodilation [92]. Unprocessed C. forskohlii extract has been proposed as a powerful phytotherapeutic antithrombotic agent [93].

9.24 Other Uses of *Coleus froskolin*

Forskolin can also help to reduce hair greying, hair loss and it also helps in bringing grey hair back to its natural hue [94]. Though the coleus plant is well-known for its medical benefits, its flowers, tubers, and stems also contains essential oil, which has an appealing odour with a spicy flavour [95]. The coleus plant's essential oil is utilized in the food flavouring business. It's also used as an antibacterial agent [96].

Forskolin must be used primarily for cosmetic purposes to help those with fair skin tan while also shielding them from UV radiation in a similar way to those with matte skin .The medication was developed by a group of American researchers and is currently undergoing testing in labs. Skin cancer is more common in people with white skin because this colour lacks melanin, the pigment that gives skin a tan. Forskolin, an active ingredient produced from the plant Coleus Forskolii, is still being studied as a cream treatment. The pigment-melanin deficient mice with white leather were able to regain system function with this therapy [94].

9.25 Medicinal Uses of 6-(3-Dimethylaminopropionyl) Forskolin Hydrochloride, or NKH477

6-(3-Dimethylaminopropionyl) forskolin hydrochloride, or NKH477 is the most powerful water soluble derivative of forskolin to date [97–100]. NKH477 is orally active, causes various cAMP-dependent actions, and show more attraction towards adenylyl cyclase type V, the predominant isoform of the enzyme in the heart, which explains its cardiovascular effects. It's the first adenylyl cyclase activator approved for use in the post-operative treatment of heart surgery patients. After repair of a

Fig. 9.7 Structure of 6-(3-Dimethylaminopropionyl) forskolin hydrochloride, or NKH477

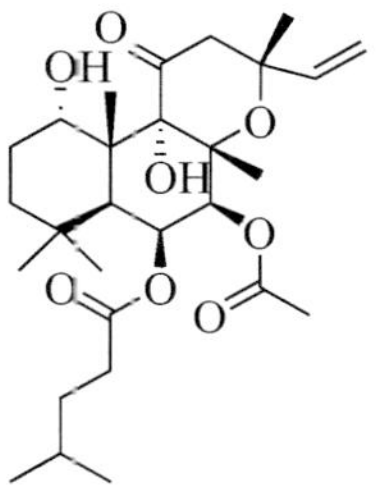

complicated congenital cardiac defect, a baby was effectively weaned off cardiopulmonary bypass with a continuous infusion of NKH477. This was made feasible by the ineffectiveness of more well-known inotropic drugs such as milrinone poor conjunction with epinephrine and isoproterenol. In Japan, NKH477 [Colforsin daropate HCl (Adehl® Inj.)] is available for heart failure treatment (Fig. 9.7) [101].

Psoriasis is a skin ailment that causes dry, rough, and dead skin to develop on the scalp, knees, groyne, and lower back. The specific reason is unknown, but experts believe a weakened immune system is one of the factors. *Coleus Forskohlii* is one of the several therapies advised for the skin disease because of its potential to improve the body's immune system.

Forskholin is an extract from the tuber of the *C. forskholi* (wild). Briq plant that has been shown to have antispasmodic and blood pressure reducing properties. It has been identified as an adenylate cyclase activator in cardiac membrane of rabbit. It works by avoiding the receptor and the G protein. The central impact of *C.forskholi* does not lower blood pressure; rather, it has a direct vasodilatory effect [102].

Despite the fact that endophytes have been demonstrated to perform a critical part in demonstrating the capability of the host plant throughout their connection, the cross-functional influence of one plant's endophytes on another plant is still poorly understood. Native endophytes of *Coleus forskohlii* (SF1), *Macrophomina pseudophaseolina* (SF2), and *Fusarium redolens* (RF1), extracted from the parts of stem as well as root effects the growth of plant and increasing the secondary metabolite in medicinal plant *Andrographis paniculata*, as well as aromatic plants Artemisia pallens and Pelargonium graveolens [103].

9.26 Coleonol (A Diterpene)

Numerous *Coleus spp.* have been characterized as remedies for painful micturition, cardiovascular problems, breathing problems, abdominal colic as well as few CNS (central nervous system) ailments including sleeplessness and seizures in old Hindu medicinal texts such as Bhau Prukash Nighantu [104] and Aurvedic materia medica. The bioactivity of a 50% ethanol extract of Coleus forskohlii, one of the Coleus species, was investigated. The findings were published in 1971 [105]. From this extract, a novel diterpene, coleonol, was identified. It was discovered to have

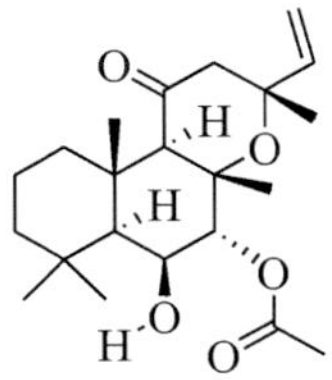

Fig. 9.8 Structure of Coleonol (A diterpene taken from Coleus forskohlii)

hypotensive and spasmolytic properties. Coleonol, a diterpene synthesized from Coleus forskohlii, was studied for its therapeutic characteristics (Fig. 9.8).

Its main potential is to relax the vascular smooth muscle, which lowers blood pressure in anaesthetized cats, rats, and rats that are naturally hypertensive. It also exhibits *in vivo* beneficial inotropic impact on separated rabbit hearts as well as cat hearts at low dosages. Coleonol has general spasmolytic action on smooth muscle of the gastrointestinal system in many species but not on guinea pig bronchial musculature. When taken in large doses, coleonol has a depressing impact on the central nervous system [52].

9.27 Biosynthesis of Forskolin

Forskolin is a labdane-type diterpenoid with the unique complex structure that acts as a cyclic AMP booster and is used for the remedy of glaucoma and heart failure.Forskolin occurs naturally in the Coleus forskohlii plant, where it collects in the root cork, and is used for commercial production. A cascade of events catalyzed by five cytochrome P450s (CfCYP76AH8, CfCYP76AH9, CfCYP76AH10, CfCYP76AH11, CfCYP76AH15) and two acetyltransferases (such as CfACT1–8 and CfACT1-6) converts the precursor of forskolin that is 13R-manoyl oxide into forskolin and other derivatives of labdane type diterpenoids. The conversion of 13R-manoyl oxide to forskolin is catalysed by at least three P450 enzymes working together with one acetyl transferase, according to transient expression in Nicotiana benthamiana (Scheme 9.1) [106].

9.28 Synthesis of Forskolin

Complete synthesis of forskolin was reported, started with reduction of compound **(1)** using excess amount of lithium aluminium hydride in the presence of diethyl ether at temperature ranging from −78 °C to 20 °C. The reaction mixture then heated from room temperature to refluxing temperature resulting in the formation of compound **(2)** (93%) in the presence of carbonyldimidazol in toluene. Then the reaction underwent further Clean desilylation and Swern oxidation to give aldehyde derivative **(3)** (79%). Adding propynyl lithium to the compound **(3)** was a tough reaction because of the two unreactive species and a simple retroaldol process. Other organometallic reagents

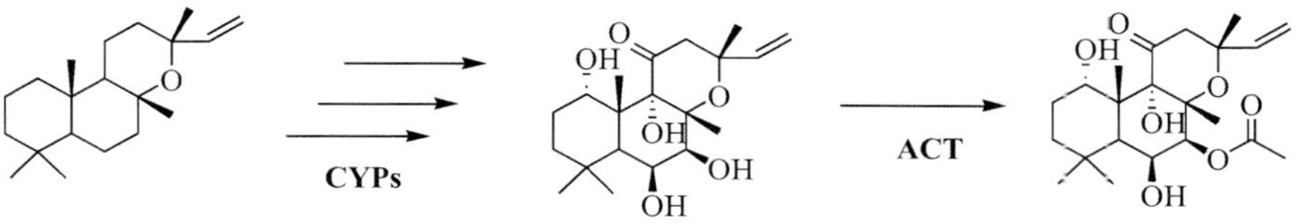

13R-manoyl oxide **Deacetyl-forskolin** **Forskolin**

Scheme 9.1 Biosynthesis of Forskolin

were ineffective with compound (**3**); retroaldolisation happened even with a premade propynyl lithium / $CeCI_3$ reagent that did not enolize $PhCH_2COCH_2Ph$, as it was tested before use. Thus, condensation of (3) at −20 °C with an additional of propynyl lithium yielded an intermediate (48%) as a mixture of 2 diastereoisomers, as well as other retroaldol products; then the intermediate (as 2 diastereoisomers) was oxidized into the alkynyl ketone (**4**) (64%) with CrO_3-py. As previously reported a similar chemical [107], conjugate addition of divinyl cuprate in the presence of BF_3- Et_2O yielded stereoselectively compound (**5**) in 61% yield. Lastly from compound (**5**), forskolin was separated in a 73% total yield (Scheme 9.2) [108].

9.29 Synthesis of 6-(3-dimethylaminopropionyl) Forskolin Hydrochloride/NHK477

6-(3-Dimethylaminopropionyl) forskolin hydrochloride/NHK477 was synthesized by reacting 7-deacetylforskolin (2.0 g), with 3-chloropropionylchloride (1 ml) in a mixture of toluene (100 ml), and dry triethylamine (4 ml) at 0 °C. For 1.5 hours, the reaction mixture was heated to 120 °C before being concentrated in a vacuum. The filtrate was placed in an acetonitrile: water (8:7, 150 ml) solution, to which 1 N sodium hydroxide (17 ml) was added. For 3 days, the reaction mixture was stirred at room temperature before aqueous dimethyl amine (30–40%, 40 ml) was added. The reaction mixture was then stirred for a further 2 h at room temperature prior to ethyl acetate extraction. The organic layer was washed, dried, and concentrated over anhydrous sodium sulphate. Purification by chromatography gave the final product from the residue. m.p. 140–142° C. Yield 1.4 g (Scheme 9.3) [109]

9.30 Synthesis of Various Analogues of Coleonol (A diterpene Isolated from Coleus Forskolin)

Coleonol (forskolin) is a highly potent phytochemical for studying enzyme control, but its limited solubility in pharmaceutically acceptable solvents prevents it from undergoing comprehensive biological investigations. After converting four hydroxyl

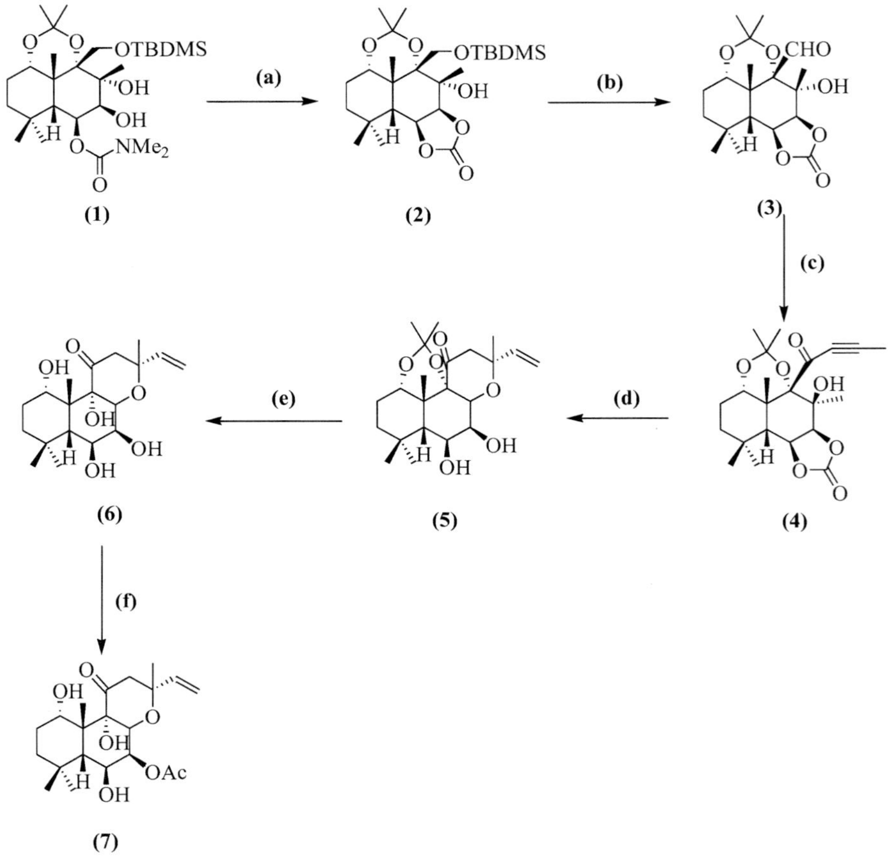

Scheme 9.2 Synthesis of Forskolin. Reaction conditions; (**a**) $LiAlH_4$/EtOH (−78 °C to R.T); CDI/toluene (R.T to reflux) (**b**) TBAF, THF, R.T; Swern oxidation; (**c**) Propynyl lithium, THF, −20 °C; (**d**) Collins, CH_2Cl_2, Room temp; Cs_2CO_3, MeCN. R.T; $(Vinyl)_2CuLi$, BF_3-Et_2O, −78 °C to RT; (**e**) 2 N NaOH, MeOH, RT; pTsOH, MeOH/CH_2Cl_2, RT; (f) Ac_2O/py, 0 °C

groups into isopropylidene groups, it was chosen to have piperazines at position C-15. The piperazine pharmacophore was selected due to its long history of usage in CVS chemotherapy, in addition to the fact that piperazine hydrochloride is hydrophilic in nature. As a result, a hydrophilic counterpart of coleonol **(12)** was synthsized and its action was documented.

As previously stated, forskolin **(7)** was extracted from Coleus forskohlii. 7-deacetyl coleonol (tetraol) **(8)** was obtained by hydrolyzing forskolin with 1 N methanolic NaOH. Intermediate **(8)** was subsequently treated with acetone in cont. H_2SO_4, resulting in 55% and 24% yields of analogues of coleonol **(9)** and **(10)**, respectively. -Analogue (9) was allowed to react with meta-perchlorobenzoic acid

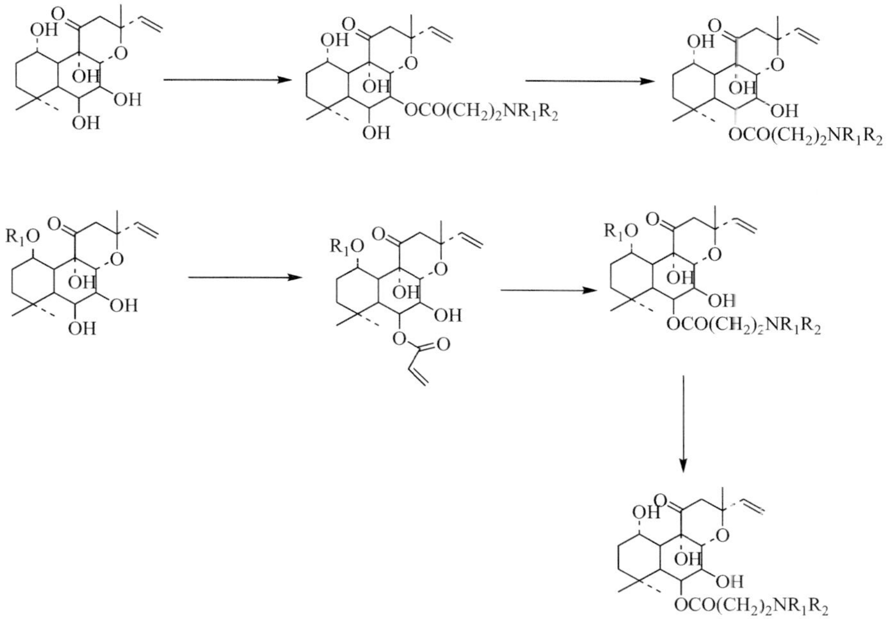

Scheme 9.3 Synthetic layout of 6-(3-dimethylaminopropionyl) forskolin hydrochloride/NHK477.

in the presence of methanol to produce a 75% yield of analogue (**11**), which further reacts with phenyl piperazine to produce a 65% yield of analogue (**12**). The hydrochloride salt of derivative (**12**) was obtained by treating it with methanolic HCl (Scheme 9.4) [110].

9.31 Future Aspects of Coleus

Despite the fact that coleus forskolin is a commercially planted medicinal plant, wild tubers are still taken, causing habitat degradation. The International Union for Conservation of Nature (IUCN) has declared that this species needs to be protected because it is in risk of extinction in the wild.. To lessen the burden on natural habitat, a procedure for rapid replication of this specie through micropropagation, should be used to create sufficient planting material for farmers' advantage. Improved plant vigour, strong root yield with large forskohlin content, resistance to leaf-eating caterpillars, mealy bugs, nematodes, bacterial infections, and other abiotic challenges are all desirable characteristics in this variety. Wide hybridization, mutation breeding, biotechnology techniques, and advanced farming elements with post-harvest technologies will give answers to the aforementioned questions. It is necessary to get knowledge of the supply and value chain, market data, and prospective opportunities in order to assist farmers and the phytopharmaceutical industry [111].

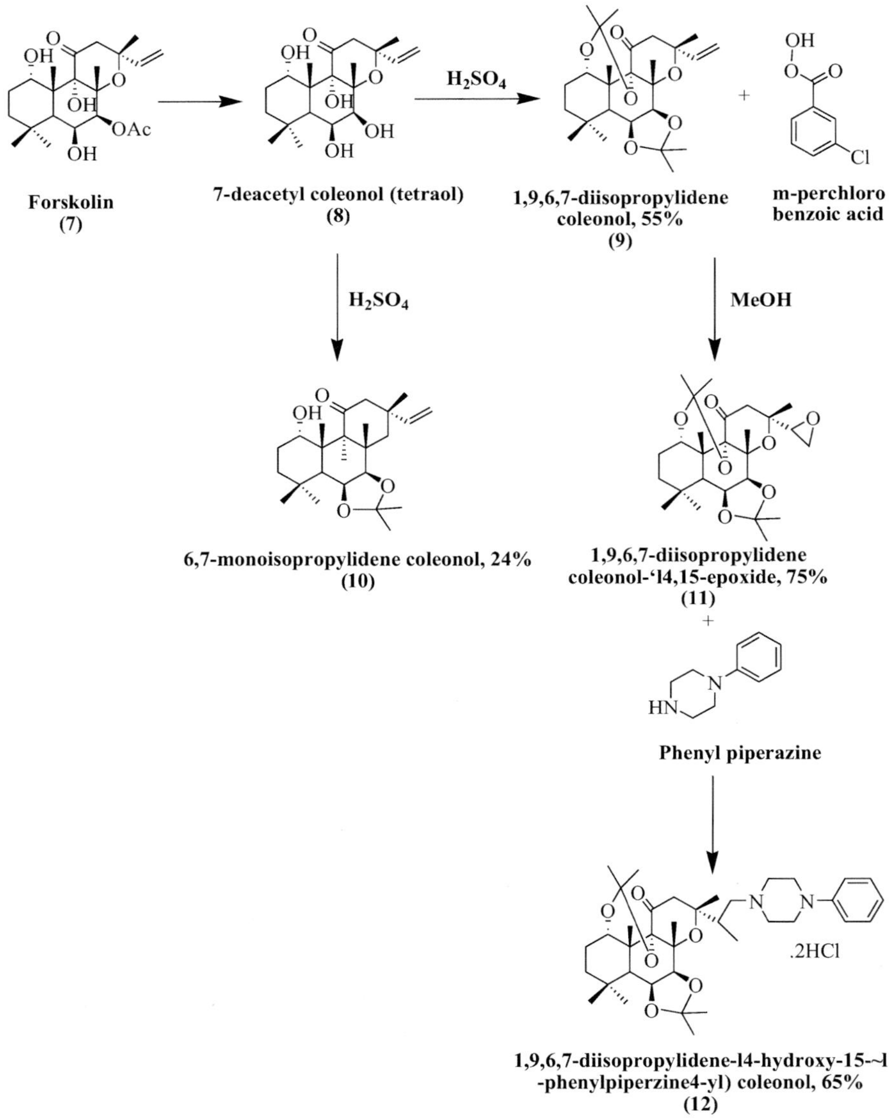

Scheme 9.4 Synthesis of analogues of Coleonol (A diterpene isolated from Coleus forskolin)

References

1. Kavitha, C., Rajamani, K., & Vadivel, E. (2010). Coleus forskohlii A comprehensive review on morphology, phytochemistry and pharmacological aspects. *Journal of Medicinal Plants Research, 4*(4), 278–285.
2. Shah, V. (1996). *Cultivation and utilization of medicinal plants (supplement)* (pp. 385–411). RRL and CSIR.

3. Prajapati, N. D., Purohit, S., Sharma, A. K., & Kumar, T. (2003). *Handbook of medicinal plants. A complete source book*. Shyam Printing Press.
4. Khan, B. A., Akhtar, N., Anwar, M., Mahmood, T., Khan, H., Hussain, I., et al. (2012). Botanical description of Coleus forskohlii: A review. *Journal of Medicinal Plants Research, 6*(34), 4832–4835.
5. Boning, C. R. (2010). *Florida's best herbs and spices: Native and exotic plants grown for scent and flavor*. Pineapple Press.
6. Morton, J. F. (1992). Country borage (Coleus amboinicus Lour.): A potent flavoring and medicinal plant. *Journal of Herbs, Spices & Medicinal Plants, 1*(1–2), 77–90.
7. Blake, S. T. (1971). *Revision of plectranthus (Labiatae) in Australasia*.
8. Paul, M., Radha, A., & Kumar, D. S. (2013). On the high value medicinal plant, Coleus forskohlii Briq. *Hygeia Journal of Drugs and Medicine, 5*, 69–78.
9. Gantait, S., & Kundu, S. (2017). Neoteric trends in tissue culture-mediated biotechnology of Indian ipecac [Tylophora indica (Burm. f.) Merrill]. *3 Biotech, 7*(3), 1–15.
10. Malik, M., Kaul, T., Yaqoob, U., & Mehta, J. (2015). High frequency and rapid in vitro plant regeneration of Coleus forskohlii Briq. *Medicinal and Aromatic Plants, 4*(193), 2167–0412.10001.
11. Vibhuti, R. K., & Kumar, D. (2019). Effect of 6-BAP on callus culture and shoot multiplication of Coleus forskohlii (syn Plectranthus forskohlli wild) briq. *Research Journal of Life Sciences, Bioinformatics, Pharmaceutical and Chemical Sciences, 5*, 574–581.
12. Suva, M. A., Patel, A. M., & Sharma, N. (2015). Coleus species: Solenostemon scutellarioides. *Inventi Rapid: Planta Activa, 2015*(2), 1–5.
13. Lebowitz, R. J. (2011). Of Coleus. *Plant Breeding Reviews, 3*(3), 343.
14. Peterson, B. J., Sanchez, O., Burnett, S. E., & Hayes, D. J. (2018). Comparison of Four Systems for Propagation of Coleus by Stem Cuttings. *HortTechnology hortte, 28*(2), 143–148.
15. Nisar, S., Hanif, M. A., Soomro, K., Jilani, M. I., & Kala, C. P. (2020). Coleus. In *Medicinal plants of South Asia* (pp. 135–147). Elsevier.
16. Veeraraghavathatham, D., Venkatachalam, R., & Sundararajan, S. (1985). Performance of two varieties of Coleus forskohlii under different spacing levels. *South Indian Horticulture, 33*, 389–392.
17. Marimuthu, T., Suganthy, M., & Nakkeeran, S. (2018). Common pests and diseases of medicinal plants and strategies to manage them. In *New age herbals* (pp. 289–312). Springer.
18. Kapur, M. L., Bhalla, S., & Verma, B. (2002). Pests of quarantine significance-some minor tuber crops. *Indian Journal of Entomology, 64*(1), 111–113.
19. Fernandes, R., & Barreto, R. (2003). Corynespora cassiicola causing leaf spots on Coleus barbatus. *Plant Pathology, 52*(6), 786–786.
20. Singh, R., Parameswaran, T., Divya, S., Puttann, K., Satyasrinivas, K., Bagyaraj, D., et al. (2009). *In management of root-rot/wilt of coleus forskohlii Briq, CIMAP golden Jubilee national symposium on medicinal and aromatic plants "fifty years of research on medicinal & aromatic plants"*. CIMAP, RC.
21. Singh, R., Gangwar, S. P., Singh, D., Singh, R., Pandey, R., & Kalra, A. (2011). Medicinal plant Coleus forskohlii Briq.: Disease and management. *Medicinal Plants-International Journal of Phytomedicines and Related Industries, 3*(1), 1–7.
22. Codd, L., Dyer, R. A., Rycroft, H., & Winter, B. d. (1963). *Flora of Southern Africa: The Republic of South Africa*. Basutoland.
23. Steenis, C. G. G. J. (1950). Flora Malesiana: Spermatophyta.(Flowering Plants). v. 1. Cyclopædia of collectors & collections, by Mrs. J. van Steenis-Kruseman. v. 2. Malesian vegetation, by CGGJ van Steenis. v. 3. Malesian plant geography, by CGGJ van Steenis.(in 2 v.) v. 4. pt. 1–5. General chapters and revisions, 1948–1954. v. 5. pt. 1–4. Bibliography, specific delimination & revisions, 1955–1958. v. 6. pt. 1–6. Systematic revisions, 1960–1972. v. 7. pt. 1–4. Systematic revisions, 1971–1976. v. 8. pt. 1–3. Cyclopædia of Collectors, Suppl. 2. Systematic revisions, 1974–1978. v. 9. pt. 1–3. Systematic revisions, 1979–1983. Noordhoff-Kolff, Vol. 1.

24. Staples, G. W., & Kristiansen, M. S. (1999). *Ethnic culinary herbs: A guide to identification and cultivation in Hawaii*. University of Hawaii Press.
25. Valdes, L., Mislankar, S., & Paul, A. (1987). Coleus barbatus (C. forskohlii) (Lamiaceae) and the potential new drug forskolin (Coleonol). *Economic Botany, 41*(4), 474–483.
26. Chandel, K., & Sharma, N. (1997). Micropropagation of Coleus forskohlii (Willd.) Briq. In *High-tech and micropropagation VI* (pp. 74–84). Springer.
27. Somanath, S., Bhaskar, S., & Sreenivasmurthy, C. (2005). Influence of FYM and inorganic fertilizer (NPK) and sources of potassium on the yield of Coleus forskohlii. *Journal of Medicinal and Aromatic Plant Sciences, 27*(1), 16–19.
28. Khatun, S., Çakılcıoğlu, U., & Chatterjee, N. C. (2011). Phytochemical constituents vis-a-vis histochemical localization of forskolin in a medicinal plant Coleus forskohlii Briq. *Journal of Medicinal Plants Research, 5*(5), 711–718.
29. Sapio, L., Gallo, M., Illiano, M., Chiosi, E., Naviglio, D., Spina, A., et al. (2017). The natural cAMP elevating compound forskolin in cancer therapy: Is it time? *Journal of Cellular Physiology, 232*(5), 922–927.
30. Revadigar, V., Shashidhara, S., Rajasekharan, P., Pradeep, N., Prakashkumar, R., & Murali, B. (2008). Variability in the chemical constituents in the roots of Coleus forskohlii from different geographical regions of India. *Acta Horticulturae, 765*, 245–254.
31. Rasineni, G. K., & Reddy, A. R. (2008). Free radical quenching activity and polyphenols in three species of Coleus. *Journal of Medicinal Plants Research, 2*(10), 285–291.
32. Malik, M., Ahmed, R., Khan, S., & Bhatty, M. (1985). Studies on the essential oil of the Coleus aromaticus plant. *Pakistan Journal of Scientific and Industrial Research (Pakistan), 53*, PMC5052183.
33. Ragasa, C. Y., Pendon, Z., Sangalang, V., & Rideout, J. (1999). Antimicrobial flavones from Coleus amboinicus. *Philippine Journal of Science, 128*(4), 347–352.
34. Yanza, Y. R., Szumacher-Strabel, M., Lechniak, D., Ślusarczyk, S., Kolodziejski, P., Patra, A. K., et al. (2022). Dietary Coleus amboinicus Lour. decreases ruminal methanogenesis and biohydrogenation, and improves meat quality and fatty acid composition in longissimus thoracis muscle of lambs. *Journal of Animal Science and Biotechnology, 13*(1), 1–19.
35. Marone, G., Columbo, M., Triggiani, M., Cirillo, R., Genovese, A., & Formisano, S. (1987). Inhibition of IgE-mediated release of histamine and peptide leukotriene from human basophils and mast cells by forskolin. *Biochemical Pharmacology, 36*(1), 13–20.
36. Wong, S., Mok, W., Phaneuf, S., Katz, S., & Salari, H. (1993). Forskolin inhibits platelet-activating factor binding to platelet receptors independently of adenylyl cyclase activation. *European Journal of Pharmacology: Molecular Pharmacology, 245*(1), 55–61.
37. Iwatsubo, K., Tsunematsu, T., & Ishikawa, Y. (2003). Isoform-specific regulation of adenylyl cyclase: A potential target in future pharmacotherapy. *Expert Opinion on Therapeutic Targets, 7*(3), 441–451.
38. Kumawat, T., & Trivedi, L. (2019). A mini review of coleus forskohlii. *World Journal of Pharmaceutical Research, 8*(7), 2324–2331.
39. Lukhoba, C. W., Simmonds, M. S., & Paton, A. J. (2006). Plectranthus: A review of ethnobotanical uses. *Journal of Ethnopharmacology, 103*(1), 1–24.
40. De Souza, N. J., Dohadwalla, A. N., & Reden, Ü. (1983). Forskolin: A labdane diterpenoid with antihypertensive, positive inotropic, platelet aggregation inhibitory, and adenylate cyclase activating properties. *Medicinal Research Reviews, 3*(2), 201–219.
41. Girish, K. (2016). Antimicrobial activities of coleus aromaticus benth. *Journal of Pharmaceutical Research, 10*(10), 635–646.
42. Khare, R. S., Banerjee, S., & Kundu, K. (2011). Coleus aromaticus Benth-A nutritive medicinal plant of potential therapeutic value. *International Journal of Pharma and Bio Sciences, 2*(3), 488–500.
43. Kirtikar, K., & Basu, B. (2005). *Text book of Indian medicinal plants* (Vol. 2, pp. 993–994). International Book Distributors.

44. Kaliappan, N. D., & Viswanathan, P. K. (2008). Pharmacognostical studies on the leaves of Plectranthus amboinicus (Lour) Spreng. *International Journal of Green Pharmacy, 2*(3), 182.
45. Shenoy, S., Kumar, H., Nayak, V., Prabhu, K., Pai, P., Warrier, I., et al. (2012). Hepatoprotective activity of Plectranthus amnoinicus against paracetamol induced hepatotoxicity in rats. *International Journal of Pharmacology and Clinical Sciences, 1*(2), 32–38.
46. Bhat, S., Bajqwa, B., Dornauer, H., do Scusa, N. d., & Fehlhaber, H.-W. (1977). Structures and stereochemistry of new labdane diterpiniods from coleus forskohlii briq. *Tetrahedron Letters, 18*(19), 1669–1672.
47. Saleem, A., Dhasan, P., & Rafiullah, M. (2005). Isolation of forskolin from stem of Coleus forskohlii. *Pharmacognosy Magazine, 1*(3), 89.
48. Seamon, K. B., & Daly, J. W. (1986). Forskolin: Its biological and chemical properties. *Advances in Cyclic Nucleotide and Protein Phosphorylation Research, 20*, 1–150.
49. Mills, I., Moreno, F. J., & Fain, J. N. (1984). Forskolin inhibition of glucose metabolism in rat adipocytes independent of adenosine 3′, 5′-monophosphate accumulation and lipolysis. *Endocrinology, 115*(3), 1066–1069.
50. Morris, D. I., Speicher, L. A., Ruoho, A. E., Tew, K. D., & Seamon, K. B. (1991). Interaction of forskolin with the P-glycoprotein multidrug transporter. *Biochemistry, 30*(34), 8371–8379.
51. Seamon, K. B., Padgett, W., & Daly, J. W. (1981). Forskolin: Unique diterpene activator of adenylate cyclase in membranes and in intact cells. *Proceedings of the National Academy of Sciences, 78*(6), 3363–3367.
52. Dubey, M., Srimal, R., Nityanand, S., & Dhawan, B. (1981). Pharmacological studies on coleonol, a hypotensive diterpene from Coleus forskohlii. *Journal of Ethnopharmacology, 3*(1), 1–13.
53. Caprioli, J., & Sears, M. (1983). Forskolin lowers intraocular pressure in rabbits, monkeys, and man. *The Lancet, 321*(8331), 958–960.
54. Lichey, J., Friedrich, T., Priesnitz, M., Biamino, G., Usinger, P., & Huckauf, H. (1984). Effect of forskolin on methacholine-induced bronchoconstriction in extrinsic asthmatics. *The Lancet, 324*(8395), 167.
55. Agarwal, K. C., & Parks, R. E., Jr. (1983). Forskolin: A potential antimetastatic agent. *International Journal of Cancer, 32*(6), 801–804.
56. Ammon, H., & Müller, A. (1984). Effect of forskolin on islet cyclic AMP, insulin secretion, blood glucose and intravenous glucose tolerance in rats. *Naunyn-Schmiedeberg's Archives of Pharmacology, 326*(4), 364–367.
57. Nourshargh, S., & Hoult, J. (1986). Inhibition of human neurophil degranulation by forskolin in the presence of phosphodiesterase inhibitors. *European Journal of Pharmacology, 122*(2), 205–212.
58. Kilmer, S. L., & Carlsen, R. C. (1984). Forskolin activation of adenylate cyclase in vivo stimulates nerve regeneration. *Nature, 307*(5950), 455–457.
59. Khatun, S., Chatterjee, N. C., & Cakilcioglu, U. (2011). Antioxidant activity of the medicinal plant Coleus forskohlii Briq. *African Journal of Biotechnology, 10*(13), 2530–2535.
60. Jagtap, M., Chandola, H., & Ravishankar, B. (2011). Clinical efficacy of Coleus forskohlii (Willd.) Briq.(Makandi) in hypertension of geriatric population. *AYU An International Quarterly Journal of Research in Ayurveda, 32*(1), 59.
61. Vibhooti, P., Ashok, D. K., Shilpa, S., & Himani, N. (2016). *Fight psoriasis naturally through ayurveda*. Shri Dev Bhoomi/Institute of Education Science & Technology.
62. Ammon, H. P., & Müller, A. B. (1985). Forskolin: From an ayurvedic remedy to a modern agent. *Planta Medica, 51*(06), 473–477.
63. Han, L.-K., Morimoto, C., Yu, R.-H., & Okuda, H. (2005). Effects of Coleus forskohlii on fat storage in ovariectomized rats. *Yakugaku Zasshi: Journal of the Pharmaceutical Society of Japan, 125*(5), 449–453.
64. De Souza, N., & Shah, V. (1988). Forskolin--an adenylate cyclase activating drug from an Indian herb. In H. Wagner & H. Hikino (Eds.), *Economic and medicinal plant research*. Norman R. Farnsworth.

65. Henderson, S., Magu, B., Rasmussen, C., Lancaster, S., Kerksick, C., Smith, P., et al. (2005). Effects of coleus forskohlii supplementation on body composition and hematological profiles in mildly overweight women. *Journal of the International Society of Sports Nutrition, 2*(2), 1–9.
66. Kamohara, S. (2016). An evidence-based review: Anti-obesity effects of Coleus forskohlii. *Personalized Medicine Universe, 5*, 16–20.
67. Kim, H. K., Song, K. S., Chung, J. H., Lee, K. R., & Lee, S. N. (2004). Platelet microparticles induce angiogenesis in vitro. *British Journal of Haematology, 124*(3), 376–384.
68. Singh, S., Painuly, P., & Tandon, J. (1984). Diterpenes from coleus-forskohlii-stereochemistry of the carbonyl chromophore. *Indian Journal of Chemistry Section B-Organic Chemistry Including Medicinal Chemistry, 23*(10), 952–955.
69. Shah, V., Bhat, S., Bajwa, B., Dornauer, H., & De Souza, N. (1980). The occurrence of forskolin in the Labiatae. *Planta Medica, 39*(06), 183–185.
70. Trivedi, A., Mehrotra, B., Tandon, R., & Jain, G. (1982). Estimation of coleonol from Coleus forskohlii Briq. *Indian Journal of Pharmaceutical Sciences, 44*, 157–158.
71. Yanighara, H., Sakata, R., Shoyama, Y., & Murakami, H. (1996). Rapid analysis of small samples containing forskolin using monoclonal antibodies. *Planta Medica, 62*(02), 169–172.
72. Mukherjee, S., Ghosh, B., & Jha, S. (1996). Forskolin synthesis in in vitro cultures of Coleus forskohlii Briq transformed with Agrobacterium tumefaciens. *Plant Cell Reports, 15*(9), 691–694.
73. Kansal, C., Srivastava, S., Dube, C., & Tandon, J. (1978). Clinical evaluation of Coleus forskohlii on hypertension. *Nagarjun, 22*, 56–58.
74. Dubey, C., Srimal, R., & Tandon, J. (1997). Clinical evaluation of ethanolic extract of Coleus forskohlii in hypertensive patients. *Sachitra Ayurveda, 49*, 931–936.
75. Suryanayanan, M., & Pai, J. (1998). Studies in micropropagation of Coleus forskohlii. *Journal of Maps, 20*(2), 379–382.
76. Baslas, R., & Kumar, P. (1981). *Phytochemical studies of the plants of Coleus genera*. Herba Hungarica.
77. Adachi, H., Ehata, S., & Hayashi, T. (1996). Antiaging cosmetics containing Coleus forskohlii root extracts and antioxidants. *Patent-Japan Kokai Tokkyo Koho-08, 176*(005), 7.
78. Majeed, M., Badmaey, V., & Rajendran, R. (1998). *Method of preparing a forskohlin composition from forskohlin extract and use of forskohlin for promoting lean body mass and treating mood disorders*. Google Patents.
79. Asolkar, L., Kakkar, K., & Chakre, O. (1992). *Second supplement to glossary of Indian medicinal plants with active principles part-I (AK)* (pp. 217–218). Council of Scientific and Industrial Research (PID).
80. Yashaswini, S., & Vasundhara, M. (2011). Coleus (Plectranthus barbatus)-A multipurpose medicinal herb. *International Research Journal of Pharmacy, 2*, 47–58.
81. Lakshmanan, G., Manikandan, S., & Panneerselvam, R. (2013). Plectranthus forskohlii (Wild) Briq.(Syn: Coleus forskohlii) – A Compendium on its Botany and Medicinal uses. *International Journal of Research in Plant Science, 3*(4), 72–80.
82. Clougherty, J. E., Levy, J. I., Kubzansky, L. D., Ryan, P. B., Suglia, S. F., Canner, M. J., et al. (2007). Synergistic effects of traffic-related air pollution and exposure to violence on urban asthma etiology. *Environmental Health Perspectives, 115*(8), 1140–1146.
83. Reader, J. R., Hyde, D. M., Schelegle, E. S., Aldrich, M. C., Stoddard, A. M., McLane, M. P., et al. (2003). Interleukin-9 induces mucous cell metaplasia independent of inflammation. *American Journal of Respiratory Cell and Molecular Biology, 28*(6), 664–672.
84. Zhu, Y., Chen, L., Huang, Z., Alkan, S., Bunting, K. D., Wen, R., et al. (2004). Cutting edge: IL-5 primes Th2 cytokine-producing capacity in eosinophils through a STAT5-dependent mechanism. *The Journal of Immunology, 173*(5), 2918–2922.
85. Cao, J., Ye, B., Lin, L., Tian, L., Yang, H., Wang, C., et al. (2017). Curcumin alleviates oxLDL induced MMP-9 and EMMPRIN expression through the inhibition of NF-κB and MAPK pathways in macrophages. *Frontiers in Pharmacology, 8*, 62.

86. Wang, Z.-D., Huang, C., Li, Z.-F., Yang, J., Li, B.-H., Liang, R.-R., et al. (2010). Chrysanthemum indicum ethanolic extract inhibits invasion of hepatocellular carcinoma via regulation of MMP/TIMP balance as therapeutic target. *Oncology Reports, 23*(2), 413–421.
87. Bruka, J. (1986). In Forskoli: Its chemical biological and medical potential. In *Proceedings of the International Symposium on Forskolin*. Hoechst India Ltd.
88. Bauer, K., Dietersdorfer, F., Sertl, K., Kaik, B., & Kaik, G. (1993). Pharmacodynamic effects of inhaled dry powder formulations of fenoterol and colforsin in asthma. *Clinical Pharmacology & Therapeutics, 53*(1), 76–83.
89. Rupp, R., De Souza, N., & Dohadwalla, A. (1986). *Proceedings of the international symposium on Forskolin – Its chemical, biological, and medical potential*. Hoechst India.
90. Siegel, A. M., Daly, J. W., & Smith, J. B. (1982). Inhibition of aggregation and stimulation of cyclic AMP generation in intact human platelets by the diterpene forskolin. *Molecular Pharmacology, 21*(3), 680–687.
91. Adnot, S., Desmier, M., Ferry, N., Hanoune, J., & Sevenet, T. (1982). Forskolin (a powerful inhibitor of human platelet aggregation). *Biochemical Pharmacology, 31*(24), 4071–4074.
92. Wysham, D. G., Brotherton, A. F., & Heistad, D. D. (1986). Effects of forskolin on cerebral blood flow: implications for a role of adenylate cyclase. *Stroke, 17*(6), 1299–1303.
93. De Souza, N. J. (1993). Industrial development of traditional drugs: The forskolin example a mini-review. *Journal of Ethnopharmacology, 38*(2–3), 167–175.
94. Ciotonea, C., & Cernătescu, C. (2010). Biological active effects of Foskolin extract. *Buletinul Institutului Politehnic DIN IASI, 4*, 95–106.
95. Misra, L. N., Tyagi, B. R., Ahmad, A., & Bahl, J. R. (1994). Variability in the chemical composition of the essential oil of Coleus forskohlii genotypes. *Journal of Essential Oil Research, 6*(3), 243–247.
96. Shivaprasad, H., Pandit, S., Bhanumathy, M., Manohar, D., Jain, V., Thandu, S. A., et al. (2014). Ethnopharmacological and phytomedical knowledge of Coleus forskohlii: An approach towards its safety and therapeutic value. *Oriental Pharmacy and Experimental Medicine, 14*(4), 301–312.
97. Hosono, M., Takahira, T., Fujita, A., Fujihara, R., Ishizuka, O., Ohoi, I., et al. (1990). Cardiovascular effects of NKH477, a novel potent water-soluble forskolin derivative. *European Journal of Pharmacology, 183*(6), 2110–2111.
98. Laurenza, A., Khandelwal, Y., De Souza, N., Rupp, R., Metzger, H., & Seamon, K. (1987). Stimulation of adenylate cyclase by water-soluble analogues of forskolin. *Molecular Pharmacology, 32*(1), 133–139.
99. Khandelwal, Y., Rajeshwari, K., Rajagopalan, R., Swamy, L., Dohadwalla, A., De Souza, N., et al. (1988). Cardiovascular effects of new water-soluble derivatives of forskolin. *Journal of Medicinal Chemistry, 31*(10), 1872–1879.
100. Tatee, T., Narita, A., Narita, K., IZUMI, G., TAKAHIRA, T., SAKURAI, M., et al. (1996). Forskolin derivatives. I. Synthesis, and cardiovascular and adenylate cyclase-stimulating activities of water-soluble forskolins. *Chemical and Pharmaceutical Bulletin, 44*(12), 2274–2279.
101. Iranami, H., Okamoto, K., Kimoto, Y., Maeda, H., Kakutani, T., & Hatano, Y. (2002). Use of corfolsin dalopate following cardiac surgery in a neonate. *The Journal of the American Society of Anesthesiologists, 97*(2), 503–504.
102. Kumari, R., Dubey, V., Mishra, S., & Singh, R. (2018). Review on: Pharmacological aspect of medicinal herb Coleus forskohlii. *Asian Journel of Pharmceutical and Educational Research, 7*(4), 16–22.
103. Mastan, A., Vivek Babu, C. S., Hiremath, C., Srinivas, K. V. N. S., Kumar, A. N., & Kumar, J. K. (2020). Treatments with native Coleus forskohlii endophytes improve fitness and secondary metabolite production of some medicinal and aromatic plants. *International Microbiology, 23*(2), 345–354.
104. KC, C. (1960). Bhav Prakasa Nighantu (Hindi translation). : Chaukhamba Vidya Bhavan, 86.

105. Bhakuni, D., Dhar, M., Dhar, M., Dhawan, B., Gupta, B., & Srimal, R. (1971). Screening of Indian plants for biological activity: Part III. *Indian Journal of Experimental Biology, 9*(1), 91–102.
106. Pateraki, I., Andersen-Ranberg, J., Jensen, N. B., Wubshet, S. G., Heskes, A. M., Forman, V., et al. (2017). Total biosynthesis of the cyclic AMP booster forskolin from Coleus forskohlii. *eLife, 6*, e23001.
107. Delpech, B., & Lett, R. (1987). Retrosynthetic studies with forskolin. *Tetrahedron Letters, 28*(35), 4061–4064.
108. Delpech, B., Calvo, D., & Lett, R. (1996). Total synthesis of forskolin Part II. *Tetrahedron Letters, 37*(7), 1019–1022.
109. Khandelwal, Y., Lal, B., & Blumbach, J. (1993). *Process for the preparation of 6-[3-substitutedaminopropionyl]-7-deacetylforskolin derivatives*. Google Patents.
110. Chauhan, P., Varma, N., & Tandon, J. (1993). The synthesis of novel, rigid and water soluble analogues of coleonol (forskolin). *Bioorganic & Medicinal Chemistry Letters, 3*(4), 677–680.
111. Lokesh, B., Deepa, R., & Divya, K. (2018). Medicinal Coleus (Coleus forskohlii Briq): A phytochemical crop of commercial significance. *Journal of Pharmacognosy and Phytochemistry, 7*, 2856–2864.

Chapter 10
Cinchona

Sana Aslam, Tooba Jabeen, Matloob Ahmad, and Arwa A. AL-Huqail

10.1 Introduction

Family	Rubiaceae
Subfamily	Cinchonoideae
Scientific name	*Cinchona officinalis*
English/Common Name	Cinchona

10.2 Plant Description

The Genus Cinchona which belongs to the *Rubiaceae* family is a large shrub or small tree that can grow to the height of 5-15meters (16 to 49 feet). It possesses 10-40 cm long evergreen leaves that are opposite and rounded to lanceolate. Flowers produced are white, pink or crimson and are present in terminal panicles. The fruit is a tiny capsule with several seeds within. The flowers have slightly hairy corolla lobes, which is a distinguishing feature of the genus. *Cinchonopsis, Jossia, Ladenbergia, Remijia, Stilpnophyllum,* and *Ciliosemina* are all members of the Cinchoneae tribe [1]. India, Vietnam, Java, Cameroon, and a few more African as

S. Aslam (✉)
Department of Chemistry, Government College Women University, Faisalabad, Pakistan
e-mail: dr.sana@gcwuf.edu.pk

T. Jabeen · M. Ahmad
Department of Chemistry, Government College University, Faisalabad, Pakistan

A. A. AL-Huqail
Department of Biology, College of Science, Princess Nourah bint Abdulrahman University, Riyadh, Saudi Arabia

M. Zia-Ul-Haq et al. (eds.), *Essentials of Medicinal and Aromatic Crops*,
https://doi.org/10.1007/978-3-031-35403-8_10

well as Asian nations also have it. It is frequently found in India's Himalayas as a result of planting. Indonesia is the greatest manufacturer of cinchona plant on the global level [2]. Countess Cinchon of spain was the one who first brought cinchona to Europe in 1638.A cinchona botanist, Carl Linnaeus, called it "Cinchona" in 1742 [2, 3].Cinchona alkaloids have been reported as the most significant natural products that have been examined, extracted, explored and investigated. Quinine itself is the most important molecule of this class due to its potential in antimalarial medication. Furthermore, many applications were reported in several fields as evidence to their potential for example medicinal, catalysis, synthetic, and beverages [4].

10.3 Agronomy of Plant

10.3.1 Soil Conditions

Cinchona enjoys undeveloped terrain that has been cleared of trees and is rich in organic debris. The tropics' densely inhabited areas, on the other hand, have no available virgin ground for planting. The volcanic northern slopes of the Preanger Regency Mountains in western Java provide the richest soils in Java. To prevent erosion, Indonesians turned the mountains into terraces. Bandung, the region's capital, was previously known as the world's quinine capital. To allow the taproot to penetrate to a large depth, the soil should be loose and deep. Cinchona cannot grow on a difficult subsoil or hardpan because it does not allow penetration. Cinchona cannot tolerate soggy soil; thus, the soil should drain properly.

Soil moisture in Cinchona-growing locations varies a lot with seasonal fluctuations in rainfall, and extended droughts happen virtually every year. During the rainy season, on the other hand, the same soils may remain saturated for lengthy periods of time. At the Toro Negro National Forest in Puerto Rico, where Cinchona is being cultivated as an experiment, the 23-year usual rainfall for the intervals of 2 weeks (from January to March) was 2.04 inches, with regular spells of famine. In January of this year, barely 0.21 inch of rain fell. From August 13 to November 18,average rainfall have been recorded as 5.69 inches for 2 week intervals.. For the past 23 years, the average annual rainfall has been 96.37 inches [5].

Cinchona grows best in deep, humus-rich, productive soils (with organic matters) over an open grainy/rough surface that has a pH of 5–6.5. Cinchona grows well on mild slopes and may even grow on steep slopes; however, soil erosion protection is required. The clay loam acidic soils of the Darjeeling hills are rich in organic matter, with organic carbon levels ranging from 1.5–3.34%, but lacking in phosphorus and potassium. The land chosen for planting had available levels of nitrogen, phosphorus and potassium of 315–414 mg/kg, 1.01–21.0 mg/kg, and 0.8–13.7 mg/kg soil, respectively [6].

An experimental design was created to see the effect of soil condition on Cinchona plant for which total 300 seeds per soil were germinated in three different types of soil. Between 13–60 days, seedlings were counted and their tallness was measured on the completion of the research. The impact of soil condition on the germination was determined using a nonparametric analysis of variance with á = 0.05. It has been shown that the natural environmental conditions (temperature humidity) and the gritty texture soil are the most suitable conditions for the seed germination of *C. officinalis*. With a $p < 0.05$ significance, it was found that the kind of soil had a substantial impact on the percentage of germination [7]

10.4 Climate

Cinchona proved to be a difficult plant to cultivate. Trees like a cool environment with minimal change between summer and winter, day and night temperatures, and do not enjoy stable humidity at their roots or nonstop sunshine. Medicinal species cannot tolerate frost. As a result, they are unable to grow on flat terrain and demand a rich soil (Watt). Seed beds in Madras were protected by "light pandals" of fern simply planted in the ground and thick enough to shade the soil. The plant also has an "exceptional tendency of breeding or hybridizing," which is "particularly conspicuous in the finest species," according to the researchers (Owen).

Cinchona loves a cold environment with 190–500 cm of annual rainfall that is evenly distributed. Showers alternate with sunshine often, resulting in vigorous growth. In compared to Darjeeling in the Eastern Himalayas, it was stated that Cinchona (with high production of quinine) ledgeriana yielded comparatively low content over the Research and Development Efforts in Cinchona highlands of South India. This might be due to high temperatures and droughts in the southern areas, where the soil is semi alluvial red loam with a medium-high carbon content. For luxuriant development, a temperature of 21–24 °C is recommended. In the summer, the temperature should not drop below 8 °C and should not rise over 30 °C. During the growth season, the temperature difference between day and night should be modest. Frost and subsurface waterlogging are both threats to all organisms [8].

10.5 Land Preparation

For plantings, forests with rich forest soil and high humus are required. They would be brushed out every month or two to admit one-fourth to one-third light around planting holes, and as the plants took hold and flourished, So, they brush out to admit lighter gradually. To avoid harming quinine trees, felling would be abandoned in favor of girdling and allowing trees to die standing until the quinine produced a cover. This method also helpful to reduce erosion during the early phases of rainstorms.

In Java, the procedure is to plant Cinchona on terrace in which a sump hole was dug on the rear end of the terrace every metre or two for aeration and to gather eroded soil and water, then dig these holes out when filled and pile the dirt on the terrace behind. In this procedure of planting, labour is numerous and inexpensive, so it is proved to be an excellent strategy [9].

10.6 Planting

The planting area should be cleaned a year ahead of time and shade trees such as Silver oak and or dedabs should be planted. Pits of 30 x 30 x 45 cm are excavated and filled with topsoil and other organic waste that has decomposed properly. When seedlings are around a year old, they are transplanted onto the main field at a spacing of 1.251.25 m. When there is enough moisture in the field, transplanting can be done at any time. The third option is to use high density planting, which involves spacing trees at 1.01.25 m or 8000 plants per hectare and gradually harvesting until only 800 plants/ha remain after 25 years.

In Darjeeling, seedlings are planted early in the monsoon season (June), with 4–6 pairs of leaves along with them. This increases the chance of their survival. On the bases of slope, pits with dimensions of 1.2 m × 1.2 m × 2.0 m × 2.0 m were dug. To achieve clean erect boles, seedlings with dirt stuck to the roots are planted and anchored. In general, 3000 plants per hectare are planted, freeing up space for shade trees. After the fourth year, trees begin to blossom. Trees that have been debudded have more laminar growth and alkaloid content [10].

10.7 Manuring

115 kg of nitrogen and potassium, as well as 15 kg of phosphorus per hectare are required each year to fertilize the plants of cinchona. Liming at 1.0–1.5 t/ha is advised every 3–4 years when the soil pH falls below 4.5.

10.8 Ecology

Some Lepidoptera species, such as the engrailed, the commander, as well as species of the genus *Endoclita*, including as *E. purpurescens*, *E. sericeus*, and *E. damor*, utilise Cinchona species as food plants. *Cinchona pubescens* has spread uncontrollably on certain islands, together with the Galapagos, posing a threat to native plant species [11]. It thrives in Ecuador's native range in volcanic soil rich in organic matter, as well as in highly rocky locations where the roots are open to the air. It thrives in disturbed environments, particularly regions where vegetation has been burned.

10.9 Pest and Diseases

Any fungal impurity that can be the reason of fast deterioration or loss of immature juicy seedlings is referred to as damping off. The illness has a global distribution and affects many people working in nurseries and greenhouses. Fungi that cause damage are ubiquitous in the soil, and when moisture and temperature circumstances are appropriate, fungal hyphae attack host plants by directly invading the host tissues. Certain damping off species target a wide range of hosts, whilst some have more restricted host range [12, 13]. Various diseases found on this plant species can be classified as the diseases caused by damping off. Fungus infection is more likely to develop in cinchona seedbeds immediately after the seeds have begun to germinate.

In French Guinea, it was observed that a severe seedling wilt of *Cinchona ledgeriana* and *Cinchona succirubra*. When the cotyledons were about 4–5 mm long, seeds in the seedbeds withered and perished in clumps. In the soil of the damaged patches, many white rhizomorphs of a tropical Clitocybe species were discovered, and when they came into contact with the cinchona rootlets, the rhizomorphs generated mycelial threads that pierced the host tissues [14].

Pestalozzia is a disease that affects cinchona plants in Pala Sumatra and is caused by Pestalozzia myristicae [15]. Infected plants should be removed and burned as soon as possible.

Copper-based fungicides are effective when applied to the soil and help in the removing and burning of the infected plants. Because cinhophyllene-type indole alkaloids along with tiny levels of 5-methoxy-tryptamine present in juvenile *Cinchona ledgeriana* plants that cause the reduction in insect damage. These chemicals give developing plants a chemical defense against herbivorous insects [16]. Caterpillars of Catalpa sphinx as well as Helopeltis spp., on the other hand, inflict harm to nursery plants. The infestation is controlled by spraying dimethyl phosphoric ester. Disphinctus humeralis, a leaf beetle that attacks fragile foliage on occasion, may be controlled using a 0.2% malathion spray.

10.10 Important Chemical Constituents of Cinchona

The genus Cinchona, comprising about 40 species [23], the widely known species include *C. calisaya*, *C. ledgeriana*, *C. officinalis*, and *C. pubescens* (*C. succirubra*) [24], which contain many biologically active substances. The bark is the primary component that is utilized for various therapeutic and other applications [3]. Previous phytochemical studies revealed a variety of secondary metabolites, including polyphenols(henolic acids, anthocyanins, and flavonoids), quinoline alkaloids, and indole alkaloids [23]. In addition, various essential oils and minerals, such as triterpene, organic and phenolic acids (quinic acid, caffeic acid), flaveniods and phytosterols are present in the bark [3]. Cinchona alkaloids are among the most familiar natural products isolated so far. They are separated from numerous species of genus cinchona such as *Cinchona succirubra*, *Cinchona ledgeriana*, and *Cinchona calisaya*, all of which are members of Rubiaceae family [4].

10.11 Cinchona Alkaloids

Alkaloids are secondary metabolites that have the ability to create a variety of medications with significant pharmacological effects. They are also utilized to investigate physiological responses and biochemical processes of action [25]. Cinchona bark of different species contains more than 36 distinct alkaloids, with quinine (**1**), quinidine (**2**), cinchonidine (**3**), and cinchonine (**4**) being the most significant. These four phytochemical components of bark are stereoisomers of one another. The quantity of alkaloids in the bark varies from 7% to 12%, with quinine being the most abundant and making up as much as 90% of it. Moreover, there is significant variety in alkaloid content and composition both within and between species. Quinine, cinchonidine, and the total amounts of the four main alkaloids, appear to be related to chronology [27]. Cinchona bark is high in quinoline alkaloids, accounting for 6–10% of total quinoline alkaloids. The species with the highest quinine contents are *Cinchona officinalis, Cinchona ledgeriana, and Cinchona calisaya*. *Cinchona pubescens* possesses a greater percentage of cinchonine (3.3–3.4%) among its higher total alkaloids (6.0–8.2%). Maximum amount of Cinchonidine (3.7–4.9%) is present in the bark of *Cinchona kartamanahs* when they are over 12 years old, and it is similarly high (2.1–2.2%) in the bark of "Hybrida" when it is approximately 6–8 years old [17] (Fig. 10.1).

In the bark of cinchona, the most essential and distinctive alkaloids comprise 16% quinine ($C_{20}H_{24}N_2O_2$), although the amount fluctuates (6–10%) depending on the species variation. A typical sample of dried cinchona or the bark of this plant was reported to have 0.4–4% of cinchona. It is often employed as an anti-malarial agent, as well as a flavouring component in carbonated beverages, a skeletal muscle relaxant, a treatment for haemorrhoids and varicose veins, and as an oxytoxic agent [3]. Quinine is separated as quinine sulphate from the total alkaloids. It's a white, sparkling, odourless substance with a strong bitter flavour. In water, it is just slightly soluble, but it shows extreme solubility towards organic solvents. In addition to quinine salts, cinchona febrifuge (powder), which is made up of residual alkaloids after quinine removal, is sold [18]. After quinine, the second most prominent alkaloid found in cinchona is cinchonine ($C_{19}H_{22}N_2O$), which is also employed as an anti-malarial medication [19] with lesser toxic effect and stronger activity than other quinine-related compounds [20]. Another major alkaloid is quinidine, a diastereoisomer of quinine, which is found in cinchona bark in concentrations ranging from 0.25% to 3.0%. It's a dextrorotatory quinine stereoisomer. Its primary role is to operate as an anti-malarial drug, but it is also effective as an anti-arrhythmic agent [21] when antiarrhythmic metabolism is achieved by membrane stability. Cinchonidine ($C_{19}H_{22}N_2O$) is found in most cinchona bark types, particularly in the

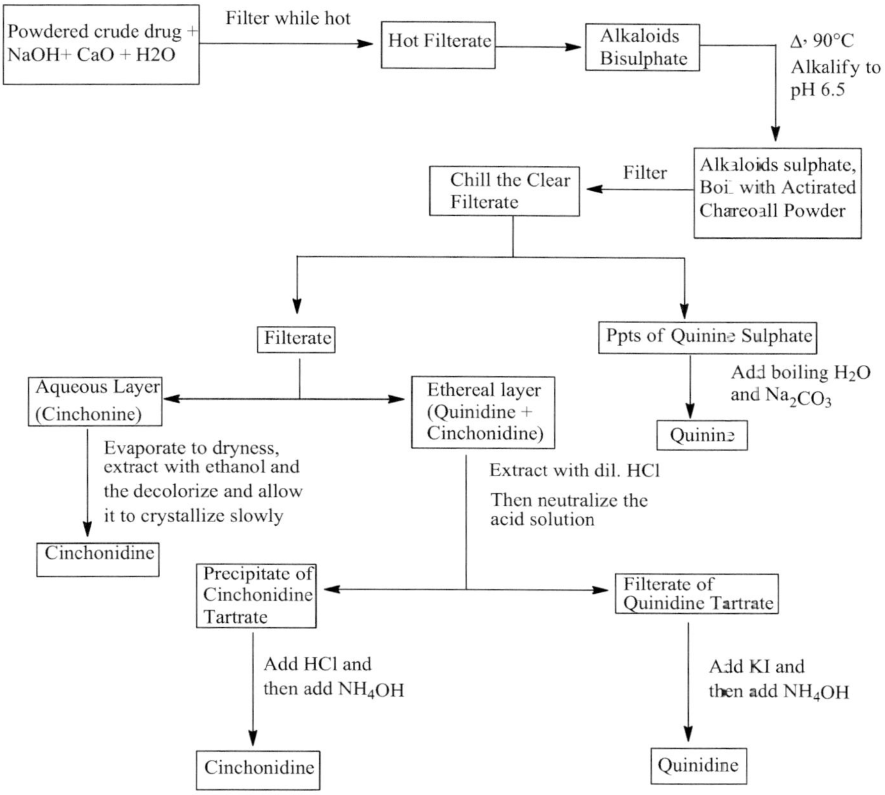

Fig. 10.1 Diagrammatic illustration of Isolation of cinchona alkaloids

bark of *C. pusescensval* and *C. pitayensis*, and is primarily utilised as an anti-malarial drug. It is a stereoisomer and pseudo-enantiomer of cinchonine that is mostly utilized as an alternative of quinine. Epicinchonidine is frequently used to treat malaria [4]. In the lack of an economically feasible synthetic technique for their synthesis, cinchona bark remains the source of these phytochemicals [22].

Quinolone and quinuclidine rings with a vinyl group are also found. In addition to alkaloids, the bark also includes a colouring substance (up to 10%), polyphenols, flavonoids, and an essential oil. [18]. Quiniarnine, cinchotine, hydrcquinine, hydro-cinchonidine, and cinchotannic acid are some of the other phytochemical constituents found in cinchona (Fig. 10.2). The methoxy group is present in quinine and quinidine, but it is absent in chichonine and cinchonidine. Cinchona also contains bitter glycosides and starch granules. Calcium oxalate and crystalline acids like quinic acid are also present. They have a total alkaloids content of at least 6.5%. Quinine accounts for 30–60% of total alkaloids.

In addition to them, a variety of other alkaloids with similar structural relationships have also been discovered such as, cinchotoxine (**5**), remijinine (**6**),

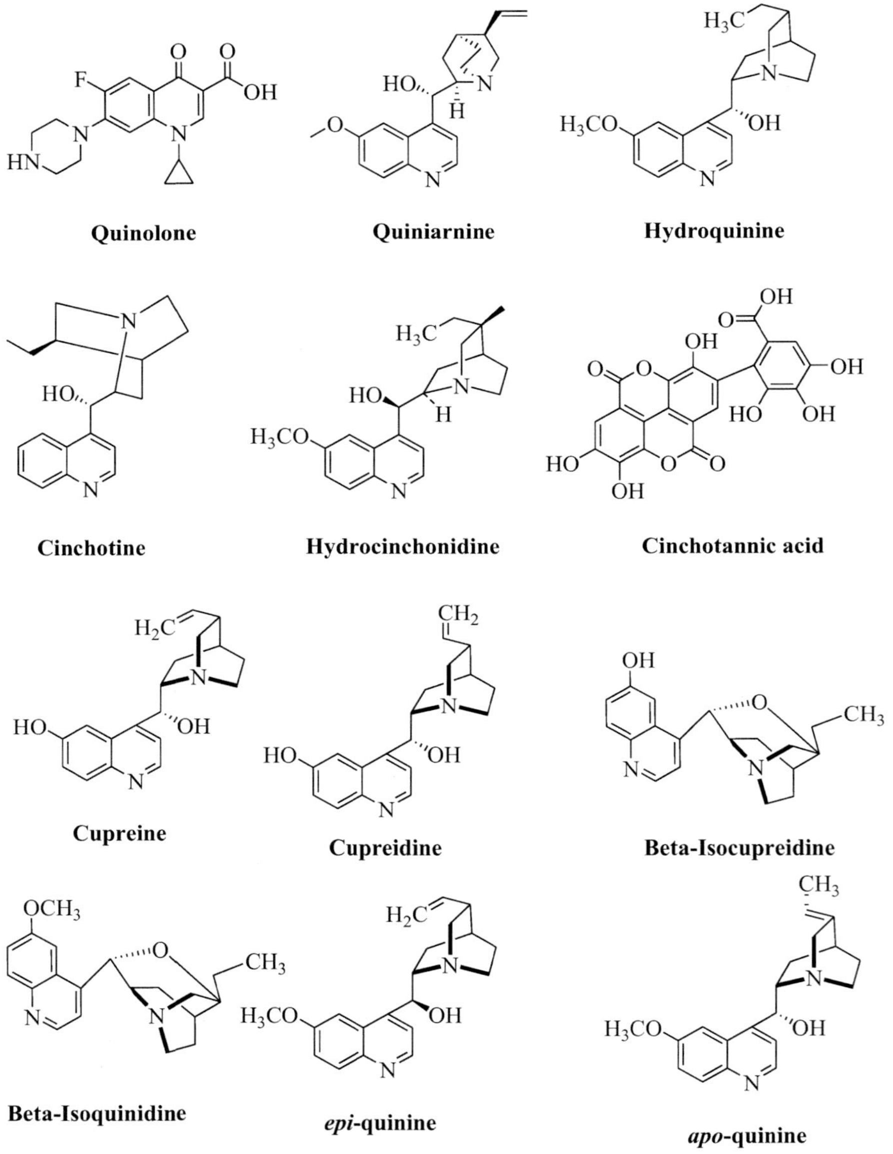

Fig. 10.2 Structures of potent phytochemicals in Cinchona Plant

cinchonamine (**7**), quinamine (**8**), liriodenine (**9**), lyscamine (**10**), cinchophylline (**11**), 9-epiquinine (**12**), 9-epiquinidine (**13**), dihydroquinine (**14**), dihydroquinidine (**15**), quinidineN(4)oxide (**16**), cinchonicinol (**17**), Epidihydrocinchonidine (**18**), Epivinylquinidine (**19**), Chloroquine (**20**), Quinotoxin (**21**) [4, 23] (Fig. 10.3).

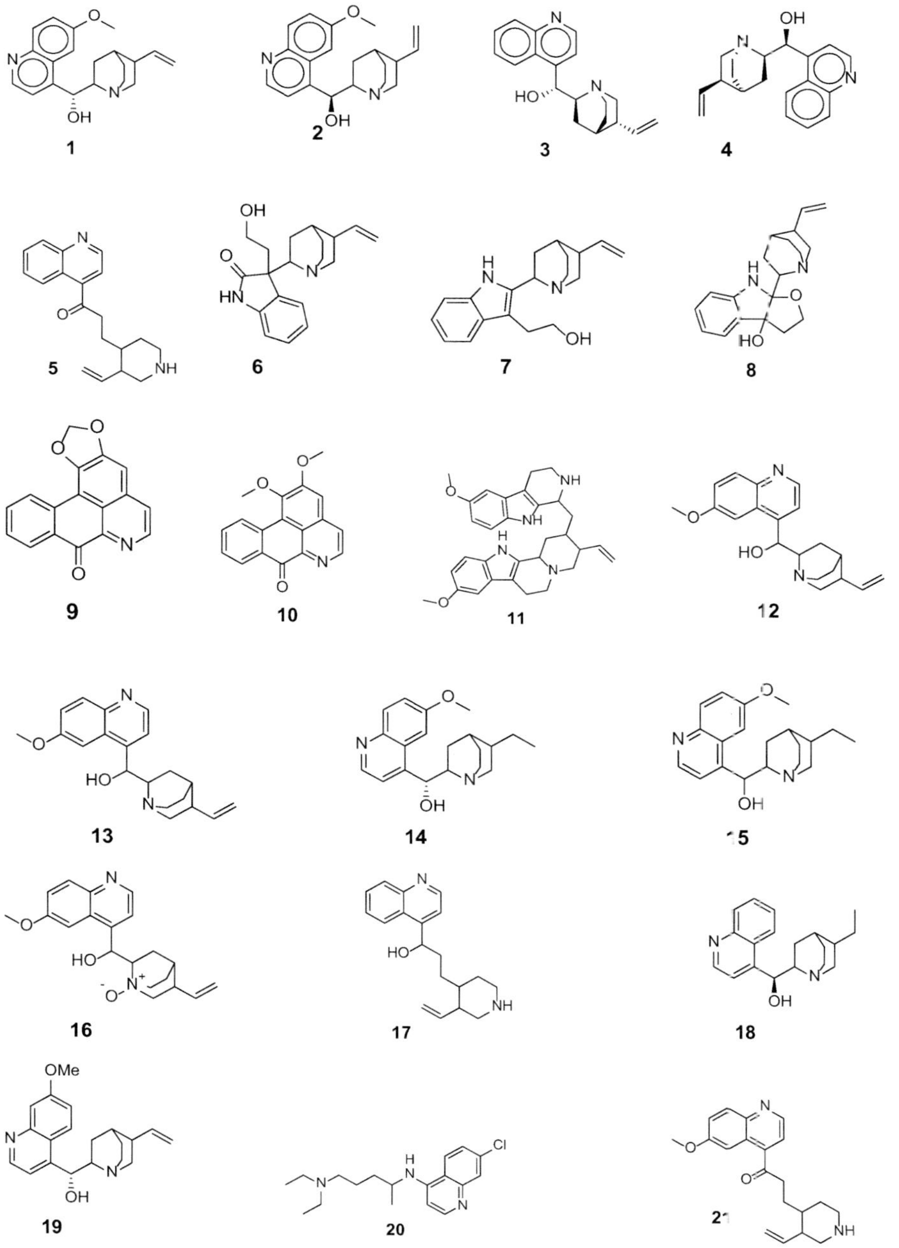

Fig. 10.3 Structures of Cinchona Alkaloids and derivatives

The bark and roots of the plant contain the majority of the cinchona alkaloids, while cinchophyllines have been found in the leaves [28]. In the bark, a portion of Cinchona alkaloids exist as the salts of acids (quinic acid, cinchotannic acid, etc.). Examination of the leaves of distinct Cinchona tree species revealed that they differ noticeably in the content of alkaloids. For instance, *C. "ledgeriana"* includes mostly indole type alkaloids, (e.g., cinchophyllamines, quinamine, and 3-epiquinamine) as compared to samples of *C. ledgeriana* from Guatemala, where quinoline alkaloids 1–4 predominate [29]. Up to 1% of the total alkaloids are found in leaves, with younger leaves having a higher concentration [26]. There are various different varieties of bark: red (*Cinchona succirubra*), grey (*C. micrantha* and *C. nitida*), yellow (*C. calisaya*), Colombian (*C. lancifolia*), or colourless (*C. officinalis*), each of which has a constant ratio of alkaloids. Compared to trees with red or grey bark, which have the lower compound content, plants with yellow bark appear to have more quinine concentration [30]. Alkaloids make up 5–7% of the "red" bark of *C. succirubra*, 4–7% of the bark of *C. calisaya*, and up to 5–14% of the bark of *C. officinalis* [29].

10.12 Other Chemical Constituents of Genus Cinchona

The findings of the phytochemical screening revealed that the tissues of Cinchona leaves and bark contain a variety of flavonoids, terpenoids, tannins, saponins, and lipophilic chemicals, with varying degrees of colour intensity that may indicate their quantity of accumulation [26, 31, 32]. Thin layer chromatography (TLC) examination of Cinchona pubescens demonstrated the presence of four flavonoids, catechin(**23**), kaempferol(**24**), apigenin(**25**), and quercetin(**26**) [31, 33]. Extracts of different cinchona species contain variable levels of these compounds. For instance, a leaf of *C. ledgeriana* extracted with 70% ethanol included flavonoids, tannins, saponins, other than alkaloids, and terpenoids. The concentration of total phenol (40.23%) and flavonoids (65.34%) was highest in the ethyl acetate fraction [26]. The C. succirubra extract contains the alkaloids, flavonoids, amino acids, glycosides, saponins, tannins (3–10%), phenolic organic acids, terpenoids and steroids [24]. In the bark of *C. calisaya*, alkaloids (6.0%), flavonoids (5.0%), saponins (2.0%), and cardiac glycosides (3.54%) all seemed to be found [34]. The anthroquinones are a different class of compounds present in *C. pubescens*. Another significant category is essential oils, which account for 0.02 to 0.08% of all substances. Some other components have been identified including glycosides, organic acids (quinotanic acid, cinconic red), monoglycosides such as quinovic acid (3β-hydroxyurea-droxibenzoico acid), [33] Coumarins and terpene compounds [35]. Typically, Cinchona plants produce all these metabolites for protection and adaptation [31] (Fig. 10.4).

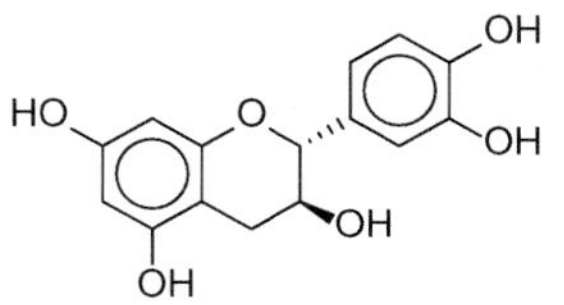

23

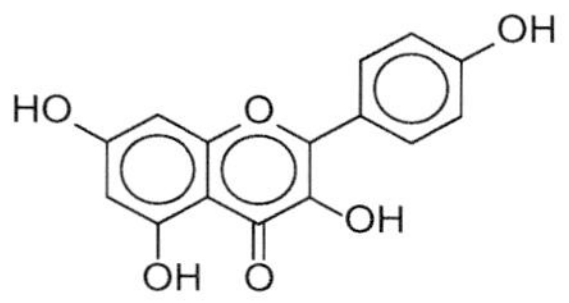

24

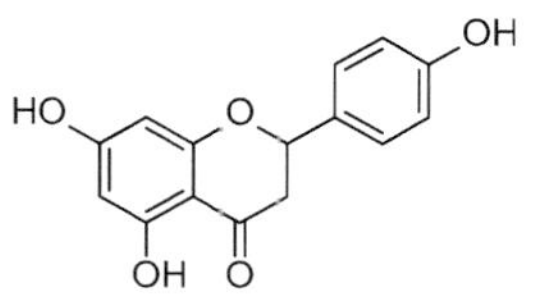

25

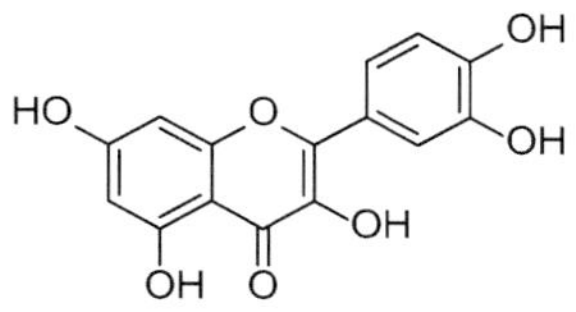

26

Fig. 10.4 Important flavonoids in Cinchona plant

10.13 Origin and Uses of Cinchona

For decades, the bark of various cinchona species has provided the febrifuge chemical quinine, which is useful against malaria. The crust of wild cinchona trees has been intensively collected in the Andes, resulting in dwindling inhabitants. Numerous varieties as well as hybrids have been grown in warm, moist climates across the world, notably in India and Southeast Asia.

Cinchona is used to stimulate hunger, promote the flow of digestive fluids, and alleviate bloating, fullness, and other stomach issues. Hemorrhoids, varicose veins, and leg cramps are among the conditions for which it is prescribed. Cinchona is used to treat moderate influenza, swine flu, the common cold, malaria, and fever in some persons. Cancer, tongue and throat illnesses, an enlarged spleen, and muscular cramps are among the other applications. Cinchona is used to dull discomfort, destroy germs, and act as an astringent in eye treatments. Cinchona extract is also used topically for haemorrhoids, ulcers, hair growth stimulation, and varicose vein management. Cinchona is used in tonic water and alcoholic drinks due to its bitter flavour.

10.14 Medicinal Uses of Cinchona

Plants have long been used to cure sickness in tropical areas, and their usage has been documented since the time when individuals from the old world first met the indigenous inhabitants of the Americas [36]. In South America, the Rubiaceae plants recognized as quina/cinchona (Cinchona spp.) and ipeca or ipecacuanha

(Carapichea ipecacuanha (Brot.) have sparked researcher's attention [37]. These plant species have been utilised by numerous groups of humans throughout history, including locals and settlers [38]. Indians from South America used cinchona bark as an anti-fever remedy before it was brought to Europe around 1640. The bark controlled malaria treatment in Europe until 1820, when pure quinine was discovered and largely substituted the natural substance [3]. Cinchona bark has a bitter flavour, was used to make beverages. Bitters have also been used as tonics and medications for recurring fevers; in the early 1800s, the impact of cinchona bark on fever therapy and prevention was demonstrated.

Cinchona bark was added as an adulterant to Jesuit's bark or Peruvian bark, which was originally thought to be connected to another fever remedy called Myroxylon peruiferum. Cinchona bark may be obtained in a variety of methods. The tree could be chopped down, however, girdling is as harmful and unsustainable as the chopping of the tree, so little pieces were cut and different treatments including "mossing," which involves applying bog to the damaged places, were utilized to let the tree to recuperate. Coppicing and cutting of side branches, which were later peeled of bark, were two other methods. Before 1820, the bark of the cinchona tree was dried, ground into fine particles, and then blended with a liquid (often wine) before being ingested. In 1820, French chemists extracted quinine from the bark before pure quinine eventually replaced the bark as the standard treatment for malarial illness. Quinine and other cinchona alkaloids, such as cinchonine, cinchonidine, and quinidine, are all antimalarial agents [39, 40]. Economically, Cinchona is the sole viable origin of quinine, and prescribed as the medicine to treat falciparum malaria [41].

10.15 Cinchona Alkaloids

Cinchona bark has historically been prized for its medicinal properties due to the presence of quinine and other alkaloids. For years, quinine and other cinchona alkaloids have been used as anti-pyretic and as treatments for malaria [71]. Besides these, cinchona alkaloids have various applications in the treatment of cardiac, muscular,cancer and other diseases [4].

10.16 Medicinal Uses of Quinine and Its Derivatives

It is the primary alkaloid found in bark of the cinchona plant, which is mostly derived from Cinchona calisaya Wedd. and Cinchona pubescens Vahl. for the treatment of malaria [42]. It is the most significant alkaloid that utilized in the form of salts, sulfate, bisulfate, hydrochloride and dichloride [26]. The first component discovered in the bark of the Cinchona tree (Cinchona officinalis) that effectively reduced some malarial fevers was quinine alkaloid [72]. It can be considered the first, unpolluted, and truly dynamic chemotherapeutic in this context. Regardless of the fact that quinine has a limited therapeutic window and the malaria vector *Plasmodium falciparum* is becoming increasingly resistant to it, it may still compete with newer antimalarial medicines, making it an essential medication for severe malaria infections [4]. The small half-life, unpleasant taste, as well as severe adverse health impacts such as headache, giddiness, seasickness, vomiting, and hemolysis all contributed to quinine's long-term usage as an antimalarial drug. Furthermore, drug resistance hindered the effectiveness of malarial infection remedy [43]. Numerous new antimalarial medicines using quinine scaffolds, such as hydroxyethylapoquinine, have been developed recently to decrease disadvantages as well as avoid drug resistance [44, 45].

Quinine is one of the most effective antimalarial alkaloids among the cinchona alkaloids [46]. Phenobarbital has the ability to partially overcome quinine resistance, according to a modest *in vitro* research [47]. It is being investigated whether calcium antagonists and additional medications (such prochlorperazine) can assist P. falciparum to restore its quinine resistance [48]. Antibiotics (such as artemisinin, artemether, clindamycin, doxycycline, and mefloquine) are being studied and utilised in conjunction with quinine to treat resistant P. falciparum strains [49, 50]. Quinine is supposed to have an antipyretic effect. Despite the fact that quinine given before acetaminophen causes a faster reduction in temperature as compared to quinine given after acetaminophen, quinine has no impact on temperature [51]. Quinine suppresses cancer cell growth by influencing apoptosis and inhibiting cell proliferation in a dosage and time dependent manner [52]. Anticancer medications such as bleomycin, cisplatin, anthracyclines, and radiation have been shown to cause an increase in the production of intracellular ROS [2].

In the nineteenth century, quinine was recognised as an ototoxic medication. Quinine was employed as an analgesic (for herpes zoster, otalgia) and as a healing mediator (for vertigo, Menière's illness, furuncles in the auditory canal, and purulent otitis media) in a variety of otologic conditions [53].

Maternal anaemia, delay in intrauterine development, low weight, premature births, and abortion are all complications of malarial infection during pregnancy. In order to minimise the above-mentioned negative consequences, it is vital to prevent and treat malaria during pregnancy. Because the use of artemisinin compounds during this period has not been proved due to its safety issues, the WHO now advises the pregnant women to use quinine with clindamycin for curing malarial infection in their first week to week 12 pregnancy (WHO 2010). Because most medical studies exclude women in their first 12 weeks, evidence on the efficiency as well as protection of anti-malarial medications in this time period of pregnancy is scarce. The majority of the evidence regarding quinine's safety in pregnancy is historical, and just a few medical studies have been available [54, 55].

For expectant women with malarial disease caused by chloroquine-resistant *P. falciparum* infection, quinine sulphate and clindamycin in a combined form is suggested. QS, in combination with other compounds such as clindamycin, doxycycline, or tetracycline, is a second therapy option for Plasmodium falciparum infections obtained in chloroquine-resistant regions. Infections obtained in Southeast Asia are treated for 7 days, while infections attained in Africa or South America are treated for 3 days. The goal of the combination is to improve quinine effectiveness, reduce treatment time, and reduce toxicity [56]. Clindamycin, instead, has an excellent pregnancy safety profile [57], and its PK parameters are typically unaffected throughout pregnancy [58].

The remedy of leg cramps produced by vascular spasm is another prominent usage of quinine. Quinine, hydroquinine, as well as quinidine have been used to avoid muscular cramps for over 50 years [59].

COVID-19, which is caused by the SARS-CoV-2 virus, has lately overloaded medical facilities and crippled economies. The unprecedented public outcry produced by this epidemic necessitated a hasty search for an efficient way to control or treat the disease. Quinine and their derivatives have been investigated as possible COVID-19 treatments due to their well-known anti-infectious and anti-inflammatory characteristics. Indeed, these molecules were first used to cure and prevent malaria, and then subsequently in the therapy of autoimmune, rheumatic, and dermatologic illnesses [60]. These antimalarial medications, on the other hand, are utilized for a number of chronic illnesses, including systemic lupus erythematosus, with minor adverse effects and are known to have anti-inflammatory and antiviral qualities. Because of its method of action, the antiviral activity of hydroxychloroquine and chloroquine has piqued the curiosity of several investigators [61]. Several in vitro and in vivo studies initially indicated encouraging findings for the use of hydroxychloroquine (HCQ) in the treatment of COVID-19 patients [62]. The daily doses of HCQ used to treat COVID-19 have ranged between 800 and 1600 mg. They did, however, establish the effective and safe dosage of HCQ based on evidence from in vitro investigations and clinical trials in one study. They looked at the link between viral load reduction and HCQ dosing in COVID-19 patients who had been treated. The results of this study indicate that an amount of HCQ taken every day should not be greater than 800 mg [62].

According to a recent research, it has been reported that Quinine Sulphate (QS) is a possible therapy option for SARS-CoV-2 infection that is well tolerated and has a predicted toxicological profile that is much better than hydroxychloroquine and chloroquine [63]. QS does, however, have a number of adverse effects that impact haematological, renal function, liver function, and cardiovascular health [64].

Quinine was used as a lead structure in the synthesis of various antimalarial medications such as chloroquine (**27**), mefloquine (**28**), pyrimethamine (**29**), proguanil (**30**), atovaquone (**31**). Quinine, is still used today to treat acute cases of severe P. falciparum, alone or in combination with doxocycline, tetracycline, or clindamycin [73].It is used for various purposes other than antimalarial such as:

- a flavour component in beverages and food industries [74].
- muscle relaxant
- hemorrhoid therapy
- an oxytoxic agent

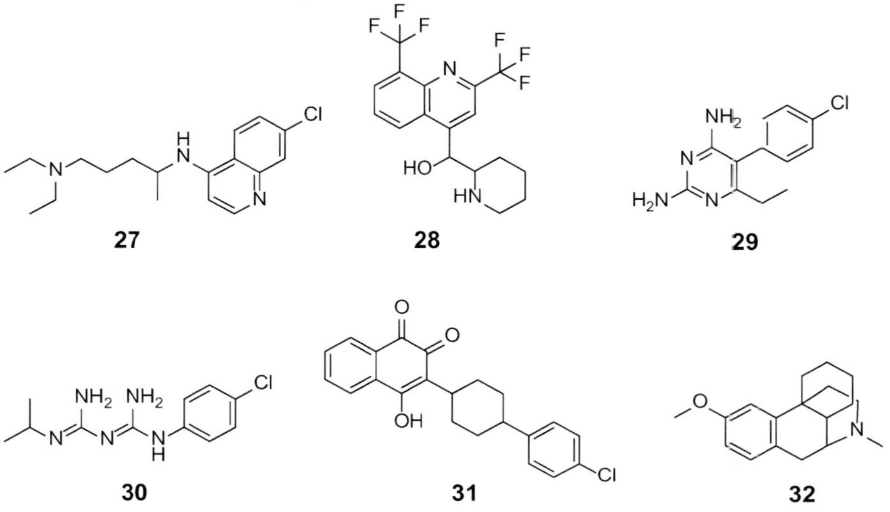

Fig. 10.5 Several Anti-malarial drugs

Cinchonism is the commonest side effect of quinine overdose. The primary site of damage by quinine overdose is the Nervous system including optic and auditory nerve [2] (Fig. 10.5).

10.17 Medicinal Uses of Quinidine and Its Derivatives

It is another important alkaloid found in the bark of cinchona plant. It works as antimalarial and antihypertensive [76].Quinidine was the first drug to be utilized for curing the cardiac arrhythmias. Quinidine, a diastereomer of the antimalarial quinine, was discovered to be the utmost effective of the antiarrhythmic compounds derived from this plant in early twentieth-century studies. Quinidine has been used as an antiarrhythmic drug since the 1920s to sustain sinus rhythm following alteration from atrial flutter or atrial fibrillation and to avoid ventricular tachycardia or ventricular fibrillation recurrence [65]. But the prevalence of gastrointestinal issues following its administration has been one of the main limiting factors in its utilization [66, 67]. A novel quinidine derivative was recently created to eliminate this adverse effect [68]. The novel derivative, quinidine polygalacturonic acidate, is created by the reaction of quinidine base and polygalacturonic acid. It has been demonstrated that the pharmacology of quinidine polygalac turonate is comparable to that of common quinidine compounds in both animals and humans. It acts on the heart to prevent fibrillation of the auricles by extending the heart muscle's refractory period, slowing conduction time, and reducing the heart muscle's excitability [69].

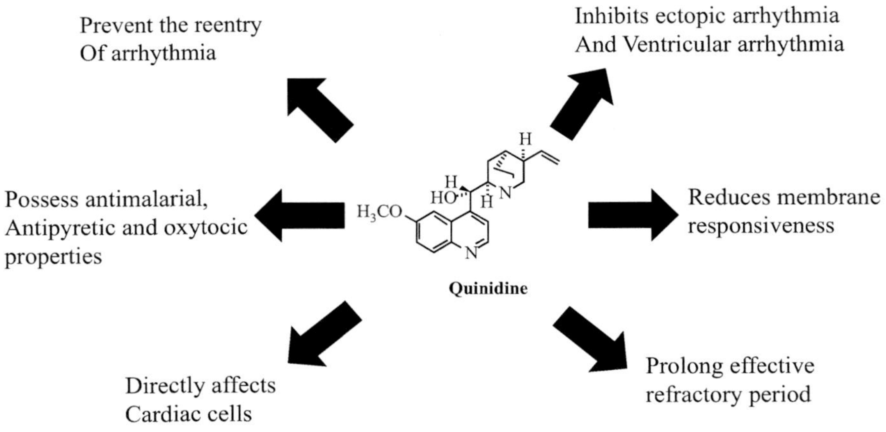

Fig. 10.6 Effects of quinidine

Quinidine has a variety of effects on the heart, many of which are paradoxical and incompletely understood. The following is a list of quinidine's effects: Quinidine works by blocking the vagus nerve's function on the heart, similar to atropine but with a less dramatic effect. When the vagus nerve is activated, its anticholinergy activity counteracts the effects of carotid sinus pressure and prevents heart rate lowering. Myocardial excitability is also reduced by using quinidine. 5′,6 The maximal rate at which the atrium reacts to repeated electrical stimulation is reduced after the heart has been exposed to quinidine. This mechanism is responsible for the drug's capacity to prevent or eliminate cardiac arrhythmias while simultaneously lowering myocardial contractility [70] (Fig. 10.6).

It is observed that relatively low doses of quinidine with dextromethorphan (**32**) is effective to treat the pseudobulbar affect disorder, and quinidine alone is rarely used to treat seizures [77].

10.18 Medicinal Uses of Cinchonine

It is also used as an antimalarial drug [3]. In comparison to Quinine, it is more active and less poisonous than other quinine-related substances [2]. It serves as an antibiotic agent and is used to treat schizonticide, amoebiasis, fever, dysentery, and fever. It causes mild aggravation of the gastric mucosa [74].Cinchonine has also a role in inhibition of osteoclast differentiation by regulating TAK1 and AKT through targeting TRAF6. It acts as a preventative measure against osteoporosis and bone problems brought on by inflammation [75].

10.19 Medicinal Uses of Cinchonidine

Cinchonidine is most frequently used as an anti-malarial drug. It can be present in the bark of most cinchona species, including *C. pusescensval* and *C. pitayensis* [3].

10.20 Other Applications of Cinchona Alkaloids

10.20.1 Anti-microbial Activity

The disc diffusion method is used to assess the antibacterial impact of cinchona alkaloids on Staphylococcus aureus, which ranges from 8 to 18 mm. Its antibacterial effectiveness increases as cinchona alkaloids concentrations increase. Herpes and disorders brought on by Plasmodium falciparum are treated with Cinchona bark [78]. Cinchona combats microorganisms that are harmful to the human body [79].

10.20.2 Anti-inflammatory Activity

For hundreds of years, cinchona bark has been used to treat inflammation and prevent malaria [80].

10.20.3 Anti-oxidant Activity

Cinchona has strong antioxidant qualities since it contains phenolic components (Phenols and flaveniods). The biological features of phenols include inhibition of lipid peroxidation, anti-HIV, anti-virus, and anticancer effects [3]. The extract of C. ledgeriana leaf has excellent antioxidant activity in the ethyl acetate fraction and water insoluble fraction, with IC_{50} values of 23.57 g/mL and 17.63 g/mL, respectively [26]. Cinchona alkaloids have the potential to be employed in anti-aging cosmetic preparations due to their strong antioxidant activity ($IC_{50} < 50$ g/mL) [24]. A possible anti-oxidant raw material for the cosmetics sector is the dry extracts from the species Cinchona pubescens Vahl [33].

10.20.4 Anti-obesity Properties

Cinchoquine which is considered a strong alkaloid is obtained from cinchona bark. Although it is mostly employed in anti-malaria activities, it can also treat the problem of obesity. In 2012,experiments have shown that cinchoquina is superior to other chemicals(derived from other plants)in the fight against obesity [3]. The body weight loss caused by chincinnovine was larger than that of EGCG and curcumin. A beneficial dietary phytochemical for preventing obesity and fatty tissue inflammation is pentachloroquine [81].

10.20.5 Anti-Cancer Activity

Cinchona alkaloids derived from C. legeriana and C. succirubra have been shown to be effective against a number of cancer cells, including lung cancer A-549, human myeloid leukaemia HL-60, breast cancer MCF-7, colon cancer SW480 and hepatocellular carcinoma SMMC-7721 [23]. Quinine affects the death of cancer cells and reduces their ability to proliferate in a dose- and time-dependent manner [2]. Apoptotic signals, the loss of apoptotic bodies and adhesion, cell contraction, membrane blistering, chromatid condensation, and nuclear fragmentation are only a few of the typical morphological alterations brought on by quinine's impact. In the future, quinine may be a successful anticancer medicine because it has a significant apoptotic action in cancer [82].

10.20.6 Anti-platelet Activity

It has been observed that Cinchona alkaloids works as an antiplatelet by preventing human platelet aggregation. Protein kinase C (PKC) and Ca2+ entrance are both inhibited, which prevents platelet aggregation [75].

10.20.7 Anti-viral Activity

Broad-spectrum antiviral properties are present in chloroquine phosphate, an analogue of quinine that was initially obtained from the bark of the Cinchona tree [83]. It has been experimented that Quinine has antiviral activity against SARS-CoV-2 in Vero cells. As evidenced by the endogenous expression of ACE2 and TMPRSS2 in Calu-3 lung cells, Quinine at concentrations over 50 M suppressed SARS-CoV-2 infection (IC_{50}: 3.7 to 50 M). Quinine thus, has the potential to be a less harmful antiviral medication for SARS-CoV-2. This demonstrates Quinine's potential to become a useful antivirus in the future [84].

10.20.8 Anti-diabetic Activity

The leaf extract of *C. ledgeriana* was screened for alpha-glucosidase inhibition activity by using Kim method at 400 nm (UV-Visible spectrophotometer) [85]. All C. ledgeriana fractions and extracts demonstrated significant enzyme inhibition activity with IC_{50} values ranging from 14.44 to 61.56 g/mL and LC_{50} values ranging from 14.79 to 120.22 g/mL Consequently, C. ledgeriana leaf extract may have anti-diabetic effects, although more isolation is required to lessen its toxicity [24]. Additionally, aqueous extracts of C. calisaya bark showed greater antidiabetic action at a lower dose of 50 mg/kg body wt when used to treat alloxan-induced diabetes mellitus [34].

10.20.9 Anti-fungal Activity

Various cinchona alkaloids and their derivatives were evaluated for activity against eight plant pathogenic fungi, including R. solani, P. zeae, B. cinerea, M. melonis, F. graminearum, M. oryzae, S. sclerotiorum, f.sp. vesinfectum and F. oxysporum by applying mycelium growth rate method. Major Quinolines (1, 2, 3, and 4) had weak antifungal activity. Derivatives, such as simplified quinone and quinotoxine molecules, on the other hand, frequently tend to boost the antifungal action [24, 86].

10.20.10 Hair Growth Stimulant

Major quinoline chemicals found in cinchona extract (1–4) may prevent hair follicle death and prevent hair loss [87]. Cinchona alkaloids have demonstrated their activity as hair growth stimulants. To exhibit the activity, their penetration into hair follicles and dermal papillae is very important to enhance the production of Vascular Endothelial Growth Factor. It is crucial for the growth of hairs and their regeneration. Cinchonine resulted in 17–43% increased hair length as compared to the controls [24].

10.20.11 Muscle Cramp

Cinchona alkaloids have neuromuscular and muscle effects [24] and has been used in the treatment of muscle cramps. Quinine (**1**) and quinine sulphate [4] have been found to reduce the occurrence and severity of muscle cramp including leg cramp [88] but use of quinine to alleviate muscle cramps is still a matter of debate regarding its effectiveness and safety [89] especially in older people, dose related side effects. Cinchona bark works as a muscle anti-cramp in low doses. It blocks the response of acetylcholine in Xenopus laevis oocytes. The response-blocking activity of acetylcholine is influenced by the concentration of cinchona alkaloids, such as Qunine (IC_{50}: 1.70 μM) and Quinidine (IC_{50}: 3.96 μM) [90]. Acetylcholine concentration has no effect on the response of blocking mechanism [24].

10.20.12 Anesthetic and Antipyretic Activity

Herbal medicines can be divided into local and general anesthetics depending on how they work. General anesthetics impair sensory and motor function by interacting with protein receptors and membranes, whereas local anesthetic herbs perform their function by interacting with voltage-gated Na+ channels [24]. In comparison to the anaesthetic effect of 2% xylocaine, the 10–20% concentration of aqueous extract of C. officinalis had an anesthetic effect ($p < 0.001$) of 72.12 and 88.08%. The *C. officinalis* extract displayed the antipyretic efficacy comparable to aspirin at single doses [91].

10.20.13 Insecticidal Agent

Acyloxy derivatives of cinchonidine and cinchonine have found to be effective insecticidal agents. This activity has been evaluated in vivo against Mythimna separate by leaf dipping method at a concentration of 1 mg/mL. The derivatives **33** and **34** demonstrated the highest insecticidal activity with final fatality rates of 75.0% and 71.4% respectively [92].

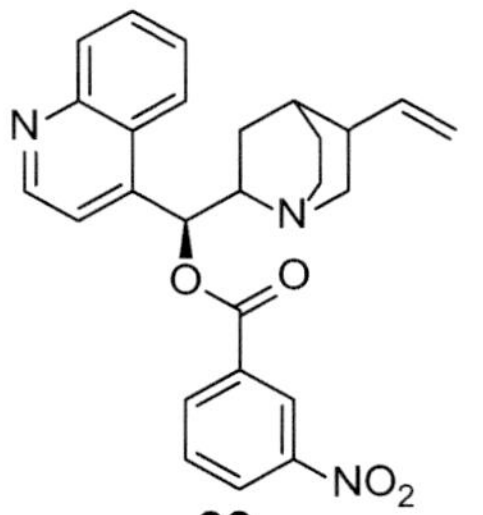

33

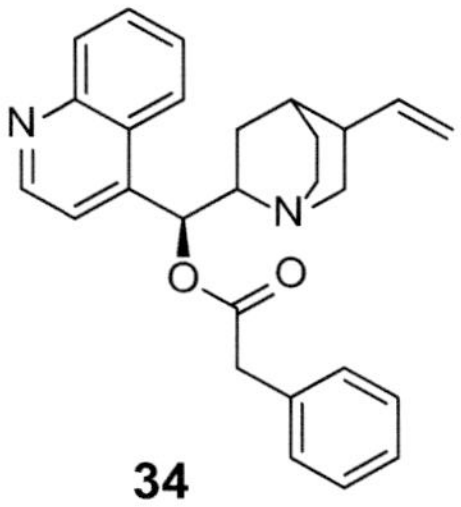

34

10.20.14 Applications in Organic Chemistry

Cinchona alkaloids have been employed as chiral catalysts, ligands, and NMR discriminating substances in organic chemistry for the past 30 years [93–96].Cinchona alkaloids have been reported as enantioselective organocatalyst under mild reaction conditions for the allylic amination of Morita–Baylis–Hillman carbonates using isatins as pronucleophiles. Quinine, Quinidine, O-benzylquinidine, (DHQD)2PHAL, (DHQ)2- PHAL, were used as a catalyst in this reaction and results have shown that (DHQD)2PHAL performed best catalytic activity afforded the required products in 81% yield with 57% enantiomeric excess [94].

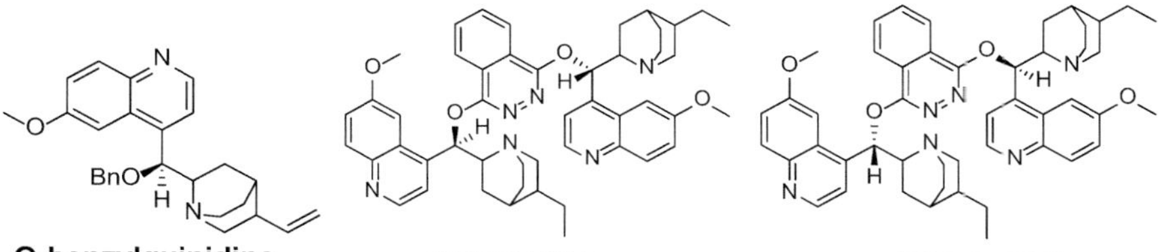

O-benzylquinidine (DHQD)2PHAL (DHQ)2- PHAL

They act as chiral resolving agents as well as catalysts or cocatalysts in both homogeneous and heterogeneous systems [97] and also works as a chiral selector in stationary phases or as a catalyst in asymmetric organic synthesis. Due to their low cost, ease of quaternization, and adaptability, cinchona alkaloids have been substantially for the synthesis of phase transfer catalysts [98, 99]. In 2016, Lajkó et al., stereoselectively separated the cyclic β-aminohydroxamic acid enantiomer pairs on the quinine or quinidine and chiral sulfonic acid-based zwitterionic chiral stationary phases ZWIX(+)™, ZWIX(−)™, ZWIX(+A) and ZWIX(−A) by high performance liquid chromatography [99] (Fig. 10.7).

10.21 Biosynthesis of Cinchona Alkaloids

The complete biosynthesis of alkaloids was suggested in the 1960s [100] and has only been slightly altered (Schemes 10.1) [101, 102]. Cinchona alkaloids belong to the terpene-indole alkaloid family, which also contains well-known natural compounds like strychnine, yohimbine, reser pine, and ajmaline. Strictosidine,which can be traced back to geraniol is formed from tryptamine and secologanine and it is a common biosynthetic intermediary for all of them.. It is converted to cinchonaminal through the Cinchona alkaloids-specific pathway as this step is mediated by strictosidine glycosidase, which involves in the formation of six membered ring containing nitrogen atom, followed by hydrolysis and decarboxylation of an intermediate molecule to produce corynantheal. In chinchonaminal, the azabicyclo[2.2.2]octane system is eventually produced via ring rearrangement through

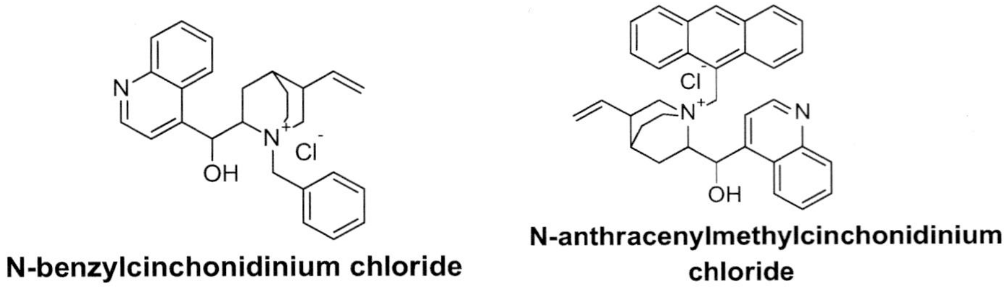

Fig. 10.7 Cinchona alkaloids as phase transfer catalysts

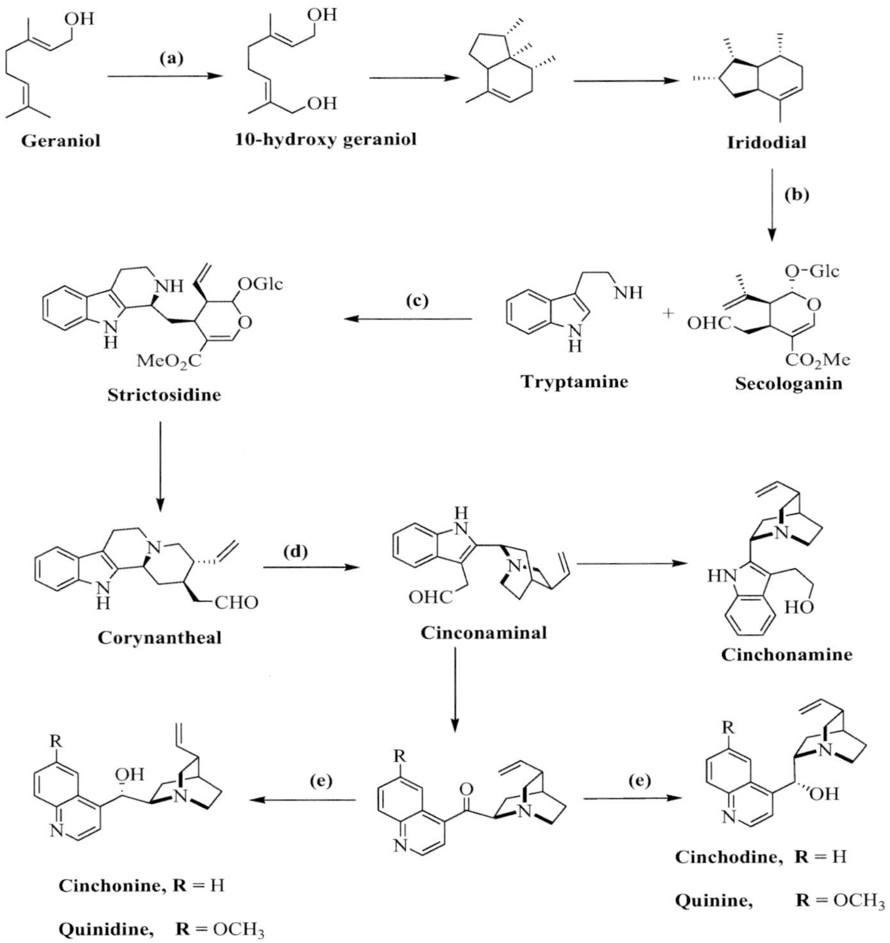

Scheme 10.1 Biosynthesis of chemical components present in Cinchona. Enzymes included in the pathway: (**a**) geraniol 10-hydroxylase; (**b**) secologanin synthase; (**c**) strictosidine synthase; (**d**) strictosidine glucosidase; (**e**) cinchoninone: NADPH oxidoreductase

iminium intermediates named as cinchonamial. Cinchonamine, a minor alkaloid, may be formed by reducing the pendant aldehyde group in cinchonamial. The primary route, on the other hand, leads to cinchonaminal by the hydrolytic breakage of the indole C–N link, which results in the dicarbonyl intermediate.

10.22 Future Prospective of *Cinchona officinalis*

Natural goods are becoming increasingly popular, and it is expected that cinchona bark will continue to be a staple of herbal treatment for a long time. Natural goods are becoming more popular in western nations. Cinchona bark is used to treat sore throats as a gargle. It's used to treat nervous ergotamines, anaemia, and convalescence in homoeopathy. Cinchona bark is a leading antiprotozoal treatment in European herbal medicine, used to treat a variety of fevers. It aids in the treatment of gall bladder, enlarged spleen, and liver problems. It's utilised as a stimulant in the products use for the growth of hair. The bark has made a resurgence in the form of "bark tea," that is utilized to treat malaria.

Throughout the previous four eras, the natural alkaloid quinine found in Cinchona has played a significant part in both the anticipation as well as handling of malarial desease. Malaria was treated before the discovery of quinine by releasing humours, which included blood loss, removal, and by employing laxatives. English physicians considered Robert Talbor, the first person to employ quinine in England, as a quack. He gained prominence only after successfully curing Charles II of France from a deadly malarial illness [103]. Even the synthetic versions of quinine that have taken its place are based on a new model of the molecule, demonstrating its uniqueness.

References

1. Andersson, L., & Antonelli, A. (2005). Phylogeny of the tribe Cinchoneae (Rubiaceae), its position in Cinchonoideae, and description of a new genus. *Ciliosemina Taxon, 54*(1), 17–28.
2. Raza, M. A., Rehman, F. U., Anwar, S., Zahra, A., Rehman, A., Rashid, E., et al. (2021). The medicinal and aromatic activities of cinchona: A review. *Asian Journal of Advances in Research, 29*, 42–45.
3. Gurung, P., & De, P. (2017). Spectrum of biological properties of cinchona alkaloids: A brief review. *Journal of Pharmacognosy and Phytochemistry, 6*(4), 162–166.
4. Kishor, C. B., Tanu, G., & Radhey, M. S. (2018). Alkaloid group of Cinchona officinalis: Structural, synthetic, and medicinal aspects. *Synthesis of Medicinal Agents from Plants, 2018*, 205–227.
5. Loustalot, A. J., Winters, H. F., & Childers, N. F. (1947). Influence of high, medium, and low soil moisture on growth and alkaloid content of Cinchona ledgeriana *Plant Physiology, 22*(4), 613.
6. Datta, M., Sah, K., Gupta, S., & Banerjee, S. (1990). Characteristics of soils supporting cinchona plants in Darjeeling hills. *Indian Agriculturist, 34*(2), 73–77.

7. Barrutia, R. R. R., Barreto, I. B., & Velásquez, T. D. M. (2020). Germination of Cinchona officinalis L. seeds in three soils types of Cajamarca, Peru. *Revista Cubana de Ciencias Forestales, 8*(1), 75–87.
8. Chatterjee, S. (1992). In domestication studies of some medicinally important exotic plants growing in India. In *WOCMAP I-medicinal and aromatic plants conference: Part 3 of 4 331* (pp. 151–158).
9. Fischer, C. A. F. (1944). Growing Cinchona under American Control. *Torreya, 44*(1), 1–5.
10. Nandi, R. (1993). *Cultivation of cinchona and production of its alkaloids in India* (pp. 1–67). S. Nandi.
11. Jäger, H., & Kowarik, I. (2010). Resilience of native plant community following manual control of invasive Cinchona pubescens in Galápagos. *Restoration Ecology, 18*, 103–112.
12. Boyce, J. S. (1938). *Forest pathology*. Forest Pathology.
13. Hartley, C. P. (1919). *Damping-off in forest nurseries*. US Government Printing Office.
14. Heim, R. (1940). Un Agaric rhizomorphique parasite des semis de Quinquina en Haute-Guinée. *Journal d'agriculture traditionnelle et de botanique appliquée, 20*(222), 77–87.
15. Hunger, F. W. T. (1907). *Proeve omtrent schaduw-cultuur met Deli-tabak op Sumatra's oost-kust*. G. Kolff & Company.
16. Aerts, R. J., Stoker, A., Beishuizen, M., Jaarsma, I., Van De Heuvel, M., Van Der Meijden, E., et al. (1992). Detrimental effects ofCinchona leaf alkaloids on larvae of the polyphagous insectSpodoptera exigua. *Journal of Chemical Ecology, 18*(11), 1955–1964.
17. Dayrit, F. M., Guidote, A. M., Jr., Generalao, M. L., & Serna, C. (1994). *Determination of the quinine content in the bark of the cinchona tree grown in Mt.* Kitanglad.
18. Nair, K. P. (2021). *Tree crops*. Springer.
19. Tracy, J. W. (1996). Drugs used in the chemotherapy of protozoal infections. In *Goodman and Gilman's the pharmacological basis of therapeutics* (pp. 965–985).
20. Genne, P., Duchamp, O., Solary, E., Pinard, D., Belon, J., Dimanche-Boitrel, M., et al. (1994). Comparative effects of quinine and cinchonine in reversing multidrug resistance on human leukemic cell line K562/ADM. *Leukemia, 8*(1), 160–164.
21. Munther, K., & Homoud, M. (2008). Tufts-new England medical center. In *Introduction to Antiarrhythmic agents*. Spring.
22. McCalley, D. V. (1990). Quantitative analysis of alkaloids from cinchona bark by high-performance liquid chromatography. *Analyst, 115*(10), 1355–1358.
23. Cheng, G. G., Cai, X. H., Zhang, B. H., Li, Y., Gu, J., Bao, M. F., et al. (2014). Cinchona alkaloids from Cinchona succirubra and Cinchona ledgeriana. *Planta Medica, 80*(02/03), 223–230.
24. Hariyanti, H., Mauludin, R., Sumirtapura, Y. C., & Kurniati, N. F. (2022). *A review: Pharmacological activities of quinoline alkaloid of Cinchona sp.*
25. Martins, D., & Nunez, C. V. (2015). Secondary metabolites from Rubiaceae species. *Molecules, 20*(7), 13422–13495.
26. Sundowo, A., Artanti, N., Hanafi, M., Minarti, M. & Primahana, G. (2017, November). Phytochemical screening, total phenolic, total flavonoids contents and antioxidant activity of Cinchona ledgeriana leaves ethanol extract. In *AIP conference proceedings* (Vol. 1904, no. 1, p. 020067). AIP Publishing LLC.
27. Canales, N. A., Hansen, T. N. G., Cornett, C., Walker, K., Driver, F., Antonelli, A., et al. (2020). Historical chemical annotations of cinchona bark collections are comparable to results from current day high-pressure liquid chromatography technologies. *Journal of Ethnopharmacology, 249*, 112375.
28. Maldonado, C., Barnes, C. J., Cornett, C., Holmfred, E., Hansen, S. H., Persson, C., et al. (2017). Phylogeny predicts the quantity of antimalarial alkaloids within the iconic yellow cinchona bark (Rubiaceae: Cinchona calisaya). *Frontiers in Plant Science, 8*, 391.
29. Kacprzak, K. M. (2013). Chemistry and biology of cinchona alkaloids. *Natural Products, 1*, 605–641.

30. Júnior, W. S. F., Cruz, M. P., dos Santos, L. L., & Medeiros, M. F. T. (2012). Use and importance of quina (Cinchona spp.) and ipeca (Carapichea ipecacuanha (Brot.) L. Andersson): Plants for medicinal use from the 16th century to the present. *Journal of Herbal Medicine, 2*(4), 103–112.
31. Pratiwi, D. R., Sulistyaningsih, Y. C., & Ratnadewi, D. (2020). Localization of alkaloid and other secondary metabolites in Cinchona ledgeriana Moens: Anatomical and histochemical studies on fresh tissues and cultured cells. *HAYATI Journal of Biosciences, 27*(1), 1–1.
32. MN, R., & Padh, H. (2003). Antioxidant activity of Cinchona officinalis stem bark extracts. *Advances in Traditional Medicine, 3*(4), 205–211.
33. Noriega, P., Sola, M., Barukcic, A., Garcia, K., & Osorio, E. (2015). Cosmetic antioxidant potential of extracts from species of the Cinchona pubescens (Vahl). *International Journal of Phytocosmetics and Natural Ingredients, 2*(1), 14–14.
34. Ezekwesili, C. N., Ogbunugafor, H. A., & Ezekwesili-Ofili, J. O. (2012). Anti-diabetic activity of aqueous extracts of Vitex doniana leaves and Cinchona calisaya bark in alloxan–induced diabetic rats. *International Journal of Tropical Disease & Health, 2*(4), 290–300.
35. Somavilla, N. S., Cosenza, G. P., Fagg, C. W., & Brandão, M. G. (2017). Morpho-anatomy and chemical profile of native species used as substitute of quina (Cinchona spp.) in Brazilian traditional medicine. Part II: Remijia ferruginea. *Revista Brasileira de Farmacognosia, 27*, 153–157.
36. Ortiz Crespo, F. (1994). La Cinchona antes y después del Virreinato del Conde de Chinchón. *Interciencia, 19*(3), 130–136.
37. de Boer, H. J., & Thulin, M. (2005). Lectotypification of Callicocca ipecacuanha Brot. and neotypification of Cephaelis acuminata H. Karst., with reference to the drug ipecac. *Taxon, 54*(4), 1080–1082.
38. Möller, M., Herzer, K., Wenger, T., Herr, I., & Wink, M. (2007). The alkaloid emetine as a promising agent for the induction and enhancement of drug-induced apoptosis in leukemia cells. *Oncology Reports, 18*(3), 737–744.
39. Christensen, S. B. (2015). Drugs and drug leads based on natural products for treatment and prophylaxis of malaria. In *Evidence-Based Validation of Herbal Medicine* (pp. 307–319). Elsevier.
40. Dobson, C. M. (2001). The structural basis of protein folding and its links with human disease. *Philosophical Transactions of the Royal Society of London. Series B: Biological Sciences, 356*(1406), 133–145.
41. Achan, J., Talisuna, A. O., Erhart, A., Yeka, A., Tibenderana, J. K., Baliraine, F. N., et al. (2011). Quinine, an old anti-malarial drug in a modern world: Role in the treatment of malaria. *Malaria Journal, 10*(1), 1–12.
42. Cosenza, G. P., Somavilla, N. S., Fagg, C. W., & Brandão, M. G. (2013). Bitter plants used as substitute of Cinchona spp.(quina) in Brazilian traditional medicine. *Journal of Ethnopharmacology, 149*(3), 790–796.
43. Shanks, G. D. (2016). Historical review: Problematic malaria prophylaxis with quinine. *The American Journal of Tropical Medicine and Hygiene, 95*(2), 269.
44. Sanders, N. G., Meyers, D. J., & Sullivan, D. J. (2014). Antimalarial efficacy of hydroxyethylapoquinine (SN-119) and its derivatives. *Antimicrobial Agents and Chemotherapy, 58*(2), 820–827.
45. Mohammadi, S., Jafari, B., Asgharian, P., Martorell, M., & Sharifi-Rad, J. (2020). Medicinal plants used in the treatment of malaria: A key emphasis to Artemisia, cinchona, Cryptolepis, and Tabebuia genera. *Phytotherapy Research, 34*(7), 1556–1569.
46. Leung, A. Y. (1980). *Encyclopedia of common natural ingredients used in food, drugs, and cosmetics*. Wiley.
47. Mangou, F., Platel, D. F., & Haumont, G. (1998). Plasmodium falciparum: Modulation by phenobarbital of sensitivity to quinine, chloroquine and mefloquine in vitro. *Transactions of the Royal Society of Tropical Medicine and Hygiene, 92*(5), 561–562.

48. Watt, G., Na-Nakorn, A., Bateman, D., Plubha, N., Mothanaprakoon, P., Edstein, M., & Webster, H. K. (1993). *Amplification of quinine cardiac effects by the resistance-reversing agent prochlorperazine in falciparum malaria* (p. 96346). Armed Forces Research Inst of Medical Sciences APO San Francisco.
49. de Vries, P. J., Bich, N. N., Van Thien, H., Hung, L. N., Anh, T. K., Kager, P. A., et al. (2000). Combinations of artemisinin and quinine for uncomplicated falciparum malaria: Efficacy and pharmacodynamics. *Antimicrobial Agents and Chemotherapy, 44*(5), 1302–1308.
50. Kremsner, P. G., Radloff, P., Metzger, W., Wildling, E., Mordmüller, B., Philipps, J., et al. (1995). Quinine plus clindamycin improves chemotherapy of severe malaria in children. *Antimicrobial Agents and Chemotherapy, 39*(7), 1603–1605.
51. Olin, B. R., & Hebel, S. (1994). *Drug facts and comparisons*. Facts and Comparisons St.
52. Jacoby, D. B., & Youngson, R. M. (2004). *Encyclopedia of family health*. Marshall Cavendish.
53. Semedo, M. G., Dias-Silva, N., Miguéis, J., & Pita, J. R. (2021). Quinine in otology and neurotology: Ototoxicity and historic role in therapy. *Otology & Neurotology, 42*(1), 145–152.
54. McGready, R., Cho, T., Villegas, L., Brockman, A., van Vugt, M., Looareesuwan, S., et al. (2001). Randomized comparison of quinine-clindamycin versus artesunate in the treatment of falciparum malaria in pregnancy. *Transactions of the Royal Society of Tropical Medicine and Hygiene, 95*(6), 651–656.
55. Adam, I., Ibrahim, M., Aelbasit, I., & El Bashir, M. (2004). Low-dose quinine for treatment of chloroquine-resistant falciparum malaria in Sudanese pregnant women. *EMHJ-Eastern Mediterranean Health Journal, 10*(4–5), 554–559.
56. Vinetz, J. M., Clain, J., Bounkeua, V., Eastman, R. T., & Fidock, D. (2011). Chemotherapy of malaria. *The pharmacological basis of therapeutics, 12*, 1383–1418.
57. Lell, B., & Kremsner, P. G. (2002). Clindamycin as an antimalarial drug: Review of clinical trials. *Antimicrobial Agents and Chemotherapy, 46*(8), 2315–2320.
58. Philipson, A., Sabath, L., & Charles, D. (1976). Erythromycin and clindamycin absorption and elimination in pregnant women. *Clinical Pharmacology & Therapeutics, 19*(1), 68–77.
59. Jansen, P., Veenhuizen, K., Wesseling, A., de Boo, T., & Verbeek, A. (1997). Randomised controlled trial of hydroquinine in muscle cramps. *The Lancet, 349*(9051), 528–532.
60. Younis, N. K., Zareef, R. O., Al Hassan, S. N., Bitar, F., Eid, A. H., & Arabi, M. Hydroxychloroquine in COVID-19 patients: Pros and cons. *Frontiers in Pharmacology, 2020, 11*, 597985.
61. Sinha, N., & Balayla, G. (2020). Hydroxychloroquine and covid-19. *Postgraduate Medical Journal, 96*(1139), 550–555.
62. Garcia-Cremades, M., Solans, B. P., Hughes, E., Ernest, J. P., Wallender, E., Aweeka, F., Luetkemeyer, A. F., & Savic, R. M. (2020). Optimizing hydroxychloroquine dosing for patients with COVID-19: An integrative modeling approach for effective drug repurposing. *Clinical Pharmacology & Therapeutics, 108*(2), 253–263.
63. Große, M., Ruetalo, N., Businger, R., Rheber, S., Setz, C., & Rauch, P., et al. (2020). *Evidence that quinine exhibits antiviral activity against SARS-CoV-2 infection in vitro.*
64. Taylor, W. R. J., & White, N. J. (2004). Antimalarial drug toxicity. *Drug Safety, 27*(1), 25–61.
65. Bozic, B., Uzelac, T. V., Kezic, A., & Bajcetic, M. (2018). The role of quinidine in the pharmacological therapy of ventricular arrhythmias 'quinidine'. *Mini Reviews in Medicinal Chemistry, 18*(6), 468–475.
66. Shaftel, N., & Halpern, A. (1958). The quinidine problem. *Angiology, 9*(1), 34–46.
67. Linenthal, A. (1955). The use of quinidine in the treatment of cardiac arrhythmias. *Modern Concepts of Cardiovascular Disease, 24*(11), 299–301.
68. Halpern, A., Shaftel, N., & Bovi, A. M. (1958). The synthesis and study of quinidine polygalacturonate. *American Journal of Pharmacy and the Sciences Supporting Public Health, 130*(6), 190–201.
69. Schwartz, G. (1959). A clinical evaluation of quinidine polygalacturonate. *Angiology, 10*(2), 115–119.

70. Sokolow, M., & Perloff, D. B. (1961). The clinical pharmacology and use of quinidine in heart disease. *Progress in Cardiovascular Diseases, 3*(4), 316–330.
71. Kumar, S., & Egbuna, C. (2019). *Phytochemistry: An in-silico and in-vitro update*. Springer.
72. Zhang, B. M., Zhi-Bin, W. A. N. G., Ping, X. I. N., Qiu-Hong, W. A. N. G., He, B. U., & Kuang, H. X. (2018). Phytochemistry and pharmacology of genus Ephedra. *Chinese Journal of Natural Medicines, 16*(11), 811–828.
73. Wink, M. (2012). Medicinal plants: A source of anti-parasitic secondary metabolites. *Molecules, 17*(11), 12771–12791.
74. Ramawat, K. G., & Mérillon, J. M. (2013). *Phytochemistry, Botany and Metabolism of Alkaloids, Phenolics and Terpenes, Natural products*. Springer.
75. Jo, Y. J., Lee, H. I., Kim, N., Hwang, D., Lee, J., Lee, G. R., et al. (2021). Cinchonine inhibits osteoclast differentiation by regulating TAK1 and AKT, and promotes osteogenesis. *Journal of Cellular Physiology, 236*(3), 1854–1865.
76. Jung, S. A., Choi, M., Kim, S., Yu, R., & Park, T. (2012). *Cinchonine prevents high-fat-diet-induced obesity through downregulation of adipogenesis and adipose inflammation*. PPAR Research.
77. Eyal, S. (2018). The fever tree: From malaria to neurological diseases. *Toxins, 10*(12), 491.
78. Tyagi, R., Sharma, G., Jasuja, N. D., & Menghani, E. (2016). Indian medicinal plants as an effective antimicrobial agent. *Journal of Critical Reviews, 3*(2), 69–71.
79. Rojas, J. J., Ochoa, V. J., Ocampo, S. A., & Muñoz, J. F. (2006). Screening for antimicrobial activity of ten medicinal plants used in Colombian folkloric medicine: A possible alternative in the treatment of non-nosocomial infections. *BMC Complementary and Alternative Medicine, 6*(1), 1–6.
80. Pap, T., Van Der Laan, W. H., Aupperle, K. R., Gay, R. E., Verheijen, J. H., Firestein, G. S., et al. (2000). Modulation of fibroblast-mediated cartilage degradation by articular chondrocytes in rheumatoid arthritis. *Arthritis & Rheumatism: Official Journal of the American College of Rheumatology, 43*(11), 2531–2536.
81. Sae-Tan, S., Grove, K. A., Kennett, M. J., & Lambert, J. D. (2011). (−)-Epigallocatechin-3-gallate increases the expression of genes related to fat oxidation in the skeletal muscle of high fat-fed mice. *Food & Function, 2*(2), 111–116.
82. Kampa, M., Alexaki, V. I., Notas, G., Nifli, A. P., Nistikaki, A., Hatzoglou, A., et al. (2004). Antiproliferative and apoptotic effects of selective phenolic acids on T47D human breast cancer cells: Potential mechanisms of action. *Breast Cancer Research, 6*(2), 1–12.
83. Sohail, M. I., Siddiqui, A., Erum, N., & Kamran, M. (2021). Phytomedicine and the COVID-19 pandemic. In *Phytomedicine* (pp. 693–708). Academic.
84. Große, M., Ruetalo, N., Layer, M., Hu, D., Businger, R., Rheber, S., et al. (2021). Quinine inhibits infection of human cell lines with SARS-CoV-2. *Viruses, 13*(4), 647.
85. Zribi, B., Roy, E., Pallandre, A., Chebil, S., Koubaa, M., Mejri, N., et al. (2016). A microfluidic electrochemical biosensor based on multiwall carbon nanotube/ferrocene for genomic DNA detection of Mycobacterium tuberculosis in clinical isolates. *Biomicrofluidics, 10*(1), 014115.
86. Yang, G. Z., Zhu, J. K., Yin, X. D., Yan, Y. F., Wang, Y. L., Shang, X. F., et al. (2019). Design, synthesis, and antifungal evaluation of novel quinoline derivatives inspired from natural quinine alkaloids. *Journal of Agricultural and Food Chemistry, 67*(41), 11340–11353.
87. Ou, H. C., Keating, S., Wu, P., Simon, J. A., Raible, D. W., & Rubel, E. W. (2012). Quinoline ring derivatives protect against aminoglycoside-induced hair cell death in the zebrafish lateral line. *Journal of the Association for Research in Otolaryngology, 13*(6), 759–770.
88. Rabbitt, L., Mulkerrin, E. C., & O'Keeffe, S. T. (2016). A review of nocturnal leg cramps in older people. *Age and Ageing, 45*(6), 776–782.
89. El-Tawil, S., Al Musa, T., Valli, H., Lunn, M. P., Brassington, R., El-Tawil, T., & Weber, M. (2015). Quinine for muscle cramps. *Cochrane Database of Systematic Reviews, 4*, CD005044.

90. Gisselmann, G., Alisch, D., Welbers-Joop, B., & Hatt, H. (2018). Effects of quinine, quinidine and chloroquine on human muscle nicotinic acetylcholine receptors. *Frontiers in Pharmacology, 9*, 1339.
91. Li, Y., & Tian, J. (2016). Evaluation of local anesthetic and antipyretic activities of cinchona alkaloids in some animal models. *Tropical Journal of Pharmaceutical Research, 15*(8), 1663–1666.
92. Che, Z. P., Yang, J. M., Zhang, S., Sun, D., Tian, Y. E., Liu, S. M., et al. (2021). Synthesis of novel 9 R/S-acyloxy derivatives of cinchonidine and cinchonine as insecticidal agents. *Journal of Asian Natural Products Research, 23*(2), 163–175.
93. Raheem, I. T., Goodman, S. N., & Jacobsen, E. N. (2004). Catalytic asymmetric total syntheses of quinine and quinidine. *Journal of the American Chemical Society, 126*(3), 706–707.
94. Zhao, M. X., Chen, M. X., Tang, W. H., Wei, D. K., Dai, T. L., & Shi, M. (2012). Cinchona alkaloid catalyzed Regio-and enantioselective allylic amination of Morita–Baylis–Hillman carbonates with Isatins. *European Journal of Organic Chemistry, 2012*(19), 3598–3606.
95. Quigley, C., Rodríguez-Docampo, Z., & Connon, S. J. (2012). Highly tunable arylated cinchona alkaloids as bifunctional catalysts. *Chemical Communications, 48*(10), 1443–1445.
96. Dijkstra, G. D., Kellogg, R. M., Wynberg, H., Svendsen, J. S., Marko, I., & Sharpless, K. B. (1989). Conformational study of cinchona alkaloids. A combined NMR, molecular mechanics and x-ray approach. *Journal of the American Chemical Society, 111*(21), 8069–8076.
97. Lai, J., Ma, Z., Mink, L., Mueller, L. J., & Zaera, F. (2009). Influence of peripheral groups on the physical and chemical behavior of cinchona alkaloids. *The Journal of Physical Chemistry B, 113*(34), 11696–11701.
98. Yeboah, E. M., Yeboah, S. O., & Singh, G. S. (2011). Recent applications of cinchona alkaloids and their derivatives as catalysts in metal-free asymmetric synthesis. *Tetrahedron, 67*, 10.
99. Lajkó, G., Orosz, T., Grecsó, N., Fekete, B., Palkó, M., Fülöp, F., et al. (2016). High-performance liquid chromatographic enantioseparation of cyclic β-aminohydroxamic acids on zwitterionic chiral stationary phases based on cinchona alkaloids. *Analytica Chimica Acta, 921*, 84–94.
100. Leete, E. (1969). Biosynthesis of quinine and related alkaloids. *Accounts of Chemical Research, 2*(2), 59–64.
101. O'Connor, S. E., & Maresh, J. J. (2006). Chemistry and biology of monoterpene indole alkaloid biosynthesis. *Natural Product Reports, 23*(4), 532–547.
102. Dewick, P. M. (2002). *Medicinal natural products: A biosynthetic approach*. Wiley.
103. Keeble, T. (1997). A cure for the ague: The contribution of Robert Talbor (1642–81). *Journal of the Royal Society of Medicine, 90*(5), 285–290.

Chapter 11
Patchouli

Muhammad Umar Ijaz, Ali Akbar, Haseeb Anwar, Sana Inam, Asma Ashraf, and Muhammad Riaz

11.1 Introduction

Pelletier-Sautelet was the person who discovered the plant (*P. cablin*) for the first time and named it Pogostemon patchouli in 1845. This plant was generally known as "cabalam" in the certain regions of Philippines. A scientist named Holmes characterized this plant as *Pogostemon cablin* in 1896 [1]. *P. cablin* is exclusively indigenous to the Philippines as well as it can also grow naturally in various South Asian regions. In South East Asia, patchouli is a significantly regarded as perennial herb that blooms efficiently in warm, humid environmental conditions [2]

P. cablin grows in soil which is considerably enriched with nutrients, shaded with other trees or certain other objects, having well-drainage system and receives frequent irrigation supply as well as moderate rainy seasons. In general, patchouli plant grows and reproduces at its best in areas having similar environmental conditions to those in the Philippines such as Southern India, Andaman and Nicobar Islands [1]. This plant can grow approximately up to 500 m above the sea level & can survive in moderate temperature of about 22–28 °C. Furthermore, humidity

M. U. Ijaz · A. Akbar
Department of Zoology, Wildlife and Fisheries, University of Agriculture, Faisalabad, Pakistan

H. Anwar
Department of Physiology, Government College University, Faisalabad, Pakistan

S. Inam
Department of pharmacy, Government College University, Faisalabad, Pakistan

A. Ashraf (✉)
Department of Zoology, Government College University, Faisalabad, Pakistan
e-mail: asmaashraf@gcuf.edu.pk

M. Riaz
Department of Pharmacy, Shaheed Benazir Bhutto University, Sheringal, Pakistan

M. Zia-Ul-Haq et al. (eds.), *Essentials of Medicinal and Aromatic Crops*,
https://doi.org/10.1007/978-3-031-35403-8_11

level of over 75% following PH level of 5.5–6.2 is considered as favorable for optimum growth and high-quality essential oils of *P. cablin* [3].

P. cablin is also regarded as an efficient and profitable crop which is widely distributed in several regions of Malaysia, Singapore Indonesia and China [3, 4]. *P. cablin* plant is well known for its potent therapeutic abilities such as antioxidant, antimutagenic, antimicrobial, antiplatelet, antidepressant, antithrombotic, aphrodisiac, analgesic, antiemetic, cytotoxic and fibrinolytic properties [5, 6].

Furthermore, essential oil of *P. cablin* is also used to treat sadness, involves in relaxing mind, regulation of appetite, reduces stress and stimulate sex attraction. Additionally, patchouli plant has antibacterial, antifungal, as well as insecticidal properties [5]. All the aforementioned characteristics enable this plant to be universally recognized as a commercial crop with significant industrial importance as well as certain biological properties [7, 8].

11.1.1 Taxonomic Position

P. cablin is a member of mint family, Lamiaceae. The taxonomic classification of this plant is given as:

Kingdom: Plantae
Division: Magnoliophyta
Class: Magnoliopsida
Order: Lamiales
Family: Lamiaceae (Labiatae)
Genus: *Pogostemon*
Species: *P. cablin*

11.1.2 Morphological Features

P. cablin is a quadrangular plant which can grow up to 1–1.2 m in length having extensive branching with upright stature. Leaves of this plant are simple, decussate, opposite. When grown in the open environment leaves of this plant are visible as purplish-green in coloure, but they can turn bright green when grown in the shady environmental conditions [9].

Using electron microscopy, the epidermal and mesophyll kinds of secretory glands are discovered in immature *P. cablin* plant leaves. Both the adaxial and abaxial epidermis of this plant give rise to glandular trichomes [10]. The micromorphological traits of patchouli flowers and its pollens collected from various populations revealed the existence of variations in morphology with alterations in environmental conditions [11].

11.2 Nutritive and Bioactive Compounds

P. cablin contains approximately 14.19% crude protein, 3.85% fat and 17.09% of crude fibers. *P. cablin* waste also contains bioactive compounds such as tannins, steroids, glycosides, anthraquinones, steroids, terpenoids, tannins, saponins, phenol, alkaloids as well as flavonoids [12].

11.2.1 Aerial Parts

Alpha-guaiene (6.4–17.9%), pogostol (0.3–5.0%), alpha-bulnesene (9.0–15.2%), and (E)-caryophyllene (2.1–3.6%), Patchouli alcohol (42.2–57.7%), seychellene (3.4–6.9%), are the main components of the essential oils of aerial parts (inflorescence, leaves & entire aerial parts) of *p. cablin* [13].

11.2.2 Roots

The larger concentrations of norpatchoulenol (5.3%), Beta-pinene (4.5%) and pogostone (70.2%) are the characteristic compounds of oils extracted from roots of *P. cablin* (Table 11.1) [13].

11.3 Chemical Composition of *P. Cablin*

Zhang et al. investigated Twenty-four molecules from *P. cablin* plant, the majority of those include volatile compounds. *P. cablin* contains 7-patchoulene, Geranium ketone, alpha-patchoulene, 5-cedrol, alpha-bulnesene, and eucalyptus oil ketene which accounts for 96% of aforementioned volatile compounds [14]. In addition, four potential patchoulene derivatives such as 2b, 12-dihydroxypatchoulol, 3a, 8a-dihydroxypatchoulol, 6 hydroxy patchoulol and 8a, 9a-dihydroxypatchoulol were isolated from aerial region of *P. cablin* [15].

Table 11.1 Nutritive and bioactive compounds in *P. cablin*

Plant part	Compound	Percentage	Reference
Aerial parts	Pogostol	0.3–5.0%	[13]
	Patchouli alcohol	42.2–57.7%	
	Alpha-bulnesene	3.4–6.9%	
Roots	Beta-pinene	4.5%	
	Pogostone	70.2%	

Pachypodol and Retusine are the two significant flavones that Guan et al. isolated from patchouli plant [16]. Furthermore, multiple other flavones of non-volatile nature are also isolated by Zhang et al. and Itokawa et al. which includes 5-hydroxy-3,3-,4-,5,7-pentahydroxyflavone (4), 5-hydroxy-3,4-,7-trimethoxyflavone (3), and 5-hydroxy-4-,7-dimethoxyflavone (3) [14, 17].

Similar to this, Ding et al. investigated ethanol-based extraction from aerial parts of *P. cablin*, obtaining a number of glycosides from it such as apigenin 7-(O-methyl glucuronide), apigenin 7-galacturonide and quercetin-7-b-D [18]. Moreover, numerous other nonvolatile chemical types have also been isolated from *P. cablin* in recent years for example 5-hydroxy 3, 4-, 7- trimethoxy flavone and 5-hydroxy-4-,7- dimethoxyflavone [19].

Additionally, Huang et al. isolated four flavonoid glycosides from the ethanol extract of the stem and leaf of *P. cablin*. This extraction resulted in the isolation of isorhamnetin-3-O-b-Dgalactoside (12) and isisolidone- 7-O-a-L-rhamn from *P. cablin* [20]. Meanwhile, Li et al. revealed the following information on several flavonoids that are isolated from *P. cablin*: 4-, 5-dihydroxy-3, 7-dimethoxyflavanone (I), 5-hydroxy-7, 3-, 4-trimethoxyflavanone (II) [11]. [Fig. 11.1 and Table 11.2]

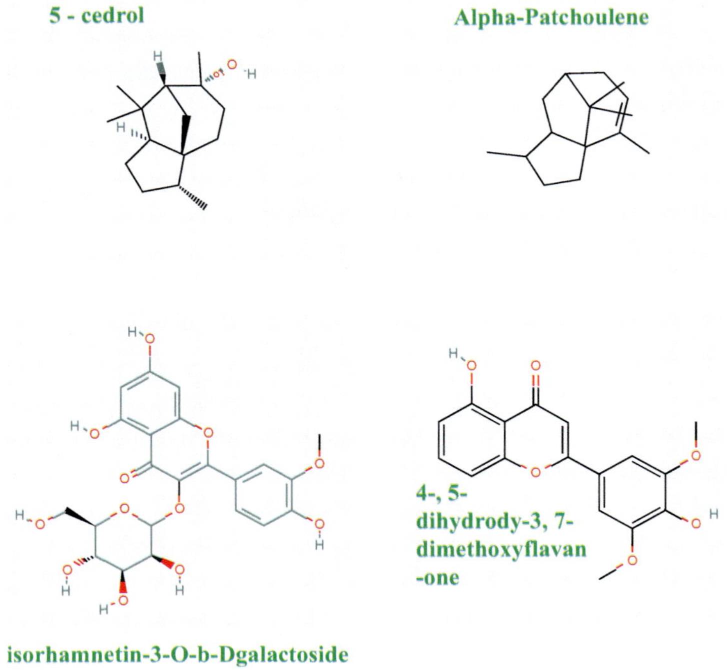

Fig. 11.1 Structure of 5-cedrole, alpha patchoulene, isorhamnetin-3-O-b-Dgalactoside (Glycoside) and 4-, 5-dihydroxy-3, 7-dimethoxyflavanone (Flavonoid)

Table 11.2 Therapeutic compounds derived from *P. cablin*

Plant	Isolated compound		References
Patchouli	5- cedrol		[14]
	Alpha-patchoulene		
	Geranium ketone		
	Derivatives of patchoulene	2b, 12-dihydroxypatchoulol	[15]
		3a, 8a-dihydroxypatchoulol	
		8a, 9a Dihydroxypatchoulol	
		6-Hydroxy Patchoulol	
	Glycosides	Isorhamnetin-3-O-b-Dgalactoside	[20]
		Isisolidone- 7-O-a-L-rhamn	
	Flavonoids	4-, 5-dihydroxy-3, 7-dimethoxyflavanone	[11]
		5-hydroxy-7, 3-, 4-trimethoxyflavanone	

11.4 Cultivation of *P. Cablin*

11.4.1 Why Vegetative Propagation is Preferred in P. Cablin*?*

P. cablin is an asexually reproducing plant and it is generally cultivated by utilizing 10- to 12-centimeter-long stem parts with three to four nodes in each. The lack of seeds has led to the adoption of this strategy for better cultivation of *p. cablin*. For vegetative propagation, stem parts containing minimum 2 or more leaves are preferred due to their greater capacity for roots and shooting [21].

11.4.2 How to Get Better Stem and Roots with Best Herbage of P. Cablin*?*

In order to promote better root & stem growth in patchouli Garbuio et al. suggested that utilizing stem parts from the apical and medium regions that contains minimum 2 leaves and nodes are favorable. Because it prefers the shady environment, *P. cablin* can be grown as an intercrop. It needs deep, loamy, fertile soil that is somewhat acidic and well-drained along with an annual rainfall between 300 and 400-centimeters with reliable frequent system of irrigation [22]. For high quality essential oil along with better crop yield, certain conditions with a soil pH between 5.5 and 6.2, humidity of 75% & a temperature between 22 and 28 °C are optimal [23].

11.4.3 Constraints During Cultivation

Presence of various pests as well as a number of diseases due to various factors such as fungus, viruses and nematodes cultivation of *P. cablin* is severely hampered. Yellow mosaic virus is the major factor that exerts adverse effects on the growth and reproduction of *P. cablin* [24, 25]. Furthermore, Peanut Strip Virus and Mottle virus are also major growth and productivity inhibitors of patchouli plant (27, 28, 29). One of the significant difficulties encountered during the commercial cultivation of *P. cablin* is nematode infection. The most destructive nematode, *Meloidogyne incognita*, considerably reduces the output of essential oils and instigate stunt growth in *P. cablin* [29].

11.4.4 Climate & Soil Profile

Patchouli grows best in fertile, shaded, well-drained soil that gets frequent irrigation or steady rainfall. Generally, Southern India following Andaman, Western Ghats and Nicobar Islands are the places where patchouli flourishes and reproduces at its best under conditions that are identical to those in the Philippines [1].

Productivity of *P. cablin* declines when grown in excessively flooded soils. It can grow up to 500 meters above sea level and can sustain in moderate temperatures of 22–28 °C and over 75% humidity. Higher-quality oil has been shown to be produced on soils with pH levels between 5.5 and 6.2 [3].

11.4.5 Genotype or Cultivars

Swamy et al. found that, in addition to environmental factors including soil and climate the genotype of the cultivars influences development and the quality of patchouli oil [6]. There are multiple varieties of patchouli plant present in current times, most commonly cultivated varieties are Java, Johore, Singapore and Indonesia, Malaysia [30].

11.4.6 Planting Time

Planting season of *P. cablin* is generally regarded as between March and May, then September and October. It is best to avoid planting in the winter (December–January) and during a time of heavy rainfall (July–August) because these months are not considered as favorable for optimum growth of *P. cablin* [31].

Vijayalalitha & Rajasekaran revealed that between the month of Nov. to Feb. with the existence of low ambient temperature, germination of *P. cablin* is significantly marvelous (80.4–100%) [32]. The rooting percentage is lower (0.8%) in the month of June when the temperature reached up to thirty-eight degrees Celsius. When compare to crops *P. cablin* planted in March, patchouli crops sown in Feb. produces considerably high yield of the essential oil (26.6 kg/ha) and herb (3.78 t/ha) [33].

11.4.7 Intercrop

P. cablin can be cultivated in combination with fruit tree, pine, coconut, arecanut, coffee and rubber crops as an intercrop [34]. In contrast to monoculture, patchouli cultivation in combination with papaya elevate oil quality by 8–11% as well as oil yield by 76%. *P. cablin* and papaya should be cultivated together for their greatest yield along with efficient utilization of valuable resources [35].

11.5 Essential Oil Extraction of *P. Cablin*

11.5.1 Major Plant (P. Cablin) *Part for Extraction of Essential Oil*

Due to prominence of patchouli alcohol as the primary component of essential oil, *P. cablin*'s essential oil is in high demand. The extensive use of essential oils is because of their excellent quality and unbeatable value. Usually, dried patchouli leaves are used to make *P. cablin* essential oil. The steam distillation method, supercritical carbon dioxide (scCO_2) [36], hydro-distillation through a Clevenger apparatus & microwave-assisted hydro-distillation mechanism is used to extract the essential oil from the dried leaf of *P. cablin*. Heat-sensitive compounds from aromatic plants could be distilled using steam distillation [37].

11.5.2 Extraction Processes of P. Cablin

In contrast to the hydro-distillation method, which packs plant materials in a particular compartment before adding water in sufficient quantity (usually about two thirds of the compartment) and heating it to a temperature of 100 °C, the procedure of steam distillation includes passing of steam (bubbles) through a heated raw material sample of particular aromatic plant. Steam distillation is superior than hydro-distillation because it requires shorter extraction time. It produces fewer chemical

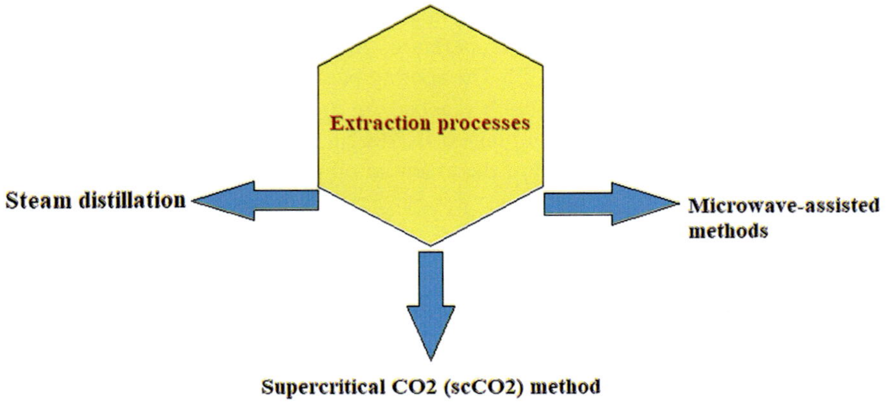

Fig. 11.2 Extraction mechanism used for extraction of essential oils of *P. cablin*

changes or alterations to the composition of essential oils because it reduces the levels of oxidation, uses less energy as well as reduces the likelihood of loss of multiple polar compounds [38].

The Mechanism of Supercritical CO_2 ($scCO_2$) distillation is primarily utilized for the extraction essential products such as essential oils from a number of plants. In contradiction to other traditional methods scCO2 involves numerous extraordinary benefits. With this procedure, natural compounds could be effectively extracted specifically without experiencing any thermal deterioration. Because it uses minimal solvents concentration, the essential oil it produces from particular plant is of a better quality (Fig. 11.2) [39].

Micro-wave assisted procedure is considered as environment friendly as well as an effective process for isolating essential oils from a considerable quantity of medicinal and herbal plants [38]. Different *P. cablin* leaf drying techniques are standardized by Pandey et al. for better quality and high yield of essential oils. The essential oil extracted from their leaves which are dried in the shade has the lowest patchouli alcohol content (1.40%) while sunlight-dried essential oil has the highest content of patchouli alcohol (36.63%) [39].

11.5.3 Components of Essential Oil

P. cablin is utilized as a base material for the synthesis of certain kinds of perfumes. Therefore, it is a necessary component for the perfume formulating industries [40]. The extraction of essential oils using steam, hydro-distillation & supercritical carbon dioxide distillation shows that the dried leaves contain more essential oils than fresh leaves [41]. The peculiar aromatic quality of *P. cablin* essential oil is mostly due to existence of patchouli alcohol which is a fundamental component of the essential oils of *P. cablin* [42].

Furthermore, the major sesquiterpene in the essential oil of *p. cablin* is patchouli alcohol. Along with patchouli alcohol, several other sesquiterpene hydrocarbons like guaiene, patchoulene & seychellene contributes to the fragrance of *P. cablin* essential oil. Numerous studies have found that the two main substance that regulate and control the quality of patchouli essential oil are patchoulol as well as patchoulene [36, 43].

Beside this, Alpha-bulnesene (8.7%), (E)-caryophyllene (2.4%), patchouli alcohol (61.6%), alpha-guaiene (10.0%), pogostone (1.0%) patchoulene (1.6%), norpatchoulenol (1.6%), b-patchoulene (1.0%), seychellene (3.1%) and gamma-patchoulene (1.3%) are the main ingredients found in the essential oil of *P. cablin* [44].

11.6 Diseases & Pests

11.6.1 Wilt (Fungal Disease)

One of the most devastating diseases of patchouli plant is wilt (fungal diseases) which is caused by *Rhizoctonia solani* [45]. *Gliocladium virens*, a biologically antagonistic agent in combination with *Glomusaggegatum* have successfully inhibited patchouli wilt disease induced by *R. solani* [46].

11.6.2 Witches Broom Disease

Witches broom disease (WBD) in *P. cablin* was discovered in 1996–97 in India (Singh et al., 2002b). Leaf shrinkage, axillary branch multiplication, diminished internodes, stunted development and a bushy appearance are all hallmarks of the disease. Phytoplasma was identified as the casual pathogen using electron microscopy [47].

11.6.3 Viruses

11.6.3.1 Yellow Scale of Cotton

P. calin viruses generally transmits mechanically, but a limited fraction can spread via vectors or grafting processes. These viruses have a greater dilution end point along with heat inactivation level than most of the other viruses [47]. A new record of the yellow scale of cotton on *P. cablin* has been detected in Mudigere, India [48].

11.6.3.2 *P. Cablin* Mild Mosaic Virus and Patchouli Mottle Virus

In Japan, Natsuaki et al. found two viruses (sap transmissible) from patchouli. Plants infected with the virus displayed minimal mosaicism, moulting or sometimes no symptoms at all. In electron microscopy, virus-like particles are observed which are either elongated or spherical in appearance. Patchouli Mild Mosaic Virus (PaMMV) is serologically similar to but distinct from Broad Bean Wilt Fabavirus. Patchouli Mottle Virus (PaMoV), an elongated virus with a limited host range was identified as a Potyvirus which has adverse effects on *P. cablin* [49].

11.6.3.3 *P. Cablin* Virus X

Meissner Filho et al. described Patchouli virus X, a novel potexvirus identified from *P. cablin* in Brazil. The plants having mosaic symptoms is infected by a Potexviral and Potyvirus [27].

11.6.3.4 Peanut Strip Virus

Patchouli strip virus was discovered in the Institute of Himalayan Bioresource Technology campus in India, with prominent viral sickness having mosaic marks on its leaves [26]. Using Enzyme Linked Immunosorbent Assay (ELISA), Reverse transcription polymerase chain reaction (RT-PCR), Immunocapture-PCR (IC-PCR) and sequencing techniques, the virus was later identified as a Peanut stripe virus (PStV) [28].

11.6.3.5 *P. Cablin* Yellow Mosaic Virus

Zaim et al. (2013) discovered golden mosaic patterns on the leaves of *p. cablin* plant in the experimental fields in Lucknow, India. Using serological and genetic data the causal virus was identified as Patchouli Yellow Mosaic Virus (PaYMV) which is a member of the Potyvirus genus.

Healthy patchouli plants have dark green leaves whereas PaYMV-infected plants contains deformed leaves with yellow mosaic symptoms including yellow blotches & discolored patches of random shapes and sizes. Infected plants also have severe growth retardation [50].

11.6.3.6 Prevention from Viruses

If insect vectors are eliminated and virus-resistant/virus-free plants are cultivated by employing meristem culture techniques, patchouli viruses can be protected more effectively. In addition to this, use of botanical or chemical insecticides can contribute in the control of insect vectors [47].

Fig. 11.3 Aphid pests of *P. cablin*

11.6.4 Aphids

Aphids such as *Aularchis milliaris*, *Graphium* sp., *Oxya chinensis*, *Myzus persicae* and *Myzus coccinellids*, *Pachyzancla stultalis* are found to damage patchouli leaves in an Indonesian study on insect attack of *P. cablin* (Fig. 11.3) [51].

11.6.5 Nematodes

Patchouli production have been found to be reduced up to 75% by the attack of parasitic nematodes [52]. Seven parasitic nematode genera of plants are discovered in a survey of sick patchouli gardens in West Java [53]. *Meloidogyne, Pratylenchus, Scutellonema, Rotylenchulus, Hemicriconemoides, Helicotylenchus* and *Xiphinema* are among these 7 lethal nematodes. The primary nematode pathogens associated with patchouli roots are thought to be three species: *Pratylenchus brachyurus, Meloigogyne incognita* and *Meloidgyne hapa. Pratylenchus brachyurus* is also isolated for the first time from patchouli roots and soils in Tamil Nadu and Bangalore [54, 55].

11.6.6 Management & Protection from Nematodes

Best cultural practices including the usage of organic matter, fertilizers, mulch & dolomite as well as the use of natural nematicides comprising of castor meal, neem powder can surely reduce damages due to attack of nematodes. Beside this, chemical pesticides and most importantly preventing the spread of nematodes from infected to uninfected areas can be used to manage these parasitic nematodes [52]. Clove, Castor, citronella plants and Neem as well as several Fungi used for trapping multiple nematodes (*Dactylaria sp., Arthrobotrys sp., and Dactylelia sp.*) as well as soil bacteria such as *Bacillus* and *micrococcus* can eradicate these nematodes up to

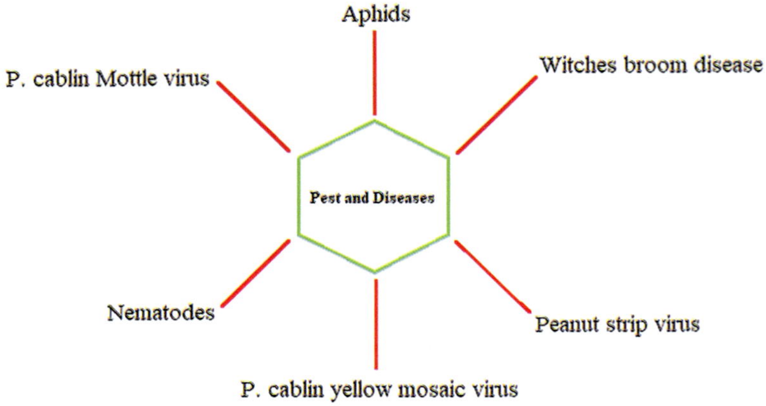

Fig. 11.4 Major Pests and Diseases of *P. cablin*

82% [56]. Furthermore, Harni et al. showed that immersing the patchouli plant roots in endophytic bacterial suspension before transplantation to the field might suppress *P. brachyurus* levels up to 80% (Fig. 11.4) [57].

11.7 Applications of *P. Cablin*

11.7.1 Antibacterial Activities

Patchouli plant is used in traditional medicine for the treatment of cold & fungal infections [58]. Patchouli oil is discovered to be more efficient in controlling 20 bacterial strains as well as 12 different fungal infections demonstrating its anti-bacterial and anti-fungal potential [59]. The essential oils of patchouli are tested in-vitro against 17 pathogenic fungus and 16 microorganisms from 3 different geographical locations comprising of China, India, and Indonesia from the nail, mucous membrane, foot, skin and armpit of different subjects. These findings revealed that, *P. cablin*'s oil expressed clear antifungal and antibacterial action to antagonize damages due to different fungal and bacterial agents [60].

The essential oil of *P. cablin* effectively suppress the growth of *Pseudomonas aeruginosa, Klebsiella pneumonia* & *Escherichia coli* [61]. Patchoulol & hexane extract of patchouli leaves (Dried) evidences higher antibacterial activity against *B. subtilis* & *S. aureus* [62, 63].

In-vitro antimicrobial & Molecular docking technology assays are significantly used to investigate the antibacterial properties of *P. cablin* [64]. Pogostone and (−) patchoulol are the main constituents with the broadest therapeutic potential for bacterial infection. Patchouli as well as geranium leaf aqueous and organic extracts demonstrated potent antimicrobial effects against *E. aerogenes, E. coli* and *S. aureus*

[65]. Staphylococcus aureus, a food-borne pathogen, is successfully prevented from forming biofilms by patchouli oils [66].

Recently, the essential oils of *P. cablin* is used in the green synthesis of nanoparticles. These nano biosystems demonstrated excellent resistance to *Klebsiella pneumoniae* and *Staphylococcus aureus* clinical strain adhesion and biofilm development [67]. Additionally, patchouli oil shown antibacterial action against the bacteria isolated from the faces of lizards [68]. Patchoulol demonstrated selective antibacterial action against *H. pylori* without damaging the natural flora of the digestive system. Patchoulol has the capacity to reduce urease, making it an effective biomolecule for treating *H. pylori* induced infections [69]. Using patchouli oil, gram-positive bacteria such as *Staphylococcus, Streptococcus* and *Bacillus* species can also inhibited successfully (Fig. 11.5 and Table 11.3) [70].

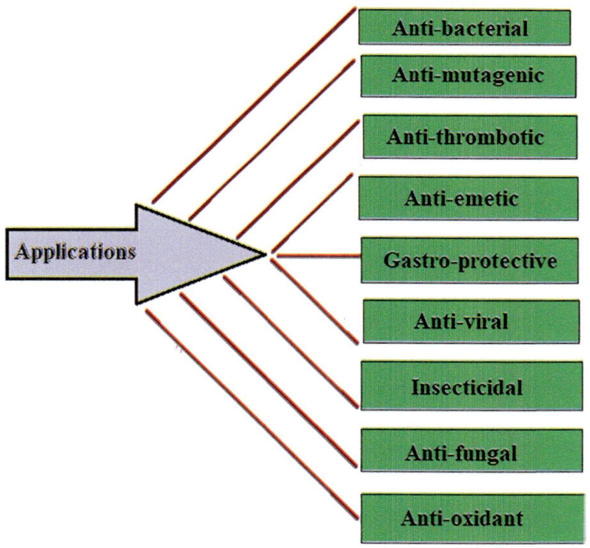

Fig. 11.5 Applications of *P. cablin*

Table 11.3 Micro-organism (Bacteria) which can be prevented sufficiently while utilizing essential oils of *P. cablin*

Plant	Micro-organism (Bacteria) which can be effectively managed by patchouli	References
Patchouli	*E. coli*	[61]
	H. Pylori	[69]
	Bacillus & staphylococcus	[70]
	K. Pneumoniae	[67]
	S. aureus	[66]

11.7.2 Antioxidant Activities

Patchouli oil demonstrated effective free radical scavenging capabilities & reduces the conversion of hexanal to hexanoic acid [71]. Patchouli herb can be used to alleviate brain cell damage brought on by ROS [72]. *P. cablin* prevents hydrogen peroxide-induced necrosis and apoptosis in the human neuroglioma cell line A172, indicating that it may be used to treat a variety of neurodegenerative diseases. Patchouli oil's antioxidative properties inhibited photoaging and preserved the structural integrity of the skin [73].

Additionally, the regeneration of UV-induced skin lesions is accelerated by patchoulol's anti-inflammatory and antioxidant characteristics along with its capacity to reduce the expression of matrix metalloproteinases [74]. According to a study by Li et al. patchouli alcohol is effective in lowering the synthesis of TNF-alpha which instigate mastitis by alterations in lipopolysaccharides located on the outer membrane of Gram-negative bacteria. Patchouli alcohol has been discovered to attenuate breast histopathologic aberrations, making it a useful treatment for preventing mastitis [75].

11.7.2.1 Mechanism of Action

Oxidative stress is described as the increased level of intracellular reactive oxygen species that further damages DNA, proteins as well as lipids [76]. This mechanism is correlated with the generation of free radicles in vivo [77]. Oxidative stress can be induced from several endogenous and external factors [78], such as alcohol, environmental pollutants as well as various sorts of drugs, which ultimately leads towards various problems such as pre-mature aging along with negative sequelae of multiple diseases [79]. Nrf-2 is a significant transcriptional factor that stimulates the antioxidant response which maintains the expression specific cytoprotective genes comprising of detoxifying, anti-apoptotic along with anti-oxidant proteins (Fig. 11.6) [80].

Patchouli oil express remarkable antioxidant properties due to its mechanism of action that promote the removal of superoxide anion free radicals along with hydroxy free radicles as well as ameliorate the process of lipid peroxidation [81]. Besides this, administration of Pachypodol, an essential compound from *p. cablin* can mitigate the production of ROS as well as protect liver cells from oxidative cell death prompted due to exposure of tert-butyl-hydroperoxide. The critical mechanism of this potential effect correlates with the amplification of endogenous antioxidant defense system through the ERK1/2- dependent activation of Nrf2 pathway [82].

Intake of alcohol can increase the damages due to generation of ROS in the hepatic tissues, as it elevates the activity of cytochrome P450 2E1 & proceeds to the production of huge quantity of ROS, this process eventually deteriorates the oxidation-reduction equilibrium regulated by GSH or GSSG [81, 83]. Ethanol-instigated acute hepatic damage, generally induced via extreme alcohol

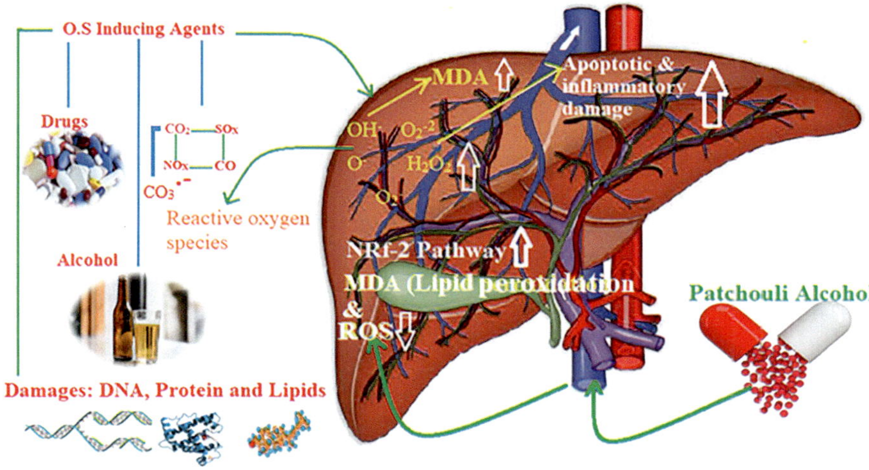

Fig. 11.6 ROS producing agents, damages and therapeutic potential of patchouli alcohol

consumption, has been correlated with multiple lethal liver disorders such as hepatitis, hepatic fibrosis, alcoholic fatty liver, seatosis along with cirrhosis [84].

Steatosis in hepatic tissues has been broadly documented as an early as well as reversible outcome of extreme consumption of alcohol [85], pre-treatment with patchouli alcohol can effectively elevates the concentration antioxidant enzyme of liver and promotes the expression of GSH/GSSG proportion. Moreover, patchouli alcohol stimulates the anti-oxidant enzyme glutathione reductase along with superoxide dismutase in exposed individuals. These mechanisms help to reduce the accumulation of ROS as well as suppresses the expression of various cytokines along with mRNA coding CYP2-E1. Treatment with patchouli alcohol could also protect degeneration of fats through its capability to accelerate metabolism of adipose tissues [86]. Heat shocks instigates oxidative stress could accelerates the damages to skin as well as apoptotic damage to the epithelial-6 cells of the intestine [87]. Pre-administration of patchouli alcohol evades the dysregulations to the cellular architecture as well as results in a reduced concentration of MDA. Extensive dosage of patchouli alcohol resulted in considerable elevation in the expression of HO^{-1} as well as Nrf2. To conclude, these results illustrated that, patchouli alcohol is capable of attenuating cell damages and ameliorating the oxidative stress actions of IEC-6 cells via the stimulation of the Nrf2-keap1 [88].

11.7.3 Antifungal Activities

The mycelial growth of *Candida albicans* can be restricted by the medium containing 100 g/mL of mixture of various oils including lemon grass, thyme, patchouli and cedarwood [89]. The essential oil of *P. cablin* contains patchouli alcohol

(44.52%) which has antifungal effects against an *Aspergillus* species population [90].

Li et al. provides evidence regarding the properties of pogostone, a naturally occurring substance derived from *P. cablin* that is useful in treating *Candida* infections, notably for vulvovaginal candidiasis [79]. Similarly, patchouli oil effectively inhibited *C. albicans* instigated damages [91]. Pogostone and its synthetic compounds also demonstrated potent antibacterial effects against *C. albicans*. According to molecular docking studies, pogostone analogues can be structurally modified to provide a promising antifungal potential against various ailments [92].

11.7.4 Insecticidal Activity

11.7.4.1 Museum Pests

Leaf ashes of *P. cablin* can be utilized to kill various pests and they are considered to be effective *Stegobium paniceum* repellent with a repellency rate of about 64% [93]. *P. cablin* oil demonstrated strong fumigation insecticidal activity against different museum insect pests such as *S. zeamais* and *F. sauteri* [94].

11.7.4.2 Prevention from Mites

To control house dust mites, *P. cablin* and *C. cassia* petroleum ether extracts can be used as a biodegradable as well as environment-friendly insecticide [95]. Patchouli alcohol is also considered as the most effective component of *P. cablin* essential oil which helps in protection against different species of urban ants and *Anopheles stephensi* [96, 97].

11.7.4.3 Mosquito Repellent

Similarly, *P. cablin*'s sesquiterpene hydroperoxides shows significant insecticidal action against *T. cruzi* [98]. Under controlled laboratory conditions, 38 essential oils are tested for their ability to repel mosquitoes. According to the findings, patchouli oil used in its pure form can completely repel mosquitoes for about 2 h [99]. *P. cablin* along with valamus (75:25) significantly involves in the protection against *A. aegypti* mosquitoes in different fields [100].

11.7.4.4 Larvicidal & Pupicidal

Progostone recently demonstrated potent larvicidal, Pupicidal and growth-inhibitory properties against *S. litura* and *S. exigua* making it a good option to manage a variety of insects that damage agricultural plants [101].

11.7.5 Aromatherapy

Aromatherapy using *P. cablin* essential oil helps to relieve stress, sleeplessness, and anxiety. Its intoxicating wine-like fragrance serves as an aphrodisiac and assists in sharpening intelligence and improving focus. It is used in incense sticks to promote relaxation on a spiritual level. By evaluating the quantity of catecholamine present in the serum & observing blood pressure variations, normal adult individuals were subjected to tests for the determination of how odor inhalation *P. cablin* affected sympathetic activity. After breathing the fragrance, normal participants displayed a decrease in sympathetic activity [102].

11.7.5.1 Dementia Treatment via Aromatherapy

An aqueous creamy mixture comprising *P. cablin* as well as marjoram oil was applied to the body of different individuals five times per day, and the findings demonstrate a reduction in the frequency of dementia-related activities. The usage of essential oils improves' mental alertness and awareness in treated individuals. Because essential oils stimulate adrenal hormone secretion, they can alleviate stress-related lethargy and anxiety [103].

11.7.5.2 Menopausal Transition

Anti-inflammatory & cooling oils including *P. cablin* and Citrus limonum are used to treat symptoms of menopause including perspiration and hot flashes. *P. cablin* and rose along with the jasmine (oil mixture) can certainly stimulate clarity and engage in the thyroid gland's activity as well as modulates hormonal discharge [104].

11.7.5.3 Psychiatric Issues

Aromatherapy can be used to treat a variety of psychiatric problems. Patchouli oil has relaxing, sedative, and elevating effects on the brain [105].

In 10 dementia patients, an essential oil vaporized mixture (patchouli, orange, ylang-ylang, chamomile, rosewood, basil, peppermint, rosemary and jasmine) reduces disturbed behavior [106].

11.7.5.4 Physiological Processes

Physiological reactions such as blood pressure and heart rate are also maintained after the inhalation of *P. cablin*'s fragrance. This indicates that inhaling the fragrance of patchouli oil improves mood and significant healing qualities. The principal ingredient, patchouli alcohol is also accountable for the stimulation of the sense of smell [107]. Aromatherapy can also help in chronic pain perception and stress management [108].

11.7.6 Antiviral Activities

Patchouli as well as other herbal traditional medicines are tested for anti-influenza viral activity. It is reported that the methanol extract of patchouli leaves at a concentration of approximately 10 g/mL could inhibit the influenza virus A/PR/8/34 (H1N1) up to 99.8%, and the IC50 values is determined as 2.6 micrometers [109].

Wu et al. indicated that patchoulol has significant anti-influenza viral properties with an IC50 of 4.0 M, making it a useful anti-influenza viral agent for the pharmaceutical industry [110]. Furthermore, patchouli alcohol also expresses an anti-influenza viral activity by stimulating immunological action as well as reducing pulmonary inflammatory response [111].

Furthermore, Wu et al. reported that patchoulol (10 g/mL) increase immunological response and restricted the expression of inflammatory cytokines to reduce their adverse responses when used to examine anti-influenza virus effects [112].

11.7.7 Protective Effect Against HIV/AIDS

Essential oil of *P. cablin* is also employed as an efficient antimicrobial factor for the treatment of certain infections brought on by methicillin-resistant *Staphylococcus aureus* and *Cryptococcus neoformans, Candida albicans,* Herpes simplex types I and type II in French hospitals for subjects with acquired immunodeficiency syndrome [108].

11.7.8 Gastrointestinal Protective Activity

The water extract of *P. cablin* plays an exceptional role in gastrointestinal protection by maintaining the amounts of nitric oxide and tumor necrosis factor in the blood as well as by maintaining the fluidity of the outer membrane of intestinal epithelial cells [113].

11.7.9 Antiemetic Activity

Hexane extracts of patchouli plant's leaves have an anti-emetic effect on young hens. The antiemetic properties of *P. cablin* are demonstrated at doses of 50–70 mg/kg [114].

11.7.10 Defecation and Constipation

A study was directed by Mikuriya et al. to examine the impacts of patchouli oil on constipation and defecation brought on by a reduced consumption of fibrous foods. It is observed that when the fragrance of the patchouli oil is inhaled by experimental animals, both models produce more and heavier waste demonstrating its therapeutic potential against constipation and defecation. Furthermore, it has been demonstrated that mice with constipation can be helped by olfactory neurotransmission systems [115].

11.7.11 Fibrinolytic & Blood Coagulation Potential

Blood coagulation & the fibrinolytic activity of many essential oils along with patchouli are studied using in vitro enzymatic activities such as fibrin production from fibrinogen using thrombin and fibrin resolution following urokinase activity. Eucalyptus, chamomile and neroli oils shows evidence of both coagulation and fibrinolytic activity whereas patchouli and frankincense oils effectively displayed hyperfibrinolysis [116].

11.7.12 Antithrombotic Activities

Sunghyangjunggisan an herbal remedy containing *P. cablin* as a significant component, dramatically reduces platelet aggregation in rats. Additionally, it demonstrated notable protection against mouse pulmonary thrombosis-related mortality [117].

11.7.13 Anti-inflammatory and Analgesic Activity

The methanol extract of patchouli plants revealed anti-inflammatory & analgesic potential in mice [118]. According to analgesic studies, patchouli extracts decrease writhing response induced due to acetic acid administration. The Chinese traditional medicine recipe CPZ contains the *P. cablin* herps in significant amount. Extracts from these herbs demonstrated a potent anti-inflammatory effect by modulating IL-1 and prostaglandin E. 2. Lipopolysaccharide-induced inflammation is present in RAW264.7 cells, patchouli essential oil successfully controls the expression of inflammatory markers concentration in these cells [119].

The ethanol extract of patchouli's root, rhizome & patchouli oil display potent anti-inflammatory effects on living organisms [120, 121]. Mouse macrophages and human colorectal cancer cells are used to study the mechanism underlying patchoulol's anti-inflammatory properties and the results are in favors of *P. cablin* [122].

Li et al. demonstrated the anti-inflammatory activities of pogostone and proposed its uses in the synthesis of a pharmaceutical drug for the treatment of septic shock [75]. By reducing pro-inflammatory marker's expression patchouli plant water extracts reduced colon inflammation in treated animals [123].

11.7.14 Anti-Tumorous and Immuno-Protective Role

Licochalcone A which is an essential component of *P. cablin* essential oils, exhibits inhibitory action against PI-PLC gamma 1 in laboratory experiments. Beside this, Licochalcone A administration effectively maintained promyelocytic leukemia cells (HL-60) generated monocytes as well as involve in prevention from terminal differentiation after the exposure to licochalcone A [124]. Patchoulol, on the other hand, increases apoptosis and decreases cell differentiation in the human colorectal cancer cell lines SW480 and HCT116 [122, 125]. Patchouli alcohol downregulated c-myc & activate the NF-kB pathway while increasing histone deacetylase 2 synthesis along with enzymatic activity. The patchouli sesquiterpenoid bulnesene efficiently suppressed the platelet dysregulating factors [126].

To treat skin irritations, Xinxiang granule (XXG) is made from the blossoms of *P. cablin* herbs [127]. Recently, animal models were used to confirm the anti-nociceptive and anti-allergic properties of patchouli oil, which exerts marvelous action against various allergic conditions of different subjects [128].

11.7.15 Amelioration of Skin Diseases

The use of essential oils having 12% patchouli oil used for the treatment of patients with ulcers, ruptured skin, skin abrasions and pressure sores effectively by managing their skin infections. Furthermore, essential oil of *P. cablin* accelerates the healing process of administrated subjects [129]. Applying patchouli essential oil protected the structural integrity of UV-irradiated skin impairments and reduces photoaging as well as enhances regeneration of skin lesions (124, 125).

11.7.16 Pharmacokinetic Activities

To examine pogostone therapeutic potential in rats following intravenous and oral administration, a quick and accurate liquid chromatography mechanism was created. In pharmacokinetic studies for preclinical research, this technique is very advantageous [132].

Li et al. employed a liquid chromatography technique to examine *in-vitro* & *in-vivo* profiles of pogostone [75]. Significant gastroprotective properties against gastric ulcers are revealed by patchouli oil through COX-mediated PGE2 activation, enhanced antioxidant and anti-inflammatory response as well as increased synthesis of stomach mucus [130].

11.7.17 Anti-mutagenic Potential

Methanolic extract of *P. cablin* reduces umu gene expression which is stimulated due to a mutagenic substance named: acrylamide in Salmonella typhimurium [131].

11.7.18 Intestinal Micro-ecological Effects

Microbiota of the gut play a very crucial activity for the maintenance of human's health; therefore, their appropriate concentration and balance is very necessary in the body of humans. These microorganisms are involved in extraction of energy and nutrients [132], defend body of animals against enteropathogens [133], prevent cancer [134], as well as might even effects behavior along with brain functionality [135]. Dysregulations of gut microbiota that is known as dysbiosis, could be an influencing factor which is correlated with IBD [136, 137], neoplastic disease [138], obesity [139, 140]. Results of the various findings have proposed that patchouli alcohol along with its derivatives including pogostone helps to regulate activity of the epithelial barrier of the gut, to accelerates the polarization of M1-M2 macrophage phenotypes, to elevates the diversity of the gut microbiota as well as overcome pro-inflammatory proteins in mouse [141] (Fig. 11.7).

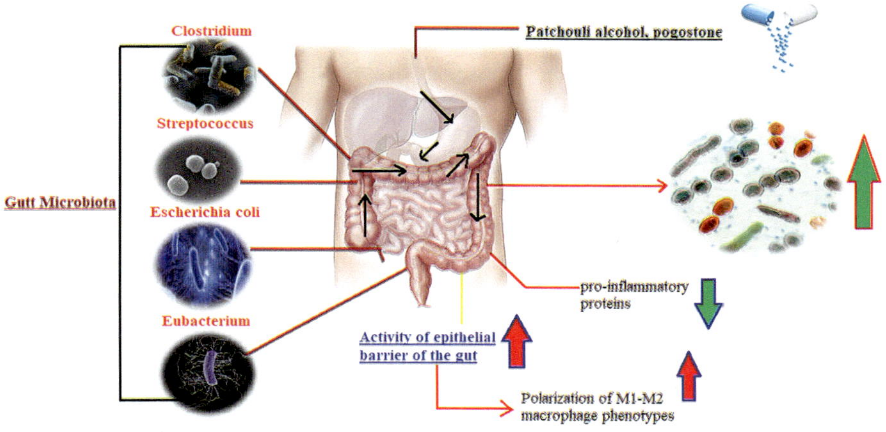

Fig. 11.7 Effect of Patchouli alcohol and its derivatives on intestinal micro-biota

11.7.19 Anti-Diabetic Properties

Obesity is extensively connected with diabetes type-2 and is considered as key risk factor for multiple other metabolic disorders. It is significant aspect contributory to the metabolic disorder, this ailment is aggravated by environmental influences such as a sedentary life style, a fat enriched diet as well as considerably via aging [142]. The treatment of patchouli alcohol expressed a net decline in weight of high fat diet instigated obese rats, Patchouli alcohol demolish adipogenesis as well as fat aggregation in adipocytes via elevating the expression along with activation of beta-catenin [143]. The chronic exposure of a high fat diet (HFD) has also been correlated with several other disorders such as nonalcoholic liver disease (NAFLD). NAFLD is a key root of hepatic disorders that disturbs approximately 30% of the United States population [144], and it is the utmost mutual chronic liver disease throughout the world [145]. Recent investigations have revealed the curative potential of patchouli alcohol when utilized to treat HFD-prompted liver steatosis in mouse, these investigations illustrated that patchouli alcohol have significant role in mitigating steatosis resulted from an HFD. Patchouli alcohol accelerated this effect via suppressing endoplasmic reticulum's stress signals as well as by maintaining the uptake, aggregation along with release of very minute concentration of lipoproteins. Furthermore, the treatment of patchouli alcohol is linked with the maintenance of aforesaid minimal level lipoprotein receptor and microsomal triglyceride transfer protein expression [146].

11.7.20 Anti-hypertensive Characteristics

Hypertension is a chronic as well as crucial condition that stimulates various other disorders & may proceeds towards premature death [146]. It is revealed that, more than 1 billion individuals throughout the world are facing the problem of hypertension, this disease is linked with approximately 9.4 million mortalities per Anum [147]. Specific factors which are able of declining the systematic blood pressure may considerably decrease the danger of actions correlated with main cardiovascular diseases comprising of coronary heart disease along with heart stroke [148]. It is fundamentally significant to regulate blood pressure within an optimum range. Patchouli alcohol accelerates considerable vasorelaxant influence due to its potential activity as a Ca^{2+} antagonizing factor in an endothelial autonomous pathway. The fundamental action comprises the obstruction of Ca^{2+} influx (extracellular) through the membrane of vascular smooth muscle cells as well as stimulate the intracellular secretion of Ca^{2+} via different Ca^{2+} channels in sarcolemma [148]. Pocahemiketal-b, an essential component of patchouli's essential oil exhibits remarkable vasorelaxant activity to antagonize vasoconstriction instigate by phenylephrine in aortal rings of rats [149].

11.8 Conclusion

P. cablin (Patchouli) is globally cultivated for its prominent therapeutic and commercial applications. Tropical and sub-tropical regions throughout the world are considered as standard environment for cultivation of *P. cablin* (Patchouli) to obtain better growth and high-quality essential oils. Chemically *P. cablin* plant consist of various essential chemical as well as phytochemical compounds such as organic acids and pogostone respectively with a number of other volatile and non-volatile compounds. These chemicals are major factors behind certain biological and pharmaceuticals properties of this plant. *P. cablin* is responsible for treatment of various skin diseases, immuno-protective, anti-tumorous, anti-thrombotic action along with various other biological application discussed during the entire chapter. Thus, we can conclude that *P. cablin* could be utilized in future for the synthesis of wide range of medicines and various cosmetics products of commercial value.

References

1. Bhaskar, S., & Kumar, T. V. (2000). Agronomic bottlenecks, genetic barriers and marketing impediments in patchouli Pogostemon patchouli production. *Journal of Medicinal and Aromatic Plant Sciences, 22*(1B), 396–403.
2. Maheshwari, M. L., Vasantha Kumar, T., Sharma, N., & Chandel, K. P. S. (1993). Patchouli-an Indian perspective. *Indian perfumer, 37*, 9–9.
3. Bhaskar, S. (1995). Growth, herbage and oil yields of patchouli (Pogostemon patchouli) as influenced by cultivars and nitrogen fertilization. *Indian Perfumer, 39*, 35–35.
4. Swamy, M. K., Balasubramanya, S., & Anuradha, M. (2010). In vitro multiplication of Pogostemon cablin Benth. through direct regeneration. *African Journal of Biotechnology, 9*(14), 2069–2075.
5. Chakrapani, P., Venkatesh, K., Singh, B. C. S., Jyothi, B. A., Kumar, P., Amareshwari, P., & Roja, A. R. (2013). Phytochemical, pharmacological importance of patchouli (Pogostemon cablin (Blanco) Benth) an aromatic medicinal plant. *International Journal of Pharmaceutical Sciences Review and Research, 21*(2), 7–15.
6. Swamy, M. K., Mohanty, S. K., Sinniah, U. R., & Maniyam, A. (2015). Evaluation of patchouli (Pogostemon cablin Benth.) cultivars for growth, yield and quality parameters. *Journal of Essential Oil-Bearing Plants, 18*(4), 826–832.
7. Xian, Y. F., Suo, J., Huang, X. D., Hou, S. Z., Chen, J. N., Ye, M. R., & Su, Z. R. (2007). A pharmacological study on anti-inflammatory effects of refined Huodan recipe. *Chinese Journal of Experimental Traditional Medical Formulae, 13*, 54–56.
8. Angadi, S. P., & Vasanthakumar, T. (1995). Patchouli. In K. L. Chadha & R. Gupta (Eds.), *Advances in horticulture: Med & Arom* (Plants) (Vol. 11, pp. 751–771). MalhotraPublishing House.
9. Chadha, K. L. (Ed.). (1995). *Advances in horticulture. 11. Medicinal and aromatic plants.* Malhotra Publishing House
10. Maeda, E., & Miyake, H. (1997). Leaf anatomy of patchouli (Pogostemon patchouli) with reference to the disposition of mesophyll glands. *Japanese Journal of Crop Science, 66*(2), 307–317.
11. Li, Y. C., Xian, Y. F., Ip, S. P., Su, Z. R., Su, J. Y., He, J. J., et al. (2011). Anti-inflammatory activity of patchouli alcohol isolated from Pogostemonis Herba in animal models. *Fitoterapia, 82*(8), 1295–1301.

12. Daud, M., Yaman, M. A., Usman, Y., & Aqmarina, Y. (2021). Potential of ciplukan extract (Physalis angulata L.) and patchouli waste (pogostemon patchouli pellet) as alternative sources of phytogenic feed additive. In *IOP conference series: Earth and environmental science* (Vol. 644, no. 1, p. 012072). IOP Publishing
13. Verma, R. S., Padalia, R. C., Chauhan, A., & Singh, V. R. (2019). Chemical composition of leaves, inflorescence, whole aerial-parts and root essential oils of patchouli {Pogostemon cablin (Blanco) Benth.}. *Journal of Essential Oil Research, 31*(4), 319–325.
14. Zhang, G., Ma, X., & Su, J. (2001). Flavonoid compounds in Pogostemon cablin. *Chinese Traditional and Herbal Drugs, 32*, 871–874.
15. Zhou, L., Xu, M., Yang, C. R., Wang, Y. F., & Zhang, Y. J. (2011). New patchoulol-type sesquiterpenoids from Pogostemon cablin. *Helvetica Chimica Acta, 94*(2), 218–223.
16. Guan, L., Quan, L. H., Xu, L. Z., & Cong, P. Z. (1994). Chemical constituents of Pogostemon cablin (Blanco) Benth. *Zhongguo Zhong yao za zhi= Zhongguo zhongyao zazhi= China journal of Chinese materia medica, 19*(6), 355–356.
17. IToKAWA, H. I. D. E. J. I., SUTO, K., & TAKEYA, K. (1981). Studies on a novel p-coumaroyl glucoside of apigenin and on other flavonoids isolated from patchouli (Labiatae). *Chemical & Pharmaceutical Bulletin, 29*(1), 254–256.
18. Ding, W., Liu, M., Wei, X., & Lin, L. (2009). Strong polarity components of Pogostemon cablin (Blance) Benth. *Journal of Tropical and Subtropical Botany, 17*(6), 610–616.
19. Ding, W. B., Lin, L. D., Liu, M. F., & Wei, X. Y. (2011). Two new sesquiterpene glycosides from Pogostemon cablin. *Journal of Asian Natural Products Research, 13*(7), 599–603.
20. Huang, L., Mu, S., Zhang, J., Deng, B., Song, Z., & Hao, X. (2009). Chemical constituents from involatile moiety of Pogostemon cablin. *China Journal of Chinese Materia Medica, 34*(4), 410–413.
21. Pandey, S. K., Bhandari, S., Sarma, N., Begum, T., Munda, S., Baruah, J., et al. (2021). Essential oil compositions, pharmacological importance and agro technological practices of patchouli (Pogostemon cablin Benth.): A review. *Journal of Essential Oil-Bearing Plants, 24*(6), 1212–1226.
22. Garbuio, C., Biasi, L. A., Kowalski, A. P. J., Signor, D., Machado, E. M., & Deschamps, C. (2007). Cutting propagation of patchouli with different number of leaves and types of cuttings. *Science in Agriculture, 8*, 391–398.
23. Yahya, A., & Yunus, R. M. (2013). Influence of sample preparation and extraction time on chemical composition of steam distillation derived patchouli oil. *Procedia Engineering, 53*, 1–6.
24. Swamy, M. K., & Sinniah, U. R. (2016). Patchouli (Pogostemon cablin Benth.): Botany, agrotechnology and biotechnological aspects. *Industrial Crops and Products, 87*, 161–176.
25. Sastry, K. S., & Vasanthakumar, T. (1981). Yellow mosaic of patchouli (Pogostemon patchouli) in India. *Current Science, 50*, 767–768.
26. Singh, M. K., Chandel, V., Hallan, V., Ram, R., & Zaidi, A. A. (2009). Occurrence of Peanut stripe virus on patchouli and raising of virus-free patchouli plants by meristem tip culture. *Journal of Plant Diseases and Protection, 116*(1), 2–6.
27. Meissner Filho, P. E., Resende, R. D. O., Lima, M. I., & Kitajima, E. W. (2002). Patchouli virus X, a new potexvirus from Pogostemon cablin. *The Annals of Applied Biology, 141*(3), 267–274.
28. Noveriza, R., Suastika, G., Hidayat, S. H., & Kartosuwondo, U. (2012). Potyvirus associated with mosaic disease on patchouli (Pogostemon cablin (Blanco) Benth.) plants in Indonesia. *Journal of ISSAAS (International Society for Southeast Asian Agricultural Sciences), 18*(1), 131–146.
29. Bhau, B. S., Borah, B., Ahmed, R., Phukon, P., Gogoi, B., Sarmah, D. K., ... & Wann, S. B. (2016). Influence of root-knot nematode infestation on antioxidant enzymes, chlorophyll content and growth in Pogostemon cablin (Blanco) Benth. 54(4): 254–261.
30. Tripathi, Y. C. P. K. (2006). *Studies on yield and quality traits of fragrant products from selected humid-tropical aromatic plant – project completion report*. Rain Forest Research Institute.

31. http://en.wikipedia.org/wiki/Patchouli
32. Vijayalalitha, S. J., & Rajasekaran, L. R. (1997). Seasonal influence of rooting of cuttings of patchouli (Pogostemon patchouli pellet). *Advances in Plant Sciences, 10*, 241–242.
33. Sarma, P. C., & Kanjilal, P. B. (2000). Effect of planting time and row spacing on growth yield and quality of patchouli (Pogostemon patchouli Benth). *Advances in Plant Sciences, 13*(1), 201–204.
34. http://www.nedfi.com/node/165
35. Ram, M., Ram, D., Singh, S., Naqvi, A. A., & Kumar, S. (1999). Studiescn intercropping of patchouli (Pogostemon patchouli) with papaya (Caricapapaya). *J Med Aromat Plant Sci, 21*(2), 358–360.
36. Donelian, A., Carlson, L. H. C., Lopes, T. J., & Machado, R. A. F. (2009). Comparison of extraction of patchouli (Pogostemon cablin) essential oil with supercritical CO2 and by steam distillation. *The Journal of Supercritical Fluids, 48*(1), 15–20.
37. Kongkathip, N., Sam-ang, P., Kongkathip, B., Pankaew, Y., Tanasombat, M., & Udomkusonsri, P. (2009). Development of patchouli extraction with quality control and isolation of active compounds with antibacterial activity. *Agriculture and Natural Resources, 43*(3), 519–525.
38. Mohammad Hosseini, M. (2017). The ethnobotanical, phytochemical and pharmacological properties and medicinal applications of essential oils and extracts of different Ziziphora species. *Industrial Crops and Products, 105*, 164–192.
39. Mohammadhosseini, M., Akbarzadeh, A., Shafaghat, A., Hashemi-Moghaddam, H., Mohammadi Nafchi, A., & Ashouri, H. (2016). Chemical composition of the essential oils from flowers and leaves of Marsdenia erecta using microwave assisted hydrodistillation technique. *Journal of Essential Oil-Bearing Plants, 19*(4), 863–874.
40. Farooqi, A. A., & Sreeramu, B. S. (2004). *Cultivation of medicinal and aromatic crops*. Universities Press.
41. Lal, M., Pandey, S. K., Dutta, S., Munda, S., Baruah, J., & Paw, M. (2018). Identification of high herbage and oil yielding variety (Jor Lab P-1) of Pogostemon cablin (Blanco) Benth through mutation breeding. *Journal of Essential Oil-Bearing Plants, 21*(1), 131–138.
42. Deguerry, F., Pastore, L., Wu, S., Clark, A., Chappell, J., & Schalk, M. (2006). The diverse sesquiterpene profile of patchouli, Pogostemon cablin, is correlated with a limited number of sesquiterpene synthases. *Archives of Biochemistry and Biophysics, 454*(2), 123–136.
43. Ramya, H. G., Palanimuthu, V., & Rachna, S. (2013). An introduction to patchouli (Pogostemon cablin Benth.)–A medicinal and aromatic plant: It's importance to mankind. *Agricultural Engineering International: CIGR Journal, 15*(2), 243–250.
44. Verma, R. S., Padalia, R. C., & Chauhan, A. (2012). Assessment of similarities and dissimilarities in the essential oils of patchouli and Indian valerian. *Journal of Essential Oil Research, 24*(5), 487–491.
45. Narayanappa, M., Chacko, C. I., & Vasanthakumar, T. (1984). Wilt of patchouli-a new disease caused by Rhizoctonia solani. In *Current Science* (Vol. 53). Current Science Association.
46. Mishra, R. C., Singh, R., Shahi, S. K., Singh, H. B., & Dikshit, A. (2000). Biological management of patchouli (Pogostemon cablin) wilt caused by Rhizoctonia solani. *Current Science, 78*(3), 230–232.
47. Noveriza, R. (2015). Patchouli viruses: Identification, biological and physical characters, and control strategy. *Jurnal Penelitian dan Pengembangan Pertanian, 34*(1), 1–8.
48. Umesha, K., Thyagaraj, N. E., & Chandrappa, H. M. (1998). New record of the yellowscale of cotton, Cerococcus hibisci green (Homoptera: cerococcidae) onpatchouli, Pogostemon cablin (Blanco) Benth. *Pest Manag. Hortic. Ecosyst, 4*(1), 51.
49. Natsuaki, K. T., Tomaru, K., Ushiku, S., Ichikawa, Y., Sugimura, Y., Natsuaki, T., et al. (1994). Characterization of two viruses isolated from patchouli in Japan. *Plant Disease, 78*(11), 1094–1094.
50. Zaim, M., Ali, A., Joseph, J., & Khan, F. (2013). Serological and molecular studies of a novel virus isolate causing yellow mosaic of patchouli [Pogostemon cablin (Blanco) Benth]. *PLoS One, 8*(12), e83790.

51. Jamalius, A., Hasan, Z., & Idris, H. (1990). Some pests of patchouli (Pogostemoncablin Benth.). *Pemberitaan Penelitian Tanaman Industri, 16*(2), 59–64.
52. Mustika, I., & Nuryani, Y. (2006). Control strategy of parasitic nematodes on patchouli.J. *Litbang Pertan, 25*(1), 7–15.
53. Djiwanti, S. R., & Momota, Y. (1991). Parasitic nematodes associated with patchouli disease in West Java. *Industrial Crops Research of Journal, 3*, 31–34.
54. Khan, M. R., & Reddy, P. P. (1991). Occurrence and distribution of potentially pathogenic endophytic nematodes associated with some ornamental, medicinal and aromatic crops. *Indian Phytopathology, 44*(4), 501–504.
55. Khan, R. M., & Reddy, P. P. (1990). An association of Pratylenchus brachyurus with Pogostemone cablin. *Indian Journal of Nematology, 20*(1), 116–119.
56. Djiwanti, S. R. (2009). Nematoda parasit dan teknologi pengendaliannya dalam budidaya nilam (Pogostemon cablin) di Indonesia. *Perkembangan Teknologi Tanaman Rempah dan Obat, 21*, 40–47.
57. Harni, R., Supramana, S., & Supriadi, S. (2014). Efficacy of endophytic bacteria in reducing plant parasitic nematode Pratylenchus brachyurus. *Indonesian Journal of Agricultural Science, 15*(1), 29.
58. Ichikawa, K. T., Kinoshita, T., & Sankawa, U. (1989). The screening of Chinese crude drugs for calcium antagonist activity: Identification of active principles from the aerial part of *Pogostemon cablin* and the fruits of *Prunus mume*. *Chemical & Pharmaceutical Bulletin, 37*, 345–348.
59. Pattnaik, S., Subramanyam, V. R., & Kole, C. (1996). Antibacterial and antifungal activity of ten essential oils in vitro. *Microbios, 86*(349), 237–246.
60. Yang, D., Michel, D., Mandin, D., Andriamboavonjy, H., Poitry, P., Chaumont, J. P., & Millet Clerc, J. (1996). Antifungal and antibacterial properties in vitro of three patchouly oils from different origins. In *Acta Botanica Gallica (France)*. Taylor & Francis.
61. Hammer, K. A., Carson, C. F., & Riley, T. V. (1999). Antimicrobial activity of essential oils and other plant extracts. *Journal of Applied Microbiology, 86*(6), 985–990.
62. Wang, D., Yin, Z., Zhang, Q., Ye, W., Zhang, X., & Zhang, J. (2010). Nonvolatile chemical constituents from Pogostemon cablin. *Zhongguo Zhong yao za zhi= Zhongguo zhongyao zazhi= China journal of Chinese materia medica, 35*(20), 2704–2707.
63. Edwards-Jones, V., Buck, R., Shawcross, S. G., Dawson, M. M., & Dunn, K. (2004). The effect of essential oils on methicillin-resistant Staphylococcus aureus using a dressing model. *Burns, 30*(8), 772–777.
64. Yang, X., Zhang, X., Yang, S. P., & Liu, W. Q. (2013). Evaluation of the antibacterial activity of patchouli oil. *Iranian Journal of Pharmaceutical Research: IJPR, 12*(3), 307.
65. Pullagummi, C., Rao, N. B., Singh, B. C. S., Bheemagani, A. J., Kumar, P., Venkatesh, K., & Rani, A. R. (2014). Comparitive studies on antibacterial activity of patchouli [Pogostemon cablin (Blanco) Benth] and Geranium (Pelargonium graveolens) aromatic medicinal. *African Journal of Biotechnology, 13*(23), 2379–2384.
66. Vázquez-Sánchez, D., Cabo, M. L., & Rodríguez-Herrera, J. J. (2015). Antimicrobial activity of essential oils against Staphylococcus aureus biofilms. *Food Science and Technology International, 21*(8), 559–570.
67. Bilcu, M., Grumezescu, A. M., Oprea, A. E., Popescu, R. C., Mogoşanu, G. D., Hristu, R., et al. (2014). Efficiency of vanilla, patchouli and ylang ylang essential oils stabilized by iron oxide@ C14 nanostructures against bacterial adherence and biofilms formed by Staphylococcus aureus and Klebsiella pneumoniae clinical strains. *Molecules, 19*(11), 17943–17956.
68. Singh, B. R., Singh, V., Ebibeni, N., & Singh, R. K. (2013). Antimicrobial and herbal drug resistance in enteric bacteria isolated from faecal droppings of common house lizard/gecko (Hemidactylus frenatus). *International Journal of Microbiology, 2013*, 340848.

69. Yu, X. D., Xie, J. H., Wang, Y. H., Li, Y. C., Mo, Z. Z., Zheng, Y. F., et al. (2015). Selective antibacterial activity of patchouli alcohol against Helicobacter pylori based on inhibition of urease. *Phytotherapy Research, 29*(1), 67–72.
70. Karimi, A. (2014). Characterization and antimicrobial activity of patchouli essential oil extracted from Pogostemon cablin [Blanco] Benth.[Lamiaceae]. *Advances in Environmental Biology, 8*, 2301–2310.
71. Wei, A., & Shibamoto, T. (2007). Antioxidant activities and volatile constituents of various essential oils. *Journal of Agricultural and Food Chemistry, 55*(5), 1737–1742.
72. Kim, H. W., Cho, S. J., Kim, B. Y., Cho, S. I., & Kim, Y. K. (2010). Pogostemon cablin as ROS scavenger in oxidant-induced cell death of human neuroglioma cells. *Evidence-based Complementary and Alternative Medicine, 7*(2), 239–247.
73. Lin, R. F., Feng, X. X., Li, C. W., Zhang, X. J., Yu, X. T., Zhou, J. Y., et al. (2014). Prevention of UV radiation-induced cutaneous photoaging in mice by topical administration of patchouli oil. *Journal of Ethnopharmacology, 154*(2), 408–418.
74. Feng, X. X., Yu, X. T., Li, W. J., Kong, S. Z., Liu, Y. H., Zhang, X., et al. (2014). Effects of topical application of patchouli alcohol on the UV-induced skin photoaging in mice. *European Journal of Pharmaceutical Sciences, 63*, 113–123.
75. Li, Y. P., Yuan, S. F., Cai, G. H., Wang, H., Wang, L., Yu, L., et al. (2014). Patchouli alcohol dampens lipopolysaccharide induced mastitis in mice. *Inflammation, 37*, 1757–1762.
76. Schieber, M., & Chandel, N. S. (2014). ROS function in redox signaling and oxidative stress. *Current Biology, 24*(10), R453–R462.
77. Rolo, A. P., Teodoro, J. S., & Palmeira, C. M. (2012). Role of oxidative stress in the pathogenesis of nonalcoholic steatohepatitis. *Free Radical Biology and Medicine, 52*(1), 59–69.
78. Kaspar, J. W., Niture, S. K., & Jaiswal, A. K. (2009). Nrf2: INrf2 (Keap1) signaling in oxidative stress. *Free Radical Biology and Medicine, 47*(9), 1304–1309.
79. Li, S., Tan, H. Y., Wang, N., Zhang, Z. J., Lao, L., Wong, C. W., & Feng, Y. (2015). The role of oxidative stress and antioxidants in liver diseases. *International Journal of Molecular Sciences, 16*(11), 26087–26124.
80. Qiang, M. (2013). Role of nrf2 in oxidative stress and toxicity. *Annual Review of Pharmacology and Toxicology, 53*, 401–426.
81. Cao, Y. W., Jiang, Y., Zhang, D. Y., Wang, M., Chen, W. S., Su, H., et al. (2015). Protective effects of Penthorum chinense Pursh against chronic ethanol-induced liver injury in mice. *Journal of Ethnopharmacology, 161*, 92–98.
82. Kim, E. K., Kim, J. H., Jeong, S., Choi, Y. W., Choi, H. J., Kim, C. Y., & Kim, Y. M. (2019). Pachypodol, a methoxyflavonoid isolated from Pogostemon cablin bentham exerts antioxidant and cytoprotective effects in HepG2 cells: Possible role of ERK-dependent Nrf2 activation. *International Journal of Molecular Sciences, 20*(17), 4082.
83. Wu, D., & Cederbaum, A. (2009). Oxidative stress and alcoholic liver disease. *Seminars in Liver Disease, 29*, 141–154.
84. Bai, T., Yang, Y., Yao, Y. L., Sun, P., Lian, L. H., Wu, Y. L., & Nan, J. X. (2016). Betulin alleviated ethanol-induced alcoholic liver injury via SIRT1/AMPK signaling pathway. *Pharmacological Research, 105*, 1–12.
85. Ding, R. B., Tian, K., Cao, Y. W., Bao, J. L., Wang, M., He, C., et al. (2015). Protective effect of panax notoginseng saponins on acute ethanol-induced liver injury is associated with ameliorating hepatic lipid accumulation and reducing ethanol-mediated oxidative stress. *Journal of Agricultural and Food Chemistry, 63*(9), 2413–2422.
86. Huang, Q. H., Wu, X., Chen, X. H., Wu, J. Z., Su, Z. R., Liang, J. L., et al. (2018). Patchouli oil isolated from the leaves of Pogostemon cablin ameliorates ethanol-induced acute liver injury in rats via inhibition of oxidative stress and lipid accumulation. *RSC Advances, 8*(43), 24399–24410.
87. Niederlechner, S., Baird, C., Petrie, B., Wischmeyer, E., & Wischmeyer, P. E. (2013). Epidermal growth factor receptor expression and signaling are essential in glutamine's

cytoprotective mechanism in heat-stressed intestinal epithelial-6 cells. *American Journal of Physiology. Gastrointestinal and Liver Physiology, 304*(5), G543–G552.
88. Liu, X., Jiang, L., Liu, F., Chen, Y., Xu, L., Li, D., et al. (2016). Effect of patchouli alcohol on the regulation of heat shock-induced oxidative stress in IEC-6 cells. *International Journal of Hyperthermia, 32*(5), 474–482.
89. Abe, S., Sato, Y., Inoue, S., Ishibashi, H., Maruyama, N., Takizawa, T., et al. (2003). Anti-Candida albicans activity of essential oils including lemongrass (Cymbopogon citratus) oil and its component, citral. *Nihon Ishinkin Gakkai Zasshi, 44*(4), 285–291.
90. Kocevski, D., Du, M., Kan, J., Jing, C., Lačanin, I., & Pavlović, H. (2013). Antifungal effect of Allium tuberosum, Cinnamomum cassia, and Pogostemon cablin essential oils and their components against population of Aspergillus species. *Journal of Food Science, 78*(5), M731–M737.
91. Liu, Q., Luyten, W., Pellens, K., Wang, Y., Wang, W., Thevissen, K., Liang, Q., Cammue, B. P., Schoofs, L., & Luo, G. (2012). Antifungal activity in plants from Chinese traditional and folk medicine. *Journal of Ethnopharmacology, 143*, 772–778. https://doi.org/10.1016/j.jep.2012.06.019
92. Yi, Y. Y., He, J. J., Su, J. Q., Kong, S. Z., Su, J. Y., Li, Y. C., et al. (2013). Synthesis and antimicrobial evaluation of pogostone and its analogues. *Fitoterapia, 84*, 135–139.
93. Kardinan, A., Wikardi, E. A., & Halid, W. (1997). *The prospect of botanical insecticides on stored food insects management* (Vol. 59, pp. 199–208). Biotrop Special Publication.
94. Chun, W., DeJun, Y., ShiLin, H., & Qun, T. (2000). Determination of toxicity of plant essential oils to museum insect pests. *Journal of Southwest Agricultural University, 22*, 494–495.
95. Wu, H. Q., Li, J., He, Z. D., & Liu, Z. G. (2010). Acaricidal activities of traditional Chinese medicine against the house dust mite, Dermatophagoides farinae. *Parasitology, 137*(6), 975–983.
96. Albuquerque, E. L., Lima, J. K., Souza, F. H., Silva, I. M., Santos, A. A., Araújo, A. P. A., et al. (2013). Insecticidal and repellence activity of the essential oil of Pogostemon cablin against urban ants species. *Acta Tropica, 127*(3), 181–186.
97. Gokulakrishnan, J., Kuppusamy, E., Shanmugam, D., Appavu, A., & Kaliyamoorthi, K. (2013). Pupicidal and repellent activities of Pogostemon cablin essential oil chemical compounds against medically important human vector mosquitoes. *Asian Pacific Journal of Tropical Disease, 3*(1), 26–31.
98. Kiuchi, F., Matsuo, K., Ito, M., Qui, T. K., & Honda, G. (2004). New sesquiterpene hydroperoxides with trypanocidal activity from Pogostemon cablin. *Chemical & Pharmaceutical Bulletin, 52*(12), 1495–1496.
99. Trongtokit, Y., Rongsriyam, Y., Komalamisra, N., & Apiwathnasorn, C. (2005). Comparative repellency of 38 essential oils against mosquito bites. *Phytotherapy Research: An International Journal Devoted to Pharmacological and Toxicological Evaluation of Natural Product Derivatives, 19*(4), 303–309.
100. Phal, D., Naik, R., Deobhankar, K., Vitonde, S., Ghatpande, N., & Lifescience, R. (2012). Laboratory evaluation of herbal mosquito coils against Aedes aegypti mosquito. *Bulletin of Environment, Pharmacology and Life Sciences, 1*(10), 16–20.
101. Huang, S. H., Xian, J. D., Kong, S. Z., Li, Y. C., Xie, J. H., Lin, J., Chen, J. N., Wang, H. F., & Su, Z. R. (2014). Insecticidal activity of pogostone against *Spodoptera litura* and *Spodoptera exigua* (Lepidoptera: Noctuidae). *Pest Management Science, 70*, 510–516. https://doi.org/10.1002/ps.3635
102. Haze, S., Sakai, K., & Gozu, Y. (2002). Effects of fragrance inhalation on sympathetic activity in normal adults. *Japanese Journal of Pharmacology, 90*(3), 247–253.
103. Bowles, E. J., Griffiths, D. M., Quirk, L., Brownrigg, A., & Croot, K. (2002). Effects of essential oils and touch on resistance to nursing care procedures and other dementia-related behaviours in a residential care facility. *International Journal of Aromatherapy, 12*(1), 22–29.

104. Smith, M. (2015). *Aromatherapy care for the caregivers research paper*. Available online: essentialoiltechniques.com/docs/Aromatherapy-for-the-care-giver.PDF. Accessed on 14 January 2015.
105. Perry, N., & Perry, E. (2006). Aromatherapy in the management of psychiatric disorders: Clinical and neuropharmacological perspectives. *CNS Drugs, 20*, 257–280.
106. Beshara, M. C., & Giddings, D. (2002). Use of plant essential oils in treating agitation in a dementia unit 10 case studies. *The International Journal of Aromatherapy, 4*(12), 207–212.
107. Pujiarti, R., Ohtani, Y., Widowati, T. B., Wahyudi, W., Kasmudjo, K., Herath, N. K., & Wang, C. N. (2012). Effect of Melaleuca leucadendron, Cananga odorata and Pogostemon cablin oil odors on human physiological responses. *Wood Research Journal, 3*(2), 100–105.
108. Buckle, J. (2002). Clinical aromatherapy and AIDS. *The Journal of the Association of Nurses in AIDS Care, 13*(3), 81–99.
109. Kiyohara, H., Ichino, C., Kawamura, Y., Nagai, T., Sato, N., & Yamada, H. (2012). Patchouli alcohol: In vitro direct anti-influenza virus sesquiterpene in Pogostemon cablin Benth. *Journal of Natural Medicines, 66*, 55–61.
110. Wu, H., Li, B., Wang, X., Jin, M., & Wang, G. (2011). Inhibitory effect and possible mechanism of action of patchouli alcohol against influenza A (H2N2) virus. *Molecules, 16*(8), 6489–6501.
111. Li, Y. C., Peng, S. Z., Chen, H. M., Zhang, F. X., Xu, P. P., Xie, J. H., He, J. J., Chen, J. N., Lai, X. P., & Su, Z. R. (2012). Oral administration of patchouli alcohol isolated from Pogostemonis Herba augments protection against influenza viral infection in mice. *International Immunopharmacology, 12*, 294–301.
112. Wu, X. L., Ju, D. H., Chen, J., Yu, B., Liu, K. L., He, J. X., et al. (2013). Immunologic mechanism of patchouli alcohol anti-H1N1 influenza virus may through regulation of the RLH signal pathway in vitro. *Current Microbiology, 67*, 431–436.
113. Xie, Y. C., & Tang, F. (2009). Protective effect of *Pogostemon cablin* on membrane fluidity of intestinal epithelia cell in ischemia/reperfusion rats after ischemia/reperfusion. *Chinese Journal of Integrated Traditional and Western Medicine, 29*, 639–641.
114. Yang, Y., Kinoshita, K., Koyama, K., Takahashi, K., Tai, T., Nunoura, Y., & Watanabe, K. (1999). Anti-emetic principles of Pogostemon cablin (Blanco) Benth *Phytomedicine, 6*(2), 89–93.
115. Mikuriya, N., Kim, Y., & Fujimura, K. (2004). The effect of the aroma of patchouli essential oil on defecation and constipation. *Aroma Res, 5*, 70–75.
116. Sumi, H. (2003). Fibrinolysis-accelerating activity of the essential oils and Shochu aroma. *Aroma Research, 4*(3; NUMB 15), 60–63.
117. Park, E. K., Yoon, H. K., & Kim, D. H. (2002). Antithrombotic activity of Sunghyangjunggisan. *Natural Product Sciences, 8*, 71–75.
118. Lu, T. C., Liao, J. C., Huang, T. H., Lin, Y. C., Liu, C. Y., Chiu, Y. J., & Peng, W. H. (2011). Analgesic and anti-inflammatory activities of the methanol extract from Pogostemon cablin. *Evidence-based Complementary and Alternative Medicine, 2011*, 671741.
119. Xian, Y. F., Li, Y. C., Ip, S. P., Lin, Z. X., Lai, X. P., & Su, Z. R. (2011). Anti-inflammatory effect of patchouli alcohol isolated from Pogostemonis Herba in LPS-stimulated RAW264. 7 macrophages. *Experimental and Therapeutic Medicine, 2*(3), 545–550.
120. Li, C. W., Wu, X. L., Zhao, X. N., Su, Z. Q., Chen, H. M., Wang, X. F., et al. (2013). Anti-inflammatory property of the ethanol extract of the root and rhizome of Pogostemon cablin (Blanco) Benth. *Scientific World Journal, 2013*, 1–12.
121. Li, K., Zhang, H., Xie, H., Liang, Y., Wang, X., & Ito, Y. (2011). Preparative isolation and purification of five flavonoids from Pogostemon cablin Benth by high-speed countercurrent chromatography and preparative high-performance liquid chromatography. *Journal of Liquid Chromatography and Related Technologies, 34*(15), 1617–1629.
122. Jeong, J. B., Choi, J., Lou, Z., Jiang, X., & Lee, S. H. (2013). Patchouli alcohol, an essential oil of Pogostemon cablin, exhibits anti-tumorigenic activity in human colorectal cancer cells. *International Immunopharmacology, 16*(2), 184–190.

123. Park, S. Y., Neupane, G. P., Lee, S. O., Lee, J. S., Kim, M. Y., Kim, S. Y., et al. (2014). Protective effects of Pogostemon cablin Bentham water extract on inflammatory cytokine expression in TNBS-induced colitis in rats. *Archives of Pharmacal Research, 37*, 253–262.
124. Park, E. J., Park, H. R., Lee, J. S., & Kim, J. W. (1998). Licochalcone A: An inducer of cell differentiation and cytotoxic agent from *Pogostemon cablin*. *Planta Medica, 64*, 464–466.
125. Yu, J., Qi, Y., Luo, G., Duan, H. Q., & Zhou, J. (2012). Extraction and analysis of the essential oil in Pogostemon cablin by enzymatic hydrolysis and inhibitory activity against Hela cell proliferation. *Zhong yao cai= Zhongyaocai= Journal of Chinese Medicinal Materials, 35*(5), 796–799.
126. Hsu, H. C., Yang, W. C., Tsai, W. J., Chen, C. C., Huang, H. Y., & Tsai, Y. C. (2006). α-Bulnesene, a novel PAF receptor antagonist isolated from Pogostemon cablin. *Biochemical and Biophysical Research Communications, 345*(3), 1033–1038.
127. Dang, H., & Huang, Q. Y. (2002). Preliminary investigation into the pharmacodynamics of Xinxiang granule. *Chinese Journal of Information on Traditional Chinese Medicine, 9*, 20–23.
128. He, J. J., Chen, H. M., Li, C. W., Wu, D. W., Wu, X. L., Shi, S. J., et al. (2013). Experimental study on antinociceptive and anti-allergy effects of patchouli oil. *Journal of Essential Oil Research, 25*(6), 488–496.
129. Kerr, J. (2002). The use of essential oils in healing wounds. *International Journal of Aromatherapy, 12*(4), 202–206.
130. Chen, H., Li, Y., Wu, X., Li, C., Li, Q., Qin, Z., et al. (2013). LC-MS/MS determination of pogostone in rat plasma and its application in pharmacokinetic studies. *Biomedical Chromatography, 27*(9), 1092–1099.
131. Miyazawa, M., Okuno, Y., Nakamura, S. I., & Kosaka, H. (2000). Antimutagenic activity of flavonoids from Pogostemon cablin. *Journal of Agricultural and Food Chemistry, 48*(3), 642–647.
132. Yatsunenko, T., Rey, F. E., Manary, M. J., Trehan, I., Dominguez-Bello, M. G., Contreras, M., et al. (2012). Human gut microbiome viewed across age and geography. *Nature, 486*(7402), 222–227.
133. Fukuda, S., Toh, H., Hase, K., Oshima, K., Nakanishi, Y., Yoshimura, K., et al. (2011). Bifidobacteria can protect from enteropathogenic infection through production of acetate. *Nature, 469*(7331), 543–547.
134. Vétizou, M., Pitt, J. M., Daillère, R., Lepage, P., Waldschmitt, N., Flament, C., et al. (2015). Anticancer immunotherapy by CTLA-4 blockade relies on the gut microbiota. *Science, 350*(6264), 1079–1084.
135. Cryan, J. F., & Dinan, T. G. (2012). Mind-altering microorganisms: The impact of the gut microbiota on brain and behaviour. *Nature Reviews Neuroscience, 13*(10), 701–712.
136. Manichanh, C., Borruel, N., Casellas, F., & Guarner, F. (2012). The gut bacteria in IBD. *Nature Reviews Gastroenterology & Hepatology, 9*, 599–608.
137. Marchesi, J. R., Adams, D. H., Fava, F., Hermes, G. D., Hirschfield, G. M., Hold, G., et al. (2016). The gut microbiota and host health: A new clinical frontier. *Gut, 65*(2), 330–339.
138. Sobhani, I., Tap, J., Roudot-Thoraval, F., Roperch, J. P., Letulle, S., Langella, P., et al. (2011). Microbial dysbiosis in colorectal cancer (CRC) patients. *PLoS One, 6*(1), e16393.
139. Boulangé, C. L., Neves, A. L., Chilloux, J., Nicholson, J. K., & Dumas, M. E. (2016). Impact of the gut microbiota on inflammation, obesity, and metabolic disease. *Genome Medicine, 8*, 1–12.
140. Ley, R. E., Turnbaugh, P. J., Klein, S., & Gordon, J. I. (2006). Human gut microbes associated with obesity. *Nature, 444*(7122), 1022–1023.
141. Leong, W., Huang, G., Khan, I., Xia, W., Li, Y., Liu, Y., et al. (2019). Patchouli essential oil and its derived compounds revealed prebiotic-like effects in C57BL/6J mice. *Frontiers in Pharmacology, 10*, 1229.

142. Kim, K. A., Gu, W., Lee, I. A., Joh, E. H., & Kim, D. H. (2012). High fat diet-induced gut microbiota exacerbates inflammation and obesity in mice via the TLR4 signaling pathway. *PLoS One, 7*(10), e47713.
143. Lee, J., Kong, B., & Lee, S. H. (2020). Patchouli alcohol, a compound from Pogostemon cablin, inhibits obesity. *Journal of Medicinal Food, 23*(3), 326–334.
144. Rinella, M. E. (2015). Nonalcoholic fatty liver disease: A systematic review. *JAMA, 313*(22), 2263–2273.
145. Younossi, Z. M., Koenig, A. B., Abdelatif, D., Fazel, Y., Henry, L., & Wymer, M. (2016). Global epidemiology of nonalcoholic fatty liver disease—meta-analytic assessment of prevalence, incidence, and outcomes. *Hepatology, 64*(1), 73–84.
146. Binti, A. A., & Binti, A. E. A. (2018). Brief overview of a decade of genome-wide association studies on primary hypertension. *International Journal of Endocrinology, 2018*, 1–14.
147. World Health Organization. (1959). *Hypertension and coronary heart disease: classification and criteria for epidemiological studies, first report of the expert committee on cardiovascular diseases and hypertension [meeting held in Geneva from 13 to 18 October 1958]*. World Health Organization.
148. Hu, G. Y., Peng, C., Xie, X. F., Xiong, L., Zhang, S. Y., & Cao, X. Y. (2018). Patchouli alcohol isolated from Pogostemon cablin mediates endothelium-independent vasorelaxation by blockade of Ca2+ channels in rat isolated thoracic aorta. *Journal of Ethnopharmacology, 220*, 188–196.
149. Zhu, H., Zhou, Q. M., Peng, C., Chen, M. H., Li, X. N., Lin, D. S., & Xiong, L. (2017). Pocahemiketals A and B, two new hemiketals with unprecedented sesquiterpenoid skeletons from Pogostemon cablin. *Fitoterapia, 120*, 67–71.

Chapter 12
Black Pepper

Muhammad Umar Ijaz, Muhammad Faisal Hayat, Asma Ashraf, and Ishrat Rahman

12.1 Introduction

Herbs impart critical role in the nutritional balance since the emergence of human era. Values of herbs are increasing day by day due to their use to enhance colors, aroma, flavors, medicinal as well as preservative properties [1]. The Piperaceae family, which comprises of 13 genera & includes piper species, is primitive plant family found in tropical climates [2]. According to Durant-Archibold et al. genus Piper contains 4166 plants out of this 1457 approved, 1333 unassessed while 1376 species are synonyms [3].

Black Pepper also known as *Piper nigrum L.* is widely regarded as the "king of spices" owing to its enormous trading shares in the international markets [4]. The word pepper came from Sanskrit word "pippali" which means "berry" [5]. Black pepper is a 50–60 cm long woody tree having strong aroma and marvelous climbing properties [6]. Black & white peppers are processed differently and are harvested at different times. Black pepper contains the pulp because it is made by dehydrating unripe fruits until wrinkles develops.

M. U. Ijaz (✉) · M. F. Hayat
Department of Zoology, Wildlife and Fisheries, University of Agriculture, Faisalabad, Pakistan
e-mail: umar.ijaz@uaf.edu.pk

A. Ashraf
Department of Zoology, Government College University, Faisalabad, Pakistan

I. Rahman
Department of Basic Dental Sciences, College of Dentistry, Princess Nourah bint Abdulrahman University, Riyadh, Saudi Arabia

M. Zia-Ul-Haq et al. (eds.), *Essentials of Medicinal and Aromatic Crops*,
https://doi.org/10.1007/978-3-031-35403-8_12

12.2 Taxonomic Rank of *P. nigrum*

Kingdom: Plantae
Class: Equisetopsida
Sub class: Magnoliidae
Super order: Magnolianae
Order: Piperales
Family: Piperaceae
Genus: Piper
Species: nigrum

Geographically, *P. nigrum* is primarily grown in humid and hot climate [7] while south Indian Peninsula's Western Ghats are the main sites for growing black pepper, which was later spread to other nations in South and Southeast Asia. Currently, Vietnam is the major contributor in black pepper production with most of it being grown in the country's southern area [8]. The top four pepper-producing nations in the globe in 2016 were Indonesia (70,000 metric tonnes), Vietnam (total production: 140,000 metric tonnes), Brazil (45,000 metric tonnes) followed by India (48,500 metric tonnes). Production of black pepper is increased in recent years due to high cultivation in Indonesia and Vietnam [9]. The actual yield of plant depends directly upon the number of leaves per stems. The dry form of seeds is designated as "peppercorns".

12.3 Stem and Leaf of Black Pepper

Stem of black pepper grows up to 10 m and is delicate, green and herbaceous which is covered by blossom, fruits and leaves. In case of pepper plant, there are three types of stems which include runner shoots, primary stem and fruit producing stem. A single axis of stem bears 20–30 spikes of fruits [10]. The plant of black pepper has alternate leaves along with 50–150 blooms and spikes. The leaves are lustrous from dorsal side while light greenish from ventral and arrange alternatively with green sparkling. Leaves are 3 to 6 cm wide and 5–20 cm long with a blade of 18 cm [11] (Fig. 12.1).

12.4 Nutritive Composition

The leaves and fruits of *P. nigrum* are rich sources of carbohydrates, proteins, minerals, fats and fibers. *P. nigrum* accessions are enriched by vitamin C, Ca, Mg, Zn while moderate level of iron. Leaves are comprising of carbohydrates (32–46%), proteins (17.46–20.30%), fat (3.59–4.04%), ash (10.30–12.94%), fibers (10.64–13.62%) and total energy 273–295 kcal/100 g. While the fruit composition

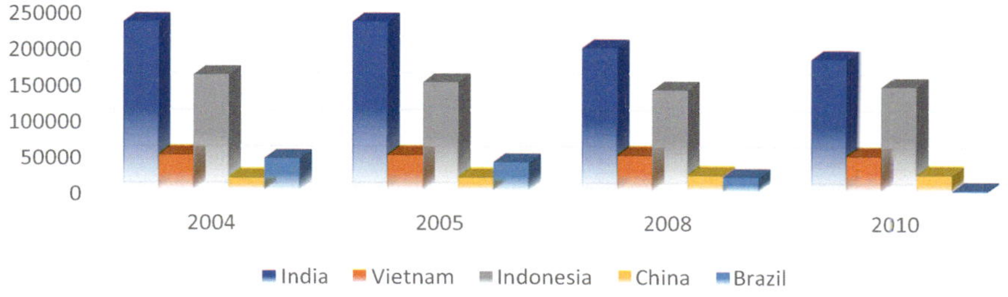

Fig. 12.1 The cultivation of black pepper in different regions throughout the world. (International community 2011)

Table 12.1 Derivatives of black pepper and their activities

Derivative	Functions	Reference
Piperine	Anti-cancerous	[13]
	Prevent fat deposition	[14]
	Anti-microbial	
	Anti-inflammatory	[15]
Essential oil	As an aromatic compound	[16]
	Prevent worm infection	[17]

includes carbohydrates (51.37–53.74%), proteins (8.44–9.18%), fibers (5.13–5.50%) and fat (12.77–15.05%) and total energy 361–384 kcal/100 g [12] (Table 12.1).

12.5 Essential Oils of *P. nigrum*

A colorless to greenish liquid is extracted from black pepper which constitutes 1 to 3% of black pepper having peppery aroma that is referred as essential oil of *P. nigrum*. The chemistry of essential oil includes gravity (0.860–0.884), boiling point (54 °C) and refractive index (1.478–1.488) at room temperature along with levorotatory optical position [18, 19]. Essential oil from black pepper can be extracted by different sorts of mechanism such as micro-extraction, liquid carbon dioxide and steam distillation. Evaluation of its nutritive value is carried out by using gas-chromatography, column chromatography and mass spectrometric analysis [20, 21].

12.6 Chemical Constituents of Black Pepper

Guenther [22], Govindarajan [23] and Parmar et al. [24], have investigated the chemical composition of pepper. Analytical information on the chemical make-up of *P. nigrum* and the anti-microbial properties of pepper oil was reported by Menon

Table 12.2 Essential oil and its chemical constituents

Chemical	Percentage	References
Cumaldehyde	15.7–63%	[28–32]
ρ-mentadien-1,3-al-7	2.2–27.4%	
ρ-mentadien 1,4-al-7	0.3–17.4%	
β-pinene	0.8–21%	
ρ-cymene	4.2–23.2%	
γ-terpinene	1.5–23.9%	

[25], Menon & Padmakumari [26] as well as Sasidharan and Menon in [27]. Oleoresin, piperine, and pepper oil are three fundamental components that are frequently synthesized by chemical processing. The extraction of pepper oil from ground pepper uses steam distillation, while the extraction of the other two components uses solvents extraction techniques. One of the most widely traded pepper products is oleoresin which is synthesize by combining the flavour (oil) and pungency (piperine) elements (Table 12.2).

The black pepper oils and terpenes are major constituents of black pepper having nitrogenous compounds [33, 34]. Other major odorants from black peppers include, butyric acid, 2-isoproplyle-3- acid, α-pinene and phellandrene as well as limonene and methyl propanal [35].

12.7 Farming of Black Pepper

Cultivation of *P. nigrum* and its management degrading the ecosystem to some extent until shortening the life cycle of pepper vines [36]. Black pepper is evolved from Miocene in the Western Ghats evergreen forests [37]. The domestication of *P. nigrum* still remain obscure, however its cultivation was started decades ago in India [38].

12.7.1 Different Varieties of Black Pepper

Only a few of *P. nigrum* cultivars are famous for their cultivation out of more than 100 varieties [39, 40]. *P. nigrum* is self-pollinating crop having high rate of variability among actual yields. Best quality and high yield cultivars were selected by using different techniques such as hybridization, clonal selection as well as open selection during pollination [40]. The first cultivar of *P. nigrum* is Panniyur-1 which was synthesized by hybridization technique. The quality of *P. nigrum* depends upon the nutrients of essential oils and pipperine. Variability in the varieties of *P. nigrum* is investigated on the basis of quality parameters [41]. Other varieties of black peppers are Chunuk, Malonan, Natar 1& 2, LDK, Petaling 1& 2, Bengkayang and ciinten [42] (Table 12.3).

Table 12.3 The production of black pepper in various countries during 2009–2010 (International pepper community 2011)

Country	2009–2010	Production (metric tons)
India		49,000–49,550
Indonesia		33,000–40,000
Brazil		32,000–38,200
Malaysia		154,000–16,450
Vietnam		88,000–101,750

12.7.2 Climate Conditions

A number of investigations have reported differences in growth and production of pepper on the basis of their genetic makeup, agronomic practices and environmental conditions [43]. Black pepper is a moist tropic plant that needs between 2000 and 3000 mm of rainfall, high levels of humidity, tropical temperatures and relatively constant day length throughout the entire year. The amount and distribution of precipitation have a substantial impact on the yield of pepper crop. The optimum average rainfall for *P. nigrum* is 2000 mm/per year. The flushing and number of flowers started by a rainfall of 70 mm over the course of 20 days in May–June, but once these processes get started, rain should remain continue till the *P. nigrum* is fully ripe [44].

12.7.3 Fertilizer Practices and Sustainable Farming

Black pepper has great demand due to its nutrients and therefore it should be enriched with these nutrients to obtain high yield [45]. Nutrients uptake of a mature *P. nigrum* plant is estimated 13 kg P, 202 kg N, 18 kg Mg, 156 kg K and 68 kg of calcium/ha [46]. The term "Natural Farming" is not only farming without the involvement of inorganic fertilizers but it directs multiple dimensions to use microorganisms for improvement of soil quality and soil fertility [47]. Sustainable farming can be defined as "utilization of resources in such a way that they do not affect the sustainability of ecosystem without any sort of compromise on human health status [48].

12.7.4 Role of Soil Microflora

The presence of microorganism is only effective when they attain optimum conditions (water, oxygen, pH and temperature) to perform their functions on their substrate [49]. During natural farming, the population of these effective microflora is made to grow to large scale and then applied to cultivated soil [50]. However, the beneficial effects of microflora depend upon the environmental conditions of soil (temperature, pH, moisture) and appear only when they establish as well as become dominant to perform their metabolic activities [36]. In case of black pepper, it takes years to fully establish the microflora in the soil. Recurrent month - to - month administrations, particularly during the initial cropping period, might significantly speed up the development of the supplied beneficial microbes [51]. It is also reported that when this microflora is isolated from soil, cultured in the lab & then reintroduced into the native soil at enormous population, it is immediately exposed to antagonistic & competitive effects from the native soil microflora and then the population density declined [52]. Therefore, the opportunities to enhance yield, quality and health status of cultivated crops become high when microflora of soil is applied in mixed form [53].

12.7.5 Harvest and Post-Harvest

The first harvest of *P. nigrum* is performed upto 2 to 3 years followed by plantation. The flowering stage of black pepper starts after 1 year of plantation but they shed them to gain more nutrient for subsequent years. The yield of black pepper reach to the peak in the seventh year followed by plantation but then decline gradually. However, the yield can remain sustained for 15 years if managed according to aforementioned conditions [54]. To achieve high quality black pepper fruits, it should be handled with appropriate management after harvesting. To make the fruits separation easy, crop should be bagged for half or full day and night period covered under heaps for complete fermentation. After this period the spikes of pepper separated by using mechanical thrashers (for large crop) or by trampling under the foot [55]. Processing and storage of black pepper and its derivatives is the crucial step in order to obtain maximum benefits. Since heat, such as that used in domestic cooking procedures, readily depletes piperine, the active principal ingredient therefore, the method used to store and handle pepper is crucial. Light-induced isomerization transforms piperine and its isomers into chavicine, isopiperine and isochavicine. The rate of isomerization rises with exposure time and light intensity. When chavicine is stored, it gradually turns into piperine, losing some of its pungency [56].

The amount of piperine lost from P. nigrum during thermal processing varied between 16% and 34%, with the maximum loss occurring during pressure cooking, suggesting decreased accessibility of pepper active ingredients from cooking when the constituents have only been exposed to high temperature such as boiling or pressure cooking for minimum time duration. Total extractable oleoresin as well as piperine of pepper are degraded according to first-order thermodynamics [57, 58]. Ionizing radiation has the capacity to purify the black pepper and its derivatives from microorganisms. A substantial reduction was reported in the viscosity as a function of radiation dose in the dose range necessary for microbiological disinfestations [59]. Antioxidant potential of black pepper is reported to reduce in response to 10–30 kGy of radiation [60].

12.8 Various Sorts of Diseases of *P. nigrum*

12.8.1 Phytopthora Infection

During the rainy season, fungus (*P. capsica*) causes the soaking of water and forms lesions on the margins of leaves which later get enlarged and spread from leaf lamina to stem. The severity of disease becomes higher if the shoots emerge from foot rot. These runner shoots carry the infectious fungus towards roots [61]. If nursery conditions are favorable for the fungus, it reproduces thousands of times and infect the roots to cause root rot disease. To prevent from aforementioned infection, the removal of runner shoot should be carried out on immediate basis from the garden or by using different fungicides [62, 63].

12.8.2 Blight & Rot Disease

This condition caused by *Pythium* sp., *Collectotrichum* sp. and *Rhizoctonia solani* with different appearance after attack. During the attack by *R. solani*, leaves developed spots of greyish color but the infected leaves remained anchored with each other. In case of *Collectotrichum* sp. Attack, the leaves develop yellowish color in the form of necrotic spots. Both of these conditions can be avoided by eliminating runner shoots from the lawns and spraying the solution of Bordeaux mixture (1%). However, the process of *Phytosanitations* also play crucial role in preventing these conditions during which infected veins are eliminated from the gardens and eradicated completely [64, 65].

12.8.3 Phyllody and Stunted Disease of Black Pepper

Phyllody is suspected to cause by a mycoplasma like organism (MLO) which has tendency to convert the spikes into leaf like structure in contrast to the anatomy of normal flower. However, the molding of spikes into such shape does not destroy rather it hamper the berries production. In case of stunted disease, the growth of vines becomes ceased thus directly affecting the actual yield. The most prominent signs include molting of leaves, narrowing lamina and reduction in the length of internode thus referred as little leaf disease [66] (Fig. 12.2).

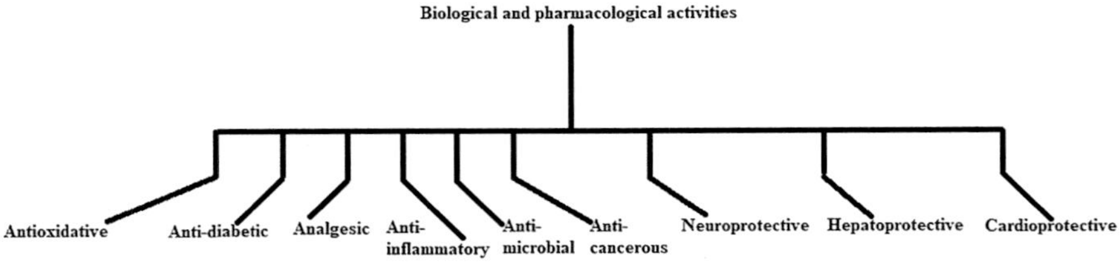

Fig. 12.2 Flowsheet showing the biological and pharmacological activities of black pepper

12.9 Biological and Pharmacological Applications of Black Pepper

Black pepper is reported to show anti-oxidant [67, 68], anti-cancerous [69] anti-apoptotic [70] immune-modulatory [69] antimicrobial [1] and anti-spasmodic [71]. Following are some important biological and pharmacological applications of black pepper:

12.9.1 Antimicrobial Activities

The antimicrobial action of any spice depends upon certain factors such as quantity, chemical constituents, type of spice, type of pathogen attacking the spices and intensity of such attack and storage conditions of spices [72]. Emergence of drug resistant pathogen led to inefficient response of antimicrobial drugs against them. The treatment of microbial infections becomes challenging and thus requires new antimicrobial drugs or therapies [73]. A wide variety of pathogens have been observed to be resistant to *P. nigrum 's* antimicrobial effectiveness. The sensitivity of the microorganisms against drug depends upon the concentrations and use of different solvents for black pepper extraction. For instance, Khan et al. investigated that water extract of black pepper fruit have the highest zone of inhibition (ZOI) against *E. coli* whereas the extract of hot water demonstrated the highest ZOI against *S. typhi* and *S. aureus* amongst multiple solvent extracts (hot water, methanol and cold water) [74]. Additionally, the extract of methanol had the highest degree of inhibitory activity against *P. aeruginosa, S. typhi* & *E. coli* but had no effect against *S. aureus.* Another investigation conducted by Karsha and Lakshmi found that use of acetone for *P. nigrum* extraction showed highest degree of inhibition against *S. aureus* and found to be more active in contrast to the extract of dichloromethane against numerous bacteria [75]. Furthermore, it was discovered that extraction of black pepper by using methanol displayed high degree of inhibition against pathogenic fungi [76]. Various pathogens including *S. aureus, B. subtilis, E. coli,* and P. aeruginosa showed resistance against n-Hexane extract while inhibited by the action of ethanol along with acetone extract of *P. nigrum.* A number of leaf extract solvents showed marvelous antimicrobial activities against above mentioned pathogens [77, 78].

12.9.2 Antioxidant Activity

Numerous in-vitro investigations were designed to examine the antioxidant abilities of *P. nigrum*. The ethanol extract exhibited highest ferric reducing abilities as compare to aqueous extract of black pepper but the aqueous extract displayed high degree of scavenging capabilities against hydrogen peroxide, superoxide anion, leading to reduced oxidative stress. A number of investigations were conducted by Shanmugapriya et al. to determine the antioxidant potential of *P. nigrum* by employing acetone, ethyl acetate and aqueous solvents [77]. Extract of ethyl acetate represented high degree of scavenging against superoxide anion, DPPH (2,2-diphenyl-1-picrylhydrazyl) and ABTS (2,2′-azino-bis (3-ethylbenzothiazoline-6-sulfonic acid) while acetone based showed excellent scavenging activities against nitrous oxide and hydrogen peroxide. Essential oils (derivatives of black pepper) exhibited inhibitory activities against lipid peroxidation, FRAP (ferric reducing ability of plasma) and DPPH in contrast to the extracts of ethanol and ethyl acetate [79]. Alcoholic extract of *P. nigrum* is documented to enhance the level of anti-oxidant level such superoxide dismutase, glutathione-S-transferase and glutathione peroxidase in rats [80]. Essential oils of *P. nigrum* is documented to diminish malondialdehyde (MDA) levels in the hepatic tissues of rabbits [81].

12.9.3 Anti-cancerous Activity

Scientific community is continuously working to explore new medicines from plants after the failure of multiple drugs in cancer therapy and plethron of its side effects on patients more than their beneficial impacts. Many researchers demonstrated the anti-cancerous effects of black pepper and its derivatives on various cancer cell lines. For example, chloroform as well as ether extracts of black pepper root demonstrated cytotoxic activities for the treatment of myeloid leukemia (HL-60 cell) in human after 3 days of continuous exposure [82]. Methanol extract from the fruit of black pepper showed excellent anti-cancerous actions against numerous cancer lines such as MDA-MB-231 and MCF-7 while the extract of dichloromethane demonstrated similar effects on MDA-MB-468 cell line of breast cancer [83]. Ethanolic extract of black pepper exhibited anti-cancerous activities against various cell lines of colorectal cancer such as HCT-15, HCT-116 and HT-29 [84].

Pellitorine, a derivative of black pepper root was documented to show cytotoxic effects in dicancer cell lines (HeLa and MCF-7) [82]. The top-notch derivative of black pepper which is well-known for its anti-cancerous activities is piperine which was found to show suppressive effects against breast cancer gene HER-2. Pipperine was reported to block the signaling pathway of ERK1/2 and inhibit the activity of FAS and SREBP-1 gene. It was also reported to suppress the expression of EGF induced gene MMP-9 by blocking the activation of AP-1 and NF-jB [85]. Cell lines of prostate cancer such as PC-3, LNCap and DU145 were also reported to be inhibited by piperine. Furthermore, it enhances the formation of LC3B and elevated the level of LC3B-II [86]. Aforementioned cell lines of prostate cancer were inhibited by dose dependent effect of piperine. It was reported to enhance the level of p27Kip1 and p21Cip1 in both DU145 and LNCaP but PC-3 remain unaffected. Several investigations demonstrated that piperine directly block the potassium channels which are the key regulator of cancerous cells proliferation. Piperine demonstrated effective results at IC50 = 49.45 μM in human prostate cancer while 39.91 μM in LNCaP. The blockage of these channels led to the arrest of G0/G1 cell cycle which ultimately cease the proliferation of cancerous cells and enhance apoptosis [87]. Administration of piperine enhance the activation of PARP-1 and caspase-3 while inhibit the expression of phosphorylated STAT-3 and NF-kB in various human prostate cancer cell lines such as PC-3, LNCaP and DU-145 [88]. Treatment of pipperine during in vivo investigation is proved to reduce tumor formation and level of serum gamma glutamyl transpeptidase and serum sialic acid in rat. This compound is also reported to decrease the level of total protein and protein bound carbohydrates which are involved in instigating the lung cancer [89].

Piperine has demonstrated anti-proliferative properties in different types of cancerous cells by arresting the cell cycle in vatious phases such as G0/G1, S or G2/M. Arrest of cell cycle in S phase only occurrs in leukemia cells [90] while G1 phase is halted in prostate [86, 87], colon [91] and melanoma cancerous cells [92]. The inhibition of G1 phase is correlated with the upregulation of p21 protein while downregulation of cyclin D. p21 protein inhibits the cyclin-dependent kinases as well as promote the blockage of phosphorylation of retinoblastoma and reduces the expression of E2F transcriptional factor 1 [93]. The over expression of p21 is reported to block the G2/M phase of cell cycle when piperine was provided in ovarian [94], osteosarcoma [95], lung [96] and breast cancer [97, 98]. Furthermore, piperine is documented to inhibit both G1/S and G2/M transition in a model of mammalian breast cancer [99].

The capability of malignant cells to invade the other tissues of body is the major hallmark. Metastasis is a complicated process which need to acquire different stages such as resistance against apoptosis, proteolysis with the help of metalloproteinases, enhancing cell mortality and invasion to surround tissues as well as reduction in cell adhesion [100]. Interestingly, piperine is found to reduce the expression of metalloproteinases in different types of cancerous cells. For example, piperine inhibits the metalloproteinases-2 and 9 in breast cancer cell lines. Inhibitory effects of pipperine are not only related with the inhibition of metalloproteinases but it also inhibits the expression of HER2 [85] which is the primary regulator of breast cancer invasion and enhance the expression of both aforementioned metalloproteinases [101]. The inhibitory actions of piperine are also relied on the inhibition of Akt signaling pathway as investigated in the study of fibrosarcoma, osteosarcoma, prostate and breast cancer cell lines [102]. All the aforementioned investigations were performed in the mammalian model of different carcinoma cell lines [97] in which piperine inhibited the expression of metalloproteinases-13 [103]. In case of osteosarcoma cell lines, piperine enhances the expression of TIMP1 and 2 which directly inhibit the regulation of metalloproteinases2 and 9 [95].

12.9.4 Anti-inflammatory Activity

The Anti-inflammatory potential of black pepper and its derivatives were determined by using plethysmometer [104]. Pipperine was exhibited anti-inflammatory response at different concentrations of doses such as 5, 10 and 15 mg/kg of body mass. The activity was observed maximum at 15 mg/kg after 2 h of continuous exposure but still this response was less than the response represented by diclofenac sodium. After administration of 500 mg/kg essential oil of black pepper, paw inflammation was reduced significantly (about 72%) in rat after 3 h of treatment in contrary to the control group while the administration of 100 mg/kg reduced the inflammation to about 66.1% in same time duration. In case of chronic inflammation of paw, a dosage of 500 mg/kg reduced the inflammation to about 50% while the standard treatment of diclofenac sodium (10 mg/kg) reduced 57.5% of paw inflammation [105].

12.9.5 Analgesic and Anticonvulsive Activity

Black pepper has been extensively studied by scientists to be used as safe painkiller drug. The pain alleviating effects of black pepper and its derivatives were evaluated in ethanolic and hexane extracts. To evaluate the pain alleviating effects of pipperine in hexane extract, tail immersion technique was used in which a dosage of 5 mg/kg was administrated in 2 h while the analgesy meter test was performed to determine the pain killing effects of pipperine in ethanol extract at 10 mg/kg for 1 h which demonstrated that ethanol extract proved to be more pain alleviating as compare to hexane extract [104]. Ethyl alcohol and hexane extract of black pepper were reported to encounter the duration of seizures in maximal electroshock seizure induced model and pentylenetetrazol (PTZ) induced model of Wistar rats. Treatment of pipperine (70 mg/kg) has been documented to prevent animals from PTZ induced seizers [106].

12.9.6 Anti-diabetic and Hypolipidemic Activity

Administration of aqueous extract of black pepper to diabetic rats for 28 days bring down the blood sugar from 270 mg/100 mL to 129 mg/100 mL [107]. Treatment of methanolic extract of *P. nigrum* with different doses 100, 200 and 300 mg/kg was reported to diminish the glucose level in alloxan induced diabetic mice [108]. As soon as the activity of polyol pathway increases, it enhances the level of aldose reductase which is directly involved in the complications during diabetic conditions, thus maintaining the level of aldose reductase can inhibit the complications such as nephropathy, retinopathy and neuropathy. Hyromethanolic extract of black pepper is documented to reduce the level of aldose reductase to reference range of 25–64 lg/mL, thus showing antidiabetic effects [109].

12.9.7 Neuroprotective Effects

Methanolic extract of black pepper is found to demonstrate anti-depressant as well as anxiolytic activity in Wistar rats. This extract has tendency to regulate amyloid beta which causes impairment of spatial memory by reducing oxidative stress in rats [110]. Pipperine was proved to enhance cognitive functions and motor coordination [111] as well as its treatment can recover the motor deficit and locomotory activity controlled by motor coordination [112]. Collected information on cognitive and brain problems is alarming on a global scale. Neural conditions and stress are estimated to impact over 450 million people [113]. Recently, the neuroprotective properties of food products were emerged as a potential treatment for a variety of diseases. In this regard, black pepper‘s bioactive constituents and antioxidant capability are considered essential for enhancing brain health. Additionally, supplementation with piperine (5, 10, and 20 mg/kg body weight/day) improves neuroprotective properties as well as enhances the cognitive function. It was documented that piperine reduced chronic mild stress brought by fluctuations in consumption of sugar among the individuals [114]. Forced swimming test was reported that piperine demonstrated efficient anti-depressant properties at the dosage of 10–20 mg/kg which is a successful way to prevent brain related disorders such as Parkinson and Alzheimer and epilepsy [115]. Piperine is also reported to fight against various sorts of neurological and psychiatric diseases. Oxidative stress is considered as the major factor behind the parthenogenesis of Alzheimer and other neurodegenerative diseases. An injection of streptozotocin was injected to a rat model of Alzheimer disease which induced oxidative stress as well as cause the malfunctioning of neurotransmission [116]. Similar group of rats were administrated with 2.5 mg/kg of piperine for 28 days. Piperine suppress the generation of oxidative stress and enhance the synatptic plasticity as well as improve mobility and chlionergic function of streptozotocin injected rats [117, 118].

Furthermore, the administration of piperine suppress the production of Ca^{2+} current [119] which ultimately cease the release of glutamate from nerve endings of hippocampal, thus improving synaptic plasticity and facilitate paired pulse. It was reported that administration of piperine expressed the acetylcholine receptors in hippocampus region which regulate the cognitive functions as well as plasticity of synaptic nerves. Piperine inhibit the production of acetylcholine esterase [120] which break down the acetylcholine into choline and acetate (Fig. 12.3).

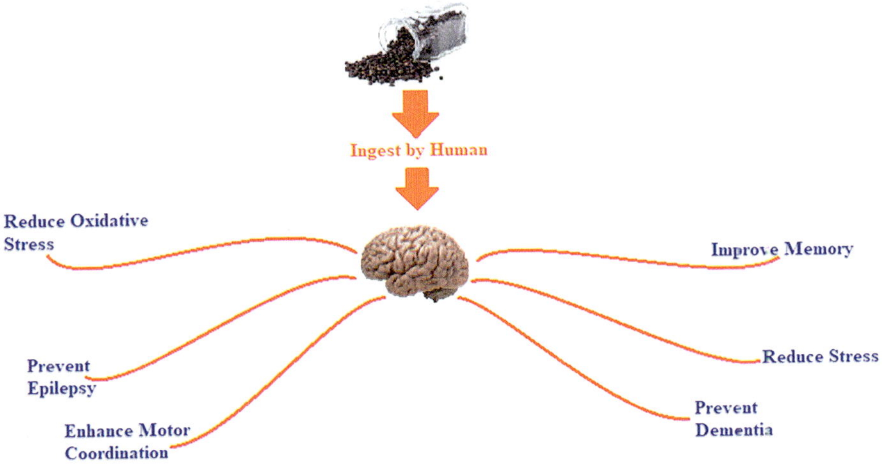

Fig. 12.3 Piperine ingestion demonstrated the neuroprotective properties

The molecular aetiology of the cognitive issues brought on by streptozotocin is still obscure, but its infusion dramatically raises the levels of synapsin, amyloid, and tau-phosphorylation [121]. Therefore, it may be assumed that piperine therapy might affect tau hyperphosphorylations and misfolded proteins. According to a study, piperine therapy inhibits secretase, which prevents amyloid and tau pathogenesis in addition to improving cholinergic activities [122, 123].

Piperine has been documented through numerous investigators for its curative action in epilepsy, a neurological disorder characterized via frequent seizures. Piperine acknowledge significant consideration in treatment of epilepsy for its remarkable monoamine oxidase A (MAO) repressive as well as neuro-protective capabilities. Piperine administration can also maintain the serotonin concentration as well as moderated MAO along with -amino butyric acid pathway [124]. It is also revealed that administration of piperine enhances the glycine, GABA along with taurine level whereas decreases nitrate in the brain & sera of a pilocarpine-instigated epileptic rats [125].

In a maximal electroshock instigated seizure model of epilepsy, piperine dosage diminishes the morbidity via negatively maintaining the NA^+ ion gates which retard the beginning of tonic seizures [126], demonstrating that piperine may control the sodium ion channel to control tonic convulsion as well as morbidity in epilepsy. For example, blockage of transient receptor potency cation channel family (V) member-1 via piperine-mediated anti-epilepsy activity [127], illustrated that TRPV1 could also be included in the piperine process, piperine can directly affect the pore-forming segment (S6) to escalate opening of these ion's channel [128]. Furthermore, co-administration of piperine along with carbamazepine in the patients of epilepsy displayed a considerable elevation in the bioavailability of carbamazepine, AUC along with absorption as well as decrease the elimination half-life [129, 130].

12.9.8 Nutrient Absorption in Gastrointestinal Track

From decades, black pepper along with its derivatives are broadly accepted and utilized in the treatment of gastrointestinal disturbances. After the ingestion, *P. nigrum* stimulates the synthesis of hydrochloric acid from the walls of stomach which aids in the digestion of food by enhancing histamine receptors. Furthermore, uptake of black peppers was reported to enhance the urination as well as diaphoretic (promote sweating) condition [131]. Following oral ingestion, pipperine gets absorbed from the mucosal layer and then enters into intestinal and serosal fluids [132]. Pipperine is found to enhance the absorption of Vit B, Ca, selenium and β-carotene throughout the gastrointestinal track [133, 134].

12.9.9 Anti-diarrheal Activity of Black Pepper

Anti-diarrheal activity of this plant is referred to its pungent pepperine derivative & numerous reserachers reported anti-diarrheal acitivty of black pepper [135]. Interestingly, a large proportion of people particularly from developing countries are still using black pepper in severe diarrheal conditions [136]. Moreover, similar anti-diarrheal effects of black pepper were observed in rats [137].

12.9.10 Cardioprotective Effects of Black Pepper

Pipperine is proved to involve in the activation of peroxisome proliferator activated receptor- γ to enhance cardiac protection against various sorts of cardiac toxicities [138]. Furthermore, pipperine has been documented to show ameliorative effects against curcumin induced cardiac toxicity in rat [139]. These investigations suggested the cardio-protective abilities of pipperine against various chemotherapeutic drugs induced toxicity.

12.9.11 Insecticidal Activity

P. nigrum has demonstrated its insecticidal activity against different insects such as *Amphimallon majal, Scarabaeidae and Coleoptera* [2]. Essential oil of black pepper (0.2% concentration) demonstrated strong repellent ability against *Tribolium castaneum* which is common insect of wheat during storage conditions [140]. Additionally, the repellent potential of *P. nigrum* was observed to be dose depend against T. castaneum [141].

12.9.12 Anti-obesity and Carminative Effects of Black Pepper

Obesity is becoming the world one of the largest health problem due to different conditions such as gastric or other metabolic disorders. *P. nigrum* is proved to be efficient in reducing obesity along with other non-pharmaceutical approaches such as exercises, yoga and diet maintenance [142]. Moreover, black pepper has potential to stimulate strong carminative responses which ultimately causes the secretion of saliva, enhance appetite and secretion of gastric juice for efficient digestion. In many countries these are employed as condiments on skin due to their ability to dilate the superficial vessels of skin which ultimately reduced the body temperature [143].

12.9.13 Anti-pyretic Effects of Black Pepper

In India, different herbal medicines such as folklore, Yunani and ayurvedic uses black pepper against various sorts of treatment or conditions such as pain, cold, neuritis, fever, throat infections and malaria therefore, known for its anti-pyretic and anti-analgesic actions [4]. The anti-pyretic actions of black pepper have been reported after successful experimentation in different medications [144].

12.9.14 Hepatoprotective and Gastrointestinal Activity

Researchers have conducted a number of investigations to determine the various biological impacts of black pepper's extracts or its active ingredient, piperine, during the past few decades to establish the scientific explanation for the pharmacological effectiveness of *P. nigrum*. Thus, a wide range of pharmacological activity and health-promoting aspects of black pepper have been described in several researches [145, 146] such as hepatoprotective, anticonvulsant, chemo preventive anti-inflammatory, hepatoprotective, antidiarrheal, immunomodulating and antioxidant properties. Although there were apparently some conflicting findings on the safety of *P. nigrum* as a food additive, it seems to be safe even at dosages much higher than the average human intake. Previous investigations have highlighted concerns about piperine's possible carcinogenic effects due to its structural resemblance to natural carcinogens such methyleugenol, safrole and estragole [147, 148].

Treatment with *P. nigrum* or its derivative (pipperine) has been documented to enhance activity of metabolic and intestinal enzymes which ultimately increase the bioavailability of therapeutic drugs in the body [4]. The uses of herbal medicines are increasing continuously, an estimation provided by WHO demonstrated that homeopathic medicines contributed about 30 to 50% of total medicines consumption in which China is consuming about 70% [149]. In the jejunal mucosa, the brush

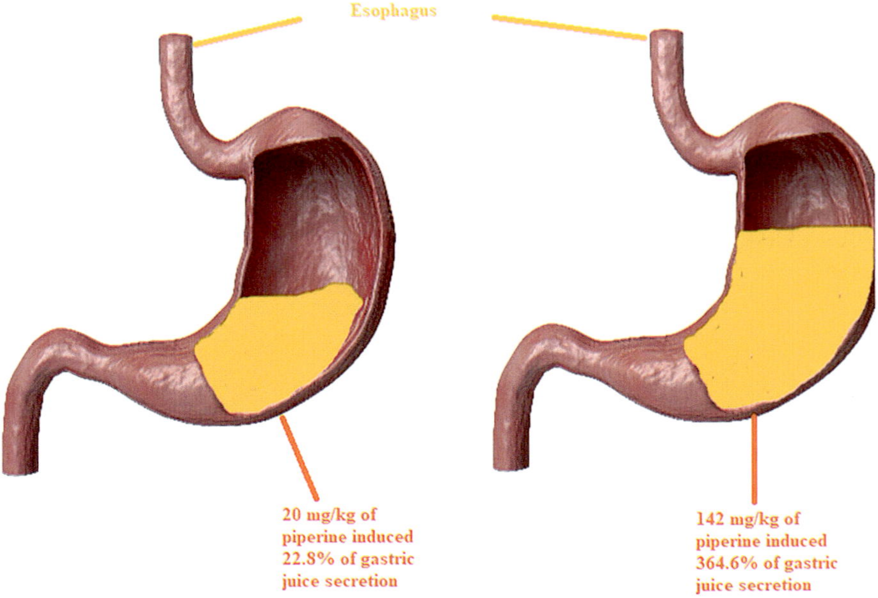

Fig. 12.4 Dose dependent effects of piperine on the secretion of gastric juice

boundary transmembrane enzymes such as glutamyl transpeptidase, glycyl-glycine dipeptidase, alkaline phosphatase and leucine amino peptidase were also activated by black pepper and piperine [150]. Additionally, piperine and supplementary black pepper increased the activity of intestinal Na^+ and K^+ ATPase by 111% and 52% respectively. All the aforementioned findings imply that piperine is capable of altering membrane dynamics because of its nonpolar nature by interacting with nearby lipids & hydrophobic protein regions, which may reduce the likelihood that membrane lipids will act as steric hindrance to enzymatic proteins and change the conformation of the aforementioned enzymes. Additionally, scanning electron microscopy ultra - structural investigations showed that piperine increased microvilli length and proposed that piperine may cause changes to membrane dynamics and permeation characteristics and it stimulates the production of various proteins which are regulators of cytoskeletal function which ultimately increases the small intestine absorptive surface and facilitates effective drug absorption through the epithelial layer [151]. Piperine is reported to elevate the level of gastric acid secretion against dose dependent effects of piperine in albino rats when the dose concentration reduces from 20 to 142 mg/kg. Piperine is documented to induce the secretion of gastric acid 22.8% and 334.6% the in response to 20 mg/kg and 142 mg/kg of dose respectively. Cimetidine 1 mg/kg reported to counteract the effects induced by piperine and thus proved to involve in cholinergic receptors. However, there is possibility that piperine is responsible for the activation of H2 receptors which ultimately elevate the acidity level [131] (Fig. 12.4).

12.9.15 *Immuno-modulatory and Anti-allergic Effects*

Piperine exhibited immune-modulatory and anti-allergic activity and documented to inhibit the proliferative response which was induced by the antibodies (lipopolysaccharide and immunoglobulin) during in vitro studies. Piperine is documented to inhibit the expression of aforementioned antibodies as well as reduced the differentiation of CD86 [152]. An investigation conducted by Lee et al. reported that combine therapy of piperine along with gamma-aminobutyric acid up-regulate the IL-10 and NF-kB by mediating the expression of JNK MAPK and p28 which ultimately enhance the activation of EPO and EPO-R [153]. In contrast to upregulation of immune response, piperine significantly reduces allergic rhinitis in rats which were provoked by ovalbumin. Piperine substantially decreased allergy symptoms such as rubbing and redness brought on by nerve ending hypersensitivity brought on by histamine released as a result to antigen-antibody reaction but it also diminishes nitric oxide (NO) thresholds because eosinophil transition into nasal epithelial tissue was less pronounced. It was discovered that piperine administration diminished swelling, redness, and disturbance of alveoli and bronchioles, much like in the histopathological portion of the nasal mucosa [154]. Piperine therapy inhibited eosinophil invasion and attenuated airway hypersensitivity in an ovalbumin-induced asthma model by inhibiting T cellular activities production of Th2 cytokine and Th2 [155].

12.9.16 *Anti-thyroidal Activity of Black Pepper*

Combine administration of piperine along with carbimazole elevated the density of lipoproteins while reduced the level of plasma proteins and lipoproteins [136]. In another investigation, piperine is reported to induced the level of testosterone while lowered the level of Apo B and TSH (88). Furthermore, another study conducted by Panda and Kar demonstrated that administration of piperine to albino rat reduces the level of T3 and T4 [156].

12.9.17 *Effects of Black Pepper and Its Derivatives on Metabolic Enzymes*

There are several studies which demonstrated the that black pepper and its derivatives directly effect on intestinal and hepatic metabolic enzymes by altering the biotransformation pathways of these enzymes. Piperine is demonstrated dose depended response on oxygenase system of liver in rats. The activity of different hepatic enzymes such as benzphetamine, aminopyrine N-demethylase and cytochrome P450 significantly reduced in response to intraperitoneal (100 mg) and intragastric (800 mg) administration of piperinein rats [157]. The substantial decrease in the activity of oxygenase system of liver is induced by several mechanisms including inhibition of D-aminoleyulininc acid synthetase, reducing hepatic heme and by binding with cytochrome P450. The levels of hepatic microsomal cytochrome P450 and cytochrome b5, NADPH-cytochrome-C reductase, benzphetamine N-demethylase, aminopyrine N-demethylase, and aniline hydroxylase were significantly increased in rats when the intragastric dose was reduced to 100 mg/kg. These findings imply that piperine, depending on the administration dose and method of administration, has both inhibitory and stimulatory effects on the cytochrome P450-mediated activities of the microsomal monooxygenases [158]. All the aforementioned results indicate that piperine is not a hepatotoxic substance because none of the treatments significantly increased the activities of aspartate aminotransferase, isocitrate dehydrogenase and serum sorbitol dehydrogenase. In another investigation, the effects of black pepper were assessed after providing the feed comprisinf of 0.5, 1.0 and 2% of *P. nigrm* for 10 and 20 days of duration [159]. In response to aforementioned dosage, the level of glutathione S-transferase and sulfhydryle content was increased in response to 0.5% of black pepper containing feed. Black pepper is reported to exhibit chemo-preventive role against different carcinogenic substances. The level of hexobarbital hydroxylase was significantly increase in piperine treated rats [160].

12.9.18 Effects of Black Pepper on Cholesterol Level

P. nigrum and its derivatives accelerated the translocation process of cholesterol transporters as well as elevated the rate of fats break down into simple and digestible molecules to prevent their accumulation into the surrounding tissues of body [161]. Use of *P. nigrum* along with highly fatty meal (40 mg/kg) substantially reduced the weight of body as well as VLDL, triglycerides, total cholesterol and LDL but enhance the value of HDL. Daily uptake of P. nigrum and piperine lowers the risk of getting arthrosclerosis [162].

12.9.19 Preservation of Orange Juice by P. nigrum Essential Oil

Citrus fruits are significant source of vitamins specifically vitamin C and A along with calcium, folic acids, iron as well as thiamine. They also have remarkable concentration of flavonoids and carotenoids. Juice of citrus fruits such as oranges is very nutritious and contains fundamental preserving capabilities for the fruits of Rutaceae family. However, without appropriate preservation fruits can decays whereas its taste, color and aroma could be affected severely [79]. Preservation of this juice may be carried out via thermal pasteurization, pulsed electric and high hydrostatic pressure mechanism. Chemicals comprising of sodium benzoate and potassium sorbate along with natural preservative including organic acids, bacteriocins, phenolic acids and essential oils could also be employed for their preservation [163]. Essential oil of *P. nigrum* can be significant if employed in industries as a result of its ability to escalates the shelf-life of orange juice. Besides this, orange juice having essential oils of *P. nigrum* exhibits considerably reduced microbial load, titratable acidity, pH, percent loss as well as nano-enzymatic browning whereas reduces the content of ascorbic acid when compare with fruit juices stored by employing ethanol along with oleosins [79].

12.9.20 Bioavailability and Potential of Piperine

Piperine, the alkaloid component of long and black peppers is currently recognized as an enhancer of bioavailability of numerous therapeutically as well as structurally diverse drugs along with various other substances. The therapeutic action of piperine to escalate the bioavailability of pharmaceutical drugs in humans is of great clinical importance. A significant mechanism accounts for its bioavailability mediating potential is still not well understood. Atal et al. have analyzed the scientific basis of the usage of the trikatu group of acrids (black pepper, long pepper as well as ginger) in the enormous prescriptions in the native ayurvedic system of medication [164]. Long pepper increases the level of blood of vasicine, test drug by approximately 233% under the potential of piperine test's blood level indicating that these acrids contains the capability to elevates the bioavailability of specific drugs. It is concluded that the trikatu group of drugs enhances the bio-availability of pharmaceuticals either by mediating the rapid absorption from the gastro-intestinal tract or via preventing the drug from being metabolized in the passage through the hepatic tissue after being absorbed. Piperine has been documented to have a significant effect on the activation as well as de-activation of exogenous constituents [165].

12.10 Other Health Advantages

Levels of LDL cholesterol and triglycerides are all significant risk factors behind coronary vascular disorders. Certain dietary suggestions can help in prevention of CVDs. Increasing the function of the peroxisome proliferator-activated receptor is one of the top approaches to enhance the lipid profile. The highest transactivation activity was demonstrated by piperine having values of EC (50) were 84 M and 49 M, respectively [166]. It has been investigated that the use of black pepper extract is proved to be successful in treating a skin disorder called vitiligo which is characterized by lesions on the skin and lack of skin pigmentation. Primary alkaloid of

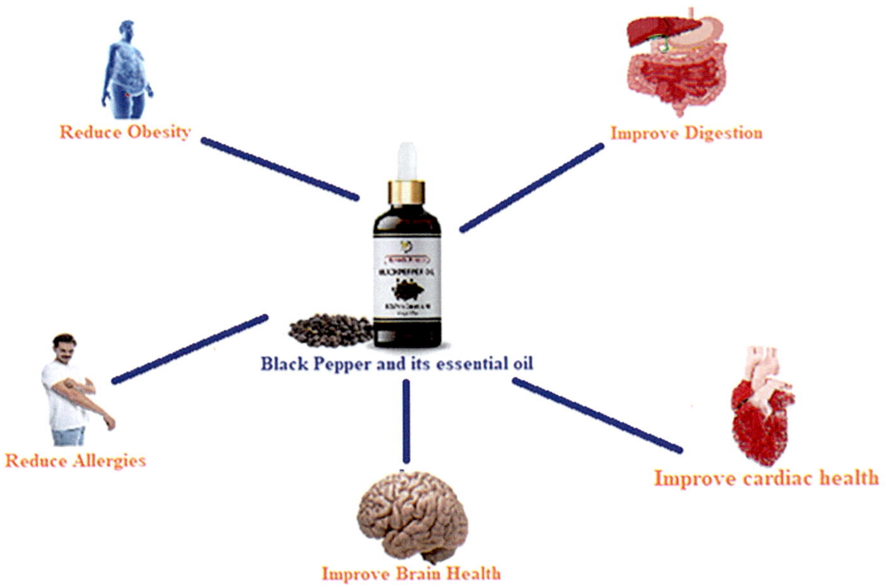

Fig. 12.5 Benefits of black pepper

black pepper promotes the growth of melanocytes as well as involved in the promotion of swallowing hence worked as strong appetizer [167] (Fig. 12.5).

12.11 Conclusions

In the current era, spices are not only confined to use in food for taste, aroma and color but become the part of ayurvedic, Unani and numerous other herbal medicines against the treatment of different disorders. Black pepper (King of Spices) and its bioactive compounds have been successfully investigated in several studies which demonstrated their role in the maintenance of nutritional balance as well as to show potential anti-inflammatory, anti-microbial, anti-cancerous, anti-diarrheal, anti-pyretic and anti-analgesic properties. Additionally, both *in-vivo* and *in-vitro* studies proved their significant role in different diseases of respiratory, digestive and circulatory system as well as against chronic conditions such as in metastatic cancer. Black pepper is reported to prevent the protein aggregation in the brain and protect from Alzheimer's conditions. Different extracts of black pepper showed multiple pharmacological and biological activities in different animal models. Pharmacological uses of black pepper are enhancing the importance of black pepper all over the world. Future studies should be conducted on different extracts of black pepper to ascertain its safety and effectiveness in detail.

References

1. Nagalingam, M., & Arumugam, G. (2011). Antimicrobial activity of some Indian folklore medicinal plants against drug resistant bacteria and fungi isolated from clinical samples. *Asian Journal of Plant Science & Research*.
2. Scott, I. M., Helson, B. V., Strunz, G. M., Finlay, H., Sánchez-Vindas, P. E., Poveda, L., et al. (2007). Efficacy of *Piper nigrum* (Piperaceae) extract for control of insect defoliators of forest and ornamental trees. *The Canadian Entomologist, 139*(4), 513–522.
3. Durant-Archibold, A. A., Santana, A. I., & Gupta, M. P. (2018). Ethnomedical uses and pharmacological activities of most prevalent species of genus Piper in Panama: A review. *Journal of Ethnopharmacology, 217*, 63–82.
4. Srinivasan, K. (2007). Black pepper and its pungent principle-piperine: a review of diverse physiological effects. *Critical Reviews in Food Science and Nutrition, 47*(8), 735–748.
5. Kumar, S., Kamboj, J., & Sharma, S. (2011). Overview for various aspects of the health benefits of Piper longum linn. fruit. *Journal of Acupuncture and Meridian Studies, 4*(2), 134–140.
6. Bui, T. T., Piao, C. H., Song, C. H., Shin, H. S., Shon, D. H., & Chai, O. H. (2017). *Piper nigrum* extract ameliorated allergic inflammation through inhibiting Th2/Th17 responses and mast cells activation. *Cellular Immunology, 322*, 64–73.
7. Ravindran, P. N., Divakaran, M., & Pillai, G. S. (2012). *Handbook of herbs and spice, 1*, 86–115.
8. Hao, C. Y., Rui, F. A. N., Ribeiro, M. C., Tan, L. H., Wu, H. S., Yang, J. F., & Huan, Y. (2012). Modeling the potential geographic distribution of black pepper (*Piper nigrum*) in Asia using GIS tools. *Journal of Integrative Agriculture, 11*(4), 593–599.
9. Hussain, M. S., Hegde, L., Sharatbabu, A. G., Hegde, N. K., Shantappa, T., Gurumurthy, S. B., et al. (2017). Evaluation of local black pepper (*Piper nigrum* L.) genotypes for yield and quality under arecanut based cropping system. *International Journal of Pure & Applied Bioscience, 5*(5), 1396–1400.
10. Thapa, A., Datta, S., Dey, A. N., & Baisare, P. (2017). Advance propagation techniques in important spice crops. *International Journal of Current Microbiology and Applied Sciences, 6*(9), 1979–1985.
11. Shango, A. J., Majubwa, R. O., & Maerere, A. P. (2021). Morphological characterization and yield of pepper (*Piper nigrum* L.) types grown in Morogoro District, Tanzania. *CABI Agriculture and Bioscience, 2*, 1–13.
12. Nwofia, G. E., Kelechukwu, C., & Nwofia, B. K. (2013). Nutritional composition of some *Piper nigrum* (L.) accessions from Nigeria. *International Journal of Medicinal and Aromatic Plants, 3*(2), 2249–4340.
13. Vellaichamy, L., Balakrishnan, S., Panjamurthy, K., Manoharan, S., & Alias, L. M. (2009). Chemopreventive potential of piperine in 7, 12-dimethylbenz [a] anthracene-induced skin carcinogenesis in Swiss albino mice. *Environmental Toxicology and Pharmacology, 28*(1), 11–18.
14. Okumura, Y., Narukawa, M., & Watanabe, T. (2010). Adiposity suppression effect in mice due to black pepper and its main pungent component, piperine. *Bioscience, Biotechnology, and Biochemistry, 74*(8), 1545–1549.
15. Bang, J. S., Oh, D. H., Choi, H. M., Sur, B. J., Lim, S. J., Kim, J. Y., et al. (2009). Anti-inflammatory and antiarthritic effects of piperine in human interleukin 1β-stimulated fibroblast-like synoviocytes and in rat arthritis models. *Arthritis Research & Therapy, 11*(2), 1–9.
16. Zachariah, T. J., Safeer, A. L., Jayarajan, K., Leela, N. K., Vipin, T. M., Saji, K. V., et al. (2010). Correlation of metabolites in the leaf and berries of selected black pepper varieties. *Scientia Horticulturae, 123*(3), 418–422.
17. Dorman, H. D., & Deans, S. G. (2000). Antimicrobial agents from plants: antibacterial activity of plant volatile oils. *Journal of Applied Microbiology, 88*(2), 308–316.

18. George, K. M., Joy, M. T., Chandran, C. V., & Verghese, J. (1988). The angular rotation of black pepper oil. *Indian Perfumer, 32*(1), 51–54.
19. Shankaracharya, N. B., Jaganmohan Rao, L., Pura Naik, J., & Nagalakshmi, S. (1997). Characterisation of chemical constituents of Indian long pepper (Piper longum L.). *Journal of Food Science and Technology, 34*(1), 73–75.
20. Tipsrisukond, N., Fernando, L. N., & Clarke, A. D. (1998). Antioxidant effects of essential oil and oleoresin of black pepper from supercritical carbon dioxide extractions in ground pork. *Journal of Agricultural and Food Chemistry, 46*(10), 4329–4333.
21. Ferreira, S. R., & Meireles, M. A. A. (2002). Modeling the supercritical fluid extraction of black pepper (*Piper nigrum* L.) essential oil. *Journal of Food Engineering, 54*(4), 263–269.
22. Guenther, E. (1952). *Essential oils of the plant family Piperaceae. The Essential Oils, 5* (pp. 135–161). Van Nostrand.
23. Govindarajan, V. S., & Stahl, W. H. (1977). Pepper—chemistry, technology, and quality evaluation. *Critical Reviews in Food Science & Nutrition, 9*(2), 115–225.
24. Parmar, V. S., Jain, S. C., Bisht, K. S., Jain, R., Taneja, P., Jha, A., et al. (1997). Phytochemistry of the genus Piper. *Phytochemistry, 46*(4), 597–673.
25. Menon, A. N. (2000). The aromatic compounds of pepper. *Journal of Medicinal and Aromatic Plant Sciences, 22*(2–3), 185–190.
26. Menon, A. N., & Padmakumari, K. P. (2005). Studies on essential oil composition of cultivars of black pepper (*Piper nigrum* L.)—V. *Journal of Essential Oil Research, 17*(2), 153–155.
27. Sasidharan, I., & Menon, A. N. (2010). Comparative chemical composition and antimicrobial activity fresh & dry ginger oils (Zingiber officinale Roscoe). *International Journal of Current Pharmaceutical Research, 2*(4), 40–43.
28. Georgiev, E., & Stoyanova, A. (2005). *Handbook for the specialist in aromatic industry*. Plovdiv, BNAEOPC. Search in.
29. Al Juhaimi, F. Y., & Ghafoor, K. (2013). Extraction optimization and in vitro antioxidant properties of phenolic compounds from Cumin (Cuminum cyminum l.) seed. *International Food Research Journal, 20*(4).
30. Iacobellis, N. S., Lo Cantore, P., Capasso, F., & Senatore, F. (2005). Antibacterial activity of Cuminum cyminum L. and Carum carvi L. essential oils. *Journal of Agricultural and Food Chemistry, 53*(1), 57–61.
31. Moghaddam, M., Miran, S. N. K., Pirbalouti, A. G., Mehdizadeh, L., & Ghaderi, Y. (2015). Variation in essential oil composition and antioxidant activity of cumin (Cuminum cyminum L.) fruits during stages of maturity. *Industrial Crops and Products, 70*, 163–169.
32. Mostafa, D. M., Kassem, A. A., Asfour, M. H., Al Okbi, S. Y., Mohamed, D. A., & Hamed, T. E. S. (2015). Transdermal cumin essential oil nanoemulsions with potent antioxidant and hepatoprotective activities: in-vitro and in-vivo evaluation. *Journal of Molecular Liquids, 212*, 6–15.
33. Jeleń, H. H., & Gracka, A. (2015). Analysis of black pepper volatiles by solid phase microextraction–gas chromatography: A comparison of terpenes profiles with hydrodistillation. *Journal of Chromatography A, 1418*, 200–209.
34. Clery, R. A., Hammond, C. J., & Wright, A. C. (2006). Nitrogen-containing compounds in black pepper oil (*Piper nigrum* L.). *Journal of Essential Oil Research, 18*(1), 1–3.
35. Jagella, T., & Grosch, W. (1999). Flavour and off-flavour compounds of black and white pepper (*Piper nigrum* L.) I. Evaluation of potent odorants of black pepper by dilution and concentration techniques: I. Evaluation of potent odorants of black pepper by dilution and concentration techniques. *European Food Research and Technology, 209*, 16–21.
36. Megir, G., & Paulus, A. D. (2011). Pepper production technology in Malaysia. In L. K. Fong & S. S. Liang (Eds.), *Malaysian pepper board*. Sarawak.
37. Sen, S., Gode, A., Ramanujam, S., Ravikanth, G., & Aravind, N. A. (2016). Modeling the impact of climate change on wild *Piper nigrum* (Black Pepper) in Western Ghats, India using ecological niche models. *Journal of Plant Research, 129*, 1033–1040.

38. Kumar, B. M. (2008). *Krishi Gita (Agricultural Verses)[A treatise on indigenous farming practices with special reference to Malayalam desam (Kerala)]* (p. 111p). Asian Agri-History Foundation (AAHF), Secunderabad, Andhra Pradesh.
39. Ravindran, P. (2000). Other economically important species of Piper. In P. N. Ravindran (Ed.), *Black Pepper–Piper nigrum* (pp. 497–509). Medicinal and Aromatic Plants–Industrial Profiles.
40. Ravindran, P. N., Babu, K. N., Sasikumar, B., & Krishnamurthy, K. S. (2000). Botany and crop improvement of black pepper. In *Black pepper* (pp. 43–164). CRC Press.
41. Gopalam, A., & Ravindran, P. N. (1987). Indexing of quality parameters in black pepper cultivars. *Indian Spice, 22*, 8–11.
42. Meilawati, N. L. W., Susilowati, M., & Bermawie, N. (2020). Phyllogenetic of nine superior black pepper (*Piper nigrum* L.) varieties based on morphological and molecular markers. In *IOP Conference Series: Earth and Environmental Science* (Vol. 418, No. 1, p. 012056). IOP Publishing.
43. Pannaga, T. S., Narayanpur, V. B., Hiremath, J. S., Hegde, L., Gandolkar, K., Rathod, V., & Chandrakala, R. (2021). Evaluation of black pepper (*Piper nigrum* L.) cultivars for yield and quality parameters under hill zone of Karnataka. *Journal of Pharmacognosy and Phytochemistry, 10*(1), 1497–1500.
44. Kandiannan, K., Krishnamurthy, K. S., Ankegowda, S. J., & Anandaraj, M. (2014). Climate change and black pepper production.
45. Zu, C., Li, Z., Yang, J., Yu, H., Sun, Y., Tang, H., … & Wu, H. (2014). Acid soil is associated with reduced yield, root growth and nutrient uptake in black pepper (*Piper nigrum* L.). *Agricultural Sciences*, 2014.
46. Yap, C. (2012). Determination of nutrient uptake characteristic of black pepper (*Piper nigrum* L.). *Journal of Agricultural Science and Technology, B, 2*(10), 1091–1099.
47. Zakaria, Z., Gairola, S., & Shariff, N. M. (2010). Effective microorganisms (EM) technology for water quality restoration and potential for sustainable water resources and management.
48. Talaat, N. B., Ghoniem, A. E., Abdelhamid, M. T., & Shawky, B. T. (2015). Effective microorganisms improve growth performance, alter nutrients acquisition and induce compatible solutes accumulation in common bean (Phaseolus vulgaris L.) plants subjected to salinity stress. *Plant Growth Regulation, 75*, 281–295.
49. Lee, C. T., Ismail, M. N., Razali, F., Muhamad, I. I., Sarmidi, M. R., & Khamis, A. K. (2008). Application of effective microorganisms on soil and maize. *Journal of Chemical and Natural Resources Engineering, Special Edition*, 1–13.
50. Keymer, D. P., & Lankau, R. A. (2017). Disruption of plant–soil–microbial relationships influences plant growth. *Journal of Ecology, 105*(3), 816–827.
51. Park, H., & DuPonte, M. W. (2008). How to cultivate Indigenous Microorganisms. *Biotechnology, 9*, 1–7.
52. Van Bruggen, A. H. C., & Finckh, M. R. (2016). Plant diseases and management approaches in organic farming systems. *Annual Review of Phytopathology, 54*, 25–54.
53. Van Der Heijden, M. G., Bardgett, R. D., & Van Straalen, N. M. (2008). The unseen majority: soil microbes as drivers of plant diversity and productivity in terrestrial ecosystems. *Ecology Letters, 11*(3), 296–310.
54. Sandford, H. (1952). Pepper in Sarawak. *Malaysian Agricultural Journal, 35*, 208–224.
55. Risfaheri, & Nurdjannah, N. (2000). Pepper processing – The Indonesian Scenario. In P. N. Ravindran (Ed.), *Black pepper* (pp. 355–366). Harwood Academic.
56. Kozukue, N., Park, M. S., Choi, S. H., Lee, S. U., Ohnishi-Kameyama, M., Levin, C. E., & Friedman, M. (2007). Kinetics of Light-Induced cis– trans isomerization of four piperines and their levels in ground black peppers as determined by HPLC and LC/MS. *Journal of Agricultural and Food Chemistry, 55*(17), 7131–7139.
57. Suresh, D., Manjunatha, H., & Srinivasan, K. (2007). Effect of heat processing of spices on the concentrations of their bioactive principles: Turmeric (Curcuma longa), red pepper (Capsicum annuum) and black pepper (*Piper nigrum*). *Journal of Food Composition and Analysis, 20*(3–4), 346–351.

58. Nisha, P., Singhal, R. S., & Pandit, A. B. (2009). The degradation kinetics of flavor in black pepper (*Piper nigrum* L.). *Journal of Food Engineering, 92*(1), 44–49.
59. De Alwis, H. M. G., & Grandison, A. S. (1992). Viscometry as a detection method for electron beam irradiation of black pepper. *Food Control, 3*(4), 205–208.
60. Suhaj, M., Rácová, J., Polovka, M., & Brezová, V. (2006). Effect of γ-irradiation on antioxidant activity of black pepper (*Piper nigrum* L.). *Food Chemistry, 97*(4), 696–704.
61. Sarma, Y. R., Premkumar, T., Ramana, K. V., Ramachandran, N., & Anandaraj, M. (1988). Diseases and pest management in pepper nurseries. *Indian Cocoa, Arecanut & Spices Journal, 11*, 123–127.
62. Sarma, Y. R., Ramachandran, N., & Anandaraj, M. (1988). Integrated disease management of 'quick wilt' (foot rot) of black pepper caused by Phytophthora palmivora MF4. *Coffee Res*, 68–72.
63. Anandaraj, M., Ramachandran, N., & Sarma, Y. R. (2017). Epidemiology of foot rot disease of black pepper (*Piper nigrum* L.) in India.
64. Mammootty, K. P., & Pillay, V. S. (1981). Studies on the chemical control of rotting disease of pepper cuttings in nursery. *Indian Phytopathology, 34(2)*, 240.
65. Mammootty, K. P., Cheeran, A., & Peethambaran, C. K. (1980). Rhizoctonia stem rot of pepper (*Piper nigrum* L.) rooted cuttings. *Indian Arecanut, Spices and Cocoa Journal, 4(2)*, 31.
66. Sarma, Y. R., Solomon, J. J., Ramachandran, N., & Anandaraj, M. (1988). Phyllody disease of black pepper (*Piper nigrum* L.). *Journal of coffee research, 18*, 61–67.
67. Gülçin, İ. (2005). The antioxidant and radical scavenging activities of black pepper (*Piper nigrum*) seeds. *International Journal of Food Sciences and Nutrition, 57*(7), 491–499.
68. Saxena, R., Venkaiah, K., Anitha, P., Venu, L., & Raghunath, M. (2007). Antioxidant activity of commonly consumed plant foods of India: contribution of their phenolic content. *International Journal of Food Sciences and Nutrition, 58*(4), 250–260.
69. Sunila, E. S., & Kuttan, G. (2004). Immunomodulatory and antitumor activity of Piper longum Linn. and piperine. *Journal of Ethnopharmacology, 90*(2–3), 339–346.
70. Pradeep, C. R., & Kuttan, G. (2002). Effect of piperine on the inhibition of lung metastasis induced B16F-10 melanoma cells in mice. *Clinical & Experimental Metastasis, 19*, 703–708.
71. Bajad, S., Bedi, K. L., Singla, A. K., & Johri, R. K. (2001). Antidiarrhoeal activity of piperine in mice. *Planta Medica, 67*(03), 284–287.
72. Zachariah, T. J., & Parthasarathy, V. A. (2008). *Chemistry of spices: Black pepper*.
73. Spaulding, C. N., Klein, R. D., Schreiber, H. L., IV, Janetka, J. W., & Hultgren, S. J. (2018). Precision antimicrobial therapeutics: the path of least resistance? *NPJ Biofilms and Microbiomes, 4*(1), 4.
74. Khan, M., & Siddiqui, M. (2007). Antimicrobial activity of Piper fruits.
75. Karsha, P. V., & Lakshmi, O. B. (2010). Antibacterial activity of black pepper (*Piper nigrum* Linn.) with special reference to its mode of action on bacteria.
76. Park, I. K., Kim, J., Lee, Y. S., & Shin, S. C. (2008). In vivo fungicidal activity of medicinal plant extracts against six phytopathogenic fungi. *International Journal of Pest Management, 54*(1), 63–68.
77. Akthar, M. S., Birhanu, G., & Demisse, S. (2014). Antimicrobial activity of *Piper nigrum* L. and Cassia didymobotyra L. leaf extract on selected food borne pathogens. *Asian Pacific Journal of Tropical Disease, 4*, S911–S919.
78. Shanmugapriya, K. S. P. S., Saravana, P. S., Payal, H., Mohammed, S., & Williams, B. (2012). Antioxidant potential of pepper (*Piper nigrum* Linn.) leaves and its antimicrobial potential against some pathogenic microbes.
79. Kapoor, I. P. S., Singh, B., Singh, G., De Heluani, C. S., De Lampasona, M. P., & Catalan, C. A. (2009). Chemistry and in vitro antioxidant activity of volatile oil and oleoresins of black pepper (*Piper nigrum*). *Journal of Agricultural and Food Chemistry, 57*(12), 5358–5364.
80. Vijayakumar, R. S., Surya, D., & Nalini, N. (2004). Antioxidant efficacy of black pepper (*Piper nigrum* L.) and piperine in rats with high fat diet induced oxidative stress. *Redox Report, 9*(2), 105–110.

81. Abdelnour, S., Alagawany, M., Abd El-Hack, M. E., Sheiha, A. M., Saadeldin, I. M., & Swelum, A. A. (2018). Growth, carcass traits, blood hematology, serum metabolites, immunity, and oxidative indices of growing rabbits fed diets supplemented with red or black pepper oils. *Animals, 8*(10), 168.
82. Ee, G. C. L., Lim, C. M., Lim, C. K., Rahmani, M., Shaari, K., & Bong, C. F. J. (2009). Alkaloids from Piper sarmentosum and *Piper nigrum. Natural Product Research, 23*(15), 1416–1423.
83. Sriwiriyajan, S., Ninpesh, T., Sukpondma, Y., Nasomyon, T., & Graidist, P. (2014). Cytotoxicity screening of plants of genus Piper in breast cancer cell lines. *Tropical Journal of Pharmaceutical Research, 13*(6), 921–928.
84. Prashant, A., Rangaswamy, C., Yadav, A. K., Reddy, V., Sowmya, M. N., & Madhunapantula, S. (2017). In vitro anticancer activity of ethanolic extracts of *Piper nigrum* against colorectal carcinoma cell lines. *International Journal of Applied and Basic Medical Research, 7*(1), 67.
85. Do, M. T., Kim, H. G., Choi, J. H., Khanal, T., Park, B. H., Tran, T. P., et al. (2013). Antitumor efficacy of piperine in the treatment of human HER2-overexpressing breast cancer cells. *Food Chemistry, 141*(3), 2591–2599.
86. Ouyang, D. Y., Zeng, L. H., Pan, H., Xu, L. H., Wang, Y., Liu, K. P., & He, X. H. (2013). Piperine inhibits the proliferation of human prostate cancer cells via induction of cell cycle arrest and autophagy. *Food and Chemical Toxicology, 60*, 424–430.
87. Ba, Y., & Malhotra, A. (2018). Potential of piperine in modulation of voltage-gated K+ current and its influences on cell cycle arrest and apoptosis in human prostate cancer cells. *European Review for Medical and Pharmacological Sciences, 22*(24), 8999–9011.
88. Samykutty, A., Shetty, A. V., Dakshinamoorthy, G., Bartik, M. M., Johnson, G. L., Webb, B., et al. (2013). Piperine, a bioactive component of pepper spice exerts therapeutic effects on androgen dependent and androgen independent prostate cancer cells. *PLoS One, 8*(6), e65889.
89. Selvendiran, K., Singh, J. P. V., & Sakthisekaran, D. (2006). In vivo effect of piperine on serum and tissue glycoprotein levels in benzo (a) pyrene induced lung carcinogenesis in Swiss albino mice. *Pulmonary Pharmacology & Therapeutics, 19*(2), 107–111.
90. Li, N., Wen, S., Chen, G., & Wang, S. (2020). Antiproliferative potential of piperine and curcumin in drug-resistant human leukemia cancer cells are mediated via autophagy and apoptosis induction, S-phase cell cycle arrest and inhibition of cell invasion and migration. *Journal of BUON, 25*, 401–406.
91. Yaffe, P. B., Power Coombs, M. R., Doucette, C. D., Walsh, M., & Hoskin, D. W. (2015). Piperine, an alkaloid from black pepper, inhibits growth of human colon cancer cells via G1 arrest and apoptosis triggered by endoplasmic reticulum stress. *Molecular Carcinogenesis, 54*(10), 1070–1085.
92. Fofaria, N. M., Kim, S. H., & Srivastava, S. K. (2014). Piperine causes G1 phase cell cycle arrest and apoptosis in melanoma cells through checkpoint kinase-1 activation. *PLoS One, 9*(5), e94298.
93. Dimri, G. P., Nakanishi, M., Desprez, P. Y., Smith, J. R., & Campisi, J. (1996). Inhibition of E2F activity by the cyclin-dependent protein kinase inhibitor p21 in cells expressing or lacking a functional retinoblastoma protein. *Molecular and Cellular Biology, 16*(6), 2987–2997.
94. Qiu, M., Xue, C., & Zhang, L. (2019). Piperine alkaloid induces anticancer and apoptotic effects in cisplatin resistant ovarian carcinoma by inducing G2/M phase cell cycle arrest, caspase activation and inhibition of cell migration and PI3K/Akt/GSK3β signalling pathway. *Journal of BUON, 24*(6), 2316–2321.
95. Zhang, J., Zhu, X., Li, H., Li, B., Sun, L., Xie, T., et al. (2015). Piperine inhibits proliferation of human osteosarcoma cells via G2/M phase arrest and metastasis by suppressing MMP-2/−9 expression. *International Immunopharmacology, 24*(1), 50–58.
96. Lin, Y., Xu, J., Liao, H., Li, L., & Pan, L. (2014). Piperine induces apoptosis of lung cancer A549 cells via p53-dependent mitochondrial signaling pathway. *Tumor Biology, 35*, 3305–3310.

97. Lai, L. H., Fu, Q. H., Liu, Y., Jiang, K., Guo, Q. M., Chen, Q. Y., et al. (2012). Piperine suppresses tumor growth and metastasis in vitro and in vivo in a 4T1 murine breast cancer model. *Acta Pharmacologica Sinica, 33*(4), 523–530.
98. Abdelhamed, S., Yokoyama, S., Refaat, A., Ogura, K., Yagita, H., Awale, S., & Saiki, I. (2014). Piperine enhances the efficacy of TRAIL-based therapy for triple-negative breast cancer cells. *Anticancer Research, 34*(4), 1893–1899.
99. Greenshields, A. L., Doucette, C. D., Sutton, K. M., Madera, L., Annan, H., Yaffe, P. B., & Hoskin, D. W. (2015). Piperine inhibits the growth and motility of triple-negative breast cancer cells. *Cancer Letters, 357*(1), 129–140.
100. Jiang, W. G., Sanders, A. J., Katoh, M., Ungefroren, H., Gieseler, F., Prince, M., et al. (2015, December). Tissue invasion and metastasis: Molecular, biological and clinical perspectives. In *Seminars in cancer biology* (Vol. 35, pp. S244–S275). Academic.
101. Pellikainen, J. M., Ropponen, K. M., Kataja, V. V., Kellokoski, J. K., Eskelinen, M. J., & Kosma, V. M. (2004). Expression of matrix metalloproteinase (MMP)-2 and MMP-9 in breast cancer with a special reference to activator protein-2, HER2, and prognosis. *Clinical Cancer Research, 10*(22), 7621–7628.
102. Hwang, Y. P., Yun, H. J., Kim, H. G., Han, E. H., Choi, J. H., Chung, Y. C., & Jeong, H. G. (2011). Suppression of phorbol-12-myristate-13-acetate-induced tumor cell invasion by piperine via the inhibition of PKCα/ERK1/2-dependent matrix metalloproteinase-9 expression. *Toxicology Letters, 203*(1), 9–19.
103. Balduyck, M., Zerimech, F., Gouyer, V., Lemaire, R., Hemon, B., Grard, G., et al. (2000). Specific expression of matrix metalloproteinases 1, 3, 9 and 13 associated with invasiveness of breast cancer cells in vitro. *Clinical & Experimental Metastasis, 18*, 171–178.
104. Tasleem, F., Azhar, I., Ali, S. N., Perveen, S., & Mahmood, Z. A. (2014). Analgesic and anti-inflammatory activities of *Piper nigrum* L. *Asian Pacific Journal of Tropical Medicine, 7*, S461–S468.
105. Jeena, K., Liju, V. B., Umadevi, N. P., & Kuttan, R. (2014). Antioxidant, anti-inflammatory and antinociceptive properties of black pepper essential oil (*Piper nigrum* Linn). *Journal of Essential Oil Bearing Plants, 17*(1), 1–12.
106. Belemkar, S., Kumar, A., & Pata, M. K. (2013). Pharmacological screening of herbal extract of *Piper nigrum* (Maricha) and Cinnamomum zeylanicum (Dalchini) for anticonvulsant activity. *Invent Rapid Ethnopharmacol, 2*, 1–5.
107. Kaleem, M., Sheema, S. H., & Bano, B. (2005). Protective effects of *Piper nigrum* and Vinca rosea in alloxan induced diabetic rats. *Indian Journal of Physiology and Pharmacology, 49*(1), 65–71.
108. Onyesife, C. O., Ogugua, V. N., & Anaduaka, E. G. (2014). Hypoglycemic potentials of ethanol leaves extract of black pepper (*Piper nigrum*) on alloxan-induced diabetic rats. *Annals of Biological Research, 5*(6), 26–31.
109. Gupta, S., Singh, N., & Jaggi, A. S. (2014). Evaluation of in vitro aldose reductase inhibitory potential of alkaloidal fractions of *Piper nigrum*, Murraya koenigii, Argemone mexicana, and Nelumbo nucifera. *Journal of Basic and Clinical Physiology and Pharmacology, 25*(2), 255–265.
110. Hritcu, L., Noumedem, J. A., Cioanca, O., Hancianu, M., Postu, P., & Mihasan, M. (2015). Anxiolytic and antidepressant profile of the methanolic extract of *Piper nigrum* fruits in beta-amyloid (1–42) rat model of Alzheimer's disease. *Behavioral and Brain Functions, 11*(1), 1–13.
111. Yang, W., Chen, Y. H., Liu, H., & Qu, H. D. (2015). Neuroprotective effects of piperine on the 1-methyl-4-phenyl-1, 2, 3, 6-tetrahydropyridine-induced Parkinson's disease mouse model. *International Journal of Molecular Medicine, 36*(5), 1369–1376.
112. Correia, A. O., Cruz, A. A. P., Aquino, A. T. R. D., Diniz, J. R. G., Santana, K. B. F., Cidade, P. I. M., … & Viana, G. S. D. B. (2015). Neuroprotective effects of piperine, an alkaloid from the Piper genus, on the Parkinson's disease model in rats.

113. Xu, C., Gong, X., Fu, W., Xu, Y., Xu, H., Chen, W., & Li, M. (2020). The role of career adaptability and resilience in mental health problems in Chinese adolescents. *Children and Youth Services Review, 112*, 104893.
114. Lindheimer, J. B., Loy, B. D., & O'Connor, P. J. (2013). Short-term effects of black pepper (*Piper nigrum*) and rosemary (Rosmarinus officinalis and Rosmarinus eriocalyx) on sustained attention and on energy and fatigue mood states in young adults with low energy. *Journal of Medicinal Food, 16*(8), 765–771.
115. Bhutani, M. K., Bishnoi, M., & Kulkarni, S. K. (2009). Anti-depressant like effect of curcumin and its combination with piperine in unpredictable chronic stress-induced behavioral, biochemical and neurochemical changes. *Pharmacology Biochemistry and Behavior, 92*(1), 39–43.
116. Nazifi, M., Oryan, S., Esfahani, D. E., & Ashrafpoor, M. (2021). The functional effects of piperine and piperine plus donepezil on hippocampal synaptic plasticity impairment in rat model of Alzheimer's disease. *Life Sciences, 265*, 118802.
117. Wang, C., Cai, Z., Wang, W., Wei, M., Kou, D., Li, T., et al. (2019). Piperine attenuates cognitive impairment in an experimental mouse model of sporadic Alzheimer's disease. *The Journal of Nutritional Biochemistry, 70*, 147–155.
118. Khalili-Fomeshi, M., Azizi, M. G., Esmaeili, M. R., Gol, M., Kazemi, S., Ashrafpour, M., et al. (2018). Piperine restores streptozotocin-induced cognitive impairments: Insights into oxidative balance in cerebrospinal fluid and hippocampus. *Behavioural Brain Research, 337*, 131–138.
119. Hsieh, T. Y., Chang, Y., & Wang, S. J. (2019). Piperine-mediated suppression of voltage-dependent Ca 2+ influx and glutamate release in rat hippocampal nerve terminals involves 5HT 1A receptors and G protein βγ activation. *Food & Function, 10*(5), 2720–2728.
120. S Figueredo, SA., G de Oliveira, M., MVV Safadi, G., de Paula da Silva, C. H., B da Silva, V., A Taft, C., & LB de Aquino, G. (2015). The natural alkaloid piperine and its acid and ester synthetic derivatives are acetylcholinesterase inhibitors. *Current Physical Chemistry, 5*(4), 294–300.
121. Ravelli, K. G., Rosário, B. D. A., Camarini, R., Hernandes, M. S., & Britto, L. R. (2017). Intracerebroventricular streptozotocin as a model of Alzheimer's disease: neurochemical and behavioral characterization in mice. *Neurotoxicity Research, 31*, 327–333.
122. Murata, K., Matsumura, S., Yoshioka, Y., Ueno, Y., & Matsuda, H. (2015). Screening of β-secretase and acetylcholinesterase inhibitors from plant resources. *Journal of Natural Medicines, 69*, 123–129.
123. Kumar, S., Chowdhury, S., Razdan, A., Kumari, D., Purty, R. S., Ram, H., et al. (2021). Downregulation of candidate gene expression and neuroprotection by piperine in streptozotocin-induced hyperglycemia and memory impairment in rats. *Frontiers in Pharmacology, 11*, 595471.
124. Pal, A., Nayak, S., Sahu, P. K., & Swain, T. (2011). Piperine protects epilepsy associated depression: a study on role of monoamines. *European Review for Medical and Pharmacological Sciences, 15*(11), 1288–1295.
125. da Cruz, G. M. P., Felipe, C. F. B., Scorza, F. A., da Costa, M. A. C., Tavares, A. F., Menezes, M. L. F., et al. (2013). Piperine decreases pilocarpine-induced convulsions by GABAergic mechanisms. *Pharmacology Biochemistry and Behavior, 104*, 144–153.
126. Mishra, A., Punia, J. K., Bladen, C., Zamponi, G. W., & Goel, R. K. (2015). Anticonvulsant mechanisms of piperine, a piperidine alkaloid. *Channels, 9*(5), 317–323.
127. Chen, C. Y., Li, W., Qu, K. P., & Chen, C. R. (2013). Piperine exerts anti-seizure effects via the TRPV1 receptor in mice. *European Journal of Pharmacology, 714*(1–3), 288–294.
128. Dong, Y., Yin, Y., Vu, S., Yang, F., Yarov-Yarovoy, V., Tian, Y., & Zheng, J. (2019). A distinct structural mechanism underlies TRPV1 activation by piperine. *Biochemical and Biophysical Research Communications, 516*(2), 365–372.
129. Ren, T., Yang, M., Xiao, M., Zhu, J., Xie, W., & Zuo, Z. (2019). Time-dependent inhibition of carbamazepine metabolism by piperine in anti-epileptic treatment. *Life Sciences, 218*, 314–323.

130. Pattanaik, S., Hota, D., Prabhakar, S., Kharbanda, P., & Pandhi, P. (2009). Pharmacokinetic interaction of single dose of piperine with steady-state carbamazepine in epilepsy patients. *Phytotherapy Research: An International Journal Devoted to Pharmacological and Toxicological Evaluation of Natural Product Derivatives, 23*(9), 1281–1286.
131. Ononiwu, I. M., Ibeneme, C. E., & Ebong, O. O. (2002). Effects of piperine on gastric acid secretion in albino rats. *African Journal of Medicine and Medical Sciences, 31*(4), 293–295.
132. Suresh, D., & Srinivasan, K. (2007). Studies on the in vitro absorption of spice principles–curcumin, capsaicin and piperine in rat intestines. *Food and Chemical Toxicology, 45*(8), 1437–1442.
133. Badmaev, V., Majeed, M., & Prakash, L. (2000). Piperine derived from black pepper increases the plasma levels of coenzyme Q10 following oral supplementation. *The Journal of Nutritional Biochemistry, 11*(2), 109–113.
134. Capasso, R., Izzo, A. A., Borrelli, F., Russo, A., Sautebin, L., Pinto, A., et al. (2002). Effect of piperine, the active ingredient of black pepper, on intestinal secretion in mice. *Life Sciences, 71*(19), 2311–2317.
135. Ahmad, N., Fazal, H., Abbasi, B. H., Rashid, M., Mahmood, T., & Fatima, N. (2010). Efficient regeneration and antioxidant potential in regenerated tissues of *Piper nigrum* L. Plant Cell. *Tissue and Organ Culture (PCTOC), 102*, 129–134.
136. Singh, A., & Duggal, S. (2009). Piperine-review of advances in pharmacology. *International Journal of Pharmaceutical Sciences and Nanotechnology (IJPSN), 2*(3), 615–620.
137. Reshmi, S. K., Sathya, E., & Devi, P. S. (2010). Isolation of piperdine from *Piper nigrum* and its antiproliferative activity. *Journal of Medicinal Plants Research, 4*(15), 1535–1546.
138. Yan, J., Xu, S. C., Kong, C. Y., Zhou, X. Y., Bian, Z. Y., Yan, L., & Tang, Q. Z. (2019). Piperine alleviates doxorubicin-induced cardiotoxicity via activating PPAR-γ in mice. *PPAR Research, 2019*.
139. Chakraborty, M., Bhattacharjee, A., & Kamath, J. V. (2017). Cardioprotective effect of curcumin and piperine combination against cyclophosphamide-induced cardiotoxicity. *Indian Journal of Pharmacology, 49*(1), 65.
140. Upadhyay, R. K., & Jaiswal, G. (2007). Evaluation of biological activities of *Piper nigrum* oil against Tribolium castaneum. *Bulletin of Insectology, 60*(1), 57.
141. Naseem, M. T., & Khan, R. R. (2011). Comparison of repellency of essential oils against red flour beetle Tribolium castaneum Herbst (Coleoptera: Tenebrionidae). *Journal of Stored Products and Postharvest Research, 2*(7), 131–134.
142. Vijayakumar, R. S., Surya, D., Senthilkumar, R., & Nalini, N. (2002). Hypolipidemic effect of black pepper (*Piper nigrum* Linn.) in rats fed high fat diet. *Journal of Clinical Biochemistry and Nutrition, 32*, 31–42.
143. Rani, S. S., Saxena, N., & Udaysree, N. (2013). Antimicrobial activity of black pepper (*Piper nigrum* L.). Global. *Journal de Pharmacologie, 7*(1), 87–90.
144. Damanhouri, Z. A., & Ahmad, A. (2014). A review on therapeutic potential of black pepper (*Piper nigrum* L.): the king of spices. *Journal of Medicinal and Aromatic Plants, 3*(3), 161.
145. D'Hooge, R., Pei, Y. Q., Raes, A., Lebrun, P., Van Bogaert, P. P., & De Deyn, P. P. (1996). Anticonvulsant activity of piperine on seizures induced by excitatory amino acid receptor agonists. *Arzneimittel-Forschung, 46*(6), 557–560.
146. Bai, Y. F., & Xu, H. (2000). Protective action of piperine against experimental gastric ulcer. *Acta Pharmacologica Sinica, 21*(4), 357–359.
147. Concon, J. M., Newburg, D. S., & Swerczek, T. W. (1979). Black pepper [*Piper nigrum*]: evidence of carcinogenicity.
148. Namiki, K., Yamanaka, M., Osawa, T., & Namiki, M. (1984). Mutagen formation by nitrite-spice reactions. *Journal of Agricultural and Food Chemistry, 32*(4), 948–952.
149. World Health Organization. (2008). *Traditional medicine: fact sheet no. 134*. World Health Organization.
150. Johri, R. K., Thusu, N., Khajuria, A., & Zutshi, U. (1992). Piperine-mediated changes in the permeability of rat intestinal epithelial cells: the status of γ-glutamyl transpeptidase activity, uptake of amino acids and lipid peroxidation. *Biochemical Pharmacology, 43*(7), 1401–1407.

151. Prakash, U. N., & Srinivasan, K. (2010). Beneficial influence of dietary spices on the ultra-structure and fluidity of the intestinal brush border in rats. *British Journal of Nutrition, 104*(1), 31–39.
152. Bernardo, A. R., da Rocha, J. D. B., de Lima, M. E. F., Ricardo, D. D., da Silva, L. H. P., Peçanha, L. M. T., & Danelli, M. D. G. M. (2015). Modulating effect of the piperine, the main alkaloid from *Piper nigrum* Linn., on murine B lymphocyte function. Brazilian. *Journal of Veterinary Medicine, 37*(3), 209–216.
153. Lee, Y. M., Choi, J. H., Min, W. K., Han, J. K., & Oh, J. W. (2018). Induction of functional erythropoietin and erythropoietin receptor gene expression by gamma-aminobutyric acid and piperine in kidney epithelial cells. *Life Sciences, 215*, 207–215.
154. Aswar, U., Shintre, S., Chepurwar, S., & Aswar, M. (2015). Antiallergic effect of piperine on ovalbumin-induced allergic rhinitis in mice. *Pharmaceutical Biology, 53*(9), 1358–1366.
155. Kim, S. H., & Lee, Y. C. (2009). Piperine inhibits eosinophil infiltration and airway hyper-responsiveness by suppressing T cell activity and Th2 cytokine production in the ovalbumin-induced asthma model. *Journal of Pharmacy and Pharmacology, 61*(3), 353–359.
156. Panda, S., & Kar, A. (2003). Piperine lowers the serum concentrations of thyroid hormones, glucose and hepatic 5′ D activity in adult male mice. *Hormone and Metabolic Research, 35*(09), 523–526.
157. Dalvi, R. R., & Dalvi, P. S. (1991). Comparison of the effects of piperine administered intra-gastrically and intraperitoneally on the liver and liver mixed-function oxidases in rats. *Drug Metabolism and Drug Interactions, 9*(1), 23–30.
158. Dalvi, R. R., & Dalvi, P. S. (1991). Differences in the effects of piperine and piperonyl butox-ide on hepatic drug-metabolizing enzyme system in rats. *Drug and Chemical Toxicology, 14*(1–2), 219–229.
159. Singh, A., & Rao, A. R. (1993). *Evaluation of the modulatory influence of black pepper.* (*Piper nigrum*, L.) on the hepatic detoxication system. *Cancer letters, 72*(1–2), 5–9.
160. Kang, A. H., Won, S. M., Park, S. S., Kim, S. G., Novak, R. F., & Kim, N. D (1994). Piperine effects on the expression of P4502E1, P4502B and P4501A in rat. *Xenobiotica, 24*(12), 1195–1204.
161. Duangjai, A., Ingkaninan, K., Praputbut, S., & Limpeanchob, N. (2013). Black pepper and piperine reduce cholesterol uptake and enhance translocation of cholesterol transporter pro-teins. *Journal of Natural Medicines, 67*, 303–310.
162. Yang, Y., Kanev, D., Nedeva, R., Jozwik, A., Rollinger, J. M., Grzybek, W., et al. (2019). Black pepper dietary supplementation increases high-density lipoprotein (HDL) levels in pigs. *Current Research in Biotechnology, 1*, 28–33.
163. Rupasinghe, H. V., & Yu, L. J. (2012). Emerging preservation methods for fruit juices and beverages. *Food additive, 22*, 65–82.
164. Atal, C. K., Zutshi, U., & Rao, P. G. (1981). Scientific evidence on the role of Ayurvedic herb-als on bioavailability of drugs. *Journal of Ethnopharmacology, 4*(2), 229–232.
165. Lambert, J. D., Hong, J., Kim, D. H., Mishin, V. M., & Yang, C. S. (2004). Piperine enhances the bioavailability of the tea polyphenol (−)-epigallocatechin-3-gallate in mice. *The Journal of Nutrition, 134*(8), 1948–1952.
166. Mueller, M., Beck, V., & Jungbauer, A. (2011). PPARα activation by culinary herbs and spices. *Planta Medica, 77*(05), 497–504.
167. Gomez-Pinilla, F., & Nguyen, T. T. (2012). Natural mood foods: the actions of polyphenols against psychiatric and cognitive disorders. *Nutritional Neuroscience, 15*(3), 127–133.

Chapter 13
Wild Marigold

Hammad Majeed, Tehreema Iftikhar, Syeda Shehwar Zahra, Muhammad Waheed, Mubashir Niaz, Samar Bashir, Faizah Altaf, and Arwa A. AL-Huqail

13.1 Introduction

Wild marigold (*Tagetes minuta* L.) is an annual aromatic herb. It is plant of sunflower family (Asteraceae) which is the largest family including about 1000 genera and over 23,000 species [1]. According to Soule, this genus have 56 species, including annuals and 29 perennials and 27 annual species [2]. The word *Tagetes* has Latin origin (minutes) which means small, this is because of its small sized capitula. It is widely grown species in several countries because of its high value. Another reason for its preferred cultivation is percentage of its essential oil [3]. Morphologically the herb has woody stem with ridges. Its leaves are pinnate compound with (4–6) pairs

H. Majeed (✉)
Knowledge Unit of Science, University of Management and Technology, Sialkot, Pakistan

T. Iftikhar (✉) · S. S. Zahra · S. Bashir
Applied Botany Lab. Department of Botany, Government College University Lahore, Lahore, Pakistan

M. Waheed
Applied Botany Lab. Department of Botany, Government College University Lahore, Lahore, Pakistan

Govt. Islamia Graduate College, Civil Lines, Lahore, Pakistan

M. Niaz
Atlas Environmental Laboratories, New York, NY, USA

F. Altaf
King Fahd University of Petroleum and Minerals, Dahran, Saudi Arabia

A. A. AL-Huqail
Department of Biology, College of Science, Princess Nourah bint Abdulrahman University, Riyadh, Saudi Arabia

M. Zia-Ul-Haq et al. (eds.), *Essentials of Medicinal and Aromatic Crops*,
https://doi.org/10.1007/978-3-031-35403-8_13

of pinnae and glossy green in texture. Its flowers have pale yellow to orange floret. Capitula are small in size [1].

Scientific name of wild marigold is *Tagetes minuta* L. (Species Plantarum 887. 1753) [4]. *Tagetes minuta* L. has many common as well as local names. In English, it is called Mexican marigold, Aztec marigold, wild marigold, Roger, stinkweed, Stinking, tall khaki weed and khaki bush. In Hindi, it is called as Jungli Gainda. Spanish people called it chinchilla, huacatay, enana whereas in Angola it is known as ekaibulo. In Argentina, it is named as chil chil while, in Brazil, it carries various common names like chinchilla, coora, margarita, rabo de rajao, voadeira and suique. Quinchihue is common name of wild marigold known in Chile. Wilde Sammetblume is its common name in German. In Kenya, its name is bhangi, nyanjaga, mubangi, muvangi, omotioku, omubazi gwemhazi, nyanjagra. In Madagascar, it is commonly known as mavoadala. In other countries like Malawi its name is khaki; In Portugal, cravo de defunct and in Paraguay it is known as agosto, and suico .In Pakistan it is locally known as Gul-e-Sad Barga [5].

13.2 Classifications [6]

Kingdom: Plantae
Subkingdom: Viridiplantae
Division: Tracheophyta
Subdivision: Spermatophytina
Class: Magnoliopsida
Order: Asterales
Family: Asteraceae
Genus: *Tagetes* L.
Species: *Tagetes minuta* L.

13.3 Crop Description

13.3.1 Origin and Distribution

Wild marigold can grow in both conditions (moist and dry areas). In tropical and subtropical areas its height can be up to 3000 m from sea level. It grows as weed in orchards, gardens, waste grounds and along roadside and causes many problems for pastures and other crops. It is grown in many areas of the world [7]. Due to its agrochemical and pharmacological properties wild marigold has been extensively grown around the globe. Due to Spanish conquest, *Tagetes minuta* L. had been introduced in different regions of the world *i.e.,* Europe, Asia, Africa, Madagascar, India, Australia and Hawaii. It has been deliberately distributed as an ornamental, medicinal or fragrant plant as well as, as a weed across the tropics, subtropics and several temperate countries. According to Maheshwari, the plant has acclimatized itself in Himalayan and sub-Himalayan regions (up to altitudes of 2000 m) [3].

13.3.2 Plant Morphology

Tagetes minuta L. is an erect herb that can grow up to the height of about 1–2 m. Its leaves are glossy green and are about 15 cm long with serrate margins. On the lower side of leaf multicellular glands can be observed that seem orange in color. Ray florets and disc florets of yellow to orange color can be observe. Flower heads are small with the length of 10–15 mm. The plant is native to South America including countries (Argentina, Chile, Bolivia, Peru and Paraguay) [8].

13.3.3 Agronomy

This species of *Tagetes* has ability to adapt to various soil conditions. Plant grown in deeper and fertile soil can be helpful for the enhanced production. When soil is well drained, friable and has high water-holding capacity it can result in the growth of a healthy plant. If soil pH is maintained up to 7.0–7.5, it ensures the better growth of plant. For the best growth of the *Tagetes minuta* L. mild climate is considered most suitable. In mild climates where temperatures ranges from 14.5 to 28.6 °C increased flowering takes place. Higher temperatures between 26.2 and 36.4 °C negatively effects production of flowers. In the same tune, cool temperature induces germination, whereas profuse vegetative, growth and flowering take place at high temperatures [9]. Human and animals are the source of seeds dispersal. Seeds are dispersed by clinging the hair of animals and human's cloths [10] (Figs. 13.1 and 13.2).

13.4 Cultivation Practices

13.4.1 Preparation of Land

To get optimum growth and yield, soil must be free from all weeds. Moreover, medium light to medium heavy soil is considered best for growth of *Tagetes minuta* L. The land ploughing is done 4–6 times. About (45–50 t/ha) of farmyard manure is applied at the time of ploughing. Due to ploughing ridges (40 cm) and channels of convenient size are formed and seedlings are transplanted on the sides of the ridges. Spacing of 30 × 45 is considered optimum. To enhance the plant growth, inoculations of soil with Rhizobacteria like *Pseudomonas fluorescens, Azospirillum brasilense* is done. It will improve the yield of essential oil (EO) [11].

Fig. 13.1 Floral view of Marigold

13.4.2 Seed Dispersal

Human and animals are the source of seeds dispersal. Seeds are dispersed by clinging the hair of animals and human's cloths [12].

13.4.3 Sowing Season

In a year, this crop can grow in three seasons (rainy, winter and summer seasons) [13].

Fig. 13.2 Lateral and aerial view of Marigold

13.4.4 Phenology

Flowering in *Tagetes minuta* L. takes place in the month of September. Crop harvesting is carried out from October till November when the crops fully bloom. At the end of November, seed harvesting is done. It is recommended to remove apical meristem during the mid of month of August, which will check the growth and production of lateral branches [14].

13.4.5 Processing and Handling of Seeds

Fruit is obtained in the form of pods having many seeds in it. Mature fruits are black to brown in color that are attached to the stalks. Husk is removed by sieving of dried fruits [15].

13.4.6 Propagation

Propagation by seed can be done by two methods *i.e.,* direct sowing and transplantation of plants.

1. **Direct sowing**

   About half kilogram of seeds is enough for one hectare land. Best time for the sowing of the seeds is before frost. Sowing is followed by germination that starts within 15 days. After about 2 months, blooming starts. Seeds are saved in dry and dark location. For annual cultivar, spring/summer season is preferred. Suitable spacing for sowing is 20 × 30 cm. Seeds of wild marigold are sown in the depth of about 0.5–1 cm to get better yield. Soil with moisture and location where sunlight is enough is considered best for sowing. It takes about 1–3 weeks to germinate [11].
2. **Transplantation**

   Preferable time for the transplantation of plant in main field is from the month of July to August (Table 13.1). Accurate time has prime importance to get maximum biomass yield of leaf and flowers. In 2008, it was concluded by researcher those old seedlings (45 days) are suitable for transplantation and it is

Table 13.1 Time period of seed sowing, transplanting, flowering, fruiting, harvesting [16]

Sowing period	Mid June	Mid-September	January–February
Transplanting period	Mid July	Mid October	February–March
Flowering period	February	June	November
Fruiting period	December	January	March–April
Harvesting period	February	May	October

suggested that seed sowing should be done in the 2nd week of June. The crop of nursery raising practice is preferred over direct sowing to maintain desired crop geometry. For successful transplantation 60–45 cm or 30–30 cm spacing is preferred. Seedling with 3–4 leaves is selected for transplantation [3].

13.5 Seed Germination

Young plants are susceptible to weed attack. So to establish the crop, 2–3 manual weeding operations are required. When plant reaches at the phase of maturity it can overcome the attack of weeds. Promotive effect on growth of branches has been observed when apical buds are nipped in 50–60 days after sowing the seeds or it can also be done about 30–45 days after transplantation. Nipping of buds also results in greater proportion of leaves and flowers. Bud nipping is also helpful to withstand the period of moisture shortage [4].

13.6 Harvesting

In developing countries, harvesting is done mechanically because low-cost labor is available in these countries. Another reason for mechanical harvesting is that this method can give efficient results [17]. The procedure of harvesting starts with the cutting of main stem. Cutting is followed by drying of individual branches and folding of plant material into a bundle. The bundles are then tied by using the twine of grasses. After that dried bundles are placed at dry places and are dried under sunlight [9].

13.7 Flowering and Fruiting

The flowering period of plant is February, June and November (Table 13.1). Whereas, fruiting can take place three times during a year that are from December to January, March to April and July to September [18] (Fig. 13.3).

13.8 Oil Distillation

Oil can be obtained through steam distillation. Plants dried under shade are considered best for distillation [19].

Fig. 13.3 Different stages in the life cycle of *T. minuta* L. (**a**) Seeds of *Tagetes minuta* L. (**b**) Plantation of crop (**c**) Stage of bud formation (**d**) Flowering stage (**e**) Full bloom stage (**f**) Mature crop view (**g**) Crop harvesting stage (**h**) Oil distillation (**i**) Essential oil extraction of *T. minuta* L.

13.9 Storage of Oil

Initially the oil extracted from wild marigold is allowed to stand for 12–24 h. This is done to separate water present in the oil. To make the oil free of moisture, it is treated with anhydrous sodium sulphate. For removing impurities, the oil is filtered. Oil should be stored in moisture free conditions by keeping it in amber colored bottle. Stainless steel utensils are also considered safe for storage [8]. It should be placed in cool area away from sunlight. For storing oil for a long period, it is advisable to keep it in refrigerator [8].

13.10 Physical and Chemical Properties of Wild Marigold Essential Oil

It is a clear liquid with pale yellow to dark yellow in color. Sometimes, it also appears orange or reddish in color. It is fruity or woody in odour. Specific gravity of essential oil is 0.8405–0.9440. Optical rotation of wild marigold EO is +1° to +5.43°.

Refractive index of oil is 1.4850–1.5896 [20]. Different factors can be regulated to get better quality of wild marigold essential oil. EO collected from plants at high altitudes are rich in ketones. So, it is preferable to gather oil from plants grown on high altitudes. Moreover, the planting and harvesting season also effects the oil quality. It is recommended to harvest wild marigold plants at full bloom stage. Plant parts also influences the quantity of oil produced. Flowers and leaves of wild marigold are considered principal parts for extraction of essential oil [21].

13.11 Fertilizers and Manures

Fertilizers are necessary for growth of plant. Different fertilizers have specific role in different plants. It is not wrong to say that fertilizers have irreplaceable role in raising crop yields. In case of wild marigold, N and S fertilizers play vital role to improve the crop's yield. When transplantation time arrives, required amount of potassium, phosphorous and FYM is mixed with the soil and same is done at the time of seed sowing. In an experiment conducted, plants were provided with 0, 60, 90 and 120 kg ha^{-1} nitrogen. Similarly different levels of S were 0, 20, 40, and 60 kg ha^{-1}. Higher biomass *i.e.* 183.9 production was seen with 120 kg ha^{-1} and 178.90 was noticed with 60 kg ha^{-1}. Whereas, 102.09 and 88.60 kg ha^{-1} increase in yield of EO was observed with the application of N and S respectively [22].

13.12 Doses and Time for Application of Fertilizers

It is recommended that crop should be treated with nitrogen in split doses at the time of sowing or after 2 months of seed germination when transplantation is to be carry out. Fertilizers are also provided at the time of bud initiation. N P and K with the ratio (120: 60: 60 kg/ha) is applied to get maximum yield. [23].

13.13 Crop Rotation

To combat the weed growth, 1–2 weeding's are required. Mostly, *Tagetes minuta* L. is inter cropped with *Brassica oleracea* var. *sabellica* and *Zea mays* L. Inter cropping has proved beneficial for enhancing the yield of oil [24].

13.14 Invitro Propagation of *Tagetes minuta* L.

Tagetes minuta L. is a source of number of secondary products such as pharmaceuticals, pesticides and flavoring agent which are used in food industry. So to improve the production of secondary metabolites production from *Tagetes minuta* L. several *invitro* propagation techniques have been opted. In research conducted by Mohamed et al. (1998) the cotyledons and hypocotyls of *Tagetes minuta* L. were cultured on MS medium containing IAA or NAA and BA. Adventitious shoots developed from the hypocotyl-derived callus. However, these adventitious shoots failed to generate further. Contrarily, cotyledon-derived callus cultured on MS medium containing IAA generated adventitious shoots which later developed into plantlets. Highest number of shoots were obtained by culturing cotyledons initially on medium containing 71 μM IAA + 44.4 μM BA and later transferring them to MS medium. From the developing shoots nodal segments were taken which were micro-propagated on half-strength MS medium containing 2.58 μM IAA. 95% of plantlets developed from these nodal segments successfully adapted to greenhouse conditions. The plantlets produced from *invitro c*ulturing of nodes had many shoots whereas those micropropagated from shoot tips consisted of a single main stem. Morphological differences were studied after 2 months growth in greenhouse. However, no significant difference was noticed in dry weight of leaf and shoot. After six times subculturing of cotyledons-derived callus on medium containing IAA and BA all explants lost their regeneration capacity except those growing on 17.23 μM IAA and 44.4 μM BA. This research developed a novel protocol for shoot regeneration from cotyledons *of Tagetes minuta* L [25].

13.15 Factors Influencing the Essential Oil Composition

Essential oil of *Tagetes minuta* L. is very rich in composition. A number of chemical constituents such as Tagetone, tran-caryophyllene, bisabolene, carvotanacetone, carvacrol, citral, B-cubebene, pclirnene, dihydrotagetone, isoeugenol, p-ionone,linalyl acetate,linalool, isolimonene, ethyl-n-heptyl ketone, B-myrcene, menthol, nerolidol, 2-nonanone, p-ocimene, ocimenone, 1-penten-3-ol, phenylacetaldehyde, o-pinene, spathulenol, cr-terpineol and y-terpineol. However, this composition of wild marigold essential oil is influenced by different factors that are discussed below.

Altitude Essential oil obtained from *Tagetes minuta* L. growing on high altitude is rich in ketone constituents while wild marigold plants growing in plains has low olfactory value.

Planting and Harvesting Season Depending upon the required number of harvests and the location of plantation wild marigold can be planted in different seasons. Growing wild marigold in different seasons significantly varies in composition of essential oil. In a research conducted by Moghaddam and Omidbiagi (2007) different composition of essential oil at different harvesting stages was found. Hydrodistillation was used to extract oil at budding, full-bloom and fruiting stage. GC/MS was used to analyze the oil content. On basis of their dry weight the oil content at these three stages were e 1.55%, 1.44% and 1.0% (w/w). Nineteen, 20 and 19 compounds have been identified in the oils of budding, full flowering and fruit set stages of *T. minuta* L., respectively. The major compounds present were limonene, dihydrotagetone, (Z)-tagetone, (E)-ocimenone and a-terpineol. The results showed that harvest times had a major effect on the constituents of the oil of *T. minuta* [26].

Plant Ontogeny Harvesting of *Tagetes minuta* L. at pre-flowering stage produces essential of poor quality as compared to the oil produced by plants harvested at full bloom stage. It is found that at vegetative stage the principal component of oil is tagetone. However, with increasing age the concentration of dihydrotagetone increases whereas the concentration of tagetone and ocimenone decreases. In full bloom stage the percentage of ketones also varies at different locations.

Plant Parts The principal parts of wild marigold containing essential oil are its flowers and leaves. In a research conducted by Moradalizadeh et al. (2013) different components of essential oil extracted from different parts of wild marigold were analyzed. Constituents of oil were studied by comparing the mass spectra and retention indices with those cited in literature. Total 26 components were determined in different plant organs. Major components isolated from leaf oils were dihydrotageton (45.9%), cis-β-ocimene (11.9%) and borneol (11.1%) and those of the seed oils included dihydrotagetone (21.0%) and benzoic acid-4-hydroxy-methyl ester (33.5%).Also, trans-ocimenone (27.0%), cis-βocimene (26.0%) and cis-ocimenone (17.6%) were the major constituents in the flower oils [27].

13.16 Impact of Different Abiotic Stresses on Yield of *Tagetes minuta* L.

Wild marigold because of their highly flavored and fragrant essential oil are in great demand by industries. They can be grown in wide range of climates but due to changing climatic conditions their growth is now becoming restricted as a result of different abiotic stresses. These abiotic stresses at elevated level results in reduced germination, growth and low quality essential oil. Wild marigold is vulnerable to following abiotic stresses.

Effect of Drought on *Tagetes minuta* L. Formation of Reactive Oxygen Species (ROS) takes place in wild marigold due to excessive drought, salinity and heavy metal stress resulting in deterioration of plant. Drought affects the physiology and biochemical responses of wild marigold and its essential oil [28]. Shortage of water results in short plants with small leaf in order to minimize the water loss. Elevated drought stress impacts both physiological and morphological characteristics of wild marigold. Under drought stress vegetative dry matter of aromatic plants reduces significantly [29]. Water limitation in wild marigold reduces its physiological parameters such as relative water content (RWC) together with its chlorophyll and carotenoid content. Photosynthesis also gets limited because of reduced stomatal and mesophyll conductance in limited water availability [30]. In drought stress stomata while acclimatizing to their environment varies in their anatomical features such as size and density. This affects the transpiration rate in wild marigold. Reduction in growth of *Tagetes minuta* L due to drought is the result of formation of ROS in cellular compartments of chloroplast, peroxisomes and mitochondria [31]. Due to accumulation of ROS in plants, degradation of membranes starts. Photosynthetic activity also reduces due to these ROS. To survive in water stress conditions wild marigold increases its root biomass and length, reduces growth of its shoot and alters leaf orientation to minimize the loss of water. Severe drought conditions prevents wild marigold from absorbing Photosynthetically Active Radiations (PAR) thus preventing it from accumulating dry matter. It is suggested that root traits should be thoroughly studied to make the crop drought tolerant. In response to dehydration abscisic acid (ABA) activates. It protects the plant by closing stomata, reducing leaf size and promoting root elongation [32].

Effect of Temperature Fluctuation In wild marigold seed germination is the major stage which is affected by fluctuations in temperature which ultimately impacts mature crops [33]. Seed germination in *Tagetes minuta* L. is inhibited at temperature above 35 °C however, if transferred to 25 °C then its achenes germinate. Pre-treating seeds by thermoinhibition enables the seeds to undergo rapid germination on return to conducive temperatures. Thermoinhibition can be used for wild marigold achenes. When wild marigold achenes are imbibed at 25 °C they continue to germinate for several days. In a research conducted it has been found that imbibing achenes at 35 °C and then transferring it to 25 °C within 24 h results in 100% germination [34].

Effect of Irradiance Stress on Biomass and Yield of *Tagetes minuta* L. Essential Oil Photosynthesis in plants is greatly influenced by solar radiation. Irradiance directly affects plant growth and adaptation, reproduction, geographical distribution and metabolite composition [35]. Yield of essential oil and biomass significantly reduces by irradiation stress. In a research conducted on impact of irradiance stress on plants it was found that wild marigold plants growing in 25% shade were taller than those growing in full sunlight. However, it was observed that increased shading reduced the number of branches in wild marigold plants. Moreover, experiment showed that highest plants with increased branches were produced when spacing between the plants was 45 × 45 cm. In contrast to open light conditions maximum biomass was attained with 45 × 30 cm plant spacing accompanied with 25% shading. Essential oil content of wild marigold reduces when provided with 50 or 70% shading. Wild marigold growing at different places contains essential oils with different grades [36].

Effect of Waterlogging on Wild Marigold In regions with high rainfall the productivity of *Tagetes minuta* L. is adversely impacted by waterlogging [10]. Due to lack of oxygen to roots in waterlogged soils wild marigold fails to uptake water and minerals thus restricting the crop yield. In order to genetically improve waterlogging tolerance in wild marigold prior knowledge about genetic characteristics and selection strategy in natural populations of wild marigold under waterlogged conditions is necessary. Under oxygen stressed conditions ethylene production changes the tolerance mechanism in plants by expression of certain genes which induces formation of aerenchyma. However, much literature is not available on oxidative stress due to waterlogging in wild marigold.

Effect of Salinity on Growth of *Tagetes minuta* L. Various physiological and yield contributing factors are effected by soil salinity. Photosynthetic as well as enzymatic activity is also affected [37]. For wild marigold to survive in saline conditions it is essential to have salinity tolerance during seed germination and early growth stages. With increased salinity the percentage of seed germination decreases. Moghaddam et al. (2019) observed that germination rate in wild marigold declined significantly with increasing percentage of sodium chloride and potassium nitrate (0–500 mM). Lowest rate of germination was recorded at 200 mM concentration of both salts. Above 200 mM the germination rate of wild marigold achenes underwent complete inhibition. This experiment showed that above tolerance limit salinity can cause complete inhibition. Moreover, this salinity stress induces enhanced accumulation of ROS which results in disruption of plant cell membranes [38].

Heavy Metal Stress in Wild Marigold Because Wild marigold produces essential oil thus it can accumulate heavy metals. These can extract harmful heavy metals from the environment together with providing economic benefits to the growers. Wild Marigold accumulates lead (Pb) from the atmosphere thus they can be employed for phytoextraction of lead. In an experiment conducted by Sosa et al. (2016) wild marigolds growing at five different sites were observed. These sites were Pb contaminated as they contained former battery plants. It was found that despite of wild marigolds growing in lead contaminated soil with their leaves accumulated with lead, their essential oil didn't contain any traces of lead [39].

Impact of Other Abiotic Stresses Upon Growth of *Tagetes minuta* L. Plant nutrients *i.e.* nitrogen and sulfur significantly influences the growth of wild marigold and its essential oil content. Increased usage of nitrogen and sulfur fertilizers results in high production of biomass and essential oil content. In contrast to yield and growth parameters, stomatal density decreases with increased use of sulfur and nitrogen fertilizers. To overcome this nutrients stress it is also essential to select lines of wild marigold that are non-responsive to both nitrogen and sulfur fertilizers [22].

Under abiotic stresses secondary metabolites accumulates in large quantity in wild marigold. The proportion of essential oil content to biomass increases under stress conditions however the overall biomass and essential oil content decreases [40]. Wild marigold contains variety of secondary metabolites such as terpenes, phenols, flavonoids and thiophenes. Primary metabolites acts as a precursor of secondary metabolites which play a pivotal role in plants defense against different abiotic stresses [41]. These secondary metabolites are produced through different pathways. For example the terpenoids produced by mevalonic acid pathway (MVA) are responsible for aromatic properties of *Tagetes minuta* L. essential oil. Under oxidative stress conditions terpenes provide protection to plants Shikimic Acid

Pathway is responsible for production of phenols and flavonoids. The phenolic compounds reduces the formation of Reactive Oxygen Species (ROS) and also contributes to HM chelation. Thus different secondary metabolites play significant role in protection of plants.

Geographical location, different crop cultivation practices and seasonal variations also effects essential oil constituents of *Tagetes minuta* L. [42]. Essential oil constituents of wild marigold varies from individual to individual within a natural population. Desired grade of essential oil can be produced by employing different crop practices. Essential oil of *Tagetes minuta* L. produced in different seasons contains different percentages of its various constituents. In winter season ocimene-rich essential oil is produced whereas in autumn and summer the produced essential oil contains high percentage of dihydrotagetone, tagetone, and ocimenone. Factors such as location, growth conditions, the plant part used, cultivation technique and oil distillation process greatly impacts the chemical profile of *Tagetes minuta* L. essential oil. At higher altitude the essential oil content of wild marigold increases whereas in dry farming it decreases [12]. So in dry conditions stress tolerant lines accompanied with advance agricultural practices can help in successful growth of wild marigold.

13.17 Diseases and Pests Control

Various pests and microorganisms infect the wild marigold which will ultimately effect the production of marigold (Table 13.2).

13.17.1 Wilt and Stem Rot

In this disease, collar portions of the plants are affected by fungus (*Sclerotiana sclerotium*). The infection results in damping-off and aggravation. As the result of infection, plants show wilting. The disease can be controlled by treating soil with Mancozeb, Fosetyl-Al and Metalaxyl. To overcome disease seed treatment with thiram is suggested [43].

13.17.2 Collar Rot

Collar rot is also caused by *S. sclerotium*. In this disease, black lesions appear on the stem of the plant. Rotting occurs at the collar regions hence leading to the death of the plant. This disease can be overcome by soil sterilization and controlled watering conditions [13, 21].

Table 13.2 Infectious agents to various diseases of Marigold with probable treatment [24]

Diseases	Infectious agents	Treatment
Wilt and Stem Rot	*Sclerotiana sclerotium*	Treatment of soil with Mancozeb, Fosetyl-Al and Metalaxyl Treatment of seed with Thirman
Collar Rot	*Sclerotiana sclerotium*	Sterilization of soil, Controlled water conditions
Leaf Spot and Blight	*Alternaria, Cercospora* and *Septoria* sp.	Fungicidal Spray
Damping off	*Pythium* sp.	Treatment of soil with 2% formaline, Treatment of seeds with Dithane
Powdery Mildew	*Oidium* sp.	Spray of Sulfex
Flower Bud Rot	*Alternaria dianthi*	Spray of Mancozeb
Black-foot disease	*Cylindrocarpon destructans.*	Spray of *Prochloraz manganese chloride*

13.17.3 Leaf Spot and Blight

This disease is caused by *Alternaria, Cercospora and Septoria* sp. development of brown necrotic spots can be observed on leaves, as the result the entire foliage gets damage and results in poor growth. Fungicidal spray is helpful in controlling the disease [11].

13.17.4 Powdery Mildew

It is caused by *Oidium* sp. Symptoms appear as white color powder growing on the aerial parts of the plant. Disease can be controlled by spraying Sulfex [4, 10].

13.17.5 Flower Bud Rot

Flower Bud Rot is caused by *Alternaria dianthi.* Young flower buds are susceptible to attack. The infected buds appear dark brown in color. The pathogen also infects leaves, causing blight in them. On margins and tips of older leaves, brown necrotic spots can be seen. Bud and leaf infections can be control by Mancozeb spray [23].

13.17.6 Damping Off

Causative agent of this disease is *Pythium* sp. The fungus attacks at the seedling stage. Symptoms of disease appear in the form of necrotic spots and rings causing collapse of the seedlings. Soil is sterilized with Formalin (2%). Beside this seed is treated with Dithane [44].

13.17.7 Thrips

It is caused by *Thrips tabaci*. This pest results in distorted and discolored plant tissue. Thrips use rapidly growing tissues for their nourishment. Resultantly, the leaves become distorted. Leaves roll and fall off. Color break is seen in the petals due to feeding by thrips before opening of buds. Most species of wild marigold are resistant to the pest however, some are affected by pests. For example, cultivated forms of *Tagetes erecta* L. has some serious pests (Red spider mite and root knot nematode) whereas, at the same site these pests have not been found on *T. minuta* L [11, 23].

Black Foot Disease This disease was first reported in 2011 in Fars province of Iran. The disease manifests itself in form of yellowing, root and foot rot, drying of leaves, and ultimately the death of plant. In a research conducted to find the causative agent of this disease. For this purpose the infected plants were collected and transferred to laboratory. The samples were then disinfected by using 0.5% NaClO solution, and cultured on Potato Dextrose Agar (PDA). The pH of medium was maintained at 4.5 with 0.5% lactic acid. Morphological studies revealed the pathogen to be *Cylindrocarpon destructans*. For further confirmation DNA was extracted from the sample and studied using PCR. Results showed 100% similarity index with *Cylindrocarpon destructans*. Further pathogenicity tests were carried out. The pathogenicity tests were done with a suspension of 1×10^6 conidia per ml homogenized in sterile water. The symptoms on infected plants were similar to that was observed previously.

13.18 Nutrient Deficiency Symptoms in Wild Marigold

The deficiency of primary and secondary nutrients results in different symptoms that are discussed below.

13.18.1 Nitrogen Deficiency

As the result of nitrogen deficiency, the older leaves become chlorotic. They may show purple to red discoloration [21].

13.18.2 Phosphorous Deficiency

Stunted growth and deep green foliage are the symptoms of this deficiency. Some plants may show purple coloration. Chlorosis followed by necrosis takes place in older leaves. Abnormal elongation of roots take place. Leaves lose their dark green color and the underside of leaves appears red and yellow in color. These symptoms spread across the leaf with older leaves falling off [18].

13.18.3 Magnesium Deficiency

Chlorosis among the veins takes place. In many leaves before chlorosis red and purple pigmentation starts appearing. Chlorosis starts from the tip and spreads across the leaf in a V-shape formation. After chlorosis necrosis takes place.

13.18.4 Sulfur Deficiency

This deficiency results in change in size and shape of flowers, turning to pale yellow in color.

13.18.5 Iron Deficiency

It results in white appearance of overall leaf blade.

13.18.6 Boron Deficiency

It has major impact on flowers. The flowers are incomplete, have fewer number of petals, the petals suddenly wilts, notches appearing in flower stems indicating missing tissues. It results in death of roots. Chlorosis of young leaves also takes place (Table 13.3).

Table 13.3 Different nutrient deficiency symptoms in *Tagetes minuta* L.

Nutrients	Deficiency symptoms
Nitrogen	Chlorosis
Phosphorous	Stunted growth, Discoloration, Abnormal elongation of roots
Magnesium	Chlorosis followed by necrosis
Sulfur	Impacts size and color of flowers
Iron	Whitish appearance of leaf blade
Boron	Fewer number of petals, chlorosis, wilting

13.19 Crop Duration

Crop has short duration whether it is planted in winter (December) or in rainy season from July to September. Crop planted in rainy seasons have to face many problems due to growth of weeds in the field. Weeds can interfere with the yield of crop. If weeds are not cleared from the crops, the crop can face the problems related to productivity and growth [8].

13.20 Annual Production

"*Tagetes* oil" in the quantity of 42 l imported by India is worth 10,028 USD. The largest supplier of EO of *Tagetes minuta* L. is Switzerland that accounts for about 9037 USD. Other countries like France exported USD 677 and United States exported USD 147 "*Tagetes* oil" in 2021. 98% of imports are accounted by Bombay Air Cargo and 2% imports are accounted by Nhava Sheva Sea [19]. Average price and average value per shipment of *Tagetes minuta* L. essential oil is estimated about USD 241.21 and USD 346 respectively [45]. 100% export is accounted by Delhi Air Cargo. Estimated Export price of (about 475 l) oil in India is about USD 54,572. Average price of "*Tagetes* oil" per unit and average value per shipment is USD 114.96 and USD 1605 respectively [9].

13.21 Chemical Composition of Wild Marigold

13.21.1 Secondary Metabolites

Wild marigold is abundant in a variety of secondary metabolites including (flavonoids monocyclic and bicyclic monoterpenes, acyclic, sesquiterpenes, thiophenes and aromatics). The main components include Beta phelandrene, Limonene, Beta ocimene, Dihydrotagetone, Tagetone, Tagetenone. Through oil extraction researchers have extracted various chemical compounds from stem, leaves and flower [18].

13.21.2 Constituents of Leaf

Thirteen compounds were investigated from leaf oil which was about 96.9%. Ddihydrotagetone is about 17.6%, cis-beta-ocimene is 11.9% and borneol is 11.1% [18, 22].

13.21.3 Constituents of Flower

As far as the composition of various constituents in flower is concerned, trans-Ocimenon is about 27%, cis- beta-ocimine is about 26% and cis-ocimenone is about 17.6% [18, 22].

13.21.4 Constituents of seed

Dihydrotagetone and benzoic acid 4- hydroxy-methylester makes about (21%) and (33.5%) are main constituents of seed respectively [18, 22].

13.21.5 Constituents of Capitula

Researchers have found that capitula contains many components *e.g.*, limonene that shares 2.1–6.2%, Beta ocimine constitutes about 15–20%, alpha aterpinolene makes 15.6% of capitula. Similarly, other compounds such as E and Z tagetone makes 2.5 and 4.62% respectively. Piperitenone and piperitone are also constituents of plant's capitula [43].

13.22 Quantitative Analysis

From quantitative point of view essential oil from flowers contain relatively greater amount of β-ocimene and tagetenone as compared to the essential oil from the leaves. Dihydrotagetone is the main component of non-bloomed plants which decreases when plant blooms contrarily quantity of tagetenone and β-ocimene increases [27, 46, 47].

13.23 Essential Oil from *Tagetes minuta* L.

Oil is the one of the most famous products made from wild marigold. Countries all over the world earn through export of its oil. Medical companies have also produced many medicines by using different parts of this plant. Many products related to food have also been produced by wild marigold. It is very profitable crop. Almost all parts of plants can be used for the extraction of essential oil. In 1984, about 1.5 tons was produced annually, which is now between 8 and 15 tons per annum with the estimated amount of (50 USD/kg). However, it had been quite hard to keep the record of production and market value of the oil due to unknown reasons [45]. Some countries have set organizations for maintaining selling price of oil *e.g.*, an organization in India named Floracopeia Aromatic Treasures, has set that price at 1.07 USD/ml. Mountain Rose a herbs company of USA has set rate at 11.75 USD for half ounce, supplying towards South Africa and France. Whereas 8.20£ is set by (Hermitage soil) that is a main supplier of special essential oils in the UK since 1979. Egypt and Zimbabwe are identified as world's major producers of *T. minuta* L. essential oil by International Trade Center. Some other countries like Nepal, Brazil, India, Madagascar, Ukraine, Australia, Chile, Peru, Bolivia, Morocco, Ecuador, Kenya and Paraguay are also the major producers of essential oil extracted from *T. minuta* L.[13].

13.24 Impact of Drying Processing on Yield of Essential Oil

Different drying methods can significantly enhance the production of wild marigold (*Tagetes minuta* L.) essential oil. Swati et al. (2020) conducted a research in which three dry processing methods were investigated. Hydrodistillation was used to obtain essential oil which was later analyzed by GC-MS. In this research different time periods (0, 12, 24, 36, 48, 60, and 72 h) were used. Results showed that sun drying for 72 h produced higher yield of essential oil whereas moisture content was lower. However, these drying treatments didn't alter the number of essential oil components which remained nine in total. Major constituents were (Z)-ocimene, dihydrotagetone, (E) & (Z)-tagetone, and (E) & (Z)-tagetenone. The amount of EO and the percentage of two of its components (Z)-ocimene and dihydrotagetone were immensely impacted by various drying methods and time. Sun drying for 60 h showed higher percentage of (Z)-ocimene and lower amount of dihydrotagetone in essential oil. However with rise in drying time the percentage of (Z)-ocimene reduced in essential oil of wild marigold (*Tagetes minuta* L.) [48].

13.25 Yield and Production of Essential Oil

Essential oil of the plant having greater percentage of *Z-beta*-ocimene *i.e.*, about 35–50% has high rate and demand internationally. According to an estimate in year 2016, the EO of the wils marigold plant showed about 15 tons annual production and it is expected that its market value will reach up to 11.8 USD million by the year 2025 with a CAGR of 7.0% from 2020 to 2025 [20].

13.26 Impact of Nitrogen and Sulfur on Yield and Quality of Essential Oil

Fertilizers plays a crucial role in enhancing yield of crops; however, using excessive amount of chemical fertilizers contaminates the atmosphere. It is hypothesized that using nitrogen and sulfur fertilizers significantly improves the yield and oil quality of essential oil of *Tagetes minuta* L. In order to prove this claim an experiment was carried out in which different concentrations of both fertilizers were employed to *Tagetes minuta* L. plants which were then studied for 2 years. Using nitrogen at 120 kg ha^{-1} and sulfur at 60 kg ha^{-1} produced higher biomass and essential oil yield in contrast to control. At 120 kg ha^{-1} of nitrogen and 40 kg ha^{-1} of sulfur it was found that stomatal density decreased whereas as density of oil glands increased significantly. Moreover, the concentration of EO component Z-B-ocimene increased significantly at this concentration. So it was concluded that 120 kg ha^{-1} of nitrogen and 40 kg ha^{-1} of sulfur is best for maximum production of *Tagetes minuta* L. essential oil.

13.27 Importance of Wild Marigold

13.27.1 Pharmaceutical Importance

The oil of *T. minuta* L. has many useful applications in the field of medicine for example, it is used to treat wounds and a variety of infections. *Tagetes* Oil (in a 5% dilution) has been used to kill maggots in open wounds. By mixing with cream or lotion, it can be used for fungal and microbial infections and particularly it is proved to be very helpful to cure wounds or athlete's foot. By drinking the boiled mixture (of boiled leaves) of the *Tagetes minuta* L., infections of respiratory system can be cured. Its vapors have showed positive effects for example vapor from its leaves have potential to treat coughs, bronchitis, headaches and chest infections [19].

In agricultural industry, it is being used as an insect repellent to avoid plants from insect related disease. The vapors also show antiseptic activity in a room. Wild marigold plant also has such compounds which exhibits topical applications [49].

Flower decoctions of plant have been found to heal wounds and sores [50]. In Kenya, its sap is being used for topical application. It is reported by Rahman et al. that wild marigold is used to treat dental related diseases e.g. to heal mouth wounds [51]. Digestive disorders can be cured by using the plant in different ways. In various countries (Bolivia, Paraguay, Argentina, Brazil and Peru) *T. minuta* L. infusions have role as digestives and appetizers. Its decoction is being used as carminatives, vermifuges and anti-diarrheic. Leaves of plant can be chewed and it is helpful to treat bad breath [52].

It is used as anti-parasites against food poisoning. Many diseases such as dyspepsia, gastritis and intestinal disorders can be cured by mixing its various constituents in various medicines [46]. The leaf constituents are found to be helpful to reduce milk secretion. Its decoction is also used as an expectorant or an anti-abortive. The infusion of the plant is helpful in regulating menstrual flow. The whole plant is diuretic and also has its applications in reducing fever and act as antipyretic. Its leaves have been proved to be effective against liver diseases [53]. Its leaves and stem both are used to make medicines to treat earache [54]. Disorders of bones, joints, muscles and nerves can also be cured by the tea made up of its flowers. This is mostly done in Morocco and Bolivia [55]. In Brazil, it is used as a drug to induce sleep [14].

13.28 Medicinal Properties for Animals

Tagetes minuta L. has also showed medicinal properties for animal for example in Ethiopia it is used to cure various diseases of animals such as blackleg, anthrax and amoebiasis. This medicine is prepared by chopping the leaves and stems of the plant and mixing the contents with water. The resultant medicine is administered orally to the cattle and sheep [18].

13.29 Antifungal Activity of *T. minuta* L. Essential Oil

The essential oil of *Tagetes minuta* L. is found to have antifungal activity due to presence of limonene, camphor, cineole and α-pinene in it. *Rhizoctonia solani, Penicillium digitatum, Trichoderma and Aspergillus niger* are some of the fungi against which the *T. minuta* L. has shown strong antifungal activity. Research has showed that the essential oil extracted from *T. minuta* L. leaves have high antifungal activity than that extracted from its flowers. It is also observed that the activity of EO increases with the increase in concentration [18, 56].

13.29.1 Antimicrobial Activity

In Argentina *Tagetes minuta* L. is commonly used in folk medicines for treating a number of gastrointestinal diseases. Using extracts of wild marigold in different fractions have been proved effective against the activity of gram positive and gram negative bacteria. Quercetagetin-7-arabinosyl-galactoside present in *Tagetes minuta* L. extract has shown significant antimicrobial activity against *Escherichia coli*. The leaf extract of *Tagetes minuta* L. completely inhibits the growth of this enteropathogen bacteria. Growth was inhibited due to total flavonoid content at low concentration [57].

13.29.2 Insecticidal Activity of **T. minuta** *L.*

Awareness about the harmful impacts of synthetic insecticides has led people towards the utilization of natural insecticides. Essential oils of various plants are used for this purpose as they are easy to extract, biodegradable and ecofriendly. The essential oil of *T.minuta* L. is found to be effective against head lice. It has also shown insecticidal activity against *Anopheles stephensi.* EO extracted from aerial parts of *T. minuta* L. is also found effective against *Aedes aegypti*, black moth and beetles as well [11].

13.30 Repellent Activity

Repellents are the substances that deters the arthropods flying or landing on human beings [58]. Researches are being carried out for production of natural repellents. Essential oil of *T.minuta* L. is being studied for its repellent activities against mites and ticks [11, 21].

13.31 Activity Against Nematodes

β-ocimene and dihydrotagetone are two compounds present in essential oil of *T.minuta* L. that have shown nematicidal activity against *Meloidogyne incognita*, a plant nematode. It has found to be highly active at 1–4% concentration against both the eggs and the juvenile nematodes [10, 11].

13.32 Activity Against Invasive Weeds

Weeds pose a serious threat to agro-ecosystems. It is found that *Tagetes minuta* L. essential oil possess allelopatic potential and can act upon invasive weeds like *Chenopodium murale* L., *Phalaris minor Retz.* and *Amaranthus viridis* L. It is observed that the concentration of volatile essential oil of wild marigold significantly reduces the reproduction, growth and chlorophyll content of the attacking weeds. The respiratory ability of weeds also diminishes by application of this essential oil. Moreover, it wild marigold essential oil completely arrests the mitotic activity within the growing weed. The dried leaf powder of *T.minuta* L. is also used to suppress growth of two rice paddy weeds.

13.33 Commercial Importance of *Tagetes minuta* L.

Essential oil (EO) is one of the main products obtained from *Tagetes minuta* L., which is further used in food products, beverages, in bakery items, deserts, as flavoring agent and in perfume industry. Oil is the one of the most famous product made from wild marigold [10]. Countries all over the world earn through export of its oil. Medical companies also have produced many medicines by using different parts of this plant. Many products related to food have also being produced by wild marigold. Thus it is very profitable crop [43].

13.34 Ethnobotanical Usage of *Tagetes minuta* L.

Since early times *T. minuta* L. has captured the interest of human beings. It is used in beverages, condiments, medicinal decoction, and in ritual/religious practices. Its leaves, stem and flowers are used as a culinary herb in South America where it is known as 'huacatay'. The huacatay paste is then used to make popular potato dish of South America called as 'Ocopa'. Since the times of Inca civilization in Peru *T.minuta* L. is used for seasoning that imparts a unique herbal flavor to the dishes. Moreover, it also used in some areas to make ethnomedicinal teas. In Northern Chile a traditional preparation of *Tagetes minuta* L. is used to make suico that enhances flavor of dishes [59]. The plant decoction and/or tea preparations may be consumed either warm or cool and may be sweetened as per individuals' taste. It has been officially recognized by UN for its use in food color and for flavor purpose in condiments. It is also used in mayonnaise, pasta, dairy products like yogurt for flavor, as well as in bakery items. Moreover, the orange pigment obtained from the petals of its flowers is in great demand in poultry[24]. It can also be grown for commercial purposes such as for sale in city markets. In India there is an increased interest in *Tagetes minuta* L. for its remarkable properties of essential oil whereas its flowers are used to decorated Hindu temples.

13.35 Arable Farming Systems

Farmers often plant wild marigold in their arable farming fields. It has been reported to retain humidity in arable farming fields. It is beneficial in several ways such as rapid growth of *Tagetes minuta* L. shades out other plants that may be of less usage to farmers. It is often an alternative host for *Sclerotinia sclerotiorum* in arable farming system [2]. *Sclerotinia sclerotiorum* is a fungal pathogen that infects variety of crops. The essential oil of *Tagetes minuta* L. is also found effective against three species of common grain pests that are *Tribolium castaneum, Rhyzopertha dominica,* and *Callosobruchus anali.* Thus the spray of wild marigold essential oil can protect the crops during storage. According to United States Environmental Protection Agency *Tagetes minuta* L. essential oil is found to be effective as an acaricide and an insecticide against variety of pathogens attacking food crops. Thus essential of wild marigold plays a significant role in food security by protecting crops in fields and during storage. Talking about livestock, so 20% concentration of *Tagetes minuta* L. essential oil shows 95% efficacy against *Rhipicephalus sanguineus, Amblyomma cajennense and Argas miniatus*. This oil has potential to control other livestock tick species that causes severe economic losses to farmers. Thus using this essential oil provides an environment friendly approach for production of safe acaricide to milk and beef in livestock industry [43].

13.36 Myths, Legends, Tales, Folklore and interesting Facts

Wild marigold is cultivated almost in every part of the world and most of the people have various myths, tales, songs, saying etc. about them. The available literature tells that Portuguese introduced marigolds in Subcontinent and since then it is extensively planted in this region to make floral chain/garlands to use them on marriages, festivals and for home decoration as well. Particularly, individuals decorate their homes with marigold on various religion occasions [60]. In some countries marigold is linked with autumn and death. Whereas somewhere it is considered special flower of October. It is featured in all saints and souls' celebrations in Europe. It is replaced with saffron to color the soul-cakes. In some parts of Europe, it is used to make garlands along with other flowers and twined about tombstones. In India, people use this flower to make offerings to their gods [13]. According to a dictionary of gardening of Marigold, In the Mexico it is known as "Flower of death" They consider the marigold flower "Mexican soldiers" who (who fought against the Spanish conquistadors) spilled their blood. They symbolize reddish brown patches on the petals with blood. It is used to symbolize a feeling of despaired love [8, 13].

In Mexico it is symbol of the annual holiday to celebrate and honor the souls of departed. In an encyclopedia of religious (rites, rituals, and Festivals) it is written that yellow is the symbol of death among the pre-hispanic people therefore marigolds are profuse at the Day of the Dead celebration. In England like other flowers marigold is also used to predict that whether he or she "loves me" or "loves me not". Some Asian singers have used marigold in their songs *i.e.,* "Mery jurray me gaindy ka Phool", "Susral gainda phool" Germans (in sixteenth century) are in belief that men should carry a marigold root along with a violet silk handkerchief with them all the time. The purpose of this myth was to be popular with the girls. Another concept that German has is that peasant girls would sow marigolds in the footsteps of those they wished to embrace. Marigolds represent power, strength, and light within a person. It is often linked with often linked to the powerful strength of the sun [51]. No doubt, Marigold remained most sacred flower for the people of Aztecs. Moreover, it was a common tradition to use marigold in religious events as well as a medicine. The Aztecs had strong belief that the flowers of marigold helped to get relieve from hiccups and it also has capacity to heal people affected from lightning [61].

References

1. Sadia, S., Khalid, S., Qureshi, R., & Bajwa, A. A. (2013). Tagetes minuta L., A useful under-utilized plant of family asteraceae: A review. *Pakistan Journal of Weed Science Research, 19*, 179–189.
2. Soule, J. (1994). *Infrageneric systematics of tagetes*. Paper presented at the Compositae: Systematics. Proceedings of the International Compositae Conference.
3. Bandana, K., Raina, R., Kumari, M., & Rani, J. (2018). Tagetes minuta: An overview. *International Journal of Chemical Studies, 6*(2), 3711–3717.
4. Maheshwari, J. K. (1972). *Tagetes Minuta Linn. in Simla Hills*.
5. Qureshi, R. A., Ghufran, M. A., Gilani, S. A., Sultana, K., & Ashraf, M. (2007). Ethnobotanical studies of selected medicinal plants of Sudhan Gali and Ganga Chotti hills, district Bagh, Azad Kashmir. *Pakistan Journal of Botany, 39*(7), 2275–2283.
6. Creative Commons Corporation, Integrated Taxonomic Information System (ITIS), an online database. (2022). (Publication no. https://doi.org/10.5066/F7KH0KBK). Retrieved March 23, 2022, https://creativecommons.org/publicdomain/zero/1.0/legalcode, http://www.itis.gov/
7. Meshkatalsadat, M. H., Safaei-Ghomi, J., Moharramipour, S., & Nasseri, M. (2010). Chemical characterization of volatile components of Tagetes minuta L. cultivated in south west of Iran by nano scale injection. *Digest Journal of Nanomaterials and Biostructures, 5*(1), 101–106.
8. Farooqi, A. A., & Sreeramu, B. (2004). *Cultivation of medicinal and aromatic crops*. Universities Press.
9. Chan, G., Towers, G. N., & Mitchell, J. (1975). Ultraviolet-mediated antibiotic activity of thiophene compounds of Tagetes. *Phytochemistry, 14*(10), 2295–2296.
10. Singh, V., Singh, B., & Kaul, V. K. (2003). Domestication of wild marigold (Tagetes minuta L.) as a potential economic crop in western Himalaya and north Indian plains. *Economic Botany, 57*(4), 535–544.
11. Muzemu, S., Mvumi, B., Nyirenda, S., Sileshi, G., Sola, P., Chikukura, L., et al. (2011). *Pesticidal effects of indigenous plants extracts against rape aphids and tomato red spider mites*. Paper presented at the African crop science conference proceedings.
12. Singh, M., Tripathi, R., Singh, S., & Yaseen, M. (2011). Influence of row spacing and nitrogen levels on herb and essential oil production and oil quality of Tagetes minuta L. *Journal of Spices and Aromatic Crops, 17*(3).
13. Vasey, E. (2008). Marigolds – A few stray petals. *The Marlowe Society Research Journal, 5*, 1–4.
14. Chamorro, E. R., Ballerini, G., Sequeira, A. F., Velasco, G. A., & Zalazar, M. F. (2008). Chemical composition of essential oil from *Tagetes minuta* L. leaves and flowers. *The Journal of Argentine Chemical Society, 96*(1–2), 80–86.

15. Salehi, B., Valussi, M., Morais-Braga, M. F. B., Carneiro, J. N. P., Leal, A. L. A. B., Coutinho, H. D. M., et al. (2018). Tagetes spp. essential oils and other extracts: Chemical characterization and biological activity. *Molecules, 23*(11), 2847.
16. Anjarwalla, P., Belmain, S., Sola, P., Jamnadass, R., & Stevenson, P. C. (2016). *Handbook on pesticidal plants* (p. 64). World Agroforestry Centre.
17. Cáceres, A., Girón, L. M., Alvarado, S. R., & Torres, M. F. (1987). Screening of antimicrobial activity of plants popularly used in Guatemala for the treatment of dermatomucosal diseases. *Journal of Ethnopharmacology, 20*(3), 223–237.
18. Walia, S., & Kumar, R. (2021). Wild marigold (*Tagetes minuta* L.) biomass and essential oil composition modulated by weed management techniques. *Industrial Crops and Products, 161*, 113183.
19. Singh, A., Khanuja, S. P., Arya, S. J., Singh, S., & Yadaw, A. (2006). Essential oil quality and yield with respect to harvest index in Tagetes minuta cultivated in sub tropical plains of North India. *Journal of Essential Oil Research, 18*(4), 362–365.
20. Babu, K. G. D., & Kaul, V. K. (2007). Variations in quantitative and qualitative characteristics of wild marigold (*Tagetes minuta* L.) oils distilled under vacuum and at NTP. *Industrial Crops and Products, 26*(3), 241–251.
21. Baser, K. H. C., & Buchbauer, G. (2009). *Handbook of essential oils: Science, technology, and applications*. CRC Press.
22. Walia, S., & Kumar, R. (2021). Nitrogen and sulfur fertilization modulates the yield, essential oil and quality traits of Wild Marigold (*Tagetes minuta* L.) in the Western Himalaya. *Frontiers in Plant Science, 11*, 631154. https://doi.org/10.3389/fpls.2020.631154
23. Njoroge, G. N., & Bussmann, R. W. (2007). Ethnotherapeautic management of skin diseases among the Kikuyus of Central Kenya. *Journal of Ethnopharmacology, 111*(2), 303–307.
24. Syamasundar, K., & Rajeswara Rao, B. R. (2013). Wild marigold (*Tagetes minuta* L.): Cultivation technology and essential oil composition. In *Natural essential oils* (pp. 228–248). Aavishkar Publisher Distributors.
25. Mohamed, M.-H., Harris, P., & Henderson, J. (1998). An efficient in vitro regeneration protocol for Tagetes minuta. *Plant Cell, Tissue and Organ Culture, 55*(3), 211–215.
26. Moghaddam, M., Omidbiagi, R., & Sefidkon, F. (2007). Changes in content and chemical composition of Tagetes minuta oil at various harvest times. *Journal of Essential Oil Research, 19*(1), 18–20.
27. Moradalizadeh, M., Mehrabpanah, M., Salajeghe, M., & Nayebli, M. (2013). Chemical constituents of the essential oils from the leaves, flowers and seeds of Tagets minuta L. By GC/MS. *International Journal of Advanced Biological and Biomedical Research, 1*(9), 1124–1128.
28. Babaei, K., Moghaddam, M., Farhadi, N., & Pirbalouti, A. G. (2021). Morphological, physiological and phytochemical responses of Mexican marigold (*Tagetes minuta* L.) to drought stress. *Scientia Horticulturae, 284*, 110116.
29. Shahri, A., Ganjali, H. R., & Fanayi, H. R. (2013). Effect of drought stress on quantitative and qualitative yield of safflower (Goldasht cultivar) in different planting densities. *International Journal of Agriculture Crop Sciences, 6*(19), 1342–1346.
30. Ouyang, W., Struik, P. C., Yin, X., & Yang, J. (2017). Stomatal conductance, mesophyll conductance, and transpiration efficiency in relation to leaf anatomy in rice and wheat genotypes under drought. *Journal of Experimental Botany, 68*(18), 5191–5205.
31. Chen, W., Li, P., & Chen, T. (2000). Glycinebetaine increases chilling tolerance and reduces chilling-induced lipid peroxidation in Zea mays L. *Plant, Cell and Environment, 23*(6), 609–618.
32. Hagar, H., Ueda, N., & Shah, S. V. (1996). Role of reactive oxygen metabolites in DNA damage and cell death in chemical hypoxic injury to LLC-PK1 cells. *American Journal of Physiology. Renal Physiology, 271*(1), F209–F215.
33. Forsyth, C., & Staden, J. V. (1983). Germination of Tagetes minuta LI Temperature effects. *Annals of Botany, 52*(5), 659–666.

34. Hills, P., Van Staden, J., & Viljoen, C. (2001). Differences in polypeptide expression in thermoinhibited and germinating achenes of Tagetes minuta L. *Plant Growth Regulation, 34*(2), 187–194.
35. Keller, P., & Lüttge, U. (2005). Photosynthetic light-use by three bromeliads originating from shaded sites (Ananas ananassoides, Ananas comosus cv. Panare) and exposed sites (Pitcairnia pruinosa) in the medium Orinoco basin, Venezuela. *Biologia Plantarum, 49*(1), 73–79.
36. Kumar, R., Sharma, S., Ramesh, K., Pathania, V., & Prasad, R. (2014). Irradiance stress and plant spacing effect on growth, biomass and quality of wild marigold (*Tagetes minuta* L.) – An industrial crop in western Himalaya. *Journal of Essential Oil Research, 26*(5), 348–358.
37. Sayyed, A., Gul, H., Ullah, Z., & Hamayun, M. (2014). Effect of salt stress on growth of *Tagetes erecta* L. *Pakhtunkhwa Journal of Life Science, 2*(3–4), 96–106.
38. Moghaddam, M., Farhadi, N., Panjtandoust, M., & Ghanati, F. (2020). Seed germination, antioxidant enzymes activity and proline content in medicinal plant Tagetes minuta under salinity stress. *Plant Biosystems-An International Journal Dealing With All Aspects of Plant Biology, 154*(6), 835–842.
39. del Carman Sosa, M., Salazar, M. J., Zygadlo, J. A., & Wannaz, E. D. (2016). Effects of Pb in *Tagetes minuta* L.(Asteraceae) leaves and its relationship with volatile compounds. *Industrial Crops and Products, 82*, 37–43.
40. Hossein, A. F., Sayed, A. V., Jahanfar, D., Amir, H. S., Mohammad, A. K., & Forestry. (2009). Medicinal and aromatic plants farming under drought conditions. *Journal of Horticulture, 1*(6), 86–92.
41. Mahajan, M., Kuiry, R., & Pal, P. K. (2020). Understanding the consequence of environmental stress for accumulation of secondary metabolites in medicinal and aromatic plants. *Journal of Applied Research on Medicinal Aromatic Plants, 18*, 100255.
42. Wasaya, A., Zhang, X., Fang, Q., & Yan, Z. (2018). Root phenotyping for drought tolerance: A review. *Agronomy, 8*(11), 241.
43. Cornelius, W. W., & Wycliffe, W. (2016). Tagetes (Tagetes minuta) oils. In *Essential oils in food preservation, flavor and safety* (pp. 791–802). Elsevier.
44. Rahman, I. U., Ijaz, F., Iqbal, Z., Afzal, A., Ali, N., Afzal, M., et al. (2016). A novel survey of the ethno medicinal knowledge of dental problems in Manoor Valley (Northern Himalaya), Pakistan. *Journal of Ethnopharmacology, 194*, 877–894.
45. Chalchat, J.-C., Garry, R.-P., & Muhayimana, A. (1995). Essential oil of Tagetes minuta from Rwanda and France: Chemical composition according to harvesting location, growth stage and part of plant extracted. *Journal of Essential Oil Research, 7*(4), 375–386.
46. Güneş, S., Savran, A., Paksoy, M. Y., Koşar, M., & Çakılcıoğlu, U. (2017). Ethnopharmacological survey of medicinal plants in Karaisalı and its surrounding (Adana-Turkey). *Journal of Herbal Medicine, 8*, 68–75.
47. Anonymous. (2017a). *Integrated taxonomic information system online database*. http://www.itis.gov
48. Walia, S., Bhatt, V., & Kumar, R. (2020). Influence of drying processing on essential oil yield and composition of Wild Marigold (*Tagetes minuta* L.) in the western Himalayas. *Journal of Essential Oil-Bearing Plants, 23*(4), 686–696.
49. Tankeu, S. Y., Vermaak, I., Viljoen, A. M., Sandasi, M., & Kamatou, G. P. P. (2013). Essential oil variation of Tagetes minuta in South Africa–A chemometric approach. *Biochemical Systematics and Ecology, 51*, 320–327.
50. Arora, K., Batish, D. R., Singh, H. P., & Kohli, R. K. (2015). Allelopathic potential of the essential oil of wild marigold (Tagetes minuta L.) against some invasive weeds. *Journal of Environmental Agricultural Sciences, 3*, 56–60.
51. Skinner, C. M. (1913). *Myths and legends of flowers, trees, fruits, and plants in all ages and in all climes*. JB Lippincott Company.
52. Bourdy, G., Chāvez de Michel, L. R., & Roca-Coulthard, A. (2004). Pharmacopoeia in a shamanistic society: The Izoceño-Guaraní (Bolivian Chaco). *Journal of Ethnopharmacology, 91*(2-3), 189–208. https://doi.org/10.1016/j.jep.2003.09.013

53. Teixidor-Toneu, I., Martin, G. J., Ouhammou, A., Puri, R. K., & Hawkins, J. A. (2016). An ethnomedicinal survey of a Tashelhit-speaking community in the High Atlas, Morocco. *Journal of Ethnopharmacology, 188*, 96–110.
54. Agra, M. D. F., Baracho, G. S., Nurit, K., Basílio, I. J. L. D., & Coelho, V. P. M. (2007). Medicinal and poisonous diversity of the flora of "Cariri Paraibano", Brazil. *Journal of Ethnopharmacology, 111*(2), 383–395.
55. Kidane, B., van der Maesen, L. J. G., van Andel, T. R., & Asfaw, Z. (2014). Ethnoveterinary medicinal plants used by the Maale and Ari ethnic communities in southern Ethiopia. *Journal of Ethnopharmacology, 153(1)*, 274–282.
56. Dulger, B., & Gonuz, A. (2004). Antimicrobial activity of certain plants used in Turkish traditional medicine. *Asian Journal of Plant Sciences, 3*, 20031215.
57. Tereschuk, M. A. L., Riera, M. V., Castro, G. R., & Abdala, L. R. (1997). Antimicrobial activity of flavonoids from leaves of Tagetes minuta. *Journal of Ethnopharmacology, 56*(3), 227–232.
58. Wanzala, W., & Ogoma, S. B. (2013). Chemical composition and mosquito repellency of essential oil of Tagetes minuta from the Southern slopes of Mount Elgon in Western Kenya. *Journal of Essential Oil-Bearing Plants, 16*(2), 216–232.
59. Sadia, S., Khalid, S., Quresh, R., & Bajwa, A. A. (2013). Tagetes minuta L., a useful underutilized plant of family asteraceae: A review. *Pakistan Journal of Weed Science Research, 19*(2).
60. Roth, N. J. (2020). *Marigolds and Munshīs: Horticultural writing and garden culture in Mughal South Asia.*
61. Grissell, E. (2020). *A history of zinnias: Flower for the ages*. Purdue University Press.

Chapter 14
Vanilla

Tehreema Iftikhar, Hammad Majeed, Muhammad Waheed, Syeda Shehwar Zahra, Mubashir Niaz, and Arwa A. AL-Huqail

14.1 Introduction

Vanilla and its relatives are the existing individuals of an ancient ancestral group of angiosperms. A large number is restricted to remote localities while others are threatened with extinction. We surely know a lot about *Vanilla planifolia*, but fundamental natural history of the entire genus Vanilla and its closest relatives is still poorly known irrespective that we are well aware of its methods of cultivation, diseases that affect the domesticated vines, and techniques of fruit processing. So, the systematic study of vanilla and its relatives is continuing to be surrounded by controversies. Due to these controversaries, it is encouraging to witness latest research and increased level of knowledge of vanilla in recent years about their evolution and classification, which due to the abundant use of DNA-based data in systematic studies [1–3].

T. Iftikhar (✉) · S. S. Zahra
Applied Botany Lab. Department of Botany, Government College University Lahore, Lahore, Pakistan

H. Majeed (✉)
Knowledge Unit of Science, University of Management and Technology, Sialkot, Pakistan

M. Waheed
Applied Botany Lab. Department of Botany, Government College University Lahore, Lahore, Pakistan

Govt. Islamia Graduate College, Civil Lines, Lahore, Pakistan

M. Niaz
Atlas Environmental Laboratories, New York, NY, USA

A. A. AL-Huqail
Department of Biology, College of Science, Princess Nourah bint Abdulrahman University, Riyadh, Saudi Arabia

M. Zia-Ul-Haq et al. (eds.), *Essentials of Medicinal and Aromatic Crops*,
https://doi.org/10.1007/978-3-031-35403-8_14

14.2 Family, Genus and Common Names

Within the genus Vanilla, the plants were formally placed into two possible sections by Rolfe (1896). Aphyllae, the first Vanilla section, appears erected to accommodate all the leafless species in the genus (e.g., *V. aphylla*, *V. barbellata*, *V. roscheri*, and others). Species of this section normally grow on the African mainland, the Caribbean islands, Madagascar, and also on Southeast Asia.

No doubt, a few of these species produce fleshy fruits but still there is no evidence that any of them are aromatic. These species together share a recent common ancestor according to Rolfe's classification, but molecular studies have demonstrated reverse of it [4]. The species of this section is not monophyletic, but an artificial grouping of species with shared vegetative morphology derived by convergent evolution. According to modern rules of natural classification, it should not be recognized formally.

Rolfe created section Foliosae for those Vanilla species that are not in section Aphyllae. All these species are leafy as the name indicates. This is a large group of vanilla species. Portères [5] divided the section into further subsections. Vanilla section Foliosae subsection Membranaceae is a small cluster of species characterized by thin stems, thin leaves, short aerial roots, and flowers in which the labellum is not fused with the column. The labellum also lacks the complex bristles, hairs, and scales characteristic of other Vanilla species, and the fruits tend to dry on the vines and split lengthwise. *Vanilla mexicana* exemplifies this section, and molecular systematic studies have demonstrated that the group is the most primitive of all Vanilla species. These plants are very difficult to cultivate, probably because they have close relationships with mycorrhizal fungi, and there is no evidence that the fruits produce aromatic vanillin.

The remaining species of the vanilla genus are classified into either subsection Papillosae or subsection Lamellosae, including *V. planifolia* and *V. pompona*. The latter group is named so because species of this section have flowers with flattened scale-like appendages (lamellae), bristles, hairs. Moreover, a complex ornamentation is present on their labella, that is always merged with column along the margins to form a floral tube [6].

Most orchids are exotic to Pakistan. They are cosmopolitan, except Antarctica, but a large number are found in the humid tropics and subtropics but only a few are reported in Pakistan. The orchids belong to family Orchidaceae and is represented by 1000 genera, 15,000–25,000 species and more than 1,25,000 hybrids, created by crossbreeding of species, and this number is being increased each year [7]. Some 3K–4K hybrids are being produced each year but only a few get recognition which are then available commercially. Furthermore, inter-generic hybrids are also being created. All new hybrids must be named and recognised by the International Orchid Commission in association with Royal Horticultural Society of Great Britain. It is the sole international authority for registration of orchids [8]. Synonym of *Vanilla planifolia* Jacks. Ex Andrews are *Vanilla fragrans* Salisb. Ames, *Epidendrum rubrum* Lam., *Notylia sativa* (Schiede) Conz., *Vanilla bampsiana* Geerinck, and *Vanilla sylvestris* Schiede [9].

Vanilla planifolia is commonly called as 'flat-leafed' vanilla, 'vanilla vine', and 'vanilla orchid'. In Castilian (Spanish); vainilla vansa, vainilla, vainilla colibri. Mayan (Mesoamerica); Zizbic [10]. ونیلا Urdu (Pakistan).

14.3 Classification [11]

Kingdom: Plantae
Division: Tracheophyta
Subdivision: Spermatophytina
Class: Magnoliopsida
Order: Asparagales
Family: Orchidaceae
Genus: *Vanilla* Mill.

14.4 Crop Description

14.4.1 Origin and Distribution

Only 47 species, all terrestrial, belonging to 26 genera, are reported in Pakistan [7]. The Orchidaceae are terrestrial, perennial but rarely have short live, epiphytic, lithophytic, autotrophic, saprophytic or sometimes mycotrophic herbs, tubers rhizomes, or rootstocks having mycorrhizal association [12].

Vanilla and its relative surviving plant members belongs to an ancient lineage of flowering plants. Large numbers are limited to isolated areas while some species are still having a threat of extinction. There are many *Vanilla* species like *Vanilla barbellata*, *V. claviculata*, *V. dilloniana*, *V. mexicana*, *V. phaeantha*, *V. planifolia*, *V. poitaei*, *V. pompna*, etc but medically and commercially most important species is *Vanilla planifolia* Jacks. Ex Andrews [13].

14.4.2 Plant Morphology

Stems are sympodial or monopodial, normally leafy but have at least swollen internodes at base. The base usually forms a pseudobulb while epiphytics are aerial with photosynthesizing adventitious roots, frequently bear velamen consisting of one or more layers of dead cells called. The leaves are usually alternate but sometimes opposite, often distichous, sometimes terete or canaliculate, glabrous or very infrequently hairy, leathery or whorled and mostly have a sheathing base and a parallel veined blade that is often fleshy. The leaves often form a false petiole, apex is often emarginate. These are sometimes reduced to bract-like scales [14].

Inflorescence is racemose that may be lateral, basal or sometimes terminal, may be erect or pendulous, sub-umbellate, spicate or paniculate. One to many flowers are present which are infrequently secund or distichously arranged. Flowers are mostly showy but small to large, typically zygomorphic, virtually actinomorphic, hermaphrodite/bisexual (sometimes monoecious and polymorphic), pedicellate, mostly resupinate with or without pedicel. Sepals are often free but rarely adnate, dorsal (median) is mostly dissimilar to laterals while laterals sometimes adnate to column foot to form a spur like mentum. Corolla has free Petals but hardly adnate to sepals, may or may not be like sepals, mostly showy; lip entire, variously lobed or 2/3 partite, ornamented, with or without a lower spur or nectary, margins entire to laciniate. The perianth usually have 6 tepals in two whorls that are similar or different. The outer most whorl of three distinctive or diversely connate tepals are sometimes sepaloid. Two members out of 3 tepals are identical and may be quite like the outer tepals. The third tepal of the same whorl usually forms a labellum that is highly modified in shape as well as in appearance. The androecium and gynoecium are almost always adnate into a column or gynostegium. Anthers are mostly 1, rarely 2 or 3, ventral or terminal on column, cap like or opening by longitudinal slits. Pollen normally form distinct pollinia, rarely loose, pollinia are 2, 4, 6, and 8 may be waxy, or horny, sessile or attached by stalks to 1 or 2 sticky viscidia. Moreover, stigma is 3 lobed, mid-lobe is mostly modified into rostellum, other lobes sunken on ventral surface of column behind the anther. Capsule fruit is present. It is sometimes berry like and opens laterally by three or six slits. Seed are numerous dust-like, infrequently winged and lack endosperm [15].

14.4.3 Agronomy

A variety of soil types can be used for growth of vanilla. Hot humid tropical climates best suited for vanilla growth having an abundant amount of organic material. To evade stagnant water condition, a gentle slope is favourably advantageous to avoid the water logging and disease. While dry soils will always need a better irrigation system to maintain adequate moisture contents in the soil. Mulch is the most used media for vanilla cultivation. The specific mulch type is not as necessary as its ability to release nutrients slowly, but it must also retain an optimal level of moisture for proper root growth [16]. Undoubtedly, hot humid climate is most favourable for cultivation of vanilla. *V. planifolia* Jacks. ex. Andrews grows best in areas where temperature ranges from 20 to 30 °C and it may also tolerate high temperatures up to 32 °C. It is sensitive to cold temperatures (below 20 °C) including freezing and may inhibit plant growth and flowering capacity. The temperature above 32 °C causes premature fruit drop and yellowing of vegetative parts. Ideal rainfall for vanilla's growth is reported around 170–280 cm/year. Extremely wet conditions during capsule ripening may cause bean rot because it requires a dry season for about 2 months to initiate flowering. Moreover, additional irrigation may be useful to establish new cuttings [17].

14.5 Cultivation Practices

14.5.1 Land orientation

If the land is predominated by slopes with an eastward orientation, then it is advisable to select an area that is mostly covered by sunlight in mornings. This is advantageous to prevent decrease in moisture contents in the soil and to save the plant, and beans from over drying due to afternoon sun [10] (Fig. 14.1).

Fig. 14.1 Aerial view of Vanilla sp.

14.5.2 Preparation of Land

Preparations are normally done keeping in mind the height and types of native trees selected for vanilla production. When trees are taller than 10 m, it is called as forest type and have a minimum diameter of 50 cm; acahual type if the area is dominated by bushes and smaller trees; if both trees and bushes are absent then the land is called as deforested land [10, 18].

14.5.3 Forest Type Land Preparation

Controlling weeds along with branches that are under 4 m from the ground is essential. Moreover, damaged branches, irrespective of their height; small trees and bushes must be eliminated if they are beneath the shadow of taller trees to enhance ventilation and distribution of shadow is up to 50%. After this activity, tutors usually planted to support the vanilla plants [18].

14.5.4 Acahual Type Land Preparation

Mulch is prepared from dried bushes and branches and weeds must be eradicated as discussed earlier. Bushes having only one stem should be selected so that they can provide shadow and support. Bushed and trees taller than 2 m can only be used as shadow contributors. Now, the land is ready for the tutors and vanilla plantation [19].

14.6 Propagation

Stem cuttings is the commonly used technique for vanilla propagation. The cuttings are generally prepared or obtained from another planter, or from a governmental agricultural entity. Cuttings are usually made from highly productive and healthy plants, which are selected earlier and marked before its harvesting. Moreover, fresh cuttings are kept under low light at room temperature for 1–2 days and let cut sites heal before it is planted. It is most crucial to note that the using longer cuttings more quickly establishes and initiate flowering [20]. Cuttings that have length under 1 m will normally take 3–4 years to flower while a meter long cutting may bloom 1–2 years earlier. It must be noted that the cutting is not a flowering shoot and should have at least 3 nodes for producing new shoots. Furthermore, cuttings must always be free from damage or of any symptoms of disease to check proliferation of the disease in future [21]. Cuttings are usually 80–120 cm long and 1 cm in diameter or it may have at least 6–8 nodes because longer cuttings are more difficult to handle during plantation and are mostly expensive, irrespective of their high productivity. Cuttings have length under 80 cm are best managed as nursery plants before their transfer to land. Cuttings may be directly planted on top of growth substrate and mulch prepared from different sources, before the healing of cut sites. The top of the longer cuttings may be attached/tied to a support and make sure that at least 1–2 nodes are in contact with the substrates. Similarly, leaves that are in contact with substrate, must be removed from nodes [22]. Moreover, tissue culture *V. planifolia* Jacks. ex. Andrews plants are also available from selected stores/companies in various countries. No doubt, tissue culture plants are very much appealing as a clean source of starting material, but it will take 3 or 4 years to mature enough to flower [23]. Vanilla is not usually cultivated by seeds because of germination challenges. The lignified thick seed cover inhibits germination as well as its seed also take considerably longer time to grow into a young and mature plant as compared to cuttings. Similarly, seed germination tends to associate with fungi and with another microorganisms. Such type of constraints has played a critical role to use cuttings as the primary propagation method [16].

14.7 Collection of Material for Propagation

It is mandatory to follow the recommendations to check the propagation of unwanted/unproductive plants *i.e.*, 'donkey ear' that falls almost 100% of its fruits after 2 months [13]. Don't plant the cuttings of unidentified origin. Parent plant should be in full production. To attain this objective, select most suitable plants, mark, and locate the design of the plot. The selected plants must have healthier phenotype and must bear at least 3 months old fruits. Right after harvest, obtain cuttings from selected plants and the portion having the fruit may be used as propagation material. The disinfection for cuttings is normally prepared with a fungicide such as benomyl at nominal concentration of 2 gram per litre along with an adherent. This preparation is then mixed with water in a large container that can immerse a bundle of at least ten cuttings [8, 13].

14.8 In Vitro Plants

Another method for producing commercial stocks of *Vanilla* plants is through *in vitro* propagation. For this purpose, the propagules must be placed under controlled conditions in green house, where they can adapt to photosynthesis. As soon as, the propagules reach a height of 30 cm, they must be transplanted. It takes around 1 year by plant to reach the fruiting stage. Plant obtained by this method are healthier. However, this method is expensive for most farmers [10, 24].

14.9 Disinfection of Cuttings

Before planting, the basal leaves should be removed by hand taking care not to break open the stem. To prevent stem rot, stem cuttings are disinfected by immersing in fungicidal solution for 2–5 min. The infected cuttings are then hanged at a tall structure of 1–1.5 m for 7–15 days. The cuttings become flexible, and calluses form on the points of leaves removal [8, 13].

14.10 Planting of Cuttings

Moist organic matter, established shadows, weed free land and tutors organised in blocks (as discussed earlier) are mandatory for the plantation of *Vanilla* cuttings. The cuttings must be healed, normally wounds arise after 1–2 days of mechanical damage, disinfected and without 2–3 base leaves [18]. Once the support trees develop enough foliage, cuttings are planted. It is done so that the support trees protect vanilla from burning. Shade cloth is used during planting of cuttings. Humid, warm, and dry months are preferable for cuttings plantation. This is done before the rainy season [25].

14.11 Planting Season

In Asian regions, particularly in Pakistan, the best season is July and August when irrigation lands can be used more freely, except beyond the favourable temperatures [7].

14.12 Planting Procedure

First, holes are made using the cuttings at a 45° angle on the ground up to 20 cm depth. The leafless portion of the cuttings is inserted and then a slight force is applied to keep the cutting fix into the land. Similarly, the lower portion of the cuttings are covered by organic matter and finally the upper portions are knotted to the tutors, it is strongly recommended to use herbal or biodegradable materials like strips of tree bark, banana leaves or some creepers [21].

14.13 Manuring

Applications of fertilizers are commonly ignored in vanilla farming because of involvement of high cost and high risk of low or loss of production due to certain disease and instability in market price. Whenever the price of vanilla, in international market, is relatively high then fertilizers are usually applied. This also considerably contributes to the instability of production level. The mostly recommended dosage of manure or fertilizers is around 50–100 g NPK with a ratio of 1:2:2/plant/year for samplings that are under 2 years, around 100–200 g NPK with a ratio 1:2:3/plant/year for those who are older than 2 years. These concentrations are normally applied 50% in the beginning and remaining after the end of rainy season. Moreover, it is also observed that high level of potash than nitrogen is believed to increase the stem tolerance against rot disease. Vanilla also requires foliar spray of nutrients, whenever it is necessary, at 5–8 g NPK/L of water at equal ratio. This is generally applied after every 1–2 weeks. The nutrient solution is generally applied in the morning before the sunrise or before evening between 5 and 8 p.m. during high relative humidity. Vanilla farmers, on the other hand, preferably use compost or manure but unfortunately its application ratio is too low around 3–5 kg/plant/year [6].

14.14 Nutrition for Vanilla

Humus is the primary source of nutrition for vanilla plant that results from decomposition of organic matter by bacteria, microorganisms, and worms. It consists of plants and animals' residues *i.e.*, mulch. Using mulch has various advantages. It keeps the humidity of soil maintained. Mulch keeps the soil aerated and does not hinder the growth of roots. Maintained temperature is available and growth of weed is inhibited. Decaying leaf mulch is best for vanilla growth. Mulch should be placed 10–20 cm deep on both sides of the root. New mulch is added 2–3 times a year [22, 26].

14.15 Compost Formation

Developing a composting system is a useful method to fulfil the nutritional requirements of vanilla plant. There are several ways of making compost but the best one is to use pine sawdust and sheep manure. *Vanilla* plants grow on orange tree supports. By mixing 70%-line sawdust and 30% sheep manure, compost could be made. For moisture purpose, 40–65% water is applied to it. The compost should be covered with plastic to protect it from rain. The temperature should not be more than 65 °C. For aeration and decomposition, the compost should be mixed after each 15 days. In 3 months, the compost gets ready for use. At this point it should contain 30% moisture. It is applied one to two times each year, after which the growers irrigate immediately so that maximum absorption of nutrients take place [10, 18].

14.16 Shade Control

In *vainillales,* shade is controlled by pruning of support trees such as *Erythrina* sp. Usually 2–3 times a year. Pruning should be done in rainy season *i.e.*, July to November to prevent disease development in vanilla. During rainy season, shade is around 30–50%. In dry season, the canopy of support trees is dense that provides 70–80% shade. This protects the vanilla plant from burning and early fruit drop. When orange trees act as support trees, young buds are also pruned that obstructs the growth of vanilla [19, 27].

14.17 Weed Control

It is done using hoe or machete. Weeds are also being pulled out by hands in order not to disturb the shallow roots of vanilla. Those weeds that are herbs can be added in mulch. Weeding is done three to four times a year. When vanilla plants are grown in shade houses, volcanic rock or ground limestone is used to prevent the growth of weeds [28, 29].

14.18 Irrigation

No doubt, vanilla needs a moist climate beside repeated rains, but excessive rainfall is responsible for high risk of rot diseases. Conversely, under protracted drought conditions, the plants may be physiologically destroyed and surely the plants may not recover. During tremendously dry conditions, irrigation must be done repeatedly after 4–5 days. The vanilla yield can be significantly increased if mulching is done with mango leaves at 25 kg/plant/year along one water cycle in 4–5 days [30]. Keeping the cultivation area in view, sprinkler/hose irrigation technique can also be used. The use of micro-sprinkler or fog irrigation system is also encouraged to keep the relative humidity moderately high to get high yield. It is mandatory to avoid the standing water around the base of plant or extreme dehydration at anywhere of vine, especially during development of pod [18].

14.19 Flowering

The flowering of vanilla is mainly depended on the size of propagules used during plantation. Inflorescence in vanilla vines appear after 2 years, normally, in the third year of plantation. Waxy green calyx is formed on axillary inflorescences. The diameter of flowers may be 5–10 cm while it may attain 6–8 cm height/length. One main petal forms a lip-shape cover, having two pollinia and stigma, while other two are similar in appearance to the sepals. Rostellum, a specialised structure situated between the stigma and pollinia, successfully prevents auto-pollination. The flowers usually bloom in the drier season of the year and are mostly triggered by low rainfall. There are 20–25 flowers within inflorescence, mostly a single flower opens each day but occasionally up to three flowers in a cluster may bloom together, normally early in the morning. Flowering may continue for about 2 or even 3 months, once a year. It worth noting that each individual flower only lasts for 1 day only. Flower wilts and drops within 1–2 days if pollination does not occur. The fruit reaches its maximum length about 6 weeks after fertilization and ripens 7–9 months after flowering [6, 31, 32].

14.20 Factors Encouraging Flowering

The flowering may be promoted by providing climatic or mechanical stress. The following examples may be adopted for the said purpose:

(i) Water Stress: Drought brings reproductively matureness to vanilla plants. The plant flowers twice a year in Uganda due to two different dry weathers [33].
(ii) Low temperatures: Low temperatures of autumn–winter are the main stress component in Mexico that induces flowering. Moreover, cold temperature, below 10 °C, raises the expectation of better flowering in the following year. the greater the expectation of a good flowering year. The low temperatures damage the apical tip, killing it, that results in breaking of apical dominance of the plant. Loss of apical dominance stimulate the lateral floral buds to develop and bloom.
(iii) Plant management: In various regions, farmers are suggested to implement a series of exercises 2 months prior to flowering:

    (a) Trimming of the apex to break apical dominance, at least 10–15 cm.
    (b) Delayed irrigation to bring stress conditions.
    (c) After opening of about 10% of flowers, use abundant irrigation.
    (d) Thinning of the support up to 75% brightness level [22], but only when sunlight strength is not too high to cause burning (light is very much crucial factor in stimulating floral buds [34].
    (e) Allow the shoots to attain a height upto 1.5 m and direct them towards a horizontal support to enhance flowering and aerial roots.

14.21 Natural Pollination

There are a few countries where vanilla beans are being collected using natural pollination. No doubt flowers pollinated very rarely by this method, accounting for only about 1% successful ratio. The people are still unclear about the natural pollinators of vanilla. Hummingbirds (*Cynniris sp.*), bees (*Melipona beechii*), and bats are considered natural pollinator for a long time. The most popular hypothesis accept that common pollinator is *Euglossa viridissima*, a shiny green orchid bee [35, 36]. Moreover, these bees have been observed visiting the flowers irregularly and their effect on pollination is too small, approximately just 1 fruit per 100 or 1000 flowers [35, 37]. Furthermore, only single fruit is normally observed per raceme with natural pollination. A few orchid insects, namely, individuals of *Eulaema sp.* (jicotes) occasionally visit the flowers of Vanilla sp in northern Veracruz, Mexico. Rarely, they may also pollinate the flowers upto 5% while searching for nectar along the base of the labellum. The mechanism through which above mentioned bees originally pollinate the vanilla flowers is not determined yet [6].

14.22 Hand Pollination

This was introduced by Charles Morren in 1836 and Edmond Albius was the first person to practice it on the island of Reunion in 1841 [25, 38]. In labellum of flower, the part that attaches to and wraps the column, is a tissue that flaps down from the column called the rostellum. The rostellum hangs between the female organ (stigma) and male organ (anther) sac and is considered in result of evolution to prevent self-fertilization. In hand pollination, pollen is manually moved, bypassing the rostellum, from the anther sac to the stigma.

Hand pollination is normally done with a thin, small wooden stick roughly the size and shape of a toothpick. This can be prepared from bone, spines, bamboo, or some other materials. Interestingly, hand pollination was similar one that is still in use today and can be done in the following steps:

(i) Pierce a longitudinal incision in the labellum, using a toothpick or similar tool, on the side opposite of the column to expose the reproductive organs.
(ii) Lift the rostellum, with the same side of the wood stick, and flip vertical so that the anther sac can hang down freely over the stigma lobes.
(iii) Carefully press the anther towards stigma until both join each other and then pull out the toothpick.

14.23 Timing for Hand Pollination

The Pollen of *V. planifolia* remains potent for a period that begins 23 h before sprouting and remains so until 16 hours after closing of flower. Similarly, the stigma is responsive 41 h before flower blooming and ending 17 h after closing of flower [39]. Hand pollination is normally done between 7 a.m. to noon, or whenever it is overcast, for practical reasons but never when the flowers are closed or withered.

It is most important that hand pollination must be performed by experienced hands. An experienced person may pollinate 1000–1500 flowers in a period of 5–7 h (about 4 flowers/min), considering that the plants are within the same vicinity. It is worth noting that the first flowers in the inflorescence that are pollinated yield straighter and longer fruits while the later flowers produce curved and smaller fruits that are less valuable. Moreover, hand pollination is a continuous activity for a period of 3–4 months. Normally, 300–600 days labour work per hectare is needed to complete pollination, keeping the abundance of flowers in mind, effectiveness of the pollinator, their location, and distance among plants.

14.24 Success in Hand Pollination

Keep noticing, fertilized flowers do not segregate, wither too early, and drop from their pedicel. On occasion, the column may separate from the fertilized ovary, but ideally column and petals stay attached to growing fruit because they keep fruit hydrated and prevent gathering of pests or fungal strains.

Unpollinated flowers, pollinated incorrectly, or have been tested by high temperature or rain fall off after 2–3 days of pollination. almost 50% pollinated flowers fall on rainy because its adhesive property is lost due to high humidity [6].

14.25 Flowers to Be Pollinated

Normally, 6–8 flowers per inflorescence must be pollinated to produce at least 4–5 fruits of good quality (pollination is not always 100% successful). There is need of at least 8–5 pollinated flowers per inflorescence to obtain 100–120 fruits per plant. No doubt, these estimations are rough because much depends on the position of flower, vigour of the plants, and environmental conditions as well as the biological traits of selected cultivar.

Moreover, vanilla producers also estimate the number of flowers to be pollinated by observing both market demand and pricing because over-pollination causes abundance of smaller fruits of less economic value. Similarly, it also increases the labour cost of pollination of inflorescence. Over-pollination also causes major variations in crop volume from year to year [34].

14.26 Development of Fruit

Immediately after hand pollination, pollen tubes germinate, begin their growth, germination and finally fertilization of the ovules. The ovary quickly after fertilization begins to enlarge and assume a strong, dark green colour keeping its orientation downward. The maximum diameter and length of fruit is normally gained 45 days after hand pollination. After this, growth of fruit ceases and enters a period of maturation that roughly lasts for 7–8 months.

14.27 Harvesting

Hand shears are used to harvest the entire raceme immediately in order to prevent dehydration. For preventing mechanical damage, the fruits are placed immediately in baskets and plastic crates. Any mechanical damage can lead to infection. Fruits are kept in shady and ventilated areas. Pruning of shoots is essential harvesting. Once cut these shoots do not grow again unless any buds are present. Pruning is done using knife or blade that is disinfected by using 1% bleach solution before use [10, 27, 40].

14.28 Curing

The curing of vanilla is done to produce an aromatically attractive and microbiologically resistant product from dark green pods harvested just before complete ripeness. Before curing it has no special aroma apart from a vague plant odor. The curing techniques are of broad range and may vary from region to region, but generally complete in four separate stages.

The first curing step is devised to stop pod dehiscence (especially in *Vanilla planifolia* G. Jackson, most widely marketed species) by inhibiting the normal metabolic pathway during maturation. "Killing" the fruit is the most common techniques that is used in inducing pod senescence. The technique used for killing the fruit is originated from Réunion. it involves soaking of vanilla pods in hot water. This is known as "scalding". In Mexico, the pods are traditionally exposed to the sun or placed in an oven.

In the second step, the heat stored during the initial step is maintained, if possible, by keeping the pods in closed wooden boxes that are used as heat insulator. This step is called "sweating". Some authors [41] report that during the second step the pods may release a blackish fluid due to drying and sweating of pods [6, 42]. The blackish fluid is released due to excess condensation.

During third step called drying, the products, vanilla pods, are stabilized after drying them.

The fourth step is termed as "conditioning". In this step, the product is packaged, stored and preserved in the heat tight environment to promote "aromatic maturation" overtime [6].

14.29 Curing and Development of Aroma

The initial heat cures and initiate the aroma development in vanilla, that is a ongoing process till the completion of curing. The breakdown of various glycosylated precursors is the most recognized phase in the improvement of the aromatic excellence of vanilla. However, it is worth noting that several aromatic chemicals, such as p-hydroxybenzoic, p-hydroxybenzaldehyde, Acid, vanillin, vanillic acid, etc. are present in green vanilla pods in glucoside type and their breakdown is necessary to make them enable to release fragrance or aroma [6].

Vanillin is the main aromatic constituent of vanilla both quantitatively and qualitatively. It is observed [6, 43] that hydrolysis of its precursor is started majorly during the scalding and sweating processes, that continued in slow drying and even in conditioning stages. A generally similar behaviour was also studied with other glucoside precursors of vanilla aroma [42, 44].

14.30 Pests Control

Insect pests are not seriously harmful for vanilla, but slugs and snails may be challenging if remains unchecked. Larval feeding can be observed on young plants. Sometimes, a butterfly, *Clysia vanillana*, can be found, whose caterpillars damage or engulf the capsules (Table 14.1). Similarly, a few snail species that cause damage to vegetative parts of vanilla are abundant in some regions of Mexico [52].

14.31 Viral Diseases

Viruses are obligatory cellular parasites that infect all the living organisms. They have most likely appeared and evolved in their hosts along the emergence of tree life [53]. Viral diseases are one of the major hurdles in increasing the vanilla production since few decades probably due to diversity in crop fields and intensified cultivation [6]. There are more than 20 virus species that infect orchids [54] and Wilser was the first person to demonstrate vanilla viral infection in French Polynesia. Sometimes, it is much difficult to identify the infected plants because some plants do not exhibit clear symptoms or are asymptomatic [55]. Currently, 10 positive-sense RNA viral species belonging to four families are reported to infect vanilla. Cucumber Mosaic Virus (CMV), Cymbidium Mosaic Virus (CymMV), Odontoglossum Ringspot Virus (ORVV) and Potyviruses are the most common viruses that infect vanilla crop. However, viruses have great ability to evolve faster and the emergence of new viral strains in vanilla fields are likely to occur probably due to speedy environmental mutations caused by climate change [56, 57].

Table 14.1 Comparative account of various pests infecting vanilla crops in various parts of world

Country & pest types		Pests name	Nature of damage
Insect-pests	Fiji, Samoa, Tonga Islands & India	Cicada, *Holotrichia serrata*	Underground stem and roots of the vines [45, 46]
	Madagascar, India	Bug (*Memmia vicina* Sign.)	Bugs damage the crop by sucking the sap from plant parts [47]
		Longicorn beetles (*Hoplia retusa* Klug. & *Enaria malanichtera* Klug.)	Damage the flowers [28, 45]
	Mexico, Madagascar, Comoro, Islands, Tamil Nadu	Coleopteran weevils (*Perissoderes oblongus* Hustache *P. ruficollis* Waterh.)	Shoot, shoot tip, Leaves damage [45, 46, 48]
	Madagascar, Reunion, Mauritius, India	Ashy-grey weevil (*Craytopus punctum* Fabr.)	Leaves, Entry holes in the flowers and destroys the column on severe cases [45, 46]
	Mexico, Puerto Rico	Black weevil (*Diorymerellus* sp.)	Damages the terminal portion of shoots [45, 48]
		Leaf tyer (*Platynota rostrana* Walker)	Infests the vine [45, 48]
		Simplica inarcualis	Caterpillar damages the shoots [45, 48]
		Wolly bear caterpillar (*Ecpantheria il1casia* Cramer.)	Damage the shoots [45, 48]
	Puerto Rico, Seychelles & Reunion	Aphid (*Cerataphis lataniae* Boisd.)	Damages the vegetative buds by sucking sap [49]
	Mexico & Reunion	*Cochylis vanilliana* de Joannis	Damage the young shoots and the rudiment of young flowers after fertilization resulting in drying up or malformed beans [45]
		Plant bug (*Trioza litseae* Giard.)	Buds and flowers are punctured and the affected parts decay, subsequently [48, 50]
	Costa Rica, Tropical Africa, Reunion, Madagascar, St. Helena, Teneriffe & South Europe	*Plusia aurifera* (Hb.), *Agrotis* sp.	Feeds on buds [45, 51]
Non-Insect pests	Puerto Rico	Slug & snails (*Thelidomus lima* Fer. & *Veronicalla kraussii* Fer.)	Damage the crop by removing the outer portions or entire and sections of bud, leaf, shoot and immature fruits during warm weather [25, 48]
		Helix sp.	Attacks and damages all parts of the plant [45]
Avian pest	Tahiti	Green dove	Eats flower buds [45]

14.32 Diseases Control

A key weakness in high vanilla production in many regions of the world is stem and root rot disease mainly caused by *Fusarium oxysporum*. Vanilla crops are badly affected under poor management conditions and does not get high degree resistance against the stem rot. *Fusarium* is a ubiquitous soil borne fungus, also transmitted by water splash that also causes rot in many other species. Under severe environmental conditions, the fungus produces spores *i.e.*, chlamydospores to extend their survival in the soil. In some conditions, the pathogen may incubate within plant body without showing any prominent symptoms. Because of general poor performance, due to adverse environmental conditions, of vanilla crop and influence of various diseases, farmers lose their interest in vanilla production and abandon their farms within 10 years [58]. Some main diseases of vanilla are Anthracnose and Rust (Table 14.2). Anthracnose is caused by fungus *Colletotrichum sp.* It attacks fruits, leaves, stems and flowers of vanilla plant. Dark brown sunken spots are the symptoms of this disease. Resultantly, the infected fruit falls off the plant thus causing a great loss. Sometimes, due to Anthracnose the yield drops by 50%. Fungicides such as copper oxy-chloride and Mancozeb could be used at a concentration of 2 g/L mixed in distilled water or Bordeaux mixture. Spraying is done immediately after nortes begin. Moreover, the plants should be well nourished to prevent disease. Yellow- orange, round pustules on the abaxial surface of leaves are the symptoms of Rust in vanilla plant. Poor ventilation, excessive shade and high precipitation are the causes of Rust. Resultantly, the productive capacity of vanilla plant is greatly reduced. It may entirely defoliate the plant. Copper containing mixtures are used to prevent rust. Infected leaves are removed from vanilla plant and buried. Black rot is another disease caused by *Phytophtora* sp. that kills the plant only in few days. Greenish or blackish coloured watery injuries appear on the infected plant. Damage starts from the apical part of the plant and then spreads in rest of the body. Poor irrigation, high rains and excess shade leads to development of this disease. It causes excessive loss due to fruit loss and rotting [12]. This disease could be prevented by maintaining proper distance between the plants. Usually, a distance of 2 m should be between plants and 2–2.5 m among the rows of plants. Plants should be pruned regularly so that they could receive maximum sunlight. Infected parts of the plants should be burnt or removed. Another problem faced in vanilla cultivation is rotting of recently planted cuttings. This is mostly done by *Fusarium oxysporum*. In this infection, rotting of lower portion takes place which then spreads in the upper portion. A white cottony mycelium is observed to be formed at the point of infection. About 5–50% damage takes place due to this. It could be controlled spraying carbendazim on plants [69].

Table 14.2 Comparative account of various fungi infecting vanilla crops in various parts of world

Sr. No	Casual fungus	Disease name	Occurrence region
1	*Atichia vanillae* (Pat.) V. Hach	Anthramose	Tahiti [27]
2	*Calospora vanillae* Massae		Seychelles, Madagascar, Reunion, Comoro, Tahiti, Antigua, Mauritius, New Granada & Thailand, India (Kerala) & Thailand [59–61]
3	*Colletotrichum gloeosporioides*	Leaf spot	
		Inflorescence rot	
4	*Colletotrichum vanillae* & *Fusarium oxysporum*	Leaf rot	
5	*Fusarium batatis* Wollen var. Vanillae	Root rot	Puerto Rico [22, 25, 47]
6	*Fusarium oxysporum* & *Colletotrichum gloeosporioides*	Shoot tip rot	India (Kerala) [62]
			South India [61]
		Stem rot	Indonesia, Ottawa, Bali, north Sulawesi and North Sumatra, India [61, 63, 64]
7	*Gloeosporium vanilla* (Zimm.) Petch & *Vermicuiaria vanillae* Delaer	Rots	Columbia, Mauritius & Sri Lanka [27]
			Mauritius [27]
8	*Macrophomina vanillae* Averna		Brazil [27]
9	*Marasmus* sp.	Horsehair blight	[61]
10	*Mycoleptodescus indicus*	Black crust	Brazil (Taperoa regions) [65]
11	*Pencillium* sp., *Aspergillus* sp., *Rhizospus* sp.	Storage spoilage	India & French [47, 61]
12	*Phiaiospora vanillae* Zimm		Java [27]
13	*Phytophthora capsici* & *P. parasitica*	Bean rot	Tahiti, Moorea [66, 67]
			South India (Kerala) [66, 67]
		Stem blight	India [66, 67]
14	*Phytophthora paimivova*	Fruit rot	French Polynesia [67]
15	*Phytophthora parasitica* Dast.		Tahaa, Raiatea & Huahine [27]
16	*Sclerotium rolfsii*	Sclerotium rot	India, French Polynesia, China, Uganda [61, 66, 68]

14.33 Recommendations

There are some recommendations, to check and reduce the severity of disease [70]. Choose a site having abundance of organic matter with no waterlogging. Always select healthy plants. Make it sure that the wounds are healed, plant is disinfected. And finally, the gaps between plants must be large enough to avoid denseness. Manage the light and shade as well as soil organic matter by trimming of tree parts. Always plant the vanilla after supporting vegetation have been fully established. Also, remove the diseased vegetation in the area. Rejuvenate the plantation periodically. Moreover, don't damage the roots of vanilla by stepping in the vicinity. Mulch having high contents of lignin may stimulate the growth of *actinomycetes* in the soil [58, 62].

14.34 Annual Production

Among several vanilla species, *Vanilla planifolia* Jacks. ex. Andrews has most of the economic importance as edible crop, *i.e.*, condiment. Most of the developed countries are working on it while major vanilla growing and exporting countries are Republic of Madagascar (1681.62 tons), France (>1000 tons), European union (>1000 tons), Jordan (856.47 tons), United States (637.72 tons), Dominican Republic (355.46 tons), Turkey (318.01 tons) and Indonesia (287.47 tons) (Table 14.3). The gross annual production of vanilla is approximately 8000 tons with a revenue of US$1121.74 million. United States of America (USA) is the main consumer (1535.85 tons) followed by European Union (1327.74 tons), United Kingdom (1014.12 tons), France (854.408 tons) and Germany (514.55 tons). It is the mostly used fragrance in perfume industry. Moreover, it also has very pleasant taste owing to the production of aromatic pod, but the vanilla pod doesn't have characteristics aroma in its green state. This special aroma, due to the presence of a chemical vanillin, develops as soon as the curing process is done, which may vary from country to country. Synthetic vanillin is also being prepared from wood because of very high market demand, although the natural production is superior in fragrance and flavour. Still, the production of synthetic vanillin is assessed at 15,000 tons that is almost two times of natural production [72].

14.35 Chemical Constituents of Vanilla

A limited amount of data is available about the chemical composition of mature green beans of vanilla. Fibers (lignin, cellulose, hemicellulose, etc) make up almost 8% of the fresh weight that is too high because most of vegetables and fruits constitute between 1% and 4%. Sucrose is the most abundant form of carbohydrate and

Table 14.3 International trade of vanilla in 2019 [71]

Sr. No	Country	Trade value (Million dollar)	Sr. no	Country	Trade value (Million dollar)
1	Madagascar	651.03	16	Romania	4.10
2	France	89.03	17	Czechia	4.05
3	Indonesia	81.66	18	United Kingdom	3.75
4	Canada	65.81	19	Belgium	2.39
5	Germany	65.77	20	Denmark	2.10
6	Papua New Guinea	37.99	21	South Africa	2.08
7	Uganda	30.15	22	Switzerland	1.77
8	Netherlands	22.34	23	Austria	1.70
9	United States	20.27	24	Australia	1.64
10	Mauritius	18.81	25	Turkey	1.43
11	Poland	13.22	26	Thailand	1.26
12	India	11.43	27	Italy	1.13
13	United Arab Emirates	10.09	28	Mexico	1.12
14	Comoros	8.06	29	Jordan	1.07
15	French Polynesia	7.69	30	China	0.90

total carbohydrates contents are much lesser than present in most of the fruits. Similarly, the ratio of proteins is much lesser and is comparable with most of the fruits and vegetables that is <1%. Due to high glucosidase activity, it is indicated that vanilla contains high level of aromatic compounds. It contains high level of lipids about 12% of fresh weight. This is too high compared to most of the vegetables and fruits (0.2–0.4%). Most abundant organic acids are malate and citrate around 80% of all organic acids. Similarly, 470 mg and 170 mg of potassium and calcium are present, respectively. These contents are much higher as compared to their contents in fruits and vegetables. *V. planifolia* Jacks. ex. Andrews has exceptionally high contents of glucovanillin around 1.7% of fresh weight. The main constituent is vanillin 4 hydroxy-3- metoxybenzaldehyde that is almost 80–90% among all volatiles. The characteristics flavour and aroma of vanilla is produced by a series of enzymatic reactions during curing process. Vanillic acid, anisic acid, anisaldehyde, anisyl alcohol, vitispiranes, caproic acid, phenol ether, phenols, eugenol, carbonyl compounds, lactones, benzyl ethers, B-complex, 15% fats, 25% carbohydrates, and almost 6% mineral salts like zinc, magnesium, calcium, iron, manganese and potassium are also reported. Moreover, 35–40% of water contents are also present in vanilla [29].

14.36 Vanilla Essence

Vanilla is a chemical extract of beans of orchid species like *Vanilla pompona*, *Vanilla tahitensis* and *Vanilla planifolia* Jacks. ex. Andrews. The flavour of their extracts shows a noticeable difference. Moreover, the aroma and flavour of a same species vary if grown in different regions of world due to climate and agronomy of the region. No doubt, it is a labour-intensive product, but it is also among the most expensive condiment after saffron and cardamon. The pure organic vanilla is rich in many fragrance and flavouring chemicals than synthetic one. Original vanilla is highly pure extract of vanilla seeds/beans with elegant flavour. It is the source of most complex taste in the world due to the presence of more than 250 organic compounds in it. It is worth noting that the original vanilla extract has dark brown shade that is characterised by sweet, fruity flavour. The synthetic vanilla is being popular in market due to shortage of organic vanilla because it is unable to meet the demand of international market. The noticeable achievements are being made to match the international requirement of organic vanilla [19].

14.37 Medicinal Importance

The usual medicinal usage of *Vanilla planifolia* Jacks. ex. Andrews has disappeared with the passage of time irrespective of its economic importance in the world due to its flavour and aromatic characteristics of its fermented pods. According to literature, it is being cultivated since centuries in various countries of the world and once it was utilized to treat more than 12 diseases in sixteenth century. Unfortunately, currently it is only being used in grave fever and abdomen pain. No doubt, it has anti-cancer, anti-cell stress, antioxidant, anti-inflammatory, anti-microbial, anti-nociceptive, anti-depressant, anti-cholesterol properties. Therefore, vanilla has huge potential in modern novel medicine, so its cultivation must be promoted to get maximum benefit from this ancient orchid [73]. Vanillin present in vanilla acts as an active agent in pain relief. It is also responsible for lowering insulin resistance. It is also used to cure lipid dysregulation [74].

14.38 Commercial Importance

The famous and major international companies use vanilla to enhance the flavour of their products *i.e.,* Pepsi, Coca Cola, Smoothie king, Starbucks, Jones Soda company, Cadbury, etc. Moreover, it is also used in men's fragrances *i.e.*, Houbigant, Faberge, New Dance and also in Women's fragrance *i.e.*, Coty, Lauder, Calvin Klein, Vera Wang, etc. [72].

14.39 Trade of Vanilla

Vanilla was ranked at 908th most traded product in 2020 across the world having a total trade of \$937M. No doubt, the exports of Vanilla decreased by −17.4%, from \$1.14B to \$937M, between 2019 and 2020. Moreover, Vanilla represents 0.0056% of total world trade [75].

There are two major forms of vanilla being traded in the international market:

- Bourbon vanilla, cured from *Vanilla planifolia* "Andrews". This is the most popular cultivated from/variety in Madagascar. It is worth noting that *V. planifolia* is the same variety started and being grown in the Mexico but has become synonymous with Madagascar. Bourbon vanilla is the most preferred form of vanilla used in Europe, although vanilla extracts are also available [75].
- Extract-grade vanilla harvested and cured from *Vanilla tahitensis*. It has a weaker vanilla with "fruity, floral, and sweet flavours", which is mostly grown in Papua New Guinea and Indonesia. Extract-grade vanilla is used to produce vanilla extract, which is the main form used in the United States of America [75].

In Europe, vanilla is mostly marketed in the following forms:

Ground or Whole natural vanilla normally used as a spice. Vanilla is most traded in pods (i.e., beans), but vanilla powder can also readily available in the market. Vanilla extract, used as a flavour in the food industry and as a fragrance in the cosmetics industry [75] (Tables 14.4 and 14.5).

Table 14.4 Export of vanilla of five major countries

Sr. No	Name of country	Export amount (Million dollar)	Share in world export of vanilla (%)	Vanilla share in total export of the country (%)
1	Madagascar	617	68	22
2	France	70	7.8	5.9
3	Germany	43	4.85	3.06
4	Indonesia	39	4.41	4.03
5	Canada	31	3.45	3.35

Table 14.5 Import of vanilla of six major countries

Sr. No	Name of country	Import amount (Million dollar)	Share in world import of vanilla (%)	Vanilla share in total import of the country (%)
1	USA	324	36	0.011
2	France	196	21	0.027
3	Germany	96	10.7	–
4	Canada	70	7.83	0.014
5	Netherlands	56	6.26	–
6	Indonesia	15.8	1.76	–

14.40 Conservation of Animals Through Vanilla

A major portion of natural habitats are being converted to agroecosystems due to increasing global demand of food to accommodate crop cultivation. it is worth noting that the agricultural expansion is more prominent in the tropical regions. there are the areas where huge rural communities are solely dependent on farming income for their livelihoods. The conversion of Such agricultural land may have severe impacts on local fauna [76]. Hending et al. (2022) compared the diversity of vertebrate species between natural forest habitat and three types of vanilla plantations maintained under varying management regimes in northeast Madagascar. They used diurnal and nocturnal transect method to survey vertebrate diversity. Natural forest habitat contained the greatest vertebrate species diversity and had proportionally more threatened and endemic species than all vanilla plantation types. They observed a greater number of species and a higher inverse Simpson index in minimally managed vanilla plantations located within or near natural forest compared to intensively managed vanilla plantations. It is worth noting that these findings are important and encouraging for animal conservation and sustainable crop cultivation in Madagascar, and suggest that newly created vanilla plantations, and already existing plantations, should endeavour to follow the more traditional, minimalistic management approach to improve sustainability and promote higher faunal diversity [76].

14.41 Production of Vanilla

There is continuous rise in the harvesting and production of vanilla since 2001. The rise in production is depicted in Table 14.6 (Figs. 14.2 and 14.3) [75].

14.42 Biotechnological Applications in Vanilla

No doubt, biotechnology has a significant role on the pattern of development and quality of life globally [77]. The end of twentieth century observed the bloom in industries due to discoveries made in the field of biological sciences and the progress made during recent years in genetic engineering, molecular biology and plant tissue culture provided a new sight in crop improvement.

In vitro culture is a significant tool of plant biotechnology, which enables us to use the totipotent nature of the cells, a concept introduced by Haberlandt (1902) and first time practically demonstrated by Steward et al. [78]. It may be used to produce disease resistant/free clones, conservation of gene pool, mass cloning of desired genotypes, selection and separation of mutants, boosting of hybrids among sexually

Table 14.6 Vanilla harvesting and production since 2001 [75]

Years	Harvested areas (Hectares)	Total production of Vanilla (Tons)
2000	38,272	3988
2001	39,451	4739
2002	39,867	5309
2003	42,393	4826
2004	42,267	5841
2005	79,455	7499
2006	83,773	9060
2007	85,038	9327
2008	89,643	9302
2009	69,793	9176
2010	86,521	6258
2011	95,252	4602
2012	99,582	8052
2013	96,628	7601
2014	96,402	7080
2015	93,030	7217
2016	93,623	7298
2017	92,840	7180
2018	95,373	7040
2019	91,335	6813
2020	91,585	6881
2021	92,066	6888

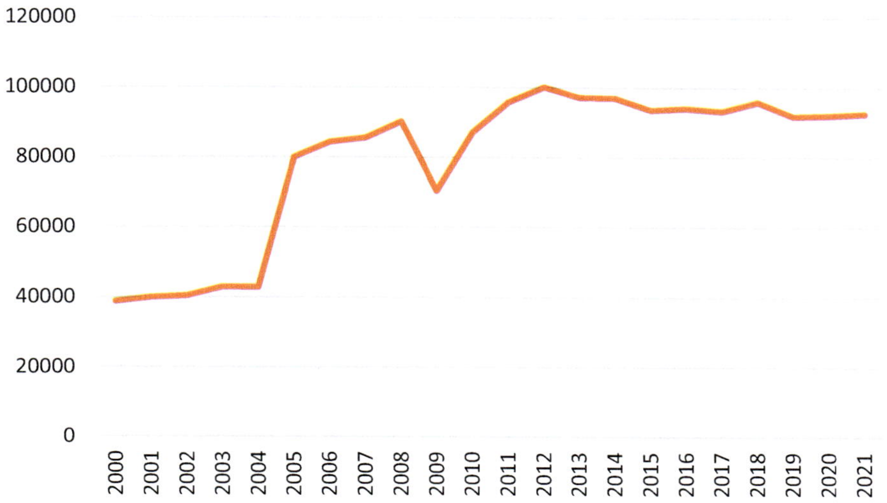

Fig. 14.2 Harvesting of Vanilla since 2001

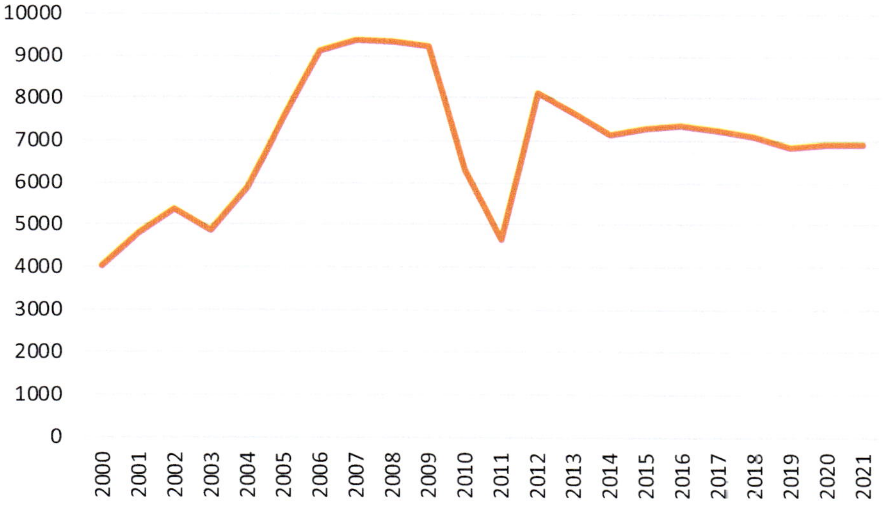

Fig. 14.3 Production of Vanilla since 2001

incompatible taxa by somatic hybridization, introduction of selected traits by genetic engineering, and in the mass production of secondary metabolites using cultured clones, i.e., cells or tissues [15].

14.43 Problems to Be Addressed

As a cash crop, vanilla is playing critical role in the economy of Comoros, Indonesia, Madagascar, and Uganda. Continuous clonal propagation of *V. planifolia* promotes the monoculture, exposing the crop to severe damage [79] because vanilla is being badly affected by many diseases (Table 14.2.) and pests (Table 14.1). Asexual propagation is a major hurdle in introduction of new genetic combination. The flowers are mostly self-pollinated, and this is a serious threat to wild populations of vanilla [80]. So, there is dire need to search for alternative methods to induce genetic variation into vanilla gene pool due to insufficient genetic variability, the continuous threat of destructive diseases that may wipe out vanilla crop, as well as the destruction of its natural habitats. Interspecific hybridization between available vanilla's primary gene pool and secondary gene pool, a close relatives of *V. planifolia*, can be used to broaden the narrow gene pool. This can Improve the quality characteristics of vanilla, *i.e.*, higher vanillin content, improved aroma and taste, larger bean size that may benefit vanilla producers and consumers. Moreover, nonviable hybrid seeds can be used to retrieve and regenerate immature embryos and develop it into a complete plant.

Cell or protoplast culture can be used for creating somatic hybrids and for transfer of desirable traits from closely related sources. Clonal propagation of elite lines, in vitro conservation, and international germplasm exchange is possible using micropropagation techniques. DNA makers such as Amplified Fragment-Length polymorphism (AFLPs), Random Amplified Polymorphic DNA (RAPD), Restriction Fragment Length Polymorphism (RFLP) and biochemical markers (isozyme, protein) can be used for the identification of germplasm and somaclonal variants [6].

14.44 Vanillin Spice

It is one of the most popular and most expensive spice in the world. It is a premium flavouring agent and perfume ingredient, with high global demand. In 2015, the market cap of synthetic vanillin was 60 tons. A vanilla bean produces 2% vanillin, with other components present that enhances its quality. The highest producer of orchid-based vanillin in the world is Madagascar, followed by China, Indonesia and Mexico. These states are also the biggest exporters of seed pods. USA is the biggest importers of vanillin extract and vanilla seed pods. It is expected that the consumption of vanillin produced from vanilla orchids will grow in future. Largest use of vanilla is in ice-creams. It is widely used in yogurt, cookies, brownies and cake. Vanilla essence is easily available in markets [24, 81].

14.45 Myths, Legends, Folklore and Tales About *Vanilla* sp.

14.45.1 Siona Necklet

In 1942, an ethnobotanist R.E. Schultes received a necklet in the department from a Siona village, Puerto Ospina, located along Putumayo River. The necklace was about 65 cm in length while hanging (140 cm total in length). It contained ten bundles of vanilla along with other components. It was worth noting that all vanilla bundles were still quite fragrant after years. This necklace, named "Siona necklace", is probably the only surviving ancient artefact that clearly reflects the usage of vanilla in a traditional perspective. Interestingly, it is also documented from Mesoamerica that in ancient times the vanilla beans were also used in necklaces. Moreover, vanilla was used to wear in an amulet around the neck, like Gujjar families wear in Pakistan, as a medicinal charm as reported in Badianus Manuscript [26]. It was once popular to use pendants of vanilla beans due to religious tradition but unfortunately, this tradition decreased significantly due to cultural erosion. Similarly, with the advancement of scientific tools and the concentration of rare skills regarding the use of vanilla in the hands of a few remaining individuals who are now hesitant to transfer it to someone outside their family/tribe the use of vanilla has become limited [72, 82].

Morning Star Another Folklore about vanilla is that it came into existence by the birth of the princess named morning star, who was killed along with her lover deer, by the priest due to misunderstanding. Prince Zkatan-Oxga known as deer fell in love with princess Tzacopontziza (Morning Star). One day deer abducted morning star and took her to a cave. However, a monster attacked them there. As soon as they ran to save their lives, they were killed by priest as morning star had devoted her life to the temple. Soon after their death, at the place they both were murdered, by the blood of deer and morning star a vine and an orchid came into existence. Both were considered as an offering to Goddess. So, this orchid was vanilla meaning "a nectar of Gods" [83, 84].

References

1. Cameron, K., & Arenas, M. (2003). Vanilloideae. *Genera Orchidacearum, 3*, 281–334.
2. Cameron, K. M. (2004). Utility of plastid psaB gene sequences for investigating intrafamilial relationships within Orchidaceae. *Molecular Phylogenetics and Evolution, 31*(3), 1157–1180.
3. Cameron, K. M. (2006). A comparison and combination of plastid atpB and rbcL gene sequences for inferring phylogenetic relationships within Orchidaceae. *Journal of Systematic Floristic Botany, 22*(1), 447–464.
4. Cameron, K. (2005). *Recent advances in the systematic biology of vanilla and related orchids (Orchidaceae: Subfamily Vanilloideae)*. Paper presented at the Vanilla: First international congress.

5. Cameron, K. M. (2018). Vanilla phylogeny and classification. In *Handbook of vanilla science technology* (pp. 375–390.
6. Odoux, E., & Grisoni, M. (2010). *Vanilla*. CRC Press.
7. News, D. (2009). *Orchids: Exotic plants*. Retrieved from https://www.dawn.com/news/print/883645
8. Ranadive, A. S. (1994). Vanilla – Cultivation, curing, chemistry, technology and commercial products. In *Developments in food science*. Elsevier.
9. Ranadive, A. S. (1992). Vanillin and related flavor compounds in vanilla extracts made from beans of various global origins. *Journal of Agricultural and Food Chemistry, 40*(10), 1922–1924.
10. Medina, J. D. L. C., Jiménes, G. C. R., & García, H. S. (2009). *VANILLA: Post-harvest operations*. Food and Agriculture Organization of the United Nations.
11. Creative Commons Corporation, Integrated Taxonomic Information System (ITIS), an online database. (2022). (Publication no. https://doi.org/10.5066/F7KH0KBK). Retrieved 23-03-2022 https://creativecommons.org/publicdomain/zero/1.0/legalcode, http://www.itis.gov/
12. Curti, D. E. (1989). *Manual para el cultivo de vainilla en la región de Papantla, Veracruz, México* (p. 33). Comisión Nacional de Fruticultura. Mimeografiado.
13. Purseglove, J. W., Brown, E., Green, C., & Robbins, S. (1981). *Spices* (Vol. 2). Longman.
14. Childers, N., Cibes, H., & Medina E. H. (1959). Vanilla-The orchid of commerce. In: C. L. Withner (ed.), The orchids, a scientific survey. The Ronald Press, New York, pp. 477–508.
15. Thorpe, T. A. (1990). The current status of plant tissue culture. In *Developments in crop science* (Vol. 19, pp. 1–33). Elsevier.
16. Uchida, J. Y. (2011). *Farm and forestry production and marketing profile for vanilla* (Vol. 1). Permanent Agriculture Resources.
17. Rahman, K. U., Thaleth, M. K. B., Kutty, G. M., & Subramanian, R. (2019). Pilot scale cultivation and production of Vanilla planifolia in the United Arab Emirates. *Bulgarian Journal of Agricultural Science, 25*(6), 1143–1150.
18. Shriver, J. (2013). *Revitalizing vanilla in Madagascar*.
19. Bhattacharjee, S. K., & Shiva, K. N. (2008). *VANILLA: The world's most flavourful spice orchid of commerce*. Aavishkar Publishers, Distributors.
20. Fay, M. F., & Krauss, S. L. J. (2003). Orchid conservation genetics in the molecular age. In *Orchid conservation, Kota Kinabalu* (pp. 91–112). Natural History Publications.
21. Wong, C., Wong, M., & Grisoni, M. (2003). Culture de la vanilla. *Fiches Techniques/Technical Leaflets, 37*.
22. Ranadive, A. S. (2005). *Vanilla cultivation*. Paper presented at the Vanilla: First international congress.
23. Tyagi, R., Yusuf, A., Jeyaprakash, P., & Dua, P. (2001). Effects of polyamines on in vitro conservation of vanilla planifoiia (Salisb.) Ames. *Indian Journal of Plant Genetic Resources, 14*(2), 300–302.
24. Gantait, S., & Kundu, S. (2017). In vitro biotechnological approaches on Vanilla planifolia Andrews: Advancements and opportunities. *Acta Physiologiae Plantarum, 39*(9), 1–19.
25. Childers, N. F., & Cibes, H. R. (1948). *Vanilla culture in Puerto Rico* (p. 28). Puerto Rico, Kansas Agricultural Experiment Station.
26. Bruman, H. (1948). The culture history of Mexican vanilla. *The Hispanic American Historical Review, 28*(3), 360–376.
27. Correll, D. S. (1953). Vanilla-its botany, history, cultivation and economic importance. *Economic Botany, 7*(4), 291–358.
28. Farooqi, A. A., Sreeramu, B., & Srinivasappa, K. (2005). *Cultivation of spice crops*. Universities Press.
29. Havkin-Frenkel, D., & Belanger, F. C. (2018). *Handbook of vanilla science and technology*. Wiley.
30. Geetha, S. P., Babu, K. N., Rema, J., Ravindran, P. N., & Peter, K. V. (2000). Isolation of protoplasts from cardamom (*Elettaria cardamomum* Maton.) and ginger (*Zingiber officinale* Rosc.). *Journal of Spices and Aromatic Crops, 9*(1), 23–30.

31. Rodolphe, G., Séverine, B., Michel, G., & Pascale, B. J. T (2011). Biodiversity and evolution in the Vanilla genus. In *The dynamical processes of biodiversity-case studies of evolution spatial distribution* (pp. 1–27).
32. Chambers, A. H., Moon, P., Edmond, V., & Bassil, E. (2019). Vanilla cultivation in Southern Florida: HS1348, 11/2019. *EDIS, 2019*(6), 7–7.
33. Hernández, H., & Lubinsky, P. (2010). *Cultivation systems* (pp. 75–95). CRC Press Taylor Francis Group.
34. Hernández Apolinar, M. (1997). *Crecimiento y reproducción de Vanilla planifolia (Orchidaceae) en Usila, Oaxaca*. Master's thesis, Facultad de Biologia, Universidad Nacional Autónoma de México.
35. Soto Arenas, M. (1999). *Conservation of the genetic resources of Vanilla*. Paper presented at the Journal of the Canadian Orchid Congress.
36. Lubinsky, P., Van Dam, M., & Van Dam, A. (2006). Pollination of Vanilla and evolution in Orchidaceae. *Lindleyana, 75*(12), 926–929.
37. Tamayo, R. G. (2006). Hágsater, E., MÁ Soto Arenas, GA Salazar Chávez, R. Jiménez Machorro, MA López Rosas y RL Dressler. 2005. Las orquídeas de México. Instituto Chinoín México, DF 304 pp. *Acta Botanica Mexicana* (75), 101–103.
38. Lecomte, H. (1902). *Le vanillier: sa culture préparation et commerce de la vanille*. C. Naud.
39. Shadakshari, Y., Madaiah, D., Dinesh Kumar, M., Shivakumar, K., & Bhagavantha Goudra, K. (2003). Pollen viability and stigma receptivity in vanilla (Vanilla planifolia Andrews). *Journal of Spices Aromatic Crops, 12*(2), 194–196.
40. Ranadive, A. (1994). Vanilla – Cultivation, curing, chemistry, technology and commercial products. *Developments in Food Science, 34*, 517–577.
41. Bourriquet, G. (1954). Altérations et défauts de la vanille préparée. In *Le vanillier et la vanille dans le monde* (pp. 623–644). Paul Lechevalier.
42. Pérez Silva, A. (2006). *Contribution à l'étude de la genèse des composés d'arôme au cours du procédé mexicain de transformation de la vanille (Vanilla planifolia G. Jackson)*. Montpellier 2.
43. Gatfield, I., Hilmer, J.-M., Weber, B., Hammerschmidt, F., Reiss, I., Poutot, G., et al. (2007). Chemical and biochemical changes occurring during the traditional Madagascan vanilla curing process. *Perfumer Flavorist, 32*, 20–28.
44. Dignum, M. J., Kerler, J., & Verpoorte, R. (2002). Vanilla curing under laboratory conditions. *Food Chemistry, 79*(2), 165–171.
45. Varadarasan, S., Gopakumar, B., & Chandrasekar, S. S. (2003). Pests and their management in vanilla. *Spice India, 16*(6), 30–37.
46. Vanitha, K., Karuppuchamy, P., & Sivasubramanian, P. (2011). Pests of vanilla (Vanilla planifolia Andrews) and their natural enemies in Tamil Nadu, India. *International Journal of Biodiversity and Conservation, 3*(4), 116–120.
47. Bouriquet, G. (1954). *vanillier et la vanille dans le monde* (Paul Lechevalier ed.).
48. Duke, J. A., & duCellier, J. L. (1993). *Handbook of alternative cash crops* (p. 434). CRC Press.
49. Duffels, J. P. (1988). The Cicadas of the Fiji, Samoa and Tonga Islands: Their taxonomy and biogeography (Homoptera, Cicadoidea). *Entomonograph, 10*, 108.
50. Ridley, H. N. (1912). *Spices*. Macmillan and Company Limited.
51. Ocampo, R. A. (1987). *Seminario sobre el cultivo de especias en Costa Rica*. Colegio de Ingenieros Agrónomos.
52. Cameron, K. M., & Carmen, M. M. (2006). Photosystem II gene sequences of psbB and psbC clarify the phylogenetic position of Vanilla (Vanilloideae, Orchidaceae). *Cladistics, 22*(3), 239–248.
53. Forterre, P. (2006). The origin of viruses and their possible roles in major evolutionary transitions. *Virus Research, 117*(1), 5–16.
54. Gibbs, A., & Mackenzie, A. (1997). A primer pair for amplifying part of the genome of all potyvirids by RT-PCR. *Journal of Virological Methods, 63*(1–2), 9–16.
55. Pearson, M., & Cole, J. (1991). Further observations on the effects of Cymbidium mosaic virus and Odontoglossum ringspot virus on the growth of Cymbidium orchids. *Journal of Phytopathology, 131*(3), 193–198.

56. Canto, T., Aranda, M. A., & Fereres, A. (2009). Climate change effects on physiology and population processes of hosts and vectors that influence the spread of hemipteran-borne plant viruses. *Global Change Biology, 15*(8), 1884–1894.
57. Garrett, K. A., Dendy, S. P., Frank, E. E., Rouse, M. N., & Travers, S. E. (2006). Climate change effects on plant disease: Genomes to ecosystems. *The Annual Review of Phytopathology, 44*, 489–509.
58. Varadarasan, S., Ali, M. A., & Chandrasekar, S. (2002). *Vanilla vine weevil, a new insect pest on vanilla (Vanilla planifolia)*. Paper presented at the Proceedings of the 15th Plantation Crops Symposium Placrosym XV, 10–13 December 2002.
59. Divakaran, M., Jayakumar, V. N., Veena, S. S., Vimala, J., Basha, A., Saji, K. V., Babu, K. N., & Peter K. V. (2008). Genetic variations and interrelationships in Vanilla planifolia and few related species as expressed by RAPD polymorphism. *Genetic Resources and Crop Evolution, 55*(3), 459–470.
60. Talubnak, C., & Soytong, K. (2010). Biological control of vanilla anthracnose using *Emericella nidulans*. *Journal of Agricultural Technology, 6*(1), 47–55.
61. Thomas, J., Suseela Bhai, R., & Vijayan, A. K. (2003). Vanilla – Diseases and their management. *Spice India, 16*(6), 19–28.
62. Philip, S. (1980). Wilt of Vanilla planifolia caused by *Fusarium oxysporum* f. sp. vanillae. *Agricultural Research Journal of Kerala, 18*(1), 139–140.
63. Hadisutrisno, B., & Rakhmadiono, S. (1976). *Epidemiology of stem rot disease (caused by pathogenic fungi, Fusarium batatatis Tucker) in Vanilla planifolia Andrews*. Paper presented at the Kongres Nasional Perhimpunan Fitopatologi Indonesia, Balai Penelitian Teh dan Kina.
64. Tombe, M., Nurawan, A., Nazarudin, S. B., Tsuchiya, K., Oniki, M., & Matsumoto, K. (1991). Experiments on the introduction of biological and cultural control of stem rot disease of vanilla. *Industrial Crop Research Journal, 4*(2), 20–26.
65. Bezerra, J. L., & Ram, A. (1986). The black crust of vanilla vine (Vanilla fragrans), caused by Mycoleptodiscus indicus (Moniliales, Hyphomycetes). *Fitopatologia-Brasilerira (Brazil), 11*(3), 717–724.
66. Suseela-Bhai, R., & Thomas, J. (2000). Phytophthora rot–a new disease of vanilla (*Vanilla planifolia* Andrews) in India. *Journal of Spices & Aromatic Crops, 9*(1), 73–75.
67. Tsao, P. H., & Mu, L. (1987). *Phytophthora blight and root rot of vanilla in French Polynesia: occurrence and causal species*. Paper presented at the International Congress of Plant Protection.
68. Qiuping, H. (1995). *Sclerotium rot infection on Vanilla fragrans and its control* (p. 24). Subtropical Plant Research Communication.
69. Grisoni, M., Pearson, M. N., & Farreyrol, K. (2010). Virus diseases of vanilla.
70. Simoes, A. J. G., & Hidalgo, C. A. (2011). *The economic complexity observatory: An analytical tool for understanding the dynamics of economic development*. Paper presented at the Workshops at the twenty-fifth AAAI conference on artificial intelligence.
71. Simoes, A. J. G., & Hidalgo, C. A. (2021). Vanilla. In *The economic complexity observatory: An analytical tool for understanding the dynamics of economic development.* Retrieved from https://oec.world/en/profile/hs92/vanilla
72. Hemming, J. (2009). *Tree of rivers: The story of the Amazon*. Thames & Hudson.
73. Niazi, J., Kaur, N., Sachdeva, R., Bansal, Y., & Gupta, V. J. D. D. T. (2014). Anti-inflammatory and antinociceptive activity of vanillin. *Drug Development and Therapeutics, 5*, 145.
74. Singletary, K. W. (2020). Vanilla: potential health benefits. *Nutrition Today, 55*(4), 186–196.
75. Simoes, A. J. G., & Hidalgo, C. A. (2021). Vanilla. The Economic Complexity Observatory: An Analytical Tool for Understanding the Dynamics of Economic Development. Retrieved from https://oec.world/en/profile/hs92/vanilla
76. Hending, D., Andrianiaina, A., Rakotomalala, Z., & Cotton, S. (2023). Conservation value of vanilla agroecosystems for vertebrate diversity in north-east Madagascar. *Oryx, 57*(1), 118–128.

77. Bhatia, C. (1996). *Biotechnology—Many application areas* (pp. 173–175). Hindu Survey of Indian Agriculture. The Hindu Daily.
78. Steward, F., Mapes, M. O., & Mears, K. (1958). Growth and organized development of cultured cells. II. Organization in cultures grown from freely suspended cells. *American Journal of Botany*, 705–708.
79. Gopinath, C. (1994). *Secret of vanilla—Farmer's notebook on vanilla* (p. 31). Indian Spice Associates.
80. Lubinsky, P. (2003). *Conservation of wild vanilla.* Paper presented at the first international congress on the future of the vanilla business.
81. Khoyratty, S., Verpoorte, R., & Kodja, H. (2020). Vanillin: Biosynthesis, biotechnology, and bioproduction. In J.-M. Merillon & H. Kodja (Eds.), *Orchids phytochemistry, biology and horticulture: Fundamentals and applications* (pp. 1–18). Springer.
82. Abreu-Runkel, R. (2020). *Vanilla: A global history*. Reaktion Books.
83. Teoh, E. S. (2019). The story of vanilla. In *Orchids as aphrodisiac, medicine or food* (pp. 109–130). Springer.
84. Hammond, P., & Lillie, S. (2015). Fruit & veg: V is for vanilla. *Child Care, 12*(12), 14–15.

Chapter 15
Tuberose

Tehreema Iftikhar, Hammad Majeed, Muhammad Waheed, Syeda Shehwar Zahra, Mubashir Niaz, Bushra Bilal, and Muhammad Riaz

15.1 Introduction

Polianthes tuberosa L. also known as tuberose is perennial plant species. Twelve species of tuberose are being cultivated, that reaches 60–120 cm in height. Its leaves are broader at upper end and narrower at posterior region. These plants have grass like light green leaves [1]. The word tuberose is derived from Latin word "Tuberosa" which means swollen. It belongs to the family Asparagaceae. The *Asparagus* family comprises of 153 genera and 2500 species. The family members are more similar in genetics and phylogeny while less similar in morphology. The members of this family include perennial herbs and shrubs with fusiform roots and branch stem. From anatomical point of view, these plants have anomocytic stomata, simple perforated vessels and raphides cells [2]. Tuberoses have tuberous roots for the storage of material during dormant stage. It is planted for its attractive appearance. The flowers are loved by gardeners due to its sweet fragrance, beauty, appearance and good

T. Iftikhar (✉) · S. S. Zahra · B. Bilal
Applied Botany Lab. Department of Botany, Government College University Lahore, Lahore, Pakistan

H. Majeed (✉)
Knowledge Unit of Science, University of Management and Technology, Sialkot, Pakistan

M. Waheed
Applied Botany Lab. Department of Botany, Government College University Lahore, Lahore, Pakistan

Govt. Islamia Graduate College, Civil Lines, Lahore, Pakistan

M. Niaz
Atlas Environmental Laboratories, New York, NY, USA

M. Riaz
Department of Pharmacy, Shaheed Benazir Bhutto University, Sheringal, Pakistan

M. Zia-Ul-Haq et al. (eds.), *Essentials of Medicinal and Aromatic Crops*,
https://doi.org/10.1007/978-3-031-35403-8_15

quality [3]. English name of *Polianthes tuberosa* L. is tuberose. In Spanish it is known as Nardo comun and Rajnigandha and Nishigandha in Hindi. Mexican names of *Polianthes tuberosa* L. are Omicochtl, Amole and Azucena.

15.2 Classification [4]

Kingdom: Plantae
Subkingdom: Viridiplantae
Super-division: Embryophyta
Division: Tracheophyta
Class: Magnoliopsida
Order: Asparagales
Family: Asparagaceae
Genus: *Polianthes* L.
Species: *Polianthes tuberosa* L.

15.3 Crop Description

Origin and Distribution Tuberose first originated in Mexico. It is cosmopolitan in distribution. Tuberose gained popularity in Europe at the end of sixteenth century. Tuberose has white flowers while its cultivars such as Double Pearl Vaib and a few others have red doted flower with brown margins. In home gardens, tuberose is being planted in bed and borders [5]. Tuberose is being cultivated in many tropical and subtropical regions including USA, South Africa, France, North Carolina and India. Worldwide there are different varieties of tuberose available such as Shrinagar, Mexican, Yellow baby, Sensation, Double pearl. Classification of tuberose is based upon rows of corolla. There are different varieties grown by gardener and commercial growers [6].

Plant Morphology Genus *Polianthes* has about 15 species, 3 varieties and few cultivars including tuberose (*Polianthes tuberosa* L.). Among these 15 species, tuberose is the only one being globally cultivated. Tuberose is bulbous plant. Leaves of tuberose are pale-green and narrow with length of 30–40 cm. Their width is 1.2–1.5 cm and are often reddish at base. White waxy textured raceme is present at the top of stem. Flowering stem is about 100 cm long. Roots are mostly adventitious and superficial. The flowers are bracteates having funnel shaped perianth with waxy white and fragrant petals [7]. They are 3–6 cm in length. The stamens are six in number with long anthers and short filaments inserted in tube. These plants are epigynous, having numerous ovules with three locular ovary and fruit is capsule which is rarely formed [8]. The seeds of tuberose are flat (Fig. 15.1).

Fig. 15.1 *Inflorescence Polianthes Tuberosa* (Tuberose)

Agronomy of Tuberose The agronomic factors play vital role in the cultivation of tuberose. Better production of crop depends upon the soil conditions. Tuberoses are mostly cultivated in warm, humid regions with temperature ranging from 16 to 30 °C (Fig. 15.2). It can also be grown in hilly areas at a height of 1200–1500 m. Plants survives best in temperature range of 40 °C, spikes length decreases and quality of flower falls if temperature increases further. Below 13 °C, leaf fall occur and frost damages the cell wall of plant. Tuberose has best growth in sunny condition while shady situation drastically reduces the flower growth. In case of photoperiodism, long day causes vegetative axis growth such as increase in length of stem and root. Flower growth and emergence is affected by light intensity [9].Tuberose requires large amount of water for survival. Plants cannot flower even in minor waterlogged soil. Tuberose can grow in wide range of soil conditions. It requires an optimum level of pH that is 6.5–7.5 with loamy or sandy soil for maximum growth. Soil should have high organic matter for more yield and increased growth. In a pot, where tuberose is planted, it must contain a mixture of leaf mold and compost [10]. Moreover, using a mixture of coconut coir and farmyard manure [FYM] significantly affects the growth of tuberose. Another mixture containing poultry manure and sand is recommendable for tuberose growth as this mixture is rich in the nutrients needed for its growth [11].

15.4 Cultivation Practices

Site Selection Tuberose requires more exposure to sunlight for proper growth. Shady temperature is also needed to some extent. In long term period of shade, the physiology of plant disturbs and its growth rate declines. The soil must have proper

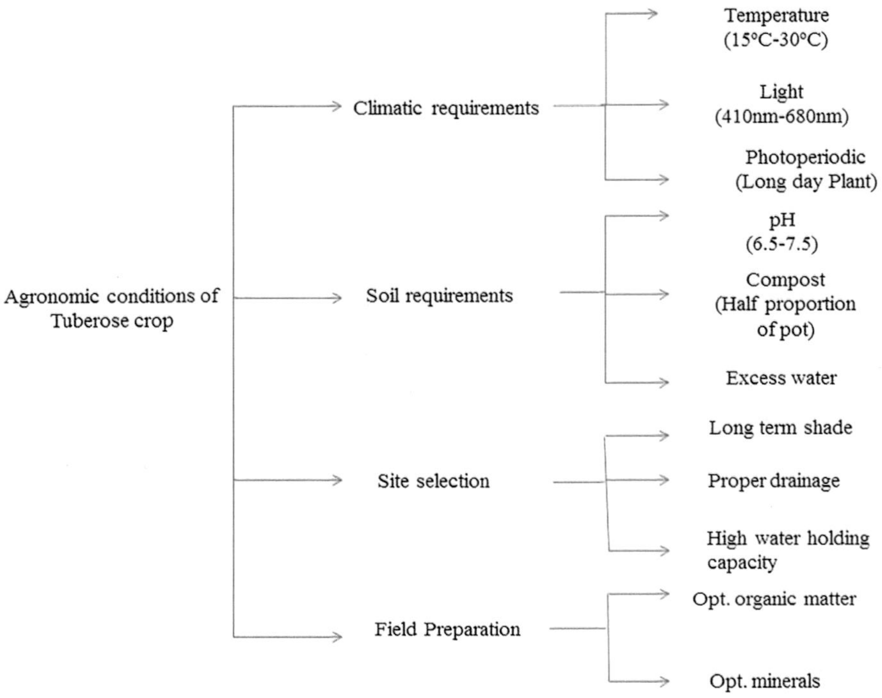

Fig. 15.2 Agronomy practice layout for Tuberose (*Polianthes tuberosa*) [11]

drainage and good water holding capacity. The waterlogged site damages root system, thereby it will decline the absorption rate of water.

Field Preparation The field must be prepared properly for healthy growth of plant. The quantity of water, soil, organic matter and mineral contents is required at optimum level. The temperature should be strictly accurate. Field must be ploughed properly at least twice for proper growth. The ploughed soil must contain manure which should be added at regular interval and left for 15 days [12].

Bulb and Seed Treatment Tuberose seed sowing requires optimum conditions for proper growth. The tuberose seeds are hard to germinate however about 60–80% seed germination is ensured. The bulbs are cleaned properly and soaked in Bavastin solution for 30 min. Treatment of bulbs with Bavistin reduces the chances of fungal infection. The growing months of plants varies in plains and hills: In plains, it is January-March whereas in hills, April-May [13]. The bulbs can be treated with Gibberellic acid [GA, A3, and GA3] and Thiourea to break the dormancy so that they can sprout easily. Storing pre-plant bulb at 8–10 °C for a period of 1 month not only increases plant growth but also enhances the yield of its spikes and flowers. Treating bulbs with GA_3 results not only in early appearance of spikes but also improves its length and number of florets on it [3].

Propagation Tuberose is mostly propagated by bulbs or division of bulbs and rarely by seeds. For 1 ha soil, 10–12 tons of bulbs are needed.

Propagation by Seeds Tuberose can be propagated by seeds under suitable climatic conditions. For this purpose, a nutrient rich growing medium is prepared by mixing molds of leaves, garden soil and vermicompost (a decomposition product created by worms). Temperature and moisture content highly effects the germination of seeds. A temperature of 25 °C is considered as an optimum temperature for seed propagation. Before plantation the beds should be properly dug and sufficient quantity of FYM should be mixed in it. The seedling is sown in rows 10–12 cm apart and 5 cm deep in heavy soil and 2.0 cm in light soil [14].

By Division of Bulbs It is another type of propagation method. Sprouting in this technique depends upon size of bulbs. Large size bulbs have high regeneration capacity. For propagation by bulbs they are cut in 2–3 segments, with each segment containing a part of bulb and a basal plate. After treatment with fungicide, each segment is planted vertically with tips protruded above. New bulbs arise from basal plates. They also have roots. At this time, these newly formed bulbs are transferred to the ground [15].

Dormancy and Sprouting of Bulbs Research has shown the presence of high amount of abscisic acid (ABA) in the bulbs of tuberose during dormancy. However, with the end of dormancy period, abscisic acid level reduced significantly which proved the involvement of ABA in induction and maintenance of dormancy in tuberose bulbs. Likewise a gradual rise was observed with the end of rest period. It was observed that the amount of ABA and phenols varied with different phases of dormancy. Moreover, it was also found that during early stages of sprouting the bulbs contained low level of Indoleacetic acid (IAA) which gradually increased as the sprouting progressed [16].

Role of Gibberellins in Corms Research has proved that gibberellins play a significant role in initiation of flowering in *Polianthes tuberosa* L. High performance liquid chromatography (HPLC), gas chromatography-mass spectrometry (GC-MS) accompanied with bioassay helped in identification and isolation of gibberellins. Four different types of gibberellins namely A_1, A_{19}, A_{20} and A_{53} were quantified in tuberose. Presence of 13-hydroxylation pathway in corms of *Polianthes tuberosa* L. is also reported. Observation showed that with the onset of flowering stage, level of GA_1 and GA_{20} increased whereas a decline was observed in GA_{19} levels. However, GA_{53} level remained constant during three different stages. Presence of GA_1 during early stages of flowering suggests involvement of this hormone in floral initiation and development. Application of all five gibberellins *i.e.* GA_1, GA_3, GA_4, GA_{20} and GA_{32} to the corms at vegetative stage initiated flowering in tuberose. GA_{32} most actively promoted flowering however it doesn't promote shoot elongation [17].

Micro-propagation In order to get nematode free plant material, micro propagation is done. It is an advance technology to get pests free plants [18].

Invitro Propagation Studies have showed that 3% BAP is best for shoot generation in tuberose. Thidiazuron (TDZ), a growth regulator is also found to be effective in shoot generation [19]. For enhancing multiple shooting in *invitro* culture of tuberose, optimized amount of BAP along with NAA and kinetin should be used. Supplementing MS media with different concentrations of NAA induces root formation. Developed mericlones are then transferred to greenhouse for acclimatization. *Invitro* propagation of tuberose is often contaminated by *Meloidogyne incognita* infection. This infection occurs in roots of tuberose culture. Infection begins 2 days post inoculation followed by root gall formation after 6 days at infected place. Cycle of this parasite lasts for 45 days. To obtain nematode free plant material regeneration technique can be used. It will produce disease free planets that can later be used for mass propagation. Indole 3 butyric acid (IBA) can also be used to generate roots.

Water and Irrigation Management The soil should be irrigated properly before sowing. It must be provided with moisture at regular intervals until bulb sprouts. In summer, weekly watering is required while in winter after 15 days the crop should be watered. If proper moisture is not provided, it will destroy all the enzymatic reactions and plant will die.

Sowing Time and Spacing Between Plants Various factors contribute for getting high yield of plants. Climatic conditions, soil requirements and size of bulbs, all have major impact on flower production in tuberose. Both, low and high planting density negatively impacts the production of flowers in tuberose. For high yield, it is advisable to plant bulbs at 30 × 20 cm or 20 × 20 cm or 30 × 30 cm. The depth of plants should be 5.0–7.0 cm. Depth also depends on size of bulbs, large bulbs require more depth and small bulbs do not need much depth. Plantation period of tuberose varies from region to region. It is from January to March in plains and April to May on hills. On a fertile clayey soil, the best month for plantation is June.

Earthing Up and Staking When tuberose plant height reaches up to 20–25 cm, earthing up should be done up to 10–15 cm. Earthing up is a process in which the soil is piled up around the base of plant. For staking, bamboo stick is used. Another method is using iron angles in beds. String or rope could be tied in three rows along the plants for lodging.

Nutrient Management In tuberose cultivation both inorganic and organic fertilizers have significant importance. For production of quality flowers, organic fertilizers must be applied whereas inorganic fertilizers must be used to get proper growth and nourishment. It is advisable to apply high percentage of N and P to get high yield of tuberose leaves and bulbs [20]. However, excess nitrogen can make flower spikes tall and soft which results in high damage from wind and pests. An ideal fertilizer for tuberose cultivation *i.e.,* 20 tons/ha, must contain 120 kg of nitrogen, 150 kg of triple super phosphate and 80 kg of potassium fertilizer [21]. Full dose of P and K should be applied during planting, whereas, half dose of N should be applied at the time of plantation and other half after 30–60 days of sowing [13].

Application of Growth Regulators In vegetative and reproductive growth of tuberose, growth regulators play a vital role by accelerating them. Application of GA3 after 40, 55 and 60 days is beneficial. For early induction of flowering, increasing length of pedicels, high number of florets and improving the quality of flowers, CCC and GA3 should be applied at 5000 ppm and 1000 ppm respectively.

Harvesting Flowering of tuberose starts after 85–100 days of planting. Flowers grows in month of July and bloom throughout the year. Spikes with bud scales are cut with sharp knife at sunrise and sunset. Its remaining parts stays in the ground. The remaining of tuberose bulb gives rise to new plants at regular intervals. The spikes are immersed in water from lower portion for longer life of spikes. Plucking of flowers is done at regular intervals for loose flower arrangement in every morning [22].

Lifting, Curing and Storage of Buds The maturation of bulbs can be traced by yellowing of leaves in February and March. The irrigation is done, and soil is left dry for proper yield in next season. The bulbs are taken off from soil and shaken properly for the removal of soil. It must be kept in mind that soil must be cleaned properly after irrigation. The internodes of stem are linked and then offset are removed thoroughly, some seeds are stocked for next growing season. The bulbs are graded according to size at a temperature from 20 to 30 °C. Artificial temperature increases the growth of bulb in shorter period of time. The size of bulb must be altered at regular intervals. For proper treatment, bulb should be soaked in Bavistin solution for few hours, it reduces infection. The mature and commercial bulbs are ready after 20–30 days of graded arrangement. Longer storage beyond 35 °C induces plant spikes growth. These spikes are of lower cost and improperly grown. The bulb size reduced and lowers the yield [23].

Yield The yield of plants depends upon spikes position, number and variety as well as climatic conditions. Planting distance or spacing has another point of importance. Each part has different yield *i.e.*, bulb 20 tons/ha in 2–3 years, spikes 4–5 lakhs in 1 year and flower yield is 11 tons/ha in 1 year [22].

Post-harvest Activities The tuberose plants are processed through grading, packing, transport and holding solution after harvesting.

Grading The lower spikes are graded according to their length and number of flowers. Pests and disease-free flowers are better to get high yield. Because tuberose cut flowers have large surface, so they immediately loose water and wilt. Thereby, post-harvest it is essential to place tuberose cut flowers immediately in water so that the xylem has direct access to water.

Packing and Transport The long spikes flowers are arranged together and transported to the long distances. The packing is done in flower basket for delivery to longer distance. The flowers are wrapped in paper or cloth. These flowers are used for ornamental purposes. For long distance travelling, these flowers are packed in square boxes. In Mexico, tuberose spikes are packed, while they are still in the ground. For this purpose, a blanket is used on which spikes are arranged in 12 rows in alternate way. Each row contains around 13–15 spikes. This packaging is known as gruesa in Mexican. Later, these packages are transported to retailers via trucks. For local delivery, tuberose delivered in dry form, however, if they are to be exported then tuberoses are placed in pulsing solution and wrapped in newspaper for delivery. Over packing or mishandling can destroy the quality of flowers. Proper packing helps in maintaining the quality of flowers. This not only maintains the freshness of flowers but reduces mechanical damage. Using polyethylene sheets or metallic paper can reduce the damage. Moreover, these sheets also slow down the wilting process. Wrapping also increases the post-harvest life of spikes which reduces in unpacked spikes. It has been observed that using newspaper sheets as wrapping material increases the water uptake by spike thus reducing wilt.

Pulsing or Holding Solution The term pulsing refers to placing freshly harvested flowers in specially formulated solution to increase the storage and vase life of the flowers. Flowers are placed in solution for a brief period of time. A solution of 10% sucrose can increase the post-harvest life of tuberose flowers for 4 days. It also increases the floral opening by 13%. The holding solution of 2% sucrose and 300 ppm of aluminum sulfate is best for preserving plant material in solution [24]. Treating flowers with 20 mM silver thiosulfate for 1 h increases the life up to 7 days. Floral opening also increases by 25% by using silver thiosulfate. Moreover, sucrose solution also reduces the percentage of damaged flowers, and the color and freshness of flowers is preserved (Fig. 15.3).

Requirements for Getting High Flower Production It depends upon various factors including size of bulbs, plantation time and the density of plants planted. In 3–4 months, flowers become ready for harvesting. In first year, 4–5 lacs/ha spikes are produced that are used as cut flowers in homes, whereas, 7–10 tons/ha loose flowers are produced that are used in making flower jewelry, in religious prayers, for decoration and rangoli making etc. However, in the next years the production of both cut flowers and loose flowers increases. In second and third year, 5–6 lacs/ha of cut flowers are produced, whereas, loose flowers production rises to 10–13 tons/ha. At the end of third, year around 25–30 tons of bulbs and bulbets could be harvested [26].

Micronutrients Requirement by Tuberose Tuberose not only requires nitrogen, potassium and phosphate for its growth and development but macronutrients like calcium, magnesium and some micronutrients are equally important in this regard. Micronutrient namely manganese, zinc, iron and boron also effects the vegetative growth of *Polianthes tuberosa.*

Calcium is one of the crucial macronutrient required by tuberose. It provides strength to plant and avoids lodging of spike. Deficiency of calcium in soil results in bud rot of tuberose flower. Depending upon the soil type this calcium deficiency can be cured by applying lime or gypsum to soil. Low level of zinc in tuberose results in stunted growth and other deformities. It has also been observed that zinc deficiency also results in small sized leaves. Significant improvement has been observed in tuberose plant after application of 20 kg of zinc per hectare. It is

Fig. 15.3 Cultivation process of Tuberose [25]

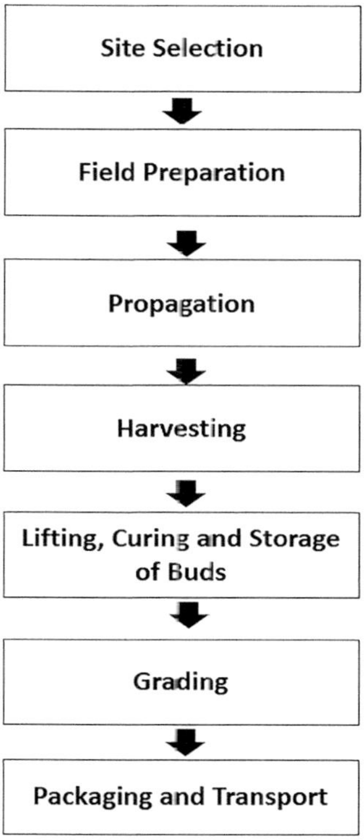

observed that adding zinc as a nutrient improves length of tuberose spike and rachis, number of bulbs per plant and number of florets per spike. Deficiency of boron in tuberose results in stunted growth of spike and plant as a whole accompanied with deformed leaves. Boron deficiency is common in area with acidic soil and high rainfall. Adding borax to soil or spraying it on standing crop can compensate boron deficiency. Foliar application of boron in amount of 100 ppm twice in a month results in increased height of tuberose. Moreover, it also enhanced the yield of spikes in tuberose. Iron is another major nutrient required by tuberose for assimilation of nitrogen and synthesis of chlorophyll. Its deficiency results in interveinal chlorosis in young leaves accompanied with poor yield of flowers. Magnesium deficiency causes chlorosis along the veins in older leaves of tuberose. Applying 0.2% of $MgSO_4$ can correct this magnesium deficiency. In new leaves, interveinal chlorosis is caused by manganese deficiency in soil. Sometimes it also results in yellowish color spikes. Spraying 0.1% of $MnSO_4$ on leaves of tuberose can remove this deficiency. Phosphorous deficiency results in purple colored leaves.

Integrated Management of Nutrients To minimize the use of artificial fertilizers, avoidance of chemical pollution, improvement of soil and for reduction in cost of cultivation, recently approaches for integrated management of nutrients in tuberose have been introduced. For this purpose several researches have been conducted till now. In these researches organic manure has been used as an alternative to inorganic fertilizers.

Karuppaiah (2019) reported that the treatment of tuberose plant with 25 tons ha^{-1} of FYM together with 200 kg ha^{-1} of NPK and 0.50% zinc sulphate and same ratio of borax improved vegetative growth of tuberose. It was also found that using 75% RDF together with farmyard manure (FYM), vermicompost, *Azospirillum* and phosphate-solubilizing bacteria increases number of spikes and bulbs per tuberose plant [28]. Salman et al. (2016) reported that application of gibberellic acid incorporated with organic manure significantly improved the yield of tuberose. Three levels of organic manure was used in this research *i.e.* control, cow dung 30 t ha^{-1} and poultry litter 20 t ha^{-1} with different concentrations of gibberellic acid. Poultry litter was found best in this regard. It significantly improved the yield of tuberose. It was found that 20 t ha^{-1} of poultry litter produced 3,50,000 t ha^{-1} of spikes and 21.72 t ha^{-1} of bulbs. In another research, Wasim et al. (2016) evaluated the impact of bio-fertilizers on growth and flowering of *Polianthes tuberosa* L. Using bulb dip method bio-fertilizers were applied at planting stage which was then followed by shade drying method. Among different bio-fertilizers, Agrobacterium significantly improved vegetative growth of tuberose such as its height, number, length and width of leaves [29]. In a research carried out by Shankar et al. (2010) it was found that using a mixture of vermicompost and PSB significantly enhance the weight of bulbs, number of spikes and their longevity.

Best Practices to Get High Tuberose Yield For getting maximum yield of tuberose bulbs, it is recommendable to treat bulbs with a solution of 0.2% Bavistin before plantation. Fertilizers should be applied in prescribed quantity. Approximately, 1.2–2.5 cm bulbs should be used for propagation. Before sprouting, no weeding should be done. Proper drainage should be done as stagnation would lead to poor quality and yield. When the leaves turn yellow and falls, it is the best time to plant bulbs. At high temperature, the crops grow again from the previously planted bulbs. This process is known as ratooning. It results in production of more spikes and light weight flowers which can be used for oil extraction process.

Enhancing Tuberose Yield by Rhizobacteria and Optimized Growth Medium Experiment has shown that using plant growth promoting rhizobacteria and optimized growing medium significantly improves impacts the quality of tuberose. It was found that among rhizobacteria, *Pseudomonas* and *Bacillus* greatly enhance growth of *Polianthes tuberosa.* Together with it, medium containing two ratios of sand and FYM in one ratio together with *Bacillus* sp. HCA 61 applied in early stages of growth proved to be best for growth of tuberose. At later stages combination of sand and vermicompost accompanied with *Pseudomonas* sp. CP 109 was found effective [30].

Impact of Organic Fertilizers on Tuberose Research has shown that the foliar application of organic fertilizers improves the vegetative and reproductive growth of tuberose. Algaren and Drin are the two organic fertilizers being used. Application of these fertilizers on the leaves results in increase of their chlorophyll content [31].

Maximum yield of flowers hence the essential oil of tuberose can also be obtain by adding nitrogen, phosphorous or potash in different concentrations to the crop. Research showed that using 80, 60 and 40 kg/ha of nitrogen, phosphorus and potassium contributes towards enhanced flowering and production of essential oil. Moreover, bio-fertilizers can also be added to tuberose plants. Bio-fertilizers accompanied with reducing level of inorganic fertilizers greatly improves different vegetative and floral characters of plants. Bio-fertilizers could be applied at the time of planting bulbs. *Azotobacter* treatment significantly enhances growth of vegetative parts. For spike production PSB is more effective. Using sand and leaf compost significantly can improve the flower number and diameter together with their shelf life.

Foliar Application of Silicon to Mitigate Salt Stress It has been found that foliar application of silicon greatly reduces salt stress in tuberose. In Asian markets tuberose comes in late summer and autumn during which the stalks possess only few flowers. During this season flower production in tuberose is affected by a number of abiotic factors. One of these factors is salt stress that significantly hampers plant growth thus significantly affecting its vase life, flower yield and quality. In order to overcome impact of salt stress on tuberose different concentrations of silicon was applied to plant. Salt stress greatly affects the morphological characters of tuberose such as its height, stalk, root and spike length, number of leaves, weight of florets etc. It was observed that foliar application of silicon greatly reduced the quantity of soluble proteins and proline in tuberose. Thus it was concluded that foliar application of silicon can alleviate salt stress in *Polianthes tuberosa* [32, 33].

Cold Storage Research have shown that the cold storage decreases the vase life of inflorescence of cut tuberose. Moreover, wet storage if continued for longer period can have detrimental impact on tuberose flowers. Researchers thereby concluded that storage procedures effects the tuberose quality and not the desiccation itself [34].

Weed Control Weed management is critical for plant growth. The plant should be cleaned from weeds at weekly basis to avoid all hindrances. For chemical control Atrazine 0.1–0.9% is sprayed on crop after sowing bulbs [35].

Pest Control There are number of insects affecting plant such as aphids, thrip and weevil. A solution of 1.75–200 mg/l of Metasystox or Rogor is sprayed on plants to control aphids and thrips while weevils are controlled by the application of Thiodon at 2 mg/l [36].

Bud Border Bud border also known as *Helicoverpa armigera* can be controlled by spray of Thiodon, monocortophos, endosulfan in 0.6%, 0.2%, 0.2% respectively (Table 15.1).

Nematodes It is one of the most noxious pests that harms tuberose. Reniform nematode, root knot nematodes and greasy streak nematodes result in stunted growth leading to excessive loss of tuberose flowering plants. Using neem 2 g/tonne hectare, carbofuran 24 kg/ha and Furadon 2 kg/ha controls nematode infestation [37].

Table 15.1 Deficiency symptoms of essential nutrients in Tuberose [27]

Macronutrients and micronutrients required by Tuberose	Deficiency symptoms
Calcium	Bud Rot in Tuberose flower
Zinc	Small sized leaves and stunted growth
Boron	Deformed leaves and stunted growth of spike
Iron	Interveinal chlorosis in young leaves
Magnesium	Interveinal chlorosis in older leaves
Manganese	Interveinal chlorosis in new leaves and yellowing of spikes
Phosphorous	Purple coloration of leaves
Nitrogen	Reduction in number of spikes and flowers, pale green foliage

Red Spider Mites These are found at lower surface of plants growing in hot and arid conditions. These red or brown insects have high multiplication capacity. They feed on plant sap, resulting in formation of yellow streaks on the leaves and stunted growth of plant. To protect the plant from mites it is recommendable to apply 1.2% Kelthane [3].

Grasshopper It severely damages tuberose plants by feeding on young leaves and flower buds. To control such attack insecticides such as 0.1% Roger or Malathion should be applied at regular intervals.

Aphids Feeding on flower buds, aphids are tiny insects with dark purple or black soft bodies. Damage from aphids could be prevented using 0.1% Malathion or dimethoate sprays at an interval of 10 days [3].

Weevils They inhabit ruined leaves and shoots. Weevils feed on border of tuberose leaves thus creating a notched effect. Their larvae feed upon roots and bulbs of tuberose. Disease caused by this pest can be treated by applying 10% BHC dust in the soil in which tuberose is planted.

Thrips These insects feed on leaves, flowers and stalks. Thrips sucks sap from the flowers of tuberose. Bunchy Top is contagious disease associated with thrips. In this disease the inflorescence of tuberose becomes malformed. Spraying 0.1% Malathion can control this disease.

Disease Control The diseases of tuberose crop can be managed through a number of ways using different techniques.

Flower Bud Rot Flower bud rot is caused by *Erwinia* sp. It results in bud rotting with brown scales and discoloration of peduncle. In such cases uprooting and destruction of diseased plants is done. Spectinomycin is used to control flower bud rot in tuberose.

Stem Rot It is caused by *Sclerotium rolfsi*. Symptoms of this disease are formation of coarse mycelial masses near base of stem. This disease can be controlled by the application of brassicol at concentration of 30 kg per hectare. The stem rot spreading can be controlled by wider spacing of plants and minimizing moisture contents of field in which plants grown.

Alternaria **Leaf Spot** *Alternaria* leaf spot disease is caused by *Alternaria polyantha* which affects the leaf of plants. It is a fungal disease. It can be controlled the spraying by Iprodione and Mancozeb in 0.2% in concentration each for 11 days.

Leaf Blight/Botrytis Bight It is caused by *Botrytis elliptica* who damages plant leaf. It can be controlled by the application of foliar spray containing Greeno and Ammonical copper at 0.5% and 2.0% concentration. This treatment is repeated after every 12 days. Sometime it takes longer period of time for complete removal of disease [38].

Blossom Blight *Fusarium equisetti* is the causative agent of this disease. Symptoms of this disease appear in form of light brown lesions which soon turns black thus making the tissue dry. It also causes infection in flower stalk resulting in its collapse. In humid atmosphere flower tips of tuberose turns brown due to accumulation of mass of brown spores. Spraying tuberose plants with 0.02% of Bavistin can control this disease [39].

Sclerotium Wilt It is caused by *Sclerotium rolfsii*, the fungus mainly affecting the roots. Other names of it are basal, foot and root tuber. This disease appears in form of patches and initially it results in drooping of leaves. At later stages it turns the plant yellow and dry. The target of this fungus is roots of tuberose which later spreads upward through the tuber and collar portion of the stem. Rotting appears in both roots and tubers. A thick cottony layer of fungus develops on the rotten stem and petioles of tuberose plant at the soil level. Adding 0.3% Zineb to soil effectively controls this disease.

Root-Knot Nematode In this disease root swelling and galling takes place. Due to heavy galling, the tuberose plant becomes stunted and its leaves turns yellow at their tips. *Meloidogyne sp,* responsible for this disease can be transferred through gall tubers or soil contaminated with this pathogen.

Greasy Streak *Aphelenchoides besseyi* Christie, a foliar nematode, is a causative agent of this disease. In wet warm weather, a greasy streak forms on leaves of tuberose that rapidly rots its foliage. Early symptoms appear in form of small water-soaked spots along the midrib. Later, these spots become elongated about 2–6 in. together with becoming black and greasy in texture. Size of these spots cause the tuberose leave to bend, wilt and dry. At the margins of these greasy spots nematodes are found in abundance [39]

Rusty Flower This disease is common in Hawaii. During warm wet weather *Thrips hawaiianses* inhabits large number of tuberose flowers and extremely damages their petals. Initially yellowish-brown spots appear on the petals and when symptoms become severe the whole flower turns rusty brown.

Table 15.2 Crop management techniques of Tuberose [41]

Management technique	Factors	Affecting agents	Control
Weed management	Weed growth	Intensive weed growth	Atrazine (0.1–0.9%)
Pest management	Bud borders	*Helicoverpa armigera*	Thiodin (0.6%) Monotrophous (0.2%) Endosulfan (0.2%)
	Nematodes	*Aphelencoids besseyi*	Carbofuran (24 kg) Furan (24 kg)
Disease management	Flower bud rot	*Erwinia* sp.	Streptomycin
	Stem rot	*Sclerotium rolfsi*	Brassicol (30 kg/ha)
	Notched leaves	Weevils	BHC Dust (10%)
	Bunchy Top	Thrips	Malathion (0.1%)
	Alternaria leaf spot	*Alternaria polyantha*	Ipordine (0.2%) Mancozeb (0.2%)
	Leaf blight	*Botrytis elliptica*	Greeno (0.5%) Ammonical copper (0.2%)
	Blossom Blight	*Fusarium equisetti*	Bavistin (0.02%)
	Sclerotium Wilt	*Sclerotium rolfsii*	Zineb (0.3%)
	Root-knot Nematode	*Meloidogyne sp.*	Zineb (0.3%)
	Greasy Streak	*Aphelenchoides besseyi*	Foliar application of carbosulfan (T1)
	Rusty Flower	*Thrips hawaiianses*	Spraying natural pyrethrins
	Peduncle Blight	*Lasiodiplodia theobromae*	Bioformulations containing bacterial and fungal isolates

Peduncle Blight Disease This disease was found to be major limiting factor for tuberose cultivation in India. It is caused by pathogen *Lasiodiplodia theobromae* which causes necrosis of peduncle and rachis. This pathogen attacks pith of peduncle thus causing redness [40] (Table 15.2).

Diversification of Tuberose Tuberose flowers are greatly used for oil extraction as well as ornamental purposes throughout the globe. Due to its diverse applications, tuberose is being planted in Africa, Asia and America. These countries produce large amount of tuberose flower and exports to USA, Europe and Japan. Tuberose is among the oldest plants cultivated in the world. In 1519, Aztecs were fond of growing tuberose as investigated by Spanish from the old world. In 1500s Indians also started growing tuberose. The fragrant flowers and showy petals made it popular. These flowers were famous from Victorian's period. In Europe [EU] the major tuberose flowers producer is Netherland which transport flower to other EU members because of its high demand. Flowering plants industry in Pakistan is at Lahore, Islamabad, Multan, Faisalabad, Quetta, Karachi and Hyderabad. The areas near Lahore such as Sahiwal and Pattoki are considered as floriculture sites for tuberose cultivation [42].

Annual Production of Tuberose The annual production of tuberose varies in different regions of world. It varies in different countries such as in Netherland (55%), Columbia (35%), Ecuador (28%), Belgium (24%), Ethiopia (10%), Italy (8%), Malaysia, Germany and Israel (1%). According to ISPF report, in 2017, annual production was maximum in Asia that was 2,30,000, in Europe 57,000, in Netherland 45,000 tons per hectare production. The demand and production increased till 2020 up to 5,35,000 from annual rate [43].

Chemical Composition of Tuberose The chemical constituents of tuberose are of great importance due to various reasons. Different methods are adopted for enhancing chemical constituents of different varieties of tuberose. The aerial parts of this plant contains cholestane glycoside along with spirostanol saponins. The bulbs of tuberose are reported to have glycosides and a long chain alcohol that is 1- tricosanol. Underground part of tuberose spirostanol saponins and different monosaccharides (Table 15.3).

Flower Preparations Tuberose panicle is placed in water for freshness of flowers. The flowers are collected regularly, and chemical constituents are extracted.

Tuberose Oil Extraction Methods Extraction methods *i.e.*, enfleurage and solvent extraction method are used for extraction of tuberose oil.

Enfleurage For extraction of tuberose oil, both cold and hot enfleurage processes are used. In cold enfleurage, palm wax is heated to 80 °C for 2 h. Later, this melted wax is poured into rectangular shaped glass trays, with each tray containing 100 mL of melted wax. After the solidification of palm wax at room temperature, tuberose flowers are being placed in the wax trays with each wax tray covered with another one. Flowers are being replaced with fresh flowers after every 24 h. Floral scents are being collected using ethanol. The ethanol evaporates leaving behind absolute de enfleurage. In hot enfleurage process, the flowers are being placed in 400 mL palm oil which is warmed at 60 °C. This process is continued for 30 min and cooled down at room temperature. After leaving overnight, the palm oil is again warmed at 60 °C and the previous flowers are replaced with new ones. The flower scent can be extracted using a pomade [44].

Table 15.3 Physiochemical properties of Tuberose oil

Property	Characteristic/value
Appearance	Liquid
Color	Orange to Brown
Odor	Sweet floral fragrance
Relative Density	0.959 kg/m^3
Refractive Index	1.482
Solubility	Insoluble in water
Storage	Stored in cool, dark and dry place
Shelf Life	24 months

Solvent Extraction For the extraction of scent from tuberose flower, petroleum and hexane are used. These are non-polar solvent for oil extraction from tuberose flower. The tuberose extract is collected from concrete samples using ethanol as a solvent. Chemical constituents present in tuberose are Pentacosane, 7-decene-5 olides, Benzyl benzoate [45].

Characteristics of Tuberose Essential Oil Tuberose essential oil is characterized by a number of parameters which are discussed below.

Color of Tuberose Oil The first parameter that is crucial in this regard is the color of tuberose essential oil. Color of essential oil is characterized by the amount of pigment present in it. Freshly extracted oil is reddish, greenish or brownish in color depending upon the species from which it is extracted [46].

Refractive Index The second significant element of immense importance for characterization of tuberose oil is its refractive index. In essential oils refractive index determines their purity. In case of tuberose oil its refractive index is 14,916 at 25 °C. If low quality of tuberose flowers are used for oil extraction then it has low refractive index. Therefore, for oil extraction fresh tuberoses should be used [47]. Moreover, the soil quality, the climate in which tuberose grows, the age in which it is harvested together with extraction method used. The amount of water present in oil also lowers its refractive index. Water present in oil reflects the incoming light which determines its lower refractive index. Therefore the tuberose oil with greater refractive index is of good quality.

Specific Gravity Specific gravity of oil is the ratio between the weight of oil and water in the same volume of water and oil. According to research the approximated specific gravity of essential oil lies in range 0,698–1,188 at 15 °C. The type and components of essential oil determines its specific gravity. Each component has its own specific gravity. Greater the concentration of chemical, greater the value of specific gravity [48]. For the extraction of oil it is advised to select inert solvents as they do not react with the oil components of the tuberose flowers.

Acid Value It is also essential to measure as it determines the quality of oil. It tells about the value of free fatty acids present on the material. Greater the value of fatty acid, higher the concentration of free fatty acid in sample [47].

Ester Value The quantity of hydroxide (Mg KOH) measured in milligram to saponify the ester present in one gram of oil is known as ester value. This value is very crucial as it determines the quality of oil. Higher ester value shows that the oil produces fragrance. Greater the ester value higher the fragrance produced.

Antioxidant Activity Any chemical compound that is capable of contributing one or more electrons for free radicals are known as antioxidants [49]. Consequently the number of free radicals reduces. To determine the antioxidant activity of plant extracts DPHH method is useful. It can be easily performed and gives quick results. This method basically measures the ability of specific compound to seize free radicals. In maceration method solvent is used as oil absorption media. Media used in

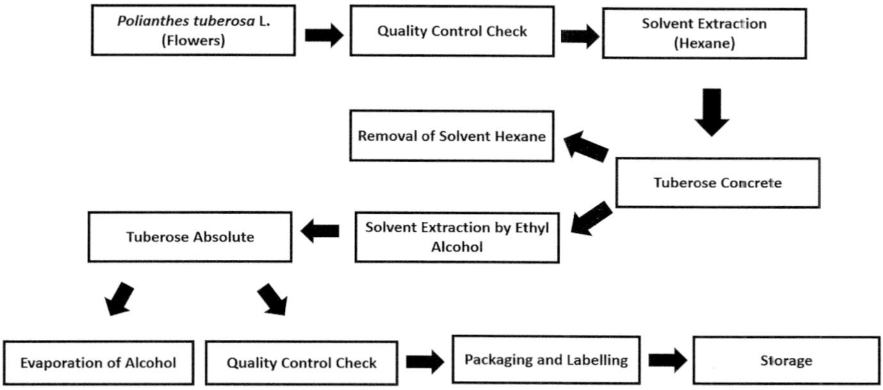

Fig. 15.4 Extraction method of Tuberose absolute

enfleurage method is fat [50]. Tuberose waste extract contains fat that is used as oil absorption media in enfleurage method. The intensity of tuberose compound is lower due to the presence of fat in it. As a result its antioxidant activity is less. The method opted for extraction of natural oxidants depends upon the type of antioxidant to be extracted. Moreover, the solvent used for extraction should match with the polarity of extracted compounds. In contrast to nonpolar solvents or those with high polarity, solvents with medium polarity should be used. Solvents like petroleum and ether are nonpolar in nature thus have low capacity to seize electrons. It is the reason when used in extraction of tuberose oil, it shows low antioxidant activity (Fig. 15.4).

Importance of Tuberose Tuberoses are used for several purposes in different fields that are discussed in detail below.

Commercial Importance

Ceremonial Uses The flower bunches are used in ceremonies because of its showy appearance in different parts of the world. The flowers are used in bouquet and presented in different ceremonies as wedding, birthdays etc.

Cut Flowers Different cut flowers are used in vases for decoration. The flower life can be increased by placing it in a preservative solution for a longer period of time. The lifetime of flower increased through immersion in solution. The flowers are activated from 12 to 15 days in vases during preservation.

As Garland Flowers of tuberose are used as garland for religious purposes in different regions of the world. The people use single flower and wear in some ceremonies [44].

In Perfumes Tuberose is often used to produce floral scents. Mostly feminine fragrances contains tuberose. Its white flowers smell like jasmine and lily. It is often used to give exotic and voluptuous effect to perfumes. Names of some female fra-

Fig. 15.5 Tuberose cut flowers

grances containing tuberose available in market are Gucci Bloom, Café Tuberosa by Atelier Cologne, Zara Tubereuse Noir and Zara Tuberose Summer.

In Cosmetics Tuberose, due to its antioxidant properties and unique chemical composition is frequently used in facial creams, lotions and masks. One such product is Coco Fiji Face and Body Lotion. It contains coconut oil and tuberose extract. Another product containing tuberose extract is Jasmine Tuberose Cream. This extract gives pleasure and beauty to costumers (Fig. 15.5).

As an Adsorbent Stick Waste tuberose sticks have been found effective in absorbing dies from wastewater. It was found that tuberose sticks have capacity to remove cationic dyes present in industrial waste water. It's a cost effective and ecofriendly approach which can be used at commercial level.

Pharmaceutical Importance

Increase Blood Circulation Tuberose has warming effect, therefore, increase blood circulation. This enhance the warm effect in cold season, increase respiratory activity and helps to avoid sexual disorders [51].

Reduce Infectious Symptoms The tuberose oil is used to prevent skin infection as well as hair fall. It prevents the entry of infectious particle in the body. It is also used in treatment of skin infection caused by *Staphylococcus aureus*. Tuberose contains lactate dehydrogenase and IMP dehydrogenase that are active agents against *Staphylococcus aureus* infection. These compounds interrupt the life cycle of *Staphylococcus* thus preventing the infection [52].

Prevention of Anemia Tuberose extract is used as most important chemical constituent. It is used against anemia to prevent blood loss in body [53].

Relaxation and Boost Stamina Tuberose oil is used as relaxant during tired situation. It is also used as stamina booster in hectic days.

Prevention from Insomnia Insomnia is the condition in which person have lack of sleep. It promotes health as the body needs are fulfilled.

Treat Cataract The cataract is the effect on eyes which makes the vision blur. The blurred vision creates problems in eyesight. The tuberose flowers are used for treatment of vision.

Influenza Treatment To cure influenza, mixture of garlic, ginger and tuberose is used. It quickly reduces the effect of cold. It helps to overcome the chilled effect of severe cold.

Prevention from Strep Throat The mixture of tuberose flower and ginger is used against strep throat. It is used thrice a day to prevent the damaging effect of sore throat [54].

Anti-tumor Property Chemical composition of tuberose shows that it contains geraniol. Geraniol possess anti-tumor properties. It has been demonstrated to have significant impact mevalonate and lipid metabolism in the human hepatocarcinoma cell line, Hep G2. Using 100 micromol/L of geraniol has been found to reduce the activity of Hep G2 cells.

Traditional Uses of Different Parts of Tuberose Tuberoses have different body parts as tubers, bulbs, flowers used for several purposes as:

Use of Bulb: Bulbs are used to cure gonorrhea in female. They are used by infants to get relief from rashes. Bulbs are rubbed with turmeric and butter. It is applied on pimple to get rid of infection. Juice of grass is mixed with bulb pieces and applied on lymph at armpit. It helps to prevent from rheumatism in joints [55].

Uses of Tubers: Tubers are used on infections as well as on burning skin. It is also used to treat malaria and healing wounds.

Uses of Flower: Flower extract is used as eye wash during eye cataract treatment. The boiled extract is used for ulcer treatment. It is also for sore throat (Table 15.4).

15.5 Other Species of Tuberose

The genus *Polianthes* comprises of 15 species of tuberose. All these species have different flower colors *i.e.*, white, red and white tinged with red. Among these only *Polianthes tuberosa* L. is grown commercially while all other species are wild [3].

Polianthes polustris Erect stem, oval to oblong bulbs, basal leaves with parallel venation, 3–5 pairs of flowers originating from single bract, short-segmented and curved perianth are the characters of *Polianthes polustris*. The plant grows up to 40 cm in height.

Table 15.4 Uses of *Polianthes tuberosa* (Tuberose)

Importance		
Commercial	Medicinal	Traditional
As garlands and bouquets in ceremonies	Increases blood circulation	Bulbs of tuberose cures gonorrhea
For production of floral scents	Reduces skin infections and prevents hair fall	Tubers prevents infections
As cut flowers for vase decorations	Prevents blood loss in body thus used to treat Anemia	Tuberose flowers extract is used in eye cataract treatment
In cosmetics industry for synthesizing facial creams, masks and lotions	Possess antitumor properties	Used in traditional Chinese medicines
As absorbent stick for absorption of dyes	Effectively controls influenza, prevents insomnia and treats strep throat	

Polianthes montana This specie has white flowers. *Polianthes montana* has oblong bulbs, long and slender stem with each pedicel possessing 12 pairs of flowers. Flowers are small in size with short lobes and rounded shape.

Polianthes durangensis Flowers in this species are small, sessile, arranged in one to six pairs. With age the flowers become purplish in color.

Polianthes graminifolia Deep red color flowers are characteristics of this species.it has close resemblance with *P. geminiflova* in regard to saponaceous roots and its usage as a substitute of soap. Stem is slender with profuse hairs at the bottom and glabrous above. Tuberous bulbs are present. Leaves are narrow, grass like in appearance. Ten to fifteen pairs of flowers are present in this species.

Polianthes geminiflora Originally it is known as *Bravoa geminiflora*. Smooth stems with numerous basal leaves are characteristic feature of this species. Seven or more pairs of light orange-red flowers are present. The Perianth is similar to that of the species *P. graminiflora*.

Polianthes longiflora These are tall plants with 3–5 pairs of flowers. It comprises of perianth that is white in color with light purplish shade.

Polianthes platyphylla In this species the flowers are white tinged with reddish. The florets are arranged in fifth to seven pairs with tube of the floret bent at the base and with short, rounded lobes.

Polianthes pringlei Small bulbs and smooth stem is the characteristic of this species. Four to five pairs of sessile flowers are present. Flowers are white in color which on drying changes to purple.

Pueraria tuberosa This specie is inhabitant of Asian regions. Since early times tubers and leaves of this species are used in Ayurveda and traditional Chinese medi-

Table 15.5 Cultivars of tuberose

Characteristics	Single type	Double type	Semi double type
Rows of petals	Single	Three	Two to three
Color of flowers	White	Creamy	White tinged with pinkish red
Usage	Loose flowers	Floral arrangements	Cut flowers

cine. Extracts of this plant possess anticancer, antitumor, antidiabetic, neuroprotective and wound healing properties.

Polianthes alboaustralis *Polianthes alboaustralis* is a new species native to State of Oaxaca, Mexico. Morphologically this species resembles to *Polianthes nelsonii*. However, it differs in its larger leaves, inflorescence, anthers and long filaments. It also contains extra-floral nectaries.

Cultivars of Tuberose Tuberose has three different types of cultivars

Single Type In this type petals of tuberose are found arranged in single row. Flowers are white in color and have more fragrance than double cultivars. About 0.08 to 0.11 percent of concrete content is found in this type. Flowers of single type are used as loose flowers in making garlands and other types of flower arrangements.

Double Type In this type of cultivar petals are arranged in three rows. It is also known as Pearl Double Tuberose. It possess small creamy flowers with tall stems. Grass like foliage is present. Its flowers are long lasting and are used in flower arrangements. These are easy to grow.

Semi Double Type In this type of cultivar two to three rows of petals are present. They possess straight spikes with white colored flowers tinged with pinkish red. These flowers are also used as cut flowers (Table 15.5).

Poisonous Nature of Tuberose Tuberose is poisonous in nature. Its fragrance has certain toxicity which can cause dizziness in people who smell it. Other symptoms like chest congestion and dyspnea. Therefore it is recommended not to plant tuberose within houses.

15.6 Myths, Legends, Folklore and Tales

Dark Legends of White Flowers The flowers are known as folklore of various nations. The sense created through fragrant white flowers that become legend of tragedy life. The darkest legends of bright life are considered these white flowers.

Love of Fashion Lovers The fashion lovers use these flowers sketched or original to beautify their dresses, boudoirs or jewelry. ***"The Mirror of Fashion"*** is a book written by great philosopher in which he writes about the beauty of flowers. He wrote, "Thousand dollars for flower, what we do with all those flowers in Valley?"

Young Girls Advised to Inhale Aroma of Tuberose Young girls are advised to inhale aroma of tuberose flowers; it prevents them from indulging in sin in some regions. It keeps them in conscious control to avoid any unethical activity, advised by many people.

Long Time Hanging of Tuberose Flowers Tuberose flowers are advised not to hang in bedrooms for longer period of time. Long term inhaling creates difficulty in breathing and result in mental illness, headache and sometimes hallucination.

Amorous Madness Some people believe that the aroma emerged from these flowers produce madness. The chemical constituents of these flowers effect brain cells.

Revengeful Ghost Attack It is believed in Mexico that tuberose is sign of revengeful attack on seaside. A woman in Mexico loved her husband by all hearts. She got ready for her. She arranged tuberoses around herself and put on some flowers on head. She wished to look like bridal to her husband as day passed. One day she realized that her husband does not love her. She got mad and killed her children from him. As he came back, he looked blood on her clothes, flowers around and killed children. He ran away. She lagged behind and killed him on seaside. The fallen tuberose flowers were evidence that murder was committed. Thus, people of region believed it the sign of dreadful ghost.

Dangerous Happiness Tuberose is considered as dangerous happiness. But why? The answer lies in the fact of its strong smell. The flowers create happiness for its beauty and fragrance. But prolonged fragrance creates difficult breathing pattern.

Legends of Tuberose In ancient times, there was a young man named Zhang. He was very good flute maker. Once, he sat near wood trunk and played the flute. A fairy came from moonlight; he invited her to play with flute. The moonlight owner asked her to know the notes of flute. He told her and said to leave marks here in the earth. The fairy moved up and threw a pearl down towards him. The pearl fell on the ground. He didn't search it. After some time, the pearl grew into the flower with beautiful fragrance called "Tuberose".

Queen of the Night Tuberose is known as queen of the night. The facts are its high fragrance. It relies in nature beauty or nature beauty relies upon it. A question always sparked everywhere. The night queen is famous for its brilliant fragrance spread all around.

Rajanigandha -Story of a Flower It is also known as white night flower. In 1974, the Indian film was produced named as Rajnidandha (Yeh hai sacha piyaar) to symbolize the beauty of plant by exploring beauty, sensuality and eroticism (Fig. 15.3).

"Itre-Nawras Shahi" Famous book written by Ibrahim Adil Shah II in seventeenth century to explore the fragrance released by tuberose have strong influence.

References

1. Mazed, H., Pulok, A. I., Rahman, H., Monalesa, N., & Subrato, G. P. (2015). Growth and yield of tuberose as influenced by different levels of manures and fertilizers. *International Journal of Multidisciplinary Research and Development, 2*(4), 555–558.
2. Chase, M. W., Reveal, J. L., & Michael, F. (2009). A subfamilial classification for the expanded asparagalean families Amaryllidaceae, Asparagaceae and Xanthorrhoeaceae. *Botanical Journal of the Linnean Society, 161*(2), 132–136.
3. Mandal, M., Maitra, S., & Mahata, D. (2018). Production technology of tuberose (*Polianthes tuberosa* L.) cultivation. *Journal of Pharmacognosy and Phytochemistry, 7*(6), 2360–2364.
4. Creative Commons Corporation, Integrated Taxonomic Information System (ITIS), an online database. (2022). (Publication no. https://doi.org/10.5066/F7KH0KBK). Retrieved 23-03-2022 https://creativecommons.org/publicdomain/zero/1.0/legalcode, http://www.itis.gov/
5. Barba-Gonzalez, R., Rodríguez-Domínguez, J. M., Castañeda-Saucedo, M. C., Rodríguez, A., Van Tuyl, J. M., & Tapia-Campos, E. (2012). Mexican geophytes I. The genus Polianthes. *Floriculture and Ornamental Biotechnology, 6*(1), 122–130.
6. Safeena, S. A., Thangam, M., & Singh, N. P. (2019). Evaluation of different cultivars of tuberose (*Polianthes tuberosa* L.) under humid agro climatic conditions of Goa. *Journal of Horticultural Sciences, 14*(2), 109–114.
7. Bakhsh, A., Khan, M. A., Ayub, C. M., Shah, M. A., & Afzal, M. (1999). Effect of various chemicals on vase life and quality of cut tuberose flowers. *Pakistan Journal of Biological Sciences, 2*, 914–916.
8. Kumar, M., Chaudhary, V., Kumar, M., Sirohi, U., & Yadav, M. K. (2021). Application of conventional and mutation approaches in genetic improvement of tuberose (Polianthes tuberosa L.): A review on recent development and future perspectives. *International Journal of Agriculture Environment and Biotechnology, 14*(3), 277–297.
9. Mandal, M., Maitra, S., Mahata, D., & Mandal, M. (2018). Production technology of tuberose (*Polianthes tuberosa* L.) cultivation. *Journal of Pharmacognosy and Phytochemistry, 7*(6), 2360–2364.
10. Asif, M., Qasim, M., & Mustafa, G. (2001). Effect of planting dates on growth, flowering and corm characteristics of tuberose (Polianthes tuberosa) cv. Single. *International Journal of Agriculture and Biology, 3*(4), 391–393.
11. Ikram, S., Habib, U., & Khalid, N. (2012). Effect of different potting media combinations on growth and vase life of tuberose (Polianthes tuberosa Linn.). *Pakistan Journal of Agricultural Sciences, 49*(2), 121–125.
12. Afifipour, Z., & Khosh, K. M. (2015). Efficacy of spraying a mixture of amino acids on the physiological and morphological characteristics of tuberose (Polianthes tuberosa L.). *International Journal of Horticultural Science and Technology*.
13. Yadav, L. P., Bose, T. K., & Maity, R. G. (1985). Response of tuberose (*Polianthes tuberosa* L.) to nitrogen and phosphorus fertilization. *Progressive Horticulture, 17*(2), 83–86.
14. Qureshi, A., Gupta, C. Y., & Abrol, A. (2018). Prospects of value addition in jasmine and tuberose. *Research article*.
15. Hussain, S. (1999). *Effect of bulb size and depth of planting on growth and flowering of tuberose (Polianthes tuberosa L.) cv. single*. M. Sc. Thesis. Department of Horticulture, University of Agriculture.
16. Nagar, P. (1995). Changes in abscisic acid, phenols and indoleacetic acid in bulbs of tuberose (Polianthes tuberosa L.) during dormancy and sprouting. *Scientic Horticulturae, 63*(1–2), 77–82.
17. Chang, S. T., Chen, W. S., Koshioka, M., Mander, L. N., Huang, K. L., & Du, B. S. (2001). Gibberellins in relation to flowering in Polianthes tuberosa. *Physiologia Plantarum, 112*(3), 429–432.

18. Vendrame, W. A., Maguire, I., & Moore, K. K. (2005). *Growth of selected bedding plants as affected by different compost percentages*. Paper presented at the proceedings of the Florida State Horticultural Society.
19. Shagufta, N., Farah, A., Saiqa, I., Kiran, S., & Amina, T. (2012). In vitro propagation of tuberose (Polianthes tuberosa). *Journal of Medicinal Plant Research, 6*(24), 4107–4112.
20. Singh, K. P., & Ha, K. S. (1990). Influence of different levels of nitrogen and phosphorus in gladiolus cv. Green Meadow cormels production. *South Indian Horticulture, 38*(4), 208–210.
21. Polara, N. D., Dhola, S. N., Khimani, R. A., Delvadia, D. V., & Viradia, R. R. (2004). Effect of different levels of inorganic fertilizers on flower quality and nutrient content of tuberose. *Current Biology, 2*, 194–197.
22. Sepaskhah, A. R., & Yarami, N. (2009). Interaction effects of irrigation regime and salinity on flower yield and growth of saffron. *The Journal of Horticultural Science Biotechnology, 84*(2), 216–222.
23. Prakash, M., & Ramachandran, K. (2000). Effects of moisture stress and anti-transpirants on leaf chlorophyll, soluble protein and photosynthetic rate in brinjal plants. *Journal of Agronomy and Crop Science, 184*(3), 153–156.
24. El-Shoura, H. A. S., & Arafa, F. F. (2018). Effect of soaking tuberose bulb (*Polianthes tuberosa* L.) in growth regulators and thiorea on growth and flowering. *Middle East Journal of Applied Sciences, 8*(1), 79–82.
25. Safeena, S. A., Thangam, M., Devi, S. P., Desai, A. R., & Singh, N. P. (2015). Ready reckoner on cultivation of tuberose. *Technical Bulletin*.
26. Qureshi, A., Gupta, C. Y., & Abrol, A. (2018). Prospects of value addition in jasmine and tuberose. *Journal of Emerging Technologies and Innovative Research, 5*(8), 263–277.
27. Datta, S., & Gupta, Y. C. (2022). *Floriculture and ornamental plants*. Springer Nature.
28. Karuppaiah, P. (2019). Effect of zinc and boron on growth, yield and quality of tuberose (Polianthes tuberosa L.) cv. Prajwal. *Horticulture International Journal, 3*(1), 7–11.
29. Wasim, M., Gupta, N., Dubey, S., & Mohanty, A. (2014). Effect of inorganic fertilizers in combination with bio-fertilizers on growth and yield of tuberose (Polianthes tuberosa). *Indian Horticulture Journal, 4*(1), 37–42.
30. Yadav, R. (2022). Effect of plant growth promoting Rhizobacteria and growing media on growth parameters of tuberose (Polianthes tuberosa L.) cv. Prajwal.
31. Sood, S., & Nagar, P. K. (2005). Alterations in endogenous polyamines in bulbs of tuberose (Polianthes tuberosa L.) during dormancy. *Scientia Horticulturae, 105*(4), 483–490.
32. Shahzad, S., Ali, S., Ahmad, R., Ercisli, S., & Anjum, M. A. (2022). Foliar application of silicon enhances growth, flower yield, quality and postharvest life of tuberose (Polianthes tuberosa L.) under saline conditions by improving antioxidant defense mechanism. *Silicon, 14*(4), 1511–1518.
33. Trujillo, E. (1968). Diseases of tuberose in Hawaii.
34. Waithaka, K., Reid, M., & Dodge, L. (2001). Cold storage and flower keeping quality of cut tuberose (*Polianthes tuberosa* L.). *The Journal of Horticultural Science and Biotechnology, 76*(3), 271–275.
35. Hamsa, K. R., & Murthy, S. P. (2017). An economic analysis of net returns from major crops in central dry zone of Karnataka under different valuation approaches. *International Journal of Agriculture Environment and Biotechnology, 10*(5), 597–601.
36. Prasad, K., Neha, P., & Gaurav, K. A. (2016). *Harvesting commercial cut flowers*. Division of Post-harvest Technology and Agricultural Research Institute.
37. Zaman, R. (2019). *Investigation and Survey on Diseases of Tuberose (Polianthes tuberosa L.) in Jashore District of Bangladesh*. Department of Plant Pathology, Sher-e-Bangla Agricultural University.
38. Salehi, H., & Bahadoran, M. (2015). Growth and flowering of two tuberose (Polianthes tuberosa L.) cultivars under deficit irrigation by saline water. *Journal of Agricultural Science and Technology, 17*(2), 415–426.
39. Pathak, B., & Khan, M. (2009). Yield loss potential of foliar nematode, Aphelenchoides besseyi in tuberose. *Indian Journal of Nematology, 39*(1), 38.

40. Durgadevi, D., & Sankaralingam, A. (2012). First report of peduncle blight of tuberose caused by Lasiodiplodia theobromae in India. *New Disease Reports, 26*(5), 2044-0588.
41. Sudagar, I. P., Sankarnarayannan, R., & Aruna, P. (2009). Effect of chemicals in increasing the vase life of tuberose cultivars. *Asian Journal of Horticulture, 4*(2), 421–423.
42. Singh, A. K., Sah, R., Sisodia, A., & Pal, A. K. I. (2017). Effect of gamma irradiation on growth, flowering and postharvest characters in tuberose varieties. *International Journal of Current Microbiology and Applied Sciences, 6*(8), 1985–1991.
43. Chaudhary, S. S. (2007). Effect of nitrogen, phosphorus and biofertilizer application on plant growth and bulb production in tuberose. *Haryana Journal of Horticultural Sciences, 36*(1/2), 82–85.
44. Nyambe, M. N., Beukes, D. R., Van De Venter, M., Swanepoel, B., Hlangothi, B. G., & Nyambe, M. N. (2021). Isolation and characterisation of altissimin: A novel cytotoxic flavonoid C-apioglucoside from Drimia altissima (Asparagaceae). *Natural Product Research, 35*(5), 717–725.
45. Naznin, A., Mofazzal Hossain, M., Anju-Man Ara, K., Islam, M., & Mokarroma, N. (2015). Effect of different preservatives on vase life of tuberose. *Journal of Ornamental Plants, 5*(2), 105–113.
46. Waste, F. (2017). Characterization of essential oils from tuberose. *Journal of Advanced Agricultural Technologies, 4*(1).
47. Guenther, E. (2014). *The essential oils-Vol 1: History-origin in plants-production-analysis*. Read Books Ltd..
48. Budiarto, R., & Sholikin, M. M. (2022). Kaffir lime essential oil variation in the last fifty years: A meta-analysis of plant origins, plant parts and extraction methods. *Horticulturae, 8*(12), 1132.
49. Sylvi, D., Syukri, D., Rozi, F., & Sari, D. P. (2022). Physicochemical properties of herbal tea bags from belimbing wuluh leaves (Averrhoa bilimbi) and gotu kola leaves (Centella asiatica) with the addition of cinnamon powder (Cinnamomun burmanii). *EKSAKTA: Berkala Ilmiah Bidang MIPA, 23*(02), 117–132.
50. Cardoso-Daodu, I. M., Fasuyi, F. P., Magaret, O., & Ilomuanya, C. P. A.. Formulation and evaluation of antioxidant cream based on markhamia tomentosa ethanolic extract and citrus sinensis oil.
51. Swarnkar, S., & Katewa, S. S. (2008). Ethnobotanical observation on tuberous plants from tribal area of Rajasthan (India). *Ethnobotanical Leaflets, 2008*(1), 87.
52. Prusty, S., Rout, K., Naik, C., Pandey, M., & Bhattacharyay, D. (2020). Prevention of *Staphylococcus aureus* causing skin disease by Polianthes tuberose. *Journal of Zoology, Uttar Pradesh*, 61–66.
53. Rahmatullah, R. N., Jannat, K., Islam, M., Rahman, T., Jahan, R., & Rahmatullah, M. (2019). A short review of Polianthes tuberosa L. considered a medicinal plant in Bangladesh. *Journal of Medicinal Plants, 7*(1), 01–04.
54. Yashaswini, S., Hegde, R. V., & Venugopal, C. K. (2011). Health and nutrition from ornamentals. *International Journal of Research in Ayurveda Pharmacy, 2*(2), 375–382.
55. Mudassir, S., Ahmad, R., & Anjum, M. A. (2021). Foliar application of micronutrients enhances growth, flowering, minerals absorption and postharvest life of tuberose (*Polianthes tuberosa* L.) in calcareous soil. *Horticultural Science and Technology, 4*, 41–47.

Chapter 16
Thyme

Tehreema Iftikhar, Hammad Majeed, Syeda Shehwar Zahra, Muhammad Waheed, Mubashir Niaz, and Naheed Bano

16.1 Introduction

Commonly known as garden thyme, *Thymus vulgaris* L. is a perennial flowering plant, member of the mint family Lamiaceae. This family comprises of 236 genera and more than 6000 species. The members of family Lamiaceae are quite diverse and cosmopolitan in distribution. Square stems, zygomorphic flowers and distichous phyllotaxis are the distinguishing characteristics of this family. Lamiaceae species are mostly aromatic and are a source of valuable essential oils [1]. Members of this family are greatly used in self-care products, flavoring, fragrance, scents, pesticide, and medicine industries [2]. Globally *Thymus vulgaris* L. has several common names. It is also known as Garden Thyme, French Thyme, rubbed Thyme, red Thyme, folia Thyme depending upon their states in which they grow. In Romania it is called as cimbru and avishan in Persian.

T. Iftikhar (✉) · S. S. Zahra
Applied Botany Lab, Department of Botany, Government College University Lahore, Lahore, Pakistan

H. Majeed (✉)
Knowledge Unit of Science, University of Management and Technology, Sialkot, Pakistan

M. Waheed
Applied Botany Lab, Department of Botany, Government College University Lahore, Lahore, Pakistan

Government Islamia Graduate College, Civil Lines, Lahore, Pakistan

M. Niaz
Atlas Environmental Laboratories, New York, NY, USA

N. Bano
Faculty of Veterinary and Animal Sciences, MNS-University of Agriculture, Multan, Pakistan

M. Zia-Ul-Haq et al. (eds.), *Essentials of Medicinal and Aromatic Crops*,
https://doi.org/10.1007/978-3-031-35403-8_16

16.2 Classification [3]

Kingdom: Plantae
Subkingdom: Tracheobionta
Super-division: Spermatophyta
Division: Magnoliophyta
Class: Magnoliopsida
Subclass: Asteridae
Order: Lamiales
Family: Lamiaceae
Genus: *Thymus* L.
Species: *Thymus vulgaris* L.

16.3 Crop Description

16.3.1 Origin and Distribution

Thyme is a common term used that encompasses all varieties of *Thymus* species found globally. *Thymus* species are native plants of Europe and Asia. They are also indigenous to the Mediterranean region, south-eastern Italy and northwest Africa (Table 16.1). *Thymus zygis* L. is native to Portugal and Spain whereas the s*erpyllum* species is indigenous to all European countries. Spain produces a total of 90% of thyme oil of world trade, In Pakistan, Thyme is being cultivated in different agro-ecological zones of Baluchistan [4].

16.3.2 Plant Morphology

The genus *Thymus is* a member of Nepetoideae subfamily of the Lamiaceae family that consists of approximately 215 species. *Thymus vulgaris* L. is a widely known aromatic herb. Thyme is a tiny perennial shrub that seldom grows 40 cm tall with

Table 16.1 Occurrence of various types of *Thymus* sp. [4]

Common name	Countries	Botanical name
Thyme (Common thyme) (Garden thyme)	Spain, Portugal, France, Germany, Italy, the UK, North Africa, Canada, and the USA	*Thymus vulgaris*
Spanish thyme	Spain, Portugal, Morocco	*Thymus zygis* L.
Wild thyme	Europe and North Africa	*Thymus serpyllum* L.

evergreen groundcover. With growing age, the stem becomes woody. *Thymus vulgaris* L. leaves are extremely small and fleshy, usually 2.5 to 5 cm long with oval to rectangular form. *Thymus vulgaris* L. have greyish-green leaves and purplish or pinkish flowers in early summers. The leaves of Spanish thyme (*T. zygis* L.) are narrower that are found clustered at the nodes. It possess an elongated inflorescence with clusters of whitish flowers spaced at regular intervals. Wild thyme (*T. serpyllum* L.) and large thyme (*T. pulegioides* L.) are entirely different in appearance. Only at the base they are woody. As a whole they are more herbaceous. They have ciliated sub-sessile leaves with flat, linear to elliptical lamina. Both these species of thyme have capitate inflorescence. Thyme prefers a dry, sunny and temperate environment for its growth. Full exposure to the sun is essential for thyme to grow to its full potential. Excessive moisture will lead to disease development in thyme. Thyme has extremely small and round seeds that retain their germination capacity for around 3 years [5].

16.3.3 Agronomy

Thymus grows well in lightweight, well-drained and aerated soil. Soil with a pH ranging from 5.0 to 8.0 is best for thyme cultivation. Coarse and rough soils that would be unsuitable for any other plant, supports thyme growth. Although calcareous, stony and dry soils are preferable for thyme growth yet they can also be grown in heavy wet soils but it reduces their aroma. It should be planted in permeable soil. Different *Thymus* species require different substrata for their growth [6]. Thyme grows well in a sunny, warm and dry climate. Full exposure to sunlight is essential for its best production. Thyme cannot grow under shade. It requires 68–86 °F (20–30 °C) temperature for best growth (Fig. 16.1). Excessive moisture leads to the development of rot disease in the thyme. The annual rainfall in the Mediterranean region, where thyme is mostly cultivated, is 500–1000 mm. They can survive in extreme climatic conditions [7].

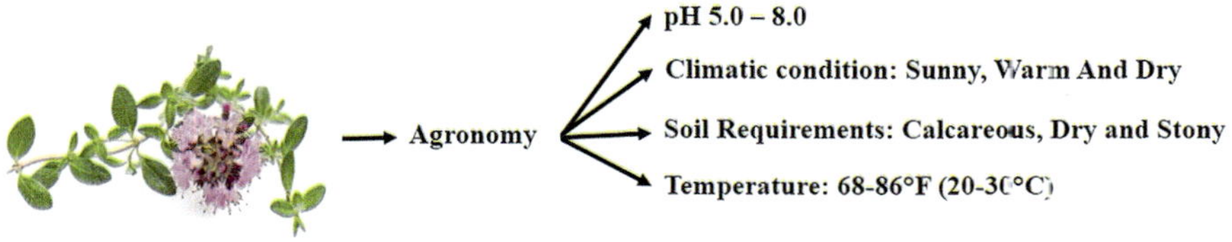

Fig. 16.1 Agronomy of *Thymus vulgaris* L. Source [8]

16.4 Cultivation Practices

16.4.1 Soil Preparation

As discussed earlier, thyme needs shallow soil for its growth. Such soil is not suitable for any other plant. Soil pH should also be in the given range. Before sowing if the pH of soil is below 5.5 it is recommended to add agricultural lime to the soil. Once established thyme can tolerate drought.

16.4.2 Using Vermicompost

Research has shown that using 50% vermicompost enhances the plant yield. It effects both the fresh and dry weight of thyme, its chlorophyll and carotenoids content along with its essential oil production. Increased percentage of vermicompost also suppresses pathogens such as *Fusarium oxysporum* and *Phytophthora infestans*.

16.4.3 Planting

It is advisable to keep a distance of 15–30 cm between the rows of plants, with a row width of 60 cm. In beds, the spacing should be 20–30 cm. In a research conducted it was found that the maximum yield of oil and thymol was obtained when spacing was of 15 cm. Maximum percentage of thymol was found at blooming stage with spacing of 45 cm. Distance of 15 cm between plants and their harvesting at beginning of blooming season was found to be best approach for gaining high yield of dry matter, thyme oil and thymol [9]. Moreover, during the first growing season spacing doesn't create much difference. But in second harvesting season spacing between the plants significantly affects its height, diameter of canopy, fresh and dry weight of its foliage and the number of leaves. Seeds can also be planted in a nursery in plant trays which can later be transplanted when ready. The season for sowing seeds and transplant cuttings is the spring season. Sometimes sowing is carried out before winter. Small seedlings can be more frost resistant than mature plants of thyme. The quality of thyme can be determined by the number of leaves present in the herb, as the glands secreting essential oil are present numerously in leaves and scarcely in stem [10].

16.4.4 Propagation

Propagation of thyme is done via cuttings, seedlings and layering. Dividing the plant at its root can increase thyme plantation. Seeds are sowed in spring at a depth of 6 mm or less. Direct sowing of seeds is often troublesome therefore seedlings are often prepared in greenhouses. Germination takes place in around 2 weeks. However, if seeds are planted in seed trays then it will take 6–8 weeks for germination. Germination rate of thyme seeds is relatively low therefore it is recommendable to use vegetative propagation. It is advisable to use cuttings for homogenous thyme plants. Cuttings of 5 to 6 cm are useful for this purpose. The vegetation period of thyme is 200–210 days. Under suitable conditions, it germinates after 4–5 days. Flowering in thyme takes place after 40–50 days. As soon as seeds are formed, thyme is harvested. *Thymus* plants are capable of surviving 2 or 3 years but in cold regions like Canada and Eastern Europe, the plants freeze thereby in these countries the thyme is treated as 1 year culture [11]. Thyme used for culinary purposes needs to be propagated after every 2–3 years as they become woody with time. Agriculture practices and environmental conditions greatly influence the yield of the *Thymus*.

16.4.5 Fertilization

Over fertilization leads to tall, spindly and weak growth of thyme. Based on annual soil analysis results, fertilizers containing nitrogen, phosphorous, potassium and sulfur should be applied. Applying nitrogen fertilizer after harvesting promotes shoot growth in thyme during the growing season. However excessive nitrogen may affect the quality of oil produced. Globally, the organic cultivation of thyme is a common practice [12].

16.4.6 Irrigation

For irrigation of thyme, drip or trickle irrigation is mostly done. This irrigation system helps to place water directly into the root zone to minimize water loss via evaporation. Another method used is overhead irrigation. It is similar to a lawn sprinkler [13]. This method is used to irrigate a larger area of land. It should be taken care that the soil doesn't dries out.

16.4.7 Harvesting

Thyme is harvested annually for obtaining oil. Generally, thyme is most aromatic during blooming period therefore it is considered as best time for harvesting. However, depending upon the different geographical conditions in which thyme exists their vegetation and blooming period vary. For example, in Spain, the major producer of thyme, there the harvesting takes place from February to August during the blooming season. Whereas in Morocco collection of plant material for oil extraction takes place at the beginning of April. And if we talk about France there harvesting is carried out twice a year, once in May and then again in September. Thus in various parts of the world harvesting period differs. The quality of thyme depends largely upon geographical location and climatic conditions. Moreover, light greatly influences the production of essential oil within thyme. Herbs growing in full sunlight showed highest production of thymol and myrcene within them. Weather conditions also matters a lot. Harvesting of thyme should be carried in morning on sunny days when there is no dew. Dew and moisture deteriorates the herb much faster by destroying its color and beneficial properties. Harvesting begins during late summer when flowering begins. The low growing habitat of thyme makes it mechanical harvesting quite difficult. Therefore, thyme plants are picked by hand. For obtaining a dried yield, the entire plant i.e. 10–15 cm above the ground is harvested [14]. For fresh yield, only tips of branches are being cut. Harvesting should be done carefully because if the crop is left too long or is cut too low, plant death can occur. Heavy cuttings should be avoided. Therefore, sharp tools should be used to prevent the splitting of stems.

16.4.8 Effect of Harvesting Time on Composition and Content of Essential Oil

The most important feature of aromatic herbs is their yield and composition of essential oil. The composition of thyme essential oil varies throughout the year. This has been proved through various researches. For example it was found that in May percentage of *p*-cymene was highest in *Thymus vulgaris* L. while the amount of thymol was lowest at the same time. Later it increased gradually and was stable at the time of vegetation. It was also found that the concentration of α-terpinene was rather low in June and in the beginning of July. However in the end of June and beginning of August its concentration increased considerably when the percentage of thymol reduced. So this shows that harvesting period significantly impacts the composition of essential oil. It is therefore possible to optimize the amount and quality of thyme essential oil by selecting right stage of development for harvesting the plant.

16.4.9 Post-harvest Treatment

Temperature is the most critical factor in maintaining the quality of post-harvest thyme crop. An optimum temperature of 0 °C provides a shelf life of 3–4 weeks. Temperature of 5 °C keeps the plant alive for 2–3 weeks. Thus an optimum period of cooling is required to extend the shelf-life of thyme. In post-harvest treatment firstly, the leaves should be removed from dried stems and then sieved to remove dirt and get the uniform product. There are several methods for drying plant material. The simplest way of drying is to keep thyme for some time and wait for it to dry naturally. However natural drying can be problematic as the leaves of thyme dries instantly whereas its stem takes time to dry. Thyme can also be dried by keeping it in sun. This solar drying maintains the green color of thyme thus making it attractive. To maintain the flavor and green color of thyme, it should be dried by hot air drying below 40 °C. The dried material should be again sieved to remove all dirt. Hot air drying is cost effective approach. However it has a major disadvantage; it results in loss of volatiles. Contrarily, freeze drying although it maintains the color and quality of thyme but it is quite expensive.

16.4.10 Grading

According to International standards, a least of 0.5% of essential oil should be present in thyme leaves, which is equal to 5 ml/kg dried herb. Likewise, ground thyme should contain 0.2% essential oil [15]. For essential oil production, there are six chemotypes present. The most frequent constituent that is found in thyme is thymol and carvacrol. Thymol has powerful disinfectant properties.

16.4.11 Packaging

The main objective of packing is to protect thyme from both external and internal conditions. Generally, dried thyme is stored in cool and dry conditions. Storing dried thyme at 5–7 °C keeps it stable for 12 months. Room temperature decreases the stability of product. Ideally plants should be stored in air-tight boxes so that it doesn't oxidize. Because of their volatile nature, essential oils need to be handled carefully. They are packed in small quantities as well as in bulk quantity. Storage at −20 °C for a period of 12 months is best for maintaining the composition of essential oil. Oil should be packaged carefully because moisture, heat, oxygen and light destroys the oil. The glass of dark color should be used for packaging [16]. Fresh thyme is packaged either in crates or in cellophane packets. Multi-layered paper sacs can also be used for this purpose.

16.4.12 Storage

An essential oil should be stored in cool and dry areas. Airtight glass bottles should be used. Oil should not be exposed to heat and heavy metals (Fig. 16.2).

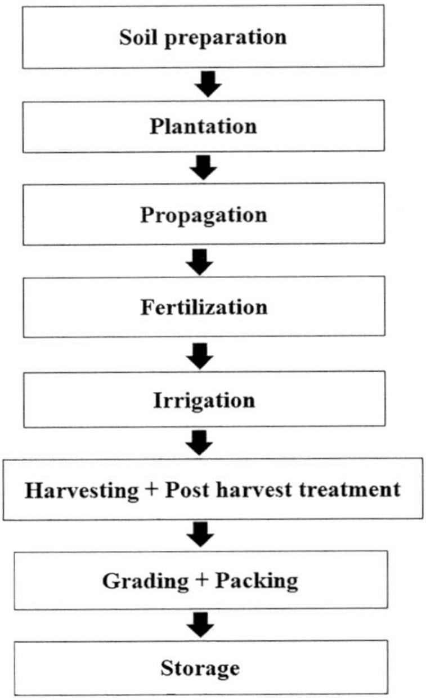

Fig. 16.2 Cultivation process of Thyme

16.4.13 Maintaining the Quality of Thyme Plant

In this regard two factors are important i.e. (a) the quantity of leaves and stem, and (b) the content of essential oil. For getting higher quantity of leaves it is advisable to harvest before blooming. Leaves present high quality. Essential oil yield is dependent upon the climatic conditions of the region. In dry places, care should be taken during drying of leaves and it shouldn't be exposed to sunlight. For this purpose plants are placed in ventilated beds that are supplied with warm air. The temperature is maintained at 30–45 °C. Plant material dries in 2–3 days. Resultantly, 10–12% of moisture is left behind. The cost of thyme cultivated in mountains is expensive because there it is impossible to provide such conditions.

16.4.14 Maintaining the Quality of Thyme Essential Oil

Different processes used for thyme oil extraction affects the quality of oil produced. Therefore, much care should be taken. Fresh thyme has a different taste. Researches showed that *Thymus vulgaris* L. dried by hot air or microwave ovens results in the loss of important constituents. It was found that the percentage of thymol content in thyme essential oil decreased on processing by microwave oven. Different methods used for thyme oil extraction are mentioned below [10].

16.4.15 Steam Distillation

It is widely used method for extraction of thyme oil. This method is advantageous because it results in lesser instability of compounds at higher temperature. Simple steam distillation and Clevenger system are two methods used for this purpose.

16.4.16 Solvent Extraction

For extraction of phenolics from the thyme leaves solvent extraction is performed. In this process soluble compounds are being separated from solid to liquid matrix. It is used to obtain antioxidants from raw material [17].

16.4.17 Supercritical Fluid Extraction

To get good yield of antioxidants this technique is used. For extraction of polyphenols from thyme, this extraction process is used. It has low extraction time and results in minimum degradation. It is the most widely used process [11].

16.4.18 Pressurized Liquid Extraction (PLE)

It is the recent solvent technique used to minimize damage. It uses solvents above normal boiling points and pressure, thus keeping the fluid in liquid state during the whole process. Resultantly, high yields are obtained from this process.

16.4.19 Extraction at Atmospheric Pressure

For this purpose raw material is prepared which is then exposed to the extraction solvent. After extraction, miscella is removed from the extracted material. And finally the solvent is removed. For this process, it is essential to have optimal particle size. The particle size shouldn't be so small that it doesn't allows the penetration of solvent. Penetration of solvent is accomplished in three stages i e. penetration of solvent in dry mass, achieving equilibrium and finally replacing the solute with new solvent. This process can be done by making batches [18].

16.4.20 Weed Control

It is necessary to keep the final product weed-free. For this purpose, organic mulches and hand weeding is used. Mulching is an effective approach for controlling weed growth. It enhances plant growth and its diameter. In case of thyme crop it is advisable to utilize organic mulch. Plastic mulch is also used to prevent weed growth. It is also an effective technique to grow a dense stand of pasture crops followed by fallowing of land before planting thyme crop. Both these steps will significantly reduce the weed.

16.4.21 Pest Control

Thyme contains volatile oils that have insect repellant properties. But still, spiders, mites and whitefly may infest thyme plants. To control pests, a pest management program should be followed. Crops should also be scouted regularly. Thyme is being attacked by aphids as well. *T. vulgaris* L. also gets infected by *Botrytis cinerea* that causes gray mold of thyme. *Rhizoctonia* also causes root rot in thyme [19]. Correct identification of these pests is also essential for their control. It is preferable that instead of chemical control, biological control should be used to eliminate pests. Moreover, plant extracts can also be used to control pests. Certain insect repellant herbs can also be used for pest control.

16.4.22 Disease Control

Wetter environment and imperfect soil drainage result in disease development in thyme plants. *Rhizoctonia* root rot, *Botrytis* blight, *Alternaria* and rust are some of the diseases that develop in thyme if not provided with a suitable environment. To prevent the diseases uncontaminated seeds and plant material should be used. Regular scouting should be followed. The earlier the disease is diagnosed, crops could be saved from large scale damage. Moreover, mulching thyme with limestone gravel layer is a common practice of suppressing unnecessary weeds and retaining moisture [20].

16.4.23 Diseases of Thyme

Thyme is susceptible to a number of plant diseases such as botrytis rot, *Rhizoctonia*, *Alternaria* blight etc. In disease caused by fungus *Alternaria* small round, black or yellow spots form on the leaves surface resulting in lesions and holes. This fungus makes the leaves dry and ultimately leads to plant death. Mostly it spreads by infected seeds. To prevent widespread of this disease infected leaves should be removed from plant. Moreover, proper spacing should be done so that air can properly circulate around the leaves. Thyme plants are also vulnerable to root rot. Treating thyme plants with hydrogen peroxide reduces the incidence of this disease. Grey mold is another pathogen damaging thyme plants. It causes discoloration of thyme leaves. A layer of grey spores is formed on the leaf surface as symptoms of this disease.

Thyme plants are also frequently attacked by aphids and spiders. Aphids use phloem sap as their nutrition thus dehydrating the plant. Using biological methods like releasing larvae of lady bug can reduce the severity of attack. Spraying neem oil is also an effective remedy. Spider mites also badly destroys thyme crop. Their presence can be witnessed by the presence of spider webs between stems. These are reddish brown insects inhabiting hot and dry environment similar to one suitable for growth of thyme. Spider mites are relatives of spiders and mites. They feed on plant juices thus rendering the leaves yellow and eventually their dropping off. Using pesticides is of no use as these insects have developed resistance for the variety of them. Another pathogen *Colletotrichum destructivum* causes Anthracnose. In this disease chlorotic lesions form on the surface of leaves which gradually extends towards the stem. People cultivating thyme in warm and humid climate should beware of this disease.

16.4.24 Annual Production

In 2020, 191.45 thousand metric tons of Thyme oil was produced globally. North America, Latin America. The Middle East and Africa, Europe and the Asia Pacific region are chief contributors of thyme oil. Red and white thyme essential oil is available in the market. It is expected that in 2027, the thyme oil market will grow by 7.8%. In Pakistan, thyme plants are grown in Baluchistan, as the environment there is feasible for its growth.

16.4.25 Major Producers of Thyme Oil

Since sixteenth century thyme oil is in use. However, the exact record of its extraction dates back to 1930s. For many years 5–10 tons of oil used to be extracted. This essential oil is of immense importance because of its rich constituents and aromatic properties. Different species of thyme produces oil rich in different compounds. For example *T. serpyllum* produces thyme oil rich in thymol, same is the case with *Thymus zygis* L. carvacrol rich thyme oil can be obtained from *T. capitates*. In beginning of twentieth century, thyme oil was extracted from *Thymus vulgaris* L. in Germany. In Spain, thyme oil is being produced by *Thymus zygis* L. [21].

16.4.26 Thyme Market

United States is the biggest market of essential oils followed by Japan and Europe. Europe is center for essential oils production, with world's seven largest oil producing firms present there. In US, soft drink manufacturers are biggest consumers of essential oils. Only world's 10% essential is used by Japan. These essential oil are also utilized in perfume and pharmaceutical industries. As France is biggest manufacturer of perfumes and Switzerland leads in pharmaceutical industries, both utilize big quantity of essential oils. For essential oil production, most countries import dried thyme from Morocco and Spain, the biggest producers of thyme [22].

16.4.27 Chemical Composition of Thyme Essential Oil

Oils are complex mixtures comprising of around 30–60 components mixed in different concentrations (Table 16.2). The property of oils is due to various components that are part of it. The essential oil of thyme is responsible for its spicy aroma. Like all other members of family Lamiaceae, the essential oil of thyme is stored in its glandular peltate trichomes that are present at both sides of the leaf. The gland head consists of 8–16 secretory cells. The oil is secreted into the sub-cuticular space by the secretory cells. Rupturing of cuticle by rubbing or grinding releases the volatile oil into the air thus spreading the spicy aroma of thyme. Around 1–2.5% essential oil is present in dried plant material.

Some of the major constituents of thyme essential oils are Thymol, Carvacrol, Linalool, Apigenin, Eugenol, p-cymene, γ-terpinene and Rosmarinic acid. Among the above mentioned 6 chemotypes, thymol is a predominant substance. When dried, it comprises 2.5% of essential oil [23]. The primary constituents of this oil are thymol, carvacrol, p-cymene, γ-terpinene, linalool, β-myrcene, terpinene-4-ol [Fig. 16.3]. Thymol is a white crystalline aromatic monoterpenoid phenol and major compound found in thyme essential oil. It is a strong antiseptic and possess antibacterial and antifungal properties. Carvacrol, another monoterpenoid and an isomer of thymol is reported to have antithrombic, anti-inflammatory and antimicrobial activity. It constitutes 1–5% of thyme oil. Moreover, p-cymene (15–20%) and γ-terpinene

Table 16.2 Chemical constituents of thyme essential oil [26]

Chemical constituents	Biological activities
Thymol	Antiseptic, antibacterial and antioxidant in nature
Carvacrol	Anti-thrombotic, anti-inflammatory antimicrobial
Linalool	Anti-viral, antioxidant and analgesic in nature
Eugenol	Neuro-protective and anti-anaphylactic
Apigenin	Anti-cancerous, antiviral and anti-inflammatory
Rosmarinic acid	Anti-mutagenic and anti-allergic

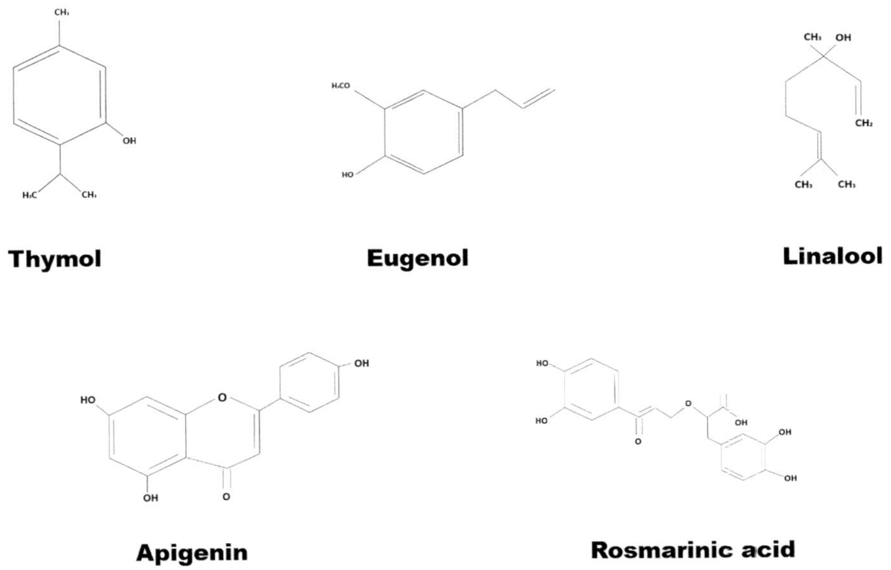

Fig. 16.3 Various chemical structures in *Thymus* sp.

(5–10%) are also present. These two are the precursor thymol biogenetic pathway. Eugenol present in *Thymus* essential oil is popular for its analgesic effects. It also possess neuro-protective and anti-anaphylactic properties. Linalool a major constituent of thyme EO and several other essential oils also possess a number of medically important properties. Rosmarinic acid and apigenin both have anti-mutagenic properties thus attacking cancerous cells. Few partly glycosidic compounds are also present.

Thymus vulgaris L. is chemically polymorphous species. Today in France seven genetically different chemotypes of *Thymus vulgaris* L. are present. They are named based upon the dominant monoterpene present in their essential oil. In Spain only two chemotypes are found namely cineole type and linalool type. However, among all the types thymol chemotype is commercially used. It contains 56.53% of oxygenated monoterpenes and a relatively low concentration of monoterpene hydrocarbons (28.69%), sesquiterpenes about 5.04% and 1.84% of oxygenated sesquiterpenes. Monoterpenes are responsible for the typical spicy aroma of thyme. The major component of *T. vulgaris* L. is thymol about 51.34% while the concentration of other components is less than 19%. According to European Pharmacopoeia, *Thymus vulgaris* L. contains 12 mL/kg of oil in it. In Spain where *Thymus zygis* is the major source of essential oil. Its chemical composition is similar to that of *T. vulgaris* but it contains high percentage of thymol. The dried herb of *T. serpyllum* contains 0.2–0.6% of essential oil. For production at commercial level plant material rich in phenolic content is needed. The percentage and composition of essential oils in *Thymus* species depends upon their genetic makeup, their surrounding environment together with their agronomic conditions and storage conditions [24].

16.4.28 Flavonoid Distribution in Thyme

Aglycones have been widely found in *Thymus* species. In *Thymus* taxa none isoflavonoids have been discovered so far. The two flavonols found in *Thymus* genus are Kaempferol and quercetin. However, flavonones are scarcely found. Anthocyanins are rarely found in this genus [25]. In *Thymus* species 16 flavonoid glycosides have been reported till date. Flavones apigenin and luteolin are most important flavonoids found in thyme. Till now 9 phenolic compounds are found to be the part of genus *Thymus*.

16.4.29 Biphenyl Compounds

Due to their antioxidant properties and deodorant effects biphenyl compounds have gained significant importance. From thyme leaves different biphenyl compounds have been isolated so far. Spectroscopic methods were used for identification of these compounds.

16.4.30 Biological Activities of the Chemical Compounds Present in T. vulgaris L.

Commonly known as a yellow oil, *Thymus vulgaris* L. oil has been found to contain 47 chemical constituents. Thyme oil has been listed among the top ten essential oils in the world [26]. However, the percentage of thymol is highest in *Thymus vulgaris* L. The percentage of these chemical compounds can vary due to the climatic conditions of the geographical region in which the plant grows. The quantity of the oil may vary because of various factors *i.e.,* plant variety and its health, plant age at the time of its collection, the methods used for its drying and later for oil extraction. Research showed that the antimicrobial activity of *Thymus* oil increases with an increase in concentration. Moreover, the antifungal activity of thyme oil is also concentration-dependent. Apart from this, thymol also has antiseptic and antioxidant properties. This antioxidant activity is due high phenolic compounds in *Thymus vulgaris* L. Compounds like carvacrol also shows anti- swelling and acetylcholinesterase inhibitory activities. The antiviral effect of linalool compound has also been reported in thyme oil. Eugenol and apigenin present in thyme oil results in their antiviral and anti-anaphylactic activities. Rosmarinic acid present in oil has anti-allergic, anti-mutagenic and astringent activities.

16.4.31 Nutritional Value of Thyme

Thyme is very rich in nutrients. These nutrients prevent various diseases and are healthy for the body. Vitamins minerals and phytonutrients that are vital for good health are found in *Thymus vulgaris* L. *Thymus* also comprises of different percentages of sodium, potassium, calcium, iron, manganese, zinc, riboflavin, pyridoxine and carotene-ß [28].

16.4.32 Vitamins

Thyme is rich in vitamin A and vitamin C [29]. Both vitamins have their nutritional importance. Vitamin A is an antioxidant that prevents night blindness, helps in keeping mucus membranes healthy and is also essential to keep skin healthy. Vitamin C helps in keeping the defense mechanism strong. It provides resistance against harmful diseases and keeps the body inflammation free. Vitamin B6 or pyridoxine is also essential for the body and it is recommended to take 27% or 0.35 mg of this vitamin on daily basis. Vitamin B6 is responsible for making antibodies and hemoglobin, breaking down proteins and for keeping blood glucose levels in the range [27]. Vitamin B6 also acts as a stress buster. Folic acid, vitamin K and vitamin E are also constituents of *Thymus vulgaris* L.

16.4.33 Minerals

Thyme is also rich in minerals that are essential to keep the body healthy. *Thymus* leaves contain K, Ca, Fe, Mn, Mg and Se that are required by the body to perform various cellular functions [30]. Selenium is a part of glutathione peroxidase which functions as an antioxidant enzyme. It converts oxygen free radicals into less toxic or non-toxic forms. Potassium is an intracellular cation that transmits nerve impulses, maintains acid-base balance, prevents osteoporosis and helps in keeping blood pressure lower. Manganese helps in keeping bones strong while iron is responsible for the formation of red blood cells.

16.4.34 Volatile Oils

Thymol is the major and significant component of essential oil found in thyme which is well-known for its activity against bacteria and fungi. Other volatile oils such as carvacrol, geraneol and borneol are also present in *Thymus vulgaris* L. Thymol and carvacrol are the major phenols of this oil. Whereas linalool and p-cymene are non-phenolic components of thyme oil [31]. In addition aromatic terpenes, terpenoids and flavonoids are also a part of thyme essential oil. Hybridization and polyploidization are responsible for the presence of diverse compounds in different species of thyme. The chemical makeup and isolation yield of essential oil are dependent upon the ecological conditions and cultivation practices. *Thymus vulgaris* L. contains 12 mL/kg of essential oil in it. The essential oil has both antioxidant and antimicrobial properties.

16.4.35 Monoterpenes and Sesquiterpenes

Most of the terpenoids found in *Thymus* genus belong to monoterpenes group. Monoterpenes constitute around 90% of thyme oil composition. Minute percentages of sequiterpenes are also present. In *Thymus* oils about 270 terpenes are present. The phenolic terpenes *i.e.* thymol and carvacrol are of significant importance. They constitute more than 10% of thyme oil composition. In plant kingdom thymol and carvacrol rarely exists. Thus genus *Thymus* is the most common source of these monoterpenoid phenols. This limited occurrence of these phenols is the major cause of interest in thyme. Another compound linalool ranks third in importance for the genus *Thymus*. It differs from other phenols in its fine sweet smell. Phenylpropanoids constitutes around 7.8% of thyme essential oil. Two compounds isoeugenol and cinnamol are significant phenylpropanoids found in genus *Thymus*.

16.4.36 Antioxidants

Flavonoid phenolic antioxidants are also present in abundance in thyme. Lutein, luteolin, thrombin and zeaxanthin are some of the flavonoids present in thyme plants. All these compounds play, important roles in the human body. Like zeaxanthin helps in building yellow pigment that protects the eye from the effects of harmful light sources. Lutein prevents cataracts and vision loss in adults. All these show the nutritional importance of Thyme.

16.4.37 Antioxidant Properties of **Thymus vulgaris** *L.*

Compounds that prevent the oxidation of other compounds are known as antioxidants. In oxidation, electrons are transferred from a compound to an oxidizing agent. As a result, free radicals are produced. If this chain reaction takes place within the cell it will cause cell death. Antioxidants keep the cell safe by stopping these chain reactions. They remove free radicals and intermediates from the cell and stop the oxidation *reactions*. Thycarvacrol and its oil is used in food for aroma and flavor. They are also used for the preservation of food. Since early times thyme leaves are used in folk medicines. Thyme oil contains the per gram concentration as arvarcrol (45 mg), Thymol (24.7 mg), β-Phellandrene (9.7 mg), essential oil (4.1 mg), Humuline (3.1 mg), α-Phellandrene (2.3 mg) and Myrcene (2.1 mg) [32]. This composition of the oil was studied in research conducted to study anti-oxidant properties of thyme oil. These compounds were not only the part of *Thymus vulgaris* L. oil but also determined how effective it is in controlling aflatoxin induce oxidative stress in rats. Aflatoxins disturb the rapid composition of blood, reduce situations activity, with increase the amount of urea and nitric acid in blood together with bringing changes in tissues of the liver. Results showed that a high dose of oil can prevent such situations [33].

Globally, *Thymus vulgaris* L. and other species of thyme are used for its remarkable benefits in medicinal and non-medicinal aspects. With time great development has been witnessed in uses of thyme hence transitioning it from a traditional herb to serious drug in phytotherapy. Numerous researches have explored the pharmacological activities of essential oil and natural extract of thyme. Essential oil of thyme possess antibacterial, antiviral, anticarcinogenic, antifungal, insecticidal and spasmolytical properties. The non-medicinal usage of thyme is also equally important because it is used in cosmetics, food and aroma industries. Moreover, raw material of thyme is also used in perfumery. Both the pharmacological and non-pharmacological uses of thyme and its essential oil are discussed in the successive paragraphs.

16.5 Importance of Thyme

16.5.1 Pharmaceutical Importance of **Thymus vulgaris** *L.*

Since early times, the *Thymus vulgaris* L. is being used for treating diseases such as respiratory diseases like asthma, bronchitis and whooping cough. It is being administered to patients in form of tea, syrup, ointment or by inhaling its steam. Atherosclerosis, toothache, infection of the urinary tract and dyspepsia could be cured via *Thymus vulgaris* L. It also has antifungal properties. It eliminates fungal infections from the stomach and intestine. Thymol, a component of the *Thymus* can kill bacteria and parasites present in the stomach or intestine, thereby increasing the appetite. Both *Thymus* plant extract and essential oil are reported to have pharmacological activities. Now Thyme is being used in drug rational phototherapy. Being an incredible source of several nutrients, it maintains blood flow and has a great impact on the overall system of the body [34].

Thymus vulgaris L. is also used in treating mental weakness. It has also been proven effective in insomnia treatment. These remedial properties of the *Thymus* are due to the presence of compounds *i.e.,* flavonoids, phenols, eugenols and most importantly thymol. Antispasmodic, bactericidal, anthelmintic are other properties of *Thymus* species. Among all *Thymus* species, *Thymus vulgaris* L. is mostly cultivated in rural areas owing to its medicinal properties. It is used to treat inflammatory diseases, swelling by insect bites, muscle swelling and pain. The reason for its high usage is the presence of a large amount of essential oil than any other species of the *Thymus* genus. It also contains a high percentage of thymol because of its expectorant effects. However, *Thymus vulgaris* L. is expensive, therefore industries prefer using other species [35]. Tablets named naturactive made from thyme are available in markets for improvement in respiratory disorders.

Thymus essential oil contains thymol that is active against various infections. The disinfectant properties of the *Thymus* make it effective for treating chronic fungal infections and chest issues such as bronchitis, asthma and whooping cough. Oil of thyme is also used as a disinfectant in mouth wash due to its antibacterial and antifungal properties. Chewing *Thymus vulgaris* L. leaves relieves throat infections. Thymol is used in chest infection syrups. Moreover, its leaves extract is used to treat worms in children. Infestations and infections caused by hookworms, ascarids, *Candida albicans*, fungi, gram +ve and gram -ve bacteria are also treated using Thyme extract and essential oil [36]. Thymol also improves liver functioning. In villages, urinary infections are also being treated by using Thyme extracts. Laryngitis can also be treated using plant extracts of *Thymus vulgaris* L. Thymol is also found to be effective against enterobacteria infections. It is also reported for resolving dermal issues such as dermatitis and bug bites. Massaging thyme oil can cure sciatica. *Thymus serpyllum* is found to show anti-thyrotropic effects in rats. They have been reported for reduction of thyroid hormone in rats. However, in humans, such results haven't been reported yet. Thyme is also used since ancient times for curing chest infection and congestion. Fresh leaves are the best remedy for treating sore throat. Before production modern antibiotics medical bandages were used. To treat liver and urinary infections, *Thymus* leaves are also used. The problem of oily skin and acne can also be cured using thyme oil. In worship places it has also been used as a fumigant. In pharmaceutical medicines, *Thymus vulgaris* L. is used mostly because it contains the highest concentration of thymol which shows high antitussive and expectorant effects. Because *Thymus vulgaris* L. is expensive thereby other species of *Thymus* are used in pharmaceutical companies. Due to its diverse properties, the essential oils of the *Thymus* have been studied in detail for use in various industries [37].

Thymus vulgaris L. also has anti-cancerous properties. It has been proved to be effective against breast carcinoma. It was found controlling cancer proliferation in the colorectal HCT116 cancer cell model. It has also shown tumor inhibitory effects in human leukemia. *T. vulgaris* has been observed to impede the growth of human oral cavity carcinoma [38]. Research showed that essential oil inhibits the growth of several tumor cells depending upon the concentration of oil. Thyme oil is also found to be effective to control parasitic mites and honeybees. As compared to lavender and laurel oil, *Thymus vulgaris* L. has the lowest LC_{50} for Varroa destructor, a parasitic mite [26]. *Thymus vulgaris* L. oil is also found to be effective in the reduction of laying eggs consequently resulting in a 10% mortality when used at 1%

concentration. Red Thyme contains 54% thymol that was found to be effective against *Dermatophagoides farina* (dust mite) [39].

Research has been carried out to determine the biological activities and chemical makeup of *T. vulgaris* L. The activity of thyme oil against bacteria was evaluated against nine cultures of gram -ve bacteria and six cultures of gram +ve bacteria [6]. Antibacterial activity was being evaluated by using bio impedance methodology, keeping detection time as an evaluation parameter. The inhibitory effect was studied via the plate counting technique. All the thyme essential oils showed antibacterial activity against tested microorganisms. However, maximum activity was seen against gram +ve cultures.

The essential oil obtained from thyme full flower showed principal effectiveness against microorganisms. Only a few species retained 50% metabolic activity while most of the cultures have been incapacitated entirely [40].The chemical constituents of essential oils play a pivotal role in determining the antimicrobial activity of essential oils [41]. The activity of thyme against microbes is linked to the thymol and terepene hydrocarbons' presence in the essential oil. P-Cymene when present alone does not exhibit antimicrobial activity whereas when present along with thymol and terpene, it is found to have antimicrobial efficacy. Research has also shown that not only the major constituents but also the minor components of essential oil also show stronger antimicrobial efficacy which shows synergistic effects. Thymol present in thyme oil has been found effective against microbes.

Research has been conducted to determine the antiviral activity of aqueous extracts of the Lamiaceae family against the Herpes simplex virus. Extracts of *Thymus vulgaris* L. showed inhibitory activity against Herpes simplex virus type 1 and 2 were tested in vitro on RC-37 cells in a plaque reduction assay.

16.5.2 Usage in Traditional Medicines

Since Romans and Greeks thyme has been cultivated and used for various purposes. It was considered as an indispensable plant. Still it is believed that consuming one cup of thyme extract provides freshness, lightness in the stomach and protects one against the morning cough. Externally, thyme infusions and essential oil has been used in folk medicines for treatment of injuries, bruises and ulcers [42]. It is believed that thyme bath has sedative properties. Massaging thyme oil is used to treat sciatica and arthritis. It also improves oxygenation. In traditional medicines thyme was used for treating baldness. It also improves blood circulation within capillaries thus preventing seborrhhoea. Moreover, it possess ability to cure ulcers and various types of dermatitis, injuries and bruises. It is also used in medicines for treatment of sleeping disorders like insomnia. Gastro-intestinal troubles like dyspepsia, gastritis, diarrhea and gastric ulcers can also be treated by using thyme oil.

16.5.3 Thyme Usage in Homeopathy

Homeopathy was introduced by German doctor Samuel Hahnemann. The underlying principle of this therapy is the internal regulation of human body by a special drug, who has different reactivity corresponding to each individual. It is also known as "Simile principle". This principle means that similar can be cured by something similar. Till date, different plants and their part have been used for this purpose. *T. vulgaris* L. *and T. serpyllum* L. for a long time have been used in homeopathy [43]. However, in 1989 this usage had been stopped due to negative monographs.

16.5.4 Thyme Oil as Antispasmodic Agent

Thyme oil has found to have antispasmodic activity on intestinal smooth muscles that have been contracted by various agents. It was observed to have antispasmodic activity against all the agents causing contractions [44]. Thymol and carvacrol both acts as antispasmodic agents.

16.6 Commercial Importance

16.6.1 Used in Mouth Wash

Boiling 5gm dried *Thymus vulgaris* L. leaves in 100 mL distilled water for 10 min and then straining (5% infusion). This remedy is effective for periodontal prophylaxis. Listerine, a mouthwash, contains thymol which is effective in reducing oral bacteria [45].

16.6.2 Oils and Hair Tonics

Thyme oil is also used in hair tonics and oils. 2 to 3 drops of a combination of thyme, lavender, and cedarwood essential oil added to grapeseed and jojoba oil, when massaged into the scalp for 7 months cures *alopecia areata*. Paronychia can be treated by adding 1 drop of thymol to chloroform and applying it to chronically affected areas 3 times a day can cure it [46]. Diluted Thyme oil is also being used in many ointments for curing skin diseases.

16.6.3 Use as a Preservative

Thyme oil is used as a botanical preservative to stop the deterioration of food during storage. Because of its antifungal and aflatoxigenic properties, *Thymus* oil is used as a preservative for storing food items. Research reported that thyme oil completely inhibited the growth of fungus *Aspergillus flavus* [47]. Thyme oil completely arrests the production of aflatoxin B1. It is also used to control postharvest fungal diseases. Usage of thyme oil increases the shelf life of the cake. It is also reported to reduce the decaying of vegetables.

16.6.4 Aromatherapy

A pleasant odor has always been desirable and it still makes people feel good. In ancient civilizations, similar to aromatherapy fumigation was carried out to treat patients. It was an effective treatment as using it disinfected the air and the sweet smell induced calmness in patients. Until nineteenth century good smelling candles were ignited by people to make the air clean. Even various essential oils were being sniffed for protection against infections.

In aromatherapy, different thyme oils are used. Thymol, red thyme oil. Linalool are used due to their soft texture and antibacterial activities. White thyme oil is used too. It has a slighter effect on the skin. It is also used to cure ringworm, athletes' foot and fungal infections.

16.6.5 Use as Garden Plants

Several *Thymus* species are used as ornamental and garden plants. A huge trade of thyme plants is carried between North America and Europe. They are being traded for their decorative foliage and aromatic properties.

16.6.6 Food Applications

Since early times, herbs have been utilized for improving the look and taste of food. Usage of thyme in food industry is restricted to preservation of meat products. It acts as an antioxidant and a preservative. Literature demonstrates that addition of thyme natural antioxidants can greatly reduce rancidity in different types of food. Thereby, different researches have been conducted to prove the preservative properties of thyme. It has been found that bioactivity of compounds and their preservative properties in foods can be greatly influenced by the addition of thyme. There are several implications of addition of thyme to food but these can be avoided by different techniques. The primary goal of using thyme in food is to enhance its shelf life however, there is one limiting factor. The usage of thyme plant extract or its essential oil develops negative organoleptic characteristics in food leading to undesiring odor and taste. These sensory limitations can be avoided by using low concentrations of thyme and other conservation methods. Moreover, nanocarriers' encapsulated natural compounds can also be used for this purpose. This encapsulation not only masks taste but also protects against the oxidative degradation by controlled release of compounds. It also increases the bioactivity of compounds present in thyme essential oil and its plant extract. Thyme essential oil and extracts are also used for preserving fish and seafood. For this purpose, the fish or seafood is immersed in essential oils or natural extract followed by drainage of product after half an hour. EO can also be used by applying on the films or adding it to the surface of fish. Due to short shelf life, fresh thyme is not used as food. Use of whole dried thyme in food processing is also limited. It is because the volatile compounds are not evenly distributed throughout the product and often additional treatments are required for processing of food. One such method of food processing is comminution. In this method leaves of thyme can be crushed, ground, chopped or sliced. When ground the oil structure becomes crushed and the oil is released. However during grinding heat is generated that vaporizes the volatile oil as a result flavor of oil reduces. To prevent destruction of volatile compounds by heat it is essential to keep the temperature low during the grinding process. Moreover, ground thyme should have optimum moisture to retain the characteristic odor and flavor. To prevent contamination of thyme by microbes in processed food, it is sterilized through different chemical and physical sterilization techniques.

16.6.7 Usage in Perfumery

Because thyme has a capacity to suppress smells it is used in deodorants. Moreover, it also acts as an antimicrobial agent so have no harm in using in cosmetics. Its essential oil is also used in soap perfumes. The oil greatly masks the tarry odor. Usage of thyme EO in trace amounts in lotions, perfumes or colognes provides freshness to body. Names of some perfumes containing thyme essential oil or its plant extract are Milano Fraganze Basilica and Acqua Colonia Lavender & Thyme.

16.6.8 Culinary Uses

In Mediterranean region, thyme is widely used for seasoning. Its strong agreeable aroma is highly recommended for seasoning of food products like pizzas etc. It also gives a pleasant touch to fatty foods like sausages, fatty meats and strong cheeses.

Moreover, it can also be used in soups, meat and fish dishes. Fresh leaves of thyme are also used in salads. Thyme is also used in liquor industry to give flavor and aroma to wine.

16.6.9 Pest's Control

Thyme essential oil has been effective against a number of pests attacking various plants. Every year in United States billions of dollars are lost due to vegetables and fruits losses during harvest and distribution process. Mostly these post-harvest losses are due to mishandling and flaws in storage procedures and diseases. *Botrytis cinerea* commonly known as gray mold fungus attacks almost 200 different plant species. It is also major pest that attacks strawberry (*Fragaria ananassa*) infecting its fresh fruit thus causing major crop loss. *Botrytis cinerea* reduces the shelf life of strawberries. In a recent research conducted it has been found that using thyme essential oil on post-harvest strawberries can significantly mitigate the impact of this fungus [48].

Thyme EO can also be used to reduce the incidence of bacterial blight in soybean plants caused by *Pseudomonas savastanoi*. Usually this pathogen is controlled by using toxic pesticides that causes pathogen resistance in plants. Using eco-friendly essential oils is beneficial option. Research conducted showed that using thyme essential oil in seed treatment greatly reduced the incidence and severity of this disease [49]. Essential oil of thyme is also beneficial in poultry industry. New Castle Virus (NDV) is the most deadly viral disease in poultry industry. Even by using vaccines this disease has not been controlled. As *Thymus vulgaris* L. contains variety of volatile and lipophilic properties such as monoterpenes, sesquiterpenes and phenylpropanoids, it has been proved effective in controlling this disease. These substances can act against different microorganisms such as fungi, yeast, bacteria and viruses. Thyme essential oil attacks envelope of New Castle Virus and its viral components that are necessary for attachment and penetration in host cells [50].

Maize is widely cultivated crop throughout the planet. Solely in China around 11 billion tons of corn has been utilized in year 2021–22. Due to its high nutrient content maize is loved globally. Its high demand in world has led to the development of techniques to prevent maize losses in fields and in warehouses. *Sitophilus zeamais,* is one of the most lethal pests for the stored grains. Particularly found in tropical regions and warm zones, this pest is the major destroyer of maize crop. It severely damages maize during storage. Using artificial pesticides have harmful impacts on consumers. Therefore, continuous researches are being conducted to find eco-friendly alternatives for this pest control. Application of thyme oil on this pests has found to significantly reduce its mortality. Research has shown that both encapsulated and non-encapsulated thyme essential oil is effective in controlling *Sitophilus zeamais* without altering the quality of crop [51].

Essential oil of *Thymus vulgaris* L. is also reported to control early blight disease in tomatoes. This disease causes severe economic losses by damaging the plants.

Infection by this pathogen starts at the bottom of plants and spreads upward thus turning the leaves yellow and creating spots upon them. It was found that thyme essential oil greatly inhibited the mycelial growth and conidial germination in tomato plants. Moreover, spraying tomato plants with thyme essential oil enhances their defense mechanism by increasing the production of defense related enzymes like peroxidase and polyphenol oxidase. Thyme EO is also reported to boost callose deposition within tomato plants infected with this fungus [52].

16.6.10 Other Uses

Thyme is effective in repelling cabbage fly, white fly and aphids. Moth can also be repelled by using thyme. For this purpose, take a muslin ribbon and fill it with dried thyme. Placing these bags within the folds of blanket and cloths and keeping them in suit cases or drawers will keep the moths away. For ornamental purposes only few species of thyme are used. It is best for garden's rock. It can also be grown as border plants. Thyme is also used in making butter, bread, thyme vinegar, cocktails and other condiments.

The essential oils of *Thymus vulgaris* L. and *Thymus serpyllum* works against *Anathoscelides obtectus,* the pest that damages *Phaseolus vulgaris* L. (kidney bean) in field and during its storage. Thyme essential oil have toxic effect on insects and stops their reproduction through ovicidal and larvicidal effects. Essential oil of *Thymus vulgaris* L. also inhibits the reproduction of mite *Tetranychus urticae*. The compound thymol present in thyme oil greatly reduces the egg laying capacity of this mite. It also possess genotoxic activity. Such effects have been visualized in a study conducted on somatic mutation and recombination test on Drosophila. Other compounds constituting thyme essential oil are found ineffective with regard to insecticidal activity. Another pest sensitive to *Thymus vulgaris* L. essential oil is *Spodoptera littoralis*. If the larvae of this pest are fed with leaves treated with *T. vulgaris* essential oil it reduces their development and egg laying capacity. Thymol is also extremely toxic against housefly. Researches have shown that aromatic plants exhibit dual insecticidal activities *i.e.* causing direct toxicity on adult insects and inhibiting their reproduction.

16.6.11 Thyme Honey

Thyme honey produced by different thyme species the most prominent among them is garden thyme (*Thymus vulgaris*) and wild thyme (*Thymus capitatus*). Thyme honey is renowned worldwide for its unique flavor and aroma. It is being produced from the mid of June till the first half of July. Color of thyme honey varies from light amber to brown. In liquid state its color varies from light amber to amber whereas in crystallized form it ranges from beige to brown color. It has strong aroma and a taste similar to tropical fruits. It is rich in flavonoids, phenols and flavanols. Often it also contains epicatechin gallate. Thyme honey, if kept outside, crystallizes within short time.

Thyme honey was traditionally used in Greece. It is widely found in Greece. Famous source of thyme honey in Greece is slopes of Mount Hymettus near Athens. A popular way of eating thyme honey there is by eating it with fresh Greek yogurt and nuts. In Sicily thyme honey is extracted from wild thyme (*T. capitatus*). There it is known as Hyblaean Honey. In Italy and New Zealand thyme honey is quite popular. It was introduced in New Zealand by gold miners back in 1800's. Thyme plant became popular in New Zealand countryside Central Otago as a source of delicious thyme honey. Every year in Central Otago Alexandra Thyme Festival is celebrated by the tourists and natives to celebrate the beauty of landscape when the wild thyme blooms and also the fine honey produced from it.

Thyme honey due to its useful ingredients is being employed in number of treatments. Several researches have been conducted to explore its miraculous properties. Thyme honey is found effective in treating oral mucositis. This oral mucositis is one of main side effect of head and neck treatment of cancer patients. Using thyme honey has found to improve the quality of life in patients suffering from head and neck cancer [53]. Thyme honey nasal spray is also found to have positive effects on patients suffering from chronic rhinosinusitis after their endoscopic sinus surgery. This nasal spray constituting thyme honey significantly reduces inflammation, polyp formation and also boosts the healing of mucosal lining within the patients suffering from chronic rhinosinusitis. However, further research is required in this regard [54].

Traditionally honey has been used as a remedy for wound treatment. Several clinical trials have been carried out to obtain super normal wound healing. To achieve this goal research has been conducted upon the efficacy of thyme honey in wound treatment. It has been found that if applied twice a day on wounds, thyme honey accelerates the healing process. Thyme honey also helps in providing relief from cough. Moreover, its viscous sticky nature soothes the throat.

16.7 Medicinal Uses of Other Species of Genus *Thymus* [55]

16.7.1 Thymus pubescens

This species is widely found in Iran and Turkey. It is a perennial plant with a woody stem and red or purple flowers. Flowering begins in spring and continues till summer. It is used as a food additive and for treating gastrointestinal disorders (Table 16.3).

Table 16.3 Physicochemical properties of Thyme [27]

Property	Characteristic/value
Molecular Formula:	$C_{50}H_{82}O_4$
Molecular Weight:	747.18368 g/mol
Components:	0.8–2.6% of thyme oil; Thyme oil from *T. vulgaris* contains 3–80% Thymol carvacrol; linalool, p-cymene, γ-terpinene, borneol, 1, 8-cineole, geraniol, various other terpenoids, alcohols and esters.
Form at 25 °C/1 Atm.	liquid in state
Color	colorless to reddish brown
Odor	Pleasant smell
Specific Gravity	0.916 g/cm^3
MP	250 °C
BP	190 °C
Solubility	Only slightly soluble in water; soluble in alcohol (80%), Thymol is soluble in glacial acetic acids, oils and fixed alkali hydroxides
VP	0.167 mm Hg at 25 °C
pH	Thymol: Solution neutral to litmus test
Water coefficient	3.3 (Log K_{ow}: −1.77 to −1.69)
Miscibility	Miscible in water
Flammability	At 60 °C it is flammable
Storage stability	Stable if stored under optimum conditions
Corrosion characteristics	Corrosive

16.7.2 Thymus caramanicus

It is also a woody plant, about 25–50 cm tall. This plant shows high antibacterial activity. Leaves of *Thymus caramanicus* are used in rheumatism treatment, allergy and dermal problems. It is also a natural food preservative. Major constituents of this plant are thymol, borneol and p-cymene.

16.7.3 Thymus kotschyanus

It is a perennial plant about 20 cm tall. It has dark green pointed leaves. Leaves of this species are used in spices and medicines. This species is mostly found in the Mediterranean region. It also grows at high altitudes in mountains. It contains 14.9% thymol. Altitude greatly affects the chemical composition of the plant.

16.7.4 Thymus transcaspicus

It is also known as Khorasan thyme. As the name indicates it is native to Iran. This species is also found in Turkmenistan. It is an aromatic plant containing thymol and carvacrol as the main constituents. It contains 60% thymol whereas around 90% of monoterpenes are also reported. This species is reported to have high antifungal activity.

16.7.5 Thymus persicus

This species is found in the West Azerbaijan province. It is also endemic to Iran. This species is characterized by small leaves and by the presence of glandular hair. Important constituents of this plant are thymol, sabinene hydrate, carvacrol, borneol, geraniol.

16.7.6 Thymus fallax *Fisch. & C.A. Mey*

It is a pleasant-smelling shrub that is found in Western Mediterranean, Southern Italy, Iran, and Turkey. It grows on rocks and grassy plains. This species is used in making soups, herbal tea for treating seasonal flu. Its leaves extract and oil is also used in pesticides, mouthwashes, fumigants etc.

16.7.7 Thymus algeriensis

It is a North African species widely grown in Tunisia, Morroco and Algeria. Both fresh and dried leaves of *Thymus algeriensis* are used as kitchen herbs. It is used for curing respiratory diseases and issues of digestive tracts. Researches indicate that it is also used as anti-abortive.

16.7.8 Thymus serpyllum

It is grown locally in Mediterranean Europe and North Africa. *T. serpyllum* is commonly known as mother of thyme. It is mostly used in home remedies. It is used as an antiseptic and stimulant in mouthwashes. It contains various compounds that have high resistance against infections. It is also used for treating ailments.

Table 16.4 Ethnobotanical uses of *Thymus* sp. [55]

Thymus species	Treatment of diseases
Thymus pubescens	Gastrointestinal disorders
Thymus caramanicus	Rheumatism
Thymus kotschyanus	Depression and cancer
Thymus persicus	Toothache reliever
Thymus fallax	Seasonal flu
Thymus algeriensis	Respiratory and digestive issues
Thymus serpyllum	Treat infections
Thymus transcaspicus	Fungal diseases
Thymus citriodorus	Bacterial disease

16.7.9 Thymus citriodorus

It is commonly known as lemon thyme. It is smaller in size than *Thymus vulgaris* L. *T. citriodorus* has deep pink flowers and dark green leaves. Its leaves have lemon smell. Out of the total 50 known species of this thyme, only 3 are cultivated for cooking purposes. It helps in curing bacterial diseases (Table 16.4).

16.7.10 Myths, Legends, Folklore and Tales About Thymus vulgaris *L.*

The word thyme is derivative of the Greek word 'Thymos' which means to fumigate. Studies showed that it was utilized by ancient Egyptians in the embalming process. Since ancient times thyme has been considered a symbol of bravery. Roman soldiers used to bathe in thyme and they believed that it gives them courage. Until medieval times it was given to soldiers upon leaving for battle as it was considered as a sign of valor. The togas of generals were been embroidered with thyme picture.

Greeks knew about thyme from early times. This was the herb the bees loved and honey produced by bees was loved by Greek Gods. In medieval Europe, when people had a strong belief in fairies, people used thyme to attract fairies during summer evenings. There are also myths that thyme was present in hay for infant Jesus and since then it has had a great aroma.

References

1. Lawrence, B. (1992). Chemical components of Labiatae oils and their exploitation. *Advances in Labiatae Science*, 399–436.
2. Özkan, M. (2008). Glandular and eglandular hairs of Salvia recognita Fisch. & Mey. (Lamiaceae) in Turkey. *Bangladesh Journal of Botany, 37*(1), 93–95.
3. Creative Commons Corporation, Integrated Taxonomic Information System (ITIS), an online database. (2022). (Publication no. https://doi.org/10.5066/F7KH0KBK). Retrieved March 23, 2022 https://creativecommons.org/publicdomain/zero/1.0/legalcode, http://www.itis.gov/

4. Ahmad, S., Koukab, S., Islam, M., Ahmad, K., Aslam, S., Aminullah, G. A., et al. (2008). Germplasm evaluation of medicinal and aromatic plants in highland Balochistan, Pakistan. *Pakistan Journal of Botany, 40*(4), 1473–1479.
5. Ahmad, N., Fazal, H., Ahmad, I., & Abbasi, B. H. (2012). Free radical scavenging (DPPH) potential in nine Mentha species. *Toxicology Industrial Health, 28*(1), 83–89
6. Saleh, N., Allam, T., El-Latif, A., & Ghazy, E. (2014). The effects of dietary supplementation of different levels of thyme (Thymus vulgaris) and ginger (Zingiber officinale) essential oils on performance, hematological, biochemical and immunological parameters of broiler chickens. *Global Veterinaria, 12*(6), 736–744.
7. Galambosi, B., Galambosi, Z., Pessala, R., Repcak, M., Hupila, I., & Aflatuni, A. (2001). *Yield and quality of selected herb cultivars in Finland.* Paper presented at the International Conference on Medicinal and Aromatic Plants. Possibilities and Limitations of Medicinal and Aromatic Plant 576.
8. Zarzuelo, A., & Crespo, E. (2002). The medicinal and non-medicinal uses of thyme. In *Thyme* (pp. 277–306). CRC Press.
9. Badi, H. N., Yazdani, D., Ali, S. M., & Nazari, F. (2004). Effects of spacing and harvesting time on herbage yield and quality/quantity of oil in thyme. *Thymus vulgaris L. Industrial Crops Products, 19*(3), 231–236.
10. Omidbaigi, R., Sefidkon, F., & Hejazi, M. (2005). Essential oil composition of Thymus* citriodorus L. cultivated in Iran. *Flavour and Fragrance Journal, 20*(2), 237–238.
11. Herrero, M., Cifuentes, A., & Ibañez, E. (2006). Sub-and supercritical fluid extraction of functional ingredients from different natural sources: Plants, food-by-products, algae and microalgae: A review. *Food Chemistry, 98*(1), 136–148.
12. Bakowski, J., & Michalik, K. (1986). *Suitability of several vegetable species for drying.* Bulletin of Vegetable Crops Research Work.
13. Cabo, J., Crespo, M., Jimenez, J., Navarro, C., & Risco, S. (1987). Seasonal variation of essential oil yield and composition of Thymus hyemalis. *Planta Medica, 53*(04), 380–382.
14. Guillen, M., & Manzanos, M. (1998). Study of the composition of the different parts of a Spanish Thymus vulgaris L. plant. *Food Chemistry, 63*(3), 373–383.
15. Moldão-Martins, M., Bernardo-Gil, M., Beirão da Costa, M., & Rouzet, M. (1999). Seasonal variation in yield and composition of Thymus zygis L. subsp. sylvestris essential oil. *Flavour Fragrance Journal, 14*(3), 177–182.
16. Tainter, D. R., & Grenis, A. T. (1993). *Spices and seasonings: A food technology handbook.* John Wiley & Sons.
17. Venskutonis, R., Poll, L., & Larsen, M. (1996). Influence of drying and irradiation on the composition of volatile compounds of thyme (Thymus vulgaris L.). *Flavour. Fragrance Journal, 11*(2), 123–128.
18. Deans, S., Svoboda, K. P., & Bartlett, M. C. (1991). Effect of microwave oven and warm-air drying on the microflora and volatile oil profile of culinary herbs. *Journal of Essential Oil Research, 3*(5), 341–347.
19. Venskutonis, P. (1997). Effect of drying on the volatile constituents of thyme (Thymus vulgaris L.) and sage (Salvia officinalis L.). *Food Chemistry, 59*(2), 219–227.
20. Jalas, J. (1971). Notes on Thymus L. (Labiatae) in Europe. I. Supraspecific classification and nomenclature. *Botanical Journal of the Linnean Society.*
21. Moldao-Martins, M., Palavra, A., Da Costa, M. B., & Bernardo-Gil, M. (2000). Supercritical CO2 extraction of Thymus zygis L. subsp. sylvestris aroma. *The Journal of Supercritical Fluids, 18*(1), 25–34.
22. Regnault-Roger, C., & Hamraoui, A. (1994). Inhibition of reproduction of Acanthoscelides obtectus Say (Coleoptera), a kidney bean (Phaseolus vulgaris) bruchid, by aromatic essential oils. *Crop Protection, 13*(8), 624–628.
23. Ballester-Costa, C., Sendra, E., Fernández-López, J., Pérez-Álvarez, J. A., & Viuda-Martos, M. (2013). Chemical composition and in vitro antibacterial properties of essential oils of four Thymus species from organic growth. *Industrial Crops and Products, 50*, 304–311.
24. Alizadeh, A., Alizadeh, O., Sharafzadeh, S., & Mansoori, S. (2011). Effect of different ecological environments on growth and active substances of garden thyme. *Advances in Environmental Biology, 5*, 780–783.

25. Kulevanova, S., Stafilov, T., Anastasova, F., Ristić, M., & Brkić, D. (1997). Isolation and identification of flavonoid aglycones from some taxa of section Marginati of genus Thymus. *Die Pramazie, 52*, 886–888.
26. Beata, K., & Kiełtyka-Dadasiewicz, A. (2015). Yield and herb quality of thyme (Thymus vulgaris L.) depending on harvest time. *Turkish Journal of Field Crops, 20*(1), 78–84.
27. Javed, H., Erum, S., Tabassum, S., & Ameen, F. (2013). An overview on medicinal importance of Thymus vulgaris. *Journal of Asian Scientific Research, 3*(10), 974–982.
28. Malik, N. R., Yadav, K., & Verma, A. (2016). Optimization of process parameters in extraction of Thyme oil using response surface methodology (RSM). *International Journal of Science, Engineering Technology, 4*(1), 248–256.
29. Dauqan, E. M., & Abdullah, A. (2017). Medicinal and functional values of thyme (Thymus vulgaris L.) herb. Journal of Applied Biology Biotechnology, 5(2), 0–2.
30. Yilmaz, A., Alibas, I., & Asik, B. B. (2021). The effect of drying methods on the color, chlorophyll, total phenolic, flavonoids, and macro and micronutrients of thyme plant. *Journal of Food Processing Preservation, 45*(11), e15915.
31. Hancı, S., Sahin, S., & Yılmaz, L. (2003). Isolation of volatile oil from thyme (Thymbra spicata) by steam distillation. *Food/Nahrung, 47*(4), 252–255.
32. Lee, S.-J., Umano, K., Shibamoto, T., & Lee, K.-G. (2005). Identification of volatile components in basil (Ocimum basilicum L.) and thyme leaves (Thymus vulgaris L.) and their antioxidant properties. *Food Chemistry, 91*(1), 131–137.
33. El-Nekeety, A. A., Mohamed, S. R., Hathout, A. S., Hassan, N. S., Aly, S. E., & Abdel-Wahhab, M. A. (2011). Antioxidant properties of Thymus vulgaris oil against aflatoxin-induce oxidative stress in male rats. *Toxicon, 57*(7–8), 984–991.
34. Bernhoft, A., Siem, H., Bjertness, E., Meltzer, M., Flaten, T., & Holmsen, E. (2010). *Bioactive compounds in plants–benefits and risks for man and animals*. The Norwegian Academy of Science Letters, Oslo
35. Amiri, H. (2012). Essential oils composition and antioxidant properties of three thymus species. *Evidence-Based Complementary Alternative Medicine, 2012.*
36. Thompson, J. D., Chalchat, J.-C., Michet, A., Linhart, Y. B., & Ehlers, B. (2003). Qualitative and quantitative variation in monoterpene co-occurrence and composition in the essential oil of Thymus vulgaris chemotypes. *Journal of Chemical Ecology, 29*(4), 859–880.
37. Sacchetti, G., Maietti, S., Muzzoli, M., Scaglianti, M., Manfredini, S., Radice, M., et al. (2005). Comparative evaluation of 11 essential oils of different origin as functional antioxidants, antiradicals and antimicrobials in foods. *Food Chemistry, 91*(4), 621–632.
38. Agili, F. (2014). Chemical composition, antioxidant and antitumor activity of Thymus vulgaris L. essential oil. *Middle-East. Journal of Scientific Research, 21*(10), 1670–1676.
39. Gill, A. O., Delaquis, P., Russo, P., & Holley, R. (2002). Evaluation of antilisterial action of cilantro oil on vacuum packed ham. *International Journal of Food Microbiology, 73*(1), 83–92.
40. Mosandl, A., Hener, U., Kreis, P., & Schmarr, H. G. (1990). Enantiomeric distribution of α-pinene, β-pinene and limonene in essential oils and extracts. Part 1. Rutaceae and gramineae. *Flavour Fragrance Journal, 5*(4), 193–199.
41. Soliman, K. M., & Badeaa, R. (2002). Effect of oil extracted from some medicinal plants on different mycotoxigenic fungi. *Food chemical toxicology, 40*(11), 1669–1675.
42. Arras, G., & Grella, G. (1992). Wild thyme, Thymus capitatus, essential oil seasonal changes and antimycotic activity. *Journal of Horticultural Science, 67*(2), 197–202.
43. Benito, M., Jorro, G., Morales, C., Peláez, A., & Fernández, A. (1996). Labiatae allergy: systemic reactions due to ingestion of oregano and thyme. *Annals of Allergy, Asthma Immunology, 76*(5), 416–418.
44. Buchbauer, G., Li, H., & Ye, Y. (2000). Methods in aromatherapy research. *Perfume Flavorist, 36*, 21–31.
45. Zarzuelo, A., & Crespo, E. (2002). *Uses of thyme* (p. 263). The Genus Thymus.
46. Haraguchi, H., Saito, T., Ishikawa, H., Date, H., Kataoka, S., Tamura, Y., et al. (1996). Antiperoxidative components in Thymus vulgaris. *Planta Medica, 62*(03), 217–221.

47. Schroder, J., & Vollmer, H. (1932). The excretion of thymol, carvacrol, eugenol and guaiacol and the distribution of these substances in the organism. *Archiv fur Experimentelle Pathologie und Pharmakologie, 168*, 331–353.
48. Vasilatis, A. A. (2022). *Inhibition of post-harvest pathogen gray mold (Botrytis cinerea) on strawberry (Fragaria ananassa) with application of thyme (Thymus vulgaris) essential oil.* Rutgers University-School of Graduate Studies.
49. Sotelo, J. P., Oddino, C., Giordano, D. F., Carezzano, M. E., Oliva, M., & J. I. M. (2021). Effect of Thymus vulgaris essential oil on soybeans seeds infected with Pseudomonas syringae. *Physiological Molecular Plant Pathology, 116*, 101735.
50. Rezatofighi, S. E., Seydabadi, A., & Nejad, S. M. S. (2014). Evaluating the efficacy of Achillea millefolium and Thymus vulgaris extracts against Newcastle disease virus in Ovo. *Jundishapur Journal of Microbiology, 7*(2).
51. Barros, F. A., Radünz, M., Scariot, M. A., Camargo, T. M., Nunes, C. F., de Souza, R. R., et al. (2022). Efficacy of encapsulated and non-encapsulated thyme essential oil (Thymus vulgaris L.) in the control of Sitophilus zeamais and its effects on the quality of corn grains throughout storage. *Crop Protection, 153*, 105885.
52. Saltos-Rezabala, L. A., Silveira, P. R. D., Tavares, D. G., Moreira, S. I., Magalhães, T. A., Botelho, D. M. D. S., et al. (2022). Thyme essential oil reduces disease severity and induces resistance against Alternaria linariae in tomato plants. *Horticulturae, 8*(10), 919.
53. Charalambous, M., Raftopoulos, V., Paikousis, L., Katodritis, N., Lambrinou, E., Vomvas, D., et al. (2018). The effect of the use of thyme honey in minimizing radiation-induced oral mucositis in head and neck cancer patients: A randomized controlled trial. *European Journal of Oncology Nursing, 34*, 89–97.
54. Hashemian, F., Baghbanian, N., Majd, Z., Rouini, M.-R., Jahanshahi, J., & Hashemian, F. (2015). The effect of thyme honey nasal spray on chronic rhinosinusitis: a double-blind randomized controlled clinical trial. *European Archives of Oto-Rhino-Laryngology, 272*(6), 1429–1435.
55. Mehran, M., Hoseini, H., Hatami, A., Taghizade, M., & Safaie, A. (2016). Investigation of components of seven Species of Thyme Essential oils and comparison of their antioxidant properties.

Chapter 17
Onion

Sara Zafar, Nazia Aslam, Abida Kausar, Shagufta Perveen, and Muhammad Riaz

17.1 Introduction

Botanical Description
Kingdom: Plantae
Division: Magnoliophyta
Class: Liliopsida
Order: Asparagales
Family: Amaryllidaceae
Genus: *Allium*
Scientific name: *Allium cepa*
English/common name: Onion

Onion is an herb, which grows to about **80** cm in height. It is developed from a pungent bulb. The bulb is almost of ovoid in shape, brownish in color, glossy, papery and solitary. Different colored bulbs are found like yellow, white and red and flavor like sweet or pungent [**1**]. The leaves are simple, linear, fistulose and entire. The sheath part of the leaves forms a pseudostem, which hides a great part of the above-ground scape [**2**]. The scape is of about **1** m in length. Inflorescence is of globose umbels. Bracteoles are present at the bases of the pedicels. The spathe is short and the flowers are campanulate [**2**]. The parianth consists of **6** petals, which are white, ovate,

S. Zafar (✉) · N. Aslam · S. Perveen
Botany Department, Government College University, Faisalabad, Pakistan
e-mail: sarazafar@gcuf.edu.pk

A. Kausar
Botany, Department, Government College Women University, Faisalabad, Pakistan

M. Riaz
Department of Pharmacy, Shaheed Benazir Bhutto University, Sheringal, Pakistan

M. Zia-Ul-Haq et al. (eds.), *Essentials of Medicinal and Aromatic Crops*,
https://doi.org/10.1007/978-3-031-35403-8_17

oblong and about **4** mm in length. The androecium consists of **6** stamens. The ovary is minute and sub-globose [**3**]. Allium show capitate stigmas. Ovary is trilocular with septal nectaries having distinct nectariferous pores. Two ovules are present in each locule, which develop into seeds [**2**]. Family is also known as Aliaceae [**4**].

17.1.1 Karyology

The basic chromosome numbers found in the onion plants are x = 8 and x = 7 with polyploids [2].

17.1.2 Morphology of Onion

17.1.2.1 Colors

The bulbs of onions are of different colors like yellow, red, brown, chartreuse and white.

17.1.2.2 Size and Shape

Bulbs are of variable shapes like flat, elongated, pyriform and round. Height and shape of the onion bulb has high inheritability as compared to the diameters and weight.

17.1.2.3 Soluble Solids Content

Onions bulbs can accumulate significant amount of carbohydrates called non-structural carbohydrates or SSC [5]. Dehydrating industries use this higher SSC content for producing onion powders or flakes for seasoning. The main soluble content found in onion bulbs are sucrose, fructose, fructans of insulin neoseries and insulin series. Fructans stored in the plants play important role in regrowth or sprouting and help in protection of plant from low-temperature and drought [6].

17.1.3 Plant Characters

17.1.3.1 Premature Flowering

Bolting (Flowering during bulb production) creates somehow useless bulbs and great loses for growers. After the emergence of true leaves, young onion plants are subjected to cooler temperatures. There appear clear cut phenotypic differences from germplasm to bolt. It is important for the plants, grow in winter to resist bolting to bring out bulb in the harvest season of spring. Planting date is also important for the plant, as the plants which are grown in the early season, faces long cold temperatures and results in the increase in the bolting [7].

17.1.3.2 Seed Yield

Seed yield is defined as the "average quantity of seeds produced from a single plant" or it can be the total number of flowering scape or umbel produced by a single plant or the number of perfect flower produced by a single umbel. Onion is considered as the poor producer of seed as compared to the many of other crops. Normally, the ovary of onion contains six carpels, expected to produce maximum six seeds per flower. However, in order to meet the maximum seed production content, seed production must be excellent. The long day storage onions produced smaller umbels and shorter scapes as compared to the short day onions. Level of inbreeding and seed yield have a correlation [8]. Reduced seed yield often results in the loss of vigor in the inbreeding results. Relatively good seeds are produced by the double haploid varieties [8, 9]. Production of these variable seeds is due to the variability of insects pollinated the plants. For example, the plants with high low sucrose content in their nectar produced lower quantities of seeds as compared to the plants, producing higher amount of sucrose containing nectar as also reported by [10]. However, no genetic basis was given for the attractiveness of insects in relation to the seed production.

17.2 Onion Agronomy

Now a days the high yield of the onion can be achieved by the use of crop-protection chemicals [11–13]. These chemicals control the weed effectively. It also works as a constituent of the costly hand labour. Experiments were performed in Netherlands to find out the total inputs of all the pesticides which were about 23 kg ha^{-1} per crop and 10.5 kg ha^{-1} of pesticide the post-transplant cropping season in Norway [14]. Along with these pesticides, onion also need some of the mineral quantities for proper growth. Large number of pesticides and nitrogen are emitted from the onion fields then from any other crop, which effect the on aquatic and soil organism [2, 15].

Integrated crop management (ICM) systems used some of the resources for the production of onion crops which are less wasteful and refined then the older and traditional crop production methods. This process deals with the usage of external resources like water [16, 17], chemical input and fuel for the production of crop along with using pesticides. Processors and consumers used onion in different forms such as fresh bulbs after harvest, dry bulbs when fresh bulbs are not available, green or salad onion with or without bulbs, some other used for pickling, freezing, for dehydration as powder and flakes, onion oils, fresh cut products etc. The distance between the farm and the factory is of great economic importance [2, 18].

Table 17.1 Pungency of different varieties of onion

Variety	Pungency
Garlic	
Spanish	Mild
Southpart white	Strong
Bessanovski	V. Strong
Hysam	Strong
Red Baron	Strong
Purple intercross	
Outer scales	Mild
Inner scales	Mild

17.2.1 Onion Pungency

Pungency of the onion is produced by the hydrolysis of S-alken/alknyl cysteine sulfoxide molecules [14, 19, 20] (Table 17.1). This pungency is performed by the allinase enzyme right after cell disruption. The disruption of the cell produces substances like, pyruvic acid, ammonia and sulfenic acid. Weston and Schwimmer method proved that, the pyruvic acid is formed by the enzymatic cleavage of those molecules which are the precursor of the flavor forming substances [21–23] (Fig. 17.1).

17.2.2 Ecology

The genus Allium shows diversification in distribution. Majority of the species grow in open sunny areas. Areas inhabited by Allium species are Central Asian high mountains, European subalpine pastures, alpine and sub-alpine grasslands of the Himalayan, different types of forests and along the river banks as well. Some taxa are able to tolerate saline and alkaline environments. Taxa with spring, summer and autumn flowering also exists. These are mostly short and long perennials, with one to many cycles of leaf formation. Many ephemeroids species of genus Allium show limited annual growth for a very short period of time in spring and early summer which complete their life cycle from leaf sprouting to seed maturation in 2–3 months. *A. cepa* is a not ephmeroid. It's leaves growth begins after the frost has ceased in the spring and in the coming autumn and winter, growth limited [2].

17.2.3 Distribution

The specie *cepa* is known as petrophytes, which usually grows in open plant formations like rock cervices, river-banks, rocks, gravelly deposits, stony slopes and many more like these. These sites mostly contain shallow soil layer. Their occurrence on

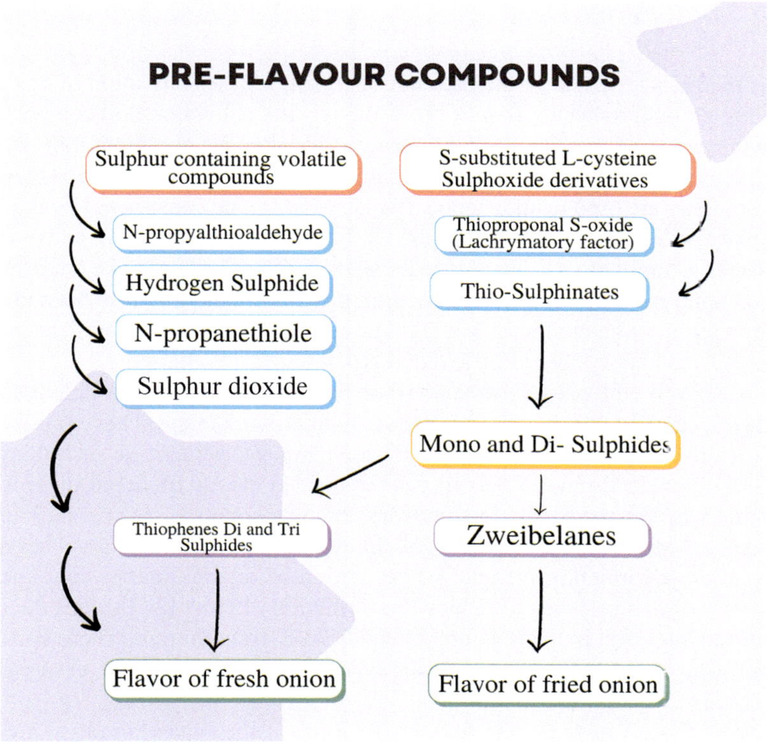

Fig. 17.1 Flavor compounds of onion

these sites are not affected by pH of soil, plant-sociological associations, minerals content or vegetation types. The distribution pattern often results in the formation of small populations. However, large populations have also been reported by various researchers [2].

17.2.4 Land Preparation and Soil Management

All the types of the soils are suitable for the cultivation of onion like heavy clay, volcanic, sandy or peat organic soils. Stony heavy soil hindered the mechanical harvesting. The homogenous, fine structured soil is required in the upper or surface layer for growth. The best or suitable soils are found to be loamy or sandy loam soils with good quantity of organic matter and soil structure [2].

17.2.5 Soil Preparation

Various methods are used for the preparation of soil for the cultivation of onion like subsoiling plough, rotavator, mould-board plough and cultivator. Soil compaction is mostly prevented in these methods. In sandy soil, after the completion of preparatory work, a roller may be used. For combating soil erosion, elevated beds of almost 15 cm or higher are used to allow good drainage and can be constructed. Sometimes, serpentine furrows are mostly set up inside large sized beds, whose purpose is to allow water to supply to a limited area at one time [2]. One thing to be keep in mind is that the land for the transplant production of onion should not have been used for the Allium production or at least 3 years. The sites with the onion disease and severe weed problems history should be avoided [24].

On acidic soil with pH almost lesser then 6.0, 2–3 months before preparation of land, lime is applied to bring the soil pH to about 6.2–6.5 range. These applications effect the storage and quality of bulb by reducing the percentage of bolting and increasing the bulb firmness. Seedling emergence is mostly hindered by crust formation, making second sowing compulsory. In some countries, to prevent crusting, onions are grown on organic or peaty soil, known as muck. Some of the other things to prevent crust formation include, incorporation of organic matter, low intensity irrigation or mulching helps to reduce the formation of crust [2–25]. For the sweet onion production, sandy loam, loamy or sandy soils are preferred. These soils contain low Sulphur, one of the fact, which enhances the capability of soil to produce sweet onions [24, 25].

Spraying of soil surface with 500×10^{-6} g ml^{-1} of polyacrylamide at a rate of 4 l m^{-2} and phosphogypsum (PG) spread after rainstorm, proved to be useful for the crust formation and stimulating emergence of onions in loamy, non-sodic soils. In some soil conditions, PG along with water of electrical conductivity of about 0.65 deci-siemens (dS) m^{-1} on average is proved to be useful for the improvement of onion emergence. Vermiculite is a source of anti crustant which can be applied over the seed row and then covered with a thin soil layer. It helps to prevent soil erosion. H_3PO_4 is also found effective in controlling erosion when banded over the row. In some cases, emulsified plastic gel is used on the soil which form the water stable soil aggregates. This helps in increasing emergence of onions bunch and in decreasing soil cracking. Rice straw mulch is also in used in countries like Thialand [2].

Before sowing of onion, sterilization is performed by solarization. For killing pathogens and weed seeds, moist soil is heated over 4 weeks by mulching with polyethylene. In some of the countries like Egypt and Florida, this method has successfully been practiced. This method results in the increase in the yield of jumbo onions (>10 cm). Onions show sensitivity to salinity and result in the decrease in the at about 1.4 dS m^{-1} treatment solution. Therefore, the salinity of the soil is kept in consideration while sowing onions. In some arid and sub humid areas, cropping of onion is performed in saline soils. This is in turn, decrease the yield of the onions even after adopting some of the preventive techniques just like planting of seedlings on the ridges. Salinity effects increases with the increase in the temperature and decrease in the humidity. For avoiding salinity, adequate drainage system is required [2, 11].

Onion storability is also affected by the soil characteristics. Onions sprouting depends upon the presence of salts in the soil just like higher quantity of sodium (more than 5.4 mg, 100 per soil) showed highest storage aptitude, resulting in fewer sprouts. In order to avoid toxicity, the soil salinity must be lower than 370 mg of salts 100 g^{-1} of soil [2, 25].

17.3 Categories of Onions

Onion is produced in three product segments. (1) Bulbs used for fresh market, these bulbs have the greatest diversity. (2) Dehydrated onions which are used for food processing. (3) Green salad onions used for fresh consumption. Bulbs of onion are produced from sets, transplants or seeds. These can survive easily in low moisture condition but still adequate supply of water is required for normal growth.

17.3.1 Green/Salad/Spring Onion

These are also known as scallions. These are immature onions grown very short period of time, at high density. These are harvested and sold after a few days of harvesting. These can be used both for salad and cooking.

17.3.2 Semi-bulbled Oninos

These onions provide larger amount of bulb tissues as compared to the scallions. These also must be harvested before maturity and sold.

17.3.3 Bulb Onions

Bulb onion are specifically produces for freezing, drying and dehydration. For dehydration, specific onion verities are selected, whose dry matter content is up to 20% or more. This reduces the energy consumption of the dehydration process. The onions growing in the high solar radiations areas are recommended for dehydration [26]. As there are most chances to obtain the maximum dry matter. Onion is one of the most traded plant, because of it's durability and storage characteristics.

17.3.4 Storability

One of the best advantage of using onion as a crop is it's storability. Upon reaching at maturity, plants are lifted [27]. So that it's roots, shoots and skin are completely dry and keep well. The bulbs have the ability to remain dormant both in low and high temperature regions. So, these are recommended both for temperate and tropical regions. In modern refrigerators, due to efficient storage ability, the onions can be stored up to 9 months [28–30].

17.4 Onion Production Methods

One of the best onion production method is direct sowing in humid temperate climates. The place should be where the irrigation can be managed well. For producing salad onions, a bed system with many rows is formed. While for bulb production, other systems like furrows and ridges are adopted. On flat surface, pneumatic seeders are used. Means or final emergence of the onion plants depends upon the soil moisture and field temperature. The temperature being varies from 10 to 20 °C and water capacities from 40 to 80% for higher emergence. The depth for onion is found to be about 1–2 cm for optimal emergence while in the drier soils, depth of about 4–5 cm is considered. A model was developed for the effect of water stress on germination of onion. According to this model, radicle growth can be initiated when the difference between the virtual osmotic potential and ambient water potential reach a certain threshold. Water potential is found to be in inverse relation with the time of onion radicle growth and it's emergence. Mechanical impedance decreases the rate of onion root development then that of shoot. So, after recovery, onion shoots develop better and elongate the non-impede seedlings [2].

Factors under consideration for the development of onions include water availability, season, price premium and soil types for early onions, jumbo (77–101 mm) and colossal (≥102 mm) diameter bulbs. Level fields are mostly recommended for irrigation layouts [2].

17.4.1 Timing

During the early part of the growing season, onions are mostly produced during long cool season and there is also a variation in temperature year to year. Experiments conducted for growing onion shows that sowing in February show high bolting in almost all the cultivars. It is possible to control the inflorescence induction by controlling the sowing date [2].

17.4.2 Plant Density

Selection of population is a very important decision; high population decrease the bulb while onion yield increase. It also increases the disease severity. Increase in the density of the onion results in increase in the %age of bolting. Distances of sowing or planting can be established according to mathematical calculations of grade and size of onions [2–23, 25–31].

17.4.3 Seed Priming

Seed quality, it's performance in the soil and its ability to pass through the difficult chemical and physical conditions of soil. The usage of prime seeds increased the rate of emergence and decrease the time spread. Seed priming is a process in which seeds before sowing, are allowed to imbibe water in a well-controlled manner. In some cases, re-drying of seeds is performed, which bring all the seeds to the same stage of germination readiness, enhancing evenness in germination. But if the storage months' increase from 6 to 12, the abnormal seedlings begin to increase, so the priming seedlings should be done only one season depending upon the usage [2].

One of the priming methods include **aerated salt solution**, but it decreases the %age of onions emergence, which is most probably due to non-tolerating behavior of the onion seeds which reached their maximum stage of development before sowing. Other method is **Solid matrix priming**. In this method, priming is done by mixing of seeds with known proportions of water and a solid material. Of all these, successful method is **Drum priming**, a commercially used priming method. The method includes the priming of seeds in bubble-columns in turn improved seeds and seedling leek performance. Drum priming method is useful as in it controlled amount of water is added to seed. The water revolves in drum, so that the seeds takes only the required percentage of water, without becoming completely wet. The imbibed seeds then dried with warm air and sown with a normal seed drill [2, 10, 32].

17.4.4 Seed Coatings

Onions seeds are coated with fungicides or insecticides, before sowing, which results in the decrease in the chemical used per unit area of land. Usually colored coating is applied. Then palletization is performed, which changes the shape and size of the seed so that it can be sown easily with a drill [2, 33].

17.4.5 Starter Fertilizers

The onion seedlings showed slow growth. It's roots branch little and seldomly develop root hairs. This makes it difficult to supply adequate amount of nutrients to the seedling at it's early growth stages for proper growth of onion. Mostly the fertilizers are applied in granular form as base dressing before sowing onions. But it's not beneficial as N fertilizers or the N content of the soil leaches out before the onion roots are able to reach out. So, an effective method to use fertilizer is in the form of liquid rather granules, in a zone close to or below the root zone of onion. This is starter fertilizer concept. In many areas, these starters help to accelerates the growth of shoot and yield of the onion plants. Increase in the bulb development takes place with the application of starters as compared to the without these starters. Starters confirms the adequate supply of nutrients to the roots, mostly of P. Ammonium Phosphate improved early growth and yield as compared to ammonium nitrate. Experiments were performed on spring sown onions in nitrogen depleted soil. Also, the use of starter helps to prevent the water contamination and leaching of nitrogen from the soil, so that the roots have better access to the nitrogen [2, 32].

17.4.6 Onion Propagation

The application of biotechnology in the onion production is more attractive due to high cost of seed production and long generation time of onion. These things are improving with the passage of time and will give significant valuable verities of onion in the coming future.

Following are the methods used by the early scientist upto near past, for the propagation of onion verities, in the labortary.

17.4.6.1 In Vitro Propagation

Firstly, the onion were produced using callus culture and micro-propagation. These were considered as the significant usage of tools for onion production. The use of this technology depends upon the cost of propagating male sterile plants in relating to the increased revenue for the hybrid seed. For reducing the occurrence of viruses, meristem culture is used for plants [34].

17.4.6.2 Molecular Based Selection

For the identification of molecular markers in onion, gel electrophoresis and recombinant technologies are being used. Polymorphism were identified using the following methods, like amplified fragment length polymorphisms AFLPs; [35], isozymes [32], randomly amplified polymorphic DNA [36] and RFLPs [37]. These marker classes helped in the identification of onion populations. For obtaining desirable traits, these genetic markers are of great benefit. The characters of plants are identified at an early stage of development by these genetic markers. It helps in crossing of the desire plants with significant characteristics. Markers linked to the genes conditioning traits with relatively high evaluation costs, such as male-sterility res-toration and flavor profiles, would reduce the number of phenotypic evaluations that must be completed for each plant or family. For obtaining desired genotypic plants, eventual sequencing of the onion plants are done followed by the resequencing of the key population. This method reveals large number of DNA polymorphism for the identification desire genes. It automatically gives desire phenotypes. Large numbers of DNA markers will also enable the genomic estimation of breeding values, allowing the performance of hybrids from crosses of male-sterile lines with various male parents to be predicted.

17.4.6.3 Doubled Haploid

To reduce the time commitment and cost of the inbrebs lines, haploid plants are subjected to chromosome doubling. The method is performed by extracting the haploid plants from the gametophyte and then the doubling of chromosome is performed. The doubled haploid produce by this way carry out similar sexual behavior as of the normal lines [38]. Male gametophyte of the onion does not produce successful haploid. Whereas the gynogenic production of haploid is successful when flower buds are grown on simple media [39–44]. Haploid plants develop from the egg and another cell of the female gametophyte [39]. Flow cytometry is used to check this haploidy.

17.4.7 Challenges to Use Double Haploid

One of the main challenge to use double haploid is that no male sterile is available for the developed double haploids. If a sterile male is used for double haploid, then it will automatically produce male sterile and will be propagating asexually.

17.4.8 Transformation and Gene Editing

Though, onion is a monocotyledonous plant, still it's transformation is being reported by Agrobacterium [45, 46] Onion can be transformed stably the following methods, like herbicide-resistance genes, anti-sense construct which helps in the removal of lachymatory (tearing) factor [45]. Many gene editing technologies are being used for the production of desired lines of onion like CRISPR/Cas system. It also helps to reduce the long generation time involved in the selection and commercialization of new onion varieties.

17.5 Pests and Diseases

Onion plants show susceptibility to root diseases. When the onions are grown as monocultures or the same field is being used by the farmers to produce onions, then there are chances of onions to become more susceptible to pathogenic fungi [2, 37, 46].

17.5.1 Onion Maggots

Delia antiqua are the onion maggots are the severe pests of onion. These are basically the descendants of houseflies. The larva infects the young plants. These plants do not cause plant mortality. These mostly grow on decaying plant material. These also effects the plants after their damage by the frost formation. For avoiding these pests, the soil should be free from high quantity of organic matter or proper insecticides should be used [28, 47].

17.5.2 Aster Leaf Hopper (Mycrosteles Fascifrons)

This insect is basically vector of the phytoplasma which results in the Aster yellows disease in onion plants. Nymphs does not feed on onion plants whereas the adult show low preferences. Symptoms caused by the leafhopper are stunting, twisting of leaves and stem and chlorosis of leaves [47].

17.5.3 Cutworms

Cutworms are the larval stage of the most of the species of moths. These mostly feeds at night and hide during day time. The damage occurs, when the plant cut off near the soil line. For their nocturnal habit, these plants are detected only during early or late day time [27, 28].

17.5.4 Wireworms

These are the larval stage of the click beetles. These mostly feed on the ground portions of the plants. Some of the species produce multiple generation in a single year, whereas some have the ability to produce adults after a period of 1–2 years [27].

17.5.5 *Thrisp*

These are the primary insect pest of onions [47, 48]. These mostly damage the leaves. These leaves are then easily attacked by other disease as well and lost the ability of photosynthesis. With the increase in temperature their number also increase. The pests mostly congregate in the leaf folds and down the plant portions.

17.5.6 *Gout and Metabolic Syndrome*

Gout and metabolic syndrome are characterized by the abnormal xanthine oxidase (OX) and high serum uric acid concentrations. These are common health problems in many diseases. These are the common symptoms of gout and metabolic disease. These make the base of precipitation of gout. Therefore, the concentration of serum uric acid and XO should be normal. In this regard, potential effects of active ingredients of onion proves to be beneficial. Onion has inhibitory effect on these oxidants and xanthine oxidase. The onion active compounds like are the potential source of anti-gout and anti-oxidants [48, 49].

17.6 Classification of Onion

Onion plants can be classified by day length sensitivity into long-day plants (LD), very long-day plants (VLD), short-day plants (SD), intermediate-day plants (ID). These lengths of days effect the following paratmeters firmness, skin, scale colors, shape, size, sweetness, juciness, storage potential and pungency. Following scale colors were observed like bronze, light yellow, dark yellow, pink, dark purplish, red and white [2].

17.6.1 *Important Chemical Constituents*

In onion, water is about 80–90% of the fresh weight of the onion. Almost 65% of the dry weight of onion is composed of non-structural carbohydrates, in the form of fructose based polymer, called fructan. In onions, in which the fructan is in lower amount as a constituent of dry matter, there, the fructose, sucrose and glucose dominates.

The two main and important component found in the onions are Alkyl cysteine sulphoxides and flavonoids. Two of the flavonoids sub-groups are also found in the onion like flavanos e.g. quercetin which impart yellow and brown color to the onion varieties And anthocyanins, which impart red or purple color to the onion. The ACSOs are the main flavor precursors, which when broken down by the enzyme allinase, gives the characteristics odor and taste of the onion.

A volatile oil is present in the onion bulb which contains Sulphur containing compounds like allin [50], allicin, flavonoids, sterols, phenolic acids and allypropyldisulphide.

Onion gypoglycaemic activity due to the presence of allicin and allylpropyldisulphide. Antihyperglycaemic activity is also shown by onion due to the presence of Diphenylamine in the mature bulbs. Allicin and alliin both involves in the inhibition of the platlet aggregation. Onion juice contains thiosulphinates, which are involved in anti-asthmatic activity [51–56].

Some of the other chemicals found in the onions are sugars like fructans, and no starch, many decomposed products of cysteine sulphoxides [2]. Phenolic hydroxyl compounds include polyphenols. These are the class of natural onion content.

17.6.2 Medicinal Uses

It is used against headache, vomiting, malaria, constipation, blood pressure, antiseptic, diuretic, cold, rheumatism etc.[3, 54]. Some of the additional uses of onions are antibiotic, antibacterial, antisclerotic, anticoagulant, anti-inflammatory, antiasthmatic, expectorant, carminative, antispasmodic, hypotensive, antidiabetic. In China, *Allium macrostemon*, is found. It is used for treating dysponea, dysentery, angina pectoris and cough [4].

17.6.3 Insulin Requirement

The regular use of onion of about 50 g/day, decrease the insulin requirement of a diabetic patient from almost 20–40 per day [4].

17.7 Onion Stem Diseases

17.7.1 Sour Skin and Bacterial Canker

The disease is caused by *Burkholderia cepacia* (*Pseudomonas cepacia*) which causes sour smell of onion bulbs due to the which the name is given. In this disease, extensive tissue maceration occurs which releases with the water. In some cases, this bacterium cause disease in the leaf axil, from which the name bacterial canker is develop. This result in the death of the infected leaves and infection spread to the bulb and also neck tissue of the plant. The onions infected with the neck-wound disease, result in the sour skin disease of the onion while in storage. Environmental conditions effect the severity of disease, as the humidity promotes the disease while the disease vanishes when the dry conditions exceed. Also, in low humidity, canker lesion dried up and the infected leaf sloughed off [2].

17.7.2 Bacterial Streak and Bulb Rot

The disease Bacterial streak and bulb rot of onion is caused by *Pseudomonas viridiflava* (Burkholder) Dowson. The disease effects the foliage leaves and bulb both in field and in storage. In this disease, lesions develop as water soaked, dark color streak which spread to the entire length of the leaf. In severe conditions, lesions blightened entire plant. These lesions are small, oval, olive green to tan in color and are surrounded by chlorotic area. These lesions go downward to the bulb [2]. Symptoms mostly develop in a circular pattern due to the restriction of the rot by scales [49].

17.7.3 Centre Rot

The disease is caused by *Pantoea ananatis* (Serrano). In this disease, the center of the leaves is mostly affected. The infected leaves become soft, bleached white and water soaked when the disease progresses. Tissues turn tan to brown. In severe condition, wilting and bleaching of leaves occur. Bulb also get infected, it become water, have foul smell and soft. It also leads to the loss of seeds heads [2].

17.7.4 Bacterial Soft Rot

This disease is caused by *Erwinia* pathogens belonging to the soft-rot '*carotovora*' group. In this disease, tissues become infected with pale yellow to light brown rot and water-soaked condition occur. Bulb disintegrates. In severe condition, wilting and whitening of onion foliage occur. Flower stalks, pedicle also get infected with lesions [2, 32].

17.7.5 Pink Root

The disease is caused by *Phoma terrestris*, a fungus. It is the most damaging disease of the fungus. This disease is greatly increased due to stresses like cold, heat, flooding, drought and nutrient deficiencies/toxicities. The fungus survives and reproduces in the soil, so continuous production of onions in the same filed result in the increased in the disease occurrence. Roots in this case turn pink or purplish. In severe cases, the roots become brown and deteriorate. Disease at early stage result in the production of unusable bulbs while in the later stages, the plants become stunted or produce small bulbs. Crop rotation, balancing temperature and fertilizers help to reduce the disease [24, 27].

17.7.6 Fusarium Basal Rot

The disease is caused by the fungus *Fusarium oxysporum f.* sp. *cepae*. This disease also effects the onion roots both in filed and in storage. The disease increased by the repetition of the onions in the field. The symptoms include yellowing of leaf tips, necrotic areas, curling or curving of leaves, infected and decaying of basal plate and white and fluffy mycelium. The control includes the use of healthy transplants, controlling insects and avoiding fertilizer injury [27, 28, 56, 57].

17.7.7 Onion Smut

Onion sumt is caused by Urocytis cepulae. Temperature is most effecting in controlling this disease as low temperature and damp places increase its occurrence while high temperature and dry weather decrease its growth. The disease mostly effects seedlings while the adult plants are somehow resistance to this disease [58].

17.7.8 Botrytis Neck Rot

The is the most damaging disease of the fungus on onions. The fungus survives in the form of sclerotia. Conidia arise from the sclerotia and spread by wind to spread the disease. The symptoms include stunted growth, splitting of leaves and leaf distortion around the neck area. A greyish sporulation appears near the neck area. Harvesting healthy mature onions, temperature balance and avoiding over usage of fertilizers are some of the management practices [27, 28].

17.7.9 Purple Blotch

The disease is caused by *Alternaria porri*, one of the commonest diseases of the onion. Older plants are more susceptible to disease then younger ones. The fungus overwinters in leaf debris of onion. The symptoms appear as elliptical, small and tan lesions which later turn purplish brown. Concentric rings appear as they enlarge.

Use of fertilizers, deep soil turning, long rotations and destruction of cull piles are some of the management practices [27, 28].

17.7.10 Downy Mildew

Downy mildew is caused by *Peronospora* sp. It is detected in early morning as velvety, violet sporulation. Later on, these lesions become yellow. Use of fungicides, destroying the cull piles and avoiding infected onions planting are some of the recommendations for controlling disease [27, 28, 59].

17.7.11 Iris Yellow Spot Virus (IYSV) and Tomato Spotted Wilt Virus (TSWV)

IYSV is spread by onion thrips (*Thrips tabaci*). This virus has also been found in Tobacco thrips (*Frankliniella fusca*). In this disease small necrotic spots with green tissues develop in the center of the leaves. Controlling thrips, proper fertilization is some of the recommendations for controlling this disease [2, 38].

17.8 Control of Diseases

The most effective treatment in the control of diseases is the rotation which also helps to reduce the pesticide and herbicide application and minimize nitrate leaching. The order of rotation must be kept in mind while doing rotation. As Lucerne reappear as weeds while maize is a high-volume residue crop which in turn effect the onion emergence. Some of the best proceeding crops are melon, cauliflower, beans, peas, pepper, cucumber, aubergine and tomato can be used as proceedings crops. The rotation sequence was experimented for 3 years to be cauliflower as the early crop, followed by onion cv. 'Babosa', chufa (*Cyperus esculentus*), early potatoes, lettuce, artichoke and melon, for the central-eastern coast of Spain. In Spain, onions–maize–wheat/lucerne–wheat rotation sequence is mostly followed. In Switzerland, rotation of onion along with carrots, cabbage, tomato and celery is used [2].

Environmental condition also plays important part while performing crop rotation. Following shallow rooted onions, sugarbeet is used to scavenge the nitrate of soil. Beet uses the soil nitrates and in turn protect the soil from nitrate pollution of groundwater. Onion growing as intercrop helps to decrease the pest damage. Beans as intercrop, reduce the attack of leaf aphids, red spider mite (*Tetranychus urticae*) etc. [2].

17.8.1 Insecticides for Onion Control

Some of the insecticides used to control insect's pests of onions are,

1. *Bacillus thuringiensis* (Xentari): These can control beet armyworms and hornworms.
2. *Lambda-Cyhalothrin* (Karate Zeon): These control leaf miners, aphids, beet armyworms, thrips, leaf hoppers.
3. Abamectin (Newmectin): These are used for broad spectrum of insect control like leaf miners, aphids, thrips and mites.
4. Indoxacarb (Avaunt 30 WG): These are used against thrips, aphids and beet armyworms [26].

17.8.2 Fungicides

1. Copper products are effective for the bacterial and fungal disease control.
2. Menab products control fungi.
3. Sulphur products effective against insect and fungal disease.
4. Chlorothalonil control fungi.
5. Copper sulphate pentahydrate and Azoxystrobin (Amistar) are effective against purple bloctch, botrytis leaf blight and Downy mildews.
6. Mefenoxam (Ridomil) products control some of the systemic diseases [26].

17.8.3 Biological Control of Onion Pests

Onion pests can also be controlled biologically.

Leaf miner can be controlled using lacewings, parasitic wasps and Eulophidae. Soldier bug can control beet army bug. Braconid control onion maggot. And onion thrips are controlled by predatory mites, lacewing and flower bug [2, 26].

17.9 Health Benefits of Onion

17.9.1 Antibacterial Effects

Onion extracts and oils have great antibiotic properties. Onion have the chemicals to be used as a remedy for diarrhea, worms, inflammatory and other infectious diseases. All these controllable effects of onions are due to the presence of saponins, phenolic compounds, proteins and organo sulphur etc. These mostly act against gram positive bacteria rather gram-negative bacteria. Onion extract is effective for oral bacterial inhibition [2, 50, 60].

In some of the experiments, an antimicrobial protein is isolated from the onion seeds, which are highly effective against plant pathogenic fungi with the concentration below 10 micrograms per milliliter. The protein isolated was Ace-AMP1 which interacts with the phospholipid membranes [2].

17.9.2 Cardiovascular Effect

Onion plant extract is used for the antiplatelet activity which in turn decrease the number of the platelet aggregation and provide a preventive effect for cardiovascular disorders, like atherosclerosis. Also, onion extract has some lipid-lowering effects. It's extract also shows some of the antithrombotic potential. In experiments, higher concentration of onion extract was used intraperitoneally and orally to rats. Boiling of extract results in decrease in it's activity. Therefore, cooking decreases the antithrombotic activity in alliums. Almost 90 mg ml- is found to cause platelet aggregation in rabbits. Human plasma also shows the same result. Heating also effects the inhibition capability. Onions with high Sulphur contents show greater antiplatelet activity as compared to the lower concentration of the Sulphur. Extract prepared from the flowering umbel show greater antiplatelet activity as compared to the same amount of the extract obtained from bulbs. Typical organo Sulphur compounds found in the onion species are (+)-S-methyl-L-cysteine sulphoxide which show antidiabetic and hypolipidaemic activity [2].

17.9.3 Effects on Respiratory System

One of the best activities shown by onion is it's antiasthmatic activity. This activity is shown by the synthesized or isolated organo Sulphur compounds found in the onion extracts, which are potentially used as medicinal drugs. Cepaenes and Thiosuphinates effects also observed in vitro. In upcoming studies, experiments are expected to perform on guinea-pigs for observing the effects of organo sulphur compounds on platelet activating factor. A single dose of onion extract of almost 100 mg kg^{-1} body weight inhibit PAF-induced bronchial hyperactivity to histamine for over 12 h. Experiments performed on the persons suffering from allergic bronchial asthma, indicated that there was remarkable reduction in the immediate and late bronchial reactions of the patients just with the use of 100 ml of 5% ethanol extract. The extract was prepared from 100 g sliced onions and was used for allergen inhalation test [2].

17.9.4 Effects on Metabolic Diseases

Some of the effects of onion extracts on blood glucose were observed. An experiment was performed on diabetic rat. It was observed that the dietary onion induced the beneficial modulation of renal lesions in diabetic rats. The ability of onion to decrease the lipid peroxidation and reduction in the blood cholesterol levels makes it's use in the amelioration of the diabetic nephropathy. Principle organo sulphur compounds of Allium species include cysteine suphoxide [2, 53, 56].

17.9.5 Anticancer Effects

Onion organo sulphur compounds help to prevent cancer. Selenium is an important component of organo Sulphur compounds. Quercetin and it's derivatives, make the typical anticarcinogenic constituents of the onions. The ethanolic extracts of onions showed the antimutagenic effect, against the mutagenicity of N-nitrosamines was observed. Allium exhibit protective effect against stomach and oesophagal cancer. It also protects from gastrointentinal cancer. One of the studies shows that the intake of the onion and fiber reduce the risk of breast cancer [2, 61]. Onion oil is beneficial in reducing the number of tumors due to its antioxidant potential [2] and rate of it's growth in experimental mice. The growth of leukemia HL60 cells is suppressed due to the onion oil.

17.9.6 Anti-inflammatory Effects

Onion is found to be a protectant for inflammation. Thiosulphinates and cepaenes are involved in anti-inflammation capacity of the onion plants. In vitro studies, the onion extracts compounds like cepaene and thiosulphinates were found to be involved in the inhibition of chemotaxis of human granulocytes [2].

17.10 Onion Annual Production and Export

The harvesting process must be start after the falling of 75% of the plant tops. Irrigation must be stop before 15 of harvest. Careful measurements should be taken while harvest to avoid any damage to bulb. While harvesting, diseased or insect infected crop should be carefully removed. Harvesting should be performed during dry weather. Harvest the plant when the bulb is in good condition [26].

In the year 2014, China and India reached almost 47.49% of the world's total onion production. Onion is almost produced on about 5.3 million hectares of the world and the bulb production is about 88.48 million matric tones for the year 2014. Almost 8% of this onion production is traded internationally. With increasing population, the world market for onion production is increasing day by day [27, 62, 63].

17.10.1 World Onion Exports

The World onion export for the year 2013 was 7.01 million tonnes [27]. It was estimated that the world export value of dry onion approached 3266 million US $ during 2013.

It is difficult to predict the trend of onion production which depends upon the economic and technological factors. Some of the factors for improving the growth, harvesting and storage process or techniques include effective research, improved mechanization, by reducing the labour input and improving the quality of crops [51, 62, 63].

17.10.2 Breeding Goals

The main objective of breeding is the production of high quality bulbs and seeds. Following characteristics are being observing while breeding, like resistance to bolting, pathogens and pest. (Bolting is the flowing of plant during the production of bulb). High quality seeds are produced by uniform flowering, straight seed stalks, conditions for cleaning and drying of seeds, vigorous bulbs, pollinating insects. Onion breeding is a very slow process. It requires almost 2 years to complete one generation e.g. seed to bulb and bulb to seed. There is a reduction in the yield of bulb production by reducing the time to 1 year.

17.11 Myths, Legends, Tales and Interesting Facts

Onions are one of the oldest vegetables in continuous cultivation dating back to at least 4000 BC. This crop was cultivated by the ancient Egyptian along the River Nile. The center of origin of the onion plants is Afghanistan. Onions can be grown from tropics to subarctic regions. Three groups of onion bulbs are found on the basis of day length. Short day bulb with the length of about 11–12 h, intermediate with the day length of about 13–14 h and long day with the day length of about 16 years or greater mostly found in the northern climes of US and Canada. Onion were brought to the US by the early European settlers. After European settlements, onions were brought to the northeast. Then the varieties move towards the south of United States. Spain and Italy verities became important part of Vidalia onion industry. After their coming from Bermuda, the first onions were given name as "Bermuda onions". After that different varieties of onion were developed including Early Grano 502, Yellow Granex and Yellow Granex hybrid, YB986. The Vidalia onion industry began in 1931 [24].

Yellow brown onions are common in Northern Europe, similar to those found in New Zealand and Tasmania. Many countries produce open-pollinated to hybrid onion cultivars. Many north-western European countries are producing hybrids. In USA, North-Eastern and Central States, large, mild yellow pungent onions are found. South East Asian countries produce large, yellow, sweet onion which are mostly used in Japan. These onions are mostly resistance to bolting [2]. Some of the factors which effect the onion production are photoperiod like longer photoperiod result in earlier maturation [64–72]. Higher temperature causes increase in the bulb development. Excess nitrogen result in the stimulation of the leaf growth. Irrigation effects the leaf blade development. Other factors include plant age, interplant competition, plant size and cultivars [73–81]. Herbal medicines made from onion extract and bulb, chemicals found in onions appear to lessen asthma-related lung stiffness and congestion [82–85]. Additionally, it includes substances that appear to lower blood sugar and cholesterol. Onion is used by people to avoid cuts and bruises [86–93]. It is also used to treat excessive blood pressure, asthma, hair loss, obesity, and insomnia [94–99]. There isn't any solid scientific proof to back up these uses, which include treating tumors and many other conditions.

The appearance of scars appears to be improved after using an onion extract gel on the skin for at least 10 weeks, either by itself or in combination with other ingredients. However, it's unclear what dosage or ingredient mix is most effective.

Those who are overweight or fat cannot lose weight by eating onions or consuming onion extract.

The negative impacts could include heartburn and abdominal pain. When applied to the skin the use of onion extract is probably harmless. Side effects might include eczema or skin inflammation. During pregnancy and while nursing, onions are frequently used in food. However, there isn't enough trustable data to determine whether using onion as medicine in higher doses while pregnant or breast-feeding is safe. Keep to the recommended food amounts to be healthy. People who are sensitive to onion may also be allergic to mugwort and celery [100–103]. Onion may lessen blood sugar and delay blood clotting [104–109]. During and after surgery, this may make blood clotting more likely or make it more difficult to regulate blood sugar levels. At least 2 weeks prior to a major surgery, stop taking onion as medicine.

Onions can cause allergies in some people [110–113]. If you have an allergy to onions, taking aspirin may make you more sensitive to them. To be safe, do not take aspirin and consume onions if you have an allergy to onions.

Blood sugar levels may be lowered by onions [114–117]. Onion might slow blood coagulation. The chance of bleeding and bruising may increase if onions are taken with medicines that also slow blood clotting.

17.12 Future Prospects

Development in the field of onion industry led to the major changes and opportunities for the onion production. Onions are the most used vegetables and staple food in the cuisine of diverse cultures worldwide. They are the most demanded ingredient in innovative products like instant noodles, grain based snacks and fish products. For examples, production of the day-length insensitive onions. Arabidopsis plant used as the basic plant for the study of flowering and day-length responses. The same genetic system as of Arabidopsis exists in onion plant. This is estimated that by using this method, the onion plants are able to produce bulbs after sprouting number of true leaves and after their maturity after a number of degree days.

The second future focus is to reduce the representation sequence of the genomic DNAs, under the heading complementing transcriptome sequencing. For this, DH will be used as line for transcriptome sequencing. Duplex-specific nuclease and Methyl-filtered libraries is of great focus to be available for the researcher [7, 118]. To study introns and promoter regions of the onions, the random sequence should be focused. But, large number of genomic sequences should be avoid. As due to onion simple genome, these will not show any significant similarities to the cDNAs [118].

References

1. Arshad, M. S., et al. (2017). Status and trends of nutraceuticals from onion and onion by-products: A critical review. *Cogent Food & Agriculture, 3*(1), 1280254.
2. Rabinowitch, H., & Kamenetsky, R. (2002). 17 shallot (Allium cepa, Aggregatum Group). *Allium Crop Science: Recent Advances*, 409.
3. Wiart, C. (2019). *Medicinal plants of Bangladesh and West Bengal: Botany, natural products, & ethnopharmacology*. CRC Press.
4. Khare, C. P. (2008). *Indian medicinal plants: An illustrated dictionary*. Springer.
5. Jaime, L., et al. (2002). Structural carbohydrate differences and potential source of dietary fiber of onion (Allium cepa L.) tissues. *Journal of Agricultural and Food Chemistry, 50*(1), 122–128.
6. Shiomi, N., Benkeblia, N., & Onodera, S. (2005). The metabolism of the fructooligosaccharides in onion bulbs: A comprehensive review. *Journal of Applied Glycoscience, 52*(2), 121–127.
7. Baldwin, S., et al. (2014). Genetic analyses of bolting in bulb onion (Allium cepa L.). *Theoretical and Applied Genetics, 127*, 535–547.
8. Villanueva-Mosqueda, E., & Havey, M. J. (2001). Genetic analyses of seed yield in onion. *Journal of the American Society for Horticultural Science, 126*(5), 575–578.
9. Caruso, G., et al. (2014). Effects of transplanting time and plant density on yield, quality and antioxidant content of onion (Allium cepa L.) in southern Italy. *Scientia Horticulturae, 166*, 111–120.
10. Dilpreet, T. (2016). Growth, yield and quality of onion as influenced by integrated nutrient management. *International Journal of Agriculture Sciences*, 0975–3710.
11. Jackson, K., et al. (1997). *Onion information kit. Agrilink, your growing guide to better farming guide.*

12. Werner, C., et al. (1990). Prediction of the performance of inbred lines derived from a population cross in autumn-sown onions (Allium cepa L.). *Theoretical and Applied Genetics, 79*, 507–512.
13. Moore, S. R. (2010). Energy efficiency in small-scale biointensive organic onion production in Pennsylvania, USA. *Renewable Agriculture and Food Systems, 25*(3), 181–188.
14. Boyhan, G. E., et al. (2009). Sowing date, transplanting date, and variety effect on transplanted short-day onion production. *HortTechnology, 19*(1), 66–71.
15. Elhami, B., et al. (2021). Analysis and comparison of energy-economic-environmental cycle in two cultivation methods (seeding and transplanting) for onion production (case study: Central parts of Iran). *Renewable Energy, 178*, 875–890.
16. Syverson, P., et al. (2001). Towards an analysis of onion routing security. In *Designing privacy enhancing technologies: International workshop on design issues in anonymity and Unobservability Berkeley, CA, USA, July 25–26, 2000 proceedings*. Springer.
17. Kumar, S., et al. (2007). Response of onion (Allium cepa L.) to different levels of irrigation water. *Agricultural Water Management, 89*(1-2), 161–166.
18. Esmaeilzadeh, S., Asgharipour, M. R., & Khoshnevisan, B. (2020). Water footprint and life cycle assessment of edible onion production – A case study in Iran. *Scientia Horticulturae, 261*, 108925.
19. Schwimmer, S., & Weston, W. J. (1961). Onion flavor and odor, enzymatic development of pyruvic acid in onion as a measure of pungency. *Journal of Agricultural and Food Chemistry, 9*(4), 301–304.
20. Lancaster, J., & Boland, M. (1990). Chapter 3: Flavour biochemistry. In *Onions and allied crops* (Vol. III). CRC Press.
21. Whitaker, J. R. (1976). Development of flavor, odor, and pungency in onion and garlic. In *Advances in food research* (pp. 73–133). Elsevier.
22. Griffiths, G., et al. (2002). Onions – A global benefit to health. *Phytotherapy Research, 16*(7), 603–615.
23. Lanzotti, V. (2006). The analysis of onion and garlic. *Journal of Chromatography A, 1112*(1–2), 3–22.
24. Boyhan, G. E., Torrance, R. L., & Hill, C. R. (2007). Effects of nitrogen, phosphorus, and potassium rates and fertilizer sources on yield and leaf nutrient status of short-day onions. *HortScience, 42*(3), 653–660.
25. Dorokhov, A., et al. (2021). *Determining the speed of soil particles during the covering of set onion with a disk-type working element featuring a soil guide*. In *E3S web of conferences*. EDP Sciences.
26. Schwartz, H. F., et al. (2014). Onion ipm PIPE: A coordinated effort to improve the management of onion thrips and Iris yellow spot virus for the US onion industry. *Plant Health Progress, 15*(4), 172–183.
27. Khokhar, K. M. (2017). Environmental and genotypic effects on bulb development in onion – a review. *The Journal of Horticultural Science and Biotechnology, 92*(5), 448–454.
28. Boyhan, G., & Kelley, W. (2007). *Onion production guide* (Bulletin 1198). The University of Georgia Cooperative Extension Service.
29. Brice, J., et al. (1997). *Onion storage in the tropics: A practical guide to methods of storage and their selection*. Natural Resources Institute.
30. Thuong, T. T. (2015). Factors affecting the storage decision of onion farmers in Kinh Mon District, Hai Duong Province, Vietnam. *Journal of Economics, Management & Agricultural Development*.
31. Tyson, C. (2021). *Onion producers guide* (pp. 1–48).
32. Díaz-Pérez, J. C., Purvis, A. C., & Paulk, J. T. (2003). Bolting, yield, and bulb decay of sweet onion as affected by nitrogen fertilization. *Journal of the American Society for Horticultural Science, 128*(1), 144–149.
33. Wall, M., Mohammad, A., & Corgan, J. (1996). Heritability estimates and response to selection for the pungency and single center traits in onion. *Euphytica, 87*, 133–139.

34. Taylor, A. G., Eckenrode, C., & Straub, R. (2001). Seed coating technologies and treatments for onion: Challenges and progress. *HortScience, 36*(2), 199–205.
35. Walkey, D., et al. (1987). Production of virus-free garlic (Allium sativum L.) and shallot (A. ascalonicum L.) by meristem-tip culture. *Journal of Horticultural Science, 62*(2), 211–220.
36. Yoo, K. S., & Pike, L. M. (1998). Determination of flavor precursor compound S-alk (en) yl-L-cysteine sulfoxides by an HPLC method and their distribution in Allium species. *Scientia Horticulturae, 75*(1-2), 1–10.
37. Wilkie, S. E., Isaac, P. G., & Slater, R. J. (1993). Random amplified polymorphic DNA (RAPD) markers for genetic analysis in Allium. *Theoretical and Applied Genetics, 86*, 497–504.
38. Fujiwara, K., et al. (2021). Real-time PCR detection of the onion downy mildew pathogen Peronospora destructor from symptomless onion seedlings and soils. *Plant Disease, 105*(3), 643–649.
39. Khan, P. S. S. V., et al. (2020). Doubled haploid production in onion (Allium cepa L.): From gynogenesis to chromosome doubling. *Plant Cell, Tissue and Organ Culture (PCTOC), 142*, 1–22.
40. Martínez, L. (2003). In vitro gynogenesis induction and doubled haploid production in onion (Allium cepa L.). *Doubled Haploid Production in Crop Plants: A Manual*, 275–279.
41. Kim, S., Yoo, K. S., & Pike, L. M. (2007). Production of doubled haploid onions (Allium cepa) and evaluation of their field performance. *Horticulture Environment and Biotechnology, 48*(3), 143–147.
42. Dhatt, A., & Thakur, P. (2014). Production of doubled haploids in onion: A review. *Journal of Horticultural Sciences, 9*(2), 107–112.
43. Khar, A., et al. (2019). Present status of haploidy research in onion (Allium cepa) – A review. *Indian Journal of Agricultural Sciences, 89*, 396–405.
44. Khar, A., et al. (2018). Genotypic response towards haploid induction in short day tropical Indian onion (Allium cepa). *Indian Journal of Agricultural Sciences, 88*(5), 709–713.
45. Bohanec, B., Jakse, M., & Havey, M. J. (2003). Genetic analyses of gynogenetic haploid production in onion. *Journal of the American Society for Horticultural Science, 128*(4), 571–574.
46. Eady, C., Weld, R., & Lister, C. (2000). Agrobacterium tumefaciens-mediated transformation and transgenic-plant regeneration of onion (Allium cepa L.). *Plant Cell Reports, 19*, 376–381.
47. Ueno, S., et al. (2009). High-pressure induced transformation of onion. *Japan Journal of Food Engineering, 10*(1), 37–43.
48. Vernon, R. S., Judd, G., & Borden, J. (1987). Commercial monitoring programme for the onion fly, Delia antiqua (Meigen) (Diptera: Anthomyiidae) in South-Western British Columbia. *Crop Protection, 6*(5), 304–312.
49. Das, A. K., Hasan, W., & Singh, S. K. (2017). Management of onion thrips, Thrips tabaci using chemical and bio-pesticide for quality onion production. *Trends in Biosciences, 10*(22), 4384–4388.
50. Black, L., et al. (2012). *Onion disease guide. A practical guide for seedsmen, growers and agricultural advisors*. Seminis Vegetable Seeds.
51. Boelens, M., et al. (1971). Volatile flavor compounds from onion. *Journal of Agricultural and Food Chemistry, 19*(5), 984–991.
52. Brewster, J. L. (2008). *Onions and other vegetable alliums* (Vol. 15). CABI.
53. George, E. S., et al. (2018). A Mediterranean diet model in Australia: Strategies for translating the traditional Mediterranean diet into a multicultural setting. *Nutrients, 10*(4), 465.
54. Gebhardt, R., Beck, H., & Wagner, K. G. (1994). Inhibition of cholesterol biosynthesis by allicin and ajoene in rat hepatocytes and HepG2 cells. *Biochimica et Biophysica Acta (BBA)-Lipids and Lipid Metabolism, 1213*(1), 57–62.
55. Glässer, G., et al. (2002). Comparison of antioxidative capacities and inhibitory effects on cholesterol biosynthesis of quercetin and potential metabolites. *Phytomedicine, 9*(1), 33–40.

56. Goldman, I. L., et al. (1996). Antiplatelet activity in onion (Allium cepa) is sulfur dependent. *Thrombosis and Haemostasis, 76*(09), 450–452.
57. Gao, C. M., et al. (1999). Protective effect of allium vegetables against both esophageal and stomach cancer: A simultaneous case-referent study of a high-epidemic area in Jiangsu Province, China. *Japanese Journal of Cancer Research, 90*(6), 614–621.
58. Warid, W., & Tims, E. (1952). Studies on the inheritance of resistance to downy mildew studies in onion incited by Peronospora destructor. *Phytopathology, 42*, 22.
59. Walker, J., Jones, H., & Clarke, A. (1944). *Smut resistance in an Allium species hybrid.*
60. Warid, W. A. B. (1952). *Inheritance studies in the onion, Allium cepa L.* Louisiana State University.
61. Goldschlag, D., Reed, M., & Syverson, P. (1999). Onion routing. *Communications of the ACM, 42*(2), 39–41.
62. Whitwell, J., & Frith, L. (1973). Experiments on the use of maleic hydrazide as a sprout suppressant on spring sown bulb onions. *Experimental Horticulture.*
63. Gulati, A., Wardhan, H., & Sharma, P. (2022). Tomato, Onion and Potato (TOP) value chains. *Agricultural Value Chains in India: Ensuring Competitiveness, Inclusiveness, Sustainability, Scalability, and Improved Finance*, 33–97.
64. Dong, J., et al. (2022). Global vegetable supply towards sustainable food production and a healthy diet. *Journal of Cleaner Production, 369*, 133212.
65. Ariga, T., et al. (2000). Antithrombotic and antineoplastic effects of phyto-organosulfur compounds. *BioFactors, 13*(1-4), 251–255.
66. Augusti, K., et al. (1975). Partial identification of the fibrinolytic activators in onion. *Atherosclerosis, 21*(3), 409–416.
67. Bajaj, J. S., Shaker, R., & Hogan, W. J. (2004). Esophageal veggie spasms: A food-specific cause of chest distress. *Official Journal of the American College of Gastroenterology | ACG, 99*(7), 1396–1398.
68. Bianchini, F., & Vainio, H. (2001). Allium vegetables and organosulfur compounds: Do they help prevent cancer? *Environmental Health Perspectives, 109*(9), 893–902.
69. Boyle, S., et al. (2000). Absorption and DNA protective effects of flavonoid glycosides from an onion meal. *European Journal of Nutrition, 39*(5), 213.
70. Briggs, W. H., et al. (2001). Administration of raw onion inhibits platelet-mediated thrombosis in dogs. *The Journal of Nutrition, 131*(10), 2619–2622.
71. Campos, K., et al. (2003). Hypoglycaemic and antioxidant effects of onion, Allium cepa: Dietary onion addition, antioxidant activity and hypoglycaemic effects on diabetic rats. *International Journal of Food Sciences and Nutrition, 54*(3), 241–246.
72. Cho, J.-W., et al. (2010). Onion extract and quercetin induce matrix metalloproteinase-1 in vitro and in vivo. *International Journal of Molecular Medicine, 25*(3), 347–352.
73. Chung, V. Q., et al. (2006). Onion extract gel versus petrolatum emollient on new surgical scars: A prospective double-blinded study. *Dermatologic Surgery, 32*(2), 193–197.
74. Draelos, Z. D. (2008). The ability of onion extract gel to improve the cosmetic appearance of postsurgical scars. *Journal of Cosmetic Dermatology, 7*(2), 101–104.
75. El-Demerdash, F. M., Yousef, M. I., & Abou El-Naga, N. (2005). Biochemical study on the hypoglycemic effects of onion and garlic in alloxan-induced diabetic rats. *Food and Chemical Toxicology, 43*(1), 57–63.
76. Gabler, N. K., et al. (2006). Dietary onion intake as part of a typical high fat diet improves indices of cardiovascular health using the mixed sex pig model. *Plant Foods for Human Nutrition, 61*, 179–185.
77. Galeone, C., et al. (2009). Allium vegetables intake and endometrial cancer risk. *Public Health Nutrition, 12*(9), 1576–1579.
78. Galeone, C., et al. (2009). Allium vegetable intake and risk of acute myocardial infarction in Italy. *European Journal of Nutrition, 48*, 120–123.
79. Graefe, E. U., et al. (2001). Pharmacokinetics and bioavailability of quercetin glycosides in humans. *The Journal of Clinical Pharmacology, 41*(5), 492–499.

80. Herman-Antosiewicz, A., & Singh, S. V. (2004). Signal transduction pathways leading to cell cycle arrest and apoptosis induction in cancer cells by allium vegetable-derived organosulfur compounds: A review. *Mutation Research/Fundamental and Molecular Mechanisms of Mutagenesis, 555*(1-2), 121–131.
81. Hister, C., et al. (2009). Evaluation of the antiproliferative effects of infusions and essential oil of Aloysia gratissima. *Pakistan Journal of Biological Sciences, 12*(24).
82. Ho, W. S., et al. (2006). Use of onion extract, heparin, allantoin gel in prevention of scarring in Chinese patients having laser removal of tattoos: A prospective randomized controlled trial. *Dermatologic Surgery, 32*(7), 891–896.
83. Hong, Y.-J., & Mitchell, A. E. (2004). Metabolic profiling of flavonol metabolites in human urine by liquid chromatography and tandem mass spectrometry. *Journal of Agricultural and Food Chemistry, 52*(22), 6794–6801.
84. Hosnuter, M., et al. (2007). The effects of onion extract on hypertrophic and keloid scars. *Journal of Wound Care, 16*(6), 251–254.
85. Huang, T.-H., et al. (2008). Onion decreases the ovariectomy-induced osteopenia in young adult rats. *Bone, 42*(6), 1154–1163.
86. Hubbard, G. P., et al. (2006). Ingestion of onion soup high in quercetin inhibits platelet aggregation and essential components of the collagen-stimulated platelet activation pathway in man: A pilot study. *British Journal of Nutrition, 96*(3), 482–488.
87. Jalili, T., et al. (2006). Quercetin-supplemented diets lower blood pressure and attenuate cardiac hypertrophy in rats with aortic constriction. *Journal of Cardiovascular Pharmacology, 47*(4), 531–541.
88. Jung, Y.-S., et al. (2002). Antithrombotic effect of onion in streptozotocin-induced diabetic rat. *Prostaglandins, Leukotrienes and Essential Fatty Acids (PLEFA), 66*(4), 453–458.
89. Karagoz, H., et al. (2009). Comparison of efficacy of silicone gel, silicone gel sheeting, and topical onion extract including heparin and allantoin for the treatment of postburn hypertrophic scars. *Burns, 35*(8), 1097–1103.
90. Kim, J. W., Kim, Y. S., & Kyung, K. H. (2004). Inhibitory activity of essential oils of garlic and onion against bacteria and yeasts. *Journal of Food Protection, 67*(3), 499–504.
91. Koc, E., et al. (2008). An open, randomized, controlled, comparative study of the combined effect of intralesional triamcinolone acetonide and onion extract gel and intralesional triamcinolone acetonide alone in the treatment of hypertrophic scars and keloids. *Dermatologic Surgery, 34*(11), 1507–1514.
92. Kook, S., Kim, G.-H., & Choi, K. (2009). The antidiabetic effect of onion and garlic in experimental diabetic rats: Meta-analysis. *Journal of Medicinal Food, 12*(3), 552–560.
93. Kumari, K., & Augusti, K. (2002). *Antidiabetic and antioxidant effects of S-methyl cysteine sulfoxide isolated from onions (Allium cepa Linn) as compared to standard drugs in alloxan diabetic rats.*
94. Kumari, K., & Augusti, K. (2007). Lipid lowering effect of S-methyl cysteine sulfoxide from Allium cepa Linn in high cholesterol diet fed rats. *Journal of Ethnopharmacology, 109*(3), 367–371.
95. Yu, C. S., et al. (2010). Quercetin inhibited murine leukemia WEHI-3 cells in vivo and promoted immune response. *Phytotherapy Research: An International Journal Devoted to Pharmacological and Toxicological Evaluation of Natural Product Derivatives, 24*(2), 163–168.
96. Zhou, Y., et al. (2011). Consumption of large amounts of Allium vegetables reduces risk for gastric cancer in a meta-analysis. *Gastroenterology, 141*(1), 80–89.
97. Zurada, J. M., Kriegel, D., & Davis, I. C. (2006). Topical treatments for hypertrophic scars. *Journal of the American Academy of Dermatology, 55*(6), 1024–1031.
98. Akash, M. S. H., Rehman, K., & Chen, S. (2014). Spice plant Allium cepa: Dietary supplement for treatment of type 2 diabetes mellitus. *Nutrition, 30*(10), 1128–1137.
99. Albanesi, M., et al. (2019). Immunological characterization of onion (Allium cepa) allergy. *Advances in Dermatology and Allergology/Postępy Dermatologii i Alergologii, 36*(1), 98–103.

100. Armentia, A., et al. (2020). Allergic hypersensitivity to garlic and onion in children and adults. *Allergologia et Immunopathologia, 48*(3), 232–236.
101. Brüll, V., et al. (2015). Effects of a quercetin-rich onion skin extract on 24 h ambulatory blood pressure and endothelial function in overweight-to-obese patients with (pre-) hypertension: A randomised double-blinded placebo-controlled cross-over trial. *British Journal of Nutrition, 114*(8), 1263–1277.
102. Burak, C., et al. (2017). Higher plasma quercetin levels following oral administration of an onion skin extract compared with pure quercetin dihydrate in humans. *European Journal of Nutrition, 56*, 343–353.
103. Campanati, A., et al. (2010). Effect of allium cepa-allantoin-pentaglycan gel on skin hypertrophic scars: Clinical and video-capillaroscopic results of an open-label, controlled, nonrandomized clinical trial. *Dermatologic Surgery, 36*(9), 1439–1444.
104. Choi, E.-Y., et al. (2015). Effect of onion peel extract on endothelial function and endothelial progenitor cells in overweight and obese individuals. *Nutrition, 31*(9), 1131–1135.
105. Conti, V., et al. (2020). Effectiveness and tolerability of a patch containing onion extract and allantoin for cesarean section scars. *Frontiers in Pharmacology, 11*, 569514.
106. Draelos, Z. D., et al. (2010). Evaluation of an onion extract, Centella asiatica, and hyaluronic acid cream in the appearance of striae rubra. *Skinmed, 8*(2), 80–86.
107. Ebrahimi-Mamaghani, M., et al. (2014). Effects of raw red onion consumption on metabolic features in overweight or obese women with polycystic ovary syndrome: A randomized controlled clinical trial. *Journal of Obstetrics and Gynaecology Research, 40*(4), 1067–1076.
108. Güngör, E. S., et al. (2020). The efficacy of onion extract in the management of subsequent abdominal hypertrophic scar formation. *Journal of Wound Care, 29*(10), 612–616.
109. Jenwitheesuk, K., et al. (2012). Role of silicone derivative plus onion extract gel in presternal hypertrophic scar protection: A prospective randomized, double blinded, controlled trial. *International Wound Journal, 9*(4), 397–402.
110. Jeong, S., et al. (2020). Effect of steamed onion (ONIRO) consumption on body fat and metabolic profiles in overweight subjects: A 12-week randomized, double-blind, placebo-controlled clinical trial. *Journal of the American College of Nutrition, 39*(3), 206–215.
111. Lee, J.-S., et al. (2016). Onion peel extract reduces the percentage of body fat in overweight and obese subjects: A 12-week, randomized, double-blind, placebo-controlled study. *Nutrition Research and Practice, 10*(2), 175–181.
112. Nakayama, Y., et al. (2020). Effects of onion extract containing concentrated cysteine sulfoxides on sleep quality: A randomized, double-blind, placebo-controlled, crossover study. *Food Science and Biotechnology, 29*(12), 1755–1762.
113. Nishimura, M., et al. (2019). Effect of daily ingestion of quercetin-rich onion powder for 12 weeks on visceral fat: A randomised, double-blind, placebo-controlled, parallel-group study. *Nutrients, 12*(1), 91.
114. Wananukul, S., et al. (2013). A prospective placebo-controlled study on the efficacy of onion extract in silicone derivative gel for the prevention of hypertrophic scar and keloid in median sternotomy wound in pediatric patients. *Journal of the Medical Association of Thailand, 96*(11), 1428–1433.
115. Yuan, X., et al. (2021). Onion extract gel is not better than other topical treatments in scar management: A meta-analysis from randomised controlled trails. *International Wound Journal, 18*(3), 396–409.
116. Zhang, J., et al. (2021). Association between dietary onion intake and subclinical hypothyroidism in adults: A population-based study from an iodine-replete area. *Endocrine, 74*, 616–624.
117. Zhou, X.-F., Ding, Z.-S., & Liu, N.-B. (2013). Allium vegetables and risk of prostate cancer: Evidence from 132,192 subjects. *Asian Pacific Journal of Cancer Prevention, 14*(7), 4131–4134.
118. Augusti, K. (1996). Therapeutic values of onion (Allium cepa L.) and garlic (Allium sativum L.). *Indian Journal of Experimental Biology, 34*(7), 634–640.

Chapter 18
Garlic

Sara Zafar, Nazia Aslam, Muhammad Zia-Ul-Haq, Shagufta Perveen, and Naeem Iqbal

18.1 Introduction

Kingdom: Plantae
Phylum: Tracheophyta
Class: Liliopsida
Order: Asparagales
Family: Amaryllidaceae
Subfamily: Allioideae
Genus: *Allium*
Scientific name: *Allium sativum* L.

Herbaceous geophytes, with bulbs or rhizomes. Leaves simple, entire, linear to broadly lanceolate or strap-shaped, sometimes sheathing at base forming a pseudostem. Peduncle central or lateral, subtended by 2 or more spathe valves. Flowers bisexual, regular or nearly so, umbellate (rarely solitary) with one to many flowers, interrupted by ephemeral hyaline bracts. Perianth segments biseriate, equal or subequal, inserted above the ovary, free or tube-like when united, sometimes bearing a false corona. Stamens 6, rarely more (*Gethyllis*), lying opposite the perianth segments, arising from the tube or segment's base. Filaments filiform or expanded at base, connate into a pseudocorona; anthers 2-locular, basifixed, introse and have longitudinal slits for opening. Ovary have 3 locules, inferior, with axile placenta-

S. Zafar (✉) · N. Aslam · S. Perveen · N. Iqbal
Department of Botany, Government College University, Faisalabad, Pakistan
e-mail: sarazafar@gcuf.edu.pk

M. Zia-Ul-Haq
Office of Research, Innovation and Commercialization, University of Engineering and Technology, Lahore, Pakistan

M. Zia-Ul-Haq et al. (eds.), *Essentials of Medicinal and Aromatic Crops*,
https://doi.org/10.1007/978-3-031-35403-8_18

tion, bearing one to many ovules per loculus; style slender, with a 3-lobed stigma. Fruit a loculicidal capsule or a berry; seeds globose or flattened, sometimes winged [1].

18.2 Morphology

Garlic is unknown from the wild but probably was derived from central Asian species *Allium longicuspis*. It is a hard clump forming perennial herb. It is usually with 15 cloves or bulbelts. Garlic forms bulb with a straight long stalk and slender leaves, at the scapes. The scape is slender and succulent at the start when the plant is young. When the plant mature and produce flower, it produce food, either processed in to pesto or sautéed. The bulb consists of many individual cloves. Each of the clove of the garlic is wrapped in it's own papery skin. For vegetative growth, this cloves is grown to produce a complete new garlic plant. It is a tricky plant to grow because the bulbs can rot and can't grow to a larger size bulb. For bulb to grow, the flowering scape must be cut in time, because the scape spent all of it's energy in growing flowers and will not grow to complete bulb [2].

18.3 Agronomy

18.3.1 Soil

Growth of garlic is very better in well irrigated fertile land and soil which is maximum in organic matter. To make the soil fertile for field, organic matter or compost is added into the fertile soils [3]. It makes the soil friable. Garlic is highly sensitive to the saline, alkali and water logged soils, same as with the case of onion [4, 5]. The electrical conductivity of a saturation extract for garlic crop is 3.9 dS/when the ECe level beyond this, the yield of crop results in decline.

18.3.2 Varieties

The selection of variety is very crucial because it directly effects the yield of the garlic.

The ICAR-DOGR has approved different varieties for different seasons and regions of the country. These are best for the cultivation on a larger area [6].

18.3.3 Types of Garlic

Garlic is also known as "Stinking rose". Garlic is found in three varieties like Tahiti, Italian and American [7]. American specie is of white skin and has the strongest flavor of all. Italian specie is pink and has a lot of cloves. Whereas, Tahiti, being the largest grows to a height of 3 inches.

18.3.4 Field Preparation

The field where we want to grow our garlic plant should be well ploughed by the use of plough mould board and tiller by the use of cultivator 3–4 times to eliminate debris and soil clods [8]. Organic matter equal to 75 kg N/ha should be added in the soil, when the soil is being ploughed at the last time. Beds of the same size must be prepared after the soil is being leveled up.

Mostly, straight beds of 1.5–2.0 m width and 4–6 m length are made. In rainy season or for the Kharif crops, preventions are carried out for water logging conditions by avoiding the formation of flats. For rainy seasons, broad bed furrows are made, consisting of 120 cm top, 45 cm width and 15 cm height. These prove to be appropriate for the sprinkler and drip irrigation [9].

18.3.5 Planting

Planting is performed after keen observation of the clover selection. Before planting, the individual cloves of the garlic from the seed of the garlic bulb should be removed. Before planting, peel off the skin of the cloves and separate the cloves without damaging the bottom plate of the cloves [10]. The damage of the bottom plate, made them waste to plant. The ideal size of the cloves for planting is >1.5 g. The clove should be healthy, it should not contain any spoiled, diseased and parasite. For preventing the cloves from fungal disease during establishment, the cloves must be dipped in the carbendazin solution of about 0.1%. For one hector area, almost 400–500 kg seed of the clove must be available [11]. Following measures should be in consideration, while planting the cloves vertically. Plantation should be 2 cm below the soil, spacing between the plants should be 10 cm and row to row space should be 15 cm [12]. Planting of garlic in Autumn gives the largest bulb because the Garlic start to grow on the shortest days and finally mature on the longest days [13].

18.3.6 Manures and Fertilizers

A sustainable agricultural system is one which provides safe, economically viable, conserve resources and enhances the environment. Garlic is a heavy feeder but after soil test results, the cloves should be recommended. Application of 75:40:40:40 kg NPKS/ha along with a doubling of two or three organic manures equivalent to 75 kg N/ha is recommended for Bihar, Chhattisgarh, Gujarat, Karnataka, Madhya Pradesh, Maharashtra, Odisha, Rajasthan and Tamil Nadu. Application of 100:50:50:50 kg NPKS +20 t FYM/ha is recommended for Haryana, Uttarkhand and Uttar Pradesh [14]. Organic matter should be mixed good in the soil and added before the last ploughing. One third of nitrogen and complete needs dose of Phosphorus, Sulphur and Potassium must be added as basal at the time of planting of crop [15]. After 30 and 45 days of planting of crop, remaining two third of the N2 must be added after equal intervals [16].

18.3.7 Irrigation

Garlic is very sensitive to moisture throughout its growing season due to the absence of deep root system. Proper care should be there for irrigation of the garlic, where it should be irrigated right after planting and at 7–10 days interval depending upon the soil moisture available [17].

18.3.8 Harvesting

Harvesting of garlic crop start when the top of the leaf start dry, change of color and also bend towards the ground surface. Another sign of the maturity of the bulb is the reduction in the thickness of the leaves of the sheath that surrounding the portion of the bulb [18]. Normally, garlic matures in between 130–180 days that totally depending upon the location, season and cultivar [19].

18.4 Pest and Disease

18.4.1 Black Aphids

Called black aphids for their separation coloration, these small, black bugs will suck the sap out of the leaves and the stem of your garlic. An organic, oil-based spray will clog their spiracle (the small openings that allow the critters to breathe), and suffocate them [20].

18.4.2 Millipedes

Millipedes are the caterpillar or worm-like creatures that have all those legs! For some people, they are the things their nightmares are made of, and they can certainly be on the creepier side of creepy crawlies [21]. Millipedes will eat potatoes, bulbs, and tubers. The millipede lives in areas that have moist mulch or compost. Therefore, to help reduce the threat, clean out all of the old mulch around a plant and swap it with new mulch. Something else that seems to work with them is the following mixture: 6 tablespoons eucalyptus oil 6 tablespoons Dawn® dishwashing liquid (original scent) 1 quart water Mix and spray on the ground at night [22].

18.4.3 Maggot

The adult onion maggot is a gray that lays long, white eggs around the base of the plant. The larvae are tiny and white, and will bore into the plant. The mature larva is about 2/5 inches long, with feeding hooks attached [23]. The onion maggot only has a 2 to 4-week lifespan, but one maggot can lay several hundred eggs during that short life. Winters do not kill them, either, as the pupae will overwinter in the soil. The damage that the onion maggot causes includes stunted and/or wilted seedlings, breakage of the plant at the soil line when you try to pull it, and deformed bulbs, which may be more susceptible to storage rot [24]. To try to get the onion maggot under control, good sanitation is key. In addition, make sure to remove all bulbs from the soil at the end of the season so there is no food available [25].

18.4.4 Wheat Curl Mite

Common to the central plains area of the United States, the wheat curl mite is a microscopic mite that transmits wheat streak mosaic. Noticeable damage from this infection usually only appears when the infestation is severe. Leaves are streaked or twisted and growth is stunted. The worst damage may be found on stored bulbs, as the cloves may dry and crumble. If erected bulbs are planted the following season, the plant can then be elected by yellow streak virus and the leaves will be damaged [25]. The risk of wheat curl mite may be lessened by a hot water treatment for the cloves, which is where they are quickly dipped in hot water, below boiling (which can kill the clove) [26].

18.4.5 Eriophyd Mite

The eriophyd mite is a very common garlic pest. The mite is very, very tiny and may not even be noticed until the damage has begun and the leaves are twisted with yellow or light-green streaks [27]. At this point, the leaves are virtually destroyed. Early on, the leaves may not sprout, and if they do, the leaves may not separate. In addition, if bulbs are stored for too long, they may come under attack as well, with withered cloves and soft bulbs being the primary after erects. Predatory mites will help keep these under control, which just goes to show the benefits of "good" pests [28].

18.4.6 Thrips

The thrips is quite common in warm weather. At only 1/25 of an inch in length, they can hide in the angles of the leaves. The adults are pale yellow to light brown, while the nymphs are lighter in color, as well as smaller [29]. The thrips feeds on the surface of the leaves, which in turn will cause them to lighten to a white or silver color. Again, there are some predatory mites that will help in the control of the thrips [30].

18.4.7 Nematodes

The nematode is a little, worm-like creature, which lives inside the host. They don't need water, so they can survive for years in the soil and will continue to survive until there are no more hosts available to them [31]. There are various types of nematodes, with specific symptoms presenting themselves once infestation has begun. However, there are a few common symptoms, including stunted plants, yellowing, swelled stems, and deformed bulbs. One type of nematode is the stem/bulb nematode. Almost invisible to the eye, it is a very tiny worm that invades the garlic's tissue, causing the base area of mature plants to swell, and can cause sponginess with long splits. The tissues, leaves, and stems will rot, with the leaves and stems also twisting [32]. The plant will also have stunted growth. As nematodes are soil borne, they are usually introduced through the movement of the soil, which is typically the result of humans. To help manage this problem, a hot water bath may help kill bulb infestation. However, once an area of the garden is infested, rotational planting for a few years with something other than garlic or onion is the only thing that will help in getting rid of these pests [33].

18.5 Diseases

18.5.1 Garlic Rust

Garlic rust is a fungal disease that not only an erects garlic, but the other alliums as well. The fungus appears as orange and black specks on the leaves. It flourishes when the weather is cool, with low sunlight and high humidity being its ideal environment [34]. As a result, when these conditions occur, do not water until late in the day. The spores are transported by the wind, and as a result, an infection in one part of the garden where your garlic or onions are could an erect the other parts of the garden where onions, garlic, and leeks are planted [35]. Unless the infection is severe, the garlic will still be edible. However, severe cases can drastically an erect bulb development, with the possibility of killing the crop, especially if it occurs early in the season. Although no resistance to garlic rust is known, there are a few things that can help keep it at bay [36]. Plant your garlic, onions, or leeks in well-drained soil, in the sunniest spot you can and, with plenty of space between the rows for good air circulation. Clip an elected leaves as soon as the rest signs of rust appear and throw the leaves in the garbage [37]. This should not elect either the stalk or the garlic. Do not compost. If you have been around garlic rust, wash your hands and clothes, and disinfect your shoes. Finally, once again, use crop rotation and do not plant anything from the *Allium* genus in that area for at least 3 years [38].

18.5.2 Downy Mildew

Downy mildew is a furry, whitish growth with yellow discoloration in the leaves. The pathogens can survive for many years as oospores. Downy mildew needs moist conditions to spread, and one stage of the spore can even swim, allowing it to spread and infect through the use of free water, as well as wind and rain, which in turn make it airborne as well [39]. Downy mildew will kill young plants and stunt the growth of older ones. If diseased bulbs are stored, the necks will turn black and shrivel, the outer scales will become wet, and sprouting may occur [30].

18.5.3 Garlic Mosaic Virus

Transmitted by aphids, symptoms of garlic mosaic virus include chlorotic mottling and stripes on the rest leaves to appear, with mature leaves having light-yellow, broken stripes. The mature leaves will also have what looks like a mosaic pattern to them, from which the virus gets its name. Symptoms are usually mild on young leaves, but growth will be stunted and bulbs will be small with few cloves. Because the virus's elects are so mild, however, there may be no symptoms at all showing [40].

18.5.4 Botrytis Rot

Botrytis rot is a fungus that attacks plants after there has been warm, wet weather. Also called neck rot, symptoms include wet stems and a gray, fuzzy fungus. It can also develop on stored bulbs. Steps for prevention include drying the garlic as quickly as possible after harvest and providing good air circulation. Store in a cool place. When planting, use healthy bulbs and allow plenty of room between rows [41].

18.5.5 Basal Rot

The signs of basal rot appear slowly, usually beginning with the leaves yellowing and eventually dying. There can also be a white fungus at the base of the bulb, which will then lead to rot either before or after the harvest, a erecting some or all of the cloves. Basal rot prefers high temperatures and will attack already diseased or pest-damaged plants. Not all infected bulbs will show symptoms, so make certain to check each plant carefully. Basal rot is spread through infected cloves, contaminated soil improvement, or contaminated tools or equipment use [42]. Tips for management include removal of infected plants immediately after detection and avoiding planting infected cloves. Do not move infected soil and disinfect tools and equipment that have come in contact with the rot. This includes shoes! A quick hot water treatment for cloves may reduce the risk of basal rot by about half as well, helping to keep your stored garlic safer [43].

According to the United Nation's Food and Agriculture Organization (FAO) almost 22.33 million metric ton of garlic is produced worldwide [44]. The largest garlic producing continent of the world is Asia and produces more than 80% of the garlic in the world. China is one of the largest garlic producing country in the world in 2010, followed by other countries like India, South Korea, Egypt, Russia, Myanmar, Ethiopia, USA, Bangladesh and Ukraine respectively [45] China contributes about 77% of the garlic production of the world output, which is equal to 18.56 MMT of garlic [46]. In case of garlic import, it is the 19th most important commodity of Pakistan. The other commodities include palm oil, rapeseed, sugar refined, cotton lint, cake of soybeans, chick peas, sunflower seed, dry onion, dry peas, jute, tea, wheat, recants, flour of wheat, tomatoes, residuals of fatty subs, lentils and fatty acids [45] From the year 2001–2010, Pakistan was in the list of top 10 countries to produce garlic [47]. Whereas in the years 2001 and 2002, Pakistan was on the 13th and 14th position to produce garlic. In just two years, 2003 and 2008, Pakistan was in the list of top 20 countries of the garlic production [48]. In these years, Pakistan was both the importer and exporter of the garlic. Whereas in all other years, Pakistan maintained it's position of being in the top 10 countries [45]. Punjab produced about 44% of the total garlic of the Pakistan to export, followed by other provinces like Khyber Pakhtunkhwa, Baluchistan and Sindh. Production of garlic and increase in area is progressing well in this field [49, 50].

Table 18.1 Garlic production in Pakistan

Province	Khyber Pakhtunkhwa (KPK)	Punjab	Baluchistan	Sindh
Production percentage	35%	44%	13%	8%
Total garlic production	32,205 tons	24,143 tons	7880 tons	6557 tons

In Pakistan, garlic production is on 7882 hectares at the present stage, with 70,925 tons, and the average yield is 8.99 tons/ha [51] (Table 18.1).

Baluchistan, overall produced 7880 tons of garlic, which is the lowest of all the provinces of Pakistan. In Baluchistan, Sibi region produce almost 40% of the total garlic production and compared to the other province, like Ziarat, which is the main growing region of the Baluchistan [50]. To fill the yield gap, the high-yielding garlic varieties are produced on the mainland. The agriculture extension workforce is working for the awareness, technology transfer and solving grower's difficulties [52].

18.6 Chemical Constituents

The most prominent quality feature of the product of the garlic is the prominent flavor of cloves, and all of this is due to of complex biochemical reactions [52, 53]. The mainstream compounds that are responsible for that flavor are compound that contain sulphur and non-volatile amino acids, most important of them are alliin or S-allyl-cysteine sulfoxide that consist of the most predominant precursors of garlic flavor [54]. Besides these from their flavor attributes, these sulfur compounds are also necessary for the renowned medicinal properties of garlic, and in addition may enhance the formation of glutathione, with the help of that we can find important antioxidant functions [55] . There are also some important volatile compound with special bioactive attributes are ajoenes [24], as also many sulfur-containing compounds, such as allicin, such as diallyl-, 1,2-vinyldithiin, methyl allyl-, allixin and S-allyl-cysteine and sulfides, and dipropyl mono-, di-, tri- and tetra-sulfides, that are made after the thiosulfinates decomposition [56]. The volatile behavior of these bioactive molecules is important for the protective mechanism of garlic in response to pathogens and pests [57]. These can also counterpart the lesions and damage of the cell [58]. When the cell membrane is damaged, the odor and volatile organosulfur compounds of the garlic are released. These damaged released the Alliin and other sulfoxides, by the enzyme alliinase, present in the vacuole [33]. Garlic is rich in many importanat vitamins like vitamins C and B, antioxidants minerals, flavonoids [59]. Besides these volatile compounds, non-volatile compounds are also produced by the garlic plants. These non-volatile compounds are important for the therapeutic and medicinal properties [60]. These important compounds include saponins, nitrogen oxides, proteins, amides, flavonoids and sapogenins [56]. For the synthesis of ACSO's, following important Γ Γ-glutamyl peptides components are produced like γ-glutamyl-S-trans-1-propenyl-cysteine (IsoGluAlC), γ-glutamyl-Smethyl cysteine and γ-glutamyl-S-2-propenyl cysteine (GluAlC) [61]. Among all regularly used vegetables, garlic is considered as one of the important source of total phenolic compounds. It is one of the main ingested material per capita in the human diet [56]. Cultivation procedure and growing conditions produced verities which are different in the quantities of the total phenolic compounds [62]. These can be detected by several ecotypes and genotypes [63]. Growing conditions also changed the other chemical makeup of the plant like pH, contents of carbohydrates and total soluble solids. It variates in different genotypes [64].

18.7 Medicinal Uses

Oral treatment of the disease gonorrhea is performed by the decoction of the above parts. A mixture of the plant plus 3 or 4 seed pods is taken. The juice of fresh leaf is calculated orally for gonorrhea, bilharziasis, stomach troubles, and as an antiemetic. Swelling and cuts are applied with powdered leaves [65]. For chest pain, decoction of the leaves is taken. Steam of the boiling leaves is used for the inflamed eyes. To cure purulent eye infection, water extract of the dried seeds are used. Before extracting, the seeds are macerated in the water. For curing aphrodisiac, fresh roots are chewed [66].

Combination of garlic with honey is also recommended for aphrodisiac as oral medicine [67]. The seeds of garlic are toxic as reported by the reports while half portion of the seed is even reported as lethal. The seed coat of the seed should be broken as it approved to be toxic [68]. The symptoms of the acute gastroenteritis are nausea, diarrhea and vomiting which are followed by convulsions, water deficiency and even death. The pulp of this plant is effective both for men and women. Like for women, it used to facilitate childbirth. Whereas for men, it used to cure aphrodisiac when taken orally. Orally used seeds are abortive and aphrodisiac. For flue and cough like diseases the oral intake of leaves is also beneficial [69]. Pine needles when taken along with garlic, used to cure spring colds and stuck congestion. When this aromatic pairing is used with delicious tea and honey, it open sinuses and loosen stuck muscles [70].

Dried leaves and roots are used to taken hot extract and applied for the eye diseases. For emmenagogue, hot water extract of the root of the garlic is used, where it is taken as oral medicine [71]. Root brew is taken orally to produce abortion. Other uses of taken hot water extract as oral medicine include abortifacient, to prevent conception and as an anti-fertility agent [72]. For females, In Unani and Ayurvedic, seeds of the garlic are used as a poultice in vagina. Here it works as an abortifacient. Whereas for males, seeds in the boiled milk is recommended for aphrodisiac, by the Unani and Ayurvedic. It is reported that the Abrin toxic action is destroyed by the boiling of seeds in milk. The seeds of garlic are also used as birth control medicine, whereas, it is used as with completely covered with aggary and swallowed during menstrual cycle [73]. It prevents the conception of the birth for at least one year. In Unani system of medicine, hot water extract is used for stimulating the sexual desire, when taken orally. Other used include, as purgative, painful swelling sand, tuberculosis and aphrodisiac. This plant juice is inserted intra-vaginally to induce abortion. For abortifacient process, dried seeds are taken orally [71].

18.7.1 Anti-Oxidant Activity

Antioxidants are the effective molecules in controlling the free radicals produced inside the cells, which damaged the cells from inside. The source of these antioxidants are herbs and spices. These have a major role in redox signaling and antioxidants defense [74]. Furthermore, the phenolic compounds are the main elements used inside these plants which help in the better performance of herbs to reduce the toxic effects of oxidants. Cloves have the highest total phenolic compounds among all the eighteen spices and herbs. Following are the polyphenols found in cloves like quercetin, acetyl eugenol, protocatechuic acid, eugenol, gallic acid,, p-coumaric acid, syringicacid and Kaempferol [75].

18.7.2 Alkaline Phosphatase Inhibition (Analgesic Activity)

For analgesic activity examination, the seed oil of the garlic was taken orally along with Petroleum ether extract [76]. It proved to be effective on the uterus of the rats. The medicine was applied intra-peritoneally to mice. An extract was made at the ratio of 1:1 of the aerial parts of the plants with ethanol/water. The dosage applied was 500.0 mg/kg.

18.7.3 Anthelmintic Activity

For diagnosing, anthelmintic activity, the 15.8 mg/mI seeds water extract was applied on *Caenorhabditis elegans* L, which showed weak activity [23]. As compared to the roots and stem of the plant was found to be active to show anthelmintic activity on on schistosomules of the trematode *Schistosoma mansoni* and cystercoids of the cestode Hymenolepis diminuta, in vitro [77].

18.7.4 Antibacterial Activity

Antibacterial activity was observed in the following bacterial species *Salmonella typhosa, Agrobacterium tumefaciens, Escherichia coli, Staphylococcus aureus and Bacillus subtilis.* The extract of the aerial parts of the plants were found to be ineffective in showing antibacterial activity [78]. Whereas the seed extract along with ether, was found to be effective in case of *Staphylococcus aureus*. The ethanol (95%) extract was effective on *Staphylococcus aureus* and *Escherichia coli.* Elements used for the antibacterial activity involve carvacrol, terpinen-4-ol, β-pinene, eugenol, isoeugenol, myrcene, 1,8-cineole and α-pinene [79].

18.7.4.1 Anticonvulsant Activity

For determining the anticonvulsant activity, ethanol extract 70% of the fresh root was applied intra-peritoneally on both the sexes of mice at different dosage, for convulsions induced by metra-zole [80]. This found to be ineffective for convulsions induced by strychnine. The extract was of ethanol and water in the ratio of 1:1n and dosage use was 500.0 mg/kg proved to be ineffective for electro-shock induced convulsions [81].

18.7.4.2 Antidiarrheal Activity

Chromatographic fraction of dried seeds, administered intragastrically to rats at a dose of 10.0 mg/kg, was active vs castor oil-induced diarrhea [77].

18.7.4.3 Antiestrogenic Effect

Root extract of the ethanol (95%) was active for the anti-fertility effect of mice, when applied at the dosage of 10.0 mg/kg. The extract of seeds was also prepared with Chloroform/methanol and applied sub-cutaneously to the mice. 50.0 mg/animal was used for this experiment and found to be active. When ethanol extract of the seeds was made and applied subcutaneously and orally to the mice, it appears to be inactive [81]. The dosage used was 1.0 mg/animal. Similarly, ethanol and water extract found to be inactive for mice as compared to the petroleum ether extract. The extract of leaves of garlic with petroleum, water and ethanol applied to mice, appeared inactive. Male rats were applied consistently with 100 mg/kg ethanol seed extract intra-gastrically for almost 2 months, appeared to be active [82]. There observed decrease in the number of the pregnant females (Table 18.2).

Dried seeds extract of the Ethanol/water in the ratio of 1:1 was found to be active when applied at the dosage of 250.0 mg/kg [82]. Consequently, no female got pregnant out of 20 when mated with 10 males for about 2 months. After the removal of treatment, pregnancies were observed. When the human females were administered with the hot water extract of the plants, the extract appeared to be active [83]. The dosage applied was 0.28 gm/person. Extract used was the mixture of *Polianthes tube* Rosa, *Piper longum* (fruit), *Piper betele*, *Abreus precatory*, *Ferula assafoetida* and *Embelia ribes* (fruit). The treatment was done for 20 days, starting from the second day of menstruation. Two doses were taken daily and sexual intercourse was prohibited during the whole treatment period. The experiment was found to be effective for 4 months. The biological activity was completely observed [84]. Seed

Table 18.2 Activity of garlic plant parts

Plant organ	Extract	Gender of rat	Dosage	Status	Effect
Diallyl trisulphide	Invitro	Rat	7.5 mg/ml	30	Inhibition of spermatozoa motility
Bulb	Crude extract	Male	500–1000 mg/kg/day	28 days	Antispermatogenic
Garlic powder	Gavage	Male	50 mg/kg/day	45–75 days	Histological alterations in somatic cells, arrest of spermatogenesis
Garlic powder	Oral	Male	0.8 g/100 g animal	28 days	Increased testosterone secretion
Raw garlic juice	Oral drinking water	Rat	60 mg/kg/day	21 days	Toxic affecting growth
Crude garlic	Oral	Male	20 g/100 g	4 months	Apoptosis of germ cells
Crude garlic	Oral	Male	15 g/100 g	30 days	Apoptosis of spermatocytes
Crude garlic	Oral	Male	30 g/100 g	30 days	Diminution of testosterone
Fresh bulb	Water extract	Female	500 mg/day	Inactive	Anti-estrogenic

oil was also used for the experimental activity and found to be active both male and female mice, when applied in the dosage of 150. Mg/animal and 25.0 mg/animal, respectively.

18.7.5 Anti-Diabetic Activity

Alpha amylase activity is reversed in the diabetic patients. Phytochemicals present in the herbs, help to alleviate the harmful effect of alpha amylase. Along with the phenolic compounds, many other bioactive compounds help to reduce the production of false alpha amylase [85]. The hemoglobin A1C (HbA1c) test measures the amount of hemoglobin in the blood. If the glucose level in the blood is high, the amount of hemoglobin A1C will also be high. To overcome this condition, the herbs are best. Herbs contains the compounds which help to reduce the blood sugar level and improve diabetes control. Sprague-Dawley rats was used as the model to search for anti-diabetic activity, whose dose of 2% w/w was effective to give results. It results in the reduction of increase blood sugar level [86].

18.7.6 Anti-Cancer Effect

Epidemiological studies have shown that consumption of herbs and spices shows a positive correlation in controlling the cancerous effect. This can only be due to the presence of antioxidants in the herbs like garlic. [87] have shown the anticancer potential of steamed, fresh and dried herbs using human Hela cancer cells. Of all these, the steamed herbs were more effective as compared to the fresh and dry herbs [88].

18.7.7 Anti-inflammatory Effect

Inflammation is a natural condition, which arises due to some defective or unusual condition inside the body. This inflammation can be acute or chronic. In this condition, three main pathways JAK-STAT, MAPK and NF-kB are activated and cause inflammation-associated diseases [89]. In controlling the anti-inflammatory activities, the herbs are more effective as these contain bioactive compounds which are effective against inflammation and produce health benefits. Cyclo-oxygenase 2 (COX-2), an inducible enzyme increases the inflammation by the production of Prostaglandin (PG) [90, 91], examined the anti-inflammatory activity of uncooked, cooked and digested of cloves. Results revealed that cloves used in different conditions inhibited the production of COX-2 which in turn reduced the inflammatory activity [92].

18.8 The Content of Trace Elements in the Medicinal Plants

18.8.1 Silicon

Silicon works as a catalyst in redox reaction and brought about the metabolism of fats, carbohydrates and proteins. Also, Silicon is directly involved in the formation of collagen of the human body. It's lesser amount increases the chances of diabetes mellitus [93].

18.8.2 Magnesium

Magnesium acts as a regulator of the functions of many organs and system, like endocrine system, which is involved in the secretion of insulin [94]. That's why magnesium deficiency causes decrease in the synthesis of insulin.

18.9 Garlic Superstitions & Folklore

According to Pliny, garlic and onions were invoked as deities by the Egyptians at the taking of oaths [95]. People of the Elysium (Lower Egypt), used to worship onions as well as took them as food. According to Egyptian, garlic was usually taken to increase endurance and strength and to reduce illness, so these were given to slaves of Egyptian routinely. According to Egyptian records, the pyramid builders were given with raw garlic, onions, beer and flatbread as food [96]. They were given more garlic, when they threatened to give up building the constructions of pyramids. It is estimated that the Pharaoh spent almost two million dollars on supplying garlic to the pyramid builders [97]. Garlic quantity was set aside to buy the prisoners at that time, especially during the era of King Tut like fifteen pounds garlic for a healthy male slave. Bulbs of garlic were obtained scattered for the room of King Tot's comb, when his tomb was excavated. In 1200 BC, during the era of Moses, when he led the Herbew slaves out of Egypt, they all complained of missing of some of the important things in life, like garlic, leeks, onions, melons and cucumbers. The Koreans use to pickled the garlic, whenever they cross a mountain path. They were of the view that, the garlic were disliked by the tigers [98]. In Mohammed's writings; he equates garlic with Satan when he describes the feet of the Devil as he was cast out of the Garden of Eden. According to the writings, garlic sprang up, where he put his left foot and onion from, where he put his right foot. In Palestinian traditions, for the successful wedding night, bride groom should had to wear garlic in their buttonhole. In Ayurvedic culture, garlic is considered best for aphrodisiac and it's ability to increase semen [99]. Ancient Greeks used to place the garlic on cross roads, as a symbol of goddess of childbirth and wilderness and to stay away from demons. The garlic was supposed to the evil spirits and causes them to lose their way. Before competition and battle, garlic was usually used to taken by the athletes and soldiers of Greek. Greek midwives used to hang garlic cloves in birthing rooms as the custom, to keep the evil spirit away. These became the common practice in most of European home, as the time passed [100]. It was considered as the sign of courage and inspiration by the Romans. Whenever the Romans conquered any country, they used to plant Garlic there, as considering the sign of courage. They believe that courage is then transferred to their battlefield. Homer reported that Ulysses owed his escape from Circe to "yellow garlic". Some of the herbalist like Culpepper considered a link between garlic and planet Mars, the planet also connected with the blood. According to the central European folk beliefs, garlic is considered as a powerful ward against werewolves, vampires and devils. Garlic should be worn by persons, rubbed on the chimneys, keyholes and hung in the window, just to avoid the devils entry. "The touch of vampire" was given to the name of disease caused by the bite of mosquitos. There, the garlic were used as mosquito repellent. Alexander Neck, recommends Garlic as a palliative of heat of the sun for labors especially [101]. Dreams of the "Garlic in the house", is considered lucky for that person. If the dream is about, "eating garlic", that man is considered to be going to discover secrets [102]. This old Welsh saying May indeed have merit as a health remedy: "Eat leeks in March and garlic in May, Then the rest of the year, your doctor can play."

The words "gar" and "leac" both refer to spears, specifically spear-shaped leaves. It's believed that garlic has both positive and negative powers. There are local folklore in central Europe that claim garlic wards off vampires, demons, werewolves, and the Evil Eye. Europeans who were frightened of the pandemic consumed whole garlic cloves during the Middle Ages to ward off the Black Death's anger. Grave robbers in eighteenth-century France used to mash whole garlic cloves in wine. They firmly believed that by ingesting this preparation, they would defend themselves against the fatal illness recognized as plague [103–107]. Garlic migrated from ancient Egypt to the Indus Valley's highly advanced ancient cultures (Pakistan and western India today). It then moved to China. It has been utilized in China since 2000 BC [108–113]. Because of its intense smell, the privileged classes in ancient India avoided garlic. Due to its stimulant properties, teenagers, widows, and nuns were prohibited from eating garlic. In India, it's still very common to hang red chilli, lemon, and garlic at the entrance or inside a store to keep off evil. Garlic has been used to cure a number of illnesses in the Middle East, East Asia, and Nepal, including asthma, high blood pressure, TB (tuberculosis), liver problems, diarrhea, bloating, constipation, intestinal worms, arthritis, diabetes, and cramps. Hippocrates, a physician in ancient Greece, supported the use of garlic to cure respiratory issues, infections, poor digestion, and fatigue. And hence, garlic has been used as a foodstuff and a medicine for thousands of years throughout the world.

In 1897, in the Romanian city of Bistritz, it was noted that the most well-known vampire, Bram Stoker's Dracula, had a definite dislike for garlic. There, garlic was handed to Johnathan Harker as protection when going to the castle of Dracula. Professor Van Helsing, a Dutch vampire hunter, uses garlic flowers to enhance Lucy Westenra's chamber, rub them all around the window frames, the door, and the fireplace, and fix her with a "wreath of garlic around her neck" to protect her from danger. Despite this, Lucy is attacked by Dracula, who then changes her into a vampire, forcing Dracula to kill her with a wooden stake and stuff garlic into her lips. According to local folklore, Dracula disliked garlic because it "so afflicted him that he has no strength" [103, 114–123].

Before the discovery of New World vampire bats, much of the original source used to inform these behaviors was thoroughly recorded from folklore that dates back to the middle times (ie, Desmodus and other genera).Moreover, there are odd similarities because the native peoples of tropical America used a similar method of defence against bites [106, 124–132].

The important effect of garlic would therefore seem to be as a repellent in both reality and fiction, leading us to think about the bioactive components in garlic and the potential physiological mechanisms that could result in such a result after eating or inhaling it [133–135]. These mechanisms include a decreased ability for vampires to quickly feed from victims with lowered blood pressure after consuming garlic and/or the increase of signs in people with porphyria as a result of the haemolytic effect of garlic (misdiagnosed vampires!).

18.10 Garlic Facts, Myths and Legends

Ancient Greek given the name "cordon" to the present day garlic. According to Folder and Blackwood, French physician Henri Leclerc derived this from scion radon which he translated as rose Puente, or "stinking rose". Garlic is being used for thousands of years for medicinal purposes [136]. Records shows that the garlic was used some 3000 year ago by Chinese people in medicines while about 5000 years ago by Sanskrit people in medicines. The Babylonians, Romans, Greeks and Egyptians used garlic for purpose of healing. Pasteur noted antibacterial activity of the garlic in 1858. Garlic was used to sure many disease like infections, snakebites, hyper-tension. In some of the culture it was used to ward off the evil spirits [137]. Garlic is of great benefit in controlling cardiovascular risk and decreasing cholesterol levels along with it's beneficial antimicrobial and antineoplastic properties. In the past, garlic was also used to get rid of many of the insects like mosquitos, slug etc. Garlic used from the prevention of mosquitos bites. Hippocrates in 300 BC, recommend garlic for various infections, leprosy, digestive disorders and cancer. Discords used I for the treatment of heart problems. Pliny used this plant as a treatment agent for epilepsy, tapeworm and leprosy. In World War 1, garlic was used to cure wounds of the soldiers which were on front line. In 1928, discovery of penicillin by Alexander Fleming replaced the use of garlic at home. But the war fought at a larger scale instead of depending on penicillin, still depended on garlic. There garlic worked as the antibiotic of choice. In Red Army Physicians, garlic was known as "Russian Penicillin", as the army was heavily dependent on the garlic.

These days, garlic is used by many of the herbalist to cure many of the disease like colds, coughs, bronchitis, ringworm, intestinal worm, fever, flue and cholesterol. It is mostly effective for gallbladder, digestive system and liver. Many reports are provided in the publish papers of the last two years, indicating that garlic is helping to prevent heart diseases and cancer, even in controlling the cancer which is spread widely or is severe.

18.10.1 Garlic Caution

There should be some safety measures to store the garlic. Garlic must not be left at the room temperature, when mixed with olive oil. It has an awesome flavor until botulism threatens it's safety [138]. Garlic mixed with vinegar, is safe to store. As the acidic level of the vinegar does not allow the growth of botulin bacteria. Therefore, the store garlic is safe from incubating and can be used easily. In Italy, to alleviate stomachaches, the garlic poultices are applied. In the early twentieth century, there was a common practice of the people to make their children to wear necklace made up of garlic cloves, just to prevent them from getting cold. This practice kept the children away from cold at that time. Ticks were also controlled by using garlic for almost 30 minutes. As, the garlic was used to kill the ticks in animals. It was also used as the repellent towards new infestations. Mouth and hoof disease of cattle was also treated with garlic. In 1955, a study was conducted in Russia. According to which the garlic extract used to eliminate heavy metals from the body and it is used as therapeutic cure. The garlic extract bind with heavy metals and eliminate through the body. Labors infected with lead poisoning, were treated with garlic extract and were observed the reduction of symptoms of poisoning.

Other experiments were performed in the japan where the persons were infected mercury and cadmium were treated with garlic and were observed improvement in their health. These metals were bounded with the garlic extract for the removal from the body. Herbs and medicinal plants provide health or therapeutic benefits. They prevent and treat several diseases. The treated diseases include cancer, intestinal, cardiovascular, uretic and metabolic diseases [82]. These medicinal plants and herbs have many applications in perfume, toothpaste, sugar, medicines, food industries, cosmetics, soap and beverages. These applications are regulated by different countries according to their own laws. Different countries may also have their own rules for labelling approved by health and nutritional authorities. The use of these medicines required several factors to be considered. Like the variations in the exposure of different things in different communities [58]. Interaction of these herbs with food or other bioactives is also considered an important element for it's use. In conclusion, these various parameters, are important to consider when applied for the application for various purposes and for various industries.

References

1. Simpson, M. G. (2019). *Plant systematics*. Academic.
2. Aburjai, T., & Natsheh, F. M. (2003). Plants used in cosmetics. *Phytotherapy Research: An International Journal Devoted to Pharmacological and Toxicological Evaluation of Natural Product Derivatives, 17*(9), 987–1000.
3. Adefegha, A., et al. (2010). Inhibitory effects of aqueous extract of two varieties of ginger on some key enzymes linked to type-2 diabetes in vitro. *Journal of Food and Nutrition Research, 49*(1), 14–20.
4. Bachmann, J. (2001). *Organic garlic production in Davis*. National Sustainable Agriculture Information Service.
5. Potgieter, J. (2006). *Verbal communication on macroelements application time*. Researcher, Limpopo Department of Agriculture.
6. Ahmad, R., et al. (2020). Quality variation and standardization of black pepper (Piper nigrum): A comparative geographical evaluation based on instrumental and metabolomics analysis. *Biomedical Chromatography, 34*(3), e4772.
7. Ahui, M. L. B., et al. (2008). Ginger prevents Th2-mediated immune responses in a mouse model of airway inflammation. *International Immunopharmacology, 8*(12), 1626–1632.
8. Kabir, M., Rahim, M., & Majumder, D. (2016). Productivity of garlic under different tillage methods and mulches in organic condition. *Bangladesh Journal of Agricultural Research, 41*(1), 53–66.
9. Aimbire, F., et al. (2007). Effect of hydroalcoholic extract of Zingiber officinalis rhizomes on LPS-induced rat airway hyperreactivity and lung inflammation. *Prostaglandins, Leukotrienes and Essential Fatty Acids, 77*(3–4), 129–138.
10. Akbari-Fakhrabadi, M., et al. (2019). Effect of saffron (Crocus sativus L.) and endurance training on mitochondrial biogenesis, endurance capacity, inflammation, antioxidant, and metabolic biomarkers in Wistar rats. *Journal of Food Biochemistry, 43*(8), e12946.
11. Akintobi, O., et al. (2013). Antimicrobial activity of Zingiber officinale (ginger) extract against some selected pathogenic bacteria. *Nature and Science, 11*(1), 7–15.
12. Alamgir, A., & Alamgir, A. (2017). *Pharmacognostical botany: Classification of medicinal and aromatic plants (MAPs), botanical taxonomy, morphology, and anatomy of drug*

plants (pp. 177–293). Therapeutic Use of Medicinal Plants and Their Extracts: Volume 1: Pharmacognosy.
13. Alvarez, M. V., et al. (2019). Valorization of an agroindustrial soybean residue by supercritical fluid extraction of phytochemical compounds. *The Journal of Supercritical Fluids, 143*, 90–96.
14. Andlauer, W., & Fürst, P. (2002). Nutraceuticals: A piece of history, present status and outlook. *Food Research International, 35*(2–3), 171–176.
15. Andrew, R., & Izzo, A. A. (2017). Principles of pharmacological research of nutraceuticals. *British Journal of Pharmacology, 174*(11), 1177–1194.
16. Diriba-Shiferaw, G. (2016). Review of management strategies of constraints in garlic (Allium sativum L.) production. *Journal of Agricultural Science, 11*, 186–207.
17. Arimalala, N., et al. (2019). Clove based cropping systems on the east coast of Madagascar: How history leaves its mark on the landscape. *Agroforestry Systems, 93*, 1577–1592.
18. Arun, N., & Nalini, N. (2002). Efficacy of turmeric on blood sugar and polyol pathway in diabetic albino rats. *Plant Foods for Human Nutrition, 57*, 41–52.
19. Azimi, P., et al. (2014). Effects of cinnamon, cardamom, saffron, and ginger consumption on markers of glycemic control, lipid profile, oxidative stress, and inflammation in type 2 diabetes patients. *The Review of Diabetic Studies: RDS, 11*(3), 258–266.
20. Bakri, I., & Douglas, C. (2005). Inhibitory effect of garlic extract on oral bacteria. *Archives of Oral Biology, 50*(7), 645–651.
21. Borek, C. (2001). Antioxidant health effects of aged garlic extract. *The Journal of Nutrition, 131*(3), 1010S–1015S.
22. Burden, A., et al. (1994). Garlic-induced systemic contact dermatitis. *Contact Dermatitis, 30*(5), 299–300.
23. Soni, K., Rajan, A., & Kuttan, R. (1992). Reversal of aflatoxin induced liver damage by turmeric and curcumin. *Cancer Letters, 66*(2), 115–121.
24. Block, E., et al. (1993). Organosulfur chemistry of garlic and onion: Recent results. *Pure and Applied Chemistry, 65*(4), 625–632.
25. Das, T., et al. (1993). Modification of cytotoxic effects of inorganic arsenic by a crude extract of Allium sativum L. in mice. *International Journal of Pharmacognosy, 31*(4), 316–320.
26. Du, C. T., & Francis, F. (1975). Anthocyanins of garlic (Allium sativum L.). *Journal of Food Science, 40*(5), 1101–1102.
27. Fletcher, R., Parker, B., & Hassett, M. (1974). Inhibition of coagulase activity and growth of Staphylococcus aureus by garlic extracts. *Folia Microbiologica, 19*, 494–497.
28. Heinle, H., & Betz, E. (1994). Effects of dietary garlic supplementation in a rat model of atherosclerosis. *Arzneimittel-Forschung, 44*(5), 614–617.
29. Hong, S., et al. (1992). Effects of garlic oil, garlic juice and allyl sulfide on the responsiveness of dorsal horn cell in the cat. *Hanyang Uidae Haksulchi, 12*(2), 621–633.
30. Mishra, R., et al. (2014). Management of major diseases and insect pests of onion and garlic: A comprehensive review. *Journal of Plant Breeding and Crop Science, 6*(11), 160–170.
31. Ross, I. A., & Ross, I. A. (2003). *Allium sativum L. medicinal plants of the world: Volume 1 chemical constituents, traditional and modern medicinal uses* (pp. 33–102). Humana Press.
32. Jain, R., & Vyas, C. (1972). Garlic in alloxan-induced diabetic rabbits. *The American Journal of Clinical Nutrition, 28*(7), 684–685.
33. Bloem, E., Haneklaus, S., & Schnug, E. (2010). Influence of fertilizer practices on S-containing metabolites in garlic (Allium sativum L.) under field conditions. *Journal of Agricultural and Food Chemistry, 58*(19), 10690–10696.
34. Kyo, E., et al. (2001). Immunomodulatory effects of aged garlic extract. *The Journal of Nutrition, 131*(3), 1075S–1079S.
35. Lee, Y., & Jang, J. (1991). Modifying effect of garlic and red pepper extracts on diethylnitrosamine-induced hepatocarcinogenesis. *Environ Mutagens Carcinog, 11*(1), 21–28.
36. Liu, J., Lin, R. I., & Milner, J. A. (1992). Inhibition of 7, 12-dimethylbenz [a] anthracene-induced mammary tumors and DNA adducts by garlic powder. *Carcinogenesis, 13*(10), 1847–1851.

37. Martin, N., et al. (1992). Experimental cardiovascular depressant effects of garlic (Allium sativum) dialysate. *Journal of Ethnopharmacology, 37*(2), 145–149.
38. Moriguchi, T., et al. (1994). Prolongation of life span and improved learning in the senescence accelerated mouse produced by aged garlic extract. *Biological and Pharmaceutical Bulletin, 17*(12), 1589–1594.
39. Devendar, P., & Yang, G.-F. (2019). *Sulfur-containing agrochemicals. Sulfur Chemistry*, 35–78.
40. Pop, A.-V., et al. (2015). Evaluation of antioxidant activity and phenolic content in different Salvia officinalis L. extracts. *Bull UASVM Food Science and Technology, 72*(2), 210–214.
41. Pruteanu, A., et al. (2018). Biochemical analysis of some vegetal extracts obtained from indigenous spontaneous species of (Thymus serpyllum L.). *Romanian Biotechnological Letters, 23*(5), 14013–14024.
42. Rakhi, N., et al. (2018). Data-driven analysis of biomedical literature suggests broad-spectrum benefits of culinary herbs and spices. *PLoS One, 13*(5), e0198030.
43. Rawat, S., et al. (2018). Hedychium spicatum: A systematic review on traditional uses, phytochemistry, pharmacology and future prospectus. *Journal of Pharmacy and Pharmacology, 70*(6), 687–712.
44. Rebey, I. B., et al. (2019). Bioactive compounds and antioxidant activity of Pimpinella anisum L. accessions at different ripening stages. *Scientia Horticulturae, 246*, 453–461.
45. Organization, W.H. (2010). *FAO/WHO expert meeting on the application of nanotechnologies in the food and agriculture sectors: Potential food safety implications: Meeting report.* World Health Organization.
46. Rezaei, M., & Ghasemi Pirbalouti, A. (2019). Phytochemical, antioxidant and antibacterial properties of extracts from two spice herbs under different extraction solvents. *Journal of Food Measurement and Characterization, 13*(3), 2470–2480.
47. Ribeiro-Santos, R., et al. (2015). A novel insight on an ancient aromatic plant: The rosemary (Rosmarinus officinalis L.). *Trends in Food Science & Technology, 45*(2), 355–368.
48. Saibabu, V., et al. (2015). Therapeutic potential of dietary phenolic acids. *Advances in Pharmacological Sciences, 2015*, 1–10.
49. Saleh, H. A.-R., et al. (2019). Plant growth, yield and bioactive compounds of two culinary herbs as affected by substrate type. *Scientia Horticulturae, 243*, 464–471.
50. Statistics, P.B.o. (2011). *Agricultural statistics of Pakistan 2010–2011.* Statics House, 21-Mauve area, G-9/1.
51. Schulze, A. E., et al. (2015). Honeybush herbal teas (Cyclopia spp.) contribute to high levels of dietary exposure to xanthones, benzophenones, dihydrochalcones and other bioactive phenolics. *Journal of Food Composition and Analysis, 44*, 139–148.
52. Sgorbini, B., et al. (2019). Evaluation of volatile bioactive secondary metabolites transfer from medicinal and aromatic plants to herbal teas: Comparison of different methods for the determination of transfer rate and human intake. *Journal of Chromatography A, 1594*, 173–180.
53. Randle, W., & Lancaster, J. (2002). *14 sulphur compounds in alliums in relation to flavour quality Allium. Crop Science*, 329.
54. Shahi, T., Assadpour, E., & Jafari, S. M. (2016). Main chemical compounds and pharmacological activities of stigmas and tepals of 'red gold'; saffron. *Trends in Food Science & Technology, 58*, 69–78.
55. Banerjee, S., Mukherjee, P. K., & Maulik, S. (2003). Garlic as an antioxidant: The good, the bad and the ugly. *Phytotherapy Research: An International Journal Devoted to Pharmacological and Toxicological Evaluation of Natural Product Derivatives, 17*(2), 97–106.
56. Lanzotti, V., Scala, F., & Bonanomi, G. (2014). Compounds from Allium species with cytotoxic and antimicrobial activity. *Phytochemistry Reviews, 13*, 769–791.
57. Sieniawska, E. (2015). Activities of tannins–from in vitro studies to clinical trials. *Natural Product Communications, 10*(11), 1934578X1501001118.
58. Hile, A. G., et al. (2004). Aversion of European starlings (Sturnus vulgaris) to garlic oil treated granules: Garlic oil as an avian repellent. Garlic oil analysis by nuclear magnetic resonance spectroscopy. *Journal of Agricultural and Food Chemistry, 52*(8), 2192–2196.

59. Rekowska, E., & Skupień, K. (2009). The influence of selected agronomic practices on the yield and chemical composition of winter garlic. *Journal of Fruit and Ornamental Plant Research, 70*(1), 173–182.
60. Skendi, A., et al. (2019). Aromatic plants of Lamiaceae family in a traditional bread recipe: Effects on quality and phytochemical content. *Journal of Food Biochemistry, 43*(11), e13020.
61. Jabbes, N., et al. (2012). Agro-morphological markers and organo-sulphur compounds to assess diversity in Tunisian garlic landraces. *Scientia Horticulturae, 148*, 47–54.
62. Sonmezdag, A. S., Kelebek, H., & Selli, S. (2016). Characterization of aroma-active and phenolic profiles of wild thyme (Thymus serpyllum) by GC-MS-Olfactometry and LC-ESI-MS/MS. *Journal of Food Science and Technology, 53*, 1957–1965.
63. Volk, G. M., & Stern, D. (2009). Phenotypic characteristics of ten garlic cultivars grown at different north American locations. *HortScience, 44*(5), 1238–1247.
64. Sreevidya, N., & Mehrotra, S. (2003). Spectrophotometric method for estimation of alkaloids precipitable with Dragendorff's reagent in plant materials. *Journal of AOAC International, 86*(6), 1124–1127.
65. Stanković, M. S., et al. (2017). Screening of selected species from Spanish flora as a source of bioactive substances. *Industrial Crops and Products, 95*, 493–501.
66. Ares, A. M., et al. (2021). Analysis of herbal bioactives. In *Aromatic herbs in food* (pp. 201–232). Elsevier.
67. Wang, B., et al. (2018). Regional variation in the chemical composition and antioxidant activity of *Rosmarinus officinalis* L. from China and the Mediterranean region. *Pakistan Journal of Pharmaceutical Sciences, 31*(1), 221–230.
68. Wu, S.-B., Long, C., & Kennelly, E. J. (2014). Structural diversity and bioactivities of natural benzophenones. *Natural Product Reports, 31*(9), 1158–1174.
69. Yamani, H., et al. (2014). Analysis of the volatile organic compounds from leaves, flower spikes, and nectar of Australian grown Agastache rugosa. *BMC Complementary and Alternative Medicine, 14*(1), 1–6.
70. Zielińska, S., & Matkowski, A. (2014). Phytochemistry and bioactivity of aromatic and medicinal plants from the genus Agastache (Lamiaceae). *Phytochemistry Reviews, 13*, 391–416.
71. Reddy, M., Reddy, K., & Reddy, M. (1989). A survey of plant crude drugs of Anantapur district, Andhra Pradesh, India. *International Journal of Crude Drug Research, 27*(3), 145–155.
72. Jain, R. (1993). Antitubercular activity of garlic oil. *Indian Drugs Bombay, 30*, 73–73.
73. Rajendran, B., & Gopalan, M. (1979). Note on the insecticidal properties of certain plant extracts. *Indian Journal of Agricultural Sciences, 49*, 295.
74. Roychoudhury, A., et al. (1993). Use of crude extract of garlic (Allium sativum L.) in reducing cytotoxic effects of arsenic in mouse bone marrow. *Phytotherapy Research, 7*(2), 163–166.
75. Sheela, C., & Augusti, K. (1992). Antidiabetic effects of S-allyl cysteine sulphoxide isolated from garlic Allium sativum Linn. *Indian Journal of Experimental Biology, 30*(6), 523–526.
76. Singh, J., Dubeyd, A., & Tripathi, N. (1994). Antifungal activity of Mentha spicata. *International Journal of Pharmacognosy, 32*(4), 314–319.
77. Dixit, S., & Dubey, A. (2017). Medicinal properties of garlic (Allium sativum L): A review. *Hort Flora Research Spectrum, 6*(1), 66–68.
78. Török, B., et al. (1994). Effectiveness of garlic on the radical activity in radical generating systems. *Arzneimittel-Forschung, 44*(5), 608–611.
79. Bag, B. B. (2018). Ginger processing in India (Zingiber officinale): A review. *International Journal of Current Microbiology and Applied Sciences, 7*(4), 1639–1651.
80. Bechtold, T., et al. (2003). Natural dyes in modern textile dyehouses—How to combine experiences of two centuries to meet the demands of the future? *Journal of Cleaner Production, 11*(5), 499–509.
81. Hammami, I., & El May, M. V. (2013). Impact of garlic feeding on male fertility. *Andrologia, 45*(4), 217–224.
82. Ross, I. A. (2001). *Medicinal plants of the world: Chemical constituents, traditional and modern medicinal uses. Totowa, New Jersey, 2*, 81–85.

83. Brower, V. (1998). Nutraceuticals: Poised for a healthy slice of the healthcare market? *Nature Biotechnology, 16*(8), 728–731.
84. Calva-Estrada, S., et al. (2018). Microencapsulation of vanilla (Vanilla planifolia Andrews) and powder characterization. *Powder Technology, 323*, 416–423.
85. Husnu, K., K. Başer, New trends in the utilization of medicinal and aromatic plants. 2005.
86. Lubbe, A., & Verpoorte, R. (2011). Cultivation of medicinal and aromatic plants for specialty industrial materials. *Industrial Crops and Products, 34*(1), 785–801.
87. Cheng, X.-L., et al. (2011). Steamed ginger (Zingiber officinale): Changed chemical profile and increased anticancer potential. *Food Chemistry, 129*(4), 1785–1792.
88. Ozdal, T., et al. (2021). Introduction to nutraceuticals, medicinal foods, and herbs. In *Aromatic herbs in food* (pp. 1–34). Elsevier.
89. Castro-Vargas, H. I., et al. (2019). Valorization of papaya (Carica papaya L.) agroindustrial waste through the recovery of phenolic antioxidants by supercritical fluid extraction. *Journal of Food Science and Technology, 56*, 3055–3066.
90. Chang, Y.-C., et al. (2003). Proinflammatory cytokines induce cyclooxygenase-2 mRNA and protein expression in human pulp cell cultures. *Journal of Endodontics, 29*(3), 201–204.
91. Baker, I., Chohan, M., & Opara, E. I. (2013). Impact of cooking and digestion, in vitro, on the antioxidant capacity and anti-inflammatory activity of cinnamon, clove and nutmeg. *Plant Foods for Human Nutrition, 68*, 364–369.
92. Chambers, A. H. (2019). *Vanilla (Vanilla spp.) breeding* (Vol. 6, pp. 707–734). Advances in Plant Breeding Strategies: Industrial and Food Crops.
93. Chantaro, P., Devahastin, S., & Chiewchan, N. (2008). Production of antioxidant high dietary fiber powder from carrot peels. *LWT-Food Science and Technology, 41*(10), 1987–1994.
94. Charles, D. J. (2012). *Antioxidant properties of spices, herbs and other sources*. Springer Science & Business Media.
95. Dugan, F. M. (2016). Seldom just food: Garlic in magic and medicine in European and Mediterranean traditions. *Digest: A Journal of Foodways and Culture, 5*(1).
96. Chauhan, B., et al. (2013). Current concepts and prospects of herbal nutraceutical: A review. *Journal of Advanced Pharmaceutical Technology & Research, 4*(1), 4–8.
97. Chen, F., et al. (2019). Endoplasmic reticulum stress cooperates in silica nanoparticles-induced macrophage apoptosis via activation of CHOP-mediated apoptotic signaling pathway. *International Journal of Molecular Sciences, 20*(23), 5846.
98. Cox, D. N., Koster, A., & Russell, C. G. (2004). Predicting intentions to consume functional foods and supplements to offset memory loss using an adaptation of protection motivation theory. *Appetite, 43*(1), 55–64.
99. Craig, W. J. (1999). Health-promoting properties of common herbs. *The American Journal of Clinical Nutrition, 70*(3), 491s–499s.
100. Crawford, P. (2009). Effectiveness of cinnamon for lowering hemoglobin A1C in patients with type 2 diabetes: A randomized, controlled trial. *The Journal of the American Board of Family Medicine, 22*(5), 507–512.
101. Danthu, P., et al. (2014). The clove tree of Madagascar: A success story with an unpredictable future. *Bois et forêts des tropiques, 320*(2), 83–96.
102. Deng, G.-F., et al. (2012). Potential of fruit wastes as natural resources of bioactive compounds. *International Journal of Molecular Sciences, 13*(7), 8308–8323.
103. Mantzioris, E., & Weinstein, P. (2021). Garlic as a vampire deterrent: Fact or fiction? *The Medical Journal of Australia, 215*, 541–543.
104. Locatelli, D. A., et al. (2015). Home-cooked garlic remains a healthy food. *Journal of Functional Foods, 16*, 1–8.
105. Rahman, M. S. (2007). Allicin and other functional active components in garlic: Health benefits and bioavailability. *International Journal of Food Properties, 10*(2), 245–268.
106. Santhosha, S. G., Jamuna, P., & Prabhavathi, S. (2013). Bioactive components of garlic and their physiological role in health maintenance: A review. *Food Bioscience, 3*, 59–74.

107. El-Saber Batiha, G., et al. (2020). Chemical constituents and pharmacological activities of garlic (Allium sativum L.): A review. *Nutrients, 12*(3), 872.
108. Petrovska, B., & Cekovska, S. (2010). Extracts from the history and medical properties of garlic. *Pharmacognosy Reviews, 4*(7), 106–110.
109. Bergamin, A., et al. (2019). Nutraceuticals: Reviewing their role in chronic disease prevention and management. *Pharmaceutical Medicine, 33*, 291–309.
110. Ried, K. (2020). Garlic lowers blood pressure in hypertensive subjects, improves arterial stiffness and gut microbiota: A review and meta-analysis. *Experimental and Therapeutic Medicine, 19*(2), 1472–1478.
111. Borrelli, F., Capasso, R., & Izzo, A. A. (2007). Garlic (Allium sativum L.): Adverse effects and drug interactions in humans. *Molecular Nutrition & Food Research, 51*(11), 1386–1397.
112. McCrindle, B. W., Helden, E., & Conner, W. T. (1998). Garlic extract therapy in children with hypercholesterolemia. *Archives of Pediatrics & Adolescent Medicine, 152*(11), 1089–1094.
113. Choi, S., Oh, D.-S., & Jerng, U. M. (2017). A systematic review of the pharmacokinetic and pharmacodynamic interactions of herbal medicine with warfarin. *PLoS One, 12*(8), e0182794.
114. Putnik, P., et al. (2019). An overview of organosulfur compounds from Allium spp.: From processing and preservation to evaluation of their bioavailability, antimicrobial, and anti-inflammatory properties. *Food Chemistry, 276*, 680–691.
115. Zhang, N.-D., et al. (2016). Traditional Chinese medicine formulas for the treatment of osteoporosis: Implication for antiosteoporotic drug discovery. *Journal of Ethnopharmacology, 189*, 61–80.
116. Ugwu, C. E., & Suru, S. M. (2016). The functional role of garlic and bioactive components in cardiovascular and cerebrovascular health: What we do know. *Journal of Biosciences and Medicines, 4*(10), 28–42.
117. Jain, M. K., & Apitz-Castro, R. (1993). Garlic: A product of spilled ambrosia. *Current Science, 65*(2), 148–156.
118. Jamir, T., Sharma, H., & Dolui, A. (1999). Folklore medicinal plants of Nagaland, India. *Fitoterapia, 70*(4), 395–401.
119. Morakinyo, A. (2008). Effects of aqueous extract of garlic (Allium Sativum). *Nigerian Journal of Health and Biomedical Sciences, 7*(2), 26–30.
120. Markey, T. (2013). 'Garlic and sapphires in the mud':'leeks' in their early folk contexts. *Leeds Studies in English*, 10–42.
121. Cherry, R. (2014). *Garlic, an edible biography: The history, politics, and mythology behind the world's most pungent food – With over 100 recipes*. Shambhala Publications.
122. Stewart, J. (2018). Blooming marvel: The garlic flower in Bram Stoker's hermeneutic garden. *Gothic Studies, 20*(1–2), 326–345.
123. Winkler, M. G., & Anderson, K. E. (1990). Vampires, porphyria, and the media: Medicalization of a myth. *Perspectives in Biology and Medicine, 33*(4), 598–611.
124. Harini, K., et al. (2013). Garlic: It's role in oral and systemic health. *Journal of Health and Allied Sciences NU, 3*(04), 017–022.
125. Loría Gutiérrez, A., et al. (2021). General aspects about Allium sativum – A review. *Ars Pharmaceutica, 62*, 471–481.
126. Singha, R., Kumara, N., & Kumarb, P. (2019). Garlic (Allium sativum): Mankind's health superstar. *Interdisciplinary Journal of Contemporary Research, 6*(6), 93–98.
127. Bhandari, P. R. (2012). Garlic (Allium sativum L.): A review of potential therapeutic applications. *International Journal of Green Pharmacy (IJGP), 6*(2), 118.
128. Singh, R., & Singh, K. (2019). Garlic: A spice with wide medicinal actions. *Journal of Pharmacognosy and Phytochemistry, 8*(1), 1349–1355.
129. Sripradha, S., et al. (2014). Garlic, its role in oral health – A review. *Research Journal of Pharmacy and Technology, 7*(6), 727–729.
130. Nagori, B., Solanki, R., & Sharma, N. (2010). Natural healing agent: Garlic, an approach to healthy life. *International Journal of Research in Ayurveda and Pharmacy (IJRAP), 1*(2), 358–366.

131. Singh, M., et al. (n.d.). *Medicinal use of garlic (Allium sativum L.)* (p. 44). isns.
132. Alam, K., Hoq, O., & Uddin, S. (2016). Medicinal plant Allium sativum. A review. *Journal of Medicinal Plants Studies, 4*(6), 72–79.
133. Parle, M., & Vaibhav, K. (2007). *Garlic – A delicious medicinal nutrient. Indian folk medicine* (pp. 210–229). Pointer Publisher.
134. Cardelle-Cobas, A., et al. (2009). *A comprehensive survey of garlic functionality*. Nova Science Publisher.
135. Fenwick, G. R., Hanley, A. B., & Whitaker, J. R. (1985). The genus Allium—Part 1. *Critical Reviews in Food Science & Nutrition, 22*(3), 199–271.
136. Devi, I. B., et al. (2016). Direction of trade and export competitiveness of chillies in India. *Agricultural Economics Research Review, 29*, 262–272.
137. AT, D. and E. Action. (1999). Scientific concepts of functional foods in Europe consensus document. *The British Journal of Nutrition, 81*, S1–S27.
138. Divakaran, M., et al. (2018). *Legacy of Indian spices: its production and processing* (pp. 13–30). Indian Spices: The Legacy, Production and Processing of India's Treasured Export.

Chapter 19
Fennel

Sara Zafar, Muhammad Kamran Khan, Shagufta Perveen, Muhammad Iqbal, and Arwa A. AL-Huqail

19.1 Introduction

The herb fennel has been used to improve both physical and mental health [1]. The perennial plant fennel, or *Foeniculum vulgare*, is a member of the Apiaceae (carrot family) and is grown for its tasty leaves, stems, and seeds. Its common name is fennel. Most important manufacturing nations of fennel are Argentina, India, Syria, Bulgaria, Egypt, Indonesia, China and Pakistan.

The carrot family's fennel is a floral plant. It's a hardy enduring yellow blossoms and fluffy leaves on a herb. The Mediterranean beaches are home to it, although it has distributed widely over the globe particularly on arid soils close to the seashore and along riversides [2].

Kingdom	Plantae
Subkingdom	Tracheobionta (Vascular plants)
Super division	Spermatophyta (Seed plants)
Division	Magnoliophyta (Flowering plant)
Class	Mangoliopsida (Dicotyledons)
Order	Apiales
Family	Apiaceae (carrot family)
Genus	Foeniculum
Species	*Foeniculum vulgare* L.

S. Zafar (✉) · M. K. Khan · S. Perveen · M. Iqbal
Department of Botany, Government College University, Faisalabad, Pakistan
e-mail: sarazafar@gcuf.edu.pk

A. A. AL-Huqail
Department of Biology, College of Science, Princess Nourah bint Abdulrahman University, Riyadh, Saudi Arabia

M. Zia-Ul-Haq et al. (eds.), *Essentials of Medicinal and Aromatic Crops*,
https://doi.org/10.1007/978-3-031-35403-8_19

This section begins with an explanation and categorization of *Foeniculum vulgare* L., a plant that is cultivated primarily as a herb or for its fruits and regarded for its enjoyable scent, nutritious value, and therapeutic benefit. The fragrance compounds found in essential oils from herbs and seeds are 50–60% anethole and 15–20% fenchone as the main ingredients of fruit volatile oil. There are numerous treated items prepared from fennel garden-fresh herb and fruits that are in high demand on the global market place. The nutritional and efficient qualities of fennel include its therapeutic uses including antimicrobial, antiflatulent, amphetamine, flatus-relieving, and expectorant as well as adulteration, harmfulness, and allergenicity. The excellence standards for various fennel manufactured goods, such as entire seed, minced seed, oleoresins and volatile oil are provided.

Fennel has been utilized for medical and cooking ideas for centuries. The whole plant has therapeutic use; its expanded ground is utilized as a vegetable; its leaves are utilized in cooking; and Its seeds are used to make spices and to extract essential oils. The blooms and leaves are often used to make brown and yellow colors.

19.2 History

Since ancient times, fennel has been one of the most consumed herbs worldwide. The name Marathon, which means "place of fennel," is where the Athenians and Persians fought, and from where the term "fennel"is derived. In recognition of their victory, the Athenians waved fennel stalks. They honored Pheidippides with fennel when he sprinted from Marathon to Athens to announce a military victory against the Persians. Fennel became a winning symbol after this incident. However, there is no evidence to suggest that fennel was ever grown there. Since mariano, which means "to become thin," is a Greek word, the plant was given the name Marathron. Now that fennel is widely recognized as a potent appetite suppressor and help in weight loss [1].

Fennel was branded as *madhurika* in early Sanskrit literature, thus it is believed that farming began in India circa 2000 BC. It was associated with success among the ancient Greeks, and the war of Marathon (490 BC) was selected after it because it took place in a field of fennel [3]. To the Romans, fennel was also a symbol of achievement, and fennel leaves were utilized to honor winners in competitions. One of the nine plants mentioned in the tenth-century description of the occult Nine Herbs Charm of Anglo-Saxon is fennel and it is one of the nine plants named in the ancient Anglo-Saxon charm known as the "Nine Herbs". Fennel was considered a regal spice in England during the thirteenth century, and it was offered to monarchs with fruit, bread, and delicacies like pickled fish spiced with the seed of fennel.

Fennel is an indigenous of southern Europe and the Mediterranean region [4], but it has been naturalized in several areas, involving northern Europe, the US Cyprus, southern Canada, and the Far East, parts of Asia and Australia. Pioneered to North America by Spanish evangelists for use in pharmaceutical gardens, it is currently seen in California as wild anise (and is frequently misbranded as anise in American superstores), where it may be noticed increasing around San Francisco and along the Pacific coast. The herb was also brought to the New England societies by English immigrants, where it turns into a component of their kitchen vegetable garden. It is now designated a wild plant in the United States, as well as in other countries [5].

19.3 Fennel Classification

The genus Foeniculum (fennel) is a representative of the Apiaceae family and the Apiales order.

(i) F. vulgare Mill. var. piperitum (Ucria) Cout. (bitter fennel),
(ii) F. vulgare Mill. var. dulce DC Batt. et Trab. (sweet fennel), and
(iii) F. vulgare Mill. var. azoricum Thell. (Florence fennel, or finocchio)

are the three principal types [6]. Sour Unlike Florence fennel, which is grown for its fruits, vital oil, leaves (which are used in cooking), and a wider leaf base, fennel is grown for its fruits and vital oil (consumed as a vegetable). For its broadened leaf base, fruits, and valuable oil produced from the fruits, sugary fennel is planted. Varieties of fennel are categorized as perennial or biennial fragrant herbs [7], although annual, biennial, and perennial kinds are described by other writers. Foeniculum is a cross-pollinated yield with the $2n = 22$ somatic chromosomes.

19.4 Chemical Composition

Fennel's chemical makeup changes depending on morphotype, resource, climate, and reaping phase. Fennel seeds contain 8.80 g water, 15.80 g protein, 14.90 g fat, 36.60 g carbohydrate, 15.70 g fiber, and 8.20 g ash per 100 g edible quantity (comprising 1.20 g Ca, 19 mg Fe, 1.70 g K, 385 mg Mg, 88 mg Na, 487 mg P and 28 mg Zn). Vitamin A (135 IU), niacin (6 mg), thiamine (0.41 mg), and riboflavin (0.35 mg) are all present in every 100 g, with an energy amount of around 1440 kJ. Mucilage, starch sugars, essential oil, tannin and stable oil (petroselenic, oleic, linoleic, and palmitic acids) are the primary elements of the fixed oil [8] are found in the seeds. Vitamin content varies in terms of type and quantity: folates, 270 mg/kg; vitamin B3, 6.40 mg/kg; vitamin C, 8.70–340 mg/kg. calcium (5.6–363 mg/kg), Potassium (4.24–5.85 g/kg) is the mainly common mineral in fennel, with phosphorus (500 mg/kg, magnesium (8.2–389 mg/kg), and sodium (7.70–512 mg/kg) being the least abundant minerals [9].

Anethole (50–60%) and fenchone (15–20%) are the two main elements of the vital oil extracted. The primary elements of the necessary oil extracted are (E)-anethole, (Z)-anethole, and α-thujone [10, 11]. Fennel essential oil contains 20 compounds, 18 of which account for 96.04% of the total indispensable oil, with anethole (68%), fenchone (3.70%), limonene (11%), and some others being the most prominent. Trans-anethole (60–65%, but up to 90%) is the most abundant ingredient in fennel seed oil.

There are also trace amounts of α-pinene, camphene,δ-α-phellandrene, and ρ-hydroxy phenyl acetone. Petroselenic, methyl chavicol oleic,, and palmitic acids, linoleic are the primary elements of the predetermined oil [12]. The unstable oil substance of dried up fennel seeds is 0.6–6%.

Table 19.1 shows the chemical makeup of sugary and sour fennel oil. In adding up to the listing of other components given, Anethole is the primary component of the essential oils obtained from the seeds and leaves of F. vulgare, accounting for 58.5% of the seed oil and 51.1% of the leaf oil, as determined by limonene (19.6% in seed oil and 22% in leaf oil) [13]. When the fruits (seeds) are fully grown, the seeds contain up to 95% of the essential oil. Hydro distillation yields 1.5–35.0%, with hydro distilling fresh or slightly wilted foliage right before flowering yielding the most amount of herbal essential oil [14].

Fennel seeds include 0.79% essential oil, 5.82% fixed oil, and 1.17 mg/g dried upmass of total phenolic compounds [16]. α-pinene (0.36%), δ-limonene (0.071%), 1,8-cineole (5.08%), anethone (86.12%), fenchone (4.12%), and estragole (methyl chavicol) are the most important components of essential oil. However, [2] the primary ingredients are trans-anethole (60–70%), methyl chavicol (estragole) (3–5%) fenchone (13–33%), α-pinene, camphene, ρ-cymene, liomonene myrcene, α-and β-phellandrene, γ-terpinene, terpineol,cis-ocimene and γ-fenchone. Fennel seeds' dry distillation residue contains 15–22% protein and 13–18% fat, making it acceptable for use as standard food [17].

Table 19.1 Comparison of Sweet and Bitter fennel oil

Component	Sweet fennel oil (%)	Bitter fennel oil (%)
Limonene	28.92	N/A
Fenchone	2.67	2.84
Fenchol	3.18	N/A
Estragole	2.53	8.31
Anethole	52.03	47.97
α-Pinene	4.03	18.10
α-Phellandrene	N/A	12.98

Source: [15]

19.4.1 Fennel Seed Oil Compounds (Table 19.2)

Table 19.2 The essential oil composition of the Bangladesh-grown *Foeniculum vulgare* Mill [13, 18]

Seed and leaf oil components	Percentage	Seed and leaf oil components	Percentage
β-Thuzaplicin in leaf oil	4.82	p-Anisic anhydride in leaf oil	0.21
β-Pinene in leaf oil	0.14	Octahydro-1-benzothipene in leaf oil	0.08
β-Phallandrene in leaf oil	0.05	N-amyl isovalerate in leaf oil	0.01
β-Ocemene in leaf oil	0.27	Myristicin in leaf oil	0.09
β-Myrcene in leaf oil	0.63	Methylisoeugenol in leaf oil	0.13
β-Bisabolene in leaf oil	0.03	Methyleugenol in leaf oil	0.07
α-Curcumene in leaf oil	0.05	Limonene-1,2-epoxide in leaf oil	0.10
Trans-p-2,8-Menthadien-1-ol in leaf oil	0.15	Limonene in leaf oil	22.9
Trans-Carvyl propionate in leaf oil	0.41	Fenchyl acetate in leaf oil	5.35
Trans-Carvyl acetate in leaf oil	0.25	Fenchone in leaf oil	1.66
Plinol D in leaf oil	0.11	Cis-Verbenol in leaf oil	0.18
Camphor in leaf oil	0.04	Terpinolene in seed oil	0.13
Camphene in leaf oil	0.07	Sabinene in seed oil	0.70
Apiol in leaf oil	0.63	Ocimene in seed oil	0.09
Anisaldehyde in leaf oil	7.55	Limonene in seed oil	19.6
Anethole in leaf oil	51.1	l-Fenchone in seed oil	7.73
Allyl-3-methoxybenzoate in leaf oil	0.06	Isopinocampheol in seed oil	0.11
4-Hexen-1-ol, acetate in leaf oil	0.22	Germacrene in seed oil	0.47
3-Methoxycinamaldehyde in leaf oil	0.14	Fenchyl acetate in seed oil	1.20
2-Methoxybenzeneethanol in leaf oil	0.10	Eugenol in seed oil	0.09
γ-Terpinene in leaf oil	0.06	Ethenyl)-2-cyclohexeneone in seed oil	1.20
β-Thujaplicine in seed oil	0.05	Cis-Sabinenehydrate in seed oil	0.09
β-Pinene in seed oil	0.23	Caryophyllene in seed oil	0.11
β-Pinene in seed oil	1.81	Camphor in seed oil	0.63
β-Camphor in seed oil	0.17	Camphene in seed oil	0.08
β-Bisabolene in seed oil	0.09	Apiol in seed oil	0.27
α-Pinene oxide in seed oil	0.19	Anisaldehyde in seed oil	0.73
α-Phallandrene in seed oil	0.31	Anethole in seed oil	58.5
Trans-Verbenol in seed oil	0.16	4-Terpinolene in seed oil	0.28
Trans-p-Mentha-2,8-dienol (S)-2-methyl-5-1-methyl in seed oil	0.30	3-Methoxycinamaldehyde in seed oil	0.27
Trans-limonene oxide in seed oil	0.84	γ-Terpinene in seed oil	1.10

19.5 International Trade, Production and Post-harvest Processing

Numerous nations, including Romania, Russia, Germany, Italy, France, Argentina, and the United States, cultivate fennel extensively. Egypt, Syria, Morocco, Denmark, Bulgaria, China, and Japan are among the countries that grow it. Gujarat, Rajasthan, and Uttar Pradesh are the major fennel-producing states in India, whereas Bihar, Karnataka, Maharashtra, Punjab, Tamil Nadu, and Jammu & Kashmir grow it on a limited scale. In India, 114,278 tons of fennel seed were manufactured on 74,150 hectares in 2008–2009, while 7250 tons of fennel seed were shipped in 2011 [19]. Anethole is produced in 1000 tons per year around the world, with China and Vietnam being the primary manufacturers. Due of the challenges involved with anise, fennel is the predominant resource of anethole in Brazil [20].

19.5.1 Cultivation and Organic Planting

Fennel is a cool period yield that produces more seeds when the weather is dry and chilly. The ideal temperature for growth is 15–20 °C, whereas high temperatures cause premature flowering and limited seed output. During the blossoming stage, the crop is vulnerable to frost damage. Fennel can be cultivated as a mixed or intercrop because it is a long-term crop with moderate initial growth. The varieties chosen must be appropriate to the local soil and climatic environments, and if possible, have pest and disease endurance. A nursery can be used to raise seedlings before they are put outdoors, or seeds can be sown straight in the ground. In a nursery, it takes roughly 2.5 kg of seed to grow adequate seedlings for 1 ha, whereas it takes 8–10 kg of seed to spread directly over the same area as a main season crop. Because of the kind and fertility of the soil, nutritional needs of plants differ from location to county. To diminish the prevalence of diseases and pests, in addition to the initial choice of resistant varieties and the application of biocontrol strategies, In an organic farming system, crop management techniques used at planting time, a healthy diet, crop rotation, and green manuring should be promoted as plant protection measures and other practices.

Fennel is a perennial plant that may be grown everywhere. It thrives in subtropical and temperate climates. Fennel is typically grown as both a Kharif and a Rabi crop. It needs a cool, dry atmosphere for fennel. To promote germination, the seed should be soaked and pre-sprouted for a number of days. Frost can damage fennel crops. Indian fennel has a sweet anise flavor, is brownish, typically smaller, straighter, and not nearly as rounded at the ends. After 12 to 18 days, seeds begin to germinate. They are harvested after 100 days and have a two-year shelf life. Every three to four years, the plants should be replanted to maintain their health [1].

Organic fennel is in high demand, and several spice firms sale certified organic fennel online. Fennel seeds grown in India are primarily from arid and partial-arid locations and are therefore organic by default because they are grown with low or no chemical inputs. In the market, such product is referred to be "near organic" and sold as such [21]. have defined the both broad and specialized standards for organic seed spice manufacturing, together with fennel. The major markets for organic spices, including fennel, are Europe, the United States, Canada, and Japan, the two new emerging markets represented by Australia and New Zealand. Spices grown organically will be in strong demand in the future.

19.5.2 Soil Condition

Black cotton soils and well-drained loamy soil are both good for growing fennel. Rain during fennel maturity ruins the color and lowers the quality. The fennel crop requires constant exposure to the sun. The ideal temperature range for seed germination is 60–90 °F.

19.6 Traditional Medicine

19.6.1 Leaves

(a) The paste made from the leaves is used to treat kidney problems, liver pain, and mouth ulcers.
(b) The leaves of the *Foeniculum vulgare* tree are used as a remedy of diabetes [22].

19.6.2 Bark

(a) Fever is cured from the bark.
(b) Blood-related disorders can be treated with tree bark [23].

19.6.3 Root

(a) Urinary tract infections and renal calculi are treated with root, glycosuria, too [24].
(b) Fever, colic, and muscle symptoms are treated with root.

19.6.4 Flowers

(a) Bloating, flatulence, and spasmodic gastrointestinal symptoms are caused by the floral paste of *Foeniculum vulgare*. It is additionally applied to treat upper respiratory catarrh.
(b) Perfumes contain its flowers [22].

19.6.5 Aerial Parts

Aerial components are also utilized to treat and improve the flow of milk breast feeding [22].

19.7 Nutrient Value of Fennel (Table 19.3)

19.7.1 Varieties

Sugary fennel and sour fennel are the two major forms of fennel grown. Bitter fennel grows wild in southern Russia, Argentina, Brazil, Czechoslovakia, Hungary France, Germany, India, Japan Italy, and other countries, as well as being farmed. Since it doesn't naturally grow there, sweet fennel is cultivate in France Bulgaria, Italy, and Macedonia [25]. The sweet fennel known as Finnochio (or Florence) is cultivated for its edible bulbous stalk. Italian carosella fennel (*F. vulgare* var. piperitum) young stems are utilized as a vegetable and for flavoring salads. Fennel vulgare var. purpureum, also known as Rubrum, Purpureum, or Nigra, is a kind of fennel with bronze leaves that is frequently planted as a blooming plant in gardens in the United Kingdom.

Persian fennel seeds are the smallest of all and have the strongest anise taste, whereas Indian fennel seeds are lesser and straighter than European fennel seeds and have a sweeter anise flavor. About 14 different types of bitter fennel are produced in India, with the most common cultivars [26]. According to harvest stage and season, the later is appropriate for growth in semi-arid circumstances and produces a high seed production with an essential oil concentration of up to 1.6 to 2.5%.

19.7.2 Harvesting and Yield

Fennel bulbs are only picked in the late fall when the stems are also used as a vegetable. Green fennel leaves can be gathered at any time during the growing season. Vibrant green bulbs that are clear and sharp color and no signs of browning are

Table 19.3 Nutrient value of Fennel [1]

Nutrients	Per 5.8 g value
Energy	31 kcal
Water	0.51098 gram
Carbohydrate	7.29 gram
Dietary fiber	3.1 gram
Fat	0.20 gram
Ash	0.47676 gram
Protein	1.24 gram
Moisture	90.21 gram
Sugar	3.93 gram

chosen. Harvesting time is determined by the sort of result being sold. Typically, before the fruits are fully developed, the crop is cut. Umbels of green fennel are gathered to be chewed approximately 30 to 40 days subsequent to flowering, when they are still green and just half their full size (if left for grow). Umbels must be harvested 4–5 times, as and when they become ready, because not every plant reaches maturity at the same time. A 2–2.5 ton/ha yield can be attained with scientific crop management.

19.7.3 Post-harvest Processing

The stem and green leaves of common fennel are utilized similarly to Florence fennel. It can be kept cooled for up to 5 days if properly wrapped in plastic bags. In contrast, harvested umbels should be dried in the shade with good airflow, especially for light-green fennel. It is certainly not a good idea to pile umbels since the quality will suffer [27] Winnowing is used to separate and clean dried umbels, removing banter, dust, and trash. Since any greater moisture level might result in fungal infection during storage, the humidity content of the seeds must be kept below 9%. Produce is dried up, cleansed, and sorted before being packaged into standardized containers and packs are suitably branded.

The seed is dried and packaged in bags that are covered with biodegradable, eco-welcoming plastic film. Packaging that generates waste ought to be avoided. Each sealed bag is maintained in a spot that is clean, dry, and well-aerated location [21].

Every step of the organic ingredient's preparation should be carefully monitored to ensure that its critical quality is maintained. Processing procedures should be used so that how many and how much processing aids and additives are needed, are kept to a minimum. To get essential oil, mature dried seeds are distilled. Extraction is usually done using either a steam or a hydro-distillation technique. The percentage of essential oil varies according on the kind and type of fennel: Indian fennel has the lowest volatile oil concentration (0.7–2.5%), while European fennel has the greatest (2–6%). Essential oils (which are commonly found in soaps toothpaste, lotionsetc.) should be placed in aluminium cans or bottles with tight lids.

Oleoresins, powder, fixed oil, and essential (volatile) oil are examples of processed goods (likewise in demand in the global market). Fennel oleoresins are extracted from using food-grade hexane, ethyl acetate ethanol, or ethylene dichloride to smash dry seeds, followed by filtering and de-solventizing under vacuum. According to ISO maximum allowed limits, the oleoresin should be completely free of any organic solvents. Fennel dried seeds are ground to create powder; to avoid the volatile oil loss, grinding at a low temperature and ahead of cooling can be utilized.

19.7.4 Nutritional Content (Table 19.4)

Table 19.4 Fennel (per 100 g) contains the following ingredients, according to the United States Department of Agriculture's National Nutrient Database: [22]

Moisture	90.21 g
Protein	1.24 g
Carbohydrates	7.3 g
Calcium	49 mg
Magnesium	17 mg
Potassium	414 mg
Zinc	0.2 mg
Vitamin B-6	0.047 mg
Thiamin	0.01 mg
Niacin	0.64 mg
Vitamin A	48 μg
Energy	31 kcal
Total lipid	0.2 g
Dietary fiber	3.1 g
Iron	0.73 mg
Phosphorus	50 mg
Sodium	52 mg
Vitamin C	12 mg
Riboflavin	0.032 mg

19.8 Major Uses of Fennel in Food

As possible providers of different nutrients, the fennel plant's bulb, leaves, and seeds are frequently used equally uncooked and cooked in salads, pasta dishes, vegetable development, and sausages, and other foods. Carbohydrates, dietary fiber, protein, vitamin C, vitamin B complex, and minerals are all found in raw fennel bulbs. Fennel is an aromatic plant that is used as a marijuana herb. It is widely used as a vegetable and a spice, with a wide range of flavoring and cooking applications. Oil, powder, and the whole seed are utilized in confectioneries and drinks as flavoring agents, antioxidants, and preservatives.

Fennel seeds are frequently used in numerous cuisines, involving pickles, soups, sauces, cakes and breads. Fennel is utilized to flavor prepared meats like hot pepperoni and sweet Italian sausages as well as non-alcoholic drinks, baked goods, sauces liqueurs and ice creams like Anisette, and as an organoleptic flavor corrector [12].

19.8.1 Fennel Bulb and Green Herb

In the Middle East and India, the bulb and green "herb" fennel are used to flavour food while it cooks or as a garnish right before serving. While getting rid of the leaves will prevent the bulb potential development bulb fennel leaves taste similar to herb fennel. grilled, braised, sautéed, stewed, or eaten raw, the bulb is a crisp, resilient root vegetable. The fennel bulb, which is high in fiber, may aid to lower cholesterol levels.

The antioxidant content of the widespread herb fennel, F. vulgare var. dulce, and its color variant Rubrum (bronze fennel), is high. Common fennel leaves and seeds (nicely flavored and analogous in form to those of dill) may also be used in salads; its seeds and leaves, which have a delicate flavor and resemble dill in form, can also be washed-out and/or the leaves cooked, or cooked marinated in risotto. Florence fennel's thicker leaf stalks are blanched and served as a vegetable [12, 28]. Florence fennel is sometimes combined with chicory and avocado in Italian and German salads, or it can be prepared as a warm side dish and served.

Green bulbs and the plant as a whole are good sources of iron, calcium vitamins B and C, carotenes and folic acid,, and are used to make herbal teas or juice mixrd with other herbs. The leaves impart their traditionally mild, anise-like flavor in all circumstances.

19.8.2 Whole Seeds

Dehydrated fennel seed has an anise flavor and is fragrant. When the seeds are young, they are brown or green in color, but as they age, they change a faded grey color. Cooking with green seeds is the finest option. Fennel seeds from time to time mixed up with anise, which has a taste and appearance similar to fennel but is smaller. Fennel seeds are used for chomping only later than meals or in betel leaves because of their particular pleasant flavor; sugar-smeared pelleted fennel seeds are as well used as a breathing refresher. Cooked fennel is used as an organoleptic flavor corrector or as digestive purposes after-meal in some parts of India and Pakistan (thus why some Indian eateries provide a fennel seed mix up after meals).

Fresh branches of green fennel seeds are commonly chewed in agrarian communities. It's a key component of the Bengali spice blend panch phoron, as well as the Chinese five-flavor powder. Fennel seed is widely used in northern European rye breads in the west and Italian sausages. Fresh or dried up fennel leaves are used in a variety of egg, seafood, and other dishes.

In many cultures, the entire seeds are utilized as a spice and seasoning (in India, China and Egypt). Fennel seeds are a classic Indian spice used in a variety of dishes, including curries. Whole fennel seeds are usually used to flavor meat dishes, soups and sauces, bread rolls, pasties, and sweets, and a small amount can totally dominate the flavor of a meal. Fennel seed is used in English-design soups, German breads, spaghetti Polish borscht, candy pickles, salads and vegetable dishes, according to [12].

The seeds are also used in the creation of several types of pickles and for flavoring liquors. Fennel seed vinegar is widely used in salad dressings and to make herb sauces, it is simply made by adding 2 tablespoons of fennel seeds to 600 ml of white wine vinegar, letting it sit in a cool, and covering the jar dark place for 2–3 weeks, shaking it periodically. The vinegar should then be filtered into clean bottles, labelled, and kept out of direct sunlight in a cold place.

19.8.3 Fennel Essential Oil

Fennel essential oil is an amusing resource of biologically active components and is utilized as a flavoring mediator in a variety of foods, pickles, and licorice candies. It has been recognized as a pure food flavored with antibacterial and antioxidant qualities that might be used alone or in combination in beverages, bread, and other food formulations.

19.8.4 Fennel Oleoresin

Fennel oleoresin is a warm, aromatic, and agreeable flavor that comes from the seeds of the fennel plant. Fennel oleoresins are utilized in processed meals, sauces, snacks and a variety of vegetable dishes.

19.8.5 Fennel Powder and Curry Powders

Food seasoning is done with finer powders, while abstraction of oleoresin, oils, and other extractives is done with coarser powders [29].

19.8.6 Fennel-Based Commercial Blends

There are a number of commercial fennel-based blends in the market [4] involving:

19.8.7 Fennel Tea

Fresh or dried up herbs are used to make this dish. The entire leaves are uncontaminated herbs that have undergone less processing than herbal tea bags, allowing the plant oil feature to be preserved and resulting in a more strenuous tea (typically made by infusion). Fennel green seed tea is also obtainable, in addition to fennel leaves and bulbs. Herbal stores and internet distributors sell organic fennel seeds for infusion and tincture production. Catnip, calendula flowers spearmint, kullcap, lemongrass, sage leaf and rosemary are common ingredients in organic herbal tea blends, as are fennel seeds and herb.

19.8.8 Cough Syrups

Syrups comprising 10% alcohol by quantity are also available, made from fennel infused with honey and herbal ingredients that are naturally grown or gathered from the wild, including elecampane root, osha root, marshmallow root, horehound, and mullein.

19.8.9 Absinthe

Fennel is one of the three important plants used as Florence in the alcoholic concoction. Absinthe is an alcoholic concoction that instigated as a medicinal tincture in Switzerland but became a common alcoholic drink in France and some other countries by the late nineteenth century.

19.8.10 Indian Panch Phoran (Five Spices)

In the Indian union states of West Bengal, Sikkim, and Bangladesh, this spice blend is particularly common, particularly for meat meals. The usual ingredients are equal parts nigella, cumin, fenugreek, black mustard seed, and fennel; ajowan is occasionally used in place of black pepper and cumin.

19.8.11 Chinese Five Spice Blend

Organic elements including black pepper, anise, fennel seeds, cloves and cinnamon are utilized to flavor a range of dishes.

19.8.12 Antimicrobial (Antibacterial and Antifungal)

Antibacterial properties of fennel important oils [30, 31], *Helicobacter pylori*, the most common stomach infection that causes gastrointestinal dysfunction, ulceration, and even cancer, has been shown to be resistant to the crude extract obtained from fennel [32]. Experiments on Listeria and Salmonella species have demonstrated that herbal extracts like fennel oil have powerful antibacterial properties when combined with benzoic acid derivatives like methyl paraben (methyl 4-hydroxybenzoic acid) [33]. Fennel essential oil exhibits antibacterial properties against the gram-negative bacterium *Acinetobacter baumannii* [34]. The antibacterial and antifungal activity of azoricum and dulce cultivars has been observed in similar measure [35], piperitum cultivars [36].

19.8.13 Antiflatulent and Antispasmodic

Fennel is a great intestine and stomach medicine for flatulence and colic, as well as for promoting a healthy hunger and digestion. In high quantities, Fennel seeds improve gastrointestinal function and have antispasmodic properties. Extract of fennel lower maximum potential contractility and reduce acetylcholine-induced contraction [37]. Fennel at a concentration of 10% weight/volume raise stomach acid release in rats from 0.12 mL (base level) to 0.42 mL in tests, albeit the specific method of this rise is uncertain. Fennel infusions are used all over the world to stimulate gastrointestinal motion because of this characteristic.

In Indian and Chinese medicine, Fennel seeds are used to make a decoctionto treat abdominal pain, colic, and stomach chills. Indigestion and abdominal distention are treated with this infusion [28]. To treat distress, stomach colic, and gastrointestinal disorders, enteric-coated hard gelatin capsules containing fennel extract, peppermint, and ginger are also available on the market. *Satapuspadi churana, Satapushpa arka*, are commercial Ayurvedic medications that are suggested by Ayurvedic consultants to promote digestion, manage colic pain, and some other gastrointestinal disorders [38].

19.8.14 Stimulant, Carminative and Expectorant

Fennel is well recognized for improving digestion and appetite; [39] found that adding various well-known dietary spices, as well as fennel, to the diet significantly shortened the time it took to finish a meal.

19.8.15 Anticarcinogenic Properties

Several dietary components, including anetholes from fennel camphor and anise have the ability to be used to avoid and treat cancer [40] [41]. demonstrated the chemo preventative efficacy of fennel against carcinogenesis. The regulation of tumor necrosis factor (TNF)-induced cellular retorts may be a mechanism through which anethole exerts its effects.

Estragole, a component of fennel, is a procarcinogen with a low risk of cancer. Estragole have to be activate by liver enzymes to reach maximal toxicity [42–44]. Fortunately, it is inactivated by other liver enzymes, minimizing liver damage.

19.8.16 Antioxidant Activity

The fennel leaf and bulb stalk and leaf, which are generally ingested raw, are strong in antioxidants and are thought to be beneficial in diseases such as coronary artery disease, carcinogenesis, fiery disease, and aging [45]. discovered that fennel fruit had analgesic, anti-inflammatory, and antioxidant properties. Fennel fruit essential oil, ethanol extracts and water have a significant antioxidant impact [10, 46]. In the linoleic acid system, ethanol and 100 grams of water extracts reduce peroxidation by 99.1% and 77.5%, respectively, this is more than the equivalent quantity of the naturally occurring antioxidant, α -tocopherol (36.1%).

According to the sample concentration, both extracts show strong free radical scavenging, superoxide anion radical scavenging, hydrogen peroxide scavenging, and metal chelating capabilities. It appears that fennel seeds might be a source of organic antioxidants.

Flavonoids, which occur as glycosides or in a free state in the fennel herb and bulb, are known for their antioxidant properties against free radicals [35]. discovered substantial quantities of flavonoids, DPPH radical scavenging activity, and overall phenolic content, which prevented peroxidation by 45–70% (the maximum antioxidant activity was seen in an 80% ethanol extract). About forty two phenolic compounds, including derivatives of hydroxycinnamic acid, flavonoid glycosides, and flavonoid aglycons, were found in fennel [47], 27 of which had never been identified before.

19.8.17 Muscle Relaxant

In a different study involving animals, acetylcholine-induced contractions of the smooth muscles in the ileum and bladde are prevented by sweet fennel. The components of fennel oil are thought to bind to calcium-binding proteins and reduce the release of calcium from intracellular reserves as the mechanism of action [48].

19.8.18 Nausea and Stress Relaxer

Anthemis nobilis (Roman chamomile), *Pimpinella anisum* (aniseed), F. vulgare var. dulce (sweet fennel), and Mentha x piperita are combined in a beneficial way was utilized on patients experiencing from the palliative care program and sign of nausea in a hospice according to [49]. Using the Bieri scale, many patients who had aromatherapy treatments felt better (a visual-numeric analogue). It was not possible to establish a solid scientific association between the aromatherapy sessions and the decrease in nausea because the patients were simultaneously receiving other therapies for their symptoms. The research did show that the essential oils used in this aromatherapy treatment were helpful in reducing the symptoms of this disease,

19.8.19 Hepatoprotective

Fennel fruit possesses liver-protective characteristics [50], in addition to be a good treatment for chest, spleen, and renal disorders [41]. Essential oil from fennel was discovered to reduce the hepatotoxicity caused by acute carbon tetrachloride-induced liver injury, as evidenced by lower levels of serum aspartate aminotransferase, alanine aminotransferase, alkaline phosphatase, and bilirubin [50]. Various spices, including fennel, generated a rise in biliary solids and a considerable increase in bile acid secretion, likely providing to the digestive stimulating activity of the test spices [51].

19.8.20 Antidysmenorrheal

It is [52] reported that fennel seeds might be utilised as a trustworthy and successful herbal treatment for primary dysmenorrhea; however, in the quantities used in this study (2% concentration), it might be less effective than mefenamic acid. Both medicines significantly alleviated menstruation pain, with a mean onset of action of 76 ± 48.8 min for fennel and 67.6 ± 46.07 min for mefenamic acid. Primary dysmenorrhea and related symptoms are caused by an increase in ectopic uterine motility (including pain).

19.8.21 Antihirsutism

It is believed that a problem with peripheral androgen metabolism is what causes idiopathic hirsutism, which is characterized by normal blood androgen levels and extreme male-pattern hair growth in females with a normal ovulatory menstrual cycle. It was [53] examined that clinical outcome of 1% and 2% fennel extract creams, which have been used as oestrogenic drugs, were applied topically to patients with idiopathic hirsutism. Hair growth speed and diameter were measured. Both the 2% and 1% fennel extract creams performed better than the utilized placebo in terms of effectiveness (measured by the mean values of hair diameter reduction for patients getting the creams containing 1%, 2%, and 0% (placebo) fennel extract, respectively, were 7.8%, 18%, and 0.5%).

19.8.22 Antiparasitic

Fleas and other parasites are kept at bay with powdered fennel seeds. Components obtained from fennel seed oils were tested for acaricidal efficacy used for *Tyrophagus putrescentiae* adults with the help of direct contact application and then compared to substances such as dibutyl phthalate, N,N-diethyl-m-toluamideand benzyl benzoate [54].

19.8.23 Harmfulness and Allergenicity

The herb, seeds, bulb and extractives of fennel do not seem to be poisonous in any way. Fennel is non-toxic in amounts that are often consumed in food. In therapeutic quantities, fennel herbal tea and some other formulations have a wide-ranging outline and essentially no side effects. Fennel oil in excess can produce nausea, vomiting, and convulsions. Because some fennel goods include estragole, a biologically occurring cancer-causing chemical, it is recommended that huge amounts of fennel be prevented and that it be used merely on the guidance of a herbal health specialist. Fennel essential oil contains bergapten (furanocoumarin) molecules, which may be carcinogenic when exposed to sunlight.

Allergic responses of the skin and respiratory tract have been reported in rare occurrences. Due to its oestrogenic effects, prenatal women should avoid using the plant, tincture, seeds, or essential oil of fennel in therapeutic therapies; modest quantities used as a food spice are deemed safe, while fennel works as a uterine stimulant in excessive concentrations. Fennel essential oil is harmful in even tiny dosages, causing skin irritation, vomiting, seizures, and respiratory difficulties. It is not recommended that the volatile oil be consumed. In sensitive people, the plant and seed oil may induce contact dermatitis [4].

It was [55] addressed the need to explain the safety of FEO use, since its use as a treatment for primary dysmenorrhea raised worries in relation to its possible teratogenicity because of its oestrogen-like activity. FEO was found to be cytotoxic at doses as small as 0.93 mg/ml, according to the findings. However, rather than a decrease in cell differentiation, this decrease was attributed to cell loss, as demonstrated by a neutral red cell sustainability experiment. These data show that FEO may be harmful to foetal cells at the quantities examined, although there was no evidence of teratogenicity.

19.8.24 Fennel as a Food Allergen

Spices are being used more frequently as a result of variations in dietary preferences and the internationalization of foods. Children with allergy symptoms had their reactions to aniseed, cinnamon, paprika, nutmeg, curry, coriander, cumin, fennel, vanilla and sesame assessed through prick tests with the essential foodstuffs, crushed or diluted in saline; [56] performed tests for certain spices (mustard, fennel). Mustard, fennel, coriander, cumin, and curry were the spices that caused sensitization (reported in 46% of cases). A case of recurring angio-oedema was caused by fennel (positive labial challenge test). Fennel and Mustard are the most commonly implicated foods, and they are as well liable for clinical symptoms. The masking of certain allergens in mixes makes it harder to avoid them in the diet.

19.8.25 Fixed Oil

The physicochemical constants of fennel seeds, which contain 9–13% fixed oil. Essential oils and fixed oils have the same basic building blocks: their structures are built by joining hydrogen atoms and carbon in different configurations. Triglycerides, which are formed of three lengthy carbon atom chains connected at one end, are the molecules that make up fixed oils. While essential oils evaporate rapidly and don't feel greasy, fixed oils are oily [4].

19.8.26 Fennel Oleoresin

It is a liquid that is brownish green in color containing 50 ml of volatile oil per 100 g. Fennel oleoresin should be made using appropriate biological solvents, then the solvent should be removed (as per importing country requirements). In terms of flavor and scent, is similar to 45.46 kg of ground fennel seed and 2.96 kg of fennel oleoresin [12].

19.8.27 Edible Plant Parts and Uses

Whole parts of the plant are safe to eat, including the roots, swelling petioles (F. vulgare subspecies var. azoricum), swollen petiole bases, fruits, seeds, flowers, leaves, shoots, stems, sprouting seedlings, essential oil, and blossoms.

Snacks, salads, soups, stews, and seasonings are made with the tender shoots, leaves, and stems. They're commonly found in grilled fish, egg meals, omelets, cooked fish dishes and bouillons, stewed with various beans and chickpeas, and in various soups and sauces. Fresh or dried, they're also utilized in brochettes and herbal drinks. Aromatize olive brines, fig preserves, and brandy using these plant parts. The young leaves can be used as a seasoning or garnish on foods that are either raw or cooked, and they are also a wonderful addition to salads.

The blossoms and inflorescences are used as a spice and to flavor alcoholic drinks and other products. The subtle anise flavor of the yellow blooms is used in sweets, cold soups, as an embellish through dinners, and in watercress soup and fennel.

Sweet fennel (F. vulgare subspecies vulgare) fruit (mericarps) are utilized to flavor fish and other seafood dishes and can also be used to formulate a tasty herbal tea. Sweets, cakes, bread, biscuits, stuffing's, everyday foods, stews, and dainties are all flavored with the fruits and seeds. In Malaysian cuisine, the fruit mericarp known as 'Jintan manis' or 'Adas pedas' is a common spice found in satay peanut sauce, dried ground prawn curry powder and chicken curries. Meatballs, Italian sausages and northern European rye bread all contain fennel seed. Salads can be made with the sprouting seeds. In the same manner that the whole seed is utilized as a culinary flavoring, an essential oil extracted from the fully matured and dried seed is employed.

In England, the large, fleshy petiole bases of Florentine fennel are eaten either cooked or raw like a cheese complement. It's a staple in various Italian and German salads, and it's frequently combined with chicory and avocado. It can also be broiled, blanched, or marinated before being added to risotto. Salad with sliced fennel, avocados, and oranges is delicious. In sandwiches, sliced fennel is frequently mixed to the standard lettuce and tomato toppings. Fennel slices are thinly sliced and served with plain yoghurt and mint leaves. Fennel goes nicely with braised fennel and salmon, is a great side dish for scallops. Fennel and onions sautéed in olive oil make a delicious side dish.

Cooking and eating the thick roots is also an option. Fennel from Florence is one of three ingredients used to make absinthe, a famous alcoholic beverage in Europe throughout the nineteenth century. The plant is grown commercially in numerous European nations to produce anethol, which is used in foods, cordials, and alcoholic beverages such as absinthe.

Uronic acid was discovered in fennel pectin, along with rhamnose, galactose, and arabinose [57]. In the deficiency and occurrence of phaseolin protein, the extracted pectins were characterized for usage as a carbohydrate resource to make biopolymer films.

19.8.28 Toxicity Studies

In mice, acute (24-h) and chronic (90-day) oral toxicity experiments on ethanolic extracts of fennel fruit revealed that the extracts produced no major severe or chronic mortality and had no spermatotoxic effects when compared to controls [58]. During prolonged therapy, the treated male mice attain considerable weight, whereas the female mice treated with the identical extracts lost weight or did not alter much.

In-vitro tests of rat embryo limb bud rnesenchymal cells revealed that fennel essential oil at the examined quantities could be harmful to foetal cells, but no indication of teratogenicity was found [55].

19.8.29 Adverse Effects

According to [59], long-standing usage of preparations like *Foeniculum vulgare*, which is used to remove gas and control intestinal purpose in infants, can promote early thelarche, and thus should be avoided. Isolated early the larche is a frequent condition defined by breast growth in children less than the age of two years and no other puberty indicators. The larche is most commonly linked to hypothyroidism, adrenal or ovarian diseases and the use of exogenous hormones or medicines, although it can also be linked to long-term herbal medicine use.

19.8.30 Traditional Medicinal Uses

The leaves, seeds and roots are all edible; however, the seeds are the most medicinally vigorous and are the most widely utilized portion [60–64].

Carminative, diuretic, aromatic, galactogogue, antispasmodic, anti-inflammatory, hepatic, and emmenagogue are all properties of fennel seed. Fennel is a great stomach and intestine medicine for flatulence and colic, as well as for increasing digestion and appetite. Children with colic and flatulence are given fennel water in India. In amenorrhoea and cases where lacteal secretion is inhibited, a hot infusion of the fruit is beneficial. Flatulence benefits from the oil. Fennel juice has been used to improve vision, and a paste made from the seeds is used to make a cooling drink for fevers and scalding urine. Fennel seed extracts have been demonstrated to be effective in treating glaucoma in animal trials.

Fennel seeds are used to treat venereal problems in Madras; Cough syrups can be flavored with it. It has been reported that it helps breastfeeding women produce more milk.

The oil relieves muscular and rheumatic problems when applied externally. For treatment of inflammation and conjunctivitis of the eyelids, the infusion can be used as a compress (blepharitis) or an eye wash. For fainting, crushed seeds are breathed.

Fennel is well-known in Egyptian traditional medicine for its dyspepsia, diuretic, immune-boosting, antioxidant, lactagogue and estrogenic -relieving properties [65]. While in Lebanese traditional communities, Fennel is used for digestive stimulant [66]. *Foeniculum vulgare* is an ancient spice and herb that was used as a carminative, a weak diuretic, and a lactation stimulant by the ancient Egyptians and Greeks [67]. For millennia, fennel (*Foeniculum vulgare*) and anise (*Pimpinella anisum*) have been utilized as estrogenic agents [68]. They are said to enhance milk secretion, induce menstruation, ease birth, relieve male climacteric symptoms, and boost libido [61].

The leaves are also said to be diuretic, boosting urine and sweat output, and the young plant's shoots are carminative and respiratory. The roots are used to treat urinary issues and as an aperitive and purgative. In India, an infusion of the roots is used to treat toothaches and difficulties associated with childbirth. Aromatherapy uses an essential oil extracted from the seed. The essential oil is antibacterial, carminative, and stimulating, however it should not be used by women who are pregnant.

Analgesic, anti-inflammatory, antispasmodic, aromatic, carminative, diuretic, emmenagogue, expectorant, galactogogue, hallucinogenic, laxative, stimulant, and stomachic are only some of the properties of this plant. Indigestion, colic, abdominal distension, and stomach cramps can all be treated with an infusion.

It is used to cure kidney stones and, when combined with other urinary disinfectants, can be used to treat cystitis effectively. It can also be used as an eyewash to treat sore eyes and conjunctivitis, as well as a gargle for sore throats.

Fennel is often used with buttermilk and honey in home beauty products, as a cleansing lotion and skin refresher, and in toothpaste. A face steam bath is made from an infusion of crushed seeds. Fennel is utilized in anticellulite massage oil in combination with other substances. Fennel is used in weight-loss recipes as well.

*19.8.31 Fennel (***Foeniculum vulgare***) in Poultry Health as an Eco-Friendly Alternative to Antibiotics*

In response to the European Union's ban on antibiotic growth promoters, research into improving gut health has advanced quickly. The poultry sector is currently dealing with problems that were formerly controlled by antimicrobial growth promoters, thus the search for the best solutions is ongoing. In parallel, there is growing social pressure to use less antibiotics and switch to other feed additives. Customers consider several readily available solutions to be secure, with phytogenics playing a key part. This review explains how using fennel seeds might be advantageous for chickens. Along with their physical and biological characteristics, a summary of the extensive chemical variety of fennel is offered. Studies show that fennel seeds improve performance, increase immune cell proliferation, decrease oxidative stress, and increase antibody titers against infectious diseases in birds, among other biological effects. It was determined that adding fennel seeds has a number of positive effects on the growth and health of chickens [69].

19.8.32 Probiotic Yoghurt Reconstituted in Aqueous Fennel Extract

With the exception of yoghurt treatments made from whole milk powder reconstituted in 6% aqueous fennel extract (FEY6) as opposed to 2%, probiotic yoghurt made from whole milk powder reconstituted in aqueous fennel extract was successfully produced with viable probiotic starter counts, a suitable chemical composition, acceptable flavour and texture characteristics, and it was equal in overall appearance (PY). It is clear that using up to 4% of aqueous fennel extract in place of water when reconstituting whole milk powder will produce functional yoghurt with desirable properties as contrasted to plain yoghurt. Natural plant-based additives require further investigation into their effects on health and technology in the manufacturing of dairy products in order to highlight their safety and higher number of bioactive components compared to synthetic additions [70].

Activated carbon derived from fennel flower waste as high-efficient sustainable materials for improving cycle stability and capacitance performance of electroactive nanocomposite of conductive polymer

In a study, black seed fennel flower (Lc-FF) biowaste, which is inexpensive, was easily chemically activated to produce activated carbon (AC). However, functionalized graphene oxide and activated fennel flower (Ac-FF) were combined to improve the electrochemical properties of the mixture, such as cyclic stability, energy density specific capacitance (Cs) and power density (FGO). Then, an incredibly thin layer of Poly-orthoaminophenol (POAP) was layered on the binary nanocomposite (FGO/Ac-FF) to produce POAP/FGO/Ac-FF (ternary nanocomposite). The coated structure of the nanocomposite surface was discernible from the SEM and TEM images, which caused the ion transfer rate to accelerate. In addition, Brunauer-Emmett-Teller (BET) study produced a fantastic definite surface area (SSA) of 2199.2 m^2 g^1. To assess and evaluate the storage energy behaviour of electrode materials, electrochemical tests comprising galvanostatic charge-discharge (GCD), electrochemical impedance spectroscopy (EIS) and cyclic voltammetry (CV) were carried out. The ternary nanocomposite electrode's Cs value in a three-electrode system was calculated to be 1400.2 $F.g^{-1}$ at 2 $A.g^{-1}$. Additionally, this material demonstrated outstanding cyclic stability, maintaining 94.4% of its capacity after 5000 continuous charge-discharge cycles. This research offers fresh perspectives on creating biowaste- electrodes for high-performance based supercapacitors (SCs) [71].

19.8.33 Pharmacological Actions Antifungal Activity

Because of the presence of anethole, methyl chavicol, fenchone and fennel seeds and oil have antifungal activity against the mycelial growth of *Alternaria alternata*, *Aspergillus flavus* (AF), *Fusarium oxysporum*, *Aspergillus niger* (AN), *Aspergillus oryzae* (AO), *Fusarium gramine* (FG) [72].

19.8.34 Anti-inflammatory Activity

Fennel extract in methanol has anti-inflammatory properties that are inhibiting of both acute and subacute inflammatory disorders. It appears to act on the lipoxygenase and cyclooxygenase pathways, dramatically raises alanine aminotransferase (ALT, serum transaminase, aspartate aminotransferase (AST), and and exhibits anti-inflammatory effect [45, 73].

19.8.35 Antibacterial Activity

Both Gram positive and gram-negative bacteria are vulnerable to the antibacterial properties of fennel [74]. *Escherichia coli* and *Bacillus megaterium* are successfully repelled by fennel.

19.8.36 Antioxidant

Fennel contains flavonoid antioxidants like kaempferol and quercetin that help the body get rid of harmful free radicals, defending against illness and aging. Due to its capacity to expand blood capillaries, rutin improves poor circulation [23, 72].

19.8.37 Respiratory Disorder

Fennel works wonders as an expectorant and a cold cure. Strong coughs, colds, respiratory chest congestion, and bronchitis can all be treated by lowering phlegm and bronchial secretions with the use of creosol and alpha-pinene. Fennel's quercitin, which has anti-inflammatory properties, can lessen asthmatic symptoms [22, 45].

19.8.38 Digestive Aid

Anethole, limonene, pinene, anisic aldehyde, myrcene,, chavicol, cineole and fenchone are examples of phytoconstituents that are thought to have anti-inflammatory, antispasmodic, digestive, carminative, and anti-flatulent properties. Additionally, fennel reduces GIT inflammation, stimulates the production of gastric and digestive juices, and improves nutritional absorption from food. Because fennel contains volatile oils that stimulate the mucous membranes in the digestive tract and boost motility and peristalsis, it has cathartic, laxative, and purgative effects [75].

19.8.39 Anti-Cancer Effects

The most significant components of fennel are anethole, vitamin C, flavonoids, and essential oils. Anethole inhibits NF-kappa B, a molecule that alters genes and triggers inflammation, as well as tumor necrosis factor (TNF), a molecule that signals the development of cancer. Because fennel is a strong source of fiber and aids in the elimination of toxins from the colon, it may be useful in preventing colon cancer.

Different types of breast and liver cancer are prevented in part by the use of fennel seed extract [76].

19.8.40 Diuretics

Fennel is a potent natural diuretic that aids in eliminating toxins and extra water from the body through frequent urination. As a result, it aids in reducing swelling and rheumatoid arthritis-related inflammation [77].

19.8.41 High Blood Pressure

Fennel seeds, a natural diuretic, aid the kidneys in eliminating extra water from the body. Fennel lowers fluid volume and lowers arterial blood pressure without changing heart and respiration rates. Potassium, a vital nutrient, is found in abundance in fennel recognized mineral that lowers high blood pressure. Stroke and heart attack risk can be decreased by lowering blood pressure [77, 78].

19.8.42 Weight Loss

In the liver and pancreas, fennel speeds up the metabolism of carbohydrates and fats. Additionally, it dissolves bloodstream fat stores, enabling the body to use them as an energy source. Its reputation as an appetite suppressant, along with these characteristics, make it an effective weight-loss solution [78].

19.8.43 Osteoporosis

Iron, magnesium, phosphorous, calcium, manganese, vitamin K and zinc are among the nutrients found in fennel that help maintain healthy bones and ward off osteoporosis and postmenopausal bone loss. Fennel reduced the differentiation and function of osteoclasts, cells that dissolve damaged bone. The bones are given protection by this [79].

19.8.44 Hair

Fennel is a good source of iron, which increases the assembly of red blood cells, stops hair loss and avoid premature graying [80].

19.8.45 Eyes

Extracts from the herb and the root of fennel are used to cleanse congested eyes. Fennel inhibits age-related macular degeneration and eye inflammation in meals. Fennel seeds are used as an eyewash to lessen eye inflammation because of its anti-bacterial properties. Additionally, fennel seeds can treat glaucoma [81].

19.8.46 Anti-thrombotic Activity

Fennel has an ingredient called anethole, which has antithrombotic, antiplatelet, clot-destabilizing, and vasorelaxant properties. Collagen-ADP, Arachidonic acid and U46619-produced aggregation are inhibited by anethole [82].

19.8.47 CNS Activity

As it demonstrated significant anxiolytic, anti-stress, and nootropic activity in studies, fennel is very beneficial for the brain. Various fennel extracts aid in the treatment of dementia, an age-related mental condition, and Alzheimer's disease [22, 83, 84].

19.8.48 Boosts Immune System Health

Vitamin C, an excellent source that supports the health of our immune system, is present in fennel. Its ability to eliminate free radicals, as well as the fact that it restores skin tissue and safeguards blood vessel walls, make it crucial.

19.8.49 Fluid Retention

Given its capacity to remove extra fluid from the body, fennel seeds can help with edoema symptoms.

19.8.50 Anemia

Fennel seeds contain iron, the main component of hemoglobin, as well as the amino acid histidine, which stimulates the production of hemoglobin and aids in the treatment of anemia.

Table 19.5 Medicinal uses of fennel [4]

Sr. No.	Medical condition	Uses of Part	References
1.	Anti-psychotic	Seed	[85]
2.	Osteogenic	Seed	[86]
3.	Osteoporosis	Seed	[79]
4.	Cardio protective	Seed	[78]
5.	Anti-obesity	Seed	[87]
6.	Anti-aging	Seed	[88]
7.	Bronchodilator	Seed	[89]
8.	Apoptotic	Seed	[90]
9.	Spasmolytic	Seed	[67]
10.	Hypoglycaemic	Seed	[91]
11.	Hypolipidemic	Seed	[92]
12.	Anti pyretic	Seed	[76]
13.	Anti tumor	Seed	[81]
14.	Cytoprotective	Seed	[41]
15.	Chemo modulator	Seed	[93]
16.	Antimutagenic	Seed	[81]
17.	Antithrombotic	Seed	[94]
18.	Oculohypotension	Seed	[77]
19.	Diuretics	Seed	[95]
20.	Antinociceptive	Seed	[96]
21.	Anticolitic	Seed	[53]
22.	Anti-hirutism	Seed	[84]
23.	Nootropic	Seed	[97]
24.	Memory enhancing	Seed	[98]
25.	Hepatoprotective	Seed	[82]
26.	Anti-stress	Seed	[83]
27.	Anxiolytic	Seed	[73]
28	Anti- inflammatory	Seed	[45]
29.	Anti- allergic	Seed	[86]
30.	Antiviral	Seed	[74]
31.	Antimicrobial	Seed	[72]
32.	Antioxidant	Seed	[72]
33.	Antifungal	Seed	[72]

19.8.51 Protects Against Aging

A natural skin care aid is fennel. All parts of the fennel plant—the seeds, stalks, bulbs, and leaves—are nutrient-dense and a great source of vitamins. These vitamins, which include vitamin C, are crucial for maintaining healthy skin and even encourage the production of collagen to keep the skintight and firm (Table 19.5).

19.9 Other Uses

Fennel seed essential oil is used in soaps, toothpastes, perfumes, and air fresheners, with some other things. The crushed leaves are useful in keeping dog's flea-free, while the dried plant repels insects. Leaves and flowers are combined to produce brown and yellow colors. Because of its camphor-like perfume, a bicyclic ketone extracted from the plant is employed as an odor-making ingredient for utilized in air fresheners. Grazing animals graze on the plant stubble in the field.

The phenylpropenes (E)-anethole and the monoterpene (+)-fenchone as well as estragole generated from fennel fruit, have been proven to be beneficial for regulating field populations of *Callosobruchus chinensis*, *Sitophilus oryzae* and *Lasioderma serricorne [60]*. The corrosion inhibitory action of fennel oil was discovered in carbon steel [99]. Fennel oil has a mixed-type inhibitory effect. At 3 mL/L, the inhibitory efficiency peaked at 76%, but dropped as the temperature rose. The fundamental oil of F. vulgare was found to have substantial insecticidal efficacy against the preserved foot insect *Sitophilus zeamais* [100].

Bitter fennel flower, unripe, and mature fruit oils (40 ppm) inhibited in-vitro mycelial development of *Fusarium oxysporum*, *Alternaria alternata* and *Rhizoctonia solani* to varied degrees [36]. In an immersion bioassay, F. vulgare essential oil showed the most powerful action for second-stage juveniles (J2) of Meloidogyne incognita worm [101]. Trans-anethole and estragole, two of its constituents, also had nematicidal action for second stage juveniles.

References

1. Arzoo, M., & P. (2017). Fennel: A brief review. *European Journal of Pharmaceutical and Medical Research, 4*(2), 668–675.
2. Pattanayak, R. *Fennel as functional food* (p. 352). New Delhi Publishers.
3. Chadwick, J. (1976). *The Mycenaean world.* Cambridge University Press.
4. Malhotra, S. (2012). Fennel and fennel seed. In *Handbook of herbs and spices* (pp. 275–302). Elsevier.
5. Bown, D. (2001). *New encyclopedia of herbs & their uses: The definitive guide to the identification, cultivation, and uses of herbs*. DK Pub.
6. Seidemann, J. (2005). *World spice plants*. Springer.
7. Weiss, E. A. (2002). *Spice crops*. CABI.
8. Bernath, J., et al. (1994). *Production-biological investigation of fennel (Foeniculum vulgare) populations of different genotype* (pp. 287–292). Atti del Convegno Internazionale.
9. Koudela, M., & Petříková, K. (2008). Nutritional compositions and yield of sweet fennel cultivars-*Foeniculum vulgare* Mill. ssp. vulgare var. azoricum (Mill.). *Tnelt HortScience, 35*(1), 1–6.
10. Mata, A., et al. (2007). Antioxidant and antiacetylcholinesterase activities of five plants used as Portuguese food spices. *Food Chemistry, 103*(3), 778–786.
11. Singh, A., et al., Performance of fennel (*Foeniculum vulgare* Mill.) Cultivars under different irrigation levels. 2021.
12. Farrell, K. T. (1998). *Spices, condiments and seasonings*. Springer.

13. Chowdhury, J. U., et al. (2009). Constituents of essential oils from leaves and seeds of *Foeniculum vulgare* Mill. cultivated in Bangladesh. *Bangladesh Journal of Botany, 38*(2), 181–183.
14. Bellomaria, B., Valentini, G., & Arnold, N. (1999). L'olio essenziale di Foeniculum vulgate mill. ssp. vulgate. *Rivista Italiana EPPOS, 9*, 43–48.
15. Coşge, B., Kiralan, M., & Gürbüz, B. (2008). Characteristics of fatty acids and essential oil from sweet fennel (*Foeniculum vulgare* Mill. var. dulce) and bitter fennel fruits (F. vulgare Mill. var. vulgare) growing in Turkey. *Natural Product Research, 22*(12), 1011–1016.
16. El-Awadi, M., & Hassan, E. A. (2010). Physiological responses of fennel (*Foeniculum vulgare* mill) plants to some growth substances: The effect of certain amino acids and a pyrimidine derivative. *Journal of American Science, 6*(7), 120–125.
17. Anthony, K. P., Deolu-Sobogun, S. A., & Saleh, M. A. (2012). Comprehensive assessment of antioxidant activity of essential oils. *Journal of Food Science, 77*(8), C839–C843.
18. Moghtader, M. (2013). Comparative survey on the essential oil composition from the seeds and flowers of *Foeniculum vulgare* Mill. from Kerman province. *Journal of Horticulture and Forestry, 5*(3), 37–40.
19. Rao, Y. S., & Mathew, K. M. (2012). Tamarind. In *Handbook of herbs and spices* (pp. 512–533). Elsevier.
20. Moura, L. S., et al. (2003). Determination of the global yields for the system fennel (*Foeniculum vulgare*)+ CO2. In *Proceedings of the 6th international symposium on supercritical fluids, Versalles, Francia.*
21. Malhotra, S., & Vashishtha, B. (2008). *Organic production of seed spices* (p. 90). Ajmer, India.
22. Badgujar, S. B., Patel, V. V., & Bandivdekar, A. H. (2014). *Foeniculum vulgare mill: A review of its botany, phytochemistry, pharmacology, contemporary application, and toxicology* (p. 2014). BioMed Research International.
23. Rather, M. A., et al. (2016). *Foeniculum vulgare*: A comprehensive review of its traditional use, phytochemistry, pharmacology, and safety. *Arabian Journal of Chemistry, 9*, S1574–S1583.
24. Khan, M., & Musharaf, S. (2014). *Foeniculum vulgare Mill. A medicinal herb* (p. 4). Medicinal Plant Research.
25. Shiva, M., Lehri, A., & Shiva, A. (2002). *Aromatic and medicinal plants. Yielding essential oil for pharmaceutical, perfumery, cosmetic industries and trade*. International Book Distributors.
26. Malhotra, S. (2011). *Breading potential of indigenous germplasm of seed spices. Vegetable crops: Genetic resources and improvement* (pp. 477–497). New India Publishing House.
27. Singh, R., & Malhotra, S. (2007). Harvesting and maturity indices in seed spices crops. In S. K. Malhotra & V. B. B. Production (Eds.), *Development, quality and export of seed spices* (pp. 195–200). NRCSS.
28. Chevallier, A. (2001). *Encyclopedia of medicinal plants*. Dorling Kindersley Limited.
29. Malhotra, S. (2010). Augmenting export of spices through value added spice products. In *3rd National seminar and exhibition on spices and herbs*. NASC Pusa.
30. Ruberto, G., et al. (2000). Antioxidant and antimicrobial activity of *Foeniculum vulgare* and Crithmum maritimum essential oils. *Planta Medica, 66*(08), 687–693.
31. Singh, G., et al. (2002). Studies on essential oils: Part 10; antibacterial activity of volatile oils of some spices. *Phytotherapy Research: An International Journal Devoted to Pharmacological and Toxicological Evaluation of Natural Product Derivatives, 16*(7), 680–682.
32. Sadeghian, S., et al. (2005). Bacteriostatic effect of dill, fennel, caraway and cinnamon extracts against Helicobacter pylori. *Journal of Nutritional & Environmental Medicine, 15*(2–3), 47–55.
33. Elgayyar, M., et al. (2001). Antimicrobial activity of essential oils from plants against selected pathogenic and saprophytic microorganisms. *Journal of Food Protection, 64*(7), 1019–1024.
34. Jazani, N., et al. (2009). Antibacterial effects of Iranian fennel essential oil on isolates of Acinetobacter baumannii. *Pakistan Journal of Biological Sciences: PJBS, 12*(9), 738–741.

35. Anwar, F., et al. (2009). Antioxidant and antimicrobial activities of essential oil and extracts of fennel (*Foeniculum vulgare* mill.) seeds from Pakistan. *Flavour and Fragrance Journal, 24*(4), 170–176.
36. Özcan, M. M., et al. (2006). Comparative essential oil composition and antifungal effect of bitter fennel (*Foeniculum vulgare* ssp. piperitum) fruit oils obtained during different vegetation. *Journal of Medicinal Food, 9*(4), 552–561.
37. Vasudevan, K., et al. (2000). Influence of intragastric perfusion of aqueous spice extracts on acid secretion in anesthetized albino rats. *Indian journal of gastroenterology: official journal of the Indian Society of Gastroenterology, 19*(2), 53–56.
38. Dhiman, A. K., & Kumar, A. (2006). *Ayurvedic drug plants*. Daya Books.
39. Platel, K., & Srinivasan, K. (2001). Studies on the influence of dietary spices on food transit time in experimental rats. *Nutrition Research, 21*(9), 1309–1314.
40. Anand, P., et al. (2008). Cancer is a preventable disease that requires major lifestyle changes. *Pharmaceutical Research, 25*, 2097–2116.
41. Singh, B., & Kale, R. (2008). Chemomodulatory action of *Foeniculum vulgare* (Fennel) on skin and forestomach papillomagenesis, enzymes associated with xenobiotic metabolism and antioxidant status in murine model system. *Food and Chemical Toxicology, 46*(12), 3842–3850.
42. De Vincenzi, M., et al. (2000). Constituents of aromatic plants: II. Estragole. *Fitoterapia, 71*(6), 725–729.
43. Iten, F., & Saller, R. (2004). Fennel tea: Risk assessment of the phytogenic monosubstance estragole in comparison to the natural multicomponent mixture. *Forschende Komplementarmedizin und klassische Naturheilkunde= Research in complementary and natural classical medicine, 11*(2), 104–108.
44. Iyer, L. V., et al. (2003). Glucuronidation of 1′-hydroxyestragole (1′-HE) by human UDP-glucuronosyltransferases UGT2B7 and UGT1A9. *Toxicological Sciences, 73*(1), 36–43.
45. Choi, E.-M., & Hwang, J.-K. (2004). Antiinflammatory, analgesic and antioxidant activities of the fruit of *Foeniculum vulgare*. *Fitoterapia, 75*(6), 557–565.
46. Oktay, M., Gülçin, İ., & Küfrevioğlu, Ö. İ. (2003). Determination of in vitro antioxidant activity of fennel (*Foeniculum vulgare*) seed extracts. *LWT-Food Science and Technology, 36*(2), 263–271.
47. Parejo, I., et al. (2004). Separation and characterization of phenolic compounds in fennel (*Foeniculum vulgare*) using liquid chromatography– negative electrospray ionization tandem mass spectrometry. *Journal of Agricultural and Food Chemistry, 52*(12), 3679–3687.
48. Çalişkan, U. K., et al. (2014). Comparison of volatile oil from herbalist and cultivar samples of *foeniculum vulgare* mill. for their compliance with european pharmacopea and antimicrobial activity. *Journal of Faculty of Pharmacy of Ankara University, 39*(3), 195–210.
49. Gilligan, N. (2005). The palliation of nausea in hospice and palliative care patients with essential oils of Pimpinella anisum (aniseed), *Foeniculum vulgare* var. dulce (sweet fennel), Anthemis nobilis (Roman chamomile) and Mentha x piperita (peppermint). *International Journal of Aromatherapy, 15*(4), 163–167.
50. Meissner, M., et al. (2003). Clinical and serological evaluation of the secondary antiphospholipid syndrome patients (SAPS) in the course of SLE and primary antiphospholipid syndrome patients (PAPS) for the presence of vasculitis or vasculopathy. In *Annals of the rheumatic diseases*. BMJ Publishing Group British MED Assoc House, Tavistock Square, London WC1H .....
51. Platel, K., & Srinivasan, K. (2000). Stimulatory influence of select spices on bile secretion in rats. *Nutrition Research, 20*(10), 1493–1503.
52. Jahromi, B. N., Tartifizadeh, A., & Khabnadideh, S. (2003). Comparison of fennel and mefenamic acid for the treatment of primary dysmenorrhea. *International Journal of Gynecology & Obstetrics, 80*(2), 153–157.
53. Javidnia, K., et al. (2003). Antihirsutism activity of fennel (fruits of *Foeniculum vulgare*) extract–a double-blind placebo controlled study. *Phytomedicine, 10*(6–7), 455–458.

54. Lee, C.-H., Sung, B.-K., & Lee, H.-S. (2006). Acaricidal activity of fennel seed oils and their main components against Tyrophagus putrescentiae, a stored-food mite. *Journal of Stored Products Research, 42*(1), 8–14.
55. Ostad, S., Khakinegad, B., & Sabzevari, O. (2004). Evaluation of the teratogenicity of fennel essential oil (FEO) on the rat embryo limb buds culture. *Toxicology In Vitro, 18*(5), 623–627.
56. Rancé, F., et al. (1994). Sensibilisation aux épices chez l'enfant. *Revue Française D'allergologie et D'immunologie Clinique, 34*(6), 475–479.
57. Giosafatto, C. V., Mariniello, L., & Ring, S. (2007). Extraction and characterization of *foeniculum vulgare* pectins and their use for preparing biopolymer films in the presence of phaseolin protein. *Journal of Agricultural and Food Chemistry, 55*(4), 1237–1240.
58. Shah, A., Qureshi, S., & Ageel, A. (1991). Toxicity studies in mice of ethanol extracts of *Foeniculum vulgare* fruit and Ruta chalepensis aerial parts. *Journal of Ethnopharmacology, 34*(2–3), 167–172.
59. Türkyılmaz, Z., et al. (2008). A striking and frequent cause of premature thelarche in children: *Foeniculum vulgare*. *Journal of Pediatric Surgery, 43*(11), 2109–2111.
60. Burkill, I. H. (1966). *A dictionary of the economic products of the Malay Peninsula* (Vol. 2, 2nd ed.). A Dictionary of the Economic Products of the Malay Peninsula.
61. Lim, T., & Lim, T. (2016). *Nymphaea odorata*. Springer.
62. Chiej, R. (1984). *The Macdonald encyclopedia of medicinal plants*. Macdonald & Co (Publishers) Ltd..
63. Bown, D. (1995). *The Royal Horticultural Society encyclopedia of herbs & their uses*. Dorling Kindersley Limited.
64. Chevallier, A., The encyclopedia of medicinal plants. 1996.
65. Ebeed, N. M., et al. (2010). Antimutagenic and chemoprevention potentialities of sweet fennel (*Foeniculum vulgare* mill.) hot water crude extract. *Journal of American Science, 6*, 831–842.
66. Jeambey, Z., et al. (2009). Perceived health and medicinal properties of six species of wild edible plants in north-East Lebanon. *Public Health Nutrition, 12*(10), 1902–1911.
67. Abou El-Soud, N., et al. (2011). Antidiabetic activities of *Foeniculum vulgare* mill. essential oil in streptozotocin-induced diabetic rats. *Macedonian Journal of Medical Sciences (Archived), 4*(2), 139–146.
68. Albert-Puleo, M. (1980). Fennel and anise as estrogenic agents. *Journal of Ethnopharmacology, 2*(4), 337–344.
69. Khan, R. U., et al. (2022). Perspective, opportunities and challenges in using fennel (*Foeniculum vulgare*) in poultry health and production as an eco-friendly alternative to antibiotics: A review. *Antibiotics, 11*(2), 278.
70. Atwaa, E. S. H., et al. (2022). Bioactivity, physicochemical and sensory properties of probiotic yoghurt made from whole milk powder reconstituted in aqueous fennel extract. *Fermentation, 8*(2), 52.
71. Bigdeloo, M., et al. (2022). Activated carbon derived from fennel flower waste as high-efficient sustainable materials for improving cycle stability and capacitance performance of electroactive nanocomposite of conductive polymer. *Journal of Energy Storage, 55*, 105793.
72. Singh, G., et al. (2006). Chemical constituents, antifungal and antioxidative potential of *Foeniculum vulgare* volatile oil and its acetone extract. *Food Control, 17*(9), 745–752.
73. Freire, R. S., et al. (2005). Synthesis and antioxidant, anti-inflammatory and gastroprotector activities of anethole and related compounds. *Bioorganic & Medicinal Chemistry, 13*(13), 4353–4358.
74. Rathore, S., Saxena, S., & Singh, B. (2013). Potential health benefits of major seed spices. *Int J Seed Spices, 3*(2), 1–12.
75. Al-Snafi, A. E. (2018). The chemical constituents and pharmacological effects of *Foeniculum vulgare*-A review. *IOSR Journal of Pharmacy, 8*(5), 81–96.

76. Devika, V., & Mohandass, S. (2014). Apoptotic induction of crude extract of *Foeniculum vulgare* extracts on cervical cancer cell lines. *International Journal of Current Microbiology and Applied Sciences, 3*(3), 657–661.
77. Beaux, D., Fleurentin, J., & Mortier, F. (1997). Diuretic action of hydroalcohol extracts of *Foeniculum vulgare* var dulce (DC) roots in rats. *Phytotherapy Research: An International Journal Devoted to Medical and Scientific Research on Plants and Plant Products, 11*(4), 320–322.
78. Garg, C., et al. (2011). Effect of *Foeniculum vulgare* mill. fruits in obesity and associated cardiovascular disorders demonstrated in high fat diet fed albino rats. *Journal of Pharmaceutical and Biomedical Research, 8*(8), 1–5.
79. Mahmoudi, Z., et al. (2013). Effects of *Foeniculum vulgare* ethanol extract on osteogenesis in human mecenchymal stem cells. *Avicenna Journal of Phytomedicine, 3*(2), 135.
80. Lewu, F. B., & Afolayan, A. (2009). Ethnomedicine in South Africa: The role of weedy species. *African Journal of Biotechnology, 8*(6).
81. Tognolini, M., et al. (2007). Protective effect of *Foeniculum vulgare* essential oil and anethole in an experimental model of thrombosis. *Pharmacological Research, 56*(3), 254–260.
82. Koppula, S., & Kumar, H. (2013). *Foeniculum vulgare* mill (Umbelliferae) attenuates stress and improves memory in wister rats. *Tropical Journal of Pharmaceutical Research, 12*(4), 553–558.
83. Mesfin, M., Asres, K., & Shibeshi, W. (2014). Evaluation of anxiolytic activity of the essential oil of the aerial part of *Foeniculum vulgare* miller in mice. *BMC Complementary and Alternative Medicine, 14*(1), 1–7.
84. Joshi, H., & Parle, M. (2006). Cholinergic basis of memory-strengthening effect of *Foeniculum vulgare* Linn. *Journal of Medicinal Food, 9*(3), 413–417.
85. Aucoin, M., et al. (2020). Diet and psychosis: A scoping review. *Neuropsychobiology, 79*(1–2), 20–42.
86. Orhan, İ. E., et al. (2012). Antimicrobial and antiviral effects of essential oils from selected Umbelliferae and Labiatae plants and individual essential oil components. *Turkish Journal of Biology, 36*(3), 239–246.
87. Rasul, A., et al. (2012). Formulation development of a cream containing fennel extract: In vivo evaluation for anti-aging effects. *Die Pharmazie-An International Journal of Pharmaceutical Sciences, 67*(1), 54–58.
88. Boskabady, M., Khatami, A., & Nazari, A. (2004). Possible mechanism (s) for relaxant effects of *Foeniculum vulgare* on Guinea pig tracheal chains. *Die Pharmazie-An International Journal of Pharmaceutical Sciences, 59*(7), 561–564.
89. Bogucka-Kocka, A., Smolarz, H., & Kocki, J. (2008). Apoptotic activities of ethanol extracts from some Apiaceae on human leukaemia cell lines. *Fitoterapia, 79*(7–8), 487–497.
90. Piccaglia, R., & Marotti, M. (2001). Characterization of some Italian types of wild fennel (*Foeniculum vulgare* mill.). *Journal of Agricultural and Food Chemistry, 49*(1), 239–244.
91. Oulmouden, F., et al. (2011). Hypolipidemic and anti-atherogenic effect of aqueous extract of fennel (*Foeniculum Vulgare*) extract in an experimental model of atherosclerosis induced by triton WR-1339. *European Journal of Scientific Research, 52*(1), 91–99.
92. Tanira, M., et al. (1996). Pharmacological and toxicological investigations on *Foeniculum vulgare* dried fruit extract in experimental animals. *Phytotherapy Research, 10*(1), 33–36.
93. Tripathi, P., et al. (2013). Investigation of antimutagenic potential of *Foeniculum vulgare* essential oil on cyclophosphamide induced genotoxicity and oxidative stress in mice. *Drug and Chemical Toxicology, 36*(1), 35–41.
94. Agarwal, R., et al. (2008). Oculohypotensive effects of *Foeniculum vulgare* in experimental models of glaucoma. *Indian Journal of Physiology and Pharmacology, 52*(1), 77–83.
95. Nassar, M. I., et al. (2010). Secondary metabolites and pharmacology of *Foeniculum vulgare* mill. Subsp. Piperitum. *Revista latinoamericana de química, 38*(2), 103–112.

96. Chakŭrski, I., et al. (1981). Treatment of chronic colitis with an herbal combination of Taraxacum officinale, Hipericum perforatum, Melissa officinaliss, Calendula officinalis and *Foeniculum vulgare*. *Vutreshni bolesti, 20*(6), 51–54.
97. El Bardai, S., et al. (2001). Pharmacological evidence of hypotensive activity of Marrubium vulgare and *Foeniculum vulgare* in spontaneously hypertensive rat. *Clinical and Experimental Hypertension (New York, NY: 1993), 23*(4), 329–343.
98. Özbek, H., et al. (2003). Hepatoprotective effect of *Foeniculum vulgare* essential oil. *Fitoterapia, 74*(3), 317–319.
99. Lahhit, N., et al. (2011). Fennel (*Foeniculum vulgare*) essential oil as green corrosion inhibitor of carbon steel in hydrochloric acid solution. *Portugaliae Electrochimica Acta, 29*(2), 127–138.
100. Bertoli, A., et al. (2012). Volatile chemical composition and bioactivity of six essential oils against the stored food insect Sitophilus zeamais Motsch.(Coleoptera Dryophthoridae). *Natural Product Research, 26*(22), 2063–2071.
101. Ntalli, N. G., et al. (2011). Synergistic and antagonistic interactions of terpenes against Meloidogyne incognita and the nematicidal activity of essential oils from seven plants indigenous to Greece. *Pest Management Science, 67*(3), 341–351.

Chapter 20
Henbane

Sara Zafar, Khalid Sultan, Shagufta Perveen, Abida Parveen, Naeem Iqbal, and Umar Farooq Gohar

20.1 Introduction

Family	:	Solanaceae
English Name	:	Henbane, Black Henbaen
Indian Nam	:	Parasikayavani, (Sanskrit), Khurasani, Ajavayan, Khurasani Jamani (Hindi), Kurassaniyoman (Tamil)
Species and Varieties		*Hyoscyamus niger* L. *H. Muticus*, IC-66 Aela
Distribution		India, America, Central Asia, Europe, Finland, Greece, Portugal, From Western Punjab to Afghanistan, to Egypt
Uses	:	Drugs

Hyoscyamus niger L. or *H. Muticus* L. also known as black henbane or common henbane is an official drug of repute. The herb and seed contain tropane alkaloids (0.03–0.06%), 90% of which are hyoscyamine and hyoscine and the rest are scopolamine, atropine, hyoscipirin, choline, fatty oil, mucilage, albumen and KNO; they are used as an anodyne, sedative, tranquilizer, antiseptic, antispasmodic, mydriatic Hyoscine is very effective at protecting against shocks generated by accidents and loud noises [1].

S. Zafar · K. Sultan · S. Perveen (✉) · A. Parveen · N. Iqbal
Department of Botany, Government College University Faisalabad, Faisalabad, Pakistan

U. F. Gohar
Institute of Industrial Biotechnology, Government College University, Lahore, Pakistan

M. Zia-Ul-Haq et al. (eds.), *Essentials of Medicinal and Aromatic Crops*,
https://doi.org/10.1007/978-3-031-35403-8_20

20.1.1 History & Myths

Mandrake was not grown as natural but its plantation in Britain was done in the eleventh century, so everyone seeking its real or imagined qualities had to look for a substitute. Those eager for mandrake's medical properties resorted to the plants of henbane. Hyoscyamus *niger* has similar features to the henbane but this plant is more found in the northern areas. This herb is important in therapeutical uses. Henbane was a great effect on witches, this plant was used as liquor by witches at midnight and in some flying oinments. This wretched henbane was poured into the ears of Hamlet's father [2].

After being served a tasty piece, a needy dog is applied to the mandrake's root (far right). Hunter kept a piece of meal near to the plant when the dog sees the meal, the dog comes for eating the meal, uprooted the plant with the meal and eat it. After some time, the dog died and was buried. Some companies are seeking to get profit from this plant but due to small fibrous roots disappointed these marketers. These marketers select another plant *Bryonia dioica* that had a large taproot native to England [2].

For millennia, black henbane was used as a therapy, and physicians were more experienced with it in the past. Henbane was employed by Dioscorides (first century A.D.) to heal insomnia and ache. Pliny (first century A.D.) claimed that henbane pertains to alcohol and decreases brain function [2]. The Homeric nepenthes is a legendary creature from Greek mythology. It was given the names Hyoscyamus and Symphonica by Benedictus Crispus in the late seventh century (681 A.D.). Jusquiasmus was a term used in the ninth century to describe some of the benefits of BH. In some areas of the United Kingdom, the seeds of henbane as food are applied to the hens, when the robbers come to rob the hens they become conscious all night due to its poisonous effects [3].

Black henbane ointment was applied to the Women of Norwegian as a result the parts of the body had thin skin and hallucinations occurs [2]. Black henbane extract was used by witches as a magic drink or to jump over the fire [4]. Black henbane was used by the soothsayers in hallucinogenic components. Henbane was used as drunken by the criminal for example in the United States doctor Crippen poisoned his wife with this herb leaving before America with MistressHenbane [5].

20.1.2 Origin and Distribution

Black henbane is found mostly in the countries of Europe and northern Africa [4, 6–8]. It can be found in practically every part of the arctic circle, covering Asia, Europe, Central America, and regions of Brazil [9]. The black henbane has eleven species found in the regions of Europe, Africa regions Asia and the Canary Islands. They all have similar features and components. It can be found near the shore areas of the seas which are chalky-like [10]. These trees grow in Iran across Northern Iran, Karaj, Tehran, Azerbaijan, Tabriz, Arak Khorasan and other countries of the world [11, 12].

Hyoscyamus niger is found in Central Asia, India, and tropical America, and is also found in the areas of Portugal, Greece, Norway and north areas of Finland. It ranges from Portugal and Greece in the south to Norway and Finland in the north. It is primarily found growing in temperate to subtropical parts of India, primarily in Kashmir and Uttar Pradesh. Its recent successful production in Indore has demonstrated that it can also be grown as a winter crop in warmer climates. The plant has also been reported to grow successfully in Bangalore settings. *H. Muticus*, on the other hand, is essentially a desert plant that grows to a height of 90 cm (with a spread of 30 cm) and may be found growing from Western Punjab through Afghanistan to Egypt [1].

20.1.3 Description of the Plant

The plant *Hyoscyamus niger* is classified as a biennial or annual It has a long flowering pattern and is often combined with a photoperiodic system to initiate a flowering [13]. The plant's one to two metre long, oval, big leaves have an extremely hairy surface and a greenish-grey colour [14]. The blooms have nearly no stems and a corolla fashioned like a glove with purple veins that is yellowish. As the fruit reaches maturity, the calyx widens and encloses the seed capsule [15]. The Solanaceae family, also known as the Nightshade family, consists of 96 genera and 3000 species. They are found in North America, Europe, Asia, and Australia. Widespread in temperate climates and difficult to cultivate in certain tropical locations [10].

There are several economical, agricultural, and medical plants in this family [16]. One of the most significant and varied plant groups for humans is Solanaceae. Alkaloids are abundant in the Solanaceae plant family as well, however only fifteen of these many species included tropane alkaloids.Atropa belladonna and Hyoscyamus niger are two of the six genera that have significant economic importance [14]. A family of herbal dicotyledons called Solanaceae. Given that certain plants prefer to grow in the shadow and some blossom at night, the Solanaceae family got its name from the Latin for "nightshade" or "sun," or "Solanuni" [17].

The Greek names (hyos), "hog and bean," and "niger" are the source of the English term "black henbanes." The Latin word for "black" was originally used to describe the colour of the seed. The herb known as Hyoscyamus has been used as a medicine since ancient Greece. Additionally, since its initial edition, the United States Pharmacopeia's leaves and flowering plumage have been formally documented [18]. At least three species, including Hyoscyamus niger, muticus, and albus, as well as the rest of the species in the genus, are significant in terms of medicine.Additionally, they are an excellent source of the tropane alkaloids hyoscyamine, scopolamine, atropine, and hyoscine that are grown on them [19].

The Anglo-Saxon terms henn (chicken) and bana (assassin), which refer to the seeds' ability to paralyse and kill birds, are the source of the name henbane. The plant is dangerous throughout, and even little doses, when mixed with other anticholinergic side effects, can result in everything from delirium to dizziness. If a person cannot fall asleep in Anglo-Saxon, shake and strain his head with the seed juice of garden mint and henbane [18]. A plant known as henbane is an annual or biennial that has a few sticky hairs. The first year, a brooch of basal leaves is visible, and the second year, a simple, erect, or slightly trimmed stone up to 80 cm high and coated in sticky hair is seen [20].

The radial leaves are arranged in a rosette shape around the neck of the root. They are long, egg-shaped, over 30 cm long, pointed, and stalky, with teethed edges. They have a greenish grey colour with sticky hairs covering them [21]. The stems' erect, thin heights range from 20 to 100 cm. The tall, egg-shaped, pinnate, pulpy, and greyish-green leaves have a stalk of bottom leaves and half of the top amplexicaul. On the stalks are clusters of blooms. The bell-shaped crown has a stunning dirty yellow hue with a reddish-violet throat and violet reticulated veins [21]. It produces a lot of seeds between 10,000 and 500,000 seeds per plant—and only 10 to 20 of these are enough to harm a youngster. An exploding capsule is a fruit. In fruit capsules, there can be up to 500 gray-brown seeds [22].

H. Niger belongs to the Solanaceae family and is an erect, viscidly hairy biennial plant that grows up to 160 cm in height and is grown in the Indian plains as a short-duration (90–100 days) winter crop. Simple, big, ovate or oblong, coarsely dentate, and pinnately lobed leaves. Flowers are yellowish or pale, sessile or subsessile, and borne on terminal cymes. The fruit is a berry with several tiny seeds. The seeds are tiny, measuring about 1.5 mm in length. 2000/g, brown to black, with fine, noticeable reticulations. *H. Muticus* L. resembles *H. Niger* but is lower in stature (90 cm). The leaves are modest in size, with a smooth dentate border and a thick, sticky feel. The seeds are slightly heavier and coated [1].

20.1.4 Varieties

The National Bureau of Plant Genetic Resources, cultivation of medicine New Delhi, produced and released IC-66, a short-duration (100-day) variety of *H. Muticus*. It is reported to yield 25 q/ha, with total tropane alkaloids ranging between 0.05 and 0.10%. The CIMAP, Lucknow, has released another variety that is both a tetraploid and a polyploid with a high seeding rate. This cultivar has a 200-day growing time and produces 645 g/ha fresh herb or 43.2 q/ha dried herb, compared to the diploid parent's 511 q/ha and 37.3q/ha. There is also a lot of alkaloid content [3].

It's possible that it'll be more profitable than the current varieties. Through mutation breeding, the CIMAP in Lucknow has released one variant of *H. Niger*. Aela is a mutant culture of *H. Niger* that was chosen from irradiation progenies and has yellow flowers with a faint purple tinge at the base of the petals. It grows quickly and produces 73 q/ha, which is 109% higher than the parent (50 q/ha). It has a high total alkaloids content of 0.545% compared to 0.167% in the parent plant [3].

20.1.5 Ingredients

All parts of black henbane like leaves, seeds and roots are used as medicine or as abused [23]. In the plant of black henbane, some alkaloid and nonalkaloid materials are found. A gas chromatography examination of black henbane extraction revealed around thirty-four alkaloids [24]. Hyoscyamine, atropine, scopolamine and tropane are tropane alkaloids present in black henbane and other plants of this family like Atropa belladonna & datura species. These chemicals have a variety of effects, including antispasmodic smooth muscle, bronchial hypersecretion decrease, and stomach discomfort alleviation [2, 25, 26].

The percentage of alkaloids in Egyptian henbane is likely more than the European henbane. Atropine and scopolamine are the chemicals or the contents which are more present in leaves [9, 27, 28]. The principal alkaloids of the root are cuscohygrine and apoatropine. Hyoscyamine and a trace of Atropine and hyoscine are the primary alkaloids that are present in black henbane seeds [9, 29]. The seeds of this plant have a very irritating smell. The seeds of this shrub have a large quantity of alkaloids [30, 31]. The percentages of black henbane alkaloids in the seeds, roots and leaves are approximately 0.17, 0.08 and 0.05, respectively [9, 27, 30].

The synthesis of the alkaloids in this plant was decreased due to some environmental problems like osmotic stress and infection of the bacteria. These pathogens attack the metabolic pathways on the plant and damage the plant [32, 33]. Animals and people will be poisoned if they consume huge amounts of the alkaloids in all areas of the black henbane [31]. Consumption of roughly 4 BH flowers, for example, is sufficient to cause clinical signs in a preschool child [30]. Under controlled conditions, the two primary BH alkaloids, hyoscyamine and scopolamine, are employed as medications. They are used as a moderate analgesic, antispasmodic, sedative, and mydriatic [6, 34]. Hyoscyamine, a BH secondary metabolite, is a levo-isomer of atropine with double the potency of atropine. Atropine is almost completely combined with hyoscyamine [35]. The central nervous system (CNS) depression field is the primary effect of the hyoscyamine [30, 36]. The plant produces hyoscyamine, however, it is promptly converted to atropine in the herbal cells and during the extraction [24, 27].

The atropine has hyoscyamine D and L has an equal quantity [24]. Tropic acid and tropine are formed when atropine is degraded. Tropane alkaloids are created from the tropane ring, which is formed by combining piperidine and a pyrrolidine ring [26]. Furthermore, non-tropane alkaloids such as Calystegia, which are weak to moderate glycosidase inhibitors, are found in black henbane [34]. Alkaloids found in henbane were the huge cause of the poison while the non-alkaloids in that plant is the focus topic of discussion.

Some secondary metabolites which are non-alkaloids like flavonoids, terpenoids, withanolides, lignans, coumarinolignans, phenolics, glycerides and glycosides are produced by several species which are anticholinergic. [34]. Non-alkaloidal chemicals found in the plant include Cleomiscosin B, Canabisin G, hyposmia, Canabisin D, Hyosciamide, Grossamide, Hyoscyamal, Balanophonin, Cleomiscosin

A B & C [9, 37]. The seeds of the henbane contain non-alkaloids like lignanamides [27, 38], lignans, withanolides [27, 39], saponin, hyoscyamine, balanophonin, pongamoside D&C [34] and coumarino-lignans [10, 34]. In methanolic extraction of henbane seed, four coumarin lignans were discovered: cleomiscosin A & B, B9-acetate and A9- acetate [34].

20.1.6 Composition

Tropane alkaloids can be found in all of the plant's vegetative parts. Alkaloids are present in 0.15–0.18% of the roots, 0.10% of the leaves, 0.02% of the stems, and 0.10% of the seeds [40]. The biennial henbane plant has more active compounds than the annul plant [40]. Henbane plant contains trace elements of volatile amines for example pyridine, choline and methylpyrroline. By extraction of the plant, some other compounds such as scopolamine, resin, atroscine, esculetin, proteins, manganese, arsenic, zinc, sodium, coumarins, mucilaginous, zinc, scopetole, copper and gamma-aminobutyric acid are present. Henbane seeds contain 34% of fat oil and certain free fatty acids (palmitic, oleinic, stearic, and myristicin, among others) [40, 41] (Table 20.1).

These non-alkaloid compounds are steroidal lactones. These are helpful in the regeneration of the brain processes. These are also antimicrobial, and anti-inflammatory and stop tumor-making in the body. These compounds are grabbed by the ethanolic extract of the henbane seeds [39]. The seeds of the henbane contain non-alkaloidal components like hyoscyamine. Two other lignans which have biological activities were also reported in seeds extraction of henbane seeds such as hyoscyamine and balanophonin [43, 44].

Phytochemical studies identified a new non-alkaloid such as coumarinolignans in seeds of the henbane. These are formed by the two compound coumarins and phenylpropanoid by the coupling process of the oxidation [43].

In seeds of *Hyoscyamus niger* phytochemical studies explore some flavonoids such as spiraeoside, rutin, pongamoside and flavonol glycoside [43, 45]. In seeds of the black henbane, steroidal saponins like hyosctamozides were identified. These

Table 20.1 Some alkaloids constituents from the seeds of Henbane [42]

Secondary metabolites	Compounds names
Lignans	Cannabisin D&G, Grossamide, Hyscyamal, Balanophnin
Coomarino-lignans	Cleomiscosin A & B, Hyosgerin, Venkatasin,
Withanlides	16 alpha-acetoxy-hyoscyamilactol, Hyoscyamilactol
Flavonoids	Rutin, spiraeoside,
Some saponins	Compounds of Hyoscyamoside A1, B2, C1, C etc
Phenolics	N- trans-feruloyltyramine, Vanillic acid,
Glycosides of steroidal	Petunioside L, Atroposide A, C, E,

hyoscyamozides are the derivatives of furostan, spirostan, pregnan, tigogenin and diosgenin [46].

Hyoscyamus niger has long been employed for ceremonial, religious, and folk medical purposes. There is a therapeutic benefit in every part of the plant. A very significant alkaloid was found in *Hyoscyamus niger* by chemical profiling and phytochemical studies, including coumarin, hyoscine, scopolamine, atropine, hyoscyamine, kaempferol, anisodamine, quercetin, linoleic acid, rutin, myristic acid, chlorogenic acid, cuscohygrine, etc. [27].

In addition, numerous significant non-alkaloidal compounds, such as lignin amides, withanolides [38], The current investigations suggested that *Hyoscyamus niger* L. will include flavonoids and tyramine compounds [39]. An earlier study discovered a lignan called hyosmin [27], Contains four coumarinolignans, including the seeds of *H. niger* venkatasin, hyosgerin, and cleomiscosins A and B. As a consequence of continued study on the lignans of *H. niger*, two novel Solanaceae family lignans, balanophonin and hyoscyamal, as well as two glucosides, pongamoside C & D, were identified and described [44].

TThe Arabic term "al-qali," which originally originated from sodas, is where the name "alkaloids"originates. The "basic constituents" of flowering plants that are pharmacologically active are largely, but not exclusively, nitrogen compounds [47]. Alkaloids are structurally unique nitrogen compounds that are present in many plants and frequently exhibit physiological function. Throughout history, people have always employed plants that generate alkaloids and extracts for their therapeutic and poisonous effects. Modern day plant-based alkaloids are frequently employed in chemotherapy, stimulants (such as caffeine and nicotine), and analgesics (such as morphine and codeine) (vincristine, vinblastine, camptothecin, and paclitaxel) [47]. Numerous alkaloids have been found to be harmful to other creatures but are nevertheless widely utilised as medications, recreational substances, or in entheogenic rituals using raw extracts obtained by acid extract-base mining.

Traditional human uses of alkaloid toxicity include hunting and combat, while modern use include the treatment of disorders [48]. Because it is a typical pyridine alkaloid, nicotine is employed as an effective natural pesticide and a potential anti-inflammatory [49, 50] Hyoscyamines, hyoscine, and scopolamines are the tropane alkaloids that have undergone the most extensive research, perhaps because of their anticholinergic effects on the parasympathetic nervous system [51]. Hyoscyamine and scopolamine are used in the treatment of motion sickness, heart problems, and gastrointestinal issues [52].

20.1.7 Lignans

Lignans are a vast class of compounds that are evidently abundant and are present in a wide variety of higher plants [53]. First discovered in wood of trees in the nineteenth century, lignans are bio-phenol chemicals that come from plants [54]. The phenylpropanoid pathway produces lignans, which are one of the primary secondary metabolites of plants. They are particularly useful in food and medicine for humans and play a significant role in plant defence [55].

The coupling of two C_6C_3 units distinguishes a large family of naturally occurring phenolic chemicals known as lignans and neoligans. When there is a link between two C_6C_3.The molecule that results when there is a link between two C_6C_3 units is called lignan. Neoligna is made up of two C_6C_3 units connected by extra carbon bonds [56, 57]. Cancer treatment and other disorders are of great pharmacological and therapeutic interest [58].

The Lignans were detected in the roots, leaves, seeds, and plant blooms of over 70 plant families from a wide range of habitats [59]. The anticancer activity in particular was among the class's several strong, substantial biological actions [60], showed antibacterial [61], antiviral [62], increase immune system, prevent swelling [61], antioxidant [60], and hepaprotective [63], work against cancer cells [64] and osteoarthritis [65] activities that contribute to the growth of interest in the production of lignans [59]. Lignans with antiviral properties in addition to cytotoxin include pelting and podophyllotoxin. Some lignans are also known to have anti-HIV, anti-microbial, anti-hepatotoxic, and anti-depressant properties [55].

20.1.8 Medicinal Importance

20.1.8.1 Henbane Species Traditionally Uses

Black Henbane (*Hyoscyamus niger*)

With a very long history of usage as a medicine, hypscyamus niger was widely employed as a sedative and painkiller [66]. Additionally, chronic dementia with insomnia, paralysis, agitation, convulsions, neuralgia, spasmodic cough, and asthma were treated with it, along with mental problems, epileptic mania, and other conditions [67].In addition, it was used to treat toothaches, pulmonary infection pain, tumour pain, abdominal colic, worm infestation pain, pain from tooth decay, and pain from the urinary tract, particularly pain from kidney stones. For external application on rheumatic, dental, and neuralgic pain, the seed oil was employed [68–71]. Along with these conditions, it was also prescribed for bleeding gums, odontalgia, dental caries, orchitis, papillitis, rheumatoid arthritis, colic, dyspepsia, worm infection, flatulence, epistaxis cardiac weakness, otal, haematemesis, whooping cough, hemoptysis, asthma, catarrh, bronchitis, conjunctivitis [72].

Hyoscyamus reticulatus

Asthma, stomach ulcers, motion sickness, and Parkinson's disease were all treated with it. It was also used as an antidote for many toxins, a spasmolytic, mydriatic, sedative, and analgesic [73].

Hyoscyamus albus

The plant extracts were employed as an antiasthmatic and an antispasmodic in traditional medicine. It was also combined with cannabis and datura to create a sedative and hallucinogen [24, 74].

As old as humanity itself, herbal medicine has been used for healing. Written records, historical monuments, and even the original plant remedies provide enough proof of the relationship between man and his hunt for pharmaceuticals in nature dating back a very long time. As a result of man's long-standing battles with disease, he has developed the ability to seek out medications in the barks, seeds, fruit bodies, and other components of plants. Current pharmacotherapy today includes a variety of plant-derived medications that were utilized for millennia by ancient civilizations and whose active effect has been acknowledged by modern research. The ability of pharmacists and doctors to respond to challenges that have emerged with the spread of professional services in the facilitation of man's life has improved thanks to their knowledge of the development of ideas related to the use of medicinal plants as well as the evolution of awareness [75].

Antipyretic, Analgesic and Anti-inflammatory Effects

Experimental studies on standardized *Hyoscyamus albus* methanolic extract looked at its analgesic (acetic acid-induced writhing response and other formalin-induced paw licking in rats) and antipyretic (brewer's yeast-induced fever in rats) characteristics. In the second stage of the formalin test, the acetic acid-induced writhing responses and the licking time were decreased by 100 and 200 mg/kg of the methanolic extract of *Hyoscyamus albus*, respectively. Additionally, it demonstrated a dose-dependent reduction in body temperature at both doses for up to 3 h; the impact was comparable to that of paracetamol [76]. In experimental animal models, the methanolic extract of *Hyoscyamus niger* seeds was tested for analgesic, anti-inflammatory, and antipyretic effects at various doses. The methanolic extract of *Hyoscyamus niger* seed generated a considerable increase in hot plate reaction time while a dose-dependent reduction in writhing response indicated analgesic efficacy. Additionally, it worked well in inflammation that was both acute and chronic, as determined by carrageenin-induced paw edoema and cotton pellet granuloma techniques. Additionally, it shown antipyretic properties in the yeast-induced pyrexia model [34]. Mice's acetic acid-mediated writhes were reduced in number by the crude extract of *Hyoscyamus niger* (Hn. Cr) in a dose-dependent manner [77]. Rats were used to study the analgesic efficacy of the alcoholic extract of *Hyoscyamus niger* seeds in both acute and chronic pain. The findings showed that injection of an alcohol extract from Hyoscyamus niger seeds greatly raised the chronic pain threshold and significantly decreased both acute and chronic pain brought on by formalin [78].

Mice were used to test the methanolic extract of *Hyoscyamus reticulates* antinociceptive properties. Acetic acid-induced writhing test and hot plate test on mice were utilized as two models to examine the effects of the extracts on nociception. With a maximal impact of 35.56% decrease, the methanolic extract significantly reduced the acetic acid-induced writhings in mice [79].

Plants have always played an important part in the treatment of numerous diseases throughout history. We simply need to go to ancient cultures to see how important plants were in medicine. The use of Unani and Hindu medicinal systems, for example, was effective in treating a variety of diseases and symptoms [80]. With the passage of time, however, those approaches proved to be less successful. Then, as a pressing human need, organic chemistry arose to meet people's rising demand for pharmaceuticals [81].

The alkaloids hyoscyamine and scopolamine have a characteristic parasympathetic action. The group of the alkaloids in the medicines like atropine sulfate which is the combination of the DL -isomers of the henbane are inactive while the L isomers are 2 times greater biological activity of the atropine. In the first case, the drugs have a peripheral postganglial cholinergic inhibiting effect at low therapeutic dosages. High doses of henbane have a bad effect on the central nervous system. It also affects the release of saliva, tears, and affects the stomach and glands of the intestine, which also affects the smooth muscles. While the use of this herb is beneficial to some heart diseases. This plant had a bad impact on the eyes, chemicals present in it cause paralysis of the eyes muscles as well as mydriasis to the muscles.

Scopolamine has higher central effectiveness. Henbane has more quantity of scopolamine as compared to other plant species which is related to that plant-like Atropa and datura species. Due to these qualities, this herb has a sedative effect on the central nervous system. The drugs are used for peace of mind, but they cause damage to the CNS. Atropine was helpful to the intoxication of these drugs. At the receptors site, atropine replaces the cholinomimetic because the atropine and cholinomimetic have antagonistic to each other. Symptoms of the atropine have the same as the scopolamine. High doses of scopolamine have been found s sedative to CNS, function as anti-Parkinson's disease and anti-seasickness. To use this plant henbane as a Parkinson's disease reliever or as a sedative by the prescription of the doctor. By rubbing henbane oil extract (oleum Hyoscyami) into the skin, it is used topically for neurologic and rheumatic problems [40, 82–84].

Henbane's physiological activity is comparable to that of Belladonna, Stramonium, Scopola, and other Solanaceae medicines. Henbane has fewer side effects, is more hypnotic and calming and also does not cause constipation. Uses from the past Henbane has been used as a medicinal since ancient times. Baron Hammer-Purgstall was convinced that " bent " (Arabic for Henbane) was Homer's Nepenthe. Hippocrates utilized white henbane as one of his "simples" (fifth century, B.C.). In terms of medical benefits, this herb is very similar to our Henbane. Dioscorides (about 60 A.D.) the plant's name hyoscyamus, Moreover, they say "dioskyamos" as an obsolete name, presumably derived from the use of medicine in temple "mysteries" [3].

Three species of this plant white, black and yellow was discovered by the Dioscorides. He singled out the white as the least risky. (In truth, it is medicinally similar to black henbane, albeit with a milder effect.). Some physician was suggested that the roots of this plant with vinegar were used as a mouthwash to relieve toothache. At the end of the seventeenth-century scientist make a toothpaste by the mixture of this plant leaves with a combination of poppy heads, red and violet rose blossoms and leaves of sage in water that was anti-odontalgic. The burned seeds of henbane were used to kill the worms in the teeth which cause discomfort to the patients. The drops of extract pour into the teeth and due to foul smell & poison worms comes out and dies. Pliny (fl. 60 A.D.) identified the drug's psychogenic properties and asked that "Henbane is of the nature of wine, and thus unpleasant to the mind, and disturbs the head (It ought) to be used with great caution and discretion. A person who consumes more than four leaves in a single drink will go insane [3].

By mixing the seeds with milk of mares and binding a paste made from a chunk of wild bull skin, an early contraceptive was created. Henbane was officially used as an anesthetic in the middle ages in combination with hemlock, mandrake, opium, aconite and juice of datura. The black henbane was used unlawfully as sleeping pills in the form of knock-out drops. In the fourteenth century, Gui de Chauliac defined the usage of narcotic inhalation as "imitating the pities of old physicians." To this last limb, cast one unconscious before they display their cut, then cut the unhealthy area [3].

The Arabians, too, were aware of Henbane's medicinal value, as evidenced by the following passage from the great anonymous classic, "The Thousand and One Nights": "Presently he filled a cresset8 with firewood on which he strewed powdered Henbane, and lighting it, went round about the tent with it till Henbane growing wild in a sheep corral and elsewhere near Barmouth, Montana". This foul smell of smoke goes into the nostrils and they fell asleep [3].

The herb henbane was found in the medicines in areas of the western it was used as a narcotic, to comfort or increase sleep, to enlarge the eye pupils, less laxative or it may carminative. Torch and another burning object holder (digestion aid), sedative (nerve quietener) and use in relaxation of the contraction of the muscles. It is also used in acute cough, hysteria, chronic mania, decreasing, hallucination, priapism epilepsy and delirium treatments [3].

Previously, root bits were hung between the baby's throat to create the so-called "anodyne necklace". The black henbane is still used as a spasmodic in the treatment of asthma, chorea, tetanus, constipation, whooping cough, phthisical coughs, croup, and other disorders. For the treatment of irritable ulcers, swellings, tumors, orchitis, and other diseases this plant can be used. Henbane and related solanaceous medications are most commonly used as a griping therapy, such as powerful purgatives, correction. It is used to treat morphine addiction and as an antidote to mercury and other poisons. In youngsters, this herb is used as a hypnotic when the derivatives of opium are not used. Henbane is the second most important plant rather than the species of Belladonna. For the treatment purpose, the dose of the henbane was given to the patients is 0.2 grams. Professional poisoners have used it for years for non-medicinal purposes. Finally, the plant's leaves are claimed to deter mice [3].

The extracts of the alkaloids from the henbane plant have a cytotoxic effect on the mice and also had reduce the spontaneous frequency of the chromosomal changes increasing the mitotic index. The extract displayed cytotoxic action against cancer cell lines at all time points and has chronic cytotoxic effects on the AMN-3 or on the RD while it has no effect on the normal cells. In the cancer cell line like A549 & PC-3, the henbane extract contains alkaloids that cause apoptosis, not cytotoxicity [20]. In cultivated cancer cell of humans, the components of cannabisins d and G that comes from the henbane have less impact on these cells [38].

Because scopolamine has a large concentration in henbane, high doses of henbane plant primarily cause psychomotor retardation, which is followed by CNS excitement for example manic episode, hallucination, lack of rest, or delirium. In animals, mostly horses caused by this plant-like dryness of the mucosa, constipation, dilation of the pupil, fluctuation in heart rates, as well as effects on the central nervous system and depression in the respiratory tract [85]. Overdose of scopolamine has the same symptoms as atropine. Scopolamine causes agitation, mydriasis, convulsions, tachydriasis etc. [86]. Toxic effects of the black henbane are difficulty talking, thirst, pyrexia, blurry vision, drowsiness, and drying of the skin and it also affects hearing and vision [6, 8, 87–89]. The severe symptoms of this plant in humans are convulsions, ulcers in the respiratory tract, anxiety and tensions [88]. Due to doses of this plant, humans may feel like in the world of dreams and shows the reactions like sleepy person [6].

Work as Anti-Diabetic

The methanolic leaf extract of *Hyoscyamus albus* was tested on diabetic rats to determine its anti-diabetic potential. For 30 days, 100 and 200 mg/Kg bw of streptozotocin were given to diabetic rats. The levels of blood glucose and glycosylated haemoglobin in diabetic rats were dramatically lowered by the oral administration of both dosages of methanolic leaf extracts from *Hyoscyamus albus*. The extract had an insulin-stimulating activity, as determined by the measurement of plasma insulin levels [90]. Extracted from Hyoscyamus are calystegines and polyhydroxylated alkaloids [90]. For their in vivo antidiabetic action on streptozotocin-induced diabetes in mice, calystegines and polyhydroxylated alkaloids were isolated from Hyoscyamus albus seeds. After 20 days of therapy with Therapeutic importance of Hyoscyamus species growing in Iraq, they significantly lowered blood glucose levels and lipid parameters of diabetic mice to normal values.

Effects as Antioxidant

The methanolic extracts of *Hyoscyamus albus* showed the highest levels of DPPH antiradical, nitric oxide scavenging, and metal chelating activities when used to evaluate the antioxidant effects of Saudi medicinal plants (*Retama raetam, Salsola inermis, Hyoscyamus albus*, and *Fagonia arabica*) [91]. Using the DPPH assay and the -carotene bleaching method, the antioxidant effect of Hyoscyamus albus leaf extracts was examined. The *Hyoscyamus albus* leaf methanolic extract had the best antioxidant activity in the -carotene bleaching test (76.00%). *Hyoscyamus albus* chloroform leaf extract's IC50 for antiradical activity changed [92]. Using 2, 2-di-phenyl-1-picrylhydrazyl as a free radical scavenger, the antioxidant activity of the crude extract of the leaves of *Hyoscyamus albus* was calculated. The total phenolic and flavonoid levels of crude extract were associated with its reducing potential [93]. Using the 2, 2-diphenyl-1-picrylhydrazyl assay, the ability of seven fractions of *Hyoscyamus niger* alkaloidal extract to scavenge free radicals was examined. Only one fraction of the alkaloidal extract demonstrated a moderate ability to scavenge free radicals when compared to the positive and negative controls [94]. The methanolic extracts of *Hyoscyamus niger* showed antioxidant activity in contrast to -tocopherol, which was used as the positive control [95]. Two methods (2, 2-diphenyl-1-picrylhydrazyl) and ferric-reducing antioxidant were used to investigate the antioxidant activity of *Hyoscyamus niger* aerial component extracts. Ascorbic acid and butylated hydroxytoluene had antioxidant concentrations of 21.68 and 4.8.32 g/ml and 377.12.21 g/ml, respectively, for methanol extract [71]. The ABTS scavenging capacity technique was used to examine the antioxidant capability of an aqueous extract of *Hyoscyamus reticulatus'* aerial parts. Significant antioxidant scavenging abilities were shown by the *Hyoscyamus reticulatus* aqueous extract (533.26 mol TE/g dry extract weight) [96]. Four independent test systems, including the radical scavenging (DPPH assay), total antioxidant capacity, ferric and cupric reducing abilities, were used to evaluate the antioxidant capacities of hexane and water extracts of *Hyoscyamus reticulates*. The findings showed that compared to hexane extract, water extract exhibited better antioxidant activity [97] (Table 20.2).

Table 20.2 Pharmacological effects of alkaloids

Body parts	Effecs	References
Skin	Reduction in sweating	[98]
Digestive	Reduction of tone, increase pancreatic juice, regulate gastric juice	[98]
Eye	Increase the dilation of the eyes, cause vision problems	[98]
Antispasmodic	Relax the tract of bladder	[98]
Urinary	Retention of the urinary	[98]
Cardiovascular	Effects on the central nervous system, peripheral nervous system, reduce the rate of cardia,	[98]
Respiratory	Reduction of secretions and cause the dilation of the bronchi	[98]
Secretions	Blockage of the salavary glands, effects on the mucous membrane	[98]
Cholinergic agonists	Use against mushrooms poisons	[98]
Central nervous system	Low memory, cause coma and hallucinations	[98]

Effective against Bacteria

Staphylococcus aureus, Pseudomonas stutzeri, Klebsiella pneumonia and *Escherichia coli* were all susceptible to the antibacterial effects of Hyoscyamus albus alkaloid extracts [99]. The antibacterial activity of the Hyoscyamus niger seeds' methanol extracts against pathogens of the urinary system was examined (*Escherichia coli, Candida albicans, Klebsiella pneumoniae Enterococcus faecalis, Pseudomonas aeruginosa, Klebsiella pneumoniae, Proteus mirabilis*). With inhibition zones of 16.0, 19.0, and 26.0 mm against *Klebsiella pneumonia, Candida albicans* and *Enterococcus faecalis* respectively, the extracts demonstrated high antimicrobial activity. However, they only shown moderate activity against the remaining test species [100]. The anti-clostridial (*Clostridium refringent*) action of the aqueous extract of Hyoscyamus niger seeds was dose-dependent (diameter of the zone of growth inhibition, 16–18 mm) [101]. *P. vulgaris, P. aeruginosa, S. aureus,* and *E. coli* were evaluated against the crude protein extract of Hyoscyamus niger. It displayed growth inhibition zone sizes of 14, 15, and 20 mm for each of these diseases, respectively [102].

Effects of Henbane Eating

Because BH contains a significant amount of scopolamine, taking a high dose of this plant primarily causes somnolence, which is followed by CNS excitation symptoms as agitation, hallucinations, delirium, and manic episode. Constipation and colic (in horses) are common symptoms of BH poisoning in livestock, as well as dryness of the upper digestive and respiratory tract mucosa, pupil dilation (mydriasis), changes in heart rate, and CNS effects like ataxia, irritability, restlessness, seizures, and respiratory depression [85].

Patients who are intoxicated display symptoms similar to atropine overdose, including mydriasis, tachycardia, arrhythmia, agitation, convulsion, and coma [86]. BH intoxication may also result in dry mouth, thirst, slurred speech, difficulty speaking, dysphagia, warm flushed skin, pyrexia, nausea, vomiting, headache, blurred vision, photophobia, urinary retention, distension of the bladder, drowsiness, hyperreflexia, auditory, visual, or tactile hallucinations, confusion, disorientation, delirium, aggressiveness, and combative behavior [6, 8, 87, 89, 103].

Convulsions, coma, respiratory arrest, hypertension, and severe intoxication are all symptoms of the condition [103]. The more frequent symptoms include hallucinations, restlessness, mydriasis, and skin flushing. Contact with BH when wearing naked skin can result in skin irritation [6].

When BH is consumed either unintentionally or on purpose, there are several case reports and case series on the subject. Hallucinations are among the most frequent symptoms of these conditions. Recorded 19 BH intoxication instances between 1984 and 1989, and every single one of them involved Bedouin children [104].

In the spring, three children, two cases, and fourteen cases were all hospitalized. The most frequent signs and symptoms were a decreased level of consciousness, restlessness (89.5%), hallucinations (89.5%), hot, dry skin, and mydriasis (94.8%). 3.5% of children) experienced a profound coma. The following less frequent symptoms were also present: ataxia, revolutionary movements, enhanced tendon reflexes, hypertension, tachycardia, vomiting, convulsions, and hyperpyrexia. Physostigmine (36.8%) and sedatives like Diaz have been administered intravenously to the kids. Upon topical contact with his contaminated hand by Datura stramonium, a child with anisocoria was recorded [105].

Additionally, a case series involving roughly 900 patients revealed that the majority of them were boys. Springtime had seen the majority of their cases admitted (92%). The predominant symptom was CNS stimulation, and the most prominent symptoms were hallucinations and convulsions [11]. Another instance of BH intoxication was found in 81.6% of 20 kids who had consumed various plant parts in Anatolia, Turkey, between 1982 and 1983. May and June saw the highest levels of drunkenness. Additionally, he disclosed that 15.8% of intoxicated children had consumed BH just to try it, 18.4% of them had accidently swallowed it, and 65.8% had utilised it to create a pleasurable experience. Some of them had given it a second or third try. Alkaloids were eliminated in the urine within 24 h in close to 85% of cases [84]. Another study found that among a group of children, the most noticeable signs of BH intoxication were slurred speech (82.6%), aggression (60.9%), dilation of pupils (87%), flushing (87%), and somnolence (82%) [106].

There are several geological evidences throughout Asia and elsewhere in the world that henbane smoke was once used as a traditional treatment for toothaches and other illnesses [107]. Henbane smoke's historical use as a folk remedy for toothaches and other diseases is supported by a number of geological findings throughout Asia and other parts of the world [85]. Through various methods, people have attempted to obtain a cure from the medicinal plant throughout history. There are numerous books and articles that describe these protocols, but the primary issue is that the majority of them lack scientific support and cannot be used in contemporary medicine as a means of disease therapy. For the first time in this study, we created and scientifically validated one of the most traditional techniques of treatment using a significant herb, black henbane. In the past, the heated release chemical from *H. niger* was utilised in numerous regions of the world to eradicate the parasitic [34].

For thousands of years, Bohemia has been a centre for this kind of treatment. In order to treat patients, they would toss henbane seeds onto hot iron plates or flaming coals and let the patients to breathe in the rising steam. Additionally, the narcotic and intoxicating properties of extract and vaping have been used in a variety of ways. Before consulting the oracle, pagan priests in Greek antiquity utilised them. Therefore, we propose that heat release from *H. niger* and other medicinal plants might be employed as an alternative natural material for antibacterial activity. These chemicals also have a significant impact on biological applications and the development of new medications [108].

Soil

Henbane, as a foliar crop, demands a level, of well-drained soil. For optimal growth, it requires sandy loamy soil rich in organic materials. The crop prefers soil pH levels between 7 and 8.5. Deep dark and wet soils are unsuitable for farming. It has been reported that it may also be cultivated as a winter crop in warmer climates. In India, the plant has been discovered to thrive well in hot weather [1].

Waterlogged soil should not be used for germination. In the winter season, chalk runs over the soil which causes damage to the plant. Drought stress cause damage to the plant and hinder its growth. Late frost and drought stress cause very early blossoms in the plants. If the environmental temperature fluctuates especially in the dry season of spring or in the summer biennial black henbane may start to flower in first season but if the soil is well manured then it stops this flowering [9].

Climate

Henbane is a cool-season crop that does not handle very cold temperatures, as the growth and development of these plants slow significantly below 20 °C and becomes essentially stagnant if the temperature continues below 10 °C. Frost kills the vegetation. Seed germination has also been shown to be affected by low temperatures. Seeds of *H. Niger* and *H. Muticus* germinate best in 8–10 days if kept at a temperature of 20–30 °C and kept wet. It is grown as a summer crop in temperate countries, but as a winter crop in subtropical ones. In any case, the plantation must be exposed to direct sunshine, which is essential for the healthy growth and development of this crop. Similarly, high temperature and humidity must be avoided because both are damaging to plant growth and alkaloids content [1].

20.2 Cultivation

20.2.1 Propagation

The crop can be grown by either immediately sowing the seeds in the main field or by raising seedlings in the nursery and transplanting them into the main field [1].

20.2.2 Direct Sowing

Early October is the ideal time for direct seeding in subtropical locations, whereas June–July is best in temperate regions. Because henbane seeds are so minute, it is critical that the soil be prepared to a fine tilth before sowing. For good germination, seeds should be planted no deeper than 2 to 2.5 cm into the soil. Soaking the seeds overnight and incubating them for 7–8 days at 25 °C for sprouting. Improves germination rate before seeding. Before sowing, the seeds should be mixed with dry, powdered soil or sand in a 1:10 ratio to ensure uniform distribution over the field. Sowing requires about 2–3 kilograms of seed [1].

The ideal sowing time for *H. Niger* in Indore is estimated to be the second fortnight of November with 30 × 30 cm spacing. In Delhi, the first two weeks of November were judged to be suitable, with 30 × 15 cm. The optimal period to raise the nursery is September in the plains and April–May in temperate zones. The seeds are blended with sand before being broadcast in 8 cm rows in raised nursery beds. The seeds are then carefully covered with a fine mixture of soil and FYM or sand, mulched, and watered promptly. To produce enough seedlings, 500 g of seeds are necessary. Germination takes roughly 8–10 days. The mulch is removed at this point, but watering and weeding continue as usual. When the temperature is low, the seeds take a long time to germinate, and the growth takes 2 months to reach a proper stage for transplantation. When seedlings are about 20 days old, a foliar spray of 1% urea has been observed to enhance their growth. The seedlings are likewise very sluggish. It takes roughly a year for the seeds to germinate [1].

20.2.3 Planting

In a well-prepared field, seedlings are transplanted 45 cm apart in rows and 15–30 cm apart from plant to plant, followed by a light irrigation. Transplanted seedlings take roughly a week to establish. Because transplanted seedlings have a high death rate, it is advised that this crop be cultivated on a commercial basis solely via direct sowing [1].

20.2.4 Manures and Fertilizers

A better yield fertilizer that contains minerals like nitrogen, potassium, and phosphorous is important. These mineral fertilizers are the main component of the growth as a result high yield obtained [109–112]. Again, and again the application of mineral fertilizers may cause problems to the environment because these nitrogen minerals are the primary source of the accumulation of nitrites and nitrates for the plants and soil. In the farming system, new techniques are now being used for lowering nitrogen fertilizers such as biofertilizers. A well management of the nitrogen mineral or nutrition is required for the production of secondary metabolites such as tropane alkaloids in henbane. Maintenance of the nitrogen for the plant to maintain the qualities of this medicinal plant [112–114].

At the time of land preparation, a well-decomposed FYM at the rate of 15–20 t/ha is added. In addition, a dose of 40–80 kg nitrogen, 20–30 kg of potassium and

30–40 kg of phosphorous /ha have been advised to achieve the highest dry-matter yield. The entire amount of P and K, as well as 20 kg N, is given as a basal dosage. During the growing season, the remainder of the nitrogen is applied as a top-dressing in two equal split doses [1].

20.2.5 Irrigation

The transplanted crop receives its first irrigation soon after transplantation. Subsequent irrigations are administered at 8–10-day intervals until the crop is about a month old, at which point it is irrigated at 15–20-day intervals. In total, 4–5 irrigations are required. However, in the case of directly planted crops, one pre-sowing irrigation is required before plowing to ensure proper seed germination. If seed germination is poor, further irrigation is required a week after sowing. Following that, 3–4 further irrigations at 15–20-day intervals are required for the standing crop [1].

20.2.6 Intercultural

Direct sowing crops should be thinned after a fortnight of germination, and rows should be spaced 30 cm apart. During the entire cropping season, two weeding-cum-hoeing operations are often required [1].

20.2.7 Pests and Diseases

There have been no reports of major insect pests or disease assaults on either henbane species. Cotton bug larvae, on the other hand, attack [1].

Egyptian henbane was reported to have been infected with PVX, tobacco rattle virus, and the cucumber mosaic virus (CMV), which causes green mosaic infestations. White spots on the leaves are the symptom, and they get bigger the worse the condition gets. Mosaic disease of the opium poppy, which has been linked to aphids and sap transmission, commonly manifests as stunting, vein banding, and malformed capsule development. According to reports, three distinct viruses that cause mosaic disease have infected Solanum species. Stunted plants with infestations yield less fruits and blooms.Tobacco leaf curl virus on Asgandh reported to produce vein clearing followed by vein banding, leaf curving, puckering, and occasionally rolling of leaves, under severe conditions flower output is much reduced and there is no seed set [115].

Several types of bugs and stinkbugs feed on cotton bolls. In southern America attacking pests are brown and green stink bugs (*Acrosternum hilare, Euschistus servus*) and the third one is the southern green stink bug (*Nezara cirıdula* L.) [116]. The combination of these insects with others causes damage to the yield and the quality of the lint. Insects mostly feed on the bolls and cause to deformation of the bolls and abscission of these bolls [117, 118]. The feeding of stinkbugs also allows germs to enter the fruits, causing physiological damage and fruit degeneration [119]. In 2005 stinkbug attacked on the plant which lost its yield in South Carolina [118]. This crop's leaves and capsule were severely damaged over the months of July and August.

Aphids are the main cause of spread the of viruses. From the ornamental plants, vegetables and other plants these aphids were to spread the viruses. Cucumber, beet, squash, pumpkin, melon, potato and beans are the crops affected by these aphids. The viruses cause plant growth to be slowed by mottling, yellowing, or curling of the leaves. The elimination of these aphids is hard from the fields. These cause the disease even in low populations and loss to yield. Insecticides are used to eliminate the aphids take more time than to disseminate the virus [120]. The crop is also infested. These two insects may be controlled by 2–3 sprayings (at 10–15-day intervals) of Endosulphon (2 ml of water), Methyl Parathion, or Metasystox (2 ml/1). In India, a virus-caused green mosaic has been reported to be wreaking havoc on the crop. The root-knot disease is caused by the worms Meloidogyne incognita and Meloidogyne Javanica. The root-knot disease is found in several crops which are caused by the nematodes [121]. These nematodes were identified as dangerous to aromatic and medicinal plants [122–124].

In Egyptian henbane, root-knot nematodes severely reduce crop yield. Patches of chlorotic and stunted growth are visible as symptoms in the crop. The production of Coleus forskohlii tubers is hindered by Meloidogyne incognita, which causes knots or galls. It has been claimed that the nematodes root-knot and root lesion inflict serious harm to mentha. In Mentha arvensis, M. incognita is a more common cause of root knot disease; symptoms include stunted plants with smaller, chlorotic leaves. Mentha spicata, Mentha piperita, and Mentha citrata are more vulnerable to Pratylenchus penetrans, P. minus, P. scribneri, and P. thornei, which cause root lesion nematodes. Plants that are infected have reduced growth, burned leaves, and lesions on the root system. *M. halpa*, M. incognita, and *Helicotylenchus dihystera* damage the rose geranium, Pelargonium graveolens crop. The root-knot nematode-infected plant exhibits slowed growth, burning of lower leaves, yellowing, and severe galling of the root system, along with a drop in the geranium yield [115].

Black henbane was known as the source of different alkaloids and is affected by nematodes like *Meloidogyne incognita*, and *Meloidogyne incognita*. These two nematodes control by the spraying of carbofuran 3 kg per hectare [124].

According to the literature, the Colorado potato beetle is the worst enemy of henbane; it is claimed to favor henbane over all other plants. The plants must be sprayed with a pesticide if they are not completely damaged by this bug. Lead arsenate was commonly utilized in the past. Another pest named Paris green is harmful to the henbane and can be controlled by using sulfur. Another control of this insect was reported to spraying of lead arsenate in plants life is sufficient for the survival of the plant. Peronospora (Phycomycetes) and the mildew Erysiphe have both been observed to attack Hyoscyamus (Ascomycetes) [3].

20.2.8 Harvesting Processing and Storage

The leaves and flower-tops of *H. Niger* are picked during the flowering stage of the plant, while the entire aerial section of *H. Muticus* can be taken at the same time. In *H. Niger*, the elder leaves of the plant's lower parts that touch the ground should be harvested first and dried separately. The herb is sun-dried for 2–3 days before being shade-dried for 6–7 days, with continual raking with sticks. Every 100 g of fresh leaves yields 14 g of dry leaves. After drying, the material is placed in gunny bags and stored in a cold, dry location. The product should not be kept for more than 2 months [1].

20.2.9 Yield

The production of the herb varies substantially depending on the spacing. An average yield of 15 q/ha of dry herb can be produced with a spacing of 45 × 30 cm. Estimation of Hyoscyamine in Leaves, for this reason, 40 g of leaf powder is weighed and put to a flask holding 200 ml of a mixture of 4:1 ratio of solvent and ethanol. It is thoroughly shaken and left to stand for 10 min. After this take 6 ml of the solution of ammonia add and the mixture is agitated repeatedly for 1 h. The mixture is then transferred to a percolator that has been sealed with cotton wool and packed tightly when the liquid no longer flows. Continue the percolation, starting with 100 ml [1].

20.2.10 Review of Literature

For ages, black henbane (BH), also known as *Hyoscyamus niger*, has been utilized as a medicinal and is mentioned in all standard therapies. Black henbane is used as a homeopathic medicine; however, it inadvertently or deliberately causes drunkenness. Alkaloids like scopolamine, tropane, atropine and hyoscyamine are present in seeds, roots and leaves. Black henbane has the properties of antisecretory, pupil dilating and spasmolytic. It also works in the relaxation of the urinary bladder. Black henbane has also the properties of sedative and anti-diarrheal.

The symptoms of this plant are increased heart rate, irregular heartbeat, irritation, muscle spasms and deep sleep, sore throat, appetite, difficulty speaking, mental confusion, swallowing disorders, warm drained skin, chills, stomach pain, vomiting, headache, loss of vision, urinary incontinence, bladder distension, tiredness, muscular rigidity, less hearing, visual problem, hearing voices, uncertainty. The primary treatment for BH inebriated people is supportive therapy, which includes stomach emptying (but not Ipecac), doses of benzodiazepines and charcoal. Emergency doctors, health care units and clinics of the drugs must know the poison of the henbane effects and its cure. These physicians should know the management of the poison in the henbane plant [42].

Some anti-inflammatory medicines contain seeds of the *Hyoscyamus niger*. Due to these properties, it is now validated in different medications. Different doses of this plant methanolic extract were indicated in animals as antipyretic, analgesic and anti-inflammatory. In the yeast-induced pyrexia model, it displayed antipyretic effects. Some investigations show that the components like coumarinolignans found in that plant are important chemical constituents of the herb and effective in anti-inflammatory medications [34].

Henbane plant belongs to the Solanaceae family containing 84 genera and 3000 species. henbane is important medicinal species of this family. Some alkaloids like Hyoscyamus and scopolamine which are tropane found in this plant in large quantities. The phytochemical research revealed that the species of henbane have components of tannins, terpenes, alkaloids, cardiac glycosides, and saponins. Henbane herb was found to be effective in cancer, diabetes, asthma, allergic, dyrrhoyea, blocking of Ca^{2+}, Parkinson's disease and depression [125].

The henbane was a narcotic used by the ancients. Originally employed as both a poison and a narcotic, it was commonly used part in flying ointments and hallucinogenic by witches, wizards, and soothsayers. In the nineteenth century, Landenburg discover his plant and called it hyoscine on the basis of this alkaloid. Notorious poisoner Madame Voisin supplied this drug to France. It was discovered to be a tropane alkaloid extremely similar to atropine. These hyoscine alkaloids characteristics were similar to the tropane alkaloids. Tropane and hyoscine alkaloids were helpful in the study of parasympathetic parts of the central nervous system [5].

Henbane (*Hyoscyamus niger*) is the source of several dangerous alkaloids and hallucinogenic tropane alkaloids like atropine, hyoscyamine and scopolamine. Potatoes and tomatoes are among the plants of the Solanaceae family. Scopolamine is a medication that is used to treat depression and nausea. Scopolamine was used in allergic and muscarinic medicines. If someone takes it more than the proper dose it causes depression. Scopolamine was hallucinogenic but mostly it causes bad results. The alkaloid atropine was used in medicines of allergy and muscarinic which may increase the heart rate, dilation of the pupil and promotes the secretion of the saliva. A toxic dose of scopolamine was around 2–4 mg while 10 mg of the atropine was also deadly. In Italy, the belladonna term was used to make women's beautiful eyes for the attraction of males. The properties of the hyoscyamine were the same as atropine and scopolamine and it was the isomer of atropine used for the production of the scopolamine alkaloid. In seeds and leaves, the presence of the hyoscyamine which is a tropane alkaloid in the henbane genus was named [126].

Hyoscyamus niger is a hallucinogenic plant that is commonly available and includes anticholinergic compounds. Ingestion, whether purposeful or unintentional, can result in catastrophic psychophysical deterioration and, in extreme cases, death. The intoxication is similar to the one caused by atropine. Usage of this plant hypertension, irregular heartbeat, restlessness, vomiting, and unconsciousness are some of the symptoms that may appear. Clinical symptomatology and history are used to make the diagnosis. The therapeutic actions of the atropine were physostigmine, therapy and washing or cleaning the stomach. Prognosis is usually a favorable [86].

The chemical constituents of the plant *Hyoscyamus niger* are alkaloids, lignans, saponins, flavonoids, and other non-alkaloidal substances which are found. Besides this coumarinolignanses are also found in the extracts of the plant. Through the techniques of thin-layer chromatography, gas chromatography and high-performance liquid chromatography *experiments were performe*d to investigate various constituents of this plant. These techniques were utilized to identify a number of chemical ingredients which are found in the species of *H. niger*. *H. niger* possesses painkiller, anti-inflammatory, antipyretic, neuropsychiatric, anti-spasmodic. It was also found helpful in the treatment of bronchodilatory, antisecretory, relaxing of the bladder. Usage medicine is seen in the cure of heart problems, hypolipidemic, cardiosuppressant, in medicines of blood dilatation, anticancer and eating repellant properties, according to modern pharmacological investigations [88].

Improper use of compounds that are present in leaves, roots and other parts of the plant poisonous to humans as well as to the animals. Deadly consumption of the henbane was reported in different areas of the world. In Turkey, 31 children were reported off the henbane poison. These children were poisoned by black henbane over the period of 3 years. The ratio of the poison was more in villages than in urban areas [127].

Secondary metabolites of alkaloids and non-alkaloids metabolites were found in the seeds of several species of the Solanaceae family plants. Important non-alkaloidal secondary metabolites are coumarinolignans, glycerides, lignans, saponins, flavonoids, glycosides and withanolides reported in henbane seeds. The seeds of the henbane were helpful against microbes. It functions against the diseases of inflammation, and analgesic [34].

All parts of the henbane have great uses in medicines. The crude extract of the henbane seeds had an anticarcinogenic effect on the relaxation of the spontaneous contraction of the rabbit jejunum by blocking the Ca^+ channel. This extract is effective in diseases of diarrhea and the accumulation of fluid in mice. The combination of the antagonistic mechanism of Ca^{2+} and anticholinergic has an antispasmodic effect of henbane extract [128].

It was stated that an old man became poisoned after drinking tea made from the plant Hyoscyamus niger. The hallucinogenic plant *Hyoscyamus niger* is commonly available contains anticholinergic compounds and is a hallucinogen. Ingestion, whether intentional or unintentional, can lead to a major deterioration of the physical condition and can result in central anticholinergic syndrome, which can show up as a wide range of indications and central or periphery symptoms. The symptoms of central manifestations range from delirium and agitation, which are excitatory signs, to coma, stupor, and depression of the central nervous system. Dry mouth, heat due to decreased sweating, tachycardia, arrhythmia, constipation, and urine retention are some of the peripheral symptoms of this illness. The history and clinical symptoms are used to make the diagnosis. Supportive therapy, gastrointestinal lavage, active choral, and the particular antidote physostigmine are examples of therapeutic procedures [129].

Alkaloids, tyramine derivatives, withanolides, lignan amides, and flavonoids are all present in plant seed, according to earlier investigations on the phytochemical examination of *Hyoscyamus niger* [100]. Additionally, numerous significant lignans were discovered in *H. niger*, including sesamol, pinoresinol, sesamin, and sesamolin. These lignans exhibit potent pharmacological and action. This strong biomedical activity may be relative to the presence of toxic metabolites [130].

In seizures brought on by pentylene tetrazole, the anticonvulsant properties of an alcoholic extract of *Hyoscyamus niger* seed were assessed at doses of 50, 100, and 200 mg/kg ip. The administration of Hyoscyamus niger seed extract, particularly in the last stages of convulsion, had an inhibitory effect on the stages, advancement, and length of seizures, according to the results. However, the administration of henbane seed extract had a significant anticonvulsant effect that peaked at the eighth injection and continued until the twelfth [130].

An alcoholic extract of the Hyoscyamus niger seed was tested for its anticonvulsant effects in pentylene tetrazole-induced seizures at dosages of 50, 100, and 200 mg/kg ip. The results showed that injection of Hyoscyamus niger seed extract had an inhibitory effect on the phases, progression, and length of seizure, particularly in the closing phases of convulsion. However, therapy with henbane seed extract resulted in a powerful anticonvulsive effect that peaked at the twelfth injection after the eighth [130].

In Europe and Asia, henbane, or *Hyoscyamus niger* L. (Solanaceae), is a widely grown plant. In Turkish folk medicine, henbane seeds have been used to get rid of eye worms. Insecticidal activity of *H. niger* seed was the focus of the current study. Alkaloid, n-hexane, ethyl acetate, and methanol extracts were made from the plant's seeds and tested for their ability to kill Lucilia sericata larvae. It was determined what the alkaloid extract's EC_{50} and EC_{90} values were, and morphological anomalies were looked into. Significant insecticidal activity was demonstrated by an alkaloid extract made from this plant's seeds. The EC_{50} values of the alkaloid extract from H. niger seeds were determined to be 8.04, 8.49, and 7.96 g/mL against the first, second, and third instars, respectively. Malformed larvae were found to include those with weak cuticles, tiny sizes, and contractions. Alkaloid extract from H. niger seeds underwent HPLC examination, and its primary constituents were identified. It was shown that injured larvae with tiny size, contraction, and weak cuticle were among the abnormalities of larvae. Furthermore, the primary components of the extract's alkaloid extract from H. niger seeds were identified using HPLC analysis [131].

References

1. Farooqi, A. A., & Sreeramu, B. (2004). *Cultivation of medicinal and aromatic crops*. Unsiversities Press.
2. Paulsen, B. S. (2010). *Highlights through the history of plant medicine* (Vol. 50). Bioactive compounds in plants-benefits and risks for man and animals.

3. Hocking, G. M. (1947). Henbane—Healing herb of Hercules and of apollo. *Economic Botany, 1*(3), 306–316.
4. Carter, A. J. (2003). Myths and mandrakes. *Journal of the Royal Society of Medicine, 96*(3), 144–147.
5. Lee, M. (2006) Solanaceae III: Henbane, hags and Hawley Harvey Crippen. *Journal-Royal College of Physicians of Edinburgh, 36*(4), 366.
6. Pokorny, M., & Mangold, J. (2010). *MontGuide: black henbane: Identification, biology and integrated management*. The US Department of Agriculture (USDA) MSUaMSUE, avalable in www. msuextension. org.
7. Liberman, A., & Mitchell, J. L. (1993). An analytic dictionary of English etymology. *Journal of Germanic Linguistics, 5*(1), 47–92.
8. Schultes, R. E., & Smith, E. W. (1976). *Hallucinogenic plants* (Vol. 35). Golden Press.
9. Grieve, M. (1931). *A modern Herbal* (Vol. 1992). Jonathan Cape Ltd.
10. Sajeli, B., et al. (2006). Hyosgerin, a new optically active coumarinolignan, from the seeds of *Hyoscyamus niger. Chemical and Pharmaceutical Bulletin, 54*(4), 538–541.
11. Daneshvar, S., Mirhossaini, M., & Balali-Mood, M. (1992). Hyoscyamus poisoning in Mashhad. *Toxicon, 30*, 501.
12. Yousefi, M. J., et al. (2009). *Genetic variation of some Iranian black henbane accessions (Hyoscyamus niger* L.) *using RAPD and SDS-PAGE of seed proteins. International Journal of Plant Breeding, 3*(2), 92–98.
13. Raghavan, V. V., & Wong, S. M. (1986). A critical analysis of vector space model for information retrieval. *Journal of the American Society for Information Science, 37*(5), 279–287.
14. Saidon, N. (2008). *The establishment of embryogenic callus culture of Hyoscyamus niger and the detection of hyoscyamine in the culture*. M.Sc. thesis, University of Sains Malaysia, Malaysia.
15. Heiser, C. B., Jr., & Pickersgill, B. (1969). Names for the cultivated capsicum species (Solanaceae). *Taxon, 18*(3), 277–283.
16. Wink, M. (2003). Evolution of secondary metabolites from an ecological and molecular phylogenetic perspective. *Phytochemistry, 64*(1), 3–19.
17. Stewart, G.W., Phyto-chemical examination of Montana *Hyoscyamus niger.* 1934.
18. Haas, L. (1995). *Hyoscyamus niger* (henbane). *Journal of Neurology, Neurosurgery, and Psychiatry, 59*(1), 114.
19. Lavania, U. C., et al. (2010). Chromosomal localization of rDNA and DAPI bands in solanaceous medicinal plant *Hyoscyamus niger* L. *Journal of Genetics, 89*(4), 493–496.
20. Ismeel, A. O. (2011). *Cytogenetic and cytotoxic studies on the effect of phytoinvestigated active compounds of Hyoscyamus niger (in vivo and ex vivo)*. Ph. D. Thesis in Philosophy of Sciencein Biotechnology, University of Al .....
21. Pudersell, K. (2006). *Tropane alkaloid production and riboflavine excretion in the field and tissue cultures of henbane (Hyoscyamus niger L.)* (Vol. 120). Tartu University Press.
22. Chevallire, A. (1996). *The encyclopedia of medicinal plants*. DK Publishing.
23. Heber, D. (2004). *PDR for herbal medicines [M]*. Thomson Health Care.
24. El Bazaoui, A., et al. (2012). Gas-liquid chromatography-mass spectrometry investigation of tropane alkaloids in Hyoscyamus albus L. from Morocco. *Zeitschrift für Naturforschung C, 67*(9–10), 461–465.
25. Bernhoft, A. (2010). A brief review on bioactive compounds in plants. *Bioactive compounds in plants-benefits and risks for man and animals, 50*, 11–17.
26. Li, R., et al. (2006). Functional genomic analysis of alkaloid biosynthesis in *Hyoscyamus niger* reveals a cytochrome P450 involved in littorine rearrangement. *Chemistry & Biology, 13*(5), 513–520.
27. Begum, A. S. (2010). *Bioactive non-alkaloidal secondary metabolites of Hyoscyamus niger Linn. seeds: A review. Research Journal of Seed Science, 3*(4), 210–217.
28. Robbers, J. E., Speedie, M. K., & Tyler, V. E. (1996). *Pharmacognosy and pharmacobiotechnology*. Williams & Wilkins.

29. Uniyal, M.R., *Medicinal Flora of Garhwal Himalayas*. 1989.
30. Frohne, D., & Pfander, H. (1983). *A color atlas of poisonous* (p. 291). Plants–Wolfe Publishing Ltd.
31. Graham, J., & Johnson, S. (2010). *Managing black henbane; fact Sheet-04-10*. The University of Nevada Reno.
32. Ghorbanpour, M., et al. (2013). Two main tropane alkaloids variations of black henbane (*Hyoscyamus niger*) under PGPRs inoculation and water deficit stress induction at flowering stage. *Journal of Medicinal Plants, 12*(45), 29–42.
33. Ghorbanpour, M., et al. (2010). Hyoscyamine and scopolamine production of black henbane (*Hyoscyamus niger*) infected with Pseudomonas putida and P. fluorescens strains under water deficit stress. *Planta Medica, 76*(12), P167.
34. Begum, S., et al. (2010). Study of anti-inflammatory, analgesic and antipyretic activities of seeds of *Hyoscyamus niger* and isolation of a new coumarinolignan. *Fitoterapia, 81*(3), 178–184.
35. Ebadi, M., & Shields, K. M. (2007). *Book review: Pharmacodynamic basis of herbal medicine*. SAGE.
36. Cooper, M. R., & Johnson, A. W. (1984). *Poisonous plants in Britain and their effects on animals and man*. HM Stationery Office.
37. Zhang, W.-N., Luo, J.-G., & Kong, L.-Y. (2012). Phytotoxicity of lignanamides isolated from the seeds of *Hyoscyamus niger*. *Journal of Agricultural and Food Chemistry, 60*(7), 1682–1687.
38. Ma, C.-Y., Liu, W. K., & Che, C.-T. (2002). Lignanamides and nonalkaloidal components of Hyoscyamus n iger seeds. *Journal of Natural Products, 65*(2), 206–209.
39. Ma, C.-Y., Williams, I. D., & Che, C.-T. (1999). Withanolides from *Hyoscyamus niger* seeds. *Journal of Natural Products, 62*(10), 1445–1447.
40. McIntyre, M. (1993). British herbal compendium. Vol 1: A handbook of scientific information on widely used plant drugs. (Companion to Vol 1 of the British Herbal Pharmacopoeia): edited by Peter R. Bradley Bournemouth: British Herbal Medicine Association, 1992. 239 pp£ 45 ISBN: 0 903032 09 0. Churchill Livingstone.
41. Duke, J. A. (1992). *Database of phytochemical constituents of GRAS herbs and other economic plants*. CRC Press.
42. Alizadeh, A., et al. (2014). Black henbane and its toxicity–A descriptive review. *Avicenna Journal of Phytomedicine, 4*(5), 297.
43. Begum, A. S., et al. (2006). Hyosmin, a new lignan from *Hyoscyamus niger* L. *Journal of Chemical Research, 2006*(10), 675–677.
44. Begum, A. S., et al. (2009). Hyoscyamal, a new tetrahydrofurano lignan from *Hyoscyamus niger* Linn. *Natural Product Research, 23*(7), 595–600.
45. Steinegger, E., & Sonanini, D. (1960). A study of the flavone of *Hyoscyamus niger*. *Pharmazie, 15*, 643–644.
46. Lunga, I., et al. (2008). Steroidal saponins from the seeds of *Hyoscyamus niger* L. *Chemistry Journal of Moldova, 3*(1), 89–93.
47. Hughes, E. H., & Shanks, J. V. (2002). Metabolic engineering of plants for alkaloid production. *Metabolic Engineering, 4*(1), 41–48.
48. De Luca, V., & St Pierre, B. (2000). The cell and developmental biology of alkaloid biosynthesis. *Trends in Plant Science, 5*(4), 168–173.
49. Isman, M. B. (2006). Botanical insecticides, deterrents, and repellents in modern agriculture and an increasingly regulated world. *Annual Review of Entomology, 51*, 45–66.
50. Mabley, J., Gordon, S., & Pacher, P. (2011). Nicotine exerts an anti-inflammatory effect in a murine model of acute lung injury. *Inflammation, 34*(4), 231–237.
51. Yun, D.-J., Hashimoto, T., & Yamada, Y. (1992). Metabolic engineering of medicinal plants: Transgenic Atropa belladonna with an improved alkaloid composition. *Proceedings of the National Academy of Sciences, 89*(24), 11799–11803.

52. el Jaber-Vazdekis, N., et al. (2008). Effects of elicitors on tropane alkaloids and gene expression in Atropa baetica transgenic hairy roots. *Journal of Natural Products, 71*(12), 2026–2031.
53. Sainvitu, P., et al. (2012). *Structure, properties and obtention routes of flaxseed lignan secoisolariciresinol, a review*. BASE.
54. Patel, D., et al. (2012). Therapeutic potential of secoisolariciresinol diglucoside: A plant lignan. *International Journal of Pharmaceutical Sciences and Drug Research, 4*(1), 15–18.
55. Hazra, S., & Chattopadhyay, S. (2016). An overview of lignans with special reference to podophyllotoxin, a cytotoxic lignan. *Chemical Biology Letters, 3*(1), 1–8.
56. Moss, G. (2000). Nomenclature of lignans and neolignans (IUPAC recommendations 2000). *Pure and Applied Chemistry, 72*(8), 1493–1523.
57. Parreira, N. A., et al. (2010). Antiprotozoal, schistosomicidal, and antimicrobial activities of the essential oil from the leaves of Baccharis dracunculifolia. *Chemistry & Biodiversity, 7*(4), 993–1001.
58. Hemmati, S. (2007). *Biosynthesis of lignans in plant species of the section Linum: Pinoresinol-lariciresinol reductase and justicidin B 7-hydroxylase*. Ph.D. Dissertation, University of Heinrich-Heine, Düsseldorf, Germany.
59. Pilkington, L. I., et al. (2015). Enantioselective synthesis, stereochemical correction, and biological investigation of the rodgersinine family of 1, 4-benzodioxane neolignans. *Organic Letters, 17*(4), 1046–1049.
60. Pan, J.-Y., et al. (2009). An update on lignans: Natural products and synthesis. *Natural Product Reports, 26*(10), 1251–1292.
61. Saleem, M., et al. (2005). An update on bioactive plant lignans. *Natural Product Reports, 22*(6), 696–716.
62. Yousefzadi, M., et al. (2010). Podophyllotoxin: Current approaches to its biotechnological production and future challenges. *Engineering in Life Sciences, 10*(4), 281–292.
63. Huang, W.-Y., Cai, Y.-Z., & Zhang, Y. (2009). Natural phenolic compounds from medicinal herbs and dietary plants: Potential use for cancer prevention. *Nutrition and Cancer, 62*(1), 1–20.
64. Webb, A. L., & McCullough, M. L. (2005). Dietary lignans: Potential role in cancer prevention. *Nutrition and Cancer, 51*(2), 117–131.
65. Habauzit, V., & Horcajada, M.-N. (2008). Phenolic phytochemicals and bone. *Phytochemistry Reviews, 7*(2), 313–344.
66. Matsuda, J., Okabe, S., Hashimoto, T., & Yamaday. (1991). Cultured roots of *Hyoscyamus niger*. *Journal of Biological Chemistry, 25*(15), 460–464.
67. Nadkarni, K. (1976). *Indian materia medica Mumbai*. Popular Prakashan Pvt.
68. Bown, D. (1995). *The Royal Horticultural Society encyclopedia of herbs & their uses*. Dorling Kindersley Limited.
69. Chevallier, A., The encyclopedia of medicinal plants. 1996.
70. Duke, J. (1985). *CRC handbook of medicinal herbs* (p. 677). CRC Press. Inc.
71. Hajipoor, K., Sani, A., & Mohammad, A. (2015). In vitro antioxidant activity and phenolic profile of *Hyoscyamus niger*. *IJBPAS, 4*(7), 4882–4890.
72. Aparna, K., Joshi, A. J., & Vyas, M. (2015). Phyto-chemical and pharmacological profiles of *Hyoscyamus niger* Linn (Parasika yavani)-a review. *Pharma Science Monitor, 6*(1).
73. Oto, G., et al. (2013). Antinociceptive activity of methanol extract of *Hyoscyamus reticulatus* L. in mice. *American Journal of Phytomedicine and Clinical Therapeutics, 1*(2), 117–123.
74. Bellakhdar, J. (1997). *Pharmacopée marocaine traditionnelle*. Ibis Press.
75. Petrovska, B. B. (2012). Historical review of medicinal plants' usage. *Pharmacognosy Reviews, 6*(11), 1.
76. Benhouda, A., & Yahia, M. (2014). Toxicity, analgesic and anti-pyretic activities of methanolic extract from Hyoscyamus albus' leaves in albinos rats. *International Journal of Pharmacy and Pharmaceutical Sciences, 6*(3), 121–127.

77. Khan, A.-U., & Gilani, A. H. (2010). In vivo studies on the bronchodilatory and analgesic activities of *Hyoscyamus niger* and Aspalathus linearis. *Latin American Journal of Pharmacy, 29*(5), 777.
78. Moghadam, M. (2012). Assessment of *Hyoscyamus niger* seeds alcoholic extract effects on acute and chronic pain in male NMRI rats. *Journal of Basic and Clinical Pathophysiology, 1*(1).
79. Oto, G., et al. (2013). Antinociceptive activity of methanol extract of *Hyoscyamus reticulatus* L. in mice. *AJPCT, 1*(2), 117–123.
80. Evans, W., & Pharmacognosy. (1989). *Brailliar Tiridel Can*. Macmillanpublishers.
81. Schuhly, W. (2004). Pharmacognosy: Phytochemistry, medicinal plants. *Phytomedicine: International Journal of Phytotherapy & Phytopharmacology, 11*(1), 90–91.
82. Allikmets, L., & Nurmand, L. (1982). *Farmakoloogia*. Valgus.
83. Blinova, K., Vandyshev, V., & Komarova, M. (1996). *Rasteniya dlya nas: Spravochnoe izdanie [Plants for Us: Reference Edition]* (p. 654). GP Yakovleva i KF Blinovoi. Saint Petersburg, Uchebnaya Kniga.
84. Tammeorg, J., O. Kook, and G. Vilbaste, Eesti NSV ravimtaimed. 1973.
85. Verstraete, F. (2010). Management and regulation of certain bioactive compounds present as inherent toxins in plants intended for feed and food. *Bioactive compounds in plants-benefits and risks for man and animals*.
86. Vidović, D., et al. (2005). Intoxication with henbane. *Lijecnicki Vjesnik, 127*(1–2), 22–23.
87. Long, D. J., et al. (1999). Black henbane (*Hyoscyamus niger* L.) in the Scottish Neolithic: A re-evaluation of Palynological Findings from Grooved Ware Pottery at Balfarg Riding School and Henge, Fife. *Journal of Archaeological Science, 26*(1), 45–52.
88. Jun, L., et al. (2011). Chemical and pharmacological researches on *Hyoscyamus niger*. *Chinese Herbal Medicines, 3*(2), 117–126.
89. Prance, G., & Nesbitt, M. (2012). *The cultural history of plants*. Routledge.
90. BenhoudaA, Y. M., et al. (2014). Hypoglycemic activity of methanolic extract of Hyoscyamus albus l. leaves in straptozotocin induced diabetic rats. *Natural Products Chemistry & Research, 2*, 5.
91. Mobin, M., Khan, M. N., & Nourabbas, Z. B. K. A. (2015). Ecotype difference in bioactive constituents and in vitro antioxidant activities of some Saudi medicinal plants. *European Journal of Medicinal Plants, 7*(3), 125.
92. Benhouda, A. et al. *Algerian Journal of Natural Products*.
93. Alghazeer, R., et al. (2012). Antioxidant and antimicrobial properties of five medicinal Libyan plants extracts. *Natural Science, 4*(5), 324–335.
94. Ismeel, A. O. (2011). *Cytogenetic and cytotoxic studies on the effect of phytoinvestigated active compounds of Hyoscyamus niger (in vivo and ex vivo)*. Al-Nahrain University-College of Science.
95. Souri, E., et al., Antioxidative activity of sixty plants from Iran. 2004.
96. Mohammad, M. K., et al. (2010). Antioxidant, antihyperuricemic and xanthine oxidase inhibitory activities of *Hyoscyamus reticulatus*. *Pharmaceutical Biology, 48*(12), 1376–1383.
97. Güneş, E., et al. (2014). *Hyoscyamus reticulatus*'un hekzan ve su özütlerinin antioksidan ve antimikrobiyal özellikleri üzerine bir çalişma. *Selçuk Üniversitesi Fen Fakültesi Fen Dergisi, 39*, 21–29.
98. Craig, C. R., & Stitzel, R. E. (2004). *Modern pharmacology with clinical applications*. Lippincott Williams & Wilkins.
99. Kadi, K., et al. (2013). In vitro antibacterial activity and phytochemical analysis of White henbane treated by phytohormones. *Pakistan Journal of Biological Sciences: PJBS, 16*(19), 984–990.
100. Dulger, G., & Dulger, B. (2015). Antimicrobial activity of the seeds of *Hyoscyamus niger* L.(henbane) on microorganisms isolated from urinary tract infections. *The Journal of Medicinal Plants Studies, 3*(5), 92–95.
101. Akhtar, M., & GilI, S. (1992). Evaluation of anticlostridial efficacy of indigenous medicinal plant drugs: Rasoot, Ajwain Khurasani Neem and Bakain. *The Pakistan Journal of Agricultural Sciences, 29*(4), 371–375.

102. Mateen, A., et al. (2015). Screening and purification of antibacterial proteins and peptides from some of the medicinal plants seeds. *International Journal of Pharma and Bio Sciences, 6*(4), 774–781.
103. Li, J., et al. (20 1). Chemical and pharmacological researches on *Hyoscyamus niger. Chinese Herbal Medicines*, 117–126.
104. Urkin, J., et a. (1991). Henbane (*Hyoscyamus reticulatus*) poisoning in children in the Negev. *Harefuah, 120*(12), 714–716.
105. Macchiaiolo, M., et al. (2010). An unusual case of anisocoria by vegetal intoxication: A case report. *Italian Journal of Pediatrics, 36*(1), 1–3.
106. Doneray, H., Orbak, Z., & Karakelleoglu, C. (2007). Clinical outcomes in children with *hyoscyamus niger* intoxication no receiving physostigmine therapy. *European Journal of Emergency Medicine, 14*(6), 348–350.
107. Fenwick, R. S., & Omura, S. (2015). Smoke in the eyes? Archaeological evidence for medicinal henbane fumigation at Ottoman Kaman-Kalehöyük, Kırşehir Province. *Turkey. Antiquity, 89*(3-6), 905–921.
108. Wagner, H. (20 3). *Rauschgift-Drogen* (Vol. 99). Springer-Verlag.
109. Attar, H., et al (2012). Relationship between phosphorus status and nitrogen fixation by common beans (Phaseolus vulgaris L.) under drip irrigation. *International journal of Environmental Science and Technology, 9*(1), 1–13.
110. Dawood, M. G. Abdelhamid, M. T., & Schmidhalter, U. (2014). Potassium fertiliser enhances the salt-tolerance of common bean (Phaseolus vulgaris L.). *The Journal of Horticultural Science and Biotechnology, 89*(2), 185–192.
111. El-Lethy, S. R., Abdelhamid, M. T., & Reda, F. (2013). Effect of potassium application on wheat (Triticum aestivum L.) cultivars grown under salinity stress. *World Applied Sciences Journal, 26*(7), 840–850.
112. Rady, M. M., et al. (2016). Growth, heavy metal status and yield of salt-stressed wheat (Triticum aestivum L.) plants as affected by the integrated application of bio-, organic and inorganic nitrogen-fertilizers. *Journal of Applied Botany and Food Quality*, 89.
113. Abdelhamid, M T., Selim, E., & El-Ghamry, A. (2011). Integrated effects of bio and mineral fertilizers and humic substances on growth, yield and nutrient contents of fertigated cowpea (Vigna unguiculata L.) grown on sandy soils. *Journal of Agronomy, 10*(1), 34–39.
114. Awad, N., et al. (2012). Ameliorate of environmental salt stress on the growth of Zea mays L. plants by exopolysaccharides producing bacteria. *Journal of Applied Sciences Research, April*, 2033–2044.
115. By, O. *Advances in Medicinal and Aromatic Plants Research.*
116. Greene, J., et al. (2009). Continued evaluations of internal boll-injury thresholds for stink bugs in the southeast. In *Proceedings of the Beltwide Cotton Conferences, 5Ð8 January* (p. 1092Ð1101)
117. Watkins, G. (1981). *Compendium of cotton diseases St. Paull, Minnesota.* The American Phytopathology Society. Cotton Disease Council. Minnesota. 87p.
118. Wene, G. P., & Sheets, L. (1964). Notes on and control of stink bugs affecting cotton in Arizona. *Journal of Economic Entomology, 57*(1), 60–62.
119. Bacheler, J., et al. (2009). Use of the dynamic threshold for stink bug management in the southeast. In *Proceedings, beltwide cotton conferences.*
120. Flint, M. (2013) *Aphids: Integrated Pest Management for Home Gardeners and Landscape Professionals.* University of California Davis.
121. Sasser, J. N., & Kirby, M. (1979). *Crop cultivars resistant to root-knot nematodes, Meloidogyne species, with information on seed sources.* Crop cultivars resistant to root-knot nematodes, Meloidogyne species, with information on seed sources.
122. Haseeb, A. (199-). Plant parasitic nematodes of medicinal and aromatic plants. In *Vistas in seed biology* (pp 98–119). Printwell Jaipur, India.
123. Haseeb, A., & Pandey, R. (1987). Incidence of root-knot nematodes in medicinal and aromatic plants-new host records. *Nematropica, 17*(2), 209–212.

124. Haseeb, A. and R. Pandey, Root-knot disease of henbane, Hyoscyamus—A new disease record. 1989.
125. Al-Snafi, A. E. (2018). Therapeutic importance of Hyoscyamus species grown in Iraq (Hyoscyamus albus, *Hyoscyamus niger* and Hyoscyamus reticulates)-A review. *IOSR Journal of Pharmacy, 8*(6), 18–32.
126. Clarke, K., Trim, C., & Hall, L. (2014). Principles of sedation, anticholinergic agents, and principles of premedication. *Veterinary Anaesthesia*, 79–100.
127. Kürkçüoğlu, M. (1970). Henbane (*Hyoscyamus niger*) poisonings in the vicinity of Erzurum. *The Turkish Journal of Pediatrics, 12*(1), 48–56.
128. Gilani, A. H., et al. (2008). Gastrointestinal, selective airways and urinary bladder relaxant effects of *Hyoscyamus niger* are mediated through dual blockade of muscarinic receptors and Ca2+ channels. *Fundamental & Clinical Pharmacology, 22*(1), 87–99.
129. Erkal, H., Özyurt, Y., & Arikan, Z. (2006). The Central Anticholinergic Syndrome after ingesting Henbane (*Hyoscyamus niger*) plant in a geriatric patient. *Turkish Journal of Geriatrics, 9*(3).
130. Huiqin, X., et al. (2019). Extraction of total lignans from Radix Isatidis and its antiviral activity. *Biomedical Research and Reviews, 2*(1), 108.
131. Küpeli Akkol, E., et al. (2020). Insecticidal activity of *Hyoscyamus niger* l. on lucilia sericata causing myiasis. *Plants, 9*(5), 655.

Chapter 21
Holy Thistle

Shagufta Perveen, Khalid Sultan, Abida Parveen, Sara Zafar, Naeem Iqbal, and Arwa A. AL-Huqail

21.1 Introduction

Family name: Asteraceae
Name in English: Holy thistle
Name in India: Pavitramullina Gida
Name of Varieties & Species: *Silybum marianum*
Distribution: South Africa, Europe, India, South America, Asia regions
Uses: Drugs & medicines

21.2 Origin and Distribution

Holy thistle is mostly found in areas of India, and central and western Europe. But now this presents in many areas of the world like Africa, India, China, Europe, America, Australia and other countries as well. The need for the seeds of this plant mostly comes from the markets of Europe from Argentine. In the markets of the U.S., these seeds are commercially supplied from cultivation in the areas of Texas. These plants tolerate the water shortage and survive in drought conditions. This plant well grows or loves the dry soil which is fully drained. These plants are found in waste places or along roads [1–3].

Originally from the Mediterranean region, milk thistle is now common in Central Europe, Western and Central Asia, North and South America, Southern Australia

S. Perveen (✉) · K. Sultan · A. Parveen · S. Zafar · N. Iqbal
Department of Botany, Government College University, Faisalabad, Pakistan

A. A. AL-Huqail
Department of Biology, College of Science, Princess Nourah bint Abdulrahman University, Riyadh, Saudi Arabia

M. Zia-Ul-Haq et al. (eds.), *Essentials of Medicinal and Aromatic Crops*,
https://doi.org/10.1007/978-3-031-35403-8_21

and North Africa [4, 5]. This plant also grows in regions of Africa, China Middle East, Australia and India. During the nineteenth century, European colonists brought the plant to North America, and it is now naturally grown there as well as in South and Central America, Australia, China, and other parts of the world [6]. During the nineteenth century, European immigrants brought the plant to North America, and it is now a native species in both South and North America. It was once grown in gardens and can now be found on vacant lots, old pastures, and along highways. In some regions, it is regarded as a troublesome invasive weed that needs to be controlled using traditional biological methods [7].

Silybum is a minor herb genus found in Asia, Africa, and Europe. The only species found in India is *Silybum marianum*. It grows wild in Punjab and the northern Himalayas, from Kashmir to Jammu. It has taken root in both North and South America [8].

21.3 Description of the Plant

The tubby, rigid, broad, and jagged herb *S. marianum* is an annual or biennial plant. In its first stage, which lies flat on the ground, unwieldy large basal leaves emerge. It is 0.9 m in width and rises to a height of 2.5 m. Up to four hollow, spherical stems filled with milky white sap are produced by each plant. It has large, broad, glossy green leaves. They stand out because of white vein-like lines and spiky borders. Young leaves feature shallow spiny lobes, whereas older leaves have deeper lobes and a wider border. *S. marianum* produces single, thistle-like flowers that are blue or white in hue. *S. marianum* is a tubby, rigid, broad, and jagged herb that grows as an annual or biennial plant. In its formative stage, which is close to the ground, awkward badge basal leaves sprout up. It has a 0.9-meter width and a 2.5-meter height. Each plant grows up to four hollow, spherical stems that are dripping with milky white sap. It has large, glossy green leaves that are broad. They stand out because to white vein-like patterns and spiky limits. Young leaves are shallowly lobed with spin, whereas mature leaves have deeper lobes and a wavier border. *S. marianum* has single, thistle-like, blue, or white blooms [9].

6–7 mm in length and 3 mm in width, the seeds are obliquely obovoid. The seeds are brown in hue, and one end of each one has a canaliculated hilum and a yellowish extrapolative expanded ring. *S. marianum* often comes in two varieties: purple flowers and white flowers, and they are both found in some parts of Asia. Magenta to purple hues can be found in the 200 florets that a single bulb can generate. The fruit has two large, opaque, white greasy cotyledons and a thin, chocolate-colored exterior region that is visible when cut from corner to corner. A plump, elaiosome-shaped cluster of fruit attached to the fruits is rich in lipids, attractive to ants, and helps disperse seeds by doing so [9].

A sample of 1000 fruits revealed that the largest fruits had the highest mass and best vigour, and that there was a substantial correlation between the shapeliness of the fruits and their mass. The seeds were substantially lighter in the secondary flower heads than in the primary flower heads. The first bloom was smaller on the plant when the sowing was postponed from October to February. The quantity of flower heads and the duration of time until the initial bloom both shcwed a similar downward tendency [9].

It's a tall perennial or biennial that can reach 120–130 cm in height. The leaves are big, speckled with white, pinnatifid into oval, triangular, sinuate-tocthed, spiny lobes, and the stem is pale green, solitary, or slightly branched. The spherical heads are 6–10 cm in diameter and concave at the base. The involucre has oblong outer scales that widen into an ovate, prickly, cellar-edge appendage that tapers abruptly into a long spine, while the inner scales are lanceolate and whole.

The hermaphrodite and fertile flowers are purple or white, with a short tube and a five-fold limb. a harvest Several fungi have been identified as causing fruit storage harm. Within the first month of storage, Aspergillus ochraceus produces significant damage to xanthotoxin, whereas Fusarium oxysporum causes damage to Aspergillus niger and Aspergillus flavus and dramatically reduces the xanthotoxin concentration during the first three months [8].

21.4 Harvesting and Yield

The crop Ammi majus lasts for 5–6 months. When planted between August and September, the crop blooms between November and December. It takes roughly a month for seeds to blossom and mature. According to studies, the mature, brown-colored fruit only contains 0.4% xanthotoxin, although it includes 0.72% when harvested when it reached maturity and one percent is found in green fruits of this plant.

Harvesting at the mature green stage is preferred because the crop's total yield is reduced significantly at the immature green stage. Another advantage of harvesting at this point is that seed-shattering, which occurs when seeds become brown, can be reduced [8].

21.5 Constituents

21.5.1 Formation of Silymarin

silydianin, isosilybin, taxifolin silychristin, and Silybin are present in the dry fruit of the holy thistle plant. These are the active compound of flavonoids. Two compounds of the flavonoids dihydro flavonol and flavonolignan are mostly used in medicine or therapeutic processes. The other names of this constituent found in this plant are silibinin, silybin and silibinin. From this plant, milk is extracted [10].

21.5.2 Some Constituents in Milk Thistle

Part of plant	Compounds	Main compound	Yiled
Dried fruits of plant	Silydianin	Flavonoids	0.4
	Silybin A &B	Flavonoids	4.0 & 7.0
	Taxifolin	Flavonoids	0.6 mg/g
	Isosilybin	Flavonoids	
	Silychristin	Flavonoids	4.0
In seeds	Dihydroflavonol	Flavonoids	
	Flavonolignan	Flavonoids	

When the seeds of this plant remove the fat then some compounds such as silychristin, silibinin, taxifolin and silidianin obtained. The yield of these compounds is 4.0, 7.0, 0.6 and 0.4 mg/g. some scientists recommended that only methanol is best for the extraction of these compounds. The extraction of silibinin is a very hard and hot water technique also being used [11].

At the temperature of 100 °C, it was seen that the yield of the taxifolin was near about 1.2 mg^{-1}, which was nearly six-fold higher than the findings obtained using ethanol by the method of Soxhlet estimation process seeds. Similarly, the output of silychristin increased by nearly four-fold to 5.0 mg/g of seed. The yield of Silibinin A & B is 1.8 and 3.3 mg/g in seeds. This yield is about 30% of Soxhlet's total yield. The concentration of the extraction of these compounds according to the time taken for their extraction it was shown that the more polar such as silychristin and taxifolin can be extracted at 85 °C. while the compound which is less polar is extracted at 100 °C.

With the exception of taxifolin, the concentrations of these substances are commonly given as silymarin content [12]. Two items from Argentina, one from Brazil, one from Spain, and one from Italy were tested using this novel method (which included tablets, sugar tablets, and capsules). According to the results, between 50% and 90% of the prescribed value was dissolved when the dose form crumbled. The products were examined using UV spectrophotometry and HPLC [13].

In addition, as the hunt for softer and "greener" solvents grows, hot water has gotten a lot of interest as an extraction solvent for recovering chemicals from plant material. It investigated a method of extracting milk thistle at temperatures above 100 °C by using hot water as a solvent. They discovered that when the temperature rose, taxifolin and every of the silymarin compounds' maximal extraction yields did not increase., indicating that the compounds were degrading. When the extraction temperature was raised from 100 to 140 °C, however, the time it took for the compound yields to reach their peak was halved, going from 200 to 55 min. At 140 °C, first-order degradation kinetics of silymarin compounds were detected [13].

21.5.3 Structure and Formation of Silymarin

A variety of lipid microsphere components, including oily core, internal cosurfactants like span 20, tween 20, tween 80, propylene glycol, and surfactants like soybean lecithin, were tested in different concentrations to optimise the final formulation's properties, including globule size range, structural integrity, sustainability, and drug-holding capacity percent. Transmission electron microscopy and the laser diffraction method were used to evaluate the size and shape of the final formulation.

In 36 h, silymarin-loaded lipid microspheres had a higher mean release percentage of 57% than the silymarin solution, which was only 19%. These findings suggested that a consistent delivery strategy for passive liver targeting may be developed, with improved hepatoprotective benefits of silymarin and soybean lecithin [14].

21.5.4 Uses in Medicine

There is a long history of milk thistles used in medicine. Theophrastus, a Greek philosopher and Aristotle's successor called milk thistle Pternix [15]. Both Pliny the Elder and Dioscorides provided descriptions of this plant and its [16]. Hepatobiliary disorders were treated with milk thistle as a preferred medication from the sixteenth of the century [17]. Milk thistle is a fantastic treatment for blockages of the liver and the gallbladder, according to famed English herbalist Nicholas Culpeper, who wrote The English Physician in 1652 [18]. The milk thistle arrived in the Americas with the first European settlers. At the beginning of the twentieth century, a group of herbalists known as the Eclectics utilised milk thistle extracts to treat menstrual disorders, liver, spleen, and kidney ailments [19]. Before the 1960s, when scientific studies, primarily carried out in Germany, reignited interest in milk thistle and its components as a treatment for acute and chronic liver disease and as a hepatoprotective agent to prevent toxic liver injury [16, 20, 21].

21.5.5 Antioxidant Effects of Silymarin

In vitro, silymarin plays a role in the antioxidant process. Silymarin, like curcumin, resveratrol, and other phytochemicals, has anti-oxidant properties. Numerous systems have demonstrated the free radical-scavenging and antioxidant abilities of silybin. Reactive oxygen intermediate production and lipid peroxidation are both triggered by TNF and are suppressed by silymarin, according to research from our lab [22]. In an investigation, the antioxidative and free-radical scavenging capabilities of compound silybin on the microsomal lipid peroxidation was found [23].

In both the pilot clinical studies and the preclinical NAFLD models, natural antioxidants have been demonstrated to have favorable benefits [24]. It is thought that silymarin's role as a natural antioxidant helps explain why MT formulations have hepatoprotective properties. Recent reviews have examined silymarin's antioxidant properties [25]. Some of the putative antioxidant mechanisms of silymarin include direct free radical scavenging, ion chelation of iron and copper in the intestine, and activation of the production of defense-enhancing molecules such heat shock proteins, thioredoxin and sirtuins [25]. Additionally, antioxidant enzymes including superoxide dismutase and nonenzymatic pathways are activated (e), primarily through Nrf2 activation. In patients with nonalcoholic steatohepatitis, silymarin, for instance, has been shown to significantly boost the expression of superoxide dismutase [26, 27]. It lowers the oxidative stress in "thalassemia sufferers [28]. Silymarin does not directly affect how ethanol is metabolised, nor does it have any effect on how quickly ethanol is removed from the body or on blood alcohol levels. Cytochrome-interaction between silymarin and silybin has not been established. Together, these findings suggest that MT's antioxidant and free radical scavenging properties are likely responsible for its ability to be antitoxic [29].

21.5.6 Silymarin Anti-inflammatory Effects

Silymarin inhibits oxidative and nitrosative immunotoxicity as well as T lymphocyte activation, which may be how it reduces autoimmune and immune-mediated liver disorders [30, 31]. In several rat/mouse models of liver disorders, such as cholestatic liver damage, silymarin or MT extracts have been found to exhibit anti-inflammatory properties [32], Hepatotoxicity caused by CCl_4 [33], the acute liver damage brought on by stress [34], limit acute liver damage brought on by stress [35], Isoniazid/zidovudine-induced liver damage [36] a model of steatohepatitis brought on by a diet low in methionine and choline, and eventually [37].

Silymarin works by preventing pro-inflammatory pathways from being activated. Silymarin has been shown to be a powerful anti-inflammatory drug in a variety of studies. Antioxidant activity, suppression of stopping of 5-lipoxygenase, inflammatory cytokines, and suppression of NF-B activation are all evidence of silymarin's anti-inflammatory properties. Silymarin's impact on leukocyte migration and acute inflammation models [38].

Oral silymarin alleviated food-pad abscesses in rats with carrageenan-induced paw edema (ED50 = 62.42). Silymarin administered orally was less effective than silymarin used externally in mice with xylene-induced ear irritation, with effects comparable to indomethacin. In mice, silymarin reduced the number of neutrophils while inhibiting carrageenan-induced leukocyte buildup in inflammatory exudates. An important step in the early stages of neurodegenerative illnesses is the activation of microglia, which cause an inflammation in the central nervous system [38].

21.5.7 Antifibrotic Activity

Another cause of the liver fibrosis is the activation of the hepatic stellate cells and the Kupffer cells that results injury of the hepatocyte cells. The conversion of hepatic stellate cells into myofibroblasts is an important step in the fibrogenesis process. Hepatic insufficiency, portal hypertension, and hepatic encephalopathy can all be caused by the modification of the liver's architecture that results from liver fibrosis [39]. Studies on animals have demonstrated that silymarin has the ability to stop the advancement of the initial liver fibrosis and to block the fibrogenesis pathways in the early stages of the fibrotic process [33, 40]. In these investigations, biliary obstruction in rats with 50 mg/kg/day silymarin caused a 30% drop in the amount of collagen and pro-collagen III [33]. Silymarin decreases the expression of pro-fibrogenic according to experiments aimed at determining the mode of action [34], limits the NF-κB [30]; slows down HCS activation [33] and changes how some genes that are important for mitochondrion electron transport chain and cytoskeleton organization are expressed [41]. A compound sitagliptin, a dipeptidyl peptidase-4 inhibitor that is clinically utilized as an oral anti-diabetic drug, and silymarin together have the ability to reduce the liver fibrosis caused by carbon tetrachloride in rats [42]. Curiously, silymarin inhibits liver fibrosis in a young adult non-alcoholic steatohepatitis model [43], which, given the rising prevalence of NAFLD in teenagers, may have clinical significance. There is evidence that silymarin's anti-fibrotic properties could be enhanced by novel nanoparticle forms of the compound. In fact, cholestasis-induced liver damage has been proven to be resolved by a unique formulation of silymarin-loaded Eudragit® RS100 nanoparticles. Fibrosis by reestablishing the hepatic regeneration capacity [44].

21.5.8 Inhibition of Prostate Antigen

Hormone-refractory of human PCA, it has been suggested that a biomarker for treatment success be a decline in blood PSA levels. Silybin in cells of the human PCA lower the prostate antigen work as [45]. It was found that silybin lowers PSA levels in both intracellular and secretory forms. A G1 cell cycle halt also causes a highly significant overall slowdown in cell proliferation. Silybin suppressed PSA expression and cell proliferation induced by 5-DHT androgens. Silybin has the capacity to decrease PSA levels when combined with the coactivator of the AR prostate epithelium-specific Ets transcription factor in PCA LNCaP cells [46]. Silybin inhibited PSA mRNA expression and secretion in conditioned media, whether it was given with or without 5-DHT. PSA was likewise reduced when silybin and 10–8 M DHT were given together.

21.5.9 Antitumor Effects of Silymarin

Many of the studies mentioned that silymarin/silybin had anticancer properties in vivo in various animal models. a mechanistic analysis of silymarin's in vivo therapeutic effectiveness against skin cancers DMBA-TPA-induced established skin papilloma-bearing SENCAR mice were given 0.5% silymarin in an AIN-93 M-purified diet (w/w) for 5 weeks, and tumor growth was inhibited (74%) and the established tumors were regressed (43%). The suppression of ERK1/2 cell proliferation and phosphorylation as well as the induction of apoptosis in malignancies accompanied these results [47].

Parts of the holy thistle were employed as a treatment for those who were punished by having their bodies dragged backward, according to Dioscorides, a prominent Greek physician who accompanied the invading Roman army in Britain in the first century AD. It is also described in Bede's writings (872–735 BC) as a plant that was a cure-all and had magical properties. Its seeds include the flavonoids silymarin and sylibin, which are thought to be the primary components responsible for the drug's antihepatotoxic (liver-damaging) properties [8].

Jaundice, intermittent fevers, dropsy, and uterine problems are all treated with this herb in Europe. The leaves are considered to be effective galactagogues whether consumed or used as an infusion. Pungent, demulcent, and antispasmodic are all qualities of the seeds. They're utilized to reduce bleeding and treat liver and gall bladder calculi. Diabetics should consume the blossoming heads. The seed extract is purgative and increases small intestinal peristalsis, according to studies. Due to its prickly nature, this plant can easily be grown in regions where cattle or other grazing animals are a hazard. In addition, the plant is naturally hardy, making cultivation simple and low-risk [8].

21.5.9.1 Liver

Silymarin, the active substance isolated from the milk thistle plant, has been proven to protect experimental mice from a variety of hepatotoxins such as acetaminophen, poisoning of fungus iron overload, and carbon tetrachloride [48].

Silybum marianum has been used as a hepatoprotective since 2000 years ago. Additionally, silymarin has been shown in numerous clinical and laboratory investigations to protect the liver from toxicity brought on by a variety of poisons, including carbon tetrachloride, acetaminophen, and tetrachloromethane [3]. According to reports, silymarin exerts its hepatoprotective effects through a variety of processes, including antioxidant activity and the removal of free radicals, a rise in cellular glutathione levels, the stimulation of DNA polymerase, and the stabilization of the hepatic membrane. Numerous studies have strongly implied that silymarin's antioxidant activity and free radical scavenging capacity are what is primarily responsible for its hepatoprotective properties. The modification of glutathione and the membrane stability that it produces are evidence of this action [49]. Silymarin stimulates DNA polymerase, which increases ribosomal RNA synthesis and promotes the regeneration of liver cells. A rise in cellular glutamine levels stabilises The enzymes glutathione peroxidase and superoxide dismutase. Silymarin reduces the size of the enlarged liver by blocking the 5-lipoxygenase cycle and preventing the liver's Kupffer cells from producing leukotrienes and free radicals. In addition, silybin suppresses the generation of peroxidation lipid and cellular damage in mouse hepatocyte cells [50]. Silymarin really alters membrane lipids like cholesterol and phospholipids, which is why it has an impact on cellular permeability. Additionally, silymarin has an effect on the liver's other lipid compartments, which may affect how lipoproteins are taken in and secreted [51]. There is a dearth of information on silymarin's impact on the liver's lipid metabolism. The rise in total lipids and triglycerides caused by carbon tetrachloride in the liver of rats can be partially countered by silibinin, and it is also likely to promote fatty acid b-oxidation. Additionally, it has been hypothesized that silymarin may reduce the liver's capacity to produce triglycerides [52]. In mice, silymarin protects against liver damages effects and lipid peroxidation of carbon tetrachloride. Silymarin is thought to reduce metabolic activation by carbon tetrachloride and act as an antioxidant to avoid chain rupture [53]. Other studies have shown that silymarin protects the liver from particular damage caused by the liver toxins paracetamol, alloxan, microcystin and halothane in a variety of laboratory animals [54]. Silymarin may shield liver cells from a variety of illnesses, including viruses, chemicals, and naturally hazardous substances, according to various studies done on living animals. Silymarin pretreatment shields experimental animals against becoming intoxicated by *Amanita muscaria*. A fatal dose of *Amanita muscaria* was administered to dogs, and even 40 h later, silybin (50 mg/kg) treatment prevented the canines from becoming intoxicated [55]. The injection of halothane, tetrachloride, thallium, acetaminophen, and carbon tetrachloride into laboratory animals also results in liver poisoning and the generation of peroxidation lipid, which is inhibited by pretreatment with silymarin [56]. Rats given silymarin or silybin showed decreased activity of hepatotoxicity as indicated by aspartate transaminase, g-glutamyl transpeptidase, and alanine transaminase [57]. Silymarin prevents the rat's alcoholic liver caused by biliary atresia [58]. Clinical studies show that taking 120 mg of silybin twice a week for two months reduced the levels of alanine transaminase and aspartate transaminase in individuals' blood serum who had liver problems [57]. In a study involving 2637 patients with chronic liver diseases, 88% of the patients who received extract of the silymarin which has a significant effects in decreasing of the enzyme level in the liver. Only 1% of patients experienced minor adverse effects from the medication [59]. Intoxication brought on by Amanita muscaria is also treated with silymarin. Up to 80% of patients who received this medication experienced moral decline. 18 Up to 48 h after *Amanita muscaria* intoxication, liver damage was fully prevented by intravenous silybin administration (for 3–4 days @ 20–50 mg/kg/day). In a study conducted on 250 people intoxicated by *Amanita muscaria*. In the untreated group, there were 46 recorded deaths. But in the group of 16 people who took silybin, there was no mortality found [60]. In a different trial, silybin was administered to 18 participants who had consumed *Amanita muscaria* poison. The findings showed that only one person who used Amanita mushrooms for suicide and went untreated for 60 h had passed away [61, 62].

Results from the use of silymarin in the treatment of alcohol-induced liver damage are conflicting. A multiple controlled double-blind experiment including 300 people with alcohol-induced liver damage found that silymarin (420 mg/day) significantly reduced enzyme levels and liver histology evaluation after 4 weeks [63]. A second study that lasted four years and employed silymarin administration involved 170 people with alcohol-related cirrhosis. The results showed that mortality was lower than in the control group [64]. In another investigation, 116 people with alcohol-induced hepatitis were given 420 mg of silymarin daily for three months. Despite the fact that 46% of the patients were successful in quitting drinking, the outcomes did not seem to significantly differ from the control group [65]. Two effects of ethanol in rats are known to be mitigated by silibinin: a reduction in the incorporation of labelled glycerol into isolated hepatocyte lipids and a blockade of the phospholipid transporter [66].

Additionally, silibinin increases the activity of choline phosphate uridylyltransferase and phosphatidylcholine synthesis in liver of rats [67]. In patients with hepatitis, silymarin administration has contradictory effects. In a double-blind research on 20 patients with active chronic hepatitis receiving 240 mg silipide twice daily for a week, the g-glutamyl transpeptidase level significantly lowered in comparison to the control group [3]. 157 patients with viral hepatitis were evaluated in a different investigation. Silymarin (140 mg, 3 times day) was administered to 29 patients, whereas placebo was given to 28 others. When compared to the control group, the silymarin-treated group's bilirubin content, aspartate transaminase level, and alanine transaminase level all significantly decreased [68]. 151 patients with liver cirrhosis were examined, but silymarin did not improve their conditions [69]. Through the promotion of liver tissue regeneration and an increase in protein synthesis, silymarin also protects against liver damage. Along with enhancing protein synthesis, silibinin also promotes the development of synthesis of the DNA and ribosomesHowever, the increase in protein synthesis only occurs in livers that are damaged, not in livers that are healthy. Silibinin increases the production of ribosomes and protein synthesis in the liver, likely through the RNA polymerase I physiological control [70]. One of silymarin's most effective properties is its ability to decrease cellular permeability, which is related to both quantitative and qualitative modifications in the membrane lipids [71]. According to this, silymarin might affect the release and absorption of lipoproteins. In this context, it has been shown that silymarin and silibinin reduce the formation and turnover of phospholipids in the rat liver. Furthermore, it has been shown that silibinin inhibits the incorporation of labelled glycerol into the lipids of isolated hepatocytes and counteracts the inhibition of phospholipid synthesis [72].

Silibinin can also boost the activity of choline phosphate cytidylyltransferase and accelerate the formation of phosphatidylcholine in rat liver [67]. It should be emphasized that there is a paucity of information on silymarin's impact on hepatic triglyceride metabolism. The rise in triglycerides and total lipids caused by carbon tetrachloride in the liver is somewhat countered by silibinin. In the liver, it might also reduce the production of triglycerides [62]. Carbon tetrachloride's hepatotoxicity and lipid peroxidation are also something that silymarin can protect you from. The hepatoprotective effects of silymarin against damage brought on by alloxan, paracetamol, microcystin and halothane have also been demonstrated in a number of experimental animals [62, 73].

Although milk thistle is frequently recommended as a therapy for alcoholic hepatitis and cirrhosis, scientific investigations have produced inconsistent outcomes. The majority of studies show that milk thistle enhances liver function and prolongs life in patients with cirrhosis or chronic hepatitis. However, issues with the study design (such as a small number of participants and variances in milk thistle dosing and duration) make it difficult to draw any definite conclusions [74].

21.5.10 Production of Protein

Hepatic healing from acute or chronic injury requires the regeneration of liver cells. Fibrosis and cellular regeneration both take place concurrently in chronic illness. In partially hepatectomized rat livers, MT extracts, silymarin, and active ingredients such silybin have been demonstrated to accelerate hepatic regeneration [75, 76]. It has been established that silybin given intraperitoneally significantly increases the production of ribosomal RNA. Evidence suggests that activation of polymerase I can be a factor, even though the exact route of action is yet unclear. Silybin-activated ribonucleic acid (RNA) polymerase I and ribosomal RNA in multiple preclinical trials, causing ribosomes to develop more quickly and speeding up protein synthesis. It may be possible to heal damaged hepatocytes and return to normal liver functioning thanks to silybin's stimulating influence on ribosome synthesis [77, 78].

Milk thistle is commonly used to treat viral hepatitis (especially hepatitis C), however, studies have yielded inconsistent outcomes. Some research discovered improved liver function, whereas others did not. Milk thistle dramatically lowered the viral load of hepatitis C in a trial of 16 individuals who had not responded to interferon and ribavirin treatment. After 14 days of treatment, the virus was undetectable in seven of the participants [79].

Milk thistle has been employed as an antidote for death cap mushroom poisoning in the past, according to folklore (Amanita phalloides). Milk thistle extract, when administered within 10 min of intake, entirely counteracts the mushroom's harmful effects, according to animal tests. It minimizes the risk of liver damage and mortality when taken within 24 h [48].

21.5.10.1 Cancer

- Silymarin and other active ingredients in milk thistle have been shown to have anti-cancer properties in laboratory experiments. It appears that these substances:
- To prevent cancer cells from dividing and replicating, use the following steps.

- Reduce the cancer cell's lifespan
- Cut off the flow of blood to malignancies.

According to certain research, silymarin can help lower skin cancer risk by supplementing sunscreen protection. According to other research, milk thistle can help chemotherapy work better. More research is needed to determine if milk thistle has any physiological effects (not just in test tubes).

If you are taking milk thistle, make sure your doctor knows. Milk thistle and cancer preventive medicines, such as Raloxifene, may interact, according to preliminary research [80].

Milk thistle is a popular herb for treating liver problems, and it also appears to help with the metabolism of lipids and glucose. Hyperinsulinemia is thought to result from reduced insulin clearance by the liver, which is caused by liver dysfunction that decreases the effectiveness of postprandial hepatic glucose storage. According to phase III clinical trials, silymarin (the active ingredient in milk thistle) is the best treatment for NAFLD since it is effective in lowering fibrosis, reducing the severity of steatosis and aminotransferase levels and liver ballooning. In trials of patients with NAFLD, milk thistle was also found to behave as an insulin sensitizer [81].

It possesses antioxidant qualities, as well as anti-inflammatory and anti-fibrotic capabilities, and protects against lipid peroxidation. Silymarin has been subjected to large-scale controlled trials in Europe, with mixed results. Ferenci and colleagues looked at 170 patients with cirrhosis who were part of a 41-month therapy regimen that included 40 mg silymarin t.i.d. Patients with alcoholic cirrhosis and those with the lesser disease showed a good beneficial effect (CTP A disease) [82].

Pares, et al., on the other hand, observed no benefit from 150 mg of silymarin taken twice daily in 200 individuals with alcoholic cirrhosis, some of whom also had hepatitis C [20] Despite mixed outcomes, silymarin has grown in popularity as a complementary and alternative medicine treatment for liver illness due to its favorable safety profile. Large multicenter studies of silymarin in Hepatitis C and NASH should shed light on the drug's efficacy in the fibrosis of the liver treatment [83].

The biennial milk thistle's dried seed extracts are widely used to treat patients with liver disease. Several diseases, such as acute alcoholic liver disease and chronic viral hepatitis and also drug-related hepatitis. It has been demonstrated that *S. marianum* is secure, well-tolerated, and improves the chemistries and signs of the liver. In these situations, *S. marianum* appears to have a dose-dependent effect on liver chemistries, with results visible within 7 days [84, 85].

Despite the fact that S. marianum is commonly utilized by HCV-infected people, there is little evidence of its effectiveness in CHC. Researchers executed a pilot double-blind, placebo-controlled crossover experiment in CHC sufferers to analyze the influence of dietary *S. marianum* on blood HCV RNA, alanine aminotransferase (ALT) levels, as well as both QOL and psychosocial variables.

Research in the lab demonstrates that particularly silybin, and silymarin, have anticarcinogenic effects on animal breast cancer, and cancer cells of the epidermal and prostate [86]. Silymarin showed cytoprotective activities on prostate and human breast cancer cells exposed to carcinogenic substances. Before being exposed to silybin, cancer cells were pre-inoculated with silybin, which boosted Adriamycin effect in the inhibition of cell proliferation [87].

21.5.10.2 Failure of Renal

According to one study, the silybin in silymarin prevents kidney damage that results from giving lab mice cisplatin. Additionally, silybin has prevented renal problems caused by the effects of cyclosporine in laboratory mice [88].

Different investigation revealed that the silymarin has positive effects on the health of liver as well as kidney [3]. More accumulation of silymarin in cells of the kidney and promotes the creation of more protein and nucleic acids, which helps the kidney cells regenerate. Due to two significant silymarin components, silybin and silychristin, Silymarin has been hypothesized to increase cell multiplication by 30%. Diabetic patients benefit from silymarin [89–91].

21.5.10.3 Osteoporosis

Significant estrogenic effects are present in the taxifolin found in *Silybum marianum*. Additionally, silymarin contains flavonoid chemicals that can influence the metaphysis of the femur by acting as an estrogenic agonist on the uterus without interacting with B estrogenic receptors [92].

21.5.10.4 Nervous System

Alcohol use has reportedly been linked to decreased learning capacity in rat pups. When it was taken along with silymarin, though, this effect was avoided. The primary factor causing severe damage to nerve cells is inflammation of the nerves. Additionally, silymarin can prevent brain damage brought on by clogged blood arteries in the brain [93]. It has been demonstrated that silymarin helps diabetes patients' nerve fibers transmit electricity more efficiently [94].

Silymarin increased the ability of diabetic rats to retain information in memory and remember it when given at a concentration of 10 mg per kilogram of weight. Thinking, memorization, and mental function are all hampered by diabetes mellitus, research has shown. It does not, however, improve short-term spatial memory in diabetic rats. Attenuation of lipid peroxidation in hippocampal tissue has been linked to silymarin's positive benefits [95].

21.5.10.5 Hematologic Effects

The effects of silymarin's antioxidants on blood components are one of its key characteristics. It has been discovered that blood molecule oxidation plays a significant role in the progression of chronic diseases such as cardiovascular. Silymarin suppresses the hemolysis of red blood cells induced by the injection of hydrogen peroxide and other oxygen-derived free radical-producing agents, as well as hemolysis induced by copper [96].

21.5.10.6 Endocrine Gland

According to findings from experiments done on animals, silymarin defends the pancreas against chemicals like alloxan and cyclosporine. Silymarin was used to treat 60 patients with diabetes and insulin resistance who had cirrhosis of the liver caused by alcoholism. Over the course of six months, the data showed considerably decreased average daily sugar levels, overnight sugar levels, low blood sugar, and demand for insulin [97].

21.5.10.7 Immune System

The chemotactic and phagocytic activities of nonstimulated neutrophils and silymarin are unaffected, according to laboratory studies. But when neutrophils are stimulated, silymarin prevents the release of myeloperoxidase. Leukocyte mobility inhibitors' activity is delayed when neutrophils are incubated with silymarin.

In a double-blind, placebo-controlled experiment including 40 patients with alcoholic liver cirrhosis, silymarin therapy increased lectin as the activator of lymphoblast deformation. Additionally, there was a significant decrease in lymphocytotoxicity in comparison to the control group [98].

21.5.10.8 Treatment of Psoriasis

Historically, silymarin has been used to treat psoriasis. In addition to suppressing the cAMP cycle and leukotriene formation, silymarin also functions for the alleviation of psoriasis by eliminating undesirable body cells' metabolites, including those from the liver. Silymarin's suppression of the cAMP cycle, which is boosted in psoriasis patients and results in increased leukotriene production, may have beneficial effects [99].

21.5.10.9 Symptoms

When milk thistle is used orally, adverse consequences can include:

- issues with the digestive system (like diarrhea nausea, flatulence, dyspepsia, abdominal fullness or pain, abdominal bloating, changes in bowel habits, and anorexia)
- Skin repercussions (eczema, urticaria, rash, and pruritus)
- Headache
- occurrences involving neuropsychology (like insomnia, malaise, and asthenia)
- Loss of the ability
- Rhinoconjunctivitis
- Arthralgia
- Anaphylaxis

But in the reports that are available, causality is rarely discussed. Both the milk thistle and control groups experienced almost the same frequency of randomized studies reporting negative effects [100].

21.5.11 Pests & Control

The infections are typically brought on by powdery mildew and ray mold (*Botrytis cinerea, Erysiphe cichoracearum*). Leaf blotch is caused by fungi Ramulariacynarae [101]. *Septoria silybi* is a disease that is also found in the american state of California [102]. It was revealed that *S. silybi* also exists in Italy [103], in Bulgaria [104], in Pakistan [105] and in Poland [106].

Except for a fungus called *Septoria silybi*, *Alternaria silybi* is the most common milk thistle pathogen [107]. Cercospora sp. has been identified as leaf blotches and abundant sporulation, but no *Alternaria silybi* spores in any live tissue. *P. tyrimni Puccinia cruchetiana, P. mariana,* and *P. laschii* are among the rusts that can be found on the milk thistle [108]. Fusarium fungus induces milk thistle fading, and *Puccinia punctiform* on leaves is one of them [109].

On the milk thistle stem, there is six Fusarium species [106]. Viruses were added to the list of pathogens discovered in milk thistles. The virus that causes spotted wilt in tomatoes is known as Cucumber mosaic virus [110], Tomato Spotted Wilt [111], and the smut fungus *Microbotryum silybum* was discovered in the flowerheads of milk thistle in Greece [112, 113]. Serious damage to seeds-heads caused by weevil larvae *Larinus latus* Herbst was seen in Egypt [114]. Aphids called *Dysaphis lappae* cynarae are to blame for plant damage in Greece [112]. While in Iran these plants are harmed by the *Aphis fabae*. At the end of flowering, Spodoptera sp. caterpillars damage the leaves. During damp weather, snails are a common problem [105].

21.5.11.1 Herbicidal

Silybum marianum is especially susceptible during the seedling and rosette stages of its development. As it ages, the plant develops greater resistance to treatment. When the plant was sprayed with the ester 2, 4-d with 80% of the active ingredients and diluted with the water at the rate of 1600 times in four parts (like 400) and it lessens the effects of the infections from the herbicides and also sprays as it grows older. Parsons recommends 1/2 pint per acre to 1 1/2 pints per acre for boom spraying. Autumn spraying will need to be followed up in the winter to deal with plants that germinate later since fall germination spans several months.

When selectively sprayed with picloram and methabenzthiazuron in combination with phenoxy acetic acid compound "at rates commonly indicated," the thistle was entirely controlled, according to research by R. Meissner and C. Mulder. They made a rosette of about three whorls and found that methabenzthiazuron alone was not sufficient to control S. marianum. They say dicamba spot therapy could be useful in the future [115].

21.5.11.2 Fertilizers

Fertilization Because this crop is tolerant of poor soil quality, its nutrient requirements are low to moderate. Pre-sowing fertilizers high in phosphorus and potassium are frequently applied to the soil and absorbed there. Because nitrogen fertilizer leaches from the root zone in high-rainfall places, it can be given on a regular basis throughout the crop's life cycle. In order to do this, half of the nitrogen is applied before planting and the remaining nitrogen is applied during the beginning of stem elongation. Poland normally administers, 58 kg ha^{-1} of potassium, 30.5 kg ha^{-1} of phosphorus, and 50 kg ha^{-1} of nitrogen prior to planting [116].

In the country of the Egypt flavonolignans in seeds contents, yield of the oil, enhancement of the seeds, and percentage of the seeds were seen when 115 kg/ha of potassium oxide and 140 kg of the nitrogen were applied [117]. Furthermore, before sowing, Bulgarian farmers applied 49.5 kg ha1 nitrogen, 138 kg ha1 P2O5, and 150 kg/ha potassium to the soil [118]. Irrigation Drought resistance is a plus for milk thistle, and regular rain is usually enough. Silymarin buildup is hampered by both an excess of water and a deficiency of it. During seed growth and filling, the crops in a Mediterranean environment should be irrigated during severe drought circumstances [116].

This experiment was conducted as a factorial based on a randomized complete blocks design with three replications in the Roudehen Research field, in 2010. In this research vermicompost in 5 levels (5, 10, 15, and 20 ton/ha and nonapplication) as the first factor and phosphate bio-fertilizer in two levels (inoculation and noninoculation) as the second factor was considered. The results showed that seed inoculation with phosphate bio-fertilizer had a significant effect on capitol number, leaf fresh weight, stem fresh weight, flower weight, total fresh weight per plant, flower dry weight, total dry weight per plant, and grain yield. The maximum grain yield was conducted with 20 ton/ha vermicompost, because of increased capitol number, capitol diameter, and 1000 grain weight. Also, the plant height, leaf fresh weight, stem fresh weight, flower fresh weight, total fresh weight per plant, flower dry weight and total dry weight per plant were highest in this treatment. With inoculation of phosphate bio-fertilizer, because of increase in capitol number and capitol diameter, the highest grain yield was observed. The maximum grain yield (2150.2 kg. ha-1) was obtained at the interaction of 20 ton/ha vermicompost and phosphate bio-fertilizer consumption. Therefore, 20 ton/ha vermicompost application and phosphate inoculation, because of a significant increase in grain yield recommend as the best treatment [119].

21.5.12 Soil and Climate of the Plant

The plant milk thistle can grow in many and different places and in a different type of soil. This plant adjusts itself in any growing conditions. The total rainfall in any area benefits to the growth of the plant. This plant grows well where the rainfall is near about 180 mm. The yield of the seeds of this milk thistle plant is approximately 550–1680 kg/ha while the yield of silymarin content was seen near about 13–35 kg/ha. These plants can be grown in water deficit area in the world. These plants has deep root system to absorb water and minerals from the soil. Milk thistle grown on any type of the soil like clay, sandy or loamy soil. It also grows on light soils [116].

The milk thistle is tolerant of various environments and climes. It may be grown in desert, southern, and northern regions, including those in Canada [120] due of its durability and adaptability [15, 106] reported that silybin is only developed in sows from subtropical climates and not those from mild ones since greater temperatures seem to make it easier for the top component to accumulate. On a variety of soil types, milk thistle grows well and produces a decent yield [15, 116, 121]. Typically thought of as a plant that grows on wastelands and along roadside ditches [122], is reportedly considered a harmful plant in many nations since it both competes with agriculture [123]. In nitrogen-rich environments, such as dairy yards, chicken coop waste, landfills, and abandoned agricultural fields, milk thistle grows best [15]. Fruits' silymarin concentration varies according to the milk thistle type, as well as the geographic and meteorological circumstances where it thrives. To the contrary, subtropical climates have higher concentrations of silibinin, the primary ingredient in silymarin, than temperate climates [15]. Ingesting milk thistle can be fatal for livestock because it accumulates nitrates, especially in the early wilting stage [105]. Growing phase depending on the climate, milk thistle can be a biannual or winter annual plant. In the fall and spring, germination takes placec [6]. Studies have revealed that temperature and light levels have an impact on milk thistle seed germination [122]. In order to germinate, fresh milk thistle seeds appear to need a post-ripening stage and perform better at low temperatures than at high temperatures. The seeds are still good for at least nine years. An experiment revealed that seeds kept at higher temperatures require a lengthy ripening period. The optimal germination temperature for *S. marianum* is stated to be between 20 and 25 °C, with 10 and 35 °C being the minimum and highest consistent germination temperatures [124]. Milk thistle plants can germinate at the temperatures as low as 15 °C and as high as 35 °C [125]. Milk thistle overwinters in a rosette after seedling establishment, and the number of basal leaves increased in the meanwhile. Low temperatures stimulate milk thistle to blossom in late winter and early spring. Between April and May, flowers anthesis occurs. July marks the maturity of the achenes (fruits). The milk thistle's whole development cycle lasted between 125 and 140 days, and it could be divided into five stages: seedling (15–20 days), vegetative (45–60 days), flowering (between 60 and 80 days), fruit-bearing (between 60 and 80 days), and withering (between 80 and 120 days). Anthesis normally lasted five days within a capitulum. About 17 days later, ripe fruits were distributed. A plant may produce an average of 55 capitulum per individual [126]. A single seed head can produce anywhere from 100 to 190 seeds. When a flower is at its late blooming stage, the flavonolignan content in seeds is at its maximum [121].

There is proof that the amount of silymarin varies genetically among milk thistle populations. The production of silymarin is also influenced by environmental factors including rainfall and average temperature, with higher temperatures and drier conditions appearing to increase output. Although it may thrive in a wide range of environmental conditions, milk thistle appears to require well-drained conditions. Site selection should take containment into account due to its weedy inclination so that its expansion can be limited if necessary. Two weeks pass before a seed sprouts after being sowed in the spring or fall. Easy to self-seed [1].

21.5.13 Preparation of the Land

Typically, the earth is ploughed to a depth of 25–30 cm [127]. If that seed is buried 6 cm or deeper, the likelihood of the plant emerging is significantly decreased. As a result, the milk thistle seedbed needs to be well-prepared. A healthy seedbed is created before to sowing using rotary hoe cultivation [122].

21.5.14 Sowing and Spacing

Direct seeding of milk thistle occurs in soils. Best germination happens between 2 and 15 °C [128]. Compared to 25 and 35 °C, 15 °C had a higher percentage of germination [125]. The milk thistle seeds have after-ripening requirements for germination temperature that restricted germination to 10–20 °C temperatures one month after harvest [128].

21.5.14.1 Seed Germination

Milk thistle (*Silybum marianum* L.) seeds (achenes) were studied for germination. The amount of time it took to meet after-ripening requirements as determined by the temperature at which the seeds were planted. The longer the after-ripening requirement is, the higher the incubation temperature during germination (up to a maximum of 5 months). Milk thistle seeds germinated at temperatures ranging from 0 to 30 °C once the post-ripening criteria were met. Germination was best when the temperature was between 2 and 15 °C. Cold periods of 16 h alternate with warm intervals of 8 h at temperatures ranging from 10 to 30 °C.

Milk thistle seedling emergence slowed as burial depth increased, although it was still significant at 8 cm. When compared to germination on the surface of the soil or litter, germination on the surface of the soil or litter was drastically reduced. At 2 and 5 °C incubation temperatures, adding 1.0 mM potassium nitrate (KNO_3) to the germination substrate improved the milk thistle seed germination [128].

21.5.14.2 Review of the Literature

It is uncertain if milk thistle has any therapeutic benefit. It is challenging to understand the evidence because of subpar research practises and/or inadequate reporting in publications. Issues that need to be addressed in the research design include heterogeneity in the origin and severity of liver disease, small sample numbers, and variations in the formulation, dose, and duration of milk thistle therapy. Liver function tests are by far the most often examined outcome measure, and improvement in aminotransferases has most frequently, but not always, been demonstrated to have a potential benefit. The least investigated clinical outcome indicators include survival and others, with inconsistent findings. The data is insufficient to determine if milk thistle is more successful for some liver diseases than for others, or whether efficacy is influenced by the length of treatment or the severity of the liver condition. Although there is no evidence linking milk thistle to specific side effects, the research that is available shows that it is connected to a few rather minor adverse effects. The mechanism of action of milk thistle is uncertain and may involve many factors despite significant in vitro and animal research. Future milk thistle mechanism research and clinical trials would benefit from a thorough analysis of the literature to clarify what is known and pinpoint knowledge gaps [100].

Milk thistle (*Silybum marianum* (L.) Gaertn.) is a useful medicinal plant for treating liver problems. Flavonolignans, also known as silymarin, contains three isomers like silydianin silycristin, and silybin, which are the very energetic ingredients in this plant. The presence of silymarin accounts for its medicinal benefits. The seeds have the largest concentration of silymarin, while the rest of the plant has a lower concentration. The amount of silymarin in milk thistle fruits varies depending on the variety of milk thistle as well as the geographical and meteorological conditions [9].

Milk thistle (*Silybum marianum*) has been used as a therapeutic plant for ages, according to folklore; its distinctive violet blooms and white-veined leaves are said to have come from the Virgin Mary's milk. It is a member of the Asteraceae family, which also contains sunflowers and daisies, and is native to the Mediterranean. Dioscorides (40–90 AD), a Greek physician and botanist, was the first to identify milk thistle's curative qualities. Milk thistle was later highlighted as "the best treatment against melancholy disorders" by John Gerard in 1597. More recently, fluoxetine or an extract produced from the leaves of the milk-thistle plant demonstrated equivalent benefits in individuals with obsessive-compulsive disorder who received either fluoxetine or an extract obtained from the leaves of the milk-thistle plant in a small randomized study from Iran [129].

Silymarin is a flavonoid that can be found in the dried fruit of the milk thistle plant *Silybum marianum* (silybin, iso-silybin, silychristin, silydianin, and taxifolin). Silymarin's antioxidant and hepatoprotective properties are well recognized, but its anticancer potential has only recently been discovered. Cell cycle arrest at the G1/S phase, induction of cyclin-dependent kinase inhibitors (such as p15, p21, and p27), down-regulation of anti-apoptotic gene products (e.g., Bcl-2 and Bcl-xL), inhibition of cell-survival kinases (AKT, PKC, and MAPK), and inhibition of infl kinases (AKT, PKC, and MAPK. Silymarin inhibits the expression of genes involved in tumor cell proliferation (cyclin D1, EGFR, COX-2, TGF-, IGF-IR), invasion (MMP-9), angiogenesis (VEGF), and metastasis (cyclin D1, EGFR, COX-2, TGF-, IGF-IR) (adhesion molecules). Silymarin inhibits NF-B-regulated gene products such as COX-2, LOX, inducible iNOS, TNF, and IL-1, resulting in anti-inflammatory actions. UV radiation, 7,12-dimethylbenz(a)anthracene (DMBA), phorbol 12-myristate 13-acetate (PMA), and other carcinogens/tumor promoters have all been shown to be chemopreventive agents in vivo by numerous research. Through downregulation of the MDR protein and other processes, silymarin has been found to make tumors more sensitive to chemotherapeutic treatments. It inhibits PSA production by binding to estrogen and androgen receptors. Silymarin has anticancer action in rodents, in addition to its chemoprotective properties. Silymarin is bioavailable and pharmacologically safe, according to multiple clinical investigations. The therapeutic efficacy of silymarin across various malignancies is now being investigated in studies [130].

Milk thistle (*Silybum marianum*) is being used in liver cancer and is also used as a remedy at home to control hepatoprotection. Silymarin crude extract and silibinin semipurified product are two commercially available formulations. Silymarin is made up of at least seven flavonolignans, with the diastereoisomers silybin A and silybin B being the most common; silibinin is made up entirely of silybin A and silybin B. The CYP2C9 inhibition properties of silybin A and silybin B, as well as their corresponding regioisomers, iso-silybin B and iso-silybin A, were assessed using human liver microsomes recombinant, CYP2C9 enzymes, and the clinically relevant probe warfarin, based on a recent clinical study showing an interaction between a silymarin product. In HLMs, silybin B was the most effective inhibitor, followed by silybin A, iso-silybin B, and iso-silybin A (IC50 values of 8.2, 18, 74, and > 100 M, respectively). Then, for further investigation, silybin A and silybin B were chosen. Silybin B was more powerful than silybin A against rCYP2C9*1 (6.7 versus 12 M), rCYP2C9*2 (9.3 against 19 M), and rCYP2C9*3 (2.4 compared 9.3 M) than silybin A was against HLMs. Both diastereoisomers inhibited (S)-warfarin 7-hydroxylation in a mixed-type inhibition model (Ki values of 4.8 and 10 M for silybin B and silybin A, respectively), using a matrix of five substrate (1–15 M) and six inhibitors (1–80 M) concentrations and HLMs. These findings, along with the high systemic silibinin concentrations (>5–75 M) obtained in a phase

I investigation including prostate cancer patients, prompted clinical evaluation of a possible warfarin-milk thistle interaction [89].

This study examines the literature on the novel and emerging applications of silybin (a chemically defined compound) and silymarin, which was primarily published in this millennium (flavonoid complex from *Silybum marianum* - milk thistle seeds). Anticancer and cancer-protective properties, as well as hypocholesterolemic activity, have been discovered in these substances, which were primarily utilized as hepatoprotectants. Prostate, lungs, CNS, kidneys, pancreas, and skin protection were all demonstrated to have these effects. Silybin's cytoprotective activity is mediated by its antioxidative and radical-scavenging capabilities, but it also has additional functions based on the unique receptor contact. These were investigated at the molecular level, and silybin was found to modulate a number of cell-signaling pathways, including NF-B, suppression of EGFR-MAPK/ERK1/2 signaling, activity on Rb and E2F proteins, and IGF-receptor signaling. Other notable findings include silybin's proapoptotic action in pre-and/or cancerogenic cells, as well as its anti-angiogenic activity. The discovery of silybin and some of its novel derivatives inhibiting and modulating drug transporters, P-glycoprotein, estrogenic receptors, and nuclear receptors adds to our understanding of silybin's molecular function. Also discussed is the use of silymarin in veterinary medicine. The crucial importance of optically active silybin, particularly in receptor studies, has been established by recent research employing optically pure silybin diastereomers. The exponential g illustrates how crucial silymarin and its constituents are to medicine. The rise in publications on the topic, with more than 800 papers published in the previous five years, amply demonstrates the significance of silymarin and its components in medicine [131].

Silybum marianum, a herbal medicine, is often used by those with chronic hepatitis C (CHC). This pilot study examined the efficacy and safety of S. marianum for treating CHC patients by lowering blood levels of hepatitis C virus (HCV) RNA, alanine aminotransferase, and general health. On 24 CHC patients, a randomised, double-blind, placebo-controlled crossover experiment was carried out. *S. marianum* (600 or 1200 mg/day) was administered for 12 weeks before being discontinued for 4 weeks. Testing for biochemical, virological, psychological, and quality-of-life factors was done at the beginning of each treatment period. Biochemical tests were repeated monthly, and tests for HCV RNA titer, psychological, and quality-of-life factors were repeated at the conclusion of each treatment period. Seventeen patients completed the study. Subjects receiving S. marianum did not vary significantly from control subjects in terms of their HCV RNA titers, serum ALT levels, or Short Form-36 scores. Mean State-Trait, Anxiety Inventory State-Anxiety ratings on *S. marianum* did not substantially deviate from the initial values. *S. marianum* and placebo both caused identical side effects. Although *S. marianum* is well tolerated in CHC patients, it has a significant impact on blood HCV RNA, alanine aminotransferase levels, quality of life, and psychological well-being [79].

Chemoprevention selective estrogen response modulators, such as raloxifene, are recommended for women who are at a high risk of breast cancer. To augment approved treatments, patients frequently turn to complementary and alternative treatment approaches, such as herbal products. Women who are using raloxifene, which largely undergoes first-pass glucuronidation in the colon, may take milk thistle formulations like silibinin and silymarin, which are popular herbal treatments. With IC50 values of less than 10 M, flavonolignans, a major component of milk thistle, have been shown to be potent inhibitors of intestinal UDP-glucuronosyl transferases (UGTs). Combining milk thistle formulations may result in negative raloxifene interactions. Using human intestinal microsomes and human embryonic kidney cell lysates overexpressing the highly expressed UGT1A10, UGT1A1, and UGT1A8, the researchers examined the inhibitory effects of several milk thistle ingredients on the intestinal glucuronidation of raloxifene. In all enzyme systems, the flavonolignans silybin B and silybin A prevented 6-glucuronidation and the effects of raloxifene 4. Kis (human intestinal microsomes, 27–66 M; UGT1A1, 3.2–8.3 M; UGT1A8, 19–73 M; and UGT1A10, 65–120 M) covered the reported intestinal tissue concentrations (20–310 M), allowing a mechanistic static model to forecast the likelihood of a clinical interaction. It was projected that silibinin and silymarin would raise raloxifene systemic exposure by four- to five-fold, indicating a considerable interaction risk that necessitates further research. This thorough examination of a potential interaction between a popular herbal supplement and a chemopreventive medication highlights how important it is to grasp natural product-drug interactions in the context of cancer [132].

The most studied herb in the therapy of liver illness is *Silybum marianum*, also known as milk thistle (MT). Silymarin, one of three isomer flavonolignans that make up MT, is the active component (silybin, silydianin, and silychristine). The most biologically active component of silymarin is silybin, which makes up 50–70% of it. By reducing free radical production and lipid oxidation, silymarin functions as an antioxidant. Silymarin has been used to treat viral hepatitis, toxin-induced liver disorders and alcoholic liver disease, and it may be helpful in the battle against hepatocellular carcinoma [133].

References

1. Barl, B., Loewen, D., & Svendsen, E. (1996). *Saskatchewan herb database*. Department of Horticulture Science, University of Saskatchewan.
2. Sewell, R., & Rafieian-Kopaei, M. (2014). The history and ups and downs of herbal medicine usage. *Journal of Herbmed Pharmacology, 3*(1), 1–3.
3. Rafieian-Kopaie, M., & Nasri, H. (2012). Silymarin and diabetic nephropathy. *Journal of Renal Injury Prevention, 1*(1), 3–5.
4. Libster, M. (2002). *Delmar's integrative herb guide for nurses*. Delmar/Thomson Learning.
5. Pharmacopoeia, B. (2009). British Pharmacopoeia herbal drugs and herbal drug preparations. *Oak Bark, 3*, 7203.
6. Das, S. K., Mukherjee, S., & Vasudevan, D. (2008). Medicinal properties of milk thistle with special reference to silymarin – An overview. *Natural Product Radiance, 7*(2), 182–192.
7. Groves, R., & Kaye, P. (1989). Germination and phenology of seven introduced thistle species in southern Australia. *Australian Journal of Botany, 37*(4), 351–359.

8. Farooqi, A. A., & Sreeramu, B. (2004). *Cultivation of medicinal and aromatic crops*. Universities Press.
9. Qavami, N., et al. (2013). A review on pharmacological, cultivation and biotechnology aspects of milk thistle (Silybum marianum (L.) Gaertn.). *Journal of Medicinal Plants, 3*(47), 19–37.
10. Wallace, S. N., Carrier, D. J., & Clausen, E. C. (2003). Extraction of nutraceuticals from milk thistle. In *Biotechnology for fuels and chemicals* (pp. 891–903). Springer.
11. Wallace, S. N., Carrier, D. J., & Clausen, E. C. (2005). Batch solvent extraction of flavanolignans from milk thistle (Silybum marianum L. Gaertner). *Phytochemical Analysis: An International Journal of Plant Chemical and Biochemical Techniques, 16*(1), 7–16.
12. Campodónico, A., et al. (2001). Dissolution test for silymarin tablets and capsules. *Drug Development and Industrial Pharmacy, 27*(3), 261–265.
13. Duan, L., Carrier, D. J., & Clausen, E. C. (2004). *Silymarin extraction from milk thistle using hot water*. Proceedings of the Twenty-Fifth Symposium on Biotechnology for Fuels and Chemicals Held May 4–7, 2003, Springer.
14. Abrol, S., Trehan, A., & Katare, O. (2004). Formulation, characterization, and in vitro evaluation of silymarin-loaded lipid microspheres. *Drug Delivery, 11*(3), 185–191.
15. Morazzoni, P., & Bombardelli, E. (1995). Silybum marianum (Carduus marianus). *Fitoterapia (Milano), 66*(1), 3–42.
16. Scott Luper, N. (1998). A review of plants used in the treatment of liver disease: Part 1. *Alternative Medicine Review, 3*(6), 410–421.
17. Schuppan, D., et al. (1999). Herbal products for liver diseases: A therapeutic challenge for the new millennium. *Hepatology, 30*(4), 1099–1104.
18. Culpeper, N., & Siderits, R. (1952). *The English physitian: Or an astrologo-physical discourse of the vulgar herbs of this nation*. Benefit of the Commonwealth of England.
19. Giese, L. A. (2001). Complementary healthcare practices. *Gastroenterology Nursing: the Official Journal of the Society of Gastroenterology Nurses and Associates, 24*(1), 38–40.
20. Wu, C.-H., Huang, S.-M., & Yen, G.-C. (2011). Silymarin: A novel antioxidant with antiglycation and antiinflammatory properties in vitro and in vivo. *Antioxidants & Redox Signaling, 14*(3), 353–366.
21. Brodniewicz, T., & Grynkiewicz, G. (2012). Plant phenolics as drug leads n what is missing? *Acta Poloniae Pharmaceutica, 69*, 1203.
22. Manna, S. K., et al. (1999). Silymarin suppresses TNF-induced activation of NF-κB, c-Jun N-terminal kinase, and apoptosis. *The Journal of Immunology, 163*(12), 6800–6809.
23. Basaga, H., et al. (1997). Free radical scavenging and antioxidative properties of 'silibin' complexes on microsomal lipid peroxidation. *Cell Biochemistry and Function: Cellular Biochemistry and Its Modulation by Active Agents or Disease, 15*(1), 27–33.
24. Salomone, F., Godos, J., & Zelber-Sagi, S. (2016). Natural antioxidants for non-alcoholic fatty liver disease: Molecular targets and clinical perspectives. *Liver International, 36*(1), 5–20.
25. Surai, P. F. (2015). Silymarin as a natural antioxidant: An overview of the current evidence and perspectives. *Antioxidants, 4*(1), 204–247.
26. Milosevic, N., et al. (2014). Phytotherapy and NAFLD-from goals and challenges to clinical practice. *Reviews on Recent Clinical Trials, 9*(3), 195–203.
27. Stiuso, P., et al. (2014). Serum oxidative stress markers and lipidomic profile to detect NASH patients responsive to an antioxidant treatment: A pilot study. *Oxidative Medicine and Cellular Longevity, 2014*, 1–8.
28. Darvishi-Khezri, H., et al. (2018). Iron-chelating effect of silymarin in patients with β-thalassemia major: A crossover randomised control trial. *Phytotherapy Research, 32*(3), 496–503.
29. Abenavoli, L., et al. (2010). Milk thistle in liver diseases: Past, present, future. *Phytotherapy Research, 24*(10), 1423–1432.
30. Esmaeil, N., et al. (2017). Silymarin impacts on immune system as an immunomodulator: One key for many locks. *International Immunopharmacology, 50*, 194–201.

31. Milić, N., et al. (2013). New therapeutic potentials of milk thistle (Silybum marianum). *Natural Product Communications, 8*(12), 1934578X1300801236.
32. Alaca, N., Ozbeyli, D., Uslu, S., Ahin, H. H., Yigitturk, G., Kurtel, H., Oktem, G., & Qaglayan Yegen, B. (2017). Treatment with milk thistle extract (Silybum marianum), ursodeoxycholic acid, or their combination attenuates cholestatic liver injury in rats: Role of the hepatic stem cells. *The Turkish Journal of Gastroenterology, 28*(6), 476–484.
33. Clichici, S., et al. (2015). Silymarin inhibits the progression of fibrosis in the early stages of liver injury in CCl4-treated rats. *Journal of Medicinal Food, 18*(3), 290–298.
34. Kim, S. H., et al. (2016). Silymarin prevents restraint stress-induced acute liver injury by ameliorating oxidative stress and reducing inflammatory response. *Molecules, 21*(4), 443.
35. Pais, P., & D'Amato, M. (2014). In vivo efficacy study of milk thistle extract (ETHIS-094™) in STAM™ model of nonalcoholic steatohepatitis. *Drugs in R&D, 14*(4), 291–299.
36. Raghu, R., & Karthikeyan, S. (2016). Zidovudine and isoniazid induced liver toxicity and oxidative stress: Evaluation of mitigating properties of silibinin. *Environmental Toxicology and Pharmacology, 46*, 217–226.
37. Aghazadeh, S., et al. (2011). Anti-apoptotic and anti-inflammatory effects of Silybum marianum in treatment of experimental steatohepatitis. *Experimental and Toxicologic Pathology, 63*(6), 569–574.
38. De La Puerta, R. et al. (1996). Effect of silymarin on different acute inflammation models and on leukocyte migration. *Journal of Pharmacy and Pharmacology, 48*(9), 968–970.
39. Seki, E., & Schwabe, R. F. (2015). Hepatic inflammation and fibrosis: Functional links and key pathways. *Hepatology, 61*(3), 1066–1079.
40. Clichici, S., et al. (2016). Beneficial effects of silymarin after the discontinuation of CCl4-induced liver fibrosis. *Journal of Medicinal Food, 19*(8), 789–797.
41. Li, C.-C., et al. (2012). Identification of novel mechanisms of silymarin on the carbon tetrachloride-induced liver fibrosis in mice by nuclear factor-κB bioluminescent imaging-guided transcriptomic analysis. *Food and Chemical Toxicology, 50*(5), 1568–1575.
42. Sokar, S. S., et al. (2017). Combination of Sitagliptin and Silymarin ameliorates liver fibrosis induced by carbon tetrachloride in rats. *Biomedicine & Pharmacotherapy, 89*, 98–107.
43. Marin, V., et al. (2017). Effects of oral administration of silymarin in a juvenile murine model of non-alcoholic steatohepatitis. *Nutrients, 9*(9), 1006.
44. Younis, N., Shaheen, M. A., & Abdallah, M. H. (2016). Silymarin-loaded Eudragit® RS100 nanoparticles improved the ability of silymarin to resolve hepatic fibrosis in bile duct ligated rats. *Biomedicine & Pharmacotherapy, 81*, 93–103.
45. Zi, X., & Agarwal, R. (1999). Silibinin decreases prostate-specific antigen with cell growth inhibition via G1 arrest, leading to differentiation of prostate carcinoma cells: Implications for prostate cancer intervention. *Proceedings of the National Academy of Sciences, 96*(13), 7490–7495.
46. Thelen, P., et al. (2004). Inhibition of telomerase activity and secretion of prostate specific antigen by silibinin in prostate cancer cells. *The Journal of Urology, 171*(5), 1934–1938.
47. Singh, R. P., et al. (2002). Silymarin inhibits growth and causes regression of established skin tumors in SENCAR mice via modulation of mitogen-activated protein kinases and induction of apoptosis. *Carcinogenesis, 23*(3), 499–510.
48. Hanje, A. J., et al. (2006). The use of selected nutrition supplements and complementary and alternative medicine in liver disease. *Nutrition in Clinical Practice, 21*(3), 255–272.
49. Yang, Z., et al. (2014). Effects and tolerance of silymarin (milk thistle) in chronic hepatitis C virus infection patients: A meta-analysis of randomized controlled trials. *BioMed Research International, 2014*, 1–9.
50. Yin, F., et al. (2011). Silibinin: A novel inhibitor of Aβ aggregation. *Neurochemistry International, 58*(3), 399–403.
51. Muriel, P., & Mourelle, M. (1990). Prevention by silymarin of membrane alterations in acute CCI4 liver damage. *Journal of Applied Toxicology, 10*(4), 275–279.

52. Heidarian, E., & Rafieian-Kopaei, M. (2012). Effect of silymarin on liver phoshpatidate phosphohydrolase in hyperlipidemic rats. *Bioscience Research, 9*(2), 59–67.
53. Lettéron, P., et al. (1990). Mechanism for the protective effects of silymarin against carbon tetrachloride-induced lipid peroxidation and hepatotoxicity in mice: Evidence that silymarin acts both as an inhibitor of metabolic activation and as a chain-breaking antioxidant. *Biochemical Pharmacology, 39*(12), 2027–2034.
54. Ramakrishnan, G., et al. (2008). Silymarin downregulates COX-2 expression and attenuates hyperlipidemia during NDEA-induced rat hepatocellular carcinoma. *Molecular and Cellular Biochemistry, 313*, 53–61.
55. Desplaces, A., et al. (1975). The effects of silymarin on experimental phalloidine poisoning. *Arzneimittel-Forschung, 25*(1), 89–96.
56. Muriel, P., et al. (1992). Silymarin protects against paracetamol-induced lipid peroxidation and liver damage. *Journal of Applied Toxicology, 12*(6), 439–442.
57. Wang, M., et al. (1996). Hepatoprotective properties of Silybum marianum herbal preparation on ethanol-induced liver damage. *Fitoterapia (Milano), 67*(2), 166–171.
58. Valenzuela, A., et al. (1985). Silymarin protection against hepatic lipid peroxidation induced by acute ethanol intoxication in the rat. *Biochemical Pharmacology, 34*(12), 2209–2212.
59. Nasri, H., et al. (2014). Turmeric: A spice with multifunctional medicinal properties. *Journal of HerbMed Pharmacology, 3*(1), 5–8.
60. Mills, S., & Hutchins, R. (2003). *European scientific cooperative on Phytotherapy (ESCOP) monographs* (pp. 345–350). Thieme Publishers.
61. Salmi, H., & Sarna, S. (1982). Effect of silymarin on chemical, functional, and morphological alterations of the liver: A double-blind controlled study. *Scandinavian Journal of Gastroenterology, 17*(4), 517–521.
62. Škottová, N., & Krečman, V. (1998). Silymarin as a potential hypocholesterolaemic drug. *Physiological Research, 47*(1), 1–7.
63. Feher, J., et al. (1989). Liver-protective action of silymarin therapy in chronic alcoholic liver diseases. *Orvosi Hetilap, 130*(51), 2723–2727.
64. Benda, L., et al. (1980). The influence of therapy with silymarin on the survival rate of patients with liver cirrhosis (author's transl). *Wiener Klinische Wochenschrift, 92*(19), 678–683.
65. Bunout, D., et al. (1992). Controlled study of the effect of silymarin on alcoholic liver disease. *Revista Médica de Chile, 120*(12), 1370–1375.
66. Platt, D., & Schnorr, B. (1971). Biochemische und elektronenoptische untersuchungen zur frage der beeinflussbarkeit der aethanolschadigung der rattenleber durch silymarin. *Arzneimittel-Forschung, 21*, 1206–1208.
67. Schriewer, H., & Weinhold, F. (1979). The influence of silybin from Silybum marianum (L.) Gaertn. on in vitro phosphatidyl choline biosynthesis in rat livers. *Arzneimittel-Forschung, 29*(5), 791–792.
68. Magliulo, E., Gagliardi, B., & Fiori, G. (1978). Results of a double blind study on the effect of silymarin in the treatment of acute viral hepatitis, carried out at two medical centres (author's transl). *Medizinische Klinik, 73*(28–29), 1060–1065.
69. Bode, J. C., Schmidt, U., & Dürr, H. (1977). Silymarin for the treatment of acute viral hepatitis? Report of a controlled trial (author's transl). *Medizinische Klinik, 72*(12), 513–518.
70. Sonnenbichler, J. (1986). Biochemical effects of the flavonolignane silibinin on mRNA, protein, and RNA synthesis in rat livers. *Progress in Clinical and Biological Research, 213*, 319–331.
71. Muriel, P., & Mourelle, M. (1990). The role of membrane composition in ATPase activities of cirrhotic rat liver: Effect of silymarin. *Journal of Applied Toxicology, 10*(4), 281–284.
72. Videla, L. A., & Valenzuela, A. (1982). Alcohol ingestion, liver glutathione and lipoperoxidation: Metabolic interrelations and pathological implications. *Life Sciences, 31*(22), 2395–2407.
73. Mardani, S., et al. (2013). Herbal medicine and diabetic kidney disease. *Journal of Nephropharmacology, 2*(1), 1–2.

74. Barve, A., et al. (2008). Treatment of alcoholic liver disease. *Annals of Hepatology, 7*(1), 5–15.
75. Sonnenbichler, J., et al. (1986). Silibinin'in kısmen hepatektomi uygulanmış sıçan karaciğerlerinde DNA sentezi üzerindeki uyarıcı etkisi: hepatom ve diğer malign hücre hatları. *Biochemical Pharmacology, 35*(3), 538–541.
76. Yormaz, S., et al. (2012). The comparison of the effects of hepatic regeneration after partial hepatectomy, silybum marinaum, propofol, N-acetylcysteine and vitamin E on liver. *Bratislava Medical Journal-Bratislavske Lekarske Listy, 113*(3), 145–151.
77. Bousserouel, S., et al. (2012). Silibinin inhibits tumor growth in a murine orthotopic hepatocarcinoma model and activates the TRAIL apoptotic signaling pathway. *Anticancer Research, 32*(7), 2455–2462.
78. Pradhan, S., & Girish, C. (2006). Hepatoprotective herbal drug, silymarin from experimental pharmacology to clinical medicine. *Indian Journal of Medical Research, 124*(5), 491–504.
79. Gordon, A., et al. (2006). Effects of Silybum marianum on serum hepatitis C virus RNA, alanine aminotransferase levels and well-being in patients with chronic hepatitis C. *Journal of Gastroenterology and Hepatology, 21*(1), 275–280.
80. Jiang, C., Agarwal, R., & Lü, J. (2000). Anti-angiogenic potential of a cancer chemopreventive flavonoid antioxidant, silymarin: Inhibition of key attributes of vascular endothelial cells and angiogenic cytokine secretion by cancer epithelial cells. *Biochemical and Biophysical Research Communications, 276*(1), 371–378.
81. Da Casa, P. D. A. A, Massa, P., & Definizione, F. E. (n.d.). Alimentazione Uomo: La Nutrizione per Alimentare il Testosterone
82. Ferenci, P., et al. (1989). Randomized controlled trial of silymarin treatment in patients with cirrhosis of the liver. *Journal of Hepatology, 9*(1), 105–113.
83. Parés, A., et al. (1998). Effects of silymarin in alcoholic patients with cirrhosis of the liver: Results of a controlled, double-blind, randomized and multicenter trial. *Journal of Hepatology, 28*(4), 615–621.
84. Buzzelli, G., et al. (1993). A pilot study on the liver protective effect of silybin-phosphatidylcholine complex (IdB1016) in chronic active hepatitis. *International Journal of Clinical Pharmacology, Therapy, and Toxicology, 31*(9), 456–460.
85. Martines, G., et al. (1979). Silymarin in pregnancy and during hormonal contraceptive treatment. Blood chemistry and ultrastructural findings in the experimental model. *Archivio per le Scienze Mediche, 136*(3), 443–454.
86. Zi, X., Mukhtar, H., & Agarwal, R. (1997). Novel cancer chemopreventive effects of a flavonoid antioxidant silymarin: Inhibition of mRNA expression of an endogenous tumor promoter TNFα. *Biochemical and Biophysical Research Communications, 239*(1), 334–339.
87. Rafieian-Kopaei, M., Baradaran, A., & Rafieian, M. (2013). Oxidative stress and the paradoxical effects of antioxidants. *Journal of Research in Medical Sciences, 18*(7), 628.
88. Zima, T., et al. (1998). The effect of silibinin on experimental cyclosporine nephrotoxicity. *Renal Failure, 20*(3), 471–479.
89. Brantley, S. J., et al. (2010). Two flavonolignans from milk thistle (Silybum marianum) inhibit CYP2C9-mediated warfarin metabolism at clinically achievable concentrations. *Journal of Pharmacology and Experimental Therapeutics, 332*(3), 1081–1087.
90. Vessal, G., et al. (2010). Silymarin and milk thistle extract may prevent the progression of diabetic nephropathy in streptozotocin-induced diabetic rats. *Renal Failure, 32*(6), 733–739.
91. Soto, C., et al. (2010). Effect of silymarin on kidneys of rats suffering from alloxan-induced diabetes mellitus. *Phytomedicine, 17*(14), 1090–1094.
92. Nasri, H., & Rafieian-Kopaei, M. (2014). Medicinal plants and antioxidants: Why they are not always beneficial? *Iranian Journal of Public Health, 43*(2), 255–257.
93. Wang, M. J., et al. (2002). Silymarin protects dopaminergic neurons against lipopolysaccharide-induced neurotoxicity by inhibiting microglia activation. *European Journal of Neuroscience, 16*(11), 2103–2112.
94. Zhang, J., Mao, X., & Zhou, Y. (1993). Effects of silybin on red blood cell sorbitol and nerve conduction velocity in diabetic patients. *Zhongguo Zhong Xi Yi Jie He Za Zhi Zhongguo*

Zhongxiyi Jiehe Zazhi= Chinese Journal of Integrated Traditional and Western Medicine, 13(12), 725–6708.
95. Roghani, M., et al. (2013). Protective effect of silymarin on learning and memory deficiency in streptozotocin-diabetic rats. *Journal of Gorgan University of Medical Sciences, 15*(2).
96. Zou, C.-G., Agar, N. S., & Jones, G. L. (2001). Oxidative insult to human red blood cells induced by free radical initiator AAPH and its inhibition by a commercial antioxidant mixture. *Life Sciences, 69*(1), 75–86.
97. Soto, C. P., et al. (1998). Prevention of alloxan-induced diabetes mellitus in the rat by silymarin. *Comparative Biochemistry and Physiology Part C: Pharmacology, Toxicology and Endocrinology, 119*(2), 125–129.
98. Láng, I., et al. (1990). Hepatoprotective and immunological effects of antioxidant drugs. *The Tokai Journal of Experimental and Clinical Medicine, 15*(2–3), 123–127.
99. Koch, H., Bachner, J., & Löffler, E. (1985). Silymarin: Potent inhibitor of cyclic AMP phosphodiesterase. *Methods and Findings in Experimental and Clinical Pharmacology, 7*(8), 409–413.
100. Mulrow, C., et al. (2000). Milk thistle: Effects on liver disease and cirrhosis and clinical adverse effects: Summary. In *AHRQ evidence report summaries*. Agency for Healthcare Research and Quality (US).
101. Brandenburger, W. (1985). *Parasitische pilze an gefässpflanzen in Europa*. G. Fischer.
102. Moscow, D., & Lindow, S. (1989). Infection of milk thistle(Silybum marianum) leaves by Septoria silybi. *Phytopathology, 77*(10), 1085–1090.
103. Zechini D'aulerio, A., Zambonelli, A., & Morara, M. (1991). Ulteriori indagini sulle malattie crittogamiche di piante officinali in Emilia-Romagna. *Informatore fitopatologico, 41*(12), 45–52.
104. Margina, A., & Zheljazkov, V. (1996). Leaf spot on milk thistle (Silybum marianum L. Gaerth.) in Bulgaria, *2*(1), 254–256.
105. Khan, M. A., Blackshaw, R. E., & Marwat, K. B. (2009). Biology of milk thistle (Silybum marianum) and the management options for growers in North-Western Pakistan. *Weed Biology and Management, 9*(2), 99–105.
106. Cwalina-Ambroziak, B., et al. (2012). The effect of mineral fertilization on achenes yield and fungal communities isolated from the stems of milk thistle Silybum marianum (L.) Gaertner. *Acta Scientiarum Polonorum Hortorum Cultus, 11*(4), 157–168.
107. Ondřej, M., Odstrčilová, L., & Dostálová, R. (2006). Phomopsis pisi – A new species causing pea stem canker. *Plant Protection Science, 42*(3), 95–98.
108. Gannibal, P. B. (2010). Taxonomic studies of Alternaria from Russia: New species on Asteraceae. *Mycotaxon, 114*, 109.
109. Kovacikova, E., & Kubinek, J. (1986). *The wilting of milk thistle Silybum marianum*. Ochrana Rostlin-UVTIZ (Czechoslovakia).
110. Sacristán, S., Fraile, A., & García-Arenal, F. (2004). Population dynamics of cucumber mosaic virus in melon crops and in weeds in Central Spain. *Phytopathology, 94*(9), 992–998.
111. Chatzivassiliou, E., et al. (2001). Differential tomato spotted wilt virus infection and vector species infestation of weeds in greenhouses and tobacco fields. *Plant Disease, 85*, 40–46.
112. Souissi, T., Berner, D., & Smallwood, E. (2005). First report of smut caused by Microbotryum silybum on ivory thistle. *Plant Disease, 89*(11), 1242–1242.
113. Vanky, K., & Berner, D. (2003). Microbotryum silybum sp. nov.(Microbotryales). *Mycotaxon, 85*, 307–311.
114. Abdel-Moniem, A. (2002). The Seed-Head Weevil, Larinus Latus Herbst (Coleoptera: Curculionidae) as a New Record in Egypt on the Milk Thistle, Silybum Marianum (L.) (Asteraceae: Compositaea). *Archives of Phytopathology and Plant Protection, 35*(2), 157–160.
115. Meissner, R., & Mulder, C. (1974). *Herbicidal control of volunteer Silybum marianum in wheat*. Agroplantae.

116. Andrzejewska, J., Sadowska, K., & Mielcarek, S. (2011). Effect of sowing date and rate on the yield and flavonolignan content of the fruits of milk thistle (Silybum marianum L. Gaertn.) grown on light soil in a moderate climate. *Industrial Crops and Products, 33*(2), 462–468.
117. Omer, E., et al. (1993). Effect of spacing and fertilization on the yield and active constituents of milk thistle, Silybum marianum. *Journal of Herbs, Spices & Medicinal Plants, 1*(4), 17–23.
118. Geneva, M., et al. (2008). Improvement of milk thistle (Silybum marianum L.) seed yield and quality with foliar fertilization and growth effector MD 148/II. *General and Applied Plant Physiology, 34*(3), 309–318.
119. Valaii, L., et al. (2015). *Effect of organic manure and bio-fertilizer on growth traits and quantity yield in milk thistle*. Silybum marianum L.
120. Wallace, S., et al. (2008). Milk thistle extracts inhibit the oxidation of low-density lipoprotein (LDL) and subsequent scavenger receptor-dependent monocyte adhesion. *Journal of Agricultural and Food Chemistry, 56*(11), 3966–3972.
121. Carrier, D. J., et al. (2003). Milk thistle, Silybum marianum (L.) Gaertn., flower head development and associated marker compound profile. *Journal of Herbs, Spices & Medicinal Plants, 10*(1), 65–74.
122. Montemurro, P., Fracchiolla, M., & Lonigro, A. (2007). Effects of some environmental factors on seed germination and spreading potentials of Silybum marianum Gaertner. *Italian Journal of Agronomy, 2*(3), 315–320.
123. Abenavoli, L., et al. (2010). *Recent progress in medicinal plants*. Spllc Press.
124. Mel'nktov, T. (1983). Morphological-biological characteristics of Silybum marianum seeds as sowing material. *Khimiko-Farmatsevticheskii Zhurnal, 17*(8), 958–963.
125. Ghavami, N., & Ramin, A. (2007). Salinity and temperature effects on seed germination of milk thistle. *Communications in Soil Science and Plant Analysis, 38*(19–20), 2681–2691.
126. Dodd, J. (1989). Phenology and seed production of variegated thistle, Silybum marianum (L.) Gaertn., in Australia in relation to mechanical and biological control. *Weed Research, 29*(4), 255–263.
127. Zheljazkov, V. D., Zhalnov, I., & Nedkov, N. K. (2006). Herbicides for weed control in blessed thistle (Silybum marianum). *Weed Technology, 20*(4), 1030–1034.
128. Young, J., Evans, R., & Hawkes, R. (1978). Milk thistle (Silybum marianum) seed germination. *Weed Science, 26*(4), 395–398.
129. Siegel, A. B., & Stebbing, J. (2013). Milk thistle: Early seeds of potential. *The Lancet Oncology, 14*(10) 929–930.
130. Agarwal, R., et al. (2006). Anticancer potential of silymarin: From bench to bed side. *Anticancer Research, 26*(6B), 4457–4498.
131. Silverman, A. L., Kumar, A., & Borum, M. L. (2018). Re:"herbal use during breastfeeding" by Anderson (breastfeed med 2017; 12 (9): 507–509). *Breastfeeding Medicine, 13*(4), 301–301.
132. Gufford, B. T., et al. (2015). Milk thistle constituents inhibit raloxifene intestinal glucuronidation: A potential clinically relevant natural product–drug interaction. *Drug Metabolism and Disposition, 43*(9), 1353–1359.
133. Abenavoli, L., & Milic, N. (2017). Silymarin for liver disease. In *Liver pathophysiology* (pp. 621–631). Elsevier.

Chapter 22
Guggul

Khalid Sultan, Shagufta Perveen, Sara Zafar, Abida Parveen, Naeem Iqbal, and Arwa A. AL-Huqail

22.1 Introduction

Family: Burseraceae
English name: Indian Bedelium tree
India name: Gugguluh, Mahisaksah (Sansakrit), Guggul, Gukkulu (Tamil)
Species and Varities: *Commiphora stockciana* Engl, *C.agollochoa*, Engl, *C. berry* (Aen). Engl, *Commiphora. mukul* Hook ex stock
Distribution: India, Asia, Africa, Australia, Bangladesh, Pacific Islands Pakistan
Uses: Drugs, perfumery

The plant name from the centuries is *Commiphora Mukul* or another name *Balsamodendron mukul*, but this name was changed by the Santapau to a new name *C. roxburghii* in 1962 [1]. The guggul tree is locally known to India. The local name of the plant is Guggal and is also known as the Bedellium India tree. The scientific name of the Guggal tree is *Commiphora mukul* or *Balsomodendron mukul.* It belongs to the Burseraceae family [2].

Purseglove (1975) suggested that Commiphora spp. originated in Africa and Asia. Species of this plant were founded in areas of Asia, Africa and Madagascar. It is also present in regions of Austria as well as the Pacific islands [3]. Commiphora spp. Found in southwest Africa [4, 5], while Lisowsky, [6] identified many species of Commiphora in Zaire. The petiole structure has been used to identify several species of Commiphora in Africa [4].

K. Sultan · S. Perveen (✉) · S. Zafar · A. Parveen · N. Iqbal
Department of Botany, Government College University, Faisalabad, Pakistan

A. A. AL-Huqail
Department of Biology, College of Science, Princess Nourah bint Abdulrahman University, Riyadh, Saudi Arabia

M. Zia-Ul-Haq et al. (eds.), *Essentials of Medicinal and Aromatic Crops*,
https://doi.org/10.1007/978-3-031-35403-8_22

Commiphora spp. can be found in Pakistan (Sindh), Baluchistan [7] and on the Indian subcontinent [7]. In the areas of the Rajputana, Bednore, Khandesh, Berar, Mysore\and Bellary (Karnataka) where it may be found. There are 185 species was found in all over the world, besides these 185 species just three species of this plant were found in India. In the Indian regions of Karnataka, Gujrat and Maharashtra *Commiphora wightii* was found. In southwest India, *Commiphora berryi* is used in medicines and this present in areas of Gwalior [8].

Guggulu kalpas is a species of the guggul plant that has a high value in Ayurvedic medicines. The Guggal plant is an important part of the Ayurvedic. More than 100 species of this species have been used in Ayurvedic medicines. This plant is also used in the faculty of Unani and Siddha. The resin part of the plant is used in the medicines. The 8-year-old plants are tapped, and the resin is obtained in the winter season. During the tapping process plants may die due to the improper tapping of the plant. Due to the high value of this plant in medicine, less knowledge of the cultivation, demand on a large scale, and improper taking care make this plant endangered [9].

From this plant guggul, the oleo gum resin was obtained by cutting on the stem bark in India. This resin or gum is yellow, pale, dull green, or brown in color with a foul smell and in taste it was bitter. In the guggul plant, the oleo gum resin was found mostly in vascular tissues. Volatile oil 1.45%, resin 61% and resin 29.3% were found in a mixture of the oleo gum. Some other compounds such as polymyrecene, myrcene and dimyrecene were also combined with this resin of oleo gum. About 12% of the ethyl acetate soluble neutral part which consists of eight steroids contains ketonic mixtures. Two steroids (Z and e-guggulsterone) are which constitute about 2% of the gum resin are responsible for lowering lipid activity in guggal plants. The biological activity of the ketonic fraction found synergetic action by the nonketotic fractions. From the guggul plant, ethyl acetate is extracted and named guggulipid [2].

A study is conducted on the trade of medicinal plants in India. It was shown that the *C. wightii* is the source of oleoresin and is highly traded in the industries of Indian herbs. In industries of medicines of India, the total demand for this oleoresin is about 500–1000 tonnes yearly. Imports from Pakistan meet more than 90% of this need. Pakistan exports 504.6 tonnes of oleoresin which obtained from the *Commiphora wightii*. So it was shown that the required amount of resin in the Indian herbal industry was 500 tonnes per year from Pakistan [10].

22.2 Origin and Distribution

There are 165 species in the genus Commiphora, which can be discovered in Africa, India, Asia, Australia, the Islands in the Pacific, Pakistan, and Bangladesh. Native to Africa is the guggul plant. The guggul plant can be found primarily in the Indian states of Rajasthan and Gujarat. Rajasthan's districts of Ajmer and Jaisalmer are well-known ecosystems [11].

There are four species in India *C. stocksiana, C. mukul C. agollochoa,* and *C. berryi*, which are the genus of guggul gum's originator and were found in areas of the Karnataka, Gujrat, Tamil Nadu, Assam and Rajasthan in India [2].

22.3 Description of the Plant

The length of the plant branches is 3–4 m high which are scented and twisted. At the end of the branches, there are spines present. The bark on the oldest parts of the stem is papery and peels away in pieces. Sessile, alternating, or fascicled leaves with 1–3 foliate leaflets. The plants are dimorphic, with female flowers with staminodes on one side and bisexual and male blooms on the other. Male flowers are sessile and 3–5 mm in length. The color of the male flower is pinkish-white, or red. Flowers emerge in two or three flower clusters. The disc and the calyx are joined at the base [2].

22.4 Uses of the Guggul Plant in Medicines

22.4.1 The Activity of Hypolipidemic

The hypolipidemic effect of crude Guggulu in rabbits was reported to be extremely encouraging [12].

Chloride retention and bile acid sequestrating activity in the oleoresin fraction are two ways that an ion exchange property can be identified. Hypercholesterolaemic activity is another method [13]. It was discovered that the crude medication and its two fractions—alcohol soluble and alcohol insoluble—significantly decreased serum turbidity and cholesterol while simultaneously speeding up the coagulation and prothrombin times. In this regard, crude Guggulu and the alcohol-insoluble fraction were both marginally more potent than the alcohol-soluble fraction [14].

For two to three weeks, eight-week-old male white-leg horn chicks were administered PE fractions A (petrol-soluble), B (alkali-washed neutral part), and C (petrol-insoluble) in an atherogenic diet-induced hypercholesterolemia. All fractions cut serum cholesterol, although A is the most effective and B the least effective [15].

The PE extract's alcohol extract and two pure fractions—a terpenoid and a steroid—showed that the steroid fraction was extremely effective as a hypolipidemic drug, reducing serum cholesterol by 69.3% and the c/p ratio. The Terpenoid reduced cholesterol by 54.3%, compared to the Alcohol extract's 59.2% reduction [16].

When administered orally to Indian domestic pigs fed a conventional atherogenic diet for six weeks, the alcohol extract of Guggulu dramatically changed the lipoprotein ratio and lowered total serum cholesterol and serum beta-lipoprotein percentage [17].

The overall lipid content (i.e., total lipids, cholesterol, TG, and phospholipids) of both hepatic and aortic tissues were decreased by the steroidal component recovered from fraction A of PE extract. The reaction was dose-dependent, and 10 mg/kg produced the greatest effect [18].

The overall lipid content (i.e., total lipids, cholesterol, TG, and phospholipids) of both hepatic and aortic tissues was decreased by the steroidal component recovered from fraction A of PE extract. The reaction was dose-dependent, and the highest impact was found at 10 mg/kg [19].

Serum cholesterol levels in healthy and hyperlipidemic rats and rabbits were decreased by alcoholic extract (25–50 mg/kg orally). On normal and triton-induced hyperlipidemic rats, a resin fraction, a pure steroid, and fraction F extracted from crude extract demonstrated the hypercholesterolaemic effect [20, 21] (Table 22.1).

Animals and people with obesity and hypercholesterolemia both demonstrated the hypolipidemic action [61].

The hypolipidemic activity of guggulu and guggulipid was confirmed by a number of clinical trials [62, 63].

A highly significant decrease in mean serum cholesterol and triglyceride levels was seen in another study in groups of animals given a high-fat diet coupled with Guggulu for a month, demonstrating the herb's hypolipidemic effect. Aortic atherosclerosis brought on by a high-fat diet was also partially reversed by the administration of guggulu [64].

13. Clinical research on C. mukul demonstrated the hypolipidemic impact of the drug and the resultant alteration in lipid profile. After taking guggulu, this study found that both total cholesterol and LDL cholesterol significantly decreased [65].

14. Animal models have also been used to study the hypolipidemic action of the isomers E- and Z-guggulsterone. Rats with either triton (WR-1339)- or cholesterol-induced hyperlipidemia received guggulsterone (Z and E), and this drastically reduced their serum lipid levels [66].

22.4.2 *Effect on Platelet Aggregation and Fibrinolytic Activity*

Serotonin, adrenaline-induced platelet aggregation were fully prevented by the isolated steroid combination from Guggulu. There was no discernible difference in the efficacy of the steroid mixture and the purified guggulsterone E or Z. The inhibitory effects of clofibrate and guggulsterone E and Z were strikingly similar. This discovery is useful for treating myocardial infarction and thromboembolism [67].

2. Guggulu's impact on fibrinolysis and platelet adhesiveness in coronary heart disease has been researched. Both healthy people (group I) and people with coronary artery disease (group II) received daily doses of 1 g of guggulu fraction A (pet ether extract) for 30 days. In both healthy people and CAD patients, there was a statistically significant increase in serum fibrinolytic activity and a drop in the platelet sticky index. The management of coronary artery disease may benefit from using guggulu fraction A in light of this [68].

Table 22.1 Pharmacological activities of Mukul plant species

Species	Locations	Native names	Uses	Chemical	References
Commiphora mukul	India	Guggal, gugglu	Antibacterial, anti-inflammatory, antidiabetic, arthritis, ulcers of the mouth, fractures of bones,	Steroids, Triterpenoids	[22–33]
Commiphora caudata	India	Kizhuvam	Anti- inflamatory, analgesic	Leaves & seeds	[34, 35]
Commiphora holtziana	Africa	Hagagr of gum	Anti- parasitic	Extract of hexane	[36]
Commiphora berryi	Jordan, India, Egypt	Mor makha, Mukiluvai, Myrrh	Antioxidant, gastric antiulcer	Stem	[37–40]
Commiphora myrrha	America,China, Britain, France,	Myrrh, Mo Yao,	Analgesic, antitumor, antioxidant, pain, skin ulcers, anti-tumor, tongue ulcers, pains of teeth	Resin, essential oil	[41–47]
Commiphora opobalsamum	Saudia Arabia, China	Balessan Mo Yao	Hypertensive, antiulcer, antiproliferative, anti-inflammatory	Triterpenoids	[48, 49]
Commiphora markeri	Areas of south Africa	Zebra-bark	Control infection	triterpenoid	[50]
Commiphora africana	Nigeria Uganda	Ekadeli	Wound's healing	Resin	[51]
Commiphora harveyi	Eastern & Southern Africa	Umnumbi	Snake bite, antibacterial	Extract of leaves	[52]
Commiphora kuva vollesen	Africa	Rahan-reb	Stomach infections, gonorrhea, snake bites		[53]
Commiphora guidottii	Areas of Somalia	Habakhadi	Antimalarial, relaxing of muscles	Sesquiterpenoids	[54, 55]
Commiphora erlangeriana	Eastern Africa	Dhunkal	Poison of arrow, antitumor	As lignans	[56]
Commiphora erythraea	Regions of Ethiopia	Hagar, agarsu	Skin infection, anti- tick	Extract of hexane	[57]
Commiphora schimperi	Regions of Kenya	Osilalei	Effective as antimalarial	Methanol extract	[58]

(continued)

Table 22.1 (continued)

Species	Locations	Native names	Uses	Chemical	References
Commiphora tenuis	Ethiopian regions		Healing of wounds, antibacterial	Essential oils	[59]
Commiphora habessinica	Uganda	Ekadeli	Effective against eye pain	Resin	[60]

22.4.3 Thyroid Stimulatory Activity

Serum triiodothyronine and ratios were increased after 15 days of administration of an ethanolic Guggulu extract to female albino mice, but serum thyroxine concentrations were not significantly changed [69].

It has been established that Z-guggulsterone is the compound in guggulu that stimulates the thyroid. The uptake of iodine by the thyroid, the activity of the enzymes involved in the manufacture of thyroid hormones, and tissue oxygen uptake were significantly increased in rats after receiving isolated Z-guggulsterone, indicating that the compound has thyroid-stimulating properties [70].

22.4.4 Anti-inflammatory and Anti-arthritic Activity

1. Oleoresin was discovered to be a highly effective anti-inflammatory drug against Brownlee's formaldehyde-induced arthritis in albino rats, as compared to hydrocortisone and butazoladin [71]

With a minimal effective dose of 12.5 mg/100 g body weight, oleoresin fraction has strong anti-arthritic and anti-inflammatory properties. The monoacid and solid portions were inert and only the acidic fraction had any discernible activity [72].

On rat paw edema caused by carrageenin, the Steroidal component isolated from PE extract exhibited considerable anti-inflammatory activity [73, 74].

The outcomes of numerous investigations support Guggulu's anti-inflammatory and anti-arthritic properties [22, 71, 75–77].

On mice with adjuvant-induced air pouch granuloma, it was discovered that the 50% aqueous methanolic extract had an anti-inflammatory effect. In lipopolysaccharide activated mouse peritoneal macrophages, the methanolic extract reduced nitric oxide synthesis [23].

Guggulosomes made by triturating and bath sonicating guggul with ibuprofen were tested for their ability to reduce inflammation. Guggulosomes were demonstrated to be more effective than ibuprofen, and guggul and ibuprofen worked in concert. The study established that guggul could operate as a carrier for medications that would allow for their continuous release [78].

The efficacy of guggul extract in common osteoarthritis (OA) models has been shown in a number of animal experiments. Prior to this study, the authors had looked at guggulu's potential for treating OA in both animal and human studies. This study's objective was to evaluate guggul's efficacy in easing OA-related discomfort, stiffness, and other symptoms [79].

22.4.5 The Activity of the Antiatherosclerosis

In atherosclerotic lesions, LDL has been observed to build up and is the main cause of cholesterol buildup in human foam cells. According to data, LDL oxidation is necessary for atherogenesis, and antioxidants that stop this oxidation may either prevent or slow down atherogenesis. Effectively inhibiting in vitro LDL oxidation was guggulsterone, the lipid-lowering component of Guggulu (as discussed under antioxidant action). As a result, Guggulu is notably effective against atherogenesis due to its combination of antioxidant and lipid-lowering characteristics [27].

22.4.6 Effects on Heart

The cardioprotective effects of guggulsterone have been demonstrated. Serum levels of glutamate pyruvate transaminase and creatine phosphokinase significantly increased after isoproterenol-induced myocardial necrosis in rats. Following the depletion of glycogen, phospholipids, and cholesterol, phospholipase, xanthine oxidase, and lipid peroxides were simultaneously increased in the ischemic heart. As determined by the reversal of blood and heart biochemical markers in ischemic rats, treatment with guggulsterone at a dose of 50 mg/kg effectively prevented cardiac damage [80].

22.4.7 Antifertility Activity

Female rats were given guggulu orally in doses of 2 and 20 mg/100 g body weight. This caused the weight of the uterus, ovaries, and cervix to decrease, while the amounts of glycogen and sialic acid in these organs rose. This implied that guggulu might function as a fertility-reducing medication [81].

22.4.8 Skin Diseases

Treatment for nodulocystic acne has been reported to be successful when guggulipid is administered. Guggulipid was found to be just as effective in treating patients as tetracycline in a trial involving 21 participants. The guggulipid therapy worked better on patients with oily faces [82].

22.4.9 Antihyperglycemic Activity

In streptozotocin-induced diabetic rats, plasma glucose levels were decreased by 200 mg/kg of an alcoholic extract of C. mukul given over the course of 60 days [83].

A study examining the impact of guggulsterone extracted from C. mukul in diabetic rats produced by high-fat diet has also been published. The hypoglycemic effect was clearly demonstrated by various biochemical parameters including GTT, glycogen content, insulin release in vivo, glucose homeostatic enzymes (like glucose-6-phosphatase and hexokinase), and expression profiles of different genes involved in carbohydrate and lipid metabolism. The findings indicated that guggulsterone has effects that are both hypoglycemic and hypolipidemic, which can help treat type II diabetes [84].

22.4.10 Antimicrobial Effects

The methanolic extract of guggul gum contained an active ingredient called (1-methyl, 1-aminoethyl)-5- methyl-2- octane, which had strong antibacterial activity against Gram-positive bacteria and moderate activity against Gram-negative bacteria [85–87].

22.4.11 Effects of Cytotoxic

In the approach for the prevention and treatment of inflammation, neoplasia, and cardiovascular disease, ferulate chemicals are utilised to stop aberrant cell development and proliferation. Significant in vitro cytotoxicity was demonstrated by ethanol acetate extract. Spectral and chemical techniques were used to identify a percentage that exhibited cytotoxic activity as a combination of two ferulates with a unique structure. The 2,2-diphenyl-1-picryl hydrazyl (DPPH) radicals were likewise moderately scavenged by this fraction [88]. With an IC50 of 1 M (24-hour treatment), guggulipid treatment significantly reduced the viability of the human prostate cancer cell line LNCaP (androgen-dependent) and its androgen-independent variant (C-81) [52]. This suggests that guggulipid may play a role in the prevention of cancer and in inducing apoptosis. A normal prostate epithelial cell line is resistant to growth suppression and apoptosis induction by this phytoconstituent, according to the study's findings, but guggulsterone decreased PC-3 cells' proliferation by inducing apoptosis in culture. These findings gave preclinical and clinical testing of guggulsterone's effectiveness against prostate cancer the justification it needed to proceed [89].

In perfumes products, oleoresin gum was used on large scale. While in medicines it was also used on large scale as a fixative. It is used as a digestive, antiseptic, carminative, stomachic, and astringent in indigenous medicine products. The resin gum stimulates phagocytosis and increases the number of leucocytes in circulation. It acts as a diuretic, expectorant, and diaphoretic. It is also useful in emmenagogues and uterine stimulants. The oleoresin gum was effective in the treatment of ulcers, pyorrhea, dyspepsia, obesity, alveolar, ulcerated throat, and diseases of chronic tonsils. After the burning of this tress, the ashes inhaling was helpful in acute diseases of bronchitis, hay fever, laryngitis and chronic catarrh. This plant was also used as an ingredient in the medicine that is used in the treatment of healing ulcers [2].

Chemical constituents of this plant guggul are inorganic chemicals, amino acids, esters, steroids, carbohydrates, and diterpenoids are identified by a complete examination of this plant guggul. Cholesterol and sesamin were isolated previously. Some other compounds such as guggulsterol I, guggulsterol II, guggulsterone-Z, guggulsterol III and guggulsterone-E were also identified [90].

Kidney failure disease is caused by increasing oxidative stress and swelling in the kidneys, resulting in reduced or stop kidney function. The combination of *B. serrata* with *Curcuma longa* was found in the treatment of inflammation of chronic kidney disease by increasing the performance of the prostaglandin [91]. Furthermore, this medicine was found not dangerous or harmful and the patients can tolerate this. This function as well as boosting inflammatory cytokine levels in chronic kidney disease patients [92].

A worm parasitic disease is caused by two liver flukes such as *Fasciola gigantica* and *Fasciola hepatica*. Myrhh, a *Commiphora molmol* gum resin, has been proven to be secure, well-practiced and effective in treating this condition. The preparation of this medicine was done by mixing 3.5 and 8 parts of the volatile oil extract and resin respectively from the myrrh. By the application of the medicine to the patients, it was seen that the patients improve their condition and fewer number eggs in their stools and the symptoms of the disease were reduced. This medicine has no negative effects on the patients' [93].

Eczema, commonly known as dermatitis and psoriasis is characterized by skin irritation. Boswellia has been shown to be useful in the treatment of two diseases psoriasis and eczema. Scientists work on the Boswellia cream and indicated that the use of this cream was helpful to minimize skin problems as well as reduce the problems of erythema. In a study to check the effectiveness of the novel Bosexil which contains components of lecithin and *B. serrata* resin against the diseases of eczema and psoriasis. Results of these medicines were compared to the placebo it was shown that Bosexil was more effective than the placebo in psoriasis erythema 50% and scales 70% . Furthermore, When eczema sufferers were given the Bosexil dosage, they improved significantly, it improved their erythema by 60% and hurt 60% without causing any side effects [94].

E-Guggulsterone, Zguggulsterone, and other guggulsterone derivatives are the most commonly reported substances. Ingredients of guggul (oleo gum resin) and their percentage are present by weight 3.50% myrcene, 4.75% alpinene, 5.40% methyl chavicol, 3.5% 1,8-cineole (eucalyptol), and additional unknown substances Guggulsterone concentrations are as follows: Guggul-2% crude gum 4.5–4.5%

ethyl acetate extract 4.2–4.7%, 35–40% of a neutral subfraction of ketonic of guggulsterone E, Z were obtained from 10% of this source [95].

Usages in Medicine Guggul literally means "disease fighter." Obesity and high cholesterol are two of the most common conditions treated with it. Other uses of this gum in medicines such for stomach pain, as a liver tonic, carminative, to increase the appetite, work against inflammation, diuretic, stimulation of thyroid glands, sedative and for rheumatoid arthritis. Oleo gum resin obtained from the guggul plant was also used as a gargle for ulcerated throats and is utilized in the treatment of indolent ulcers in the form of a lotion. Guggul is a natural substance that is used as a primary or secondary ingredient in a variety of medications and treatments. Hay fever, laryngitis, and chronic bronchitis are among the conditions for which it is indicated. It's also utilized in the treatment of gout and cardiac problems. Chemical Components and Their Use Guggul's steroid content has been linked to anti-inflammatory and hypolipidemic effects. Ethanol extraction of the stem of *Commiphora wigrtii* is a chemical found which is antifungal name flavone or in past, it was named muscanone or naringenin. In microbiological assays, Muscanone was reported to be effective against *Candila albicans* [96].

Guggultetrol ferulate was obtained from the cytotoxic fraction of a guggul extract extracted with ethyl acetate. E-Guggulsterone and Z-Guggulsterone (ketonic component) have a hypolipidemic (blood lipid reducing) effect [97]. Naringenin has anti-bacterial, anti-inflammatory, and anti-viral effects in addition to preventing lipoprotein buildup [98]. Cembranoids regulate the absorption of cholesterol and fats in the gastrointestinal tract [99]. Myrrhanol, a guggul gum triterpenoid, is an anti-inflammatory and pain reliever for osteoarthritis sufferers. Alpha pinene has antifungal and antibacterial properties [100]. Eugenol (monoterpenoid) is an antioxidant that also helps tumor cells grow. It has antimicrobial properties as well [101]. Anti-inflammatory and antibacterial properties of mansumbinoic acid [102]. Alpha terpineol is an antimicrobial compound with a lot of power [103].

Beta-sitosterol suppresses cholesterol production in the body and lowers cholesterol levels. 8-cineole has anti-inflammatory and anti-nociceptive properties [104]. For anticarcinogenic activity, quercetin has the most potent inducer-impacted [105]. Diayangambin is used to minimize ear edema and has immunomodulatory and anti-inflammatory properties [106]. Ellagic acid possesses anti-mutagenic, anti-inflammatory, and cancer-preventative properties [107]. It attaches to cancer cells, immobilizing them. Although L-Arabinose has no biological function, it is an excellent sugar source [98].

22.5 Phytoconstituents of Guggul

By steam distillation, the gum resin of C. wightii produces around 0.4% of essential oil, with myrcene, dimyrcene, and polymyrcene as its main constituents. The oil also contains eugenol, d-limonene, -pinene, linalool, cineole, -terpineol, d—phellandrene, methylheptanone, bornyl acetate, geraniol, and a few more unknown substances [108].

22.5.1 Diterpenoids

Cembrene-A, cembrene, and other cembrenoids are among the diterpenoid components of guggulu [109]. One of the most fundamental tetraenes, cmembrene-A is produced by cycling C1 to C14 of geranylgeranyl pyrophosphate. A novel cembrane alcohol called mukulol (allylcembrol) was discovered in Guggulu's aerial parts and resin [110, 111]. Through spectrum analysis and a slight dehydration that produced cembrene, the structure of the allylcembrol was determined. Isocembrol and 4-epiisocembro are two more identified cembrane-type diterpenes [22].

22.5.2 Sesquiterpenoids

According to reports, cadinene and bicyclic sesquiterpene are present in the gum resin of C. wightii [108].

22.5.3 Triterpenoids

From the gum resin, myrrhanol A, B, and C, myrrhanone A, myrrhanone B, myrrhanone A acetate, commipherol, commipherin, and epimansumbinol, an octanordammarane triterpenoid, have been identified. Mansumbinone and mansumbinoic acid have been named as the two additional triterpenoidal components that have been isolated [40, 76–78]. Myrrhanol A's complete stereochemistry was [23, 112–114]. The (3S, 5S, 8R, 9R, 10S)-3, 8, 30-trihydroxypolypoda-13E, 17E, 21E-triene was shown to be the absolute stereochemistry of myrrhanol A. Myrrhanol B is the 30-oic acid form of myrhanol A with a different C-5 stereostructure (5R as opposed to 5S in myrhanol A). The 3-keto analogues of myrhanols A and B are called myrrhanone A and B, respectively [23]. A derivative of myrrhanone, (13E, 17E, 21E)-8-hydroxypolypoda-13, 17, 21-trien-3-one, and a derivative of myrrhanol, (13E, 17E, 21E), Additionally, polypoda-13, 17, 21-trien-3, and 18-diol have been isolated [22].

Three new and recently isolated steroids are guggulsterone-M, dihydro guggulsterone-M and guggulsterone-Y28. The steroidal constituents have been related to hypolipidemic and anti-inflammatory activities of the drug [22].

Flavonoids a novel antifungal flavone called muscadine and the well-known naringenin were produced from an ethanolic extract of the trunk of C. wightii. A microbiological sensitivity experiment revealed that muscadine was effective against C. albicans. According to research conducted on C. wightii flowers, the main flavonoid compounds include quercetin, quercetin-3-O-L-arabinose, quercetin-3-O-Dglucuronide, quercetin-3-O-Dgalactoside, quercetin-3-O-L-rhamnoside, and pelargonidin-3, 5, di-O-glucoside [79].

22.5.4 Guggultetrols

The saponified gum resin yielded a crystalline substance that was isolated and identified as a mixture of octadecan-1,2,3,4-tetrol, nonadecan-1,2,3,4-tetrol, and eicosan-1,2,3,4-tetrol with trace amounts of other components, possibly lower (C-16 and C-17) and higher (C21 and C-22) homologous tetrols. These substances make up the guggu tetrols, a novel class of lipids that occur naturally. They are long-chain linear aliphatic tetrols with hydroxyl functions at the locations of C-1, C-2, C3, and C-4. The pure forms of guggultetrol-18 and guggultetrol-20 were produced through derivatization and preparative GLC. The cytotoxic effects of the medicine were discovered to be caused by a combination of two ferulates with an unique skeleton. They were separated from the cytotoxic component of the guggulu ethyl acetate extract. Based on homologous long-chain tetrols and acid, it was determined to be an ester combination [41, 115].

22.5.5 Lignans

Two lignans, sesamin [116] and diayangambin [24] have been reported from guggulu. The compound 5,5-tetrahydro-1H,3H-furo[3,4-c] Methanolic extract of Guggulu has been reported to contain the compound furan-1,4-diylbis [7- (methoxy)-1,3-benzodioxole] [22].

22.5.6 Sugars

L-arabinose, D-galactose, L-fructose (traces), and 4-Omethyl-D-glucuronic acid were all produced after the gum portion of the resin underwent complete hydrolysis. A graded hydrolysis of the gum produced aldobiouronic acid [6-O-(4-O-methyl-Dglucopyranosyluronic acid)-D-galactose]. 2,3,4,6-tetra-O-methyl-D-galactose, 2,3-di-O-methyl-L-arabinose, 2,3,4-tri-Omethyl-D-galactose, 2,4-di-O-methyl-D-galactose, and 2,3,4-tri-Omethyl-D-glucuronic acid were produced by hydrolysis of methylated gum in the ratios of 1: 1: 1: 2: 1. According to the preliminary structure, the gum is a highly branched polysaccharide with 1–6, 1–3, and 1–5 type of linkages [117, 118].

22.5.7 Amino Acids

After the solvent had been removed, the alcohol extract of C. wightii was divided between water and ether. The presence of different amino acids was detected when the aqueous fraction was chromatographed. The amino acids that were discovered included cystine, histidine, lysine, arginine, aspartic acid, serine, glutamic acid, threonine, alanine, proline, tyrosine, tryptophan, valine, leucine, and isoleucine [119].

In addition to tiny levels of sesamin and other unidentified compounds, the guggul plant contains steroids, polysaccharides, diterpenoids, ferulates, inorganic ions, triterpenoids, long-chain aliphatic esters & tetrols and lignans.

The steam distillation of the oleoresin obtained from the *Commiphora wightii* plant extract has about 0.4% of the essential oils with a combination of some other compounds such as polymyrcene, myrcene and di myrcene [120]. Some other compounds which were found in oil constituents such as bornyl acetate, cineole, pinene, geraniol, methyl heptanone and eugenol [108]. A sesquiterpene was found in the oleoresin gum which was obtained from the extract of the *Commiphora wightii* plant named cadinene which is a bicyclic [108]. Some other diterpenes were identified in Guggulu plant extracts such as cembrenoids, cembrene-A and camphene. Basic tetraenes such as cembrene-A and geranylgeranyl pyrophosphate were also found and cembrane alcohol named Mukulol has been extracted from the resin and upper parts of the plant guggul [110, 111].

Triterpinoids components such as myrrhanol A as well as myrrhanol B& C polypodane-type triterpenes [23, 112]. Two more triterpenoid components, mansumbinone and mansumbinoic acid are also discovered [114].

22.5.8 Steroids

The gum resin has been used to isolate a number of steroidal components. Guggulsterone E& Z, guggul sterol I, II, III anI guggul sterol VI [116] are the most important ingredients [90]. Cholesterol has been mentioned as well. Guggulsterone-M, dihydro guggulsterone-M, and guggul sterol-Y are three new and newly discovered steroids [114]. The drug's hypolipidemic and anti-inflammatory properties have been linked to the steroidal components [22].

22.5.9 Flavonoids

A flavone named muscanone was extracted from the bark of the *Commiphora wightii* by the methanolic extract with silica gel. Another flavonoid was also identified in this process known as naringenin. In a microbiological sensitivity assay, this muscanone was effective against the candida species [121].

22.5.10 Phenolics

Phenolics are natural compounds found in plants that have powerful antioxidant and anti-inflammatory properties. In plants, phenolic substances such as gallic acid, protocatechuic acid, gentisic acid, vanillic acid, p-hydroxybenzoic acid, syringic acid, ellagic acid, and cinnamic acid derivatives such as caffeic acid, chlorogenic acid, ferulic acid, sinapic acid (SA), and p-coumaric acid are abundant. These phenolic chemicals are abundant in guggul, which adds to its enormous biological activity in the treatment of a variety of chronic illnesses in humans [122].

22.5.11 Guggultetrols

The saponified gum resin yielded a crystalline compound that was identified as a mixture of octadecan-1,2,3,4-tetrol, nonadecan-1,2,3,4-tetrol, and eicosan-1,2,3,4-tetrol with a little amount of other components, likely lower (C-16 and C-17) and higher (C-21 and C-22) homologous tetrols. Guggultetrols are a new type of naturally occurring lipid that includes these chemicals. They are linear aliphatic tetrols with hydroxyl functionalities at locations C-1, C-2, C3, and C-4. Guggultetrol-18 and guggultetrol-20 were isolated in their pure form via derivatization and preparative GLC [90].

L-arabinose, D-galactose, L-fructose (traces), and 4-O-methyl-D-glucuronic acid were all produced after the gum component of the resin was completely hydrolyzed. By hydrolysis of the gum aldibiouronic acid was obtained which is a D-galactose [117] Some other galactose is also obtained by the hydrolysis of the methylated gum. The gum which is obtained from the plant is a highly branched polysaccharide and this gum has 1–5, 1–3 and 1–6 linkage [118].

22.5.12 Varieties

Since 1981, over 26 cultures have been assessed in Anand, Gujarat, and the top yielding variety, 'Marusudha', has been released for the cultivation [2].

22.5.13 Soil

Guggal is not cultivated on a large scale in any commercial plantation. The guggul plants were growing in wild areas of the arid regions of western India. The soil of these wild areas was silty that is low in organic matter but high in other minerals. Plants grow more quickly in damp soils. It may be grown in a wide range of soil types [2]. Though these trees may grow in a variety of soil types, they thrive on sandy or sandy loam soils [123].

The plant is adaptable and can grow in arid environments under a variety of conditions. The strong bark and short leathery leaves. It is protected from arid conditions by a white waxy layer on the stem. For a high yield of oleo-gum-resin, it prefers a warm, dry climate [2]. Warm, dry climates and heme-suitable arid locations are ideal for the Guggul plant to thrive. Frost sensitivity is a problem with these plants [123].

22.6 Land Preparation

Three to four ploughings should be done on the land well before the rainy season. Remove any previous crops' weeds and divide the field into manageable pieces. With a spacing of 3 × 3 in meters dig up the 50 by 50 × 50 cm plots for the sowing. Farmyard manure (FYM) or compost, along with topsoil, should be used to fill the pits. 1000 plants per acre or 2500 plants per hectare are required. During the wet season, the rooted cuttings should be placed in trenches. Cutting the side branches as the plant grows will help to train it appropriately [123].

2–3 ploughings are done well before the rainy season to prepare the ground, which is then divided into plots of manageable size. Pits measuring 0.5 × 0.5 × 0.5-meter cube are dug in rows with a 3–4 m space between them. To protect the plants from termite infestations, the pits are filled with FYM and topsoil, which has been mixed with Aldrin (5%) [2]. The soil is prepared by plowing 2–3 times and mixing animals' waste or manure to increase the yield. A little amount of Aldrin is also used to prevent the attack of the termites. The beds are prepared with the size of 6 × 3 to 10 × 4. Climate conditions have a significant effect in the tree's establishment. The highest temperature for optimal growth is 37 °C, the minimum temperature is 27 °C, and the best humidity is 76% [9].

22.7 Cultivation

For the purpose of the plantation rooted cuttings are planted during the rainy season in the pits. When the plants grow properly the side branches are cut down to trim the plant. During the early phases of growth, intercultural is limited to weeding and hoeing. However, surrounding the shrubs, the soil must be churned twice a year. Plants grow faster when the soil is stirred [2].

22.8 Propagation

Seeds and stem-cuttings can both be used to grow Guggal. In this plant, air-layering has also worked. The meristem culture protocol is also available in the literature [2].

22.9 Propagation by Seeds

Because of the existence of the hard seed coat, seed germination is delayed and poor (5%). This is not a typical technique of reproducing this plant. Seeds of the plant were clean from the soil and kept in running water for a proper and higher rate of germination. During the Kharif season, seedlings can be maintained in polythene bags and then transplanted in the main field after strengthening [2]. The guggul plant's seed process gradually is quite weak, and the plant may not grow correctly as a result. The seeds produce just a small number of seedlings. Seeds are dispersed to germinate seedl ngs, and the thickest of the Euphorbia plant germinates these seeds [9]. Seedlings may die if the seeds are buried deeply because of the deep root system. If the seeds germinate from the cuttings, they will grow quickly and thrive [9].

22.10 Propagation Through Stem Cuttings

For propagation of the plant around 15–20 long and 10-millimeter-thick stem, cutting is required. These cuttings are semi-hardwood. For the purpose of high growth rate, the cuttings are treated with growth hormones such as IBA and NAA. During the months of June and July, these cuttings are planted in well manured and prepared soil. After the plantation of the cuttings, the soil should be irrigated on regular basis. In the next rainy season when the cuttings properly grow are prepared to transfer to the soil. In stem-cuttings, the rate of rooting varies between 80% and 94% [2].

In the late summer season cuttings of the guggal leafless plant were selected for the plantation. Here the disease-free branches are selected. Plant starts foliation or growth when the season of monsoon starts and plant become physiologically active. The second and third week of June is best for the plantation of cuttings at the depth of 15–20 cm [9].

22.11 Through Air Layering

The mortality rate is very low when the gugal plants are propagated through the seeds and a higher rate of surviving when they are planted through the cuttings. Besides the cuttings, air layering is also successful. A 60–90 cm long healthy branch is selected under the leaf bud or node for the production of an individual plant and a ring is marked. On the ring, the bark is removed and is covered by the soil and bandaged it tightly [9].

22.11.1 Fertilizer Application

Except for a minimal degree of nitrogen fertilization, this crop hasn't shown a satisfactory reaction to fertilizers. As a result, urea or ammonium sulfate is applied twice a year before irrigation, at a rate of 25/50 g bush [2]. The requirement of the farmyard manure for plantation is about ten tonnes per acre or twenty-five tonnes per hectare [123]. From the industries, the dung of the livestock and plant residues are the sources of organic fertilizers. Plants take their nutritional requirements from organic fertilizers and plant pests populations can be decreased by using organic fertilizers. Besides this, organic fertilizers play an important role to increase the activity of microorganisms. These fertilizers also increase the activity of the cation and anion exchange. To increase the carbon contents of the soil and organic matter this fertilizer plays a role. As inorganic fertilizer increases, the quality, quantity and yield of the crops the organic fertilizer also enhance the yield [124]. For the uptake of the nitrogen, potassium, phosphorous, magnesium and calcium increased with using agro-industrial wastes as soil fertilizers, farmyard manure contains soluble phosphoric acid lime and potash. Due to the use of cow manure for a long time, increased aggregate stability, and macropores [125].

22.11.2 Irrigation

Plant growth is aided by light irrigation during the summer. Because it is an arid-zone plant, irrigation is not required until the middle of November, unless there is a severe drought. Plants aged one to five years must be irrigated if there is winter precipitation. Irrigation is only required in the summer for plants aged 6–7 years. For irrigation purposes, a water tank or the loading of the head can be used [126].

22.11.3 Pests and Diseases

During the process of the guggul cultivation leaf-eating termites, caterpillars and whiteflies were found as the harmful insects of this plant. The pits should be filled with the farmyard to prevent the attack of the termites. Pesticides such as chlorpyriphos should be used as 0.5 ml per liter of the water. The solution of the metcoid 0.2% to control the whiteflies and caterpillar must be sprayed to the crop. Some other diseases such as leaf blight and leaf spot caused by bacteria are also identified. For the prevention of these diseases attack spraying of the agromycin and blitox 0.1 and 4 grams per liter [123].

During the rain grasses and *Ziziphus nummularia* are the harmful weeds. They damage the crop and loss the yield. In the months of September and December, these weeds must be removed by hand or by some metal thing. In the drier months of the year, these plants are attacked by termites' insects. These insects commonly attack 3–4-year-old plants. These plants are attacked by the termites when a small amount of water is present. In the root zone and in the stem these termites attack and cause severe damage. Through the buried stem a hole is made, and they make the hole large like a tunnel into the cutting. These termites exit from the stem by making another hole or from this tunnel. When plants are attacked by the termites, they cause yellowing of the leaves or wilting and cause the death of the plants. Some control measures are taken against the termite's [127]. Some of these measures are as follows.

1. An insect termitorium when attacked then it causes damage to the plant. It controls by the spraying of kerosine or carbon disulfide.
2. *Garcenia gumnifera* and Asafoetida, Aloes, and Guggal resin 1:2:2:2 ration aqueous solution (called Gondal) are effective in the control of the termites.
3. Three percent of haptafan given to the soil is helpful in the killing of the pests.
4. An aqueous solution of copper sulfate and mercuric chloride 55 and 25% respectively are deterrents to the termites.
5. During the process of the infestation, 5% of aldrin in water can destroy the termites when pour into the pits.
6. 250 g of the benzene hexachloride powder is mixed in the soil to prevent infestation during the transplantation of the plants.
7. Treating a fungicide (Tofasan) to the cuttings can control the disease's spreading.
8. Waterlogging should be avoided
9. Destruction of infected plants should be done [127].

22.12 Review of the Literature

Guggul is gaining popularity in the natural drug market. Over fifty Guggul formulations from well-known pharmaceutical businesses such as Himalaya, Baidyanath, and Patanjali have been released in India. The *Commiphora wightii* plant produces guggul, a gum resin. Some diseases such as liver problems, arthritis, diabetes, inflammation, obesity, anemia and constipation are treated with this plant herb. Because of its medicinal properties, it is used as a strong adhesion agent and is mixed in many herbal treatments, among other things. To conduct a thorough bibliographic search utilizing multiple scientific engines and databases to review the primary phytochemical, pharmacological characteristics, and analytical procedures of Guggulu detection. There are 66 phytochemicals found in the guggul plant such as guggulsterone E&Z, quercetin and gallic acid. Pharmacological effects of these chemicals in the diseases of cancer, mutagenic, venom and function as an antioxidant. The mechanism of action of guggulsterone is still unknown. There aren't a lot of pharmacology and toxicology studies. This research has revealed a significant gap in the literature that will be filled by in-depth in-vivo and in-vitro studies [128].

The oleo-gum-resin (Guggul or bdellium) is produced by incisions on the stem and major branches of the *Commiphora wightii*, a shrub found in dry areas of the Indian subcontinent. Despite its importance in traditional Indian medicine, guggul has only recently been shown to be beneficial in clinical trials of hyperlipidemia, rheumatism, hypercholesterolemia and arthritis. This resulted in a sustainable study of the gum's structure, growth practices, tapping and isolating procedures, and chemical contents. More research is needed, however, to establish a desirable plant ideotype, create a suitable gum exudate, and standardize forest felling cycles [123].

Guggul is a flowering plant that belongs to the family of Burseraceae with the Indian name bdellium and produces oleoresin. It provides a wide range of therapeutic benefits. For many chemical constituents, it is a gold mine. The guggul plant can be found predominantly in arid and rocky areas around the world. It's a wonder drug for inflammation, obesity, rheumatism, and problems with lipid metabolism, to name a few. Chemical components discovered in this plant include terpenoids, sugars, guggul tetrols, amino acids, flavonoids and sterols [123].

Guggul is a blooming plant that produces a gum resin, *Commiphora wightii*, often known as Indian bdellium. It provides a wide range of therapeutic benefits. For many chemical constituents, it is a gold mine. The guggul plant can be found predominantly in arid and rocky areas around the world. It is a medicinal gem for a variety of ailments, including inflammation, obesity, rheumatism, and lipid metabolism issues. Terpenoids, steroids, sterols, amino acids, sugars, guggul tetrols, and flavonoids are among the chemical elements found in this plant [129].

If the plant of the guggul plant was injured, then the oleo-gum resin was obtained from its bark. In Indian Ayurveda industries, it was used to treat obesity, lipid metabolism, gout and inflammation. The oleo gum resin is a mixture of monoterpenoids, flavonoids, volatile oil, steroids, lignans and sugars. It also contains amino acids and guggultetrols [96].

Only the dry portions of India's Rajasthan and Gujarat states, as well as Pakistan's borderlands, are home to this species. The Indian herbal industries consume large amounts of oleo-gum resin, Guggulu, which was obtained by the tapping of this plant. Due to the more harvesting, loss of the habitat and quick degradation of this plant as well as irregular tapping of oleoresin the natural population of this species has declined during the previous several decades. As a result, this species is listed as near threatened on the red list of threatened species [96].

The Burseraceae family's Guggulu (*Commiphora mukul*) is a 2–3.5 m tall plant. The plant can be found growing wild in the desert, rocky locations as well as low-rainy, hot climates. Resin, which is collected by tapping the barks, is used in medicinal preparation. In Ayurvedic medicine, Guggulu has a great value. Guggulu is a Rasayana also known as Vatakaphaghna and is used to treat various ailments. The quantity of plants has greatly decreased as a result of high values and excessive demand, incorrect gathering methods, unchecked forest degradation, and insufficient cultivation knowledge. It is now considered to be a threatened species. As a result, it is critical to cultivate and conserve this plant. Seeds and vegetative propagation are both options for propagating Guggulu. The rate of seed germination is quite low. The most popular and successful method for vegetative propagation is stem cutting. For appropriate growth, farming care is also required. Knowledge of collection methods and awareness can both help with the conservation [9].

In the industries of the Unani Ayurvedic and Siddha, the gum of this plant was used in Pakistna and India. Refined *Commiphora wightii* oleo-resin commodities are exported from India to 42 countries, including Pakistan, for anti-inflammatory and anti-obesity therapy, which is considered to reduce heart disease and cholesterol levels. Countries such as France, Hungary, Belgium, Germany, Italy, the Netherlands, and the United Kingdom are all importers, so *C. wightii* exports are extremely significant to the European Union. The demand for *C. wightii* oleoresin is increasing, whereas the supply of wild *C. wightii* is declining. Overexploitation of *C. wightii* following extraction of its economically valuable oleoresin is not a new issue. It has existed for more than 50 years. The implementation of felling and pruning trees for the purpose of collecting *C. wightii* oleoresin has had a disastrous impact on the species' populations [130].

Guggul, an oleo-gum resin derived from *Commiphora wightii*, is well-known for its anti-inflammatory, antioxidant, thyroid stimulatory, platelet aggregation, fibrinolytic, and cytotoxic properties. The main ingredients responsible for their pharmacological use are guggulsterone, specifically E and Z guggulsterones. It has traditionally been used as an antimalarial, antidysenteric, anticholesterolemic, antihypertensive, and anti-rheumatic agent for a variety of clinical diseases including dysmenorrhea, dyspepsia, impotence, leprosy, leucoderma, and anemia, among others. Guggul is now available as a marketed formulation for the treatment of a variety of clinical problems, and it can be mixed with a variety of other components.

Though traditional dosage forms have advantages such as patient compliance and ease of access, they have been proven to have drawbacks such as dose variation, peak-valley effect, non-adjustment of the delivered drug, invasiveness, and so on. Due to its limited bioavailability and water solubility, guggul fails to produce the desired effect. As a result, it is only a partial or ineffective treatment for a variety of signs and symptoms. A new technique in the pharmaceutical industry, the novel drug delivery system (NDDS), has eliminated many of the limitations of traditional dosage forms. Nanoparticles, nanovesicles, gugglusomes, and proteasomal gel are examples of innovative guggul dosage forms. However, innovative guggul formulations are less common in the market. Guggul can be used to make a profit [131].

Guggulsterone is a steroidal ketonic chemical produced by the *Commiphora wightii* vertical rein channels and water channels. Guggulsterone is in great demand in the pharmaceutical, aroma, and spice sectors due to its wide range of medicinal and therapeutic properties, including its other major bioactivities. A resin gum guggulsterone is becoming increasingly popular in the pharmaceutical and fragrance industries, necessitating quantitative analysis in extant wild populations of *C. wightii*. Traditional and biotechnological methods can be used to identify top germplasm with greater guggulsterone levels [132].

The purpose of this study was to look at two guggulsterone variations, E and Z guggulsterone, in crude secretions of *C. wightii* accessions from three states in Southeast India: Gujarat, Rajasthan, and Haryana. In a stem sample by using high-performance liquid chromatography a steroid was identified at the wavelength of 250 nm. The HPLC examination of the two stem samples from the Indian regions of Rajasthan and Haryana two guggulsterone E& Z was found. The quantity of these isomers of guggulsterone was higher in areas of Rajasthan as compared to Haryana and Gujrat. While the samples were taken from the areas of the Palana and Devikot has the amount of the guggulsterone E was 487.68 g/g and the amount of the guggulsterone Z was 487.45 g/g. Some parameters such as flow rate, selectivity, recoveries, linearity and specificity are derived from the sample of guggul plant which contains E and Z guggulsterone and a graph was made to quantify the quantity assay. Factors of the environment play an important role in the production of the guggulsterone isomers [132].

In regions of Asia, the resin extracted from the guggul plant was used as a drug for the lowering of cholesterol. In the United States and Western Europe, its popularity for this purpose is rising. It was shown that guggulsterone has two nuclear hormones receptors which have the hypolipidemic effects [133].

Guggulipid is formed by the guggulsterone steroidal hormone present in the oleo gum resin in *Commiphora wightii* plant. Guggul resin is extracted from the guggul plant used in medicine to cure different diseases such as problems of the lipids in the body like arteriosclerosis and obesity. Two guggulsterone E & Z are closely related steroidal ketones responsible for the extracts' hypolipidemic action, which dates back to ancient times. Appetite regulation hormones such as leptin, cholecystokinin & ghrelin and dopamine & serotonin doses of 100, 200, 400 mg according to the body weight. The guggulsterone hormones' effects on lipids of the plasma membrane weight of the body and uptake of the food in rats as well as glucose level were determined. A dose of 400 mg per kilogram of body weight was given for 15 days of guggulsterone was effective increasing body weight and lowering the consumption of the food. The guggulsterone has no effect or little effect on the cholecystokinin. The impact of this hormone was shown that it decrease the glucose level as well as ghrelin and triglyceride. This hormone enhances the plasma of serotonin, dopamine and leptin [134].

When the bark of the Bhandari (*Commiphora wightii*) is wounded, it produces Guggulu, an oleogum resin. Guggulu is used to make a variety of compound medications in Ayurveda, the majority of which include the suffix 'Guggulu'. The oleoresin gum contains the constituents such as carbohydrates, aliphatic esters, inorganic molecules and amino acids as well as steroids. The oleoresin obtained from the guggul plant was used to reduce the triglycerides and cholesterol levels in humans. The oleo gum resin which is obtained from the bark of the guggul plant or *Commiphora wightii* injured. In India, this plant was grown from the exudates of the myrrh plant in semi-arid and arid areas of Pakistan, India and Bangladesh. "One who defends against sickness," according to the Sanskrit definition of the term "guggul." This reflects the vast esteem and therapeutic Ayurvedic applications for this herb, which is thought to be the most significant for the clearance of "ama," harmful compounds that build up by the regulation and digestion of the sluggish through which the metabolism slow [135].

Herbal treatments are among the oldest remedies known to mankind. Ayurvedic treatments are well-known in India. A small tree belibgs to the Burseraceae family known has the species of the *Commiphora wightii*. Guggulipid, guggulsterone, guggulsterone, mukolol, and other steroids are found in C. wightii. In Ayurvedic medicines, Guggulu is used as an astringent, antimicrobial, expectorant, aphrodisiac, carminative, antispasmodic, and emmenagogue [126].

It is the most effective herb for Medoroga and Vata problems according to Ayurveda. It is commonly used for obesity and is recognized all over the world as a fat-burning agent. It aids in the reduction of triglyceride levels. Rheumatoid arthritis, gout, and sciatica can all benefit from it. Ayurveda's Rasayana is also one of the most significant. It also helps with sluggish livers, libido stimulation, neurological disorders, heart problems, bronchitis and blood circulation problems, healing of the wounds, enhancing the process of digestion, women's problems and different diseases of the skin in humans. In Indian medication industries, the *Commiphora wightii* plant has an important role and these industries are dependable to this plant for the oleoresin [126].

According to an investigation, it was shown that the sesquiterpene extract from the *Commiphora molmol* has antifungal and antimicrobial activities. It was most effective against the species of the *Staphylococcus aureus*, *Pseudomonas aeruginosa* and *E. coli*. The concentration of the sesquiterpene from 0.18 mg to 2.80 mg/ml has a growth inhibitory effect on the bacteria and fungi. In the membranes of mammals, the extract has the activity of sodium blocking [136].

The extract of the myrrh was effective to control the leafworm in cotton. This extract was separately or with other insecticides used to control this leafworm in cotton [137].

In an investigation, it was shown that this myrrh extract obtained from the Commiphora species was effective to kill or control the mosquito. When this extract was given to the mosquito it was shown that it affects the gut, nervous tissues, fat and muscles of the body [138].

An extract from the Commiphora species of myrrh was used for the making of the capsules of the soft gelatin, emulsion and suppositories which are used against the microbes and bacteria. This extract was commercially used and named Mirazid. This was used against two species of microbes *Schistosoma mansoni* and *Schistosoma haematopium* [55].

Commiphora molmol was used against the snails. An investigation shows that the myrrh obtained from the *C. molmol* has molluscicidal activities against the *Biommphalaria alexandrina*. This activity is due to the presence of the oil in the extract. It was also shown that the oil extract has more effective than the oleoresin extract [138].

References

1. Bhandari, M. (1964). Notes on Indian desert plants 4-new names and combinations. *Nelumbo – The Bulletin of the Botanical Survey of India, 6*(2–4), 327–328.
2. Farooqi, A., & Sreeramu, B. (2004). *Cultivation of medicinal and aromatic crops. Hyderabad.* India: University Press (India) Ltd.
3. Good, R. (1953). *The geography of the flowering plants* (2nd ed.). The Geography of the Flowering Plants.
4. Van Der Walt, J. (1974). A preliminary report on the genus Commiphora in South West Africa. *Madoqua, 1974*(8), 5–23.
5. Van der Walt, J. (1975). The south west African species of Commiphora. *Mitteilungen der Botanischen Staatssammlung Muenchea, 12*, 195–266.
6. Lisowski, S., Malaisse, F., & Symoens, J. J. (1972). *Genre Commiphora Jacq.(Burseraceae) au Zaire (Ex-Congo-Kinshasa).* Boletim.
7. Hooker, J. (1872). *The Flora of British India Part I* (p. 514). L. Reeve.

8. Bhatnagar, L., Singh, V., & Pandey, G. (1973). On the occurrence of Guggulu yielding Commiphora with reference to Commiphora berryi Engl. Madhya Pradesh. *Indian Journal of Medical Research, 8*, 69–75.
9. Thosar, S. L., & Yende, M. R. (2009). Cultivation and conservation of Guggulu (Commiphora mukul). *Ancient Science of Life, 29*(1), 22–25.
10. Ved, D., & Goraya, G. (2008). Demand and supply of medicinal plants. *Medplant-ENVIS Newsletter on Medicinal Plants, 1*(1), 2–4.
11. Phogat, P., et al. (2010). Introduction to hyperlipidemia and its management: A review. *Pharmacology, 2*, 251–266.
12. Satyavati, G. (1966). *Effect of an indigenous drugs (Guggulu) disorders of lipid metabolism with special reference to atherosclerosis and obesity (Medorog)* (p. 77). Thesis submitted of the degree of D. Ay. M.(BHU) Brochure PGIM BHU.
13. Satyavati, G., Dwarakanath, C., & Tripathi, S. (1969). Experimental studies on the hypocholesterolemic effect of Commiphora mukul Engl.(guggul). *Indian Journal of Medical Research, 57*, 1950–1962.
14. Shastry, V., & Tripathi, S. (1968). Experimental and clinical studies on effect of Guggulu (C. mukul) in hyperlipidaemia and thrombosis. *Indian Journal of Medical Research, 2*(1), 195.
15. Mehta, V., Malhotra, C., & Katrah, N. (1968). The effects of various fractions of gum guggulu on experimentally produced hypocholesterolaemia. *Indian Journal of Physiology and Pharmacology, 12*, 87.
16. Malhotra, C., et al. (1970). The effect of various fractions of gum guggul on experimentally produced hypercholesteraemia in chicks. *Indian Journal of Medical Research, 58*(3), 394–395.
17. Khanna, D., et al. (1969). A biochemical approach to anti-atherosclerotic action of Commiphora-mukul: an Indian indigenous drug in Indian domestic pigs (Sus scrofa). *The Indian Journal of Medical Research, 57*(5), 900–906.
18. Malhotra, S., & Ahuja, M. (1972). Effect of steroidal compound isolated from fraction A of Commiphora mukul on hepatic and aortic lipid content in rats fed on atherogenic diet. *Indian Journal of Pharmacology, 4*, 110.
19. Das, D., Sharma, R., & Arora, R. (1973). Antihyperlipidaemic activity of fraction A of Commi-phora mukul in monkeys. *Indian Journal of Pharmacology, 5*, 285.
20. Kapoor, N., & Nityanand, S. (1971). *Hypocholesterolaemic effect of the fraction isolated from the C. mukul (Guggulu)*. A Seminar on Disorders of Lipid Metabolism, held in New Delhi, (India) October.
21. Nityanand, S., & Kapoor, N. (1971). Hypocholesterolemic effect of Commiphora mukul resin (guggal). *Indian Journal of Experimental Biology, 9*(3), 376–377.
22. Francis, J. A., Raja, S. N., & Nair, M. G. (2004). Bioactive terpenoids and guggulusteroids from Commiphora mukul gum resin of potential anti-inflammatory interest. *Chemistry & Biodiversity, 1*(11), 1842–1853.
23. Kimura, I., et al. (2001). New triterpenes, myrrhanol A and myrrhanone A, from guggul-gum resins, and their potent anti-inflammatory effect on adjuvant-induced air-pouch granuloma of mice. *Bioorganic & Medicinal Chemistry Letters, 11*(8), 985–989.
24. Matsuda, H., et al. (2004). Absolute stereostructures of polypodane-and octanordammarane-type triterpenes with nitric oxide production inhibitory activity from guggul-gum resins. *Bioorganic & Medicinal Chemistry, 12*(11), 3037–3046.
25. Singh, S. V., et al. (2005). Caspase-dependent apoptosis induction by guggulsterone, a constituent of Ayurvedic medicinal plant Commiphora mukul, in PC-3 human prostate cancer cells is mediated by Bax and Bak. *Molecular Cancer Therapeutics, 4*(11), 1747–1754.
26. Ulbricht, C., et al. (2005). Guggul for hyperlipidemia: a review by the Natural Standard Research Collaboration. *Complementary Therapies in Medicine, 13*(4), 279–290.
27. Wang, X., et al. (2004). The hypolipidemic natural product Commiphora mukul and its component guggulsterone inhibit oxidative modification of LDL. *Atherosclerosis, 172*(2), 239–246.

28. Ojha, S. K., et al. (2008). Effect of Commiphora mukul extract on cardiac dysfunction and ventricular function in isoproterenol-induced myocardial infarction. *Indian Journal of Experimental Biology, 46*, 646.
29. Singh, R. B., Niaz, M. A., & Ghosh, S. (1994). Hypolipidemic and antioxidant effects of Commiphora mukul as an adjunct to dietary therapy in patients with hypercholesterolemia. *Cardiovascular Drugs and Therapy, 8*, 659–664.
30. Szapary, P. O., et al. (2003). Guggulipid for the treatment of hypercholesterolemia: a randomized controlled trial. *JAMA, 290*(6), 765–772.
31. Panda, S., & Kar, A. (2005). Guggulu (Commiphora mukul) potentially ameliorates hypothyroidism in female mice. *Phytotherapy Research: An International Journal Devoted to Pharmacological and Toxicological Evaluation of Natural Product Derivatives, 19*(1), 78–80.
32. Saeed, M. A., & Sabir, A. (2004). Antibacterial activities of some constituents from oleo-gum-resin of Commiphora mukul. *Fitoterapia, 75*(2), 204–208.
33. Saxena, G., et al. (2007). Gugulipid, an extract of Commiphora whighitii with lipid-lowering properties, has protective effects against streptozotocin-induced memory deficits in mice. *Pharmacology Biochemistry and Behavior, 86*(4), 797–805.
34. Annu, W., et al. (2010). *Anti-inflammatory, analgesic and anti-lipid peroxidation studies on leaves of Commiphora caudata (Wight & Arn.) Englera*. Indian Journal of Natural Products and Resources.
35. Vikneshwaran, D., Viji, M., & Rajalakshmi, K. (2008). Ethnomedicinal plants survey and documentation related to Paliyar community. *Ethnobotanical Leaflets, 2008*(1), 146.
36. Chikamai, B., & Gachathi, N. (1994). Gum and resin resources in Isiolo district, Kenya: ethnobotanical and reconnaissance survey. *East African Agricultural and Forestry Journal, 59*(4), 345–351.
37. Gowrishankar, N., et al. (2004). A preliminary study on gastric antiulcer activity of Commiphora berryi (Arn) Engl in rats. *Indian Drugs-Bombay, 41*(2), 97–100.
38. Shankar, N. G., et al. (2008). Hepatoprotective and antioxidant effects of Commiphora berryi (Arn) Engl bark extract against CCl4-induced oxidative damage in rats. *Food and Chemical Toxicology, 46*(9), 3182–3185.
39. Abdul-Ghani, A.-S., & Amin, R. (1997). Effect of aqueous extract of Commiphora opobalsamum on blood pressure and heart rate in rats. *Journal of Ethnopharmacology, 57*(3), 219–222.
40. Alzweiri, M., et al. (2011). Ethnopharmacological survey of medicinal herbs in Jordan, the Northern Badia region. *Journal of Ethnopharmacology, 137*(1), 27–35.
41. Su, S.-L., et al. (2009). Isolation and biological activities of neomyrrhaol and other terpenes from the resin of Commiphora myrrha. *Planta Medica, 75*(04), 351–355.
42. Shoemaker, M., et al. (2005). In vitro anticancer activity of twelve Chinese medicinal herbs. *Phytotherapy Research: An International Journal Devoted to Pharmacological and Toxicological Evaluation of Natural Product Derivatives, 19*(7), 649–651.
43. Bradley, P. R. (1992). *British herbal compendium. Volume 1. A handbook of scientific information on widely used plant drugs. Companion to Volume 1 of the British Herbal Pharmacopoeia*. British Herbal Medicine Association.
44. Omer, S., Adam, S., & Mohammed, O. (2011). Antimicrobial activity of commiphora myrrha against some bacteria and candida albicans Isolated from Gazelles at King. *Research Journal of Medicinal Plant, 5*(1), 65–71.
45. Racine, P., & Auffray, B. (2005). Quenching of singlet molecular oxygen by Commiphora myrrha extracts and menthofuran. *Fitoterapia, 76*(3–4), 316–323.
46. Schilcher, H. (1997). *Phytotherapy in paediatrics—Handbook for physicians and pharmacists*. 2nd German ed. translated by AR Meus. Medical Pharmacy Publishers, Stuttgart.
47. Tyler, V. E. (1992). *The honest herbal: a sensible guide to the use of herbs and related remedies*. Pharmaceutical Products Press.

48. Al-Howiriny, T., et al. (2004). Hepatoprotective properties of Commiphora opobalsamum ("Balessan"), a traditional medicinal plant of Saudi Arabia. *Drugs under Experimental and Clinical Research, 30*(5–6), 213–220.
49. Shen, T., et al. (2009). A triterpenoid and sesquiterpenoids from the resinous exudates of Commiphora myrrha. *Helvetica Chimica Acta, 92*(4), 645–652.
50. Fourie, T. G., & Snyckers, F. O. (1989). A pentacyclic triterpene with anti-inflammatory and analgesic activity from the roots of Commiphora merkeri. *Journal of Natural Products, 52*(5), 1129–1131.
51. McGuffin M. J. T., Kartesz, A. Y., & Leung, A. O. (2000). Tucker editors. Herbs of Commerce, 2nd edition. Silver Spring (MD): American Herbal Products Association.
52. Watt, J. M., & Breyer-Brandwijk, M. G. (1962). *The medicinal and poisonous plants of southern and eastern Africa being an account of their medicinal and other uses, chemical composition, pharmacological effects and toxicology in man and animal* (2nd ed.). The Medicinal and Poisonous Plants of Southern and Eastern Africa being an Account of their Medicinal and other Uses, Chemical Composition, Pharmacological Effects and Toxicology in Man and Animal.
53. Kokwaro, J. O. (1976). *Medicinal plants of east Africa*. East African Literature Bureau.
54. Claeson, P., Andersson, R., & Samuelsson, G. (1991). T-cadinol: A pharmacologically active constituent of scented myrrh: Introductory pharmacological characterization and high field 1H-and 13C-NMR data. *Planta Medica, 57*(4), 352–356.
55. El Ashry, E., et al. (2003). Components, therapeutic value and uses of myrrh. *Die Pharmazie – An International Journal of Pharmaceutical Sciences, 58*(3), 163–168.
56. Habtemariam, S. (2003). Cytotoxic and cytostatic activity of erlangerins from Commiphora erlangeriana. *Toxicon, 41*(6), 723–727.
57. Carroll, J., Maradufu, A., & Warthen, J., Jr. (1989). An extract of Commiphora erythraea: A repellent and toxicant against ticks. *Entomologia Experimentalis et Applicata, 53*(2), 111–116.
58. Koch, A., et al. (2005). Evaluation of plants used for antimalarial treatment by the Maasai of Kenya. *Journal of Ethnopharmacology, 101*(1–3), 95–99.
59. Asres, K., et al. (1998). Terpenoid composition of the wound-induced bark exudate of Commiphora tenuis from Ethiopia. *Planta Medica, 64*(05), 473–475.
60. Gradé, J. T., Tabuti, J. R., & Van Damme, P. (2009). Ethnoveterinary knowledge in pastoral Karamoja, Uganda. *Journal of Ethnopharmacology, 122*(2), 273–293.
61. Satyavati, G. (1988). Gum guggul (Commiphora mukul) – The success story of an ancient insight leading to a modern discovery. *The Indian Journal of Medical Research, 87*, 327–335.
62. Nityanand, S., Srivastava, J., & Asthana, O. (1989). Clinical trials with gugulipid. A new hypolipidaemic agent. *The Journal of the Association of Physicians of India, 37*(5), 323–328.
63. Verma, S., & Bordia, A. (1988). Effect of Commiphora mukul (gum guggulu) in patients of hyperlipidemia with special reference to HDL-cholesterol. *Indian Journal of Medical Research, 87*, 356–360.
64. Baldwa, V., et al. (1981). Effects of Commiphora mukul (Guggul) in experimentally induced hyperlipemia and atherosclerosis. *The Journal of the Association of Physicians of India, 29*(1), 13–17.
65. Hasani-Ranjbar, S., et al. (2010). The efficacy and safety of herbal medicines used in the treatment of hyperlipidemia; a systematic review. *Current Pharmaceutical Design, 16*(26), 2935–2947.
66. Chander, R., Khanna, A., & Kapoor, N. (1996). Lipid lowering activity of guggulsterone from Commiphora mukul in hyperlipaemic rats. *Phytotherapy Research, 10*(6), 508–511.
67. Mester, L., Mester, M., & Nityanand, S. (1979). Inhibition of platelet aggregation by "Guggulu" steroids. *Planta Medica, 37*(12), 367–369.
68. Bordia, A., & Chuttani, S. (1979). Effect of gum guggulu on fibrinolysis and platelet adhesiveness in coronary heart disease. *Indian Journal of Medical Research, 70*, 992–996.

69. Panda, S., & Kar, A. (1999). Gugulu (Commiphora mukul) induces triiodothyronine production: possible involvement of lipid peroxidation. *Life Sciences, 65*(12), PL137–PL141.
70. Tripathi, Y. B., Malhotra, O., & Tripathi, S. (1984). Thyroid stimulating action of Z-guggulsterone obtained from Commiphora mukul. *Planta Medica, 50*(01), 78–80.
71. Gujral, M., et al. (1960). *Antiarthritic and antiinflammatory activity of gum guggul* (pp. 267–273). Balsamodendron Mukul.
72. Santhakumari, C., Gujral, M., & Sareen, K. (1964). Further studies on the anti-arthritic and antiinflammatory activities of gum guggul. *Indian Journal of Physiology and Pharmacology, 8*(2), 36–37.
73. Arora, R., et al. (1971). Isolation of a crystalline steroidal compound from Commiphora mukul & its anti-inflammatory activity. *Indian Journal of Experimental Biology, 9*(3), 403–404.
74. Arora, R. (1972). Anti-inflammatory studies on a crystalline steroid isolated from Commiphora mukul. *Indian Journal of Medical Research, 60*(6), 929–931.
75. Chaudhary, G. (2012). Pharmacological properties of Commiphora wightii Arn. Bhandari – An overview. *International Journal of Pharmacy and Pharmaceutical Sciences, 4*(3), 73–75.
76. Khanna, D., et al. (2007). Natural products as a gold mine for arthritis treatment. *Current Opinion in Pharmacology, 7*(3), 344–351.
77. Karan, M., et al. (2012). Effect of traditional ayurvedic purification processes (sodhanvidhi) of guggulu on carrageenan-induced paw oedema in rats. *Journal of Pharmaceutical and Biomedical Sciences, 21*(5), 1–5.
78. Verma, S., Jain, A., & Gupta, V. (2010). Synergistic and sustained anti-inflammatory activity of guguul with the ibuprofen: A preliminary study. *International Journal of Pharma and Bio Sciences, 1*(1).
79. Singh, B. B., et al. (2003). The effectiveness of commiphora mukul fir osteoarthritis of the knee: An outcomes study. *Alternative Therapies in Health & Medicine, 9*(3), 74–79.
80. Chander, R., et al. (2003). Cardioprotective activity of synthetic guggulsterone (E and Z-isomers) in isoproterenol induced myocardial ischemia in rats: a comparative study. *Indian Journal of Clinical Biochemistry, 18*(2), 71–79.
81. Amma, M., et al. (1978). Effect of oleoresin of gum guggul (Commiphora mukul) on the reproductive organs of female rat. *Indian Journal of Experimental Biology, 16*(9), 1021–1023.
82. Thappa, D. M., & Dogra, J. (1994). Nodulocystic acne: oral gugulipid versus tetracycline. *The Journal of Dermatology, 21*(10), 729–731.
83. Bellamkonda, R., et al. (2011). Antihyperglycemic and antioxidant activities of alcoholic extract of Commiphora mukul gum resin in streptozotocin induced diabetic rats. *Pathophysiology, 18*(4), 255–261.
84. Sharma, B., et al. (2009). Effects of guggulsterone isolated from Commiphora mukul in high fat diet induced diabetic rats. *Food and Chemical Toxicology, 47*(10), 2631–2639.
85. Goyal, P., Chauhan, A., & Kaushik, P. (2010). Assessment of Commiphora wightii (Arn.) Bhandari (Guggul) as potential source for antibacterial agent. *Journal of Medicine and Medical Sciences, 1*(3), 71–75.
86. Kalpesh, B. I., & Yogesh, N. M. (2010). In vitro assessments of antibacterial potential of Commiphora wightii (Arn.) *Bhandari. gum extract. Journal of Pharmacognosy and Phytotherapy, 2*(7), 91–96.
87. Romero, C. D., et al. (2005). Antibacterial properties of common herbal remedies of the southwest. *Journal of Ethnopharmacology, 99*(2), 253–257.
88. Zhu, N., et al. (2001). Bioactive constituents from gum guggul (Commiphora wightii). *Phytochemistry, 56*(7), 723–727.
89. Xiao, D., et al. (2011). Reactive oxygen species-dependent apoptosis by gugulipid extract of Ayurvedic medicine plant Commiphora mukul in human prostate cancer cells is regulated by c-Jun N-terminal kinase. *Molecular Pharmacology, 79*(3), 499–507.

90. Bajaj, A. G., & Dev, S. (1982). Chemistry of ayurvedic crude drugs – V: Guggulu (resin from commiphora mukul) – 5 some new steroidal components and, stereochemistry of guggulsterol-I at C-20 and C-22. *Tetrahedron, 38*(19), 2949–2954.
91. Shelmadine, B. D., et al. (2017). A pilot study to examine the effects of an anti-inflammatory supplement on eicosanoid derivatives in patients with chronic kidney disease. *The Journal of Alternative and Complementary Medicine, 23*(8), 632–638.
92. Moreillon, J. J., et al. (2013). The use of an anti-inflammatory supplement in patients with chronic kidney disease. *Journal of Complementary and Integrative Medicine, 10*(1), 143–152.
93. Massoud, A., El Sisi, S., & Salama, O. (2001). Preliminary study of therapeutic efficacy of a new fasciolicidal drug derived from Commiphora molmol (myrrh). *The American Journal of Tropical Medicine and Hygiene, 65*(2), 96–99.
94. Togni, S., et al. (2015). Clinical evaluation of safety and efficacy of Boswellia-based cream for prevention of adjuvant radiotherapy skin damage in mammary carcinoma: a randomized placebo controlled trial. *European Review for Medical and Pharmacological Sciences, 19*(8), 1338–1344.
95. Mesrob, B., et al. (1998). High-performance liquid chromatographic method for fingerprinting and quantitative determination of E- and Z-guggulsterones in Commiphora mukul resin and its products. *Journal of Chromatography B: Biomedical Sciences and Applications, 720*(1–2), 189–196.
96. Sarup, P., Bala, S., & Kamboj, S. (2015). Pharmacology and phytochemistry of oleo-gum resin of Commiphora wightii (Guggulu). *Scientifica, 2015*, **1–14**.
97. Macha, M. A., et al. (2010). 14-3-3 zeta is a molecular target in guggulsterone induced apoptosis in head and neck cancer cells. *BMC Cancer, 10*(1), 1–12.
98. Kay, M. A. (1996). *Healing with plants in the American and Mexican West*. University of Arizona Press.
99. Yu, B.-Z., et al. (2009). Effect of guggulsterone and cembranoids of Commiphora mukul on pancreatic phospholipase A2: Role in hypocholesterolemia. *Journal of Natural Products, 72*(1), 24–28.
100. Jasuja, N. D., et al. (2012). A review on bioactive compounds and medicinal uses of Commiphora mukul. *Journal of Plant Sciences, 7*(4), 113–137.
101. Nagababu, E., & Lakshmaiah, N. (1992). Inhibitory effect of eugenol on non-enzymatic lipid peroxidation in rat liver mitochondria. *Biochemical Pharmacology, 43*(11), 2393–2400.
102. Duwiejua, M., et al. (1993). Anti-inflammatory activity of resins from some species of the plant family Burseraceae. *Planta Medica, 59*(01), 12–16.
103. Park, S.-N., et al. (2012). Antimicrobial effect of linalool and α-terpineol against periodontopathic and cariogenic bacteria. *Anaerobe, 18*(3), 369–372.
104. Santos, F., & Rao, V. (2000). Antiinflammatory and antinociceptive effects of 1, 8-cineole a terpenoid oxide present in many plant essential oils. *Phytotherapy Research: An International Journal Devoted to Pharmacological and Toxicological Evaluation of Natural Product Derivatives, 14*(4), 240–244.
105. Ghosh, B. (1999). Quercetin inhibits LPS-induced nitric oxide and tumor necrosis factor-α production in murine macrophages. *International Journal of Immunopharmacology, 21*(7), 435–443.
106. De León, E. J., et al. (2002). Diayangambin exerts immunosuppressive and anti-inflammatory effects in vitro and in vivo. *Planta Medica, 68*(12), 1128–1131.
107. Thresiamma, K., George, J., & Kuttan, R. (1996). Protective effect of curcumin, ellagic acid and bixin on radiation induced toxicity. *Indian Journal of Experimental Biology, 34*(9), 845–847.
108. Saxena, V., & Sharma, R. (1998). Constituents of the essential oil from Commiphora mukul gum resin. *Journal of Medicinal and Aromatic Plants Science, 20*, 55–56.
109. Rücker, G. (1972). Monocyclic diterpenes from Indian gugul resin (Commiphora mukul). *Archiv der Pharmazie, 305*(7), 486–493.

110. Prasad, R., & Dev, S. (1976). Chemistry of ayurvedic crude drugs – IV: guggulu (resin from commiphora mukul – 4 absolute stereochemistry of mukulol). *Tetrahedron, 32*(12), 1437–1441.
111. Patil, V., Nayak, U., & Dev, S. (1973). Chemistry of ayurvedic crude drugs – II: Guggulu (resin from Commiphora mukul)-2: Diterpenoid constituents. *Tetrahedron, 29*(2), 341–348.
112. Matsuda, H., et al. (2004). Absolute stereostructures of polypodane-type triterpenes, myrrhanol A and myrrhanone A, from guggul-gum resin (the resin of Balsamodendron mukul). *Chemical and Pharmaceutical Bulletin, 52*(10), 1200–1203.
113. Xu, J., et al. (2011). Neuroprotective cadinane sesquiterpenes from the resinous exudates of Commiphora myrrha. *Fitoterapia, 82*(8), 1198–1201.
114. Hanuš, L. O., et al. (2005). Myrrh-commiphora chemistry. *Biomedical Papers, 149*(1), 3–28.
115. Kumar, V., & Dev, S. (1987). Chemistry of ayurvedic crude drugs – VII guggulu (resin from Commiphora mukul) – 6: absolute stereochemistry of guggultetrols. *Tetrahedron, 43*(24), 5933–5948.
116. Patil, V., Nayak, U., & Dev, S. (1972). Chemistry of ayurvedic crude drugs – I: Guggulu (resin from Commiphora mukul) – 1: steroidal constituents. *Tetrahedron, 28*(8), 2341–2352.
117. Bose, S., & Gupta, K. (1966). Structure of commiphora mukul gum. 3. methylation+ periodate oxidation studies. *Indian Journal of Chemistry, 4*(2), 87.
118. Ali, M., & Hasan, M. (1967). Chemical investigation of Commiphora mukul Engl. (Burseraceae). *Pakistan Journal of Scientific and Industrial Research, 10*, 21–23.
119. Satyavati, G. (1991). *Guggulipid: A promising hypolipidaemic agent from gum guggul (Commiphora wightii)*. Academic Press.
120. Bhati, A. (1950). Essential oil from the resin of Commiphora mukul, Hook. EX. Stocks. *Journal of the Indian Chemical Society, 27*, 436–440.
121. Fatope, M. O., et al. (2003). Muscanone: a 3-O-(1″, 8″, 14″-trimethylhexadecanyl) naringenin from Commiphora wightii. *Phytochemistry, 62*(8), 1251–1255.
122. Hazra, A. K., et al. (2018). HPLC analysis of phenolic acids and antioxidant activity of some classical ayurvedic Guggulu formulations. *International Journal of Ayurveda and Pharmaceutical, 9*(1), 112–117.
123. Kumar, S., & Shankar, V. (1982). Medicinal plants of the Indian desert: Commiphora wightii (Arnott) Bhand. *Journal of Arid Environments, 5*(1), 1–11.
124. Arancon, N. Q., et al. (2003). Effects of vermicomposts on growth and marketable fruits of field-grown tomatoes, peppers and strawberries: The 7th International Symposium on Earthworm Ecology Cardiff·Wales·2002. *Pedobiologia, 47*(5–6), 731–735.
125. Olayinka, A., & Ailenubhi, V. (2011). Influence of combined application of cow dung and inorganic nitrogen on microbial respiration and nitrogen transformation in an alfisol. *Nigerian Journal of Soil Science, 2*, 15–20.
126. Bhardwaj, M., & Alia, A. (2019). Commiphora wightii (Arn.) Bhandari. Review of its botany, medicinal uses, pharmacological activities and phytochemistry. *Journal of Drug Delivery and Therapeutics, 9*(4-s), 613–621.
127. Bhatt, G., & Dixit, R. (1974). A preliminary study on extensive cultivation of guggul at Mangliawas Hebal Farm, Ajmer, Rajasthan. *Indian Journal of Medical Research, 9*, 51–58.
128. Kumar, V., Singh, S., & Singh, R. (2020). Phytochemical constituents of guggul and their biological qualities. *Mini-Reviews in Organic Chemistry, 17*(3), 277–288.
129. Mruthunjaya, K., et al. (2019). Guggul – A treasure of chemical constituents. *International Journal of Pharmacognosy and Phytochemical Research, 11*(2), 49–52.
130. Cunningham, A., et al. (2018). Rising trade, declining stocks: The global gugul (Commiphora wightii) trade. *Journal of Ethnopharmacology, 223*, 22–32.
131. Singh, H., et al. (2019). Novel drug delivery approaches for guggul. *Plant Archives, 19*(2), 983–993.
132. Kulhari, A., et al. (2015). Quantitative determination of guggulsterone in existing natural populations of Commiphora wightii (Arn.) Bhandari for identification of germplasm having higher guggulsterone content. *Physiology and Molecular Biology of Plants, 21*(1), 71–81.

133. Nohr, L. A., Rasmussen, L. B., & Straand, J. (2009). Resin from the mukul myrrh tree, guggul, can it be used for treating hypercholesterolemia? A randomized, controlled study. *Complementary Therapies in Medicine, 17*(1), 16–22.
134. Mithila, M., & Khanum, F. (2014). The appetite regulatory effect of guggulsterones in rats: A repertoire of plasma hormones and neurotransmitters. *Journal of Dietary Supplements, 11*(3), 262–271.
135. Rout, O. P., Acharya, R., & Mishra, S. K. (2012). Oleogum resin Guggulu: A review of the medicinal evidence for its therapeutic properties. *Int J Res Ayurveda Pharm, 3*, 15–21.
136. Dolara, P., et al. (2000). Local anaesthetic, antibacterial and antifungal properties of sesquiterpenes from myrrh. *Planta Medica, 66*(04), 356–358.
137. Shonouda, M., Farrag, R., & Salama, O. (2000). Efficacy of the botanical extract (myrrh), chemical insecticides and their combinations on the cotton leafworm, Spodoptera littoralis boisd (Lepidoptera: Noctuidae). *Journal of Environmental Science & Health Part B, 35*(3), 347–356.
138. Massoud, A., & Labib, I. (2000). Larvicidal activity of Commiphora molmol against Culex pipiens and Aedes caspius larvae. *Journal of the Egyptian Society of Parasitology, 30*(1), 101–115.

Chapter 23
Glory Lily

Khalid Sultan, Shagufta Perveen, Sara Zafar, Abida Parveen, Naeem Iqbal, and Muhammad Riaz

23.1 Introduction

Phylum: Tracheophyta
Class: Liliopsida
Order: Liliales
Family: Liliaceae
Genus: Gloriosa
Specie: Superba
English Name: Glory Lily
Common Name: Langali, Visalya, Karihari
Species and Varieties: *Gloriosa lutea, G. superba, G. magnifica, G. longifolia, G. virescens, G. plantii, G. latifolia*
Distribution: Africa, India, Asia, Sri- Lanka, America, Madagascar
Uses: drugs

Gloriosa is also known as glory lily, gloriosa lily, or the lily flower. It belongs to the Liliaceae family, subfamily Wurmbacoidae [1]. It's a fascinating genus with a lot of research going on it. Agnishike, Akka-thangiballi, Karadikannina gadde, or Nangulika are some of the Kannada names for it, while Karihari is the Hindi name. Since time immemorial, the plant has been employed in traditional Indian medicine. Its tubers have been used as a tonic, antiperiodic, anti-helminthic, and

K. Sultan · S. Perveen (✉) · S. Zafar · A. Parveen · N. Iqbal
Department of Botany, Government College University, Faisalabad, Pakistan

M. Riaz
Department of Pharmacy, Shaheed Benazir Bhutto University, Sheringal, Pakistan

M. Zia-Ul-Haq et al. (eds.), *Essentials of Medicinal and Aromatic Crops*,
https://doi.org/10.1007/978-3-031-35403-8_23

anti-snakebites and scorpion stings, according to reports. The medication is a gastrointestinal irritant that can cause vomiting and bowel movements [2].

It can be used to induce labor or as an abortifacient. Colic, chronic ulcers, piles, and gonorrhea are all conditions that it is thought to help with. It's employed as a cataplasm for urological symptoms and local treatments against parasitic skin infections. Children's asthma has been claimed to be cured by applying the leaves as a paste to their foreheads and necks. Head lice can be treated with juice from the leaves. Alkaloids, particularly colchicine and gloriosine, are responsible for the drug's therapeutic benefits. Gout is a prevalent disorder in temperate areas of the world, and colchicine is used to treat it [2].

23.2 Origin and Distribution

Gloriosa superba Linn. is found in Assam (India), the Himalayas Northwest, and the Deccan Field Islands [3–5]. This plant grows natively throughout Asia and Africa [6, 7] India [4, 8] and Sri Lanka are examples of equatorial nations where it can be found [9–14]. Glory lily, flame lily, tiger claw climbing lily and creeping lily are some of the popular names given to this species [9].

This magnificent lily, which is Zimbabwe's national flower, is native to Africa. It is quite common throughout India and Tamil Nadu's official flower [15]. Gloriosa is now a common pot plant in gardens because of people's love of floral beauty [11, 16].

Gloriosa is a tropical Asian and African plant native to the region. The name of the genus comes from the Latin word glorious, which means "glory" in Latin. This plant grows in Indian Assam, the Deccan peninsula, and the Himalayas reaching elevations of 2120 meters. It grows along the Western Ghats in Karnataka and is rather common. It can also be found on Madagascar, Sri Lanka, and Indochina islands. In India, roughly 2000 acres are planted with this crop [2].

23.3 Description of the Plant

This plant is herbaceous in nature and perennial that can reach a height of 1.5 m above ground level and grows about a length of 3.5 to 6 m. The vines are tall, with weak stems and tuberous roots that rely on cirrhosis tips to support themselves [17, 18]. The leaves are oblong, lanceolate, and acuminate, with twisted tips that function as tendrils. At the bud stage the petals of this plant hang on the ovary and when they mature, they stand tall, exposing the stigma at right angles [2, 19, 20]. Each stamen has a lengthy anther with a lot of orange-yellow pollen. The ovary is made up of three cells and is shaped like an ellipsoidal capsule. A capsule contains many seeds, each of which is warty and compressed dorsally.

The glory lily plant is a very beautiful creeper and has a perennial hollow stem that is near about 6 centimeters. This plant emerges in the rainy season from the underground tuber per year. This plant belongs to the monocot family and is widely distributed throughout India. Several species of glory lily plants were found in India like *G. grandiflora, G. superba, G. planti G. simplex, G. rotheschildiana, G. lutea, G. longifolia, G. sudanica*, etc. leaves of this plant are sessile, etiolated, lanceolate, spear-shaped and alternate with curve end that helps the plant for creeping and climbing [21, 22]. In the months of November and March, the flower blooms of this plant the color yellow-edged brilliant and red [23]. Colored, large and solitary from green to yellow to orange to scarlet to crimson, the flowers are found at the ends of the branches. Small insects are unable to pollinate the enormous flowers' unique architecture, which includes At the point of the ovary where it is attached to the style, the curve is almost 90 degrees, six radiating anthers, and six perianth lobes twisted backward [24]. Oblong, ellipsoid capsules make up fruits. Many spherical seeds are produced. A single fleshy, cylindrical, V-shaped tuber produces one to four stems. Once they have grown to a critical size, daughter corms only bloom after 2–3 years [25]. 3–4 years are needed for an economically viable tuber crop. Although corms have been the traditional method of propagation, the rate of proliferation is still quite modest [21, 26, 27]. Only long-beaked birds like *Nectarinia Asiatica* and *Nectarinia zeylonica* and big insects like bumble bees have been observed visiting these blooms [28]. Although the wind is another aspect that would be assisting in its pollination, this reduces the likelihood of effective cross-pollination.

23.4 Crop Improvement

A crossbreeding program including *G. superba, G. lutea*, and *G. planti* was attempted in 1979. (diploids). Tetraploids of *G. carsonii, G. virescens*, and *G. richmondensis* in various combinations. *G. virescens x G. carsonii* and its reciprocal, *G. virescens x G. richmondensis* and *G. carsonii x G. richmondensis*, were the only successful hybrids. All of the hybrids' morphological and cytogenetic behaviors have been determined. According to these investigations, the tetraploid Gloriosa species had a higher cross-ability than the diploid Gloriosa species. Furthermore, despite being involved in several hybridization combinations, *G. superba* has always failed to create hybrids [2].

23.5 Soil

It grows best in acidic sandy loam soils with good drainage. This plant best grows in black and red loamy soils that have medium water holding capacity and rich in humus in regions of south India. This crop can be grown in soil with a pH of 6–7 [2].

23.6 Climate

The glory lily is a tropical flower that thrives in hot, humid climates. It may reach a height of 600 meters above sea level when growing in natural settings. This crop thrives in a climate with an average annual rainfall of 373 cm, evenly spread throughout the year. It cannot resist prolonged moisture stress and, during dry conditions, requires repeated irrigations till flowering. 15–20 °C during the day and 10–15 °C at night are ideal for its growth and flowering. It's important to have a lot of relative humidity.

Continuous cloudy weather, on the other hand, is ideal for the disease Curvularia, a severe hazard that causes 75–80% vine mortality. In the event that the crop must be cultivated in such conditions, appropriate preventive precautions must be taken [2].

23.7 Cultivation

23.7.1 Propagation

This plant commercially can be grown from the rhizomes, from seed or from the subterranean. Flowering takes about 3–4 years for plants to develop from seeds. As a result, seed propagation is not preferred by growers, save in the case of research. During the growth season, Gloriosa grows a bifurcated tuber with only one developing bud on each fork. Tubers must be handled with caution because they are brittle and can shatter. The tuber will not sprout if the budding bud suffers any harm. Because the size of the tubers affects the vine's vigor, flowering, and fruiting capabilities, it should weigh at least 50–60 g [2].

Very few seeds are produced by glory lilies [29]. The traditional method of propagation has a number of drawbacks: It is necessary to reserve 50% of the produce in order to prevent from spreading of soil-borne illnesses, from one area to another and crop-to-crop spreading during the 2–3 months of storage between crop harvest and crop planting [30]. Gloriosa is frequently propagated vegetatively in horticulture, but the growth is very slow [31]. A sophisticated developmental process thought to be influenced by physiological genetic and environmental factors results in the creation of underground stems like corms and tubers [32, 33]. Chances are unaffected by the low seed set. *G. superba* reproduces through its tubers, but it can also reproduce through its seeds, tubers, and flowers. Cross-pollinated seeds ensure fresh gene combinations that allow colonizing the plant species in different climatic regions and in the areas of tropical, subtropical, and temperate regions [24].

During their first year, plants grown from smaller tubers do not blossom. By slicing them in half, large tubers can be split in half. From the month of May, the latent tubers begin to grow. Planting in July and August has been observed to support good growth and production under Bangalore circumstances. Planting requires approximately 2.5 to 3.0 tha of tubers. To prevent the tubers from decaying before they sprout. Planting should only be done with healthy tubers. Suitable fungicides, such as Emisan-6 (methoxyethyl mercury chloride) @ 0.08%, should be applied to chosen tubers or tuber parts [2].

23.7.2 Field Preparation and Planting

Plow and harrowed the land multiple times till it is finely tilted. Remove all of the grass stubs and roots. To prevent waterlogging during rainstorms, the field must be properly leveled, and drainage measures installed. After then, the field is separated into manageable subplots. Incorporate 15–20 t/ha of FYM or compost into the soil. At 45–60 cm intervals, one-foot-deep furrows are opened. Plant the treated tubers at a depth of 6–8 cm, with a plant-to-plant distance of 30 to 45 cm depending on the soil type. Cross-pollination has been shown to improve fruit sets when plants are spaced closer together [2].

Glory lily needs a trellis or standards for support because it is a climber. The sprouted tubers are trained 1.5 metres over a support constructed of galvanised iron wire. Additionally, farmers use dead wood from Dodonaea viscosa, cashew, neem, and other locally accessible trees as well as live fences like Balsmodendron (Kiluvai) to support the vines [34].

23.7.3 Micropropagation

Plants used as raw materials in herbal treatment compositions are in high demand right now. The resurgence of herbal medicine as an alternative health care system has the potential to reduce allopathic drug adverse effects. As a result, developing a strategy for mass cultivation of a number of valuable medicinal plants while increasing yield is unavoidable. One of the manufacturing tools for increasing productivity is micropropagation.

Many challenges in the cultivation of medicinal plants can be solved using a biotechnological approach, particularly in large-scale planting material multiplication [35]. The stages of micropropagation plant production were described. Several tissue culture approaches for bulk plant multiplication have been established and explored by numerous researchers [36–39].

Liliaceae plants have recently been cultivated for medicinal and ornamental purposes [40]. *G. superba* was shown to be a major source of the natural alkaloid 'colchicine' among them. [41].

23.7.4 Irrigation

During the sprouting period, frequent irrigation is essential to keep the surface moist and prevent the creation of a hard pan, allowing for easy sprouting and the emergence of the growing tip from the soil. To keep the tubers from decaying, don't irrigate them until after they've finished flowering. Excessive watering causes yellow or brown areas on the leaves, which fall off prematurely [2].

23.7.5 Crop Monitoring

It is required to provide some form of assistance in order to grow gloriosa successfully. Because the stem is so fragile, the plants should be staked or fastened to wires or allowed to climb on some form of the frame until they are around 30–40 cm tall [2].

23.8 Constituents

Medicinal plants contain chemicals that are biologically important and play a significant function in the medication development [42]. Several phytochemical elements are obtained from various sections such as the stem, root, leaf, fruit, park, and seed [43]. Flavonoids are abundant in medicinal plants, and each category of flavonoids possesses antioxidant properties [44]. Catechin components are the most potent flavonoids when it comes to protecting the organism from reactive oxygen species (ROS [45]. Flavonoid contents like kaempferol, rutin, quercetin and myricetin are anti-inflammatory, anticancer, antioxidant, anti-allergic and antiviral in properties [46–48]. Thirty plants from Indian Ayurveda medicine exhibit anticancer properties [49].

Researchers looking for medicinally essential bioactive chemicals are now focusing their efforts on herbal items. *Gloriosa superba*, sometimes known as Malabar glory, is a Liliaceae-family perennial creeper endemic to Africa and Southeast Asia. The glory lily is both Zimbabwe's national flower and the state flower of India's Tamil Nadu province. It has a thin stem that grows 20 feet every year. Plant leaves are 6–8 inches long with an oval shape at the tip of the leaves with a thread-like apex that aids in tree climbing. Among the medicinal plant species, *G. superba* is an endangered species [50] and Colchicine and gloriosine are two of the plant's poisonous alkaloids [51, 52].

In cytological and plant breeding studies, colchicine is sometimes used to double chromosomes. External application of *G. superba* tuber paste is used to treat parasite skin infections, and this plant has also been utilized as an ayurvedic medicinal herb to treat inflammatory diseases, gout, ulcers, and bleeding [11, 53]. This plant is

Table 23.1 Pharmacological properties in different parts of the glory Lilly

Extract	Parts	Activities	References
Water and Ethanol	All parts	Anthelmintic	[55]
n-Butanol and chloroform	Seeds, tubers, leaves	Anti- cancer and antimicrobial	[56]
Methanolic	Stem and leaf	Antimicrobial and antioxidant	[57]
Petroleum ether, methanolic and aqueous	Tubers	Mutagenic, antibacterial and antifungal	[58]
Water, methanolic and acetone	Tubers	antioxidant	[59]
Alcoholic	Tubers	Antihemolytic	[60]
Aqueous	Leaves	Anticoagulant	[61]
Acetone	Stem and tubers	Antifungal	[62]
Alcoholic	Tubers	Antimicrobial	[63]
acetone	Leaves and tubers	Antibacterial	[64]
Alcoholic	Tubers	Anthelmintic	[63]
Hexane, methanol and chloroform	Seeds and tubers	Against bacteria	[64]
Methanolic	Leaves, stem	Antimicrobial and antioxidant	[57]

being used in suicide attempts, abortions, and murders as the presence of the colchicine [54]. The roots are said to be safe for African porcupines and some moles to eat [55] (Table 23.1).

G. superba is an effective abortifacient, causing a fetus to be expelled while it is present in the womb. Astringent acrid, purgative, germicidal, anthelmintic, bitter, and cholagogue characteristics are all present in root extracts. Snakebite paste is an antidote, and the plant's extract is a CNS depressant [63, 65]. When the tuberous root of *G. superba* is heated in sesame oil and applied twice a day to arthritis-affected joints, it lowers inflammation [66]. Glory lily can treat infection on the skin, fever, wounds, contraction of the uterine, piles, abnormalities of the blood, poisoning and inflammation [67].

Gloriosa species' seeds and tubers are the main sources of colchicine. The alkaloid colchicines are thought to be abundant in G. superba tubers [68, 69]. The tubers can be used to treat ulcers, bleeding piles, inflammation, scrofula, white discharge, leprosy, dyspepsia, skin conditions, helminths, intermittent fever, snake bites, sluggishness and baldness [70]. It is also thought to be helpful in encouraging labour and placenta ejection [71]. Sesame oil is used to cure the *G. superba* tuber [66].

Given the foregoing, it's no surprise that *Gloriosa Superba* pharmacological properties have piqued people's curiosity. As a result, the current review includes updated information on *Gloriosa superba* phytochemical and pharmacological qualities, as well as its various applications.

23.8.1 Phytochemical Properties

Colchicines, 1,2-dimethyl colchicine, gloriosine, 3-demethyl colchicine, tannins, 2,3-dimethyl colchicine, N-diacetyl colchicines, superbine, N-formyl and colchicocide are all found in the tubers of *G. superba* [72]. This plant's seed and rhizome contain the main component colchicine [73] and Gloriosine is another crucial molecule [51, 52]. *G.superba* tubers contain 0.25% colchicine, while flowers include N-formyl de-Me-Colchicine and luteolin [74, 75] it was found that in the seeds of this plant, 3–0-dimethyl colchicine was present.

23.8.2 Presence of Polyphenols

G. superba seeds and tubers have total phenolic concentrations of 0.561 mg/g and 0.975 mg/g, respectively, according to a phytochemical study. The contents of the carotenoids presen in tubers were near about 22.74 mg per 100 g while it presents in seeds was about 25.62 g. while the concentration of the ascorbic acid was found in seeds 23.34 and in tubers was 21.06 g/100 g of weight [76].

Glycosides, Carbohydrates, flavonoids, alkaloids, phenolics, terpenoids, and steroids have all been discovered in *G. superba* seeds. Flavonoids, carbohydrates, alkaloids, terpenoids, and steroids were also found in the leaves of *G. superba*. In tubers of the glory lily, other compounds like vitamin C, alkaloids, phenols, flavonoids glycosides, carbohydrates, saponins, minerals, and vitamin E were found [77]. Saponins, flavonoids, alkaloids, phenols, glycosides, tannins, and steroids were discovered in the leaves and tubers of *G. superba*. Furthermore, the *G. superba* plant is said to be high in a number of physiologically active chemicals that could be exploited as a source of crude medications to supplement traditional therapies [77–80].

23.8.2.1 Pharmacological Activities

It's a long-held belief that plants have therapeutic properties, and even though the twentieth century saw amazing advancements in synthetic organic chemistry, more than 25% of medicines prescribed in industrialized nations come either directly or indirectly from plants. However, there is still little research on plants used in traditional medicine, especially in the field of clinical microbiology [81]. The role of biological control in chemical antimicrobials has greatly increased recently [82], primarily as a result of synthetic product adverse effects. Even commonly consumed foods contain several naturally occurring antimicrobial substances that aid in preventing food from degrading [83]. The recent emergence of bacterium strains with lower resistance to antibiotics and the rising incidence of multidrug-resistant bacteria enhance the possibility of incurable bacterial illnesses and make it more urgent to find alternative infection-control tactics [84]. Comparatively speaking to synthetic products, plant-based antimicrobials have a high safety profile and few negative effects. They also possess tremendous therapeutic promise for treating a variety of infectious disorders [85].

The pharmacological activity of the various components of *G. superba* was distinct. Gout is characterized by the deposition of uric acid microcrystals in the joints, which is produced by a malfunctioning endogenous inactive synthesis regulating mechanism. Colchicine is supposed to stop the formation of new crystals, which appears to be necessary for acute gout to continue. In polyploid breeding in agricultural research, these alkaloids are also used as polyploidizing agents.

Gloriosa superba Linn. has a colchicine concentration that ranges from 0.15% to 0.3% in tubers and 0.7% to 0.9% in seeds. Until recently, only the tubers were harvested; however, after learning that the seeds contain a significant amount of alkaloids, the crop is now farmed primarily for its seeds, which are in high demand both domestically and internationally [2].

Scleroderma is an autoimmune illness with fibrosis, vascular changes, and autoantibodies. Individual organ system issues are treated rather than scleroderma itself, which has no therapy options [86, 87]. The treatment is tailored to the individual patient and focuses on the disease's signs and symptoms. Colchicine, thalidomide, extracorporeal photophoresis, IFN-alpha, and D-penicillamine have all been demonstrated to be effective in the treatment of aggressive disease [88].

23.8.2.2 Anticancer Effects

Plant secondary metabolites are recognized as bioactive substances for both primary and secondary prevention of cancer and are well-classified and recognized as chemo preventive agents [89–91]. As a result, a lot of data show that consuming more secondary metabolites can reduce the risk of developing cancer [92, 93]. These substances influence metabolic and signalling pathways, which in turn regulate angiogenesis, restrict the creation of microtubule assemblies in cells, and prevent apoptosis [94]. Flavonoids, alkaloids, terpenes, steroids, lignans,, curcumins, glucosides, saponins and phenolics are some of the secondary metabolites that can be generally categorised [95].Because these plant-based bioactive chemicals have the necessary geno-protective actions, such as preventing DNA damage in healthy cells, these secondary metabolites can be employed to create customised cancer prevention regimens, either individually or together [96–98].

Methanol leaf extract of the glory lily plant is effective against DLA tumor cells, a novel green strategy that has been used to reduce the negative effects of the silver nanoparticles from the environment and the human. From the methanolic extract of this plant encapsulated silver nanoparticles with active ingredients of the colchicines are poisonous while the without cap particles are not poisonous. *G. superba* leaves make good starting materials for the synthesis of AgNPs because they are stable, eco-friendly, economically viable, and produce particles with lower sizes. The anticancer effect of the silver nanoparticles are apoptotic inducing agents to the tumor cells and cytotoxic against the tumor cells. It was also observed to extend the average lifespan of DLA tumor-induced mice compared to methanol extract. By adding *G. superba* leaf to the AgNs, then the silver nanoparticles are less toxic as compared to others [99].

Both the rhizome and the seeds of this plant were being used in traditional medicine to cure the disease of the gout [100, 101]. The leaf juice is used to kill head lice and is also used in arrow poisons. It also has anti-mosquito effects [102]. Glory lily rhizome is used as antidotal against snake bites, and it is widely used to discourage snakes on Indian windowsills. Many cultures see the species as possessing a variety of magical characteristics [103–105]. Lipoxygenase inhibition was excellent in the crude extract and fractions. *Gloriosa superba* includes alkaloids such as N-Formyldesacetyl-colchicine, 3-desmethyl colchicine, 2-desmethyl colchicine, and lumicolchicine, as well as chemicals such as salicylic acid and chelidonic acid [106].

23.9 Antimicrobial Activity

The phytochemicals derived from *G. superba* tubers have antibacterial activity. This plant has strong effects against the microbes like gram-negative bacteria *Escherichia coli* [58]. *G. superba* extracts were tested for antimicrobial activity against *Candida glabrate, Candida albicans, Microsporum canis, Staphylococcus aureus,* and *Trichophyton longifusus* [107]. The antibacterial effects of acetone, ethanol, hexane and methanol extracts from the stem and root of the glory lily were effective against *A. flavus, E. coli, S. aureus*, and *A. niger*. The plant's acetone extract, on the other hand, demonstrated the most potent antifungal action against *E.coli* [62]. Significant antibacterial activity against *C. albicans*, a fungus strain, and gram-negative bacteria was found in the alcoholic extract of *G. superba* tubers (Table 23.2).

23.10 Anthelmintic Activity

The alcoholic extracts of *G. superba* tubers were found to be effective against the earthworm *Eisenia fatida* [63]. The activity of ethanol and water extracts of the entire *G. superba* plant against the Indian earthworm *Pheretima posthuma* was studied. These two extracts were when compared then it was seen that it has produced significant activity and exhibited a strong anthelmintic action [55].

Table 23.2 Pharmacological activities of the Glory lily plant

Uses	Countries	References
Abortifacient	India, Sri Lanka, Tanzania, Nepal, Zambia, Bangladesh, Uganda	[11, 104, 108–114]
Antiparasitic and anthelminthic	South Africa, india	[103, 109]
Wounds healing	Tanzania, India	[104, 108, 109, 111, 115–117]
Dislocation and arthritis	Sri lanka, india, Nigeria	[104, 108, 109, 118]
Baldness	India	[58]
As cough relief	Sierra, India, Ivory	[104, 108, 109]
Ulcers	India	[119, 120]
As female sterility	South Africa, India, Ivory Coast, Zambia, Congo,	[103, 108–110, 121]
Hemorrhoids	India	[119, 120, 122, 123]
Malaria & Fever	Tanzania, Bangladesh, India	[11, 108, 109, 111]
Inflammation	India	[119]
Tumor & Gout	Thailand, Ethiopia, India	[119, 120, 124, 125]
As succidal poison	India, Burma, Cambodia, Kenya, Tanzania, Sri Lanka, Zambia, Nigeria	[104, 109, 126, 127]
Indigestion	India	[58]
As muscle relaxant	India	[128]
Bites of scorpions	Zambia and Sri Lanka	[108–110]
Worms of intestine	India	[117]
Stomach pain relief	Kenya, Congo, Nepal, Mozambique	[103, 104, 109, 113, 129]
Pains of teeth	Zimbabwe	[104, 130, 131]
Human hair lice killer	South Africa, Indonesia, Cameron, india, Senegal, Guyana	[103, 104, 108, 126, 132]
General and abdominal pain	Thialand, Kenya, Ivory Coast, Nepal	[104, 108, 109, 113, 133]

23.11 Lily Plant Antioxidant Effects

It was suggested that the plant extract from methanol can be used as a strong antioxidant and has antimicrobial activity. It is also used as a natural source of antimicrobial and antioxidant [57, 134]. It was also reported that the extracts of tubers, leaves, and seeds of the glory lily plant have significant antioxidant activities.

23.11.1 Anti-inflammatory Activity

The G. superba tuber has been demonstrated to have potent anti-inflammatory properties in alcoholic, hydroalcoholic, and aqueous extracts [135, 136]. These studies revealed that the highest anti-inflammatory effect was displayed by an aqueous extract of 250 mg/kg *G. superba* tubers.Colchicine was administered orally at doses of 2, 4, and 6 mg/kg body weight. These doses produced respective inhibitions of 48.[illegible], 68.7, and 79.1%, while the once-daily, 4-day treatment with 100 mg/kg of phenylbutazone produced a 30.9% inhibition. These findings unequivocally show that colchicine is more potent as an anti-inflammatory drug [137].

23.12 Other Benefits

When a mouse was treated with an alcoholic extract of the glory lily plant, it increases its immunity like as venom dose [60]. In an analysis of hyposaline-induced hemolysis, it was seen that the glory lily plant dose impacts the human red blood cell membrane stabilization. A result of 2.97 mg/ml for the IC50 indicated that *G. superba* leaf extracts have clotting effects by reducing the thrombin-induced clotting [60].

In the plant of lily colchicine is an important marker. It was also seen that chemicals like 3-demethylcolchicine and lumicolchicine were present [138]. This chemical was most effective against Familial Mediterranean Fever and gout disease [139], as well as being applied to treat other illnesses like scrofula, cancer, and anthelmintic, and acts as an antipyretic, antiabortion, and purgative. Naturopaths frequently utilize everything for a variety of treatments, including the relief of chronic back problems. It is also applied to the long-term management of Behcet's illness as an anti-inflammatory drug [140].

The use of colchicine for the cure of pericarditis has previously been suggested. Colchicine may be helpful for patients with recurrent pericarditis, according to emerging research, despite the fact that doctors are frequently dubious about its potential for treating this condition [141]. In order to create polyploids, it is utilized in genetic investigations, particularly in biological studies on breeding. and tubulin-binding assays as a positive control, given that colchicine has a strong connection with the tubulin [142]. This chemical was being used in the treatment of cancer cells [139] but it is also a poison for the human who uses it to treat as anti-tumor medicine and it has more effects on the cells of the animals as compared to the plant cells. When low amount of this is given to the animals it affects them by dividing the cells hence it is very lethal to the animal's [140]. On the other hand, plant cells can endure a state of mitotic arrest brought on by colchicine (referred to as C-mitosis) for a number of days before restarting their division after the colchicine is withdrawn [141].

Additionally, the ED50 (effective dosage) value for the *G. superba* extract against the p388 cell line was extremely low. They claimed that 5- fluorouracil is similar to 3-dimethyl-N-formyl-N-deacetylcolchicine which is a popular drug against cancer, indicating the compound's potent toxicity to the cell properties as well [143]. The number of colchicines in glory lilies ranges from 0.1% to 0.9%, however, in Indian corms, it is only 0.02% [144]. In plant breeding, colchicine has been employed to cause polyploidy and mutation [26]. Colchicine showed a 100- percent result in the metaphase through the fuchsia breeding, when a mosquito applied this when it was making the chromosome during the process of the metaphase [53]. The polyploids were prepared by using the glory lily seeds extracted from the colchicine [145]. A report claims that giving fresh extracts from glory lily tubers to seeds of maize plants before planting led to the development of tetraploid roots. Colchicine is a cytotoxic drug that has been used to treat cancer that is incurable [146].

It was claimed that *G. superba* had hepatoprotective properties [24]. It assessed the ability of its rhizomes' crude extract to block the enzymes lipoxygenase, acetylcholinesterase, butyrylcholinesterase, and urease as well as its successive fractions of rhizomes' extracts in chloroform, ethyl acetate, n-butanol, and water. The chloroform fraction had the highest enzyme inhibition potency (90%) out of all the studied fractions for butyrylcholinesterase, lipoxygenase and acetylcholinesterase. While in the fraction urease was not inhibited [147].

Both the seed and the rhizome are used to cure gout in traditional medicine, and its main therapeutic purpose is as a source of colchicine [100, 101, 148, 149]. The juice from the leaves is used to eradicate head lice and has anti-mosquito effects in addition to being a component of arrow poisons [80]. In India, the rhizomes are frequently placed on window sills to ward off snakes since they contain antidotal qualities for snakebites. Many cultures think the species has a variety of magical abilities [103, 104, 108].

23.13 Effects

Tubers of glory lily consumption can result in colchicine toxicity, [150] which frequently happens when someone ingests the plant (*Ipomoea batatas*),which is similar to the glory lily plant [151]. The primary symptoms in adults are gastric, however the symptoms in children are more serious which would include heat, edema, hypoglycemia, irregular heartbeats, and seizures [152].

Gloriosa superba roots were consumed by a 21-year-old who mistook them for yams [51]. The quantity of colchicine consumed was estimated to somehow be 350 mg based on the roots' 0.3% colchicine composition.She began feeling sick after approximately 2 h and began to have diarrhea 8 h later that persisted all night. The menstrual cycle of this woman lasted further 20 days after it had already begun. She experienced extensive alopecia after 12 days, but it soon cleared up.

An attempt at suicide by consuming Gloriosa superba tubers resulted in a 29-year-old man experiencing stomach distress, intense dryness, scorching inside the throat and mouth, headache, and stomach pain [54]. The acute renal failure resulted from dehydration. ST-segment elevation was seen on the electrocardiogram. White blood cell and platelet counts decreased along with hemoglobin concentration.

Colchicine is a natural compound obtained from the tuber of *Gloriosa superba*. It was introduced into clinical practice more than 200 years ago and is still used to treat spontaneous gout and cyclosporin A-induced gout in organ transplant recipients. The major neurological side effect of colchicine is toxic, vacuolar myopathy with lysosomal accumulation, but peripheral neuropathy may also occur, probably secondarily to impaired microtubular assembly [153]. The risk of developing peripheral neuropathy is increased by concomitant renal impairment, a rather common complication of gout, and by chronic administration of the drug, which is generally required [153].

Colchicine neuropathy is uncommon, and even when it occurs, the distal sensory damage is minimal. In the affected patients, axonal damage affecting both myelinated and unmyelinated fibers is seen, and recovery is usually complete after the treatment is stopped. Sural nerve studies have, however, occasionally revealed intracytoplasmatic inclusions in endothelial, perineurial, and Schwann cells, as well as segmental demyelination. At the neurophysiological level, axonopathy is usually coupled with an aberrant spontaneous activity that is more visible in the proximal muscles on an electromyographic test, resulting in a colchicine mononeuropathy [153].

Colchicine injection in rats and monkeys resulted in two distinct forms of brain injury. Tetradentate colchicine, at certain doses, preferentially damaged DGC in rats, but damage in monkeys was less selective and more severe. To examine the structure-function link between tubulin binding and DGC death, experiments were conducted using several tubulin-binding agents. These and other medications' tubulin-binding properties were documented in the literature, but this did not indicate that they could harm DGC. By monitoring the EEG in rats and monkeys given phenobarbital treatment, the role of seizure-induced cell death was examined. According to the statistics, seizures are a rare epiphenomenon of colchicine's effect. Colchicine, according to our theory, damages the brain by causing an unfocused inflammatory response and is not a selective neurotoxic. Both the dose and the species have an impact on this reaction. The medical ramifications of the current and suggested usage of colchicine were covered as our final point [154].

There is evidence that 10 mg of colchicine is a hazardous amount that can have fatal effects on people [155]. Colchicine may actually increase female fertility and pregnancy outcomes, according to this study. Colchicine is not linked to lower female fertility rates, a greater likelihood of miscarriages, or stillbirths. A discovery that is in direct opposition to [156], who discovered colchicine had early abortifacient and oxytocic effects on the reproductive system of female rats. Humans that consume more than 40 mg of colchicine die within 3 days of doing so [157].

the negative effects of using it to treat FMF. Patients who are elderly or who suffer from liver or renal impairment experience more side effects [155]. The symptoms appear 2 h after ingesting lethal amounts of colchicine. The initial symptoms of poisoning include vomiting, numbness, acute throat discomfort, and diarrhea that causes dehydration. In the 2–3 weeks following poisoning, the two main symptoms that manifest are alopecia and dermatitis [155, 158, 159]. Twenty-four to seventy-two hours after consumption, multi-organ failure may appear. These include hemolytic anemia, arrhythmias, paralysis, disseminated intravascular coagulation, liver injury, respiratory distress syndrome, and bone marrow depression [155, 158, 159]. In many cases, overdosage results in a cholera-like syndrome that includes dehydration, shock, acute renal failure, alopecia, hyperthermia, hepatocellular failure, epileptic convulsions, coma, and death [155].

23.14 Manures and Fertilizers

The *Gloriosa superba* plant is a demanding crop, requiring additional soil nutrients to plant growth, blooming, seeds release, and maturity of the pod, as well as to supply the nutritional needs for tuber growth and multiplication for the succeeding generation. The tubers' shallow roots are largely made of fibrous roots, which may gather water and nutrients in one place. For these tiny fibrous roots to absorb the nutrients in increasing amounts, the nutrients must be given in several divided doses. The amount of fertilizer used should be adequate to meet nutrient needs during various critical phases of crop growth.

The percentage or ratio of the (100 K, 150 N & 50 P/ha) of NPK was given to the soil before sowing. While the amount of vermicompost was given at the rate of 5 t/ha with a combination of zinc sulfate at the rate of 25 kg/ha sodium molybdite @ 0.5 kg/ha and iron sulfate @ 50 kg/ha and Borax with the quantity of 10 kg/ha at the time of plantation. While the foliar application of the iron sulfate 1% boric acid at 0.56% and zinc sulfate at 0.5% was given [34].

23.15 Intercultural Practice

Drought tolerance is established in the crop. The majority of glory lily producers in Tamil Nadu use drip irrigation. Irrigation should be done twice or three times a week throughout the first 3 months. Watering to the plant was given two times within a week in the months of October and December. Because flowering, pollination, pod set, and pod maturity all happen at the same time, irrigation must be equally applied to all stages of growth. Weeding is necessary frequently throughout the early stages of crop growth to control weeds. If this wedding is not done then plants compete with other plants for their survival as a result the growth of the plants might slow down. Chemical weed control is only possible when the rows and plants are separated by a large distance [34].

23.16 Pests and Their Control

23.16.1 Weeding

The gloriosa plantation requires continuous weeding in the early stages to keep weeds from competing for moisture and nutrients, limiting the plant's growth. When weeding, take special care not to damage the growing tip, as it will not sprout again during the season of damaged. Only considerable space between rows and between plants allows for chemical weed control. [2].

23.16.2 (Polytela gloriosae) Caterpillar of Lily

It is a harmful and common pest of glory lily that can be found from seedling to maturity at any time during the cropping season (August to February) [160].

23.16.3 Semilooper (Plusia signata)

On the undersides of the leaves, eggs are placed singly. The green-colored larvae that live beneath the leaves defoliate the leaves extensively. The larvae, on the other hand, are not attracted to the developing tips. In addition to lily caterpillars, these semiloopers are considered pests. Glory should be managed with foliar application of neem seed kernel extract (5%) or neem oil (3%). When the number of pests was most on the plant (0% effect on the leaf) then the insect killers like chlorpyriphos, quinalphos, azadiractin, and the bacillus thuringiensis was given to the plant. After the spray, the plant might grow properly and can repeat the spray [160].

23.16.4 The Disease of Thrips (Thrips tabaci)

Glory lily plants are infected with necrosis by the thrips which is a viral disease. Plants infected with the virus take on a bronze or purple hue. Curled and deformed leaves On the leaves and leaf stems, many minute dark dots appear. Wilting and death of affected leaves is possible. The foliar spray of the spinosad or the fipronil might be applied on the plant for 15 days, with a gap of 7 days to control the thrips disease [160].

23.16.5 The Disease of Root Rot (Macrophomina phaseolina)

Gloriosa superba suffers from root rot, which reduces yield by 20 to 30%. In patches, plants begin to wilt suddenly and completely. The signs of root rot illness include yellowing foliage, discoloration, and decomposing roots. Because of severe root rot, such plants are eventually dead [161].

23.16.6 *Leaf Blight (*Alternaria alternata*)*

It's a dangerous disease that affects *Glory lily* and can be seen between October and December. The first symptoms of the infection are circular to oval, small, light brownish spots, 2–6 per leaf, scattered throughout the tip, border, and midrib. The center necrotic lesion in each place is surrounded by concentric rings. The spots turn dark brown to blackish in hue as they progress, and consolidate. They start to change shape, and finally the damaged leaves completely blight. The ideal conditions for *A. alternata* growth were a pH range of 6.6–6.6 and a temperature range of 25–30 °C. [162].

23.16.7 *The Tuber Rot (*Rhizoctonia bataticola*)*

Gloriosa superba is susceptible to tuber rot, which can result in yield losses of up to 20%. The infection wreaks havoc on the plant's underground tuber, killing it. Infected tubers become mushy and the foliage becomes yellow in the early stages of the infection. The entire tuber becomes affected in advanced stages, resulting in a discolored mass and the plant's death. It is possible to manage the sickness. By removing affected plants, ensuring appropriate drainage, and other measures, diseases can be controlled. *Pseudomonas fluorescens* is applied to the soil at a rate of 2.5 kg per hectare coupled with neem cake at a rate of 250 kg per hectare, and tubers are dipped in *P. fluorescens* (2 g/L) or carbendazim (0.1%) for 20 min prior to planting and sprayed with tebuconazole (0.1%) [163].

23.16.8 *Yield*

Seed output varies substantially depending on the plant's vigor and age, which are in turn determined by the tuber's size. The first year's yield will be poor, but it will progressively rise with time. We may harvest roughly 200–250 kg/ha of dried seeds after 3 years from a well-managed area. Around 75% of the seed production comes from the pericarp (husk). [2].

23.17 Review of the Literature

Colchicine poisoning is recorded in a patient who ate *Gloriosa superba* tubers. The most common adverse signs included acute renal failure,, cardiotoxicity, abnormalitiesof haematology, and gastroenteritis. No hypotension or neurological symptoms were present. Changes in electrocardiography were notable and had not been previously documented [54].

Suicide attempts with poisonous extracts from the *Gloriosa superba* creeper vine are common in this area. A family involved in commercial dealings with this plant was poisoned after consuming a crude liquid extract of its root, according to this story. Gastrointestinal problems were the most common. Sweating, hypotension, jaundice, bradycardia, and convulsions were among the symptoms that the children experienced. With only symptomatic treatment, the traits could be reversed [152].

One of the poisonous plants native to Japan is *Gloriosa superba*. It contains strong alkaloids like colchicine, which binds to tubulin and stops it from forming microtubules in human cells. *Gloriosa superba* tubers contain toxic substances that can be lethal if consumed. We present a case of *Gloriosa superba* tuber-induced colchicine poisoning. *Gloriosa superba* tubers were accidentally swallowed by a 58-year-old man. It was wild yam, he thought. 30 min after intake of the plant, abdominal pain, vomiting, and diarrhoea occurred [151].

Gloriosa superba, often known as the glory or superb lily, is a tropical climbing plant with a stunning red blossom. Because of high levels of colchicine in all parts of the plant, it is poisonous. It is produced commercially in India and Africa for use in Ayurvedic medicine and as a cash crop for extracting colchicine. In Sri Lanka, it is a wild species, and commercial production is uncommon. Gloriosa tuber poisonings, both accidental and suicidal, are well-known and reported. In Sri Lanka, there have been no cases of Gloriosa seed poisoning. In other parts of the world, Google and PubMed searches revealed no known examples of seed poisoning or use with homicidal intent [164].

In Asian countries, there are numerous deadly plants. This instance emphasizes the dangers of using Gloriosa seeds or extracts to create potentially lethal poisoning, whether by accident or on purpose. Because of many consequences that can mimic a systemic infection, it would be difficult to diagnose Gloriosa as the source of poisoning without any previous information. This story demonstrates how plants can be used as biological weapons [164].

Gloriosa superba (L.) is an African and Southeast Asian perennial creeper belonging to the Liliaceae family. The glory lily is Zimbabwe's national flower and the state flower of India's Tamil Nadu state. Colchicines and Gloriosine, two poisonous alkaloids found in the plant, are used to cure gout and rheumatism. Use in traditional medicine, similar to many poisonous plants, and it has been used to treat, numerous allied species that contain colchicine. *G. superba* has antioxidant, antibacterial, antimicrobial, and anthelmintic effects throughout the entire plant. *G. superba* is also an effective abortifacient, causing the fetus to leave the womb. As a result, this paper examines the most recent knowledge on the qualities of pharmacological and phytochemical implications of glory lily extract and its various applications [165].

Plants have always been important to humans. Secondary metabolites are prevalent in certain plants, and they serve an important role in developing treatments for various illnesses. In this field, researchers are extracting various plants and extracts and then experimenting with them to see if they may provide positive results. *Gloriosa superba*, popularly known as Glory Lily, is one of these plants. This plant, which belongs to the Liliaceae family, is high in active principal substances such as alkaloids, flavonoids, terpenes, and saponins, which are mostly collected from the plant's seeds and tubers. Colchicine is the most important alkaloid being studied for its medicinal potential [79].

The antibacterial and antifungal properties of a methanol extract of *Gloriosa superba* Linn (Colchicaceae) rhizomes and its subsequent fractions in various solvent systems were investigated. The n-butanol fraction showed excellent antifungal sensitivity against *Candida albicans* and *Candida glaberata* (up to 90%) and *Trichophyton longifusus* (78%) followed by the chloroform fraction against Microsporum canis (80%). In an antibacterial bioassay, the crude extract and subsequent fractions demonstrated modest to moderate antibacterial activity. The crude extract has the highest antibacterial sensitivity against Staphylococcus aureus (88%) followed by the chloroform fraction (59%). The antibacterial activity of the crude extract and fractions of the plant had no significant association with the overall phenol concentration [56].

Gloriosa superba is a climber with v-shaped tubers that grows herbaceous or semi-woody. The plant's therapeutic qualities, including colchicines, colchicosides, and alkaloids, are highly recognized for their use in treating arthritis, rheumatism, cancer-related disorders, impotency, and gout. Colchicine inhibits mitotic cell division and hence functions as an anti-mitotic agent. TCC is a semi-synthetic derivative of natural colchicoside that has anti-inflammatory and analgesic properties. Cross-pollination is encouraged by the beautiful brightly colored blossoms and hercogamous character. Genotypes with higher seed yield, alkaloid content, and field resistance to key pests and diseases are crucial to meet the needs of the phytopharmaceutical industry. Glory lily needs a trellis or standards for support because it is a climber. The pressure on wild-harvested tubers has been relieved by the rapid proliferation of micro tubers created from seed material [34].

By using plant extract from Gloriosa superba L. as fuel, the inquiry seeks to synthesise copper oxide nanoparticles (CuO Nps), characterise them, and conduct tests on their antibacterial properties against particular harmful bacteria. The particles' monoclinic nature was shown by X-ray diffraction experiments.The blue shift with an increase in plant extract concentration is seen by the UV-visible absorption spectra of CuO Nps. The particles' spherical form may be seen in SEM pictures.The present study reveals how glory lily. extract may be easily used as a fuel for the effective production of CuO nanoparticles using a green synthesis technique to produce material that is considerably active against bacteria [7].

Gloriosa superba L., a perennial climber, is utilized in Southeast Asia and different regions of Africa as an ayurvedic medicine to treat illnesses. The plant was listed as endangered because it was being recklessly harvested from the wild while being heavily used in the pharmaceutical industry for its colchicine content. It also has a low seed set issue, but it is presently being grown because of industrial demand. The plant is used to treat a variety of conditions including gout, rheumatism, inflammation, ulcers, bleeding piles, leprosy, impotence, and snakebites. Gloriosine, colchicine, N-formyl deacetylcolchicine, lumicolchicine, and colchicoside are just a few of the chemicals that have been isolated from plant parts, primarily tubers and seeds [53].

Due to the overuse of this glory lily plant, it was listed in Red Data Book. In vitro, mass multiplication techniques and other biotechnology applications have been used to conserve this plant. The in vitro regeneration of this plant utilizing nodal explants and shoot cuttings has been the subject of numerous investigations [166–168], explants shoot tips [169] culturing in invitro [25], the embryo of the leaf tissues [170–172], from axillary and apical buds [120], more efficient in vitro colchicine synthesis [173, 174].

The ability to inhibit enzymes was tested in an alcoholic extract taken from the rhizomes of *Gloriosa superba* Linn (Colchicaceae). The crude extract and its succeeding fractions, which included n-butanol, ethyl acetate, chloroform, and aqueous, were tested for the presence of urease, butyrylcholinesterase, acetylcholinesterase, and lipoxygenase. On lipoxygenase, a remarkable inhibition was seen. Out of all the studied fractions on lipoxygenase, the chloroform fraction demonstrated the highest enzyme inhibition potency (90%). Overall, there was a 67–90% inhibition of lipoxygenase, a 46–69% inhibition of acetylcholinesterase, and a 10–33% inhibition of butyrylcholinesterase. Urease was not suppressed [147].

The last 10 years have seen a tremendous increase in interest in green chemistry methods for designing therapeutically important nanomedicine. Here, we report for the first time on the anticancer potential of naturally occurring palladium and platinum nanoparticles (PdNPs) utilising an extract from the medicinal plant *Gloriosa superba* tuber (GSTE). The development of dark brown and black colours for PtNPs and PdNPs, respectively, along with an increase in peak intensity in the UV-visible spectra served as evidence that the nanoparticles' synthesis was finished in less than 5 h at 100 °C. These results demonstrate the effectiveness of producing nanoscale platinum and palladium medicines by phytogenic processes for treating and managing breast cancer [175].

Initial phytochemical analysis of extracts revealed the presence of reducing sugars, proteins, amino acids, steroids, flavonoids, terpenoids, saponins, alkaloids, tannins, and phlorotannins. Comparing tuber extracts to shoot and flower extracts against all of the studied bacteria and fungi, tuber extracts shown effective antibacterial and antifungal activity. Comparing the tuber's ETOH extract to the shoot and flower extracts revealed that it had the strongest antibacterial effects on Staphylococcus aureus (19 mm), Escherichia coli (18 mm), Micrococcus luteus (17 mm), Pseudomonas aeruginosa (17 mm), and Salmonella abony (16 mm). When compared to shoot and flower extracts, G. superba tuber extracts showed more potent antifungal activity against all of the fungi tested. Comparing the tuber's ETOH extract to shoot and flower extracts, Rhizopus oryzae (20.17 mm), Mucor Sp. (19.87 mm), Aspergillus niger (18.02), Candida krusei (17.98 mm), and Candida albicans (16.88 mm) showed the strongest antifungal activity [176].

We looked into the effectiveness of the whole plant extracts of *Gloriosa superba* Linn. (Liliaceae) against the Indian earthworm *Pheretima posthuma*. The period at which the worms became paralyzed and when they died were timed during tests using various concentrations (20–60 mg mL–1) of each extract. Significant anthelmintic activity was detected in both extracts (ethanol and aqueous). When compared to piperazine citrate (15 mg mL-1), which is included as a standard reference, and normal saline as a control, both extracts (aqueous and ethanol) at the tested dose level (20–60 mg mL-1) demonstrated substantial activity (p0.01). The current study suggests that the complete plant of *Gloriosa superba* could be useful as an anthelmintic [55].

Methanol extracts from the rind, rhizomes, leaves, flowers, and seeds of *G. superba* were tested in vitro for their antifungal properties. The mycelial growth of *F. oxysporum* was inhibited by all of the extracts examined, with seed extract (82.45%) having the highest mycelial inhibitory action at 100% concentration when compared to the others, followed by leaves (77.45%), rind (82.33%), flower (76.39%) extracts and rhizome (82.06%). The biggest mycelial growth reduction was seen with flower extract at a concentration of 25%, followed by seed, rind, leaves, and rhizomes. The findings of this study show that seed and rhizome extracts have the highest potential to stop *F. oxysporum* from growing mycelial [177].

References

1. Vaishnavi, B., Khanm, H., & Bhoomika, H. (2019). Review on pharmacological properties of glory lily (Gloriosa superba Linn.)-an endangered medicinal plant. *Journal homepage: http://www.ijcmas.com, 8*(02), 2019.
2. Farooqi, A. A., & Sreeramu, B. (2004). *Cultivation of medicinal and aromatic crops*. Universities Press.
3. Sastri, B. (1950). The Wealth of India, A dictionary of raw material and industrial products. *Publ. Inf. Dir, 5*, 285–293.
4. Chopra, R.N., Glossary of Indian medicinal plants. 1956.
5. Chandel, K., G. Shukla, and N. Sharma, Biodiversity in medicinal and aromatic plants in India. 1996.
6. Kavina, J., Gopi, R., & Panneerselvam, R. (2011). Gloriosa superba Linn—A medicinally important plant. *Drug Invention Today, 3*(6).
7. Naika, H. R., et al. (2015). Green synthesis of CuO nanoparticles using Gloriosa superba L. extract and their antibacterial activity. *Journal of Taibah University for Science, 9*(1), 7–12.
8. Clewer, H. W. B., Green, S. J., & Tutin, F. (1915). XCI.—The constituents of Gloriosa superba. *Journal of the Chemical Society, Transactions, 107*, 835–846.
9. Gooneratne, I. K., et al. (2014). Toxic encephalopathy due to colchicine—Gloriosa superba poisoning. *Practical Neurology, 14*(5), 357–359.
10. Srivastava, S., et al. (2014). A validated over pressured layered chromatography (OPLC) method for separation and quantification of colchicine in Gloriosa superba (L.) tubers from different geographical regions. *RSC Advances, 4*(105), 60902–60906.
11. Ghani, A. (1998). *Medicinal plants of Bangladesh: chemical constituents and uses*. Asiatic Society of Bangladesh.
12. Martin, J. S., & Martin, M. M. (1982). Tannin assays in ecological studies: lack of correlation between phenolics, proanthocyanidins and protein-precipitating constituents in mature foliage of six oak species. *Oecologia, 54*(2), 205–211.
13. Yadav, R., & Pandya, I. (2014). First record of critical endangered plant (Gloriosa superba L.) at verge of Sai river, Jaunpur, Uttar Pradesh, India. *Current Trends in Life Sciences, 3*(2), 23–26.

14. Shukla, R. (2009). Patterns of plant species diversity across Terai landscape in north-eastern Uttar Pradesh, India. *Tropical Ecology, 50*(1), 111.
15. Acharya, D., et al. (2006). Rare herb of patalkot-Gloriosa superba. *Traditional Herbs and Medicines*, 10–24.
16. Arumugam, A., & Gopinath, K. (2012). In vitro micropropagation using corm bud explants: an endangered medicinal plant of Gloriosa superba L. *Asian Journal of Biotechnology, 4*(3), 120–128.
17. Singh, A. (2006). *Flower crops: Cultivation and management*. New India Publishing.
18. Jayaweera, D. (1982). *Medicinal Plants Used in Ceylon. vol* (Vol. 5, p. 201). National Science Council of Sri Lanka, Colombo.
19. Pulliah, T. (2002). Medicinal plants in India. *Regency Publ New Delhi India*, 137–139.
20. Finnie, J., & Staden, J. V. (1994). *Gloriosa superba L.* (flame lily): Micropropagation and in vitro production of colchicine. *Medicinal and aromatic plants, VI*, 146–166.
21. Pullaiah, T. (2002). *Medicinal Plants in India* (Vol. 1). Regency Publication.
22. Finnie, J., & Staden, J. V. (1994). *Gloriosa superba L.* (flame lily): micropropagation and in vitro production of colchicine. In *Medicinal and aromatic plants VI* (pp. 146–166). Springer.
23. Rajak, R., & Rai, M. (1996). Herbal medicines, bio-diversity, and conservation strategies. In *National seminar on herbal medicines, bio-diversity, and conservation strategies (1994: Chhindwara, India)*. International Book Distributors.
24. Gupta, L., & Rana, R. (2001). Significance of sequential opening of flowers in Gloriosa superba L. *Current Science, 80*(10), 1266–1267.
25. Sivakumar, G., Krishnamurthi, K., & Rajendran, T. (2003). In vitro corm production in Gloriosa superba L., an Ayurvedic medicinal plant. *The Journal of Horticultural Science and Biotechnology, 78*(4), 450–453.
26. Ambasta, S. S. (1986). *The useful plants of India*. Publications & Information Directorate.
27. Warrier, P., Nambiar, V. P. K., & Ramankutty, C. (1994). *Indian medicinal plants* (Vol. 3, p. 84). Orient Longman.
28. Subramanya, S., & Radhamani, T. (1993). Pollination by birds and bats. *Current Science*, 201–209.
29. Mamatha, H., et al. (1992). Pollen studies in *Gloriosa superba Linn*. In *WOCMAP I-Medicinal and Aromatic Plants Conference: Part 3 of 4 331*.
30. Shirgurkar, M. V., John, C., & Nadgauda, R. S. (2001). Factors affecting in vitro microrhizome production in turmeric. *Plant Cell, Tissue and Organ Culture, 64*(1), 5–11.
31. Krause, J. (1985). Production of Gloriosa tubers from seeds. In *IV International Symposium on Flower Bulbs '77*.
32. Villafranca, M., et al. (1998). Effect of physiological age of mother tuber and number of subcultures on in vitro tuberisation of potato (Solanum tuberosum L.). *Plant Cell Reports, 17*(10), 787–790.
33. Ewing, E., & Struik, P. (2010). Tuber formation in potato: induction, initiation, and growth. *Horticultural Reviews, 14*, 89–198.
34. Padmapriya, S., Rajamani, K., & Sathiyamurthy, V. (2015). Glory lily (Gloriosa superba L.)-A review. *International Journal of Current Pharmaceutical Review and Research, 7*(1), 43–49.
35. Hughes, K., Henke, R., & Constantin, M. (1978). *Propagation of higher plants through tissue culture: A bridge between research and application*. Dept. of Botany; UT-ERDA Comparative ....
36. Murashige, T. (1974). Plant propagation through tissue cultures. *Annual Review of Plant Physiology, 25*(1), 135–166.
37. Thorpe, T. A. (1993). In vitro organogenesis and somatic embryogenesis: physiological and biochemical aspects. In *Morphogenesis in plants* (pp. 19–38). Springer.
38. Dodds, J. H., & Roberts, L. W. (1985). *Experiments in plant tissue culture*. International Potato Center.
39. Dixon, R. A., & Gonzales, R. A. (1994). *Plant cell culture: a practical approach*. IRL Press.

40. Sharma, M. (1980). Cytogenetical investigations into some garden ornamentals II. The genus Aloe L. *Cytologia, 45*(3), 515–532.
41. Bellet, P., & Gaignault, J. C. (1985). [Gloriosa superba L. and the production of colchicinic substances]. *Annales Pharmaceutiques Françaises, 43*(4), 345–347.
42. Devi, K. S., Annapoorani, S., & Ashokkumar, K. (2011). Hepatic antioxidative potential of ethyl acetate fraction of Cynodon dactylon in Balb/c mice. *Journal of Medicinal Plants Research, 5*(6), 992–996.
43. Ashokkumar, K., Selvaraj, K., & S.D. KM. (2013). Reverse phase-high performance liquid chromatography-diode array detector (RP-HPLC-DAD) analysis of flavonoids profile from curry leaf (Murraya koenigii. L). *Journal of Medicinal Plants Research, 7*(47), 3393–3399.
44. Muthukrishnan, S. D., Kaliyaperumal, A., & Subramaniyan, A. (2015). Identification and determination of flavonoids, carotenoids and chlorophyll concentration in Cynodon dactylon (L.) by HPLC analysis. *Natural Product Research, 29*(8), 785–790.
45. De Groot, H. (1994). Reactive oxygen species in tissue injury. *Hepato-Gastroenterology, 41*(4), 328–332.
46. Fraga, C. G., et al. (1987). Flavonoids as antioxidants evaluated by in vitro and in situ liver chemiluminescence. *Biochemical Pharmacology, 36*(5), 717–720.
47. Halliwell, B. (1994). Free radicals, antioxidants, and human disease: curiosity, cause, or consequence? *The Lancet, 344*(8924), 721–724.
48. Murugesan, A. K., et al. (2021). Facile green synthesis and characterization of Gloriosa superba L. tuber extract-capped silver nanoparticles (GST-AgNPs) and its potential antibacterial and anticancer activities against A549 human cancer cells. *Environmental Nanotechnology, Monitoring & Management, 15*, 100460.
49. Adhvaryu, M. R., Reddy, N., & Parabia, M. H. (2008). Anti-tumor activity of four Ayurvedic herbs in Dalton lymphoma ascites bearing mice and their short-term in vitro cytotoxicity on DLA-cell-line. *African Journal of Traditional, Complementary and Alternative Medicines, 5*(4), 409–418.
50. Aldam, C. (2002). Endangered medicinal plant species in Himachal Pradesh. *Current Science, 83*, 797–798.
51. Gooneratne, B. (1966). Massive generalized alopecia after poisoning by Gloriosa superba. *British Medical Journal, 1*(5494), 1023.
52. Angunawela, R., & Fernando, H. (1971). Acute ascending polyneuropathy and dermatitis following poisoning by tubers of Gloriosa superba. *The Ceylon Medical Journal, 16*(4), 233–235.
53. Jana, S., & Shekhawat, G. (2011). Critical review on medicinally potent plant species: Gloriosa superba. *Fitoterapia, 82*(3), 293–301.
54. Mendis, S. (1989). Colchicine cardiotoxicity following ingestion of Gloriosa superba tubers. *Postgraduate Medical Journal, 65*(768), 752–755.
55. Pawar, B. M., et al. (2010). Anthelmintic activity of Gloriosa superba Linn (Liliaceae). *International Journal of PharmTech Research, 2*(2), 1483–1487.
56. Budhiraja, A., et al. (2012). Antimicrobial and cytotoxic activities of fungal isolates of medicinal plant Gloriosa superba. *The International Journal of Advanced Research in Pharmaceutical, 2*, 1–9.
57. Moteriya, P., et al. (2014). In vitro antioxidant and antibacterial potential of leaf and stem of Gloriosa superba L. *American Journal of Phytomedicine and Clinical Therapeutics, 2*(6), 703–787.
58. Hemaiswarya, S., et al. (2009). Antimicrobial and mutagenic properties of the root tubers of Gloriosa superba Linn.(Kalihari). *Pakistan Journal of Botany, 41*(1), 293–299.
59. Jagtap, S., & Satpute, R. (2014). Phytochemical screening, antioxidant, antimicrobial and flavonoid analysis of Gloriosa superba Linn. Rhizome extracts. *Journal of academia and industrial research, 3*(6), 247–254.

60. Kumarapppan, C., Jaswanth, A., & Kumarasunderi, K. (2011). Antihaemolytic and snake venom neutralizing effect of some Indian medicinal plants. *Asian Pacific Journal of Tropical Medicine, 4*(9), 743–747.
61. Kee, N. L. A., et al. (2008). Antithrombotic/anticoagulant and anticancer activities of selected medicinal plants from South Africa. *African Journal of Biotechnology, 7*(3).
62. Kamna, B., & Anirudha, R. (2012). Antimicrobial efficacy of an endemic plant species (Gloriosa superba L.). *International Journal of Pharma and Bio Sciences, 3*(4), 353–359.
63. Suryavanshi, S., Rai, G., & Malviya, S. (2012). Evaluation of anti-microbial and anthelmintic activity of Gloriosa Superba tubers. *Advance Research in Pharmaceuticals and Biologicals, 2*(1), 45–52.
64. Banu, R., & Nagarajan, N. (2011). Anti bacterial potential of glory lily, Gloriosa superba linn. *International Research Journal of Pharmacy, 2*, 139–142.
65. John, J., et al. (2009). Analgesic and anti-inflammatory activities of the hydroalcoholic extract from Gloriosa superba Linn. *International Journal of Green Pharmacy, 3*(3), 215.
66. Singh, V. (1993). Selected Indian Folk medicinal claims and their relevance in primary health care programme. *Glimpses Plant Research, 10*, 147–152.
67. Haroon, K., et al. (2008). Antimicrobial activities of Gloriosa superba extracts. *The Journal of Enzyme Inhibition and Medicinal Chemistry, 22*(6), 722–725.
68. Srivastava, U., & Chandra, V. (1977). Gloriosa superba Linn. (kalihari) an important colchicine producing plant. *Indian Journal of Medical Research, 10*, 92–95.
69. Prajapati, N. D., et al. (2003). *Medicinal plants* (3rd ed., p. 353). Agrobios Published Company.
70. Gupta, L., et al. (2005). Colchicine content in Gloriosa superba L. *J Res, 4*, 238–241.
71. Samy, R. P., et al. (2008). Ethnobotanical survey of folk plants for the treatment of snakebites in Southern part of Tamilnadu, *India. Journal of Ethnopharmacology, 115*(2), 302–312.
72. Capraro, H. (1984). In the alkaloids.(Ed.): A. Brossi. *Academic Press, 23*(1984), 1–70.
73. Sarin, Y., et al. (1974). Colchicine from the seeds of Gloriosa superba. *Current Science*.
74. Veeraiah, S., & Reddy, J. (2012). Current strategic approaches in ethnomedicinal plants of Tinospora cordifolia and Gloriosa superba—A review. *International Journal of Pharma and Bio Sciences, 3*(2), 320–326.
75. Suri, O. P., et al. (2001). A new glycoside, 3-O-demethylcolchicine-3-O-α-D-glucopyranoside, from Gloriosa superba seeds. *Natural Product Letters, 15*(4), 217–219.
76. Megala, S., & Elango, R. (2012). In Vitro Antibacterial Activity Studies of Tuber and Seed Extracts of Gloriosa Superba Linn. against some Selected Human Pathogen. *International Journal of Pharmaceutical Sciences and Research, 3*(11), 4230.
77. Muthukrishnan, S., & Annapoorani, S. (2012). Phytochemical constituents of Gloriosa superba seed, tuber and leaves. *Research Journal of Pharmaceutical, Biological and Chemical Sciences, 3*(3), 111–117.
78. Senthilkumar, M. (2013). Phytochemical screening of Gloriosa superba L.-from different geographical positions. *International Journal of Scientific and Research Publications, 3*(1), 1–5.
79. Kaur, R. (2021). Phytochemical investigations and therapeutic potential of Gloriosa superba: A review. *Plant Cell Biotechnology and Molecular Biology*, 15–19.
80. Rehana, B., & Nagarajan, N. (2012). Phytochemical screening for active compounds in Gloriosa superba leaves and tubers. *Journal of Pharmacognosy and Phytochemical Research, 4*(1), 17–20.
81. Whitfield, P. (1996). Medicinal plants and the control of parasites. *Royal Society of Tropical Medicine and Hygiene, 90*, 596–600.
82. Von der Weid, I., et al. (2003). Antimicrobial activity of Paenibacillus peoriae strain NRRL BD-62 against a broad spectrum of phytopathogenic bacteria and fungi. *Journal of Applied Microbiology, 95*(5), 1143–1151.
83. Shelef, L., Naglik, O., & Bogen, D. (1980). Sensitivity of some common food-borne bacteria to the spices sage, rosemary, and allspice. *Journal of Food Science, 45*(4), 1042–1044.

84. Sieradzki, K., et al. (1999). The development of vancomycin resistance in a patient with methicillin-resistant Staphylococcus aureus infection. *New England Journal of Medicine, 340*(7), 517–523.
85. Iwu, M., Duncan, A., & Okunji, C. (1999). In J. Janik (Ed.), *Perspectives on new crops and new uses* (pp. 457–462). ASHS Press.
86. Gabrielli, A., Avvedimento, E. V., & Krieg, T. (2009). Scleroderma. *New England Journal of Medicine, 360*(19), 1989–2003.
87. Klippel, J. (1997). Systemic lupus erythematosus: demographics, prognosis, and outcome. *The Journal of Rheumatology. Supplement, 48*, 67–71.
88. Frank, P. J. (2001). Scleroderma. *Dermatology Online Journal, 7*, 1.
89. Astani, A., Reichling, J., & Schnitzler, P. (2012). Melissa officinalis extract inhibits attachment of herpes simplex virus in vitro. *Chemotherapy, 58*(1), 70–77.
90. Hossain, R., et al. (2021). *Lasia spinosa chemical composition and therapeutic potential: A literature-based review*. Oxidative Medicine and Cellular Longevity 2021.
91. Sharifi-Rad, J., et al. (2022). Ethnobotany, phytochemistry, biological activities, and health-promoting effects of the genus Bulbophyllum. *Evidence-based Complementary and Alternative Medicine, 2022*.
92. Sharifi-Rad, J., et al., Genistein: an integrative overview of its mode of action, pharmacological properties, and health benefits. Oxidative Medicine and Cellular Longevity 2021. **2021**.
93. Kato, Y., et al. (2018). Cancer metabolism: new insights into classic characteristics. *Japanese Dental Science Review, 54*(1), 8–21.
94. Sharifi-Rad, J., et al. (2021). Cinnamomum species: bridging phytochemistry knowledge, pharmacological properties and toxicological safety for health benefits. *Frontiers in Pharmacology, 12*, 600139.
95. Sharifi-Rad, J., et al. (2022). Hyssopus essential oil: an update of its phytochemistry, biological activities, and safety profile. *Oxidative Medicine and Cellular Longevity, 2022*.
96. Islam, M. T., et al. (2021). A literature-based update on Benincasa hispida (Thunb.) Cogn.: traditional uses, nutraceutical, and phytopharmacological profiles. *Oxidative Medicine and Cellular Longevity*, 2021.
97. Hossain, R., et al. (2022). Biosynthesis of secondary metabolites based on the regulation of microRNAs. *BioMed Research International, 2022*.
98. Salehi, B., et al. (2018). Epibatidine: a promising natural alkaloid in health. *Biomolecules, 9*(1), 6.
99. Muthukrishnan, S., Vellingiri, B., & Murugesan, G. (2018). Anticancer effects of silver nanoparticles encapsulated by Gloriosa superba (L.) leaf extracts in DLA tumor cells. Future. *Journal of Pharmaceutical Sciences, 4*(2), 206–214.
100. Anon. (1956). *The wealth of India: A dictionary of raw material and industrial products*. CSIR Delhi.
101. Bhakuni, D., & Jain, S. (1995). *Advances in horticulture, 11*. Malhotra Publishing House.
102. Choochote, W., et al. (2004). Evaluation of the colchicine-like activity of Gloriosa superba-extracted fractions for mosquito (Diptera: Culicidae) cytogenetic study. *Journal of Medical Entomology, 41*(4), 672–676.
103. Watt, J. M., & Breyer-Brandwijk, M. G. (1962). *The medicinal and poisonous plants of Southern and Eastern Africa being an account of their medicinal and other uses, chemical composition, pharmacological effects and toxicology in man and animal*. The Medicinal and Poisonous Plants of Southern and Eastern Africa being an Account of their Medicinal and other Uses, Chemical Composition, Pharmacological Effects and Toxicology in Man and Animal. (Edn 2).
104. Neuwinger, H. D. (1996). *African ethnobotany: poisons and drugs: chemistry, pharmacology, toxicology*. CRC Press.
105. Burkill, H. M. (1995). *The useful plants of West Tropical Africa. Vol 3: Families JL. ed. 2*. Royal Botanic Gardens.
106. Duke, J. A. (2002). *Handbook of medicinal herbs*. CRC Press.

107. Khan, H., et al. (2008). Antimicrobial activities of Gloriosa superba Linn (Colchicaceae) extracts. *Journal of Enzyme Inhibition and Medicinal Chemistry, 23*(6), 855–859.
108. Burkill, H. M. (1994). *The useful plants of west tropical Africa. Volume 2: Families EI*. Royal Botanic Gardens.
109. Dounias, E. (2006). *Gloriosa superba L.* PROTA.
110. Fowler, D. G. (2007). *Zambian plants: their vernacular names and uses*. Royal Botanic Gardens.
111. Haerdi, F. (1964) The Indigenous Medicinal Plants of Ulanga District, Tanzania, East Africa. *Acta Tropica*, (Suppl. 8), 1–278.
112. Jain, A., et al. (2004). Folk herbal medicines used in birth control and sexual diseases by tribals of southern Rajasthan, India. *Journal of Ethnopharmacology, 90*(1), 171–177.
113. Manandhar, N. P. (2002). *Plants and people of Nepal*. Timber Press.
114. Maurya, R., et al. (2004). Traditional remedies for fertility regulation. *Current Medicinal Chemistry, 11*(11), 1431–1450.
115. Katewa, S., Chaudhary, B., & Jain, A. (2004). Folk herbal medicines from tribal area of Rajasthan, India. *Journal of Ethnopharmacology, 92*(1), 41–46.
116. Chopda, M., & Mahajan, R. (2009). Wound healing plants of Jalgaon district of Maharashtra state, India. *Ethnobotanical Leaflets, 2009*(1), 1.
117. Bhargava, N. (1983). Ethnobotanical studies of the tribes of Andaman and Nicobar Islands, India. I. Onge. *Economic Botany, 37*(1), 110–119.
118. Singh, V. K., et al. (2007). Anti-inflammatory activity of Gloriosa superba Linn. *Pharmacist, 2*(1), 19–20.
119. Kala, C. P., Farooquee, N. A., & Dhar, U. (2004). Prioritization of medicinal plants on the basis of available knowledge, existing practices and use value status in Uttaranchal, India. *Biodiversity and Conservation, 13*, 453–469.
120. Hassan, A., & Roy, S. K. (2005). Micropropagation of Gloriosa superba L. through high frequency shoot proliferation. *Plant Tissue Culture, 15*(1), 67–74.
121. Bryant, A.T., Zulu medicine and medicine-men. 1966.
122. Sahu, S., Dhal, N., & Mohanty, R. (2010). Potential medicinal plants used by the tribal of Deogarh district, Orissa, India. *Studies on Ethno-Medicine, 4*(1), 53–61.
123. Lather, A., et al. (2011). Pharmacological potential of the plants used in treatment of piles-A review. *Journal of Natura Conscientia, 2*(1), 255–265.
124. Saralamp, P., et al. (1996). *Medicinal plants in Thailand (Vol. 1). bangkok: Faculty of Pharmacy*. Mahidol University.
125. Yineger, H., & Yewhalaw, D. (2007). Traditional medicinal plant knowledge and use by local healers in Sekoru District, Jimma Zone, Southwestern Ethiopia. *Journal of Ethnobiology and Ethnomedicine, 3*(1), 1–7.
126. Dalziel, J. M. (1937). *The useful plants of west tropical Africa*. The Useful Plants of West Tropical Africa.
127. De Padua, L. S., Bunyapraphatsara, N., & Lemmens, R. H. M. J. (1999). *Plant resources of South-East Asia 12:(1) medicinal and poisonous plants 1*. Backhuys Publishers.
128. Nadkarni, A. (2002). *Nadkarni's Indian materia medica*. Popular Prakashan pvt Ltd. Mumbai Reprint.
129. Yamada, T. (1999). A report on the ethnobotany of the Nyindu in the eastern part of the former Zaire. *African Study Monographs, 20*(1), 1–72.
130. Gelfand, M., et al. (1985). *Zambeziana: a new series on culture and society in Central Africa. 17. The traditional medical practitioner in Zimbabwe: his principles of pratice and pharmacopoeia*. Mambo Press.
131. Mavi, S. (1996). Medicinal plants and their uses in Zimbabwe. *Indigenous knowledge and its uses in Southern Africa, 67*, 73.
132. Maradjo, M., & Soediarto, A. (1977). *Kacang-kacangan*. Karya Nusantara.
133. Saralamp, P. (1996). *Medicinal plants in Thailand* (Vol. 1). Department of Pharmaceutical Bota.

134. Ashokkumar, K. (2015). Gloriosa superba (L.): A brief review of its phytochemical properties and pharmacology. *International Journal of Pharmacognosy and Phytochemical Research, 7*, 1190–1193.
135. Li, Z., et al. (1996). Inhibition of LPS-induced tumor necrosis factor-α production by colchicine and other microtubule disrupting drugs. *Immunobiology, 195*(4–5), 624–639.
136. Ding, A. H., et al. (1990). Downregulation of tumor necrosis factor receptors on macrophages and endothelial cells by microtubule depolymerizing agents. *The Journal of Experimental Medicine, 171*(3), 715–727.
137. Maroyi, A., Van der Maesen, L., & Gloriosa superba L. (2011). (family Colchicaceae): Remedy or poison. *Journal of Medicinal Plants Research, 5*(26), 6112–6121.
138. Sapra, S., et al. (2013). Colchicine and its various physicochemical and biological aspects. *Medicinal Chemistry Research, 22*(2), 531–547.
139. Akbulut, S. (2009). Minimum leaf width as an indicator of Colchicum speciosum Steven (Liliaceae) suitable for collection. *Journal of Medicinal Plants Research, 3*(5), 377–381.
140. Imazio, M., et al. (2009). Colchicine for pericarditis: hype or hope? *European Heart Journal, 30*(5), 532–539.
141. Bharathi, P. (2006). Antimitotic effect of colchicine from six different species of Gloriosa in onion roots (Allium cepa). *Journal of Medical Sciences, 6*, 322–327.
142. Labib, S., et al. (2014). Fatal colchicine intoxication. *Saudi Journal of Anaesthesia, 8*(3), 394.
143. Mahidol, C., et al. (1998). Biodiversity and natural product drug discovery. *Pure and Applied Chemistry, 70*(11), 2065–2072.
144. Thakur, R., Potešilova, H., & Šantavý, F. (1975). Substances from plants of the subfamily wurmbaeoideae and their derivatives. Part LXXIX1. *Planta Medica, 28*(07), 201–209.
145. Kumar, L. (1953). Doubling of Chromosomes induced by Gloriosine isolated from Gloriosa superba, Linn. *Nature, 171*(4357), 791–792.
146. Brown, W., & Seed, L. (1945). Effect of colchicine on human tissues. *American Journal of Clinical Pathology, 15*(5), 189–195.
147. Khan, H., Ali Khan, M., & Hussan, I. (2007). Enzyme inhibition activities of the extracts from rhizomes of Gloriosa superba Linn (Colchicaceae). *Journal of Enzyme Inhibition and Medicinal Chemistry, 22*(6), 722–725.
148. Joshi, C., Priya, E. S., & Mathela, C. (2010). Isolation and anti-inflammatory activity of colchicinoids from Gloriosa superba seeds. *Pharmaceutical Biology, 48*(2), 206–209.
149. Shah, K. K., & Sagar, G. V. (2015). Phytochemical and pharmacological evaluation of Gloriosa superba. *Journal of Drug Delivery and Therapeutics*, 27–42.
150. Nagaratnam, N., De Silva, D., & De Silva, N. (1973). Colchicine poisoning following ingestion of Gloriosa Superba tubers. *Tropical and Geographical Medicine, 25*(1), 15–17.
151. Ide, N., et al. (2010). Case of colchicine intoxication caused by tubers of Gloriosa superba. *Chudoku Kenkyu: Chudoku Kenkyukai jun Kikanshi= The Japanese Journal of Toxicology, 23*(3), 243–245.
152. Aleem, H. (1992). Gloriosa superba poisoning. *The Journal of the Association of Physicians of India, 40*(8), 541–542.
153. Cavaletti, G. (2007). Toxic and drug-induced neuropathies. In *Neurobiology of Disease* (pp. 871–883). Elsevier.
154. Dasheiff, R. M., & Ramirez, L. F. (1985). The effects of colchicine in mammalian brain from rodents to rhesus monkeys. *Brain Research Reviews, 10*(1), 47–67.
155. Rigante, D., et al. (2006). The pharmacologic basis of treatment with colchicine in children with familial Mediterranean fever. *European Review for Medical and Pharmacological Sciences, 10*(4), 173.
156. Malpani, A. A. (2011). Effect of the aqueous extract of Gloriosa superba Linn (Langli) roots on reproductive system and cardiovascular parameters in female rats. *Tropical Journal of Pharmaceutical Research, 10*(2).
157. Bruneton, J. (1999). *Toxic plants dangerous to humans and animals*. Intercept Limited.

158. Cerquaglia, C., et al. (2005). Pharmacological and clinical basis of treatment of Familial Mediterranean Fever (FMF) with colchicine or analogues: An update. *Current Drug Targets-Inflammation & Allergy, 4*(1), 117–124.
159. Maxwell, M., Muthu, P., & Pritty, P. (2002). Accidental colchicine overdose. A case report and literature review. *Emergency Medicine Journal, 19*(3), 265–266.
160. Suganthy, M., & Sakthivel, P. (2012). Field efficacy of biopesticides against Plusia signata (Fabricius) on Gloriosa superba. *Madras Agricultural Journal, 99*(4/6), 368–370.
161. Meena, B., & Rajamani, K. (2016). Biological management of root-rot disease in Gloriosa superba. *International Journal of Noni Research, 11*(1), 82–85.
162. Maiti, C. K., et al. (2007). First report of leaf blight disease of Gloriosa superba L. caused by Alternaria alternata (Fr.) Keissler in India. *Journal of General Plant Pathology, 73*(5), 377–378.
163. Meena, B., et al. (2015). Integrated disease management of root rot (Macrophomina phaseolina) in Gloriosa superba. In *III International Symposium on Underutilized Plant Species 1241.*
164. Kande Vidanalage, C. J., Ekanayeka, R., & Wijewardane, D. K. (2016). Case report: a rare case of attempted homicide with Gloriosa superba seeds. *BMC Pharmacology and Toxicology, 17*(1), 1–4.
165. Ashokkumar, K. (2015). Gloriosa superba (L.): a brief review of its phytochemical properties and pharmacology. *International Journal of Pharmacognosy and Phytochemical Research, 7*(6), 1190–1193.
166. Somani, V., John, C., & Thengane, R. (1989). In vitro propagation and corm formation in Gloriosa superba. *Indian Journal of Experimental Biology, 27*, 578–579.
167. Samarajeewa, P. (1993). Clonal propagation of Gloriosa superba L. *Indian Journal of Experimental Biology, 31*, 719–720.
168. Custers, J., & Bergervoet, J. (1994). Micropropagation of Gloriosa: Towards a practical protocol. *Scientia Horticulturae, 57*(4), 323–334.
169. Sivakumar, G., & Krishnamurthy, K. (2000). Micropropagation of Gloriosa superba L.-an endangered species of Asia and Africa. *Current Science, 78*(1), 30–32.
170. Finnie, J., & Van Staden, J. (1989). In vitro micropropagation of Sandersonia auranicaceae and Gloriosa superba. *Plant Cell, Tissue and Organ Culture, 19*, 151–158.
171. Sivakumar, G., Krishnamurthy, K., & Rajendran, T. (2003). Embryoidogenesis and plant regeneration from leaf tissue of Gloriosa superba. *Planta Medica, 69*(05), 479–481.
172. Sivakumar, G., & Krishnamurthy, K. (2004). In vitro organogenetic responses of Gloriosa superba. *Russian Journal of Plant Physiology, 51*(5), 713–721.
173. Sivakumar, G., et al. (2004). Enhanced in vitro production of colchicine in Gloriosa superba L.—an emerging industrial medicinal crop in South India. *The Journal of Horticultural Science and Biotechnology, 79*(4), 602–605.
174. Ghosh, B., et al. (2002). Enhanced colchicine production in root cultures of Gloriosa superba by direct and indirect precursors of the biosynthetic pathway. *Biotechnology Letters, 24*(3), 231–234.
175. Rokade, S. S., et al. (2018). Gloriosa superba mediated synthesis of platinum and palladium nanoparticles for induction of apoptosis in breast cancer. *Bioinorganic Chemistry and Applications, 2018.*
176. Chimahali, J., et al. (2019). Phytochemical analysis and evaluation of antimicrobial activity in the whole plant extracts of Gloriosa superba. *Phytochemical Analysis, 12*(6), 245–249.
177. Ashokkumar, N., et al. (2017). Studies on antifungal activity of different plant parts of Glory lily (Gloriosa superba L.) against fungal wilt pathogen, Fusarium oxysporum. *International Journal of Current Microbiology and Applied Sciences, 6*(9), 428–433.

Chapter 24
Aniseed

Huma Umbreen, Razia Noreen, Mahr Un Nisa, Hamna Saleem, and Umar Farooq Gohar

24.1 Introduction

Common herb called Aniseed (Fig. 24.1) also named as Anis, Anise, Anis seed, or Sweet cumin, is a medicinal flowering plant which produces fruit commonly known as seeds [1]. The scientific name of the plant is called *Pimpinella anisum* L and aniseed belongs to the family Apiaceae or Umbelliferae [2, 3]. Pimpinella is a broad spectrum genus including many herbaceous plants which are produced annually [4]. This genus consists of almost 150 species, but only a few are important economically. Aniseed is worthwhile for its industrial and economic benefits and its taxonomic description of *Pimpinella anisum* L is given in Fig. 24.2. Being herb aniseed plant reaches to a height of 12–27 inches. These plants mostly have compound leaves; the flowers are white and smaller in size and are arranged in whorls [5]. Fruit i.e. seeds are small, curved, greyish brown to greyish green in color and have a sweet fragrance and flavor [6].

The plant is widely grown worldwide, used as spice as well as its seeds are used to extract the essential oil. The plant is instinctive to the South Asia and Eastern Mediterranean region and is widely cultivated in these areas to flavor the foods and drinks. The synonym species include *Anisum odoratum, Anisum officinale, Anisum officinarum, Anisum vulgare*, which are often used as adulterant in products of aniseed [2]. Its use in industry is mainly linked to its fragrant and flavoring properties.

H. Umbreen (✉) · M. U. Nisa · H. Saleem
Department of Nutritional Sciences, Government College University Faisalabad, Faisalabad, Pakistan

R. Noreen
Department of Biochemistry, Government College University Faisalabad, Faisalabad, Pakistan

U. F. Gohar
Institute of Industrial Biotechnology, Government College University, Lahore, Pakistan

M. Zia-Ul-Haq et al. (eds.), *Essentials of Medicinal and Aromatic Crops*,
https://doi.org/10.1007/978-3-031-35403-8_24

Fig. 24.1 Aniseed plant

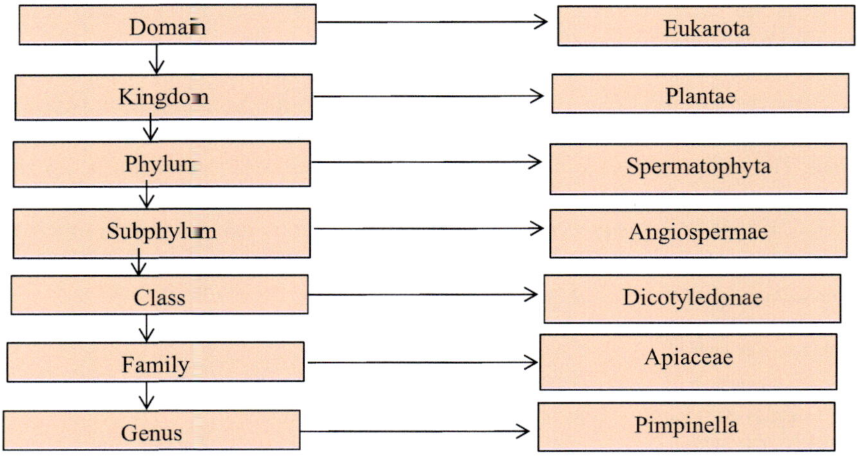

Fig. 24.2 Taxonomic description of Aniseed (*Pimpinella anisum* L). (Derived from Refs. [2, 3, 9])

Furthermore, many folk medicines related to aniseed since the ancient times are still in use for treatment of gastric issues (antispasmodic and carminative), pulmonary problems (treatment of cough, asthma and bronchitis), anti-inflammatory agent and as ameliorative agent to renal issues (kidney stones) [7, 8].

As described earlier Aniseed is famous for its food related, economic and commercial properties throughout the world and the common names for the plant in different countries are as follows;

• France	Anis vert
• USA	Anise and Star anise
• Japan	Anise seed
• England	Anise
• Italy	Annesella
• North Africa	Boucage anis, Petit anise
• Iran, India, Pakistan	Anisa, Badian, Kuppi, Muhuri, Saunf and Sop [9, 10].

24.2 Agronomy

24.2.1 Cultivation Conditions

Aniseed has a slow growth which needs dry, warm and sunny environment for cultivation and for better production of seed and essential oil. The cultivation conditions are also important for next season crop as the seeds from previous year are the best to get maximum yield. After cultivation care should be taken to protect the field from wind so as to avoid the lodging of the plants [11]. Owing to the favorable conditions the growth is enhanced resulting in production of more umbels from the nodes, however these umbels are smaller in size than the previous one and also get mature later [12]. Further cultivation conditions (Table 24.1) in detail are described as under;

(a) **Soil and Sowing:** The soil required for plantation of Aniseed should be fertile yet light, loam sandy soil is more appropriate due to its drainage properties. The field should be rich in nutrients, moist and free of weeds. The seed is planted almost half inch deep in the soil at the rate of one to two seeds/inch and rows are kept about 1.5–2.5 feet apart from each other. It requires comprehensive cultivation all through the development season. Before cultivation it is also important to manage the weeds present in the field. Furthermore, overnight soaking of the seeds in water is recommended for better germination and protection from molds. The suitable pH of the soil is in between 6.3 and 7.3 for the plant to thrive [2, 13, 14].

(b) **Climate**: The market value and quality of seed and oil derived from the Aniseed plant is very much affected by seasonal conditions. It requires warm and frost free season for at least 4 months to thrive and for the seeds to get ripen, whereas growing season needs even type of rainfall. The temperature requirements

Table 24.1 Cultivation conditions for Aniseed

Parameter	Description
Weather	Warm frost free
Soil	Sandy
Soil pH	6.3–7.3
Soil temperature	18–21 °C
Climate temperature	8–25 °C
Rainfall	1000–1200 mm
Germination time	14 days
Germination time (with prior soaking)	8–10 days
Environment type for germination	Dark
Fertilizers	Phosphorus pentoxide Potassium oxide

demand for uniformity throughout growing season. The most appropriate temperature for the plant to germinate and grow is 8–25 °C while, optimum soil temperature requirement is between 18–21 °C and the bed prepared for seeds needs to be warmed before planting. Although it is grown in different seasons in different countries, but crops cultivated at the end of October and start of November have been found to increase the productivity of the crop and essential oil as compared to late November [15–17]. The rainfall of 100–1200 mm is the most suitable to give the maximum yield. However, upper limit of tolerable rainfall is 2000 mm [12].

(c) **Germination:** The time required for germination after sowing of seeds is 14 days. Ripe fruit seeds are preferred due to its quick germination; however, germination can be obtained in 8–10 days if prior soaking in water is done as described earlier. Furthermore, 3–4 irrigations are also recommended within 20 days' time period [15]. Seeds stored for a longer period of time lose its germinating power and gradually it will even vanish, hence it is recommended to use ripe last year seeds. The cultivation season may vary depending upon the area and it can be carried out in autumn or spring. The seeds are planted at a rate of 20–25 Kg per hectare area and once germinated, the plants grow gradually and weeds control is an important step at this stage. The seeds of aniseed prefer a dark environment for germination, therefore it is necessary that aniseed fruit is sown 1–1.5 cm below the surface of the ground, it should be slightly pressed and covered with further soil [18].

To enhance growth, it is suggested to add fertilizers before the cultivation of aniseed, however the requirement of fertilizers is also dependent upon requirement of the crop i.e.

- Uptake of the plant
- Condition of the soil (contribution towards mineral provision)

Under normal conditions it has been demonstrated that the use of phosphorus pentoxide (50–75 Kg/ha) and Potassium oxide (80–100 Kg/ha) for few weeks is recommendable. Nitrogen is added at the rate of 50–100 Kg/ha, however its excessive amount may reduce the yield due to overproduction of weeds. Moreover, bio-fertilizers and ammonium nitrate have been found to enhance the yield of the crop as well as also contribute towards increased production of essential oils from the plant [2, 17].

24.2.2 Harvest Conditions

The aniseeds can be harvested after a month of flowering, at this stage the flower turn brownish grey in color. The umbels of aniseed don't ripen at once rather gradually, even seeds also get ripe randomly, therefore harvesting becomes challenging. Harvesting is done usually in August and September [3], whereas in some areas it is harvested from the end of July till September [2]. According to the study conducted by Ozel [19] it was concluded that harvesting upon complete maturation of the main umbel (called 4th stage) results in better yield therefore, it is more recommendable to harvest when the umbel is mature enough. If during maturation of the flowers there is alternate change in weather as dry and wet, the color of the seeds become dark which affects the quality of the seed and its extracted products. Moreover, the presence of weeds while harvesting may cause adulteration and distresses the economic and commercial value of the crop. In some countries the top of the plant is cut down at green stage of umbel, tied in form of bundles and arranged in conical piles upside down until ripen. This results in natural drying and after just shaking aniseed can be obtained, further it also prevents the discoloration of the fruit, while in other countries the seeds are whirled and can be threshed out when dry using machines [20].

24.3 Diseases and Pests

Diseases related to aniseed are mostly fungal in nature, the occurrence of diseases and attack of pests are caused by critical environmental conditions and growth of weeds. The crop of aniseed is subtle to weeds for having a slow growth as described earlier. On the other hand sound environment results in better growth and prevents the attack of diseases and pests [21]. The important diseases and pests of aniseed (Table 24.2) are discussed below;

24.3.1 Mycoflora

The less information is available about the mycoflora and associated mycotoxins of aniseed but knowledge about it is important specially if it attacks on seeds. Seeds may be important means to transmit disease to the crops and may result in critical infections, further it also reduces the medicinal properties of the due to compromise

Table 24.2 Common diseases of Aniseed and their details

Sr. No.	Common Diseases	Causes	Management	References
1	Alternaria blights	Fungi	Hot water treatment of seeds before harvest Fungicide spray	[34]
2	Passalora blight	Fungi	Fungicide spray Seed treatment with Azoxystrobin and carbendazim + chlorothalonil	[29]
3	Downy mildew	Fungi	Don't overcrowd plants Rotate plants with other species Use of Eco-friendly fungicides	[35]
4	Powdery mildew	Fungi	Avoid excess fertilization Fumigation Use of Eco-friendly fungicides	[35]
5	Rust	Fungi	Plant in well-drained soil Fungicide spray	[7]

on quality and usefulness [22, 23]. However, aniseed is less affected by fungal attack due to its strong antifungal properties [24]. Different strains of mycoflora have been identified in different countries as follows,

- The leading species attacking on the aniseeds in Saudi Arabia belong from Aspergillus genus [25].
- The predominant and damaging disease for anise is the rust; it affects more than 26% of seeds. This rust was formerly known to be *Puccinia pimpinellae* [7].
- Furthermore, Ghoneem et al. [26](2012) reported that several species from almost twenty one fungal have been reported to be linked with aniseed seeds in Egypt the including the leading species such as
    - *Alternaria alternata*
    - *Drechslera tetramera*
    - *Cladosporium sp.*
    - *Stemphiulium sp.*
- It has also been reported that species of genus Fusarium also affect the crop but at a very low percentage [26].
- Bulajić et al. [27] reported that in Serbia most pronounced effect on aniseed was observed by *Alternaria alternata*.
- Pavlović et al. [23] observed that that genus Fusarium is also abundantly present with aniseed in Serbia and identified *F. tricinctum* and *F. sporotrichoides* for the first time in the world.
- In Egypt the emerging issue is of rust fungus such as obligate parasite fungus (*Puccinia pimpinellae*). The infection of rust affects all parts of plant including stem, flower buds, inflorescence and fruits. This infection may result in curled leaves, which get brown and ultimately drop. Moreover, flower and umbels number and fruit size decrease [7].
- Furthermore, concerning the destructive effects of *Puccinia pimpinellaeI*, Ghoneem et al. [28] identified severe action of this rust on germination, quality of aniseed and declared it as a major threat to aniseed plant in Egypt.

- The studies from Turkey have reported Passalora blight in aniseed which is caused by *Passalora malkoffii* and deteriorates all parts of the plant above the ground especially flowers. Pathogen transmission has been observed in pathogen free soil where infected seeds were grown [29].
- Srivastava and Chandra [30] described that genus of Aspergillus and Fusarium are the major culprits to cause diseases in aniseed and some other spice plants of India.

Once identified the attack of pathogen it becomes easier to take necessary measures to prevent transmission of the pathogen through seeds. The use of chemical fungicide may be an effective treatment but it has its own adverse effects on the consumers due to its potent teratogenic and carcinogenic effects. Planting the treated seeds (hot water treatment) in rotation with other plants from non-Umbelliferae family may prevent the spread of these fungal infections [22, 31]. Furthermore, spray of chitosan (1000 ppm) has been observed to be strong agent in decreasing severity and incidence of rust, it is also found to improve plant characteristics and yield of aniseed [7].

24.3.2 Mycotoxins

Mycotoxins are the poisonous secondary metabolites produced by fungi and these are present on all substrates where is fungi outbreak. Concerning the mycotoxins of aniseed, these belong the most commonly to fungi including Aspergillus, Penicillium, and Fusarium genera. The aflatoxins produced by these fungi include citrinin, aflatoxins and ochratoxin. These toxins are usually carcinogenic, hepatotoxic and nephrotoxic. Aflatoxins are naturally occurring secondary metabolites from some species of Aspergillus and they are carcinogenic [32]. Globally, the mycotoxins adulteration has been demonstrated in different countries including Egypt, Turkey, Ethopia, Italy, Portugal and Morocco. Though usually the concentration of aflatoxin on spices is negligible but it might be of concern in case of aniseed as it is being immensely used in colic issues, as carminative, for expectorant properties and flatulence for children [25]. Therefore, some of the countries have customized rules and regulations to deal with these mycotoxins, which are more specifically related to aflatoxins found in spices and its upper limit has been set as 10 μg of aflatoxin/Kg of spice [33].

24.3.3 Pests

The common pests of Aniseed plant are insects including Aphids (*Cavariella aegopodii*), Armyworm (*Pseudaletia unipuncta*), and Cutworm (*Peridroma saucia*). These insects and their larvae suck the juice of plant at flowering stage and may cause destruction of plant through loss of flower and fruit (seed). These may produce 21 generations in a year and cause huge loss. Butterfly caterpillars cause severe destruction to the leaves and also damage umbels of the plant. The chemical insecticide may be harmful for aniseed plant as it is used as food. Furthermore, application of chemicals also may harm honeybee activation during the flowering phase. Therefore, it is a better strategy to use essential oils and extract of plants as natural insecticides. These extracts/ essential oils are usually mixture from different plants and help to kill or repel the insects [36, 37].

(a) **Aphids**

The aphids (*Cavariella aegopodii*) are smaller in size but result in huge loss if favorable conditions are available for their growth (grow very rapidly) that results in formation of colonies. These feed on sap of the plant and in larger quantities remove so much of sap that the plant becomes stunted due to lack of nutrients. Infestation with Aphid may result in leaves to get yellow, necrotic spots on leaves and development of mold. Moreover, while sucking they may also transmit viral diseases to the plant. Washing the plant with strong pressure helps to give away the aphid, however if the attack is severe insecticidal soap spray may be useful. Moreover, pretreatment of the soil with reflective mulches as silver colored plastics can prevent the aphid to feed on plant.

(b) **Armyworm**

Armyworms (*Pseudaletia unipuncta*) are also smaller pests, which cause single or grouped holes in the foliage. Eggs are present in form of clusters and are covered with whitish scales. Larva is generally dark green in color and feeds not only on seeds but also causes damage to the aniseed fruit. These insects have 3–5 generations in a year. The management of army worm has been recommended through organic methods using biological controls, the chemical controls are also available but these don't provide enough remedy against larva.

(c) **Cutworm**

Cutworms as *Peridroma saucia*, *Nephelodes minians* and others have wide range of host. The infestation results in irregular holes on fruit, larvae are usually more active at night. Larva may be of many pattern and colors but one thing common is turning to C shape on disturbing. As the management before plant the aniseed if any other host of cutworm is planted I the land, then it is necessary to remove all previous plant residues from the soil. As a preventive measure add diatomaceous earth around the plant. On infestation suitable insecticide can be applied to deal with cutworms.

Table 24.3 Top producers of aniseed, badian, fennel and coriander

Country	Production (tonnes)
Syria	27,700.00
Egypt	28598.00
Turkey	39277.00
Spain	4229.00
North Africa	68,212
Central America	129446.00
China	56,395
Iran	65792.00
India	678,938
Mexico	127,244
World	**1146610.00**

24.4 Description of Crop

24.4.1 Origin and Production

The origin of aniseed production is not properly known, however perhaps it started from the Eastern Mediterranean area. It was used by the primordial Egyptians near about 1500 BC, similarly its roots are also found from the Greeks and the Romans from where it was transported to India through Persia. It is narrated that aniseed was observed to be presented in China (1200 AD) from Java which was also not a producer of aniseed rather has been brought there probably from West. Now it is frequently present and used in Syria, Egypt, Cyprus, Turkey and Greece [38, 39]. Although, it is extensively produced in Spain, Turkey, Central and South America, China, Iran and India but in tropical low lands of South East Asia, (as the crop needs warm climate) it does not grow well, so care is taken concerning this when growing there. If look into history about the use of aniseed fruit as spice, it was grown and used by primitive Egypt and later in Greece, Rome and Arabia [40]. According to FAOSTAT [41] statistics about anise, badian, fennel, coriander (aniseed and synonyms of aniseed) the total world production is 1146610.00 tonnes, irrespective of the area of production and yield. The top producing countries have been mentioned in Table 24.3.

24.4.2 Products of Aniseed

The aniseed fruits are broadly used as flavoring agent (spice) in baked items, household cookery, toothpaste (as tooth polishing agent), mouthwashes and for extraction of essential oils which are used in food items, confectionary and drinks to enhance organoleptic properties of the food. The fruit extract has its roots in ancient Chinese medicines since earlier in fifth century. The seed as well as essential oils have carminative, antiseptic, expectorant, diuretic, and antispasmodic activities (Will be discussed later in this chapter). It is also added in tobacco and alcoholic beverages due to its flavoring properties. Aniseed has a sugary and aromatic taste, which produces a characteristic flavor and aroma on crushing. The major ingredient which causes this pleasant flavor and odor is anethole, other active components are methyl chavicol, p-methoxyphenol, terpenes and acetone [1, 4, 42, 43]. Moreover, aniseed and its oil is important part of Italian sausages, pizza, pepperoni, processed meat products and cookies [2].

Essential oil of aniseed acts as imperative ingredient in formulation of soaps, skin creams and perfumes due to its properties to act as masking substance against unpleasant odor [44]. It is used with licorice in sweets as a long time tradition and thus has confused the flavor with licorice [45]. Furthermore, aniseed has long been used for its properties in skin care and to improve complexion [3]; for example aniseed tea is known to reduce the oiliness of skin and fruit is recommended to be used as an active ingredient in face packs [46]. Moreover, along with the use of flower/seed in food items at household level, the leaves of this plant are being used in salads, salad dips and cheese spreads [2].

(a) **Essential Oil**

Pimpinella anisum essential oil is colorless to pale yellowish and has characteristic strong aromatic flavor of licorice (spicy sweet) [47]. The essential oil obtained from the genus Pimpinella consists of different essential components including alkenes, sesquiterpenes and phenolic compounds. Schizogenetic oil ducts present in fruit, leaves and shoot are responsible for production of essential oil. Depending upon the origin of the plant the concentration of essential oil varies in plants, however, to act as drug it must contain at least 2% of the essential oil [48]. It is reported that most of the plants of Aniseed grown in Europe produce an ample amount of essential oil (1–5%). The concentration of the essential oil has been found to be higher in fruit (2–6%) as compared to the roots (0.05%), whereas the shoot and leaves also contain minute quantities of this oil. It is not only the genetics of the plant which determines the concentration of essential oil but it is also dependent upon the stage of development and the maximum production has been observed at waxy stage [49]. Furthermore it is demonstrated that treatment with GA_3 (@ 50 ppm) at appropriate times, may result in significant increase in production of the oil from the plant. Likewise, similar trend has also been observed with Kinetin treatment (@50 ppm) [50].

24.4.3 Important Chemical Constituents

Aniseed constitute a good amount of lipids i.e. 8–11%, consisting of both saturated (mainly palmitic acid, stearic acid and lauric acid) as well as unsaturated fatty acids (mainly Petroselinic acid, lioleic acid and oleic acid). Other macronutrients levels include 4% carbohydrates (lignin) and protein (18%) [3, 51]. Sousa et al. [52] has demonstrated that aniseed contain 1.6–6% of essential oil, however the production of oil and content of important constituents may be different conditions based upon genotype of the plant, the environmental factors, the method and time of irrigation, fertilizers used, plantation and harvesting conditions as described earlier in this chapter [9, 53]. The essential oil is colorless to pale yellow and give characteristic sweet-spicy flavor and aromatic odor (licorice like) related to aniseed. It oil is derived from schizogenetic ducts present in its roots (0.05%), leaves and stem (minutes quantities) and fruits (2–6%) [54].

The essential oil obtained from aniseed plant consists of various constituents broadly classified as phenolic compounds, sesquiterpenes and alkenes. The core constituent responsible for specific flavor and odor is trans-anethole (75–95%), which also imparts its role, as therapeutic and medicinal purposes (discussed later in this chapter). The other components include umbelliferone, umbelliprenine, scopoletin, fat based β-armyrin and stigmasterol and flavonoids including flavones, flavonol, glycosides, rutin, isovitexin and isoorientin [9, 55]. Furthermore, it also comprises estragole (adds in odor but not taste), anisaldehyde, α and γ himachalene, cis-anethole and others described in Table 24.4. Sesquiterpenes hydrocarbon present in essential oil of aniseed include germacrene-D, β- bisbolene, γ-himachalene, βhimachalene, α-zingiberene and ar-curcumene [56]. Tabanca et al. [57] detected 140 different chemical compound including mono-, sesqui- and trinorsesquiterpenoids, propenylphenols and pseudoisoeugenols through GC and GC-MS (Table 24.4).

24.5 Medicinal Uses

Aniseed plant parts and oils have been proved by different research studies to be an excellent strategy to manage and treat various medical conditions. It is used not only as antimicrobial, anti-insecticidal, antioxidant and anti-inflammatory agent but is also effective managing different conditions of tissues and organ systems which have been shown in Fig. 24.3 and are explained as under;

24.5.1 Antimicrobial Properties

Infectious diseases have become a major health concern and the side effects (especially anti-microbial resistance) and cost of synthetic anti-microbial agents has derived the focus toward natural agents which are effective as well as give no harm to the recipient. The versatile medicinal plant the aniseed has marvelous anti-microbial properties. The most effective compound found to be involved in antimicrobial effect is anethole, which shows its efficacy against bacteria, fungi and virus; however the levels of antimicrobial activities vary in accordance with species, and plant properties [3]. The aqueous and ethanolic extracts of aniseed have been used for

Table 24.4 Major chemical constituents present in aniseed oil and extract

Sr. no.	Chemical constituent	Sample used	Method used	References
1	Trans-anethole	Aniseed oil	GC/GC –MS	[58]
2	Estragole			
3	(E)-methyleugenol			
4	β-bisabolene			
5	p-anisaldehyde			
6	α-cuparene			
7	α-himachalene			
8	Cis-anethole			
1	Methylchavicol	Aniseed Oil	GC/GC –MS	[47]
2	γ-himachalene			
3	Trans-pseudoisoeugenyl 2-methylbutyrate			
1	Carvone	Aniseed Oil	GC/GC –MS	[59]
2	B-caryophyllene			
3	Dihydrocarvyl acetate			
4	Estragole			
1	Anethole	Aniseed extract	Supercritical Extraction with CO_2/GC-MS	[60]
2	γ-himachalene			
3	p-anisaldehyde			
4	Methylchavicol			
5	Cis-pseudoisoeugenyl 2-methylbutyrate			
6	Trans-pseudoisoeugenyl 2-methylbutyrate			
1	Glucosides	Methanolic fruit extract	^{13}C NMR spectroscope	[61]
2	Alkyl glucoside			
3	α-glucide			
1	Oleic acid	Aniseed Oil	Silver Ion HPLC	[62]
2	Petroselinic acid			
3	Cisvaccenic acid			
4	Catechein	Ethanolic extract of seed	HPLC	[9]
5	Chlorogenic compound			
6	Pyrogallo			
7	Syrinic compounds			
8	Salycilic compound			
9	P. coumaric			
10	Kampferol			
11	Cinnamic acid			
12	Hypersoid			
13	Luteolin			
14	Quercetin			

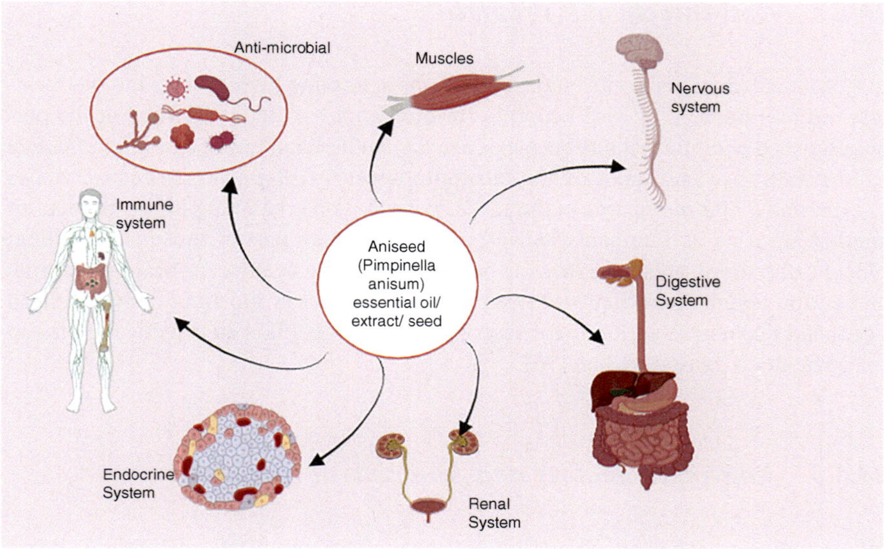

Fig. 24.3 Effectiveness of Aniseed

their anti-bacterial properties against genus Staphylococcus, Streptococcus, Micrococcus, Mycobacterium, Bacillus and *E. coli* by different studies and effective results are shown [63–65]. Similar outcomes have also been observed by using a combination of methanolic extract and essential oil against nine different species of bacteria [66]. In a study conducted by Mahady et al. [67] it was proved that that aniseed fruit extract shows positive effect against the Helicobacter pylori (primary cause for higher gastric acid production resulting in gastritis and peptic ulcer disease.

Along with having antibacterial activities, extract of aniseed has also been reported to show its effect as antifungal agent against genus Candida. Furthermore, extract also showed positive antimycotic activity against dermatophytes and yeast [68]. Likewise, antifungal activity of aniseed essential oil was also investigated for antifungal activities and fruitful results were obtained against *A. niger, A. parasiticus and Aternaria alternata* showing most prominent effects on *Aspergillus parasiticus* species [58]. Similar finding have also been obtained from ethanolic extracts used against dermatophytes and yeast [69]. Moreover, the aniseed essential oil has been found effective to cure the lesions caused by potato virus X, tobacco mosaic virus, and tobacco ring spot virus [70].

Amer and Aly [71] observed the anti-bacterial activities of aniseed extract against the invasion of four strains of most common disease causing bacteria including *S. aureus*, *B. cereus, S. typhimurium* and *E. coli* and observed potential effects of extracts as antibacterial agents. However, the effect was more pronounced by the extract of seed as compared to aerial parts. Moreover, ethanolic extract showed better effects as compared to the water extracts.

24.5.2 Anti-insecticidal Properties

Insects attack is major contributor towards the economic losses caused to the farmers; further insect biting also becomes furious if not handled properly. For the purpose aniseed essential oil has been extensively studied and has been found effective against both ova and larva of the mosquito species (*Anopheles stephensi, Aedes aegypti and Culex quinquefasciatus*) [72]. Essential oil containing higher amount of anethol are used as fumigants as well as are important part of mosquito repellent [73]. Furthermore, aniseed extract is proved to be toxic against carmine spider mite and cotton aphid. Anise has also been found beneficial in finishing body and head lice, head lice and other itching insects [74, 75] and its oil is an effective treatment for pediculosis caused by lice [76].

24.5.3 Anti-inflammatory and Antioxidant Properties

Inflammation may be caused by tissue damage, invasion of microbes, and/or toxic compounds, develop to chronic inflammation and may result in even death [77]. Therefore, aniseed can be of much importance because it possesses anti-inflammatory properties. Tas et al. [78] has observed that oil of aniseed shows anti-inflammatory activities which are comparable to indomethacin a known Non-steroidal anti-inflammatory drug (NSAID) to treat arthritis. Similarly, Okuyama et al. [79] found out that topical use of ethyl acetate and hexane based extract of aniseed, acted as an anti-inflammatory agent to treat chemically induced inflammation in a mouse model.

Antioxidant properties of aniseed (ethanol and water extracts) have been found comparable to butylated hydroxyanisole (BHA), butylated hydroxytoluene (BHT), and α-tocopherol, with water extrct showing higher antioxidant activity compared with ethanol one [63]. Similarly, Rajeshwari et al. [5] reported that aniseed oil and its methanol oleoresin depict the highest antioxidant properties even better that BHA and BHT. There is a strong relationship between flavonoids content and antioxidant properties of the aniseed extract, so aniseed shows higher DPPH (2,2-diphenyl-1-picrylhydrazyl) free radical scavenging properties and has been approved by different studies [55, 80]. Whereas, Speisky [81] observed the antioxidant properties of different herbal teas in terms of Troloxequivalent antioxidant capacity, hypochlorite quenching assay and peroxynitrite quenching assay and found out that aniseed tea has weak antioxidant properties as compared to others.

24.5.4 *Muscular System*

Aniseed extract has been observed to cause muscle relaxing effects through an experiment on guinea pig tracheal chain. The results showed a better effect with aqueous and ethanol extract as compared to essential oil. This effect was comparable to the one caused by theophylline used as bronchodilator for the treatment of chronic obstructive pulmonary disease and asthma. This relaxing effect in muscles of trachea is due to inhibition of muscarinic receptors, which are G-coupled protein receptors involved in the parasympathetic nervous system [82]. Further studies on relaxing effect of aniseed was conducted by Tirapelli et al. [83] using hydroalcoholic extracts of aniseed which showed positive outcomes on anococcygeus smooth muscle of rat model. This effect was found to be dose dependent and was attributed to stimulation of the NO-cGMP pathway.

24.5.5 *Nervous System*

Aniseed shows its effect on nervous system through its properties

- Analgesic effect
- Anticonvulsant effect
- Causing aversion to Morphine dependence

Aniseed has been found to show analgesic properties in thermal test [84]; and effect comparable to aspirin and morphine [85]. Similarly, Tas et al. [78] observed that fixed oil of aniseed has analgesic effect comparable to that of aspirin (100 mg Kg^{-1}) and morphine (10 mg Kg^{-1}). Pourgholami et al. [86] reported that aniseed fruit essential oil can be used to treat tonic convulsions, the study was done on male mice, which were induced seizures using pentylenetetrazole or maximal electroshock. Similar results were demonstrated by using aniseed extract which delayed the onset of seizures and also caused a reduction in death rate [87]. Furthermore, Sahraei et al. [88] conducted a research on rats to check the effect of aniseed essential oil on conditioned place preference induced by morphine. The results showed that essential oil of aniseed has ability to cause conditioned place aversion.

24.5.6 *Digestive System*

Aniseeed has been in use traditionally to soothe the gastrointestinal tract (GIT) related issues. Aqueous extract of aniseed has cytoprotective and anti-ulcer properties against gastric lesions. As has been described earlier in this chapter that aniseed has anti-bacterial activities against *H. pylori* (a major causative agent for acid production and gastric ulcer), Al Mofleh et al. [89] has described ethanol based extract suspension of aniseed helps to get rid of ulcer in ruminants. The extract is useful to replace exhausted gastric mucosal Non-Protein Sulfhydryl (NP-SH) as well as gastric wall mucus concentration. The anti-ulcer properties of aniseed may be linked to the mediation of prostaglandin and also via anti-secretory and anti-oxidtaive properties of aniseed described earlier.

Aniseed has been in use for effective treatment of nausea through aroma therapy. The study was conducted on 25 patients having different symptoms of nausea [90]. Furthermore, aniseed has been proved to be a commendable remedy to deal with constipation. Picon et al. [55] observed that laxative effect of aniseed not only causes a reduction in colonic transit time but also increases number of daily evacuation. Another, plus point of this therapy is absence of any side effect; therefore it can be used effectively for the treatment of constipation, which otherwise causes lots of other health problems.

24.5.7 Other Systems

Along with the specific action described above aniseed plays important function in some other systems. Kreydiyyeh et al. [1] used aniseed oil in water to check its effect on renal system and observed a volume reduction in urine due to enhanced activity of renal Sodium-Potassium ATPase activity. Further this study also demonstrated that enhanced activity of Na^+-K^+ pump results in efficient glucose absorption. Aniseed has also a tremendous effect on endocrine system in terms of its effect in lowering blood glucose and cholesterol concentration. Aniseed powder is not only involved in correcting the metabolism of macromolecules but also due to its antioxidant power in dealing with peroxidation reactions [91]. Moreover, aniseed acts as expectorant thus easing the breathing process, Haggag et al. [92] has demonstrated that aniseed helps to reduce the sleep discomfort, cough frequency and intensity related to pulmonary disease and attributed this effect to anethole. Owing to having antimicrobial, anti-inflammatory and antioxidant properties, aniseed has immunomodulatory effect [3]. It is also thought that aniseed can help to stimulate sexual desire especially in females, whereas it is also reported that it also does the same in males too and lessen signs of male climacteric [45].

24.6 Myths, Legends, Tales, Folklore, and Interesting Facts

The most customarly herbs have been in use as herbal teas in tales and folklores but the concept still exits, however these have also been proven through research studies described earlier. One of such example is the use of aniseed as “nursing tea” which has efficiency to enhance lactation but now has been proved to be linked with hormone production. Similarly, the use of its herbal tea in digestive issues provides versatile methods to deal with mild digestive problems including lack of appetite, stomach ache, and flatulence also because these ease acid production and have antimicrobial properties. Among some of the myth about aniseed one is the golden age of absinthe i.e. in almost end of nineteenth century. The recipe of Absinthe contains fruit of aniseed and some other plants extracted with concentrated alcohol and are thought to change the mental state of the consumer. It is supposed to act as brain stimulant, an aphrodisiac and a hallucinogen; it is also famous to enhance psychic abilities and to protect against the Evil Eye. Furthermore, aniseed is burned as incense to help in the management of spiritual dreams [93].

24.7 Conclusion

The data collected from different sources has demonstrated that Aniseed a medicinal as well as culinary herb, has been in use for ancient times due to its valuable properties. The nutrients present in it are of much importance not only with respect to macro and micronutrients but also its phytochemical profile is versatile and provides it with characteristic odor, flavor, and medicinal properties known for long time. Although the biochemical composition varies with respect to area where it is grown, however, overall the benefits are almost same. It is being used in raw form, as product (food, cosmetics, nutraceutical, pharmaceutical) and as essential oil. Worldwide it is considered as productive crop that is demand of many industries, therefore research work on the plant yield and productivity are ongoing to get maximum benefits out of it. Not only at modern and industrial scale it is in demand but also some myths and folklores are related to it and it's considered as sign of good luck and well being.

References

1. Kreydiyyeh, S. I., Usta, J., Knio, K., Markossian, S., & Dagher, S. (2003). Aniseed oil increases glucose absorption and reduces urine output in the rat. *Life Sciences, 74*(5), 663–673.
2. Özgüven, N. (2001). Aniseed. In K. V. Peter (Ed.), *Handbook of herbs and spices* (Vol. 1). Woodhead Publishing Limited.
3. Shojaii, A., & Fard, M. A. (2012). Review of pharmacological properties and chemical constituents of Pimpinella anisum. *International Scholarly Research Network ISRN Pharmaceutics, 2012*, 8.
4. Andallu, B., & Rajeshwari, C. U. (2011). Aniseed (Pimpinella anisum L.) in health and disease. In V. R. Rreedy, R. R. Watson, & V. B. Patel (Eds.), *Nuts and seeds in health and disease prevention*. Academic.
5. Rajeshwari, U., Shobha, I., & Andallu, B. (2011). Comparison of aniseeds and coriander seeds for antidiabetic, hypolipidemic and antioxidant activities. *Spatula DD, 1*(1), 9–16.
6. Figueiredo, A. C., Barroso, J. G., Pedro, L. G., & Scheffer, J. J. C. (2008). Factors affecting secondary metabolite production in plants: volatile components and essential oils. *Flavour and Fragrance Journal, 23*, 213–226.
7. Saber, W. E. I. A., Ghoneem, K. M., & ElMetwally, M. M. (2009). Identification of newly detected Puccinia pimpinellae on anise plant in Egypt and its control using biotic and abiotic

elicitors in relation to growth and yield. *African Journal of Microbiology Research, 3*(4), 153–162.

8. Zheljazkov, V. D., Astatkie, T., O'Brocki, B., & Jeliazkova, E. (2013). Essential oil composition and yield of Anise from differet distillation times. *Horticulture Science, 48*(11), 1393–1396.
9. Sun, W., Shahrajabian, M. H., & Cheng, Q. (2019). Anise (Pimpinella anisum L.), a dominant spice and traditional medicinal herb for both food and medicinal purposes. *Cogent Biology., 5*, 1673688.
10. Khare, C. P. (2007). *Indian medicinal plants: An illustrated dictionary*. Spinger.
11. Poss, E. (1991). *Studies on the influence of a seed treatment on germination of selected medicinal and Spice plant species, Thesis*. Technical University Munich-Weihenstephan.
12. Stephens, J. M. (1997). *Cooperative extension service, institute of food and agricultural sciences, University of Florida*, Doc. HS 54,1Gainesville FL. 32611.
13. Ross, I. A. (2001) Medicinal plants of the world. In *Chemical Constituents, Traditional and Modern Uses* (Vol. 2). Humana Press Inc.
14. Barnes, J., Anderson, L. A., & Phillipson, J. D. (2002). Aniseed. In *Herbal Medicines e A guide for healthcare professionals* (2nd ed.). Pharmaceutical Press.
15. Maheshwari, S. K., Gangrade, S. K., & Tarivedi, K. C. (1989). Effect of date and method of sowing on grain and oil yield and oil quality of anise. *Indian Perfumer., 33*(3), 169–173.
16. Tainter, D. R., & Grenis, A. T. (1993). *Spices and Seasonings: A Food Technology Handbook*. VCH Publishers.
17. Zidan, M. A., & Elewa, M. A. (1995). Effect of salinity on germination, seedling growth and some metabolic changes in four plant species (Umbelliferae). *Indian Journal of Plant Physiology., 38*(1) 57–61.
18. Fazecas, I., Borcean, I., Tabara, V., Lazar, S., Samaila, M., & Nistoran, I. (1985). Studies on the effects of fertilizer and sowing date on the yield and essential oil content in Pimpinella anisum in the year 1978–1980. *Horticulturae Abstracts, 55*, 387.
19. Ozel, A. (2009). Anise (pimpinella anisum): changes in yields and component composition on harvesting at different stages of plant maturity. *Explanatory Agriculture, 45*, 117–126.
20. Stephens, J. M (1997). *Anise—Pimpinella anisum L. one of a series of the Horticultural Sciences Department*. UF/IFAS Extension. http://edis.ifas.ufl.edu
21. Heeger, E. F. (1956). *Handbuch des arznei- und gewürzpflanzenbaues* (pp. 578–583). Deutscher bauernverlag.
22. Essono, G., Ayodele, M., Akoa, A., Foko, J., Olembo, S., & Goskowski, J. (2007). Aspergillus species on cassava chips in storage in rural areas of southern Cameroon: Their relationship with storage duration, moisture content and processing methods. *African Journal of Microbiology Research., 1*, 1–8.
23. Pavlović, S., Ristic, D., Vučurović, I., Stevanović, M., Stojanović, S., Kuzmanović, S., et al. (2016). Morphology, pathogenicity and molecular identification of Fusarium spp. associated with anise seeds in Serbia. *Notulae Botanicae Horti Agrobotanici Cluj-Napoca, 44*(2), 411–417.
24. Azzoune, N., Mokrane, S., Riba, A., Bouras, N., Verheecke, C., et al. (2015). Contamination of common spices by aflatoxigenic fungi and aflatoxin B1 in Algeria. Quality assurance and safety of crops & foods. Wiley. 8(1), 137–144.
25. Bokhari, F. M. (2007). Spices mycobiota and mycotoxins available in Saudi Arabia and their abilities to inhibit growth of some toxigenic fungi. *Mycobiology., 35*(2), 47–53.
26. Ghoneem, K. M., Al Sahli, A. A., & Rashad, Y. M. (2012). Detecting of Verticillium dahliaon anise seeds using a new seed health testing technique. *African Journal of Microbiology Research., 6*, 1171–1177.
27. Bulajić, A., Djekic, I., Lakić, N., & Krstić, B. (2009). The presence of Alternariaspp. on the seed of Apiaceae plants and their influence on seed emergence. *Archives of Biological Sciences., 61*, 871–881.
28. Ghoneem, K. M., Elwakil, M. A., & El-Sadek Ismail, A. (2009). Puccinia pimpinellae, a new pathogen on anise seed in Egypt. *Plant Pathology Journal., 8*(4), 165–169.

29. Erzurum, K., Demirci, F., Karakaya, A., Caklr, E., Tuncer, G., & Maden, S. (2005). Passalora Blight of Anise (Pimpinella anisum) and its control in Turkey. *Phytoparasitica, 33*(3), 261–266.
30. Srivastava, R. K., & Chandra, S. (1985). Studies on seed mycoflora of some spices in India: Qualitative and quantitative estimations. *International Biodeterioration, 21*, 19–26.
31. Desjardins, A. E. (2015). *Fusarium Mydotoxins: Chemistry, Genetics and Biology*. American Phytopathological Society (APS Press).
32. Jeswal, P., & Kumar, D. (2015). Mycobiota and natural incidence of aflatoxins, ochratoxin a, and citrinin in Indian spices confirmed by LC-MS/MS. *International Journal of Microbiology, 2015*, 8.
33. Commission Regulation (EC). (2006). Commission Regulation No. 1881/2006 of 19 December 2006. *Official Journal of European Communities*. L 364/5, 20/12/2006.
34. Meena, M., Ghasolia, R. P., & Sharma, J. (2020). Efficacy of fungicides against Alternaria alternata causing alternaria blight of fennel. *International Journal of Chemical Studies., 8*(2), 723–726.
35. Zaker, M. (2016). Natural plant products as eco-friendly fungicides for plant diseases control–A Review. *The Agriculturists, 14*(1), 134–141.
36. Abramson, C. I., Wanderley, P. A., Maria, J., & Wanderley, A. (2006). Effect of essential oil from Citronella and Alfazema on Fennel Aphids Hyadaphis foeniculi Passerini (Hemiptera: Aphididae) and its Predator Cycloneda sanguinea L. (Coleoptera: Coccinelidae). *American Journal of Environmental Sciences, 3*(1), 9–10.
37. Kocak, E., Kesdek, M., & Yildirim, E. (2007). A new anise (Pimpinella anisum L.) pest: Carterus dama (Rossi, 1792) (Coleoptera: Carabidae). *Journal of the Faculty of Agriculture, 21*(42), 1–3.
38. Melchior, H., & Kastner, H. (1974). *Gewürze botanische und chemische untersuchung verlag paul parey* (pp. 83–88). Berlin und Hamburg.
39. Ceylan, A. (1997). *Tıbbi Bitkiler II*. Ege Üniversitesi Ziraat Fakültesi Tarla Bitkileri Bölümü Yayını No: 481, 305s, Bornova İzmir.
40. Boztaş, G., & Bayram, E. (2020). Foreign Trade and Production of Anise (Pimpinella anisum L.) in Turkey. *Ziraat Fakültesi Dergisi Türkiye, 13*, 103–108.
41. FAOSTAT. (2018). http://data.un.org/Data.aspx?d=FAO&f=itemCode%3A711. Accessed on 22 May 2022.
42. Buchgraber, K., Frühwirt, P., Köppl, P., & Krautzer, B. (1997). Produktionsnischen im Pflanzenbau Ginseng. In *Kümmel* (pp. 35–37). Hanf & Leopold stocker Verlag.
43. Ozguven, M., Sekin, S., Gurbuz, B., Sekeroglu, N., Ayanoglu, F., Ekren, S. (2005). Tobacco, medicinal and aromatic production and trade. In *Proceedings of sixth technical congress of Turkish agricultural engineers* (Vol. 1, pp. 481–501).
44. Harry, R. G. (1963). *The principles and practice of modern cosmetics* (2nd ed.). Leonard Hill.
45. Leung, A. Y., & Foster, S. (1996). *Encyclopedia of common natural ingredients used in food, drugs and cosmetics* (2nd ed.). Wiley.
46. Bremness, L. (1991). *The complete book of herbs, colour library Books*. Dorling Kindersley.
47. Orav, A., Raal, A., & Arak, E. (2008). Essential oil composition of Pimpinella anisum L. fruits from various European countries. *Natural Product Research., 22*(3), 227–232.
48. European Pharmacopoeia. (2000). *Dritter Nachtrag* (3rd ed, pp. 499–500). Council of Europe.
49. Omidbaigi, R., Hadjiakhoondi, A., & Saharkhiz, M. (2003). Changes in content and chemical composition of Pimpinella anisum L. oil at various harvest time. *Journal of Essential oil Bearing Plants, 6*(1), 46–50.
50. El Hady, S. (2005). Enhancement of the chemical composition and the yield of anise seed (Pimpinella anisum L.) oils and fruits by growth regulators. *Annals of Agriculture and Science., 50*(1), 15–29.
51. Besharati-Seidani, J., & A. & Yamini, Y. (2005). Headspace solvent microextraction: a very rapid method for identification of volatile components of Iranian Pimpinella anisum seed. *Analytica Chimica Acta, 530*(1), 155–161.

52. Sousa, M. P., Matos, E. O., Matos, J. J. A., Machado, M. I. L., & Craveiro, A. A. (1991). *Constituintes quimicos ativos de plantas medicinais Brasileiras*. Edic¸oes UFC/Laborat ˜ orio de Produtos Natu ais, Fort- ´ aleza.
53. Acimovic, M. G., Korac, J., Jacimovic, G., & Oljaca, S. (2014). Influence of ecological conditions on seeds traits and essential oil contents in anise (Pimpinella anisum L.). *Notulae Botanicae Horti Agrobotanici ClujNapoca., 42*(1), 232–238.
54. Lee, J. G., Knowr, Y. J., Jang, H. J., Kwag, J. J., Kim, O. C., & Choi, Y. H. (1997). A comparison of different extraction methods for volatile components of anise (Pimpinella anisum L.). *Agricultural Chemistry and Biotechnology., 40*, 144–147.
55. Picon, P. D., Picon, R. V., Costa, A. F., et al. (2010). Randomized clinical trial of a phytotherapic compound containing Pimpinella anisum, Foeniculum vulgare, Sambucus nigra, and Cassia augustifolia for chronic constipation. *BMC Complementary and Alternative Medicine, 10*(17).
56. Santos, P. M., Figueiredo, A. C., Oliveira, M. M., Barroso, J. G., Pedro, L. G., & Youns, S. G. (1998). Essential oils from hairy root culture and from fruits and roots of Pimpinella anisum. *Phytochemistery, 48*, 455–460.
57. Tabanca, N., Demirci, B., Kirimer, N., Baser, K. H. C., Bedir, E., Khan, I. A., et al. (2006). Gas chromatographic-Mass spectrometric analysis of essential oil from Pimpinella species gathered from Central and Northern Turkey. *Journal of Chromatography, 1117*, 194–205.
58. Ozcan, M. M., Chalchat, J. C., & Shojaii, A. (2006). Chemical composition and antifungal effect of anise (Pimpinella anisum L.) fruit oil at ripening stage. *Annals of Microbiology, 56*(4), 353–358.
59. Embong, M. B., Hadziyev, D., & Molnar, S. (1997). Essential oils from spices grown in Alberta: Anise oil (Pimpinella anisum). *Canadian Journal of Plant Science, 57*, 681–688.
60. Rodrigues, V. M., Rosa, P. T. V., Marques, M. O. M., Petenate, A. J., & Meireles, M. A. A. (2003). Supercritical extraction of essential oil from aniseed (Pimpinella anisum L) using CO2: Solubility, kinetics, and composition data. *Journal of Agricultural and Food Chemistry, 51*(6), 1518–1523.
61. Fujimatu, E., Ishikawa, T., & Kitajima, J. (2003). Aromatic compound glucosides, alkyl glucoside and glucide from the fruit of anise. *Phytochemistry, 63*(5), 609–616.
62. Denev, R. V., Kuzmanova, I. S., Momchilova, S. M., & Nikolova-Damyanova, B. M. (2011). Resolution and quantification of isomeric fatty acids by silver ion HPLC: fatty acid composition of aniseed oil (Pimpinella anisum, Apiaceae)," Journal of AOAC International. 94(1), 4–8.
63. Gulcin, I., Oktay, M., Kirecci, E., & Kufrevioglu, O. I. (2003). Screening of antioxidant and antimicrobial activities of anise (Pimpinella anisum L.) seed extracts. *Food Chemistry, 83*(3), 371–382.
64. Ates, D. A., & Erdogrul, O. T. (2003). Antimicrobial activities of various medicinal and commercial plant extracts. *Turkish Journal of Biology, 27*(157–162), 2003.
65. Akhtar, A., Deshmukh, A. A., Bhonsle, A. V., et al. (2008). In vitro Antibacterial activity of Pimpinella anisum fruit extracts against some pathogenic bacteria. *Veterinary World, 1*(9), 272–274.
66. Al-Bayati, F. A. (2008). Synergistic antibacterial activity between Thymus vulgaris and Pimpinella anisum essential oils and methanol extracts. *Journal of Ethnopharmacology, 116*(3), 403–406.
67. Mahady, G. B., Pendland, S. L., Stoia, A., Hamill, F. A., Fabricant, D., Dietz, B. M., et al. (2005). In vitro susceptibility of helicobacter pylori to botanical extracts used traditionally for the treatment of gastrointestinal disorders. *Phototherapy Research, 19*(11), 988–991.
68. Kosalec, I., Pepeljnjak, S., & Kuatrak, D. (2005). Antifungal activity of fluid extract and essential oil from anise fruits (Pimpinella anisum L., Apiaceae). *Acta Pharmaceutica, 55*(4), 377–385.
69. Yazdani, D., Rezazadeh, S., Amin, G., Zainal Abidin, M. A., Shahnazi, S., & Jamalifar, H. (2009). Antifungal activity of dried extracts of anise (Pimpinella anisum L.) and star anise (Illicium verum Hook, f.) against dermatophyte and saprophyte fungi. *Journal of Medicinal Plants, 8*(5), 24–29.

70. Shukla, H. S., Dubey, P., & Chaturvedi, R. V. (1989). Antiviral properties of essential oils of Foeniculum vulgare and Pimpinella anisum L. *Agronomie, 9*(3), 277–279.
71. Amer, M. A., & Aly, U. I. (2019). Antioxidant and antibacterial properties of anise (*Pimpinella anisum* L.). *Egyptian. The Pharmaceutical Journal, 18*(1), 68–73.
72. Prajapati, V., Tripathi, A. K., Aggarwal, K. K., & Khanuja, S. P. (2005). Insecticidal, repellent and ovipositiondeterrent activity of selected essential oils against Anopheles stephensi, Aedes aegypti and Culex quinquefasciatus. *Bioresource Technology, 96*(16), 1749–1757.
73. Erler, F., Ulug, I., & Yalcinkaya, B. (2006). Repellent activity of five essential oils against Culex pipiens. *Fitoterapia, 77*(7–8), 491–494.
74. Spoerke, D. G. (1980). *Herbal medications* (p. 83). Woodbridge Press Publishers Co.
75. Buchman, D. D., (1987). *Herbal medicine: The natural way to get well and stay well, century Hutchinson.*
76. Newall, C. A., Anderson, L. A., & Phillipson, J. D. (1996). *Herbal medicines- A guide for health-care professionals*. The Pharmaceutical press.
77. Michels da Silva, D., Langer, H., & Graf, T. (2019). Inflammatory and molecular pathways in heart failure-ischemia, hfpef and transthyretin cardiac amyloidosis. *International Journal of Molecular Science, 20*(9), 10.
78. Tas, A., Ozbek, H., Atasoy, N., Altug, M. E., & Ceylan, E. (2006). Evaluation of analgesic and antiinflammatory activity of Pimpinella anisum fixed oil extract. *Indian Veterinary Journal, 83*(8), 840–843.
79. Okuyama, T., Matsuda, M., Matsuda, Y., Baba, M., Masubuchi, H., Adachi, M., et al. (1995). Studies on cancer biochemoprevention of natural resources. X. Inhibitory effect of spices on TPA- enhanced 3H-choline incorporation in phospholipid of C3 H 10 T1/2 cells and on-TPA induced ear edema. *Zhonghua Yaoxue Zashi, 47*, 421–430.
80. Nickavar, B., & Abolhasani, F. A. S. (2009). Screening of antioxidant properties of seven Umbelliferae fruits from Iran. *Pakistan Journal of Pharmaceutical Sciences, 22*(1), 30–35.
81. Speisky, H., Rocco, C., Carrasco, C., Lissi, E. A., & Lopez-´Alarcon, C. (2006). Antioxidant screening of medicinal herbal teas. *Phytotherapy Research, 20*(6), 462–467.
82. Boskabady, M. H., & Ramazani-Assari, M. (2001). Relaxant effect of Pimpinella anisum on isolated Guinea pig tracheal chains and its possible mechanism(s). *Journal of Ethnopharmacology, 74*(1), 83–88.
83. Tirapelli, C. R., de Andrade, A. O., Cassano et al., (2007). Antispasmodic and relaxant effects of the hidroalcoholic extract of Pimpinella anisum (Apiaceae) on rat anococcygeus smooth muscle, Journal of Ethnopharmacology 110(1), 23–29.
84. Twaij, H. A. A., Elisha, E. E., Khalid, R. M., & Paul, N. J. (1988). Analgesic studies on some Iraqi medicinal plants. *International Journal of Crude Drug Research., 25*(4), 251–254.
85. Tas, A. (2009). Analgesic effect of Pimpinella anisum L. essential oil extract in mice. *Indian Veterinary Journal, 86*(2), 145–147.
86. Pourgholami, M. H., Majzoob, S., Javadi, M., Kamalinejad, M., Fanaee, G. H. R., & Sayyah, M. (1999). The fruit essential oil of Pimpinella anisum exerts anticonvulsant effects in mice. *Journal of Ethnopharmacology, 66*(2), 211–215.
87. Abdul-Ghani, A. S., El-Lati, S. G., & Sacaan, A. I. (1987). Anticonvulsant effects of some Arab medicinal plants. *International Journal of Crude Drug Research., 25*(1), 39–43.
88. Sahraei, H., Ghoshooni, H., Hossein, S., et al. (2002). The effects of fruit essential oil of the Pimpinella anisum on acquisition and expression of morphine induced conditioned place preference in mice. *Journal of Ethnopharmacology, 80*(1), 43–47.
89. Al Mofleh, I. A., Alhalder, A. A., Mossa, J. S., AlSoohalbani, M. O., & Rafatullah, S. (2007). Aqueous suspension of anise "Pimpinella anisum" protects rats against chemically induced gastric ulcers. *World Journal of Gastroenterology, 13*(7), 1112–1118.
90. Gilligan, N. P. (2005). The palliation of nausea in hospice and palliative care patients with essential oils of Pimpinella anisum (aniseed), Foeniculum vulgare var. dulce (sweet fennel), Anthemis nobilis (Roman chamomile) and Mentha x piperita (peppermint). *International Journal of Aromatherapy, 15*(4), 163–167.

91. Rajeshwari, C. U., Abirami, M., & Andallu, B. (2011). In vitro and in vivo antioxidant potential of aniseed (Pimpinella anisum L.). *Asian Journal of Experimental Biology and Science, 2*(1), 80–89.
92. Haggag, E. G., Abou-Moustafa, M. A., Boucher, W., & Theoharides, T. C. (2003). The effect of a herbal water-extract on histamine release from mast cells and on allergic asthma. *Journal of Herbal Pharmacotherapy, 3*(4), 41e54.
93. Kovács, K., Lente, D. C. G., & Gunda, T. (2014). *100 chemical myths misconceptions, misunderstandings, explanations*. Springer.

Chapter 25
Sacred Basil

Huma Umbreen, Kainat Khalid, Aqsa Khalid, Razia Noreen, and Romina Alina Marc

25.1 Introduction

Scientific name:	*O. sanctum/O. tenuiflorum*,
English name:	Sacred basil/Holy basil
Common name:	Tulsi/Rama Tulsi/Krishna Tulsi
Family:	Lamiaceae
Taxonomy:	

Sacred basil is one of the most famous herbs grown worldwide, is locally grown in Asia (India, Pakistan, Iran, Thailand, and other countries) and can be seen maturing abundantly in tropical and sub-tropical regions. Due to the popularity and beneficial health effects, Sacred basil is represented as the "king of the herbs." As the word basil is derived from a Greek word *Basileus* which means "king" or *Basilikon* meaning "royal". A Latin word, *Basiliscus*, indicates to "basilisk" a mythical fire-breathing dragon so repulsive that a glimpse could kill [2]. Approximately 30–160 annual and perennial herbs and shrubs are grown indigenously to the tropics in addition to subtropical areas of Africa, Asia and South America [3, 4]. Different species (Table 25.1) and forms of Ocimum sanctum differ in growth pattern, color,

H. Umbreen (✉) · K. Khalid · A. Khalid
Department of Nutritional Sciences, Government College University Faisalabad, Faisalabad, Pakistan

R. Noreen
Department of Biochemistry, Government College University Faisalabad, Faisalabad, Pakistan

R. A. Marc
Department of Food Engineering, Faculty of Food Science and Technology, University of Agricultural Sciences and Veterinary Medicine Cluj-Napoca, Cluj-Napoca, Romania

M. Zia-Ul-Haq et al. (eds.), *Essentials of Medicinal and Aromatic Crops*,
https://doi.org/10.1007/978-3-031-35403-8_25

Table 25.1 Commonly found species of basil

Common name	Scientific name	Important uses and benefits	References
Bush or Greek basil	*Ocimum basilicum* L.	Ornamental and floral plant	[9]
Camhpor basil	*Ocimum Kilimandscharicum* Guerke	Culinary herb Spicy pungent aroma due to linalool	[10]
Genovese basil	*Ocimum basilicum* Genovese	Antioxidant Aromatic compound with volatile oil property.	[11]
Sweet basil	*Ocimum basilicum* L.	Volatile oil Flavoring agent in food products	[12]
Purple basil	*Ocimum basilicum* Purpurascens	Purple varieties, due to accumulation of anthocyanins Ornamental plant Antioxidant	[13]
Holy basil (Kemangi)	*Ocimum sanctum* Linn	Antipyretic, anti-inflammatory, anti-cancer, and neuroprotective agent Phytochemical compound	[14]
Peruvian basil	*Ocimum micranthemum*	Culinary herb in spite of medicinal and sweet flavor	
Scented basil	*Ocimum africanum* Lour	Lemon aroma Diverse ethno-botanical and pharmacological properties	[15]
Lettuce leaf basil	*Ocimum basilicum* Crispum	Ornamental crop Raw material in the perfumery, food, and pharmaceutical industries	[16]
Thrysiflora basil	*Ocimum basilicum* var. Thrysiflora	Used as food seasoning Enriches products with unique flavor and aroma Beneficial in cosmetics, medicines, food, beverages and household products	[17]

structure and aromatic conformation, which make the botanical identity of basil much difficult (Fig. 25.1).

The genus Ocimum belongs to the family Lamiaceae, consists of about 68 species native to tropical regions of Asia, Africa and, Central and South America. The aromatic herb *O. sanctum* is commonly known as 'Sacred Basil' or 'Holy Basil' as it is worshipped by the Hindus. Commonly known as *Tulsi* (in India) is also identified as two common species, *Rama Tulsi* with green leaves and *Krishna Tulsi* having purple leaves [5]. This aromatic herb is of great importance in the field of medicine since the time of ancient civilization. As a result of its polymorphism and being economical, the herb is considered as a rich source of various naturally available essential oils and aromatic chemicals [6].

The height of this short-lived perennial shrub is 30–60 cm having hairy stems and sporadically hairy leaves, allocated in the Himalayas up to an altitude of 6000 ft.

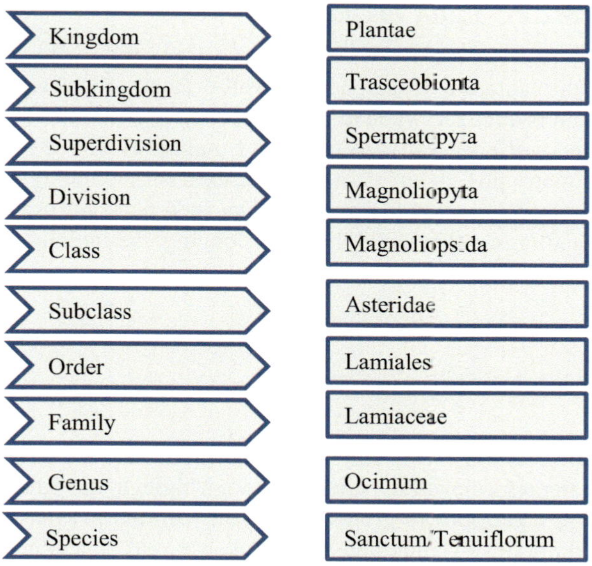

Fig. 25.1 Taxonomic description of Basil plant. (Derived from Ref. [1])

[7]. During trade, usually exported in the dried form, having much-broken leaves along with dispersed pieces of small stems and a few brownish flowers; varying number of small brown fruits are also observed. It has a strong, aromatic and characteristic odor and is slightly bitter in taste [8].

25.2 Agronomy

25.2.1 Climate

It grows under diverse climatic conditions on various types of soils with specific pH ranges from 5 to 8.5. Moreover, it is also raised as a wild plant over hills. In the subtropical areas this plant nurtures in winter and in the grasslands of North India, holy basil grows in summer. *Ocimum basillicum* L. is unable to grow properly in areas where possibilities of snow falling or frost are higher [2]. The suitable growing temperature for basil is around 80 to 90 degrees Fahrenheit, however, the plant can still do well starting from 50 up to 90 degrees Fahrenheit. The important point to keep in mind is that once the temperature drops below 50 degrees Fahrenheit and there is cold environment, the growth of the plant can be affected; most of the plants grow stunted and can't reach to the maximum potential under such environmental conditions. Therefore, this implies that basil plants are sensitive to cold and they can rapidly show signs of damage or even die off once the temperature goes down. The addition of mulch to the base of the plant is another way to minimize the cold effect and add a layer of insulation [18].

25.2.2 Land Preparation

For plantation of sacred basil, well-drained soil is required. The maximum production is obtained when the farm is properly ploughed and harrowed for several times and soil is added with farm yard manure (FYM @15 tonnes/hectare). Both local country plough and tractor can be used for ploughing the field but it should be until the soil becomes highly tilth. Fine seed beds are used for transplantation of basil seedlings and to avoid the water logging, the internal drain system should be properly developed [19].

25.2.3 Soil

Basil will grow the best in an area where there is 6–8 h of full sun daily; however it can perform well in partial sun too. Although, the Tulsi plant shows a wide range of soil conditions to grow but the soil with the pH range of 4.3–8.2, gives the most fruitful results. It is well cultivated on humus-rich or moderately fertile, well drained loamy or sandy-loam soils. The best soils for basil to grow are those, which are in good physical condition having appropriate water holding capacities along with well-prepared internal water drainage system, therefore occasionally waterlogged lands should be prevented. In Egypt, the domestic soil has been used for Basil production by enhancing the soil quality with 30–40 tetra hydro furfuryl alcohol (THFA) organic matter. In greenhouses Sacred Basil is cultivated in flower beds or pots on special substrates, such as fertilizers, different ready made mixtures or even artificial soils. Specifically, for the production of fresh basil, the plants are grown in beds or in different tank farming systems such as growing plants without soil and nutrient film technique or using deep flow technique [20, 21].

25.2.4 Cultivation

This plant can be cultivated by two methods:

(1) Seed propagation
(2) Vegetative propagation

25.3 Seed Propagation

Seeds are used to be sown directly in soil by broadcasting method. The soil is blended with sand to ensure an even distribution. Before propagating, the field is distributed into rows 50–60 cm apart from each other. Seeds are planted at the depth of 2 cm in the soil. If seeds are planted deeper than 2 cm in the soil it results in germination failure. Depending on the moisture content of soil, the field is irrigated after almost 24 h. The process of germination starts in 10–15 days. First thinning and weeding of plants is performed after 20–25 days when plants attain a height of 10–15 cm [22].

25.4 Vegetative Propagation

For a period from October to December this plant is usually propagated vegetatively. This is done by cutting 8–10 nodes from a length of 10–15 cm, leaving behind the first 2–3 pairs of leaves, the rest are trimmed off to prepare the plant. Afterwards, these are planted in polythene bags or well prepared nursery farming beds. The time of rooting gets completed in about 4–6 weeks and they are all set for transplanting into the main field. In order to attain the required growth of plant, manuring is a mandatory step. For the preparation of the soil, farmyard manure i.e. nitrogen, potassium and phosphorus are given as pre planting dose, and rest of the nitrogen is applied in two stages as after first and second cuttings (discussed under manuring) [2].

25.4.1 Light

Holy Basil grows the best during long days and in hot or sunny weather. Light directly affects the weight and other properties of plant and helps it to flourish. The standard plant weight of *Ocimum sanctum* ranges between 90–93 g/pot under 1.5–2 h of light. The flowering stage appears promptly after 18 h of light exposure, and the highest yield obtained is found to be 102 g/pot under 24 h of light exposure [23]. The most favorable day/night temperature for seed germination is 19–22 °C under laboratory conditions [24].

25.4.2 Transplanting

It is appropriate to plant the seeds in the nursery from end of February to March in already grown beds. Germination is usually generous under favorable conditions and gets completed in about 10 days. A seedling with 4–6 leaves becomes ready for transplantation in about 6–7 weeks. Seedlings should be transplanted at distance of 60 × 60 cm. Irrigation of fields is compulsory both before and after transplantation [22].

25.4.3 Manuring

Farmyard manure (FYM) is applied to the site during the land preparation process and it is incorporated it into the soil thoroughly to be more effective. The level of application is as, nitrogen fertilizer in doses of 48 Kg, potash in doses of 24 Kg, and phosphorus in doses of 24 Kg/acre in the form of Urea, Muriate of potassium (MOP), and Single superphosphate (SSP) (Table 25.2). The baseline dosage of nitrogen should be reduced to half and the entire dose of phosphate pentoxide be provided at the time of transplantation. At three different times throughout a plant's growth cycle, a certain amount of nitrogen is administered and micronutrients needed are Manganese (Mn) at 50 ppm concentration and Cobalt (Co) at 100 ppm concentration. After the first and second cuts, the remaining Nitrogen dosage is delivered in two splits [25].

25.4.4 Irrigation

In the lands of India, *Ocimum basilimum* L. is cultivated mainly in the months of July–August, where heavy rains are expected. Therefore, due to the life span of nearly, 65–70 days and expected weather, the crop sown at this time of July–August does not need irrigation. But normally it requires only 2–3 irrigations [26].

25.4.5 Harvesting

The Basil crop requires 65–70 days to get mature, and on maturation nearly 80% of the plants, change their color from green to golden and cluster of flower appears on the branches, hence this is the right time for harvesting. The basil leaves are picked as soon as the plants are 6–8 inches tall. Once the temperature hit 80 °F (27 °C), basil will really start leafing out [27]. Harvesting is mostly done in the early morning, when leaves are at their juice extreme to get maximum benefits. The crop is harvested at full blooming stage and cutting is done at 15 cm from ground level to make sure good reproduction for later harvests. First harvesting is done after 90 days and later on harvesting is carried out after every 75 days [19].

Table 25.2 Manuring for basil crop

Fertilizers	Kg/acre
Urea	104
SSP	150
Muriate of potassium	40
Nitrogen	48
Phosphorus	24
Potassium	24

25.4.6 Distillation of Oil

Harvested plant is dried for 6–8 h on the ground before distillation. This results in reduction in moisture content. Later on, oil is distilled using water vapor method. Oil is extracted from the plant. One acre of plant yields about 20–25 kg of oil. Sweet basil possesses a clove like scent with an aroma and is somewhat saline in taste. It yields evaporative essential oil (oil of Basil) that is used as a flavoring agent and also in perfumes. Characteristic and composition of oil from different regions vary and is also dependent upon the soil and environmental conditions but these are mostly rich in phenylpropanoids [28].

Four major types of oils have been recognized and described as under:

1. Methyl cinnamate type (methyl ester formed through condensation of methanol and methyl cinnamic acid)
2. European type (having good quality aroma, contain linalool and methylchavicol)
3. Reunion/Exotic type (having coarse-herbaceous odor, which gives sweetness of the estragole)
4. Eugenol type (It is allyl chain-substituted guaiacol, and is a member of chemical compound called allylbenzene) [29].

25.5 Pests and Diseases

As discussed already in this chapter, the basil plant doesn't require much of care and are easy to look after; however there are many diseases which may result in low yield and further, attack of disease and pests is more common when there is increased humidity. The sample collection identification is usually done with 70% alcohol and can be preserved in it. This is done in order to get a better understanding of the behavior of the pests as well as to find the damages caused by insects, which is necessary in order to develop effective control strategies. The important pests and diseases have been described as under;

25.5.1 **Cochlochila Bullita *and* Syngamia Abruptalis**

The Indian species *Cochlochila bullita* is widely found on *O. sanctum* in areas like Southeast Asia and Africa. Under heavy infection with the pest the leaves may get yellow and fall off. It results in dark brown dirty spot on lower side of the leaves. If the attack is repeated it may result in death of the plant.

Similarly, *Syngamia abruptalis* (Lepidoptera) is also common pest of basil plant and is found in Southeast Asia and is also observed in Thailand in the plants of basil. The Burla and Hirakud regions of Sambalpur have lately been infected by a new sucking bug that feeds on Tulsi (Odisha). As a treatment the heavy spray with water may provide helpful results and prevent from nymphs, moreover removal of leaves and twigs with heavy infestation may also be useful. The insecticide may also be helpful to avoid further damage to the yield [30, 31].

25.5.2 *Nipaecoccus viridis*

In a research study Atanu Seni found for the first time that Tulsi from Odisha, India, was infected with *Nipaecoccus viridis* (Newstead). First and second instar nymphs on the delicate area of the plants indicate an infestation has begun, and is visible after a careful inspection. Later, these spread the whole tulsi plant's stalk and leaves. Purple bodily fluid is observed to spill from their crushed corpses. Seni (2022) found that the leaves of heavily affected plants become yellow, wither off, and eventually fall to the ground. Holy basil plant is progressively infested after the arrival of nymphs in their first instar on fragile parts of plants. It takes between a month and two months to completely dry out to severely affected plants. In addition, it has been shown that plants in the shadow are more vulnerable than those in the open. When the temperature is between 26 and 30 degrees Celsius and the relative humidity is between 70% and 85%, insects are raised at a higher rate. In light yellow to white ovisacs, the female deposits 107–138 purple eggs. Both *Paratrechina longicornis* and *Solenopsis geminata* have been shown to be connected with the mealybug. It is more common in shady places, and affected plants dry up within a month or two after the insect invasion. Besides *O. sanctum* and *Euphorbia hirta*, Citrus species and Rosa species are also affected. When a pest is discovered on this vital medicinal plant, proper care and control must be implemented immediately [31].

Most common species of pests

- *Paratrechina longicornis* (Latreille)
- *Paratrechina longicornis* (Latreille)
- Hymenoptera (Formicidae)
- *Nematode* (Meloidogne incognita)
- *Solenopsis geminata* (Fabricius)

The plants that are less harmed by mealybugs may be salvaged by brushing them off the plant and then eliminating them. Mealybugs may be prevented from spreading to new plants by keeping away the ants [32]. After the seedlings have been planted, they need to be irrigated appropriately. Two to three control treatments are needed during the life of the plant to keep it free of weed infestation [33].

25.5.3 Pest Management

Pests are more common in regions with higher average temperatures. As areas in the temperate zone do not supply a great deal of information about the pests that affect sweet basil grown by manual cultivation, therefore it can be assumed that the manually cultivated basil crop is less susceptible to infestation. In addition to crop rotation, a variety of nematicides, including aldicarb, carbofutan, and bavistin, or oilseed cakes may be used to combat the most prevalent nematode (*Meloidogne incognita*). Spraying the plants with pesticides is required because the larvae of some pests may cause considerable harm to the plants by attaching themselves to the underside of the leaves, folding them in half lengthwise, and webbing them [20].

25.5.4 Curl Leaf Disease

Holy basil is one of the most vital economic crops that has a significant impact on export earnings. However, the holy basil prices could be dropped due to a curl leaf disease caused by pests. Several previous studies focused on plant leaf disease detection based on the leaf color. Unfortunately, the leaf curl disease sometimes changes a shape of the leaf. However, monitoring the basil leaves manually is a difficult task since it requires a lot of time and effort. From previous studies it is obvious that the plant diseases are usually examined by pre-processing color images, extracting the image feature and by classifying leaf disease. Dhingra et al. proposed a novel computer vision to identify and classify leaf disease based on the Neutrosophic approach [34]. In their work, various species of basil leaves were used as a dataset and they classified five classes of diseases which are downy mildew, aphids, gray mold bacterial leaf spot and Fusarium wilt [34–36].

25.5.5 Damping-off

Basil wilts in the presence of *Pythium ultimum* which is especially harmful to young plants since it can lead to wet rotting or traditional dampening down. Near the soil's surface, it enters the host stem and spreads, resulting in a soft, colorless to dark brown rot. While the disease does not cause any noticeable damage to young plants, it causes the stems to collapse about 5–10 cm above the ground. In later phases, the spread of the infection is slowed considerably, yet it still results in the death of the plants. Damping-off is a typical cause of stunted development in basil and is commonly linked to *P. ultimum*. Conditions such as soil moisture, temperature, pH, cation content of the medium, and light intensity, all have a role in promoting *P. ultimum* infection. It becomes contagious at temperatures below 20 degrees Celsius once it thrives [37].

Unfortunately, plants that have already experienced damping off cannot be saved. However, it can be prevented since it spreads through the soil. Greenhouses can avoid these problems by using sterile soil and clean containers. For optimal growth, basil needs well-drained soil and prevention from overwatering is of much importance [38].

25.5.6 *Fusarium Wilt and Crown Rot*

Basil is susceptible to a vascular wilt disease brought on by the fungus *Fusarium oxysporum* The onset of basil fusarium wilt in the United States was first documented in the early 1990s. Fusarium wilt manifests initially as stunted growth, bent young leaves, and then full-blown rot. Affected plants exhibit symptoms such as asymmetric growth, epinasty, curling, chlorosis, and wilting, starting at the apical leaves. In the apical area of the plant, discoloration of the xylem is more noticeable and is linked to the plant's outward manifestations. As the illness progresses, the plants wilts further, the shoots die back, the leaves fall off and eventually the plant dies from a black stem rot that has moved upward. Following this, necrosis spreads basipetally from the vegetative apex to the rest of the plant, while the roots and base of the stem show no signs of infection until it has progressed to its final stage [39].

Although older plants may survive longer, young ones typically dry out within 4–7 days of the first signs of distress. If basil plants suddenly cease receiving water, they will wilt very fast. Crown and root rot symptoms have been observed in infected plants. Macroconidia of *Fusarium oxysporum* F. appear as a thin, pale pink or orange film on the stems of diseased plants [38]. It can spread through the air and soil. Airborne propagations are easily distributed due to the higher number of conidia produced and the consistent air flow created by fan operation in the greenhouse. Macroconidial masses on stem surfaces serve as a primary source of airborne inoculum, which is further dispersed by soil particles and the harvesting process. Epidemiological evidence suggests that it is responsible for the long-distance transmission of Fusarium wilt and crown rot of sacred basil through seed borne inoculum, while its rapid local proliferation produces its devastating effects on the plant itself. It has been found that the *Ocimum basilicum* var. minimal variety is especially vulnerable to the fungus *Fusarium oxysporum* F. sp. Basilica [40].

To stop the spread of disease, you should get rid of the infected plant or portions of plants and dispose off them properly. The farmers should plant only healthy seedlings or seeds. The basil plants need to have their soil wetted rather than having their leaves sprayed with water. Instead of using fast-acting synthetic nitrogen fertilizers, which may increase plant disease, try a slow-release organic fertilizer. If the disease does not abate through treatment of the soil with solar radiation, the plant should be thrown out entirely. Moreover, give the plants a good dose of the organic supplement to make them flourish. Furthermore, using the soaking techniques for the ground with non-toxic fungicide may also be helpful [38].

25.5.7 *Leaf Spots*

(a) Black Spot

Basil black spot is caused by *Colletotrichum gloeosporioides*. Its symptoms manifest themselves primarily on the plant's stems and leaves. On leaves, lesions manifest as necrotic spots ranging from irregular to circular in shape, and these frequently expand and join with surrounding lesions. Older lesions appear dried out, with torn tissue at their core [41].

Similar to stem lesions brought on by *R. solani* or Sclerotinia sp., these can girdle the entire stem and ultimately kill the plant. Colletotrichum spp. thrives in conditions of high relative humidity and temperature between 15 and 20 °C, which leads to a prolonged period of leaf wetness. It usually manifests itself more severely in thickly planted crops [42].

(b) Cercospora Leaf Spot

The symptoms are typically little dark spots with lighter centers on leaves, but their size and shape might fluctuate. *Cercospora ocimicola* is the causative agent while the ideal growth takes place in wet/humid environments. Fungal spores can survive the winter in nearby infected plant debris or on the soil's surface, from where they can be carried by the wind and spread to new areas [38].

Management

Management is to prevent water from splashing all over the garden, water the plants from below and cover the ground around them with mulch. Leaves exhibiting the symptoms should be plucked off the plant and be burned. If a fungicide is sprayed with potassium bicarbonate once a week, one can keep mild illnesses at bay [38].

(c) Bacterial Leaf Spot

Another name for basil leaf spot caused by bacteria is "basil shoot blight." The pathogen responsible for this is called *Pseudomonas cichorii*. Bacterial leaf spot is a common problem for basil hybrids such Purple basil, African Blue Basil, and Holy Basil. Initially, the leaves show little water-soaked lesions around the leaf margin or over the entire leaf. Their leaf shapes and veining can range from angular to irregular. These wet lesions can go from brown to black and have a yellow halo around them. These lesions frequently combine to cause severe leaf blight when environmental conditions are favorable to bacteria. Dried diseased leaf spots may appear brittle and break [43].

It has been hypothesized that bacteria can spread from plant to plant via the medium of seeds. Extreme conditions of heat and humidity are ideal for its growth. Infected soil water can further reach to healthy plant tissues, and infected hands can also spread disease to healthy plants. This disease may not be very harmful in the field, but it can have a devastating effect on the number of seedlings you harvest.

Management

Seeds and cuttings should be checked for diseases before planting. Eliminating disease-ridden leaves and plants can help reduce spread of disease. It's best to water the plants in the morning so that the leaves can dry when the sun is out. Instead of using overhead sprinklers, you should water the soil directly. Hygiene standards are also very important. Everyone who comes into contact with the sick leaves, including workers, should wash their hands and dispose off the plant debris in a safe location, away from unaffected areas. Take care to only use clean and sterile tools. The condition can be treated with copper or antibacterial chemicals, but these aren't always successful strategies in stopping the spread of disease when the weather is wet [38].

25.5.8 Basal Rot

In basil, *Rhizoctonia solani* and *Sclerotinia sclerotiorum* are the most common pathogens, causing basal rots throughout the plant's life cycle. *Microdochium tabacinum*, however, can cause infections at any stage of growth and development. The pathogen that causes damping-off is particularly dangerous for the young plants. At the base of the stem, *R. solani* quickly colonizes, forming large, necrotic, dry, sunken zones [44]. The entire stem is commonly girded by lesions. Small, circular lesions usually appear in cultivated fields and grow up to 1 m in diameter. Infested soil colonizes rapidly with the pathogen as it spreads through the soil. There has been a lot of moisture in the patch. *Sclerotinia sclerotiorum* and *Sclerotinia minor* are the two most common causes of basil stem rot and are often responsible for basil stem rot [44].

Management
As the use of excessive water, overcrowding, or excessive fertilization are responsible factors for the development of basal rot therefore, the simple management is to avoid and keeping control on these factors.

25.5.9 Downy Mildew

A fungus called *Peronospora belbahrii* causes basil Downy mildew. Symptoms of the infection include yellowing of leaves, which is indicative of a nutritional deficit, once the pathogen has attacked the plant. Yellowing starts at the main vein of the leaf and spirals outward, eventually turning into dark brown spots. The lower surface of the leaf shows gray fuzzy or downy growths caused by fungus [45]. It spreads by seeds contaminated with disease, leaf pieces contaminated with disease and wind-blown spores from another sick basil plant. A humid environment is the most often conducive to the disease's growth [45].

Management
The unhealthy plant should be removed and destroyed so that proper spacing of plants can be achieved and good air circulation is ensured. The foliage should not be wet when irrigation is done with drips. The use of fungicides for the treatment of basil downy mildew is not recommended. Rather than starting with seedlings when growing basil, start with dormant seeds. Moreover, instead of planting basil outdoors, you should grow it in pots.

25.5.10 Gray Mold

Basil growing systems are highly susceptible to *Botrytis cinerea*, which causes gray mold symptoms on leaves and stems. A lot of dark conidia are produced by organs infected with mycelia. As a result of wind currents or rain splashing from plant to plant, conidia can easily be dispersed. Crop residues are a source of saprophytic mycelia and sclerotia during unfavorable conditions. In commercial greenhouses, gray mold develops immediately after harvesting on stem cuttings. All of the leaves and secondary buds of the plant are killed as the disease spreads throughout the plant parts. Once the disease reaches the main stem at the bottom, the entire plant dies. Despite the fact that most infected stem cuttings remain alive and well 48 h after harvest, stem cuttings are most susceptible to infection right after harvest. Similarly, *M. cinerea* can also cause rot in bunches when transported to markets [46].

Management

Maintaining a clean growth environment is crucial to preventing botrytis gray mold. Drip irrigation rather than overhead irrigation is recommended for proper air circulation, heating, and ventilation. Some control can be achieved with the bio-fungicide "Actinovate" when applied to foliar. Furthermore, it is suggested that field work should be avoided during rainy weather.

25.6 Description of Crop

Normal agro-climatic conditions are more likely to result in crop output being negatively affected by significant nutrient deficits such as nitrogen, phosphate, and potassium. Sweet basil is indigenous to India's northwestern states. A number of countries throughout the world produce this plant including United States, Bulgaria, South Africa, and France. An exceptionally diverse genus, the Ocimum family has over 160 species that may grow from sea level to over 550 m in height, making it one of the most diverse groups in the plant kingdom. Tropical and rain forest habitats are the most common areas to find this species. In India's mountainous areas, it grows wild [7] (Fig. 25.2).

The mature plant grows to a height of 30–60 cm and is upright with many branches and a strong fragrance. By whole or sub-serrate edges and up to 5 cm in length, the fragrant leaves of this plant are simple and contrary. The annual herb Ocimum Spp. often known as aromatic basil grows to a height of 80–100 cm. Tulsi blossoms are tiny, purple to reddish in hue, and clustered together on tubular spikes in relatively compact bunches. At the foot of each flower cluster, there are heart-shaped bracts without stalks. The inside of the sepal cup is not hairy. The calyx tube is bearded at the base, and the flower is seldom higher than 5 mm. Flowers have hairy stems. There are little fruits with yellow to reddish seeds [47].

Different varieties have leaves that are yellowish green to dark green in hue. Essential oils are made from the plant's blooming tops and leaves, which are used in the perfume and pharmaceutical industries. Foliage and branching abound on the stems of this plant. *Allium basilicum, Ocimum sanctum*, and other types of basil may be found in nature, including *Ocimum gratissimum* and *Ocimum sanctum*. Basil grows on soils with a pH range of 5.0–8.30 and have great water retention capacity. Intercropping basil-chamomile and mint, basil-mustard and mint and

Fig. 25.2 Sacred basil plant

basil-potato mint in the subtropics is the most ideal method. There must be an average temperature of 27 °C in order to thrive [6]. Basil is grown from seeds and summer is the time of year when seedlings are transferred from the nursery to the field area. Nursery-grown saplings are spaced 30–35 cm apart from one another and 45–50 cm apart from one another in rows [33].

25.6.1 Products of Basil Plant

- Basil, one of the most popular culinary herbs,
- Basil herb have many applications as ornamental crops
- Its raw material used in the perfumery
- It is used in food industries, beverages and household products
- It is also used in pharmaceutical industries for centuries.
- Its secondary metabolites used in the form of volatile oily liquids
- It is mainly used in cosmetics industry
- Its dried leaves are used commonly as a flavoring agent in many food products
- Basil plant has many antioxidants, which are linked rosmarinic acid [13, 17].

25.6.2 Importance of Chemical Constituents

Volatile oil, phenolics and flavonoids are found in abundance in *Os* leaves, as well as fatty acid metabolites and terpenes. In the unsaponifiable composition of the seeds, essential oil, mucilaginous, polysaccharide, and -sitosterol are all present. *Os* seed oil has a higher percentage of triglycerides fat (94–98%), most of which is linolenic-acid (43.8%) [48].

Ocimum sanctums fresh leaf and stalk extract revealed bioactive compounds and antioxidants such cirsilineol, rosameric acid, apigenin, circimaritin, isothymusin as well as significant amounts of eugenol. Its leaves produce 0.7% volatile oil, mostly composed of eugenol (71%) and methyl eugenol (20%). Carvacrol and sesquiterpene hydrocarbon caryophyllene are also found in the oil. *Ocimum sanctum* leaf extract yields the flavonoids orientin and vicenin, which are further purified [49, 50].

The fragrant leaves may be utilized fresh or dried, while the seeds are a valuable source of essential oils and commercial value. Eugenol is a component of the volatile oil (about 71%). Compounds including nerol, caryophyllene and selinene are also found in eugenol, as well as cineole and carvacrol (3%). Volatile oil derived from various sources has a wide range of concentrations of these components. Ascorbic acid (83 mg/100 g) and β -carotene (2.5 mg/100 g) are both present in the leaves [51]. *O. sanctum* and *O. gratissisum* contain 1,8-cineole, 1,8-bisabolene, eugenol, methyl eugenol, and linalool, caryophyllene, and caryophyllene oxide and is extracted for various purposes in different countries like India, Nigeria and Northeastern Brazil. The plant has been in use in many herbal products and medicines for years in India. This plant's blossoms and leaves may be used to make teas and infusions also [52].

For its antibacterial powers, *O. gratissisum* may rely on its volatile oil, which has a higher content of thymol and eugenol. *O. kilim* and *O. scharicum* are generally known as Kapur tulsi in India contain limonene and camphene found in the leaves' aqueous extract of carbonyl 1,8-cineole and 1,8-cineole carbonyls [53]. Additionally, linalool and endo-borneol are also present in it. In addition to polyphenols and saponins, the tulsi plant is a source of steroid compounds. Moreover, nutrients and non-nutients compounds present include proteins, carbohydrates and triterpenes. The majority of the essential oil is composed of camphor (64%), but also have other fragrant components i.e. mono-terpenoids such as limonene (8.7%) and ocimene (6.4%) [54, 55] (Table 25.3).

25.7 Medicinal Uses

The ancient writings of ayurveda identify *Ocimum sanctum* as a Rasayana medication for the treatment of cough, respiratory problems, poisoning, impotence, and arthritis [56]. It's utilized as an adaptogen, a nerve tonic and to improve health during cancer treatment [57]. Anticancer, antioxidant, leishmanicidal, anti-inflammatory, radiation-protective and mosquitocidal properties of *Os* chemical components are primarily explored for their therapeutic potential. Additionally, this

Table 25.3 Major chemical constituents present in basil oil and extract

Sr. No	Chemical Constituents	Sample Used	References
1	Cirsilineol	Fresh leave and stalk extract	[49]
2	Rosameric acid		
3	Circimaritin		
4	Isothymusin		
5	Apigenin		
6	Eugenol		
1	Carvacrol	Volatile oil	[55]
2	Sesquiterpene hydrocarbon caryophyllene		
3	Thymol		
4	Ocimene		
1	Orientin	Leaf extract	[50]
2	Vicenin		
1	Ascorbic acid	Leaf	[51]
2	Beta carotene		
1	Triglycerides	Seed	[48]
2	Linolenic acid		
3	Polysaccharides		
4	Sitosterol		
1	Linalool	Leaves aqueous extract	[54]
2	Endo-borneol		
3	Carbonyl		
4	1,8-cineole		
5	Limonene		
6	Camphene		
1	Tannins	Plant	[54]
2	Saponins		
3	Steroid		
4	Triterpene		

plant is mostly used for its medical characteristics since it has a broad variety of therapeutic agents [58, 59] as described under;

- "The elixir of life" name given to tulsi because it promotes longevity.
- In Ayurveda & Siddha systems of medicine, various parts of plant are used for prevention and cure of many diseases
- The leaves are used to enhance the memory
- Chewing of leafs helps to treat ulcers and infections of mouth [60].

25.7.1 Anticancer Activity

Anticancer and antitumor qualities may be found in a variety of plants in Ayurveda including basil. In mice with Sarcoma-180 solid tumors, the ethanolic extract of *Ocimum sanctum* has been found to decrease tumor cell size and resulted in better survival. Similar effects were also observed when leaf extract (200 mg/kg) was supplied orally to mice with cancer and resulted in healthier weight gain and lifespan of mice. Furthermore, the plant extract has also been found to protect cell's DNA from the harmful radiations [50, 61].

A study was conducted on anti-proliferative effects of Vicenin-2 (a flavonoid component of *Ocimum sanctum*), on human colorectal cancer cells. It has been observed that Vicenin-2 causes a significant cell cycle arrest in the G2 M phase of the cell cycle as a consequence of this compound. Caspase-3 and Bcl-2 are also raised in the presence of Vicenin-2, however cytochrome C, Bax, and caspase-3 are found to be reduced. Holy basil's active ingredient vicenin-2 has been studied alone and in combination with docetaxel to determine its effectiveness against prostate cancer. Vicenin-2 effectively promots anti-proliferative, anti-angiogenic, and pro-apoptotic effects in the cell. Vicenin-2 and docetaxel have been found to work together synergistically in order to slow down the development of prostate cancers [62]. Following these findings, researchers have concluded that vicenin-2 is an effective anti-prostate cancer treatment and found that co-administration of vicenin-2 with docetaxel is more effective than either of the solo agents in androgen-free prostate cancer.

Furthermore, Vicenin-2, alone or in conjunction with radiation, has been demonstrated to significantly reduce the number of cancer cells and to increase the survival. In addition, DNA fragmentation is enhanced and caspase-3 activity is raised whereas it lowers the levels of MMP-2 and p21 proteins. Anti-metastatic activity of holy basil extracts has been observed in case of lung cancer and is attributed to its antioxidant properties. The extract is found to be cytotoxic to lung cells. It dramatically reduces cell adherence and invasion in cancer cells matrix metalloproteinase-9 activity and, moreover, extract results in the management of tumor nodules, lung mass, and shows anti-metastasis action [63, 64].

25.7.2 Anti-diabetic Activity

The anti-diabetic effects of *O. sanctum* have been well documented. Hydro-alcoholic extract of basil is reported to be effective against streptozotocin-induced diabetes in rats, and this effect is comparable with that of gliben-clamide at 250 and 500 mg/Kg body weight. Similrly, ethanol-extract of *O. sanctum* has been found to manage hyperglycemia in alloxan diabetic rats in both acute and long-term feeding experiments. In rats, the ethanol extract and three different phases (butanol, ethyl acetate, and aqueous) of *O. sanctum* have been shown to have significant insulin-secretory effects. Similar results are also reported in investigations using isolated rat islets for acute insulin release [65]. According to findings of another study, on treating with this extract there is 17.6% reduction in fasting blood glucose concentration, while 7.3% reduction has been observed in postprandial blood sugar compared to the blood sugar levels of placebo [66, 67].

25.7.3 Anti-lipidemic Activity

These days, hyperlipidemia, atherosclerosis, and associated disorders have become a major public health issue. An aqueous extract of *O. basilicum* with the triton WR 1339 is found to significantly reduce total cholesterol, triglyceride, and LDL levels in rats models [68]. Likewise, in a study conducted on mice, it has been reported that a diet containing 1–2% fresh Tulsi leaves for 28 days reduced total lipid levels in the blood. Another research found that treatment with fixed oil of basil herb (during the last three weeks of high fat diet induced dyslipidemia) reduced the increased level of blood lipids and showed cardio-protective effects against hyperlipidemia. Further, it is demonstrated that hypolipidemic properties of holy basil fixed oils can be attributed to combined effects linolenic and linoleic acids [62, 69].

25.7.4 Antibacterial Activity

Ocimum sanctum leaf extracts and oils have been analyzed for antibacterial activity against *Escherichia coli*, *S. typhimurium*, *P. aeruginosa*, and *S. aureus* in aqueous, alcoholic and chloroform extract forms. Both gram+ve and gram-ve pathogenic bacteria are shown to be equally susceptible to *O. sanctum* extracts. When compared to the tulsi essential oil made from dry leaves, fresh leaves essential oils show stronger antibacterial activity, while the opposite is found true for anti-fungal properties [70]. Besides having anti-fungal, anti-septic and anti-viral properties, essential oils also inhibit the growth of a wide range of bacteria, including *E. coli*, *B. anthracis* and *M. tuberculosis* etc. Moreover, basil extract also significantly reduces the etiology, signs and symptoms and biochemical indications of many types of viral infections in human subjects [71].

25.7.5 Diseases of the Eyes

The basil extracts have anti-oxidant properties that prevent eyes from free radicals and oxidative damage. Along with prevention from eye diseases, it also helps to recover from macular degeneration and treatment of cataracts. Moreover, its leaf juice and triphala are both used in ayurvedic eye drop formulations for the treatment of glaucoma, chronic conjunctivitis, and other eye disorders. Three drops of tulsi oil mixed with honey used on a daily basis have been demonstrated to enhance one's vision [72].

25.7.6 Anti-fertility Activity

Tulsi leaf extracts in benzene and petroleum ether have been proven to cause anti-fertility by 80% and 60% in female rats, respectively [73]. Furthermore, Tulsi leaves have been claimed to show anti-fertility effects by both native women and ayurvedic practitioners in Kerala (India) [74]. Tulsi leaves contain ursolic acid, which have an anti-fertility impact. The mechanism of action shows that in males, its anti-estrogenic action is linked to the cessation of spermatogenesis, whereas in females, it has been linked to the inhibition of ovulation. An anti-fertility drug that does not cause adverse effects may be found in this component. Tulsi leaves slow down the activity of the sertoli cells in men, reducing spermatogenesis [75]. Moreover, ursolic acid has also been shown to possess contraceptive effects in rat model when it was tested to check its impact on fertile animals. This result has been attributed to its anti-estrogen effects, where estrogen is necessary for spermatogenesis while in females required for production and maturation of ovum [71].

25.7.7 Antioxidant Activity

Tulsi contains polyphenol called as rosmarinic acids as major non-nutritive component that is well known for its antioxidant properties along with having anti-inflammatory and antimicrobial activities. This acid is responsible for safeguarding the human body's cells from damage caused by free radicals and thus prevents oxidative stress in the body. Due to its effectiveness as antioxidants it works very well in degenerative diseases of lungs including asthma. More oxidation in the body causes cell damage as well therefore; it becomes further important for body through its protection from excessive oxidation in the body [76].

25.7.8 Adaptogenic Activity

Tulsi has been found to have adaptogenic properties that are meant for increasing the resistance against stress by reduction in sensitivity for stressors. The basil plant contains quite a lot of Rasayana properties which provide the body with a calm and clear mental state and thus helps to cure the common mood-changing activity of the body. Eugenol and caryophyllene are the two major components of the tulsi that are responsible to show these properties [47].

25.7.9 Mosquitocidal Activity

The plant has also been found to involve as having anti-insecticidal properties especially with respect to mosquito. Tulsi's triglyceride and eugenol (derived from tulsi's hexane extract) have been tested on fourth instar *Aedes Aegyptus* larvae to see whether they have any mosquitocidal action. As soon as Tulsi seeds were submerged in water, these began to release polysaccharides (a mucilaginous material) within an hour. This suffocated the larvae that had come in touch with the seeds, resulting in their death by drowning. So, it is concluded that the Tulsi plant is effective against mosquitos bite and growth [1]. Several studies have shown that Tulsi is both prophylactic/preventative and curative for insect stings or bites. Intake of leaf juice and administration of the juice again after a few hours is the best remedy. Moreover, the paste prepared from clean roots is also used to treat insect bites [71].

25.7.10 Immunomodulatory Agent

It has been reported that the anti-inflammatory activities of *Ocimum sanctum* oils are helpful for rats with paw edema caused by carrageenan or other mediators (Labiatae). *Ocimum sanctum* may be used to decrease inflammation since it has been shown to inhibit both cyclooxygenase and lipoxygenase. Tulsi oil dramatically decreases the arthritis and edema symptoms in rats subjected to Freund's adjuvant, formaldehyde, and turpentine oil. Both cell and humoral immunity are strengthened, resulting in a stronger immunological response. Contrary to aspirin, it has no negative effects on the body. Therefore it is recommended as supplement in the management of osteoarthritis-related pain and inflammation [54]. As a result of its outstanding immunity enhancing properties, tulsi helps to strengthen the immune system in order to build immunity against unfamiliar elements such as bacteria, viruses, microbes, allergens etc. Thus, it helps to maintain the balance within the body as well [77].

25.7.11 Fever and Common Cold

Various types of fevers are being treated using basil leaves and helps to reduce the symtoms by acting as anti-pyretic agent. During the rainy season, one of the most prevalent methods is to boil the leaves with tea and then serve it to the patients. This is widely used to treat dengue and malaria fevers too [71]. Furthermore, chewing tulsi leaves is also used as remedy to get relief from viral common colds and flu [78].

25.7.12 Coughs and Sore Throat

This plant is a main ingredient in many ayurvedic cough syrups and other kinds of dosage forms because it helps to expectorate mucus in respiratory disorders [71]. Moreover, the extract from holy basil is used as healing agent in gargles. Likewise, boiled leaves of the plant are given to people with sore throats as a treatment for sore throats [13]. In addition to its uses as a gargle, this water can also be used for other issues of throat and cough [79].

25.7.13 Respiratory Diseases

Tulsi is a marvellous herb to cure the asthma and other diseases of respiratory tract. Kehwah (water infusion) of leafs with ginger and honey has been reported as an effective home remedy used for all kind of respiratory disorders and cold. This extract with the mixture of lavang and lavana (cloves and salt) shows instant aid in influenza. According to the results of another study by Puri et al. 2002, this herb has been shown to be useful for treating respiratory system disorders, as well as remedy for cough and colds. Decoction of the leaves, with honey and ginger, can be used to treat bronchitis, asthma, influenza, cough and cold. In addition to providing immediate relief in the case of influenza, a decoction of leaves, cloves, and common salt is also effective. These should be boiled until only half of the water remains, and then is taken as remedy [80].

25.7.14 Snake and Insect Bites

Tulsi oil has been reported to have natural antiseptics and anti-inflammatory properties. Therefore it may be considered as an effective combating strategy to treat snake bites, including those caused by poisonous snakes. For the purpose, all parts of the plant are consumed or mixed with other plants to form a paste for topical application. Tulsi leaves are often placed in bowls of water outside homes and in bathwater in order to repel insects that are attracted to their odor and thus prevent the entry inside the house [71].

25.7.15 Stress and Headaches

The leaves of basil are considered 'adaptogens' and provide significant stress relief. Researchers have found that they offer significant protection against stress. In addition to purifying the blood and protecting several common elements, basil makes a good headache medicine. Its leaves can be given as a decoction and medicinal essence has been found as effective remedy to headache. Moreover, for relief from heat, headaches, and for general cooling effect, pounding leaves with sandalwood paste can be applied to the forehead.

25.7.16 Other Common Health Issues

Several studies have shown that tulsi leaves have a great effect on kidney function, and if the juice of the leaves is given along with honey for six months, the renal stone will be removed by the kidneys. Moreover, it helps to lower blood cholesterol levels, relieves the "weakness" of the heart and cures all types of heart diseases. There are several common diseases that can be treated by the juice of tulsi leaves, such as common cold, upset body temperature, loose stools, and vomiting. If blisters of chicken pox are late to appear, tulsi leaves mixed with kesar (Safforn yellow) are worth the effort [76]. Tulsi juice is able to effectively treat fungal infections on the skin, as well as other types of skin disorders like leukoderma. Additionally, it is also reported to be effective in treating fungal infections [76]. Dry leaves compressed into a powder and soaked in a hot water can be used to clean teeth and treat the dental diseases. Some of the power of this plant is derived from mixing it with mustard oil and applying it as a dental cream for the treatment of pyorrhoea and other dental problems [71].

25.8 Nutritional Facts

Ocimum sanctum is a highly nutritive in composition, vitamins such as tocopherol, beta-carotene, A, K, B1, B2, B3, B5, B6, B9, Vit-C and choline, as well as minerals like iron, calcium, magnesium, phosphorus, manganese, sodium, potassium, and zinc are abundantly present in this plant. Additionally, it has a positive impact on digestion and absorption of the nutrients. It provides 30 Kcal, 4.2 g protein; 0.5 g fat; 2.3 g carbohydrate; 25 mg calcium; 287 mg phosphorus; 15.1 mg iron; 25 mg vitamin C per 100 grams [71].

Many other metabolites are also present in basil such as bioactive compounds phenols, anthocyanins, tannins, flavonoids, essential oil, steroids as described earlier in this chapter [81]. Minerals play an essential role in the food and nutraceutical industries since they are often consumed in the diet. It has been used for centuries to enhance the taste of food and as a home treatment for a variety of ailments. *Ocimum sanctum*'s nutraceutical benefits have recently attracted more attention because of the abundance of vitamins, minerals, fats, proteins, polysaccharides, dietary fiber, colors, and mucilage found in this plant [82, 83].

The therapeutic properties of the leaves are the most common. Confectionery, baking, sauces, ketchups, tomato paste, pickles, gourmet vinegars, spiced meat and sausage, and drinks all use basil oil as a flavoring agent in the form of a flavoring ingredient in the oil. A tincture of the plant may be used as a flavor enhancer in liqueurs. This oil has methyl chaviocol, methyl cinnamate, and linalool as its main components. Moreover, transanethole is also present as one of the most important components [84].

25.9 Myths, Tales and Folklores

Basil plant has its roots as divine plant in many cultures and ethnic groups as under;

- In India, Tulsi is revered as a holy plant, in Christianity, tulsi also consider as divine gift, and is grown at site of Christ crucifixion.
- In Hindu homes, Tulsi is a must-have plant since it is believed to ward off bad spirits.
- Tulsi is revered in Hinduism as a manifestation of Goddess.
- During religious ceremonies, the leaves are infused into holy water.
- Charnaamrit, an offering to Lord Vishnu, has leaves as a component [85].

Hindu legend says goddess Tulsi raised it from her own fire ashes. It is described that a sage with tangled hair and a sparkling face blocked Indra and Bhaspati's approach to Mount Kailash to visit Shiva. Shiva transformed to test Bhaspati and Indra. Indra was outraged when the yogi wouldn't move despite his efforts to identify him. Indra's thunderbolt scared him away. Shiva's eyes became red as Indra's actions stunned him. Bhaspati asked Shiva to pardon Indra when his third eye opened to kill him. Shiva shot fire from his eye into space, where it met water and became Jalandhara.

Sukra appointed Jalandhara demon king after he grew up. Vrinda is Klanemi's daughter. Sage Bhgu's stories turned Jalandhara against Visnu and other gods. Jalandhara and Visnu fought to draw. Inspired by Jalandhara's courage, he asked him to grant any desire. Visnu granted Jalandhara's request to live in Sgara, destroying the gods and making Jalandhara King of the three worlds (heaven, earth and hell). Lord Iva's son Devas refused to submit to Jalandhara's reign and grew sad. He visited Jalandhara after consulting the Devas. It was to show Jalandhara how gorgeous Kaila was [86]. It is also considered as sign of both hate and love, in Italy it is symbol of love while in Sicilian as sign of love and death. Furthermore, in Moldavian folklores it is assumed that if a young man accepts basil from a young woman, he will fall in love with her.

25.10 Conclusions and Future Perspectives

The above mentioned data provides the information that components of *Os* have been isolated and investigated using bioassays and resulted in discovery of new compounds in various extracts. It has been known to contain several chemical components, including as phenolics and flavonoids, phenylpropanoids, coumarins, terpenoids and fatty acid derivatives, as well as a variety of essential oils and fixed oils. Due to its high concentration of eugenol, *Os* essential oil has been extensively studied in analytical, chemical, and biological contexts.

It is widely used for treating a wide range of disorders, including antiviral agents, antiseptics, inflammation reduction, antioxidants, ulcers, injuries, diabetes mellitus, bacillary dysentery loose motions, purgatives, vermifuges, astringents, leprosy, prevention of goitre, treatment of tumors, remedy for pest bites and venom, indigestion and relief of flatulence, as well as treating diseases of intestine. The studies show that the above mentioned positive effects of basil plant and its product are without any side effects, therefore, it should be further verified for its effects at extensive levels and to create the scientific basis of whole properties and knowledge.

References

1. Verma, S. (2016). Chemical constituents and pharmacological action of Ocimum sanctum (Indian holy basil-Tulsi). *The Journal of Phytopharmacology, 5*(5), 205–207.
2. Makri, O., & Kintzios, S. (2008). Ocimum sp.(basil): Botany, cultivation, pharmaceutical properties, and biotechnology. *Journal of Herbs, Spices & Medicinal Plants, 13*(3), 123–150.
3. Simon, J. E., Quinn, J., & Murray, R. G. (1990). *Basil: A source of essential oils* (pp. 484–489). Timber Press.
4. Paton, A., Harley, R., & Harley, M. (1999). *Ocimum: An overview of classification and relationships* (pp. 11–46). Basil.
5. Vani, S. R., Cheng, S., & Chuah, C. (2009). Comparative study of volatile compounds from genus Ocimum. *American Journal of Applied Sciences, 6*(3), 523.
6. Ahmad, N., et al. (2010). Rapid synthesis of silver nanoparticles using dried medicinal plant of basil. *Colloids and Surfaces B: Biointerfaces, 81*(1), 81–86.
7. Verma, A., et al. (2018). Programmatically interpretable reinforcement learning. In *International conference on machine learning*. PMLR.
8. Jackson, B. P., & Snowdon, D. W. (1990). *Atlas of microscopy of medicinal plants, culinary herbs and spices*. Belhaven Press.
9. Koutsos, T., Chatzopoulou, P., & Katsiotis, S. (2009). Effects of individual selection on agronomical and morphological traits and essential oil of a "Greek basil" population. *Euphytica, 170*(3), 365–370.
10. Simon, J. E., et al. (1999). Basil: A source of aroma compounds and a popular culinary and ornamental herb. *Perspectives on New Crops and New Uses, 16*, 499–505.
11. Pennisi, G., et al. (2019). *Unraveling the role of red: Blue LED lights on resource use efficiency and nutritional properties of indoor grown sweet basil* (p. 10). Frontiers in Plant Science.
12. Antonescu, A.-I., et al. (2021). Perspectives on the combined effects of Ocimum basilicum and Trifolium pratense extracts in terms of phytochemical profile and pharmacological effects. *Plants, 10*(7), 1390.
13. Prinsi, B., et al. (2019). Insight into composition of bioactive phenolic compounds in leaves and flowers of green and purple basil. *Plants, 9*(1), 22.
14. Kustiati, U., Wihadmadyatami, H., & Kusindarta, D. L. (2022). Dataset of phytochemical and secondary metabolite profiling of holy basil leaf (Ocimum sanctum Linn) ethanolic extract using spectrophotometry, thin layer chromatography, Fourier transform infrared spectroscopy, and nuclear magnetic resonance. *Data in Brief, 40*, 107774.

15. Kumar, A., et al. (2021). The pre-eminence of agro-parameters and chemical constituents in the influence of harvest interval by traits× environment interaction over the years in lemon-scented basil (Ocimum africanum Lour.). *Industrial Crops and Products, 172*, 113989.
16. Złotek, U., et al. (2020). Potential acetylcholinesterase, lipase, α-glucosidase, and α-amylase inhibitory activity, as well as antimicrobial activities, of essential oil from lettuce leaf basil (Ocimum basilicum L.) elicited with jasmonic acid. *Applied Sciences, 10*(12), 4315.
17. Łyczko, J., et al. (2020). Chemical determinants of dried Thai basil (O. basilicum var. thyrsiflora) aroma quality. *Industrial Crops and Products, 155*, 112769.
18. D'Alessandro, A., et al. (2021). Characterization of basil volatile fraction and study of its agronomic variation by asca. *Molecules, 26*(13), 3842.
19. Smitha, G., et al. (2019). Nutrient management through organics, bio-fertilizers and crop residues improves growth, yield and quality of sacred basil (Ocimum sanctum Linn). *Industrial Crops and Products, 128*, 599–606.
20. Putievsky, E., & Galambosi, B. (1999). Production systems of sweet basil. *Basil: the genus Ocimum, 10*, 39–65.
21. Yamamoto, A., & Takano, T. (1996). *Effects of anion variations in a nutrient solution on the basil [Ocimum basilicum] growth, essential oil content and composition.* Scientific Reports of the Faculty of Agriculture-Meijo University (Japan).
22. Mandal, J., Pattnaik, S., & Chand, P. K. (2000). Alginate encapsulation of axillary buds of Ocimum americanum L. (hoary basil), O. Basilicum L. (sweet basil), O. Gratissimum L.(shrubby basil), and O. Sanctum. L. (sacred basil). *In Vitro Cellular & Developmental Biology-Plant, 36*(4), 287–292.
23. Skrubis, B., & Markakis, P. (1976). The effect of photoperiodism on the growth and the essential oil of Ocimum basilicum (sweet basil). *Economic Botany, 30*(4), 389–393.
24. Werker, E., Putievsky, E., & Ravid, U. (1985). The essential oils and glandular hairs in different chemotypes of Origanum vulgare L. *Annals of Botany, 55*(6), 793–801.
25. Anwar, M., et al. (2005). Effect of organic manures and inorganic fertilizer on growth, herb and oil yield, nutrient accumulation, and oil quality of French basil. *Communications in Soil Science and Plant Analysis, 36*(13–14), 1737–1746.
26. Thakur, N., & Verma, K. (2012). Financial flows from sacred basil (Ocimum sanctum) based agroforestry land-use systems in mid hills of Western Himalayas. *Indian Forester, 138*(7), 638–645.
27. Rihan, H. Z., et al. (2020). A novel new light recipe significantly increases the growth and yield of sweet basil (Ocimum basilicum) grown in a plant factory system. *Agronomy, 10*(7), 934.
28. Kothari, S., Bhattacharya, A., & Ramesh, S. (2004). Essential oil yield and quality of methyl eugenol rich Ocimum tenuiflorum Lf (syn. O. sanctum L.) grown in South India as influenced by method of harvest. *Journal of Chromatography A, 1054*(1–2), 67–72.
29. Charles, D. J., & Simon, J. E. (1990). Comparison of extraction methods for the rapid determination of essential oil content and composition of basil. *Journal of the American Society for Horticultural Science, 115*(3), 458–462.
30. Das, U., Pal, S., & Ponnusamy, N. (2020). Biology and seasonality of lace bug Cochlochila bullita (Stal)(heteroptera: Tingidae) on tulsi Ocimum sanctum L. L. *International Journal of Bio-resource and Stress Management, 11*(2), 114–118.
31. Seni, A. (2022). Nipaecoccus viridis (Newstead)(Hemiptera: Pseudococcidae): An emerging threat of Tulsi; Ocimum sanctum L. *National Academy Science Letters, 45*(2), 135–137.
32. Ahmed, M. Z., & Deeter, L. (2022). Rapid species-level hemolymph color test for all life stages of Nipaecoccus viridis (Newstead)(Hemiptera: Pseudococcidae), an invasive and regulatory pest in the United States. *Journal of Applied Entomology, 146*(4), 454–460.
33. Srivastava, R., Kumar, S., & Sharma, R. (2018). Ocimum as a promising commercial crop. In *The Ocimum Genome* (pp. 1–7). Springer.
34. Dhingra, G., Kumar, V., & Joshi, H. D. (2019). A novel computer vision based neutrosophic approach for leaf disease identification and classification. *Measurement, 135*, 782–794.

35. Thorat, A., Kumar, S., & Valakunde, N. D. (2017). An IoT based smart solution for leaf disease detection. In *2017 International conference on big data, IoT and data science (BID)*. IEEE.
36. Dhakate, M., & Ingole, A. (2015). Diagnosis of pomegranate plant diseases using neural network. In *2015 fifth national conference on computer vision, pattern recognition, image processing and graphics (NCVPRIPG)*. IEEE.
37. Matta, A. (1978). *Fusarium tabacinum (Beyma) W. Gams patogeno in natura su basilico e pomodoro* (pp. 119–125). Rivista di Patologia Vegetale.
38. Singh, P. S. (2022). *Traditional Medicinal Plants*. (Volume 5). Akinik Publication, New Delhi, India
39. Dutky, E., & Wolkow, P. (1994). First report of fusarium wilt of basil in Maryland. *Plant Disease, 78*, 1217
40. Katan, T., Gamliel, A., & Katan, J. (1996). Vegetative compatibility of fusarium oxysporum from sweet basil in Israel. *Plant Pathology, 45*(4), 656–661.
41. Gullino, M., Garibaldi, A., & Minuto, G. (1995). First report on 'black spot' of basil incited by Colletotrichum gloeosporioides in Italy. *Plant Dis, 79*, 539.
42. Garibaldi, A., et al. (1995). Colletotrichum gloeosporioides Penz. – A new pathogen of basil in Italy. *Informatore Fitopatologico, 45*(2), 34–35.
43. Holcomb, G., & Cox, P. (1998). First report of basil leaf spot caused by Pseudomonas cichorii in Louisiana and cultivar screening results. *Plant Disease, 82*(11), 1283–1283.
44. Matta, A., & Garibaldi, A. (1981). *Malattie delle piante ortensi*. Edagricole.
45. Jakhi, P. S. (2021). *Traditional Medicinal Plant*. Ed. Singh, P. S. Vol. (3). Akinik Publication Delhi, India.
46. Sharabani, G., et al. (1996). Development of gray mold in sweet basil. *Phytoparasitica, 24*, 140.
47. Buddhadev, S. G. Buddhadev, S. S., & Mehta, N. D. (2014). A review article on Ocimum Sanctum Linn. *International Journal of Research in Ayurveda and Pharmacy, 2*(2), 1–6.
48. Naji-Tabasi, S., & Razavi, S. M. A. (2017). New studies on basil (Ocimum bacilicum L.) seed gum: Part III–steady and dynamic shear rheology. *Food Hydrocolloids, 67*, 243–250.
49. Devi, P. U. (2001). Radioprotective, anticarcinogenic and antioxidant properties of the Indian holy basil, Ocimum sanctum (Tulasi). *Indian Journal of Experimental Biology, 39*, 185.
50. Gupta, S., Prakash, J., & Srivastava, S. (2002). *Validation of traditional claim of Tulsi, Ocimum sanctum Linn. as a medicinal plant*. NISCAIR-CSIR.
51. Wongsen, W., et al. (2013). Relationship between leaf position and antioxidant properties in three basil species. *International Food Research Journal, 20*(3).
52. Aparna, D., & Musarraf, M. (2013). *In vitrocytotoxic effect of methanolic crude extracts of Ocimum sanctum*. IJPSR.
53. Arivuchelvan, A., et al. (2012). Immunomodulatory effect of Ocimum sanctum in broilers treated with high doses of gentamicin. *Indian Journal of Drugs and Diseases, 1*(5), 109–112.
54. Kumar, R., et al. (2022). A systemic review of Ocimum sanctum (Tulsi): Morphological characteristics, phytoconstituents and therapeutic applications. *International Journal for Research in Applied Sciences and Biotechnology, 9*(2), 221–226.
55. Anuradha, B., & Murugesan, A. (2001). *Immunotoxic and haematotoxic impact of copper acetate on fish Oreochromis mossambicus and modulatory effect of Ocimum sanctum and Valairasa chendhuram*. M. Sc., thesis submitted to Manonmaniam Sundaranar University.
56. Bano, N., et al. (2017). Pharmacological evaluation of Ocimum sanctum. *J Bioequiv Availab, 9*(3), 387–392.
57. Chulet, R., & Pradhan, P. (2009). A review on rasayana. *Pharmacognosy Reviews, 3*(6), 229.
58. Purushothaman, B., et al. (2018). A comprehensive review on Ocimum basilicum. *Journal of Natural Remedies, 18*(3), 71–85.
59. Machado, M. M., de Oliveira, L. F. S., & Zuravski, L. (2019). *Ocimum basilicum L.: Antiinflammatory actions and potential usage for arthritic conditions, in bioactive food as dietary interventions for arthritis and related inflammatory diseases* (pp. 481–487). Elsevier.

60. Sharma, K., et al. (2019). Efficacy of chlorhexidine, hydrogen peroxide and tulsi extract mouthwash in reducing halitosis using spectrophotometric analysis: A randomized controlled trial. *Journal of Clinical and Experimental Dentistry, 11*(5), e457.
61. Monga, J., et al. (2011). Antimelanoma and radioprotective activity of alcoholic aqueous extract of different species of Ocimum in C57BL mice. *Pharmaceutical Biology, 49*(4), 428–436.
62. Suanarunsawat, T., et al. (2010). Anti-hyperlipidemic and cardioprotective effects of Ocimum sanctum L. fixed oil in rats fed a high fat diet. *Journal of Basic and Clinical Physiology and Pharmacology, 21*(4), 387–400.
63. Almatroodi, S. A., et al. (2020). Ocimum sanctum: role in diseases management through modulating various biological activity. *Pharmacognosy Journal, 12*(5).
64. Kim, S.-C., et al. (2010). Ethanol extract of Ocimum sanctum exerts anti-metastatic activity through inactivation of matrix metalloproteinase-9 and enhancement of anti-oxidant enzymes. *Food and Chemical Toxicology, 48*(6), 1478–1482.
65. Vats, V., Yadav, S., & Grover, J. (2003). Effect of T. foenumgraecum on glycogen content of tissues and the key enzymes of carbohydrate metabolism. *Journal of Ethnopharmacology, 85*(2–3), 237–242.
66. Prabhakar, P. K., et al. (2013). Synergistic interaction of ferulic acid with commercial hypoglycemic drugs in streptozotocin induced diabetic rats. *Phytomedicine, 20*(6), 488–494.
67. Mahajan, N., et al. (2013). A phytopharmacological overview on Ocimum species with special emphasis on Ocimum sanctum. *Biomedicine & Preventive Nutrition, 3*(2), 185–192.
68. Tan, P. V., et al. (2005). Healing effect on chronic gastric ulcers and short-term toxicity profile of the leaf methanol extract of Ocimum suave Wild (Lamiaceae) in rats. *African Journal of Traditional, Complementary and Alternative Medicines, 2*(3), 312–325.
69. Surkar, A., et al. (1994). Changes in the blood lipid profile after administration of Ocimum sanctum (Tulsi) leaves in the normal albino rabbits. *Indian Journal of Physiology and Pharmacology, 38*, 311–311.
70. Mirdha, B., Naik, S., & Mahapatra, S. (2007). *Antimicrobial activities of essential oils obtained from fresh and dried leaves of Ocimum sanctum (L.) against enteric bacteria and yeast*. International Symposium on Medicinal and Nutraceutical Plants 756.
71. Bhadra, P. (n.d.). A review paper on the Tulsi plant (Ocimum sanctum). *Indian Journal of Natural Sciences, 10*(60), 20854–20860.
72. Patil, R., et al. (2011). Isolation and characterization of anti-diabetic component (bioactivity—Guided fractionation) from Ocimum sanctum L.(Lamiaceae) aerial part. *Asian Pacific Journal of Tropical Medicine, 4*(4), 278–282.
73. Nagarajun, S., Jain, H., & Aulakh, G. (1989). In C. K. Atal & B. M. Kapoor (Eds.), *Indigenous plants used in fertility control. Cultivation and utilization of medicinal plants* (p. 558). PID CSIR.
74. Batta, S., & Santhakumari, G. (1971). The antifertility effect of Ocimum sanctum and Hibiscus rosa sinensis. *The Indian Journal of Medical Research, 59*(5), 777–781.
75. Prakash, P., & Gupta, N. (2005). Therapeutic uses of Ocimum sanctum Linn (Tulsi) with a note on eugenol and its pharmacological actions: A short review. *Indian Journal of Physiology and Pharmacology, 49*(2), 125.
76. Kumar, V., et al. (2018). Comparative study on antimicrobial activity of tulsi (Ocimum sanctum) and neem (Azadirachta indica) methanol extract. *Asian Journal of Pharmaceutical and Clinical Research, 6*, 514–517.
77. Anderson, E. N. (2014). *Everyone eats: Understanding food and culture*. NYU Press.
78. Staples, G. W., & Kristiansen, M. S. (1999). *Ethnic culinary herbs: A guide to identification and cultivation in Hawaii*. University of Hawaii Press.
79. Kuhn, M. A., & Winston, D. (2012). *Winston & Kuhn's herbal therapy and supplements: A scientific and traditional approach*. Lippincott Williams & Wilkins.
80. Puri, H. S. (2002). *Rasayana: Ayurvedic herbs for longevity and rejuvenation*. CRC Press.
81. Filip, S. (2017). Basil (Ocimum basilicum L.) a source of valuable phytonutrients. *International Journal of Clinical Nutrition & Dietetics, 3*, 118.

82. Mandal, A. K., et al. (2022). Phytochemistry, pharmacology, and applications of (Tulsi). In *Edible plants in health and diseases* (pp. 135–174). Springer.
83. Pachkore, G., & Dhale, D. (2012). Phytochemicals, vitamins and minerals content of three Ocimum species. *IJSID, 2*(1), 201–207.
84. Nakar, R. N., Dhaduk, H. L., & Chovatia, V. P. (2017). *Medicinal plants: Cultivation and uses* (pp. 460–464). DAYA.
85. Verma, N. (2018). Medicinal as well as sacred plants of Bilaspur town, district Rampur (Uttar Pradesh). *Bulletin of Pure & Applied Sciences-Botany, 2.*
86. Mowahhedian Attar, A., Alimardi, M. M., & Rohani, S. M. (2021). Investigating the possibility of Hinduism being. *Religious Research, 9*(17), 27–53.

Chapter 26
Khus

Sadia Zafar, Inam Mehdi Khan, Muhammad Muddasar, Rehman Iqbal, and Umar Farooq Gohar

26.1 Introduction

Chrysopogon zizanioidesis a member of the Poaceae family, which has 11,337 species and 707 genera [1]. In tropical and subtropical Asia, including India (USDA-ARS 2020), Cambodia, Pakistan, Sri Lanka, Myanmar, Thailand, and Vietnam, the genus Chrysopogon is locally widespread [2]. There are 21 species of Chrysopogon in India, and they go by the names *C. aciculatus, C. asper, C. aucheri, C. castaneus, C. copei, C. fulvus, C. gryllus, C. hackelii, C. lancearius, C. lawsonii, C nodulibarbis, C. orientalis, C. polyphyllus, C. pseudoze*. Later, this genus expanded to 23 species with the addition of *C. festucoides* [3, 4] as a new record of enlarged distribution [5, 6] and the transfer of *Vetiveria zizanioides* syn. *Chrysopogon zizanioides* [7]. Later, its reach widened to include areas including Asia, Africa, North and South America, the Caribbean, Australia, and the Pacific [2, 8–11]. In Asia, it is introduced to Bangladesh [2], China [12, 13] Fujian [14], Guangdong [13], Hainan, Jiangsu, Sichuan, Yunnan, Zhejiang Japan, Bonin Islands [10], Nepal [2], Philippines [10], Sri Lanka [2], Singapore, and Taiwan [15]. In Europe and Oceania, it is introduced to Spain [15], Australia [11], Cook Island, Fiji, French Polynesia, New Caledonia, Palau, Samoa, Tonga, Wallis, and Futuna [10].

Later, it was also expanded into various North and South American countries and territories, including Antigua and Barbuda, Barbados, Sint Eustatius, Cuba, Dominica, Grenada, Guadeloupe, Haiti, Jamaica, Martinique, Netherlands Antilles, Saint Martin, Puerto Rico, Saint Kitts and Nevis, Saint Lucia, Saint Martin, Saint

S. Zafar (✉) · I. M. Khan · M. Muddasar · R. Iqbal
Department of Botany, Division of Science and Technology, University of Education Lahore, Lahore, Pakistan
e-mail: sadia.zafar@ue.edu.pk

U. F. Gohar
Institute of Industrial Biotechnology, Government College University, Lahore, Pakistan

M. Zia-Ul-Haq et al. (eds.), *Essentials of Medicinal and Aromatic Crops*,
https://doi.org/10.1007/978-3-031-35403-8_26

Vincent and the Grenadines, Sint Maarten, Trinidad and Tobago [16]. Kerala, Karnataka, Andhra Pradesh, Tamil Nadu, Kashmir, Rajasthan, Orissa, Bihar, Gujarat, Madhya Pradesh, Panjab, and Uttar Pradesh are among the Indian states where it has been grown [10].

The World Bank first encouraged its production in India in the middle of the 1980's as a means of conserving land and water. The World Bank then released a manual on vetiver grass and its function in land and soil conservation in 1989 to emphasize the potential uses of vetiver [17]. Since then, approximately 100 nations have begun cultivating plants for a variety of environmental problems, including the stabilization of steep slopes, wastewater disposal, phytoremediation of toxic soil, and others [8, 18]. In addition to this, the main factor is its essential oil, which is produced mostly in Haiti, India, Indonesia, West Java, and Réunion [8, 18].

Evergreen perennial grass *Chrysopogon zizanioides* grows in clusters that are typically between one and three meters tall [8, 19]. When growing conditions are good, its roots expand quickly and can reach a depth of 3–4 m in the first year of growth, enabling them to withstand drought and flooding as well [8, 18–20]. The root portion of the plant is the component that is most sought after for oil extraction for use in food and drink, spa treatments, household appliances, and medicinal purposes. Vetiveria oil sold on the global market in 2019 was 408.8 tons, and from 2020 to 2027, that market is projected to grow by 7.8% [21]. Because grass must be completely destroyed in order to collect the roots that are essential for oil production. The quality of the land is being continuously deteriorated by the removal of grasses for their roots, which poses a serious risk of soil erosion and flooding. A review study like this one is a first step in preserving the conventional wisdom about the uses of this aromatic and medicinal plant, which can then be used to find novel pharmacological molecules.

26.2 Botanical Description

Mostly found in India's plains, vetiver is a densely tufted, scented grass that can reach heights of 1200 m. This grass is mostly developed with the aid of the pruning technique, which further encourages strong growth of roots and leaves, which is of utmost importance [22]. The roots have a 2 mm diameter, range in colour from cream to light yellow, and occasionally even brown, and have a slightly bitter, pungent, clustered, and aromatic texture. They also grow densely in all types of soil. These roots have cells that produce oil deep within the roots that are difficult to access [23–25]. The inner bark's cortical layer and the lysigen lacunae are where the secreted oil is kept [26]. The leaves are six or more on each side, tapering, straight, pointed, and have rough or scaly borders. They measure between 30 and 90 cm in length and 4 to 10 mm in width. The inflorescence is composed of whorls with around 6–10 flowers, each with up to 20 rays, profuse slender racemes measuring 7.5 cm in length, and whorls that are branching slant to straight. The spikelets are

Fig. 26.1 Field view of Khus Plant (Chrysopogon zizanioidesis) (**a**), inflorence of Khus plant (**b**), Leaves arrangements (**c**)

sessile, 6 mm long, and grey to purplish in colour. The callus is small and paired with thorns [22] (Fig. 26.1).

Under a microscope, it was discovered that the roots of Indian *C. zizanioides* included parenchymal cells that were replete with starch grains and oil globules. The outside layer is made up of trichomes and fibrous tissues, and the inner layer is made up of bordered pored, pitted, and reticulated capillaries. The lacunar collenchyma cells in the hypodermis layer are covered in calcium oxalate crystals in large amounts [27].

The figure's pointer indicates a variety of things, including the following: Parenchymatous cells with one or more starch granules, 2- parenchymatous cells containing oil globules, 3- parenchyma cell slice, Four: lignified permeable parenchyma cells, five: trichomes, six: a fiber structure supply of 7-pored vessels, tracheids and vessels with pitted borders on them, reticulate vessels, and 10-section illustration of cork cells collenchyma with a lacuna, calcium oxalate crystals, sclereids, phloem, and cork cells are listed in that order.

26.3 Traditional Uses

Different portions of vetiver contain a variety of medicinal characteristics that have been employed since ancient times, in addition to being used as a species for soil preservation and wastewater control. Vetiver is classified as Mutrala (which improves urine excretion), Shukrashodhana (which cures genital problems such urinary calculi, dysuria, spermatorrhoea, etc.), and Dahaprashamana in the ancient ayurvedic text "CharakaSamhita" meaning reduces fatigue [27].

In contrast to other parts of the plant, the vetiver roots have received the most attention in Ayurveda, which has made use of its properties as a blood purifier to remove excess toxins from the body, acting as a cardioprotective; gastrointestinal system strengthener by reducing indigestion and excessive acid production; in treating issues with the digestive system (such as anorexia and diarrhoea); respiratory system (such as asthma, tuberculosis, and cough); and CNS and vertig [28]. It is also used as an energizer, sedative, nervine tonic, antidepressant, and stimulant of sexual drive. Additionally, it improves blood flow, which helps with anaemia [29]. It is also used to alleviate the problem of jet lag [30]. Additionally, the plant's combo of root and leaf relieves pain from sprains, lumbago, and rheumatoid arthritis while the distilled root oil functions as a refresher in an oil diffuser. It is also used as anti-inflammatory, detoxifier, cardiac disorders, skin related problems, fever as well as for pest repellent (Table 26.1).

Tribal students, who depend on neighbouring regions for their livelihood, make up the majority of the Indian population. The tribal people use every plant part in a variety of ways to treat a variety of ailments. For example, the Santhal tribes of Bihar and West Bengal apply fresh roots for burns, snakebite, and scorpion stings; the West Bengal Lodhas use root paste to treat headaches and rheumatoid arthritis; the Mandla and Bastar tribes of Madhya Pradesh use leaf juice as an anthelmintic; and the tribes of the Himalayas use [35, 36] Sastry [37] also discussed the effective use of this plant to help ladies who are childless owing to uterus-related problems.

In addition to its biological functions, it has a wide range of other practical applications, such summertime doormats made of roots, window screens for cooling effects, eco-friendly soil binders, perfumers for clothing, and the production of paper and strawboard.

Table 26.1 Traditional and ethno-medical uses of the Khus plant

Sr. No	Uses	References
1	Pest repellent	[31]
2	As cooling agent in fever	[32, 33]
3	Skin beautification	[18]
4	Anti-rheumatoid	[23, 24]
5	Cardiac debility	[18]
6	Prevent from adverse effect of air pollution	[23, 24]
7	Refrigerants and scents	[31]
8	Aphrodisiac in emotional stress	[24]
9	Detoxifier, Anti-inflammatory	[34]

26.4 Formulations and Preparations

As one of the ingredients, *C. zizanioides* is used in the preparation of many traditional ayurvedic medicines, including those listed below: Fighter Hand Sanitizer, Yograj Guggul, Ushira Powder, Shadanga Paneeya, Nilavembu Churna, Sri SriTattva Ushira Syrup, Ushirasava, Trichup Herbal Hair Pack Powder, Prosman, and Balar are some examples of herbal hair products. Detail information about the mentioned formulation that contains *C. zizanioides* in mentioned in Table 4.

26.5 Phytochemistry

Due to differences in the soil, environment, and location in which it grows, vetiver plants exhibit variation in their chemical composition. When grown in various types of soil, including regular soil, normal soil with added microbes, and semi-hydroponic soil, a distinctive variation in its chemical composition is seen. The best results were obtained by vetiver and normal soil with additional bacteria in the form of sesquiterpenes and their ketones, aldehydes, and alcohols [38]. These oxygenated substances give vetiver oil a scent, enhancing its use in the food and fragrance industries. In addition, the age of the roots and the length of their distillation time have an impact on the oil quality. While the oil extracted from young, fresh roots has a pungent smell, the oil from older, dried roots has a significant amount of sesquiterpenes, which have a more pleasant scent.

Contrarily, it is discovered that the oil from cleaned roots is rich in alkanes with a carbon chain ranging from 19 to 29 and devoid of fungus-related metabolites such as -amorphene, −vetispirene, etc. On the other hand, when studied using the random amplified polymorphic DNA approach and gas chromatographic-mass spectrometric (GC-MS) analysis, environment and location also generate diversity in the chemical composition. Thus, the smell of essential oils varies greatly depending on their components and location [39].

Like the majority of essential oils, vetiver oil has a complicated chemical structure made up of more than 100 sesquiterpene molecules and their derivatives [40]. But C15 sesquiterpenoids, which boil at or above 200 °C, make up the majority of the oil (40).

In addition to hydrocarbons, alcohol derivatives, carbonyl derivatives, and ester derivatives, these sesquiterpenoids also appear in other forms. Cadenene, clovene, apomorphine, aromadendrin, and junipene are among the hydrocarbons, and khusimol (3.4–13.7% from 45–80%), epiglobulol, spathulenol, and khusinol are among the alcohol derivatives. Khusimone, −vetivones, and ester derivatives, such as khusinol acetate, are included in the carbonyl derivatives (1.3–7.8%). In addition to these, the vetiver oil also contains trace amounts of benzoic acid, vetivene, furfurol, khusemene, khusimone, ß-humulene, valencene, selinene, etc. [40]. According to Bhatwadeka et al. [41] of these, −vetivone, −vetivone, and khusimone are mostly accountable for the scent of the vetiver oil and are consequently referred to as the oil's "fingerprint" [42, 43]. In comparison to the other two, -Vetivone has the best scent, while its main isomer, nordihydro -Vetivone, has a substantial, woody-peppery aroma. Each of these elements, both singly and collectively contributes tothevetivers distinctive scent [40].

26.6 Chemical Composition

By getting vetiver oil from reliable sources, Curtis, [29] in 2011 compared the vetiver oil of nine distinct nations, including Brazil, China, Haiti, India, Java, Madagascar, Mexico, Reunion, and Salvador. A total of 114 compounds were found in the quantitative and qualitative examination, with a mixture of hydrocarbons, alcohols as main fractions, ketones as minor fractions, and a few additional unrelated chemicals. Around 70–90% of oil is made up of these chemical combinations, with local variations in the proportions of hydrocarbons, aldehydes, alcohols, esters, ketones, and acids When compared, it was found that the vetiver oil from Java, India, and China had the highest percentage of oil components, with varying amounts of elements such hydrocarbons, aldehydes, alcohols, esters, ketones, and acids making up 89.4%, 89.3, and 89.1% of the oil components, respectively. With the biggest variety in their hydrocarbons, Salvador, Reunion, Madagascar, and Mexico also have significant oil percentages, at 88.2%, 87.3%, 86.2%, and 85.2%, respectively. The most prevalent hydrocarbons found in oil are khusimene, amorphene, β-vetivenene, and β-vetispirene in varied concentrations between 2 and 3%. The two main alcohols, khusimol and vetiselinenol, make up 4–9% of the entire oil composition. Additionally, it includes three significant bioactive ketones, with estimated percentages of 4.1%, 3.2%, and 0.8%, respectively: khusimone, α-vetivone, and β-vetivone 3. Analyzing the composition differences statistically showed that there was no difference between the oil from different places [44].

Along with the aforementioned, it was discovered that Haitian vetiver oil only had a small amount of the other oil components, 73%, which included 32.6% hydrocarbons, 2.1% aldehydes, 23.6% alcohols, 0.2% esters, 11.8% ketones, and 3.4% acids. These substances divided into groups based on how polar and non-polar they behaved. Flash chromatography was used to first convert the polar components into methyl esters for separation, and then several analytical techniques, including GC-MS and nuclear magnetic resonance, were used to characterize the polar components and validate their structural identity (NMR). Sesquiterpene hydrocarbons were the first compounds that the gas chromatography method detected. These were then followed by secondary and tertiary alcohols, and finally, aldehydes. This chromatogram's final fraction was particularly rich in khusimol, vetivones, isovelencenol, and zizanioic acid, which are thought to be responsible for a number of biological activities as well as the distinctive scent. The majority of oils also contain other derivatives such Khusinol acetate, Epiglobulol, Spathulenol, Khusol, and Khusimone as well as small amounts of nootkatone acid and Khusinol [29].

More than 100 different compounds have been found in Indian vetiver oil, including terpenoids, sesquiterpenoids, hydrocarbons, alcohols, ketones, and acids as major constituents. Phenols and nitrogenous compounds have occasionally been found as minor constituents and have been identified using gas chromatography-mass spectroscopy (GC-MS) analysis. The main constituents of north Indian Khus oil were also abundant in the south Indian type at the same time [42–45]. The identified substances were also divided into different subgroups based on their structural characteristics, including Eremophillanes, Vetiverols, Zizaanes and Prezizaanes, Eudesmanes, Amorphanes, Murrolanes, and Cadinanes, which are described below:

1. **Eremophillanes**

The initial components were α and β-Vetivones, which were separated from the vetiver by Plattner and Pfau in the past [46], and then by Naves and Perrottet [47]. Due to the similarities in their physical and chemical properties, differing only in the C3 -carbon structure, α-vetivone was previously thought of as an epimer of β-vetivone. It was later renamed "isonootkatone" in accordance with its constituents' nomenclature, but the word "α-vetivone" is still used in the literature. At first, it was suggested that vetivones were absent from freshly extracted oil. The non-detachable component of vetiver, known as vetivones, is responsible for a variety of biological actions, according to GC-MS analysis of several research. Another important derivative of eremophillane is β-vetivene, which was first proposed as γ-vetivenene but was later corrected due to the real location of double bond 1(10). Nearly 10 years later, Takahashi also announced the existence of isovalencenol, a crucial oil component with high similarities to Takahashi's alcohol [48]. Valencene, nor-isovalencenone, isovalencenal, α and β-isovalencenol, α and β-isonootkatol, nootkatol, Kusunol (Valerianol), isovalencenic acid, methyl isovalencenate, and other key eremophillanes are just a few of the important ones.

2. **Vetiverols**

Vetiverols are mixes of non-reactive carbonyl compounds and ethers with varying amounts of primary, secondary, and tricyclic sesquiterpene alcohols. α and β-vetispirene, β-acoradiene, β-chamigrene, β-acorenone B, vetiverol, and β-vetivol, as examples. It swiftly turns into vetiveryl acetate and is primarily used in perfumery. However, because mixes are so complicated, it is difficult to give any vetiverol a single name or a clear structure [49].

3. **Zizaanes and Prezizaanes**

The two most significant compounds in this class are khusenic acid, also known as zizanoic acid, which was discovered in Japanese vetiver oil, and Khusimol, which was identified from the vetiver oil of South India. Later, it was confirmed that zizanoic acid reduction produces the same end product as khusimol. The high boiling portion of vetiver essential oil also contained another structure that was separated and recognised as Khusimone, one of the essential components of vetiver oil. Khusimene, Khusenol, isokhusimol, allo-khusiol, prezizanol, zizanoic acid/

khusenic acid, epizizanoic acid, isokhusenic acid, methyl khusenate, khusimylformate, khusimyl acetate, etc. are some more substances that fall under this group [50, 51].

4. **Eudesmanes**

Numerous eudesmane derivatives are present in vetiver oil, but junenol was the first to be isolated and recognized [52] as a substantial constituent of khus oil. Later, it was discovered among other eudesmane sesquiterpenoids in Haitian vetiver oil. Vegetselinenol was the first alcohol discovered shortly after eudesmanes' investigation [53]. Selina-4,7-diene, β-Selinene, γ-Selinene, Cascarilladiene, γ-eudesmol, β-cyperone, β-agarofun, camphor, β-eudesmol, and α-eudesmol etc. are just a few of the numerous compounds that were later discovered and classified.

5. **Amorphanes, Murrolanes and Cadinanes**

Khusinol has been recognised as the first cadnanes in khus oil, one of the crucial components of khus oil [54]. Although it can't be guaranteed for other vetiver oils, this component is unquestionably present in vetiver oil from north India. In addition to khusinol, other essential substances include 1-epi-cubenol, -cadinol, -veticadinol, and -calacorene. -amorphene, -cadinene, -murrolene, -amorphene, and khusinol/ khosinol are also important.

6. **Bisabolanes and Elemanes**

The two key chemicals found in practically all vetiver oils according to GC-MS analysis are -bisabolol and bisabola-3(15),10-dien-7-ol [55, 56]. Elemol is a compound known as elemane that is typically found in vetiver essential oil from Haiti, Japan, and Reunion [53, 57]. Sesquicineole, deoxy-saussurea lactone, β-bisabolene, zingiberene, α-curcumene, γ-elemene, β-elemol and others are some further instances.

7. **Patchoulanes and Cedranes**

Kaiser discovered the first cedranes derivative in 1972 [55, 58], and Paknikar and Zalkow redesigned its structure in 1975. Based on X-ray and chemical synthesis, this refinement of the structure was made [59, 60]. Cyperene, α-funebrene, α and β-cedrene, α-funebren-15-al, isocedranol, α-funebren-15-oic acid etc., were among the additional structures found and characterised.

8. **Cyclocopacamphanes and Nigritanes**

Tetracyclic sesquiterpenes called cyclocopacamphanes are found as alcohols, acids, and aldehydes. Cyclocopacamphanic acid was the first substance in each category to be fully defined, and 2D-NMR studies provided a detailed analysis of 6-Epi-nigitene [51, 51] and its structural makeup. Later, numerous additional compounds were investigated, including 6-epi-nigritene, nigritene, khusian-2-one, cyclocopacamphan-12-al, yclocopacamphan-12-ol, khusien-14-ol, khusian-2-ol, and methyl cyclocopacamphanoate.

9. **Phenols and nitrogenous compounds**

Vanillin, 4-ethylphenol, and 4-vinylguaiacol are among the phenolic chemicals that are more prevalent in vetiver oil and that may be extracted using standard extraction techniques. Nitrogenous substances, however, such as pyridines and pyrazines, are only trace amounts and are biologically inert [62, 63]. Other than the categories mentioned above, the remaining chemicals are found in miscellaneous.

26.6.1 Pharmacological Activities

In contrast to other plant components, this plant's roots are equipped with a variety of elements. Numerous studies used fractions, oils, isolated compounds from the roots, extracts in various solvents, and extracts from the roots on various animal models to examine its diverse biological activities. Out of all the experimental research, many of them have produced amazing results, raising the possibility of promising outcomes in the future.

1. **Antimicrobial activity:**

Plants and their derivatives have long been used as medicinal agents to speed up the recovery process in the majority of disease cases [64]. There is still a great need for identifying the precise molecule or molecules responsible for the biological activity [65], despite the fact that numerous scientific articles have described the antibacterial activity of various plants [64, 66, 67].

The antibacterial activity of various fractions of the essential oil from the root was examined by using TLC-bioautography analysis against the MRSA (Methicillin-resistant Streptococcus aureus) and VREF (Vancomycin-resistant Enterococcus faecalis) MDR Gram-positive bacteria. By separating the active fractions of oil, the application of this analysis technique allowed for the identification of the most pertinent chemicals accountable for the biological activity. The MIC values calculated for active fractions like for CC-F15 $\geq$ 250 μg/ml for both MRSA and VREF, CC-F15–7/8/9/10, MICs $\leq$125 μg/ml and in the case of VREF were CC-F15-8/9 which presented MICs $\leq$250 μg/ml, CC-F5-8 MICs for both MRSA 62.5 μg/ml and VREF 125 μg/ml and non-fractionated essential oil MIC value show 510 μg/ml [68]. The variations in the concentrations of the active compounds and the presence or absence of the bioactive compounds were the causes of the variations in the MIC values of the subfractions. Furthermore, stronger effects were seen against Gram-positive, non-MDR bacteria, particularly S. aureus [69]. According to the results of the other investigation by Barros et al. (2009a, 2009b), the MIC values of the methanol/water (7:3) fraction were less than 25,000 g/ml and as high as 150 g/ml for non-fractionated CZR-EO (*C. zizanioides* roots essential oil) against *S. aureus*.

Additionally, David et al. [70] contrasted the antibacterial properties of vetiver oil produced using four distinct techniques. These processes include hydrodistillation (HD), carbon dioxide expanded ethanol extraction (CXE), supercritical fluid extraction (SFE), and indirect vapour distillation (IVD) extraction. Using a micro-dilution method, these oils were tested for their antibacterial properties against gram-positive (*S. aureus, B. subtilis*) and gram-negative bacteria (*P. aeruginosa, E. coli*). SFE oil shown enhanced effectiveness only against *S. aureus* (MIC = 78 g/ml), while other oils, such as CXE, had varying degrees of potency. The HD oil, in contrast, demonstrated strong suppression of *S. aureus* (MIC = 39 g/ml) and moderate effectiveness against *B. subtilis, P. aeruginosa*, and *E. coli*, each with a MIC value of 312.5 g/ml. The IVD extracted oil had MIC values of 78 and 156 g/ml, respectively, and shown better effectiveness against gram-positive than gram-negative bacteria. This variance in chemical makeup, which was validated by GC-MS analysis, may be the cause of the discrepancy in MIC value and oil potency. With no other oil present, GC-MS analysis of the CXE oil revealed the existence of three additional compounds: valerenol, valerenal, and -Cadinene at 18.48%, 10.21%, and 6.23%, respectively [70]. A comparable investigation using root and leaf extract fractions against *S. aureus*, *E. coli*, two fungi, *C. albicans,* and *C. neoformans* was published by Jayashree et al. in [71]. Based on the zonal inhibition, the comparison analysis showed that roots have stronger antibacterial potency than leaves. The root extract had an average IC50 value of around 6 mg/ml, with maximum inhibition zones against *S. aureus* and *Candida albicans* of about 30 mm and 33 mm and minimum inhibition zones against *E. coli* and *Candida neoformans* of about 22 mm and 32 mm, respectively [71]. Cedr-8-en-13-ol, δ-Selinene, γ-Gurjunenepoxide-2, and Solavetivone were the key phytochemicals that were in charge of the extracts stated above and the oil fractions' antibacterial activity. These compounds can be further separated and studied for a variety of different biological effects [72].

The ethanolic extract of *C. zizanioides* and its various solvent fractions, including hexane, ethyl acetate, and methanol, were examined for their anti-mycobacterial activity against both strains of *M. tuberculosis* in addition to their antibacterial and antifungal activities (virulent and non-virulent). Only the hexane fraction (MIC 50 g/ml) of all the fractions prevented the growth of both *M. tuberculosis* strains 7 days after inoculation. The potential of this non-polar bioactive chemical as a future anti-tubercular molecule still needs to be determined. Therefore, the investigation requires additional active moiety extraction and identification from the hexane fraction [73].

2. **Anticonvulsant activity:**

Using animal models for seizures generated by pentylenetetrazole (PTZ) and maximal electroshock (MES), Gupta et al. [74] revealed the anticonvulsant efficacy of the *C. zizanioides* root [74]. To estimate the LD50, the ethanolic extract was diluted in 1% Tween 80 and administered orally to rats at dosages of 100, 200, 300, 400, 500, and 600 mg/kg. Following the completion of the acute toxicity research in accordance with OECD (Organization for Economic Co-operation and Development) guideline 425, the LD50 was found to be 600 mg/kg. In MES-induced seizures, the extract, when administered in safe quantities, delayed the onset of convulsions without causing any deaths, whereas only 83% of animals survived PTZ-induced seizures. The drugs that alleviate or lessen seizures must increase the GABA inhibitory neurotransmitter or block sodium ion channels in order to work. The generalized tonic-clonic seizures caused by MES imitate grandmal epilepsy. Additionally, studies on vetiver extracts against convulsions indicated that they significantly increased GABA levels in generalised tonic-clonic and partial seizures. Alkaloids, flavonoids, saponins, terpenoids, tannins, and phenolics are among the many phytochemicals that are rich in the roots, and the extracts showed a notable amount of activity [75]. According to Raj et al. [76], triterpenes and flavonoids, which have a larger affinity for GABA-A receptors and anti-convulsant properties, are the most significant chemicals to which the action is attributed [76].

The phytochemical component that gives plants their anticonvulsant properties needs to be identified for this investigation. Additionally, the GABA-A receptor's binding effectiveness and affinity can be suggested as the mechanism of action.

3. **Anti-inflammatory:**

Chou et al. [77] described the anti-inflammatory properties of VZ-EO (*Vetiveria zizanioides*-essential oil) in murine macrophage cell line RAW 264.7. In 12 well-plates with 3 × 105 RAW cells plated for 20 h, the essential oil was tested at four different doses of 0, 5, 7.5, 10, and 12.5 g/ml and 1 g/ml lipopolysaccharide (LPS) (control). Following incubation, essential oil demonstrated an impressive anti-inflammatory effect that gets stronger with dose. Nitric oxide generation is reduced by it at concentrations of 7.5, 10 and 12.5 g/ml, which is a good sign that inflammation has been controlled without any evidence of harm. It also demonstrated a protective effect by increasing the number of LPS-stimulated RAW 264.7 macrophages through an increase in Heme oxygenase-1 and mRNA expression and decreasing the NO (nitric oxide) release from LPS-stimulated RAW 264.7 macrophages through inhibition of cyclooxygenase-2, mRNA, and nitric oxide synthase [77, 78]. Apoptosis is suppressed as a result of the inhibition of iNOS, which is directly related to the biochemical modification of the process [79]. A distinct picture of the anti-inflammatory effects was produced by a sharp decline in TNF-, mRNA expression, IL-1, and IFN- mRNA at all doses. Oil-treated rats also shown a dosage-dependent reduction in the production of superoxide anion, MDA, and SOD activity at a dose of 12.5 g/ml, exhibiting an antioxidant and anti-inflammatory effect. The amounts of cedr-8-en-13-ol, amorphene, vatirenene, and gurjunene are 12.5%, 7.80%, 5.91%, and 5.94% respectively, are key components of this oil and are responsible for its numerous processes. Future treatments for inflammatory-related illnesses can make extensive use of this feature.

4. **Anticancer activity:**

Regarding conventional plant usage, scientists looked into the cytotoxic and protective properties of roots using in vitro cancer models. However, just a small number of scientists have identified the chemical components with anticancer potential; others must to look into and validate this. Powers et al. [80] examined the effects of *C. zizanioides* root oil on two different breast cancer cell lines, MCF-7 and MDA-MB-231, using doses of 1.44 x 106 and 1.44 x 104 cells per well plate. The essential oil was utilized as the test sample with a concentration of 0.01% w/v, while tingenone (100 g/ml) and DMSO were employed as the positive and negative controls for the MTT experiment. With an IC50 value of 23.9.0 g/mL and 36.2 g/mL, respectively, the inoculation of essential oil resulted in a considerable reduction in cell viability and demonstrated better anticancer efficiency against both breast cancer cell lines. The inclusion of a significant portion of sesquiterpene alcohols, such as (E)-isovalencenol (13.5%), khusimol (12.1%), and α-vetivone (5.4%), may be the cause of this substance's increased potency [80].

In a second study, the cancer cell lines HeLa (cervical cancer cell line) and MCF-7 were used to test the cytotoxic potential of the polyherbal extracts *Vetiveria zizanioides*, *Trichosanthescucumerina*, and Mollugocerviana (breast cancer cell line). To create the polyherbal extract, equal parts of each plant were used in a 1:1:1 ratio, and a ratio of 1:10 was used for solvent methanol extraction. When this polyherbal extract was evaluated on cancer cell lines, the LC50 values were 467 2.9 mg/ml for HeLa and > 300 mg/ml for MCF-7 cell lines. This efficacy may be attributed to the presence of phytochemicals such as polyphenols and flavonoids, among others. The enzymatic alterations, induction of apoptosis, redox reactions, hormone regulation, and activation of the immune system are all associated to the formation of cancer cells, carcinogenesis, and these flavonoids may play a part in any of these signaling pathways [81]. Additionally, polyphenols have an antioxidant effect that lowers thrombogenesis, atherosclerosis, inflammation, and the development of cancer [82].

A second study that used the MTT ((4,5-dimethylthiazol-2-yl)-2,5-diphenyltetrazolium bromide) assay to detect cancer cells called MCF-7 also found that vetiver's aqueous root extract has anticancer properties. In addition, cell morphology was examined using the Acridine Orange and Ethidium Bromide staining techniques [83]. The IC50 values were obtained between 5 and 50 g/ml, while the anti-proliferative impact was observed between 31 and 37 g/ml (MIC).

5. **Antioxidant activity**:

The antioxidant activity of vetiver was evaluated using the -carotene bleaching technique [84]. This approach depends on the free linoleic acid radicals' quick decolorization of -carotene in the absence of antioxidants. The absorbance decline at 470 nm was, however, considerably slowed from 0 to 180 min when *C. zizanioides* essential oils (10 g/mL) were administered. In a test for lipid peroxidation, it also demonstrated antioxidant efficacy on par with that of conventional butylated-hydroxyanisole (BHA). Another crucial factor to take into account is the oil's ability to boost the antioxidant enzyme system by increasing intracellular levels of glutathione (GSH), superoxide dismutase (SOD), and glutathione peroxidase (GPx), which reduces oxidative stress [85]. Cedr-8-en-13-ol (12.4%), β-amorphene (7.80%), β-vatirenene (5.94%), β-gurjunene (5.91%), and dehydroaromadendrene (5.45%) were the chemical ingredients that were studied by chromatographic methods. However, research studies showed that cedr-8-en-13-ol was crucial for this biological function. Cedr-8-en-13-ol, an essential oil from the *Peucedanum longifolium* plant, was previously shown to be a powerful inhibitor of lipid peroxidation in a study published by Tepe et al. [86]. Cedr-8-en-13-ol, the chemical, also showed some insecticidal efficacy against mosquitoes [9].

Through various experiments, Chou et al. [77] also documented the antioxidant and anti-inflammatory properties of C. zizanioides essential oil. According to Chou et al. [77], the assays used were the SOD (superoxide dismutase) assay, MDA (malondialdehyde) generation, and superoxide assay. When compared to RAW 264.7 macrophages treated with lipopolysachharide (LPS), these cells exhibit a reduction in superoxide production of about 12% and 20% at 7.5 and 12.5 l g/ml, respectively. Further suppressing the lipid peroxidation in LPS-stimulated macrophages was the MDA levels' decrease at these concentrations. Although many other factors also support the antioxidant potency of VZ-EO, the decline in the level of the SOD enzyme unfortunately cast doubt on its certainty. Apocyanin and natural antioxidants (NAO) are examples of naturally occurring antioxidants that demonstrate their efficacy by not changing the concentration of antioxidant enzymes [114]. Cedr-8-en-13-ol (12.4%), −amorphene (7.80%), −vatirenene (5.94%), and -gurjunene (5.91%) make up the majority of an essential oil's chemical makeup. AlpiniaoxyphyllaMiq essential oil also contains a significant amount of -amorphene, a potent inhibitor of Staphylococcus aureus [115]. On the other hand, −vatirenene is a chelator of ferrous ions and a DPPH free radical scavenger (80, 88).

Luqman et al. [72] also reported on the antioxidant activity of hexane extracts of *C. zizanioides* roots using reducing power (RP), total phenolic component (TPC), ferric reducing antioxidant power (FRAP), and DPPH (1-1diphenyl-2-picrylhydrazyl). This study showed that the antioxidant potential increased with extract concentration, which may be because concentration increases lead to an increase in phenolic content. Due to its ability to prevent ageing and reduce oxidative stress, the aforementioned activity can suggest its use as a dietary/extra antioxidant in nutraceuticals and cosmetics.

6. **Anti-melanogenesis:**

Melanogenesis, the process of producing melanin, results in the emission of reactive oxygen species (ROS), melanocyte-stimulating hormone (-MSH), and hydrogen peroxide (H2O2) [22, 116]. Natural antioxidants including resveratrol, catechin, and epicatechin, curcumin, and ascorbic acid [117] and others can be used to battle this stress, which will have an anti-melanogenic impact.As a result of VZ-EO's excellent antioxidant potential and ability to lessen oxidative stress, its anti-melanogenic effects in -MSH-stimulated B16 cells were also investigated. It is examined based on the levels of the antioxidant enzymes glutathione (GSH), superoxide dismutase (SOD), glutathione peroxidase (GPx), and catalase, which produce MDA (malondialdehyde). When administered in doses ranging from 2.5 to 20 g/mL, it significantly lowers MDA levels, bringing them to levels that are comparable to those of unstimulated cells at the highest dose. Another factor is the GSH level, which is important in redox reactions and can have an anti-melanogenic effect [116]. The levels of GSH enzymes were considerably increased by VZ-EO, and they continue to be elevated long after the oil has stopped being used. In addition, VZ-EO boosts the activity of antioxidant enzymes, which can quickly overcome oxidative stress. Previous studies on the protective effects of gallic acid on UV-mediated melanogenesis through increased levels of GSH and amikacin's control of antioxidant enzymes like SOD, GPX, and CAT further supported these findings [118]. Cedr-8-en-13-ol (12.4%), a significant percentage of which is found in oil, is the chemical element proposed for this activity. In order to treat foods, cosmetics, cutaneous infections, and pigmented skin, this essential oil may be useful.

7. **Mosquito repellent:**

In 2019, Khater and Gaden investigated the ability of essential oils of vetiver, cinnamon, and lavender to deter mosquitoes from biting both larval and adult house flies, Muscadomestica L. Pure vetiver oil was diluted in 5% Tween 20 for topical application and fumigation at a concentration of 0.6%, which guaranteed 100% adult mortality via ovicidal impact but was regrettably insignificant for the virucidal activity against RFV-T48 and entrance assay [119]. Even though earlier studies also claimed that it was 100% effective against L. sericata, its applicability is restricted by its expensive cost, which is somewhat offset by the fact that it is less volatile [121]. Additionally, a mixture of vetiver and sunflower oil was tested. Unfortunately, these blended mixes did not prove to be useful; rather, they only added to the cost. However, since these oils can be used in places where insecticides are contraindicated, cinnamon and vetiver were used to control flies and flies' products [120]. Additionally, a lot of work still needs to be done to create better formulations for essential oils by adding carrier molecules or modifying them into nanoparticles, microsomes, liposomes, etc. [122]. There is only one essential oil product on the market, and it has an insecticidal effect. It is a 10:5:2 combination of rosemary, geraniol, and peppermint oils [123].

8. **Effect on anxiety:**

The effects of essential oils on CNS activity, which in turn impacts how the brain works, are well known. In order to assess the anxiolytic qualities of vetiver essential oil and analyse changes in the expression of the c-fos protein, the study used an elevated plus-maze model of anxiety. In this experiment, 2.5% vetiver essential oil was administered to rats, who were subsequently placed in an elevated plus-maze apparatus to evaluate their behaviour. When compared to the closed arm, the visit and time spent in the open arm considerably increased shortly after oil inhalation, which is a true evidence of the anxiolytic effect. Additionally, the central amygdaloid nucleus' hyperactivity, which may be a marker of the anxiolytic effect, is evaluated for c-fosexpression. The prefrontal cortex and hippocampus are two more anxiety-related brain regions that still need to be studied under stress.

The study still needs further examination in order to identify the bioactive molecule responsible for the same effects, even though vetiver oil's effect is similar to that of the well-known anxiety medication Diazepam [124].

Cheaha et al. [125] conducted a further study in 2016 to investigate the effects of vetiver oil inhalation on the sleep-wake cycle and EEG. This study also dealt with the central nervous system. The three states that the brain can be in depending on changes in the sleep-waking cycle are:

(i) Non-synchronized waves were used to depict the awake state.
(ii) High levels of low-frequency waves that are associated with sleep.
(iii) Rapid eye movement (REM) sleep was manifested by the dominant parietal theta wave.

When vetiver oil is inhaled, a sizable increase in overall awake duration is seen, although slow-wave sleep is shown to be lowered and REM sleep is unaffected [126]. The anxiolytic and tranquil effects of the stimulant were confirmed by enhanced gamma waves and decreased slow EEG waves [127]. The augmentation of inhibitory neurotransmitter, GABAergic transmission, is the suggested mechanism behind these benefits. Additionally, it boosts the synaptic knob's release of acetylcholine, which works as a short-term memory enhancer when learning or memorizing Vetiver essential oil can be used to enhance learning, alertness, and task performance [125].

9. **Acaricidal activity:**

Numerous researchers have examined the acaricidal (mites-killing) properties of vetiver essential oil [69, 128], hypothesizing Khusimol as the primary causative component [129]. It has a variety of mechanisms at its disposal, including:

1. Reduction of the reproduction process by depriving female arthropods of nutrients such as proteins and lipids, which is more severe than what is found in commercial products [130, 131].
2. Increase in acetylcholine levels through inhibition of cholinesterase-induced hydrolysis [132].

As a result of the rise in acetylcholine levels, arthropods become paralysed and eventually die [133].

The presence of phytochemicals that are particular to the target species may be the cause of the oil's action against *A. cajannense, R. microplusticks*, and their larvae. Maximum sesquiterpene-containing Drymisbrasiliensis has demonstrated similar potencies against R. microplus [134]. Against *A. cajannense*, however, Eucalyptus citridodora and Cymbopon going tordus essential oils demonstrated substantially lower mortality while being used in considerably larger concentrations [135]. Even though *A. cajannense* required a higher percentage of oil, vegetable oil was effective against both ticks at the same concentration, independent of whether it was high in zizanioic acid or khusimol acid [136]. In contrast, ethanolic extracts of vetiver roots demonstrated little toxicity in rats [137] because they lacked octopamine receptors, which are primarily found in insects [138]. Future views on cosmeceutical use can greatly benefit from its efficacy and low toxicity in vertebrates.

10. **Hypoglycaemic activity:**

With the use of an alloxan-induced diabetes model, Karan et al. [113] examined the hypoglycaemic impact of *Vetiveria zizanioides* roots [113]. All albino Wistar rats (male and female) received an intravenous injection of the diabetes-inducing drug alloxan at a concentration of 150 mg/kg. After 48 hours, blood sugar levels were measured to identify diabetic rats. After then, for a total of 28 days, ethanolic root extract and the medication glibenclamide were given at various doses of 100 mg/kg, 250 mg/kg, 500 mg/kg, 750 mg/kg, and 10 mg/kg.The blood sugar level was regularly checked during the trial to compare the anti-diabetic effects of the various groups at various doses. The measured sugar values showed that vetiver extract had equivalent action to the common medication glibenclamide on days 7, 21, and 28 showed flavonoids, sterols, saponins, and polyphenolic substances when they were examined for phytochemicals. Among these, flavonoids and saponins may possess an anti-diabetic effect. However, additional study is still required to isolate and define the pure chemical, demonstrating a breakthrough in the treatment of diabetes in a more palatable manner [113].

11. **Antidepressant effect:**

The in vivo tests of the ethanolic extract of *V. zizanioides* were utilised to determine the antidepressant activity using the tail suspension test and the forced swim model of depression. In the study by Josephine et al. [126] on previously-induced depressed rats, a 100 mg/kg dose of the ethanolic extract of *V. zizanioides* roots alone as well as in combination with a 10 mg/kg dose of Fluoxetine was found to be beneficial. When compared to the immobility time data, the Fluoxetine and 100 mg/Kg vetiver extract combination in all mice showed the best antidepressant efficacy.

Due to the substantial role that oxidative stress plays in the emergence of depression or severe depression, the antioxidant properties of both are proposed as the underlying mechanism for this action [112]. As a result, any decrease in oxidative stress or increase in antioxidant enzymes like glutathione, catalase, and superoxide dismutase directly stops the lipid peroxidation process. This significant action is mediated by the major flavonoid/polyphenolic compounds or components, such as -vetivone, –vetivenene, and -vetivone [78, 87].

12. **Protective effect against Cisplatin-induced toxicity:**

In Swiss albino mice, cisplatin therapy has been shown to cause nephrotoxicity, mutation, and myelosuppression, with symptoms including increased weight loss, decreased appetite, GIT disruption, dehydration, renal atrophy, and oedema that can be seen right away [88]. (Chirino and Pedraza-Chaverri [89], Elevated creatinine levels, blood urea nitrogen (BUN), oxidative stress, and reduced antioxidant enzymes all support the harmful consequences of cisplatin therapy. Furthermore, it has been found that antioxidants help prevent these harmful outcomes by reducing oxidative stress [89]. Amifostine is the only antioxidant medication that has been licenced for use in conjunction with Cisplatin, but its price and associated adverse effects also restrict its use.

Additionally, vetiver essential oil has stronger antioxidant and free radical scavenging properties that can be used as a preventative in cisplatin and vetiver oil combo therapy. Prior to the administration of Cisplatin, vetiver oil was given three times daily for 7 days in doses of 5, 10, and 20 mg/kg. By reducing lipid peroxidation and raising the levels of the antioxidant enzymes GSH and GST, their treatment has been successful in reducing the harmful effects of oxidative stress associated with Cisplatin at 5 mg/kg. Other indicators like the animal's body weight, biochemical markers (Hb, leucocytes, thrombocytes, BUN, creatinine), and histological investigations all having normal values further support their protective function (cellular integrity). Vetiveria oil functions as a chemo-preventive drug due to all of these factors combined, although additional research is needed to confirm these results [90].

13. **Antidiuretic activity:**

Rao et al. [91] isolated the Khusimol for the first time from the vetiver roots in 1994. Several analytical methods, including nuclear magnetic resonance (NMR), infrared spectroscopy, mass spectrometry, and molecular weight determination, were used to confirm the structure of the isolated sesquiterpene alcohol khusimol (functional group determination). Khusimol was discovered to be a competitive inhibitor of vasopressin V1a receptors after the structure had been obtained and tested utilizing the [3H]-Vasopressin-Binding assay [91]. When the osmolality is high, this blockage of the V1a receptor may be used for its antidiuretic properties.

14. **Sedative activity:**

Thubthimthed et al. [92] investigated the sedative properties of vetiver oil in male Wistar rats weighing approximately 124 g. This study used two oils to test their effectiveness after being inhaled by rats for an hour: vetiver (5% w/v) and lavender (5% w/v).After an hour of continuous inhalation, they have delivered the final 30 minutes in a square block with a camera attached. After their rest period was through, their activity was continually recorded for 5 min in order to observe their rearing behaviours in the open field test. After viewing every recorded videotape, the collected data values were statistically evaluated to arrive at the conclusion. They discovered that vetiver oil lowers rearing motility more effectively than lavender oil, which is completely consistent with its traditional use [92].

15. **Antifungal activity:**

Pathogens were subjected to a broad range of vetiver oil's organic fungicidal properties. The antifungal efficacy of sesquiterpenoids conversion products identified in vetiver oil against two phytopathogenic fungi, *A. alternata* and *F. oxysporium*, was evaluated using the spore germination inhibition technique. One of the several compounds studied, husinodiolmonobrosylate, was found to be an effective antifungal agent against both fungi [93]. The sesquiterpenoid Schiff bases N-(Khusilidene)-p-methoxy aniline and N-(Khusilidene)-p-bromo-aniline were produced through reactions with p-methoxy aniline and p-bromo-aniline. N-(Khusilidene)-p-bromo-aniline was shown to diminish the growth of F. oxysporium by up to 74.5% at a concentration of 1 mg/mL, while N-(Khusilidene)-p-methoxy aniline inhibited the growth of *A. alternata* by up to 84.7% [94]. Additionally, vetiver oil's antifungal effects against Rhizoctoniabataticola and Sclerotiumrolfsii were noted [95]. When evaluated, South Indian vetiver oil has a somewhat higher fungal toxicity than North Indian oil [96].

In 2012, Sangeetha and Stella carried out a new study utilising several vetiver root and leaf extracts. This activity was performed using disc diffusion and the minimum inhibitory concentration (MIC) method utilising *A. niger, A. flavus, C. albicans,* and *S. cerevisiae* as test fungi. The antifungal activity of vetiver extracts at a dosage of 10 mg/ml was further investigated using fluconazole as a positive control. Of the tested fungal cultures, vegetable leaf and root extracts produced *Aspergillus niger* inhibition zones with mean values of 30 and 32 mm [97].

26.7 Agricultural Uses

1. **Insecticidal activity:**

Additionally, vetiver oil has insecticidal properties that are generally effective against ants, ticks, and cockroaches. Zizanal and epizizanal, two tricyclic sesquiterpenoids found in vetiver oil, were previously mentioned as potential culprits for this activity [98].However, Babprasert and Karintayakit [99] later confirmed that at least six compounds, including khusimone, zizanal, epizizanal, and (C)-(1S,10R)-1,10-dimethylbicyclo[4.4,0]-dec-6-en-3-one, were discovered to be responsible for the insect repellent effect [99]. Additionally, Ceratitiscapitata treatment with oil extracts from this plant showed anti-oviposition effects. In the humid tropics of Western Africa, Ndemah et al. [100] described the insecticidal action of an aqueous extract (vetiver oil) derived flavonoids (concentration 0.07 mg/mL) against lepidopterus stem borers (on maize) [100].

2. **Termicidal activity:**

The oil from the roots of *V. zizanioides* showed a considerable decrease in the leaf damage brought on by subterranean termite larvae [101]. Although vetiver oil is a complex mixture of more than 100 different chemicals, only three of them nootkatone, zizanol, and bicyclovetivenol—showed repellant action toward Formosan subterranean termites (Coptotermesformosanus).Therefore, vetiver oil can undergo chemical reactions that result in oxidation or reduction to produce a variety of chemicals that can be used for pest management [102, 103]. Numerous studies have shown that vetiver oil can also act as an arrestant, repellent, and feeding deterrent against *C. formosanus* and its symbiotic fauna when combined with nootkatone and disodium octaboratetetrahydrate. These characteristics might be seen by specific criteria, including as a sharp decline in their tunnelling activity, wood consumption, and survival [104, 105].

3. **Pesticidal activity:**

The efficiency of vetiver oil against specific pests such as adults of *Tribolium castaneum*, the XSM, SMC, SKS, and JTC strains of *Tribolium castaneum*, the Sitophilusoryzae, the *Callosobruchus maculates*, etc., has also been investigated. Treatment of *Tribolium castaneum* with vetiver dust (2 and 5 g) [106].The red rust flour beetle *T. castaneum* XSM, SMC, SKS, and JTC strains were all susceptible to the petroleum ether, ethyl acetate, acetone, and methanol extract of vetiver roots. The petroleum ether extract showed the greatest amount of toxicity among them (LD50 = 0.051 g/cm2).

While the methanolic extract (LD50 = 11.351 g/cm2) showed the lowest toxicity in the larval bioassay against the XSM strain [107]. Similar to this, four nonadjacent bis-tetrahydrofuranacetogenins found in the pet ether fraction, designated squamostatins B through E, exhibited insecticidal efficacy against Sitophilusoryzae (a seed-infesting species of wheat) [106].Furthermore, adult Callosobruchus maculates were shown to be harmful to acetone extracts of fresh and preserved leaves, but not to ethanol extracts [108].

4. **Anti-plasmodial (antimalarial) and larvicidal activity:**

Additionally, this oil was discovered to be efficient against the Anopheles stephensi malarial vector, which is thought to be responsible for about 85% of deaths. According to the observed death rate, the extracts may be applied as biopesticides. Second, third, and fourth instar larvae of A. stephensi had LC50 values of 0.276%, 0.285%, and 0.305%, respectively [109].

5. **Herbicidal effect:**

In 1994, Techapinyawat looked into the herbicidal effects of vetiver root and stem extracts on soybean seeds. The fact that the extracts and certain of its phytochemical components prevented soybean seeds from germinating further supports their herbicidal effect. This characteristic might also be useful for preventing the spread of undesirable weeds. Another investigation was done to confirm the herbicidal effects of vetiver oil and its minor constituent nootkatone on six common weed species, including velvetleaf, gigantic ragweed, pitted morning glory, redroot pigweed, and common lamb's quarters. The germination test, carried out in Petri dishes, was the experiment employed for the study. During the test, it was observed that this oil acted as a useful herbicide by suppressing the growth of redroot pigweed and common lamb's quarters seedlings as well as the germination of these weeds. In addition, they inhibited the growth of *Pisum sativum* L. plants in the lab and the field, respectively [109]. The non-edible plants Ricinuscommunis L. and Jatrophacurcas L. interacted allelopathically with vegetable oil, which increased the growth of the Jatropha seedlings while limiting the growth of the *R. communis* seedlings. This result suggests that jatropha and vetiver are suitable fence plants for plant-plant interaction [110].

26.7.1 Other Uses

While the dried roots are utilized as a scent in linen and clothing, vetiver roots are commonly employed as a coolant in the walls of desert coolers. To assist tribal people in making a living, they have produced woven screens, mats, shades (chik), hand fans, broom hangers, and baskets. Root mats are the traditional way that vetiver roots were used to build huts and cottages because they had a cooling effect during the hot summers. Even plant pulp used in paper and strawboard manufacturing [40].

26.7.2 Safety of Vetiver Essential Oil

The European Flavour and Fragrance Association was asked to amend the substance's name and intended use after receiving vetiver oil in the form of vetiveryl acetate, according to the International Fragrance Association's submission II dated June 2013. The Scientific Committee on Consumer Safety approved vegetable acetate with 1% alpha-tocopherol as a safe concentration for regular usage as a fragrance ingredient (SSCS). The amount of acetylated vetiver oil in typical products can range from 0.05 to 0.90 depending on the type of product, such as hydroalcoholic-based scents. (0.90, deodorants, eye makeup, hand cream, body lotion, 0.10 for face cream, 0.10 for body lotion, and 0.20 for bath items). Although not in the concentration described above, the aldehydes and ketones found in the oil can damage DNA and proteins.

According to the Scientific Committee on Consumer Safety, topical application of acetylated vetiver oil to the skin can produce minor discomfort. However, the HRIPT (Human Repeat Insult Patch Test) study defended the same because no person using vetiver oil-containing cosmetics reported experiencing an adverse reaction. The spray products cannot be created without the inhalational safety report, which is still pending [111].

26.7.3 Limitations

Vetiveria cultivation and use, despite their widespread use, have a few drawbacks that should be seriously discussed before they are expanded. The main goal of vetiver cultivation is to get vetiver oil from the roots, which can extend several feet into the ground. The deeper root, when removed from the ground, increases the likelihood of soil erosion, endangering neighbouring residents. As a result, the authors suggest putting in place certain stringent regulations for the cultivation, preservation, and harvesting of vetiver roots. It is difficult for an analytical chemist to extract and identify the chemical constituent in pure form to investigate its biological potential due to the complexity of oil, which contains a mixture of over 100 chemicals. Even the commercial unavailability or high cost of its components (if available) restricts its potential uses.

26.8 Conclusion

Vetiveria has a strong potential to serve people all over the world in terms of health, economy, and the environment, as shown by the rising number of research studies and conservation initiatives on vetiver cultivation. Long, deep roots have a number of traditional functions that are described in the ayurvedic pharmacopoeia and other works of literature, but reality regrettably reveals the bitter truth. Due to a lack of proper policies, rules, money, and international cooperation, the vetiver plant is still not extensively investigated. The cultivated portion part is only used for the utilities that have already been created. Further benefits are hindered by small institutions' limited access to and difficulties in extracting root oil. With the help of both developed and developing nations, the execution of vetiver conservation programmes might thus be shown to be a win-win situation in the future given its wide range of applications. To improve the vetiver's quality, accessibility, and usefulness, more research projects and innovative extraction methods are needed.

References

1. Stevens, P. F. (2017). *Angiosperm phylogeny website* (Version 14). Missouri Botanical Garden St. Louis.
2. Clayton, W. D., Vorontsova, M. S., Harman, K. T., & Williamson, H. (2015). *GrassBase-the online world grass flora. GrassBase-The Online World Grass Flora.*
3. Presl, K. B. (1830). *Reliquiae Haenkeanae, seu Descriptiones et icones plantarum, quas in America Meridionali et Boreali, in insulis Philippinis et Marianis collegit Thaddaeus Haenke* (Vol. 1). apud JG Calve.
4. Veldkamp, J. F. (1999). A revision of Chrysopogon Trin. Including Vetiveria Bory (Poaceae) in Thailand and Malesia with notes on some other species from Africa and Australia. *Austrobaileya, 5*, 503–533.
5. Kottekkattu, T., & Pradeep, A. K. (2014). Chrysopogon festucoides (Poaceae): A new record for South India. *Rheedea, 24*(1), 56–59.
6. Sunil, C. N., Narayanan, M. R., Sivadasan, M., Shaju, T., Kumar, V. N., & Alfarhan, A. H. (2017). A new species of Chrysopogon (Poaceae: Andropogoneae) from India. *Phytotaxa, 307*(4), 245–253.
7. Linnaeus, C. (2017). Phalaris zizanioides. In *Mantissa plantarum Altera* (Vol. 947, p. 190). Laurentii Salvii/Holmiae.
8. De Guzmam, C. C., & Oyen, L. P. A. (1999). Vetiveria zizanioides (L) Nash. In *Plant resources of South-East Asia 19 essential oil plants* (pp. 167–172). Backhuys Publishers.
9. Adams, R. P., McDaniel, C. A., & Carter, F. L. (1988). Termiticidal activities in the heartwood, bark/sapwood and leaves of Juniperus species from the United States. *Biochemical Systematics and Ecology, 16*(5), 453–456.
10. PIER. (2018). *Pacific Island ecosystem at risk. IOP publishing Pacific Island ecosystem at risk Honolulu Hawaii*. HEAR/University of Hawaii.
11. USDA-ARS. (2020). *Germplasm Resources Information Network (GRIN)*. Online Database. National Germplasm Resources Laboratory.
12. Weber, E., Sun, S. G., & Li, B. (2008). Invasive alien plants in China: Diversity and ecological insights. *Biological Invasions, 10*(8), 1411–1429.

13. Zhang, W., Liu, J. X., & Huo, P. H. (2017). Phoma herbarum causes leaf spots and blight on vetiver grass (Vetiveria zizanioides L.) in southern China. *Plant Disease, 101*(10), 1823.
14. Jian, O. U., Changyi, L. U., & O'TOOLE, D. K. (2008). A risk assessment system for alien plant bio-invasion in Xiamen, China. *Journal of Environmental Sciences, 20*(8), 989–997.
15. Chong, K. Y., Tan, H. T., & Corlett, R. T. (2009). *A checklist of the total vascular plant flora of Singapore: Native, naturalised and cultivated species*. National University of Singapore.
16. Zuloaga, F. O., Morrone, O., Davidse, G., Filgueiras, T. S., Peterson, P. M., Soreng, R. J., & Judziewicz, E. J. (2003). *Catalogue of new world grasses (Poaceae): III. Subfamilies Panicoideae, Aristidoideae, Arundinoideae, and Danthonioideae* (pp. 34–62).
17. Greenfield, J. C. (1989). *Vetiver grass (Vetiveria spp.): the ideal plant for vegetative soil and moisture conservation*. World Bank; Del Giudice, L., Massardo, D. R., Pontieri, P., Bertea, C. M., Mombello, D., Carata, E., ... & Alifano, P. (2008). The microbial community of vetiver root and its involvement into essential oil biogenesis. *Environmental Microbiology, 10*(10), 2824–2841.
18. Truong, P., Van, T. T., & Pinners, E. (2008). *Vetiver system applications technical reference manual* (pp. 89–96). The Vetiver Network International.
19. Plants, U. T. (2020). *Useful tropical plants database*. K Fern.
20. Joy, R. J. (2009). *"SUNSHINE" VETIVERGRASS Chrysopogon zizanioides (L.) Roberty*. USDA NRCS Plant Materials Center.
21. Grand view Research (2020) *Vetiver oil market size, share & trends analysis report by application (Medical, food & beverage, spa & relaxation), by region (North America, Europe, Asia Pacific, central & South America, Middle East & Africa), and segment forecasts, 2020–2027*. Grand view Research.
22. Paillat, L., Périchet, C., Pierrat, J. P., Lavoine, S., Filippi, J. J., Meierhenrich, U., & Fernandez, X. (2012). Purification of vetiver alcohols and esters for quantitative high-performance thin-layer chromatography determination in Haitian vetiver essential oils and vetiver acetates. *Journal of Chromatography A, 1241*, 103–111.
23. Chomchalow, N. (2001a). *The utilization of vetiver as medicinal and aromatic Plants with special reference to Chomchalow N (2001b)*. Office of the Royal Development Projects Board.
24. Chomchalow, N. (2001b). *The utilization of vetiver as medicinal and aromatic Plants with special reference to.Chomchalow N (2012)*. Office of the Royal Development Projects Board
25. Peyron, L. (1989). Vetiver in perfumery. *La Quintessenza, 13*, 4–14.
26. Viano, J., Gaydou, E., & Smadja, J. (1991). On the presence of intracellular bacteriums in the roots of Vetiveria zizanioides (L.). *Staph Rev Cytol Biol Végét-Bot, 14*, 65–70.
27. Putiyanan, S., Nanthachit, K., & Kittipongpatana, N. (2006). Chrysopogon zizanioides (L.) Roberty (Gramineae) part I. Pharmacognostic identification of roots. *Chiang Mai University Journal (Thailand) CMU, 5*(2), 179–198.
28. Chauhan, M. (2019). Ushira, Vetiver (Vetiveria zizanioides). *Planet Ayurveda, 75*(1), 12–34.
29. Curtis, S. (1996). *Essential oils (Neal's yard remedies)*. Aurier Press.
30. Shealy, C. N. (1998). *The illustrated encyclopedia of healing remedies*. Element Books Limited.
31. Lingga, P. (2001). *Recipes of traditional medicine*. PenebarSwadaya.
32. Usmanghani, K., Saeed, A., & Alam, M. T. (1997a). *Indusyunic medicine* (pp. 363–364). Faculty of Pharmacy, University of Karachi, Pakistan.
33. Usmanghani, K., Saeed, A., & Alam, M. T. (1997b). *Indusyunic medicine: Traditional medicine of herbal animal and mineral origin in Pakistan*. Department of Pharmacognosy, Faculty of Pharmacy, University of Karachi.
34. Ibrahim, A. U.(1996). Uses of Vetiveria nigritana grass species inNorthern Nigeria:ACase of Bauchi State. In *Abstracts of papers presented at ICV-1* (pp 1–124).
35. Jain, V., & Jain, S. K. (2016). *Compendium of Indian folk medicine and ethnobotany (1991–2015)*. Deep Publications.
36. Singh, K. K., & Maheshwari, J. K. (1983). Traditional phytotherapy amongst the tribals of Varanasi district, Uttar Pradesh. *Journal of Economic and Taxonomic Botany, 4*, 829–838.

37. Sastry, K. N. R. (1996). Socio-economic dimensions of vetiver in rainfed areas of Karnataka (India). In *Vetiver: A miracle grass* (pp. 4–8).
38. Pripdeevech, P., Wongpornchai, S., & Promsiri, A. (2006). Highly volatile constituents of Vetiveria zizanioides roots grown under different cultivation conditions. *Molecules, 11*(10), 817–826.
39. Dowthwaite, S. V., & Rajani, S. (2000). Vetiver: Perfumer's liquid gold. In *Proceedings of ICV-2 held in Cha-am* (pp. 478–81).
40. Lavania, U. C. (2003a). Other uses and utilization of vetiver: Vetiver oil. In *The third international vetiver conference, Guangzhou* (p. 475).
41. Bhatwadekar, S. V., Pednekar, P. R., Chakravarti, K. K., & Paknikar, S. K. (1982). Survey of sesquiterpenoids of vetiver oil. *Cultivation and utilization of aromatic plants/edited by CK Atal and BM Kapur.*
42. Demole, E. P., Holzner, G. W., & Youssefi, M. J. (1995). Malodor formation in alcoholic perfumes containing vetiveryl acetate and vetiver oil. *Perfumer & Flavorist, 20*(1), 35–40.
43. Sinha, S., Jothiramajayam, M., Ghosh, M., Jana, A., Chatterji, U., & Mukherjee, A. (2015). Vetiver oil (Java) attenuates cisplatin-induced oxidative stress, nephrotoxicity and myelosuppression in Swiss albino mice. *Food and Chemical Toxicology, 81*, 120–128.
44. Mallavarapu, G. R., Syamasundar, K. V., Ramesh, S., & Rao, B. R. R. (2012). Constituents of South Indian vetiver oils. *Natural Product Communications, 7*(2), 1934578X1200700228.
45. Sellier, N., Cazaussus, A., Budzinski, H., & Lebon, M. (1991). Structure determination of sesquiterpenes in chinese vetiver oil by gas chromatography—Tandem mass spectrometry. *Journal of Chromatography A, 557*, 451–458.
46. Pfau, A., & Plattner, P. (1940). Studies on volatile plant materials XI on the constituents of the β-Vetivone. *Helvetica Chimica Acta, 23*, 768–792.
47. Naves, Y. R., & Perrottet, E. (1941). Volatile plant materials XIII. Alpha-and beta-vetivones. *Helvetica Chimica Acta, 24*, 3–29.
48. Pinder, A. R. (1980). Further synthetic investigations in the eremophilane sesquiterpene group. Synthesis of (±)-isovalencenic acid,(±)-isovalencenol, and their (Z)-isomers, and experiments directed towards the synthesis of tessaric acid. *Journal of the Chemical Society, Perkin Transactions, 1*, 1752–1755.
49. Arctander, S. (1969). *Perfume and flavor chemicals:(aroma chemicals)* (Vol. 2). Allured Publishing Corporation.
50. Umarani, D. C., Gore, K. G., & Chakravarti, K. K. (1966). Terpenoids xc: Khusimol, a new sesquiterpene alcohol. *Tetrahedron Letters, 7*(12), 1255–1261.
51. Hanayama, N., Kido, F., Sakuma, R., Uda, H., & Yoshikoshi, A. (1968). Minor acidic constituents of vetiver oil. *Tetrahedron Letters, 9*(58), 6099–6102.
52. Nguyen, T. A., & Feizon, M. (1965). Sesquiterpenes of vetiver oils. *Am Perfum Cosmet, 80*(3), 40–50.
53. Homma, A., Kato, M., Wu, M. D., & Yoshikoshi, A. (1970). Minor sesquiterpene alcohols of vetiver oil. *Tetrahedron Letters, 11*(3), 231–234.
54. Del Giudice, L., Massardo, D. R., Pontieri, P., Bertea, C. M., Mombello, D., Carata, E., et al. (2008). The microbial community of vetiver root and its involvement into essential oil biogenesis. *Environmental Microbiology, 10*(10), 2824–2841.
55. Kaiser, R., & Naegeli, P. (1972). Biogenetically significant components in vetiver oil. *Tetrahedron Letters, 13*(20), 2009–2012.
56. Weyerstahl, P., Marschall, H., Splittgerber, U., & Wolf, D. (1997). New cis-eudesm-6-ene derivatives from vetiver oil. *Liebigs Annalen, 1997*(8), 1783–1787.
57. Andersen, N. H. (1970). The structures of zizanol and vetiselinenol. *Tetrahedron Letters, 11*(21), 1755–1758.
58. Naegeli, P., & Kaiser, R. (1972). A new synthetic approach to the acorane-, daucane-and cedrane skeleton. *Tetrahedron Letters, 13*(20), 2013–2016.

59. Paknikar, S. K., Bhatwadekar, S. V., & Chakravarti, K. K. (1975). Biogenetically significant components of vetiver oil: Occurrence of (−) X-funebrene and related compounds. *Tetrahedron Letters, 16*(34), 2973–2976.
60. Zalkow, L. H., & Glower, M. G., Jr. (1975). The absolute configuration of a vetiver acoradiene. The conversion of carotol to acoradienes. *Tetrahedron Letters, 16*(1), 75–78.
61. Filippi, J. J. (2014). Norsesquiterpenes as markers of overheating in Indonesian vetiver oil. *Flavour and Fragrance Journal, 29*(3), 137–142.
62. Shibamoto, T., & Nishimura, O. (1982). Isolation and identification of phenols in oil of vetiver. *Phytochemistry, 21*(3), 793.
63. Clery, R. A., Hammond, C. J., & Wright, A. C. (2005). Nitrogen compounds from Haitian vetiver oil. *Journal of Essential Oil Research, 17*(6), 591–592.
64. Mollik, M. A. H. (2013). Combination of trichosanthes cucumerina L. compounds: An analysis for novel effects of anticancer cell activites as probes for pharmacological studies. *Journal for Immunotherapy of Cancer, 1*(1), 1–1.
65. Sharifi-Rad, M., Fokou, P. V. T., Sharopov, F., Martorell, M., Ademiluyi, A. O., Rajkovic, J., et al. (2018). Antiulcer agents: From plant extracts to phytochemicals in healing promotion. *Molecules, 23*(7), 1751.
66. Silva, N. C. C., & Fernandes Júnior, A. J. J. O. V. A. (2010). Biological properties of medicinal plants: A review of their antimicrobial activity. *Journal of Venomous Animals and Toxins Including Tropical Diseases, 16*, 402–413.
67. Pareek, A. R. C. H. A. N. A., & Kumar, A. S. H. W. A. N. I. (2013). Ethnobotanical and pharmaceutical uses of Vetiveria zizanioides (Linn) Nash: a medicinal plant of Rajasthan. *International Journal of Pharmacy and Life Sciences, 50*, L12–L18.
68. Ramírez-Rueda, R. Y., Marinho, J., & Salvador, M. J. (2019). Bioguided identification of antimicrobial compounds from Chrysopogon zizaniodes (L.) Roberty root essential oil. *Future Microbiology, 14*(14), 1179–1189.
69. Hammer, K. A., Carson, C. F., & Riley, T. V. (1999). Antimicrobial activity of essential oils and other plant extracts. *Journal of Applied Microbiology, 86*(6), 985–990.
70. David, A., Wang, F., Sun, X., Li, H., Lin, J., Li, P., & Deng, G. (2019). Chemical composition, antioxidant, and antimicrobial activities of Vetiveria zizanioides (L.) Nash essential oil extracted by carbon dioxide expanded ethanol. *Molecules, 24*(10), 1897.
71. Jayashree, S., Rathinamala, J., & Lakshmanaperumalsamy, P. (2011a). Antimicrobial activity of Vetiveria zizanoides against some pathogenic bacteria and fungi. *Medicinal Plants-International Journal of Phytomedicines and Related Industries, 3*(2), 151–156.
72. Luqman, S., Srivastava, S., Darokar, M. P., & Khanuja, S. P. (2005). Detection of antibacterial activity in spent roots of two genotypes of aromatic grass Vetiveria zizanioides. *Pharmaceutical Biology, 43*(8), 732–736.
73. Saikia, D., Parveen, S., Gupta, V. K., & Luqman, S. (2012). Anti-tuberculosis activity of Indian grass KHUS (Vetiveria zizanioides L. Nash). *Complementary Therapies in Medicine, 20*(6), 434–436.
74. Gupta, R., Sharma, K. K., Afzal, M., Damanhouri, Z. A., Ali, B., Kaur, R., et al. (2013). Anticonvulsant activity of ethanol extracts of Vetiveria zizanioides roots in experimental mice. *Pharmaceutical Biology, 51*(12), 1521–1524.
75. Luqman, S., Kumar, R., Kaushik, S., Srivastava, S., Darokar, M. P., & Khanuja, S. P. (2009). Antioxidant potential of the root of Vetiveria zizanioides (L.) Nash. *Indian Journal of Biochemistry and Biophysics, 46*(1), 122–125.
76. Rajasekhar, C. H., Kokila, B. N., & Rakesh, R. B. (2014). Potentiating effect of vetiveria zizanioides root extract and essential oil on phenobarbital induced sedation-hypnosis in swiss albino mice. *International Journal of Experimental Pharmacology, 4*, 89–93.
77. Chou, S. T., Lai, C. P., Lin, C. C., & Shih, Y. (2012). Study of the chemical composition, antioxidant activity and anti-inflammatory activity of essential oil from Vetiveria zizanioides. *Food Chemistry, 134*(1), 262–268.

78. Kim, H. J., Chen, F., Wang, X., Chung, H. Y., & Jin, Z. (2005a). Evaluation of antioxidant activity of vetiver (Vetiveria zizanioides L.) oil and identification of its antioxidant constituents. *Journal of Agricultural and Food Chemistry, 53*(20), 7691–7695.
79. Brüne, B., von Knethen, A., & Sandau, K. B. (1998). Nitric oxide and its role in apoptosis. *European Journal of Pharmacology, 351*(3), 261–272.
80. Powers, C. N., Osier, J. L., McFeeters, R. L., Brazell, C. B., Olsen, E. L., Moriarity, D. M., et al. (2018). Antifungal and cytotoxic activities of sixty commercially-available essential oils. *Molecules, 23*(7), 1549.
81. Aron, P. M., & Kennedy, J. A. (2008). Flavan-3-ols: Nature, occurrence and biological activity. *Molecular Nutrition & Food Research, 52*(1), 79–104.
82. Tapiero, H., Tew, K. D., Ba, G. N., & Mathe, G. (2002). Polyphenols: Do they play a role in the prevention of human pathologies? *Biomedicine & Pharmacotherapy 56*(4), 200–207.
83. Chitra, T., Jayashree, S., & Rathinamala, J. (2014). Evaluation of anticancer activity of Vetiveria zizanioides against human breast cancer. *International Journal of Pharmaceutical Sciences, 6*, 164–166.
84. Elzaawely, A. A., Xuan, T. D., & Tawata, S. (2005). Antioxidant and antibacterial activities of Rumex japonicus H OUTT. Aerial Parts. *Biological and Pharmaceutical Bulletin, 28*(12), 2225–2230.
85. Villarama, C. D., & Maibach, H. I. (2005). Glutathione as a depigmenting agent: An overview. *International Journal of Cosmetic Science, 27*(3), 147–153.
86. Tepe, B., Akpulat, H. A., & Sokmen, M. (2011). Evaluation of the chemical composition and antioxidant activity of the essential oils of Peucedanum longifolium (Waldst. & Kit.) and P. palimbioides (Boiss.). *Records of Natural Products, 5*(2), 108.
87. Tarai, D. K., Nayak, S., & Karan, S. (2010). In vitro free radical scavenging activity of Vetiveria zizanioides. *Journal of Pharmacy Research, 3*(4), 681–683.
88. Al-Kharusi, N., Babiker, H. A., Al-Salam, S., Waly, M. I., Nemmar, A., Al-Lawati, I., et al. (2013). Ellagic acid protects against cisplatin-induced nephrotoxicity in rats: a dose-dependent study. *European Review for Medical and Pharmacological Sciences, 17*(3), 299–310.
89. Chirino, Y. I., & Pedraza-Chaverri, J. (2009). Role of oxidative and nitrosative stress in cisplatin-induced nephrotoxicity. *Experimental and Toxicologic Pathology, 61*(3), 223–242.
90. An, Y., Xin, H., Yan, W., & Zhou, X. (2011). Amelioration of cisplatin-induced nephrotoxicity by pravastatin in mice. *Experimental and Toxicologic Pathology, 63*(3), 215–219.
91. Rao, R. C., Gal, C. S. L., Granger, I., Gleye, J., Augereau, J. M., & Bessibes, C. (1994). Khusimol, a non-peptide ligand for vasopressin V1a receptors. *Journal of Natural Products, 57*(10), 1329–1335.
92. Thubthimthed, S., Thisayakorn, K., Rerk-am, U., Tangstirapakdee, S., & Suntorntanasat, T. (2003, October). Vetiver oil and its sedative effect. In *The 3rd international vetiver conference, Guangzhou* (492–494).
93. Dikshit, A. (1984). Antifungal action of some essential oils against animal pathogens. *Fitoterapia, 55*, 171–176.
94. Kaushal, S., & Chahal, K. K. (2008). Schiff bases of khusilal: Synthesis and their antifungal activity. *Pestology, 32*, 47–49.
95. Sharma, P. K., Raina, A. P., & Dureja, P. (2009). Evaluation of the antifungal and phytotoxic effects of various essential oils against Sclerotium rolfsii (Sacc) and Rhizoctonia bataticola (Taub). *Archives of Phytopathology and Plant Protection, 42*(1), 65–72.
96. Dubey, N., Raghav, C. S., Gupta, R. L., & Chhonkar, S. S. (2010a). Chemical composition and antifungal activity of vetiver oil of North and South India against Rhizoctonia solani. *Pesticide Research Journal, 22*, 63–67.
97. Sangeetha, D., & Stella, D. (2012). Screening of antimicrobial activity of vetiver extracts against certain pathogenic microorganisms. *International Journal of Pharmacy and Biological Sciences, 3*(1), 197–203.
98. Jain, S. C., Nowicki, S., Eisner, T., & Meinwald, J. (1982). Insect repellents from vetiver oil: I. Zizanal and epizizanal. *Tetrahedron Letters, 23*(45), 4639–4642.

99. Karintanyakit, P., & Babpraserth, C. (1996). Vegetables pest management by using essential oil from vetiver grass (Vetiveria zizanioides Nash). In *Vetiver: A Miracle Grass, Chiang Rai (Thailand)* (pp. 4–8).
100. Ndemah, R., Gounou, S., & Schulthess, F. (2002). The role of wild grasses in the management of lepidopterous stem-borers on maize in the humid tropics of western Africa. *Bulletin of Entomological Research, 92*(6), 507–519.
101. Chomchalow, N., Lekskul, S., Pichitakul, N., & Wasuwat, S. (1970). Researches on essential oils at ASRCT. *ASST Newsletters, 3*(5–6), 49–63.
102. Zhu, B. C., Henderson, G., Chen, F., Maistrello, L., & Laine, R. A. (2001). Nootkatone is a repellent for Formosan subterranean termite (Coptotermes formosanus). *Journal of Chemical Ecology, 27*(3), 523–531.
103. Nix, K. E., Henderson, G., & Laine, R. A. (2003). Field evaluation of nootkatone and tetrahydronootkatone as wood treatments against Coptotermes formosanus. *Sociobiology, 42*(2), 413.
104. Maistrello, L., Henderson, G., & Laine, R. A. (2001). Efficacy of vetiver oil and nootkatone as soil barriers against Formosan subterranean termite (Isoptera: Rhinotermitidae). *Journal of Economic Entomology, 94*(6), 1532–1537.
105. Maistrello, L., Henderson, G., & Laine, R. A. (2003). Comparative effects of vetiver oil, nootkatone and disodium octaborate tetrahydrate on Coptotermes formosanus and its symbiotic fauna. *Pest Management Science: Formerly Pesticide Science, 59*(1), 58–68.
106. Ibrahim, S. A., Henderson, G., Zhu, B. C., Fei, H., & Laine, R. A. (2004). Toxicity and behavioral effects of nootkatone, 1, 10-dihydronootkatone, and tetrahydronootkatone to the formosan subterranean termite (Isoptera: Rhinotermitidae). *Journal of Economic Entomology, 97*(1), 102–111.
107. Sujatha, S. (2010). Essential oil and its insecticidal activity of medicinal aromatic plant Vetiveria zizanioides (L.) against the red flour beetle Tribolium castaneum (Herbst). *Asian Journal of Agricultural Sciences, 2*(3), 84–88.
108. Pangnakorn, U. (2009). Efficiency of vetiver grass extracts against Cowpea Weevil (Callosobruchus maculatus Fabr.). *American-Eurasian Journal of Agricultural & Environmental Sciences, 6*, 356–359.
109. Aarthi, N., & Murugan, K. (2010). Larvicidal and repellent activity of Vetiveria zizanioides L, Ocimum basilicum Linn and the microbial pesticide spinosad against malarial vector, Anopheles stephensi Liston (Insecta: Diptera: Culicidae). *Journal of Biopesticides, 3*(1), 199–204.
110. Vimala, Y., Anuj, K. A., & Gupta, M. K. (2005). Physico-chemical interpretation of allelopathic interaction of vetiver with two non-edible oil yielding fence plants. *Journal of Experimental Botany, 2*, 141–150.
111. Bernauer, U., Bodin, L., Chaudhry, Q., Coenraads, P., Dusinska, M., Ezendam, J., et al. (2019). Opinion of the scientific committee on consumer safety (SCCS)–final opinion on the safety of fragrance ingredient acetylated vetiver oil (AVO)-(Vetiveria zizanioides root extract acetylated)-submission III. *Regulatory Toxicology and Pharmacology, 107*, 104389.
112. Bilici, M., Efe, H., Köroğlu, M. A., Uydu, H. A., Bekaroğlu, M., & Değer, O. (2001). Antioxidative enzyme activities and lipid peroxidation in major depression: Alterations by antidepressant treatments. *Journal of Affective Disorders, 64*(1), 43–51.
113. Karan, S. K., Pal, D., Mishra, S. K., & Mondal, A. (2013). Antihyperglycaemic effect of Vetiveria zizanioides (L.) Nash root extract in alloxan induced diabetic rats. *Asian Journal of Chemistry, 25*(3), 1555.
114. Ben-Shaul, V., Lomnitski, L., Nyska, A., Zurovsky, Y., Bergman, M., & Grossman, S. (2001). The effect of natural antioxidants, NAO and apocynin, on oxidative stress in the rat heart following LPS challenge. *Toxicology Letters, 123*(1), 1–10.
115. Xin, C., Xiaojing, L., Jiao, W., Haofu, D., & Wenquan, W. (2010). Study on chemical composition and antifungal activity in volatile oil of Alpinia oxyphylla Miq fruits. *Chinese Agricultural Science Bulletin, 26*, 366–371.

116. Munoz-Munoz, J. L., García-Molina, F., Varón, R., Tudela, J., García-Cánovas, F., & Rodríguez-López, J. N. (2009). Generation of hydrogen peroxide in the melanin biosynthesis pathway. *Biochimica et Biophysica Acta (BBA)-Proteins and Proteomics, 1794*(7), 1017–1029.
117. Liu, G. S., Peshavariya, H., Higuchi, M., Brewer, A. C., Chang, C. W., Chan, E. C., & Dusting, G. J. (2012). Microphthalmia-associated transcription factor modulates expression of NADPH oxidase type 4: a negative regulator of melanogenesis. *Free Radical Biology and Medicine, 52*(9), 1835–1843.
118. Panich, U., Tangsupa-a-nan, V., Onkoksoong, T., Kongtaphan, K., Kasetsinsombat, K., Akarasereenont, P., & Wongkajornsilp, A. (2011). Inhibition of UVA-mediated melanogenesis by ascorbic acid through modulation of antioxidant defense and nitric oxide system. *Archives of Pharmacal Research, 34*(5), 811–820.
119. Khater, H. F. (2012). Prospects of botanical biopesticides in insect pest management. *Pharmacologia, 3*(12), 641–656.
120. Khater, H. F., & Geden, C. J. (2018). Potential of essential oils to prevent fly strike and their effects on the longevity of adult Lucilia sericata. *Journal of Vector Ecology, 43*(2), 261–270.
121. Khater, H. F., & Geden, C. J. (2019). Efficacy and repellency of some essential oils and their blends against larval and adult house flies, Musca domestica L.(Diptera: Muscidae). *Journal of Vector Ecology, 44*(2), 256–263.
122. Murugan, K., Priyanka, V., Dinesh, D., Madhiyazhagan, P., Panneerselvam, C., Subramaniam, J., et al. (2015). Predation by Asian bullfrog tadpoles, Hoplobatrachus tigerinus, against the dengue vector, Aedes aegypti, in an aquatic environment treated with mosquitocidal nanoparticles. *Parasitology Research, 114*(10), 3601–3610.
123. Roni, M., Murugan, K., Panneerselvam, C., Subramaniam, J., Nicoletti, M., Madhiyazhagan, P., et al. (2015). Characterization and biotoxicity of Hypnea musciformis-synthesized silver nanoparticles as potential eco-friendly control tool against Aedes aegypti and Plutella xylostella. *Ecotoxicology and Environmental Safety, 121*, 31–38.
124. Saiyudthong, S., Pongmayteegul, S., Marsden, C. A., & Phansuwan-Pujito, P. (2015). Anxiety-like behaviour and c-fos expression in rats that inhaled vetiver essential oil. *Natural Product Research, 29*(22), 2141–2144.
125. Cheaha, D., Issuriya, A., Manor, R., Kwangjai, J., Rujiralai, T., & Kumarnsit, E. (2016). Modification of sleep-waking and electroencephalogram induced by vetiver essential oil inhalation. *Journal of Intercultural Ethnopharmacology, 5*(1), 72.
126. Nirwane, A. M., Gupta, P. V., Shet, J. H., & Patil, S. B. (2015). Anxiolytic and nootropic activity of Vetiveria zizanioides roots in mice. *Journal of Ayurveda and Integrative Medicine, 6*(3), 158.
127. Velmurugan, C., Shajahan, S. K., Kumar, A., Kumar, V., & Thomas, S. (2014). Memory and learning enhancing activity of different extracts of roots of Vetiveria zizanioides. *International Journal of Novel Trends in Pharmaceutical Sciences, 4*(6), 174–182.
128. Bizzo, H. R., Hovell, A. M. C., & de Rezende, C. M. (2009). Essential oils in Brazil: General aspects, production and perspective. *Quim Nova, 32*(3), 588–594.
129. Facey, P. C., Porter, R. B., Reese, P. B., & Williams, L. A. (2005). Biological activity and chemical composition of the essential oil from Jamaican Hyptis verticillata Jacq. *Journal of Agricultural and Food Chemistry, 53*(12), 4774–4777.
130. Williams, L. A. (1993). Adverse effects of extracts of artocarpus altilis park, and azadirachta indica (A. juss) on the reproductive physiology of the adult female tick, boophilus microplus (canest.). *Invertebrate Reproduction & Development, 23*(2–3), 159–164.
131. WILLIAMS, L. A., GARDNER, M. T., SINGH, P. D., THE, T. L., FLETCHER, C. K., CALED-WILLIAMS, L. I. S. A., & KRAUS, W. (1997). Mode of action studies of the acaricidal agent, epingaione. *Invertebrate Reproduction & Development, 31*(1–3), 231–236.
132. Loizzo, M. R., Tundis, R., Conforti, F., Menichini, F., Bonesi, M., Nadjafi, F., et al. (2010). Salvia leriifolia Benth (Lamiaceae) extract demonstrates in vitro antioxidant properties and cholinesterase inhibitory activity. *Nutrition Research, 30*(12), 823–830.

133. Ribeiro, V. L. S., Vanzella, C., dos Santos Moysés, F., Dos Santos, J. C., Martins, J. R. S., von Poser, G. L., & Siqueira, I. R. (2012). Effect of Calea serrata less. N-hexane extract on acetylcholinesterase of larvae ticks and brain Wistar rats. *Veterinary Parasitology, 189*(2–4), 322–326.
134. Ribeiro, V. L. S., Avancini, C., Gonçalves, K., Toigo, E., & von Poser, G. (2008). Acaricidal activity of Calea serrata (Asteraceae) on Boophilus microplus and Rhipicephalus sanguineus. *Veterinary Parasitology, 151*(2–4), 351–354.
135. Clemente, M. A., de Oliveira Monteiro, C. M., Scoralik, M. G., Gomes, F. T., de Azevedo Prata, M. C., & Daemon, E. (2010). Acaricidal activity of the essential oils from Eucalyptus citriodora and Cymbopogon nardus on larvae of Amblyomma cajennense (Acari: Ixodidae) and Anocentor nitens (Acari: Ixodidae). *Parasitology Research, 107*(4), 987–992.
136. Arrigoni-Blank, M. D. F., Santos, A. V., & Blank, A. F. (2011). Direct organogenesis and acclimatization of patchouli. *Horticultura Brasileira, 29*, 145–150.
137. Barros, G. C., Tresvenzol, L. M., Cunha, L. C., Ferri, P. H., Paula, J. R., & Bara, M. T. F. (2009). Composição química, atividade antibacteriana e avaliação da toxicidade aguda de Vetiveria zizanoides L. Nash (Poaceae). *Latin American Journal of Pharmacy, 28*, 531–537.
138. Rattan, R. S. (2010). Mechanism of action of insecticidal secondary metabolites of plant origin. *Crop Protection, 29*(9), 913–920.

Chapter 27
Isabgol

Zainab Maqbool, Zubaida Yousaf, Arusa Aftab, Zainab Shahzadi, and Umar Farooq Gohar

27.1 Introduction

Isabgol is an annual herbaceous plant of the Plantaginaceae family. Since ancient times, isabgol has been used to treat a variety of illnesses, including pain, injury, and cancer. The Isabgol seed's epidermal coat, known as the husk, is a particularly strong component when used medicinally. Isabgol is a natural purgative made up of beneficial natural mucilaginous chemicals. Isabgol is a thickening agent used in the pharmaceutical sector. Diarrhea, constipation, and dysentery are all conditions it is traditionally used to treat.

Mucilage, a polysaccharide-rich gel that forms when Plantago seeds are moistened, is useful as a food supplement and can bulk out dietary fibre. Plantago ovata seeds, the only commercially important variety of Plantago, are ground into the dry husk layer that produces mucilage, known as psyllium, while the remaining inner seed tissues are either discarded or utilised as low-quality animal feed. Significant interspecies variations in mucilage yield and macromolecular characteristics, according to research, are mostly the result of variations in heteroxylan and pectin composition and likely reflect a wide range of hydrocolloid functioning that can be used in industry.

Z. Maqbool · Z. Yousaf (✉) · A. Aftab (✉) · Z. Shahzadi
Department of Botany, Lahore College for Women University, Lahore, Pakistan
e-mail: Zubaida.yousaf@lcwu.edu.pk; Arusa.Aftab@lcwu.edu.pk

U. F. Gohar
Institute of Industrial Biotechnology, Government College University, Lahore, 54000, Pakistan

M. Zia-Ul-Haq et al. (eds.), *Essentials of Medicinal and Aromatic Crops*,
https://doi.org/10.1007/978-3-031-35403-8_27

Plant Profile	
Isabgol	
Kingdom:	Plantae
Family:	Plantaginaceae
Genus:	Plantago
Species:	*P. ovata*
Binomial name:	*Plantago ovata* Forssk.
English/Common name	

Ispaghula, Isabgol, Indian Plantago, Psyllium, Sand Plantain, Flea Seed, Isabgula, spogel seed, Snigdhabijab, etc.

Approximately 200 species of Plantaginaceae family are under the genus of *Plantago*. Most of them are predominantly cross pollinated in nature. Central Asia is considered as the center of diversity of *Plantago*, some species have now spread in temperate regions of the world. *Plantago* species are herbaceous, mostly grown as weeds. Because of their diversive properties *Plantago* species are used in traditional medicinal system and modern medicine system (Table 27.1). *Plantago ovata* and *P. psyllium* L. are economically most important species. Under the genus Plantago, there are about 200 species of plants belonging to the Plantaginaceae family. In nature, the majority of them are cross-pollinated. Both are farmed primarily for their mucilaginous seed husks, which are the most potent laxatives in terms of pharmacology. Isabgol is significant because its seeds and husk are traditionally used as medicines all over the world. Isabgol is a plant with enormous, narrow leaves and no stems (7–20 cm long and 0.6 cm broad). The taproot is fully grown and has a few thin, fibrous secondary roots. After 60 days from the time of planting, flowering shoots appear from the plant's base. Flowers are bracteate, 12.5–37.5 mm long, and white in colour. Isabgol has ripened when its flower spikes have turned reddish brown. Upper leaves of the plant turn yellow during the ripening period, and lower leaves become dried up. The 1.2–4 cm long and 0.5 cm wide spikes have a cylindrical to ovoid shape and can support 69–45 blooms. The blooms are tetramerous anemophilous, protogynous, and hermaphrodite. The fruit of Isabgol is an 8 mm long, boat-shaped, spheroid capsule with smooth, white-rosy seeds. The seeds' epidermis is composed of polyhedral cells with mucilage depositing on their walls to make the cells' walls thicker. At maturity, the fruits start to open. Fruit and seed husk are separated during milling [1].

The phylogenetic links of the distinct group of plants known as Plantago are yet unknown. Plantago is a tribasic genus, with four, five, and six as its three base numbers. Six is regarded as the basic base number among these, and the other numbers are secondary derivatives. The study was unable to locate a sister group for this monophyletic family based on morphological, embryological, or chemical evidence.

Since ages Isabgol has been used in medicines, now a day it is cultivated as a medicinal plant. Seeds of husk consists of mucilage, fatty oil, albuminous matter, inactive glucoside (Aucubin) and plantinose sugar. Isabgol seeds husk has the potential to stop diarrhea as it can absorb and retain water. It is a laxative, reduces

Table 27.1 Genus *Plantago* applications in traditional medicine system and modern medicine system

System	Traditional medicinal system			Modern system of medicine
	Part of plant	Mode of application	Ailment	
Sensory/ neurologic	Leaves	Food with lentil Plaster on Forehead	Epilepsy	Effects *Gamma-aminobutyric acid* (GABA) system Treat seizure
Urinogenital	Seed, leaves	Vaginal cleaning Plaster on the pubic bone	Menorrhagia (excess menstrual bleeding)	In the presence of E, phytoextracts demonstrated that [H]-17-estradiol from both receptor subtypes boosted the expression of all pS2, PR, and PTGES mRNAs Indicating an estrogen agonist effect
	Leaves, seed	Used in Vaginal washing	Uterus pain due to congestion	Tonus-raising reaction on the uterus
	Seed, Root, leaves	Oral infusion	Uterine congestion Obstruction in Kidney's ducts obstruction	Diuretic effects of Iridoid (monoterpenoids) compounds
	Seed, Root and leaves	Oral infusion	Erythrocyturia	Kidney stones can be cured by using iridoid glycosides to dissolve the stone
	Seed and leaves	Oral decoction, ointment	Bladder pain	Reliefs from pain by hindering the synthesis of prostaglandins
	Leaves	Oral extract	Dysuria	Catalpol and Aucubin exhibited diuretic effect
Ophthalmic	Leaves	Eye ointment, Liniment	Conjunctivitis	Plants have the ability to combat the herpes simplex viruses (HSV-1 and HSV-2) and adenoviruses that cause conjunctivitis (ADV-3, ADV-8 and ADV-11)
ENT2	Leaves	Nasal drop	Epistaxis	Not found
	Leaves	Ear drop	Otitis	Prostaglandin synthesis is inhibited to produce analgesic effects
	Leaves, Root	Mouth wash	Canker sores, Mouth ulcers, toothache, Gingivitis	Antifungal, oral antimicrobial, and antinociceptive properties, as well as analgesic and anti-inflammatory actions. Ehrlich ascites cancer, (tannin)umor
	Root	Hanging	Scrofula	Not found
	Leaves	Gargle	Tonsillitis	Anti-inflammatory potential due to presence of flavonoid and tannin

(continued)

Table 27.1 (continued)

System	Traditional medicinal system			Modern system of medicine
	Part of plant	Mode of application	Ailment	
Pulmonary	Leaves	foodstuff with lentil	Asthma	Mast cell density, alveolar epithelium consistency, and glycoprotein proliferation in airways all increase
	Leaves	Oral extract, plaster	Tuberculosis	Effect on bacteria (*Mycobacterium tuberculosis*) in agar-dilution (Minimum Inhibition Concentration of extract greater than 1 mg/ml)
	Leaves	foodstuff with lentil	Asthma	Increase number of mast cells, thickness of alveolar epithelium and accumulation of glycoprotein in airways
Gastrointestinal		Oral extract	Hematemesis	Not found
	Leaves	Oral extract	Hemorrhoid	Cure anus and rectum diseases by 10% plantain ointment
	Seed, leaves	Oral extract, rectal clyster	Gastroenteritis, Dysentery	Reducing the average ulcer index, Cure diarrhea
	Leaves	Oral extract	Tonic in liver and spleen, Obstruction in liver ducts	Exhibit hepatoprotective activity
	Leaves	food with lentil	Ascites	hindering effect against (ascites tumor) Ehrlich ascites carcinoma
Dermatologic	Leaves	Plaster	Deep wound	Alkaloids in the herb can heal incisions
	Leaves	Plaster	Burn by fire	Reepithelialization (cure wound) and good granulation tissue organization
	Leaves	Plaster	Carbuncle, Eczema, Herpes, Urticaria	Aqueous extract showed anti-allergic effect and anti-inflammatory potential (cAMP4, PDEI5)
	Leaves	Plaster by salt	Rabid dog's bite	Not found
	Leaves	Plaster	Chronic furunculosis (disease of hair follicle)	According to German commission E it was confirmed
	Leaves	Plaster	Deep wound	Isabgol exhibit Neoepithelium and skin appendages formation. The herb's alkaloids increase the tensile strength of the scar tissue, which aids in the healing of incisions

Keynote: Minimum inhibitory concentration (MIC), cyclic adenosine monophosphate (cAMP), phosphodiesterase inhibitor, gamma-aminobutyric acid (GABA), ENT (PDEI) kidney and bladder ailments, syphilis, arthritis, and hemorrhoids. It is diuretic and mainly used as a dietary fiber. Mild to moderate hypercholesterolemia and blood glucose level can be reduced by the use of soluble fiber cereals in diet. Isabgol husk has a soluble content that is almost eight times higher than oat bran's. The Isabgol plant's nutritional fibres can be used to make low-calorie foods and have medicinal potential [2].

27.2 Agronomy (Like Soil Conditions, Climate, Land Preparation, Planting, Manuring, Irrigation, Ecology Etc.)

27.2.1 Soil Conditions and Climate

Isabgol is a 119–130 day annual herbaceous crop with short stems (10–45 cm). It is a member of the Plantaginaceae family. It is indigenous to West Asia and the Mediterranean region. It thrives in dry, chilly climates. It thrives in light, sandy loam soils that have good drainage. This crop should not be grown in soil with inadequate drainage. With a pH of 7.2–7.9, it can withstand mild levels of soil salinity. Isabgol is a crop of Rabi season. The paper like seeds can suck water, swell, and drop off as a result of the weight gain. The biggest dangers to Isabgol agriculture include unseasonal rain, excessive humidity, and significant dew deposition, which can result in a complete loss of yield. It cannot be cultivated in places that experience winter rain. *P. ovata* needs relatively little in the way of nutrients. The plant needs little nitrogen, which is easily provided by growing a leguminous crop before growing psyllium. The swelling factor of seeds is reduced when the nitrogen content is enhanced from 0 to 50 kg/ha. Increased nitrogen application was found to boost seed yield. As a result, the basal dose is typically 25 kg/ha of nitrogen and 25 kg/ha of phosphorus [1].

27.2.2 Propagation

Isabgol is propagated through seeds. Use high standard seeds free from diseases gathered during the previous season [1]. Dormancy was observed in freshly harvested seeds of *Plantago,* dormancy can be overcome by placing at lower storage and by a period of dry storage. No significant difference in percentage of seeds grown in light and dark conditions has been observed [2].

Freshly harvested seeds when dipped in water the germination rate was observed as 17%. The germination rate of freshly harvested seeds can be enhanced to 100% when seeds were soaked in 1000 ppm potassium gibberellate consisting of 0.5% glucose, or weakening the seeds with fine sand paper. The seed dormancy of freshly harvested seeds could be broken down by aging or by cultivating seeds at a temperature of 5–15 °C. The seeds stored for over 2 years are not viable [3]. Seed dormancy is the barrier in effective propagation of Isabgol. In a study seed germination and breakage of dormancy was studied. Different methods to break the seed dormancy and their effects on seed yield were observed. KNO_3, acetone, alcohol, gibberellic acid (GA) and prechilling techniques were used. It was observed that KNO_3, acetone, alcohol and prechilling techniques had no positive effect on seed dormancy. Gibberellic acid application was observed to have significant effect in breaking dormant seeds. GA also increases the seed germination rate and seedling growth [3]. In a lab and greenhouse experiment, the rate of Isabgol seed germination and emergence in response to salinity stress (NaCl), temperature, and planting depth was noted. The evaluations for the ideal, ceiling, and base temperatures were 3.35, 21.24, and 35.04 °C. Isabgol seed is very resilient to low water potential and high salinity stress, or sensitive to it. The germination of Isabgol seeds was unaffected by a salt stress of 200 mM. With rising salt concentration, a decrease in germination rate was observed. Salinity levels of 328 mM and −1.24 MPa were necessary for 50% suppression of maximal germination, respectively. At pH = 4–6, maximum seed germination (>93%) was seen, while at pH = 7–9, it fell to 52–58%. Maximum seedling growth crop up when the seeds were planted on the soil surface and reduced with increasing the depth of planting; no seed emerged from depth of 3 cm.

27.3 Land Preparation and Planting

The fine tilth field free from clods and weed, is good for germination. The field is well plowed to make the soil in fine condition. Field is subdivided into different plots according to soil types and slopes for irrigation. A plot size of 8–12 m × 3 m is convenient for light soil [3].

27.4 Manuring and Irrigation

Isabgol seed needed irrigation as soon as it was sown. After a week, Isabgol typically begins to germinate. A second irrigation was carried out 3–4 weeks after sowing, and a third irrigation was carried out when spikes began to form. The Isabgol is watered 8–10 times during its growth phase [4]. N, P_2O_5, and K_2O are effective inorganic fertilisers for the Isabgol crop. It needs 20:10:12 kg/acre of N, P_2O_5 and K_2. According to reports, the growth and productivity of isabgol are significantly impacted by animal excrement. Application of animal dung has been reported to generate 5 tonnes per acre [4, 5].

27.4.1 Ecology

Isabgol is a winter annual plant. It is indigenous to Europe and Asia. It has naturalized throughout the world in temperate and tropical regions of the world. In the winter, it often grows in dry and semi-arid regions of the world [6].

The genetic diversity and performance of eight Isabgol genotypes (PO-001, PO-002, PO-003, PO-004, PO-005, PO-006, PO-007, and PO-008) were tested in an experiment. To assess the performance, morphological, physiological, and yield aspects were looked at. The germplasm came from different parts of Bangladesh. During the Rabi season in 2019–2020, an experiment was conducted at the Regional Spices Research Center in Magura. Three duplicates of an experiment with a random complete block design were set up. Plant height, tillers per plant, leaves per plant, length of a spike, weight of seeds, and seed yield were all highest in germplasm PO-001, whereas they were all lowest in genotype PO-007 [7].

27.4.2 Pests and Diseases

White grub is reported to be most common pest found in Isabgol cultivation. Downy mildew, powdery mildew, damping off seedlings and rhizoctonia wilt are the most common issues with Isabgol crop. It is best to cultivate in areas free of illness and insect pests. It is recommended to use cultural, biological, and mechanical strategies to control insect pests and illnesses in medicinal crops. It is preferred to apply biological pesticides on seeds. Chemical/synthetic pesticides should only be used if there is no other option or if there is enough time between application and harvest to ensure that the chemical cannot be detected in the finished product [3].

Three important fungus diseases include reportedly wilt, damping off, and downy mildew. Downy mildew, which *Peronospora plantaginis* induces when spikes first appear, is the most dangerous of all. Two symptoms are ashy-white, frost-like mycelial development on the underside of the leaf and chlorotic spots on the upper leaf surface. The problem can be effectively treated with Dithane M-45 or Dithane Z-78 at 2.0–2.5 g/l or Bordeaux combination 6:3:100. Psyllium wilt is caused by *Alternaria* sp., *Fusarium oxysporum*, and *F. solani*. Wilting begins in the outer leaves and eventually affects the entire plant. Silverish hues appear on the leaves. Wilt disease can be managed by treating seeds with 2.5 g/kg of Bavistin or Benlate. Pythium ultimum causes psyllium to become moist. Treatment with CGA-48 988 and fenaminosulf have been suggested as *P. ultimum* defences. Occasionally, powdery mildew infects psyllium. Karathane W.D. (0.2%) is in charge of battling the illness. Aphids typically attack the crop when it is nearly mature, causing significant losses. Spraying the aphids with Endosulfan (0.5%) or Dimethodafe (0.2%) twice weekly will control them [108].

27.5 Description of Crops (Like Countries Where Produced More, Annual Production, Products Made)

Plantago is a genus with around 200 species. Three significant species are grown in Pakistan, India, and numerous European nations: Plantago ovata, P. psyllium, and P. indica. The Isabgol is produced and exported mostly by India. About 120,000 tonnes of isabgol were manufactured in India each year [8, 9]. A byproduct of Isabgol called "Industrial Kha Kha" Powder is mostly used to stop soil erosion. Rice cakes, noodles, health drinks, and other bakery products may have acceptable levels of isabgol. Spraying the aphids with Endosulfan (0.5%) or Dimethodafe (0.2%) twice weekly will control them [10].

The mucilage powder from Isabgol offers properties that facilitate more rapid medicine suspension and absorption. The isabgol seed husk is used in the cosmetics industry. It performs two functions in the cosmetics industry: base and sizing agent. In trade, isabgol husk is occasionally referred to as Bhusi or Sat Isabgol [11]. Isabgol seeds can be sold whole or only with the husk still on. Companies that manufacture drugs and pharmaceuticals bought isabgol husk. The top international purchasers are Dr. Morepen (USA), Procter & Gamble (USA), and Al Parigo (USA). Isabgol seeds and husk exports generate more than 35 million annually.

A coagulant extract from Isabgol was used in a study to treat the bacteria and murky water. Utilizing FeCl3-indued crude extract, the coagulant was isolated from the Isabgol extract (FCE). An experiment was run to determine the content of humic acid, the amount of coagulant, the pH of the water, and the concentration of turbidity. The greatest reduction in turbidity was seen in water with a pH of 8. at the recommended FCE concentration of 0.8 mg/L. The level of humic acid in the water had a big impact on how cloudy it was. FCE also enhanced the water's bacterial purity. According to the study's findings, FCE can be applied in water treatment facilities [12].

Isabgol husk, gum katira, and their mixture's rheological characteristics were investigated. Different media, including pure water, phosphate buffer (pH 7.0), and 0.1 N HCl, were used to study rheological properties. To ascertain their compatibility in dispersion form at 1% mass gel strength, the blending effects of isabgol and gum katira were evaluated at four distinct concentrations (00:100, 25:50, 50:50, 75:25, and 100:00). The Krigbaum and Wall equation was used to calculate the solubility of gum katira and isabgol husk. To examine further rheological characteristics, Bingham, Power, Casson, Casson chocolate, and IPC paste analyses were carried out. Isabgol husk, gum katira, and their mixes exhibited shear thinned or pseudoplastic behaviour, as determined by the power flow index "p". In all concentrations of Isabgol husk, gum katira, and their mixtures, the power flow index was calculated to be less than "1". All blends exhibited pseudoplastic behaviour under thermal settings of 298.15, 313.15, and 333.15 K and in dispersion mediums of distilled water, o.1N HCl, and phosphate buffer. According to the study's conclusions, the mixes can be applied to the production of food as well as to in vitro and in vivo medication delivery systems [13].

Isabgol is shipped to the USA and Western Europe at a rate of about 90%. There are many industrial applications for Isabgol husk [14]. It serves as the primary ingredient in many laxative preparations used in contemporary medicine that also contain sodium bicarbonate and a variety of additional tastes. It is a stabilising agent for ice creams and a component of chocolates and other food items. Isabgol husk generates jelly when combined with hot caustic soda, which is used to replace agar-agar. To make tooth cleaning powder and germicidal lubricating gels, isabgol seeds husk gum is employed. For the acidification of petroleum wells, isabgol is employed in composition [15]. Isabgol seed husk can be used alone or in conjunction with other resins to provide a coating that is water resistant when applied to explosives. When combined with gurr, isabgol seed husk are also utilized as cattle fodder. 69% of the entire seed crop is seed without husk, which is utilised as bird feed [16]. In place of agar, isabgol husk is employed as a gelling agent to culture microbial strains [17].

27.5.1 Important Chemical Constituents and Medicinal Uses

Isabgol has a high hemicellulose content and is made up of rhamnose, arabinose, and galacturonic acid units connected to a xylan backbone (arabinoxylans). Iridoids, phenols, polysaccharides, sterols, alkaloids, and cumatines are among the phytochemicals found in isabgol [10, 11]. The presence of secondary metabolites such as saponins, tannins, phenols, glycosides, flavonoids, carbohydrates, terpenoids, steroids, alkaloids, and others in isabgol has been confirmed [18]. On top of common minor components like rhamnogalacturonan-I and cellulose, plantago mucilage contains a significant amount of heteroxylan of high complexity, a form of xylan-type polysaccharide with a b-(1?4)-linked xylose backbone. A typical element of plant cell walls, xylan is also present in the seed mucilage of species that are not related [18].

Phenolics, flavonoids, terpenoids, and iridoid glycosides make up isabgol. The fruiting body of Isabgol is a well-known source of natural antioxidants and has excellent antioxidant activity. Many flavonoids have been reported from Isabgol, including scutellarein 7-glucoside, scutellarein 7-glucuronide, scutellarein, apigenin, luteolin, hispidulin, and 5,7,4′,5′-tetrahydroxyflavanone-3′-C-glucoside [18].

It is known that flavonoids play a role in other antioxidant activities as well as defensive mechanisms against cardiovascular diseases. The majority of the 6000 different kinds of flavonoids that have been found come from the Isabgol's flowers, fruits, and leaves. Different enzyme inhibitors, such as cyclo-oxygenase and lipoxygenase, react with flavonoids. Both cyclo-oxygenase and lipoxygenase are engaged in a variety of biological processes, including apoptosis, anti-ulcer, anti-proliferative, anti-hepatotoxic, and anti-microbial actions. Due to their antioxidant properties, polyphenols and flavonoids are both employed as dietary supplements or pharmaceuticals [19].

Isabgol has a significant amount of hemicellulose, which has a xylan backbone linked to rhamnose, arabinose, and arabinoxylans. It has iridoids, phenols, polysaccharides, sterols, alkaloids, cuminatives, and other phytoconstituents that are documented. Alkaloids, carbohydrates, saponins, tannins, terpenoids, phenols, glycosides, steroids, and glycosides are among the substances found in isabgol. Isbagol is composed of 85% of a single polysaccharide made up of D-xylose (62%), L-arabinose (20%), L-rhamnose (9%), and D-galactouronic acid (9%), and only 15% of non-polysaccharide components. First, Laidlaw and Percival identified the sugar molecules in Isabgol and their composition.

From the quiet mucilage, two polysaccharides units were fractionated. While the second is soluble in hot water (4000 grammes per equivalent weight Uronic acid 3%), the first is soluble in water (700 grammes per equivalent weight Uronic acid 20%). The polysaccharide has a disaccharide side chain with D-galactouronic acid coupled with O_2 of an L-rhamnose and a linear backbone of D-xylose residues in the pyranose ring. Isabgol is a naturally occurring polymer that primarily consists of polysaccharide chains with xylan systems (1–3) and (1–4). It is used to treat diabetes, ulcerative colitis, haemorrhoids, colon cancer, hypercholesterolemia, and severe constipation [20].

The metabolic and scavenging profiles of the developing Isabgol fruit were examined. The findings demonstrated that -3 and -6 fatty acids, amino-acids (primary metabolites), secondary metabolites, and antioxidants are abundant in developing Isabgol fruits. Anatomical and morphological characteristics demonstrated that the Isabgol fruit has five developmental stages. Differentially, total lipids and fatty acids were produced. With each stage of growth, saturated fatty acids rise while total polyunsaturated fatty acids fall. For the unsaturation index and degree, a bending curve was seen. By using Principle component analysis, a significant change in fatty acid profile was seen from the bud's beginning to its maturation stage. At various developmental phases, a constant level of total amino acids was seen. The research showed concurrent decreases in total phenolic and flavonoid contents.

Twenty-two distinct metabolites in all were found, and metabolic alterations were also found as the fruit developed. At the flowering stage, six metabolites were found, and two more were found at the early and mature stages of development. All developmental stages had detectable levels of apigenin and kaempferol. Time-dependent metabolomics were validated by a change in metabolite production. Fruit growth involved progressive changes in metabolite production that were coordinatedly connected to one another [19].

27.5.2 Medicinal Uses

Isabgol has been shown to be effective in the treatment of hypercholesterolemia, colon cancer, irritable bowel syndrome, inflammatory bowel disease, constipation, and diabetes [21]. There is evidence that the Isabgol has antioxidant and antibacterial properties [22]. In the Pak-Indo subcontinent, Isabgol's seeds and husk are utilised as a traditional medicine. Both are frequently employed in the treatment of diabetes, obesity, ulcerative colitis, constipation, intestinal irritant bowel syndrome, diarrhoea, high cholesterol, and laxative use [23].

Isabgol has been used in food and for health reasons, according to a study. Obesity, diabetes, and dyslipidemia are a few of the chronic conditions that can be reduced by consuming more fibre. Isabgol husk consumption can have positive health effects. Isabgol, a non-digestible carbohydrate made up of arabinose and xylose, can be classified as a useful fibre. The active component of arabinoxylans produced by arabinose and xylose is what causes viscous gel to form. Fibers that produce viscous gels are good for your health. Isabgol and water combine to make a viscous gel. Isabgol's thick gel decreases blood cholesterol levels by preventing the absorption of fat and cholesterol [24].

27.6 Constipation

Isabgol husk can alter the pharmacological effects and stomach motility [25]. Isabgol, depending on the increase in intestinal volume caused by water and the decrease in viscosity of the luminal contents, might mechanically stimulate the gut wall. Isabgol raises the volume of intestinal contents because of its bulk-producing property when enough liquid (30 mL per 1 gramme of husk) is consumed. This causes a stretch stimulation, which eases bowel movement. A lubricating layer is formed by a swollen mucilaginous mass, which facilitates the passage of intestinal contents. The moisture content of faeces is similarly increased by isabgol husk [25].

The efficiency and rate of action is Isabgol husk, lactulose and other laxative to treat simple constipation was checked in 394 patients by 65 general practitioners. 224 patients used Isabgol husk and 170 patients used other laxatives. After 4 weeks of treatment Isabgol husk was evaluated superior by the general practitioners as compared to other laxatives. Patients administered with Isabgol has a higher proportion of normal stools. Diarrhea, belly pain and gripes were less common with Isabgol [26].

27.6.1 Hypocholesterolemia

Although the particular process by which Isabgol husk lowers serum cholesterol is uncertain, it has been discovered that it does so by 5% [27]. According to in-vivo studies, isabgol husk increases the activity of cholesterol-7-hydroxylase more than cellulose or oat bran, but less than cholestyramine [28]. In animals given a high-fat diet, isabgol increases the activity of cholesterol-7-hydroxylase and HMG-CoA-reductase [29]. Isabgol and pectin both reduced Apo B secretion, according to research on mice, and Isabgol-fed animals had LDL catabolic rates that were 100 times higher. In studies on humans, it was discovered that isabgol husk decreased LDL cholesterol, decreased cholesterol absorption, and increased the metabolism of chenodeoxycholic and cholic acids [30].

According to a study, consuming 5.1 grammes of Isabgol husk twice day for 8 weeks reduces LDL levels by 5.1% and total cholesterol by 3.5% [31]. Isabgol fibres, both soluble and insoluble, have a propensity to lower total blood and LDL cholesterol, which lowers the risk of heart disease. It was discovered that an enzymatic process might produce different gelling qualities as well as absorption capacity. A diet that has the right amount of soluble fibre decreases cholesterol [32, 33].

27.6.2 Hemorrhoids

Isabgol intake to cure hemorrhoids was tested. Fifty patients with internal bleeding haemorrhoids received either 11.6 grammes of Metamucil® or a placebo of vitamin B for 40 days. Individuals in the Isabgol husk group showed a significant decrease in bleeding when compared to the control group [34].

27.6.3 Ulcerative Colitis

Through anaerobic fermentation of soluble non-starch polysaccharides from Isabgol husk, short chain fatty acids, specifically acetate, propionate, and butyrate, were generated in the gut. The epidermis of the isabgol seed is the husk, and because the isabgol seed contains a substantial quantity of fermentable fibre, its degradation rate is slower than that of pectin, leading to the production of significant amounts of butyrate and acetate [35]. Because it has anti-cancer potential, butyric acid is advised as a colonocyte's oxidative substrate. It can be beneficial in curing ulcerative colitis. In a study colorectal cancer patient were treated with Isabgol seeds, 20 gm of seeds given daily for 3 months 42% increase in butyric acid production was observed. 10 gm of Isabgol seeds is as effective as mesalazine to treat Ulcerative colitis [36].

27.6.4 Diabetes Mellitus

Isabgol hush was found to have anti-diabetic effects in 34 patients with hypercholesterolemia and type 2 diabetes. For 8 weeks, the patients received 5.1 grammes of Isabgol twice a day and a placebo. Up to 8.9% of total cholesterol and 1.0% of LDL were reduced. Additionally, a notable decrease in the postprandial rise of glucose was seen [37].

27.6.5 Colorectal Cancer

Approximately 55,000 people in the USA died cause of colorectal cancer in 1995 [38]. Dietary fiber is considered to prevent colorectal cancer. fiber can reduce the transit rate of colorectal cancer. Isabgol causes changes in microbial bile acids metabolism after fibre fermentation, a drop in pH, and the creation of short chain fatty acids in addition to decreasing bile metabolism by the stomach microflora and concocting bile acids by stool bulking [39]. Studied documented that bile acids, deoxycholate and lithocholate are responsible for promoting colorectal cancer. anti-carcinogenic activity of Isabgol has been checked in rats [40]. It was reported that the significant reduction in ratio of lithocholic and deoxycholic acid observed during cure with Isabgol husk. Isabgol acts as a substratum for colonic microflora [41]. 7 µdehydroxylase activity is decreases by fiber fermentation. Reduction in ratio of secondary bile acids reduces the hydrophobicity of the bile acids feeding back to the liver [42]. T For the production of cholic acid and chenodeoxycholic acid (primary bile acids), this reduction in ratio changes the balance between two pathways (hepatic 7-hydroxylation and blood vessel 27-hydroxylation) of cholesterol [43], leading to a decrease in the relative proportion of chenodeoxycholic acid to cholic acid being produced [44]. Lithocholic acid is produced in smaller quantities by colonic bile acids. The gut flora produces isolithocholic acid from lithocholic acid, and a decrease in lithocholic acid will result in a drop in isolithocholic acid [45]. Dietary fibres that are not digested improve the capacity of the stool to absorb water and volume, which lowers concentrations [46]. Lithocholic acid removes the protective mucous layer, promotes the rapid growth of typically non-replicating mucosal surface cells, damages DNA in human cell lines, inhibits rat mutagenesis, decreases toxifying enzyme activity, and weakens the mucosa's oxidative defence system. Patients with big adenomas and colorectal cancer have had increased levels of lithocholic acid compared to deoxycholic acid in their faeces [47–50].

27.7 Treatment of Metabolic Disorders

Zuker rats were given 3.5% Isabgol husk as food for 25 weeks as part of a study. According to the findings, this diet can help people avoid a variety of metabolic problems, including obesity, dyspipidemia, hypertension, and endothelial dysfunction. Adiponectin and TNF- plasma concentrations are decreased by isabgol intake. In hypertensive overweight adults, supplementing with isabgol fibre for 6 months considerably lowers blood pressure. By consuming less saturated fat, one can lessen their chance of developing cardiovascular disease (CVD) by lowering their serum LDL cholesterol levels. According to epidemiology, eating foods high in water-soluble fibres, such as isabgol husk, decreases serum LDL cholesterol levels without affecting triacylglycerol or HDL levels [51].

27.7.1 Pharmacokinetic Potential of Isabgol Husk

Isabgol husk absorbs water, which causes it to swell and produce mucilage. In the stomach, over 10% of the mucilage is hydrolyzed. The stomach is where free arabinose is primarily absorbed [52].

Isabgol husk mostly begins to work 12–24 h after a single intake. The Isabgol's maximum effect won't last longer than 2 or 3 days [52].

The polysaccharides are broken down by human gut flora [52].

27.7.2 Pharmaceutical Effect

Isabgol is a herb that has traditionally been used to treat inflammatory bowel disease. A study found that rats' colonic inflammation was reduced by isabgol seeds. Reformation of intestinal cytoarchitecture, notable generation of short-chain fatty acids, and a decrease in pro-inflammatory mediators are all indicators of decreased colonic inflammation [53]. Alcoholic preparations from isabgol seeds demonstrated cholinergic activity. It has been noted that sedated cats and dogs have lower blood pressure. Rat, rabbit, and guinea pig intestinal movement was induced whereas isolated and perfused hearts of frogs and rabbits were suppressed. The effect of the extract on smooth muscle is blocked by atropine. The seed oil caused a decrease in blood cholesterol levels in rabbits [54, 55].

In substitution of corn oil, linoleic acid-enriched oil made from embryos has been suggested as a dietary hypocholesterolemic agent. Administration of embryo oil as a supplement decreases serum cholesterol levels in animal trials [56]. Patients with atheromatous heart conditions who added Isabgol husk to their diets experienced a significant drop in anginal attack [57]. A glucoside found in seeds called acubin is physiologically inactive. Isabgol tannins don't have much of an impact on bacteria or entamoeba [58]. Escherichia coli can be effectively eliminated by the ethanol and acetone extract of isabgol [59]. There is evidence that isabgol lowers cholesterol. Today, the pharmaceutical industry uses seed husk primarily [60, 61].

Isabgol seeds were found to have a considerable impact on rabbits' immune systems [62]. White blood cell and spleen leukocyte counts rise in response to isabgol seeds extract, which also lowers anti-HD antibodies. Rats with type 1 and type 2 diabetes were shown to have less hyperglycemia when given an aqueous extract of isabgol seeds. This extract can be utilised to treat diabetes, as evidenced by a decrease in hyperglycemic content [63]. Isabgol seed coat lowers the risk of cardiovascular disease by raising and stabilising HDL cholesterol levels [64]. Use of Isabgol (Isabgol) and *Foeniculum vulgare* (fennel) frequently can slow the development of obesity. In rats with high-fat diet-induced obesity, the effect was measured by improvements in body mass index (BMI), dyslipidemia, hyperinsulinemia, and hyperleptinemia, decreases in lipid accumulation, and changes in glycemic status.

The Fennel and Isabgol seeds were identified by the observed results as antioxidant, anti-inflammatory, and anti-hyperlipidemic agents. By lowering hypercholesterolemia and hyperglycemia, methanolic extract of fennel and Isabgol seeds can help cure obesity [65]. It has been proven that Isabgol methanolic extracts are more effective against gram-positive bacteria [66]. Isabgol husk was given to African Green Monkeys that were tired of eating high cholesterol-containing foods at a 10% concentration. The outcomes revealed a preventative effect [67]. When given orally to patients with mild hypercholesterolemia, isabgol seeds decreased LDL cholesterol by 8% and total cholesterol by 6% [68].

After consuming isabgol-rich cereal as part of a low-fat diet, blood lipid concentration in hypercholesterolemic patients improved [69]. Aqueous dried Isabgol seed extract was shown to have weak antibacterial activity when used to treat Sterptococcus pyrogenes [70]. Patients with type II diabetes who consumed isabgol fibre before meals did not see any appreciable weight loss. When isabgol husk is consumed by mice who are genetically predisposed to diabetes, a significant amount of insulin is produced [71]. After therapy, there was a reduction in fasting plasma glucose, total cholesterol, LDL (low-density lipoprotein), and triglyceride levels, but an increase in HDL (high-density lipoprotein) cholesterol levels [72].

After eating cereal rich in isabgol as part of a low-fat diet, hypercholesterolemic patients' blood lipid concentration improved [69]. Aqueous dried Isabgol seed extract was reported to have only marginal antibacterial activity against Sterptococcus pyrogenes [70]. Before meals, type II diabetes patients who consumed isabgol fibre showed no discernible impact on their weight. When isabgol husk is consumed in the diet of genetically diabetic mice at a rate of 2.5%, a lot of insulin is produced [71]. High-density lipoprotein (HDL) cholesterol levels increased after therapy, but fasting plasma glucose, total cholesterol, LDL (low-density lipoprotein), and triglyceride levels decreased [72].

A 1:1 ratio of wheat bran and isabgol contains 18% of the recommended daily fibre intake, preventing the growth of colon tumours [77]. It has been noted that seed bran inhibits N-,ethynitroso-induced oncogenesis in the breast. Isaptent, a cervical dilator made from Isabgol seeds, has been reported to be effective in females regardless of age, parity, or gestational phase, with no obvious harm to the cervix or vaginal flora [78]. Adult human weight decreases as a result of mucilage oral ingestion [79]. The husk's mucilaginous polysaccharide was isolated, and studies of its liquid absorption, bacterial suppression, biocompatibility, promotion of macrophage irritancy, and allergenicity supported the positive clinical outcomes in wound healing [77].

Isabgol's dried seeds and husk are utilised in traditional medicine as an emollient, demulcent, and risk-free laxative (Fig. 27.1). It is employed to treat dysentery, severe diarrhoea, and constipation [79, 80]. Isabgol has been used in Asian and European herbal therapy to treat persistent constipation since the sixteenth century. The gut was not irritated by isabgol. Isabgol seeds are recommended for usage in feverish conditions as well as to cure kidney, bladder, and urethra infections since they are thought to be cooling and diuretic. The decoction of seeds is used to treat colds and coughs. Rheumatic and glandular swellings can be treated using ointment

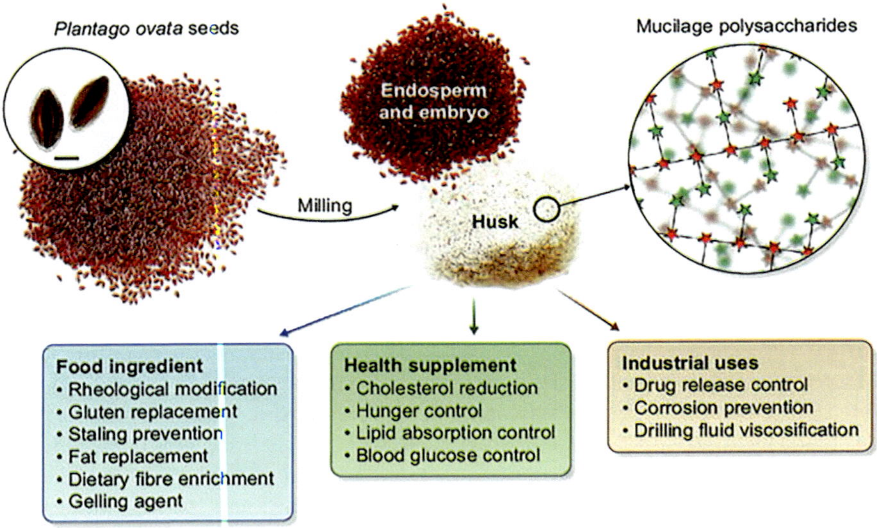

Fig. 27.1 A list of the benefits and uses of psyllium husk. Plantago ovata seeds are milled to remove the dry outer covering to create psyllium husk (Bar: 1 mm). Approximately 86% of the dry weight of the psyllium husk is made up of mucilage polysaccharides, with the remaining lignin, protein, and minerals making up the remaining 9%

produced from crushed seeds [81]. In Yemen, soaked seeds are applied topically to the hair and used as an ointment to cure boils and ulcers [82].

Isabgol seeds are claimed to be used as a therapy to treat chronic diarrhoea in children from European and local cultures when other medications have failed [83]. Isabgol seeds are made astringent and tonic by application of moderate degree heat. It is reportedly used to treat a variety of illnesses in traditional medical systems all over the world. Isabgol seeds have a positive impact on the treatment of gastritis, enteritis, diarrhoea, and other gastrointestinal conditions [84]. In India, dried seed decoction is used as a demulcent and a treatment for diarrhoea [85]. Constipation and digestive issues might be treated with seeds ointment [86]. Dry seeds are utilized as an anti-inflammatory in Iran. Dried seeds have a diuretic effect when combined with coconut juice. They are also given orally for indigestion and diarrhoea brought on by irregularities in bile secretion. As a bulk laxative, dried seed coat is taken orally [87].

Dried Isabgol seed extract in acetic acid is applied externally to treat gout and rheumatoid arthritis. Inflammations of the urinary system are treated with an infusion of dried seeds [88]. While hot water extracts of dried husk are given orally in Thailand to alleviate irritation and diarrhoea, leaf infusions are taken orally in Spain to cure colds [89]. Since it is grown in desert and semiarid areas and is utilised as a medicine in South Asia, it is known as Desert Indian wheat in India [73]. Desert Indian wheat can thrive in a broad variety of agro-climates because it is native to Mediterranean regions, including North Africa, Europe, and Pakistan. Isabgol is only available in dry areas of the world because to low water demand [72].

Isabgol's non-targeted metabolomics and antioxidant activities were investigated. Isabgol is rich in natural antioxidants, omega-3 and omega-6 fatty acids, vital amino acids, and sulfur-rich amino acids, according to the study [74, 75]. Isabgol is made up of flavonoids and phenolic compounds. The reducing and antioxidant potential of flavonoids and phenols is evident. In leaves, seeds, and husks, there were around 76%, 78%, 58% polyunsaturated, 21%, 15%, 20% saturated, and 3%, 7%, 22% monounsaturated fatty acids. Fatty acids in the C12 to C24 range were found in several locations around Isabgol [90]. In leaves, there was found to be a high concentration of -3 alpha-linolenic acid (57%) and -6 linoleic acid (18%). From Isabgol, a total of 36 metabolites were found. Ten of these 26 metabolites were found in seeds and leaves, while the remaining from Isabgol husk. The majority of the metabolites are known as natural supplements because they are phenolics, flavonoids, alkaloids, or natural antioxidants [76]. The study showed that metabolites have therapeutic potential. Saponins, a type of reactive oxygen species, were also discovered. Isabgol leaves can also be consumed as a salad, according to the study's findings [91].

27.8 Myths, Legends, Tales, Folklore, and Interesting Facts

The popular name Isabgol, which refers to the seed's form, is derived from the Persian words "isap" and "ghol," which mean "horse ear." Isabgol has a propensity to absorb 14 times its weight in water. Plantago seeds can absorb three times as much water as they weigh. The gelatinous fluid found in plantago seeds is formed of polysaccharides or soluble fibre. Mucilage makes up nearly 30% of the overall weight. About 40% of the linoleic acid (LA) in plantago seeds is present [92].

27.8.1 Traditional Therapeutic Benefits

The dried seeds and husk of Isabgol are used as emollients, demulcents, and mild laxatives in traditional medicine. Dysentery, severe diarrhoea, and constipation can all be treated with it [93, 94]. Isabgol has been used in Asian and European herbal therapy to treat persistent constipation since the sixteenth century. The gut was not irritated by isabgol. Isabgol seeds are advised for usage in feverish conditions as well as to cure infections of the kidney, urinary bladder, and urethra since they are thought to be cooling and diuretic. The decoction of seeds is used to treat colds and coughs. Rheumatoid arthritis and glandular inflammation can be effectively treated using ointment produced from crushed seeds [95]. It is reportedly used to treat a variety of illnesses in traditional medical systems all over the world. Isabgol seeds are well-known for helping to treat gastritis, diarrhea, enteritis, and diarrhea [96]. Indian uses a decoction of dried seeds to cure diarrhoea and as a demulcent [97]. Seed ointment may be used to treat stomach issues and constipation [72]. In Iran, dry seeds are used as an anti-inflammatory. Adding dried seeds to coconut juice causes a diuretic effect. Additionally, they are taken orally to treat indigestion and diarrhoea brought on by bile secretion abnormalities. Dry seed coat is administered orally as a bulk laxative [98].

Dried Isabgol seed extract in acetic acid is applied externally to treat gout and rheumatoid arthritis. Inflammations of the urinary system are treated with an infusion of dried seeds [99]. While hot water extracts of dried husk are given orally in Thailand to alleviate irritation and diarrhoea, leaf infusions are taken orally in Spain to cure colds [100, 101]. Since it is grown in desert and semiarid areas and is used to treat illnesses in South Asia, it is known as Desert Indian wheat in India. Desert Indian wheat can thrive in a variety of agro-climates because it is native to Mediterranean regions including North Africa, Europe, and Pakistan. Isabgol can only be found in desert areas of the world due to low water requirement [101]. Dried Isabgol seed extract in acetic acid is applied externally to treat gout and rheumatoid arthritis. Inflammations of the urinary system are treated with an infusion of dried seeds [99]. While hot water extracts of dried husk are given orally in Thailand to alleviate irritation and diarrhoea, leaf infusions are taken orally in Spain to cure colds [100, 101]. Since it is grown in desert and semiarid areas and is used to treat illnesses in South Asia, it is known as Desert Indian wheat in India. Desert Indian wheat can thrive in a variety of agro-climates because it is native to Mediterranean regions including North Africa, Europe, and Pakistan. Isabgol can only be found in desert areas of the world due to low water requirement [101].

27.8.2 Tradition

To make a decoction, cook 12 cup of fresh or dried leaves in 3–4 cups of water for 30–40 min. Decoction is used one or two times each day in cups. Cough, ulcers, constipation, colitis, cyctitis, and uncomfortable urination are all conditions that this tea is used to cure [102].

27.8.3 Psyllium Applications in Food Systems

The demand for functional foods has increased with the rising trend of health consciousness. The surge in demand for diets that are nutrient-rich has prompted the food sector to look at functional food products with significant health benefits. In countless studies carried out all around the world, isabgol has been added to foods like ice cream, gluten-free bread, biscuits, pizza, noodles, cake, and spaghetti. Additionally, adding Isabgol has been found to improve the mechanical characteristics of dough, create a softer crumb in high fibre wheat bread, and extend the shelf life of gluten-free bread. Isabgol's use in various food products is restricted since it contains significant levels of fibre, which may adversely affect texture and colour. It was noted that biscuits containing Isabgol husk had a darker colour than those without it among baked food products [103].

In order to thicken the chyme and inhibit the way that digestive enzymes interact with complex carbohydrates, psyllium forms a viscous gel that thickens the chyme [104]. Additionally, the enhanced viscosity lowers peak postprandial blood glucose levels and reduces glucose absorption [105]. The small intestine is frequently where nutrients are absorbed first. This delay in nutritional absorption may cause the small intestine's distal ileum, where nutrients are typically lacking, to obtain nutrients. One metabolic process that nutrients in the distal ileum can initiate in mucosal receptors is the release of glucagon-like peptide 1 into the bloodstream. Reduced hunger and increased insulin output are two advantages of glucagon-like peptide 1 in terms of health [104].

When considered as a whole, the viscosity/gel-related mechanisms for better glycemic control include lowering the glycemic index of consumer foods and increasing chyme viscosity to slow glucose absorption and starch degradation in the small intestine. This results in a gel-dependent improvement in glycemic control for those with T2DM and those at risk for developing the disease (e.g , metabolic syndrome [Met Syn]). It is important to keep in mind that the surface area of the small intestine's mucosa is comparable to a tennis court, providing ample of space for nutrition absorption. Nutrients are still absorbed before they reach the large intestine even though psyllium inhibits absorption [106] (Fig. 27.2).

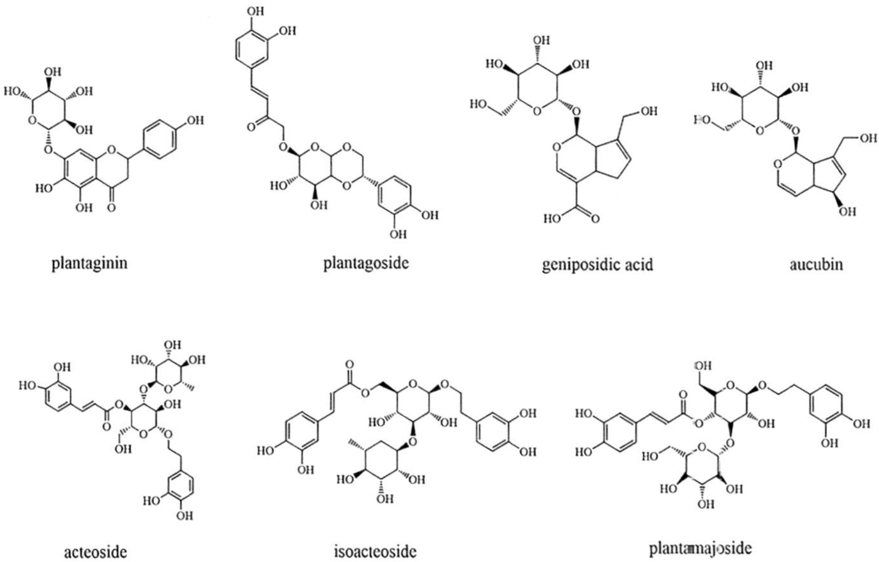

Fig. 27.2 Chemical structure of phytoactive compounds present in Pysillum husk

27.8.4 Summary

Plantago ovata is one of the most popular medicinal plants with widespread use and commercial value. There has been little progress in raising the yield and quality of isabgol despite great effort. Lack of actual diversity is the main obstacle to obtaining higher goals. In order to micro-identify heterogeneity at the molecular level, considerable work is required. This may make it easier to identify novel genotypes useful for developing better cultivars. It is necessary to conduct genomic and functional genomics investigations to find novel genes in *P. ovata*. There are 200 wild allies, and they may have a wealth of beneficial genes. It is necessary to use these wild species' potential for genetic improvement of Isabgol [107].

A natural edible, biodegradeable and renewable polymer is isabgol seed husk. It has been known to treat constipation, ulcerative colitis, diabetes and hemorrhoids. A part its traditional use in constipation, it can also lower the LDL (low density lipoprotein) Cholesterol level up to the normal value., which may cause hypercholesterolemia, hypertension, low body working efficiency, etc. The medicinal applications of Isabgol has or no negative effects. It contains dietary fiber that helps in curing obesity and consists of specific flavonoids that helps in inhibiting the formation of cancer cells.

References

1. Shahriari, Z., Heidari, B., & Dadkhodaie, A. (2018). Dissection of genotype × environment interactions for mucilage and seed yield in Plantago species: Application of AMMI and GGE biplot analyses. *PLoS One, 13*. https://doi.org/10.1371/journal.pone.0196095
2. Gupta, R. (1991). Agrotechnology of medicinal plants. In R. O. B. Wifesekera (Ed.), *The medicinal plant industry*. CRS Press.
3. Gupta, A., Parihar, S. S., Choudhary, V. K., Naseem, M., & Maiti, R. K. (2008). Germination, dormancy and its removal in Isabgol(Isabgol Forssk). *International Journal of Agriculture Environment and Biotechnology, 1*(3), 117–124.
4. Ghaderi, F. F., Alimagham, S. M., Kameli, A. M., & Jamali, M. (2012). Isabgol(Isabgol) seed germination and emergence as affected by environmental factors and planting. *International Journal of Plant Production, 6*(2), 185–194.
5. Das, M. (2011). Growth, photosynthetic efficiency, yield and swelling factor in Plantago indica under semi-arid condition of Gujarat, India. *International Journal of Plant Physiology and Biochemistry, 3*(12), 205–214.
6. Chandler, C. (1954). Improvement of Plantago for mucilage production and growth in the United States. *Contributions from the Boyce Thompson Institute, 17*, 495–505.
7. Grey, T., Eason, K., Wells, L., & Basinger, N. (2019). Effects of temperature on seed germination of *Plantago lanceolata* and management in Carya illinoinensis production. *Plants, 8*(9), 308.
8. Koocheki, A., Tabrizi, L., & Mahallati, M. (2007). The effects of irrigation intervals and manure on quantitative and qualitative characteristics of Isabgol and *Plantago psyllium*. *Asian Journal of Plant Sciences, 6*, 1229–1234.
9. Islam, M. R., Mehedi, M. N. H., Moniruzzaman, M., Obaidullah, A. J. M., Fahim, A. H. F., & Karim, M. R. (2020). Evaluation of eight Isabgol(Isabgol.) germplasm performance grown

under different climatic conditions in Bangladesh. *Archives of Agriculture and Environmental Science, 5*(4), 447–451.
10. Chatterjee, S. K. (2002). Cultivation of medicinal and aromatic plants in India a commercial approach. *Acta Horticulturae, 576*, 191–202.
11. Clapham, A. R., Tutin, T. G., & Moore, D. M. (1989). *Flora of the British Isles*. Cambridge University Press.
12. Raissi, A., Galavi, M., Ramroudi, M., Mousavi, S. R., & Rasoulizadeh, M. N. (2012). Effects of phosphate bio-fertilizer, organic manure and chemical fertilizers on yield, yield components and seed capabilities of Isabgol(*Plantago ovate*). *International Journal of Agriculture and Crop Sciences, 4*, 1821–1826.
13. Yadav, N., Sharma, V., Kapila, S., Malik, R. K., & Arora, A. (2016). Hypocholesterolaemic and prebiotic effect of partially hydrolysed psyllium husk supplemented yoghurt. *Journal of Functional Foods, 24*, 351–358.
14. Ramavandi, B. (2014). Treatment of water turbidity and bacteria by using a coagulant extracted from *Plantago ovata*. *Water Resources and Industry, 6*, 36–50.
15. Sharma, V. K., Mazumder, B., & Nautiyal, V. (2014). Rheological characterization of Isabgolhusk, gum katira hydrocolloids, and their blends. *International Journal of Food Science*. 2014, https://doi.org/10.1155/2014/506591
16. Modi, J. M., Mehta, K. G., & Gupta, R. (1974). Isabgol, a dollar earner of North Gujarat. *Indian Farming, 23*, 17–19.
17. Sahay, S. (1999). The use of psyllium (isubgol) as an alternative gelling agent for microbial culture media. *World Journal of Microbiology and Biotechnology, 15*(6), 733–735.
18. BeMiller, J. N., Whistler, R. L., & Barkalow, D. G. (1993). Aloe, chia, flaxseed, okra, psyllium seed, quince seed and tamarind gums. In R. L. Whistler & N. BeMiller James (Eds.), *Industrial gums; polysaccharides and their derivatives* (p. 239244). Academic.
19. Anon. (1989). *The wealth of India – Raw materials* (Vol. VIII, pp. 146–154). Council of Scientific and Industrial Research.
20. Sally, M. (2019). *Isabgolbased products to cost more*. Retrieved April 3, 2022, from https://economictimes.indiatimes.com/news/economy/agriculture/Isabgol-based-products-to-cost-more/articleshow/71286527.cms?from=mdr
21. Patel, M., Mishra, A., Jaiswar, S., & Jha, B. (2020). Metabolic profiling and scavenging activities of developing circumscissile fruit of psyllium (Isabgol Forssk.) reveal variation in primary and secondary metabolites. *BMC Plant Biology, 20*(1), 1–15.
22. Hillier, K. (2007). Methylcellulose. In *Xpharm: The comprehensive pharmacology reference* (pp. 1–3). Elsevier.
23. Williams, J. (2022). *Foods with psyllium fiber* | Livestrong.com. (2022). Retrieved April 3, 2022, from https://www.livestrong.com/article/288536-foods-with-psyllium-fiber/
24. Fons, F. A. G., & Repoir, S. (2008). Culture of Plantago species as bioactive components resources: A 20-year review and recent applications. *Acta Bota Gallica, 155*, 277–300. 22.
25. Ronsted, N., Gobel, E., Franzyk, H., Jensen, S. R., & Olsen, C. E. (2000). Chemotaxonomy of Plantago. Iridoid glucosides and caffeoyl phenylethanoid glycosides. *Phytochemistry, 55*, 337–348.
26. Franco, E. A. N., Sanches-Silva, A., Ribeiro-Santos, R., & de Melo, N. R. (2020). Psyllium (Isabgol): From evidence of health benefits to its food application. *Trends in Food Science & Technology, 96*, 166–175.
27. Arzneimittelregister. (2003). Klinikleitfaden Gynäkologie, Geburtshilfe, 833–882. https://doi.org/10.1016/b978-343722211-5.50032-2
28. Mehmood, M. H., Aziz, N., Ghayur, M. N., & Gilani, A. H. (2011). Pharmacological basis for the medicinal use of psyllium husk (Ispaghula) in constipation and diarrhea. *Digestive Diseases and Sciences, 56*(5), 1460–1471.
29. Childs, N. M. (1999). Marketing functional foods: What have we learned? An examination of the Metamucil, benefit, and heartwise introductions as cholesterol-reducing ready-to-eat cearls. *Journal of Medicinal Food, 2*, 11–19.

30. Matheson, H. B., Colon, I. S., & Story, J. A. (1995). Cholesterol 7-a-hydroxylase activity is increased by dietary modification with psyllium hydrocolloid, pectin, cholesterol and cholestyramine in rats. *The Journal of Nutrition, 125*, 454–458.
31. Vergara-Jimenez, M., Conde, K., Erickson, S. K., & Fernandez, M. L. (1998). Hypolipidemic mechanisms of pectin and psyllium in guinea pigs fed high fat-sucrose diets: Alterations on hepatic cholesterol metabolism. *Journal of Lipid Research, 39*, 1455–1465.
32. Everson, G. T., Daggy, B. P., McKinley, C., & Story, J. A. (1992). Effects of psyllium hydrophilic mucilloid on LDL-cholesterol and bile acid synthesis in hypercholesterolemic men. *Journal of Lipid Research, 33*, 1183–1192.
33. Sprecher, D. L., Harris, B. V., & Goldberg, A. C. (1993). Efficacy of psyllium in reducing serum cholesterol levels in hypercholesterolemic patients on high-or low-fat diets. *Annals of Internal Medicine, 119*, 45–554.
34. Aygustin, J., & Dwyer, J. (1999). Coronary heart disease: Dietry approaches to reducing risks. *Topics in Clinical Nutrition, 10*, 1–13.
35. Yu, L., Devay, G. E., Lai, G. H., Simmons, C. T., & Neilsen, S. R. (2001). *Enzymatic modification of psyllium*. US Patent. 6, 48,373.
36. Yu, L., & Perret, J. (2003). Effects of solid-state enzyme treatments on the water-absorbing and gelling properties of psyllium. *Lebensmittel-Wissenschaft & Technologie, 36*, 203–208.
37. Perez-Miranda, M., Gomez-Cedenilla, A., & Leon-Colombo, T. (1996). Effect of fiber supplements on internal bleeding hemorrhoids. *Hepato-Gastroenterology, 43*, 1504–1507.
38. Mortensen, P. B., & Nordgaard-Andersen, I. (1993). The dependence of the in vitro fermentation of dietary fibre to short-chain fatty acids on the contents of soluble non-starch polysaccharides. *Scandinavian Journal of Gastroenterology, 28*, 418–422.
39. Fernandez-Banares, F., Hinojosa, J., & Sanchez-Lombrana, J. L. (1999). Randomized clinical trial of Isabgol seeds (dietary fiber) as compared with mesalamine in maintaining remission in Vipin K. Sharma et al., IsabgolHusk: A Herbal Remedy for Human Health Source of support: Nil, conflict of interest: None declared ulcerative colitis. *American Journal of Gastroenterology, 94*, 427–433.
40. Anderson, J. W., Allgood, L. D., & Turner, J. (1999). Effects of psyllium on glucose and serum lipid responses in men with type 2 diabetes and hypercholesterolemia. *The American Journal of Clinical Nutrition, 70*, 466–473.
41. Marteau, P., Flourie, B., & Cherbut, C. (1994). Digestibility and bulking effect of ispaghula husks in healthy humans. *Gut, 35*, 1747–1752.
42. Reddy, B. S. (1993). *Foods, nutrition and chemical toxicity* (D. V. Parke, C. Ioannides, & R. Walker, Eds.). Smith- Gordon, pp. 325–336.
43. Wilpart, M., Mainguet, P., Maskens, A., & Roberfroid, M. (1983). Mutagenicity of 1, 2-dimethylhydrazine towards salmonella typhimurium, co-mutagenic effect of secondary bile acids. *Carcinogenesis, 4*, 45–48.
44. Hill, M. J. (1991). The ratio of lithocholic to deoxycholic acid in faces; a risk factor in colorectal carcinogenesis. *European Journal of Cancer Prevention, 1*(2), 75–78.
45. Owen, R. W., Henly, P. J., Day, D. W., & Thomson, M. H. (1985). *Fecal steroids and colorectal cancer: Bile acid profiles in low and high risk groups Human Carcinogenesis* (J. V. Joossens, M. J. Hill, & J. Geboers, Eds.). Elsevier, pp. 165–170.
46. Chaplin, M. F., Chaudhury, S., Dettmer, P. W., & Sykes, J. (2000). Effect of ispaghula husk on the faecal output of bile acids in healthy volunteers. *The Journal of Steroid Biochemistry and Molecular Biology, 72*, 283–292.
47. Kritchevsky, D. (1991). Bile acids: Biosynthesis and functions. *European Journal of Cancer Prevention, 1*(2), 23–28.
48. Biasco, G., Paganelli, G. M., Owen, R. W., & Hill, M. J. (1991). Faecal bile acids and colorectal cell proliferation. *European Journal of Cancer Prevention, 1*(2), 63–68.
49. Pool-Zobel, B. L., & Leuchl, U. (1997). Induction of DNA damage by risk factors of colon cancer in human colon cells derived from biopsies. *Mutation Research, 375*(2), 105–115.

50. Schnieder, H., Fiander, H., Latta, R. K., & Ross, N. W. (1993). Bile acid inhibition of xenobiotic-metabolizing enzymes is a factor in the mechanism of colon carcinogenesis: Tests of aspects of the concept with glucoronosyl transferase. *European Journal of Cancer Prevention, 2*(2), 393–400.
51. Arvind, P., Papavassiliou, E. D., Tsioulis, G. J., Duceman, B. W., & Lovelace, C. I. (1994). Lithocholic acid inhibits the expression of HLA class1 genes in colon adenocarcinoma cells, differential effects on HLA, -A, -B and -C loci. *Molecular Immunology, 31*(8), 607–614.
52. Owen, R. W., Dolo, M., Thomson, M. H., & Hill, M. J. (1983). The fecal ratio of lithocholic acid to deoxycholic may be an important aetiological factor in colo-rectal cancer. *European Journal of Cancer & Clinical Oncology, 19*, 1307.
53. Owen, R. W., Henly, P. J., Thomson, M. H., & Hill, M. J. (1986). Steroids and cancer: Fecal bile acid screening for early detection of cancer risk. *The Journal of Steroid Biochemistry and Molecular Biology, 24*, 391–394.
54. Imray, C. H. E., Radley, S., Davis, A., & Barker, G. (1992). Neoptolemos, faecal unconjugated bile acids in patients with colorectal cancer or polyps. *Gut, 33*, 1239–1245.
55. Abdel-Moaty, A., Helal, H. A., Elkhatib, B. R., & Assar, R. H. (2018). Therapeutic effect of Isabgol and Sage on obese rats. *Journal of Home Economics, 28*(1), 2–3.
56. Sharma, V. K., & Bhattacharya, A. (2009). Isabgolhusk: A herbal remedy for human health. *Journal of Pharmacy Research, 2*, 296–301.
57. Gitanjali Deokar, S. K. (2016). Pharmaceutical benefits of Plantago ovate (Isabgolseed): A review. *Pharmaceutical and Biological Evaluations, 3*(1), 9.
58. Rodríguez-Cabezas, M. E., Gálvez, J., Camuesco, D., Lorente, M. D., Concha, A., MartinezAugustin, O., Redondo, L., & Zarzuelo, A. (2003). Intestinal anti-inflammatory activity of dietary fiber (Isabgol seeds) in HLA-B27 transgenic rats. *Clinical Nutrition, 22*(5), 463–471.
59. Miettinen, T. A., & Tarpila, S. (1989). Serum lipid and cholesterol metabolism during guar gum, Isabgol and high fibre treatment. *Clinica Chimica Acta, 183*, 253–262.
60. Agarwal, O. P. (1985). Prevention of atheromatous heart diseases. *Angiology, 36*, 485–492.
61. Sharma, U., Velpandian, T., Sharma, P., & Singh, S. (2009). Evaluation of antileishmanial activity of selected Indian plants known to have antimicrobial properties. *Parasitology Research, 105*, 1287–1293.
62. Chan, J. K. C., & Wypyszyk, V. (1988). A forgotten natural dietary fiber: Psyllium mucilloid. *Cereal Food World, 33*, 919–922.
63. Dhar, M. K., Kaul, S., Sareen, S., & Kou, A. K. (2005). *Plantago ovata*: Genetic diversity, cultivation, utilization and chemistry. *Plant Genetic Resources, 3*(2), 252–263.
64. Rezaeipoor, R., Saeidnia, S., & Kamalinejad, M. (2000). The effect of Isabgol on humoral immune responses in experimental animals. *Journal of Ethnopharmacology, 72*(1-2), 283–286.
65. Kordošová, A., & Machová, E. (2006). Antioxidant activity of medicinal plants polysaccharides. *Fitoterapia, 77*, 367–373.
66. Solà, R., Godàs, G., Ribalta, J., Vallvé, J. C., Girona, J., Anguera, A., et al. (2007). Effects of soluble fiber (Isabgol husk) on plasma lipids, lipoproteins and apolipoproteins in men with ischemic heart disease. *American Journal of Clinical Nutrition, 85*, 1157–1163.
67. Abdelaaty, A. S., Hanaa, H. A., Faiza, M. H., & Haider, G. (2012). Regulation of obesity and liquid disorders by *Foeniculum vulgare* extracts and *Plantago ovate* in high-fat diet-induced obese rats. *American Journal of Food Technology*, 1–11. https://doi.org/10.3923/ajft.2012
68. Motamedi, H., Darabpour, E., Gholipour, M., & Seyyed-Nejad, S. M. (2010). Antibacterail effect of ethanolic and methanolic extracts of *Plantago ovate* and *Oliveria decumbens* endemic in Iran against some pathogenic Bacteria. *International Journal of Pharmacology, 6*(2), 117–122.
69. Mc Call, M. R., Mehta, T., Leathers, C. W., & Foster, D. M. (1992). Psyllium husk I. Effect on plasma lipoprotein, cholesterol metabolism, and artherosclerosis in African green monkeys. *American Journal of Clinical Nutrition, 56*(2), 376–384.

70. Everson, G. T., Daggy, B. P., McKinley, C., & Story, J. A. (1992). Effect of psyllium hydrophilic mucilloid on LDL cholesterol and bile acid synthesis in hypercholesterolemic men. *Journal of Lipid Research, 33*(8), 1183–1192.
71. Olson, B. H., Anderson, S. M., Becker, M. P., Anderson, J. W., Hunninghake, D. B., Jenkins, D. J., et al. (1997). Psyllium-enriched cereals lower blood total cholesterol and LDL cholesterol, but not HDL cholesterol, in hypercholesterolemic adults: Results of a meta-analysis. *Journal of Nutrition, 127*(10), 1973–1980.
72. Zagari, A. (1992). *Medicinal plants* (Vol. 4, 5th ed., p. 969). Tehran University Publications, No 1810/4.
73. Watters, K., & Blaisdell, P. (1989). Reduction of glycemic and lipid levels in DB/ DB diabetic mice by psyllium plant fiber. *Diabetes, 38*(12), 1528–1533.
74. Rodriguez-Moran, M., Guerrero-Romero, F., & Lazcano-Burciaga, G. (1998). Lipid- and glucoselowering efficacy of Plantago psyllium in type II diabetes. *Journal of Diabetes and Its Complications, 12*(5), 273–278.
75. Chiang, L. C., Chiang, W., Chiang, M. Y., Ng, L. T., & Lin, C. C. (2003). Antileukemic activity of selected natural products in Taiwan. *The American Journal of Chinese Medicine, 31*(1), 37–46.
76. Vijayalakshimi, K., Mishra, S. D., & Prasad, S. K. (1979). Nematocidal properties of some indigenous plant materials against second stage juveniles of Meloidogyne incognita (koffoid and white) chitwood. *Indian Journal of Entomology, 41*(4), 326–331.
77. Turnbull, W. H., & Thomas, H. G. (1995). The effect of a Isabgol seed containing preparation of appetite variables, nutrient, and energy intake. *International Journal of Obesity, 19*(5), 338–342.
78. Matheson, H. B., & Story, J. A. (1994). Dietary psylium hydrocolloid and pectin increase bile acid pool size and change bile acid composition in rats. *Journal of Nutrition, 124*(8), 1161–1165.
79. Alabaster, O., Tang, Z. C., Frost, A., & Shivapurkar, N. (1993). Potential synergism between wheat bran and psyllium: Enhanced inhibition of colon cancer. *Cancer Letters, 75*(1), 53–58.
80. Cohen, L. A., Zhao, Z., Zhang, E. A., Wynn, T. T., Simi, B., & Rivenson, A. (1996). Wheat bran and psyllium diets: Effects on N-methylnitrosourea-induced mammary tumorigenesis in F344 rats. *Journal of the National Cancer Institute, 88*(13), 898–907.
81. Khanna, N. M., Sarin, J. P. S., Nandi, R. C., Singh, S., Setty, B. S., & Kamboj, V. P. (1980). Isaptent – A new cervical dilator. *Contraception, 21*(1), 29–40.
82. Enzi, G., Inelmen, E. M., & Crepaldi, G. (1980). Effect of hydrophilic mucilage in the treatment of obese patients. *Pharmacotherapeutica, 2*(7), 421–428.
83. Montague, J. F. (1932). *Psyllium seeds: The latest laxative*. Montague Hospital for Intestinal Ailments.
84. Kirtikar, K. R., & Basu, B. D. (1988). *Indian medicinal plants* (Vol. III, 2nd ed., pp. 2039–2042). Bisen Singh Mahendra Pal Singh.
85. Evans, W. C. (1989). *Trease and Evan's pharmacognosy*. Bailliere Tindall.
86. Schopen, A. (1983). *Traditionelle Heilmittel in Jemen*. Steiner.
87. Koedam, A. (1977). *Plantaginaceae Pharmacologie Weekbland, 112*, 246.
88. Dagar, J. C., Kumar, Y., & Tomar, O. S. (2006). Cultivation of medicinal Isabgol(Plantago ovata) in alkali soils in semiarid regions of northern India. *Land Degradation and Development, 17*, 275–283.
89. Gupta, S., Yadava, J. N. S., & Tandon, J. S. (1993). Antisecretory (antidiarrheal) activity of Indian medicinal plants against *Escherichia coli* enterotoxin-induced secretion in rabbit and Guinea pig ileal loop models. *International Journal of Pharmacognosy, 31*(3), 198–204.
90. Shah, G. L., & Gopal, G. V. (1985). Ethnomedical notes from the tribal inhabitants of the north Gujarat (India). *Journal of Economic and Taxonomic Botany, 6*(1), 193–201.
91. Martinez-Lirola, M. J., Gonzalez- Tejero, M. R., & Molero-Mesa, J. (1996). Ethnobotanical resources in the province of Almeria, Spain: Campos de Nijar. *Economic Botany, 50*(1), 40–56.

92. Patel, M., Mishra, A., & Jha, B. (2016). Non-targeted metabolite profiling and scavenging activity unveil the nutraceutical potential of psyllium (Isabgol). *Frontiers in Plant Science, 7*. https://doi.org/10.3389/fpls.2016.00431
93. Jewvachdamrongkul, Y., Dechatiwongtse, T., Pecharaply, D., Bansiddhi, J. & Kanchanapee, P. (1982). Identification of some Thai medicinal plants. Mahidol Univ. *Journal of Pharmaceutical Sciences, 9*(3), 65–73.
94. Jamal, Z., Ahmad, M., Zafar, M., Sultana, S., Khan, M., & Shah, G. (2012). Medicinal plants used in traditional folk recipes by the local communities of Kaghan valley, Mansehra, Pakistan. *Indian Journal of Traditional Medicine, 11*(4), 634–639.
95. Abbas, J. A., El-Oqlah, A. A., & Mahaneh, A. M. (1992). Herbal plants in the traditional medicine of Bahrain. *Economic Botany, 46*, 158–163.
96. Shah, G. L., & Gopal, G. V. (1985). Ethnomedical notes from the tribal inhabitants of the north Gujarat (India). *The Journal of Economic and Taxonomic Botany, 1*(6), 193–201.
97. Zahoor, A., Ghafor, A., & Muhammad, A. (2004). *Plantago ovata– A crop of arid and dry climates with immense herbal and pharmaceutical importance; Introduction of medicinal herbs and spices as crops*. Ministry of Food, Agriculture and Livestock.
98. Martinez-Lirola, M. J., Gonzalez- Tejero, M. R., & Molero-Mesa, J. (1996). Ethnobotanical resources in the province of Almeria, Spain: Campos de Nijar. *Economic Botany, 1*(50), 40–56.
99. Jewvachdamrongkul, Y., Dechatiwongtse, T., Pecharaply, D., Bansiddhi, J., & Kanchanapee, P. (1982). Identification of some Thai medicinal plants. Mahidol Univ. *Journal of Pharmaceutical Sciences, 3*(9), 65–73.
100. Solà, R., Godàs, G., Ribalta, J., Vallvé, J. C., Girona, J., & Anguera, A. (2007). Effects of soluble fiber (Isabgol husk) on plasma lipids, lipoproteins and apolipoproteins in men with ischemic heart disease. *The American Journal of Clinical Nutrition, 85*, 1157–1163.
101. Rezaeipoor, R., Saeidnia, S., & Kamalinejad, M. (2000). The effect of Isabgol on humoral immune responses in experimental animals. *Journal of Ethnopharmacology, 1–2*(72), 283–286.
102. Montonen, J., Knekt, P., Jarvinen, R., Aromaa, A., & Reunane, A. (2003). Whole-grain and fiber intake and the incidence of type 2 diabetes. *The American Journal of Clinical Nutrition, 77*, 622–629.
103. Chen, C., Shang, C., Xin, L., Xiang, M., Wang, Y., & Shen, Z. (2022). Beneficial effects of psyllium on the prevention and treatment of cardiometabolic diseases. *Food & Function, 13*(14), 7473–7486.
104. Gonçalves, S., & Romano, A. (2016). The medicinal potential of plants from the genus Plantago (Plantaginaceae). *Industrial Crops and Products, 83*, 213–226.
105. McRorie, J., & Fahey, G. Fiber supplements and clinically meaningful health benefits: Identifying the physiochemical characteristics of fiber that drive specific physiologic effects. In T. C. Wallace (Ed.), *The CRC handbook on dietary supplements in health promotion* (pp. 161–206). CRC Press/Taylor & Francis Group.
106. Gibb, R. D., McRorie, J. W., Jr., Russell, D. A., Hasselblad, V., & D'Alessio, D. A. (2015). Psyllium fiber improves glycemic control proportional to loss of glycemic control: A meta-analysis of data in euglycemic subjects, patients at risk of type 2 diabetes mellitus, and patients being treated for type 2 diabetes mellitus. *The American Journal of Clinical Nutrition, 102*(6), 1604–1614.
107. Kawasaki, N., Urashima, M., Odaira, H., Noro, T., & Suzuki, Y. (2010). Effects of gelatinization of enteral nutrients on human gastric emptying. *Gastroenterology Research, 3*(3), 106–111.
108. Dhar, M. K., Kaul, S., Sareen, S., & Koul, A. K. (2005). *Plantago ovata* genetic diversity, cultivation, utilization and chemistry. *Plant Genetic Resources: Characterization and Utilization, 3*(2), 252–263.

Chapter 28
Kalonji

Zainab Shahzadi, Zubaida Yousaf, Arusa Aftab, Mehwish Riaz, and Shadma Wahab

28.1 Introduction

N. sativa, belongs to family Ranunculaceae, is also referred to as fennel flower, black cumin, kalonji, roman coriander, black caraway and black seeds. It is an essential herbal medicine used in conventional medical systems like Unani and Ayurveda. The Southern Europe, Mediterranean region, Syria, Pakistan, Turkey, Saudi Arabia and India are among the nations where this plant is grown. It is originated from Europe, Africa, and Asia [1].

In-depth biological study, *N. sativa* has exposed a variety of pharmacological activities, including antibacterial, anticancer, hypotensive, purgative, immuno protective, antidiabetic, antifungal, anti-inflammatory, sedative, antihelmintic, antispasmodic, bronchodilator, gastroprotective, anti-hepatotoxicity and free radical savaging activity [2]. Thymoquinone, a key phytochemical present in *N. sativa* seeds oil which have potential application in several biological activities. *N. sativa* has a low toxicity and is used in food as a flavouring component in bakeries items like prickle and cakes [3, 4].

Historical and religious context has made *N. sativa* a valuable herb. Hence used in a wide range of pharmacological formulations. *N. sativa* was reported as the cure for every illness except for death in one of the Prophetic hadiths, it is regarded

Z. Shahzadi · Z. Yousaf (✉) · A. Aftab · M. Riaz
Department of Botany, Lahore College for Women University, Lahore, Pakistan
e-mail: Zubaida.yousaf@lcwu.edu.pk

S. Wahab
Department of Pharmacognosy, College of Pharmacy, King Khalid University, Abha, Saudi Arabia

Complementary and Alternative Medicine Unit, College of Pharmacy, King Khalid University, Abha, Saudi Arabia

M. Zia-Ul-Haq et al. (eds.), *Essentials of Medicinal and Aromatic Crops*,
https://doi.org/10.1007/978-3-031-35403-8_28

among Muslims as one of the greatest forms of healing medicine available. Tibb-e-Nabwi is another site where regular use of it is advised (Prophetic Medicine) [5].

28.2 Taxonomic Position

Kingdom: Plantae
Clade: Tracheophytes
Clade: Angiosperms
Order: Ranunculales
Family: Ranunculaceae
Genus: *Nigella*
Species: *N. sativa*

28.3 English/Common Name

Black cumin, Kalonji, black seed, fennel flower, black caraway or roman coriander (Fig. 28.1).

28.4 Nutritive Values

N. sativa has a high nutritional value because of its considerable amounts of vegetable protein, fibre, minerals, and vitamins. The breakdown of nutrients is as follows: 31.94% total carbohydrates, 25–85% protein, 7–94% fibre, 38% fat. When it comes to the other amino acids that have been linked to arginine, aspartate and glutamate are the major ones, while methionine and cysteine are the minor ones. Additionally, *N. sativa* seeds contain high quantities of thiamin, niacin, pyridoxine, copper, folic acid, zinc, calcium, and phosphorus [6].

28.5 Agronomy

28.5.1 Soil Conditions

N. sativa should be cultivated in a sunny, well-drained area of the garden. They prefer well-drained soil that has been amended with manure or compost before planting. It does well in sandy soils with a pH 6–7 [7, 8]. Loamy and sandy soil having rich microbial activity is most suitable for farming. Dense rainfall areas

Fig. 28.1 Seeds and two different types of *Nigella sativa* L. flowers (**a**) Seeds (**b**) Flower with pale blue color petals (**c**) Flower with white color petals (**d**) Complete flower

having sloppy soils and temperate rainfall areas having well-drained soils are most favorable soils for cultivation. Soil with pH 7.0–7.5 is suitable for cultivation. One of the essential components for seed germination is soil moisture. It is suggested that soil moisture about 60% is the best condition for sowing seeds. When soil moisture increases to 80% and decreases to 40% do not helps to improve the rate of seed germination [8].

28.5.2 Climate

N. sativa seeds are usually cultivated in early spring. It thrives in climates with yearly daytime temperatures of 14–26 °C, but can endure temperatures of 5–30 °C. It prefers 600–800 mm of yearly rainfall, but can endure 400–1000 mm. It is a subtropical plant that may also be grown in the temperate zone and at higher elevations in the tropics between 1500 and 2500 m [9].

It is suggested that the temperature ranges from 5 to 8 °C is the best condition for sowing seeds. When the temperature rises to 18 °C then the germination of seeds in the field also increases to 48%. In such temperature plants grow well and results worthy yield. But by increasing temperature to 25–30 °C would be worsen to seed germination. Plantation of *N. sativa* seedlings in the middle of April resulting abundant rooting 75% and plant shows improved growth rate, flowering and development of valuable seeds. It shows effective growth in cool dry areas having light snowfall. Cool and moist climate favors seed setting and flowering [10].

28.5.3 Land Preparation

Before planting the seed, the field for *N. sativa* is prepared. Depending on the soil characteristics, the ground should be ploughed twice or more. The beds were made with a 120–130 cm spacing to make it easier to drain excess water from the field and to reduce the favourable circumstances for potential illnesses like wilt and damping-off. It is possible to use one ploughing together with two to three harrowing and leveling operations [11].

28.5.4 Planting

N. sativa seeds should be planted at a depth of 1/8 inch, three to four seeds together. Until seeds germinate in 7–14 days, keep the soil evenly damp—but not soggy. Thin seedlings when two or more leaves sprout, spacing them 6–9 inches apart to keep the best examples [11, 12].

Seed drill method or line sowing method is used for plantation. In line sowing method space between lines should be 30–50 cm at the distance of 2 cm. After the 20 days of planting pruning of plant is completed at a distance of 20 cm. Sowing through bulb is only possible when the soil moisture condition in the field allows for shallow ploughing (neither too dry nor too wet) [12].

28.5.5 Manuring

Despite the fact that *N. sativa* requires little fertilizer, poor fertilization can lower yields. To achieve the best yield, you will need to fertilize moderately. Fertilizers including potassium, nitrogen, and phosphorus are necessary for the growth of *N. sativa*. It is recommended that 30 kg ha^{-1} of potassium, nitrogen and phosphorous can be applied. After the plant has established, nutrients are sprayed near the root system [13].

In the soil increasing nitrogen doses raises the development of vegetative growth, number of capsule and branches per plant. Application of increases doses of nitrogen up to 60 kg ha^{-1} results in highest yield of seeds and components. But increasing doses of nitrogen showed negative impact on seed yield [13, 14].

By the addition of microorganisms, biofertilizer, cattle manure level (at 30 m^3/fed) and combination of nitrogen fixing bacteria with phosphate solubilizing bacteria to the soil increases the seed yield components, volatile oil content, chlorophyll content, carbohydrate composition in the seeds of *N. sativa*. Application of NPK (5:3:2) alongside the planted bulb is more suitable for every year [14].

28.5.6 Irrigation

N. sativa is a low-water-demanding crop, it is important to provide sufficient water during the growth period to promote flower development, seed setting, and seed yields. Irrigating the seeds at the budding stage may increase essential oil, carvone, and thymoquinone content, but not overall yield. As a result, complete irrigation is required to increase *N. sativa* seed yield. Irrigation is beneficial throughout the blooming and seed development stages to boost the oil content and seed size [15].

28.5.7 Ecology

Climate and soil conditions have an impact on plant growth and development, physiological growth, the synthesis of key phytochemicals, and the amount and quality of essential oils. A growing season with rainfall ranging from 120 to 400 mm may be suitable to maximize *N. sativa* productivity. Temperatures ranging from 0 to 25 °C are suitable for *N. sativa*, with 12–14 °C being optimal [16]. *N. sativa* requires a warm environment to grow fast, and full sun is ideal for flowering. *N. sativa* can tolerate moderate shade, although it will not bloom as profusely as it would in bright sunshine. Despite its low water requirements, the availability of water throughout the growing season is critical for the early onset of flowering and seed development in this semi-arid plant. *N. sativa* grows best in healthy, well-drained soil, but it may also survive in a variety of other settings. Growing on sandy loam soil with significant microbial activity is good. A pH range of 7.0–7.5 is required for the plant [17].

28.5.8 Pests and Diseases

The major diseases caused in *N. sativa* are wilt (which can cause yield losses of up to 72%), blight (which can cause production losses of up to 88%), and powdery mildew (cause yield losses that can amount to 60%). Aphids and mites are two of the most common insect pests that effect *N. sativa*. To acquire the best output, all diseases and insect pests of *N. sativa* should be treated with appropriate control strategies [18].

- Caterpillar creates holes inside the bulb and reduces sprouts.

Pest and disease control: Sow or dig the soil with 10% BHC and 5% Aldrine in 25 kg/ha^{-1}. Utilization of well decayed animal manure and other residue components.

- Fall Armyworn and Soyabean looper depends on the seeds, flowers and also harm the crop.

Pest and disease control: Spraying with 0.05% methyl parathion in combination with 1 ml/l Endosal 35EC or water or Thiodian, at 10-15 days interval [19].

28.5.9 Description of Crop

There are 20 species of annual plants in the genus *Nigella*. The most well-known member of this genus is *N. sativa*. This plant is grown all over the world but is native to Southwest Asia, North Africa and Southern Europe. India produces more than 1,939,000 tonnes of *N. sativa* annually, followed by Turkey (199,018 tonnes), Bangladesh (180,993 tonnes), China (113,359 tonnes), Indonesia (110,387 tonnes), Pakistan (73,472 tonnes), and Ethiopia, which produces around 36,754 tonnes [20, 21]. Moreover, *N. sativa* is vertical, straight annual herbaceous plant which attains 30.0–67.6 cm at maturity stage. Primary branches range from 4 to 10 per plant, leaves with alternate arrangement and pinnae of extensive leaves. Flowers are hermaphrodite with specific flowering pattern [21].

There are a variety of *N. sativa* products available in market, including non-GMO soft gel capsules providing by Healths Harmony, Bioferm eye balm, meadow foam, and avocado to protect the eyes. *N. sativa* seeds unprocessed cold organic oil extracted by organic pure oil (OPO) and by Ryaal. The supplements of *N. sativa* are also available in the form of an extract [22].

Additional products of *N. sativa* that are available in the marketplace are: capsules of soft gelatin by Al-Barkah, sedative creams and body soaps provided by Hermani Corporation. The Immunoviva firm uses Immuno-viva core, a powerful antioxidant product, in liquid or pill form. Bergmeister is the supplier of the *N. sativa* skin cream [22].

Vartika Nature also offers hair oil that has been supplemented with *N. sativa* extracts, which is used to reduce hair-related issues, help to whole hair care, and improve the shine and thickness of hair [22].

28.6 Important Chemical Constituents

28.6.1 Volatile Compounds

Several research conducted on *N. sativa* seeds oil have discovered the existence of numerous complexes of different compounds, including ketone, mono-terpene, monoterpenoid alcohols, sesquiterpenes and di-terpene. The structures of some of

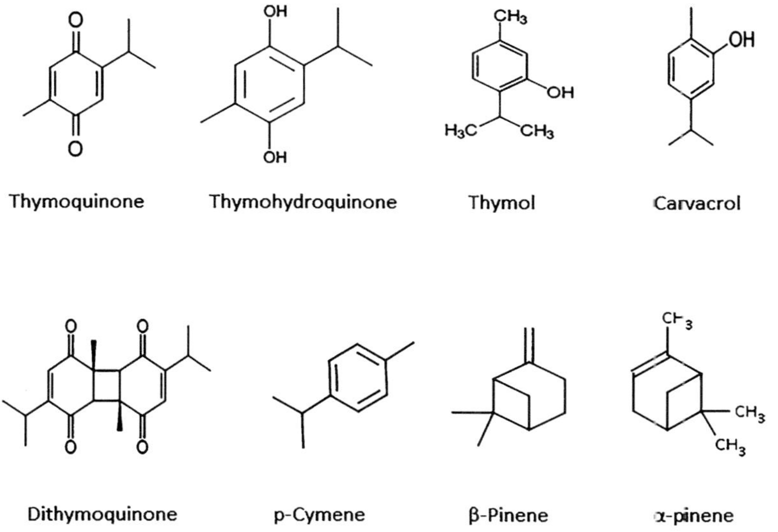

Fig. 28.2 Essential oil compounds present in *N. sativa*

compounds (Thymohydroquinone, carvacrol, thymol, thymoquinone (TQ), Phellandrene, β-pinene and α-pinene) are given in Fig. 28.2 [23].

28.6.2 Phenolic Acids and Flavonoids

Flavonoids have antioxidant characteristics that enable them to shield the tissues from free radical damage. Additionally, flavonoids reduce the adherence of inflammatory cells to the sub-endothelium, which results in a reduction in the inflammatory response. Another advantage of flavonoids is their ability to inhibit peroxidase activity, which enables the protection against oxygen reactive species. Vanillic acid, p-coumaric acid, catechin, gallic acid, chlorogenic acid, ferulic acid, nigel flavonoside B, quercetin, apigenin, flavone and rutin are some of the phenolic chemicals that are identified from the *N. sativa* seeds extract (Fig. 28.3) [24].

28.6.3 Alkaloids

Various *N. sativa* alkaloids, including nigellicine, which has nigellimine, N-oxide, isoquinoline molecule, indazole nucleus, and nigellimine ultimately nigellidine, an indazole molecule, were isolated and discovered between the years 1985 and 1995. In contrast, the aerial portion of the *N. sativa* plant magnoflorine is recently discovered molecule (Fig. 28.4) [25].

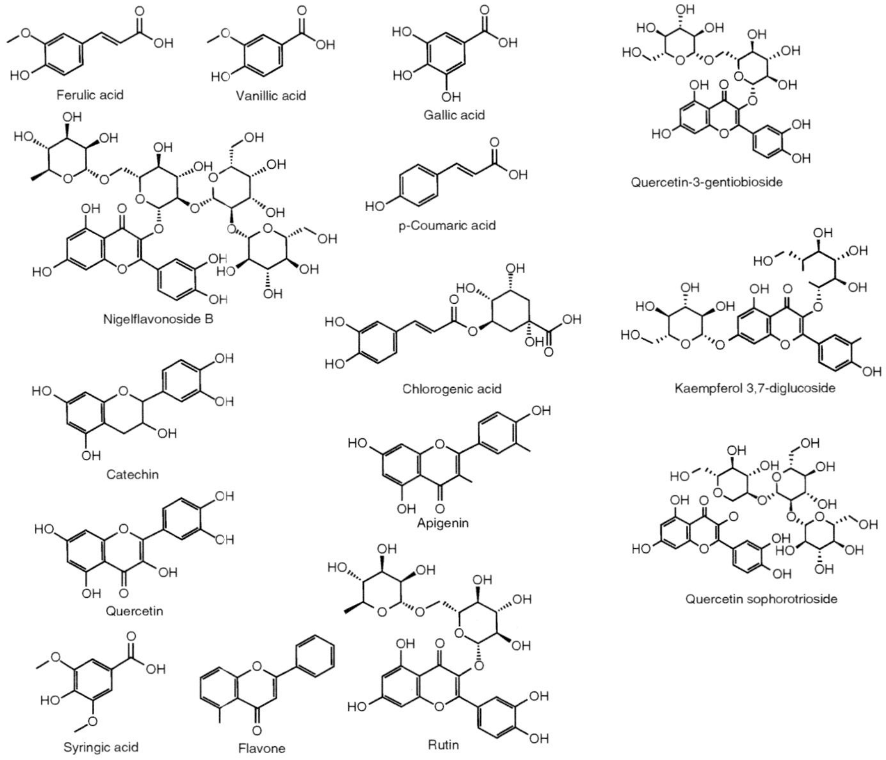

Fig. 28.3 Phenolic compounds present in *N. sativa*

28.6.4 Saponins

Saponins are the secondary metabolites found in *N. sativa* which have glycosidic bonds connecting one or more oligosaccharides to either steroids or aglycone triterpenes. Saponins have a strong ability to attach to the cell surface and the components of its membrane because they include carbohydrates which polar group and triterpenes or steroids which are non-polar groups. Several saponins, including 3-O-L-rha-(1-2)-L-ara-28-O-Lrha(1-4)—D-glu(1-6)—D-glu-hederagenine and 3-O-D-xyl(1-3)—L-rha-(1-2)—L-ara-hederagenine, were discovered in a research on the *N. sativa* methanolic extract. In a different investigation, various saponins from the plant's aerial portion were extracted and identified. Among these are Flaccidoside, Kaempferol 3-O-rutinoside and nigelloside (Fig. 28.5) [25, 26].

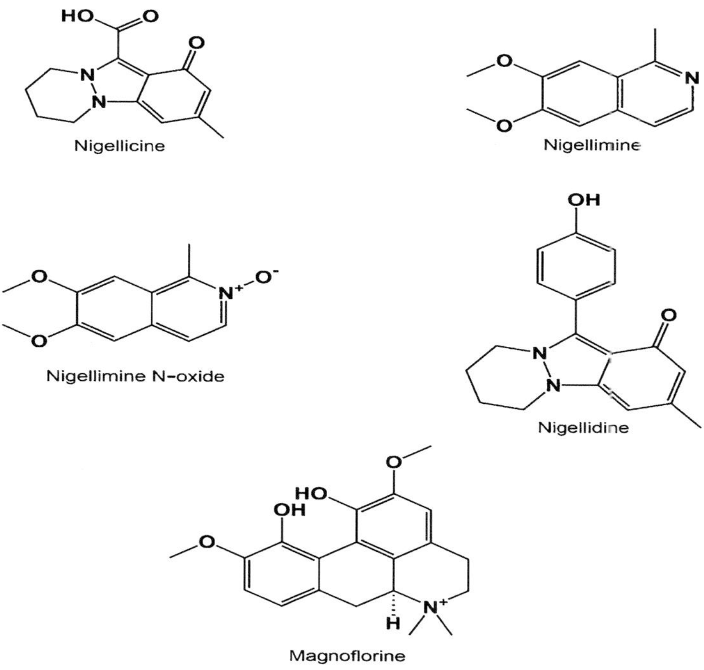

Fig. 28.4 Alkaloid present in *N. sativa*

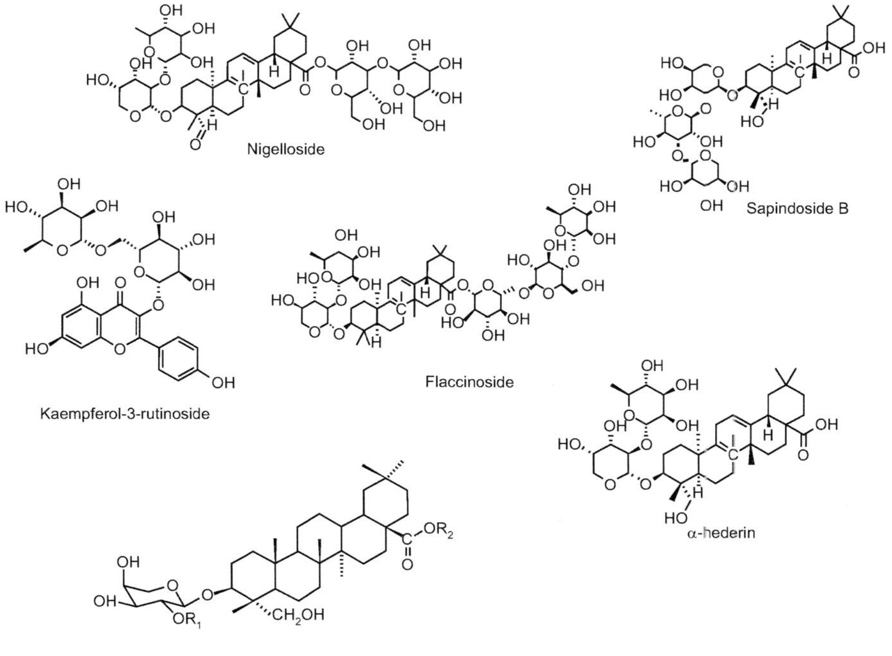

Fig. 28.5 Saponins present in aerial parts of *Nigella sativa*

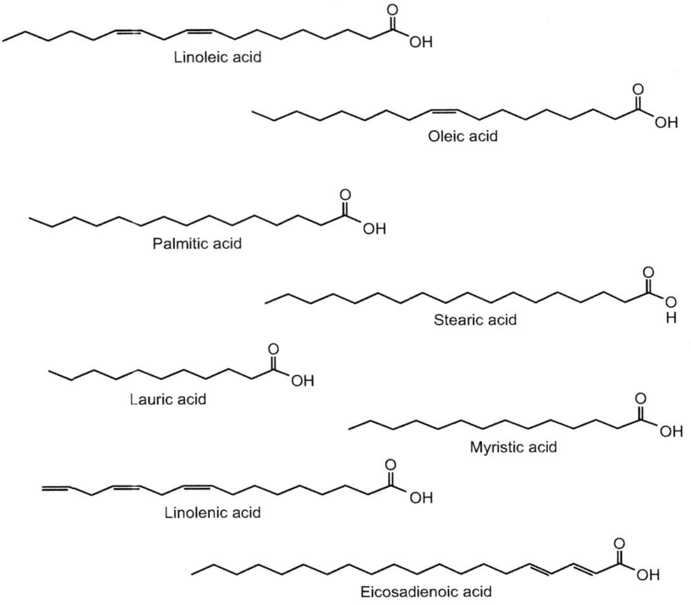

Fig. 28.6 Fatty acid present in *N. sativa*

28.6.5 Fatty Acids

The seeds of *N. sativa* are reported to have crude fibre; minerals such as Copper, sodium, phosphorus, Calcium and zinc; and vitamins including niacin, thiamine and folic acid. The *N. sativa* also contains several kinds of fatty acids, which were identified using the Gas Chromatography-Mass Spectroscopy method. The fatty acids present in *N. sativa* are linolenic acid, oleic acid, stearic acid, eicosadienoic acid, lauric acid, linoleic acid, myristic acid and palmitic acid, all are present in trace amount (Fig. 28.6) [27].

28.6.6 Terpenes and Terpenoids

Terpenes and terpenoids account for more than half of the phytocompounds discovered in *N. sativa*. Monoterpene, diterpene, sesquiterpenes, monoterpenoid alcohols, and ketone are some of them. Thymoquinone, -pinene, p-cymene, carvone and trans-anethole have been identified as the major terpenes. TQ is the main component of *N. sativa* with concentration is 388.61 g/ml (Fig. 28.7) [17].

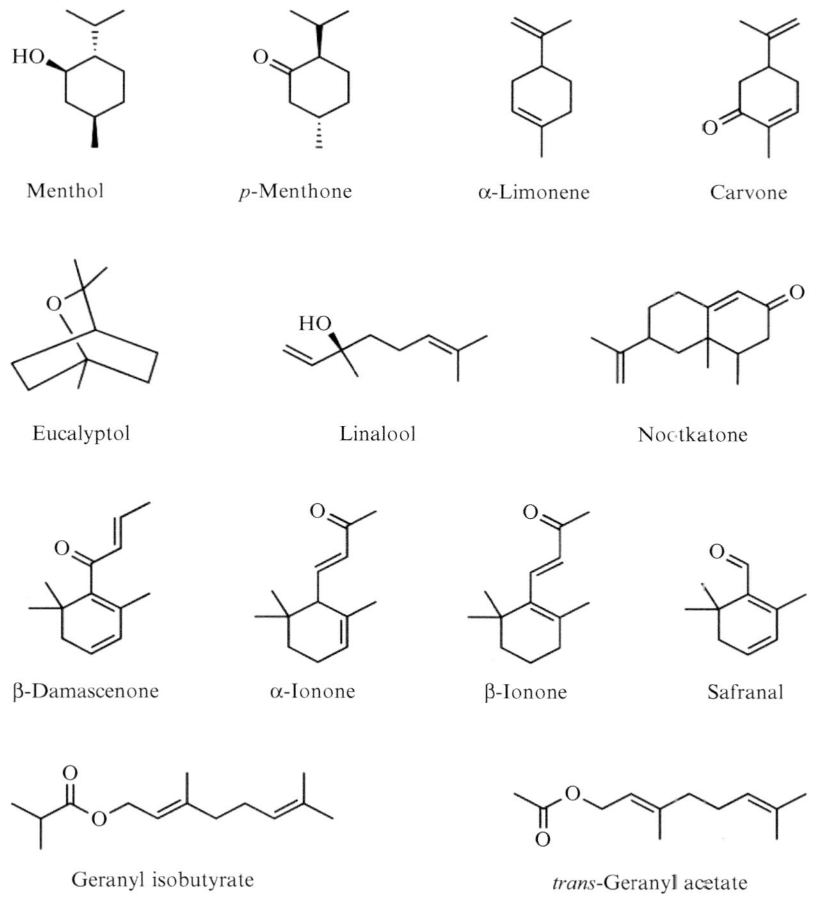

Fig. 28.7 Terpenes and terpenoids present in *N. sativa*

28.6.7 Phytosterols

In seed oil derived from *N. sativa*, the total sterol content ranges from 1993.07 to 2887.28 mg/kg. In *N. sativa* seed oil, -sitosterol (44–54%) dominates the sterol composition. The second primary sterol is stigmasterol (16.57–20.92%), which is followed by -7-stigmasterol, -7-avenasterol, campesterol, and cholesterol [27] (Fig. 28.8).

28.6.8 Other Compounds

A study revealed that *N. sativa* contains ash, carbohydrate, protein, fat, arabinose, xylose, and rhamnose. Additionally, it comprises alkane hydrocarbons, phenylpropanoids, and the triterpene saponin (171) (-hederin) (n-nonane, dodecanal) [17].

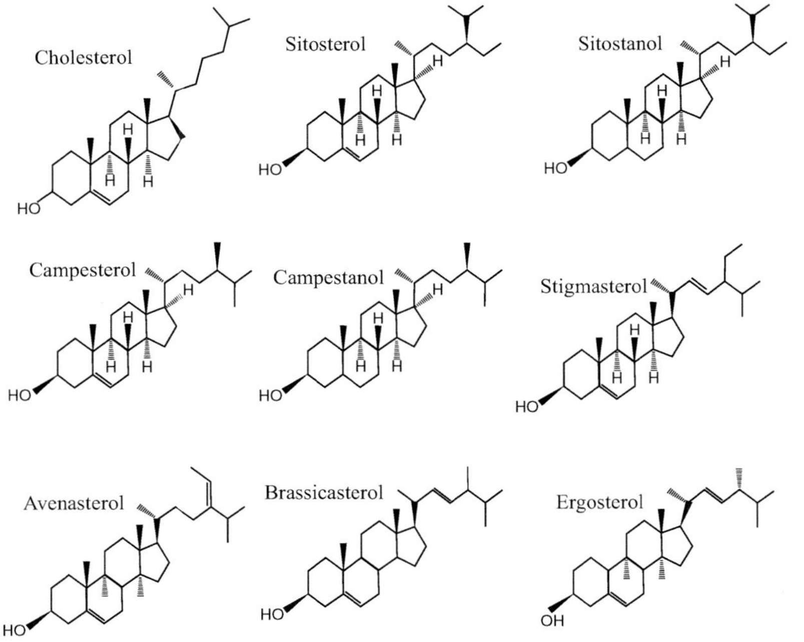

Fig. 28.8 Phytosterols present in *N. sativa*

28.6.9 Medicinal Uses

For centuries, in Asia, Central East and Africa, *N. sativa* has been widely used as an oil and herb. *N. sativa* has traditionally been used to cure a variety of diseases, including those influencing the digestive system, respiratory system, liver, renal, intestinal system, immunological system, circulatory system, and general health. It is regarded as one of the most effective methods of remedial medicine in Islam. *N. sativa* seeds, according to the Prophet Muhammad (PBUM), can cure all illnesses except for death, including how it boosts a person's energy and aids in recovery from fatigue and melancholy [28]. It is also mentioned in the "Medicine of the Prophet (Muhammad)" or the "Tibb-e-Nabavi" list of natural cures, according to the tradition, to cure all diseases. It is also recognized as a helpful medicine in the Unani medical system for a variety of disorders. *N. sativa* have historically been used to treat inflammatory diseases, rheumatism, bronchitis, and asthma in the Central East and Southeast Asian nations. Along with stimulating digestion and protecting against parasitic infections, it is also used to increase milk production in nursing mothers [29]. *Nigella* has earned the Arabic name "Habbatul barakah," which translates to "the seeds of blessing," due to its many uses. It is used in Arabic and Indian culture as food flavouring, a medication, and a way to stay warm in the winter. *N. sativa* seeds are scattered in the folding of woollen fabrics to protect them from insect damage. In Indian customs, seeds are used to increase perspiration, ward off insects, stimulate indigestion, and lessen turgidity [29, 30].

28.7 Pharmacological Activities

28.7.1 Antimicrobial Activity

It has been discovered that *N. sativa* antibacterial effects are caused by its active compounds, including melanin and thymohydroquinone. Strong antibacterial activity was shown in *N. sativa* against *Salmonella typhi* and *Pseudomonas aeruginosa*. *N. sativa* seeds methanolic extract has a potent antifungal activity. Both Gram −ive and Gram +ive bacteria were resistant to the essential oil. However, it was shown to be difficult to have compassion towards Gram positive microorganisms like *Staphylococcus aureus* and *Vibrio cholera*. *N. sativa* is particularly susceptible to some bacteria, including *Staphylococcus viridans*, *S. pyogenes*, *S. aureus*, *Corynebacterium renale*, *E. coli*, *Pasteurella multocida*, *Listeria monocytogenes*, *Trueperella* (*Arcanobacterium*) *pyogenes*, *Mannheimia haemolytica*, *Corynebacterium pseudotuberculosis*. In an *in vitro* test, volatile oil outperformed ampicillin in terms of antibacterial activity. The volatile oil was also shown to have synergistic effects with ampicillin, doxycycline, terbinafine, cephalexin, and chloramphenicol against strains which are resistant to drug including *Shigella* species, *Vibrio cholera*, and *Escherichia coli*. Additionally, it was discovered that *N. sativa* extract administered subcutaneously could successfully cure mice of a non-lethal *staphylococcal* infection [31, 32].

28.7.2 Antioxidant Activity

N. sativa antioxidant properties make it a valuable substance for treating and preventing ischemic disease and central nervous system diseases. In male rats, co-administration of NSO and cisplatin reverses the oxidative stress that cisplatin had caused in the testicles. Antioxidant substances found in *N. sativa* extract include 4-terpineol, TQ (thymoquinone), t-anethole, and carvacrol. TQ predominate antioxidant properties reduce reactive oxygen species production in an indirect manner, which prevents lipid peroxidation. Rats receiving TQ or NSO injections during an ischemia phase experience reperfusion, which boosts their levels and activity of superoxide dismutase and glutathione peroxidase. Thymoquinone restores the antioxidant non-enzymatic (vitamin C and GSH) and enzymatic (GPX, SOD, GST, and CAT) activities. However, it also brings malondialdehyde (MDA) levels in the mouse brain back to normal. Pre-treatment of TQ reduces the addition of enzyme activities like SOD, CAT, and glutathione peroxidase as well as restores the elevated levels of conjugated diene and MDA to normal. Other NSO complexes, such 4-terpineol, carvacrol, and quinone, are outstanding in bringing free radicals together to neutralize each other [33]. Iron combined with *N. sativa* prevents oxidation. Antioxidant properties of *N. sativa* help eliminate free radicals in conditions like cirrhosis or liver injury. Alkaloids, flavonoids, and tannin are present in *N. sativa* species from Khorasan hydro alcoholic extract. *N. sativa* contains flavonols glycosides and aglycones, which are flavonoids. These flavonoids function as a revealer of superoxide radicals in the blood to avoid oxidation and eliminate free radicals in cells because they have stronger antioxidant qualities and higher anti-radical effects [34].

28.7.3 Anti-inflammatory Activity

Arthritis and asthma are two chronic inflammatory diseases that are connected to a number of other inflammatory diseases and distinct pathways. Thymoquinone and *N. sativa* seed oil were reported to inhibit leucocyte eicosanoid synthesis and peroxidation of lipid membrane. Additionally, a fall in rat paw oedema and granuloma pouch weight was observed. Low concentrations of nigellone are efficient in inhibiting the production of histamine from mast cells, which continue to have an anti-asthmatic role in plants [34].

The oil from *N. sativa* seeds has long been used to treat inflammatory disorders including rheumatism, skin rashes, and back discomfort. TQ and *N. sativa* seeds oil have anti-inflammatory effects by preventing the formation of these substances. Inflammation conditions in the body are produced by cytokines, neutrophils, eosinophils, inflammatory macrophages, and oxidants. It has been discovered that administration of an aqueous extract of TQ or *N. sativa* to neutrophils activated by calcium ionophore inhibits the formation of 5-lipoxigenase [34–36].

In an *in vivo* study of peritonitis, edoema, encephalomyelitis, arthritis, and colitis, anti-inflammatory actions was observed with thymoquinone including suppression of inflammatory leukotrienes, mediators, and prostaglandins. Thymoquinone suppression of iNOS is thought to be the cause of the anti-inflammatory actions of *N. sativa* extract and TQ on LPS-induced inflammation in macrophages and mixed glial cells. This reduction in NO generation by these cells is likely what causes these effects (TQ) [35, 36].

28.7.4 Anti-Cancer Activity

A variety of neoplasms have been shown to respond favourably to *N. sativa* seeds extract, some of its energetic compounds, particularly -hederin and TQ, and its essential oil. The anti-cancerous properties of *N. sativa* have been ascribed to a variety of mechanisms, including the reduction of cell growth, the initiation of apoptosis in cancer cells via a pathway independent and dependent of p53, alterations in intracellular glutathione levels, influences on enzyme activity, and the capture of free radicals. Increased cancer cell apoptosis, growth inhibition, and morphological changes on colon cancer cells are the effects of TQ anti-cancer mechanism. TQ causes apoptosis by decreasing the expression of the target mRNAs gene for p21WAF and P53 and by inhibiting anti-apoptotic proteins (BCL-2). *N. sativa* methanol extract exhibits weak cytotoxicity to healthy lymphocytes but strong cytotoxicity to Ehrlich ascites carcinoma and Dalton's ascitic lymphoma cells. When compared to normal L929 cells, the ethanol extract of *N. sativa* has been shown to have apoptotic and cytotoxic effects on the ACHN line of renal cancer cells. With less toxicity on healthy lymphocytes, *N. sativa* seed extract has the potential to have anti-cancer effects on lymphoma-180 cells and sarcoma cells. The modern use of *N. sativa* extract inhibits the early stages of skin cancer in mice. TQ induces apoptosis in the human osteosarcoma cell line $SaOS^{-2}$, a higher percentage of growth inhibition, and blocks composition of human umbilical vein endothelial cell tube in a dose-dependent manner. These effects of TQ on osteosarcoma *in vivo* and *in vitro* have been investigated. In $SaOS^{-2}$ cells, TQ appears to normalize VEGF, XIAP, surviving, and NF-B DNA-binding activity. In $SaOS^{-2}$ cells, TQ treatments increase the levels of caspase-3 and cleaved SMAC. TQ effectively prevents tumor angiogenesis and tumor growth both *in vivo* and *in vitro* [37].

28.7.5 Anti-hyperlipidemic Activity

Greater than the palliative treatment within 2 months of involvement, *N. sativa* significantly contributes to improved lipid profiles in menopausal women increased HDL-c, LDL-c, TG, and decreased total cholesterol. After 1 month of treatment discontinuation, the profiles of lipid in the *N. sativa* treatment groups have returned to their pre-treatment levels. The potential hypolipidemic effects of *N. sativa* oil on healthy volunteers were investigated and their findings revealed a significant decrease in triglyceride, LDL, and HbA1C levels in NSO-treated volunteers compared to the placebo group. The effects of *N. sativa* seeds aqueous extract on some biochemical variables, body weight, and food intake in rats were also studied. It was concluded, *N. sativa* reduces calorie intake and body weight without reducing water intake. Low density lipoproteins (LDL), blood sugar, and serum cholesterol all decreased significantly while HDL increased [38].

28.7.6 Anti-diabetic Activity

Several traditional medicine experts have suggested *N. sativa* as a treatment for diabetes. The essential oils present in *N. sativa* seeds showed noticeable impact on lowering sugar levels of blood [39]. The anti-diabetic properties of *N. sativa* have been induced, which affects cellular uptake of proteins with hypolipidemic effects, by activating adenosine monophosphate kinase (AMPK) [40]. With oral administration, volatile *N. sativa* oil (2 $mg\bullet^{-1}$ $kg\bullet^{-1}$ BW) significantly lowers blood sugar levels in Balb/c mice. A significant hypoglycemic effect was produced by intraperitoneal injection of NSO (50 $mg\bullet^{-1}$ kg^{-1}) in both alloxan-diabetic rabbits and fasting normal rabbits [41]. *N. sativa* seed extract increases muscle Glut4 content and ACC phosphorylation, acting to increase insulin sensitivity which is a main constituent of the insulin independent AMPK signaling pathway. Additionally, *N. sativa* extract reduces the generation of free radicals, relative cell proliferation, and beta cell regeneration in streptozotocin diabetic rats. In rats, treatment with *N. sativa* seeds oil significantly lowers serum glucose and tissue MDA levels while raising tissue SOD and serum insulin levels. In the clinical treatment of diabetes, TQ may be useful in preventing oxidative stress on cells. It has been shown that rats were given 25 days treatment of TQ (in drinking water) and *N. sativa* powder (mixed with edible food) and analysis was performed by studying hematologic factors, confirmed that TQ caused a substantial decrease in level of blood sugar in control group rats. Administration of *N. sativa* extract for 2 months in rabbits improves histological and biochemical symptom of liver damage and significantly lowers ceruloplasm and blood glucose levels. Fructose 1.6 bisphosphatase, a component of the pentose phosphate pathway, and glucose 6-phosphatase, a component of gluconeogenesis, are both inhibited by *N. sativa* in living things [42].

28.7.7 Gastro-protective Activity

N. sativa exhibits anti-ulcer effects due anti-secretory, anti-oxidant and prostaglandin-mediated activities. Aqueous *N. sativa* extract is said to reduce aspirin-induced stomach ulcer indexes in rats by 36%. Histamine levels are reduced for 2 weeks when NSO is used at a dose of 0/88 $g•kg•^{-1}d^{-1}$ while glutathione and mucin levels are elevated in the stomach. In Wistar rats, it is observed that doses of 50 and 100 $mg•^{-1}$ kg^{-1} of TQ have protective effects against gastric ulcer by reducing the effects of ischemia-reperfusion, gastric ulcer, and antioxidants [43–45].

28.7.8 Cardiovascular-protective Activity

Administration of oral dosing of *N. sativa* extract after a long period of time lowers the incidence of cardiac issues and vascular contractile reactivity in diabetics [45]. The influence of *N. sativa* extract on diabetic rabbit cardiovascular actions revealed considerable asymmetrical heart activity in diabetic rats. The acute effects of diesel exhaust particles (DEP) at 4 and 18 h and protective effects of TQ on cardiopulmonary factors were studied in mice, and the results showed that pretreatment of TQ showed hypotensive effect, lowers plasma SOD activity, increases IL-6, and prevents leukocytosis. It also avoids a drop in prothrombotic events and platelet counts. Furthermore, the influence of *N. sativa* oil on blood glucose, homeostasis, and cholesterol levels have been studied, and the results show that NSO reduces TG, glucose, and serum cholesterol while increasing hematocrit, haemoglobin, platelets, and leukocytes in rats. *N. sativa* extract at 800 $mg•kg^{-1}$ dosage for 12 weeks reduced heart tissue damage induced by ischemia reperfusion. Dose of essential oil constituent in anaesthetized rats at 432 g•mL^{1} generated a dosage dependent drop in heart rate and blood pressure that was prevented by anticholinergic. 0.6 mL•kg^{-1} •d^{-1}Dichloromethane extract of *N. sativa* was shown to showed hypotension effect in hypertensive spontaneously rats [46].

28.7.9 Immuno-protective Activity

Estearidonic acid, linolenic acid and other compounds of *N. sativa* stimulate the immunological response, particularly T cells. *N. sativa* stimulates natural killer cells and raises the percentage of helper T cells to suppress T cells. In comparison to immunotherapy alone, supplementing 2 g of *N. sativa* extract for 30 days with immunotherapy leads in an enhancement in phagocytic action of macrophages. In rats, NSO usage for 6 weeks increases the quantity of blood lymphocytes and the phagocytic activity of peritoneal macrophages. *N. sativa* regulates Th1/Th2 cytokine production as well as anti-inflammatory and pro-inflammatory cytokine secretion. The effects of *N. sativa* on immuno-hematological processes on rainbow trout are researched, leading in a considerable rise in serum immunoglobulin levels. In BALB/c mice, the immunomodulatory effects of a collection of medicinal herbs comprising *N. sativa* are studied. *N. sativa* immunomodulatory activity has therapeutic significance in the supportive and preventive treatment of opportunistic illnesses [45].

28.7.10 Neuro-protective Activity

Thymoquinone (TQ) slows the degeneration of neurons in *N. sativa* extract. TQ causes the central nervous system's opioid receptors to get stimulated, which has an analgesic effect. Treatment with intracranial thymoquinone reduces seizures induced with PTZ and performed this action by raising the tone of the gabaergic and opioid receptor sites. In one study LPS was used to induce depression in an animal model and effect was analyzed by forced swim and open field test. *N. sativa* (200–400 mg•kg^{1}) results in a loss of depression. NSO has therapeutic potential by preventing oxidative stress overproduction and drug-induced nitric oxide [47, 48].

28.7.11 Wound Healing Activity

It has been discovered that *N. sativa* seeds oil in farm animals promote wound healing. When applied topically to mice with skin infected with *staphylococcus*, *N. sativa* seed ether extract promotes healing by reducing tissue damage, inflammation, and absolute differential WBC numbers as well as local infection [49, 50]. The positive effects of *N. sativa* extracts and silver sulfadiazine (SSD) cream were examined on burn wounds in another *in vivo* study. Histopathological analyses performed on the

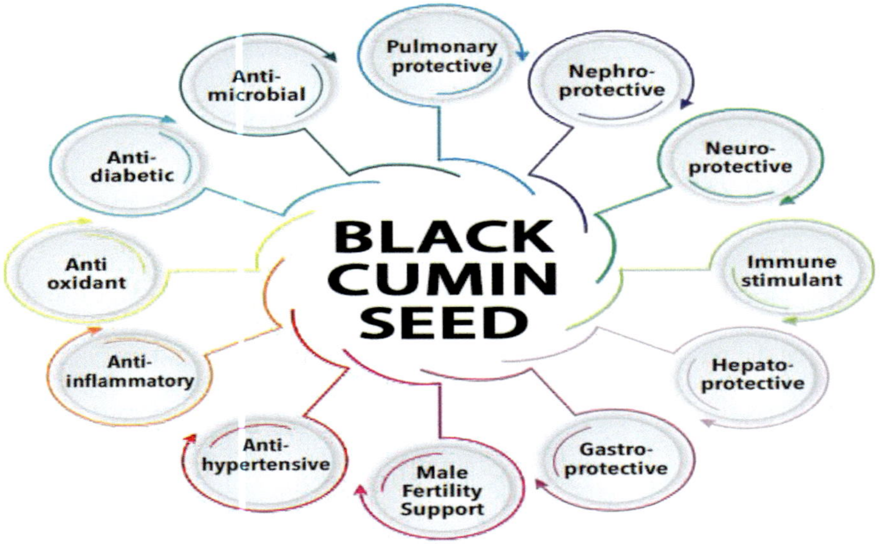

Fig. 28.9 Therapeutic applications of *N. sativa*

burns in the treatment (*N. sativa* extract) and standard control (SSD) groups on the fourth, ninth, and fourteenth days demonstrated that burn healing was superior compared to the control group. The capacity of NSO to contract for wound healing is comparable to that of Baneocin (antibiotic). In contrast to *Pluchea indica, Piper sarmentosum, and Melastoma malabathricum* extracts, *N. sativa extract* exhibits a potential improvement in human gingival fibroblast proliferation through rapid wound healing. In addition, up to 15% rise in basic Fibroblast growth factor at 100 g•mL^1 of *N. sativa* extract, a somewhat greater impact was seen on the expression of TGF. Consequently, *N. sativa* has beneficial effects in healing wounds, supporting its traditional usage for the treatment of oral wounds (Fig. 28.9 and Table 28.1) [50].

28.8 Prevention and Side Effects

There are no known side effects or health risks associated with the proper management of prescribed therapeutic levels [51].

28.9 Safety Profile

N. sativa seeds have a long antiquity of use for medicinal purposes and food. When used within the recommended dosage no contrary or side effects have been reported, though dermatitis has been reported (Table 28.2) [51].

Table 28.1 Pharmacological effects of different extracts of *N. sativa* and their mechanism of action in different studies

Sr. No.	Pharmacological effect	Type of extracts	Extract dosage	Study type	Mechanism of action
1	Antibacterial activity	*N. sativa* etheric extract	25–400 µg/ mL^{-1}	*In-vitro* test	Combining antibacterial drugs with herbal products would increase their effectiveness.
			2 mg/kg	*In-vivo* test	Ether extract cured mice with a localized *Staphylococcus aureus* infection
		N. sativa ethanolic extract	3–200 µg/mL	*In vitro* test	Ethanolic extract prevent the progression of all pathogenic bacteria.
		N. sativa seeds oil (volatile)	120–600 µg/ mL	*In vitro* test	All of the species studied were extremely sensitive to volatile oils.
2	Anti-fungal activity	*N. sativa* aqueous extract	6.6 mL/kg	*In vivo* test	Aqueous extract decreases five-fold candida in kidney. Aqueous extract decreases eight-fold candida in liver. Aqueous extract decreases11-fold candida in spleen.
		N. sativa etheric extract	2.5–40 mg/ mL	*In vitro* test	Etheric extract inhibit growth in eight isolates of dermatophytes.
3	Anti-schistosomal	*N. sativa* aqueous extract	0.2 mg/kg	*In vivo* activity	Aqueous extract protected rats from the bulk of schistosomiasis-induced hematological and biochemical changes. Schistosomiasis rats with aqueous extract had better anti-oxidant capabilities.
		N. sativa powder extract	10 mg	*In vitro* test	The extract of *N. sativa* decreased the quantity of *S. mansoni* worms and ova development in the liver.
4	Antioxidant activity	*N. sativa* decoction	3 mg/kg	*In vivo* assay	Decoction of *N. sativa* decreased blood LPO. And increases plasma TTM.
		N. sativa thymoquinone	5 mg/kg	*In vivo* test	Thymoquinone raised levels of conjugated diene and malondialdehyde. Thymoquinone also boosted the activities of catalase, glutathione, and super oxide dismutase enzyme.

(continued)

Table 28.1 (continued)

Sr. No.	Pharmacological effect	Type of extracts	Extract dosage	Study type	Mechanism of action
5	Anti-inflammatory activity	*N. sativa* thymoquinone	3 mg/kg	*In vivo* activity	Thymoquinone raises the neutrophil count in lung lavage fluid and EC50. Thymoquinone reduced the tracheal sensitivity of LLF lymphocytes to methacholine.
		N. sativa ethanolic extract	0–2000 μg/mL	*In vitro* test	Ethanolic extract decreased the proportion of living cells. ACHN cells showed maximal apoptosis at the concentration of 1000-1250 g/mL as compared to L929 cells, which displayed maximum apoptosis at 1500 g/mL.
6	Antihyperlipidemic activity	*N. sativa* powder extract	500 mg	Clinical study	*N. sativa* powder lowers density of lipoprotein, cholesterol, and triglycerides. Additionally, high density lipoprotein and cholesterol was raised by *N. sativa*.
		N. sativa powder extract	1–3 g	Clinical	*N. sativa* powder increased HDL-c/LDL-c and lowered TC, TG, and LDL-c.
		N. sativa seeds oil	2.5 mL	Clinical	*N. sativa* seed oil lowered fasting blood cholesterol, LDL, triglyceride (TG), glucose, and HbA1C levels.
		N. sativa seeds oil	50–100 μg/mL	*In vitro* study	*N. sativa* seeds oils increased type 1 plasminogen activator inhibitor.
7	Anticancer activity	*N. sativa* thymoquinone	20–80 μmol/L	*In vitro* assay	Thymoquinone increased the expression of cleaved caspase-3 and SMAC in $SaOS^{-2}$ cells while decreasing the expression of nuclear factor- κB DNA-binding activity, vascular endothelial growth factors, survivin, and X-linked inhibitor of apoptosis protein.
		N. sativa aqueous extract	50 mg/kg	*In vivo* test	Aqueous *N. sativa* extract reduced the genotoxicity and ultrastructural alterations brought on by CCl_4 dosages.
		N. sativa seeds oil	0.20 mL/kg	*In vivo* activity	The oil of *N. sativa* seeds induced partial proliferation of pancreatic-cells and lowered serum glucose levels while simultaneously decreasing serum insulin levels.

Table 28.1 (continued)

Sr. No.	Pharmacological effect	Type of extracts	Extract dosage	Study type	Mechanism of action
8	Antidiabetic activity	*N. sativa* aqueous extract	2 mL/kg	*In vivo* activity	*N. sativa* reduced tissue malondialdehyde and blood glucose, increased tissue super oxide dismutase and serum insulin.
		N. sativa seeds oil	0.2 mL/kg		
		N. sativa thymoquinone extract	3 mg/mL		
9	Cardiovascular protective activity	*N. sativa* powder extract	*N. sativa* mixed pelleted food (6.25%)	*In vivo* activity	*N. sativa* powder reduced maximal contractile in the and also stopped leukocytosis and SBP decline brought on by DEP.
		N. sativa thymoquinone	6 mg/kg	*In vivo* activity	Thymoquinone increased interleukin-6 (IL-6) concentration.
		N. sativa aqueous extract	20 mL/kg	*In vivo* assay	An aqueous extract of N. sativa reduced high glucose concentrations, red blood cell counts, white blood cell counts, packed cell volume, neutrophil count, and heart rate. Aqueous *N. sativa* extract also inhibited the development of stomach ulcers brought on by necrotizing agents.
10	Gastroprotective activity	*N. sativa* thymoquinone	5–25 mg/kg	*In vivo assay*	Thymoquinone lowered body weight, malondialdehyde levels and colonic myeloperoxidase activity. Thymoquinone raised levels of glutathione.
		N. sativa seeds oil	2.5 and 5 mL/kg	*In vivo* test	The oil from *N. sativa* seeds normalized the levels of lipid peroxide, glutathione, superoxide dismutase and lactate dehydrogenase. The glutathione content was reduced by *N. sativa* seeds oil at high dosages.
		N. sativa thymoquinone	5–100 mg/kg		
		N. sativa powder	0.01–1 g/kg	*In vivo* test	High dose of *N. sativa* were effective for serum creatinine.

(continued)

Table 28.1 (continued)

Sr. No.	Pharmacological effect	Type of extracts	Extract dosage	Study type	Mechanism of action
11	Nephron protective activity	*N. sativa* thymoquinone	2.5–5.0 mL/kg	*In vivo* activity	The oil from *N. sativa* seeds reduced TG, hepatic MDA, and plasma transaminase activity.
		N. sativa thymoquinone	13 μg/kg	*In vivo* activity	Thymoquinone reduced the severity of the diarrhea. Thymoquinone decrease significantly both total plasma IgE and OVA-specific IgE.
		N. sativa seeds oil	2 mL/kg	*In vivo* assay	MDA levels in the liver and erythrocytes were reduced by thymoquinone seed oil. Thymoquinone seed oil reduced the levels of superoxide dismutase, catalase and glutathione peroxidase.
12	Hepatoprotective activity	*N. sativa* hydroalcoholic extract	200–400 mg/kg	*In vivo*	In both the forced swim test and the open field test, a hydroalcoholic extract of *N. sativa* resulted in elimination of depression.
		N. sativa seeds oil	4 mL/kg	*In vivo* assay	The oils from *N. sativa* seeds decreased tramadol dependency. Additionally, stop the overproduction of nitric oxide. *N. sativa* oil also raised the levels of malondialdehyde.
13	Neuro-protective activity	*N. sativa* essential oil (thymoquinone)	5 mL/kg	*In vivo* test	Thymoquinone increased levels of malondialdehyde while reducing the amount of dead hippocampus neuronal cells. Thymoquinone boosted the activities of GSH, catalase, and SOD.
		Ethanolic	2 g	*In vivo* test	Ethanolic extract increased peritoneal macrophages phagocytic activity.
14	Effect on reproductive system	Alcoholic	200–400 mg/kg	*In vivo* assay	Alcoholic extract of *N. sativa* increased the testes, sperm count weight, epididymidis, concentration of blood testosterone, fertility index and luteinizing hormone.
15	Immunoprotective activity	Ethanolic	200 mg/mL	*In vivo* activity	Ethanolic extract altered Th1/Th2 cytokine production and modified secretion of inflammatory cytokine.

Table 28.2 Pharmacological properties and efficacy mechanism of *N. sativa*

Sr. No.	Pharmacological effect	Potential phytochemicals	Mechanism of efficacy
1	Anti-microbial activity	Thymohydroquinone and melanin	*N. sativa* contains active phytochemicals that have anti-microbial effects.
2	Anti-oxidant activity	Thymoquinone, carvacrol, t-anethole and 4-terpineol	Antioxidant constituents remove and capture free radicals. Lower MDA levels. Suppress lipid peroxidation.
3	Anti-inflammatory activity	Thymoquinone	Inhibits mediators which cause inflammation such as inteleukin-1 (IL-1), interleukin (IL-6), LOXs (lypo-oxygenase), and nitric oxide synthase enzymes.
4	Anti-hyperlipidemic	Sterol, mesalazine, thymoquinone and HMG-CoA reductase	Affect cholesterol production. Improves the ability of liver cells to eliminate low density lipoproteins from the bloodstream.
5	Immuno-protective activity	Thymoquinone, alpha-linoleic acid and stearic acid	Increase macrophage activity and lymphocyte counts to boost the immune system. Inhibits inflammatory processes Some seed components fortify the immunological response, particularly in T cells.
6	Anti-cancer activity	Saponins, thymoquinone and alpha-hederin	Suppress the cell growth Activate process of apoptosis via P53 independent and dependent pathways. Helps in glutathione alteration.
7	Anti-diabetic activity	Thymoquinone	Regulate the activity of liver enzymes involved in glucose metabolism Decreasing gluconeogenesis Increase antioxidant capacity Maintain and promote the growth of pancreatic beta cells Activate AMP- activated protein kinase via thymoquinone.
8	Neuro-protective activity	Thymoquinone	Slows down neuronal deterioration.
9	Hepato and nephro-protective activity	Thymoquinone	Decrease glutathione. Antioxidant effects of thymoquinone have a protective impact on the liver and kidneys.
10	Wound healing activity	Thymoquinone	Increase the production of beta-FGF Enhance fibroblast growth.
11	Protection of reproductive system	Thymoquinone	Inhibit the enzymes cyclooxygenase and lipo-oxygenase, which helps in minimizing the negative effects of free radicals. Sexual cells are protected against death and loss of tissue weight or volume.

28.10 Myths, Legends, Tales, Folklore and Interesting Facts

- *N. sativa* is a herb that originated in the Mediterranean and Asian countries and has been cultivated for over 3000 years for a variety of purposes.
- *N. sativa* has been labelled as the "Secret of the Pharaohs" because of its wide spread use in ancient Egypt; it was even discovered in Tutankhamen's tomb.
- The ancient Egyptians employed *N. sativa* to cure stomach issues as well as inflammatory skin diseases such as rashes and bites. Its potential advantages were also noted in other parts of the world, where *N. sativa* was used to cure headaches, congestion, and pain by ancient Greek physicians [52].
- Similarly, a Persian physician praised *N. sativa* for human medicine, noting that it could be used as an antifungal and anti-inflammatory ointment [52].
- *N. sativa* was utilized in Ayurvedic medicine to improve longevity and metabolism.
- Throughout ancient times, *N. sativa* seeds have been effectively used to maintain people's health. Numerous researches have shown that has a wide range of pharmacological potential. The actual importance of *N. sativa* to Muslims comes from the holy word of the Prophet Mohammed, "*N. sativa* is the treatment for every ailment except death," which is supported by both religious and scientific data. *N. sativa* is also cited in both the Old and New Testaments of the Bible, where it is believed that *N. sativa* was used as a currency to pay tithe to priests in ancient Rome and Greece [52].
- Avicenna mention *N. sativa* in his renowned book "The Canon of Medicine" like seeds help to recover *N. sativa* from depression, fatigue and also stimulate the energy of the body.
- The usage of *N. sativa* in Arabian and Indian nations as medicine and food has a long history of folktale.
- In the Southeast Asian and Central East countries *N. sativa* seeds have been used traditionally to cure many diseases and ailments like rheumatism, bronchitis, asthma and correlated inflammatory diseases [52].

28.11 Conclusion

Medicinal herbs have attracted a lot of interest since they are less expensive, easier to get, and have fewer side effect profiles than synthetic drugs. Religious and cultural traditions employ various healing plants and their constituents. Many human communities across the world, especially Muslim people, have used *N. sativa* for millennia to treat a wide range of illnesses. Numerous studies have confirmed that *N. sativa* constituents, such as thymoquinone, have a spectacular natural remedy for the treatment of a variety of diseases, including non-communicable diseases such as neurological disorders, diabetes, high blood pressure, cholesterol, inflammation, and tumor and communicable diseases such as parasitic, bacterial, viral and fungal.

Research on both humans and animals has shown that *N. sativa* may treat various diseases, and their antioxidant benefits have recently come to light since they are used as nutritional supplements with little or no negative effects. Thus, *N. sativa* is an excellent therapeutic medicinal plant that may be used to treat a variety of illnesses.

References

1. Hannan, M. A., Rahman, M. A., Sohag, A. A. M., Uddin, M. J., Dash, R., Sikder, M. H., & Kim, B. (2021). Black cumin (*Nigella sativa* L.): A comprehensive review on phytochemistry, health benefits, molecular pharmacology, and safety. *Nutrients, 13*(6), 1784.
2. Jamshidi-Kia, F., Lorigooini, Z., & Amini-Khoei, H. (2017). Medicinal plants: Past history and future perspective. *Journal of Herbmed Pharmacology, 7*(1), 1–7.
3. Hassanien, M. F., Assiri, A. M., Alzohairy, A. M., & Oraby, H. F. (2015). Health-promoting value and food applications of black cumin essential oil: An overview. *Journal of Food Science and Technology, 52*, 6136–6142.
4. Anwar, F., Muhammad, G., Hussain, M. A., Zengin, G., Alkharfy, K. M., Ashraf, M., & Gilani, A. H. (2016). *Capparis spinosa* L.: A plant with high potential for development of functional foods and nutraceuticals/pharmaceuticals. *International Journal of Pharmacology, 12*(3), 201–219.
5. Tariq, M. (2008). *Nigella sativa* seeds: Folklore treatment in modern day medicine. *Saudi Journal of Gastroenterology: Official journal of the Saudi Gastroenterology Association, 14*(3), 105.
6. Ansary, J., Giampieri, A., & J., Giampieri, F., Forbes-Hernandez, T. Y., Regolo, L., Quinzi, D., Gracia Villar, S., & Cianciosi, D. (2021). Nutritional value and preventive role of *Nigella sativa* L. and its main component thymoquinone in cancer: An evidenced-based review of preclinical and clinical studies. *Molecules, 26*(8), 2108.
7. Ansary, J., Giampieri, F., Forbes-Hernandez, T. Y., Regolo, L., Quinzi, D., Gracia Villar, S., & Cianciosi, D. (2021). Nutritional value and preventive role of *Nigella sativa* L. and its main component thymoquinone in cancer: An evidenced-based review of preclinical and clinical studies. *Molecules, 26*(8), 2108.
8. Shariq, I. M., Israil, A. M., Iqbal, A., & Brijesh, P. (2015). Morpho-physiological characterization of seeds and seedlings of *Nigella sativa* Linn.: Study on Indian germplasm. *International Research Journal of Biological Sciences, 4*(4), 38–42.
9. Nimet, K. A. R. A., Katar, D., & Baydar, H. (2015). Yield and quality of black cumin (*Nigella sativa* L.) populations: The effect of ecological conditions. *Turkish Journal of Field Crops, 20*(1), 9–14.
10. Kifelew, H., Getachew, W., Luleseged, T., Mitiku, H., Bekele, D., & Fikere, D. (2017). Seed spices production guideline.
11. Mahmood, T. (2013). Growth and yield attributes of black cumin (*Nigellla sativa* L.) as affected by sowing dates and methods. *Mycopathologia, 10*(2).
12. Zapotoczny, P., Żuk-Gołaszewska, K., & Ropelewska, E. (2019). Impact of cultivation methods on properties of black cumin (*Nigella sativa* L.) seeds. *Journal of central European. Agriculture, 20*(1), 353–364.
13. Hadi, M. R. H. S., Ghanepasand, F., & Darzi, M. T. (2015). Evaluation of biofertilizer and manure effects on quantitative yield of *Nigella sativa* L. *International Journal of Agricultural and Biosystems Engineering, 9*(8), 890–893.
14. Maleki, J., Sharifi Ashourabadi, E., Mirza, M., Heydari Sharifabad, H., & Lebaschy, M. H. (2021). Improving the productivity and quality of black cumin (*Nigella sativa* L.) by using soil fertility management practices. *Journal of Plant Nutrition, 44*(12), 1741–1757.

15. Ariafar, S., & Forouzandeh, M. (2017). Evaluation of humic acid application on biochemical composition and yield of black cumin under limited irrigation condition. *Bulletin de La Société Royale Des Sciences de Liège, 86*, 13–24.
16. Haque, M., Singh, R., Nadeem, A., Rasool, S., Wani, J. A., Khan, A., & Zehra, U. (2022). *Nigella sativa*: A promise for industrial and agricultural economic growth. In *Black seeds (Nigella sativa)* (pp. 439–460). Elsevier.
17. Majeed, A., Muhammad, Z., Ahmad, H., Hayat, S. S. S., Inayat, N., & Siyyar, S. (2021). *Nigella sativa* L.: Uses in traditional and contemporary medicines – An overview. *Acta Ecologica Sinica, 41*(4), 253–258.
18. Wako, F. L. (2020). Black cumin (*Nigella sativa* L.) production: A mini review.
19. Ndirangu, E. G., Opiyo, S., & Ng'ang'a, M. W. (2020). Chemical composition and repellency of *Nigella sativa* L. seed essential oil against Anopheles gambiae sensu stricto. *Trends in Phytochemical Research, 4*(2), 77–84.
20. Teshome, W., & Anshiso, D. (2019). Assessment of production and utilization of black cumin (Nigella sativa) at the Oromia Regional State, Ethiopia. *Asian Journal of Agricultural Extension, Economics & Sociology*, 1–12.
21. Merga, J., Wakjira, G., Habetewold, K., & Abukiya, G. (2018). Survey and identification of black cumin (*Nigella sativa* L.) disease in Ethiopia. *International Journal of Research in Agriculture and Forestry, 5*(11), 31.
22. Ainane, T., Askaoui, Z., Elkouali, M., et al. (2015). Chemical composition and antibacterial activity of essential oil of *Nigella sativa* seeds from BeniMellal (Morocco): What is the most important part, essential oil or the rest of seeds? *Journal of Material and Environmental Science, 5*(6).
23. Pachaiappan, R., Nagasathiya, K., Singh, P. K., Gopalakrishnan, A. V., Velusamy, P., Ramasamy, K., & Gopinath, S. C. (2022). Phytochemical profile of black cumin (*Nigella sativa* L.) seed oil: Identification of bioactive anti-pathogenic compounds for traditional siddha formulation. *Biomass Conversion and Biorefinery*, 1–13.
24. Atta-ur-Rahman, M. S., Cun-Heng, H., & Clardy, J. (1992). Isolation and structure determination of Nigellicine, a novel alkaloid from the seeds of *Nigella sativa*. *Journal of Natural Products, 55*(5), 676–678.
25. Badalamenti, N., Modica, A., Bazan, G., Marino, P., & Bruno, M. (2022). The ethnobotany, phytochemistry, and biological properties of Nigella damascena – A review. *Phytochemistry, 113165.*
26. Matthaus, B., & Ozcan, M. M. (2011). Fatty acids, tocopherol, and sterol contents of some nigella species seed oil. *Czech Journal of Food Science, 29*(2), 145–150.
27. Salehi, B., Quispe, C., Imran, M., Ul-Haq, I., Živković, J., Abu-Reidah, I. M., & Sharifi-Rad, J. (2021). Nigella plants–traditional uses, bioactive phytoconstituents, preclinical and clinical studies. *Frontiers in Pharmacology, 12*, 625386.
28. Yimer, E. M., Tuem, K. B., Karim, A., Ur-Rehman, N., & Anwar, F. (2019). *Nigella sativa L. (black cumin): A promising natural remedy for wide range of illnesses*. Evidence-Based Complementary and Alternative Medicine.
29. Ahmad, M. F., Ahmad, F. A., Ashraf, S. A., Saad, H. H., Wahab, S., Khan, M. I., & Athar, M. T. (2021). An updated knowledge of Black seed (Nigella sativa Linn.): Review of phytochemical constituents and pharmacological properties. *Journal of Herbal Medicine, 25*, 100404.
30. Nazarparvar, M., Shakeri, A., & Ranjbariyan, A. (2020). Chemical composition and antimicrobial activity against food poisoning of alcoholic extract of *Nigella sativa* L. *Biointerface Research in Applied Chemistry, 10*, 6991–7001.
31. Sowmya, R. K. (2021). Antibacterial Activitiy and time-kill assay of Terminalia catappa L. and *Nigella sativa* L. against selected human pathogenic bacteria. *Journal of Pure Applied Microbiology, 15*(1), 285–299.
32. Jan, K., Ahmad, M., Rehman, S., Gani, A., & Khaqan, K. (2019). Effect of roasting on physicochemical and antioxidant properties of kalonji (Nigella sativa) seed flour. *Journal of Food Measurement and Characterization, 13*, 1364–1372.

33. Pop, R. M., Sabin, O., Suciu, Ş., Vesa, S. C., Socaci, S. A., Chedea, V. S., & Buzoianu, A. D. (2020). *Nigella sativa's* anti-inflammatory and antioxidative effects in experimental inflammation. *Antioxidants, 9*(10), 921.
34. Amin, B., & Hosseinzadeh, H. (2016). Black cumin (*Nigella sativa*) and its active constituent, thymoquinone: An overview on the analgesic and anti-inflammatory effects. *Planta Medica, 82*(01/02), 8–16.
35. El Mezayen, R., El Gazzar, M., Nicolls, M. R., Marecki, J. C., Dreskin, S. C., & Nomiyama, H. (2006). Effect of thymoquinone on cyclooxygenase expression and prostaglandin production in a mouse model of allergic airway inflammation. *Immunology Letters, 106*(1), 72–81.
36. Randhawa, M. A., & Alghamdi, M. S. (2011). Anticancer activity of *Nigella sativa* (black seed)—A review. *The American Journal of Chinese Medicine, 39*(06), 1075–1091.
37. Bano, F., Wajeeh, M., Baig, N., Naz, H., Akhtar, N., & Hajra, N. (2009). Antiobesity, antihyperlipidemic and hypoglycemic effects of the aqueous extract of *Nigella sativa* seeds (Kalongi). *Journal of Biochemistry and Molecular Biology, 42*(4), 136–140.
38. Kaatabi, H., Bamosa, A. O., Lebda, F. M., Al Elq, A. H., & Al-Sultan, A. I. (2012). Favorable impact of *Nigella sativa* seeds on lipid profile in type 2 diabetic patients. *Journal of Family & Community Medicine, 19*(3), 155.
39. Benhaddou-Andaloussi, A., Martineau, L. C., Vallerand, D., Haddad, Y., Afshar, A., Settaf, A., & Haddad, P. S. (2010). Multiple molecular targets underlie the antidiabetic effect of *Nigella sativa* seed extract in skeletal muscle, adipocyte and liver cells. *Diabetes, Obesity and Metabolism, 12*(2), 148–157.
40. Khazdair, M. R., Ghafari, S., & Sadeghi, M. (2021). Possible therapeutic effects of Nigella sativa and its thymoquinone on COVID-19. *Pharmaceutical Biology, 59*(1), 694–701.
41. Benhaddou-Andaloussi, A., Martineau, L., Vuong, T., Meddah, B., Madiraju, P., Settaf, A., & Haddad, P. S. (2011). The in vivo antidiabetic activity of Nigella sativa is mediated through activation of the AMPK pathway and increased muscle Glut4 content. *Evidence-Based Complementary and Alternative Medicine, 2011.*
42. Zaoui, A., Cherrah, Y., Lacaille-Dubois, M. A., Settaf, A., Amarouch, H., & Hassar, M. (2000). Diuretic and hypotensive effects of Nigella sativa on the spontaneously hypertensive rat. *Thérapie, 55*(3), 379–382.
43. Al Mofleh, I. A., Alhaider, A. A., Mossa, J. S., Al-Sohaibani, M. O., Al-Yahya, M. A., Rafatullah, S., & Shaik, S. A. (2008). Gastroprotective effect of an aqueous suspension of black cumin Nigella sativa on necrotizing agents-induced gastric injury in experimental animals. *Saudi Journal of Gastroenterology: Official Journal of the Saudi Gastroenterology Association, 14*(3), 128.
44. Akhtar, A. H., Ahmad, K. D., Gilani, S. N., & Nazir, A. (1996). Antiulcer effects of aqueous extracts of Nigella sativa and Pongamia pinnata in rats. *Fitoterapia (Milano), 67*(3), 195–199.
45. Fararh, K. M., Atoji, Y., Shimizu, Y., Shiina, T., Nikami, H., & Takewaki, T. (2004). Mechanisms of the hypoglycaemic and immunopotentiating effects of Nigella sativa L. oil in streptozotocin-induced diabetic hamsters. *Research in Veterinary Science, 77*(2), 123–129.
46. El Tahir, K. E., Ashour, M. M., & Al-Harbi, M. M. (1993). The cardiovascular actions of the volatile oil of the black seed (Nigella sativa) in rats: Elucidation of the mechanism of action. *General Pharmacology: The Vascular System, 24*(5), 1123–1131.
47. Tayarani-Najaran, Z., Sadeghnia, H. R., Asghari, M., & Mousavi, S. H. (2009). Neuroprotective effect of Nigella sativa hydro alcoholic extract on serum/glucose deprivation induced PC12 cells death. *Physiology and Pharmacology, 13*(3), 263–270.
48. Al-Majed, A. A., Al-Omar, F. A., & Nagi, M. N. (2006). Neuroprotective effects of thymoquinone against transient forebrain ischemia in the rat hippocampus. *European Journal of Pharmacology, 543*(1-3), 40–47.
49. Ahmed, I. H., Awad, M. A., El-Mahdy, M., Gohar, H. M., & Ghanem, A. M. (1995). The effect of some medicinal plant extracts on wound healing in farm animals. *Assiut Veterinary Medical Journal, 32*(64), 236–244.

50. Yaman, I., Durmus, A. S., Ceribasi, S., & Yaman, M. (2010). Effects of Nigella sativa and silver sulfadiazine on burn wound healing in rats. *Veterinární Medicína, 55*(12), 619–624.
51. Dereli, F. T. G., Ilhan, M., & Belwal, T. (Eds.). (2022). *Novel drug targets with traditional herbal medicines: Scientific and clinical evidence*. Springer Nature.
52. Siddiqui, S., Khatoon, A., Ahmad, K., Upadhyay, S., Srivastava, A., Trivedi, A., & Arshad, M. (2022). Traditional Islamic herbal medicine and complementary therapies.

Chapter 29
Licorice

Zainab Maqbool, Mahnoor Amir, Arifa Zereen, Ghufrana Abid, and Shadma Wahab

29.1 Introduction

The shrub known as licorice, *Glycyrrhiza glabra* L., is a member of the Leguminosae family. It can grow as tall as 2.5 m. The leaves are imparipinnate and alternating, with a mean height of 7–15 cm and a range of 9–17 leaflets per leaf. Roots that are stoloniferous are well-developed. The roots are robust, branching, and coloured from red to yellow. Flowers are small, campanulate, zygomorphic, stipulate, and born in axillary spikes. Flowers range in hue from faint white blue to purple. The fruit is a 1.5 cm long compressed legume or pod that occasionally has glandular hairs and typically has three to five brown coloured reinform seeds inside [1].

Z. Maqbool · A. Zereen (✉)
Department of Botany, Government Graduate College for Women Samanabad, Lahore, Pakistan

M. Amir
College of Earth and Environmental Sciences, University of Punjab, Lahore, Pakistan

G. Abid
Department of Chemistry, Government College University, Lahore, Pakistan

S. Wahab
Department of Pharmacognosy, College of Pharmacy, King Khalid University, Abha, Saudi Arabia

Complementary and Alternative Medicine Unit, College of Pharmacy, King Khalid University, Abha, Saudi Arabia

M. Zia-Ul-Haq et al. (eds.), *Essentials of Medicinal and Aromatic Crops*,
https://doi.org/10.1007/978-3-031-35403-8_29

Kingdom:	Plantae
Order:	Fabales
Family:	Fabaceae
Genus:	*Glycyrrhiza*
Binomial name:	*Glycyrrhiza glabra* L.
English/ common name:	Liquorice, Persian Liquorice, common Liquorice, Liquorice-root, Liquorice, and Rhizoma *Glycyrrhiza*e Spanish juice, Spanish licorice, Si-Pei licorice, Sinkiang licorice, Russian licorice, and Russian Liquorice true licorice, sweet root, sweet wood, and sweet wood Liquorice
Urdu name:	Mulathi

29.2 Agronomy (Like Soil Conditions, Climate, Land Preparation, Planting, Manuring, Irrigation, Ecology Etc.)

29.2.1 Soil Conditions and Climate

Licorice, also known as *Glycyrrhiza*. It is a perennial herb from family Leguminosae. Globally, there are about 200 species that are highly valuable economically [1]. It does well in warm, subtropical, and temperate climates. It thrives in deep, dry, well-limed soil that is exposed to full sun [2]. On subtropical areas, licorice grows in deep, fertile sandy soils. The ideal licorice soil is sandy loam with no stones, which grows well in dry seasons and in warmer places with an annual rainfall of little more than 50 cm. According to a study, *G. glabra* can withstand drought and is found in the floodplains of the River Amudarya [2] (Fig. 29.1).

Fig. 29.1 *Glycyrrhiza glabra* (**a**) Plant (**b**) root

29.2.2 *Propagation*

Rhizome cuttings or harvested crowns are used to grow the crop. Since the plant's vegetative parts take years to develop, vegetative part propagation is not advantageous. As planting material, crowns and 15–25 cm long stolons with two to three eye buds are acceptable [3]. Traditional seed propagation is not possible due to inadequate flowering, a lack of seed supply, loss of seed viability during storage, and slow growth rates [4]. In a study, different stages of the *G. glabra* seed's germination potential were detected. *G. glabra* seeds had the highest rate of germination in the waxy-ripe stage. The greatest quality of licorice was produced when blooms were removed right away as they appeared. Licorice root production is increased by adding 0.0025% succinic acid, followed by nitrogen and phosphorus [2].

29.2.3 *Land Preparation and Planting*

The ground is laboriously ploughed. Rhizome cuttings are sown in fertile soils 6–8 cm deep at a distance of 90 45 cm in ridges 40–65 cm high in February, March, and July, August. The distance could be shortened to 90 45 cm in areas with medium soil fertility. Within 15–20 days of seeding, cuttings begin to germinate. One can anticipate a 60–70% sprouting rate. After 40 days, the gaps are filled [5].

29.3 Manuring and Irrigation

Except in fertile soil, manuring is not recommended [6]. 10 metric tonnes of farmyard manure are applied each hectare. A base dose of 40 kg of nitrogen and phosphorous per hectare is added, and then 20 kilogramme of nitrogen per hectare is added as a top covering. A basal dose of 20 kg of potassium per hectare is applied to soil that is low in the mineral [5]. Until the cuttings take root, only a little water is sprinkled. After 30–45 days in the dry summer season, irrigation is required for the crop [6].

29.3.1 *Ecology*

Central Asia, Europe, South Western Asia, and the Mediterranean region are the original home of licorice. Effectively growing licorice include temperate, warm, and subtropical climates. It thrives best in deep fertile sandy loam soil that has been composted and is in full sun. It thrives in regions with average annual temperatures between 5.7 °C and 25 °C, mean annual precipitation between 400 mm and 650 mm, and soil pH between 5.5 and 8.2. Frosts have no effect on it. Winter dormancy causes the transfer of characteristics to the rhizomes [7].

29.3.2 Pests and Diseases

Although there are no pests on the plant, aphids, leaf-miners, spider mites, slugs, snails, and white flies could still exist. Under conditions when there is standing water, Pythium (root rot) is conceivable. On moist vegetation and in environments with high humidity, botrytis can develop [8].

Gentrospora acerina is the culprit for licorice rot. Both Heterodera *Glycyrrhiza*e and Acanthoscelides aureolus have been identified as licorice pests in the USSR and the USA. Myllocerus undecimpustulatus, a weevil, was found to be infecting G. glabra plants, according to CIMAP [9].

29.3.3 Description of Crops

The production period of licorice takes minimum 2–3 years. Licorice producing countries are Afghanistan, Azerbaijan, China, Iran, Iraq, Pakistan, Turkmenistan, Turkey, Uzbekistan etc. Fig. 29.2. The annual production of licorice been reported to be around 40,000–50,000 tons in 1991 and 19,500–22,500 tons during 1997–2016 [2, 10]. The creation of food additives (flavouring and sweetening agents), foaming agents, skin depigmentation (in the field of cosmetics), and the manufacture of pharmaceuticals are just a few of the industrial uses of *Glycyrrhiza* spp. [11]. The Indian, Egyptian, Chinese, Greek, and Roman cultures all used licorice roots and rhizomes as carminatives [12]. In addition, licorice is commonly used in the production of biomass, bioenergy, and pulp [13].

29.3.4 Important Chemical Constituents and Medicinal Uses

According to [14], the principal chemical components of are glycyrrhizin, glycyrrhetinic acid, glabridin, quercetin, liquiritigenin, isoliquiritigenin, licochalcone C, formononetin, licopyranocoumarin, glabrocoumarin, kanzonol Y, paratocarpin B, and glycyglabrone (Fig. 29.3),

1. Sucrose is 50% less sweet than glycyrrhizin. The major components of glycyrrhizin include triterpene, saponins, and flavonoids. It prevents the production of prostaglandins, particularly prostaglandin E2, cyclooxygenase activity, and the formation of platelets [13].

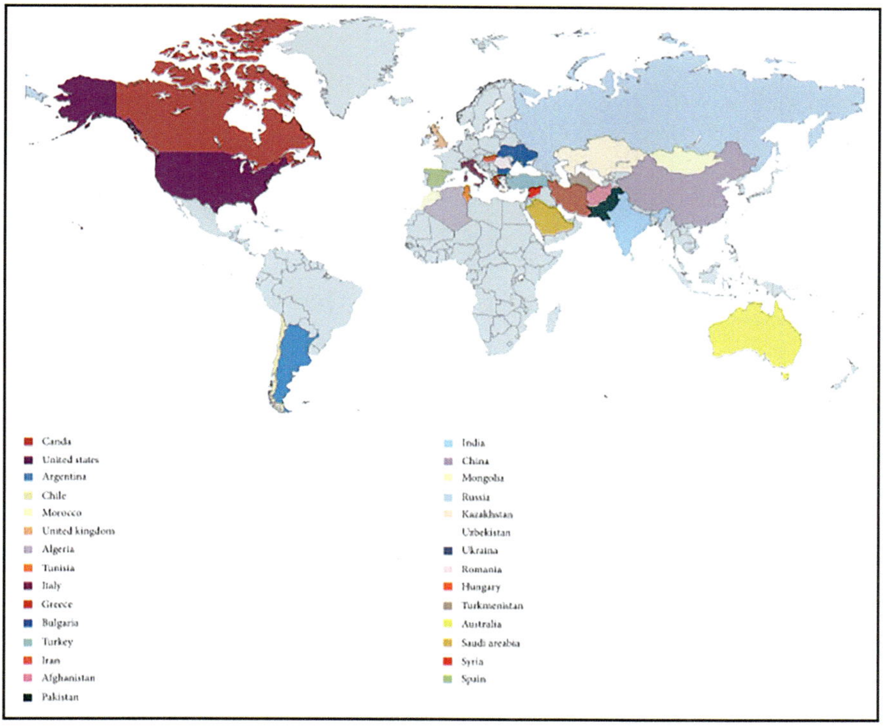

Fig. 29.2 Map of the world showing the biggest producers of *Glycyrrhiza* spp.

2. Glabrin A and B, isoflavones, 18-glycyrrhetinic acid, and glycyrrhizin are the phytoactive components of glycyrrhetinic acid. It inhibits 11-hydroxysteroid dehydrogenase and has anti-inflammatory properties [13].
3. The isoflavone glabridin. The item is a natural phenol, a more well-known chemical obtained from plants. Tyrosine and reactive oxygen species (ROS) synthesis are both hampered by glabridin, which also slows melanogenesis [14].
4. The flavonoid quercetin is obtained from plants. It lessens the generation of inflammatory metabolites and inhibits the activities of cyclooxygenase and lipoxygenase [15].
5. The chemical liquiditigenin is a phenol. It inhibits both NF- and the NLRP3 inflammasome [16].
6. The substance isoliquiritigenen is a phenol. By preventing AP-1 and NF- activity, it lessens macrophages' inflammatory activity [17].
7. The phenolic substance licochalcone C inhibits electron transport in the bacterial respiratory chain [18].
8. A bioactive isoflavone is called formononetin. Apoptosis is induced, and metastasis is stopped [19].

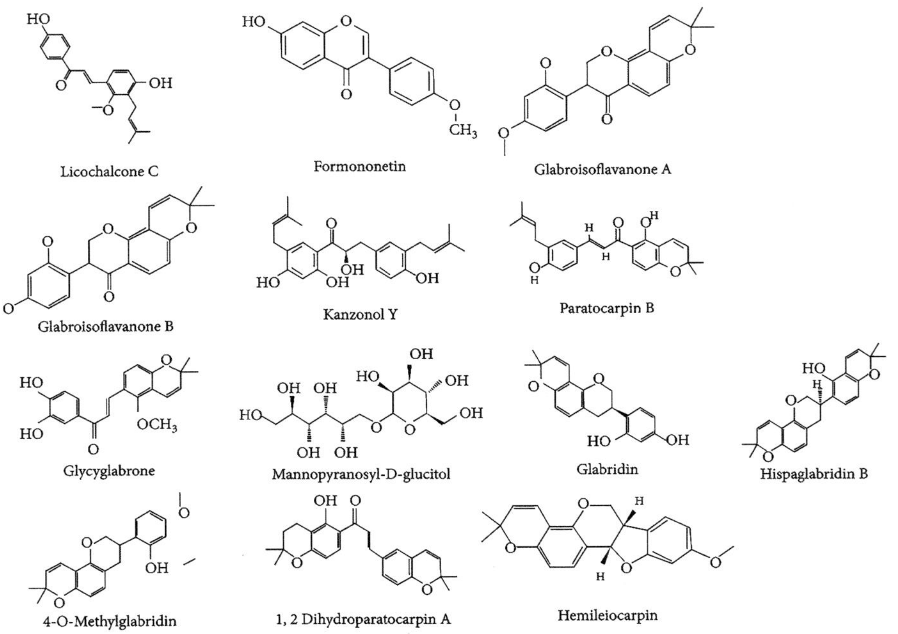

Fig. 29.3 Phytochemicals present in Licorice [124]

9. Coumarins include glabro-coumarin and licopyrano-coumarin. Both inhibit cell growth in HIV-infected cell cultures [20, 21].
10. Kanzonol Y is a chalcone. It inhibits the activity of *Bacillus subtilis* H17 [22].
11. Glycyglabrone and paratocarpin B are chalcone. Both act as powerful antioxidants [23, 24].
12. A mannose is mannopyranosyl-D glucitol [25].
13. The isoflavones glabridin and hispaglabridin B. Tyrosine activity is reduced by glabridin. Strong antioxidants include hispaglabridin [26, 27].
14. The isoflavan 4-O-Methylglabridin is. In-vitro antibacterial activity is present [28].

29.3.5 Medicinal Uses

Licorice roots and rhizomes are commonly used alone or in conjunction with other herbs to cure a variety of ailments, including colitis, ulcerative colitis, diarrhea, hyperdipsia, prostate cancer, fever, sciatica, arthritis, leucorrhoea, psoriasis, hemorrhagic disorders, malaria, and jaundice [29]. According to reports, licorice is used to treat inflammation, dermatitis, oral problems, liver conditions, stomach ulcers, lung conditions, dermatitis, and allergies [30].

29.3.6 *Antitussive Activity of Licorice*

Licorice has antitussive properties, and its powder and extracts are used to treat bronchial catarrh, coughs, and sore throats [13]. The bioactive substance glycyrrhizin in licorice is what gives it its antitussive properties. It promotes tracheal mucus secretion and lessens upper respiratory tract obstruction [31]. Liquiritinapioside, an active ingredient in licorice's methanolic extract, can block the cough-inducing chemical capsaicin [13]. Licorice's ethanolic extract was found to be able to prevent Sulfur Dioxide Gas-Induced Cough in experimental animals [32]. A study showed that licorice soothes inflammation and acts as a mild form of codeine for sore throats. By using carbenoxolone, a semi-synthetic substance produced from *Glycyrrhiza glabra*, gastric mucus secretion is increased [33].

29.4 Antiulcerogenic Activity

Since the 1070s, *G. glabra* have been employed as an antiulcerogenic due to their propensity to block the two enzymes delta 13-prostaglandin reductase and 15-hydroxyprostaglandin [34]. According to reports, 15-hydroxyprostaglandin converts prostaglandins (E2 and F2 alpha) into inactive 15-ketoprostaglandins and delta 13-prostaglandin reductase into 13,14-dihydroxy, 15-ketoprostaglandin [35]. *Glycyrrhiza* extract-derived carlbenoxolone exhibits antiulcerogenic action [36]. Taking 100 mg of licorice three times each day can treat duodenal and stomach ulcers. By increasing the secretion of prostaglandins from the digestive system and mucus from the stomach, it increases the lifetime of stomach surface cells [37].

29.5 *Glycyrrhiza glabra* Anti-cancer Activity

According to reports, *Glycyrrhiza* has the ability to fight cancer both in vivo and in a lab setting [31]. The bioactive substances glycyrrhizic and 18-glycyrrhetinic acids found in G. glabra are what give it its anticancer properties. Both substances start the mitochondrial permeability transition, which kills tumour cells [38]. The brine shrimp lethality experiment was used to test the cytotoxicity of methanolic extract of *Glycyrrhiza.* According to the study, *Glycyrrhiza* extract has an LC50 value of 0.77 g/ml [39]. Different licorice methanolic extract concentrations (0, 12.5, 25, 50, and 100 g/ml) were tested for their anticancer effects against the intestinal carcinoma cell line Caco 2 and the prostate carcinoma cell line PC-3. The findings demonstrated that methanolic extract has the ability to suppress the growth of Caco-2 and PC-3. According to the study, licorice has an excess of phytoestrogen chemicals, which gives them chemopreventive properties. It was determined that it is beneficial against stomach tumour, breast cancer, and ovarian cancer. Licorice extract is used in conjunction with cisplatin to lessen its negative effects [40]. By triggering AKT/mTOR signalling, glycyrrhizinic acid can stop breast and endometrial cancer cells from proliferating [41]. *Glycyrrhiza* extracts showed in vivo antitumor efficacy against cancer cells [43]. Adipocyte-dependent growth inhibition of MCF-7 is caused by licorice ethanolic extract. Since it stopped bone marrow cells from developing micronuclei and chromosomal abnormalities in albino mice, licorice hydromethanolic extract was revealed to have antimutagenic potential. Additionally, licorice hydromethanolic extracts inhibit thromboxane A2 in lung cancer [44].

Licorice can cause BCL2 to get phosphorylated by preventing the cell cycle's G2/M transition by employing the anti-microtubule drug paclitaxel. Human monoblastic leukaemia U937 cell lines can undergo apoptosis when exposed to licorice extract that is 70% methanolic. Glycycoumarin can bind to and inactivate the cancer-causing TOPK, which stimulates the p53 pathway and inhibits hepatocellular carcinoma cells, according to in-vivo and in-vitro studies [45]. Inhibiting IAPs and the butyrate-induced mitochondrial pathway makes the licorice component glycerol effective against human colon cancer cells (HCF116 and HT-29) [46]. The fourth most common cancer-related cause of death worldwide is colorectal cancer [47].

Oncogene mutations, tumour inhibitor inactivation, the presence of different signalling pathways, apoptotic downregulation, and morphological progression have all been documented [48]. Studies conducted both in vitro and in vivo have shown that the pentacyclic triterpenoid 18-glycyrrhetinic acid, produced from licorice root, can inhibit the growth of colorectal cancer cells [49]. Glycyrrhetinic acid lowers the levels of the proteins p-PI3K, p-AKI, p-STAT3, p-JNK, p-P38, and p-NFkBP65. After 2 hours of glycyrrhetinic acid administration, P13K and STAT3 phosphorylation in LoVo, SW620, and SW480 cells reduces [49]. Stimulation of the P13K or AKI pathways encourages cell migration, proliferation, and survival in cancer cells [50]. By misregulating the expression of MMP 1,2,3,9,12,13 and MTI-MMP, deactivation of these pathways slows down the fast growth and invasion of cancer cells [51]. The actin cytoskeleton is also disrupted by glycorylrhetinic acid, which also prevents metastasis by decreasing the p38-MAPK-AKI signalling cascade [52, 53], and promotes DNA fragmentation-mediated apoptosis [54]. Matrix metalloproteinase (MMP) expression is also decreased [49].

29.6 Antidiabetic Activity of Licorice

Diabetes mellitus type II is a long-term metabolic condition characterised by elevated blood sugar levels brought on by insulin inactivity. The metabolism of lipids and glucose is regulated by a number of transcription factors [35]. PPARs, or peroxisome proliferation activated receptors, have a role in lipid and glucose metabolism. The liver, muscle, and kidney tissues are the principal tissues that have PPAR receptors. PPAR receptors have alpha, gamma, and delta names. The PPAR gamma receptor is impacted by medications that boost insulin. Blood sugar levels were dramatically lowered by a number of compounds, including glycycoumarin, glycyrin, glyasperin D, dehydroglyasperin, glyasperin B, and iso-glycyrol ethyl solution produced from *Glycyrrhiza glabra* crude extracts. It has been demonstrated that these bioactive compounds bind to PPAR gamma receptors [35].

Two advantageous compounds produced from *Glycyrrhiza*, chalcone and amorfrutin, enhanced lipid and glucose metabolism [56]. Glabridin inhibits glucose intolerance and offers the greatest amount of glucose absorption by transferring GLUT-4 via AMPK (Adenosine Monophosphate Kinase) [57]. Glycyrrhizin reduces the level of blood insulin and the quantity of pancreatic islet cells by increasing the levels of glycohaemoglobin, cholesterol, and triglycerides [55].

29.7 Hypolipidemic Potential of Licorice

Acetyl CoA carboxylase and dehydrogenase both play important roles in lipid metabolism, and *Glycyrrhiza glabra* increases their levels while lowering their activity [40]. Laboratory tests have demonstrated that *Glycyrrhiza glabra*'s glabridin ethanolic extract, ethyl-acetate (EA), water soluble, and hexane soluble fractions increase blood High Density Lipoprotein (HDL) and lower total serum cholesterol and triglyceride levels [57].

29.7.1 Hormonal Action of Licorice

Glycyrrhiza glabra may affect cortisol levels, and estrogenic action inhibits testosterone production [59]. Glycyrrhizin and 18-glycyrrhetinic acid, the two major components of licorice, have mineralocorticoid properties and can thus block cortisol metabolism. In PCOS patients, licorice helps decrease the unwanted effects of spironolactone's diuretic activity [60]. The substance 18-glycyrrhetinic acid inhibits the enzyme 11-HSD. Higher cortisol levels in humans result from lower 11-HSD levels, and these higher cortisol levels eventually interact with mineralocorticoid receptors to boost sodium ion absorption. The enzymes 3-hydroxysteroid dehydrogenase, 17-hydroxysteroid dehydrogenase, and 17,20-lyase can all be inhibited by glycyrrhizin. Each of these enzymes is crucial for the production of androgen and oestrogen [61]. Licorice extracts can reduce blood testosterone levels by reducing the activity of the 11-HSD enzyme, which catalyses the conversion of androgen steroids into testosterone hormone [62]. When taken as a 25 mg extract, *Glycyrrhiza glabra* exerts estrogenic effects on uterine retention and vaginal opening. Licorice contains isoflavones, which can impact the function of the pituitary, ovary, and hypothalamus as well as the development of sexuality and the estrus cycle [63]. The active ingredient in licorice, glabridin, is used to relieve menopausal symptoms [64]. According to a study, the sperm can be stimulated by isoliquiritugenin and formononetin during fertilisation [65].

29.8 Anti-asthmatic Activity

Lower airways are affected by the chronic inflammatory disease known as asthma. The inflammation of these airways is treated with various corticosteroid medications. The negative effects of these corticosteroids can vary. A compound known as licochalcone A, which is derived from the licorice plant's root, has been shown to have anti-asthmatic properties. By decreasing IKB kinase complex activation, licochalcone A prevents TNF- from activating nuclear factor kappa B (NF-KB) [65]. Licorice includes extractable flavonoids that can decrease eosinophilic lung inflammation, Ig levels, IL-3, IL-5, and IL-13 levels and boost INF-gamma activity [66].

29.9 Antihepatotoxic Potential of Licorice

Chronic hepatitis is a progressive liver condition that results in cirrhosis, severe hepatocellular carcinoma, or liver failure [67]. *Glycyrrhiza* has been used to treat chronic hepatitis for more than 50 years. In a study, it was discovered that *Glycyrrhiza*, when compared to a placebo, enhances liver tissue and lowers serum levels of the enzyme aminotransferases. Swiss albino mice liver tissues were exposed to hydroethanolic extracts of *Glycyrrhiza* root at dose concentrations of 300 and 600 mg/kg for 7 days to evaluate CCl4-induced oxidative stress [68]. According to reports, 18-glycyrrhetinic acid protects the liver by lowering the expression of P450E1 [69]. Glycyrrhetinic acid can reduce oxidative stress and aflatoxin-induced liver damage [70]. A single dose of 2 mg/kg body weight of licorice was found to have a noticeable impact on acute liver conditions [71]. Aqueous and methanolic extracts of *Glycyrrhiza* inhibit the activity of aspartate and aminotransferase in a rat model [72].

29.9.1 Effect of Licorice on Fertilization

Licorice water extract enhanced the success rate of in-vitro fertilisation (IVF) [12]. Artificial insemination is the procedure of artificially injecting semen into the female vaginal tract using a catheter. Treatment of infertility is crucial for both human and cattle reproduction [73]. It was discovered that formononetin and isoliquiritigenin, which were isolated from licorice water extracts, increased the success rate of IVF in the rodent experimental model. Following maturation, the sperm cell in this process triggers an acrosome reaction, which fertilises the egg [12]. Estrogen influences sperm stimulation and response activation. There was significant estrogen-like action in isoliquiritigenin. Pubertal timing, ovarian function, impaired estrous cycle, and pituitary and hypothalamic gland activity are all aided in sexual development by formononetin [62].

29.10 Anti-atherogenic Effect

There is potential for *Glycyrrhiza glabra* to be anti-atherogenic. Studies conducted in vivo and in vitro established *Glycyrrhiza glabra's* anti-atherogenic efficacy [13]. Patients with hypercholesterolemia exhibited antioxidant potential from its alcoholic extract. When given 0.1 g/day for a month to animals with atherosclerotic apolipoprotein-E deficiency, alcoholic extracts also demonstrated antioxidant action [69].

29.11 Analgesic Activity

An unpleasant feeling called pain is connected to tissue damage [74]. There are several levels of pain. Analgesics are medications that work on the peripheral and central nervous systems to reduce pain [75]. A natural pain reliever is *Glycyrrhiza glabra* [76]. Formalin and the light tail-flick test have been used to confirm the analgesic efficacy of *Glycyrrhiza glabra*. Licorice root's aqueous extract prevents neutrophil and inflammatory mediator production as well as white globule migration [77].

29.12 Anti-allergic Activity

The three most prevalent allergy diseases are asthma, rhinitis, and atopic dermatitis [78]. Allergy problems are brought on by mast cells and basophils. These cells and cell surface-bound IgE release histamine, cytokines, and prostaglandins. The cytokines that trigger neutrophil and macrophage chemotaxis and phagocytosis generate tissue inflammation. The effectiveness of *Glycyrrhiza*'s antiallergic properties was proven by in vitro and in vivo studies [79]. The compounds glycyrhizin, 18-glycyrrhetinic acid, isoliquiritin, and liquiritigenin, which were isolated from *Glycyrrhiza*, show anti-allergic and IgE production inhibitory characteristics [13]. Methanolic extracts of licorice include liqueritigenin and 18-glycyrrhetinic, which inhibit the degranulation of RBL-2H3 cells induced by IgE and (DNP-HSA) antigen [80].

Glycyrrhizic acid, liquiritigenin, and liquiritin suppress proinflammatory mediators in BV2 cells, including inducible nitric oxide synthase (iNOS), cyclooxygenase 2 (COX2), TNF-, IL-1, IL-4, IL-5, and IL-6 [81]. Glycyrrhizic acid has been shown to increase cell permeability by up to 60% and decrease the elasticity modulus of the cell membrane in a human RBC investigation [82]. In cultured NCI-H292 cells, TNF- induces MUC5AC protein and mRNA expression. Glycyrrhizin inhibits the MUC5AC gene's transcription, weakening the excessive mucus production [83]. According to reports, glycyrrhizic acid can treat allergic asthma brought on by ovalbumin by slowing the OX40-OX40L and p38 MAPK activity that regulates the Th1/TH2 balance [84].

29.13 Anti-oxidant Effect

Glycyrrhiza has been found to have antioxidant activity in both in vitro and in vivo investigations [85]. The DPPH radical scavenging experiment was used to evaluate the *Glycyrrhiza*'s in vitro potential. At a concentration of 500 g/ml, methanolic extract possesses the highest level of scavenging action, or 67.22% [86]. The determined IC50 value for this was 359.45 g/ml [55]. The phenolic chemicals glabridin, hispaglabridin A, and 30-hydroxy-40-methylglabridin are present in licorice's ethanolic extract. Through tests for free radical scavenging, hydrogen-donating activity, metal ion chelating potential, and reducing ability, all of these phenolic compounds are demonstrated to be antioxidants. Antioxidant action of glabridin was seen against LDL oxidation. *Glycyrrhiza* contains lipidochalcones B and D that can delay microsomal lipid breakdown and shield biological systems from oxidative stress [31]. Inducible nitric oxide synthase (iNOS) activity and superoxide radical generation can both be reduced by licochalcone C [87]. Retrochalcone, a compound isolated from *Glycyrrhiza glabra*, can stop the oxidative hemolysis of red blood cells. It has been determined that *Glycyrrhiza glabra* flavonoids have 100 times more antioxidant potential than vitamin E [32].

29.13.1 Immunostimulatory Effect

Influenza, an H1N1 is a virus [88] that cause Swine flu. A study observed that N-Acetylmuramyl (analog of glycyrrhizin) possess the ability to retards the replication of the virus [89]. Polysaccharide extracts of licorice increases the immune stimulation by activating the macrophages. Phagocytic activity of neutrophil increases when mixed with *Glycyrrhiza* alcoholic extract [90]. It was noticed in a study that 100 μg/ml concentration of licorice possess immunostimulatory effect [32]. In human granulocytes *Glycyrrhiza* can stimulate the production of TCD69 lymphocytes and macrophages [32].

29.14 Antimalarial Activity

In Asia, Africa, and Latin America, malaria is a significant public health issue [91]. According to reports, *Glycyrrhiza glabra* helps to prevent malaria. 9.95 g/ml of water-methanol and 13 g/ml of ethyl acetate, which were extracted from the root extract of licorice, were shown to have outstanding anti-plasmodial activity against the P. falciparum strain and minimal HeLa toxicity, according to an in vitro investigation. Oral administration of licorice root extracts was used to test its in vivo activity, and the results showed that it slows P. berghei development in mice by 65% and 72.2%, respectively [92]. Licochalcone is responsible for licorice's antimalarial properties [93].

29.15 Antidepressant Activity

The primary inhibitory neurotransmitter in the mammalian CNS is gamma amino butyric acid (GABA) [94]. Glabridin may activate GABA induce receptors, which would have an antidepressant effect [95]. N265 and M286 amino acids, which are located in the second and third transmembrane domains on the -subunits of GABA receptors, are implicated in the same mechanism of action as general anaesthetics [96].

29.16 Hair Growth Stimulatory Effect

Licorice hydroalcoholic extract promotes hair growth [97]. Alopecia is treated by *Glycyrrhiza glabra.* Licorice was found to have a better start to hair development than 2% minoxidil [97].

29.16.1 Bleaching of the Skin

Due to its anti-oxidant and anti-inflammatory properties, glycyrrhizin helps protect skin [98]. It has bleaching properties. Tyrosinase, an enzyme necessary for the synthesis of melanin, is blocked by glabridin, licochalcone A, and isoliquiritin in cultured murine melanoma cells [99]. An in-vitro investigation found that licorice extract in the form of ethanol can reduce the tyrosinase enzyme's activity by up to 50% [99]. The skin's moisture content is maintained by *Glycyrrhiza* ethanolic extract [32].

Human keratinocytes have been able to prevent DNA damage and apoptosis activation induced by UVB radiation by blocking the effects of DNA damage and apoptosis stimulation due to UVB radiation [97]. Additionally, it is employed to treat dermatitis, pruritus, cysts, and eczema [100].

29.17 Anti-fungal Activity

Chaetomium funicola M002 had been shown to be resistant to the antifungal effects of a methanolic extract of *Glycyrrhiza glabra.* Active substances like glabridin, glabrol, and their derivatives have the potential to be antifungal [100].

29.17.1 Bacterial Resistance

Licorice root hydromethanolic extract has antibacterial properties since it contains saponins, alkaloids, and flavonoids [101]. Both aqueous and ethanolic licorice extracts have been shown in in vitro tests to have inhibitory effect against *Staphylococcus aureus* and *Streptococcus pyogenes* [100].

29.18 Memory Enhancing Activity

Mice were used to test the effects of *Glycyrrhiza glabra* on memory and learning. Licorice's impact on memory and learning was examined using the elevated plus-maze and passive avoidance paradigm. For a subsequent week, three separate doses—70 mg/kg PO, 150 mg/kg PO, and 300 mg/kg PO—were given orally. Significant improvements in mice's memory and learning were seen at 150 mg/kg p.o [101].

29.18.1 Anticoagulant

Glycyrrhizin is the first thrombin inhibitor found in plants. It was found to speed up the time that thrombin and fibrinogen took to clot. Additionally, it speeds up plasma recalcification. Glycyrrhizin inhibits thrombin-induced platelet accumulation but not platelet aggregation caused by collagen, PAF (Platelet Aggregating Factor), or convulxin [102, 103].

29.19 Anti-viral Effects

Herpes simplex, varicella zoster, Japanese encephalitis, influenza virus, vesicular stomatitis virus, and type A influenza virus have all been shown to be less active when licorice extract is used [104–109]. Glycyrrhizin prevents the virus from attaching to cells. According to reports, it has been tested for chronic hepatitis C and HIV-1. Glycyrrhizin was found to have considerable antiviral efficacy against two clinical isolates of the SARS virus (FFM-1 and FFM-2) [110–113].

29.19.1 Anticancer

Twelve licorice flavonoids inhibited the cell cycle at different phases and caused apoptosis in cancer cells, which slowed their growth. The putative anticancer properties of *Glycyrrhiza glabra* are shown in Fig. 29.4. Licochalcone A (LA), an

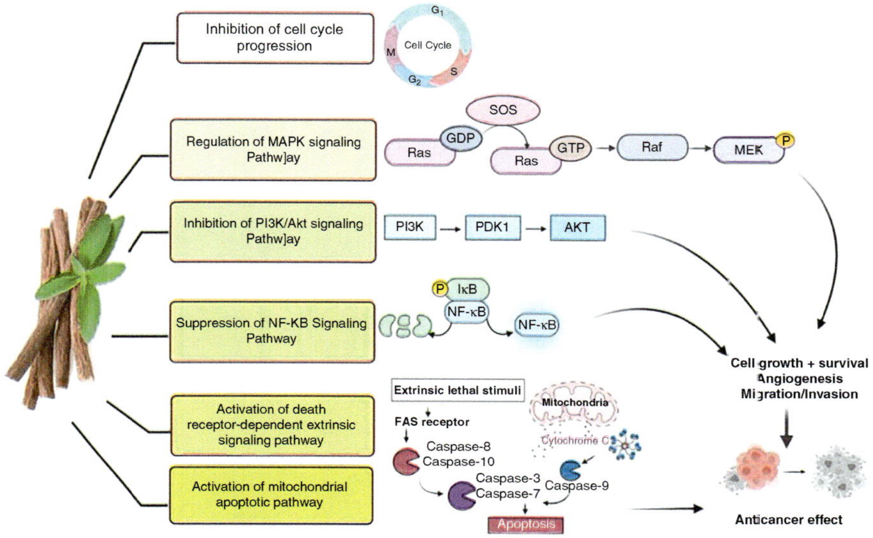

Fig. 29.4 Possible *Glycyrrhiza glabra* anticancer mechanism of action

anti-cancer flavonoid, is found in licorice. Through a rise in the LC3-II protein, which creates autophagosomes, licochalcone A displayed anticancer effectiveness [124]. According to a different study, licochalcone A increased LC3-II signalling while suppressing PI3K/RAC-serine-threonine protein kinase (Akt)/mammalian target of rapamycin (mTOR) signalling [125]. A licochacone stops the cell cycle from progressing at the G1/S and G2/M stages. Two cyclins and cyclin-dependent kinases (CDKs) whose protein levels are lowered as part of this mechanism include cyclin B1 and CDK1 [124–130]. Using flow cytometry, the number of RNA-seq. Licochalcone T (CD3e (+), B (CD45R/B220 (+), and B cells in the spleen and whole blood were determined to assess a related mechanism. The MWM test demonstrated that LA increased CBF levels and enhanced cognitive function in treated mice. Therefore, LA may improve cognition by managing the immune system [130]. In a study, the anti-cancer activity of licorice was investigated using ethanolic and water extracts. A range of cancers, including breast, colon, and liver cancers, have been successfully treated with licorice ethanolic extract. The ethanolic extract considerably decreased the incidence of breast cancer and hepatic cancer at 100 g/mL and 16.1 g/mL, but had no impact on colon cancer [131]. Another study examined the effects of ethyl acetate extracts on the licorice plant's dried leaves and roots and its potential medical applications. Using LC-MS-MS, 40 bioactive chemicals were measured, and it was discovered that there were significant variations between the extracts. The likelihood that the fresh root of the plant under analysis contains therapeutic value is strong. The results of this study suggest that licorice aids in the prevention and treatment of oxidative stress-related disorders, such as cancer [132]. Glabridin, another important flavonoid of *G. glabra*, has demonstrated anticancer potential by reducing the expression level of proteins including p-AKT, p-ERK1/2, cyclin D1, and other p-epidermal growth factor receptor-related molecules [133, 134] (Table 29.1).

Table 29.1 Chemical elements in licorice that give it its medicinal value

Compound name	Phytochemistry	Mechanism of action	Chemical structure
4-O-Methylglabridin	Isoflavanoid	Exhibit notable in vitro antibacterial activity.	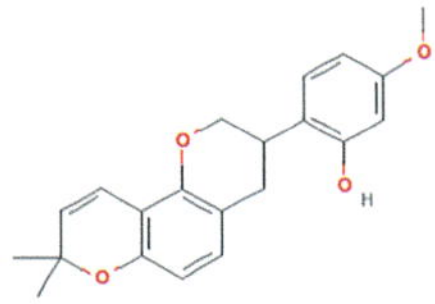
Formononetin (biochanin b)	Bioactive isoflavones	Through the targeting of various pathways, they were able to inhibit metastasis, induce apoptosis, and pause the cell cycle.	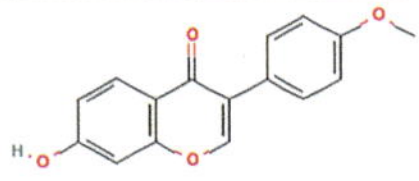
Glabridin	An isoflavane, or isoflavonoid, is glabridin.	Glabridin suppressed melanogenesis by inhibiting tyrosine and ROS generation, respectively.	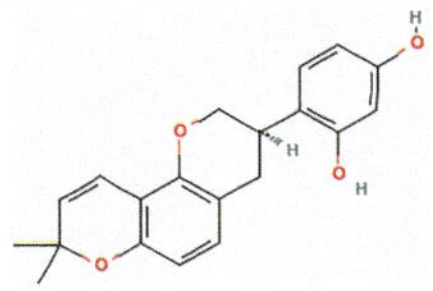
Glabridin	Isoflavones	Antagonist of tyrosine.	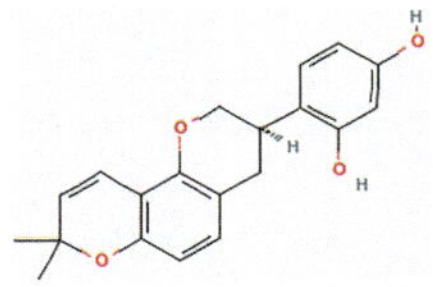
Glabrocoumarin	Coumarins	It prevented the growth of cells in HIV-infected cell cultures without cytotoxicity.	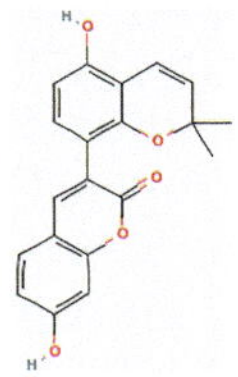
Glycyglabrone	Chalcone	It showed strong anti-free radical action.	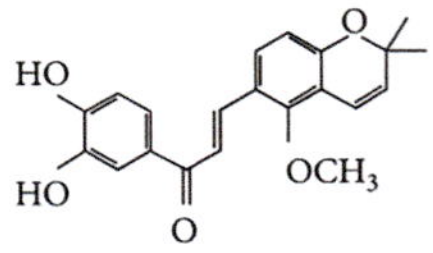
Glycyrrhetinic acid	Active phytoconstituents are 18β-glycyrrhetinic acid, isoflavones, glabrin A and B, and glycyrrhizin.	Glycyrrhetinic acid inhibited 11-hydroxysteroid dehydrogenase and displayed anti-inflammatory properties.	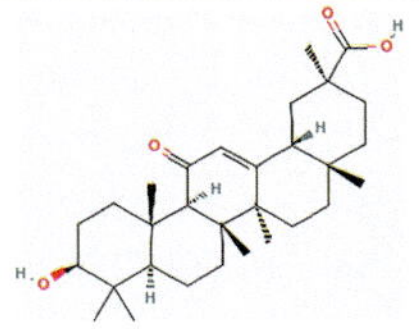
Glycyrrhizin	Triterpene, saponins, and flavonoids are the active components.	Cyclooxygenase activity, prostaglandin E2 in particular, and platelet aggregation were all inhibited.	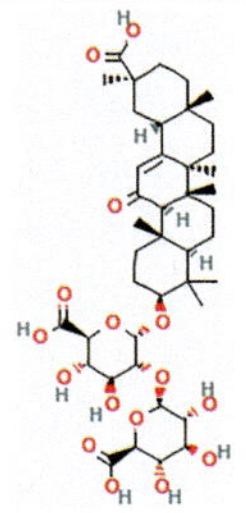

(continued)

Table 29.1 (continued)

Compound name	Phytochemistry	Mechanism of action	Chemical structure
Hispaglabridin B	Isoflavones	The strongest antioxidant agent is it. Transcription of FoxO1. Activity was hampered by lower expression of the muscle-specific E3 ubiquitin ligases MuRF1 and Atrogin1.	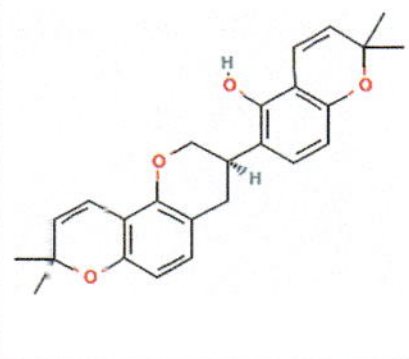
Isoliquiritigenin	Phenolic compounds	By preventing AP-1, NF-κ, and AP-1 activation, you can lessen the inflammatory response of macrophages.	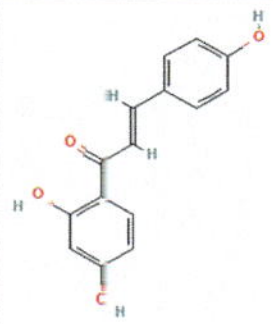
Kanzonol Y	Chalcone	Efficacy as an inhibitor of Bacillus subtilis H17	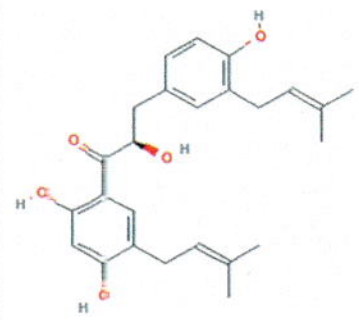
Licochalcone C	Phenolic compounds,	The bacterial respiratory chain is hindered in terms of electron transport.	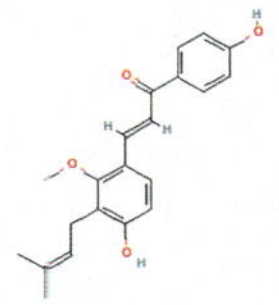
Licopyrano coumarin	Coumarins	It prevented the growth of cells in HIV-infected cell cultures without causing sny cytotoxicity.	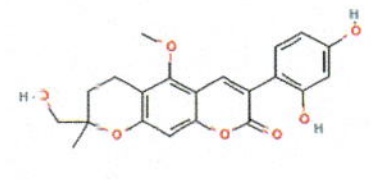
Liquiritigenin	Phenolic compounds,	Through the NLRP3 and NF-κ pathways, it is suppressed.	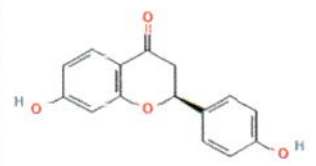
Mannopyranosyl-D glucitol	Mannose	Not disclosed	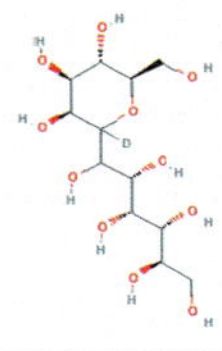
Paratocarpin B	Chalcone	The antioxidant property has been demonstrated using the peroxynitrite assay. The strongest antioxidant agent is it.	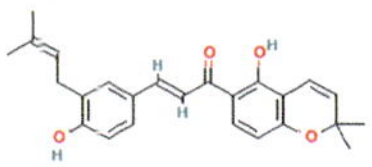
Quercetin	Plant-derived flavonoid	It is unknown how flavonoids prevented enzyme activity. It lessens the synthesis of inflammatory metabolites and inhibits the activities of cyclooxygenase and lipoxygenase.	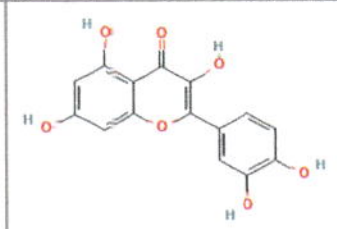

29.20 Myths, Legends, Tales, Folklore, and Interesting Facts

Egyptian use licorice root as an all-rounder herb, and there is no doubt that licorice addict throughout history have included kings of ancient Egypt and prophets [114].

Licorice was found in the tomb of King Tut, amongst many of its treasurers. He thinks that he wouldn't journey into the next world without his favorite drink, Maisus [115].

The beneficial properties of licorice are endorsed by Alexander the Great, Julius Caesar and Brahma India's great prophet. Warriors used licorice to diminish the thirst [116].

In northern European countries licorice is used to get relief from sore throat [117, 118].

29.21 Traditional Uses

Licorice has a number of traditional applications. It is used to heal burns and wounds when mixed with butter. Lactation is aided by *Glycyrrhiza* when combined with cow's milk. Its blended root is used to make a decoction and a shampoo that are used to treat erysipelas [119]. To treat raspy voice, combine rice milk and *Glycyrrhiza* [119]. When combined with honey, licorice is used as a tonic to boost intelligence [119]. It is also used to treat haemorrhage and anaemia. In India, *Glycyrrhiza* root extract is used to cure conjunctivitis. *Cardiotonic Glycyrrhiza* and *Picirrhiza kurroa* mixed with sugar water; haematemesis-curing *Glycyrrhiza* and *Santalum album* [119]. In Pakistan, licorice root, wheat, and oil are fed to cows, goats, buffaloes, and sheep to boost milk output and reproductive rates [120]. In Turkey, licorice root sap is used to make wine [121]. Root decoction is a common laxative in Italy [122]. Licorice root and stem juice is used as an astringent, tonic, and stimulant in Nepal [123]. In Egypt, it is used with tea to treat sore throats [35].

29.22 Summary

Glycyrrhiza species have been utilised in food and medicinal all throughout the world. It is notable that licorice roots and rhizomes have long been valued for their therapeutic effects, the bulk of which come from phytoconstituents. Glycyrrhizin, liquiritigenin, isoliquiritigenin, 4-O-methylglabridin, isoprenyl chalcone, formononetin, glabridin, and hispaglabridins A and B are the primary phytochemicals present in *Glycyrrhiza* spp. Throughout time, further new compounds have been found. Via in vivo and in vitro research, isolated chemicals and extracts from *Glycyrrhiza* spp. showed notable antioxidant, antibacterial, anti-inflammatory, anti-proliferative, and cytotoxic activities. The plant species and chemical glycyrrhizin (glycyrrhizic acid) have been the subject of the most study because of their biological activity. Clinical trials utilizing *Glycyrrhiza* spp. have demonstrated that its extracts can lessen pain in cancer, ischemia, and neurological patients. Clinical trials are necessary to verify *Glycyrrhiza* plant extracts as potential pharmacological and food additives, and further in-depth study is needed to close several research gaps on the safety and toxicological properties of licorice.

References

1. Blackburn, T. (2019). Depressive disorders: Treatment failures and poor prognosis over the last 50 years. *Pharmacology Research & Perspectives, 7*(3), e00472.
2. Singh, A. K., Sharma, A., Kumar, P., & Virmani, O. P. (1984). Cultivation and utilization of liquorice (*Glycyrrhiza glabra* L.): A review. *Current Research on Medicinal and Aromatic Plants, 6*(2), 98–105.
3. Hayashi, H., & Sudo, H. (2009). Economic importance of licorice. *Plant Biotechnology, 26*, 101–104.
4. Lim, T. (2016). Glycyrrhiza glabra. *Edible Medicinal And Non-Medicinal Plants, 22*, 354–457.
5. CIMAP Newsletter. (1995). In vitro propagation of licorice through multiple shoot formation. *Journal of Spices and Aromatic Crops, 21*(4), 4–5.
6. Gupta, V. K., Fatima, A., Faridi, U., Negi, A. S., Shanker, K., Kumar, J. K., et al. (2008). Antimicrobial potential of *Glycyrrhiza glabra* roots. *Journal of Ethnopharmacology, 116*, 377–380. https://doi.org/10.1016/j.jep.2007.11.037
7. Medicinal Plants. (2022). Retrieved 24 March 2022, from https://books.google.com.pk/books?
8. Yashtimadhu (*Glycyrrhiza glabra* Linn.) A Potent Medicinal Herb. (2022). Retrieved 24 March 2022, from https://books.google.com.pk/books?
9. Rao, K. V. S. (1993). A review on licorice. *Ancient Science of Life, 13*(1–2), 57–88.
10. Whipker, B. E., Owen, G. W., McCall, I., & Cleveland, B. (2014). Licorice plant (Helichrysum): Disorder diagnostics. *e-Gro Alert, 3*(12), 1–4.
11. Gupta, B., Ahmed, K., & Gupta, R. (2018). *Glycyrrhiza glabra* (medicinal plant) research: A Scientometric assessment of global publications output during 1997–2016. *Pharmacognosy Journal, 10*(6), 1067–1075. Retrieved from https://www.phcogj.com/article/742
12. Tung, N. H., Shoyama, Y., Wada, M., & Tanaka, H. (2015). Two activators of in vitro fertilization in mice from licorice. *Biochemical and Biophysical Research Communications, 467*(2), 447–450.
13. Visavadiya, N. P., Soni, B., & Dalwadi, N. (2009). Evaluation of antioxidant and anti-atherogenic properties of *Glycyrrhiza glabra* root using in vitro models. *International Journal of Food Sciences and Nutrition, 60*(sup2), 135–149.
14. Rizzato, G., Scalabrin, E., Radaelli, M., Capodaglio, G., & Piccolo, O. (2017). A new exploration of licorice metabolome. *Food Chemistry, 221*, 959–968.
15. Yu, J. Y., Ha, J. Y., Kim, K. M., Jung, Y. S., Jung, J. C., & Oh, S. (2015). Anti-inflammatory activities of licorice extract and its active compounds, glycyrrhizic acid, liquiritin and liquiritigenin, in BV2 cells and mice liver. *Molecules, 20*(7), 13041–13054.
16. Akamatsu, H., Komura, J., Asada, Y., & Niwa, Y. (1991). Mechanism of anti-inflammatory action of glycyrrhizin: Effect on neutrophil functions including reactive oxygen species generation. *Planta Medica, 57*, 119–121.
17. Simmler, C., Pauli, G. F., & Chen, S.-N. (2013). Phytochemistry and biological properties of glabridin. *Fitoterapia, 90*, 160–184.

18. Obolentseva, G. V., Litvinenko, V. I., Ammosov, A. S., Popova, T. P., & Sampiev, A. M. (1999). Pharmacological and therapeutic properties of licorice preparations (a review). *Pharmaceutical Chemistry Journal, 33*, 427–434.
19. Graebin, C. S. (2018). *The pharmacological activities of glycyrrhizinic acid ("glycyrrhizin") and glycyrrhetinic acid* (pp. 245–261). In Reference Series in Phytochemistry; Nature Publishing Group.
20. Gottlieb, D., & Shaw, P. D. (Eds.). (1967). *Mechanism of action*. Springer. ISBN 978-3-642-46053-1.
21. Anand David, A., Arulmoli, R., & Parasuraman, S. (2016). Overviews of biological importance of quercetin: A bioactive flavonoid. *Pharmacognosy Reviews, 10*, 84.
22. Zhu, X., Shi, J., & Li, H. (2018). Liquiritigenin attenuates high glucose-induced mesangial matrix accumulation, oxidative stress, and inflammation by suppression of the NF-κβ and NLRP3 inflammasome pathways. *Biomedicine and Pharmacotherapy, 106*, 976–982.
23. Glioblastoma Multiforme; Potential for the Dandenong Ranges. (2020). *Advances and avenues in the development of novel carriers for bioactives and biological agents* (pp. 383–422). Elsevier.
24. Haraguchi, H., Tanimoto, K., Tamura, Y., Mizutani, K., & Kinoshita, T. (1998). Mode of antibacterial action of retrochalcones from *Glycyrrhiza inflata*. *Phytochemistry, 48*, 125–129.
25. Jiang, D., Rasul, A., Batool, R., Sarfraz, I., Hussain, G., Mateen Tahir, M., et al. (2019). Potential anticancer properties and mechanisms of action of formononetin. *BioMed Research International, 2019*, 5854315–5854311.
26. De Simone, F., Aquino, R., De Tommasi, M. N., Piacente, S., & Pizza, C. (2001). Chapter 8: Anti-HIV aromatic compounds from higher plants. In C. Tringali (Ed.), *Bioactive compounds from biological sources* (pp. 305–336). CRC Press. 704p.
27. Wang, D., Liang, J., Zhang, J., Wang, Y., & Chai, X. (2020). Natural chalcones in Chinese Materia Medica: Licorice. *Evidence Based Complementary Alternative Medicine, 2020*, 3821248–3821214.
28. Li, W., Asada, Y., & Yoshikawa, T. (2000). Flavonoid constituents from *Glycyrrhiza glabra* hairy root cultures. *Phytochemistry, 55*, 447–456.
29. Sharifi-Rad, J., Quispe, C., Herrera-Bravo, J., Belén, L. H., Kaur, R., Kregiel, D., et al. (2021). *Glycyrrhiza* genus: Enlightening phytochemical components for pharmacological and health-promoting abilities. *Oxidative Medicine and Cellular Longevity, 2021*, 7571132–7571120.
30. Chen, J., Yu, X., & Huang, Y. (2016). Inhibitory mechanisms of glabridin on tyrosinase. Spectrochim. Spectrochimica. *Acta Part A: Molecular and Biomolecular Spectroscopy, 168*, 111–117.
31. Sharma, V., Katiyar, A., & Agrawal, R. C. (2018). *Glycyrrhiza glabra*: Chemistry and pharmacological activity. *Sweeteners, 87*.
32. Damle, M. (2014). *Glycyrrhiza glabra* (Liquorice)-a potent medicinal herb. *International Journal of Herbal Medicine, 2*(2), 132–136.
33. Pandey, S., Verma, B., & Arya, P. (2017). A review on constituents, pharmacological activities and medicinal uses of *Glycyrrhiza glabra*. *Pharmaceutical Research, 2*, 26–31.
34. Baker, M. E. (1994). Licorice and enzymes other than 11β-hydroxysteroid dehydrogenase: An evolutionary perspective. *Steroids, 59*(2), 136–141.
35. Tanaka, A., Horiuchi, M., Umano, K., & Shibamoto, T. (2008). Antioxidant and anti-inflammatory activities of water distillate and its dichloromethane extract from licorice root (*Glycyrrhiza* uralensis) and chemical composition of dichloromethane extract. *Journal of the Science of Food and Agriculture, 88*(7), 1158–1165.
36. Malek, J. M., & Ghazvini, K. (2007). In vitro susceptibility of Helicobacter pylori to licorice extract. *Journal of Pharmacy Research, 6*(1), 69–72.
37. Ram, H. A., Lachake, P., Kaushik, U., & Shreedhara, C. S. (2010). Formulation and evaluation of floating tablets of liquorice extract. *Pharmacognosy Research, 2*(5), 304–308.
38. Lee, C. S., Kim, Y. J., Lee, M. S., Han, E. S., & Lee, S. J. (2008). 18β-Glycyrrhetinic acid induces apoptotic cell death in SiHa cells and exhibits a synergistic effect against antibiotic anti-cancer drug toxicity. *Life Sciences, 83*(13–14), 481–489.

39. Sultana, S., Haque, A., Hamid, K., Urmi, K. F., & Roy, S. (2010). Antimicrobial, cytotoxic and antioxidant activity of methanolic extract of *Glycyrrhiza glabra*. *Agriculture and Biology Journal of North America, 1*(5), 957–960.
40. Icer, M. A., Sanlier, N., & Sanlier, N. (2017). A review: Pharmacological effects of licorice (*Glycyrrhiza glabra*) on human health. *International Journal of Basic and Clinical Studies, 6*(1), 12–26.
41. He, S. Q., Gao, M., Fu, Y. F., & Zhang, Y. N. (2015). Glycyrrhizic acid inhibits leukemia cell growth and migration via blocking AKT/mTOR/STAT3 signaling. *International Journal of Clinical and Experimental Pathology, 8*(5), 5175–5181.
42. Sheela, M. L., Ramakrishna, M. K., & Salimath, B. P. (2006). Angiogenic and proliferative effects of the cytokine VEGF in Ehrlich ascites tumor cells is inhibited by *Glycyrrhiza glabra*. *International Immunopharmacology, 6*(3), 494–498.
43. Deng, Q. P., Wang, M. J., Zeng, X., Chen, G. G., & Huang, R. Y. (2017). Effects of glycyrrhizin in a mouse model of lung adenocarcinoma. *Cellular Physiology and Biochemistry, 41*(4), 1383–1392.
44. Song, X., Yin, S., Zhang, E., Fan, L., Ye, M., Zhang, Y., & Hu, H. (2016). Glycycoumarin exerts anti-liver cancer activity by directly targeting T-LAK cell-originated protein kinase. *Oncotarget, 7*(40), 65732–65743.
45. Lu, S., Yin, S., Zhao, C., Fan, L., & Hu, H. (2020). Synergistic anti-colon cancer effect of glycyrol and butyrate is associated with the enhanced activation of caspase-3 and structural features of glycyrol. *Food and Chemical Toxicology, 136*, 110952.
46. Favoriti, P., Carbone, G., Greco, M., Pirozzi, F., Pirozzi, R. E. M., & Corcione, F. (2016). Worldwide burden of colorectal cancer: A review. *Updates in Surgery, 68*(1), 7–11.
47. Arvelo, F., Sojo, F., & Cotte, C. (2015). Biology of colorectal cancer. *Ecancermedicalscience, 2015*(9), 520.
48. Wang, S., Shen, Y., Qiu, R., Chen, Z., Chen, Z., & Chen, W. (2017). 18 β-glycyrrhetinic acid exhibits potent antitumor effects against colorectal cancer via inhibition of cell proliferation and migration. *International Journal of Oncology, 51*(2), 615–624.
49. Manning, B. D., & Cantley, L. C. (2007). AKT/PKB signaling: Navigating downstream. *Cell, 129*(7), 1261–1274.
50. Honore, P. M., Hoste, E., Molnár, Z., Jacobs, R., Joannes-Boyau, O., Malbrain, M. L., et al. (2019). Cytokine removal in human septic shock: Where are we and where are we going? *Annals of Intensive Care, 9*(1), 1–13.
51. Wang, X. F., Zhou, Q. M., Lu, Y. Y., Zhang, H., Huang, S., & Su, S. B. (2015). Glycyrrhetinic acid potently suppresses breast cancer invasion and metastasis by impairing the p38 MAPK-AP1 signaling axis. *Expert Opinion on Therapeutic Targets, 19*(5), 577–587.
52. Yamaguchi, H., Noshita, T., Yu, T., Kidachi, Y., Kamiie, K., Umetsu, H., & Ryoyama, K. (2010). Novel effects of glycyrrhetinic acid on the central nervous system tumorigenic progenitor cells: Induction of actin disruption and tumor cell-selective toxicity. *European Journal of Medicinal Chemistry, 45*(7), 2943–2948.
53. Pirzadeh, S., Fakhari, S., Jalili, A., Mirzai, S., Ghaderi, B., & Haghshenas, V. (2014). Glycyrrhetinic acid induces apoptosis in leukemic HL60 cells through upregulating of CD95/CD178. *International Journal of Molecular and Cellular Medicine, 3*(4), 272–278.
54. Huang, Z. Y., Wang, L. J., Wang, J. J., Feng, W. J., Yang, Z. Q., Ni, S. H., et al. (2019). Hispaglabridin B, a constituent of liquorice identified by a bioinformatics and machine learning approach, relieves protein-energy wasting by inhibiting forkhead box O1. *Br. Journal of Pharmacology, 176*, 267–281.
55. Lee, K. K., Omiya, Y., Yuzurihara, M., Kase, Y., & Kobayashi, H. (2013). Antispasmodic effect of shakuyakukanzoto extract on experimental muscle cramps in vivo: Role of the active constituents of *Glycyrrhizae* radix. *Journal of Ethnopharmacology, 145*(1), 286–293.
56. Sawada, K., Yamashita, Y., Zhang, T., Nakagawa, K., & Ashida, H. (2014). Glabridin induces glucose uptake via the AMP-activated protein kinase pathway in muscle cells. *Molecular and Cellular Endocrinology, 393*(1–2), 99–108.

57. Armanini, D., Fiore, C., Mattarello, M. J., Bielenberg, J., & Palermo, M. (2002). History of the endocrine effects of licorice. *Experimental and Clinical Endocrinology & Diabetes, 110*(06), 257–261.
58. Maurya, S. K., Raj, K., & Srivastava, A. K. (2009). Antidyslipidaemic activity of *Glycyrrhiza glabra* in high fructose diet induced dsyslipidaemic Syrian golden hamsters. *Indian Journal of Clinical Biochemistry, 24*(4), 404–409.
59. Armanini, D., Castello, R., Scaroni, C., Bonanni, G., Faccini, G., Pellati, D., et al. (2007). Treatment of polycystic ovary syndrome with spironolactone plus licorice. *European Journal of Obstetrics & Gynecology and Reproductive Biology, 131*(1), 61–67.
60. Armanini, D., Mattarello, M. J., Fiore, C., Bonanni, G., Scaroni, C., Sartorato, P., & Palermo, M. (2004). Licorice reduces serum testosterone in healthy women. *Steroids, 69*(11–12), 763–766.
61. Josephs, R. A., Guinn, J. S., Harper, M. L., & Askari, F. (2001). Liquorice consumption and salivary testosterone concentrations. *The Lancet, 358*(9293), 1613–1614.
62. Kim, S. H., & Park, M. J. (2012). Effects of phytoestrogen on sexual development. *Korean Journal of Pediatrics, 55*(8), 265–271.
63. Su Wei Poh, M., Voon Chen Yong, P., Viseswaran, N., & Chia, Y. Y. (2015). Estrogenicity of glabridin in Ishikawa cells. *PLoS One, 10*(3), e0121382.
64. Tung, N. H., Shoyama, Y., Wada, M., & Tanaka, H. (2014). Improved in vitro fertilization ability of mouse sperm caused by the addition of licorice extract to the preincubation medium. *The Open Reproductive Science Journal, 6*(1), 1–7.
65. Kim, S. H., Yang, M., Xu, J. G., Yu, X., & Qian, X. J. (2015). Role of licochalcone A on thymic stromal lymphopoietin expression: Implications for asthma. *Experimental Biology and Medicine, 240*(1), 26–33.
66. Kao, T. C., Wu, C. H., & Yen, G. C. (2014). Bioactivity and potential health benefits of licorice. *Journal of Agricultural and Food Chemistry, 62*(3), 542–553.
67. Tsochatzis, E. A., Bosch, J., & Burroughs, A. K. (2014). Liver cirrhosis. *The Lancet, 383*(9930), 1749–1761.
68. Sharma, V., & Agrawal, R. C. (2014). In vivo antioxidant and hepatoprotective potential of *Glycyrrhiza glabra* extract on carbon tetra chloride (CCl4) induced oxidative-stress mediated hepatotoxicity. *International Journal of Research in Medical Sciences, 2*(1), 314–320.
69. Jeong, H. G., You, H. J., Park, S. J., Moon, A. R., Chung, Y. C., Kang, S. K., & Chun, H. K. (2002). Hepatoprotective effects of 18β-glycyrrhetinic acid on carbon tetrachloride-induced liver injury: Inhibition of cytochrome P450 2E1 expression. *Pharmacological Research, 46*(3), 221–227.
70. Saxena, S. (2005). *Glycyrrhiza glabra*: medicine over the millennium. *Natural Products Radiance, 19*(6), 358–367.
71. Al-Razzuqi, R., Al-Jawad, F. H., Al-Hussaini, J. A., & Al-Jeboori, A. (2012). Hepatoprotective effect of *Glycyrrhiza glabra* in carbon tetrachloride-induced model of acute liver injury. *Journal of Physiology and Pharmacology Advances, 2*(7), 259–263.
72. Abd-Al-Sattar-Sadiq Layl, L. (2016). Hepatoprotective effect of *Glycyrrhiza glabra* L. extracts against carbon tetrachloride-induced acute liver damage in rats. Extracts against carbon tetrachloride-induced acute liver damage in rats (June 30, 2016). *TJPRC: International Journal of Veterinary Science, Medicine & Research (TJPRC: IJVSMR), 1*, 1–8.
73. Ombelet, W., & Van Robays, J. (2015). Artificial insemination history: Hurdles and milestones. *Facts, Views & Vision in ObGyn, 7*(2), 137–143.
74. Treede, R. D. (2018). The International Association for the Study of Pain definition of pain: As valid in 2018 as in 1979, but in need of regularly updated footnotes. *Pain Reports, 3*(2), e643.
75. Kirkpatrick, D. R., McEntire, D. M., Smith, T. A., Dueck, N. P., Kerfeld, M. J., Hambsch, Z. J., et al. (2016). Transmission pathways and mediators as the basis for clinical pharmacology of pain. *Expert Review of Clinical Pharmacology, 9*(10), 1363–1387.
76. Thombre, N. A., Gaikwad, S. M., & Chaudhari, K. S. (2019). A review on analgesic herbals. *PharmaTutor Journal, 7*(4), 37–41.

77. Kim, K. R., Jeong, C. K., Park, K. K., Choi, J. H., Park, J. H. Y., Lim, S. S., & Chung, W. Y. (2010). Anti-inflammatory effects of licorice and roasted licorice extracts on TPA-induced acute inflammation and collagen-induced arthritis in mice. *Journal of Biomedicine and Biotechnology, 2010*, 2010–2018.
78. Shoormasti, R. S., Pourpak, Z., Fazlollahi, M. R., Kazemnejad, A., Nadali, F., Ebadi, Z., et al. (2018). The prevalence of allergic rhinitis, allergic conjunctivitis, atopic dermatitis and asthma among adults of Tehran. *Iranian Journal of Public Health, 47*(11), 1749.
79. Shin, Y. W., Bae, E. A., Lee, B., Lee, S. H., Kim, J. A., Kim, Y. S., & Kim, D. H. (2007). In vitro and in vivo antiallergic effects of *Glycyrrhiza glabra* and its components. *Planta Medica, 73*(03), 257–261.
80. Fouladi, S., Masjedi, M., Hakemi, M. G., & Eskandari, N. (2019). The review of in vitro and in vivo studies over the glycyrrhizic acid as natural remedy option for treatment of allergic asthma. *Iranian Journal of Allergy, Asthma and Immunology, 18*, 1–11.
81. Curreli, F., Friedman-Kien, A. E., & Flore, O. (2005). Glycyrrhizic acid alters Kaposi sarcoma–associated herpesvirus latency, triggering p53-mediated apoptosis in transformed B lymphocytes. *The Journal of Clinical Investigation, 115*(3), 642–652.
82. Selyutina, O. Y., Polyakov, N. E., Korneev, D. V., & Zaitsev, B. N. (2016). Influence of glycyrrhizin on permeability and elasticity of cell membrane: Perspectives for drugs delivery. *Drug Delivery, 23*(3), 848–855.
83. Nishimoto, Y., Hisatsune, A., Katsuki, H., Miyata, T., Yokomizo, K., & Isohama, Y. (2010). Glycyrrhizin attenuates mucus production by inhibition of MUC5AC mRNA expression in vivo and in vitro. *Journal of Pharmacological Sciences, 113*(1), 76–83.
84. Wu, Q., Tang, Y., Hu, X., Wang, Q., Lei, W., Zhou, L., & Huang, J. (2015). Regulation of Th1/Th2 balance through OX40/OX40L signalling by glycyrrhizic acid in a murine model of asthma. *Respirology, 21*(1), 102–111.
85. Dirican, E., & Turkez, H. (2014). In vitro studies on protective effect of *Glycyrrhiza glabra* root extracts against cadmium-induced genetic and oxidative damage in human lymphocytes. *Cytotechnology, 66*(1), 9–16.
86. Chopra, P. K. P. G., Saraf, B. D., Inam, F. A. R. H. I. N., & Deo, S. S. (2013). Antimicrobial and antioxidant activities of methanol extract roots of *Glycyrrhiza glabra* and HPLC analysis. *International Journal of Pharmacy and Pharmaceutical Sciences, 5*(2), 157–160.
87. Franceschelli, S., et al. (2011). Licocalchone-C extracted from *Glycyrrhiza glabra* inhibits lipopolysaccharide-interferon-γ inflammation by improving antioxidant conditions and regulating inducible nitric oxide synthase expression. *Molecules, 16*, 5720–5734.
88. Jilani, T. N., Jamil, R. T., & Siddiqui, A. H. (2018). *H1N1 influenza (swine flu)*. StatPearls Publishing.
89. Baltina, L. A. (2003). Chemical modification of glycyrrhizic acid as a route to new bioactive compounds for medicine. *Current Medicinal Chemistry, 10*(2), 155–171.
90. Vikhe, G. P., Vikhe, P. P., Naik, S. S., Gavhane, A. J., & Gaikar, R. B. (2013). In vitro effect of G. glabra and T. cordifolia plant extracts on phagocytosis by human neutrophils. *Pravara Medical Review, 5*(2), 12–15.
91. Chen, M., Theander, T. G., Christensen, S. B., Hviid, L., Zhai, L., & Kharazmi, A. (1994). Licochalcone A, a new antimalarial agent, inhibits in vitro growth of the human malaria parasite plasmodium falciparum and protects mice from P. yoelii infection. *Antimicrobial Agents and Chemotherapy, 38*(7), 1470–1475.
92. Ramazani, A., Tavakolizadeh, M., Ramazani, S., Kheiri-Manjili, H., & Eskandari, M. (2018). Antiplasmodial property of *Glycyrrhiza glabra* traditionally used for malaria in Iran: Promising activity with high selectivity index for malaria. *Journal of Arthropod-Borne Diseases, 12*(2), 135–140.
93. Schwikkard, S., & van Heerden, F. R. (2002). Antimalarial activity of plant metabolites. *Natural Product Reports, 19*(6), 675–692.
94. Hasan, M. K., Ara, I., Mondal, M. S. A., & Kabir, Y. (2021). Phytochemistry, pharmacological activity, and potential health benefits of *Glycyrrhiza glabra*. *Heliyon, 7*(6), e07240.

95. Jin, Z., Kim, S., Cho, S., Kim, I. H., Han, D., & Jin, Y. H. (2013). Potentiating effect of glabridin on GABAA receptor-mediated responses in dorsal raphe neurons. *Planta Medica, 79*(15), 1408–1412.
96. Hanrahan, J. R., Chebib, M., & Johnston, G. A. (2011). Flavonoid modulation of GABAA receptors. *British Journal of Pharmacology, 163*(2), 234–245.
97. Roy, S. D., Karmakar, P. R., Dash, S., & Biswajit, D. (2013). Hair growth stimulating effect and phytochemical evaluation of hydro-alcoholic extract of *Glycyrrhiza glabra*. *Education, 2019*.
98. Yang, R., Yuan, B. C., Ma, Y. S., Zhou, S., & Liu, Y. (2017). The anti-inflammatory activity of licorice, a widely used Chinese herb. *Pharmaceutical Biology, 55*(1), 5–18.
99. Al-Snafi, A. E. (2018). *Glycyrrhiza glabra*: A phytochemical and pharmacological review. *IOSR Journal of Pharmacy, 8*(6), 1–17.
100. Sarkar, S., Arora, P., & Garg, K. V. (2013). Cosmeceuticals for hyperpigmentation: What is available? *Journal of Cutaneous and Aesthetic Surgery, 6*(1), 4–11.
101. Sharma, V., Agrawal, R. C., & Pandey, S. (2013). Phytochemical screening and determination of anti-bacterial and antioxidant potential of *Glycyrrhiza glabra* root extracts. *Journal of Environmental Research and Development, 7*(4A), 1552–1558.
102. Alonso, J. (2004). *Tratado de Fitofármacos y Nutracéuticos. www.fitoterapia.net* (pp. 905–911). Corpus.
103. Sianne, S., & Fanie, R. V. H. (2002). Antimalarial activity of plant metabolites. *Natural Product Reports, 19*, 675–692.
104. Revers, F. E. (1956). Clinical and pharmacological investigations on extract of licorice. *Acta Medica Scandinavica, 154*, 749–751.
105. Dhingra, D., Parle, M., & Kulkarni, S. K. (2004). Memory enhancing activity of *Glycyrrhiza glabra* Linn in mice. *Journal of Ethnopharmacology, 91*(2–3), 361–365.
106. Jeong, H. G., You, H. J., Park, S. J., Moon, A. R., Chung, Y. C., Kang, S. K., et al. (2002). Hepatoprotective effects of 18βglycyrrhetinic acid on carbon tetrachloride-induced liver injury: Inhibition of cytochrome P450 2E1 expression. *Pharmacological Research, 46*, 221–227.
107. Xu-ying, W., Ming, L., Xiao-dong, L., & Ping, H. (2009). Hepatoprotective and anti hepatocarcinogenic effects of glycyrrhizin and matrine. *Chemico-Biological Interactions, 181*(1), 15–19.
108. Alaaeldin, A. H. (2007). Curcuma longa, *Glycyrrhiza glabra* Linn and Moringa oleifera ameliorate diclofenac-induced hepatotoxicity in rats. *American Journal of Pharmacology and Toxicology, 2*(2), 80–88.
109. Mauricio, I., Francischett, B., Monterio, R. Q., & Guimaraeas, J. A. (1997). Identification of Glycyrrhizin as thrombin inhibitor. *Biochemical and Biophysical Research Communications, 235*, 259–263.
110. Mendes-Silva, W., Assafim, M., Ruta, B., Monteiro, R. Q., Guimaraes, J. A., Zingali, R. B., et al. (2003). Antithrombotic effect of glycyrrhizin, a plant-derived thrombin inhibitor. *Thrombosis Research, 112*, 93–98.
111. Pompei, R., Flore, O., Marcialis, M. A., Pani, A., & Loddo, B. (1979). Glycyrrhizic acid inhibits virus growth and inactivates virus particles. *Nature, 281*(5733), 689–690.
112. De-Clercq, E. (2000). Current lead natural products for the chemotherapy of human immunodeficiency virus (HIV)infection. *Medicinal Research Reviews, 20*, 323–349.
113. Cinatl, J., Morgenstern, B., Bauer, G., Chandra, P., Rabenau, H., & Doerr, H. W. (2003). Glycyrrhizin, an active component of liquorice roots, and replication of SARS-associated coronavirus. *Lancet, 361*(9374), 2045–2046.
114. Mitscher, L. A., Park, Y. H., Clark, D., & Beal, J. L. (1980). Antimicrobial agents from higher plants. Antimicrobial isoflavanoids and related substances from *Glycyrrhiza glabra* L. var. typica. *Journal of Natural Products, 43*, 259–269.
115. Wahab, S., Annadurai, S., Abullais, S. S., Das, G., Ahmad, W., & Ahmad, M. F. (2021). *Glycyrrhiza glabra* (licorice): A comprehensive review on its phytochemistry, biological activities, clinical evidence and toxicology. *Plants, 10*, 2751.
116. Noreen, S., Mubarik, F., Farooq, F., Khan, M., Khan, A. U., & Pane, Y. S. (2021). Medicinal uses of licorice (*Glycyrrhiza glabra* L.): A comprehensive review. *Open Access Macedonian Journal of Medical Sciences, 9*(F), 668–675.

117. Zogoolas, J. (2017). *The myths and uses of liquorice – Raw essentials tea*. Retrieved March 24, 2022, from https://rawessentialstea.com/journal/the-myths-and-uses-of-liquorice
118. Highly Effective Folk Medicine Remedies From Around the World. (2022). Retrieved March 24, 2022, from https://www.readersdigest.ca/health/healthy-living/folk-medicine-remedies/
119. Fuhrman, B., Volkova, N., Kaplan, M., Presser, D., Attias, J., Hayek, T., & Aviram, M. (2002). Antiatherosclerotic effects of licorice extract supplementation on hypercholesterolemic patients: Increased resistance of LDL to atherogenic modifications, reduced plasma lipid levels, and decreased systolic blood pressure. *Nutrition, 18*(3), 268–273.
120. WHO, World Heart Day 2017, World Health Organization. (2017). Available https://www.who.int/cardiovascular_diseases/world-heart-day-2017/en/. Accessed 21 April 2021.
121. Ibrahim, R. S., Mahrous, R. S., Fathy, H. M., Omar, A. A., & Abu EL-Khair, R. M. (2020). Anticoagulant activity screening of an in-house database of natural compounds for discovering novel selective factor Xa inhibitors; a combined in silico and in vitro approach. *Medicinal Chemistry Research, 29*(4), 707–726.
122. De Caterina, R., Husted, S., Wallentin, L., Andreotti, F., Arnesen, H., Bachmann, F., et al. (2013). General mechanisms of coagulation and targets of anticoagulants (section I). *Thrombosis and Haemostasis, 109*(04), 569–579.
123. Saeedi, M., Morteza-Semnani, K., & Ghoreishi, M. R. (2003). The treatment of atopic dermatitis with licorice gel. *Journal of Dermatological Treatment, 14*(3), 153–157.
124. Tang, Z. H., Chen, X., Wang, Z. Y., Chai, K., Wang, Y. F., Xu, X. H., et al. (2016). Induction of C/EBP homologous protein-mediated apoptosis and autophagy by licochalcone A in non-small cell lung cancer cells. *Scientific Reports, 6*, 26241.
125. Wang, J., Zhang, Y. S., Thakur, K., Hussain, S. S., Zhang, J. G., Xiao, G. R., et al. (2018). Licochalcone A from licorice root, an inhibitor of human hepatoma cell growth via induction of cell apoptosis and cell cycle arrest. *Food Chemical. Toxicology, 120*, 407–417.
126. Bortolotto, L. F. B., Barbosa, F. R., Silva, G., Bitencourt, T. A., Beleboni, R. O., Baek, S. J., et al. (2017). Cytotoxicity of trans-chalcone and licochalcone A against breast cancer cells is due to apoptosis induction and cell cycle arrest. *Biomedicine and Pharmacotherapy, 85*, 425–433.
127. Qiu, C., Zhang, T., Zhang, W., Zhou, L., Yu, B., Wang, W., et al. (2017). Licochalcone A inhibits the proliferation of human lung cancer cell lines A549 and H460 by inducing g2/M cell cycle arrest and ER stress. *International Journal of Molecular Sciences, 18*, 1761.
128. Lu, W. J., Wu, G. J., Chen, R. J., Chang, C. C., Lien, L. M., Chiu, C. C., et al. (2018). Licochalcone A attenuates glioma cell growth in vitro and in vivo through cell cycle arrest. *Food and Function Journal, 9*, 4500–4507.
129. Lin, X., Tian, L., Wang, L., Li, W., Xu, Q., & Xiao, X. (2017). Antitumor effects and the underlying mechanism of licochalcone A combined with 5-fluorouracil in gastric cancer cells. *Oncology Letters, 13*, 1695–1701.
130. Fu, Y., Hsieh, T. C., Guo, J., Kunicki, J., Lee, M. Y. W. T., Darzynkiewicz, Z., & ei al. (2004). Licochalcone-A, a novel flavonoid isolated from licorice root (*Glycyrrhiza glabra*), causes G2 and late-G1 arrests in androgen-independent PC-3 prostate cancer cells. *Biochemical and Biophysical Research Communications, 322*, 263–270.
131. Chen, R., Wang, M., Liu, Q., Wu, J., Huang, W., Li, X., et al. (2020). Sequential treatment with aT19 cells generates memory CAR-T cells and prolongs the lifespan of Raji-B-NDG mice. *Cancer Letters, 469*, 162–172.
132. Morsi, M. K., El-Magoli, B., Saleh, N. T., El-Hadidy, E. M., & Barakat, H. A. (2008). Study of antioxidants and anticancer activity licorice Glycyrrhiza glabra extracts. Egypt. *Egyptian Journal of Nutrition and Feeds, 2*, 177–203.
133. Vlaisavljević, S., Šibul, F., Sinka, I., Zupko, I., Ocsovszki, I., & Jovanović-Šanta, S. (2018). Chemical composition, antioxidant and anticancer activity of licorice from Fruska Gora locality. *Industrial Crops and Products, 112*, 217–224.
134. Zhu, K., Li, K., Wang, H., Kang, L., Dang, C., & Zhang, Y. (2019). Discovery of glabridin as potent inhibitor of epidermal growth factor receptor in SK-BR-3 cell. *Pharmacology, 104*, 113–125.

Chapter 30
Brahmi

Hina Qaiser, Roheena Abdullah, Mehwish Iqtedar, Afshan Kaleem, and Bayan Hussein Sajer

30.1 Introduction

Bacopa monnieri is a widely known medicinal herb which is aquatic, perennial and non-aromatic. It is a member of Scrophulariaceae family (Fig. 30.1), a succulent plant with oblong, thick leaves which are present in alternate arrangements on a delicate green/purplish-green stem. Its actinomorphic flowers are small (4–5 petals) and white, purple, pink or violet in in color. It has the potential to grow in salty waters and can be propagated via cuttings. Fruiting bodies appear purple and resembles ovoid containers (5 mm in length). The general characteristics of the brahmi plant has been summarized in Table 30.1.

30.2 Origin and Distribution

Bacopa monnieri has been deliberately grown by mankind for its medicinal benefits and aesthetic value. It originated from the entirety of the areas in Asia with a tropical climate and now it is widely present in tropical and subtropical countries (Table 30.2). It has been designated as nonnative plant species in Japan, Singapore,

H. Qaiser
Department of Biotechnology, Lahore College for Women University, Lahore, Pakistan

Department of Biology, Lahore Garrison University, Lahore, Pakistan

R. Abdullah (✉) · M. Iqtedar · A. Kaleem
Department of Biotechnology, Lahore College for Women University, Lahore, Pakistan

B. H. Sajer
Department of Biological Sciences, King Abdul Aziz University, Jeddah, Saudi Arabia

M. Zia-Ul-Haq et al. (eds.), *Essentials of Medicinal and Aromatic Crops*,
https://doi.org/10.1007/978-3-031-35403-8_30

Fig. 30.1 Classification of Brahmi

Table 30.1 Characteristic features of various plant parts of *Bacopa monnieri*

Plant part	Features
Stem	Herbaceous, branched, solid with prominent nodes and internodes, cylindrical, green
Root	Tap and adventitious root
Leaf	Opposite, decussate, sessile, estipulate simple
Flower	Solitary, stalked, ebracteated, zygomorphic, bisexual, hypogunos
Calyx	Gamopetalous, inferior, persistent, irregular
Corolla	Gamopetalous, deciduous, inferior, irregular, white with purple tinge
Androecium	Stamen-4, epipetalous, antisepalous, anterbiolocular, black dorsifixed
Gynoecium	Syncarpous, carpels, ovoid
Fruit	Capsule

Table 30.2 World wide distribution of *Bacopa monnieri*

Region wise distribution	Countries wise distribution
Africa	Eswatini, Madagascar, Mozambique, Nigeria, Somalia, South Africa
Asia	Bahrain, Bhutan, Cambodia, China, Guangdong, Guangxi, Hainan, Yunnan, India, Indonesia, Japan, Kuwait, Laos, Malaysia, Nepal, Oman, Pakistan, Philippines, Saudi Arabia, Singapore, Sri Lanka, Taiwan, Thailand, Yemen
Europe	Hungary, Portugal, Spain
North America	Antigua and Barbuda, British Virgin Islands, Bahamas, Barbados, Belize, Cayman Islands, Costa Rica, Cuba, Dominican Republic, El Salvador, Grenada, Guadeloupe, Guatemala, Panama. Puerto Rico, Saint Lucia, Trinidad and Tobago, US Virgin Islands, United States of America
Oceania	Australia, New South Wales, Queensland
South America	Argentina, Bolivia, Brazil, Chile, Colombia, Ecuador, French Guiana, Paraguay, Peru, Uruguay, Venezuela

Spain, Portugal and the Cayman Islands. As an aquatic plant, Brahmi has been found to grow in wet soil, shallow water, and swamps. Generally distributed among warmer countries, it occurs in Nepal, India, Srilanka, China, Taiwan, Pakistan, Vietnam, Florida and the Southern region of the USA.

30.3 Vernacular Names in Different Languages

Brahmi has been assigned various names in different languages. As it comes from India, it is commonly known as Indian pennywort in English language. Other names given to it in English are Water hyssop and Thyme leaved Gartiola. It is called 'Brahmi' in Hindi, Urdu and Sanskrit dialects. Its Chinese, French, German and Japanese names are Jia ma chi xian, Petite bacopa, Kleine fettblatt and Bakopa respectively. As it is known to enhance the memory, it is referred as 'Brahmibuti' in Punjabi.

30.4 Climate and Soil

The succulent leaves remain evergreen and tiny white flowers bloom for protracted periods. Nevertheless, unlike different Bacopa species, the leaves don't give off aroma aren't after crushing. The plant will develop in moist to wet soils, alongside streams and ponds or in marshlands in full sun exposure to partly shady conditions (Fig. 30.2). Water Hyssop can be planted in water gardens and hanging baskets or draping over the brink of a pot. It may even progress in moist soil or shallow water as a ground cover.

30.5 Chemical Composition of Brahmi

Bacopa monnieri has a moisture content of about 88% with deposits of carbohydrates, proteins and almost negligible fats. In addition to these, brahmi also has minerals. The major chemical composition of the herb is show below (Fig. 30.3) [1].

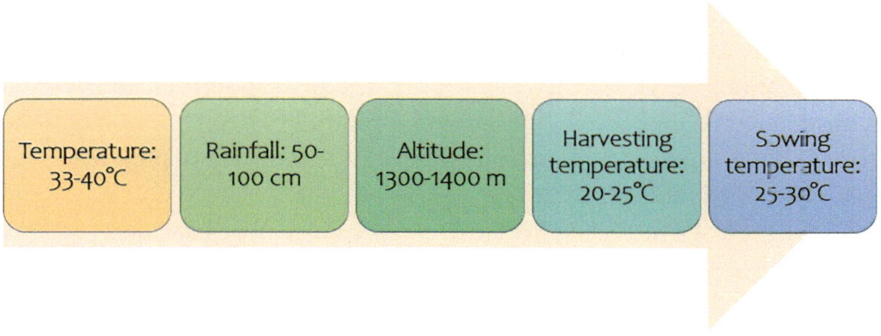

Fig. 30.2 Cultivation conditions of *Bacopa monnieri*

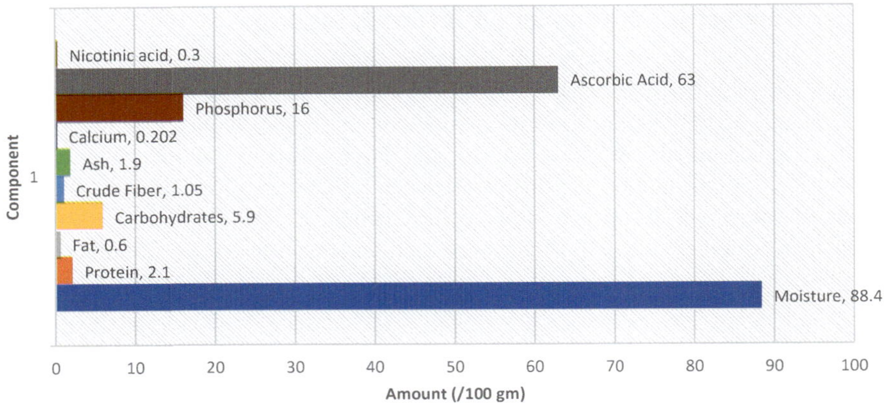

Fig. 30.3 Chemical composition of *Bacopa monnieri*

30.6 Functional Phytochemicals of Brahmi

Important phytochemicals of Brahmi are alkaloids, saponins and sterols (Fig. 30.4). The first alkaloid isolated from this herb was named 'brahmine' [2] while later on nicotine and herpestine were reported to be present [3]. Afterwards, studies confirmed the presence of D-mannitol, hersaponin and potassium salts [4]. The key compound responsible for cognitive effects of brahmi is bacoside A (3-(a-L-arabinopyranosyl)-O-b- D-glucopyranoside-10, 20-dihydroxy-16-keto-dammar-24-ene) [5]. Bacoside A is generally present with bacoside B, its optical isomer [6]. The acid hydrolysis of bacosides results in the production of aglycones, bacogenin A1, A2, A3, [7–9], saponins such as jujubogenin and pseudojujubogenin and bacogenin, A4, identified as ebelin lactone pseudojujubogenin [10]. Sequentially, another saponin; minor bacoside A1 and a novel triperpenoid saponin, bacoside A3, were discovered [10]. As per the escalated interest in the medical value of brahmi, further investigations discovered another two pseudojujubogenin glycosides named as bacopaside I and II [11]. Afterwards, three novel saponins; bacopasides III, IV and V [12], three novel phenylethnoid glycosides, viz. monnierasides I–III along with analogue plantainoside B were isolated [13]. Bacopasaponin G and phenylethyl alcohol have also been reported to present in Brahmi [14].

The two dimentional structures of some of the important phytochemicals of Brahmi are provided in Fig. 30.5.

It was found out from research that *B. monniera* extract has many active compounds [16–18]. In order to authenticate BME effects on behavior, Charles et al. [35] established that oral BME has higher chances of systemic circulation. Bacoside A, an active entity, was detected through HPLC in the serum of rats that were given BME Such active components affect the neurotransmitter system by increasing cognitive abilities. BME are nonpolar glycosides so they can pass through the blood brain barrier through diffusion [25–27, 52]. The presence of medicinal radio-compounds in brain further attest to BME crossing the BBB [53].

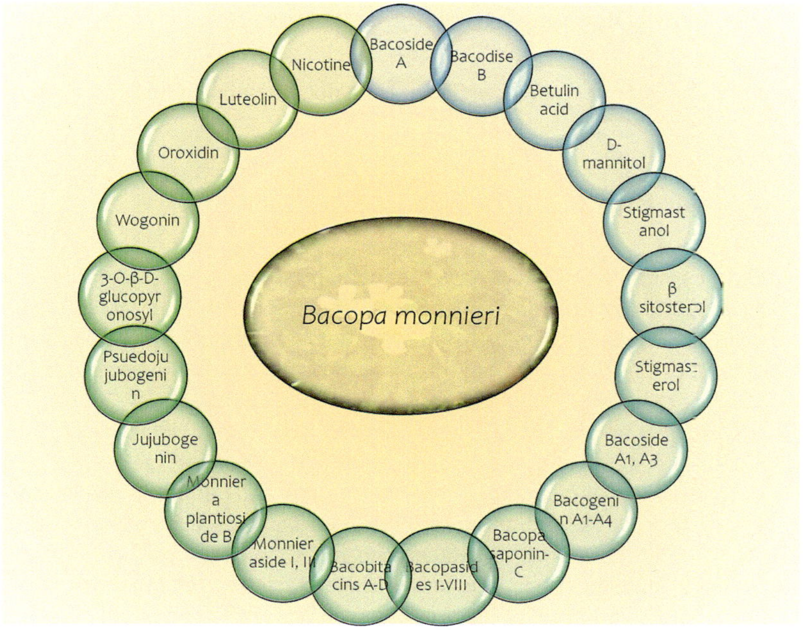

Fig. 30.4 Phytochemicals isolated from *Bacopa monnieri*

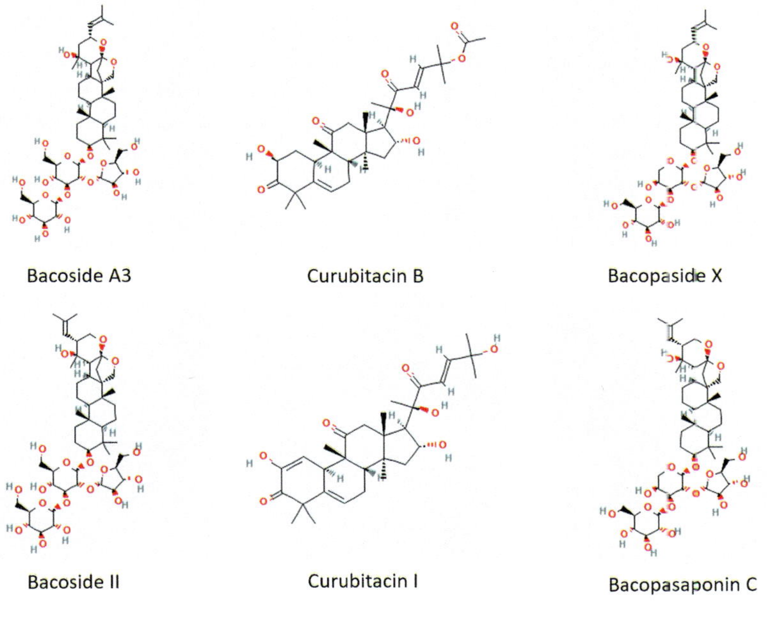

Fig. 30.5 2D structures of bacosides isolated from Brahmi

30.7 Pest and Diseases Affecting Brahmi

No known diseases or pest problems have been documented so far.

30.8 Role of Brahmi in Traditional Medicine

Since old times, traditional and ayurvedic medicines have played a pivotal role in general therapeutics treatments many human ailments which many include cancer, diabetes and skin diseases. *Bacopa monnieri* holds an eminent position in ethnomedicine. It is extensively used as a possible cure for treating neurological disorders, as a memory booster agent and tonic for brain (Fig. 30.6).

Where there is scant amount of proper health care, conventional therapy is mostly used [15–17]. *Bacopa monnieri,* as per its medicinal properties, makes an essential component of traditional medicine in various cultures and ethnic groups around the globe. Its leaf powder is used as blood purifying agent by the people residing in Nara Desert, Sindh, Pakistan [18]. Rajasthani folks make use of this herb to treat stomach related illnesses, bone regeneration and fractures, inflamed urinary duct, rheumatoid arthritis and swollen legs [19]. It is also given to patients having hoarse voice. *Bacopa monnieri* along with *Camellia sinesis* (Joyawake tea) can serve as a nervine tonic as well. Brahmi root extract has the ability to relief the symptoms of constipation and asthma and can be an effective antivenom. It can also be used as an eye ointment for treating cataracts. The oil-based preparations of Brahmi are useful for relieving headaches. It has strong antiseptic properties as well [20, 21]. In South Western Ghats, Virudhunagar district, Tamil Nadu, Brahmi is used for treating dysentery and memory enhancement [22]. Brahmi roots and leaves, as per their therapeutic value, are used against disorders of nervous system in the villages of Daksin Dinajpur, West Bengal, India [23]. Some Southern Kerala tribes exploits its leaves for urine related problems and as a growth stimulant of pubic abdominal region [24]. Bangladeshi people also consume its leaf for blood purification [25].

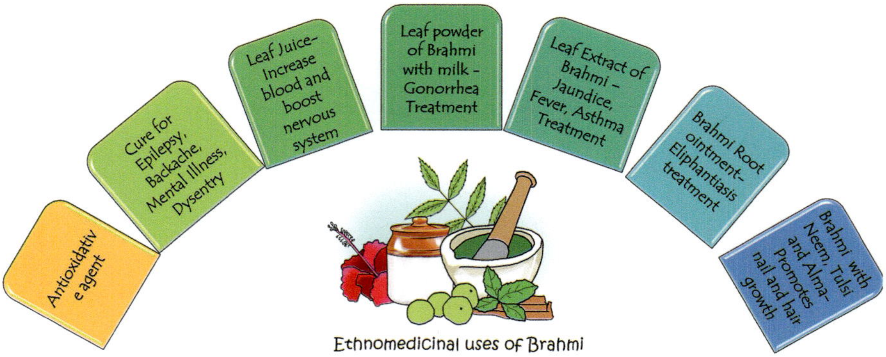

Fig. 30.6 Benefits of *Bacopa monnieri* as a traditional medicine

30.9 Pharmacological Benefits of Brahmi

Studies have revealed many neuropharmacological properties of Brahmi extract. Pharmacologically active compounds such as saponins and bacosides are responsible for most of the medicinal benefits of Brahmi (Fig. 30.7). Bacosides primarily act upon nervous system and upregulate the transmission of nerve impulses and also help in neuron damage repair by stimulating the kinase activity, neuronal synthesis and restoration of synaptic activity. Animal studies confirmed the relaxant properties of brahmi on pulmonary arteries, aorta, trachea, and ileal and bronchial tissue, perhaps facilitated by restriction of calcium-ion influx through plasma membranes into the cell (Fig. 30.8). Many clinical investigations have been conducted to evaluate the nootropic properties *Bacopa monnieri* (Table 30.3).

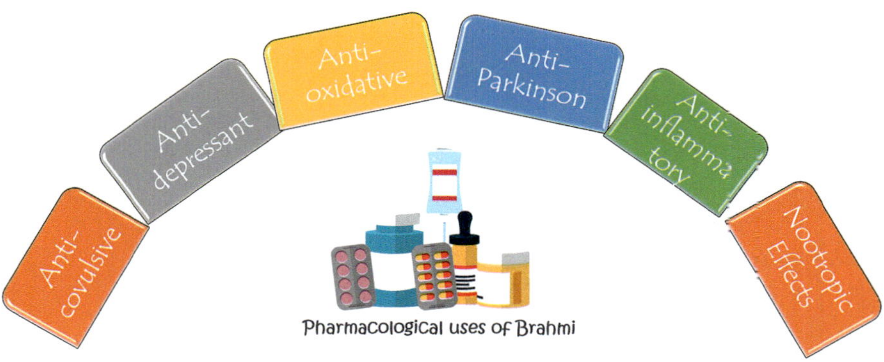

Fig. 30.7 Pharmacological benefits of *Bacopa monnieri*

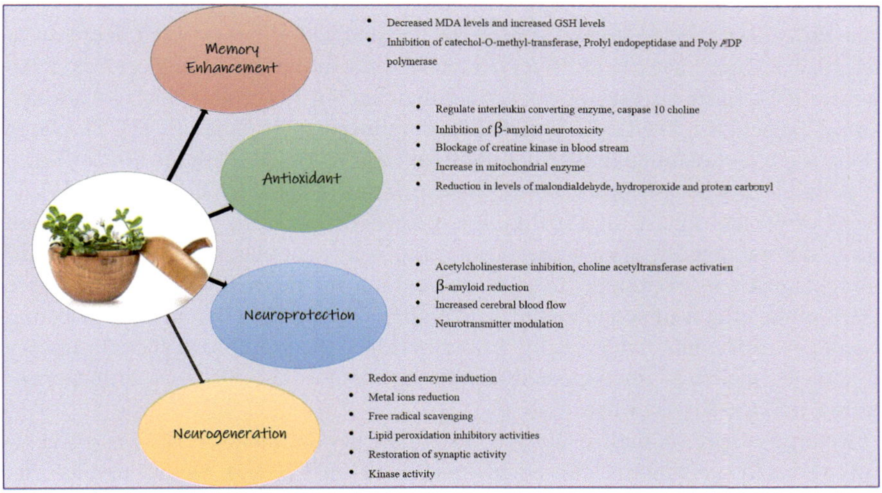

Fig. 30.8 Chief pharmacological properties of *Bacopa monnieri* with their particular target molecules

Table 30.3 Clinical investigations performed on *Bacopa monnieri* [26–30]

Substance	Treatments	Subjects	Findings
Brahmi crude extract	Chronically for 4 weeks	Patients suffering from anxiety neurosis	Memory enhancement
	Chronically for 12 weeks	Children	Memory and learning enhancement
Brahmi standardized extract	–	Mentally retarded children	Learning enhancement and control of abnormal behaviour
	Chronically for 12 weeks	Healthy adults	Early information processing improvement, verbal learning and memory consolidation
	Acute treatment for 2 hrs	Healthy adults	No significant changes found
	1st trial after chronic treatment for 3 months, 2nd trial for 6 weeks after the completion of 1st trial	Healthy adults	Retention of new information potential improved, Tasks assessing attention, verbal and visual short term memory and the retrieval of pre-experimental knowledge unaffected

30.9.1 Role in Alzheimer's Disease and Schizophrenia Management

To fully comprehend the reversing effect of Brahmi in amnesia, many experiments were performed. For this, amnesia causing drugs such as benzodiazepines, scopolamine, quinoline derivatives and phenytoin, were given to animals. Their mechanism is to interfere long-term potentiation (LTP) via gamma-aminobutyric acid-benzodiazepine pathway. Saraf et al. documented a monumental decrease in diazepam induced anemia (1.75 mg/kg) when oral Brahmi (120 mg/kg) was given to mice [31]. Later on, they also recorded the effect of Brahmi on downstream signaling molecules related to LTP in diazepam induced amnesic mice [32]. The same group used Scopolamine in place of diazepam and recorded the same effect [33].

Results showed a suppression in protein kinase C and iNOS amounts. While levels of protein kinase A, MAP kinase, cAMP, calmodulin, pCREB, CREB and nitrite did not change when amnesia was caused by scopolamine. Brahmi resulted in higher amounts of protein kinase C, calmodulin and pCREB levels and therefore reversed the effects of scopolamine. The findings highlighted the memory boosting capability of Brahmi. Additionally, Brahmi is also able to improve spatial memory in amnesic model of mice created using scopolamine [34]. This experiment was designed to use Morris water maze test.

Rotarod test in animals evaluated their Muscle coordination. Brahmi successfully reversed scopolamine caused anemia. (Fig. 30.9). Hence, we can deduce that Brahmi has a significant position in treating Alzheimer's disease.

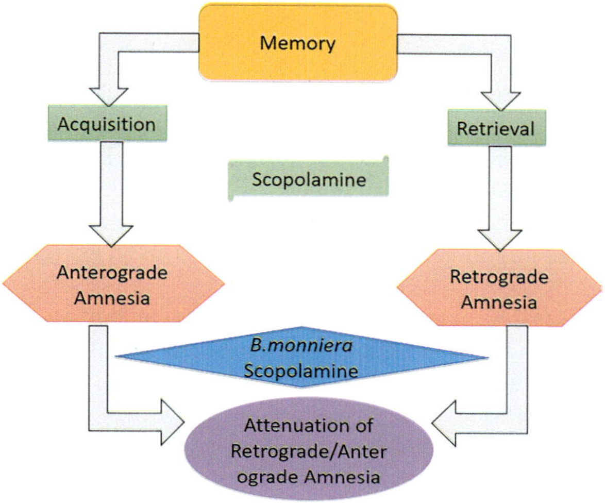

Fig. 30.9 Illustrative depiction of Brahmi extract on memory acquisition and retrieval

In another study, Brahmi was investigated as a neuroprotective agent in a schizophrenic rat [35]. The findings focused on discrimination ratio and thus indirectly cognitive ability. Moreover, the levels of N-methyl-D- aspartate receptor subtype 1 (NMDAR1) were recorded in brain mainly the prefrontal cortex, striatum, cornu ammonis fields I (CA1) and 2/3 (CA2/3) of hippocampus and dentate gyrus (DG). There was a notable decrease in discrimination ratio in schizophrenia induced group in contrast to control.

30.9.2 Anti-Asthmatic Activity

Brahmi extract has shown to possess relaxant attributes owing to (beta)-adrenoreceptor a prostaglandin [36]. It has also been able to dilate the bronchioles in the lungs of anesthetized rats [37] which makes it suitable for treatment of respiratory diseases [38]. The antagonistic action of carbachol on respiratory system (inspiratory and expiratory pressure) is responsible for the above property. Ethanolic extracts of brahmi have also exhibited the calcium antagonistic activity [27] whereas the methanolic preparations produced a mast cell stabilizer effect showing its usefulness in controlling allergic reactions [39].

30.9.3 Anti-cancer Activity

Bacosides present in Brahmi have anticancer properties. The use of Brahmi has shown to decrease the escalated values of ulcer index, adrenal gland weight, plasma glucose, aspartate aminotransferase (AST), and creatine kinase (CK) related to acute stress in cancer patients [40]. Methanolic extract exhibited potent mast cell stabilizer [41] activity. It can also accumulate heavy metals such as cadmium, chromium, lead and mercury in vast quantities, therefore can be exploited for bioremediation [42].

B. monnieri possesses the potential to fight cancer. Palethorpe et al. [43] worked to find out that bacopaside I and bacopaside II, a terpenoid from B. monnieri, are able to synergistically inhibit the membrane transport system aquaporin. AQP1 can increase the spread of tumors. Therefore, its blockage will prevent proliferation, migration and invasion in breast cancer cell lines.

30.9.4 Anticonvulsive Activity

Bacopa can be effectively used for the prevention of seizures and its role in epilepsy management has been established in traditional medicine. Scientific experiments investigating the anticonvulsive activity of brahmi and its crude water extract has been performed using animal models [44]. It also possesses sedative attributes and can effectively lengthen the soporific action of phenobarbitone. Brahmi can activate gamma-aminobutyric acid production which have been shown to display anticonvulsive, pain killing and sedative behavior [45]. It implies the indirect influence of Brahmi over the central nervous system [46]. Experimental studies on mice have been performed to elucidate the locomotors activity and maximal electroshock seizures in response to brahmi extract alone and combination with phenytoin (PHT) [47].

Acquiring the information and its retention has been shown to increase without any effect on PHT anti-convulsive activity. Furthermore, use of brahmi in conjugation with various other antiepileptic medication has produced fruitful results in the treatment of epilepsy.

30.9.5 Antidepressant

Brahmi extracts in prepared in methanol have shown depression receding proprieties in rodents and mice. Oral administration at 20 and 40 mg/kg dosage for a 5 days period exhibited profound anti-depressing effects [48]. CDRI-08 (KeenMind) was injected in mice for tail suspension test (TST) and forced swimming test (FST) resulting in antidepressant activity [49]. Oral administration of these drugs for 5 days elicited reduced immobility time span both in FST and TST. It was deduced that Bacosides A and B, bacopasaponin C, bacopasides I and II were responsible for anti-depressant ability in Brahmi with the exception of bacopaside VII [48, 50, 51].

30.9.6 Anti-inflammatory

Brahmi also display inhibitory effects when it comes to inflammation management. It does so by modulating the release of pro-inflammatory mediators [52]. Its use as an anti-inflammatory compound has been well established in traditional medicine [53]. It also negativity impacts It 5-lipoxygenase (5-LOX), 15-LOX and cyclooxygenase-2 (COX-2) activities [54]. The inhibitory effect may be attributed to the occurrence of triterpenoids and bacosides in it.

30.9.7 Anti-nociceptive Activity

Brahmi extracts can also be used as an analgesic owing to its influence on β1-adrenergic, α2-adrenergic receptors and 5-HT receptors, which are involved in analgesic pathways and hence it can be useful in pain management as well [55]. It has also been reported that aqueous preparations of brahmi used along with naloxone was not able to enhance the analgesic potential which implied the influence of opioid receptors.

30.9.8 Anti-oxidant Activity

Alcoholic and hexane extract of Brahmi are free-radical scavengers. They stop oxidation by hindering the oxidative degradation of lipids [42]. The antioxidative process of Brahmi can also be due to inhibition of enzymes like superoxide dismutase (SOD), catalase (CAT) and glutathione per-oxidase (GPX) [56]. A study revealed the ability of the hydro alcoholic extract of the whole Brahmi plant to stop the excretion of superoxide released from polymorphonuclear cells in nitro blue assay [57]. Sumathy et al. (2001) conducted experiments on rats which were given morphine. He observed the antihepatotoxic effects of Brahmi alcoholic extract on them [58]. The reason might lie in lessened mitochondrial enzyme activity in rat brains [59]. Methanolic extract Brahmi can interfere with the production of the superoxide anion by stopping it in a dose-dependent manner. Resultantly, a decrease in the concentrations of nitric oxide (NO), generated (enzymatic and non- enzymatic) by activated astrocytes is noted. A variety of neurodegenerative diseases, such as AD, ischemia and epilepsy may be linked to this [59, 60].

30.9.9 Anti-stress Activity

Brahmi extract works like an adaptogen. If low doses of Brahmi extract are given as treatment beforehand, a notable reverse in ulcer index and plasma AST can be recorded. On the other hand, if higher doses are given then significant reversed changes in ulcer index, adrenal gland weight, CK, and AST are recorded [40].

30.9.10 Anti-Spasmodic Activity

Brahmi extract acts as an anti-spasmodic in smooth muscles. The mechanism behind this is blockage of calcium influx via calcium channels of the cell membrane. Nevertheless, if no modification is observed in contractions produced by nor--adrenaline or caffeine in the presence of Brahmi extract, it proves that Brahmi extract is not involved in influx of efflux of intracellular calcium.

30.9.11 Anxiolytic Effect

Brahmi extract acts as anti-anxiolytic when given in high dose in contrast to Lorezepam [61]. Nevertheless, it is noteworthy to observe the upper hand of Brahmi over Lorazepam as it do not initiate amnesia, rather it works like a memory stimulator [62]. Shanker and Singh observed these effects and also highlighted the anxiolytic role of Brahmi extract [62].

30.9.12 Cardiovascular Activity

The ethanol-based preparation of Brahmi also restores and upregulate the cardiac health. It is able to sustain a smooth blood flow within normal pressure range through left ventricle [63]. However, Brahmi also exert positive effect on aortic and pulmonary artery cardiac flow [64].

30.9.13 Endocrine Effects

Some animal experiments document the effect of Brahmi on endocrine system. CDRI-08 (KeenMind; 200 mg/kg orally) influence thyroid hormone in male mice [65]. When noticed, it was seen that rate of production of T4 raised by 41% if Brahmi was administered. It had no effect on T3 synthesis. Some studies highlight the anti-fertility activity of Brahmi male mouse [66].

30.9.14 Gastroprotective Activity

The ulcer preventing and healing properties Brahmi plant extract may be attributed to its interactive effects on the mucosa membranes [67]. It also show positive effects on the irregular spasmodic movements of intestines [38]. It does so by restricting the flow of calcium ions through the transmembrane protein channels residing in the smooth muscle cells of small intestine. Fresh juice extracts of Brahmi has shown efficient potential of abating and treating ulcer formations [68, 69]. This may be credited to its positive association with protective agents and processes of mucosa such as aggravated mucin and glycoprotein production and down regulated cell shedding. The brahmi fresh juice also impedes stomach acidity and pepsin secretion.

30.9.15 Hepatoprotective Activity

Studies recorded the antihepatotoxic potential of Brahmi extract. Experiments were conducted on rats where they were given morphine. They were administered Brahmi extract beforehand. The result showed a positive effect on morphine stimulated liver and kidney physiology [70]. It was seen that when pre-treatment was done with Bacoside A, the lipid peroxidase levels and the activity of serum markers enzymes decreased. The antioxidant system was maintained and in return protects the rats from Diethyl nitrosamine-induced hepatic toxicity [71].

30.9.16 Learning and Memory

A range of experiments were done to prove the neuro-pharmacological ability of Plant extracts and isolated Bacosides. The results establish the nootropic action of the plant. It was observed in initial experiments that plant [72] or alcoholic extract of Brahmi [73] acted as a stimulator in learning process in rats. Further research found out that nootropic activity was due to Bacosides A and B, in ethanol extract [45]. Other functions of these active compounds were the inhibited amnesic effects of scopolamine, electroshock and immobilization stress.

The underlying pharmacology of active compounds is speculative. The recorded mechanism suggests the role of bacosides in membrane dephosphorylation, with concomitant increase in protein and RNA turnover in specific brain areas [3]. Brahmi boosts the activity of protein kinase in hippocampus further proving its nootropic activity [74]. In cases when Brahmi was given for 2 weeks: acetylcholine was completely depleted, acetyl cholinesterase catalytic activity was down regulated and colchicines induced binding of muscarinic cholinergic receptors present in frontal cortex and hippocampus [75]. It can be safely envisaged that behavioral changes induced by cholinergic deterioration could be assuaged by lessening noradrenergic activity [76].

30.9.17 Management of Diabetes Nephropathy

In diabetes mellitus, there is a relative or total absence of insulin, which causes reduced glucose uptake by insulin-sensitive tissues and has serious consequences. According to world's estimates, about 415 million people are affected by diabetes, increasing to 642 million by 2040 [77]. Diabetes is managed by drug regimens but to-date, complete prevention has not become possible.

In addition to chemotherapy, various medicinal plants have been used conventionally [78–80]. Gosh et al., used ethanolic extract of B. monnieri to check antioxidant and antihyperglycemic activity in mice. The study revealed a decrease in glycosylated hemoglobin thus making it comparable to control drug α-tocopherol [81].

An isolate of BM, stigmasterol is effective in streptozotocin-nicotinamide-induced diabetic nephropathy (DN) by decreasing glycation products and by improving free radical damage (66). It was found out that B. monnieri was also responsible for decreased reduced serum glucose and increased diabetic rat body weight [82].

30.9.18 Antimicrobial Effects

Some anti-microbial action is seen in methanolic extracts of CDRI-08 (KeenMind) [1, 83]. A study revealed antimicrobial activity of different extracts i.e., hexane and petroleum ether stopped microbial growth but methanolic extracts had greater capacity. On the other hand, aqueous extract of CDRI-08 (KeenMind) failed to exhibit any effects [84]. Rohini et al., 2004 showcased the suppression of microorganisms like Staphylococcus aureus to be more than *Salmonella typhi* and *Escherichia coli. Klebsiella pneumonia* was seen to unaffected [85].

30.10 Dosage, Safety and Toxicity

The safe use of Brahmi has been documented since centuries in ayurvedic medicine (Table 30.4). The pharmacological use of Brahmi bacosides at safe dosages has been investigated in a clinical trial performed on adult healthy individuals for a period of 4 weeks. Individuals consuming 20–30 mg of bacosides once a day and those taking daily multiple doses of 100-200 mg, both, showed no side effects [86]. The median lethal dose (LD_{50}) of Brahmi extracts (aqueous and alcoholic) has been recorded as 1000 mg and 15 g/kg respectively, in rat models [87]. The aqueous extract at 5 g/kg dosage administered through oral route also showed no toxic effects. In case of alcoholic extract given orally, the LD_{50} was 17 g/kg. Moreover, there were no behavioral changes associated with the consumption of bacosides [88].

Table 30.4 Traditional daily dosage of Brahmi [3]

Traditional daily dose of *Bacopa monnieri*	
Powder form	5–10 g
Infusion	8–16 ml
Syrup	30 ml
1:2 fluid extract	
Adults	5–12 ml/day
Children (6–12 years)	2.5–6 ml/day
20% standardized *Bacopa monnieri*	
Adults	200–400 mg/day in divided doses
Children	100–200 mg/day in divided doses

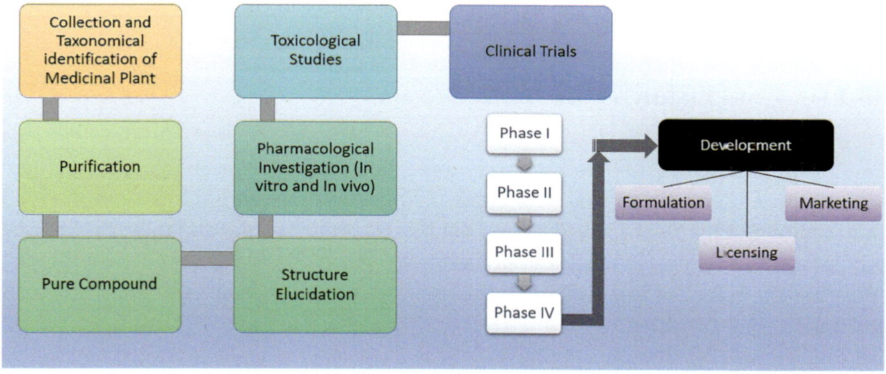

Fig. 30.10 Good manufacturing practice for formulating herbal medicine

Nevertheless, the herbal formulation is capable of enhanced pharmacokinetics and dynamics this lessening dose. In this way drugs with high potency can be easily managed. (114). So, for best results work should be done to make the herbs more bioavailable and safer. Good manufacturing practices should rule over the whole procedure. Following flow chart explains the procedure for herbal formulations (Fig. 30.10).

B. monnieri has zero toxicity. It can be used to revert any adverse reaction caused by toxic compounds. Thus, making it a very safe option in pharmaceutical domain. Nevertheless, deep research should be conducted to discover the full potential. Table 30.5 showcases toxicity and prevention findings.

Table 30.5 Efficiency of Brahmi against adverse effects of different toxic substances [70, 89–93]

Toxicity	BME dosage	Effects of BME
Sodium fluoride	300 mg/kg	Ameliorate the cholinergic system. Reduce the oxidative stress. Subdued neuropathological changes
Opioid	40 mg/kg	Restored serum ALT, AST and creatinine elevations
Lead	10 mg/kg/day	Reduced brain lead level in comparison to conventional treatment. Reduce the oxidative stress
Aluminum	40 mg/kg/day	Protect brain from oxidative damage
Methyl mercury	250 mg/ml	Prevented mitochondrial damage
Glutamate	5 mM	Prevented mitochondrial damage Prevented oxidative stress in cultured neuronal cells
Trimethylin	50 mg/kg	Ameliorates TMT-induced cognition dysfunction mainly via protecting the hippocampal neurons

30.11 Conclusion

Bacopa monnieri is recognized as a very important traditional herb in ayurvedic and folk medical literature. It can be a game changer in managing many neurological and cardiovascular disorders. Different plant extracts (ethanolic and methanolic) are used as traditional medicines for many ailments. Bacoside A, being the most important phytochemical of this herb, exhibits anticancerous, antidiabetic, antimicrobial and antioxidant properties.

References

1. Azad, A., Awang, M., & Rahman, M. (2012). Phytochemical and microbiological evaluation of a local medicinal plant Bacopa monnieri (l.) Penn. *International Journal of Current Pharmaceutical Review and Research, 3*(3), 66–78.
2. Russo, A., & Borrelli, F. (2005). Bacopa monniera, a reputed nootropic plant: An overview. *Phytomedicine, 12*(4), 305–317.
3. Deo, Y. K., & Reddy, K. (2013). Critical review on pharmacological properties of Brahmi. *International Journal of Ayurvedic Medicine, 4*(2), 92–99.
4. Sastri, M., Dhalla, N., & Malhotra, C. (1959). Chemical investigation of Herpestis monniera Linn (Brahmi). *Indian Journal of Pharmacology, 21*(1959), 303–304.
5. Chatterjee, N. (1965). Chemical examination of Bacopa monniera Wettst. Part II: The constitution of Bacoside A. *Indian Journal of Chemistry, 3*(1965), 24–29.
6. Patil, D., Bakliwal, S., Rane, B., & Pawar, S. (2012). A review on Bacopa monniera: Brahmi. *Pharma Science Monitor, 3*(4), 3196–3211.
7. Kulshreshtha, D., & Rastogi, R. (1973). Bacogenin-A1: A novel dammarane triterpene sapogenin from Bacopa monniera. *Phytochemistry, 12*(4), 887–892.
8. Kulshreshtha, D., & Rastogi, R. (1974). Bacogenin A2: A new sapogenin from bacosides. *Phytochemistry, 13*(7), 1205–1206.

9. Chandel, R., Kulshreshtha, D., & Rastogi, R. (1977). Bacogenin-A3: A new sapogenin from Bacopa monniera. *Phytochemistry, 16*(1), 141–143.
10. Rastogi, S., Pal, R., & Kulshreshtha, D. K. (1994). Bacoside A3: A triterpenoid saponin from Bacopa monniera. *Phytochemistry, 36*(1), 133–137.
11. Chakravarty, A. K., Sarkar, T., Masuda, K., Shiojima, K., Nakane, T., & Kawahara, N. (2001). Bacopaside I and II: Two pseudojujubogenin glycosides from Bacopa monniera. *Phytochemistry, 58*(4), 553–556.
12. Chakravarty, A. K., Garai, S., Masuda, K., Nakane, T., & Kawahara, N. (2003). Bacopasides III—V: Three new triterpenoid glycosides from Bacopa monniera. *Chemical and Pharmaceutical Bulletin, 51*(2), 215–217.
13. Chakravarty, A. K., Sarkar, T., Nakane, T., Kawahara, N., & Masuda, K. (2002). New phenylethanoid glycosides from Bacopa monniera. *Chemical and Pharmaceutical Bulletin, 50*(12), 1616–1618.
14. Hou, C.-C., Lin, S.-J., Cheng, J.-T., & Hsu, F.-L. (2002). Bacopaside III, bacopasaponin G, and bacopasides A, B, and C from Bacopa monniera. *Journal of Natural Products, 65*(12), 1759–1763.
15. Sheldon, J. W., Balick, M. J., Laird, S. A., & Milne, G. M. (1997). Medicinal plants: Can utilization and conservation coexist? *Advances in Economic Botany, 12*(1997), 101–104.
16. Shrestha, P. M., & Dhillion, S. S. (2003). Medicinal plant diversity and use in the highlands of Dolakha district. *Nepal. Journal of ethnopharmacology, 86*(1), 81–96.
17. Tabuti, J., Dhillion, S., & Lye, K. (2003). Traditional medicine in Bulamogi county, Uganda: Its practitioners, users and viability. *Journal of Ethnopharmacology, 85*(1), 119–129.
18. Qureshi, R., & Bhatti, G. R. (2008). Ethnobotany of plants used by the Thari people of Nara Desert, Pakistan. *Fitoterapia, 79*(6), 468–473.
19. Verma, M. (2014). Ethno medicinal and antimicrobial screening of bacopa monnieri (l.) pennell. *Journal of Phytomedicine, 6*(2014), 1–6.
20. Khan, A. V., Ahmed, Q. U., Shukla, I., & Khan, A. A. (2010). Antibacterial efficacy of Bacopa monnieri leaf extracts against pathogenic bacteria. *Asian Biomedicine, 4*(4), 651–655.
21. Choudhary, S., Kumari, I., Thakur, S., Kaurav, H., & Chaudhary, G. (2021). Brahmi (Bacopa monnieri) – A potential ayurvedic cognitive enhancer and neuroprotective herb. *International Journal of Ayurveda and Pharma Research, 9*(5), 41–49.
22. Hossain, H., Howlader, M. S. I., Dey, S. K., Hira, A., & Ahmed, A. (2012). Evaluation of analgesic, antidiarrhoeal and cytotoxic activities of ethanolic extract of Bacopa monnieri (L). *British Journal of Pharmaceutical Research, 2*(3), 188.
23. Banerjee, S., Anand, U., Ghosh, S., Ray, D., Ray, P., Nandy, S., et al. (2021). Bacosides from Bacopa monnieri extract: An overview of the effects on neurological disorders. *Phytotherapy Research, 35*(10), 5668–5679.
24. Rani, S., Rana, J., & Rana, P. (2013). Ethnomedicinal plants of Chamba district, Himachal Pradesh, India. *Journal of Medicinal Plants Research, 7*(42), 3147–3157.
25. Rahmatullah, M., Mollik, M. A. H., Islam, M. K., Islam, M. R., Jahan, F. I., Khatun, Z., et al. (2010). A survey of medicinal and functional food plants used by the folk medicinal practitioners of three villages in Sreepur Upazilla, Magura district, Bangladesh. *American Eurasian Journal of Sustainable Agriculture, 4*(3), 363–373.
26. Singh, R., & Singh, R. (1980). Studies on the antioxidant anxiety effect of the Medhay Rasayan drug Brahmi (Bacopa monniera Linn)—Part II (experimental studies). *Journal of Research in Indian Medicine, 14*(1980), 1–6.
27. Sharma, R., Chaturvedi, C., & Tewari, P. (1987). Efficacy of Bacopa monniera in revitalizing intellectual functions in children. *Journal of Research Medicine in Indian Education, 1*(2), 15–20.
28. Dave, U. P., Chauvan, V., & Dalvi, J. (1993). Evaluation of br-16 a (mentat) in cognitive and behavioural dysfunction of mentally retarded children—A placebo-controlled study. *The Indian Journal of Pediatrics, 60*(3), 423–428.

29. Stough, C., Lloyd, J., Clarke, J., Downey, L., Hutchison, C., Rodgers, T., et al. (2001). The chronic effects of an extract of Bacopa monniera (Brahmi) on cognitive function in healthy human subjects. *Psychopharmacology, 156*(4), 481–484.
30. Roodenrys, S., Booth, D., Bulzomi, S., Phipps, A., Micallef, C., & Smoker, J. (2002). Chronic effects of Brahmi (*Bacopa monnieri*) on human memory. *Neuropsychopharmacology, 27*(2), 279–281.
31. Prabhakar, S., Saraf, M. K., Pandhi, P., & Anand, A. (2008). Bacopa monniera exerts antiamnesic effect on diazepam-induced anterograde amnesia in mice. *Psychopharmacology, 200*, 27–37.
32. Saraf, M., Prabhakar, S., Pandhi, P., & Anand, A. (2008). Bacopa monniera ameliorates amnesic effects of diazepam qualifying behavioral–molecular partitioning. *Neuroscience, 155*(2), 476–484.
33. Saraf, M. K., Anand, A., & Prabhakar, S. (2010). Scopolamine induced amnesia is reversed by Bacopa monniera through participation of kinase-CREB pathway. *Neurochemical Research, 35*, 279–287.
34. Saraf, M. K., Prabhakar, S., Khanduja, K. L., & Anand, A. (2011). Bacopa monniera attenuates scopolamine-induced impairment of spatial memory in mice. *Evidence-Based Complementary and Alternative Medicine, 2011*, 236186.
35. Piyabhan, P., & Wetchateng, T. (2013). Cognitive enhancement effects of Bacopa monnieri (Brahmi) on novel object recognition and VGLUT1 density in the prefrontal cortex, striatum, and hippocampus of sub-chronic phencyclidine rat model of schizophrenia. *Journal of the Medical Association of Thailand= Chotmaihet Thangphaet, 96*(5), 625–632.
36. Dar, A., & Channa, S. (1997). Bronchodilatory and cardiovascular effects of an ethanol extract of Bacopa monniera in anaesthetized rats. *Phytomedicine, 4*(4), 319–323.
37. Gairola, G., Nikhate, S. P., Raole, V. V., Bagul, A., & Kumar, S. (2021). A literary study on bramhi and its physiological action over human memory-A medhya rasayana. *International Journal of Botany Studies, 6*(1), 552–553.
38. Dar, A., & Channa, S. (1999). Calcium antagonistic activity of Bacopa monniera on vascular and intestinal smooth muscles of rabbit and guinea-pig. *Journal of Ethnopharmacology, 66*(2), 167–174.
39. Negi, K., Singh, Y., Kushwaha, K., Rastogi, C., Rathi, A., Srivastava, J., et al. (2000). Clinical evaluation of memory enhancing properties of Memory Plus in children with attention deficit hyperactivity disorder. *Indian Journal of Psychiatry, 42*(4), 2002.
40. Rai, D., Bhatia, G., Palit, G., Pal, R., Singh, S., & Singh, H. K. (2003). Adaptogenic effect of Bacopa monniera (Brahmi). *Pharmacology Biochemistry and Behavior, 75*(4), 823–830.
41. Samiulla, D., Prashanth, D., & Amit, A. (2001). Mast cell stabilising activity of Bacopa monnieri. *Fitoterapia, 72*(3), 284–285.
42. Tripathi, Y. B., Chaurasia, S., Tripathi, E., Upadhyay, A., & Dubey, G. (1996). Bacopa monniera Linn. as an antioxidant: Mechanism of action. *Indian Journal of Experimental Biology, 34*(6), 523–526.
43. Palethorpe, H. M., Smith, E., Tomita, Y., Nakhjavani, M., Yool, A. J., Price, T. J., et al. (2019). Bacopasides I and II act in synergy to inhibit the growth, migration and invasion of breast cancer cell lines. *Molecules, 24*(19), 3539.
44. Shanmugasundaram, E., Akbar, G. M., & Shanmugasundaram, K. R. (1991). Brahmighritham, an Ayurvedic herbal formula for the control of epilepsy. *Journal of Ethnopharmacology, 33*(3), 269–276.
45. Shanker, G., & Singh, H. (2000). Anxiolytic profile of standardized Brahmi extract. *Indian Journal of Pharmacology, 32*(152), 5 pages.
46. Singh, H., Shanker, G., & Patnaik, G. (1996). Neuropharmacological and anti-stress effects of bacosides: A memory enhancer. *Indian Journal of Pharmacology, 28*, 47.
47. Vohora, D., Pal, S., & Pillai, K. (2000). Protection from phenytoin-induced cognitive deficit by Bacopa monniera, a reputed Indian nootropic plant. *Journal of Ethnopharmacology, 71*(3), 383–390.

48. Sairam, K., Dorababu, M., Goel, R., & Bhattacharya, S. (2002). Antidepressant activity of standardized extract of Bacopa monniera in experimental models of depression in rats. *Phytomedicine, 9*(3), 207–211.
49. Shen, Y.-H., Zhou, Y., Zhang, C., Liu, R.-H., Su, J., Liu, X.-H., et al. (2009). Antidepressant effects of methanol extract and fractions of Bacopa monnieri. *Pharmaceutical Biology, 47*(4), 340–343.
50. Sheikh, N., Ahmad, A., Siripurapu, K. B., Kuchibhotla, V. K., Singh, S., & Palit, G. (2007). Effect of Bacopa monniera on stress induced changes in plasma corticosterone and brain monoamines in rats. *Journal of Ethnopharmacology, 111*(3), 671–676.
51. Zhou, Y., Shen, Y.-H., Zhang, C., Su, J., Liu, R.-H., & Zhang, W.-D. (2007). Triterpene saponins from Bacopa monnieri and their antidepressant effects in two mice models. *Journal of Natural Products, 70*(4), 652–655.
52. Viji, V., & Helen, A. (2011). Inhibition of pro-inflammatory mediators: Role of Bacopa monniera (L.) Wettst. *Inflammopharmacology, 19*(5), 283–291.
53. Channa, S., Dar, A., Anjum, S., & Yaqoob, M. (2006). Anti-inflammatory activity of Bacopa monniera in rodents. *Journal of Ethnopharmacology, 104*(1-2), 286–289.
54. Viji, V., & Helen, A. (2008). Inhibition of lipoxygenases and cyclooxygenase-2 enzymes by extracts isolated from Bacopa monniera (L.) Wettst. *Journal of ethnopharmacology, 118*(2), 305–311.
55. Bhaskar, M., & Jagtap, A. (2011). Exploring the possible mechanisms of action behind the antinociceptive activity of Bacopa monniera. *International Journal of Ayurveda Research, 2*(1), 2.
56. Bhattacharya, S., Bhattacharya, A., Kumar, A., & Ghosal, S. (2000). Antioxidant activity of Bacopa monniera in rat frontal cortex, striatum and hippocampus. *Phytotherapy Research, 14*(3), 174–179.
57. Pawar, R., Gopalakrishnan, C., & Bhutani, K. (2001). Dammarane triterpene saponin from Bacopa monniera as the superoxide inhibitor in polymorphonuclear cells. *Planta Medica, 67*(08), 752–754.
58. Sumathi, T., Nayeem, M., Balakrishna, K., Veluchamy, G., & Devaraj, S. N. (2002). Alcoholic extract of 'Bacopa monniera'reduces the in vitro effects of morphine withdrawal in guinea-pig ileum. *Journal of Ethnopharmacology, 82*(2-3), 75–81.
59. Sumathy, T., Govindasamy, S., Balakrishna, K., & Veluchamy, G. (2002). Protective role of Bacopa monniera on morphine-induced brain mitochondrial enzyme activity in rats. *Fitoterapia, 73*(5), 381–385.
60. Russo, A., Borrelli, F., Campisi, A., Acquaviva, R., Raciti, G., & Vanella, A. (2003). Nitric oxide-related toxicity in cultured astrocytes: Effect of Bacopa monniera. *Life Sciences, 73*(12), 1517–1526.
61. Deo, Y. K., & Reddy, K. (2015). Nephroprotective effect of Brahmi Ghrita. *Journal of Medicinal Plants, 3*(2), 05–07.
62. Bhattacharya, S., & Ghosal, S. (1998). Anxiolytic activity of a standardized extract of Bacopa monniera: An experimental study. *Phytomedicine, 5*(2), 77–82.
63. Rashid, S., Lodhi, F., Ahmad, M., & Usmanghani, K. (1990). Cardiovascular effects of Bacopa monnieri (L.) pennel extract in rabbits. *Pakistan Journal of Pharmaceutical Sciences, 3*(2), 57–62.
64. Dar, A., & Channa, S. (1997). Relaxant effect of ethanol extract of Bacopa monniera on trachea, pulmonary artery and aorta from rabbit and guinea-pig. *Phytotherapy Research: An International Journal Devoted to Medical and Scientific Research on Plants and Plant Products, 11*(4), 323–325.
65. Kar, A., Panda, S., & Bharti, S. (2002). Relative efficacy of three medicinal plant extracts in the alteration of thyroid hormone concentrations in male mice. *Journal of Ethnopharmacology, 81*(2), 281–285.
66. Singh, A., & Singh, S. K. (2009). Evaluation of antifertility potential of Brahmi in male mouse. *Contraception, 79*(1), 71–79.

67. Dorababu, M., Prabha, T., Priyambada, S., Agrawal, V., Aryya, N., & Goel, R. (2004). Effect of Bacopa monniera and Azadirachta indica on gastric ulceration and healing in experimental NIDDM rats. *International Journal of Experimental Niology, 42*(4), 389–397.
68. Rai, K., Gupta, N., Dharamdasani, L., Nair, P., & Bodhankar, P. (2017). Bacopa monnieri: A wonder drug changing fortune of people. *International Journal of Applied Sciences and Biotechnology, 5*(2), 127–132.
69. Sairam, K., Rao, C. V., Babu, M. D., & Goel, R. (2001). Prophylactic and curative effects of Bacopa monniera in gastric ulcer models. *Phytomedicine, 8*(6), 423–430.
70. Sumathi, T., & Devaraj, S. N. (2009). Effect of Bacopa monniera on liver and kidney toxicity in chronic use of opioids. *Phytomedicine, 16*(10), 897–903.
71. Janani, P., Sivakumari, K., & Parthasarathy, C. (2009). Hepatoprotective activity of bacoside A against N-nitrosodiethylamine-induced liver toxicity in adult rats. *Cell Biology and Toxicology, 25*(5), 425–434.
72. Devendra, P., Patel, S., Birwal, P., Basu, S., Deshmukh, G., & Datir, R. (2018). Brahmi (Bacopa monnieri) as functional food ingredient in food processing industry. *Journal of Pharmacognosy and Phytochemistry, 7*(3), 189–194.
73. Singh, H., & Dhawan, B. (1982). Effect of Bacopa monniera Linn.(Brāhmi) extract on avoidance responses in rat. *Journal of Ethnopharmacology, 5*(2), 205–214.
74. Singh, H., & Dhawan, B. (1997). Neuropsychopharmacological effects of the Ayurvedic nootropic Bacopa monniera Linn.(Brahmi). *Indian Journal of Pharmacology, 29*(5), 359–364.
75. Pase, M. P., Kean, J., Sarris, J., Neale, C., Scholey, A. B., & Stough, C. (2012). The cognitive-enhancing effects of Bacopa monnieri: A systematic review of randomized, controlled human clinical trials. *The Journal of Alternative and Complementary Medicine, 18*(7), 647–652.
76. Sara, S. J. (1989). Noradrenergic-cholinergic interaction: Its possible role in memory dysfunction associated with senile dementia. *Archives of Gerontology and Geriatrics, Suppl, 1*(1989), 99–108.
77. Fatima, U., Roy, S., Ahmad, S., Ali, S., Elkady, W. M., Khan, I., et al. (2022). Pharmacological attributes of Bacopa monnieri extract: Current updates and clinical manifestation. *Frontiers in Nutrition, 9*, 972379.
78. Al-Attar, A. M., & Alsalmi, F. A. (2019). Influence of olive leaves extract on hepatorenal injury in streptozotocin diabetic rats. *Saudi Journal of Biological Sciences, 26*(7), 1865–1874.
79. Rao, M. U., Sreenivasulu, M., Chengaiah, B., Reddy, K. J., & Chetty, C. M. (2010). Herbal medicines for diabetes mellitus: A review. *Interntional Journal of PharmTech Research, 2*(3), 1883–1892.
80. Kooti, W., Farokhipour, M., Asadzadeh, Z., Ashtary-Larky, D., & Asadi-Samani, M. (2016). The role of medicinal plants in the treatment of diabetes: A systematic review. *Electronic Physician, 8*(1), 1832.
81. Ghosh, T., Kumar, M. T., Sengupta, P., Dash, D. K., & Bose, A. (2008). Antidiabetic and in vivo antioxidant activity of ethanolic extract of Bacopa monnieri Linn. aerial parts: A possible mechanism of action. *Iranian Journal of Pharmaceutical Research, 7*(1), 61–68.
82. Pandey, S. P., Singh, H. K., & Prasad, S. (2015). Alterations in hippocampal oxidative stress, expression of AMPA receptor GluR2 subunit and associated spatial memory loss by Bacopa monnieri extract (CDRI-08) in streptozotocin-induced diabetes mellitus type 2 mice. *PLoS One, 10*(7), e0131862.
83. Katoch, M., Singh, G., Sharma, S., Gupta, N., Sangwan, P. L., & Saxena, A. K. (2014). Cytotoxic and antimicrobial activities of endophytic fungi isolated from Bacopa monnieri (L.) Pennell (Scrophulariaceae). *BMC Complementary and Alternative Medicine, 14*(1), 1–8.
84. Hosamani, R., Krishna, G., & Muralidhara. (2016). Standardized Bacopa monnieri extract ameliorates acute paraquat-induced oxidative stress, and neurotoxicity in prepubertal mice brain. *Nutritional Neuroscience, 19*(10), 434–446.
85. Rohini, G., Sabitha, K. & Devi, C. (2004). Bacopa monniera Linn. extract modulates antioxidant and marker enzyme status in fibrosarcoma bearing rats.

86. Singh, H., & Dhawan, B. (1997). Neuropsychopharmacological effects of the Ayurvedic nootropic Bacopa monniera Linn.(Brahmi). *Indian Journal of Pharmacology, 29*(5), 359.
87. Martis, G., Rao, A., & Karanth, K. (1992). Neuropharmacological activity of Herpestis monniera. *Fitoterapia, 43*(5), 399–404.
88. Allan, J. J., Damodaran, A., Deshmukh, N., Goudar, K., & Amit, A. (2007). Safety evaluation of a standardized phytochemical composition extracted from Bacopa monnieri in Sprague–Dawley rats. *Food and Chemical Toxicology, 45*(10), 1928–1937.
89. Balaji, B., Kumar, E. P., & Kumar, A. (2015). Evaluation of standardized Bacopa monniera extract in sodium fluoride-induced behavioural, biochemical, and histopathological alterations in mice. *Toxicology and Industrial Health, 31*(1), 18–30.
90. Velaga, M. K., Basuri, C. K., Robinson Taylor, K. S., Yallapragada, P. R., Rajanna, S., & Rajanna, B. (2014). Ameliorative effects of Bacopa monniera on lead-induced oxidative stress in different regions of rat brain. *Drug and Chemical Toxicology, 37*(3), 357–364.
91. Jyoti, A., Sethi, P., & Sharma, D. (2007). Bacopa monniera prevents from aluminium neurotoxicity in the cerebral cortex of rat brain. *Journal of Ethnopharmacology, 111*(1), 56–62.
92. Ayyathan, D. M., Chandrasekaran, R., & Thiagarajan, K. (2015). Neuroprotective effect of Brahmi, an ayurvedic drug against oxidative stress induced by methyl mercury toxicity in rat brain mitochondrial-enriched fractions. *Natural Product Research, 29*(11), 1046–1051.
93. Pham, H. T. N., Phan, S. V., Tran, H. N., Phi, X. T., Le, X. T., Nguyen, K. M., et al. (2019). Bacopa monnieri (L.) ameliorates cognitive deficits caused in a trimethyltin-induced neurotoxicity model mice. *Biological and Pharmaceutical Bulletin, 42*(8), 1384–1393.

Chapter 31
Buckwheat

Hina Qaiser, Roheena Abdullah, Afshan Kaleem, Mehwish Iqtedar, and Bayan Hussein Sajer

31.1 Introduction

Buckwheat or common buckwheat was presumably brought to use and cultivated in Southeast Asia, probably around 6000 B.C.E. It then entered the regions of Central Asia, Tibet and Europe. It is possible that the domestic use of buckwheat originated in the western Yunnan region of China [1]. In Europe, the information on Buckwheat is reported in the Balkans by at least the Middle Neolithic (c. 4000 B.C.E.). The most ancient traces are tracked in China up to Circa 2600 B.C.E. According to a study, the existence of buckwheat pollen in Japan dated back to 4000 B.C.E.

31.2 Classification of Buckwheat

Buckwheat exists as a pseudocereal crop. It is from family *Polygonaceae* and genus *Fagopyrum* (Fig. 31.1). The family name calls attention to the swollen nodes of a few species in the stem region. The term stems from Greek; poly meaning 'many' and gony meaning 'joint'. Scientifically, Buckwheat is called *Fagopyrum esculentum* and its designated plant symbol is FAES. It also goes by *Polygonum fagopyrum,*

H. Qaiser
Department of Biotechnology, Lahore College for Women University, Lahore, Pakistan

Department of Biology, Lahore Garrison University, Lahore, Pakistan

R. Abdullah (✉) · A. Kaleem · M. Iqtedar
Department of Biotechnology, Lahore College for Women University, Lahore, Pakistan

B. H. Sajer
Department of Biological Sciences, King Abdulaziz University, Jeddah, Saudi Arabia

M. Zia-Ul-Haq et al. (eds.), *Essentials of Medicinal and Aromatic Crops*,
https://doi.org/10.1007/978-3-031-35403-8_31

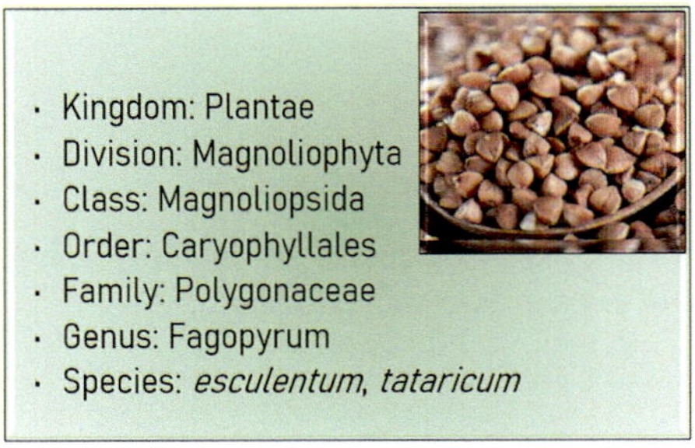

Fig. 31.1 Classification of Buckwheat

Table 31.1 Vernacular names of common buckwheat in different countries [3]

Country	Common buckwheat name
India	Ogal
Nepal	Mite Phapar
Pakistan	Jawas
Mandarin	Tian qiao mai
Japan	Soba
Bhutan	Jare
France	Sarrasin
Russia	Grecicha kul'furnaja
Italy	Fagopiro
Germany	Buchweizen

Fagopyrum fagopyrum Karst., *Fagopyrum vulgare Hill* and *Fagopyrum sagittatum Gilib.*

Around 15–16 species of plants sum up the genus Fagopyrum. It contains the common buckwheat or the Japanese (*Fagopyrum esculentum*) and the Tartary buckwheat (*Fagopyrum tataricum*). Both of them are of same use and categorized as pseudocereals. They are not a part of the grass family.

31.3 Vernacular Names in Major Languages

Throughout the evolution course of Buckwheat, it was given various names [1]. In the province of Yunnan, it is known as buckwheat er, common buckwheat er chi and tartary buckwheat er ka [2]. The sequence in the names is vital to find out the migration path in Europe and Asia. It has been assigned various names in different countries (Table 31.1).

31.4 The Buckwheat Plant

The plant produces brown seeds which do not have a defined shape. The seeds are seen with four triangular surfaces. The germination begins in 3–4 days, given the seeds are sown in warm soil. If all conditions are met, the plants grow at a very fast pace. The leaves are shaped like a heart and the stems are thinner and hollow. The next step is when flowering starts in 3 weeks after planting. It is in full motion for some weeks but as plant ages, the flowers production declines. When the flowers are in full bloom, a buckwheat field resembles an ocean due to the white petals. Then comes the pollination of the flowers which follow the development of a full-sized seed within 10 days. The seeds complete their maturity cycle in 1–2 weeks. First the seeds are seen and mature on the lower stems and then they grow up on the stem along with the growth of plant. Plant height and how fast the plant will grow is based on planting date. The plant can grow to be up to 1 m height when planted in starting of summer and provided with optimum fertility conditions. Approximately 11–12 weeks are required for maturation. If it is grown at the end of July, it has maturation time of 9–10 weeks. Hot and dry environment affects the size and development of the crop.

31.5 Climate and Soil

Common buckwheat is a facultative short-day plant [4]. It develops stably where temperatures are low and the air is moist. It is to note that the seeds are able to germinate in very dry regions [5]. Normally, it is a undemanding crop [6] but the factors affecting the total biomass production, grain yield and quality of buckwheat are conditioned to temperature, rainfall and sunlight [7]. High temperatures and hot dry winds affect the crop. Tartary buckwheat is robust in contrast to the common buckwheat; It can bear poor soil and extreme weather situations. Temperatures between 5–42 °C are good for germination and growth of Buckwheat. But the optimum temperature is 24–26 °C. When flowering is taking place temperature more than 30 °C causes loss of flower (blasting), fruit desiccation and poor grain yield [8]. Temperature is the utmost significant environmental factor affecting flavonoid and rutin accumulation in buckwheat [9]. A notable production is the result of appropriate soil moisture. When the plant is provided with moisture, the growth sets off but at the same time the maturity process is deterred. In contrast, cereals mature forcefully under low soil moisture condition. It was seen that the mass production of common buckwheat improved when the moisture in the soil rose. However, no changes were recorded in the seed set [10]. Buckwheat is able to endure a lot of environmental changes specially its ability to handle different kinds of soil including infertile, rocky and poorly tilled lands [11] puts it into limelight because other grain crops cannot handle those soils. But the favorable soil is well drained sandy soils and tolerates acidic soil conditions (pH as low as 4.8) [7]. Buckwheat does not grow well in saline and semi-arid regions [12].

31.6 Chemical Composition of Buckwheat

There is a wide array of nutrients in the Buckwheat grains e.g. proteins, rutin, polysaccharides, dietary fibre, lipids, polyphenols and micro nutrients (minerals and vitamins) [13]. Buckwheat holds good nutritive value in comparison to other cereals and can serve as an exceptional source of macronutrients (Fig. 31.2).

Factors like the type of Species and environment decide which compound will be of what percentage [15]. Buckwheat grouts mainly contain starch (550 mg/g), proteins (120 mg/g), dietary fibre (70 mg/g), lipids (mg/g) and various compounds such as organic acids, phenolics tannins and nucleic acids (about 180 mg/g) [16, 17]. Ample quantities of dietary fiber can be gained from Buckwheat bran, especially bran containing hulls holds about 400 mg/g of total dietary fibre whereas bran devoid of hulls possess about 160 mg/g of total dietary fibre. It is reported that the maximum quantity of protein is obtained from Bran fractions [18, 19]. The composition of Buckwheat resembles to that of Cereals and pseudo cereals. Figure 31.3 shows a proximate composition of buckwheat in comparison to other cereals. An interesting thing about buckwheat fibre is the lack of a key anti-nutritional factor in common wheat [18]. Buckwheat bran contains more soluble dietary fibre compared to the wheat bran or oat bran. The content of indigestible starch is around

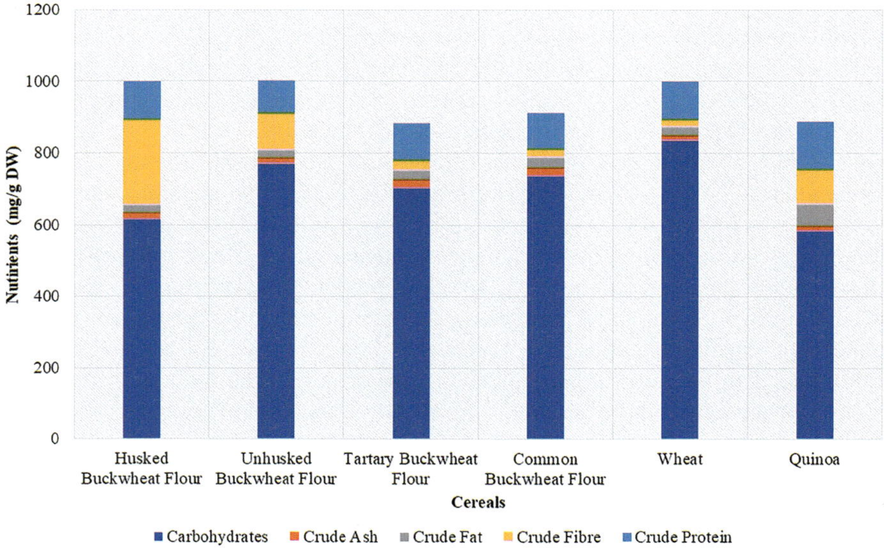

Fig. 31.2 Nutrient profile of buckwheat and other important cereals [14]

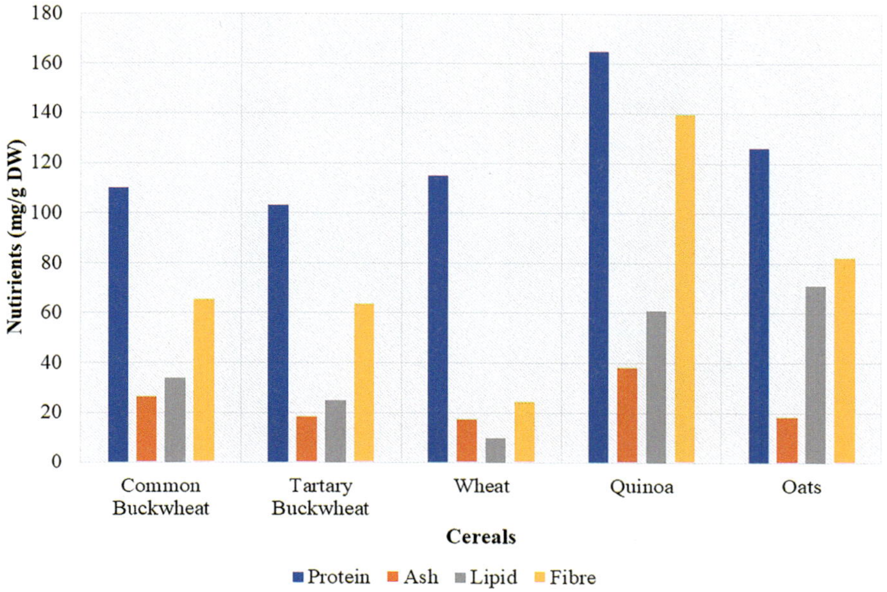

Fig. 31.3 Proximate analysis comparison of Buckwheat with other cereals

303–380 g/100 g of total starch in uncooked form of buckwheat groats while it decreases to 70–100 mg/100 g after cooking [20]. Studies showed the presence of soluble carbohydrates in the embryo with maximum concentration present in the endosperm [18].

Increased awareness regarding the nutrient profile of Buckwheat has led to its wide spread use in commercial food products (raw and processed) (Fig. 31.4).

31.7 World Buckwheat Production, Exports and Imports

The genetic resource spread of buckwheat is significantly pronounced in central Asia especially in the south-western China [21] making China the originator of buckwheat). It is easy to find the common buckwheat (*Fagopyrum esculentum*) in all the global continents except the Tartary buckwheat (*Fagopyrum tataricum*) which can be found in the hilly and mountain region of China and the Himalayas. On a global scale, buckwheat cultivation covers an area of 2.4 million hectares, typically having an average production of 2.4 million tones and 1000 kg/ha productivity. Among the buckwheat growing countries in world in 2021, Russia was the most notable country in buckwheat area and production followed by China and Ukraine (Fig. 31.5). Analysis of Buckwheat production over the past 10 years showed random fluctuations with the highest in the year 2017 (more than 3,000,000 tonnes)

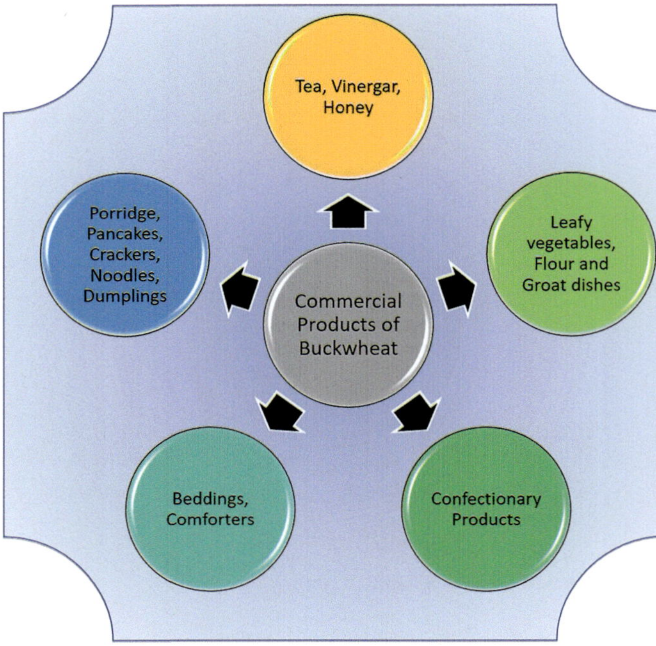

Fig. 31.4 Uses of Buckwheat in commercial products

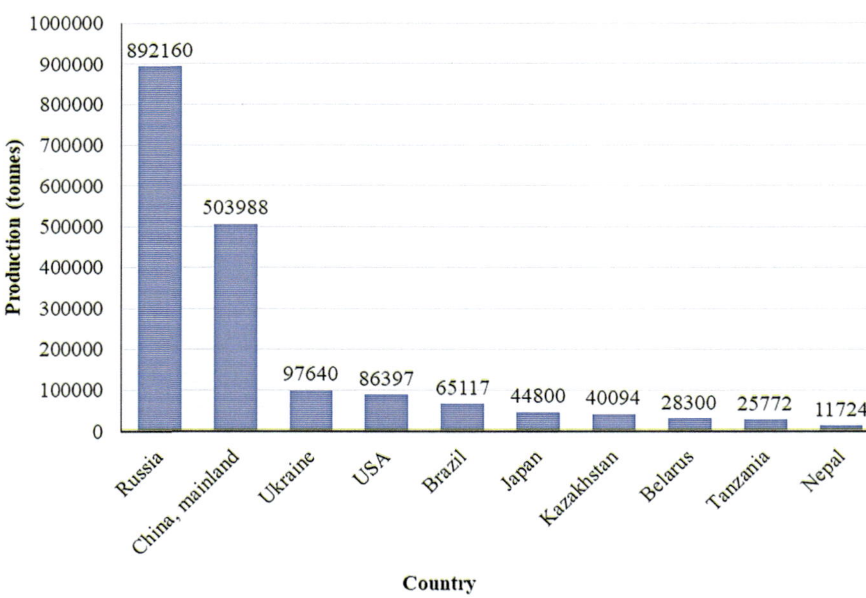

Fig. 31.5 Top ten producers of Buckwheat for the year 2021

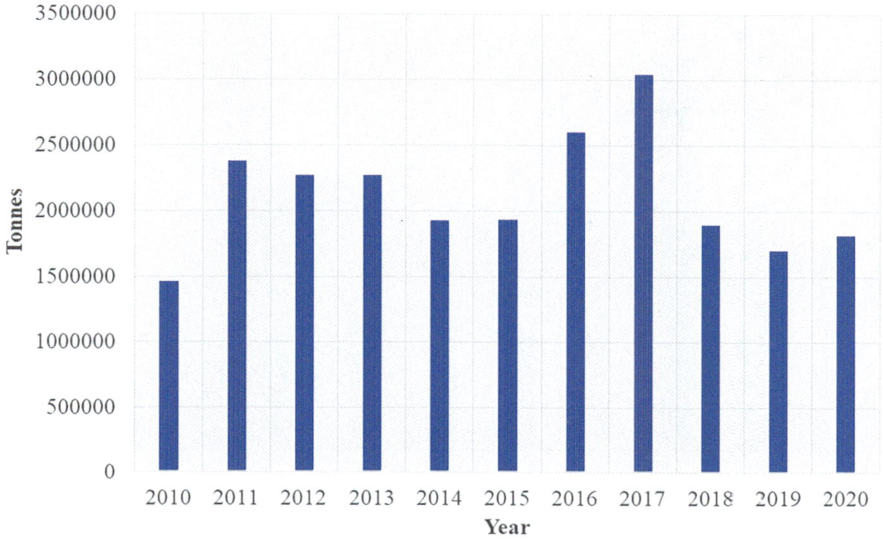

Fig. 31.6 Worldwide Buckwheat production ranging from 2010 to 2020

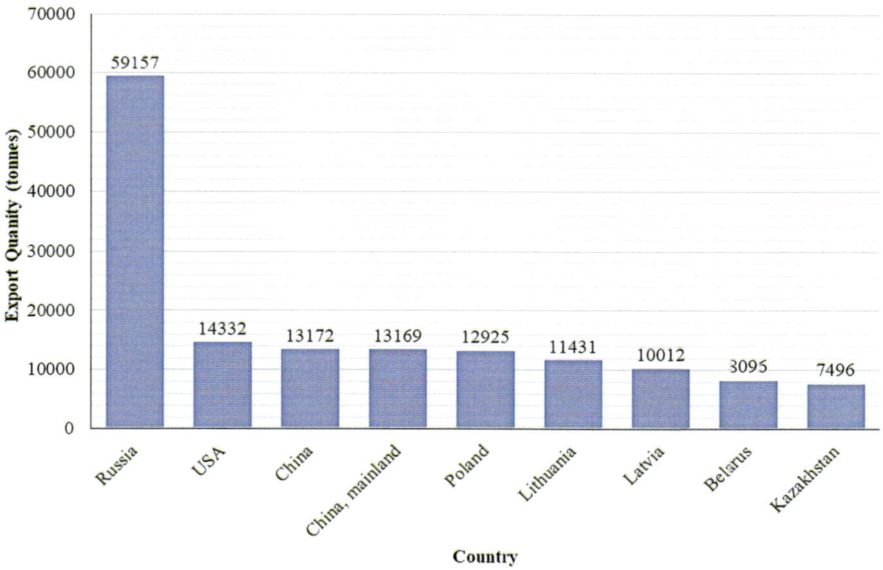

Fig. 31.7 Top ten Buckwheat exporting countries (FAO 2021)

(Fig. 31.6). The production for year 2020 was around 180,000 tonnes. In 2019, among the list of most traded goods around the world, Buckwheat secured 897th place. In 2021, Russia was the largest exporter of buckwheat followed by USA and China (Fig. 31.7). As far as the imports are concerned, Ukraine and Japan imported significant quantities of Buckwheat (Fig. 31.8).

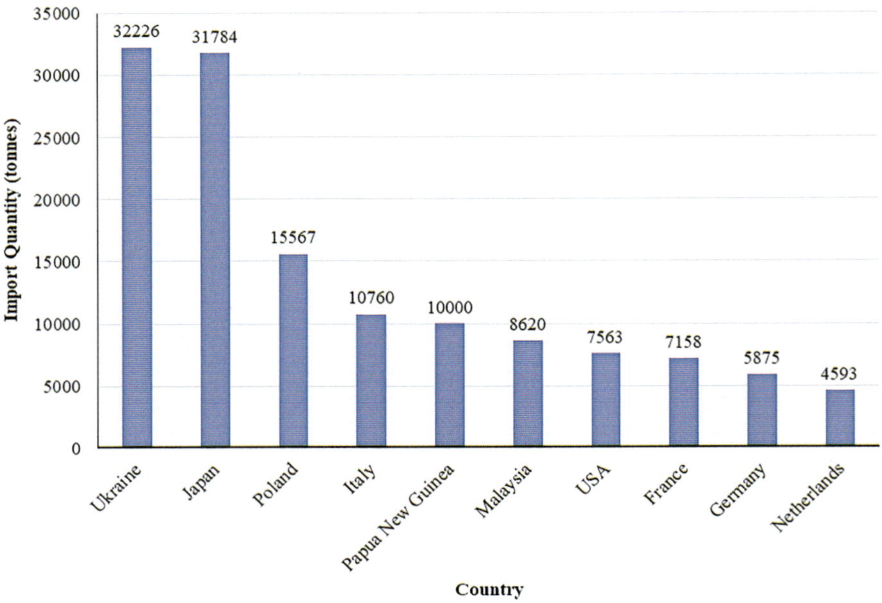

Fig. 31.8 Top ten Buckwheat Importing countries (FAO 2021)

31.8 Functional Phytochemicals of Buckwheat

Various compounds having functional importance have been discovered from a variety of buckwheat species. They include flavonoids, phenolic acids, tannins, fagopyrin, triterpenoids, steroids, stilbenes, and so on (Fig. 31.9).

31.8.1 Anti-oxidative Compounds

The anti-oxidative activity of Buckwheat grain is greater than cereal grains [22]. Antioxidants secure the body against the damaging oxidative activity of free radicals. The anti-oxidative activity corresponds significantly to the total polyphenol content [23]. Common buckwheat hulls and groats have many different anti-oxidative compounds like vitamins B1, B2, and E, and phenolic compounds (including polyphenols such as catechins, rutin, quercetin and proanthocyanidins). In common buckwheat groats, it was seen that different catechins ((−)-epicatechin and (−)-epicatechin gallate) inhibit oxidation Watanabe [24]. It is noteworthy that rutin can be called as a functional flavonoid. In addition the notable anti-oxidant activity of rutin has been taken into account as well [25]. Another study records zero mutual relation between anti-oxidative activity and rutin. Moreover, Morishita et al. [26] found out rutin contributed only to 2% of the total anti-oxidative process.

Fig. 31.9 Medicinal and health benefits of bioactive compounds present in Buckwheat

Morishita et al. [27] experimented to form different buckwheat strains having a greater inhibition to oxidation. He used recurrent individual selection and was able to create enhanced anti-oxidative breeds like 'Gamma no irodori', 'Cobalt no chikara', and 'Ruchiking'.

31.8.2 Anti-hypertensive and Anti-hyperglycemic Functional Compounds

Christa and Soral-Śmietana experiments over animal models in 2008 helped to discover the risks of buckwheat flour in relation to diseases like diabetes, obesity, hypertension, and hypercholesterolemia. Angiotensin-I converting enzyme (ACE) is responsible for regulating blood pressure. Buckwheat is known to have a compound which stops ACE activity [28]. The compound is 2″-hydroxynicotianamine, the hydroxy derivative of nicotianamine. It is present in both buckwheat flour and plant body. This compound has a great potential to block the function of ACE with the inhibitory concentration (IC50) of only 0.08 μM [29]. This is similar to that of nicotianamine [30]. Moreover, 2″-hydroxynicotianamine lowers blood pressure and nicotianamine in spontaneously hypertensive rats (SHR) [29]. The compound is a special addition to polygonaceae. The relative percentage of 2″-hydroxynicotianamine is 20–100% in Tartary buckwheat leaves [28].

D-Chiro-Inositol (DCI) in buckwheat affects high blood pressure. It works like insulin by acting as a part of a putative media- tor of insulin action [31]. It increases insulin effects therefore exerting a positive effect on hyperglycemia and hypertension [32]. DCI is among rare naturally existing isomers. Buckwheat flour is known to lower glucose in diabetic patients [33]. Such effect matches to that of synthesized DCI. Kawa et al. (2003) researched that DCI is one of the main reasons why buckwheat lowers glucose levels. Buckwheat contains a soluble carbohysrate fagopyritol which has DCI in it. Fagopyritol consists of at least six molecular species consisting of 2–4 oligomers; fagopyritol A1, A2, A3, B1, B2, B3. Fagopyritol A1 finds its use in putative insulin mediators [34]. The almost alike Fagopyritols structure can make a mark in making an innovative plant-based compound. This compound can treat insulin response disorders such as Non-Insulin Dependent Diabetes Mellitus (NIDDM) and Polycystic Ovary Syndrome (PCOS) [35]. It was seen that fagopyritol synthase might have some enzymatic reactions in buckwheat. The reaction may be the interaction with galactinol synthase or stachyose synthase.

31.8.3 Rutin

Rutin belongs to polyphenols [17]. Rutin in both the seeds and leaves of buckwheat is one of the most notable characteristics of Buckwheat. About hundred times greater, rutin is present in the seeds of Tartary buckwheat than common buckwheat. It confers many health benefits (Fig. 31.10). Rutin adds to the strength of human capillaries. It has anti-hypertensive properties anti-oxidative properties, anti-inflammatory properties [38], and alpha- glucosidase inhibitory activities [36]. Rutin is a potential major anti-oxidant [37–39]. After genetic examination of common buckwheat grains, rutin content found was 10–30 mg/100 g DW [40, 41]. Kitabayashi et al. [42] worked to find the heritability of the seed rutin content. The highest heritability comes during the days preceding to first flowering. Other roles of rutin in buckwheat include its function in the plant defense mechanism against ultra violet light, low temperatures, drought conditions, and infestation [43]. Up till now, rutin rich common buckwheat cultivars such as SunRutin and Toyomusume have been established [44].

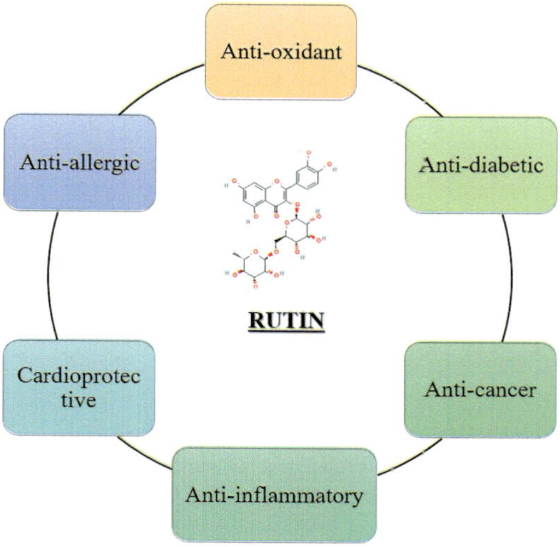

Fig. 31.10 Health benefits of Rutin

31.8.4 Polysaccharides and Dietary Fiber

The digestive enzymes of small intestine cannot break down Dietary fiber but its bacterial fermentation occurs in large intestine. One of the main uses of Dietary fiber is helping to get rid of type 2 diabetes, heart disease, and obesity [45]. Non-starch polysaccharides (i.e., cellulose, hemicellulose, and pectin) make up the main portion of dietary fiber. It has two types, based on water solubility: insoluble dietary fiber (IDF), and soluble dietary fiber (SDF) [46]. Buckwheat is a significant source of dietary fiber but not much research has been conducted on this. Normally, the total dietary fiber (TDF) content of buckwheat groats is 5–10% [18, 33]. Approximately 32% of the TDF is represented by the IDF fraction in whole buckwheat groats. However, an IDF ranging from 53.9% to 81.1% of the TDF have been reported for ten different buckwheat cultivars. Buckwheat husks possess 10 times greater TDF content than the groats [33].

31.9 Pests and Diseases Affecting Buckwheat

The occurrence of diseases and insect-pest infections in Buckwheat seems to be comparatively low. Infections due to cut worms and aphids are prominent but the yield rate is rarely affected. Melon aphid (Aphis gossypii) has been found to frequently infest buckwheat leaves [47]. In cases when aphid attack is at a large scale, the leaves turn yellow, distorted necrotic spots on leaves can be seen and sometimes stunted shoots. The Honeydew of aphids is an optimum environment for the growth of sooty mold on the plants. Petroleum servo Agrospray @ 7 ml/l or Neem oil

Table 31.2 Diseases of Buckwheat and their associated pathogens [49]

Disease/disorder	Pathogen
Aster yellow	*Mycoplasma* spp.
Root rots	*Fusarium* spp., *Botrytis* spp., *Rhizoclonia* spp.
Stern rot	*Botrytis cinerea*
Chlorotic leaf spot	*Alternaria allernals*
Stipple spot	*Bipolaris sorokiniana*
Blight	*Phyloplhora parasilka*
Downy mildew	*Peronospora* spp.

(1500 ppm) @ 3 ml/l are used for managing of aphids [48]. Some of the microbial diseases of buckwheat have been summarized in Table 31.2.

Storage conditions of buckwheat include Prevention, thorough drying of grains, proper cleaning. Treatment of floors and wall of the storage bin should be done with Neem oil (1500 ppm) @ 5 ml/l. Grains should be stored in thoroughly dried bins [50].

31.10 Benefits and Uses of Buckwheat

31.10.1 Buckwheat as a Soil Conditioner

Buckwheat crop interferes with the life cycle of certain insects, increases the nutrient profile of the soil and suppresses the growth of weeds, protects the main crops from water logging, erosion and wild animals. We get 5–7 tons of biomass/ha yield of buckwheat [9]. How and what nutrients can be extracted depend on growth stage at which crop takes up the soil. Mulch on the surface helps in maintaining soil stability, and suppresses weeds. Green manuring of buckwheat boosts soil aggregate stability and scavenges nutrients especially phosphorus and calcium (Clark, 2007). Buckwheat residues had a narrow C:P ratio (175) with high P concentration (2.7 g/kg), which enhances P availability in residue amended soil [51]. Buckwheat had high carbon: nitrogen ratio (C:N = 34), hence causes immobilizing of nitrogen during decomposition and lessens soil N availability. When buckwheat unifies with soil, a speedy breakdown starts and releases nutrients for uptake by the subsequent crop especially potassium [52]. Moreover, buckwheat can prevent the development of root pathogens [53] and insect-pests cycle.

31.10.2 Buckwheat as a Scavenger of Phosphorus

The remarkable capacity of buckwheat to scavenge phosphorus, calcium and some minor nutrients from the soil, which otherwise are unavailable to other crop, is noteworthy. Therefore, it is called as a "natural phosphorus pump". Phosphorus (P) uptake of buckwheat is about ten times higher than wheat [54]. When there is a reduction in phosphorus level, Buckwheat produces exudates with a lower pH [55], acidify rhizospheric soil and absorb P beyond its metabolic requirement [56], showed luxury consumption of P [57]. Similarly, [54] reported that buckwheat quickly absorbs phosphorus in calcareous soils but when there is an abundance of Fe and Al phosphates, lesser absorption was noted. The roots of buckwheat can store inorganic phosphorus to a great extent. So, when buckwheat plants come in contact with the soil, a speedy decomposition takes place. Thus, phosphorus and other nutrients become available to the next crop [58]. Under phosphorus deficient condition, buckwheat root increases the release of protons and P solubilizing substances and enhances P uptake [54]. Buckwheat roots secrete phosphatases and mild acids to extract phosphorous from the organic content [5] and providing it to the succeeding crops [59]. This aids in the gradual release of phosphates from organic fertilizers, making it available to the plant [50].

31.10.3 Buckwheat as a Superfood

The place of buckwheat in economy is mainly because of its high nutritive value of its grains and the presence of rutin in foliage (Tables 31.3, 31.4 and 31.5). In general buckwheat grain contains about 10–14% protein which is more than most of the cereals [60]. Buckwheat protein is also superior in quality than cereals, as it is enriched with essential amino acids [61], with about 80% digestibility [62]. According to a study, the elevated levels of lysine in buckwheat makes it very valuable supplement for those diets or cereals or which do not have enough lysine. Buckwheat flour is free from gluten and used in treating the celiac disease in human being [63]. The biological value of Tartary buckwheat is lower than common buckwheat due to high tannin content and a high hull percentage, [47]. Buckwheat grains contain a high quality slowly digested starch, helps to treat diabetics. Buckwheat flour also contains a good proportion of high-quality fat. Palmitic, oleic and linolenic acids constitute about 95% of the total fatty acids in buckwheat grains [64]. Buckwheat grain is a reserve for trace elements like Zinc, Copper and manganese as compared to cereal crops [65].

Table 31.3 Minerals and trace elements profile of buckwheat, wheat and rice

Minerals and trace elements (mg/100 g grain)	Buckwheat	Wheat	Rice
Calcium	110	10	30
Iron	4.0	0.7	3.5
Magnesium	390	65	138
Phosphorus	330	160	298
Manganese	3.4	0.5	2.3
Zinc	0.8	1.3	2.7
Potassium	450	268	284

Table 31.4 Essential amino acids present in buckwheat, wheat and rice

Essential amino acids (% of total protein)	Buckwheat	Wheat	Rice
Lysine	5.9	3.8	2.6
Methionine	3.7	3.0	3.5
Tryptophan	1.4	1.0	1.2
Leucine	6.7	8.2	6.3

Table 31.5 Vitamins present in buckwheat, wheat and rice

Vitamins (mg/100 g grain)	Buckwheat	Wheat	Rice
Thiamine	3.3	0.06	0.5
Riboflavin	10.6	0.06	0.2
Niacin	18.0	1.9	5.5
Tocopherol	40.0	–	–
Pantotehnic acid	11.0	–	–
Choline	440	–	–

31.10.4 Buckwheat as a Medicinal Plant

Buckwheat grain has very pleasant aroma because of salicylaldehyde (2-hydroxybenzaldehyde) [66]. Rutin (quercetin-3-rutinosid) constituent of buckwheat is used in preventing edema, haemorrhagic diseases, and stabilizing high blood pressure [67]. Rutin content in buckwheat varies with genotypes and growing condition. Rutin has some medicinal uses. It is used in treating the growing weakening of capillaries in hypertension leading to hemorrhage, purpurea and bleeding from kidney. Rutin also has potent anti-carcinogens properties. Fagopyrin is a naphthodianthrone substance with photo-sensitizing effect, isolated by Brockmann and Lackner for the first time in 1943 from the foliage of common buckwheat (*Fagopyrum esculentum*), however it exclusively found in cotyledons [68]. The photosensitizing effect of fagopyrin is recently used in photodynamic therapy for the treatment of microorganism and cancer cells [69]. Due to the copper abundance in buckwheat, function of iron is enhanced and this also helps to stop hypothermia in human beings. Various teas can be made out from Tartary buckwheat and they are used to lower blood pressure, sugar and lipids. Farmers of remote locations of north-east India traditionally use Tartary buckwheat to treat livestock suffering from foot and mouth disease.

31.10.5 Buckwheat as a Staple Crop in Hilly Areas

Buckwheat is a multipurpose crop, primarily grown for its grains. However, tender shoots and leaves are also used as leafy vegetables. The seed is used in a number of culinary preparations as well as alcoholic drinks; the husk is used in stuffing pillows. The grain is generally used as human food, the flour used in the preparation of chapattis, pancakes, biscuits and noodles etc. People in main land consume buckwheat during religious fasting days. It is also used as a livestock, poultry and piggery feeds. Tartary buckwheat also used in place of tea in hill region. The straw is fed to cattle immediately after threshing, when it is still green/fresh. Buckwheat flower produces a good quality of honey and can also cultivate for honey crop. Buckwheat crops can also be grown as an insect trap crop with other economically important crops to reduce the insect-pests infestation in main crops.

31.11 Conclusion

Buckwheat, as per its nutritive content, can play a significant role in improving the health of individuals. It can serve as a rich source of amino acids and important minerals. The pharmacological properties of buckwheat can be further exploited so that we can utilize its health benefits to the fullest.

References

1. Ohnishi, O. (1993). Population genetics of cultivated common buckwheat, Fagopyrum esculentum Moench. VIII. Local differentiation of land races in Europe and the silk road. *The Japanese Journal of Genetics, 68*(4), 303–316.
2. Li, S.-q., & Zhang, Q. H. (2001). Advances in the development of functional foods from buckwheat. *Critical Reviews in Food Science and Nutrition, 41*(6), 451–464.
3. Unander, D. (2002). Buckwheat. Fagopyrum esculentum Moench. Promoting the conservation and use of underutilized and neglected crops, Vol. 19. *Economic Botany, 56*(1), 110–110.
4. Quinet, M., Cawoy, V., Lefevre, I., Van Miegroet, F., Jacquemart, A.-L., & Kinet, J.-M. (2004). Inflorescence structure and control of flowering time and duration by light in buckwheat (Fagopyrum esculentum Moench). *Journal of Experimental Botany, 55*(402), 1509–1517.
5. Gardner, W., & Boundy, K. (1983). The acquisition of phosphorus by Lupinus albus L. IV. The effect of interplanting wheat and white lupin on the growth and mineral composition of the two species. *Plant and Soil*, 391–402.
6. Radics, L., & Mikóházi, D. (2010). Principles of common buckwheat production. *The European Journal of Plant Science and Biotechnology, 4*(1), 57–63.
7. Jung, G. H., Kim, S. L., Kim, M. J., Kim, S. K., Park, J. H., Kim, C. G., et al. (2015). Effect of sowing time on buckwheat (Fagopyrum esculentum Moench) growth and yield in Central Korea. *Journal of Crop Science and Biotechnology, 18*(4), 285–291.
8. Dražić, S., Glamočlija, D., Ristić, M., Dolijanović, Ž., Drazić, M., Pavlović, S., et al. (2016). Effect of environment of the rutin content in leaves of Fagopyrum esculentum Moench. *Plant Soil and Environment, 62*(6), 261–265.
9. Sobhani, M. R., Rahmikhdoev, G., Mazaheri, D., & Majidian, M. (2014). Influence of different sowing date and planting pattern and N rate on buckwheat yield and its quality. *Australian Journal of Crop Science, 8*(10), 1402–1414.
10. Gubbels, G. (1978). Yield, seed weight, hull percentage, and testa color of buckwheat at two soil moisture regimes. *Canadian Journal of Plant Science, 58*(3), 881–883.

11. Khanh, T., Chung, M., Xuan, T., & Tawata, S. (2005). The exploitation of crop allelopathy in sustainable agricultural production. *Journal of Agronomy and Crop Science, 191*(3), 172–184.
12. Horie, T., Karahara, I., & Katsuhara, M. (2012). Salinity tolerance mechanisms in glycophytes: An overview with the central focus on rice plants. *Rice, 5*(1), 1–18.
13. Qin, P., Wang, Q., Shan, F., Hou, Z., & Ren, G. (2010). Nutritional composition and flavonoids content of flour from different buckwheat cultivars. *International Journal of Food Science & Technology, 45*(5), 951–958.
14. Nalinkumar, A., & Singh, P. (2020). An overview of buckwheat Fagopyrum spp an under-utilized crop in India-nutritional value and health benefits. *International Journal of Medical Research and Health Sciences, 9*(7), 39–44.
15. Bárta, J., Kalinová, J., Moudrý, J., & Čurn, V. (2004). Effects of environmental factors on protein content and composition in buckwheat flour. *Cereal Research Communications, 32*(4), 541–548.
16. Im, J.-S., Huff, H. E., & Hsieh, F.-H. (2003). Effects of processing conditions on the physical and chemical properties of buckwheat grit cakes. *Journal of Agricultural and Food Chemistry, 51*(3), 659–666.
17. Bonafaccia, G., Gambelli, L., Fabjan, N., & Kreft, I. (2003). Trace elements in flour and bran from common and tartary buckwheat. *Food Chemistry, 83*(1), 1–5.
18. Steadman, K., Burgoon, M., Lewis, B., Edwardson, S., & Obendorf, R. (2001). Buckwheat seed milling fractions: Description, macronutrient composition and dietary fibre. *Journal of Cereal Science, 33*(3), 271–278.
19. Krkošková, B., & Mrazova, Z. (2005). Prophylactic components of buckwheat. *Food Research International, 38*(5), 561–568.
20. Skrabanja, V., Kreft, I., Golob, T., Modic, M., Ikeda, S., Ikeda, K., et al. (2004). Nutrient content in buckwheat milling fractions. *Cereal Chemistry, 81*(2), 172–176.
21. Gong, G., Qin, Y., Huang, W., Zhou, S., Yang, X., & Li, D. (2010). Rutin inhibits hydrogen peroxide-induced apoptosis through regulating reactive oxygen species mediated mitochondrial dysfunction pathway in human umbilical vein endothelial cells. *European Journal of Pharmacology, 628*(1–3), 27–35.
22. Zieliński, H., & Kozłowska, H. (2000). Antioxidant activity and total phenolics in selected cereal grains and their different morphological fractions. *Journal of Agricultural and Food Chemistry, 48*(6), 2008–2016.
23. Morishita, T., Hara, T., Suda, I., & Tetsuka, T. (2002). Radical-scavenging activity in common buckwheat (Fagopyrum esculentum Moench) harvested in the Kyushu region of Japan. *Fagopyrum, 19*(3), 89–93.
24. Watanabe, M. (1998). Catechins as antioxidants from buckwheat (Fagopyrum esculentum Moench) groats. *Journal of Agricultural and Food Chemistry, 46*(3), 839–845.
25. Terao, J. (2009). Dietary flavonoids as antioxidants. *Food Factors for Health Promotion, 61*, 87–94.
26. Morishita, T., Yamaguchi, H., & Degi, K. (2007). The contribution of polyphenols to antioxidative activity in common buckwheat and tartary buckwheat grain. *Plant Production Science, 10*(1), 99–104.
27. Morishita, T., Shimizu, A., Yamaguchi, H., & Degi, K. (2019). Development of common buckwheat cultivars with high antioxidative activity—'Gamma no irodori', 'Cobalt no chikara'and 'Ruchiking'. *Breeding Science, 69*(3), 514–520.
28. Nakamura, K., Naramoto, K., & Koyama, M. (2013). Blood-pressure-lowering effect of fermented buckwheat sprouts in spontaneously hypertensive rats. *Journal of Functional Foods, 5*(1), 406–415.
29. Aoyagi, Y. (2006). An angiotensin-I converting enzyme inhibitor from buckwheat (Fagopyrum esculentum Moench) flour. *Phytochemistry, 67*(6), 618–621.
30. Shimizu, E., HaYASHI, A., Takahashi, R., Aoyagi, Y., Murakami, T., & Kimoto, K. (1999). Effects of angiotensin I-converting enzyme inhibitor from Ashitaba (Angelica keiskei) on blood pressure of spontaneously hypertensive rats. *Journal of Nutritional Science and Vitaminology, 45*(3), 375–383.

31. Ortmeyer, H. K., Bodkin, N., Lilley, K., Larner, J., & Hansen, B. C. (1993). Chiroinositol deficiency and insulin resistance. I. Urinary excretion rate of chiroinositol is directly associated with insulin resistance in spontaneously diabetic rhesus monkeys. *Endocrinology, 132*(2), 640–645.
32. Wang, L., Li, X., Niu, M., Wang, R., & Chen, Z. (2013). Effect of additives on flavonoids, d-chiro-inositol and trypsin inhibitor during the germination of tartary buckwheat seeds. *Journal of Cereal Science, 58*(2), 348–354.
33. Lu, L., Murphy, K., & Baik, B. K. (2013). Genotypic variation in nutritional composition of buckwheat groats and husks. *Cereal Chemistry, 90*(2), 132–137.
34. Berlin, W. K., Wang, S.-N., & Shen, T. (1990). Glycosyl-inositol deivatives II. Synthesis of 2-amino-2-deoxy-D galactosyl-α-1, 3-D-chiro-inositol. *Tetrahedron Letters, 31*(8), 1109–1112.
35. Asplin, I., Galasko, G., & Larner, J. (1993). Chiro-inositol deficiency and insulin resistance: A comparison of the chiro-inositol-and the myo-inositol-containing insulin mediators isolated from urine, hemodialysate, and muscle of control and type II diabetic subjects. *Proceedings of the National Academy of Sciences, 90*(13), 5924–5928.
36. Jiang, P., Burczynski, F., Campbell, C., Pierce, G., Austria, J., & Briggs, C. (2007). Rutin and flavonoid contents in three buckwheat species Fagopyrum esculentum, *F. tataricum*, and *F. homotropicum* and their protective effects against lipid peroxidation. *Food Research International, 40*(3), 356–364.
37. Awatsuhara, R., Harada, K., Maeda, T., Nomura, T., & Nagao, K. (2010). Antioxidative activity of the buckwheat polyphenol rutin in combination with ovalbumin. *Molecular Medicine Reports, 3*(1), 121–125.
38. Afanas'eva, I. B., Ostrakhovitch, E. A., Mikhal'chik, E. V., Ibragimova, G. A., & Korkina, L. G. (2001). Enhancement of antioxidant and anti-inflammatory activities of bioflavonoid rutin by complexation with transition metals. *Biochemical Pharmacology, 61*(6), 677–684.
39. Ishiguro, K., Morishita, T., Ashizawa, J., Suzuki, T., & Noda, T. (2016). Antioxidative activities in rutin rich noodles and cookies made with a trace rutinosidase variety of Tartary buckwheat (Fagopyrum tataricum Gaertn.), 'Manten-Kirari'. *Food Science and Technology Research, 22*(4), 557–562.
40. Brunori, A., & Végvári, G. (2007). Rutin content of the grain of buckwheat (Fagopyrum esculentum Moench. and Fagopyrum tataricum Gaertn.) varieties growtn in southern Italy. *Acta Agronomica Hungarica, 55*(3), 265–272.
41. Morishita, T., & Tetsuka, T. (2002). Varietal differences of rutin, protein and oil content of common buckwheat (Fagopyrum esculentum) grains in Kyushu area. *Japanese Journal of Crop Science, 71*(2), 192–197.
42. Kitabayashi, H., Ujihara, A., Hirose, T., & Minami, M. (1995). On the genotypic differences for rutin content in Tartary buckwheat, Fagopyrum tataricum Gaertn. *Japanese Journal of Breeding, 45*(2), 189–194.
43. Gullón, B., Lú-Chau, T. A., Moreira, M. T., Lema, J. M., & Eibes, G. (2017). Rutin: A review on extraction, identification and purification methods, biological activities and approaches to enhance its bioavailability. *Trends in Food Science & Technology, 67*(2017), 220–235.
44. Ohsawa, R. (2020). Current status and prospects of common buckwheat breeding in Japan. *Breeding Science, 70*(1), 3–12.
45. Anderson, J. W., Baird, P., Davis, R. H., Ferreri, S., Knudtson, M., Koraym, A., et al. (2009). Health benefits of dietary fiber. *Nutrition Reviews, 67*(4), 188–205.
46. Elleuch, M., Bedigian, D., Roiseux, O., Besbes, S., Blecker, C., & Attia, H. (2011). Dietary fibre and fibre-rich by-products of food processing: Characterisation, technological functionality and commercial applications: A review. *Food Chemistry, 124*(2), 411–421.
47. Tahir, I., & Farooq, S. (1985). Grain composition in some buckwheat cultivars (Fagopyrum spp.) with particular reference to protein fractions. *Plant Foods for Human Nutrition, 35*(2), 153–158.

48. Babu, S., Kalita, H., Singh, R., Gopi, R., Kapoor, C., & Das, S. (2014). Buckwheat (Fagopyrum spp.). In R. K. Avasthe, Y. Pradhan, & K. Bhutia (Eds.), *Handbook on organic crop production in Sikkim*. Published by Sikkim Organic Mission, Govt. of Sikkim and ICAR RC Sikkim Centre. ISBN, 9, 788193.
49. Sindhu, R., & Khatkar, B. S. (2016). Physicochemical and functional properties of starch and flour of tartary buckwheat (F. tataricum) grains. *International Journal of Engineering Research and Technology, 5*(6), 315–320.
50. Babu, S., Yadav, G., Singh, R., Avasthe, R., Das, A., Mohapatra, K., et al. (2018). Production technology and multifarious uses of buckwheat (Fagopyrum spp.): A review. *Indian Journal of Agronomy, 63*(4), 415–427.
51. Arcand, M. M., Lynch, D. H., Voroney, R. P., & van Straaten, P. (2010). Residues from a buckwheat (Fagopyrum esculentum) green manure crop grown with phosphate rock influence bioavailability of soil phosphorus. *Canadian Journal of Soil Science, 90*(2), 257–266.
52. Kumar, V., Brainard, D. C., & Bellinder, R. R. (2008). Suppression of Powell amaranth (Amaranthus powellii), shepherd's-purse (Capsella bursa-pastoris), and corn chamomile (Anthemis arvensis) by buckwheat residues: Role of nitrogen and fungal pathogens. *Weed Science, 56*(2), 271–280.
53. Altieri, M. A., & Nicholls, C. I. (2003). Soil fertility management and insect pests: Harmonizing soil and plant health in agroecosystems. *Soil and Tillage Research, 72*(2), 203–211.
54. Zhu, Y., He, Y., Smith, S., & Smith, F. (2002). Buckwheat (Fagopyrum esculentum Moench) has high capacity to take up phosphorus (P) from a calcium (Ca)-bound source. *Plant and Soil, 239*(1), 1–8.
55. Simpson, R. J., Oberson, A., Culvenor, R. A., Ryan, M. H., Veneklaas, E. J., Lambers, H., et al. (2011). Strategies and agronomic interventions to improve the phosphorus-use efficiency of farming systems. *Plant and Soil, 349*(1), 89–120.
56. Amann, C., & Amberger, A. (1989). Phosphorus efficiency of buckwheat (Fagopyrum esculentum). *Zeitschrift für Pflanzenernährung und Bodenkunde, 152*(2), 181–189.
57. Bekele, T., Cino, B., Ehlert, P., Van der Maas, A., & Van Diest, A. (1983). An evaluation of plant-borne factors promoting the solubilization of alkaline rock phosphates. *Plant and Soil, 75*(3), 361–378.
58. Tolaini, V., Del Fiore, A., Nobili, C., Presenti, O., De Rossi, P., Procacci, S., et al. (2016). Exploitation of tartary buckwheat as sustainable ingredient for healthy foods production. *Agriculture and Agricultural Science Procedia, 8*(2016), 455–460.
59. Ragaee, S., & Abdel-Aal, E. S. M. (2006). Pasting properties of starch and protein in selected cereals and quality of their food products. *Food Chemistry, 95*(1), 9–18.
60. Nicholson, J., McQueen, R., Grant, E., & Burgess, P. (1976). The feeding value of tartary buckwheat for ruminants. *Canadian Journal of Animal Science, 56*(4), 803–808.
61. Eggum, B. O., Kreft, I., & Javornik, B. (1980). Chemical composition and protein quality of buckwheat (Fagopyrum esculentum Moench). *Plant Foods for Human Nutrition, 30*(3), 175–179.
62. Thacker, P., Anderson, D., & Bowland, J. (1983). Chemical composition and nutritive value of buckwheat cultivars for laboratory rats. *Canadian Journal of Animal Science, 63*(4), 949–956.
63. Tummaramatti, S., Laxminarayana, H., Hosamani, R., & Sampaganvi, S. (2016). Effect of bio-fertilizers on growth, yield and quality of buckwheat (Fagopyrum esculentum Moench). *Environment and Ecology, 34*(3B), 1258–1261.
64. Joshi, D. C., Sood, S., Hosahatti, R., Kant, L., Pattanayak, A., Kumar, A., et al. (2018). From zero to hero: The past, present and future of grain amaranth breeding. *Theoretical and Applied Genetics, 131*(9), 1807–1823.
65. Christa, K., & Soral-Śmietana, M. (2008). Buckwheat grains and buckwheat products–Nutritional and prophylactic value of their components–A review. *Czech Journal of Food Sciences, 26*(3), 153–162.
66. Janeš, D., & Kreft, S. (2008). Salicylaldehyde is a characteristic aroma component of buckwheat groats. *Food Chemistry, 109*(2), 293–298.

67. Omidbaigi, R., & Mastro, G. (2004). Influence of sowing time on the biological behaviour, biomass production, and rutin content of buckwheat (Fagopyrum esculentum Moench). *Italian Journal of Agronomy, 8*(1), 47–50.
68. Kreft, S., Janeš, D., & Kreft, I. (2013). The content of fagopyrin and polyphenols in common and tartary buckwheat sprouts. *Acta Pharmaceutica, 63*(4), 553–560.
69. Dai, T., Huang, Y.-Y., & Hamblin, M. R. (2009). Photodynamic therapy for localized infections—State of the art. *Photodiagnosis and Photodynamic Therapy, 6*(3–4), 170–188.

Chapter 32
Tianma

Laiba Ahmed, Maham Saeed, Khaqan Zia, Sahar Nazeer, Ayoub Rashid Ch, Shahzad Sharif, and Saima Muzammil

32.1 Introduction

Gastrodia elata belongs to the family Compositae and genus Gastrodia. In the genus Gastrodia, about 22 species are present and mostly these species grow in china. Among all of them, one which is used as medicinal tool is G. elata listed in Chinese pharmacopeia. Due to the characteristics fleshy tuber or coralloid stem present underground, it raises in forests at about 400–3200 m above the sea level. By establishing a symbiotic association with fungi called *Armillaria mellea*, nutrients and energy requirements of G. elata is satisfied. Isolation and identification of about 81 compounds of different types from this plant like phenolics, polysaccharides, organic acids and sterols have been taken place [1].

According to traditional Chinese medicine, G. elata has shown its medicinal activity in suppressing of hyperactive liver and to stop tetany. While in clinical practice, this plant species is mainly useful in insomnia, dizziness, Alzheimer disease, neurasthenia, convulsions, epilepsy, hypertensive and many other diseases. In outdated medicines, G. elata is used as tonic and aphrodisiac and by adding to alcoholic drinks as a purposeful food which is beneficial in improving sexual potency, prevention of abortion and to improve vision [2].

L. Ahmed · M. Saeed · K. Zia · S. Nazeer · A. R. Ch · S. Sharif (✉)
Department of Chemistry, Government College University, Lahore, Pakistan
e-mail: mssharif@gcu.edu.pk

S. Muzammil
Department of Microbiology, Government College University, Faisalabad, Pakistan

M. Zia-Ul-Haq et al. (eds.), *Essentials of Medicinal and Aromatic Crops*,
https://doi.org/10.1007/978-3-031-35403-8_32

32.2 Scientific Classification

Kingdom: Plantae
Phylum: Tracheophyta
Class: lilopsida
Order: Asparagales
Family: Orchidaceae
Genus: Gastrodia
Species: *G. elata*
Botanical name: *Phyllanthus emblica* or *Emblica officinalis*
Common name: Tianma

At the present time, G. elata is the part of number of prescriptions as a part of granule, injection, pills and capsules. In a study conducted to treat 53 patients having vertebral artery insufficiency, Tianma injection was used as a treatment which showed significant efficiency in comparison to control group.

In order to treat diseases or improving the therapeutic effect, Tianma preparations can be used in recombinant with other medicines. In addition to the presence of phenolics and polysaccharides as pharmacological agents, various trace elements and amino acids are also present in G. elata, due to which its nutritious and palatable value is high. G. elata can be used to prepare medicinal food in folk as "Tianma steamed egg" which is applied in treatment of dizziness and neurasthenia. Other such products include "Tianma sleeping Poridge" having a medical role in brain and body functions, sleeping effects, insomnia, forgetfulness and dizziness. Another "Tianma simmered pig brain" having core and marrow effect has shown amazing medicinal effects in dizziness, headache, palpitation and insomnia [3].

In vivo studies have also revealed that spatial memory can be improved to a great extent in Alzheimer's patients by increasing the concentration of choline acetyltransferase and inhibition of AChE in brain. Long term use of G. elata has also been reported to increase the learning and memory capabilities in scopolamine treated rats by improving the cholinergic functions to mediate cognitive effects [4].

Some of the important chemicals in G. elata are Gastrodin, 4-hydroxybenzylalcohol, 4-hydroxybenzaldehyde, vanillin and vanillyl alcohol which have anti-inflammatory, anti-oxidant and neuroprotective effects (Fig. 32.1) [5].

32.3 Nutritional Composition

The nutrient composition of G. elata has shown that it is rich in crude protein, crude fat and vitamin C. The total amino acids (TAA) composition is 10.034 g/100 g, and the ratio of essential amino acids to TAA is 41.71% and essential amino acids to nonessential amino acids is 71.17%. All essential amino acids are present in balanced proportion and the first limiting amino acid is lysine. The content of K is found to be 41,234.43 mg/kg as macro element while that of Mn is found to be as

Fig. 32.1 *Gastrodia elata* [6]

Table 32.1 Components of G. eleta with nutritional value [6]

Components	Nutritive value
Total amino acids (TAA)	10.034 g/100 g
Essential amino acids: TAA	41.71%
Essential amino acids: nonessential amino acids	71.17%
K contents	41,234.43 mg/kg
Mn contents	20.5 mg/kg
Starch concentration	14.02–15.02%

20.5 mg/kg as micro element. The concentration of starch in nutritional component vary from 14.02% to 15.02%. On the basis of these nutritional values, it can be concluded that G. elata is an excellent plant for high level research and applications (Table 32.1) [6].

32.4 Chemical Constituents

The biological active components present in G. elata are Gastrodin, 4-hydroxybenzylalcohol, 4-hydroxybenzaldehyde, benzyl alcohol, bis-(4-hydroxyphenyl) methane, vanillin, 4-(4′-hydroxybenzyloxy) benzyl methylether, vanillic acid and vanillyl alcohol.

The total phenolic contents in G. elata has been measured to be 0.0485%. Phenolic sulfuric method has been employed to measure the total polysaccharides contents with highest percentage of 21.6%. Moreover, prime percentage of Gastrodin, amino acid and total flavonoids are 0.24%, 1.92% and 0.24% respectively as given by their detection result. The phenolic compounds which are present in G. elata are 4-butoxyphenylmethanol, 4,4′-methylenediphenol, 4-O-glucopyranosyl-benzaldehyde, 4-hydroxybenzyl ethyl ether, Gastrodigenin,

Parishin, and 4-hydroxy 3-4-hydroxybenzyl) benzyl methyl ether. These phenolic constituents have been reported for playing a role in neuroprotective, anti-inflammatory and anti-oxidant action (Fig. 32.2) [7].

The polysaccharides constituents of G. elata contains AGEW, 61WGEW, Adenosine glucoside, 4-(methoxymethyl)phenyl-1-O-β-D-glucopyranoside, Gastrodinisome, Trimethylcitryl-b-galactopyranoside, 5-(hydroxymethyl)-furfural, 5-(hydroxymethyl)-2-furaldehyde. Some of these constituents have been proven to act as bioactive against dengue virus [8]. Organic acids include 2-hydroxypropane-1,2,3-tricarboxylic acid, butanedioic acid, palmitic acid, 6-methylcitrate, 1,5-dimethylcitrate and proto- catechuic acid. The sterols components are β-sitosterol, and 3-O-(4′-hydroxybenzyl)-β-sitosterol, 4-hydroxybenzyl-β-sitosterol, daucosterol. Some other components of G. elata are adenine, adenine nucleoside and many varieties of amino acids. Among trace elements, iron has highest percentage in G. elata, and chromium, zinc, manganese, and copper are the other elements present. Delay of aging and inhibition of diseases are main responsibilities of these elements (Fig. 32.3) [9].

32.5 Cultivation

G. elata is present in temperate and arctic northernmost region of the East Asia and distinctively in Korea, Japan and china's mountainous areas. The plant spends its whole life cycle underground except for florescence. G. elata is a saprophytic herb and do not have any green part and is completely dependent on a fungus called Armillaria for its nutrient requirement by a symbiotic relationship. So, it is very difficult to cultivate this plant outside of its native environment. As well as a sheltered woodland position with damp humus rich soil is also necessary for its cultivation. It grows at height of 1300–10500 feet at the edges of forests. The plant can tolerate as low as −15 °C temperature. The height of plant is in between 30 and 150 cm. Due to over-collection of this plant for the purpose of medicine, the plant is becoming rare. This plant is a shallow rooting plant of a well-drained low fertility soil so it fulfils it nutrients requirements from fungus relationship and also compete with other plants. June to July is the flowering season of this plant and this plant produces fruit from July to August [10, 11].

32.6 Cultivars

G. elata usually grows on the edge or within the forests. G. elata being achlorophyllous and aphyllous fulfils its nutrients requirements by symbiotic relationship with fungi. Three years are almost required for obtaining a developed tuber from the seeds of G. elata. Germination of seed takes place in the first year by symbiotic relationship with fungus called M. osmundicola and protocorms are formed by

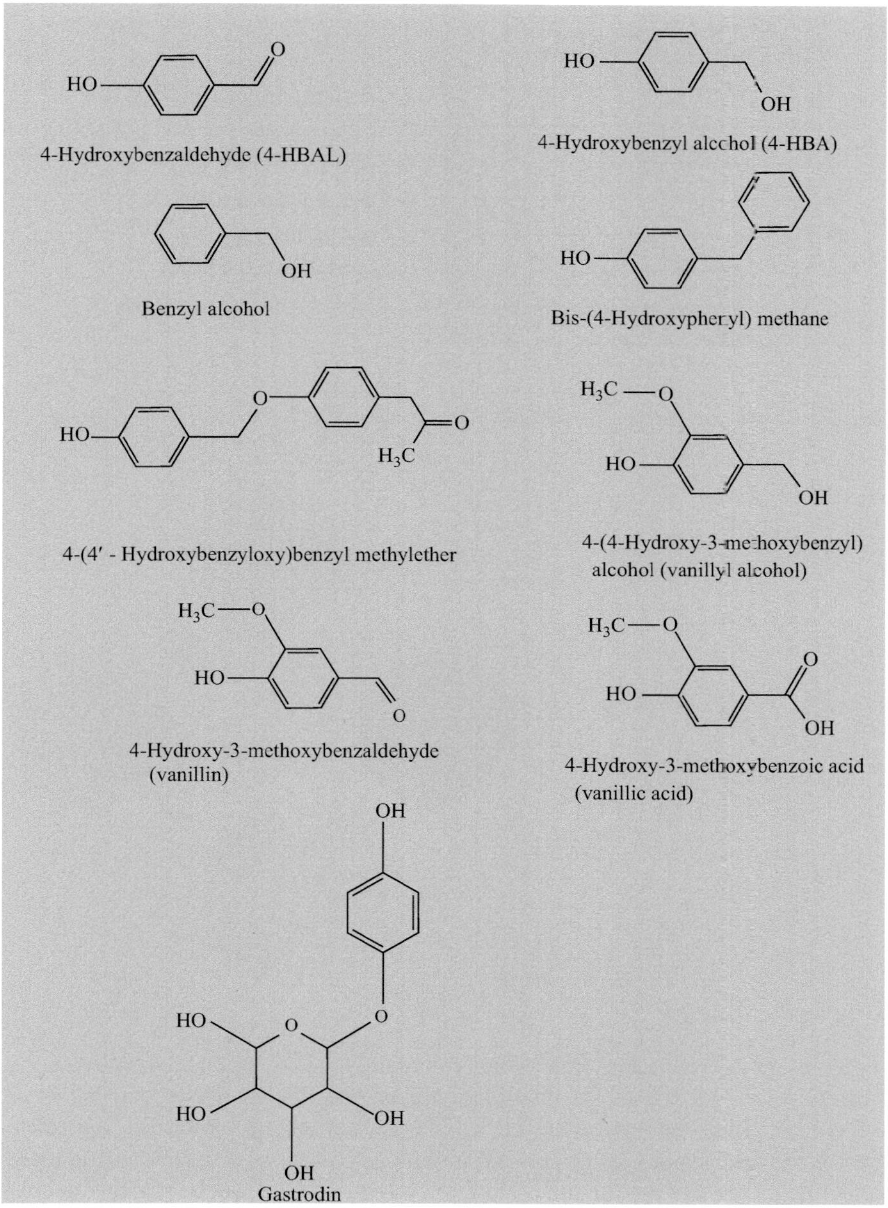

Fig. 32.2 Chemical composition of representative compounds in G. elata [5]

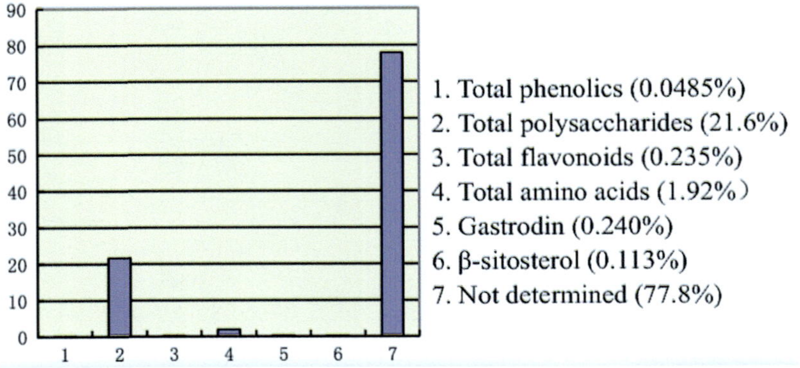

Fig. 32.3 Contents of chemical composition from G. elata [10]

Fig. 32.4 G. elata cultivation cycle in steps [12]

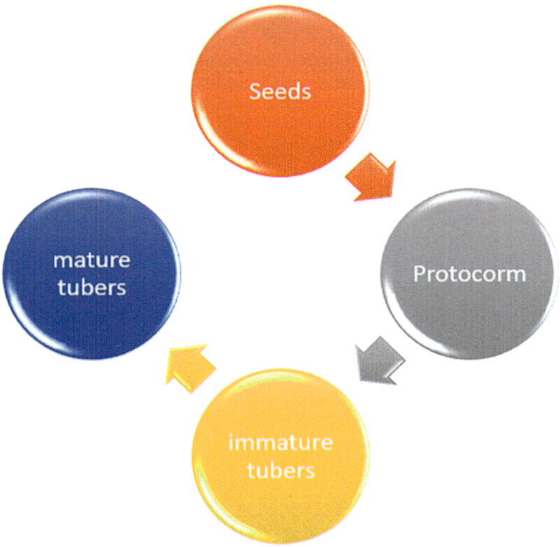

stimulation of germination, from which immature tubers are obtained cultivating it with Armillaria for 1 year. The major function of Armillaria is the decomposition of wood from dead trees and then releasing these nutrients into soil for vegetative growth. An additional year of cultivation with Armillaria is done to develop these immature tubers to a mature tube. The life cycle of G. elata can be completed only in the presence of these fungi. Instead of waiting for tubers to grow and mature from seed, there is a need of commercially developed immature tubers which can be propagated and can be used for cultivation of mature tubers (Fig. 32.4).

Traditionally, G. elata is produced in less amount very low because of deficiency of proper methods for cultivation.

A recent method which has been used for its cultivation is placing of several layers of fresh wood within a hole in the ground. Inoculation and cultivation of naturally collected tree bark and branches infected with Armillaria take place for 1 year. As a result, immature tuber of G. elata are planted and cultivation take place after an extra year to obtain mature tubers [12].

32.7 Controlling Factors in G. elata Growth

- Growth of G. elata can be varied a lot in forest and open grounds because of temperature, humidity and sunlight difference at both places.
- Another important factor which may also be a deciding factor is the optimal quantity of Armillaria needed for inoculation with G. elata to increase the yield and reduction of production cycle.
- Type of wood used for cultivation is also very important factor as almost 600 plant species are used by farmers as source of wood which includes trees, herbs and shrubs. So it is very important to know that how different wood may affect the growth cycle [11].

32.8 Pollination in G. elata Flowers

In flowers of G. elata, cross pollination with lower ovary by insects and epigynous is done and have three fused carpels in which several small ovules are present which mature into seeds after fertilization. Normally, for seed maturation around about 21 days are considered. Ovary developed naturally into dehiscent capsule. As it is an orchid in which photosynthesis is absent and most of the life cycle of G. elata completed below soil so cultivation can be done indoors or in the wild as no sunlight is required. Even though, more than two million seeds are produced by a single plant of Gastrodia, but the rate at which seed germinate is much less, and production is also not much effective because the seeds of G. elata have structure which is small and simple so it cannot store food [13].

32.9 Seed Germination

G. elata exhibit indigent seed germination rate as they are not only small but also do not contain endosperm so seed germination depends on fungi. Although Armillaria are effective in development of G. elata but some species of Armillaria are not beneficial as they inhibit the germination. An enzyme called esterase isozyme has been proved to be effective in germination of seed. The mechanism of germination of seed in G. elata by symbiotic relationship has been reported in literature. According to which, during the germination of seed in G. elata the hyphae of *M. osmundicola* attacks the seed coat. In order to obtain necessary nutrients the plant seed is cultured with some fungi which are helpful in protocorms formation as size of seed is small and no endosperm is present. Seed germination is most importantly controlled by maintaining the optimal temperature [14].

32.10 Different Form of Harvested G. elata

Harvesting of G. elata takes place either in spring or in winter. The form which is bloomed during winter is titled as Gastrodia hiemalis and it is of excellent quality while the form which is excavated in spring is of inferior quality and is called as Gastrodia fontinalis. After digging out, stem and fibrous roots are removed and coarse skin is removed by soaking in water. Further on it is boiled until the disappearance of white dots in the center [15].

32.11 Tuber of G. Elata

The dried tubers have a shape of long, oval-shaped, a bit of fat, and show symptoms of shrinking and bending. The red stem at the base which is residue has a color from red to reddish brown while rounded root marks are present at the other end. The length of tuber is 6–10 cm long, while diameter is 2–5 cm and they are 0.9–2 cm in thickness. The surface color of tuber is from yellowish white to yellowish brown. Remaining scales of light color skin and many longitudinal wrinkles are present on it. Thin wrinkles are the characteristics of Gastrodia hiemalis while Gastrodia fontinalis has thick wrinkles (Fig. 32.5) [16].

Fig. 32.5 Tuber of G. elata [17]

32.12 Obstacles in Cropping of G. elata Can Be Controlled by Rotation Planting

G. elata planting lowers stability of bacterial and fungal community, harmful fungus species and abundance in soils. And it is much noticeable during second year of planting, even if for another 2 years' soil is left, the normal level of microbial community structure could not be obtained After G. elata is planted in soils, the most momentous change is the high manufacturing of Ilyonectria cyclaminicola. The high level of the Ilyonectria fungus is significant cause of G. elata diseases. The confirmation was done by inoculation of I. cyclaminicola in the sterile soil which causes increase rate of Rot in G. elata. After harvesting 1 crop of G. elata, the community of fungi structure in soils gradually can be improved then by harvesting 1–3 crops of P. impudicus, which lowers richness of I. cyclaminicola. As a result, the Ilyonectria fungi becomes dominant flora in soils [18].

32.13 Ecological Cultivation of G. elata

Due to the exceptional biotic features, in modern agricultural production Gastrodia elata undergoes high depletion of resources and less consumption rate, due to which the green and healthy progress of this industry is pointedly lowered. Some suggestions for ecological cultivation of G. elata are recommended as under:

- Follows the code of environmental protection and no pollution.
- Binding to managing of pest and diseases by green way
- Promotion of industrial organization to capitalize profits
- Improving the reprocessing and consumption of fungus materials
- Implementing design based on science of fungus resources for releasing pressure of forest [19].

32.14 In Vitro Micro-propagation in G. elata

G. elata has always been cultivated as major therapeutic source for cure of several human diseases. The field production of G. elata has a problem is the degeneration of seed tubers, the major cause of which is soil borne pathogens. Micro-propagation has been employed to produce disease free seed tubers. Before culturing, the tuber is allowed to treat with $HgCl_2$ in vitro. Water agar medium is the most effective medium for the growth and induction of vegetative stem. Vegetative stem has shown better growth and induction in dark as compared to light. So, following the in vitro micro-propagation can be an efficient way to enhance the revenue and worth of G. elata tubers [20].

32.15 Effect of Cultivation Conditions on Functional Components of G. elata

Some of the major pharmacological substances in G elata are Gastrodin, ergothioneine and vanillyl alcohol. Cultivation conditions and time have a great impact on them. With change of cultivation areas, the contents of ergothioneine was different. The higher concentrations of Gastrodin is present in smaller tubers while concentrations of ergothioneine is low than that of larger tubers. The concentrations of both the gastrodin and vanillyl alcohol is increased when cultivation is done in sun shade screen but the levels of ergothioneine is decreased. The harvesting time also has an influence on functional substances i.e. concentration of Gastrodin becomes high in October as compared to April, while the level of ergothioneine increased in April in comparison to October. By sexual propagated seed use, the level of Gastrodin is produced by 1.9 times in high quantity as compared to asexually propagated seed [13].

32.16 Nutrient Management

G. elata is a terrestrial and perennial orchid and have no root and leave with fleshy, thickened rhizomes at apex by having buds as both apical and lateral with diverse nodes. Generally, the G. elata has their rhizomes deeply sunken in the ground, with just inflorescence stalk only visible to the ground and pollination is carried out by sunlight by help of bees. When reproductive growth of Gastrodia has been completed in 2 months, mature rhizomes completely fulfil the nutrients requirement of G. elata. The whole process from bolting to seed maturation is fulfilled by stored nutrients. In vitro germination of seeds of orchid of temperate regions take place very difficult and Gastrodia is not terrestrial but also has orchids with no leave and root and not photosynthetic.

All the nutrients requirements for lifetime of Gastrodia is fulfilled by relationship with different fungi; out of which 1 is important for germinating the seed, and another is important for the growing and developing rhizomes. Humic substances increases the immersion of numerous nutrients in plants. Absorption of humic acid by plants with low molecular weight can be done. The cell membrane permeability can be increased by these components and these components also show hormone like capabilities Amino acids possess assistance in growth factor as well as in morphogenesis, with increasing formation of callus, browning prevention, protocorm-like bodies and seedlings are increased and also yield of orchids is increased. Chitosan treatment can proliferate the rate of germination in corn seeds and also the length and dry weight of shoots and roots can be increased and is helpful for the seedlings growth under conditions of low temperature. The total weight, vitamin C content can be increased by chitosan. Seed germination can be increased by hormones and natural ingredients alongside of natural additives like potato, humic acid and coconut water. The survival time required for culture is maintained by potato and vitrification in tissue culture plants can be treated, because of which the variety and concentration of the formula content causes unaccommodating of the osmotic pressure [21].

32.17 Steaming Effect on Gastrodia elata

Steaming effects, the quality of *G. elata* to a great extent. The *G. elata* central temperature is around 62.8–67.2 °C which is close to gelatinization temperature. Steaming results in lowering the weight and shear hardness but adhesion force and elongated drying time is increased. From microstructure of steamed *G. elata*, detection of polysaccharides hydrolysis and starch gelatinization has been done due to which changes in weight, texture and drying rate has been shown. Furthermore, steaming boosted contents of gastrodin and let down *p*-hydroxybenzyl alcohol contents [22].

32.18 Pests and Diseases

32.18.1 Pests

32.18.1.1 Botrytis cinerea

Botrytis cinerea is a fungus also known as necrotrophic fungus has a role in effecting many species of plants. In G. elata it causes a disease called flower gray mold. Flowers with such symptoms were identified to have Botrytis cinerea on basis of their morphological characteristics and phylogenetic analysis [23].

32.18.1.2 Sclerotium rolfsii Sacc

It is a fungus which fits to deuteromycotina, hyphomycetes, agonomycetales, of Sclerotium genus. This pathogen causes white silk disease in Gastrodia elata. This fungus has a poor direct invasion ability, so this pathogen usually invades through wound infection.

32.18.1.3 Fusarium solani

It is complex specie of 26 closely related filamentous fungi of division Ascomycota. It causes root rot disease in G. elata at the rate of 25% [24].

32.18.2 Diseases

32.18.2.1 Black Rot Disease

Black Rot disease occurs in G. elata planting base. It is caused by Fusarium redolens. The rate of disease varies from 10% to 25%. The major symptoms of this disease occurs in the root of plant which results in necrosis, tuber rot and pungent odour. Initially the rots occur locally, then it expands gradually and covering the plant completely which results in formation of rotten nest. This disease effects a lot on yield of G. elata and its medicinal effect [25].

32.18.2.2 Tuber Rot Disease

The tuber rot disease in G. elata is caused by Fusarium oxysporum. This disease has a detrimental effect on production of plant. It is known as the first reported G. elata tuber rot disease caused by Fusarium oxysporum [26].

32.18.2.3 Root Rot Disease

It is caused by Fusarium solani, a filamentous fungus. It results in rotten nest and then empty nest, due to which the yield and medicinal value of G. elata is reduced. The affected parts of the plant become white in color, start decaying and as a result produced pungent odor started.

32.19 Medicinal Uses

Gastrodiaelata also known as tianma is a Chinese herbal medicine used to cure a number of diseases as headache, dizziness, epilepsy, spasm, stoke, amnesia, cramps, high blood pressure and many other neurological disorders. Many compounds are present in the extract of G.elata which play their role to cure a disease. While working for the active ingredient of G.elata, many small molecules were found. Among them gastrodin, phenolic compounds, perishin and 4-hydroxy benzyl alcohol were also present. All these compounds were found to have the ability to protect neuron cells against damage and inflammation. This can be shown by following figure (Fig. 32.6).

Gastrodin is an important compoment of G. elata and is used to cure Parkinson disease, Alzheimer's disease and other neurological diseases. It can be shown by following figure (Fig. 32.7).

Different compounds have been isolated from the rhizome of G.elata which can be depicted by following figure (Fig. 32.8) [27].

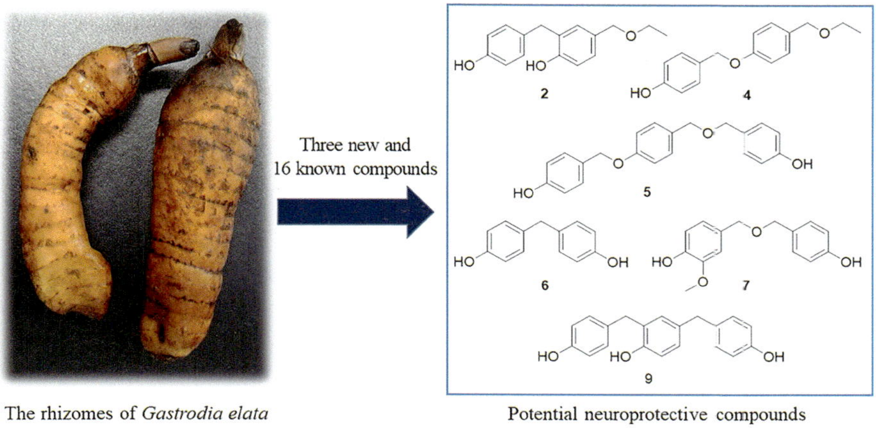

Fig. 32.6 Neuro protective componds of G. elata [27]

Fig. 32.7 Gastrodin [29]

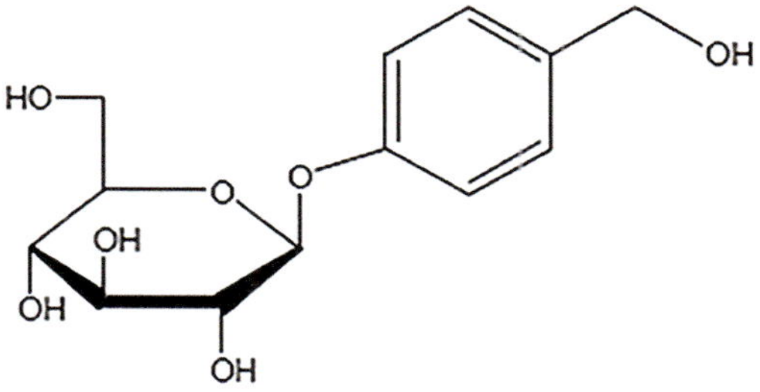

Medicinal uses of G.elata are as under;

- Immunostimulating
- Analgesic action
- Uses for skin and body
- Neuroprotective effect
- Applications in cardiovascular system
- For Anxiety
- Vascular Dementia

32.19.1 Immunostimulating

Cellular immune functions can be strengthening by the gastrodin polysaccharides. Gastrodin does so by having an inhibitory effect on the tissue culture of vesicular stomatitis virus.

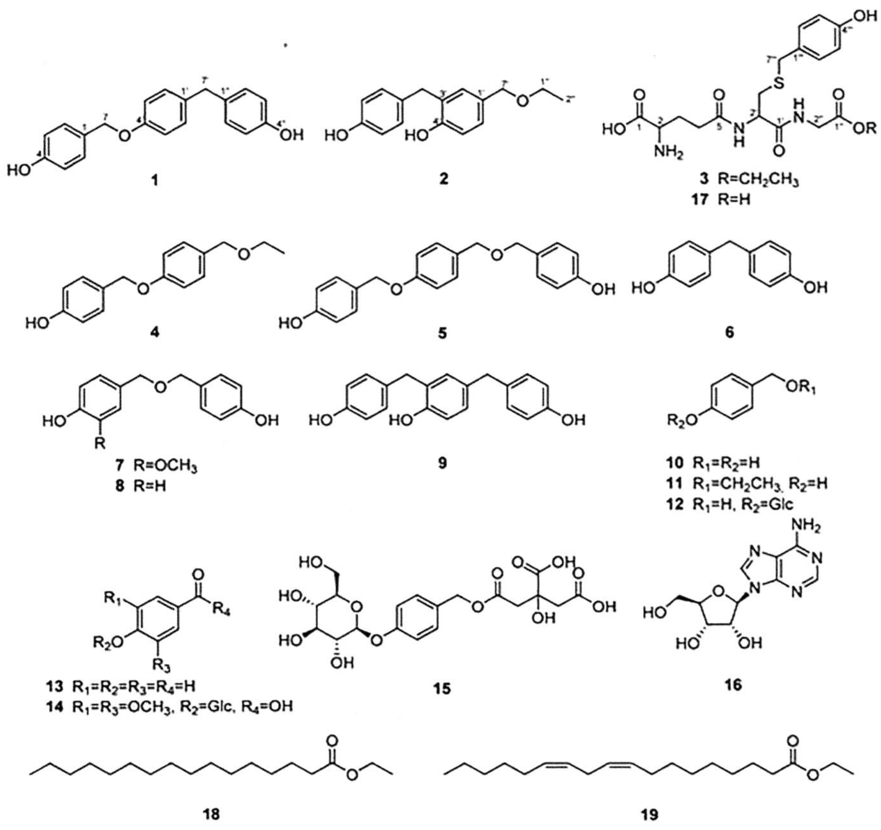

Fig. 32.8 Various isolated compounds of G. elata [27]

32.19.2 Analgesic Action

In case of neuralgia, headache and cerebrovascular diseases, an effective analgesic action can be produced by the injection of G. elata. Studies have shown that wild plants of Gastrodiaelata can produce better analgesic effect as compared to the artificially cultivated plants of this specie [28].

32.19.3 Uses of G. elata for Skin and Body

G.elatacan serve the following functions for skin and body.

- Antioxidant
- Anti-inflammatory
- For skin whitening

32.19.3.1 Antioxidant

In order to reduce the appearance of photo ageing in human dermal fibroblasts exposed to UVA radiation, G. elata increases antioxidant activity with an adequate amount of flavonoid and polyphenols [28].

32.19.3.2 Anti-inflammatory

Redness and irritation of the skin can be reduced by the anti-inflammatory properties of G. elata rhizome [28].

32.19.3.3 For Skin Whitening

Melanin synthesis is inhibited by G. elata by the repression of tyrosinase activity. Even a lower dose of g. elata extract can suppress the tyrosinase activity effectively. Therefore, G.elata can act as a natural skin whitening and anti-aging product.

32.19.4 Neuroprotevtive Effects of G.elata

G. elata is used extensively for curing many central nervous system diseases. The diseases include epilepsy, Alzaheimer's disease, dimensia and Parkinson's disease. Effects of gastrodin for curing various diseases of central nervous system can be given as follows (Fig. 32.9).

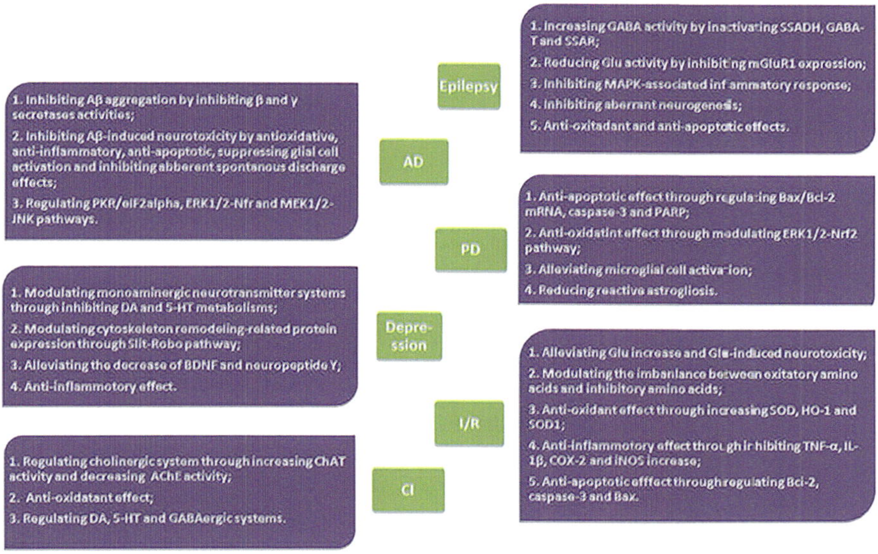

Fig. 32.9 Effect of gastrodin on the diseases of CNS [29]

32.19.4.1 Epilepsy

Epilepsy belongs to a condition in which frequent seizures occur. These seizures affect the functioning of brain and central nervous system. Though a number of drugs are available to cure epilepsy but no improvement has been observed in the cure rate of the disease. This leads to the need of discovering a novel drug with effective outcomes. It was observed that gastrodin can reduce the severity of seizures and shorten their duration as well. The anomalies in the central nervous system are caused by an imbalance in the interaction between inhibitory and excitatory neurotransmitters. The primary inhibitory brain neurotransmitter is gamma-amino butyric acid. Increased levels of gamma-amino butyric acid can significantly lower the frequency of seizures and epilepsy.. Studies have revealed that Gastrodin can inhibit the activity of all those enzymes which degrade Gamma-amino butyric acid. These enzymes include succinic semialdehyde dehydrogenase of bovine brain, succinic semialdehydereductase and GABA transaminase. Thus, gastrodin can reverse the effect of all the enzymes which have a degenerative effect on Gamma-amino butyric acid. Gastrodin treats epilepsy by lowering the excitatory neurotransmitter glutamate's activity. Unfortunately, gastrodin is ineffective when it comes to the glutamate receptor N-Methyl-D-Aspartate receptor. Hence, gastrodin cannot treat NMDA-receptor mediated seizures.

32.19.4.2 Temporal Lobe Epilepsy

Drug-resistant epilepsy, such as that of the temporal lobe, is a major source of medical issues. A study was done to evaluate the therapeutic value of gastrodin. During this, a number of rats were taken and divided into two groups. First group was experimental group in which seizures were induced by lithium-pilocarpine injection and the second group was control group (Dimethyl sulphoxide, vehicle group). Results revealed that a dosage of gastrodin reduced the severity and duration of seizures largerly. Moreover, Western blotting revealed that Gastrodin prevented GABA receptor degeneration following the occurrence of lithium-pilocarpine-induced seizures (Fig. 32.10) [29].

32.19.4.3 Alzheimer's Disease

According to 60% of instances, AD is the primary cause of dementia. The disease's progression resulted in significant neuron loss, intracellular neurofibrillary tangles, and senile plaques. As till the date the mechanism of this disease is unknown, we are lacking an effective curing method or drug for Alzheimer's disease. It has been observed by the in vitro and in vivo studies that gastrodin is quite affective to treat AD. The main component of senile plaque is Amyloid beta peptide (Aβ). It is

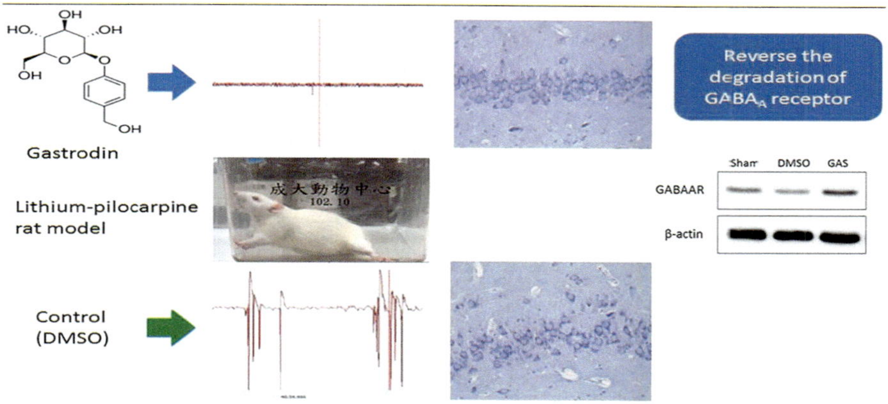

Fig. 32.10 Effect of gastrodin on temporal lobe epilepsy [29]

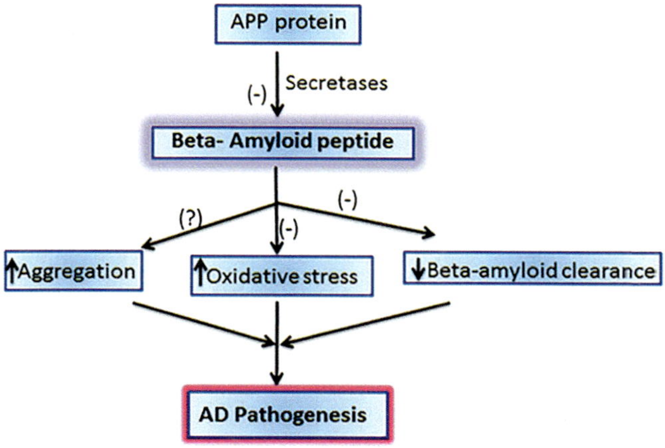

Fig. 32.11 Mechanism of AD [29]

regarded as the most important toxicant of Alzheimer's disease. From amyloid precursor protein, the amyloid beta peptide is produced. The enzymes that degrade this protein molecule into amyloid beta peptide are called secretases such as beta and gamma secretases. The mechanism can be shown by following figure (Fig. 32.11).

It was revealed by the studies that gastrodin can reduce the level of amyloid beta peptide. Gastrodin does so by reducing the level of β and γ secretases and thus reduces the risk of AD. Oxidative stress, inflammatory responses, synaptic loss, neuronal death are the pathological changes induced by Aβ. Aβ-induced neurotoxicity can be reduced by gastrodin [29]. The summary of Alzheimer's disease is as follows (Fig. 32.12).

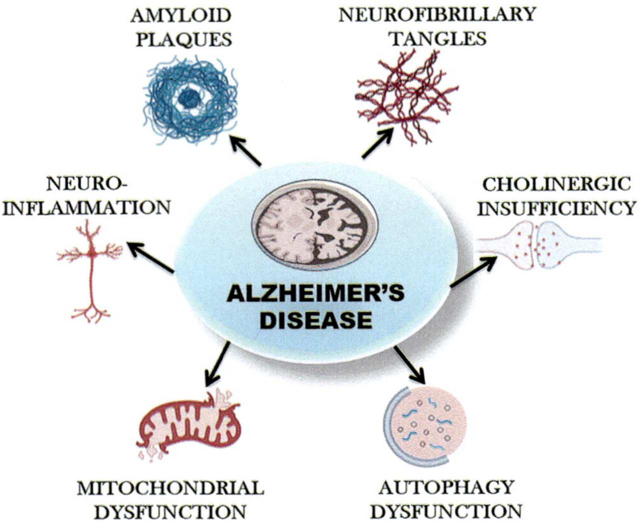

Fig. 32.12 Symptoms of AD [29]

32.19.4.4 Parkinson's Disease

Parkins' illness is the name of the neurological condition that causes uncontrollable movement. The patient also has to deal with functional handicap, low quality of life, and quick cognitive deterioration. The primary cause of Parkinson's disease is the death and degradation of dopaminergic neural cells in the mid brain's area substatia nigra. The reduced amount of neurotransmitter dopamine gives rise to various motor symptoms. The drugs used for the cure of PD can enhance the dopamine level but cannot lessen the loss of dopaminergic neutrons. In addition, the long term use of these drugs causes many side effects. Research have shown that Gastrodia elata as well as its biologically active compounds can significantly lessen the loss of doaminergic neurons, synuclein buildup, and neuroinflammation (Fig. 32.13) [30].

32.19.5 Applications in Cardiovascular System

Heart and brain health are interconnected. Given the substantial correlations among heart disease risk factors, Alzheimer's disease, and vascular dementia, it has been hypothesised that type 2 diabetes, Alzheimer's disease, and these conditions are connected.. The chance of acquiring these disorders appears to be increased by hypertensive, dyslipidemia, high cholesterol, reduced physical activity, and an unhealthy diet. Blood artery tonicity is regulated by myocytes, which also regulate tension and relaxation. The activity of contractile and morphological proteins

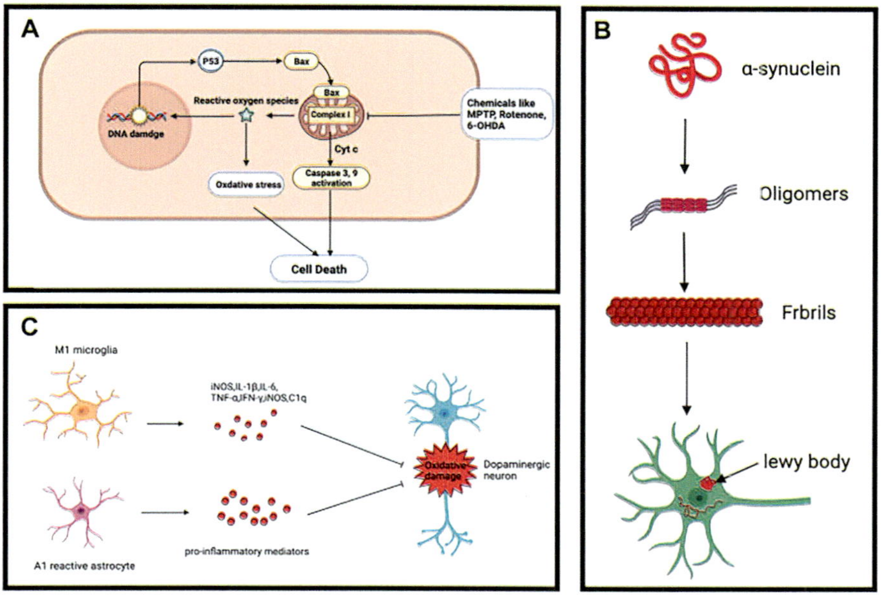

Fig. 32.13 Mechanism of PD [30]

including actin alpha 2, desmin, tubulin protein 4, and vinculin, in addition to matrix proteins like elastin, was changed by tianma treatment in low doses over time, as per functional aortic tissue proteomic findings. Atherosclerosis and cardiovascular disorders may also be prevented by elevated ANXA2 and decreased FABP4 levels. Through promoting vasodilatory effects that improve arterial structure, tianma may be capable of avoiding a range of cardiovascular disorders, including migraine, atherosclerosis, and stroke. By illuminating the processes by which tianma reduces these aberrant cardiovascular responses, research of all the bioactive molecules in tianma may thus aid in its implementation as a successful therapeutic herbal medication for curing cardiovascular illnesses (Fig. 32.14) [31].

32.19.6 Anxiety

Research using rats with post-traumatic stress disorder that had been exposed to augmented single sustained stress discovered that gastrodin could revert the anxiety-like behaviour, suggesting that it has anxiolytic effects in vivo. In order to better understand the underlying mechanism, researchers found that the anxiolytic effect of gastrodin was connected to iNOS/p38 cascades in the hippocampus of Post traumetic stress disorder rat model. Furthermore, they found that gastrodin could counteract the incline of IL-6 and IL-1 rates and increased expression of iNOS and phospho-p38 MAPK [29].

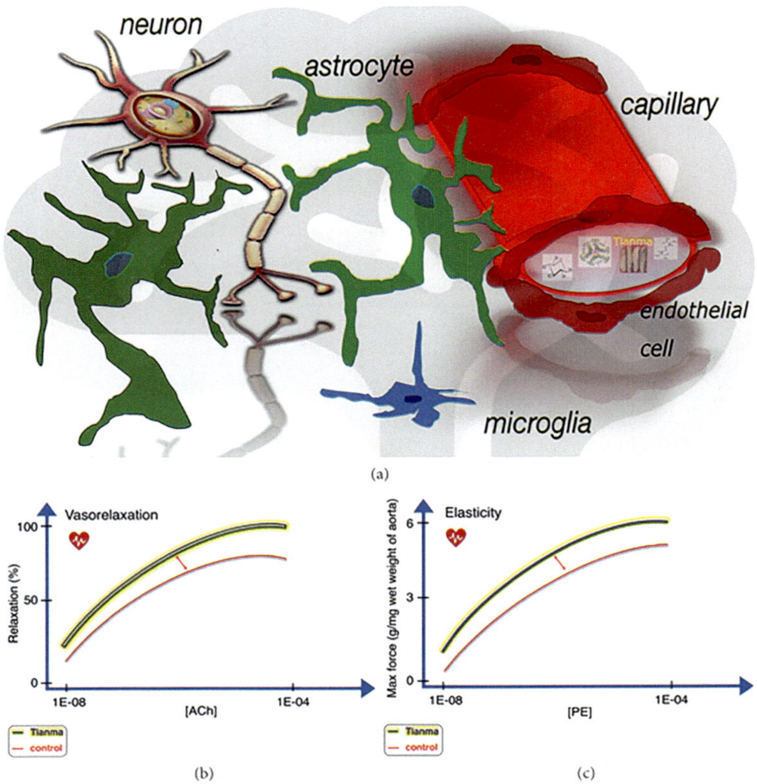

Fig. 32.14 Review of the CCV-related processes mediated by tianma in a schematic form. (**a**) Because astrocytes act (**b**) A quantitative representation of vasorelaxation which got increased in rats due to three months of treatment by Tianma (**c**) A quantitative analysis of enhanced Vascular contraction and elasticity due to tianma treatment in rats [31]

32.19.7 Vascular Dementia (VD)

In association with cerebrovascular brain damage, vascular dementia is a combination of functional and cognitive dysfunction. Almost 15% of all occurrences of dementia are caused by it, making it the second most common cause after Alzheimer's disease. Apart for a few medications with numerous side reactions and only little clinical effects, there are currently no established treatments to halt or cure the disease's course. The possibility for using gastrodin in the cure of vascular dimentia has been intriguing. Researchers created a rat model of vascular dementia and found that gastrodin could greatly boost the model's capacity for learning and memory. They also found that gastrodin could decrease the activity of acetylcholinesterase as well as Glu while increasing the action of choline acetyltransferase, implying that the mechanism underlying the memory-improving influence was connected to the upregulation of brain dopaminergic system function [29].

32.20 G. elata's Pharmacological Effects and Mechanisms

Gatrodiaelata shows medicinal and pharmacological effects through various ways. Due to presence of flavonoids, polyphenol like components it is used as an anti-oxidant agent. It also cures many neurological disorders due to presence of gastrodin, phenolic compounds, perishin and 4-hydroxy benzyl alcohol etc. The following table depicts the mechanism of G. elata to cure various diseases (Table 32.2).

Table 32.2 Pharmaceutical uses of G. elata with mechanism

Medicinal uses of gastrodiaelata	Description	Mechanism
Immunostimulating	Cellular immune functions can be strengthening by the gastrodin polysaccharides.	Gastrodin does so by having an inhibitory effect on the tissue culture of vesicular stomatitis virus.
Analgesic action	In case of neuralgia, headache and cerebrovascular diseases, an effective analgesic action can be produced by the injection of G. elata.	G. elata does so by modulating neurotransmitters.
Antioxidant	G. elata can be used as an oxidant.	In order to reduce the appearance of photo ageing in human dermal fibroblasts exposed to UVA radiation, G. elata increases antioxidant activity with an adequate amount of flavonoid and polyphenols
Anti-inflammatory	It can be used as an anti-inflammatory drug.	Redness and irritation of the skin can be reduced by the anti-inflammatory properties of G. elata rhizome.
For skin whitening	Gastodiaelata have the potential to act as a skin whitening agent.	Melanin synthesis is inhibited by G. elata by the repression of tyrosinase activity. Even a lower dose of g. elata extract can suppress the tyrosinase activity effectively. Therefore, G.elata can act as a natural skin whitening and anti-aging product.
Epilepsy	Epilepsy belongs to a condition in which frequent seizures occur. Gastrodin can reduce the severity of seizures and shorten their duration as well.	Gamma-amino butyric acid is the key inhibitory neurotransmitter in brain. Increase in the concentration of this neurotransmitter can reduce the occurrence of seizures, epilepsy to a great deal. Studies have revealed that Gastrodin can inhibit the activity of all those enzymes which degrade Gamma-amino butyric acid.
Temporal lobe epilepsy	Drug-resistant epilepsy that causes numerous medical issues is temporal lobe epilepsy. Gastrodin can reduce the severity and duration of seizures largely.	Gastrodin reverse the degradation of GABA receptor to reduce the severity of seizures.

(continued)

Table 32.2 (continued)

Medicinal uses of gastrodiaelata	Description	Mechanism
Alzheimer's disease	Gastrodin is quite affective to treat AD.	Amyloid beta peptide levels can be lowered by gastrodin. In order to do this, Gastrodin lowers the levels secretases, which lowers the risk of Alzheimer's. The neurotoxic effects of Aβ can be mitigated by gastrodin. Consequently, it is possible to state that gastrodin can prevent cell damage brought on by Aβ stimulated injury.
Parkinson's disease	The midbrain dopaminergic neurons' loss and degradation is what causes Parkinson's disease. G. elata ia an effective drug to treat PD.	Gastrodia elata and its bioactive components play an important role to reduce the death of doaminergic neurons, α synuclein accumulation and neuroinflammation
Anxiety	Gastrodin is effective to treat anxiety	Gastrodin has the ability to lower IL-6 and IL-1 levels.
Vascular Dimentia	Gastrodin can be used to treat vascular dimentia.	Gastrodin improves choline acetyltransferase activity while decreases acetylcholinesterase and glutamate activity, demonstrating the mechanism underlying the effect on memory.

32.21 Conclusions

Belonging to Compositae family, Gastrodia species are mostly grown in China among which "Gastrodiaelata" is widely applied as medicinal tool. The nutritional value of different medicinal components has made G.elata an excellent plant for high level research and applications. Some of the major pharmacological substances of it are gastrodin, ergothioneine and vanillyl alcohol. Cultivation conditions and time have a great impact on them. Its crop can be significantly reduced by various pests. Many neurodevelopmental problems and mental illnesses, such as narcolepsy, Alzheimer's disease, low mood, anxiety, and other types of cognitive loss, have been successfully treated with G elata. It influences the brain's central nervous system in both vivo and in vitro in a wide variety of biological ways. Modifying neurotransmitters, preventing neuroinflammation, regulating mitochondrial cascades, boosting neurotrophins, and preventing oxidation, inflammatory, and apoptosis are a few of its known mechanisms of action.. By conducting more research on the mechanism involved in curing diseases, Gastrodiaeleta medicinal applications can be further explored.

References

1. Ojemann, L. M., Nelson, W. L., Shin, D. S., Rowe, A. O., & Buchanan, R. A. (2006). Tian ma, an ancient Chinese herb, offers new options for the treatment of epilepsy and other conditions. *Epilepsy & Behavior, 8*(2), 376–383.
2. Perry, E., & Howes, M. J. R. (2011). Medicinal plants and dementia therapy: Herbal hopes for brain aging? *CNS Neuroscience & Therapeutics, 17*(6), 683–698.
3. Zhang, Z. C., Su, G., Li, J., Wu, H., & Xie, X. D. (2013). Two new neuroprotective phenolic compounds from Gastrodia elata. *Journal of Asian Natural Products Research, 15*(6), 619–623.
4. Shao, Z. Q. (2015). Comparison of the efficacy of four cholinesterase inhibitors in combination with memantine for the treatment of Alzheimer's disease. *International Journal of Clinical and Experimental Medicine, 8*(2), 2944.
5. Jang, J. H., Son, Y., Kang, S. S., Bae, C. S., Kim, J. C., Kim, S. H., et al. (2015). Neuropharmacological potential of Gastrodia elata Blume and its components. *Evidence-Based Complementary and Alternative Medicine, 2015.*
6. Bulpitt, C. J., Li, Y., Bulpitt, P. F., & Wang, J. (2007). The use of orchids in Chinese medicine. *Journal of the Royal Society of Medicine, 100*(12), 558–563.
7. Chen, P. J., Hsieh, C. L., Su, K. P., Hou, Y. C., Chiang, H. M., & Sheen, L. Y. (2009). Rhizomes of Gastrodia elata BL possess antidepressant-like effect via monoamine modulation in subchronic animal model. *The American Journal of Chinese Medicine, 37*(06), 1113–1124.
8. Chen, X., Xiao, F., Wang, Y., Fang, J., & Ding, K. (2012). Structure-activity relationship study of WSS25 derivatives with anti-angiogenesis effects. *Glycoconjugate Journal, 29*, 389–398.
9. Huang, T., Danaher, L. A., Brüschweiler, B. J., Kass, G. E., & Merten, C. (2019). Naturally occurring bisphenol F in plants used in traditional medicine. *Archives of Toxicology, 93*, 1485–1490.
10. Zhan, H. D., Zhou, H. Y., Sui, Y. P., Du, X. L., Wang, W. H., Dai, L., et al. (2016). The rhizome of Gastrodia elata Blume–an ethnopharmacological review. *Journal of Ethnopharmacology, 189*, 361–385.
11. Zhang, D., Mei, L., He, H., Wang, Y., Cao, A., Guo, J., et al. (2014). Study of factors for cultivating the orchid species Gastrodia elata, a traditional Chinese medicine. *Plant Diversity and Resources, 36*(2), 254–260.
12. lan ping Long, L., & fu lai Luo, L. (2021). *Effects of different years of natural recovery of Gastrodia elata on the community structure of bacteria and fungi in rhizosphere soil.*
13. Hsieh, C. H., Liang, Z. C., Shieh, W. J., Chang, S. L., & Ho, W. J. (2022). Effects of nutrients and growth regulators on seed germination and development of juvenile rhizome proliferation of Gastrodia elata in vitro. *Agriculture, 12*(8), 1210.
14. Jiang, L., Wan, S., Wang, S., & Yu, C. (2001). Isozyme analysis of Gastrodia elata f. elata and G. elata f. glaucca and their hybrid. *Zhong yao cai= Zhongyaocai= Journal of Chinese Medicinal Materials, 24*(8), 547–548.
15. Lei, Y. C. (2015). Correlation between contents of gastrodin and polysaccharides with grade of Gastrodia tuber. *Chinese Traditional and Herbal Drugs*, 418–423.
16. Tian, H., Li, H., & Yang, C. (2010). Gastrodia R. Br., a newly recorded genus of Orchidaceae in Guangdong Province. *Journal of Tropical and Subtropical Botany, 18*(5), 488–490.
17. Yu, L., Shen, Y., & Miao, H. (2006). Study on the anti-vertigo function of polysaccharides of Gastrodia elata and polysaccharides of Armillaria mellea. *Chinese Journal of Information on Traditional Chinese Medicine.*
18. Zhang, J. Q., Tang, X., Guo, L. P., Yang, Y., Wang, Y. H., Wei, Y., et al. (2022). Correlation analysis between continuous cropping obstacle of Gastrodia elata and Ilyonectria fungi and relieving strategy. *Zhongguo Zhong yao za zhi= Zhongguo Zhongyao Zazhi= China Journal of Chinese MateriaMédica, 47*(9), 2296–2303.

19. Jiang, W. K., Zhang, J. Q., Guo, L. P., Yang, Y., Xiao, C. H., Yuan, Q. S., et al. (2022). Thoughts and suggestions on ecological cultivation of Gastrodia elata. *Zhongguo Zhong yao za zhi= Zhongguo Zhongyao Zazhi= China Journal of Chinese Materia Médica, 47*(9), 2277–2280.
20. Kim, H. T., Kim, S. T., Lee, W. Y., & Park, E. J. (2013). Induction and growth of vegetative stems through in vitro culture of Gastrodia elata. *Korean Journal of Medicinal Crop Science, 21*(2), 142–147.
21. George, E. F., Hall, M. A., & Klerk, G. J. D. (2008). The components of plant tissue culture media I: Macro-and micro-nutrients. In *Plant propagation by tissue culture: Volume 1. The background* (pp. 65–113). Springer.
22. Xie, Y. K., Li, X. Y., Zhang, Y., Zheng, Z. A., Huang, L. Q., Liu, D. H., et al. (2021). Effects of high-humidity hot air impingement steaming on Gastrodia elata: Steaming degree, weight loss, texture, drying kinetics, microstructure and active components. *Food and Bioproducts Processing, 127*, 255–265.
23. Li, J., Zhang, M., Yang, Z., & Li, C. (2022). Botrytis cinerea causes flower gray mold in Gastrodia elata in China. *Crop Protection, 155*, 105923.
24. Li, J., & Li, C. (2022). Fusarium solani causing root rot disease on Gastrodia elata in Shaxi, China. *Plant Disease, 106*(1), 320.
25. Li, C., Zhang, M., Li, J., Huang, M., Shao, X., & Yang, Z. (2022). Fusarium redolens causes black rot disease in Gastrodia elata grown in China. *Crop Protection, 155*, 105933.
26. Panwar, V., Aggarwal, A., Paul, S., Singh, V., Singh, P. K., Sharma, D., & Shaharan, M. S. (2016). Effect of temperature and pH on the growth of Fusarium spp. causing Fusarium head blight (FHB) in wheat. *South Asian Journal of Experimental Biology, 6*, 186–193.
27. Kim, H. M., Kwon, J., Lee, K., Lee, J. W., Jang, D. S., & Kwon, H. C. (2020). Constituents of Gastrodia elata and their neuroprotective effects in HT22 hippocampal neuronal, R28 retinal cells, and BV2 microglial cells. *Plants, 9*(8), 1051.
28. Liu, Y., & Huang, G. (2017). The chemical composition, pharmacological effects, clinical applications and market analysis of Gastrodia elata. *Pharmaceutical Chemistry Journal, 51*, 211–215.
29. Liu, Y., Gao, J., Peng, M., Meng, H., Ma, H., Cai, P., et al. (2018). A review on central nervous system effects of gastrodin. *Frontiers in Pharmacology, 9*, 24.
30. Lu, C., Qu, S., Zhong, Z., Luo, H., San Lei, S., Zhong, H. J., et al. (2022). The effects of bioactive components from the rhizome of gastrodia elata blume (Tianma) on the characteristics of Parkinson's disease. *Frontiers in Pharmacology, 13*.
31. Heese, K. (2020). Gastrodia elata Blume (Tianma): Hope for brain aging and dementia. *Evidence-Based Complementary and Alternative Medicine, 2020*.

Chapter 33
Chili Pepper

Sahar Nazeer, Tayyaba Tur Rehman Afzal, Sana, Maham Saeed, Shahzad Sharif, and Muhammad Zia-Ul-Haq

33.1 Introduction

Capsicum annum, commonly known as the chili pepper, is used all over the world. Apart from giving taste and flavor to food, spices also play a significant role in medicine [1, 2]. Capsicums are mostly used in the food industry, giving color and flavor to sauces, snacks, processed meat, drinks and different beverages. *Capsicum annum* is one of the five major species of the genus Capsicum [3]. Among others in the family, it is the most common and is popularly cultivated in countries with temperate and semi-tropical ranges [4]. Vitamins A and C, which are significant dietary antioxidants, are present in bell peppers in substantial amounts. These chemicals' concentrations may be affected by maturation, genetic makeup, and processing [5]. Fruits and vegetables' organic compounds may be crucial in the chemoprevention of cancer [6]. Widespread in plants, flavonoids have pharmacological and biochemical effects that include antioxidant, anti-inflammatory and anti-allergic properties [7].

33.1.1 Scientific Classification and Origin

Kingdom: Viridiplantae
Phylum: Streptophyta

S. Nazeer · T. T. R. Afzal · Sana · M. Saeed · S. Sharif (✉)
Department of Chemistry, Government College University, Lahore, Pakistan
e-mail: mssharif@gcu.edu.pk

M. Zia-Ul-Haq
Office of Research, Innovation and Commercialization,
University of Engineering and Technology, Lahore, Pakistan

M. Zia-Ul-Haq et al. (eds.), *Essentials of Medicinal and Aromatic Crops*,
https://doi.org/10.1007/978-3-031-35403-8_33

Class: Magnoliopsida
Subclass: Asteridae
Order: Solanales
Family: Solanaceae
Genus: Capsicum
Specie: annuum
Common name: Chili pepper

Known evidence of the origin of the chili pepper dates back to **6000** B.C [**8**]. The reason for its naming "pepper" is because in the early period, people would give fruits in exchange for black "pepper," which was very similar in taste, and hence it was given the name "pepper," which was incorrect [**9**, **10**]. The term "capsicum" was coined by Fusch in **1543**, derived from the Greek word "capsa," which indicated the typical fruit shape, and formally came into use by Linneo in **1753**. After Christopher Columbus' incidental discovery of America later in the fifteenth century, he traveled to Europe for the first time, but this time there was no incident. After it was spread into Europe, it soon reached Asia and Africa and later, around the globe, courtesy of Portuguese and Spanish exports. After its introduction to other regions, it soon replaced other spices because it was quite cheaper than the others. It is still known as "chile" in Mexico and Central America, where it was used in the ancient period locally after its introduction to all the regions of the world. Currently, fruits with a smaller size but a more spicy flavor are referred to as "chili pepper," whereas those with a larger size but less spice or no spice at all are referred to as "sweet pepper" [**11**].

33.1.2 Fruit Morphology

Fruit morphology shows that there are different types of *Capsicum annuum*. Fruit lengths range from 40 to 140 mm, and fruit breaths range from 19 to 65 mm. The fruit stalk length is almost 20–50 mm. The shape of all the fruits is cordate, and the fruit surface is flexuous. All of the variants' fruits are red, yellow, or green. The color varies depending on the variety, with most being green before turning red (Fig. 33.1).

The plants having different morphologies based on the type of the plant, have different colors and even taste of the peppers which are having many effects including biological, antioxidant and also used worldwide as a spice in cooking for taste [13] (Table 33.1).

All the Variants have different epidermal characters like length, breadth, size, shape and anticlinal wall. The shape is mostly polygonal and the anticlinal walls are mostly straight and curved. The size, length and breadth of the cells are very small

Fig. 33.1 Examples of pod types. (Ref. [12])

Table 33.1 Morphological parameters in *Capsicum annuum* [14]

Taxa	Fruit length (mm)	Fruit breath (mm)	Fruit stalk length	Fruit shape	Fruit surface	Fruit colour
C. annuum var. *abbreviatum*	42.65	39.67	27.94	Cordate	Flexuous	Red
C. annuum var. *annuum*	71.50	20.34	36.59	Cordate	Flexuous	Green
C. annuum var. *accuminatum*	72.25	45.37	36.58	Cordate	Flexuous	Green
C. annuum var. *grossum*	115.10	65.32	43.78	Attenuate	Flexuous	Reddish
C. annuum var. *glabriusculum*	–	19.51	45.95	Cordate	Flexuous	Reddish

mostly in micrometers. All the variants are having the same shape and anticlinal wall of the epidermal cells but the variation in the size, length and breadth of the cells exists [15] (Table 33.2).

The Fruit interior in *Capsicum annuum* also varies. The locule is trilocular and placentation is axile and may be hollow or less hollowed. The no of seeds also varies in different variants [16] (Table 33.3).

Table 33.2 Epidermal characters in *Capsicum annuum*

Taxa	Size (μm)	Length (μm)	Breath (μm)	Anticlinal wall	Shape
C. annuum var. *abbreviatum*	328.51	15 (23) 31	13 (18) 25	Curve and striaght	Polygonal
C. annuum var. *annuum*	235.74	11 (20) 30	10 (14) 20	Curve and striaght	Polygonal
C. annuum var. *accuminatum*	212.17	13 (18) 24	10 (14) 21	Curve and striaght	Polygonal
C. annuum var. *grossum*	554.44	38 (41) 45	13 (16) 21	Curve and striaght	Polygonal
C. annuum var. *glabriusculum*	282.47	38 (41) 45	15 (16) 28	Curve and striaght	Polygonal

Table 33.3 Fruit interior in the *Capsicum annuum*

Taxa	Number of seeds	Locule	Placentation	Hollow
C. annuum var. *abbreviatum*	43	Trilocular	Axile	Less hollow
C. annuum var. *annuum*	39	Bilocular	Axile	Hollow
C. annuum var. *accuminatum*	58	Bilocular	Axile	Hollow
C. annuum var. *grossum*	61	Trilocular	Axile	Less hollow
C. annuum var. *glabriusculum*	97	Tetralocular	Axile	Hollow

33.2 Properties of Pepper

The occurrence of capsaicinoids in the pepper results in its being pungent, making it unique. Capsaicinoids are derived from phenylpropanoids, which are produced in the cells of the placental epidermis and are said to be secondary metabolites. The surface of the placenta is their location, where they are piled up in structure [17]. When engulfed, it comes in contact with vanilloid receptors, which sense it to be very hot. Capsaicinoids are a blend of two main compounds: capsaicin and dihydrocapsaicin. Other than these two, the other compounds account for nearly 10% of the total. Capsaicinoids aid in inflammation can fight against cancerous activity and can be considered beneficial against obesity. Peppers also fight against oxidants and play an important role in the color of fruits. In chromoplasts, the assembly of carotenoids when ripening gives rise to different colors in the fruits that are in the process of maturing. Capsanthin and capsorubin (responsible for red pigment), violaxanthin and neoxanthin (responsible for yellow pigment), lutein, and b-carotene (responsible for orange pigment) are among these carotenoids [18]. Capsicum species, including chili peppers, contain ascorbic acid in excessive amounts, which helps meet the requirement for daily consumption. B-carotene and B-group vitamins (niacin, thiamine, and riboflavin) constitute an excess quantity of vitamin A, which is an important one. When peroxyl radicals and molecules of oxygen come into contact, all the previously mentioned compounds act as anti-oxidants, giving safety to cells. The industries use colors obtained from fruits (mainly because of crytopcapsin, capsanthin, and capsorubin) and use them in the processing of food in a large number of goods [19].

33.2.1 Chemical Composition

Capsaicin content in pepper is the highest (69%). It also contains 22% dihydrocapsaicin and 7% nordihydrocapsaicin [20] (Fig. 33.2).

The chemical composition of two cultivars can be determined by mixing them with a solvent, typically organic solvent acetone or a mixture of acetone and water and then extracting and analyzing the pepper components using GC/MS. The result shows that the two cultivars, Adorno and Etna, have different concentrations of these chemicals. The peppers having oleic acid, octadecane, 5-eicosene, hexadecane, heptadecane, 4-hydroxy-4-methyl-2-pentanone, pentadecanoic acid, cycloeicosane, 1-octadecanamine, 9-octadecanamide, heneicosane, nonivamide, nordihydrocapsaicin, capsaicin, squalene, docosane, tetracosane. pentacosane, Vitamin E, alpha and beta amyrin and many other chemicals which are having specific health benefits and biological effects [21].

This is the composition of selected varieties of C.annuum which is extracted from the fruit with the help of acetone solvent. The amount of **4-hydroxy-4-methyl-2-pentanone** in Adorno is 0.98% and in etna is 2.84%. **1-hexadecene** amount in Adorno and etna is 0.06% and 0.05%. **2-methylpentadecane** having concentration of 0.04% in Adorno while **Heptadecane** amount in Adorno is 0.15% and in etna is 0.10%.

Similarly **1-Heptadecanol acetate** amount in Adorno is 0.10% while etna having 0.07%. **Octadecane** concentration in Adorno (0.21%) and etna has 0.08%.The amount of **5-eicosene** in Adorno and Etna is 0.10% and 0.03%.

Red pepper (chili)

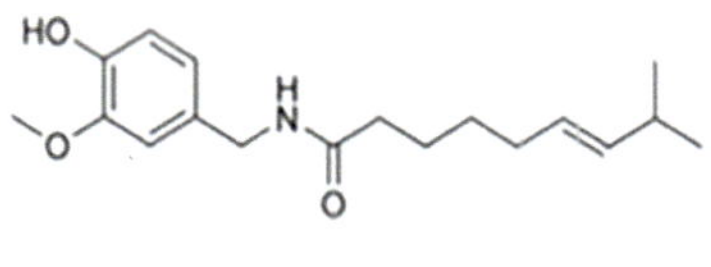

Capsaicin (69%)

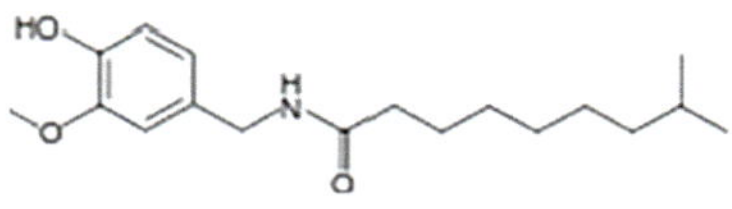

Dihydrocapsaicin (22%)

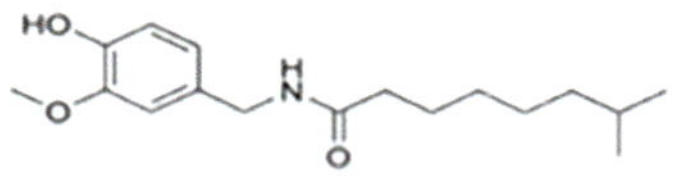

Nordihydrocapsaicin (7%)

Fig. 33.2 Red pepper composition

Eicosane value in Adorno is 0.04% and in Etna 0.09%.
Heneicosane amount in in Adorno (0.36%) and in Etna is 0.48%.
Nonivamide in Adorno (7.66%) and Etna in (13.20%).
Capsaicin in Adorno (37.22%) and in Etna (40.85%).
Dihydrocapsaicin in Adorno (28.68%) and Etna (19.87%).
Docosane in Adorno (0.52%) and Etna having (0.52%).
Homocapsaicin in Adorno (0.74%) and Etna has (0.24%).
Homocapsaicin II in Adorno (0.42%) and in Etna (1.35%).
Homodihydrocapsaicin in Adorno (1.85%) and in Etna (4.03%).
Tetracosane in Adorno 0.23% and in Etna 0.20%.
Pentacosane in Adorno 0.58% and in Etna 0.46%.
Hexacosane in Adorno 0.14% and in Etna 0.22%.
Vitamin E in Adorno 2.10% and in Etna 1.98%.
β-amyrin in Adorno 0.46% and Etna having 0.47%.

So the results indicates that both cultivars Adorna and Etna have highest amount of **Capsaicin** and second highest amount of **Dihydrocapsaicin.**

33.2.2 Nutritive Value

33.2.2.1 Phenolic Content in Fruits During Maturity

The quantity of flavonoid aglycones produced by following acid hydrolysis is used in pepper flavonoid analyses. Research of Materska et al. discovered a high figure for each polyphenol in their analysis. Six flavonoid groups and five extracts of hydroxycinnamic acid were measured in peppers at various stages of ripening (Table 33.4). To eliminate quantification mistakes and provide a more representative sample, freeze-dried specimens were engaged.

The flavonoid concentration of raw green peppers is typically 4–5 times greater than other ripening phases, yet a significant drop was noted from young to mature green as it developed. At every stage, pepper contains more O-glycosyl flavones than other flavonoids. More than 85% less green fruit was present when the pepper was raw. The size of unripe green fruit is somewhat less than that of more mature fruit therefore; this reduction is caused by both the growth in fruit size with maturation and the depletion of flavonoids. One of the key elements affecting phenolic component levels in fruits and vegetables is ripeness. Pepper has more O-glycosyl

Table 33.4 Ascorbic acid, dehydroascorbic acid, and total vitamin C of *C. annuum* L

Ripening	Vitamin C	Ascorbic acid	Dehydroascorbic acid
Red before maturation	63 (2.8)	61 (2.8)	2 (0.1)
Red	93(9.1)	90.6(9)	2.3(0.2)
Green before maturation	42.2 (0.8)	31.2 (0.09)	11 (0.8)
Green	54.2 (4)	51.2 (3.6)	3 (0.3)

flavones than other flavonoids at every stage. When the pepper was raw, there was more than 85% less green fruit. Unripe green fruit is somewhat smaller than more mature fruit therefore; this diminution is the result of both the enlargement of the fruit with maturation and the loss of flavonoids. One of the most important factors influencing the amount of phenolic components in fruits and vegetables is ripeness.

33.2.2.2 Vitamin C in Fruits During Ripening

In line with earlier research, this study discovered that the ascorbic acid concentration of fresh peppers increased as the fruits developed. Brightly adapted leaves have been shown to have greater ascorbate concentrations than leaves grown at lower densities and to act as photo- sensitizers. The cultivar examined in this study exhibited vitamin C levels that ranged from 63 to 243 mg/100 g less than other reported pepper cultivars.

The vegetable having the greatest vitamin C concentration among other significant plant foods, such as broccoli and spinach, was said to be red bell peppers. In contrast to unripe green peppers, red peppers have 45% more total vitamin C, and as the peppers develop, the pH levels become more stable. Vitamin C levels increased significantly in other fruits, such as tomatoes, up until the red stage, after which fruit senescence occurred.

33.2.2.3 Carotenoid Content in Fruit During Ripening

Chlorophyll and carotenoids decrease during pepper ripening, while fatty acids produce and esterify the carotenoid chromoplast colors. According to our findings, lutein and cis-lutein levels in the immature and mature red ripening phases fell throughout ripening to undetectable levels. Neoxanthin concentration in young green and red fruits was comparable but non-existent in mature red fruits. Their primary pigment was capsanthin, and the red phase contains a lot of provitamin A because of the high levels of beta-cryptoxanthin and carotene. The skin of peppers contains the majority of the phenolic chemicals, including C-glycosyl, flavones, luteolin derivatives, and quercetin O-glycosides, a branched-chain fatty acid thought to be added to vanillyl amine and 8-methyl-6-non-enoyl Co-enzyme A in the inner lobule of C.annuum where it is then converted into capsaicin [22].

33.2.3 Antioxidant Constituents

Numerous studies have linked eating flavonoids to a lower risk of developing cancer [23, 24]. Enzymes like prostaglandin synthase, lipoxygenase, and cyclooxygenase, which are linked to tumorigenesis, are inhibited by flavonoids [25]. Furthermore, flavonoids can activate detoxifying enzyme mechanisms like glutathione

S-transferase [26]. Fruits and vegetables are found to contain a wide range of flavonoids, with different types and concentrations depending on the variety and level of ripeness. Onions (284–486 mg/kg), cabbage (110 mg/kg), broccoli (30 mg/kg), green beans (32–45 mg/kg), and kidney beans were among the 30 fruits and vegetables that were examined. The greatest quercetin content is between 28 and 30 mg/kg [27]. Flavone levels except for fresh beans (25 mg myricetin) and red peppers (13–3 mg luteolin/kg; 31), were low in comparison to flavonols. Flavonoids exist in plant cells as glycosides, which typically have sugars attached at the C position. Following consumption, the human intestinal flora breaks down flavonoid glycoside conjugations into aglycones. Add vitamin C and flavonoids to your diet. There is little available information on the flavonoid content of various pepper species and cultivars. The concentrations of quercetin, luteolin, and ascorbic acid in several pepper types were measured, and the link between flavonoid content and antioxidant activity was investigated. When completely mature, peppers change color from green to yellow, orange, red, or purple, and variations in ripeness can impact phytonutrient content, which is significant in dietary antioxidant consumption. Fresh pepper carotenoid pigments have been researched to increase color retention throughout processing and storage and red bell peppers have the most vitamin A (beta-carotene and beta-cryptoxanthin). Chili peppers contain flavonoids [27, 28]. It was discovered that the concentration of these antioxidant substances rose as the peppers matured. To determine the contribution of these antioxidants to contemporary nutritional demands, the nutritional content of sweet peppers was studied.

33.2.4 Genetic Resources and Breeding

The capsicum genome is 3.5 Gb in size and is diploid with 12 chromosomes (2n = 2 h = 24). In it, there is also a recognized one having 13 chromosomes (2n = 2 h = 26), also tetraploid species (with 2n chromosomes), namely *C. annuum* var. gabriusculum. Fresh research has categorized Capsicum into eleven taxa respective to their central morphological characteristics, origins, and genetic relationships. The species most interesting for consumption and reproduction are found in three main groups: the Annuum, which includes three domesticated species (*C. annuum*, C. frutescens, and C. chinense) and two wild species (*C. annuum* var. glabriusculum and C. galapagoense); the Baccatum which includes three forms of C. baccatum (var. baccatum, var. pendulum, and var. umbilicatum); the wild Chacoense and C. Praetermisssum; and the domesticated species of the same name; *C. annuum* is the most commercially valuable species in the world. This species features thorny, tart inserts with herbaceous or shrubby growth and fruits of various sizes, shapes, and colours when ripe. C.frutescens and C. Chinense are primarily grown in North America, Asia, and Africa. The first includes tart accessions with mainly small fruits less than 2 cm long; the second, on the other hand, includes very tart accessions with fruits of irregular shape. Also C. baccatum and C. pubes are sown in Central and South America having certain phenotypic traits. Each one is characterised by or

Fig. 33.3 Mature fruits of wild Capsicum species: (**a**) C. chacoense, (**b**) C. praetermissum, (**c**) C. eximium, (**d**) *C. annuum* var.glabriusculum, and e C. flexuosum

fruits with distinguishing features to flower colour (white, yellow, or purple types of spots), seed color (brown or black), and flower (star, round, or bell-shaped) [29].

Specifically, wild and domesticated animals have been utilized for disease resistance, and the outcomes have been well described in the literature. The four main goals of pepper breeding are: (a) major agronomic traits like yield and fruit characteristics like color, shape, plant habit, and fruit set; (b) resistance to abiotic stresses like drought and salinity that limit cultivation in certain areas; (c) resistance to a large number of bacterial, fungal, and viral diseases that cause significant crop damage and loss of production quality; and (d) quality, for which peppers are grown (Fig. 33.3).

33.2.5 Economic and Culinary Importance

2–4.5 million tonnes of dehydrated varieties and 17–36 million tonnes of fresh varieties, the global production of black pepper has expanded dramatically during the past 20 years. The same thing can be seen in the cultivated land, which has increased by around 35% over the past 20 years to reach almost 3.8 million hectares. On all seven continents, there are 126 nations where fresh peppers are grown. Mexico comes in second place with about 3.5 million tonnes produced per year, just behind China, which produces 18 million tonnes annually [30]. There are 70 nations where dried chilies are farmed; however, Oceania has not been listed as one of them. Thailand is the second-largest producer, behind India, with over two million tonnes (349–615 metric tons). Peppers are cultivated practically everywhere on the globe and are very simple to grow in greenhouses and open fields in a variety of climatic and environmental circumstances. About 10–12% of the world's total fresh pepper production is produced in Africa, Europe, and the Americas, respectively. Asia and Africa are the two biggest producers of dried chili, with 70.3 and 21.2 percent of global production, respectively. Since 1991, pepper production has become more economically valuable, a reliable source of revenue for producers in many nations, and a contributor to world trade. Dried peppers are currently worth \$3.8 billion, while fresh peppers are worth \$30.208 million. Fresh peppers have shown a six-fold rise in observed growth over the past 25 years compared to dried peppers' four-fold increase. The astonishing diversity of many traits, including the architecture of the

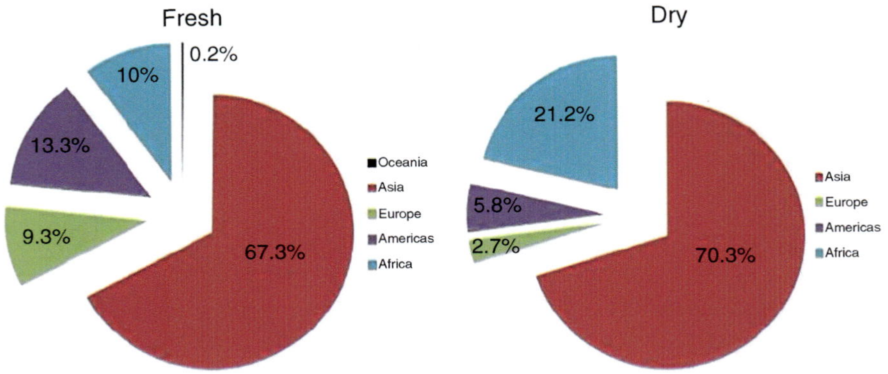

Fig. 33.4 Production of dry and fresh pepper by region. (Ref. [30])

plants, the morphologies of the blossoms, the type of fruit, the color, the permeability, and the quality attributes that distinguish this crop has sparked a growing interest in the genus Capsicum. It has high adaptability and suitability for a wide range of uses. The existence of sweet and spicy kinds ensures that there is a diversity of dishes on the food front. The former are mostly found in the arid and semi-arid parts of North America and Europe, where they are consumed raw or prepared as vegetables. The latter however, are more common in tropical areas of the Americas, Africa, and Asia, where they are used in a variety of recipes, mostly as a fresh or dried spice powder or as a flavoring in sauces. The following groups' best describe how black pepper is used nutritionally: Fresh, immature green fruits, ripe red fruits, and leaves should all be consumed. (b) Ripe fruits, dry spices, and pastes for fresh processing of sauces, pastes, pickles, brews, etc. [31] (Fig. 33.4).

More than 20 market types (including sweet peppers, chili peppers, anchovies, jalapenos, snap peas, Hungarian, mobile, and Thai waxwings) are produced commercially based on the size and form of the pod. Additionally, there may be several varieties of each of these markets; for instance, cherry vines may have little or big pods and vines may have stubby, conical, or small pods.

33.2.6 Biological Effects of Capsicum anuum

33.2.6.1 Hypocholesterolemic and Hypolipidemic Effects of Capsaicin

Researchers have looked at the biological activity of red pepper, which is well-known for its spiciness and capsaicin content [32]. Researchers looked at how cayenne pepper and capsaicin affected rats given a diet high in hydrogenated fat (40%). It is found that 5% cayenne pepper or its equivalent in capsaicin (15 mg%) added to the diet has a tendency to reduce liver and blood cholesterol levels [33].

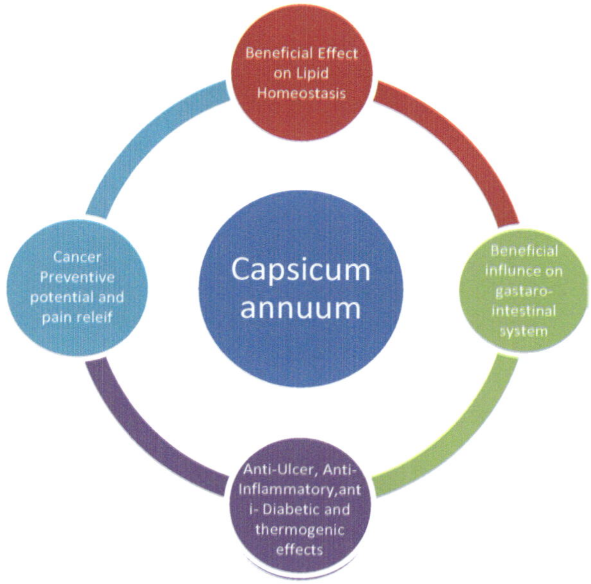

Fig. 33.5 Biological effects of *Capsicum annuum*

Also, rats given peanut butter containing 1.5, 3.0, or 15 mg of capsaicin had a 10% decrease in blood total cholesterol [34] (Fig. 33.5).

33.2.6.2 Anti-diabetic Potential

It has been demonstrated that the neuropeptide Substance P, which is released by capsaicin, may cure diabetes mellitus in rats [35], but the effect on insulin secretion seems to vary depending on the species. Capsaicin produces the neuropeptide substance P, which suppresses the release of insulin and alters blood sugar levels [35]. Additionally, research is being done on capsaicin as a possible treatment for type 1 diabetes prevention.

33.2.6.3 Thermogenic and Weight Reducing Influence of Capsaicin

Studies on animals and people show that ingesting capsaicin temporarily boosts heat production in the body. Capsaicin, according to claims, causes a shift in substrate oxidants from carbs to fats. Your risk of cardiovascular disease can be decreased by using this spice to substitute fat and salt in your diet (thereby making your food more pleasant). Therapeutic red pepper, or more specifically, its spiciness principle, which operates on capsaicin satiety and has a positive thermogenic effect, may have a significant impact on the prevalence and severity of obesity [36]. However, additional evidence is required to support this claim. Capsaicin increases satiety and lowers calorie and fat consumption when it is consumed orally and gastro-intestinally [37].

33.2.6.4 Capsaicin in Pain Relief

Capsaicin 0.025% or 0.075% was proven to be efficient and secure in two studies of individuals with osteoarthritis and rheumatoid arthritis. Additionally, post-herpetic neuralgia has been advised to be first treated with capsaicin [38]. An effective and risk-free substitute for systemically administered analgesics, which are frequently linked to possible adverse effects, is capsaicin 0.025% or 0.075% [39].

33.2.7 Medicinal Uses of Capsicum annum

33.2.7.1 Antioxidant Potential

Red Sweet or Bell pepper (Capsicum annum), among vegetables have become extremely known for their rich mixture of antioxidant content [40]. Exceptionally, have a high concentration of ascorbic acid as well as others also such as carotenoids, polyphenols and flavonoids (Fig. 33.6).

Among the other antioxidants, phytochemicals have a special place because of their free radical scavenging property [41]. Capsanthin, cryptocapsin and capsorubin (oxygenated carotenoids) are exclusive in this genus that is shown to be effective free radical scavengers. Capsaicin has various functions to show antidiabetic effects by reducing insulin resistance, decreasing total cholesterol and obesity as well as acting as a cancer-suppressing agent by its anti-inflammatory and antioxidant action and also by blocking various signaling pathways [42]. They are beneficial due to their protective role against several diseases such as cardiovascular, anemia, cancer and diabetes by performing their function on lipids by scavenging free radicals [43, 44]. It protects the body from oxidative damage and has an anticoagulant that helps to prevent blood clotting that can cause heart attacks [45]. The resulting benefits by the use of natural products that contain bioactive-rich substances have promoted growth in food industries.

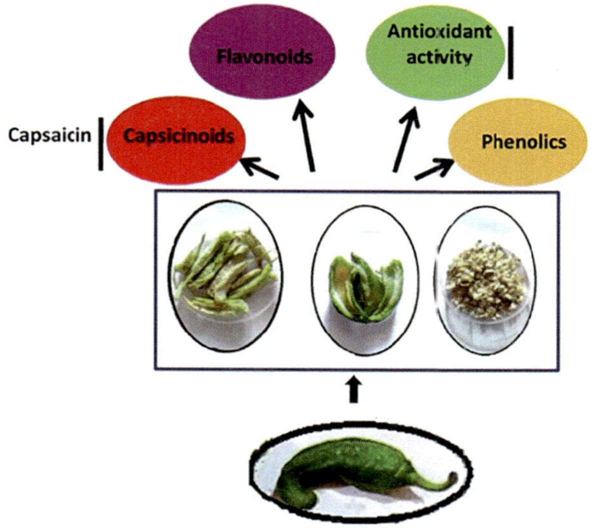

Fig. 33.6 Antioxidant component of capsaicin

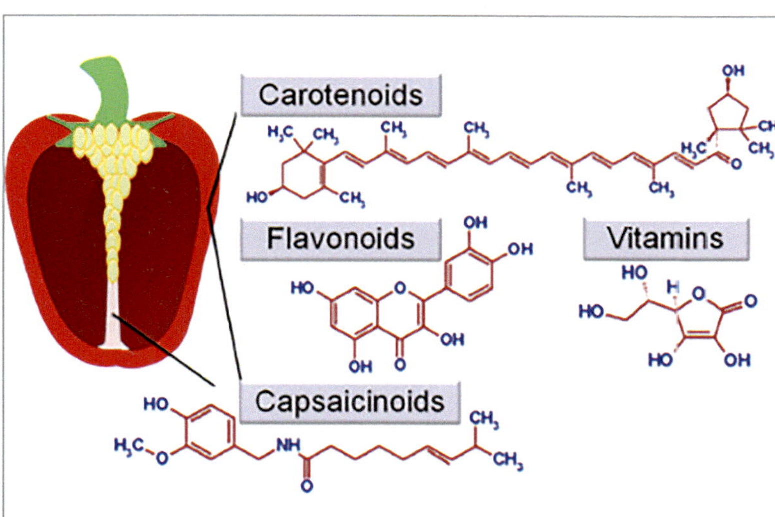

Fig. 33.7 Secondary metabolites of capsicum species

33.2.8 Source of Vitamins

Among the mechanism of protection against free radicals, antioxidant vitamins A, C, E and β -carotene for its richness, have a special place in pharmaceutical industries, in the prevention of illness [46] (Fig. 33.7).

Within the cell, there are two lines of antioxidant defenses. In fat-soluble cellular membranes, the first line is found which consists of vitamin E, coenzyme and β-carotene [45, 47]. Among these, vitamin E is known as the most potent chain-breaking antioxidant within the cell membrane. Inside the cell, various water-soluble scavengers are present which include vitamin C, superoxide dismutase (SD) and glutathione peroxidase [48]. It is used to treat osteoarthritis [49]. Extremely exceptional levels of vitamin C in capsicum annum are helpful for the immune system and also increase the production of white blood cells to protect against infection.

33.2.8.1 Antidiabetic Activity

The inhibition of intestinal glucose found by the crude extraction of fruit is considered to lower the level of blood sugar [50]. The regular intake of chili reduces postprandial insulin. The inhibitory activities of α-amylase, α-glucose of the capsicum were determined to evaluate the antidiabetic activity of the Capsicum annum against the α-glucosidase, α-amylase and as well as ACE (angiotensin-converting-enzyme) inhibitor [51, 52] (Fig. 33.8).

Baek J, Lee et al., (2013) evaluated capsicum annum's antidiabetic activity and highlighted the antidiabetic activity of chili in review [53].

Tundis Rosa et al., (2013), for the four Capsicum annum species, screened the hypoglycemia potential and how the maturity stages influence the biological activity. They screened the lipophilic protein and the total extract separately in terms of identification of the phytochemicals which are responsible for this activity. Among these extracted samples, the most promising activity was observed of the immature Capsicum annum fiesta with an IC50 value of 47.8 mg/ml [54].

On the contrary, in case of the Capsicum annum orange, interesting activity was observed. According to Loizzo, Tundis, and Menichini (2008), Capsicum annum Acuminatum in its immature stage of ripening exhibited good inhibitory activity against both enzymes with IC50 in various ranges for a-amylase and a-glucosidase, respectively. On the other side, the mature stage was observed inactive on both enzymes [55, 56]. Anthon OE et al., (2013) investigated the improvements in blood glucose levels in Wistar rats that were suffering from diabetes and also observed improvements in their body weight [57].

33.2.8.2 Anti-cancer Activity

By investigation in various types of cancer, capsaicin showed anticancer activity [58, 59]. In both vitro and vivo study, capsaicin has been investigated to show an effectual response in growing prostate cancer cells that results in apoptosis in androgen receptors of both negative as well as positive prostate cancer cells that connected with elevation of antibodies.

The anti-cancer effect is also shown by the capsaicin in colorectal carcinoma and studied this effect on SW480, Colo 205, HCT 116 and LoVo cells that induced

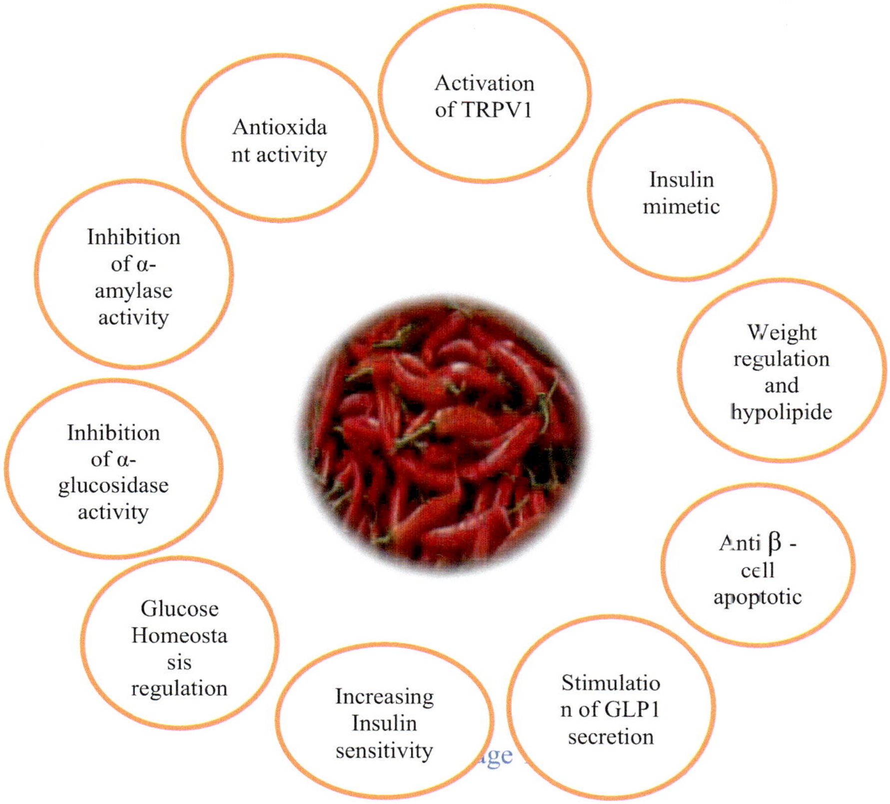

Fig. 33.8 Antidiabetic effect of *Capsicum annum*

anti-tumorigenesis, autophagy, TCF dependent signaling and deregulation of B-catechin [60, 61]. Capsaicin also shows anti-cancer activity in human breast cancer which was investigated on T47D, MDA-MB231, and BT-474 cell lines to result in apoptosis and into the G2/M phase to arrest the cell cycle and mitochondrial dysfunction [62, 63]. The anticancer activity was also studied in the recovery of human myeloid leukemia on U937, THP-1 cells line to elevate the apoptosis by regulating the calcium-MKII-Sp1 route [64] (Fig. 33.9).

In the case of melanoma, capsaicin enhances the apoptotic effect and inhibited the growth of the cells on A375 cell lines [65]. It was also found that capsaicin arrests the cell cycle at the phase of G0/G1 and results in apoptosis to cure esophageal epidermoid cancer [66]. On the PANC-1 cell line, the anticancer activity was successfully investigated against pancreatic cancer where it arrests the growth of tumor cells by inducing apoptosis as well as to arrest the G0/G1 phase of the cell cycle [67, 68].

On the HepG2 cell line, capsicum also inhibited the human hepatoma with the decline in cell viability, inducing autophagy, activated caspase-3, apoptosis and by

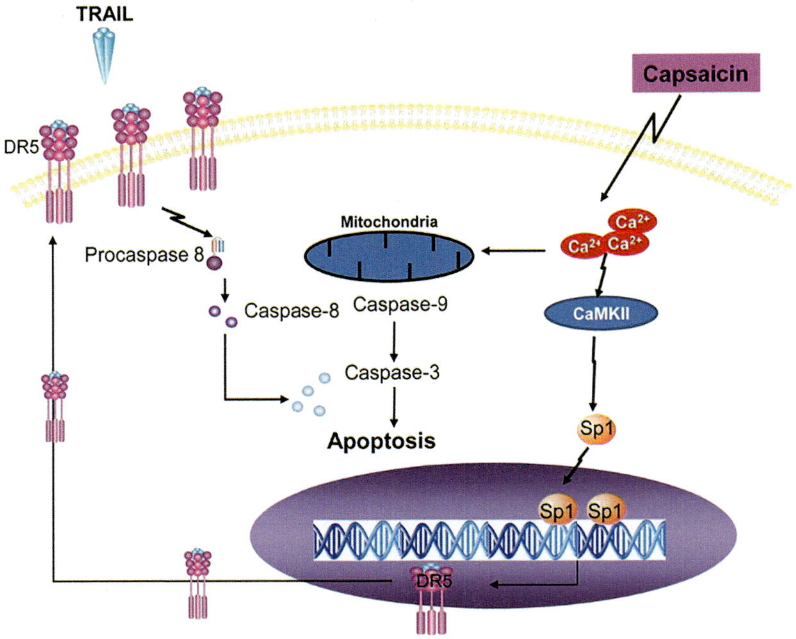

Fig. 33.9 Calcium-MPKII-SP1 route

generating ROS. By regulating the pathway of calciumCaMKII-Sp1 on Hep3B cell lines, capsaicin enhances the apoptotic effects [71]. On the cell lines NPC-TW 039, capsicum helped in the recovery of human nasopharyngeal carcinoma by inducing apoptosis, to arrest the cycle at the phase of G0/G1 and elevated the level of cytosolic calcium [69].

The study by Lin CH et al., (2013), investigated the treatment of epithelial carcinoma cells KB cells by capsaicin to result in a decline in cell death and as well as in cell proliferation through a dose-dependent manner. By the analysis of the cell cycle, it was represented that the exposure of capsaicin to KB cells, results in arrest at the G2/M phase of the cell cycle. The effectiveness of the capsicum was also studied against gastric cancer where it modulates MAPK signaling and induces apoptosis [70, 71].

33.2.8.3 Antiviral Activity

Capsicums are known to be prosperous in chemicals that show an influential response as opposed to various viruses, such as Cis-Capsaicin which shows potent response to HSV infection. A study confirmed the potent response of Cis-Capsaicin in the inhibition of the viral replication cycle. However, in another study, capsaicin showed effects on sensory neurons which play role in the persistence, and spreading of the herpes simplex virus (HSV) infection. Pereeira JAP et al., (2016) also determined the antiviral activity by capsicum [72].

33.2.8.4 Anti-fungal Activity

From the C. frutescens, a novel saponin as CAY-1 was isolated and shown to be an effective response against 16 various fungal strains that performed function by disrupting the membrane and as well as the cell integrity of the fungi [73]. An antifungal response of capsicum leaf (aqueous) was studied in opposition to various types of fungi such as Rhizopus, Aspergillus flaus, Penicillium species. The minimum fungicidal concentration (MFC) and minimum inhibitory concentration of C. frutescens extracts were evaluated. By comparing the value, it was observed that the value of MIC (minimum inhibitory concentration) is lowered in comparison to the leaf extract [74].

Another study reveals, isolated peptides and extracts have shown an effective response against various microbes such as fungi [75]. A study also reported the *Capsicum annuum* activity of anti-fungal in response to Arbuscular mycorrhizal [72] and also investigated the anti-fungal activity of *Capsicum annuum* by another study [57].

33.2.8.5 Respiratory Agents

In clinical and human pharmacological research, capsaicin is treated as cough reflex sensitivity in a mechanism of testing [76]. Capsaicin reduced allergic symptoms of nasal and as well as desensitized the nasal mucosa and also reduced pain in humans.

The Capsaicin, respiratory effects include the stimulation of the cough reflex through the broncho constriction and sensory neurons. Capsaicin causes pulmonary edema and laryngeal, chemical pneumonitis that is infrequent [77].

33.2.8.6 Effect on Cornea and Conjunctiva

An oily extract such as oleoresin was isolated in spray form by capsicum that results in burning or stinging, an increase of tear secretion, temporary blindness, eye pain, burning in mouth nose, corneal, sneezing abrasion, choking sensation, difficulties in the breathing, runny nose, and as well as asthma with bronchoconstriction in patients.

The other effects of it include eczema, rashes, erythema and dermatitis on the skin irritating area, blisters, headaches, dizziness and vesicles for the exposure, vomiting, pulmonary edema, chest pain, critical respiratory failure, motor control loss and hypotension.

Oleoresin spray of capsicum affects the cornea structure formation as well as sensitivity and reduces the production of aqueous humor [78].

33.2.8.7 Anti-arthritis Activity

The arthritis development is effectively controlled by the ethanolic extract of C. annum. It was observed that the decline in AIA (Adjuvant-Induced Arthritis) in the leaf of Capsicum Annum (hot pepper), when treated in mice represented through the decrease in CRP, level of cytokines, and ESR [79].

A topical formulation of nanovesicle was formed through the semi-purified extract of capsaicinoids extract which indicated better activity of anti-arthritis in the model of rat and also a decline in joint pain and swelling. The formulation of nanovesicle represented the acceptance and better tolerability of both human as well as animal models [80]. For arthritis treatment, a paste of leaves is locally applied [81].

33.2.8.8 Hepatoprotective Activity

Hepatoprotective activity was shown by the capsaicin against the CCl4 which cause injury to the liver in rats. It shows a defensive effect in the liver as well as in the lungs by nourishing the defense system of the pulmonary antioxidant enzyme [82].

33.2.8.9 Analgesic Response

An aqueous extract of capsicum frutescens L. and capsaicin has dose-dependent, significant central analgesic and peripheral analgesic properties that exhibit anti-inflammatory characteristics on induced pain either thermally, mechanically or chemically [83].

Capsaicin demonstrated a potent response at vanilloid receptors to show benefits in uncontrolled pain in diabetic neuropathy and trigeminal neuralgia [84].

A literature survey revealed the analgesic effect of the use of capsaicin in the mechanism of spinal. In pre-treated rats, Ojewole JA, (2002) described the protection from thermal stimuli [85].

33.2.8.10 Anthelminthic Activity

Kamal ATMM et al., (2015) carried out the activity of anthelmintic by using the Capsicum frutescens L. methanolic extract. At two different stages, the anthelmintic activity was observed in the worm in their 'death time' and 'paralysis time' A paralysis time was determined when zero moment observation of any sort happens except the worm's vigorous shaking. The conclusion was ended to death, with the passage, fading of their body color results in loss of their motility. Similarly, the paralyzing time of Tubifex was found in the case of C. frutescens.

33.2.8.11 Anti-obesity Effect

Capsaicin shows effective potential in terms of anti-obesity activity by various mechanisms such as reducing the fat intake that results from a decrease in weight, induction of energy expenditure, sensation, and lipolysis in the adiponectin.

Adiponectin receptor and its expressions are also elevated by the capsaicin (dietary) in diabetic mice which decreased metabolic dysregulation. Due to its dual functionality, capsaicin represents effects in the liver and the adipose tissue. The adipose tissues are served with capsaicin and the protein which is involved in lipid metabolism is altered.

Some effects are also shown by the Capsaicin to influence apoptosis, and adipogenesis in adipocytes. With capsaicin consumption, energy expenditure increases by activating the brown adipose tissue which results in increasing the oxidation of fats and improving lipolysis [86]. The various varieties of *Capsicum annuum* L. aqueous extracts showed an anti-obesity effect by evaluating the lipoprotein lipase (LPL) messenger RNA expression level in 3T3-L1 cells (mouse pre-adipocytes).

33.2.8.12 Cardiovascular Effects

Many studies have shown the beneficial effects of capsaicinoids on the cardiovascular system, to cure atherosclerosis, hypertension, and ischemic heart disease [87]. In those rabbits which are anesthetized, in them, capsaicin is introduced by the intravenous injection leads to hypotension whereas in dogs results in an elevation in blood pressure [88]. Capsaicin prevented the elevated level of cholesterol in the liver when female albino rats are treated with meals rich in cholesterol.

33.2.8.13 Anti-ulcer Activity

In the gastric mucosa of rats, capsaicin showed a defensive response to lesion formation. Long-term chili intake protects against hemorrhagic shock induced by gastric mucosal injury in rats [89]. Capsaicin also protects against mucosal micro-bleeding and is potent for Helicobacter pylori elimination which results in damage to the mucosa [90].

Liquid capsicum frutescens extract response was investigated on the rats which are healing from the acute gastric ulcer. The results proved that the oral administration of aqueous extract results in the gastric ulcer undergoing a decline in length and gastric juice decrease in volume, also showing improvement in the histopathological alteration [91].

Capsaicin pretreatment results in to decrease in the stomach acid secretion level which is brought out by histamine. A study reveals capsaicin in little amounts, which is provided through intragastric intubation that impedes stomach acid secretion in the human [92].

33.2.8.14 Anticoagulant Activity

Therapeutically, without any variations, capsaicin is treated to cure thromboembolism in platelets [93]. It impeded the accumulation of platelets, as well as regulated the clotting activity of its factors such as VIII and IX as characteristics to decrease cardiovascular diseases. It has proved that the ability of capsaicin can move across the platelet's plasma membrane and can change the membrane fluidity [94].

Previous studies have revealed that from intracellular platelet stores, capsaicin influences the release of calcium and is put up towards thrombin-induced-platelet-activation because of TRPV1, present in the platelets of humans [95]. An extract of Capsicum frutescens extracts showed synergistic activity with streptokinase on the thrombolysis [96]. Other work also supported the anti-coagulant response of capsaicin [97].

33.2.8.15 Dermatological Conditions

Topical application of capsaicin is known as an inhibitor of cutaneous vasodilation which provides eases and comfort on either average or serious psoriasis. Importantly a decline in erythema and scaling was observed.

In the medication study, it was observed that almost half of the patients showed symptoms such as skin burning, redness, stinging, and itching at the beginning of application of capsaicin but applying continually lessened the symptoms and results beneficial for the cure of psoriasis [98].

Acute lipodermatosclerosis was also cured by capsaicin and also used in pregnant women to treat acute lobular panniculitis [99].

33.2.8.16 Pruritus

Effectively, capsaicin is used in the treatment of skin lesions. Topical capsaicin is effectively used to serve the pruritus having association with psoriasis.

By treating with capsaicin for 24 h, it was found that lesional skin showed a decline in perfusion by 15%, also attributed by the PRP (Pityriasis Rubra Pilaris) as skin redness, scaling, also variations in pruritus [100].

Also observed, is a patient having PRP at a high level, capsaicin is used for the treatment purpose which results in comfort. The prurigo nodularis treatment is also done by capsaicin, which results due to the discharge of nodules either by lichenified or scratched [101].

33.2.8.17 Rhinitis

Snider M. (1992) conducted a study in which it was found that rhinitis, congestion and sneezing attenuated in those patients that are taking capsaicin nasal sprays repeatedly. An effective response was shown by the intranasal capsaicin, in a placebo-controlled study and the nasal symptomatology was reduced in non-allergic, non-infectious, perennial rhinitis after treatment of 9 months without any response in cellular homeostasis [102].

Another study showed the activity of capsicum against rhinitis but for an elderly patient, it was not effective [103]. Another study also investigated capsicum activeness in response to rhinitis [104].

33.2.9 Cough Challenge

In clinical research, the widespread use of capsaicin has been shown by red pepper pungent extract due to its influence on the cough in either a reproducible or dose-dependent manner [105]. Red pepper of the Solanaceae family is a fruit spice, rich in lipids, proteins, carbohydrates, and vitamins, and also has healthy phytochemicals as such flavonoids, carotenoids, and capsaicinoids which show effective responses in the prevention from coughs, sore throats and asthma. Some studies also investigated the capsicum response activity [106].

33.2.9.1 Memory Enhancing Activity

Green pepper is also referred as a favorable memory enhancer. A green pepper mechanism depends on,

1. Free radicals scavenging
2. Memory improvement
3. Inhibition of acetylcholinesterase enzyme
4. Ireversal of memory deficits

There was a significant rise in glutathione levels in the brains [107]. Another study reveals that Capsanthin showed remarkable improvement in memory acquisition [108].

33.2.9.2 Immuno-Modulation

The extract of *Capsicum annuum* showed immunological effects and its pungent capsaicin was investigated in vitro and as well as in ex vivo in PP (Peyer's patch) cells. By directly provide of the capsicum extract, capsaicin resulted in the suppression of interferon-gamma, interleukin (IL)-2, and the production of IL-4, and IL-5.

After extract oral administration in ex vivo, by using PP cells removed for four successive days, IFNgamma, IL-2,-5 elevated in oppose Con A (concanavalin A). Providing extract orally also enhanced the production INFgamma, IL-2, -4 in opposition to Con A, while no response was shown on IL-5 production. It revealed dendritic cells have the receptor for immune response and show powerful immune consequences. Additionally, the administration of CAE elevates the activity Nuclear factor kappa light chain (NF-kB) in the lungs [109].

Table 33.5 Hypoglycemia activity IC50(mg/ml) of pepper fruits at two ripening stages [53]

		Total extract		Lipophilic fraction	
	Maturity stage	α-Amylase	α-glucosidase	α-Amylase	α-glucosidase
C. annuum cultivars					
Fiesta	I	47.8 ± 1.5**	109.2 ± 1.2**	9.1 ± 0.2**	>1000
	M	>1000	>1000	20.6 ± 0.8**	>1000
Orange Thai	I	98.7 ± 3.9**	102.5 ± 3.5**	28.6 ± 0.7**	>1000
	M	437.9 ± 7.2**	166.5 ± 3.4**	30.1 ± 1.4**	>1000
Acuminatum	I	>1000	>1000	16.1 ± 0.9**	>1000
	M	118.1 ± 1.4**	71.5 ± 1.4**	28.8 ± 1.2**	>1000
Cayenne	I	129.6 ± 1.8**	81.1 ± 1.2**	15.3 ± 0.8**	>1000
Golden	M	222.8 ± 2.6**	63.6 ± 1.1**	23.5 ± 1.3**	>1000
Acarbose		50.0 ± 0.9	35.5 ± 1.2		

** p-value<0.05

33.2.9.3 Pharmacological Action of *Capsicum annum*

The oil of pepper seed holds a rich quantity of antioxidants such as vitamin E, unsaturated fatty acids and carotenoids [110]. The main constituent of capsicum annum is capsaicin which is responsible for the intense sensation of burning whenever it comes in contact with eyes, mucous membranes or skin [111]. Peppers are the most consumed spices in the world as they are used all over the world in salads, curries and sauces. Powdered dry chilies are used in culinary preparations while the capsicum leaves are added in soups and are much popular in Southeast Asia [112]. C. annum is much important concerning human health as it prevents heart diseases, high cholesterol, and excessive blood clotting [113]. Pharmaceutical activities of Capsicum are listed with mechanism and description in Table 33.5.

Pharmaceutical actions of Capsicum	Description	Mechanism
Anticancer	Useful in the treatment of Breast cancer Lung cancer Liver cancer	Suppresses cell multiplication and ROS production through modulating FOXO3a-mediated pathways In lung cancer, restriction in cell proliferation occurs through E2F pathway In liver cancer, Inhibition of migration, invasion and adhesion occurs to combat this disease
Antidiabetic	Has good antidiabetic effect as it contains α-glucosidase and α-amylase inhibitors, which are important for the degradation of disaccharides and polysaccharides	Red pepper seed extraction (RPSE) has much potentail to improve glycemic control by inhibiting hepatic gluconeogenesis by the phosphorylation of AMPK and FOXO1 in mice [114]

Pharmaceutical actions of Capsicum	Description	Mechanism
Antimicrobial	Much helpful against fungal and bacterial infections	Carbopol-based capsaicin increases the antifungal and antibacterial effects against Bacillus cereus, Escherichia coli, Salmonella typhi, Staphylococcus aureus, Candida albicans, Candida parapsilosis, Candida tropicalis and Candida albicans [115]
Antioxidant	It inhibits the peroxidation of lipids in erythrocyte membranes in humans	Hypotriglyceridemic effect produce by the 0.015% capsaicin in the diet plays an effective role in reducing oxidant stress which occurs due to countering of depleted antioxidant enzymes in erythrocytes. It is also effective in the reduction of lipid peroxide content Reduced hepatic total thiols in the situation of hypercholesterolemic were partially improved by dietary capsaicin treatment Decreased activities of hepatic antioxidant enzymes – GSH-transferase, GSH-reductase, superoxide dismutase and catalase in hypercholesterolemic rats were effectively opposed by the dietary capsaicin [20]
Neuroprotective	Capsicum has high potential to prevent neurodegenerative diseases in the medical and food industry	*Capsicum annuum* var. abbreviatum extract (CAAE) has high neuroprotective activity as it restrains the signaling of mitochondrial-related apoptosis, includes the modulation of bcl-2 families, activates the signaling pathway of Nrf2/HO-1 and inhibits caspase cascades [116]
Hepato-protective	Prevents liver injury	Capsaicin in CA provides hepatic protection against lipopolysaccharide, CCl4, and iron induced hepatotoxicity The release of IL-6 and TNF-α is promoted by kupffer cells when they are treated with ETOH. Decreased level of TNF-α and IL-6, when co-treatment with the extract of CA serves as hepatoprotective agent against hepatic lesions [117]
Ulcer-protective	Prevents gastric mucosal injury Inhibits gastric acid secretion in stomach	RCP extract is much helpful for the cure of gastric ulcer disease as it lowers the pH of gastric juice Capsaicin has an anti-ulcer effect on such stomachs which are infected by H. pylori. Capsaicin shows an active response towards the chemicals that are secreted by the stomach when it is infected by H. pylori infection [118]

Pharmaceutical actions of Capsicum	Description	Mechanism
Hypoglycemic	Capsicum annum is much efficient in the resistance of hyperglycaemia due to the action of phenolic group	The hypoglycaemic capacity of Cabsicum. annuum extracts was assessed by in vitro α-glucosidase and α-amylase inhibition analysis α-glucosidase and α-amylase inhibitors present in capsicum are much helpful for anti-hypoglycemic activity [119]
Antibacterial	Increases the percentage of wound healing Suppresses infection	Antibacterial properties of *C. annuum* extract were attributed to its total tannin, phenolic and flavonoid contents The inhibition of the H+ proton/ATPases pump, biofilm formation, and dehydrogenase activity serves as a protection against bacteria The cell membrane of bacteria is altered via the leakage of reducing sugars, nucleic acids, and proteins. This alteration also helps in the antibacterial activity of C. annum [120]
Antiaging	Highly used in cosmetic industry as it prevents aging Have good resistance against damaging effects of free radicals Widely used in cosmetic industry due to its color, flavor, fragrance, and skin care qualities	Anthocyanins present in peppers act as UV protectors and protects the skin from wrinkles and aging Fight against solar radiations and free radicals Vitamin C in capsicum is an efficient guard against reactive oxygen species (ROS) which causes oxidative damage of skin Inhibits the production of melanin [19]

33.2.10 Uses of **Capsicum annum** *as a Home Remedy*

Home remedy	Uses
Antiobesity effect	Reduces fat intake Gives sensation of satiety Reduces the metabolic deregulation in obese Reduces energy intake Capsaicin triggers TRPV1 which increases intracellular Ca^{2+} levels and activates the sympathetic nervous system which in result helps in the weight loss
Analgesic activity	Capsaicin helps in the relief of pain caused by dysfunction of CNS Much helpful in thermoregulation [121]
Boosts immunity	Vitamin A in chili maintains the health of mucus membrane Capsaicin enhances the defense mode of body β- Carotene boosts the immune system

Home remedy	Uses
Anti-rhinitis agent	Combat chronic sinus infections Tackles nasal congestion and clears the mucus of nose
Aromatherapy	Essential oil of chili seed is used in aromatherapy which improves physical and psychological well-being [122]
Cough sensitivity	Capsaicin is highly useful in the analysis of cough suppressants, like guaifenesin and benzonatate Capsaicin effects the nasal mucosa and lowers the allergic signs of nasal pain caused by agents [123]

33.3 Conclusions

The chili pepper, with its diverse morphological traits and significant properties plays an extremely important role in preventing and treating many diseases and also functions in several health benefits, specifically antidiabetic, anticancer, and antiviral activities, as well as having anti-obesity, cardiovascular, and immunomodulatory effects. After detailed research on the potential of capsaicin which is the main flavoring component and capsaicinoids which has significance in medical field, the importance of this crop and its commercial value has been increased many folds. The extracted components of chili which are used as pain reliever, attained great attention in medical field. Many researchers suggest that capsaicin because of having antioxidant, anticancer and anti-inflammatory properties are providing enormous therapeutic benefits. However for validating the effectiveness of capsaicinoid, there is a need of in vitro and in vivo analysis. Moreover, the components which give pungency and flavor will be controlled by using molecular agriculture which in turn provides more attractive benefits to human as well as for industrial consumption. Apart from its use in food, researchers are trying to add components extracted from this crop, in pharmaceutical and nutaceuticals products.

References

1. Davis, C. B., et al. (2007). Determination of capsaicinoids in habanero peppers by chemometric analysis of UV spectral data. *Journal of Agricultural and Food Chemistry, 55*(15), 5925–5933.
2. Mueller-Seitz, E., Hiepler, C., & Petz, M. (2008). Chili pepper fruits: Content and pattern of capsaicinoids in single fruits of different ages. *Journal of Agricultural and Food Chemistry, 56*(24), 12114–12121.
3. Onyenekwe, P. C., & Ogbadu, G. H. (1995). Radiation sterilization of red chili pepper (Capsicum frutescens). *Journal of Food Biochemistry, 19*(2), 121–137.
4. Belletti, P., Marzachi, C., & Lanteri, S. (1998). Flow cytometric measurement of nuclear DNA content in capsicum (Solanaceae). *Plant Systematics and Evolution, 209*, 85–91.
5. Howard, L. R., et al. (1994). Provitamin A and ascorbic acid content of fresh pepper cultivars (Capsicum annuum) and processed jalapenos. *Journal of Food Science, 59*(2), 362–365.

6. Huang, M. T., Ferraro, T., & Ho, C. T. (1994). Cancer chemoprevention by phytochemicals in fruits and vegetables: an overview. ACS Publications, 546.
7. Havsteen, B. (1983). Flavonoids, a class of natural products of high pharmacological potency. *Biochemical Pharmacology, 32*(7), 1141–1148.
8. Davenport, L. J. (2004). Genera Solanacearum: The genera of Solanaceae illustrated, arranged according to a new system by Armando T. Hunziker. *Systematic Botany, 29*(1), 221–222.
9. Basu, S. K., & De, A. K. (2003). Capsicum: Historical and botanical perspectives. In *Capsicum* (pp. 21–35). CRC Press.
10. Davenport, W. A. (1970). Progress report on the domestication of Capsicum (chili peppers). In Proceedings of the Association of American Geographers (Vol. 2, pp. 46–47). Association of American Geographers.
11. Walsh, B. M., & Hoot, S. B. (2001). Phylogenetic relationships of capsicum (Solanaceae) using DNA sequences from two noncoding regions: The chloroplast atpB-rbcL spacer region and nuclear waxy introns. *International Journal of Plant Sciences, 162*(6), 1409–1418.
12. Tripodi, P., & Kumar, S. (2019). The capsicum crop: An introduction. In *The capsicum genome*, Springer, Cham, pp. 1–8.
13. Radford, A. E. Dickson, W. C., Massey, J. R., Bell, C. R. (1974). Vascular Plant Systematic. Harper and Row Publishers. New York: P 891.
14. Zhigila, D. A., et al. (2014). Fruit morphology as taxonomic features in five varieties of Capsicum annuum L. Solanaceae. *Journal of Botany, 2014*, 1.
15. Zhigila, D. A., et al. (2014). Palynomorphs and floral bloom as taxonomic characters in some species of the genus Amaranthus l. (Aamaranthaceae). *Bayero Journal of Pure and Applied Sciences, 7*(2), 164–168.
16. Olorode, O. (1984). Taxonomy of West African flowering plants. Longman group limited, 117.
17. Stewart, C., Jr., et al. (2007). Genetic control of pungency in C. chinense via the Pun1 locus. *Journal of Experimental Botany, 58*(5), 979–991.
18. del Rocío Gómez-García, M., & Ochoa-Alejo, N. (2013). Biochemistry and molecular biology of carotenoid biosynthesis in chili peppers (Capsicum spp.). *International Journal of Molecular Sciences, 14*(9), 19025–19053.
19. Baenas, N., et al. (2019). Industrial use of pepper (Capsicum annum L.) derived products: Technological benefits and biological advantages. *Food Chemistry, 274*, 872–885.
20. Srinivasan, K. (2016). Biological activities of red pepper (Capsicum annuum) and its pungent principle capsaicin: A review. *Critical Reviews in Food Science and Nutrition, 56*(9), 1488–1500.
21. Wesołowska, A., Jadczak, D., & Grzeszczuk, M. (2011). Chemical composition of the pepper fruit extracts of hot cultivars Capsicum annuum L. *Acta Scientiarum Polonorum Hortorum Cultus, 10*(1), 171–184.
22. Davies, B. H., Matthews, S., & Kirk, J. T. O. (1970). The nature and biosynthesis of the carotenoids of different colour varieties of Capsicum annuum. *Phytochemistry, 9*, 797–805.
23. Verma, A. K., et al. (1988). Inhibition of 7, 12-dimethylbenz (a) anthracene-and N-nitrosomethylurea-induced rat mammary cancer by dietary flavonol quercetin. *Cancer Research, 48*(20), 5754–5758.
24. Yoshida, M., et al. (1990). The effect of quercetin on cell cycle progression and growth of human gastric cancer cells. *FEBS Letters, 260*(1), 10–13.
25. Yoshimoto, T., et al. (1983). Flavonoids: Potent inhibitors of arachidonate 5-lipoxygenase. *Biochemical and Biophysical Research Communications, 116*(2), 612–618.
26. Smith, T. J., & Yang, C. S. (1994). *Effects of food phytochemicals on xenobiotic metabolism and tumorigenesis*. ACS Publications.
27. Hertog, M. G. L., Hollman, P. C. H., & Katan, M. B. (1992). Content of potentially anticarcinogenic flavonoids of 28 vegetables and 9 fruits commonly consumed in The Netherlands. *Journal of Agricultural and Food Chemistry, 40*(12), 2379–2383.

28. Howard, L. R., et al. (2000). Changes in phytochemical and antioxidant activity of selected pepper cultivars (capsicum species) as influenced by maturity. *Journal of Agricultural and Food Chemistry, 48*(5), 1713–1720.
29. Barboza, G. E., & De Bem Bianchetti, L. (2005). Three new species of capsicum (Solanaceae) and a key to the wild species from Brazil. *Systematic Botany, 30*(4), 863–871.
30. Marini, E., et al. (2015). Antimicrobial and anti-virulence activity of capsaicin against erythromycin-resistant, cell-invasive group a streptococci. *Frontiers in Microbiology, 6*, 1281.
31. Poulos, J. M. (1994). Pepper breeding (Capsicum spp.): achievements, challenges and possibilities. In Plant Breeding Abstracts (No. AVRDC Staff Publication).
32. Surh, Y.-J., et al. (1995). Chemoprotective effects of capsaicin and diallyl sulfide against mutagenesis or tumorigenesis by vinyl carbamate and N-nitrosodiinethylamine. *Carcinogenesis, 16*(10), 2467–2471.
33. Sambaiah, K., Satyanarayana, M. N., & Rao, M. V. L. (1978). Effect of red pepper (chillies) and capsaicin on fat absorption and liver fat in rats. *Nutrition Reports International, 18*(5), 521–529.
34. Srinivasan, M. R., et al. (1980). Influence of red pepper and capsaicin on growth, blood constituents and nitrogen balance in rats. *Nutrition Reports International, 21*(3), 455–467.
35. Tsui, H., et al. (2007). 'Sensing' autoimmunity in type 1 diabetes. *Trends in Molecular Medicine, 13*(10), 405–413.
36. Kawada, T., Hagihara, K.-I., & Iwai, K. (1986). Effects of capsaicin on lipid metabolism in rats fed a high fat diet. *The Journal of Nutrition, 116*(7), 1272–1278.
37. Westerterp-Plantenga, M. S., Smeets, A., & Lejeune, M. P. G. (2005). Sensory and gastrointestinal satiety effects of capsaicin on food intake. *International Journal of Obesity, 29*(6), 682–688.
38. Bernstein, J. E., et al. (1989). Topical capsaicin treatment of chronic postherpetic neuralgia. *Journal of the American Academy of Dermatology, 21*(2), 265–270.
39. McCarthy, G. M., & McCarty, D. J. (1992). Effect of topical capsaicin in the therapy of painful osteoarthritis of the hands. *The Journal of Rheumatology, 19*(4), 604–607.
40. Simonne, A. H., et al. (1997). Ascorbic acid and provitamin a contents in unusually colored bell peppers (Capsicum annuum L.). *Journal of Food Composition and Analysis, 10*(4), 299–311.
41. Nadeem, M., et al. (2011). Antioxidant potential of bell pepper (Capsicum annum L.)-a review. *Pakistan Journal of Food Science, 21*(1–4), 45–51.
42. Sanati, S., Razavi, B. M., & Hosseinzadeh, H. (2018). A review of the effects of Capsicum annuum L. and its constituent, capsaicin, in metabolic syndrome. *Iranian Journal of Basic Medical Sciences, 21*(5), 439.
43. Ivorra, M. D., Paya, M., & Villar, A. (1989). A review of natural products and plants as potential antidiabetic drugs. *Journal of Ethnopharmacology, 27*(3), 243–275.
44. Zhang, S., et al. (2017). Capsaicin reduces blood glucose by increasing insulin levels and glycogen content better than capsiate in streptozotocin-induced diabetic rats. *Journal of Agricultural and Food Chemistry, 65*(11), 2323–2330.
45. Matsufuji, H., et al. (2007). Anti-oxidant content of different coloured sweet peppers, white, green, yellow, orange and red (Capsicum annuum L.). *International Journal of Food Science & Technology, 42*(12), 1482–1488.
46. Anaya-Esparza, L. M., et al. (2021). Bell peppers (Capsicum annum L.) losses and wastes: Source for food and pharmaceutical applications. *Molecules, 26*(17), 5341.
47. Kono, N., & Arai, H. (2015). Intracellular transport of fat-soluble vitamins A and E. *Traffic, 16*(1), 19–34.
48. Wheeler, C. R., et al. (1990). Automated assays for superoxide dismutase, catalase, glutathione peroxidase, and glutathione reductase activity. *Analytical Biochemistry, 184*(2), 193–199.
49. Elwan, H., et al. (2020). Efficacy of dietary supplementation with Capsicum annum L on performance, hematology, blood biochemistry and hepatic antioxidant status of growing rabbits. *Animals, 10*(11), 2045.

50. Gulati, V., Harding, I. H., & Palombo, E. A. (2012). Enzyme inhibitory and antioxidant activities of traditional medicinal plants: Potential application in the management of hyperglycemia. *BMC Complementary and Alternative Medicine, 12*(1), 1–9.
51. Oboh, G., et al. (2013). Inhibition of α-amylase and α-glucosidase activities by ethanolic extract of Amaranthus cruentus leaf as affected by blanching. *African Journal of Pharmacy and Pharmacology, 7*(14), 1–7.
52. Oboh, G., Akinyemi, A. J., & Ademiluyi, A. O. (2012). Antioxidant properties and inhibitory effect of ethanolic extract of struchium sparganophora (ewuro odo) leaf on α-amylase and α–glucosidase activities. *African Journal of Traditional, Complementary and Alternative Medicines, 9*(3), 342–349.
53. Baek, J., et al. (2013). Inhibitory effects of Capsicum annuum L. water extracts on lipoprotein lipase activity in 3T3-L1 cells. *Nutrition Research and Practice, 7*(2), 96–102.
54. Tundis, R., et al. (2013). Antioxidant and hypoglycaemic activities and their relationship to phytochemicals in Capsicum annuum cultivars during fruit development. *LWT-Food Science and Technology, 53*(1), 370–377.
55. Loizzo, M. R., et al. (2008). In vitro inhibitory activities of plants used in Lebanon traditional medicine against angiotensin converting enzyme (ACE) and digestive enzymes related to diabetes. *Journal of Ethnopharmacology, 119*(1), 109–116.
56. Menichini, F., et al. (2009). The influence of fruit ripening on the phytochemical content and biological activity of Capsicum chinense Jacq. cv Habanero. *Food Chemistry, 114*(2), 553–560.
57. Parvez, G. M. M. (2017). Current advances in pharmacological activity and toxic effects of various capsicum species. *International Journal of Pharmaceutical Sciences and Research, 8*(5), 1900–1912.
58. Luo, X.-J., Peng, J., & Li, Y.-J. (2011). Recent advances in the study on capsaicinoids and capsinoids. *European Journal of Pharmacology, 650*(1), 1–7.
59. Mori, A., et al. (2006). Capsaicin, a component of red peppers, inhibits the growth of androgen-independent, p53 mutant prostate cancer cells. *Cancer Research, 66*(6), 3222–3229.
60. Yang, J., et al. (2013). Low-concentration capsaicin promotes colorectal cancer metastasis by triggering ROS production and modulating Akt/mTOR and STAT-3 pathways. *Neoplasma, 60*(4), 364–372.
61. Lee, S.-H., et al. (2012). Capsaicin represses transcriptional activity of β-catenin in human colorectal cancer cells. *The Journal of Nutritional Biochemistry, 23*(6), 646–655.
62. Chang, H. C., et al. (2011). Capsaicin may induce breast cancer cell death through apoptosis-inducing factor involving mitochondrial dysfunction. *Human & Experimental Toxicology, 30*(10), 1657–1665.
63. de Sá-Júnior, P. L., et al. (2013). RPF101, a new capsaicin-like analogue, disrupts the microtubule network accompanied by arrest in the G2/M phase, inducing apoptosis and mitotic catastrophe in the MCF-7 breast cancer cells. *Toxicology and Applied Pharmacology, 266*(3), 385–398.
64. Moon, D.-O., et al. (2012). Capsaicin sensitizes TRAIL-induced apoptosis through Sp1-mediated DR5 up-regulation: Involvement of Ca2+ influx. *Toxicology and Applied Pharmacology, 259*(1), 87–95.
65. Kim, M. Y. (2012). Nitric oxide triggers apoptosis in A375 human melanoma cells treated with capsaicin and resveratrol. *Molecular Medicine Reports, 5*(2), 585–591.
66. Ip, S. W., et al. (2012). Capsaicin mediates apoptosis in human nasopharyngeal carcinoma NPC-TW 039 cells through mitochondrial depolarization and endoplasmic reticulum stress. *Human & Experimental Toxicology, 31*(6), 539–549.
67. Datta, P., et al. (2014). Capsaicin mediated oxidative stress in pancreatic cancer. In *Cancer* (pp. 241–246). Elsevier.
68. Zhang, J.-H., et al. (2013). Involvement of the phosphoinositide 3-kinase/Akt pathway in apoptosis induced by capsaicin in the human pancreatic cancer cell line PANC-1. *Oncology Letters, 5*(1), 43–48.

69. Lin, C.-H., et al. (2013). Capsaicin induces cell cycle arrest and apoptosis in human KB cancer cells. *BMC Complementary and Alternative Medicine, 13*(1), 1–9.
70. Sarkar, A., Bhattacharjee, S., & Mandal, D. P. (2015). Induction of apoptosis by eugenol and capsaicin in human gastric cancer AGS cells-elucidating the role of p53. *Asian Pacific Journal of Cancer Prevention, 16*(15), 6753–6759.
71. Park, S.-Y., et al. (2014). Capsaicin induces apoptosis and modulates MAPK signaling in human gastric cancer cells. *Molecular Medicine Reports, 9*(2), 499–502.
72. Pereira, J. A. P., et al. (2016). Effects of arbuscular mycorrhizal fungi on Capsicum spp. *The Journal of Agricultural Science, 154*(5), 828–849.
73. Soumya, S. L., & Nair, B. R. (2012). Antifungal efficacy of Capsicum frutescens L. extracts against some prevalent fungal strains associated with groundnut storage. *Journal of Agricultural Technology, 8*(2), 739–750.
74. Ashouri, S., & Shayesteh, N. (2009). Insecticidal activities of black pepper and red pepper in powder form on adults of Rhyzopertha Dominica (F.) and Sitophilus granarius (L.). *Pakistan Entomologist, 31*(2), 122–127.
75. Cruz, L. P., et al. (2010). Isolation and partial characterization of a novel lipid transfer protein (LTP) and antifungal activity of peptides from chilli pepper seeds. *Protein and Peptide Letters, 17*(3), 311–318.
76. Millqvist, E. (2000). Cough provocation with capsaicin is an objective way to test sensory hyperreactivity in patients with asthma-like symptoms. *Allergy, 55*(6), 546–550.
77. Smith, C. G., & Stopford, W. (1999). Health hazards of pepper spray. *North Carolina Medical Journal, 60*(5), 268–274.
78. Gerber, S., Frueh, B. E., & Tappeiner, C. (2011). Conjunctival proliferation after a mild pepper spray injury in a young child. *Cornea, 30*(9), 1042–1044.
79. Tag, H. M., et al. (2014). Potential anti-inflammatory effect of lemon and hot pepper extracts on adjuvant-induced arthritis in mice. *The Journal of Basic & Applied Zoology, 67*(5), 149–157.
80. Sarwa, K. K., Das, P. J., & Mazumder, B. (2014). A nanovesicle topical formulation of Bhut Jolokia (hottest capsicum): A potential anti-arthritic medicine. *Expert Opinion on Drug Delivery, 11*(5), 661–676.
81. Baruah, S., et al. (2014). A review on recent researches on Bhut jolokia and pharmacological activity of capsaicin. *International Journal of Pharmaceutical Sciences Review and Research, 24*(2), 89–94.
82. Hassan, M. H., et al. (2012). Antioxidant and antiapoptotic effects of capsaicin against carbon tetrachloride-induced hepatotoxicity in rats. *Toxicology and Industrial Health, 28*(5), 428–438.
83. Jolayemi, A. T., & Ojewole, J. A. O. (2014). Analgesic effects of Capsicum frutescens Linn. [Solanaceae] fruit aqueous extract in mice. In *Food sources, medical uses and health implications* (p. 75). Nova Main Publishers, Springer, Cham, New York, USA.
84. Davis, J. B., et al. (2000). Vanilloid receptor-1 is essential for inflammatory thermal hyperalgesia. *Nature, 405*(6783), 183–187.
85. Ojewole, J. A. O. (2008). Anticonvulsant activity of Hypoxis hemerocallidea Fisch. & CA Mey. (Hypoxidaceae) corm ('African potato') aqueous extract in mice. *Phytotherapy Research: An International Journal Devoted to Pharmacological and Toxicological Evaluation of Natural Product Derivatives, 22*(1), 91–96.
86. Leung, F. W. (2014). Capsaicin as an anti-obesity drug. In *Capsaicin as a therapeutic molecule* (pp. 171–179). Springer.
87. Peng, J., & Li, Y.-J. (2010). The vanilloid receptor TRPV1: Role in cardiovascular and gastrointestinal protection. *European Journal of Pharmacology, 627*(1–3), 1–7.
88. Kwon, Y. I., Apostolidis, E., & Shetty, K. (2007). Evaluation of pepper (Capsicum annuum) for management of diabetes and hypertension. *Journal of Food Biochemistry, 31*(3), 370–385.
89. Teng, C. H., et al. (1998). Protective action of capsaicin and chilli on haemorrhagic shock-induced gastric mucosal injury in the rat. *Journal of Gastroenterology and Hepatology, 13*(10), 1007–1014.

90. Mózsik, G. (2014). Capsaicin as new orally applicable gastroprotective and therapeutic drug alone or in combination with nonsteroidal anti-inflammatory drugs in healthy human subjects and in patients. *Capsaicin as a Therapeutic Molecule*, 209–258.
91. Maramag, R. P. (2013). Diuretic potential of Capsicum Frutescens Linn., Corchorus Oliturius Linn., and Abelmoschus Esculentus Linn. *Asian Journal of Natural & Applied Sciences, 2*(1), 60–69.
92. Mózsik, G., et al. (1999). Small doses of capsaicin given intragastrically inhibit gastric basal acid secretion in healthy human subjects. *Journal of Physiology-Paris, 93*(5), 433–436.
93. Ezekiel, J. A. T., & Oluwole, O. J. A. (2014). Effects of capsaicin on coagulation: Will this be the new blood thinner. *Clinical Medicine & Research, 3*, 145–149.
94. Hogaboam, C. M., & Wallace, J. L. (1991). Inhibition of platelet aggregation by capsaicin. An effect unrelated to actions on sensory afferent neurons. *European Journal of Pharmacology, 202*(1), 129–131.
95. Harper, A. G. S., Brownlow, S. L., & Sage, S. O. (2009). A role for TRPV1 in agonist-evoked activation of human platelets. *Journal of Thrombosis and Haemostasis, 7*(2), 330–338.
96. Arifuzzaman, M., et al. (2011). Evaluation of thrombolytic properties of Nigella sativa, Capsicum frutescens and Brassica oleracea. *International Journal of Research in Pharmaceutical Sciences, 2*(3), 483–487.
97. Al-Snafi, A. E. (2015). Therapeutic properties of medicinal plants: A review of plants with hypolipidemic, hemostatic, fibrinolytic and anticoagulant effects (part 1). *Asian Journal of Pharmaceutical Science & Technology, 5*(4), 271–284.
98. Bernstein, J. E., et al. (1986). Effects of topically applied capsaicin on moderate and severe psoriasis vulgaris. *Journal of the American Academy of Dermatology, 15*(3), 504–507.
99. Yosipovitch, G., et al. (2005). Topical capsaicin for the treatment of acute lipodermatosclerosis and lobular panniculitis. *Journal of Dermatological Treatment, 16*(3), 178–180.
100. Neess, C. M., et al. (2000). Treatment of pruritus by capsaicin in a patient with pityriasis rubra pilaris receiving RE-PUVA therapy. *Clinical and Experimental Dermatology, 25*(3), 209–211.
101. Lysy, J., et al. (2003). Topical capsaicin—A novel and effective treatment for idiopathic intractable pruritus ani: A randomised, placebo controlled, crossover study. *Gut, 52*(9), 1323–1326.
102. Rijswijk, V., & Van Wijk, G. (1998). The long-term effects of capsaicin aqueous spray on the nasal mucosa. *Clinical & Experimental Allergy, 28*(11), 1351–1358.
103. Fokkens, W., Hellings, P., & Segboer, C. (2016). Capsaicin for rhinitis. *Current Allergy and Asthma Reports, 16*, 1–5.
104. Bernstein, J. A., et al. (2011). A randomized, double-blind, parallel trial comparing capsaicin nasal spray with placebo in subjects with a significant component of nonallergic rhinitis. *Annals of Allergy, Asthma & Immunology, 107*(2), 171–178.
105. Dicpinigaitis, P. V., & Alva, R. V. (2005). Safety of capsaicin cough challenge testing. *Chest, 128*(1), 196–202.
106. Ikegbunam, M., Ukamaka, M., & Emmanuel, O. (2016). Evaluation of the antifungal activity of aqueous and alcoholic extracts of six spices. *American Journal of Plant Sciences, 7*(1), 118–125.
107. Parle, M., & Kaura, S. (2013). Green chilli: A memory booster from nature. *Annals of Pharmacology and Pharmaceutics, 4*(1), 17–21.
108. Suganuma, H., et al. (2004). *Ameliorative effects of dietary carotenoids on memory deficits in senescence-accelerated mice (SAMP8)*. Elsevier.
109. Takano, F., et al. (2007). Capsicum ethanol extracts and capsaicin enhance interleukin-2 and interferon-gamma production in cultured murine Peyer's patch cells ex vivo. *Life Sciences, 80*(17), 1553–1563.
110. Pinto, D., et al. (2021). Cosmetics—Food waste recovery. In *Food waste recovery* (pp. 503–528). Elsevier.
111. Cockburn, N., et al. (2017). Oral health impacts of medications used to treat mental illness. *Journal of Affective Disorders, 223*, 184–193.

112. Padmanabhan, P., et al. (2016). *Encyclopedia of food and health*. Elsevier.
113. Dini, I. (2018). Spices and herbs as therapeutic foods. In *Food quality: Balancing health and disease* (pp. 433–469). Elsevier.
114. Kim, H. K., et al. (2020). Red pepper (Capsicum annuum l.) seed extract improves glycemic control by inhibiting hepatic gluconeogenesis via phosphorylation of foxo1 and ampk in obese diabetic db/db mice. *Nutrients, 12*(9), 2546.
115. Ткачева, Н., & Елисеева, Т. (2021) Еда для разжижения крови. *Журнал здорового питания и диетологии, 16*, 30–34.
116. Byun, E.-B., et al. (2018). Neuroprotective effect of Capsicum annuum var. abbreviatum against hydrogen peroxide-induced oxidative stress in HT22 hippocampus cells. *Bioscience, Biotechnology, and Biochemistry, 82*(12), 2149–2157.
117. Das, M., et al. (2018). Hepatoprotective effects of green Capsicum annum against ethanol induced oxidative stress, inflammation and apoptosis in rats. *Journal of Ethnopharmacology, 227*, 69–81.
118. Omar, O. A. S., et al. (2013). Efficacy of capsicum frutescens in curing the peptic ulcer. *International Journal of Pure and Applied Sciences and Technology, 15*(1), 43.
119. Tundis, R., et al. (2011). Comparative study on the chemical composition, antioxidant properties and hypoglycaemic activities of two Capsicum annuum L. cultivars (Acuminatum small and Cerasiferum). *Plant Foods for Human Nutrition, 66*, 261–269.
120. Ekom, S. E., Tamokou, J.-D.-D., & Kuete, V. (2021). Antibacterial and therapeutic potentials of the Capsicum annuum extract against infected wound in a rat model with its mechanisms of antibacterial action. *BioMed Research International, 2021*, 1–17.
121. Hernández-Pérez, T., et al. (2020). Capsicum annuum (hot pepper): An ancient Latin-American crop with outstanding bioactive compounds and nutraceutical potential. A review. *Comprehensive Reviews in Food Science and Food Safety, 19*(6), 2972–2993.
122. Chakrabarty, S., Islam, A. K. M. M., & Islam, A. K. M. A. (2017). Nutritional benefits and pharmaceutical potentialities of chili: A review. *Fundamental and Applied Agriculture, 2*(2), 227–232.
123. Batiha, G. E.-S., et al. (2020). Biological properties, bioactive constituents, and pharmacokinetics of some Capsicum spp. and capsaicinoids. *International Journal of Molecular Sciences, 21*(15), 5179.

Chapter 34
Kewda

Amna Rana, Shagufta Kamal, Muhammad Zia-Ul-Haq, Ismat Bibi, Saima Rehman, and Maryam Rehman

34.1 Introduction

Pandanus odoratissimus Lam. (Syn. *Pandanus odorifer*), traditionally known as "Kewda", is quite a popular medicinal plant of great significance since it is widely utilized as a cure for a variety of diseases as well as in the perfumery industry. Each part of the plant is valuable for a wide range of purposes [1]. Also, its leaves and roots are employed in the cottage industry to manufacture baskets and mats. Other commercial applications involve food additives, soaps, hair oils, incense sticks, aromatherapy, lotions, and cosmetics [2]. Kewda plant favorably grows in saline soil and is dioecious [3]. Apart from being an ornamental plant, kewda exhibits a greatly dense and branched growing pattern making a sustainable canopy, which further gives rise to the development of unique microfauna. It also constitutes a diversified variety of phytochemicals that confer profound nutritional value [4]. It not only grows wildly but is also cultivated in its specific habitats [4, 5].

A. Rana · S. Kamal (✉)
Department of Biochemistry, Government College University, Faisalabad, Pakistan
e-mail: shaguftakamal@gcuf.edu.pk

M. Zia-Ul-Haq
Office of Research, Innovation and Commercialization, University of Engineering and Technology, Lahore, Pakistan

I. Bibi
Department of Chemistry, Islamia University of Bahawalpur, Bahawalpur, Pakistan

S. Rehman
Department of Chemistry, Government College University, Faisalabad, Pakistan

M. Rehman
Department of Botany, Government College University, Faisalabad, Pakistan

M. Zia-Ul-Haq et al. (eds.), *Essentials of Medicinal and Aromatic Crops*,
https://doi.org/10.1007/978-3-031-35403-8_34

Kewda has found tremendously valuable uses in different areas of life. It is often planted against water bodies as a primary barricade for checking sand drifting as well as soil erosion. The hydrodistillation of kewda plant has widespread applications on the grounds of high demand in international market. It can be cultivated via propagation during the months of June, July, and August. With an average of 100 flowers yearly, the flowering stage initiates followed by 3–4 years of cultivation [6]. The distillation of kewda flowers (male spadices) is performed to attain essential oil and perfumery agents. The use of kewda water in soft drinks, flavouring syrups, and food preparations, in addition to use of kewda oil for its curative effect against rheumatism and spasms make them quite valuable products. Moreover, kewda attar has also found a great number of uses in hair oils, serums, cosmetics, soaps, lotions, incense sticks as well as scenting [1]. The essential oil extract of the kewda plant refers to a truly valuable commercial product in demand internationally; however, its yield and quality are greatly affected by poor plantation techniques. Its quality is also declined when propagators use the same branch cuttings generation after generation. It ultimately leads to oil adulteration and low-quality products sold on a commercial scale. The estimated life span of a kewda tree ranges over 50–80 years. Each sexually-mature plant contributes 8–12 fruits every second year. The flowering season of kewda is at its peak during the rainy season (60%) with a gradual decline in hot (30%) and cold weathers (10%) [7]. Nonetheless, kewda has been considered an essentially remarkable plant concerning both the medical and socio-economic aspects. Kewda is classified as

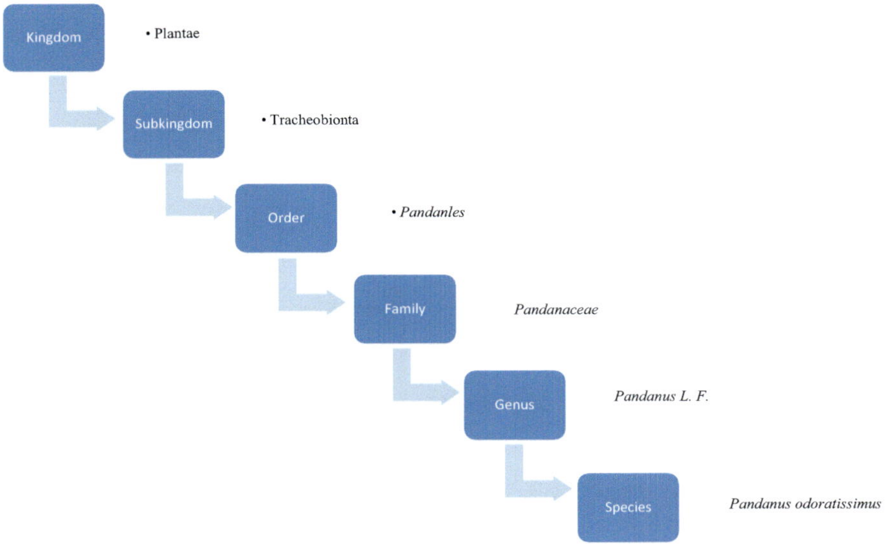

Plant name: *Pandanus odoratissimus Linn*
Kingdom: Plantae-plants
Subkingdom: Tracheobionta
Order: Pandanales
Family: *Pandanaceae*
Genus: *Pandanus L. F.*
Species: *Pandanus odoratissimus*

34.1.1 Common Names

- ajbwirōk, anewetāk (Pingelap Atoll, Pohnpei)
- binu (Kapingamarangi Atoll, Pohnpei)
- bōb (Marshall Islands)
- choy, fach, far (Yap)
- deipw, fach, far (Chuuk)
- deipw, kipar (Pohnpei)
- épo (Nauru)
- fa, fafa, laufala, falahola, kukuvalu, lou'akau (Tonga)
- fala, lau fala (Samoa, Tuvalu)
- hala (Nukuoro Atoll, Pohnpei)
- hala, pū hala (Hawai'i)
- kafu (Guam)
- mweng (Kosrae)
- ongor (Palau)
- pandanas (Vanuatu: Bislama)
- pandanus, vacouet (French)
- pandanus, screw pine (English)
- te kaina (Kiribati)
- vadra, voivoi (Fiji) [1]

Classification of Kewda

34.2 Botanical Description

Male flowers are large, pointed, concave pendulous comprising a single thyrsus of pointed anthers in the axil of each flower. Whereas female flowers exhibit a terminal appearance with equally spaced three rows of white floral leaves around the base of young fruit. Fruits are angular or oval-shaped with a rough appearance, rich-orange in color, 5–8 inches in diameter and 6–10 inches in length, and 4–8 pounds in weight having exterior ends detached from each other when ripe [5]. Nuts are characterized by wedge-shaped drupe with a hard appearance comprising numerous cells having perforations on both sides. Single in quantity, oblong in shape, and smooth in

Fig. 34.1 A fully grown tree of *Pandanus odoratissimus* Linn (Kewda) [9].

appearance, the seed adheres to whitish fibers longitudinally passing through the cell perforation. The trunk is single, erect, and long appearance with a ramous round head that gets covered with hardwood while getting matured. Leaves have a sharp spiny and smooth appearance, 3–5 ft. long, elongated towards a triangular point since densely packed into three spiral rows and stem-clasping. An oval stigma, yellow in color, is attached over the apex of the germ outside a two-lipped umbilicus. Moreover, the subterranean roots cover the surface soil layers [1, 8] (Figs. 34.1, 34.2 and 34.3).

34.3 Origin and Distribution

There are about 600 species of the genus Pandanus (Pandanaceae) that are widely located in different tropical and subtropical regions of the world, particularly including South Asia, Western Africa, Northern, Australia, Polynesia, and the Pacific Island [8]. Amongst about 36 Pandanus species found in India, Pandanus odorifer (Forssk.) Kuntze. (Pandanus odaratissimus) has been predominantly found along the extensive coast of the Indian peninsula and the Andamans [10]. Geographical distribution of Kewda is presented in Fig. 34.4.

Besides, kewda is also native to certain coastal areas of Mauritius, Taiwan, Iran, Java, Malaysia, Myanmar, China, and the south islands of Japan. However, the highly abundant growth of this plant in India denotes the reason behind profound research on kewda native to this area. The major Indian regions harboring kewda

Fig. 34.2 Top left: Male inflorescence. Top right: Female inflorescence. Bottom: Fruit heads consisting of a collection of various tightly packed wedge-shaped drupes, also known as "keys" [9]

Fig. 34.3 Left: Characteristic ring pattern on bark left followed by falling of older leaves. Right: Prop roots descending into a thin substrate over lava rock [9]

Fig. 34.4 The geographical points of sampling sites of Pandanus alongside the coast of Odisha [4]

refer to coastal regions of Kerala, Tamil, Odisha, Nadu, Gujarat, and Andhra Pradesh, as well as some districts of Uttar Pradesh [2, 4].

34.4 Habitat

Kewda is primarily native to forest habitats with an average sea-level elevation of 66 ft., however, it can reach over 1970 ft. It is acclimatized to grow in tropical and subtropical regions alongside water bodies like sea coastlines, as well as banks of rivers, canals, and ponds, etc. [5, 11]. Moreover, its dense features enable it to tolerate high saline conditions, and water scarcity in addition to strong gusts of wind and storms. Yet, highly fertile and well-irrigated soil is required for the proper cultivation of kewda plants to be used in the perfumery industry [1].

34.5 Climatic Conditions for Cultivation of Kewda

There are certain favourable conditions for proper cultivation of kewda plant, including optimum temperature range, light, rainfall, as well as nature of soil. Usually, direct or spotted sunlight would make the kewda plants grow well. However, excessive sunlight during winters and prevention of direct and prolonged sunlight exposure during summers is necessary to maintain optimum growth. As far as soil is concerned, the soil used for kewda cultivation must be fertile and well-drained whereas it can be grown in light to heavy textured soils, such as sands, loams, clays, clay loams, thin soil over limestones, saline, acidic or alkaline soils (pH 6–10). These plants are dense enough to become drought-tolerant. The average optimum temperature for proper growth of kewda plant is 24–28 °C with annual rainfall of 1000–2000 mm [7].

34.6 Phytochemistry

34.6.1 Phytochemical Extracts

A variety of phytochemical extracts of kewda leaves have been prepared. Alcoholic extract as well as extracts of n-Butanol, ethyl acetate, and methanol exhibit sticky consistency with the color appearance of blackish green, brown, brownish yellow, and reddish brown respectively. The dark brown extract of petroleum ether maintained at 40–60 °C gives off wax-like consistency. However, the chloroform extract exhibits dark green color with powdery consistency [11].

34.6.2 Chemical Composition

The essential oil of the kewda plant is extracted from the inflorescences and it comprises various valuable chemical constituents yielded after the GC-MS (Gas chromatography-mass spectrometry) analysis of kewda oil extract as depicted in the table below [2, 11] (Table 34.1).

Table 34.1 Major chemical constituents of kewda essential oil derived from [4, 11]

Major chemical constituents of kewda oil	
PEME (Phenyl ethyl methyl ether)	65–80%
Ether	37.7%
Terpene-4-ol	18.6%
α-terpineol	8.3%
2-phenyl ethyl alcohol	7.5%
Benzyl benzoate	11%
Viridine	8.8%
Germacrene-B	8.3%
Thujene	4%
Cymene	0.3–3%
Pinene	0.05–2.5%
1,8 cineole	0.1–0.2

Moreover, kewda oil also comprises of minor amount of benzyl acetate, benzyl salicylate, and benzyl alcohol [1]. Chemical structures of some of the major phytochemicals (Fig. 34.5) are given below:

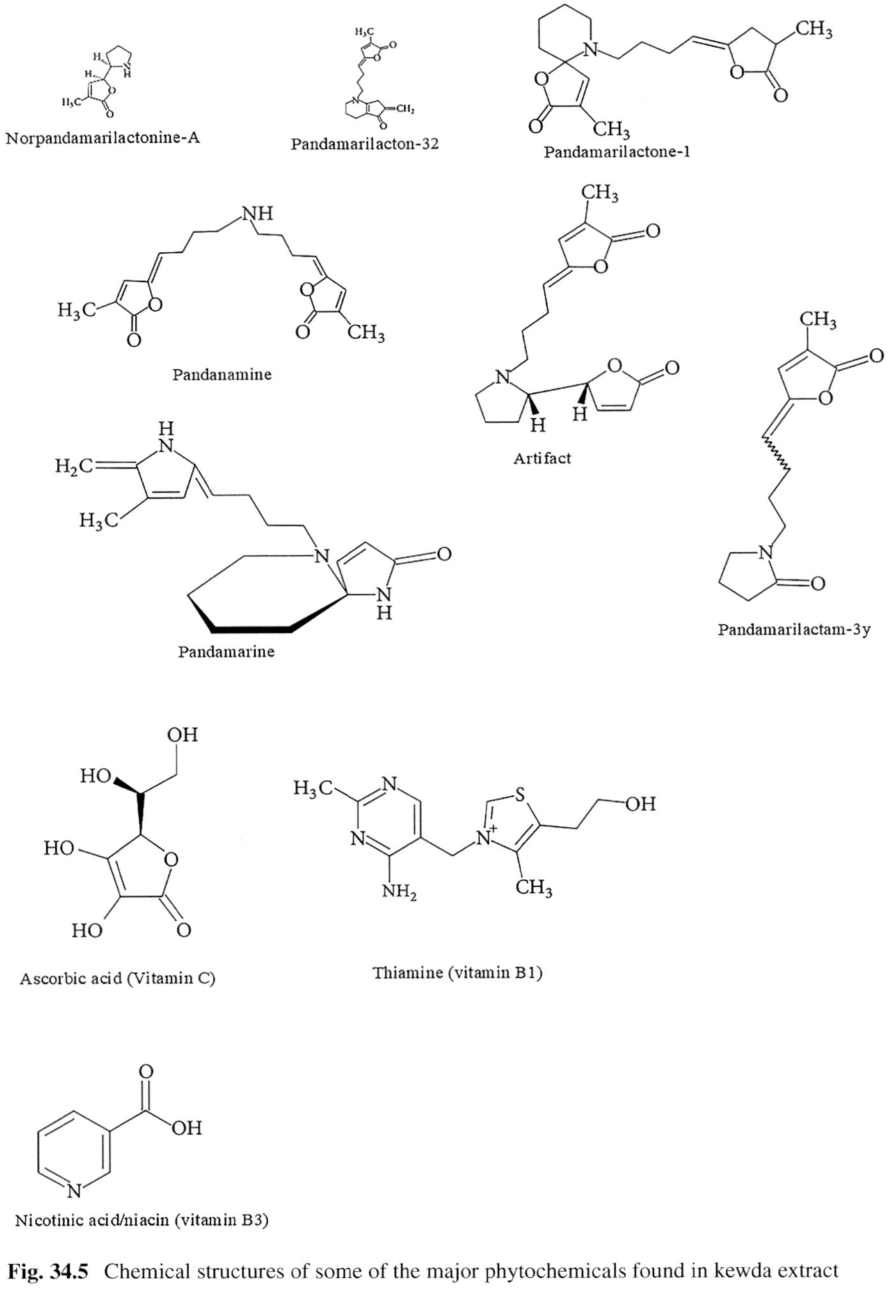

Fig. 34.5 Chemical structures of some of the major phytochemicals found in kewda extract

34.7 Nutritional Aspects of Kewda

Being a staple food in Micronesia, the adults' consumption of this area usually reaches up to 1 kg of kewda fruit daily; therefore, they also tend to preserve the fruit pulp for their utilization in different ways. They often preserve it by boiling, and baking, followed by further processing and drying [4]. The fruit pulp is a source of various major macronutrients like protein, calcium, and phosphorus as well as micronutrients like iron, thiamin, riboflavin, niacin, and beta-carotene. However, vitamin C is a predominant nutrient found in fresh *Pandanus* fruits. The fruits of orange and deep-yellow colored varieties of kewda fruits have been considered to be quite a beneficial food to prevent vitamin A deficiency in the community since these varieties are rich in provitamin A carotenoids [1, 4] (Table 34.2).

The essential oil extracted from kewda flowers is partially more soluble in water due to which it is essential to precisely examine the distillation rate required for proper isolation [1].

34.8 Cultivation and Collection

Apart from propagation via seeds, kewda can also be cultivated through branch-cutting. The propagation can be proceeded by division of suckers or by using the offsets method. The flowering stage initiates followed by 3–4 years of plantation and is limited to the monsoon season, usually referring from July to October. The morning period is preferably suitable for flower harvesting whereas it may require a fortnight for the spadices' maturation. Annually, a completely matured kewda tree yields about 30–40 spadices [1, 12].

Apical and axillary shoots or meristems, as well as adventitious shoots, can be employed for kewda regeneration. Moreover, micropropagation is also a modern propagation technique lately used for successful in-vitro growth of large pools of kewda plant, which was previously a challenge faced during propagation of kewda via conventional means. Micropropagation of the kewda plant has also been reported to be performed by direct and indirect organogenesis [13].

Table 34.2 Nutritional value of *Pandanus* derived from [1, 4]

Nutritional value of *Pandanus* (Edible pericarp) per 100 g (228 Kcal)		
Water	80 g	[6–8]
Carbohydrates	17 g	[6, 7]
Beta-carotene	19–19.000μg	[8]
Vitamin C	5 mg	[6, 8]
Protein	1.3 mg	[8, 9]
Fat	0.7 mg	[9]
Fiber	3.5 mg	[9]
Calcium	134 mg	[7]
Phosphorus	108 mg	[7]

34.9 Propagation

The branch cuttings are utilized to regrow kewda trees since the plant species propagated via seeds do not often match the qualitative aspects of the parent plant. There is a wide range of traditional varieties of kewda plants being grown selectively from their cuttings on the grounds of their ultimate fruit qualities. However, the wild varieties of kewda trees retain much longer and straighter boles; and hence, are greatly used for their timber as they propagate well from seeds in their respective native habitats [9, 13].

34.10 Propagation by Branch Cuttings

34.10.1 Propagule Collection

In most cases, the branches of shorter laterals, and about 30-40 cm in length, associated with a few aerial roots are selected with great care for propagation. The former fruits or flowers and other bearings of the trees are excluded before the plantation of cut branches. If early fruiting is required, the kewda branches already bearing fruits are selected [7] (Fig. 34.6).

34.10.2 Propagule Processing

The leaves section is limited to almost 30% as leaves are often trimmed and cut before the propagation of cut branches.

Fig. 34.6 Branch-cutting for propagation [7]

34.10.3 Propagule Storage

Preferably, the cut branches ought to be processed immediately followed by collection. In case of any need of storing the freshly-cut branches, they can be sealed in sample plastic bags under cool and shady conditions for a certain period [13].

34.11 Extraction of kewda Essential Oil

To extract the essential oil, flowers are plucked in quite early hours for immediate transportation and processing towards the distillation units. The delay in transportation time causes a reduction in the aroma and quality of kewda oil [1]. Firstly, manual plucking of the green spathe is performed to enhance the yield of essential oil leading to oil extraction by Clevenger's apparatus. In this procedure, the chopped flowers are weighed equivalent to 1000 g, which are then added to the flask along with 3000 mL of distilled water. The subsequent heating of the flask at 60–70 °C is continued for 3–4 h after which the kewda essential oil is obtained in glass vials and stored at 4 °C [6, 14].

34.12 Life Span

The average life span of a kewda tree ranges over 20–80 years or in some cases, it can live over 100 years. However, vegetative propagation is practiced to extend the life span of kewda trees with desirable qualities. In the case of fruiting abilities, the productivity and longevity of kewda trees are limited to 20–25 years; therefore, replantation and propagation are quite necessary for fruit yield maintenance [10, 14].

34.13 Seasonality of Leaf Flush, Flowering, Fruiting

The kewda trees keep growing the whole year. As far as flowering is concerned, flowering can be observed even during the off-season; however, male plants are more inclined towards once-a-year flowering, whereas female plants flower predominantly every second year. Moreover, the fruit ripening of kewda trees spans over a couple of months [5, 14].

34.14 Industrial Application of Kewda

Kewda plant has retained an exceptional place in the field of perfumery industry as it is a traditional bioresource of fragrance. Various pharmacological activities of kewda extract make it quite useful for different disorders. Apart from medical uses, a diversified range of industrial uses is also associated with kewda plants, such as the food industry, paper and crafting industry, as well as fiber industry [4, 13].

34.15 Commercial Products

The major commercially available products of the kewda plant refer to woven products like mats. In Tonga, mats, being a highly valuable gift, are woven out of leaf strips with detailed designs (fala). However, floor mats for daily use are designed rather simply via wider strips of leaves (papa & lotaha) [15]. There is a vast range of home-made products woven from hands using kewda accessories and these are widely used for regular gifts as well as general home use since 80% of women in the Ha'apai group are associated with handicraft production involving kewda plant in addition to paper mulberry (*Broussonetia papyrifera*) [6]. These handicraft products are easily accessible and can be processed in numerous ways being produced on the local level. These commercial products are long-lasting; therefore, these are an essential resource of income for the concerned population [16].

34.16 Kewda Perfumery Industry

Keeping in view the widespread perfumery industry based on kewda plants, kewda water, kewda attar, and kewda oil have been considered as quite essential aromatic products widely used in many regions of the world, especially in India. The distillation of male spadices of kewda plant is carried out to develop these perfumery products [6]. The delicate white-colored spathes around the kewda flowers are attributed to the aromatic value of the kewda plant. The process of hydrodistillation must be performed carefully owing to high solubility of the oil extract in water. According to a rough estimate, nearly 35 millions of flowers contribute to the flavouring and perfumery industry worth 400 million rupees in an year [17].

34.16.1 Kewda Oil

The essential oil of kewda plant extract finds various curative properties as it is used in ear drops formulation for ear-ache. It is also used to treat headaches, rheumatism, spasms and many other ailments [18].

34.16.2 Kewda Attar

Kewda attar has been famously used for scents production as it blends well with almost all sorts of perfumes. It is also used in hair oils, lotions, soaps, and incense sticks [6, 7].

34.16.3 Kewda Water

Kewda water is often used as a flavoring agent in syrups, sweets, soft drinks, and other food items [1, 7].

34.17 Ethnomedicinal Uses

Different parts of the kewda plant are used for medical purposes. Roots and basal white parts of leaves are often utilized for the treatment of different diseases. The use of roots as a decoction for hemorrhoids and leaves for treating cancer, asthma, hepatitis, boils, flu, and dysuria is common in Kiribati. However, male flowers, fruits, and aerial roots, either individually or combined, are utilized for treating certain respiratory and digestive health issues in different regions of Hawaii [18]. Other uses of kewda roots include relieving stomach cramps and syphilis while leaves are also used to cure vomiting and nausea conditions [7, 19]. Fruits have a wide range of applications as preparative agents of Ayurvedic formulations like rheumatism, leucorrhea, etc. [18, 19].

34.18 Pharmacological Activities of the Kewda Plant

In the case of the medical utilization of the kewda plant, it shows a wide variety of biological activities, majorly including anti-inflammatory activity, antioxidant activity, analgesic and cytotoxic activities, anticancer activity, as well as anti-diabetic and cardioprotective activities [9]. These activities make kewda extract useful in different diseases like cardiac health problems, sterility, headaches, leprosy, arthritis, scabies, spasms, colic infections and leukoderma [20].

Kewda essential oil plays a remedial role in oral inflammation that may be linked to the inflammatory cytokines, such as IL-1β that further instigates the prostaglandins and Il-6 release from fibroblasts [21]. The anti-inflammatory activity of kewda extract is caused by stopping the synthesis of serotonin, histamine and prostaglandins [22]. There are certain phytochemicals exhibiting antioxidant activity naturally with high reducing and scavenging capacities. Moreover, methanolic leaf extracts of

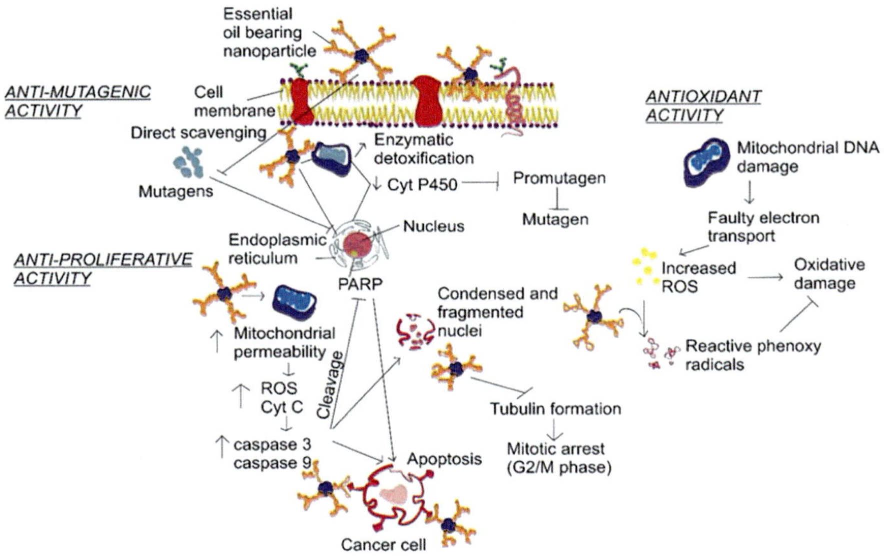

Fig. 34.7 Possible mechanism of action of kewda essential oil owing to its anti-mutagenic, antioxidant, and anti-proliferative activities. (*Cyt* cytochrome, PARP Poly-ADP ribose polymerase, ROS reactive oxygen species) [24]

the kewda plant have also been reported to comprise distinctive antioxidant enzymes like superoxide dismutase, lipid peroxidase, glutathione, and catalase [23] (Fig. 34.7).

34.18.1 Antioxidant Activity

It has been evident from various in vitro studies that methanolic extracts of leaves of kewda plant exhibit great antioxidant activity as they showed the scavenging potential against the radicals of both DPPH (1,1-diphenyl, 2-picryl hydrazyl) as well as nitric oxide. When vitamin E was taken as a reference standard, the methanolic extract of kewda leaves retained similar results of IC50 as that of vitamin E, hence, proving its antioxidant potential [1].

Superoxide Dismutase (SOD) and catalase are considered to play a quite significant role in natural defense system of living body as these enzymes are responsible for reduction of peroxides and reactive oxygen species, which would otherwise lead to cellular damage and defective biological activity. The methanolic extracts of kewda leaves have shown to increase the level of these enzymes, hence, providing a protective mechanism of action [23] (Fig. 34.8).

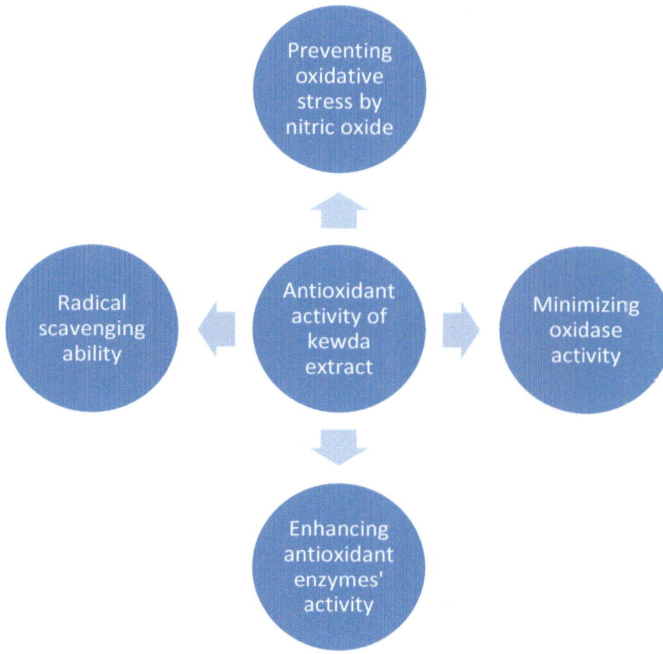

Fig. 34.8 Overview of the antioxidant activity of the kewda extract, derived from [25]

34.18.2 Anti-Inflammatory and Analgesic Activity

The methanolic extract of kewda leaves has proved to express its anti-inflammatory potential most predominantly in an in-vitro rat model study where it was administered in the concentration of 100 mg/Kg. Moreover, the analgesic activity of its aqueous extract showed comparable results to that of aspirin via using both peripheral and central mechanisms of actions, when the kewda leaves extract was administered in the dosage forms of 400 and 800 mg/Kg. Different diseases involving inflammation in body like earache, rheumatism, arthritis, fever and diarrhea have been observed to be treated via using methanolic extract of kewda leaves owing to its high anti-inflammatory, and analgesic activity [1]. Followed by the preliminary screening, the aqueous extract of kewda leaves in addition to the methanolic extract has exhibited significant anti-inflammatory activity in certain in-vivo rat models as well [26].

34.18.3 Antidiabetic Activity

In an in-vitro study, rats were made developed diabetes via alloxan and then ethanolic extract of kewda roots were examined for their antihyperglycemic activity. The long-term dosage delivery of kewda extracts exhibited more significant potential against diabetes. The occurrence of phytochemicals like tannins, 4-hydroxybenzoic acid and flavonoids majorly contribute to antidiabetic activity of kewda extracts [1, 27].

34.18.4 Antitumour Activity

The alcoholic extract of kewda plant has shown considerable level of antitumor activity in a dose-dependent manner, when it was administered orally in a Dalton's ascitic lymphoma (DAL) tumor case study. The extract was majorly involved in minimizing the mass development of solid tumor mass in addition to reducing the average body weight of the mice suffering from DAL tumor [1].

34.18.5 CNS-Depressant Action

The methanolic extract of kewda leaves have proved to exhibit CNS (Central nervous system) depressant activity via distinctive mechanisms of action like reducing dopaminergic and serotonergic transmission. The presence of flavonoids and saponins confer sedative impact and muscle relaxant potential in various in-vitro as well as in-vivo models [28].

34.18.6 Hepatoprotective & Hepatocurative Activity

In a recent study, the ethanolic extract of kewda plant has shown highly significant hepatoprotective potential with respect to membrane-stabilizing activity. It minimizes the hepatic necrosis in the range of certain dosage level (200 and 400 mg/Kg). In this study, the paracetamol-induced liver damage in rats was examined thoroughly via treating it with kewda ethanolic extract. The raised levels of liver enzymes like SGPT, SGOT, and ALP as well as concentration of triglycerides and direct bilirubin were effectively mediated owing to the presence of certain phytochemicals found in kewda extract, viz. flavonoids, and alkaloids [29] (Fig. 34.9).

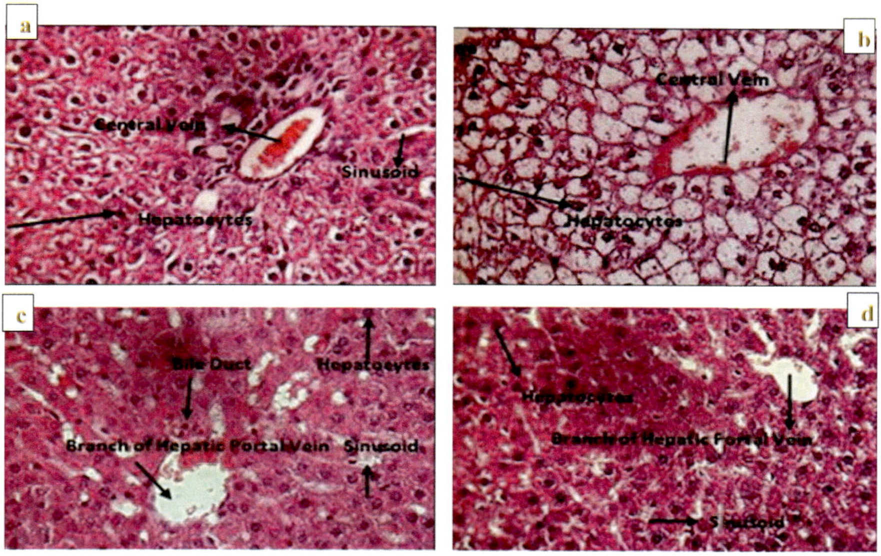

Fig. 34.9 Photomicrographs of (**a**) Normal liver cells in control group of animals indicating healthy hepatocytes; (**b**) Liver cells showing high level of degeneration in central vein in animal group developing paracetamol-induced hepatic necrosis; (**c**) & (**d**) Ethanolic kewda extracts exhibiting hepatoprotective potential via minimizing cellular damage in concentration of 200 mg/Kg and 400 mg/kg respectively [29]

34.18.7 Antimicrobial Activity

The antifungal potential of kewda oil extract was examined against keratinophilic fungi, in which kewda extract exhibited quite moderate antifungal action against *Aspergillus flavus* and *Trycophyton tonsurans* with zone of inhibition measuring up to 11 mm and 17 mm respectively. This antifungal potential can be predominantly attributed to presence of aldehydes, alcohols, phenols, and terpenes in the kewda extract [30].

Besides, alkaloids and flavonoids confer antibacterial activity to kewda extract as well. The ethanolic extract of kewda plant has indicated high antibacterial potential against gram-positive bacteria, including Staphlococcus aureus, and Bacillus subtilis, whereas it showed null effect against gram-negative Escherichia coli. Moreover, pandanin is a lectin derived from the leaf extract of kewda plant after undergoing a series of processing, such as ammonium sulphate precipitation, dialysis, affinity chromatography, and gel filtration. It shows significant level of hemagglutinating activity as well as strong antiviral action against a wide range of human pathogenic viruses like influenza virus (H1N1), and herpes simplex virus type-1 (HSV-1) [1].

34.19 Pests and Diseases

There is a distinctive number of pests and diseases affecting the kewda plant in adverse ways. In the central regions of the Pacific, sap-sucking mealybugs are the most common pests that weaken the kewda trees [25]. While hermit crabs and rats target the flourishing fruits of kewda trees. In some regions like Australia, an infestation of planthopper (*Jamella australiae*) causes dieback of kewda trees as mold grows on kewda leaves when this insect releases honeydew which is a sticky substance. The growing mold inhibits the growing areas of kewda trees which finally leads to the death of a tree. However, natural predators like wasps keep the growth of this mold in control in their native habitats. Bacterial infection caused by *Erwinia carotovora* subsp. *carotovora* and nematodes like *Helicotylenchus dihystera* also badly affect kewda trees [4].

The arthropod pests that are most commonly observed to destroy kewda trees include:

- *Aspidiotus nerii* (oleander scale)
- *Aspidiotus destructor* (coconut scale)
- *Oryctes rhinoceros* (coconut rhinoceros beetle)
- *Graeffea crouanii* (coconut phasmid)
- *Pinnaspis strachani* (hibiscus snow scale)
- *Pseudococcus perforatus* (mealybugs)
- *Pseudococcus giffardi*

Whereas the most commonly observed fungal species targeting kewda trees refer to:

- *Melanconium* sp.
- *Coniothyrium pandani*
- *Asteromella* sp.
- *Glomerella* sp.
- *Oxydothis pandani*
- *Macrophoma pandani*
- *Dothidella pandani*
- *Volutellaria fuliginea*
- *Lembosia pandani*
- *Melanconium pandani*
- *Meliola juttingii*
- *Microcyclus pandani*
- *Phomatospora cylindrotheca*
- *Phomatospora pandani* [31]

34.20 Conclusion

With an exceptional status and demand in the international market, kewda plant has successfully retained its worth as a traditional plant since it has been widely used for various medical, pharmaceutical, and industrial purposes. Kewda trees tend to grow in certain tropical and subtropical regions with their characteristic densely-growing branches. Owing to its wide range of phytochemicals, kewda extracts of aqueous and alcoholic nature have proved to exhibit a great number of pharmacological activities, majorly including analgesic, antioxidant, antitumor, antidiabetic, CNS depressant, antimicrobial, hepatoprotective, and anti-inflammatory activities. One of the most predominantly occurring phytochemical, PEME, is responsible for the extensive use of kewda essential oil extract in the perfumery industry. Moreover, kewda essential oil and its extract has also found eminent applications in the food and fiber industries. Its propagation is, therefore, consistently performed by cultivating cut branches of kewda trees. However, different sorts of pests, insects, and certain fungal species tend to attack the native kewda plants. Hence, the sustainable growth of kewda in its native regions must be ensured in order to fully optimize its applicable features while exploring the novel applications with the ever advancing scientific research.

References

1. Adkar, P.-P., & Bhaskar, V.-H. (2014). *Pandanus odoratissimus* (Kewda): A review on ethnopharmacology, phytochemistry, and nutritional aspects. *Advances in Pharmacological and Pharmaceutical Sciences, 2014.*
2. Rout, P.-K., Naik, S., & Rao, Y.-R. (2011). Liquid CO_2 extraction of flowers of *Pandanus fascicularis* Lam. And fractionation of floral concrete and comparative composition of the extracts. *Journal of Food Biochemistry, 35*(2), 500–512.
3. Misra, R., Dash, P.-K., & Rao, Y.-R. (2000). Chemical composition of the essential oils of Kewda and Ketaki. *Journal of Essential Oil Research, 12*(2), 175–178.
4. Padhy, S., Dash, S.-K., Panda, B.-B., Misra, M.-K., Padhi, N.-P., Sahu, D., et al. (2016). *Pandanus fasicularis* Lamk. (Kewda): The prime vegetation in the hinterland biodiversity of coastal Odisha, with unique ethnic utility, genetic variation and economics-a review. *Journal of Biodiversity, 7*(1), 33–49.
5. Nasim, N., Behera, J.-K., Sandeep, I.-S., RamaRao, V.-V., Kar, B., Mishra, A., et al. (2018). Phytochemical analysis of flower from *Pandanus odorifer (Forssk.)* Kuntze for industrial application. *Natural Product Research, 32*(20), 2494–2497.
6. Dutta, P.-K., Saxena, H.-O., & Brahmam, M. (1987). Kewda perfume industry in India1. *Economic Botany, 41*(3), 403–410.
7. Gurmeet, S., & Amrita, P. (2016). Unique pandanus-flavour, food and medicine. *Journal of Pharmacognosy and Phytochemistry, 5*(3), 08–14.
8. Englberger, L. (2003). Are Pacific islanders still enjoying the taste of *pandanus*? *Pacific Islands Nutrition, 58*, 10–11.
9. Adkar, P.-P., Jadhav, P.-P., Ambavade, S.-D., Bhaskar, V.-H., & Shelke, T. (2014). Adaptogenic activity of lyophilized hydroethanol extract of *Pandanus odoratissimus* in Swiss albino mice. *International Scholarly Research Notices*, 2014.

10. Englberger, L., Aalbersberg, W., Dolodolotawake, U., Schierle, J., Humphries, J., Iuta, T., et al. (2006). Carotenoid content of *pandanus* fruit cultivars and other foods of the Republic of Kiribati. *Public Health Nutrition, 9*(5), 631–643.
11. Englberger, L., Aalbersberg, W., Schierle, J., Marks, G.-C., Fitzgerald, M.-H., Muller, F., et al. (2006). Carotenoid content of different edible *pandanus* fruit cultivars of the Republic of the Marshall Islands. *Journal of Food Composition and Analysis, 19*(6–7), 484–494.
12. Murai, M., Pen, F., & Miller, C.-D. (1958). *Some tropical South Pacific Island foods: Description, history, use, composition, and nutritive value.*
13. Nasim, N., Sandeep, I.-S., Sahoo, A., Das, S., Panda, M.-K., Acharya, L., et al. (2020). Population genetic structure and diversity analysis in economically important *Pandanus odorifer (Forssk.)* Kuntze accessions employing ISSR and SSR markers. *Industrial Crops and Products, 143*, 111894.
14. Jagadev, P.-N., Naik, B., Sahoo, S.-C., Beura, S., & Maharana, T. (1993). Variability, selection and planting material in Kewda (*Pandanus fascicularis* Lam.). *Indian Journal of Genetics and Plant Breeding, 53*(03), 310–314.
15. Rao, Y.-R. (2000). Kewda (*Pandanus fascicularis Lam*), an economically important aromatic shrub in Ganjam district, Orissa, India. *Journal of Medicinal and Aromatic Plant Sciences, 22*(1B), 377–395.
16. Ghane, S.-G., & Zanan, R.-L. (2023). Ethnopharmacology and Phytochemistry of kewda [*Pandanus odorifer (Forssk.)* Kuntze; family: *Pandanaceae*]. In *Bioactives and pharmacology of medicinal plants* (pp. 451–463). Apple Academic Press.
17. Nasim, N., Sandeep, I.-S., Nayak, S., & Mohanty, S. (2021). *Cultivation and utilization of Pandanus odorifer for industrial application* (pp. 435–456). Domestication, Biotechnology and Regional Importance.
18. Kumar, N.-R., Padmalaxmi, D., Udupa, A.-L., Subir, G., Ojeh, N., Rathnakar, U.-P., et al. (2011). Acute anti-inflammatory activity of *Pandanus fasicularis Lam. Journal of Pharmacy Research, 4*(4), 1234–1236.
19. Nadaf, A., Zanan, R., Nadaf, A., & Zanan, R. (2012). Economical importance of Indian *Pandanus* species. *Indian Pandanaceae-an Overview*, 127–137.
20. Jagadev, P. N., & Beura, S. (2021). Kewda: Cultivation and perfume production.
21. Kida, Y., Kobayashi, M., Suzuki, T., Takeshita, A., Okamatsu, Y., Hanazawa, S., et al. (2005). Interleukin-1 stimulates cytokines, prostaglandin E2 and matrix metalloproteinase-1 production via activation of MAPK/AP-1 and NF-κB in human gingival fibroblasts. *Cytokine, 29*(4), 159–168.
22. Ramesh, L., Abhaykumar, K., & Reddy, V.-C. (2010). Anti-inflammatory activity of *Pandanus odoratissimus* extract. *IJP-International Journal of Pharmacology, 6*(3), 311–314.
23. Londonkar, R., & Kamble, A. (2011). Hepatotoxic and invivo antioxidant potential of *Pandanus odoratissimus* against carbon tetrachloride induced liver injury in rats. *Oriental Pharmacy and Experimental Medicine, 11*, 229–234.
24. Sharma, M., Grewal, K., Jandrotia, R., Batish, D.-R., Singh, H.-P., & Kohli, R.-K. (2022). Essential oils as anticancer agents: Potential role in malignancies, drug delivery mechanisms, and immune system enhancement. *Biomedicine & Pharmacotherapy, 146*, 112514.
25. Moon, J.-K., & Shibamoto, T. (2009). Antioxidant assays for plant and food components. *Journal of Agricultural and Food Chemistry, 57*(5), 1655–1666.
26. Chandrasekar, R., & Chandrasekar, S. (2017). Natural herbal treatment for rheumatoid arthritis-a review. *International Journal of Pharmaceutical Sciences and Research, 8*(2), 368.
27. Kumari, S., Wanjari, M., Kumar, P., & Palani, S. (2012). Antidiabetic activity of *Pandanus fascicularis* Lamk-aerial roots in alloxan-induced hyperglycemic rats. *International Journal of Nutrition, Pharmacology, Neurological Diseases, 2*(2), 105.
28. Raju, S., Subbaiah, N.-V., Reddy, K.-S., Das, A., & Murugan, K.-B. (2011). Potential of *Pandanus odoratissimus* as a CNS depressant in Swiss albino mice. *Brazilian Journal of Pharmaceutical Sciences, 47*, 629–634.

29. Mishra, G., Khosa, R.-L., Singh, P., & Jha, K.-K. (2015). Hepatoprotective potential of ethanolic extract of *Pandanus odoratissimus* root against paracetamol-induced hepatotoxicity in rats. *Journal of Pharmacy & Bioallied Sciences, 7*(1), 45.
30. Singh, P.-R.-A.-B.-A.-L., Bundiwale, R.-U.-C.-H.-I., & Dwivedi, L.-K. (2011). In-vitro study of antifungal activity of various commercially available itra (Volatile plant oil) against the keratinophilic fungi isolated from soil. *International Journal of Pharma and Bio Sciences, 2*(3), 178–184.
31. Jagadev, P.-N., Beura, S., & Maharana, T. (2001). Insect pests and diseases of Kewda (*Pandanus fascicularis Lam.*). *Indian Perfumer, 45*(4), 265–268.

Chapter 35
Jasmine

Sheeza Shoukat, Shagufta Kamal, Ismat Bibi, Naheed Akhter, Saima Rehman, and Mohammad Khalid

35.1 Introduction

The word "Jasmine" has its origin in the Arabic Persian word "Yasmyn" meaning fragrance [1, 2]. Jasmine is an ornamental plant belonging to the Oleaceae family nurtured predominantly for its fragrant flowers, industrial, and cultural impacts. The fragrant flowers are the symbol of profound love, fidelity, and merriness. It is used for hair embellishment, making garlands, and bouquets, religious purposes, extraction of oil, and perfumery [3–5]. Jasmine species are widely used in the cosmetic industry for making skin toners, perfumes, soaps, conditioners, and creams [6]. These ornamental shrubs are usually planted in public buildings and private gardens due to their soothing and delicate smell from the flowers [7, 8]. Its fragrance has medicinal importance, reducing stress, and nervous breakdown. In some countries, its flowers are also used for making jasmine tea which aids in boosting immunity [9, 10].

They are extensively cultivated in gardens which is why they are also known as garden plants. With their pink, creamy, white, or yellow fragrant or fragrance-free flowers, jasmines have broad genetic diversity and are present as wild and domesticated, deciduous or evergreen, climbing, or upright shrubs [5]. The immense genetic

S. Shoukat · S. Kamal (✉) · N. Akhter
Department of Biochemistry, Government College University, Faisalabad, Pakistan
e-mail: shaguftakamal@gcuf.edu.pk

I. Bibi
Department of Chemistry, Islamia University of Bahawalpur, Bahawalpur, Pakistan

S. Rehman
Department of Chemistry, Government College University, Faisalabad, Pakistan

M. Khalid
Department of Pharmacognosy, College of Pharmacy,
Prince Sattam Bin Abdulaziz University, Al-Kharj, Saudi Arabia

M. Zia-Ul-Haq et al. (eds.), *Essentials of Medicinal and Aromatic Crops*,
https://doi.org/10.1007/978-3-031-35403-8_35

mutation affects the jasmine flowering tithe, the flowering period, and the jasmine crop's overall yield. Appropriate selection of planting material has massive importance to produce a high-quality crop. For the selection of quality planting material complete information about the mother plant, maturity of cuttings, and age of rooted cuttings should be known [4].

The soothing fragrance of *Jasminum* helps in reducing depression. Additionally, *Jasminum* tea has several positive health effects [11]. More than 30 countries produce various types of tea. After water, tea is the second most used beverage in the world. It originates from China and besides flavoring drinks, it also has several health benefits because it helps in treating or preventing all types of disorders [12].

Chinese jasmine tea, a kind of floral-scented tea, is made by continuously combining the base tea with the fragrant *Jasminum sambac* flowers, which release the aromatic chemicals into the base tea. Aroma is one of the significant factors in determining the quality of jasmine tea. The preprocessing of the base tea, keeping the fresh flowers layering with the base tea, spreading out the mixture for heat dissipation, re-heaping up and scenting, separating the flowers from the tea, heating, and cooling are the steps that are involved in jasmine tea scenting processes. The last scenting round's resulting tea is combined with the right quantity of fresh jasmine blossoms to create the finished tea, which intensifies the jasmine smell. The quality of jasmine tea increases with increasing rounds of scenting. But, the greater the rounds of scenting process higher is the cost [13, 14].

One of the most popular reprocessed teas, jasmine-scented tea is produced from tea dhool by scenting it with jasmine flowers. People in both the north and southwest of China adore this tea. The quality of flowers, tea dhools, scenting techniques, and many other things might influence the quality of scented tea. Appearance, odor, color, and flavor are the indicators in determining the quality of tea [15].

Due to its antioxidant activity jasmine oil aids in preserving food and avert the phenomenon of degradation [16]. As a secondary metabolite, the *Jasminum* plant also produces essential oils (EOs) which are a blend of naturally occurring, highly complex volatile compounds. EOs are present abundantly in seeds, flowers, leaves, rhizomes, and bark. *Jasminum* oil is thick, transparent yellowish-brown liquid with a pleasant odor generated from concrete flower extract [2, 17]. Hydro distillation, simultaneous steam distillation-solvent extraction and carbon dioxide extraction are the different methods used to prepare essential oils [18] [19]. EOs are used in medicines, massage therapies, cosmetics, aromatherapy, and perfumery as well as in the food industries. EOs act as antimicrobials and thus play important role in plant defense. They also play a role as a repellent for herbivores and aid in pollination by acting as attractive agents for insects. They are also used majorly as antibacterial, antifungal, and insecticidal agents. Aromatic compounds, terpenes/terpenoids, and aliphatic molecules, predominantly having low molecular weight are the components of EOs [17].

35.2 Classification

Oleaceae (the olive family) comprises 28 genera and about 900 species, scattered on all the continents apart from Antarctica. This family includes commercially important genera cultivated for their fruit, oil, timber, or as ornamentals [20, 21]. Most of the species of Oleaceae are trees and it includes small shrubs (*Menodora* spp.), lianas (*Jasminum* spp.), and herbaceous plants (*Dimetra craibiana*). Oleacea (Fig. 35.1) is divided into five tribes i.e., Jasmineae, Fontanesiease, Myxopyreae, Oleeae, and Forsythieae [22].

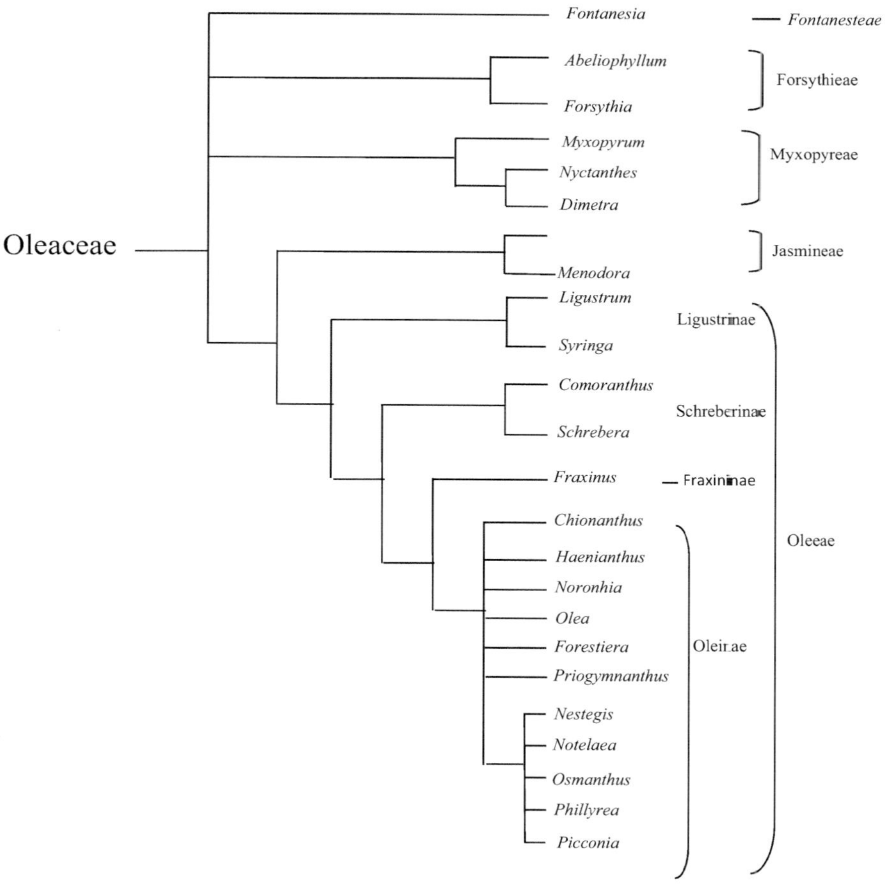

Fig. 35.1 Phylogenetic tree of Oleaceae genera's

35.2.1 Species

35.2.1.1 *Jasminum sambac*/Arabian Jasmine

Arabian jasmine is the common name of *Jasminum sambac* [23]. It is also known as the "Queen of the night" or "Moonlight of the grove" because to spread aroma the flower opens at midnight. *J. sambac* is the national flower of the Philippines it is also called Sampaguita which means "I promise you" [24, 25]. All over the year, it produces flowers while the maximum flower production is noticed during March to April. From October to February most of the religious events and wedding ceremonies take place which in turn increases the number of flowers demand [26]. From ancient times, it is considered a sacred flower. It is extremely profitable and due to its pleasant smell, its demand is increasing globally [27].

35.2.1.2 *Jasminum grandiflorum*/Spanish Jasmine

Jasminum grandiflorum also called Spanish jasmine or Royal jasmine is an alluring flower and is important industrially [27, 28].

35.2.1.3 *Jasminum multiflorum*/Winter Jasmine

Jasminum multiflorum is also known as "Winter jasmine, Downy jasmine, and Indian jasmine". It is used as folk medicine from old times.

35.2.1.4 *Jasminum gracile*/Wax Jasmine

Jasminum gracile known as wax jasmine commonly grown in Florida. *J. gracile* is a well-known ornamental shrub used for the embellishment of private lawns, parks, and public buildings. Its plants are strong have a long life, and possess subtle fragrant flowers [7].

35.2.1.5 *Jasminum auriculatum*/Needle Flower Jasmine

Jasminum auriculatum is commonly known as needle flower jasmine found in India, Nepal, Sri Lanka, and Western Penisula [29].

35.2.1.6 *Jasminum parkeri*/Dwarf Jasmine

Jasminum parkeri commonly known as dwarf jasmine is an ornamental shrub that is about to extinct. It possesses attractive flowers that are brilliant yellow and fragrant. No attempts have been made to expand this specie. This specie was discovered in India, but its ornamental potential has not yet been exploited in India. *J. parkeri* seeds have a low germination rate and slow seedling potential, but they can also be used for multiplication. Preserving its population's asexual reproduction is a successful alternate strategy. Plant growth regulators (PGRs) are essential for regulating growth, propagation, and physiological and metabolic processes when developing ornamental plants from cuttings. Asexual reproduction through stem cuttings is an efficient technique for conserving plant species that are at risk of extinction or have insufficient natural regeneration [30].

35.2.1.7 *Jasminum polyanthum*/Pink Jasmine

Jasminum polyanthum also called pink jasmine is a fast-growing creeper. This is the native plant of China hence also known as Chinese jasmine. At the beginning of spring, pinkish-white flower buds are produced which grow into flowers with five petals. This plant grows exceptionally fast and may result in a closely packed bush proliferated through stem cuttings, seeds, and suckers [6, 31].

Following are the botanical names of *Jasminum polyanthum*:

- *Jasminum* excellent
- *Jasminum* blini
- *Jasminum* delafieldii [31].

35.2.1.8 *Jasminum mesnyi*

Jasminum mesnyi is commonly known as Primrose. Jasmine is an evergreen shrub that produces bright yellow flowers from March to April. Its leaves are oval shaped oppositely attached to branches. It is native to China and also cultivated in Indian gardens [32].

35.3 Agronomy

35.3.1 Soil Conditions

In comparison to conventional cultivation, organic cultivation of jasmine increases soil fertility and reduces the chances of eutrophication. Organic cultivation increases carbon/nitrogen, carbon/phosphorous, carbon/potassium, and nitrogen/phosphorous ratios while reducing phosphorous and nitrogen concentrations in soil. *Jasminum* plants grow best at alkaline pH under organic conditions. Soil is rich in clay and less sandy in texture [33].

35.3.2 Climate

The Jasmine crop grows best in coastal climates in both plain as well as undulated land. This is a sun-loving crop and nurtures best in a comparatively dry environment. Those plants with exposure to sunlight all over the day produce a high yield. Optimum growth of this plant is observed in sandy soil, red loamy soil, and laterite soil. Soil having poor drainage is not appropriate for its production. Prolonged water-logging results in the decomposition of plant roots which results in wilting [24, 26].

35.4 Land Preparation

35.4.1 Special Cultural Practices

35.4.1.1 Watering and Weed Management

In *Jasminum* cultivation, during the period of rapid plant growth water is critically important. For better growth, the soil must be drenched with moisture [24]. Few types of jasmine flowers are planted throughout the year. The plants are watered depending on the climate and soil conditions. Every individual plant needs 20–30 l of water through drip irrigation from December to May. Extra weeds should be removed twice a year by a light digging around the basin. Enrichment with leaf litter or polythene cover limits weed growth and conserves the moisture content of soil [26].

35.4.1.2 Soil and Moisture Conservation

In undulated areas, the cultivation of *Jasminum* causes the leaching of nutrients and soil. Crescent-shaped bunds should be made to prevent leaching losses. In summer dispersing the leaf litter surrounding the plant basin aids in protecting the humidity. The dispersed leaf litter thus changes into organic manure that facilitates the plant to grow. The pH of coastal soil is acidic in nature and ranges from 4.5 to 5.0. To neutralize soil pH, agricultural limestone should be added about 15 to 20 days before the addition of fertilizer. Neutral pH facilitates the uptake of nutrients by plants from soil [26].

35.5 Planting

The spacing between the plants of about 2.0 m × 2.0 m or 2.4 m × 2.4 m accommodates up to 1000 plants per acre. Two cubic feet of pit size is needed, and the pits are occupied with the blend of topsoil, compost, and neem cake to minimize soil-borne pests. For coastal areas planting in August is suggested [26].

35.6 Fertilizers

In *Jasminum* cultivation, the biggest cost for farmers is the use of chemicals like fertilizers and pesticides. To produce the finest bud farmers are utilizing both organic and inorganic fertilizers. Animal manure and decayed plant parts are used commonly as organic fertilizers, but these are far less efficient than inorganic fertilizers. Urea, potash, and ammonium sulfate are the most used inorganic fertilizers, which increase *Jasminum* yield by up to 50%. Before application, the blending of various types of fertilizers aids in boosting the quantity and quality of *Jasminum* flowers. The side effects of these chemical fertilizers include discoloration of vegetation, yellowing of buds, burning effects, and leaf twirling. These side effects may be due to wrong application methods such as improper quantity and timing of the fertilizer. Inappropriate application methods may affect farmers' health, causing skin allergies, vomiting, and dizziness. Precautionary measures must be taken when dealing with these hazardous fertilizers [24].

35.7 Weeding

Weeding is contemplated as an arduous maintenance activity. The existence of weeds in the field provides the area for the reproduction of pests. The farmers remove weeds from *Jasminum* farm on average twice or thrice a week. For reducing weed growth, the farmers cover the basal portion of *Jasminum* with a thick polyethylene sheet or rice stalk [24].

35.8 Manuring

During the month of September and October, well-decomposed compost should be added. Application of fertilizers should also be done twice a year. Nitrogen, phosphorous, and potassium are applied in 90:570:130 g/plant during the first year, while their quantity increases in the second year up to 175:1150:260/plant. From the third year onward N: P: K of 270:1710:400 g/plant should be applied [26].

35.9 Plant Growth Hormones

Phytochromes are the hormones that are naturally produced in plants. They accelerate plant growth and influence in minute amounts. They are crucial for the growth of different organs such as stems, buds, roots, and fruits. They are also involved in the expansion of both the length and width of plant cells. Plant growth regulators

(PGRs) are synthetic organic molecules that either increase or decrease the growth of plants [34].

The seasonal blossoming of all *Jasminum* species is a severely limiting factor that significantly affects both jasmine flower producers and consumers and is expected to have an impact on commercial output. The application of Naphthalene acetic acid (NAA) reported the maximum plant height among the various treatments. Cycocel plays a useful role in the dwarfing process because it suppresses the duration of green growth and inhibits it. As a result, the internodes get shorter and the stems become thicker [1].

Although synthetic fertilizers assist farmers in increasing crop yield, they also have disastrous health effects. Excessive use of fertilizer lessens the amount of oxygen in the water and causes the emission of carbon dioxide, which is a potent greenhouse gas. Chemical fertilizer also minimizes soil pH thus making the soil acidic [35].

35.10 Removal of Flower Bud

Some plants start producing flowers right after planting as new shoot develops. Trimming of flower buds up to 6 months after planting boosts vegetative growth [26].

35.11 Pruning

Farmers follow various defoliation and trimming techniques to start flowering and development of buds. Pruning after 6 months aids in flower development and usually, these processes are performed when plants grow enough to be pruned. At the beginning of the rainy season, various types of maintenance activities are utilized by farmers that avert the existence of pests and diseases and boost plant growth [24].

Optimum sunlight, aeration, and pruning of plants at a suitable time aid in minimizing the pest population in *Jasminum*. Removing thickly grown, dead and watery shoots throughout the year results in excess flower yield. The plant covering should be preserved up to 4 feet from the ground for an effortless collection of flowers. September to October is the perfect time for pruning.

After pruning, the trim ends should be treated with a mixture of copper sulfate ($CuSO_4$), quick lime (CaO), and water. For the ubiquitous production of flowers staggered pruning is advised while pruning the entire plant at the same time results in minimal yield [26].

35.12 Harvesting

Jasmine has the main flowering season from March to October and the off-season from November to February. There is either no blooming at all or very little flowering during this off-season. Development regulators have become crucial tools in floriculture in regulating growth and blooming [1, 36].

Jasminum flower blooms for a very short interval of about 12–20 h. In the evening at about 6–8 when the temperature drops flower opening is at its peak. At dawn, the farmers usually pick the mature unopened flowers while the unopened buds are harvested from dusk until nightfall. One of the most crucial steps in *Jasminum* production is harvesting which usually begins as early as 3 in the morning. Collecting a sufficient quantity of buds in the short calculated time of morning requires active labor [24]. Economic yield might be anticipated in the first year of planting with adequate management practices [26].

Harvesting flowers at the proper time is essential because it influences the storage life of flower buds. Right after harvesting the flower buds were delivered immediately because the completely opened flowers are not commercially important [37].

35.13 Ecology

Pests and diseases

Jasminum diseases were believed critical because they reduce the plant flowers largely. Some common diseases include witches' broom caused by phytoplasma. Tobacco mosaic virus infection, cercosporoid leaf spots, and various other folicolous taxa [9].

Witches broom

In witches broom disease (Fig. 35.2), plant pathogenic phytoplasmas parasitize the phloem tissue of the diseased plant host and are spread by insect vectors. The anatomy of phytoplasma shows that it consists of a ribosome and DNA [38]. Stunting, floral deformity, excessive shoot proliferation, yellowing of leaves, and reduction in plant strength were observed in phytoplasma-infected plants that may lead to the death of the plant presented in (Fig. 35.2) [23, 39].

Viral infections

Arabic mosaic virus (ArMV) in *Jasminum mesney* and *Jasminum officinale*, Cucumber mosaic virus (CMV) in *Jasminum polyanthum*, and Tobacco ringspot virus (TRSV) in *Jasminum nudflorum*, *Jasminum mesney,* and *Jasminum officinale* were observed. *Jasminum* chlorotic ring spot virus (JCRSV) was observed in *J. sambac* and has been claimed to be transmitted by whitefly (*bemisia tabaci*) shown in the (Fig. 35.3a, b) [7].

Fig. 35.2 *Jasminum* infected with witches' broom [23]

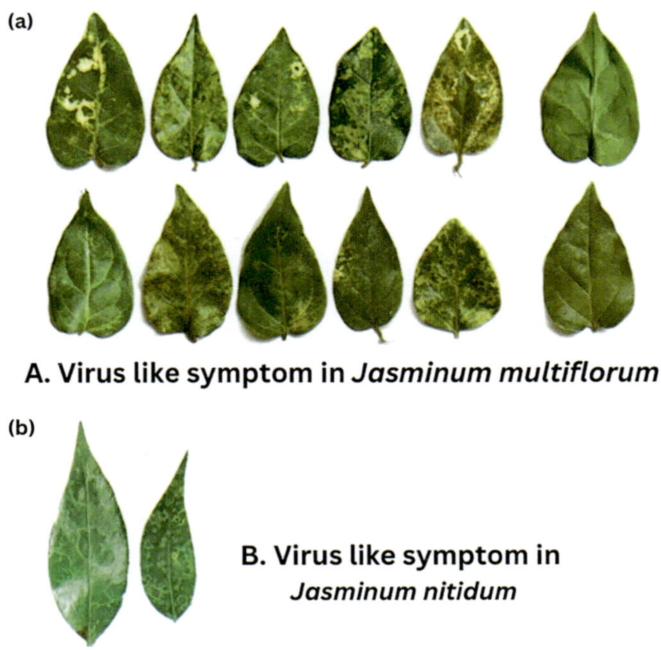

Fig. 35.3 *Jasminum* infected with viral infections [40]

Cercospora leaf spot

Cercospora is the fungi that cause cercospora leaf spot disease. Early signs of cercospora illness include small minute brown patches with grey-colored borders, which later develop into a filthy grey mass with a brown margin. Later, the disease may influence pods, petioles, and stems. The pathogen results in defoliation,

Fig. 35.4 Yellow mosaic virus-infected *Jasminum* leaves (**a**). Vein clearing (**b**). severe mosaic infected leaf [44]

premature defoliation in the severe stage, leaf form wrinkles, and pod size reduction. This disease spread widely in monsoons. The development of hazardous compounds, such as cercosporin based on perylene quinone, which can harm plants, microorganisms, and animals, makes pathogens more severe [41].

Tobacco mosaic virus infection

The tobacco mosaic virus (TMV) is the simplest. It is easily accessible and widely studied [42]. The spreading of the virus from the host cell is due to numerous processes including the multiplication of viral nucleic acids, protein formulation targeting, and dissemination of disease-causing entities [43].

Yellow Vein Mosaic Disease

In yellow vein mosaic disease (Fig. 35.4a, b) initially, the veins of the leaves turn yellow. Then, the whole leaf becomes yellow or cream-colored in a drastic situation. In the beginning, the growth of the affected plant stopped and the fruits become out of shape and hard in appearance (Fig. 35.4a, b).

35.14 Description of Crop

China is one of the largest producers of Jasmine and produces almost 60% of the world's *Jasminum* [45]. Jasmine flowers are extensively distributed all over the world but mostly in Asia. Almost 200 species of jasmine are known [6]. It is present largely in tropical areas of the world but the diversity of this specie is present in Southeast and Southwestern Asia. Jasmine is one of the profit-oriented crops cultivated both on hills and in plains, especially in China, the United States, Japan, Kashmir, Russia, Hong Kong, Afghanistan, Senegal, Persia, Africa, Europe, the Pacific Islands, Formosa, Tahiti, Australia, Sri Lanka, India and Egypt [5, 26, 45, 46]. Some of its varieties like *Jasminum grandiflorum* are exported to Morocco, Spain, Italy, and France for oil extraction shown in Fig. 35.5 [47].

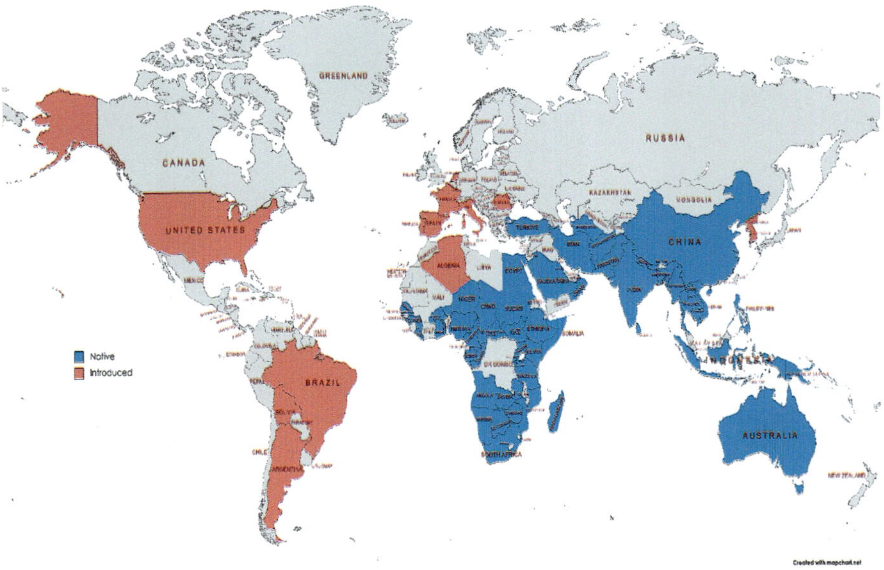

Fig. 35.5 *Jasminum* distribution map. (Created with World Map – Simple I MapChart)

1. Important chemical constituents

Jasminum leaves contain numerous constituents including leaves that are rich in phenols, polyphenols, alkaloids, flavonoids, steroids, tannins, coumarins, anthocyanins, phlobatinins, leucoanthcyanins, emodine, teroinoids, saponins, phenylethanoid glycoside, three glucosides (Inerminoside A1, C, and D) and neo-clerodance diterpinoids antiviral proteins (CIP-29 and CIP-34). While β-sitosterol, α-amyrin, β-glucoside flavonoids, rutin, and secoiridoid glucosides (9-hydroxyjasminoside, 9-hydroxy jasminosidic acid, Jasmoside, and jasminoside) are also important components [31, 32].

Flavonoids, glycosides, saponins, terpenoids, and triterpenes are present in *Jasminum* root bark, while its leaves are rich in alkaloids, phenols, phenylethanoids, polyphenols, flavonoids, tannins, saponins, and sterols. Moreover, its flowers also contain (Table 35.1) alkaloids, phenolics, flavonoids, tannins, saponins, and terpenoids (Table 35.1) [48].

The *Jasminum* floral fragrance is due to the volatile organic compounds or aromatic compounds produced by flowers. The flower's aroma captivates pollinators, which avert herbivores and flower-eating insects. Isoprene-derived terpenoids are plentiful in floral fragrances and manufactured via the mevalonate pathway. Fatty acid derivatives synthesized from acetyl-CoA and benzenoids/phenylpropanoids are the main chemical components of floral fragrance. Additionally, the fragrance (Table 35.2) may be due to benzyl alcohol, linalool, methyl benzoate, jasmine lactone, and farnesene [49, 50].

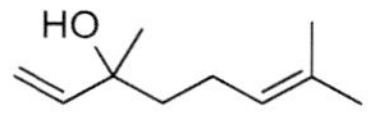 Linalool	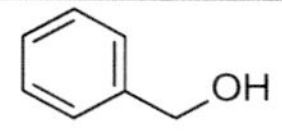 Benzyl alcohol
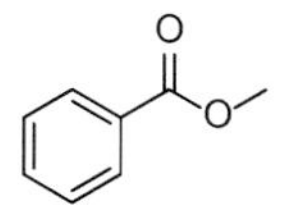 Methyl benzoate	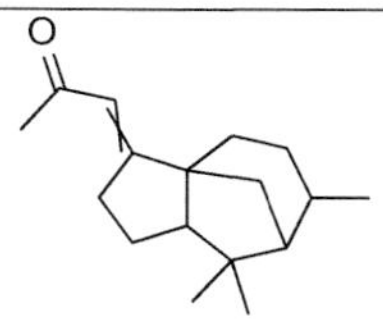 Terpenoids
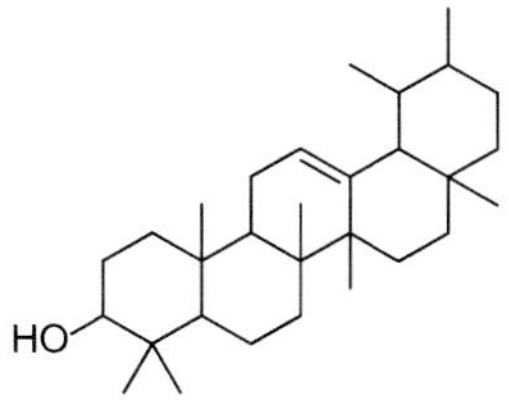 α-amyrin	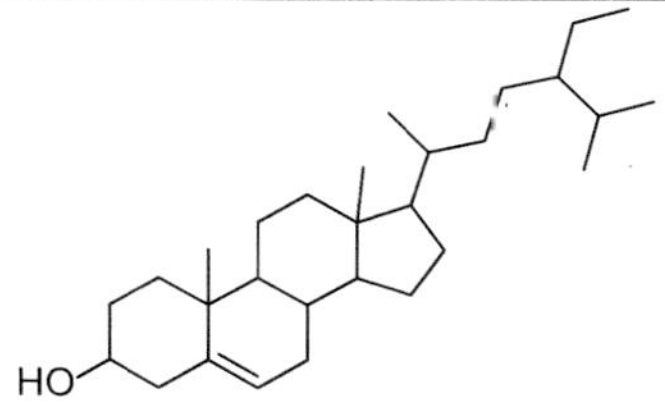 β-sitosterol
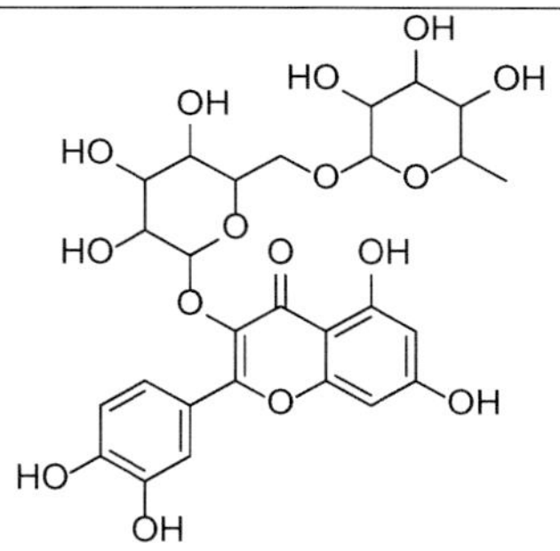 Rutin	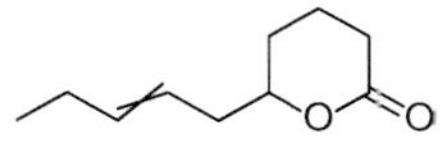 Jasmine lactone
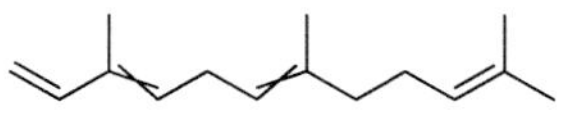 Alpha-farnesene	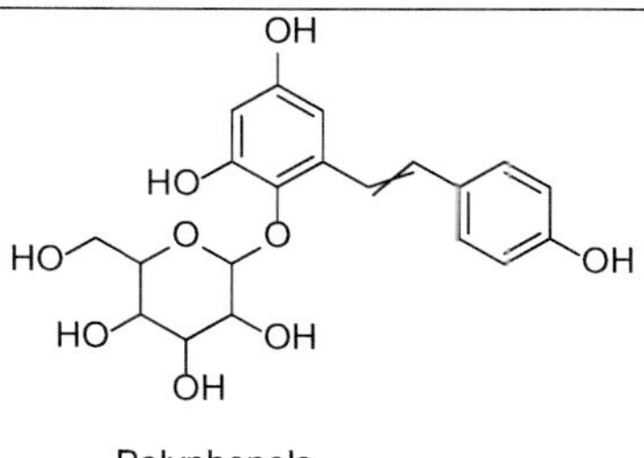 Polyphenols

Table 35.1 Chemical structures of some constituents present in *Jasminum* leaves, roots, and flowers

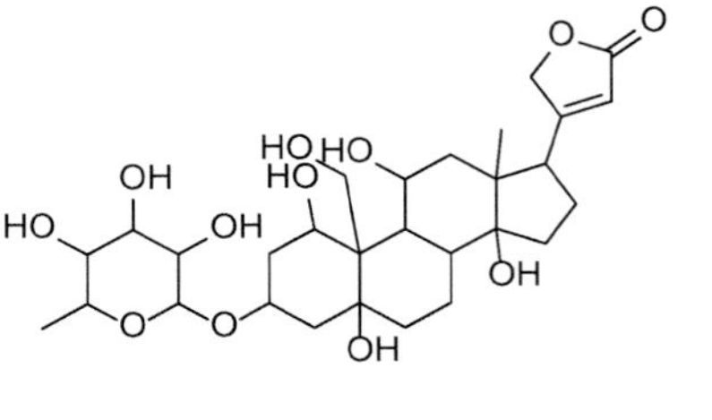 Astrobain	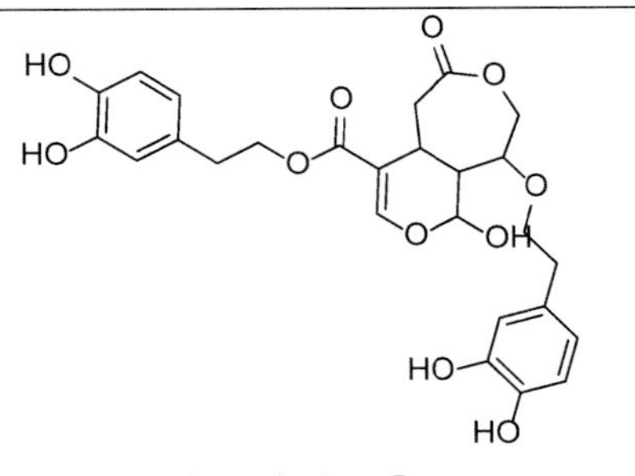 Jasmolactone B
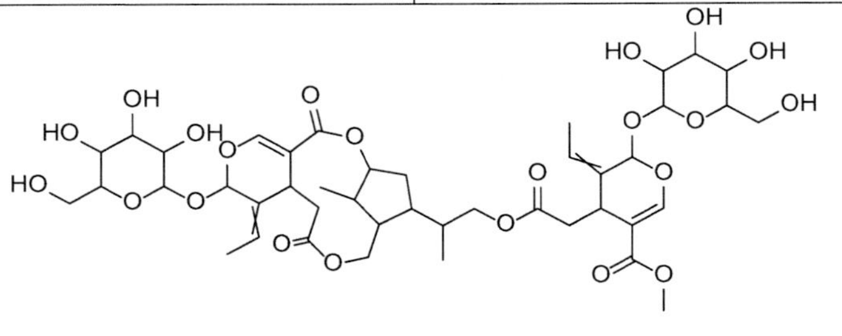 Jasmoside	
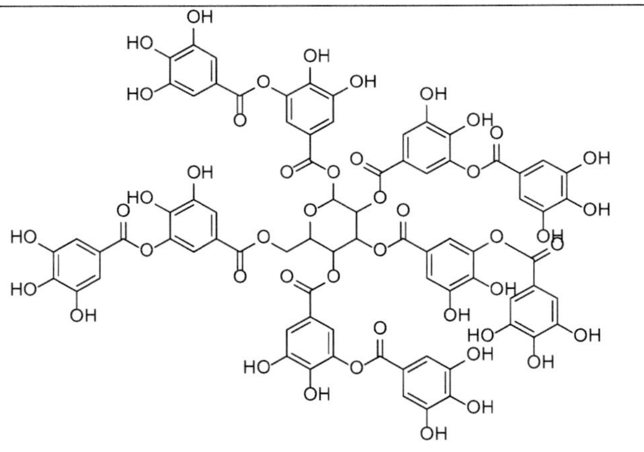 Tannins	

Tab. 35.1 (continued)

Through different tests, the qualitative bioactive compounds were assessed i.e., alkaloids by Mayer's test, saponins by Foam test, terpenoids by concentrated sulphuric acid, sugars by Fehling's test, flavonoids by concentrated HCl, qunines by NaOH test, phenol by $FeCl_3$, protein, and steroid by chloroform and sulphuric acid [31].

Table 35.2 Chemical constituents and their odor description [19, 51, 52]

Compound	Odor
Linalool	Fresh-floral, sweet fruity woody, lavender notes
Benzyl acetate	Floral fruity, fresh sweet notes
***cis* -3-Hexanol**	Green, fresh grass
6-Methyl heptan-2-one	Green, herbaceous, weak oily
Ethyl acetate	Fruity
Ethyl butanoate	Pineapple
Methyl α-methylbutanoate	Sweet, apple
Hexan-3-ol	Alcoholic
Styrene	Balsam floral
Benzyl alcohol	Floral, rose
Ethyl benzenecarboxylate	Fruity, musty
β-elemen	Sweet
Cinnamyl acetate	Floral, balsam
Benzyl tiglate	Mushroom, rose
δ-cadinene	Woody
Methyl methanthranilate	Fruity, phenolic
Geranyl formate	Fruity, green
Eugenol	Green
β-pinene	woody
Trans-linalool oxide	Fresh floral, light sweet citric
cis-Linalool oxide	Fresh floral, citric notes
Methyl benzoate	Fragrant fruity
Phenylethyl alcohol	Floral, rose like, fragrant honey notes
Phytyl acetate	Faint floral balsamic
Phytol	Weak floral balsamic, sticky side notes
Isophytol	Weak floral balsamic
Methyl linoleate	Fatty oily
***p* -Cresol**	Medicinal aromatic
Methyl oleate	Fatty oily
Geranyl linalool	Floral, rose geraniol lavender notes, weak fruity
***α*- Terpineol**	Fragrant floral, weak fruity with lilac notes
Methyl salicyclate	Fragrant minty, sweet spicy, wintergreen notes
Nerol	Floral, weak rose notes, sweet fruity
Phenylethyl acetate	Sweet fruity, rose honey notes
Indole	Unpleasant-fecal, musty cadaverous floral in dilution
Chavicol	Sweet phenolic, anise like, green minty notes
Geraniol	Sweet floral, rose notes, weak fruity
Methyl palmitate	Fatty oily
***cis* -Jasmone**	Intense jasmine like, warm floral, weak fruity
***trans* -Jasmone**	Floral jasmine like, weak spicy, warm fruity
Eugenol	Spicy, strong clove cardamon notes
Isoeugenol	Spicy, intense clove like, sweet and woody undertones

(continued)

Table 35.2 (continued)

Compound	Odor
Methyl anthranilate	Fruity, grape like
δ-Jasmine lactone	Jasmine like, warm floral
α-(*E,E*)-Farnesene	Weak floral
***trans* -Nerolidol**	Floral fruity, rose apple and green citrus notes, woody waxy
***cis* -3-Hexenyl benzoate**	Green herbal, weak woody
***cis* -Methyl jasmonate**	Floral sweet, herbaceous
Benzyl benzoate	Sweet balsamic, faint floral
Benzyl salicyclate	Faint sweet, weak herbaceous medicinal

35.15 Different Species of *Jasminum* and Their Medicinal Uses

Infectious diseases are the substantial reason of morbidity and mortality. Synthetic drugs used in treatment of these diseases have numerous side effects. Tannins, alkaloids, terpenoids and flavonoids are the secondary metabolites present abundantly in plants. These secondary metabolites exhibit therapeutic potential thus used as a safe, cheap, pollution free and effective alternative for treating different diseases and have lesser side effects than allopathic system [53].

Plants have been believed a potent medicinal source, over the past centuries. Presently, numerous people worldwide even now use plants, as a substitute for traditional/indigenous medicine. In comparison to synthetic drugs, plants and their components are preferred. Nature blesses human beings with diverse medicinal plants, which in a true sense form warehouse for treating nearly all disorders. Currently, in Africa, more than 80% of the population still uses herbal products as a treatment. Under the term of complementary medicine, even developed countries now use herbal products. Many plants are still under examination. Biological and phytochemical analysis alleviates the therapeutic effects and additionally, this may lead to the finding of a novel compounds with new mechanisms [20].

The two main issues that will influence society's health in the present are the emergence of new diseases and the development of antibiotic resistance among microorganisms. Microbial illnesses are the cause of millions of deaths. One of the most favorable antimicrobial medicines that combat resistance in the current environment may come from plants and their bioactive components. They are pioneers in developing novel medications that are either exploited as a starting point for synthesizing bioactive components [48].

Chemical constituents in medicinal plants cause physiological effects in the body, such as the removal of reactive oxygen species (antioxidant), inhibition of blood glucose-raising enzymes (anti-diabetic), inhibition of inflammatory action (anti-inflammatory), eradication of pathogens (anti-microbial) and retardation in cell multiplication (anti-cancer), etc. Considering that, bioactive plant components contribute to their therapeutic characteristics and are used to treat human ailments. These substances belong to the same class as phytochemicals such as flavonoids, alkaloids, steroids, saponins, and saponin analogs [31].

Jasminum species are used extensively for curing various diseases including ulcers, intestinal worms, boils, eye disorders, and skin and venereal diseases. Its leaf extracts are efficiently used for curing toothache, urinary tract infections, breast carcinomas, headaches, muscle pain, and ulcer in the throat, gums, and mouth [6, 54]. *Jasminum* possessess emmenagogue properties (Table 35.3) that is why its use is prohibited in pregnancy [19].

Jasminum has countless medicinal effects (Table 34.5). The role of *Jasminum* in combating different diseases, the plant parts, solvents, models and biomarkers are briefly described in table 35.4:

Table 35.3 *Jasminum* spp. and their therapeutic uses

Name of specie	Therapeutic use	References
Jasminum sambac	Antimicrobial and antioxidant effects, odontalgia, skin diseases, otalgia, leprosy, anticancer, strangury, fixing loose teeth, ringworm, dysmenorrhea, antiulcer, anti-inflammatory, spasmolytic, chemopreventive anti-acne, wound healing activities, vasodilator, antiseptic, analgesic, \antidepressant, sedative, cytotoxic activity towards Brine Shrimp Artemia, Dalton's ascites lymphoma, and Hep-G2, headache and pain due to cracked bones and joint displacement	[20]
Jasminum officinale	Anti-inflammatory, antiulcers, colitis, arthritis, enteritis, iridoid, sesquiterpene, antiseptic, expectorant, anti-depressant, tranquilizer, inflammation in the eye, cough, fever, gastrointestinal disorders, and oligomenorrhea	[55]
Jasminum mesnyi	Anorexia, central nervous system disorders, gastric issues, diabetes, myalgia, nocturnal emission, migraine, spinal pain, hepatic disorders and joint diseases	[56]
Jasminum grandiflorum	Antiulcer, odontalgia, skin diseases, otalgia, leprosy, strangury, anti-anthelmintic activity, cytotoxic effects and dysmenorrhea	[20]
Jasminum humile	Astringent, for curing hard lumps and chronic fistulas, ringworms and as a tonic, anticancer and cytotoxic activities	[20]
Jasminum multiflorum	Fever, cough, abdominal distention, diarrhea, decreasing blood glucose levels, controlling menstrual flow, anticancer, swollen eyes, conjunctivitis, cardiotropic activities and control HCV	[57]
Jasminum auriculatum	Antioxidant, antibacterial, antiseptic, suppurative, leprosy, deobstruant, antihelmintic, aphrodisiac, emollient, anti-ulcer, anti-acne, chemoprotective, wound healing and cytoprotective activity	[29]

Table 35.4 Effect of *Jasminum* spp. on different diseases

Botanical Name	Plant Parts	Solvent/Dose	Activity	Model	Biomarkers Affected	References
	Flower	Ethanol extract (300 mg/kg p.o.)	Chemopreventive	7,12-dimethylbenz(a) anthracene (DMBA)-induced rat mammary carcinogenesis	**Increase:** vitamin E, vitamin C, reduced glutathione, glutathione peroxidase, superoxide dismutase, catalase **Decrease:** thiobarbituric acid reactive substance	[48, 58]
Jasminum grandiflorum	Leaf	The methanol extract (100–800 μL/ml)	Anti-inflammatory	Lipopolysaccharides induced nitric oxide in rat peritoneal macrophage	**Decrease:** No production	[48, 59]
	Leaf	Ointment	Wound healing	Dermal wound healing in diabetic rats	**Increase:** wound contraction, total hydroxyl proline, hexosamine, protein, DNA content, tensile strength, collagen & fibrous tissue, no. of blood vessels, superoxide dismutase, catalase **Decrease:** lipid peroxidation	[48, 60]
Jasminum sambac	Root	Ethanol extract	Anti-inflammatory	Carrageenan-induced rat paw edema model and cotton pellet-induced granuloma in rats	**Increase:** superoxide dismutase, catalase **Decrease:** paw edema, granuloma formation, aspartate aminotransferase, alanine aminotransferase, lipid peroxidation	[48, 61]
	Flower	*Lactobacillus rhamnosus* fermented extract	Anti-aging	Ultraviolet B or H2O2-induced HS68 dermal fibroblast cells	**Increase:** phosphor-nuclear respiratory factor 2, antioxidant gene expression **Decrease:** reactive oxygen species, aging markers	[48, 62]
	Aerial part	Hydro-alcoholic extract (100 mg/kg)	Analgesic	Subcutaneous naloxone injection in rats	**Decrease:** pain, morphine withdrawal	[63]
	Aerial part	Hydro-alcoholic extract	Cardioprotective effect	Rabbit ventricle section from ADR-induced group in heart tissues	**Increase:** oxidative stress **Decrease:** adrenaline	[64]

Jasminum abyssinicum	Leaves	Water extract	Anthelmintic activity	Naturally infected sheep lamb with nematode	**Decrease: in** fecal egg count	[65]
	Root	Methanol extract	Anti-inflammatory activity	Sub plantar injection in mice	**Increase:** kinin synthesis, neutrophil-derived mediators	[66]
	Root	Methanol extract	Analgesic activity	Acetic acid induced writhing in mice	**Increase:** stress tolerance capacity of animal **Decrease:** no of writhing	[66]
Jasminum polyanthum	Aerial parts	Aqueous extract	Anthelmintic activity	Pinworms in naturally infected mice	**Decrease:** parasites in the body	[67]
	Leaf	Methanol extract	Antidiabetic	NA	**Decrease:** α-amylase and α-glucosidase	[68]
	Flower	Methanol extract	Antidiabetic	NA	**Decrease:** α-amylase and α-glucosidase	[68]
Jasminum auriculatum	Leave extract	Ethanol extract	Wound healing	Rats skin cut off from the dorsal back	**Increase:** wound contraction, total hydroxyl proline, tensile strength, collagen fibers, fibroblast cells, no. of blood vessels **Decrease:** epithelialization period, wound area	[69]
Jasminum officinale	Aerial part	Ethanol extract	Pentobarbital	Albino mice	**Increase:** sleeping time, γ-aminobutyric acid,	[70]
	Flower	Ethanol extract	Anti-nociceptive effect	Acetic acid induced writhing in mice	**Increase:** soothing **Decrease:** pain stimulus	[71]
	Flower	Ethanol extract	Anti-inflammatory effect	Xylene-induced ear edema in mice and cotton pellet granuloma test in rats	**Increase:** superoxide dismutase, catalase **Decrease:** granuloma formation, aspartate aminotransferase, alanine aminotransferase, lipid peroxidation	[71]
	Flower	Aromatherapy oil	Analgesic	Pregnant women	**Decrease:** pain, and labor in nulliparous women	[72]

35.15.1 Antimicrobial effect of *Jasminum* spp.

The antimicrobial analysis (Table 35.5) was carried out to study the antimicrobial activity of *Jasminum* samples. The zone of inhibition was assessed which tells about the antimicrobial sensitivity of the sample towards different pathogens is presented in Table 35.5 [73].

Table 35.5 Antimicrobial potential of different *Jasminum* spp.

Botanical name	Microbes	ZOI (mm)	References
Jasminum arborescens	*Escherichia coli*	2.8	[48, 74]
	Pseudomonas aeruginosa	3.1	
	Staphylococcus aureus	3.7	
	Bacillus subtilis	3.6	
Jasminum azoricum	*Staphylococcus aureus*	20	[48]
	Bacillus cereus	24	
	Bacillus subtilis	9	
	Escherichia coli	14	
	Pseudomonas sp.	17	
	Salmonella typhi	22	
	Vibrio cholera	18	
	Streptococcus sp.	17	
	Corynebacterium sp.	14	
	Proteus vulgaris	18	
Jasminum grandiflorum	*Bacillus subtilis*	25	[48, 58–60, 73, 75]
	Escherichia coli	21	
	Staphylococcus epidermidis	15	
	Shigella flexneri	10	
	Alternaria alternata	27	
	Candida albicans	10	
Jasminum polyanthum	*Escherichia coli*	8	[48, 73]
	Klebsiella pneumonia	9	
	Staphylococcus aureus	13	
	Pseudomonas aeruginosa	13	
	Aspergillus flavus	8	
Jasminum syringifolium	*Escherichia coli*	21.33	[48]
	Pseudomonas aeruginosa	16.67	
	Staphylococcus aureus	21.67	
	Bacillus cereus	22.33	
	Staphylococcus epidermidis	16.33	
	Vibrio cholera	18.67	
	Proteus mirabilis	15.33	
	Shigella flexneri	22.67	
	Salmolnella enterica	19.33	
	Klebsiella pneumonia	18.33	
	Aspergillus niger	17.33	
	Candida albicans	15.33	

(continued)

Table 35.5 (continued)

Botanical name	Microbes	ZOI (mm)	References
Jasminum sambac	*Aspergillus niger*	9	[73, 76]
	Aspergillus flavus	13	
	Microsporum nanum	17	
	Trichophyton floccosum	21	
	Trichophyton tonsurans	30	
	Serratia marcescens	25	
	Pseudomonas aeruginosa	18	
Jasminum aungustifolium	*Serratia marcescens*	25	[48, 73]
	Salmonella spp.	8	
	Pseudomonas sp.	6	
	Lactobacillus sp.	6	
	Yersinia sp.	5.5	
	Enterococcus sp.	5	
Jasminum humile	*Pseudomonas aeruginosa*	17	[48]
Jasminum officinale	*Klebsiella pneumonia*	18	[48]

The antimicrobial activity of *Jasminum* extracts were compared with standard antibiotics. The results display that in comparison to standard antibiotic some plant extracts exhibit more growth inhibitory activity [53].

35.15.2 *Antiviral Efficacy of Jasminum Officinale L. var. Grandiflorum Extracted Oleuropein Against Hepatitis B*

From the flower buds of *Jasminum officinale L. var. grandiflorum* oleuropein (Ole) was extracted. Serum from the congenitally infected duck with hepatitis B virus (DHBV) was used as the source of DHBV (Fig. 35.6). The isotonic saline liquid diet containing Ole was administered orally to intravenously infected ducks with HBV. Ole was effectively able to block HBsAg release in HepG2 2.2.15 cells in a dose-dependent manner with an IC50 value of 23.2 g/ml. ole (80 mg/kg, twice daily) was able to suppress DHBV DNA level in infected ducks. However, in the HepG2 2.2.15 cells the mechanism of how Ole block HBsAg gene expression is not yet clear [77].

35.16 Myths, Legends, Tales, Folklore, and Interesting Facts

In Indian Mythology trees have predominant significance as seen in the written and oral Indian traditions. Trees have exceptional admiration in Indian culture. The gods were believed as branches of trees. The trees that provide shade and produce fruits

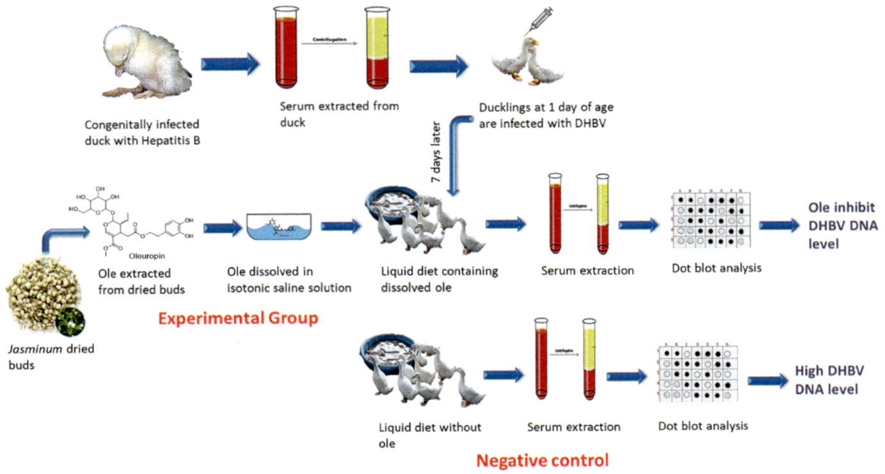

Fig. 35.6 Oleuropein in treating Hepatitis B virus

were associated with life and fertility. One of the finest attempts to sustain natural resources is the idea and faith in sacred trees and forest groves. Sacred groves also called church forests, sacred forests and fetish forests are scattered ubiquitously in the world [78, 79].

Evergreen trees, that stay green all over the year, signify eternity. While deciduous trees that shed their leaves in the autumn season and produce new leaves in the spring symbolized rebirth after death or immortality. *Jasminum* flowers have universal practice in religious worship. All flowers enriched with colors and aromas were considered good for religious purposes [79].

References

1. Dhanasekaran, D. (2019). Augmentation of flowering in Jasmine (Jasminum sambac. Ait.) through growth hormones. *Annals of Plant and Soil Research, 21*(2), 116–120.
2. Rao, Y. R., & Rout, P. K. (2003). Geographical location and harvest time dependent variation in the composition of essential oils of Jasminum sambac.(L.) Aiton. *Journal of Essential Oil Research, 15*(6), 398–401.
3. Joulain, D. (2021). Jasminum grandiflorum flowers—Phytochemical complexity and its capture in extracts: a review. *Flavour and Fragrance Journal, 36*(5), 526–553.
4. Sobhana, A. (2012). Standardisation of planmting of material production in Bush Jasmine (Jasminum sambac). In *Quality seeds and planting material in horticultural crops* (p. 487).
5. Akhtar, N., et al. (2021). Comparative palynology and taxonomic implication of Jasminum L.(Oleaceae) species from Pakistan on the bases of scanning electron microscopy. *Microscopy Research and Technique, 84*(10), 2325–2336.
6. Saraswathi, R., & Palayyan, M. (2022). Systemic Review on Jasminum Polyanthum. *World Journal of Pharmaceutical Sciences, 100117*, 151–100154. https://doi.org/10.54037/WJPS

7. Kamenova, I., Adkins, S., & Achor, D. (2006). Identification of tomato mosaic virus infection in jasmine. *Acta Horticulturae, 722*, 277–284.
8. Xiao, W., et al. (2017). Chemistry and bioactivity of Gardenia jasminoides. *Journal of Food and Drug Analysis, 25*(1), 43–61.
9. Wikee, S., et al. (2011). Colletotrichum species from Jasmine (Jasminum sambac). *Fungal Diversity, 46*(1), 171–182.
10. Chaitanya, H., Nataraja, S., & Krishnappa, M. (2019). Review on Propagation Techniques of Jasmine (Jasminum sambac (L.)). *Journal of Pharmacognosy and Phytochemistry, 8*(1), 593–596.
11. Wang, S., et al. (2020). Comparison of volatiles in different jasmine tea grade samples using electronic nose and automatic thermal desorption-gas chromatography-mass spectrometry followed by multivariate statistical analysis. *Molecules, 25*(2), 380.
12. Xing, L., et al. (2019). Recent advances in the understanding of the health benefits and molecular mechanisms associated with green tea polyphenols. *Journal of Agricultural and Food Chemistry, 67*(4), 1029–1043.
13. Chen, M., et al. (2017). Changes in the volatiles, chemical components, and antioxidant activities of Chinese jasmine tea during the scenting processes. *International Journal of Food Properties, 20*(3), 681–693.
14. Zhang, Y., et al. (2022). Analysis of Volatile Components of Jasmine and Jasmine Tea during Scenting Process. *Molecules, 27*(2), 479.
15. Li, H., et al. (2018). Characterization of volatile compounds and sensory analysis of jasmine scented black tea produced by different scenting processes. *Journal of Food Science, 83*(11), 2718–2732.
16. Wei, A., Shibamoto, T., & Chemistry, F. (2007). Antioxidant activities and volatile constituents of various essential oils. *Journal of agricultural and food chemistry, 55*(5), 1737–1742.
17. Ayaz, M., et al. (2017). Neuroprotective and anti-aging potentials of essential oils from aromatic and medicinal plants. *Frontiers in Aging Neuroscience, 9*, 168.
18. Mourya, N. M. N., Bhopte, D. B. D., & Sagar, R. S. R. (2017). *A review on Jasminum sambac: A potential medicinal plant. International Journal of Indigenous Herbs and Drugs*, 13–16.
19. Jirovetz, L., et al. (2007). Chemical composition, olfactory evaluation and antimicrobial activities of Jasminum grandiflorum L. absolute from India. *Natural Product Communications, 2*(4), 1934578X0700200411.
20. Mansour, K. A., et al. (2022). Cytotoxic effects of extracts obtained from plants of the Oleaceae family: bio-guided isolation and molecular docking of new secoiridoids from Jasminum humile. *Pharmaceutical Biology, 60*(1), 1374–1383.
21. Wallander, E., & Albert, V. A. (2000). Phylogeny and classification of Oleaceae based on rps16 and trnL-F sequence data. *American Journal of Botany, 87*(12), 1827–1841.
22. Dupin, J., et al. (2020). Resolving the phylogeny of the olive family (Oleaceae): Confronting information from organellar and nuclear genomes. *Genes, 11*(12), 1508.
23. Al-Zadjali, A., Natsuaki, T., & Okuda, S. (2007). Detection, identification and molecular characterization of a phytoplasma associated with Arabian jasmine (Jasminum sambac L.) witches' broom in Oman. *Journal of Phytopathology, 155*(4), 211–219.
24. Sanchez, F. C., Jr., Santiago, D., & Khe, C. P. (2010). PRODUCTION MANAGEMENT PRACTICES OF JASMINE (Jasminum sambac [L.] Aiton) IN THE PHILIPPINES Fernando C. Sanchez, Jr., Dante Santiago2 and Caroline P. Khe1. *The International Society for Southeast Asian Agricultural Sciences, 16*, 126.
25. Ahmed, N., et al. (2016). Jasmine (Jasminum sambac L., Oleaceae) oils. In *Essential oils in food preservation, flavor and safety* (pp. 487–494). Elsevier.
26. Chaitanya, H., et al. (2018). Review on production techniques of GI Crop, Udupi Mallige (Jasminum sambac (L.) Aiton). *Journal of Pharmacognosy and Phytochemistry, 7*(3S), 50–52.
27. Nivedha, M., et al. (2019). In vitro antifungal evaluation of various plant extracts against leaf blight disease of Jasminum grandiflorum caused by Alternaria alternata (Fr.) Keissler. *Journal of Pharmacognosy and Phytochemistry, 8*(3), 2143–2147.

28. Radha, R., et al. (2016). Pharmacognostical, phytochemical and anthelmintic activity on flowers of Jasminum grandiflorum Linn.(Oleaceae). *International Journal of Pharmacognosy, 3*(10), 455–460.
29. Gupta, D., Saini, S. L., & Sharma, T. Antimicrobial Potential of Polyherbal Formulation Tiktadya Ghrutam-A Review. *International Journal of Research and Review, 10*, 6.
30. Kashyap, U., et al. (2021). Propagation of Jasminum parkeri: a critically endangered wild ornamental woody shrub from Western Himalaya. *Agronomy, 11*(2), 331.
31. Prakkash, M. J., Ragunathan, R., & Jesteena, J. (2019). Evaluation of bioactive compounds from Jasminum polyanthum and its medicinal properties. *Journal of Drug Delivery and Therapeutics, 9*(2), 303–310.
32. Bhushan, B., Sardana, S., & Bansal, G. (2014). Acute and sub-acute toxicity study of Clerodendrum inerme, Jasminum mesnyi Hance and Callistemon citrinus. *Journal of Acute Disease, 3*(4), 324–327.
33. Wang, W., et al. (2016). Organic cultivation of jasmine and tea increases carbon sequestration by changing plant and soil stoichiometry. *Agronomy Journal, 108*(4), 1636–1648.
34. Sajjad, Y., et al. (2017). Application of plant growth regulators in ornamental plants: a review. *Pakistan Journal of Agricultural Sciences, **54***(2), 333.
35. Abdullah, N. H., et al. (2020). The Application of Waste Mussel Shell and Nitrogen-Phosphorus-Potassium (NPK) Fertilizer on The Growth Components of Jasmine (Jasminum). *Multidisciplinary Applied Research and Innovation, 1*(1), 1–12.
36. Wang, C., et al. (2019). Optimal coupling of straw and synthetic fertilizers incorporation on soil properties, active Fe dynamics, and greenhouse gas emission in Jasminum sambac (L.) field in southeastern China. *Sustainability, **11***(4), 1092.
37. Abdullah, T., Ahmad, S., & Rejab, N.. (1992). Determination of floral stages and packaging methods for prolonged storage of Jasminum multiflorum. In *WOCMAP I-Medicinal and aromatic plants conference: Part 3 of 4 331*.
38. Lee, I.-M., Davis, R. E., & Gundersen-Rindal, D. E. (2000). Phytoplasma: phytopathogenic mollicutes. *Annual Reviews in Microbiology, 54*(1), 221–255.
39. Jones, P. (2001). Phytoplasma plant pathogens 1 2. In *Plant pathologist's pocketbook* (p. 126).
40. Dey, K. K., et al. (2018). Detection of Jasmine virus H and characterization of a second pelarspovirus infecting star jasmine (Jasminum multiflorum) and angelwing jasmine (J. nitidum) plants displaying virus-like symptoms. *Archives of Virology, **163***(11), 3051–3058.
41. Gupta, P. K. (2022). Current status of Cercosporoid fungi in India, effective management strategies and future directions. *Indian Phytopathology, 75*, 1–12.
42. Klug, A. (1999). The tobacco mosaic virus particle: Structure and assembly. *Philosophical Transactions of the Royal Society of London Series B: Biological Sciences, 354*(1383), 531–535.
43. Kawakami, S., Watanabe, Y., & Beachy, R. N. (2004). Tobacco mosaic virus infection spreads cell to cell as intact replication complexes. *Proceedings of the National Academy of Sciences, 101*(16), 6291–6296.
44. Sudheera, Y., et al. (2014). Characterization of a potyvirus associated with yellow mosaic disease of jasmine (Jasminum sambac L.) in Andhra Pradesh, India. *Virus, 25*(3), 394–397.
45. Wu, Q. and J. Yang. Comparative advantage analysis of production of jasmine tea in China. in Journal of Physics: Conference Series. 2021. IOP Publishing., 012044
46. Jeyarani, J. N., et al. (2018). Phylogenetic analysis and evolution of morphological characters in the genus Jasminum L.(Oleaceae) in India. *Journal of Genetics, 97*(5), 1225–1239.
47. Rajamohan, S. & Sathish, A. (2019). *Cultivation and marketing constraints of Jasmine in Tamilnadu. International Journal of Current, 9*(7), 426–429.
48. Balkrishna, A., et al. (2021). Mechanistic insight into antimicrobial and antioxidant potential of Jasminum species: a herbal approach for disease management. *Plants, 10*(6), 1089.
49. Xu, S., et al. (2022). A high-quality genome assembly of Jasminum sambac provides insight into floral trait formation and Oleaceae genome evolution. *Molecular Ecology Resources, 22*(2), 724–739.

50. Knudsen, J. T., et al. (2006). Diversity and distribution of floral scent. *The Botanical Review, 72*(1), 1–120.
51. Arctander, S. (1969). *Perfume and Flavor Chemicals (Aroma Chemicals) I and II Published by the author: Montclair.*
52. Zhang, J., et al. (2021). Characterization of aroma-active compounds in Jasminum sambac concrete by aroma extract dilution analysis and odour activity value. *Flavour and Fragrance Journal, 36*(2), 197–206.
53. Nitha, B., Remashree, A., & Balachandran, I. (2012). Antibacterial activity of some selected Indian medicinal plants. *International Journal of Pharmaceutical Sciences and Research, 3*(7), 2038.
54. El-Hawary, S. S., et al. (2021). Phenolic profiling of different Jasminum species cultivated in Egypt and their antioxidant activity. *Natural Product Research, 35*(22), 4663–4668.
55. Lu, Y., et al. (2019). Four new sesquiterpenoids with anti-inflammatory activity from the stems of Jasminum officinale. *Fitoterapia, 135*, 22–26.
56. Kumar, M., & Randhava, N. K. (2014). Jasminum mesnyi Hance: Review at a glance. *Journal of Drug Delivery and Therapeutics, 4*(5), 44–47.
57. El-Hawary, S. S., et al. (2022). The role of active metabolites isolated from Jasminum multiflorum flowers against hepatitis C virus infection and related hepatocellular carcinoma. *Natural Product Research, 36*(10), 2625–2629.
58. Kolanjiappan, K., & Manoharan, S. (2005). Chemopreventive efficacy and anti-lipid peroxidative potential of Jasminum grandiflorum Linn. on 7, 12-dimethylbenz (a) anthracene-induced rat mammary carcinogenesis. *Fundamental & Clinical Pharmacology, 19*(6), 687–693.
59. Chaturvedi, A. P., & Tripathi, Y. B. (2011). Methanolic extract of leaves of Jasminum grandiflorum Linn modulates oxidative stress and inflammatory mediators. *Inflammopharmacology, 19*(5), 273–281.
60. Chaturvedi, A. P., Kumar, M., & Tripathi, Y. B. (2013). Efficacy of Jasminum grandiflorum L. leaf extract on dermal wound healing in rats. *International Wound Journal, 10*(6), 675–682.
61. Sengar, N., et al. (2015). Anti-inflammatory, analgesic and anti-pyretic activities of standardized root extract of Jasminum sambac. *Journal of Ethnopharmacology, 160*, 140–148.
62. Ho, C. C., et al. (2021). Extracts of Jasminum sambac flowers fermented by Lactobacillus rhamnosus inhibit H2O2-and UVB-induced aging in human dermal fibroblasts. *Environmental Toxicology, 36*(4), 607–619.
63. Moloudi, M. R. (2018). et al. *Effect of Hydro-alcoholic extract of Jasminum sambac on morphine withdrawal symptoms in rats., 23*(1), 1–7.
64. Khan, I. A., et al. (2021). Jasminum sambac: A Potential candidate for drug development to cure cardiovascular ailments. *Molecules, **26***(18), 5664.
65. Komen, C., Wanjala, F. M., & Kiprono, P. C. (2005). Efficacy of Jasminum abyssinicum treatment against Hemonchus contortus in sheep. *African Journal of Traditional, Complementary and Alternative Medicines, 2*, 3.
66. Tadiwos, Y., Nedi, T., & Engidawork, E. (2017). Analgesic and anti-inflammatory activities of 80% methanol root extract of Jasminum abyssinicum Hochst. ex. Dc.(Oleaceae) in mice. *Journal of Ethnopharmacology, **202***, 281–289.
67. Esteban-Ballesteros, M., et al. (2019). In vitro anthelmintic activity and safety of different plant species against the ovine gastrointestinal nematode Teladorsagia circumcincta. *Research in Veterinary Science, **123***, 153–158.
68. Prakkash, M. J., et al. (2019). Evaluation of bioactive compounds from Jasminum polyanthum and its medicinal properties. *Journal of Drug and Delivery & Therapeutics, 9*(2), 303–310.
69. Arun, M., Satish, S., & Anima, P. (2016). Evaluation of wound healing, antioxidant and antimicrobial efficacy of Jasminum auriculatum Vahl. Leaves. *Avicenna Journal of Phytomedicine, **6***(3), 295.
70. Al-Khazraji, S. M. J. P. A. (2018). Potentiating effects of alcoholic extract of the aerial parts of jasminum officinalle on pentobarbital-induced sleep. *Plant Archives, 18*(1), 697–702.

71. Atta, A., & Alkofahi, A. (1998). Anti-nociceptive and anti-inflammatory effects of some Jordanian medicinal plant extracts. *Journal of Ethnopharmacology, 60*(2), 117–124.
72. Kaviani, M., et al. (2014). Comparison of the effect of aromatherapy with Jasminum officinale and Salvia officinale on pain severity and labor outcome in nulliparous women. *Iranian Journal of Nursing and Midwifery Research,* ***19***(6), 666.
73. Shekhar, S., & Prasad, M. (2015). Evaluation of antimicrobial activity of Jasminum species using solvent extracts against clinical pathogens. *World Journal of Pharmaceutical Sciences, 4*, 1247–1256.
74. Arun, M., Satish, S., & Anima, P. (2016). Evaluation of wound healing, antioxidant and antimicrobial efficacy of Jasminum auriculatum Vahl. leaves. *Avicenna Journal of Phytomedicine,* ***6***(3), 295.
75. Upadhyaya, C., et al. (2021). Investigation of Mobile Communication Radio Frequency Exposure on the Medicinal Property of Jasminum Grandiflorum L. In *2021 5th International conference on computing methodologies and communication (ICCMC)*. IEEE.
76. Singh, P., Bundiwale, R., & Dwivedi, L. (2011). In-vitro study of antifungal activity of various commercially available itra (Volatile plant oil) against the keratinophilic fungi isolated from soil. *International Journal of Pharma and Bio Sciences, 2*(3), 178–184.
77. Zhao, G., Yin, Z., & Dong, J. (2009). Antiviral efficacy against hepatitis B virus replication of oleuropein isolated from Jasminum officinale L. var. grandiflorum. *Journal of Ethnopharmacology, 125*(2), 265–268.
78. Kandari, L. S., et al. (2014). Conservation and management of sacred groves, myths and beliefs of tribal communities: a case study from north-India. *Environmental Systems Research, 3*(1), 1–10.
79. Lal, H. S., Singh, S., & Mishra, P. (2014). Trees in Indian Mythology. *Discovery, 12*(29), 16–23.

Chapter 36
Opium Poppy

Muhammad Tahir Hayat, Uzma Hameed, and Muhammad Zia-Ul-Haq

36.1 Introduction

Scientific Name: ***Papaver somniferum***
Taxonomy

Synonyms

Papaver album Mill	*Papaver glabrum* Gilib.
Papaver album-nigrum Crantz	*Papaver hortense* Hussenot
Papaver amoenum Lindl.	*Papaver indehiscens* Dumort.
Papaver amplexicaule Stokes	*Papaver nigrum* Garsault
Papaver somniferum subsp. *nigrum* (DC.) Thell.	*Papaver officinale* C. C. Gmelin
Papaver polycephalum Hort. ex E.Vilm.	*Papaver opiiferum* Forssk.
Papaver somniferum subsp. *hortense* Arcangeli	

M. T. Hayat · U. Hameed (✉)
Institute of Industrial Biotechnology, Government College University, Lahore, Pakistan
e-mail: uzmahameed@gcu.edu.pk

M. Zia-Ul-Haq
Office of Research, Innovation and Commercialization, University of Engineering and Technology, Lahore, Pakistan

M. Zia-Ul-Haq et al. (eds.), *Essentials of Medicinal and Aromatic Crops*,
https://doi.org/10.1007/978-3-031-35403-8_36

Common Names:

Papaver somniferum, known as breadseed or opium poppy, belongs to the flowering plant family Papaveraceae. The genus name Papaver is the Greek word for "poppy" and the species name *somniferum* is the Latin word for "sleep-inducing [1]. *P. somniferum* is a herbaceous and highly glaucous plant that is rarely branched. Leaves are simple, with pinnate venation. The main stem bears terminal actinomorphic flowers, which can be of variable color depending upon varieties. The fruit is dry and hairless, bearing tiny, globose and kidney-shaped seeds ranging from white, pale gray, dark, or black in color.

Opium poppy cannot withstand extreme cold and requires fertile soil with appropriate moisture contents, nutrient retention ability, and suitable climatic conditions. The optimum growth and development temperature for the plant is between 16 and 20 °C. Disease management is crucial in opium poppy cultivation, as it is susceptible to several pests and diseases; therefore, control measures, including weed management, crop rotation, pesticide applications and disease-resistant varieties, are required [2]. The opium poppy has had economic, therapeutic, and symbolic value for thousands of years [3]. Opium is extensively used for its analgesic and euphoric properties [4]. Opium is obtained from the opium poppy capsules and is known as poppy tears, scientifically termed Lachryma papaveris [5]. The only *Papaver* species that produce opium today is *Papaver somniferum* [1]. Opium is dry latex that contains a mixture of alkaloids (morphine, codeine, noscapine, thebaine, etc.), phenolic compounds, and essential oils like linalool, alpha-pinene, methyl salicylate and cadinene, which provide therapeutic benefits. The opioids from opium poppy modulate numerous emotional, motivational, sensory and cognitive functions by binding specific surface receptors of neurons. Morphine and its derivatives are extensively used for myocardial infarction, surgeries and pain management in patients with long-term chronic disorders and late-stage cancer [4]. In addition, poppy seeds are healthful and edible and are used in essential oils production, culinary preparation as spices and condiments, baking and diagnostics, and pharmaceutics [6].

The opium poppy has been a double-edged sword for human civilization. The pharmaceutical products of its alkaloids offer immeasurable benefits but still are the most commonly abused illicit drugs after cannabis. The side effects of these drugs include opioid-induced tolerance, addiction, physical dependence, constipation, nausea, vomiting, drowsiness, lethargy, hyperalgesia and addiction behavior [7]. Opium has contributed to historical and current epidemics of opioid addiction and overdose deaths [8], which led to imposing bans on poppy cultivation and opium distribution and sale regulations to prohibit illegal use. Still, opium and opioid-derivative addictions contribute to significant social and health problems.

This chapter highlights the history, morphology, agroecology, pests and diseases of *P. somniferum* plant. In addition, the chemical constituents of opium and various applications of opium and poppy seeds are also reviewed.

36.2 Plant Description

36.2.1 Morphological Characters of Papaver somniferum

P. somniferum is a member of the cosmopolitan family Papaveraceae. It is about 3–5 ft high and a robust annual small shrub or herb (Fig. 36.1) that bears about 5–8 capsules [9, 10]. The opium poppy is a highly glaucous plant with a greyish-green appearance. The plant has simple and rarely branched, cauline bases clasping stem without or with sparsely distributed course hairs. Leaves are simple with undulating serrate-dentate and lobed margins. The venation is pinnate, and leave are alternately arranged on the stem. The upper larger leaves are cordate-amplexicaul and clap the stem, while the lower leaves have very short petioles. The flowers are actinomorphic and single terminal.

The flower (Fig. 36.1b) color varies within varieties and can be white, mauve, pink, red, white, pale or violet, sometimes with dark markings at the base. The fruit is a dry, hairless capsule (Fig. 36.1c), with 12–18 stigmatic rays at the top that develops from the superior ovary with a parietal placenta. The seeds (Fig. 36.1d) are tiny (0.3 mm), globose and kidney-shaped, and white, pale grayish, dark grey to black in color. After flowering, seeds are fully ripened after 20–25 days [11, 12].

The various varieties of *P. somniferum* are recognized with characters based on the number and components of carpels or the absence of an aperture. For example, the opium poppies cultivated in India and Persia have capsules without pores, which are more or less ovate or oblong (var. *album*, Boiss.). Similarly, the variety grown in Germany has capsules that are large and strongly ribbed along the lines of the union of components of the carpel [13]. The detailed morphological characters are described in Table 36.1.

Fig. 36.1 *Papaver somniferum* (opium poppy). (**a**). Field view (**b**) Flower (**c**) fruit (Capsule) (**d**) Seeds

Table 36.1 Morphological features of Papaver somniferum

Stem	Simple, erect, rarely branched, and up to 150 cm high glabrous, glaucous, lactiferous, rarely sparsely setose
Roots	Tap root, thick, erect, slender, conical, and much branched
Leaves	Simple, alternate, lanceolate, ovate-oblong, deciduous, both surfaces glabrous, glaucous, and rather waxy and 5–25 × 2–7 cm across Margins undulate, serrate-dentate, often with larger teeth alternating with smaller ones, sometimes shallowly pinnatifid Apex obtuse to acute, base obtuse, or rounded Base cordate-amplexicaul, upper cauline leaves, except the lowest, which usually narrows into a short stalk Lateral veins and midrib impressed above and prominent beneath
Flowers	Large solitary, terminal, actinomorphic, bisexual, 5–12 cm in dia. Showy, of variable colors: Pinkish or reddish, white, rarely pale violet, mauve or crimson red with or without a basal dark blotch
Flower bud	Glabrous ovoid-oblong and 1–3.5 cm long
Peduncle	Glabrous or sparsely bristly and 5–25 cm long
Sepals	2 sepals, deciduous, glabrous, ovate-orbicular, caducous & 2–3.5 cm
Ovary	1 pistil, green, superior, ovoid-globose, glabrous, unilocular, and about 10 mm long, 1–2 cm dia. Many ovules and 4–18 parietal placentas 5–18 united stigmas to form a compressed stigma disk with 7–18 stigmatic rays (deeply crenulate lobes)
Stamen	Numerous, about the size of the ovary, about 5–10 mm long Filaments yellowish, filiform, and slender Anthers oblong-elliptic, about 1–2 mm long
Fruit (capsule)	Hairless, glabrous, not ribbed, subglobose to oblong-ellipsoid, apex flat-topped, about 2–7 × 5–6 cm across Stipitate (when present) very short stipe 1-2 mm, abruptly ending round base Apex flat-topped, crowned by the stigmatic disk with stigma rays and is somewhat shorter than the capsule breadth Pale green, brown when mature Poricide, dehiscing by subapical pores or persistent disc Seeds small (about 0.3 mm dia.), globose to reinform, white, pale grayish, dark-grey to black, rich in oil
Seeds	Small (about 0.3 mm dia.), globose to reinform White, pale grayish, dark-grey to black, Rich in oil
Seeds dispersal	Autochory, anemochory, zoochory, anthropochory

36.2.2 Seed Dispersion, Reproduction and Growth

The opium poppy seeds dispersal may involved any of the four modes, i.e., anemochory (dispersal by wind), zoochory (dispersal by animals or birds), anthropochory (dispersal by humans) or autochory (self-dispersion). *P. solaniferum* are angiosperms with complete and bisexual flowers, i.e., the female (gynoecium) and functional male (androecium) are in the same flower. Therefore, opium poppy is considered predominantly self-pollinated. It can carry out entomophilous pollination, i.e., cross-pollination by insects. The pollens of the *P. somniferum* are heavy and sticky, so wind pollination is considered negligible. Flies and honeybees have been identified as the main potential cross-pollinator [14].

P. somniferum is a terrestrial plant. The tiny seeds of opium poppy germinate quickly when proper temperature, humidity and moisture contents are present.

The main stem of mature *P. somniferum* ranges between 2–5 ft in height with oblong, toothed, and lobed green leaves that sometimes reach upto15 inches in diameter. The flower bud develops on the peduncle (the top leafless part of the stem or tiller). During the bud development, the peduncle elongates to form a hook turning the bud upside down. But during flower development, it straightens, pointing the buds upward. A day or two after the peduncle straightening, the sepals fall off, revealing the four flower petals that appear initially crinkled and crushed. The petals eventually enlarge and smooth out in the sunlight. The flowers can be single or rarely double, with different varieties of white, crimson red, reddish-purple, pink, or variegated petals.

Opium poppy flowers typically appear after 90 days of growth and flowering last for 2–3 weeks. After 2–4 days, the petals fall off to reveal a tiny, round, green developing fruit. Southeast Asian opium poppy capsules are commonly oblate-shaped. The capsule is the only opium-producing part of the opium poppy. Capsule is scored while it is still attached to the plant. About 95% of the opium is secreted by the cell of the middle layer, which is scraped and collected. The capsules are cut from the stem and left to dry. After drying, these are cut open, and seeds are collected, sun-dried, and stored for planting next year. Alternatively, seeds for planting are collected from intentionally unscored capsules, as scoring pods can diminish the seed quality.

36.3 Origin and Distribution

Some other scientists proposed *P. glaucum* as its ancestor. Another hypothesis proposes the origin of *P. somniferum* as a triploid hybrid, which is also supported by the cytogenetic study [15, 16].

However, cultivating opium poppies for their psychoactive compounds is illegal in many countries [8, 15–19].

36.4 History

Opium has been known to ease pain and for its use as a surgical analgesic. Sumerians, who used the words hul "joy" and Gil as symbols for the poppy as a plant of joy, may have been aware of the psychological effects of opium as early as 4000 B.C. Based on the discovery of fossil remnants of poppy pods and poppy seed cakes in the Neolithic Swiss Lake Dwellings, the plant was known in Europe at least 4000 years ago. The first known record of opium use is thought to be the Sumerian clay tablet from around 2100 BC. The poppy goddess, with closed eyes and hairs decorated with poppy seeds and implying sedation, appeared in the Minoan culture of ancient Greece from around 1500 BC. In Egypt and Cyprus, jugglers imitating poppy capsules were discovered during the same era. Egyptian Ebers Papyrus described the use of opium to sedate children in 1552 BC. Theophrastus provided the first authentic record of using the milky juice of poppy at the beginning of the third century BC. The father of medicine, Hippocrates (460–357 BC), recommended poppy juice as a narcotic and purgative for treating leucorrhea. Dioscorides, Pliny, Celsus, and Galen also knew how to use poppy for surgeries [16–18, 20–23].

Around the year 1000 AD, opium was extensively used by Arabic doctors. Ibn Sina recommended it for ocular diseases and diarrhea. Polypharmacy, using a mixture of nonsensical medications, was common and many treatments contained opium. For example, the famous panacea "Theriaca", an ancient multi-ingredient preparation, contained up to sixty medications, including opium. Opium poppy most likely made it to China around the seventh century A.D. through Arab traders who suggested its use for therapeutic purposes. However, early records of its use are also present in Chinese literature [20]. The Swiss alchemist Paracelsus tested various opium concoctions and advised opium to relieve pain in the sixteenth century. One of his formulations, archanum or laudanum, is considered to contain opium [24]. In the early 1800s, Serturner, the founder of alkaloid research, isolated morphine from opium, which became the most common analgesic [20].

36.5 Opium and Poppy Plants: Legal Issues

Opium has been a friend and foe of human civilization throughout history [25]. Opium addictions have become a significant social and health problem. In the eighteenth century, opium addiction was widespread in *China*. In an effort to stop the devastation brought on by the widespread addiction, opium cultivation and import from Europe were forbidden during the Qing Dynasty. Opium smuggling became a common industry in China, during the nineteenth century, opium was cultivated and imported to the USA due to indiscriminative medicinal use and myriad tonics. Opium smoking dens were also frequent and unprohibited. It resulted in an opium addiction epidemic in the late 1800s. The frequent use of morphine during the Civil War in the USA for treating wounded soldiers also resulted in many addicts. It was referred to as army disease" or "soldier's disease.

In 1874, Heroin, a derivative of morphine, was synthesized in an attempt to make a less addictive and more effective alternative. It was used to treat chest pain, cough, and tuberculosis. However, the heroine later proved to be much more addictive. The USA banned opium in 1905 and approved the *Pure Food and Drug Act* in 1906, which required accurate labelling of all opium-containing patent medicine. Later on, in 1914 Harrison Narcotics Act was imposed for the sale regulation of narcotics and to prohibit illegal use.

Opioid misuse can have serious and potentially life-threatening effects on a person's physical and mental health, such as depression, anxiety, and psychosis. Opioid overdoses can lead to respiratory depression, coma, and even death [26]. Only in the USA, about 49,860 were opioid overdose death in 2019, among which 14,019 deaths were due to heroin overdoses [27]. Morphine and other opioid analgesics are crucial in managing severe pain due to being effective and safe. Nevertheless, it is essential to prescribe these medications based on each patient's needs and ensure their access whenever required [28]. Internationally, the regulation of opioid use and misuse is primarily overseen by WHO [29].

Today, opium poppy is cultivated in many parts of the world, including Southeast Asia, China, Thailand, Buma, Vietnam, Laos, India, Afghanistan, Pakistan, Iran, Tasmania, Argentina, Mexico, and some parts of Europe and America. However, it is cultivated in Turkey, Spain, India, Hungary, France, Poland and Australia for legal pharmaceutical purposes. India is the largest legal producer of opium for the world's pharmaceutical market. However, opium poppy is cultivated worldwide illegally for heroin manufacturing [8, 16, 18]. Afghanistan is the world's largest poppy producer, with approximately 224,000 hectares of cultivation area, followed by Myanmar with 29,500 ha.

36.5.1 Biomarkers for Opium Illicit Use

Biomarkers are measurable indicators of biological processes or conditions within an organism. 6-MAM (6-monoacetylmorphine) is a biomarker so the presence of 6-MAM in urine or blood samples directly indicates recent heroin abuse [30]. The detection of 6-MAM is considered a more specific and sensitive biomarker for heroin use than morphine or codeine, which can also be present in the body from other sources, such as prescribed painkillers. 6-MAM can be used to detect heroin abuse in both urine and blood with 100% and 95% sensitivity, respectively [31].

36.6 Agroecology

36.6.1 Climate

Although the poppy is a sturdy plant, it cannot withstand extremely cold temperatures. It can grow best in a temperate area [8, 17, 32].

Extreme arid and excessive moisture conditions adversely affect poppy growth and can reduce the alkaloid contents. Very little or no rain is required for poppy plants during later development periods and harvesting [8]. The opium poppy best tolerates annual rainfall from 300–17,300 mm. Heavy rainfall on poorly drained soils can cause waterlogged soils, leading to plant death. Similarly, heavy rainfall during second and third months of growth can reduce plant growth and alkaloid contents [8, 17].

36.6.2 Soil

Poppy plants can grow in a variety of fertile soils with high organic contents but grow best in sandy loam soil. These soils have good moisture and nutrient retention and provide good penetration for the roots of young poppy plants. Young poppies cannot grow well in sandy or clay soils. The pH of the soil is the principal factor that affects plant growth. The poppy plant thrives well in moderate to alkaline soils (pH 6–8). The capsule and morphine yields have been reported at a pH increase from 5.6 to 6.1 in the surface soil [8, 17].

36.6.3 Cultivation

The poppy needs rich soil and quickly depletes it. Therefore, a rotational system must be used, or the plant must be sown on heavily fertilized soil or fellow land to avoid frequently employing chemical fertilizers [33]. In Southeast Asia, maize is used for crop rotation with poppy plants each year. Maize is grown from April to August, and it helps to keep the excessive weeds in control and provides feed for animals. After harvesting, the maize stalks are left in the field, and poppy seeds are broadcasted after weeding and pulverizing ground. The maize stalks protect the young poppy plants from heavy rains. Spinach, cabbages, beans, cotton, parsley tobacco, or squash are usually panted with the poppy as cash crops [8].

36.6.3.1 Sowing

The poppy is an annual plant. The sowing time depends on the variety and climate. The alternating rainy and dry seasons affect the opium yield. Therefore, in poppy-growing regions (typically warm temperate zones), the sowing season may start in September and last until April. Before sowing, the soil must be well prepared by tilling and irrigation (Fig. 36.2) [33].

The poppy seeds can be sown in several ways, tossed or broadcast by hand, or using a dibble tool. There is no agricultural technology for the dissemination of poppy seeds. Thus, the manual dissemination requires roughly 20 kilograms of heads per hectare instead of 3–4 kg, which would be required by a mechanical shower [33].

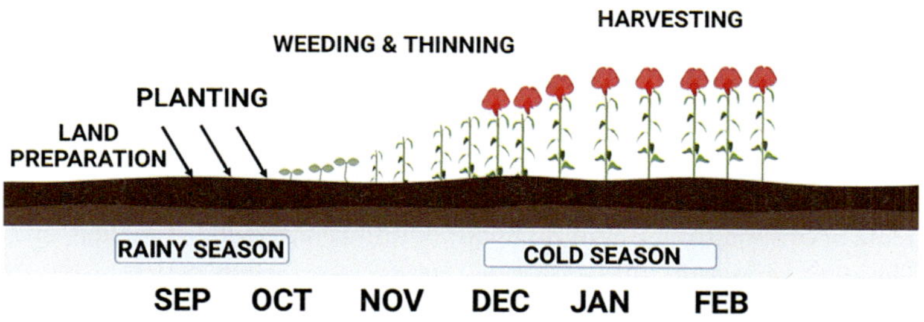

Fig. 36.2 Cultivation of opium poppy in south Asian region

36.6.3.2 Germination

The seed germinates, and the appearance of the first four leaves takes about 1 month. The stem develops after 2–3 weeks, and the plant fully develops after 2 months. The flowering season varies depending on climate, sowing time, and variety. Typically plant blooms during the day, and flowers last up to 2 days. After the petals fall off, capsules grow and ripen in about 2 weeks [33].

36.6.3.3 Thinning

After 1 month of growth, thinning is performed in the field when plants are about 1 foot high. In this process, some plants are removed to provide more space for the proper growth of other plants. The optimal space is considered to be 20–40 cm between two plants or about 8–12 plants/inch^2 [8].

36.6.4 Harvesting

Opium is a dried form of latex obtained from ripened poppy capsules. It is used to extract morphine and other alkaloids. The harvesting time can vary depending on climatic conditions but generally starts approximately 2 weeks after the petals fall off. The ripened capsules swell in size and become stiff and dark green, but the capsule's color can vary depending on the variety. The crown of the pod points downward in a mature capsule. Not all the capsules are ripened simultaneously in the same field. Therefore, harvesting can last for about 2 weeks. The individual capsules are needed to examine before harvesting [8, 33].

There are two phases in opium harvesting: capsule incision and latex collection. During incision, cuts are made on the capsule surface, and milky liquid latex oozes out. After drying, the latex (opium) is collected from the surface of the capsule. The incision is also known as tapping, lancing, or scoring. It is a critical step and must be performed without damaging or killing the capsule in a highly skilled manner.

The incision process can affect the quality and quantity of opium. The latex is present between the mesocarp and epicarp of the capsule, which large numbers of juice channels running from bottom to top. Thus, wall of capsule wall must not be cut right through or deep; otherwise, latex will flow down and waste. The desired incision depth is approximately 1 mm, made with a specially designed blade tool. It takes about 8–14 h for latex to exude and solidify. Thus incision is ideally performed late afternoon. In the cool night air, opium is oxidized, thickened, and darkened. The dark brown opium (*apeen* or *taryak*) is scrapped off from the capsule surface the next morning with a short-handled flat iron blade tool carefully without damaging the capsules. Scored capsules continue to secrete latex for many days [8, 33].

36.6.5 Yield

The yield of opium per hectare is very challenging to estimate as it depends on several factors such as climate, nature of the soil, state of seed and annual rainfall, etc. A typical opium poppy field may contain 6×10^3–12×10^4 plants per hectare, with 12×10^4 –2.75×10^5 capsules. The actual opium yield varies depending on the measures taken by individual farmers and weather conditions [8, 33]. Production loss between 12–65% can be induced due to climate conditions such as cold, drought, lack of precipitation and extreme temperatures [34].

Afghanistan is responsible for producing almost 80% of the world's opium. However, in 2022, the country experienced a decline in opium yield per hectare due to drought conditions. This decrease in yield offset the increase in cultivation, resulting in an estimated opium production of 6200 tons in 2022, which is 10% less than the previous year. The average opium yield per hectare dropped from 38.5 kg/ha in 2021 to 26.7 kg/ha in 2022 [35].

36.7 Pests and Diseases

P. somniferum is susceptible to various pests and diseases that can affect its growth, yield, and quality [36]. Significant economic losses can be attributed to *Stenocarus ruficornis* as it is recognized to harm the roots and leaves of poppies. Additionally, Hungary has documented the presence of two other types of insects, *Dasyneura papaveris* and *Carpodiplosis papaveris*, which are responsible for creating galls on poppy [37]. Among the fungal parasites most prominent are *Perenospora arborescens* which causes poppy mildew and kills the plants. The diseases of opium poppy and their causative agents are documented in Table 36.2.

Table 36.2 Pests and diseases of *P. somniferum*

Causing agent	Disease symptoms	Vector	Reporting country	Reference
Viruses				
Bean yellow mosaic virus		Aphid	Poland and Australia	[38, 39]
Beet yellows virus		Aphid	Netherlands	[40]
Opium poppy mosaic virus	Mosaic and mottling symptoms on the leaves	Aphid	Auckland (New Zealand)	[41]
Poppy mosaic virus		Aphid vector, *Myzus persicae*	India	[42]
Tomato leaf curl New Delhi virus	Severe curling of leaves	Whitefly vector, *Bemisia tabaci*	India	[43]
Tomato spotted wilt orthotospovirus	No symptoms	Thrips vectors	Tasmania (Australia)	[39]
Turnip mosaic virus		Aphid vectors	Australia, Hungary, and the Czech Republic	[39, 44, 45]
Bacteria				
Pectobacterium carotovorum	Severe symptoms of soft rot and darkening	No known vector, direct contact between infected and healthy plant tissues	Spain	[46]
Fungi				
Peronospora arborescens	Downy mildew with symptoms including dark brown 'oilspot' lesions on leaves	No known vector, Patghogen is transmitted by spores	Tasmania (Australia)	[47]
Nematodes				
Basirolaimus saccharis	Stunting, yellowing of leaves, and wilting	*Fusarium oxysporum*	India	[48]
Tylenchorhynchus latus	Stunted growth, yellowing of leaves, and reduced yield.	No unknown vector, causes direct damage to plant	Turkey	[49]
Zygotylenchus guevarai	Stunted growth, yellowing of leaves, and reduced pod formation	No unknown vector, causes direct damage to plant	Turkey	[49]

36.7.1 *Disease Control*

Effective pest and disease management in *P. somniferum* cultivation is essential to ensure healthy crops and productive yield. Several steps can be taken to minimize the risk of disease and protect the plants, such as crop rotation, seed selection, sanitation, chemical control, use of resistant cultivars and good cultural practices, including proper irrigation, fertilization, and weed control [2].

36.8 Genome of Opium Poppy

Papaveraceae are dicotyledonous, dialypetalous superovaried plants. *P. somniferum* is a member of the *Papaver* genus from the Papaveraceae family of Ranunculales order and has 22 diploid chromosome (2n). The Papaveraceae family has 43 genera and ~500 species. In 2018, the University of York led a research team in collaboration with the Wellcome Sanger Institute in the UK and other international partners to publish the genome sequence of *P. somniferum*, shedding light on how the opium poppy developed the ability to produce important pharmaceutical compounds. The study uncovered the origins of the genetic process that results in the creation of the painkillers morphine and codeine as well as the cough suppressant noscapine. By creating a high-quality assembly of the 2.7 GigaBase genome sequence, the researchers annotated the genome that predicted 51,213 protein-coding genes and 9494 noncoding RNAs (Table 36.3).

About 81.6% of protein-coding genes in the poppy genome are distributed in 11 chromosomes, while the remaining 18.4% are found in unplaced scaffolds. The long terminal repeat retro-transposons are the most prevalent repeating elements, making up 70.9% of the genome [25].

36.9 Chemical Constituents of Opium Poppy and Their Applications

As described earlier opium poppy is one of the world's oldest medicinal plants containing several different chemical constituents, including alkaloids, phenols, essential oils, and various other compounds (Fig. 36.3) [50]. Opium is obtained from

Table 36.3 Assembly and annotation of *P. somniferum* genome [25]

Parameter	Value
N50	1.77 Mb
N90	590 kb
Largest contig	13.8 Mb

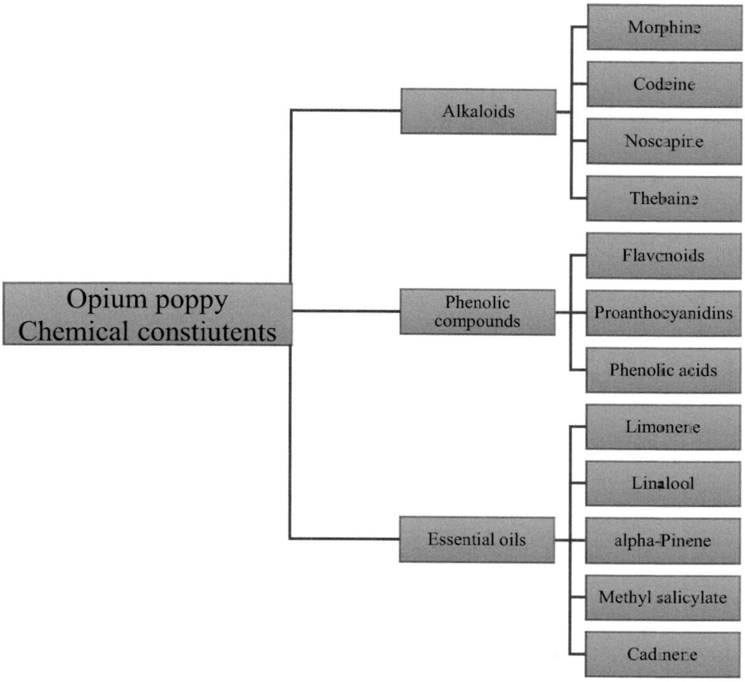

Fig. 36.3 Important chemical constituents of Opium poppy

mature capsules or seed pods. However, the exact composition of opium can vary depending on the growing conditions and the methods used to extract the alkaloids. The cultivation, harvesting, and processing of the plant and its alkaloids are highly regulated in most countries [51].

36.9.1 Alkaloids

The alkaloids produced by the poppy plant are responsible for the plant's psychoactive and medicinal properties [17]. Opioids are collectively referred to as a class of drugs that derive from, or mimic, natural substances derived from opium and can bind opioid receptors [52]. Opioids work in the brain to produce a variety of effects by binding to specific opioid receptors (Table 36.4). The most common opioid receptors in the brain and spinal cord are delta (δ), kappa (κ), Mu (μ) opioid, and nociception receptors [7].

Table 36.4 Opioid receptors, their location and functions [7, 53]

Receptor nomenclature	Common location in the CNS	Common roles & Functions
MO, μ, mu,	Cerebral cortex, thalamus, locus coeruleus, dorsal horn, amygdala, striatum, hippocampus	Analgesia, cardiovascular function, feeding, mood, hormone, secretion, thermoregulation, intestinal transit
DOP, δ, delta	Cortex, caudate putamen, thalamus, olfactory bulb, amygdala, dorsal horn, nucleus accumbens (NAc)	Analgesia, cardiovascular Regulation, gastrointestinal motility, Mood, behaviour
KOP, κ, kappa	NAc, hypothalamus, amygdala, cerebral cortex, olfactory bulb, claustrum, caudate nucleus, corpus callosum, thalamus, subthalamic nucleus	Neuroprotection, neuroendocrine functions, Analgesia in inflammation, Diuresis, feeding
N/OFQ, NOP	Amygdala, substantia nigra, hippocampus, lateral septum, hypothalamus, dorsal horn	Locomotor activity, spinal analgesia, feeding, anxiety, mood, memory

36.9.1.1 Major Types of Benzylisoquinoline Alkaloids (BIAs)

Benzylisoquinoline-derived alkaloids are predominant in opium poppy. This is a structurally diverse group of plant metabolites responsible for the pharmacological properties of the opium poppy [54]. Benzylisoquinoline alkaloids are characterized by a benzyl and an isoquinoline ring in their structure. Examples of BIAs include morphine, codeine, papaverine, berberine, noscapine and sanguinarine, among many others (Fig. 36.4).

Generally, morphine is the most abundant alkaloid in raw opium, followed by noscapine, codeine, thebaine, and papaverine. These compounds exhibit various biological activities, including analgesic, antitumor, antimicrobial, and antiparasitic properties (Table 36.5). Plants often produce BIAs as a defense mechanism against predators or pathogens. They are synthesized through complex pathways involving multiple enzymes. Various factors, including light, temperature, and stress, regulate their biosynthesis. Because of their diverse biological activities and potential therapeutic applications, BIAs have been the subject of extensive research. The biosynthesis and regulation of these compounds and their potential use in medicine and other fields are of particular interest [55, 56].

Morphine Morphine is the most well-known BIAs in the opium poppy. It was the first isolated activated ingredient from poppy in 1804 by Friedrich Serturner and named after Morpheus, the Greek god of dreams, due to its sedative effects [57]. It is a potent painkiller for cancer, surgeries, or serious injuries. It is the major constituent that makes up about 10–16% of raw opium.

The chemical structures of morphine and its metabolites, such as morphine-3-glucuronide and morphine-6-glucuronide, are closely related to the clinical effects of drugs and their capability to cross the Blood Brain Barrier (BBB) to reach the brain [58]. Morphine is extensively absorbed from the gastrointestinal (GI) tract and is distributed throughout the body. Morphine and some related compounds are

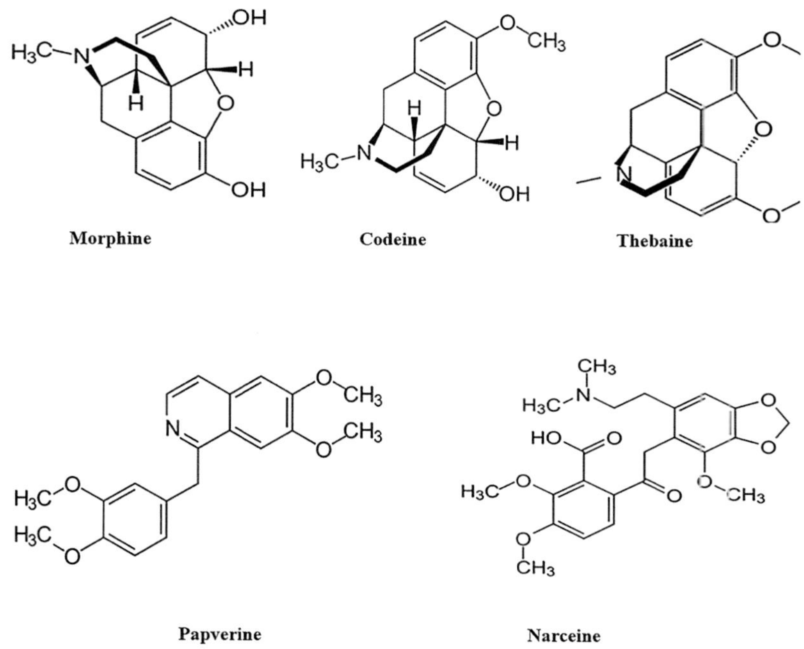

Fig. 36.4 Chemical structure of common benzylisoquinoline alkaloids

Table 36.5 Applications of selected opium alkaloids

Alkaloid	Plant part	Application	Adverse effects	References
Morphine	Unripe seed pods	Pain management	Central nervous system depression, respiratory depression, constipation, nausea, urinary retention, vomiting, dry mouth, sedation, euphoria, lightheadedness, dizziness, dysphoria, skin rashes, agitation, biliary tract, spasm anorexia	[66]
Codeine	Latex	Relieve mild to moderate pain and reduce coughing	Adrenal gland problems	[67]
Thebaine	Latex		Dizziness, urinary retention	[68]
Noscapine	Latex	Cough suppressant, Treating carcinoma cells in the breasts, skin, ovaries, colon, lungs, etc.	Headache, dizziness, and even coma	[69]
Papaverine	Latex	Improves blood flow in patients with circulation problems	Arterial hypotension, ventricular tachycardia, flushing of the face, tachycardia, and paradoxical aggravation of cerebral vasospasm.	[70, 71]

naturally synthesized in the human body in small amounts known as endogenous opioids. Therefore, these can be metabolized and converted into morphine 6-glucuronide or morphine 3-glucuronide, followed by renal excretion. Direct action of morphine on CNS (central nervous system) results in analgesia induction and pain perception alteration [59, 60]. However, it is highly addictive and can also cause side effects like respiratory and CNS depression, constipation, etc. (Table 36.5).

Codeine Codeine is another important alkaloid often used as a pain reliever and cough suppressant [61].

Thebaine It is also used to make painkillers but is not a direct analgesic [62].

Noscapine Noscapine is a cough suppressant and is also being studied for its potential anti-cancer properties [63].

Papaverine Papaverine has a smooth muscle-relaxing effect and treats certain cardiovascular conditions [64].

Narceine Narceine is similar to morphine and codeine and has similar pharmacological effects, including analgesic and sedative properties but is less potent. Narceine is a minor constituent of opium typically found in concentrations of less than 1%. It is usually extracted from opium using various chemical and chromatographic techniques [65].

36.9.1.2 Mechanism of Action

Morphine, A typified opioid, has a more binding affinity with the Mu receptor. Most of its analgesic effects are by binding this receptor within the central nervous system (CNS) and the peripheral nervous system (PNS). μ receptor is G protein-coupled receptor (GPCR). The G protein is a heterodimer and composed of α, β, and γ subunits. The binding of morphine with μ receptor induces several structural and molecular changes that lead to the activation of coupled G protein. The GDP (guanosine diphosphate) bounded to the α subunit is exchanged with GTP (guanosine triphosphate). The βγ subunits dissociate from α subunits. These subunits interact with several proteins in cell and modulate different downstream signaling pathways [72, 7, 73–75]. In presynaptic neurons, the G protein βγ subunits interact with the nearby voltage-gated calcium channels (CaV), inhibiting its opening. These channels cannot open even in the presence of action potential. Without Ca^+ ions influx, no neurotransmitters are released, resulting in decreased neuroexcitlibity and pain. In postsynaptic inhibition, morphine binds to receptors on the postsynaptic neuron, and activated G protein induces the potassium channels (Kv) opening, and positively charged potassium ions flow out, resulting in hyperpolarization of the neuron and subsequent pain decrease.

Gα protein also inhibits adenylyl cyclase (AC) and decreases cAMP production. The decrease in cAMP levels results in structural, enzymatic, and molecular changes in the neurons leading to decrease neurotransmitter release and pain (Fig. 36.5) [7, 76].

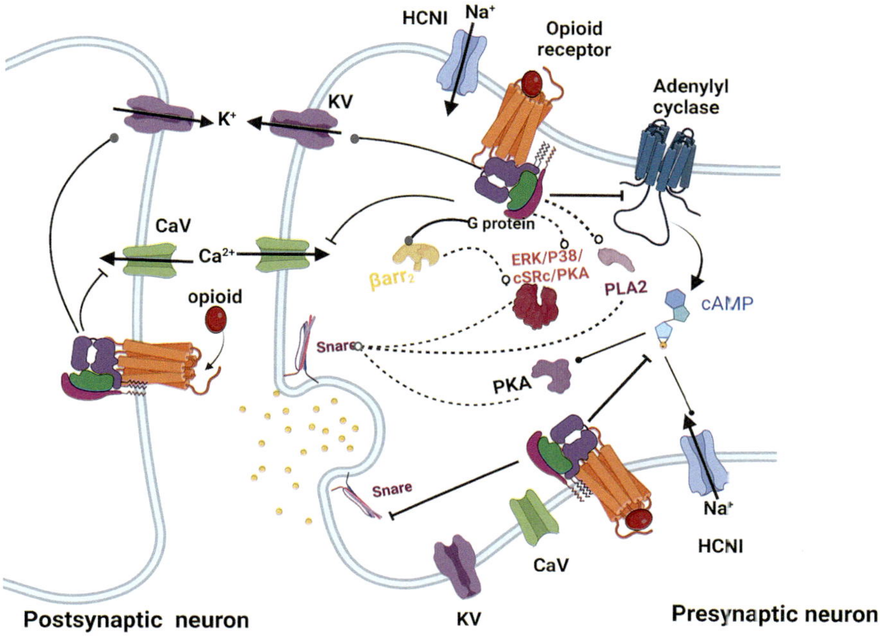

Fig. 36.5 Mechanism of action of morphine (a typified opioid)

36.9.1.3 Metabolic Pathway for the Synthesis of Alkaloids

The biosynthesis of BIAs in opium poppy involves a complex network of enzymatic reactions, which are tightly regulated at multiple levels to ensure the production of specific BIAs at different stages of plant development (Fig. 36.6). The first committed step in the BIA biosynthesis is dopamine and 4-hydroxyphenyl acetaldehyde condensation to form (S)-norcoclaurine, which is catalyzed by the enzyme norcoclaurine synthase. (S)-orcoclaurine is then converted to (S)-reticuline through a series of enzymatic reactions, including methylation, reduction, and dehydration. (S)-Reticuline is a branching point in BIAs biosynthesis and can be further modified by different sets of enzymes to produce different BIA subclasses For example, converting (S)-reticuline to salutaridine involves the steps of reduction, acetylation, and oxidation that are catalyzed by salutaridinol acetyltransferase, salutaridine reductase, and codeinone reductase (COR), respectively [77].

- **Morphine** is synthesized from thebaine through a series of enzymatic reactions, including demethylation, oxidation, and O-methylation steps. The key enzymes involved in these steps are thebaine 6-O-demethylase (T6ODM), codeinone reductase (COR), and O-methyltransferase (OMT).
- **Codeine** is synthesized from (S)-reticuline through a similar pathway to morphine, except that thebaine is not formed. Codeine is synthesized by the methyla-

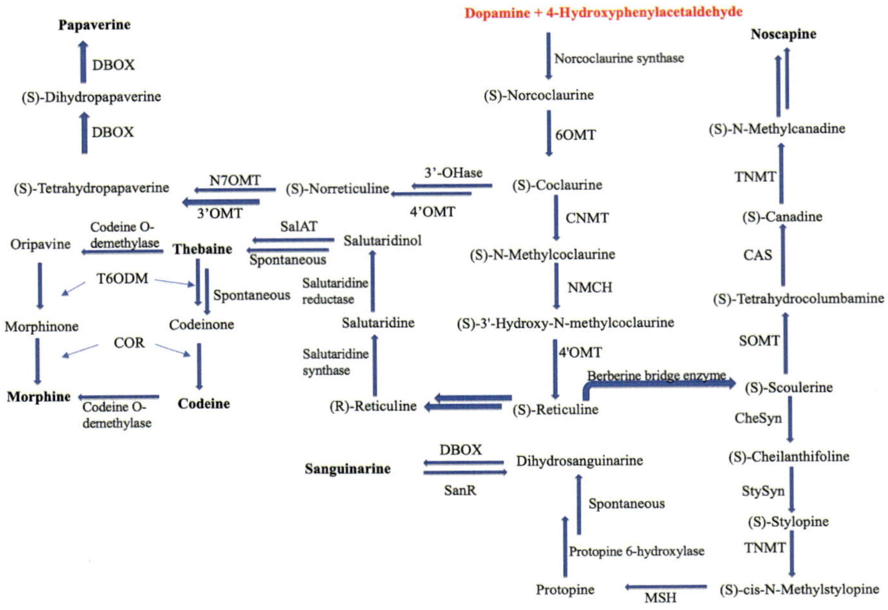

Fig. 36.6 Biosynthesis of Benzylisoquinoline alkaloids. *DBOX* dihydrobenzophenanthridine oxidase, *N7OMT* (R,S)-norreticuline 7-O-methyltransferase, *NMCH* (S)-N-methylcoclaurine 3′-hydroxylase, *3′OMT* 3′-O methyl transferase, *CNMT* (S)-coclaurine N-methyltransferase, *COR* codeinone reductase, *T6ODM* thebaine 6-O-demethylase, *SalAT* salutaridinol 7-O-acetyltransferase, *TNMT* tetrahydroprotoberberine N-methyltransferase, *SOMT* scoulerine 9-O-methyltransferase, *CheSyn* cheilanthifoline synthase, *StySyn* stylopine synthase, *SanR* Sanguinarine reductase, *BBE* berberine bridge enzyme, *MSH* methylstylopine hydroxylase [77]

tion of (S)-reticuline by OMT and the subsequent reduction of codeinone by COR.

- **Noscapine** is synthesized from (S)-reticuline through a series of enzymatic and chemical transformations involving various enzymes such as berberine bridge enzyme, scoulerine 9-O-methyltransferase, and tetrahydroprotoberberine N-methyltransferase (TNMT).
- **Sanguinarine** is a benzophenanthridine alkaloid that can be synthesized from (S)-reticuline through several steps with the help of various enzymes, including TNMT, cheilanthifoline synthase (CheSyn), stylopine synthase (StySyn), protopine-6-hydroxylase and Sanguinarine reductase (SanR).

36.9.1.4 Alkaloid Derivatives

Several derivatives of alkaloids, such as the derivatives of morphine, including heroin, codeine, and hydromorphone (Fig. 36.7), are also used as pain relievers. About 70% of morphine produced is used as a precursor for synthesizing other derivatives such as oxycodone, hydromorphone, codeine, etc. [78].

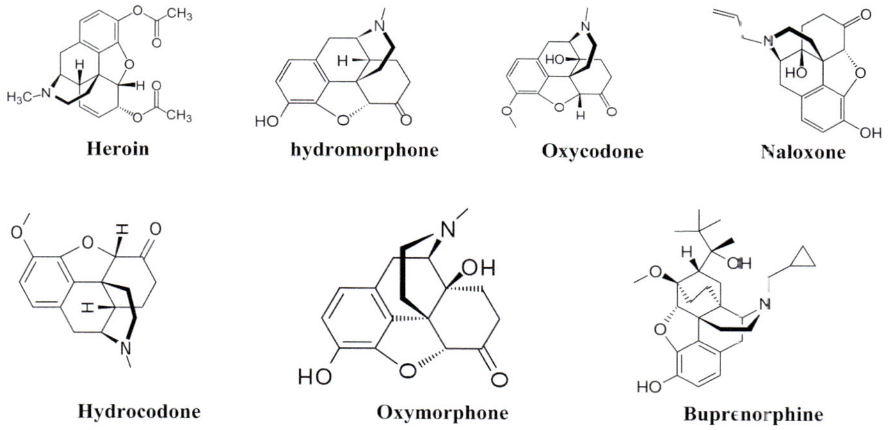

Fig. 36.7 Chemical structures of alkaloid derivatives

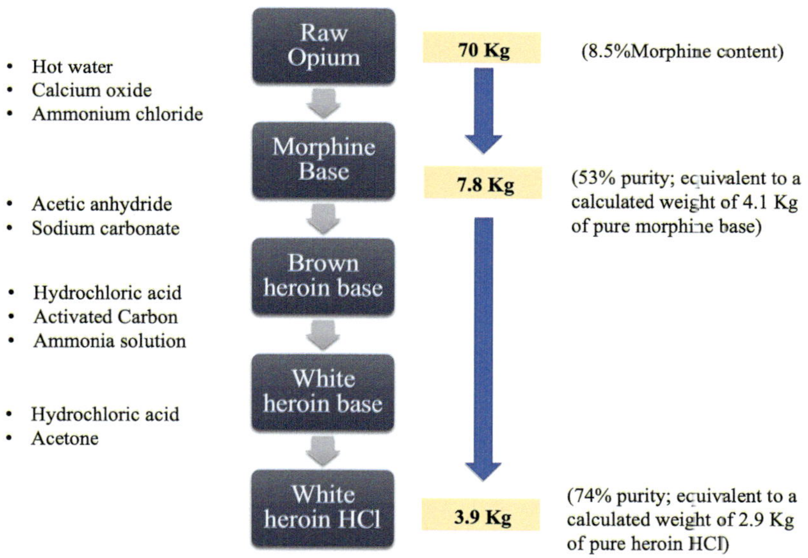

Fig. 36.8 Heroin manufacturing process [82]

- **Heroin** is an opioid drug derived from morphine (Fig. 36.8) and was first synthesized in 1874. It is used in several countries to relieve pain, such as during heart attack or childbirth, as well as in opioid replacement therapy. However, a heroin overdose can cause slow and shallow breathing, coma, and death [79].
- **Oxycodone** is derived from thebaine and is used to treat moderate to severe pain, such as pain from surgery, injury, or chronic conditions like cancer [80].

Table 36.6 Alkaloid derivatives and their applications

Derivative	Precursor	Application
Heroin	Morphine	Pain management
Hydromorphone	Morphine	Used to treat moderate to severe pain
Oxycodone	Thebaine	Treatment of pain due to surgery, injury, and cancer
Oxymorphone	Thebaine	Alleviation of moderate-to-severe pain
Naloxone	Thebaine	Used to rapidly reverse the effects of opioids
Hydrocodone	Codeine	Moderate to severe pain treatment, cough suppressor
Buprenorphine	Thebaine	Sever pain treatments: During or after an operation, a serious injury, or cancer

- **Hydrocodone** is derived from codeine or thebaine and is also used to treat moderate to severe pain and as a cough suppressor [81].
- The detail of the selected alkaloid derivatives is given in Table 36.6.

36.9.2 Phenolic Compounds

Phenolic compounds are a diverse group of secondary metabolites found in various plants, including the opium poppy. These compounds are known for their antioxidant properties and potential health benefits [83]. Some of the phenolic compounds found in opium poppy include:

Isoquinoline Alkaloids These are the most abundant phenolic compounds in the opium poppy, including morphine, codeine, and thebaine, known for their analgesic properties, as described earlier [84]. These are generally classified as BIAs but, due to the presence of phenolic groups, also come in the category of isoquinolines.

Flavonoids Opium poppy contains various flavonoids, including quercetin, kaempferol, apigenin, and isorhamnetin. These compounds have antioxidant properties and potential health benefits [84].

Proanthocyanidins These are a type of flavonoid found in the opium poppy, and they have been shown to have antioxidant, anti-inflammatory, and antimicrobial properties. The primary examples of proanthocyanidins include catechin, procyanidin B1 and procyanidin B2 [85].

Phenolic Acids Opium poppy contains various phenolic acids, including caffeic acid, ferulic acid, and sinapic acid. These compounds have been shown to have antioxidant properties and may have potential health benefits [86].

Overall, the phenolic compounds in opium poppy are the subject of ongoing research due to their potential health benefits. However, the primary focus of opium poppy cultivation and use is the production of alkaloids [87].

36.9.3 *Essential Oils and Other Components*

Opium poppy essential oils include various chemical compounds, including terpenes, sesquiterpenes, and phenylpropanoids (Fig. 36.9) [88]. Some of the compounds found in opium poppy essential oils include:

Limonene: This terpene is commonly found in citrus fruits and possesses anti-inflammatory, antioxidant, and antimicrobial properties [89].

Linalool: This terpene has a floral fragrance and calming and relaxing properties [90].

α-Pinene: This is a terpene found in various coniferous trees, and it has been shown to have anti-inflammatory, antioxidant, and antimicrobial properties [91].

Methyl salicylate: This phenylpropanoid is found in various plants, including wintergreen and birch trees. It has been used for its analgesic and anti-inflammatory properties [92].

Cadinene: This is a sesquiterpene found in various plants, including cypress trees, and it shows anti-inflammatory and antimicrobial properties.

Overall, opium poppy essential oils compounds may have potential health benefits, including anti-inflammatory, antioxidant, and antimicrobial properties. However, it is essential to note that the International Narcotics Control Board strictly regulates the cultivation and use of opium poppy due to the potential for abuse and addiction associated with its alkaloids [93].

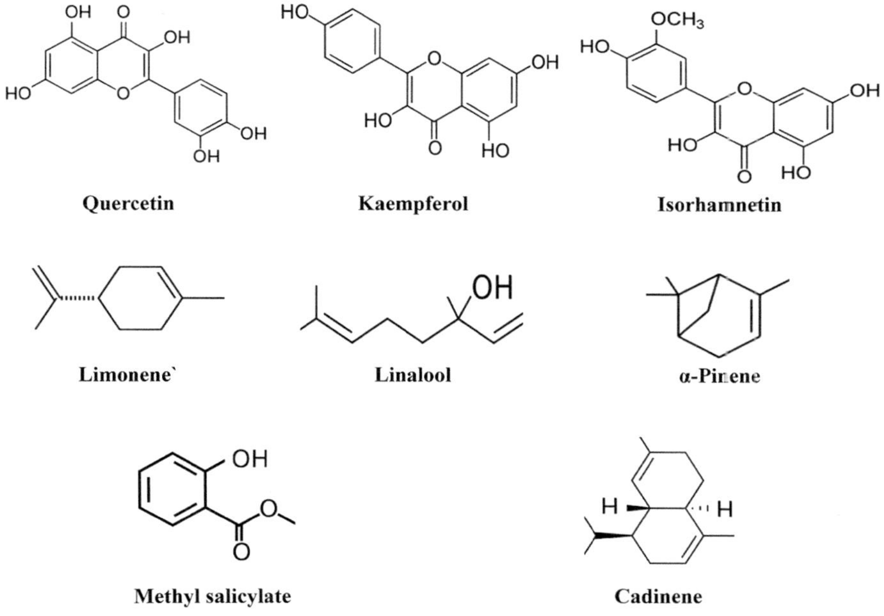

Fig. 36.9 Chemical structures of compounds of essential oils

36.9.4 *Applications of* P. somniferum *L Extracts*

P. somniferum L extracts have also shown many applications, including anti-inflammatory, antidiabetic, analgesic, and remedy for cough and lung infection. Morphine, codeine, and noscapine (antitussive) are used directly in the therapy [94].

36.9.4.1 Cytotoxicity

The alkaloids produced by *P. somniferum* L have been shown to have cytotoxic activity against various cancer cell lines in various studies (Table 36.7). For instance, morphine and codeine are known to have immunosuppressive effects, which can lead to an increased risk of infections and cancer. They can also induce apoptosis, or programmed cell death, in certain types of cells, including immune cells and cancer cells. Noscapine has been shown to induce apoptosis in various types of cancer cells, including breast, lung, and prostate cancer cells. Papaverine can also inhibit the growth of colorectal cancer cells [95].

36.9.4.2 Antimicrobial Activity

P. somniferum extracts have also been reported for antimicrobial properties in several studies. Extracts from different parts of the plant, such as the leaves, stems, seeds, and latex, have been evaluated for their antimicrobial activity against a range of microorganisms and could potentially be used to develop new antimicrobial agents. The antimicrobial activity of extracts from *P. somniferum* L is against selected species of bacteria and fungi is shown in Table 36.8.

36.9.4.3 Antioxidant Activities

P. somniferum L has been studied *in vitro* for its antioxidant properties due to the presence of phytochemicals such as alkaloids, flavonoids, and phenolic compounds. For example, morphine is a potent antioxidant and free radical scavenger that has been shown to protect cells and tissues against oxidative stress-induced damage.

Table 36.7 Cytotoxicity of *P. somniferum* L.

Extract name	Model	Cytotoxicity	Reference
Hexane	C6 cells, and Vero cell lines	Strong anticancer activity on all cell lines	[96]
Methanol			
Ethyl acetate			
Lead and iron oxide nanoparticles using extract of *P. somniferum* from the pods of the plant	HepG2 cell lines	Higher cytotoxicity with PbO NPs as compared to Fe_2O_3 NPs.	[97]

Table 36.8 Antimicrobial activities of *P. somniferum* L. [10]

Extract name	Microorganisms	Zone of Inhibition (ZOI)
Hexane methanol, ethanol, and ethyl acetate extracts	*Bacillus cereus*	0.14 mm
	Staphylococcus aureus	2.00 mm
	Escherichia coli	0.10 mm
	Salmonella typhi	0.13 mm
	Aspergillus Niger	3.00 mm
	Aspergillus oryzae	3.00 mm
	Aspergillus flavus	1.50 mm
	Penicillium chrysogenum	2.00 mm

Codeine also possesses antioxidant activity and like morphine provide protection to cells and tissues protects against oxidative damage. The antioxidant activity of the ethanolic extract, determined by total reducing power assay, was found to be 3592.56 mg/mL [98]. Similarly, in another study, the extract of *P. somniferum* L. prepared in methanol showed 49.75 IC50 (μg/mL) antioxidant activity when the Linoleic acid peroxidation assay method was used [99].

36.10 Poppy Seeds and Their Applications

They are legally sold and widely used in culinary preparations such as baked goods, dressings, and condiments. For example, poppy seeds are frequently used in muffins, bread, and other food [100]. Poppy seeds are a good source of nutrients, including protein, fiber, healthy fats, minerals, and vitamins (Table 36.9).

Mature poppy seeds do not contain latex but can become contaminated with opium alkaloids due to pest damage and during harvest [55]. Analysis of poppy seed contents by using HPLC showed that it contained alkaloids Morphine (7.65–25.15%), Noscapine (4.03–15.22%), Codeine (1.21–6.37%), Thebaine (0.97–6.38%) Papaverine(0.51–5.33%) [101]. Eating poppy seeds can have several effects on the body. Poppy seeds are a good source of dietary fiber, which can aid in digestion and help regulate bowel movements. The presence of small amounts of natural opiates such as morphine and codeine in poppy seeds can induce mild relaxation. Because of the presence of these natural opiates, consumption of poppy seeds in large amounts can lead to a positive result on a drug test for opioids. In some cases, it may lead to feelings of drowsiness or even sedation due to the presence of these natural opiates. But the chances of experiencing opiate overdoses from consuming large amounts of poppy seeds are rare unless seeds are contaminated with higher concentrations of opiates [102]. Another study showed high morphine levels of 13,857 ng/ml due to the ingestion of unprocessed poppy seed [103].

Table 36.9 Nutrient composition of poppy seed [17]

Nutrient	Composition (per 100 g)	Nutrient	Composition (per 100 g)
Water	5.95 g	Betaine	0.9 mg
Energy	525 kcal	Vitamin E (α-tocopherol)	1.77 mg
Carbohydrate	28.13 g	Total saturated fatty acids	4.517 g
Total dietary fibre	19.5 g	Total monounsaturated fatty acids	5.982 g
Total sugars	2.99 g	Total polyunsaturated fatty acids	28.569 g
Zinc	7.90 mg	Cystine	0.297 g
Copper	1.627 mg	Phenylalanine	0.758 g
Manganese	6.707 mg	Tyrosine	0.727 g
Selenium	13.5 μg	Valine	1.095 g
Vitamin C	1 mg	Arginine	1.945 g
Thiamene	0.854 mg	Histidine	0.471 g
Riboflavin	0.1 mg	Alanine	0.839 g
Niacin	0.896 mg	Aspartic acid	2.365 g
Pantothenic acid	0.324 mg	Glutamic acid	4.299 g
Vitamin B-6	0.247 mg	Glycine	0.952 g
Total folate	82 μg	Proline	2.754 g
Total choline	8.8 mg	Serine	0.952 g

36.11 Poppy Seeds and Their Applications

36.11.1 Food and Culinary

Poppy seeds have various food and culinary uses across multiple cultures, including cooking, baking, and confectionery some native tribes eat the young leaves as vegetables [104].

36.11.2 Pharmaceutical Applications

Poppy seeds contain opium alkaloids, which have been used for pain relief and sedation for thousands of years. These alkaloids also make opioid painkillers, such as morphine and codeine. Poppy seeds can be used to treat diarrhea because they contain astringent compounds that help to reduce the amount of fluid in the intestine [105].

36.11.3 Colovesical Fistula Diagnostics

Colovesical fistula diagnostics involves expensive and invasive test procedures and is very challenging. The success rate of diagnostic techniques also varies. The poppy seeds are exceptional for colovesical fistula diagnostics and can be used for the inexpensive, accurate and convenient diagnostic test with 100% accuracy [106].

36.11.4 Hepatocellular Cancer Imaging

Poppy seed oil tends to aggregate in hepatocellular cancer cells. The iodized poppy seed oil (Lipiodol) is employed for imaging vascular hepatocellular cancer and used as a vehicle or adjuvant to deliver chemotherapeutic drugs (such as cyclosporine A, cisplatin) to malignant sites [107].

36.11.5 Skincare

Poppy seed oil is often used in skin care products due to its high linoleic acid concentration, which helps hydrate and nourish the skin. The oil is also used in massage therapy to help relieve muscle tension and promote relaxation [108].

36.11.6 Industrial Uses

Poppy seeds have many applications in the paint and biofuel industry [109]. They are also used as lubricants in the machinery [110].

36.12 Conclusion

The opium poppy has been known for centuries for its medicinal properties and its psychoactive effects. It is widely being cultivated across the world for opium production that is a natural source of pharmaceutically important drugs like morphine, codeine and their derivatives. Poppy seeds are also consumed widely and used in baking and cooking. The cultivation and use of opium poppy are subject to legal restrictions due to worldwide opium misuse and derived-drug abuse and addiction. In conclusion, while the opium poppy has been a controversial plant throughout history, however, its potential for medicinal and culinary purposes cannot be ignored. Proper legal regulations can ensure its benefits are harnessed while minimizing its adverse effects.

References

1. Chouvy, P.-A. (2010). *Opium: uncovering the politics of the poppy*. Harvard University Press.
2. Alam, M., et al. (2011). Disease incidence and its management on opium poppy: A global perspective. In *International Symposium on Papaver 1036*. ISHS.
3. Kritikos, P. G., & Papadaki, S. (1967). *The history of the poppy and of opium and their expansion in antiquity in the eastern Mediterranean area*. UN.
4. Najafipour, H., & Beik, A. (2016). The impact of opium consumption on blood glucose, serum lipids and blood pressure, and related mechanisms. *Frontiers in Physiology, 7*, 436.
5. Krames, E. S. (2012). A history of intraspinal analgesia, a small and personal journey. *Neuromodulation: Technology at the Neural Interface, 15*(3), 172–193.
6. Özcan, M. M., & Atalay, Ç. (2006). Determination of seed and oil properties of some poppy (*Papaver somniferum* L.) varieties. *Grasas y Aceites, 57*(2), 169–174.
7. Paul, A. K., et al. (2021). Opioid analgesia and opioid-induced adverse effects: A review. *Pharmaceuticals, 14*(11), 1091.
8. NCJRS. (1992). *Opium poppy cultivation and heroin and processing in Southeast Asia*. U.S. Department of Justice/Drug Enforcement Administration/Office of Intelligence.
9. Kapoor, L. (1995). *Opium poppy: botany, chemistry, and pharmacology*. CRC Press.
10. Butnariu, M., et al. (2022). Papaver plants: Current insights on phytochemical and nutritional composition along with biotechnological applications. *Oxidative Medicine and Cellular Longevity, 2022*, 1–23.
11. Blamey, M., Fitter, R., & Fitter, A. (2003). *Wild flowers of Britain and Ireland: The complete guide to the British and Irish Flora*. A. & C. Black.
12. Stace, C. (2019). *New Flora of the British Isles*. CABI.
13. Bentley, R. A. H. T. (2002). *Medicinal plants*. Omsons Publications.
14. Miller, J. A. C., et al. (2005). Pollination biology of oilseed poppy, *Papaver somniferum* L. *Australian Journal of Agricultural Research, 56*, 483.
15. Salavert, A., et al. (2020). Direct dating reveals the early history of opium poppy in western Europe. *Scientific Reports, 10*, 20263.
16. Bernáth, J., & Németh, É. (2010). Poppy. In *Oil crops* (pp. 449–468). ResearchGate.
17. Lim, T., & Lim, T. (2013). Papaver somniferum. *Edible Medicinal And Non-Medicinal Plants, 5*, 202–217.
18. Dasgupta, A. (2019). Analytical true positive: Poppy seed products and opiate analysis. In *Critical issues in alcohol and drugs of abuse testing* (pp. 449–462). Academic.
19. Hong, U. V. T., et al. (2022). Insights into opium poppy (Papaver spp.) genetic diversity from genotyping-by-sequencing analysis. *Scientific Reports, 12*, 111.
20. Norn, S., Kruse, P. R., & Kruse, E. (2005). History of opium poppy and morphine. *Dansk medicinhistorisk arbog, 33*, 171–184.
21. Salavert, A., et al. (2018). The opium poppy in Europe: exploring its origin and dispersal during the Neolithic. *Antiquity, **92***, 364.
22. Krikorian, A. D. (1975). Were the opium poppy and opium known in the ancient Near East? *Journal of the History of Biology, 8*, 95–114.
23. Tétényi, P. (2010). Opium poppy (*Papaver somniferum*): Botany and horticulture. In *Horticultural reviews* (pp. 373–408). Wiley.
24. Sigerist, H. E. (1941). Laudanum in the works of Paracelsus. *Bulletin of the History of Medicine, 9*(5), 530–544.

25. Guo, L., et al. (2018). The opium poppy genome and morphinan production. *Science, 362*, 343–347.
26. Mégarbane, B., et al. (2006). Does high-dose buprenorphine cause respiratory depression? Possible mechanisms and therapeutic consequences. *Toxicological Reviews, 25*(2), 79–85.
27. Shahbandeh, M. (2022). *Main countries for opium cultivation 2020, based on acreage*. cited 2023; Available from: https://www.statista.com/statistics/264744/top-countries-for-oppium-cultivation-based-on-acreage/
28. Rajagopal, M. R., Joranson, D. E., & Gilson, A. M. (2001). Medical use, misuse, and diversion of opioids in India. *Lancet, 358*(9276), 139–143.
29. Cherny, N., et al. (2010). Formulary availability and regulatory barriers to accessibility of opioids for cancer pain in Europe: A report from the ESMO/EAPC opioid policy initiative. *Annals of Oncology, 21*(3), 615–626.
30. von Euler, M., et al. (2003). Interpretation of the presence of 6-monoacetylmorphine in the absence of morphine-3-glucuronide in urine samples: Evidence of heroin abuse. *Therapeutic Drug Monitoring, 25*(5), 645–648.
31. Borriello, R., et al. (2015). Clinical and forensic diagnosis of very recent heroin intake by 6-acetylmorphine immunoassay test and LC-MS/MS analysis in urine and blood. *Annals of Clinical & Laboratory Science, 45*(4), 414–418.
32. Acock, M., Pausch, R., & Acock, B. (1997). Growth and development of opium poppy (*Papaver somniferum* L.) as a function of temperature. *Biotronics, 26*, 47–57.
33. UN. (1953). *The opium poppy*. cited 2023; Available from: https://www.unodc.org/unodc/en/data-and-analysis/bulletin/bulletin_1953-01-01_3_page004.html
34. Yazici, L. (2022). Influence of different sowing times on yield and biochemical characteristics of different opium poppy (*Papaver somniferum* L.) genotypes. *Journal of King Saud University – Science, 34*, 102337.
35. UNODC. (2022). *Opium cultivation In Afghanistan: Latest findings and emerging threats*. cited 2023.
36. Bečka, D., et al. (2014). Poppy root weevils (Stenocarus ruficornis, Stephens 1831) control in opium poppy (*Papaver somniferum* L.). *Plant, Soil and Environment, 60*(10), 470–474.
37. Barnes, H. F. (1949). *Gall midges of economic importance. Volume VI. Gall midges of miscellaneous crops*. C. Lockwood.
38. Miciński, B., & Przybylska, M. (1983). Preliminary experiments on virus diseases of poppy (*Papaver somniferum* L.) in Poland. *Zeszyty Problemowe Postępów Nauk Rolniczych, 291*, 221–225.
39. Pethybridge, S., et al. (2005). Survey for viruses in oilseed poppy in Tasmania, Australia. *Australasian Plant Pathology, 34*, 611–613.
40. Cleij, G. (1961). Beet yellows in poppy. *Euphytica, 10*, 225–228.
41. Tang, J., et al. (2016). Opium poppy mosaic virus, a new umbravirus isolated from *Papaver somniferum* in New Zealand. *Archives of Virology, 161*, 197–201.
42. Zaim, M., Joseph, J., & Lal, R. (2011). Identification and characterization of a potyvirus associated with severe mosaic disease of opium poppy (*Papaver somniferum* L.) in India. In *International Symposium on Papaver* (p. 1036). ISHS.
43. Srivastava, A., et al. (2016). First report of tomato leaf curl New Delhi virus on opium poppy (*Papaver somniferum*) in India. *Plant Disease, 100*(1), 232.
44. Horvath, J., & Besada, W. (1975). Opium poppy (*Papaver somniferum* L.), a new natural host of turnip mosaic virus in Hungary/Mohn (*Papaver somniferum* L.) als neue natürliche Wirtspflanze für das Kohlschwarzring-Virus in Ungarn. *Zeitschrift für Pflanzenkrankheiten und Pflanzenschutz/Journal of Plant Diseases and Protection, 82*, 162–167.
45. Špak, J., & Kubelková, D. (1990). Occurrence of turnip mosaic virus in opium poppy (*Papaver somniferum* L.) in Czechoslovakia. *Sborník ÚVTIZ, Ochrana Rostlin, 26*(4), 257–261.
46. Aranda, S., et al. (2008). First report of Pectobacterium carotovorum causing soft rot of opium poppy in Spain. *Plant Disease, 92*(2), 317–317.

47. Cotterill, P., & Pascoe, I. (1998). Downy mildew of Papaver somnifmum in Tasmania. *Australasian Plant Pathology, 27*(4), 263–264.
48. Shamsi, M. A. (1979). Basirolaimus gen. n.(Nematoda: Hoplolaimidae) with the description of Basirolaimus sacchari, n. sp. from India. *Nematologia Mediterranea, 7*, 15–19.
49. Akgül, H., & Ökten, E. (2001). A list of Tylenchida associated with poppy crops (*Papaver somniferum* L.) in Afyon region, Turkey. *Nematology, 3*(3), 289–291.
50. Alcantara, J., et al. (2005). Sanguinarine biosynthesis is associated with the endoplasmic reticulum in cultured opium poppy cells after elicitor treatment. *Plant Physiology, 138*(1), 173–183.
51. Koleva, I. I., et al. (2012). Alkaloids in the human food chain–natural occurrence and possible adverse effects. *Molecular Nutrition & Food Research, 56*(1), 30–52.
52. Hemmings, H. C., & Egan, T. D. (2012). *Pharmacology and physiology for anesthesia E-Book: Foundations and clinical application*. Elsevier Health Sciences.
53. Gibula-Tarlowska, E., & Kotlinska, J. H. (2020). Crosstalk between opioid and anti-opioid systems: An overview and its possible therapeutic significance. *Biomolecules, 10*(10), 1376.
54. Hagel, J. M., & Facchini, P. J. (2013). Benzylisoquinoline alkaloid metabolism: A century of discovery and a brave new world. *Plant and Cell Physiology, 54*(5), 647–672.
55. Chain, E., et al. (2018). Update of the scientific opinion on opium alkaloids in poppy seeds. *EFSA Journal, **16***(5), e05243.
56. Menéndez-Perdomo, I. M., & Facchini, P. J. (2018). Benzylisoquinoline alkaloids biosynthesis in sacred lotus. *Molecules, 23*(11), 2899.
57. Behr, A., et al. (2020). *Nature's pharmacy-natural pharmaceuticals* (pp. 277–293). An Introduction.
58. De Gregori, S., et al. (2012). Morphine metabolism, transport and brain disposition. *Metabolic Brain Disease, 27*(1), 1–5.
59. Gennari, C. (2002). Analgesic effect of calcitonin in osteoporosis. *Bone, 30*(5), 67–70.
60. Stefano, G., et al. (2012). Endogenous morphine: Up-to-date review 2011. *Folia Biologica, 58*(2), 49–56.
61. Ræder, J. C., Steine, S., & Vatsgar, T. T. (2001). Oral ibuprofen versus paracetamol plus codeine for analgesia after ambulatory surgery. *Anesthesia & Analgesia, 92*(6), 1470–1472.
62. Yaksh, T. L., & Wallace, M. S. (2011). Opioids, analgesia, and pain management. In *Goodman and Gilman's the pharmacological basis of therapeutics* (pp. 481–526). McGraw-Hill.
63. Mahmoudian, M., & Rahimi-Moghaddam, P. (2009). The anti-cancer activity of noscapine: A review. *Recent Patents on Anti-Cancer Drug Discovery, 4*(1), 92–97.
64. Rahmanian-Devin, P., et al. (2021). Noscapine, an emerging medication for different diseases: A mechanistic review. *Evidence-based Complementary and Alternative Medicine, 2021*, 8402517.
65. Guo, Q., et al. (2013). Determination of five alkaloids of Pericarpium papaveris in hot pot broth using ultra-performance liquid chromatography coupled to triple quadruple mass spectrometry. *Food Analytical Methods, 6*, 698–704.
66. Pouraghaei, M., et al. (2021). Intranasal ketamine versus intravenous morphine for pain management in patients with renal colic: A double-blind, randomized, controlled trial. *World Journal of Urology, 39*, 1263–1267.
67. Eltantawy, W., Mohyieldin, M., & Mohammed, A. (2020). Predictive accuracy of three dimensional ultrasound measurement of fetal adrenal gland volume in preterm labor cases. *QJM: An International Journal of Medicine, 113*(Supplement_1), hcaa056. 002.
68. Tolstikova, T., et al. (2009). Thebaine as a precursor of opioid analgesic agents. *Chemistry for Sustainable Development, 17*, 109–126.
69. Singh, H., et al. (2013). A review on noscapine, and its impact on heme metabolism. *Current Drug Metabolism, 14*(3), 351–360.
70. Lawrence, P. F. (2011). *Pharmacologic Adjuncts to Endovascular Procedures, in Endovascular Surgery* (pp. 807–813). Elsevier.
71. Yavuz, S., et al. (2001). Effect of papaverine applications on blood flow of the internal mammary artery. *Annals of Thoracic and Cardiovascular Surgery, 7*(2), 84–88.

72. Haghjooy-Javanmard, S., et al. (2018). Influence of morphine on TLR4/NF-kB signaling pathway of MCF-7 cells. *Bratislavske lekarske listy, 119*(4), 229–233.
73. Bogacka, J., et al. (2020). CCR4 antagonist (C021) influences the level of nociceptive factors and enhances the analgesic potency of morphine in a rat model of neuropathic pain. *European Journal of Pharmacology, 880*, 173166.
74. Listos, J., et al. (2019). The mechanisms involved in morphine addiction: An overview. *International Journal of Molecular Sciences, 20*(17), 4302.
75. Chahl, L. A. (1996). Opioids-mechanisms of action. *Australian Prescriber, 19*, 63–65.
76. Reeves, K. C., et al. (2022). Opioid receptor-mediated regulation of neurotransmission in the brain. *Frontiers in Molecular Neuroscience, 15*, 919773.
77. Hao, D. C., Gu, X.-J., & Xiao, P. G. (2015). Phytochemical and biological research of Papaver pharmaceutical resources. *Medicinal Plants: Chemistry Biology and Omics, 6*, 217–251.
78. Singh, A., Menéndez-Perdomo, I. M., & Facchini, P. J. (2019). Benzylisoquinoline alkaloid biosynthesis in opium poppy: An update. *Phytochemistry Reviews, 18*, 1457–1482.
79. Boyd, S. (2021). Heroin and the illegal drug overdose death epidemic: A history of missed opportunities and resistance. *International Journal of Drug Policy, 91*, 102938.
80. Schmidt-Hansen, M., et al. (2022). Oxycodone for cancer-related pain. *Cochrane Database of Systematic Reviews, 6*(6), CD003870.
81. Chua, K.-P., & Conti, R. M. (2021). Prescriptions for codeine or hydrocodone cough and cold medications to US children and adolescents following US food and drug administration safety communications. *JAMA Network Open, 4*(11), e2134142.
82. Zerell, U., Ahrens, B., & Gerz, P. (2005). Documentation of a heroin manufacturing process in Afghanistan. *Bulletin on Narcotics, 57*(1–2), 11–31.
83. Olszowy, M. (2019). What is responsible for antioxidant properties of polyphenolic compounds from plants? *Plant Physiology and Biochemistry, 144*, 135–143.
84. Sharopov, F., et al. (2018). Alkaloid content, antioxidant and cytotoxic activities of various parts of *Papaver somniferum. Pharmaceutical Chemistry Journal, 52*, 459–463.
85. Haslam, E. (1998). *Practical polyphenolics: from structure to molecular recognition and physiological action*. Cambridge University Press.
86. Anouar, E., et al. (2009). New aspects of the antioxidant properties of phenolic acids: A combined theoretical and experimental approach. *Physical Chemistry Chemical Physics, 11*(35), 7659–7668.
87. Davis, C. (2014). Evolutionary and neuropsychological perspectives on addictive behaviors and addictive substances: relevance to the "food addiction" construct. *Substance Abuse and Rehabilitation, 5*, 129–137.
88. Croteau, R., Kutchan, T. M., & Lewis, N. G. (2000). Natural products (secondary metabolites). *Biochemistry and molecular biology of plants, 24*, 1250–1319.
89. Vuuren, S. V., & Viljoen, A. M. (2007). Antimicrobial activity of limonene enantiomers and 1, 8-cineole alone and in combination. *Flavour and Fragrance Journal, 22*(6), 540–544.
90. Aprotosoaie, A. C., et al. (2014). Linalool: A review on a key odorant molecule with valuable biological properties. *Flavour and Fragrance Journal, 29*(4), 193–219.
91. da Silva Rivas, A. C., et al. (2012). Biological activities of α-pinene and β-pinene enantiomers. *Molecules, 17*(6), 6305–6316.
92. Zhang, D., et al. (2011). Anti-inflammatory activity of methyl salicylate glycosides isolated from Gaultheria yunnanensis (Franch.) Rehder. *Molecules, 16*(5), 3875–3884.
93. Board, I. N. C. (2005). *Report of the International Narcotics Control Board 2004: 2005*. United Nations Publications.
94. Labanca, F., Ovesnà, J., & Milella, L. (2018). *Papaver somniferum* L. taxonomy, uses and new insight in poppy alkaloid pathways. *Phytochemistry Reviews, 17*(4), 853–871.
95. Gomes, D. A., Joubert, A. M., & Visagie, M. H. (2022). The biological relevance of Papaverine in cancer cells. *Cells, 11*(21), 3385.
96. Güler, D. A., et al. (2016). Anticancer activity of *Papaver somniferum. Journal of the Turkish Chemical Society Section A: Chemistry, 3*(3), 349–366.

97. Muhammad, W., et al. (2019). *Papaver somniferum* L. mediated novel bioinspired lead oxide (PbO) and iron oxide (Fe2O3) nanoparticles: In-vitro biological applications, biocompatibility and their potential towards HepG2 cell line. *Materials Science and Engineering: C, 103*, 109740.
98. Kačániová, M., et al. (2010). Antioxidant, antimicrobial actiyity and heavy metals content in pollen of *Papaver somniferum* L. *Ecological Chemistry and Engineering. A, 17*(1), 97–106.
99. Souri, E., Amin, G., & Farsam, H. (2008). Screening of antioxidant activity and phenolic content of 24 medicinal plant extracts. *DARU Journal of Pharmaceutical Sciences, 16*(2), 83–87.
100. Luqman, S. (2011). The saga of opium poppy: Journey from traditional medicine to modern drugs and nutraceuticals. In *International Symposium on Papaver* (p. 1036). ISHS.
101. Krenn, L., Glantschnig, S., & Sorgner, U. (1998). Determination of the five major opium alkaloids by reversed-phase high-performance liquid chromatography on a base-deactivated stationary phase. *Chromatographia, 47*, 21–24.
102. Hayes, L. W., Krasselt, W. G., & Mueggler, P. A. (1987). Concentrations of morphine and codeine in serum and urine after ingestion of poppy seeds. *Clinical Chemistry, 33*(6), 806–808.
103. Hill, V., et al. (2005). Multiple aspects of hair analysis for opiates: methodology, clinical and workplace populations, codeine, and poppy seed ingestion. *Journal of Analytical Toxicology, 29*(7), 696–703.
104. Lachenmeier, D. W., Sproll, C., & Musshoff, F. (2010). Poppy seed foods and opiate drug testing-where are we today? *Therapeutic Drug Monitoring, 32*(1), 11–18.
105. Azizi, M., Hasam, S., & Qarizada, D. (2022). Production of poppy seed in Afghanistan and its composition effects: A review. *Asian Journal of Chemical Sciences, 12*(1), 12–20.
106. Kwon, E. O., et al. (2008). The poppy seed test for colovesical fistula: Big bang, little bucks! *The Journal of Urology, 179*(4), 1425–1427.
107. Ronzoni, S., et al. (1999). Synthesis and NMR characterization of a novel class of thienomorphinans. *Organic Letters, 1*(3), 513–516.
108. Zahoor, F., Forristal, D., & Gillespie, G. (2015). *CropQuest: Minor crops report*. Teagasc.
109. Šerá, B., et al. (2013). New physicochemical treatment method of poppy seeds for agriculture and food industries. *Plasma Science and Technology, 15*(9), 935.
110. Brühl, L. (2016). Occurrence, determination, and assessment of mineral oils in oilseeds and vegetable oils. *European Journal of Lipid Science and Technology, 118*(3), 361–372.

Chapter 37
Lavender

Hammad Salahudin, Shagufta Kamal, Naheed Akhter, Ismat Bibi, Kanwal Rehman, Muhammad Sajid Hamid Akash, and Umar Farooq Gohar

37.1 Introduction

Lavender is a flowering plant that has been identified due to the sweet smell of its flowers. In past times, lavender has also been used as a holy herb. The genus Lavandula includes above 30 species of which some are used in aromatherapy. Aromatherapy is a practice in which essential oils are used for therapeutic purposes. Lavender which is a member of the Lamiaceae family; is one of the conventional and worldwide cultivated scented plant species with numerous applications in a different fields. Lavender leaves, flowers, essential oils, and other products can be industrially used in many ways such as crisped, scorched, or processed products. This plant and its products are conventionally believed to be used as a bactericidal, antifungal, carminative, relaxant, treat migraine, antidepressant, antidiabetic, and effectively potent for burns and insect bites. Lavender is also used for different food flavorings such as jams, biscuits, jelly, ice cream, coffee, and honey [1]. The essential oil that is obtained from the flower is included in the list of the 15 most essential volatile oils that are traded in the world on large scale [2]. Due to the presence of

H. Salahudin · S. Kamal (✉) · N. Akhter
Department of Biochemistry, Government College University, Faisalabad, Pakistan
e-mail: shaguftakamal@gcuf.edu.pk

I. Bibi
Department of Chemistry, Islamia University of Bahawalpur, Bahawalpur, Pakistan

K. Rehman
Department of Pharmacy, The Women University, Multan, Pakistan

M. S. H. Akash
Department of Pharmacy, Government College University, Faisalabad, Pakistan

U. F. Gohar
Institute of Industrial Biotechnology, Government College University, Lahore, Pakistan

M. Zia-Ul-Haq et al. (eds.), *Essentials of Medicinal and Aromatic Crops*,
https://doi.org/10.1007/978-3-031-35403-8_37

several volatile oils, lavender has an extreme concentration for perfumes and is also used for cosmetic preparations (such as whitening creams). Due to its efficacy in sedation, the essential oil of lavender is also commercially used for bathing products such as detergents, shampoo, oil, liniment, and salt. Besides industrial uses, lavender is commonly used as an ornamental and landscape plant because lavender can tolerate drought, it is used for decoration as green-roofing which is becoming very popular for contemporary buildings in many rural areas [3]. Lavender is also one of the most reasonable plants used in public parks and house gardens as well as indoors because of its fragranced, evergreen leaves and attractive flowers [4]. It is used for aromatic purposes in soaps, detergents cosmetics, and perfumes. The quality of essential oils derived from lavender plants directly relates to the concentration of terpenic alcohols as well as acetates [1, 4]. Although the genus Lavandula comprises more than 30 flowering species but only three of them are cultivated industrially for manufacturing essential oil. Bulgaria is the world's top lavender oil producer country growing over 6000 different varieties of lavender. One of the common lavender, Lavandula angustifolia is used as a medicinal agent in ointments, balms, perfumes, and cosmetics and is considered a model plant for isoprenoid studies. Due to the enrichment of terpenes, the essential oil of lavender is also well known for its germicidal effect and it has antiseptic, anti-inflammatory, and painkiller properties [2, 4]. One of the researchers reported the presence of many bioactive compounds which have strong antioxidant properties in living systems which include polyphenols, anthocyanins, and carotenoids. Camphor, terpene, ocimene, linalool, and linalyl acetate are the main components of lavender responsible for its typical aroma. Monoterpenes (C10 class of isoprenoids) are the main constituents present in the essential oil of lavender that is generated and accumulated in the glandular trichomes. The surface of aerial parts of the lavender plant is covered by glandular trichomes [5]. In ancient times, the Lavandula species has been used as a holy herb. The genus Lavandula includes above 30 species of which some are used in aromatherapy. Aromatherapy is a practice in which essential oils are used for therapeutic purposes. Besides aromatherapy, Lavandula species are also used for different medical purposes. One of the studies revealed that lavender oil can be efficient in combating the occurrence of antifungal-resistant infections. In a research, it was found that lavender's essential oil was deadly to all skin pathogenic strains. It kills the fungal cell by destroying its cell membrane [6]. In some studies, it is reported that essential oils derived from lavender plants can be used in the healing process of wounds. In a test carried out at Celal Bayar University, Turkey, researchers compared the effects of saline solution, tincture of iodine, and lavender oil for wound healing and they found that wound healing progressed more rapidly in lavender oil groups than in the control and other study groups [7]. Studies showed that applying lavender oil in combination with oil from thyme, rosemary, and cedarwood will enhance hair growth by up to 44% after 7 months of treatment [8]. Taking lavender orally for 2–3 months improves depression and sleep and prevents anxiety recurrence in people [9].

37.2 Scientific Name/English Common Name

Lavenders are a member of the plant family called Lamiaceae, the mint family which contains some species and a few are helpful in aromatherapy. Plants of this family often consist of 5 combined petals that are with 2 lobes on the upper side and 3 on the lower side forming lips. The leaves are on the opposite side of one another at the stem and usually, the stem is squared shape. In Latin the name used for the lavender genus is *Lavandula,* hence the lavenders start with *Lavandula.* Two widely used "lavenders" are *Lavandula angustifolia* and *Lavandula latifolia.*

Lavandula angustifolia is often known as *L. vera* or *L. officinalis* even though the exact name is *L. angustifolia* (Lawrence 1989). It also has numerous other names: English lavender, French lavender, and true lavender. *Lavandula latifolia* is routinely named *L. spica,* and its common name is spike lavender or spike. Spike is entirely different from spikenard which has a resemblance to valerian and is a member of the family Valerianaceae. *Lavandula angustifolia* and *Lavandula latifolia* were both included in the British formularies that are provided to hospitals in vats labeled simply "lavender." However, these plants have different therapeutic functions. *L. angustifolia* is used as a sedative, relaxant, and hypotension. *L. latifolia* is used as a stimulant and expectorant. *Lavandula intermedia* also called *L. hybrida.* It is a naturally existing hybrid and firstly it was discovered in 1828. Nowadays, this hybrid is known as a manufactured cultivar with the market name Lavandin. It is intercourse between *L. angustifolia and L. latifolia.* As Lavandin is a member of *Lavandula* genus also known "lavender". It confuses as people think they got the original lavender but they have lavandin. There is also another less-frequently used species of lavender that is *L.stoechas* that is used for clinical purposes.

37.3 Agronomy

37.3.1 Soil Conditions

The lavender plant loves to grow in soils where pH is slightly basic or neutral but it can also grow in a moderately acidic environment. Lavender grows well in sandy or chalky soil with good drainage. In terms of plant growth and yield, mineral uptake and accumulation are among the most essential factors. Nitrogen is the master element required by the plant to improve the yield and quantity by affecting biosynthetic compounds such as proteins, amino acids, chlorophyll, and nucleic acid [10]. If nitrogen is provided in optimum doses, the cell number of the leaf along with leaf size is increased. Moreover, it also increases leaf production and yield. In aromatic plants such as lavender, nitrogen is highly effective for the yield of essential oils. So soil should be enriched with nitrogen. But if the soil is deficient in nitrogen sources, it decreases plant growth as well as its yield and losses in quality may also occur. Nitrogen is also an essential element required for cell growth hence due to its depletion, cell division undergoes reduction as a result stunted growth may occur [11]. Depending upon the gravity of N deficiency, chlorosis could cause the demise or dropping of aged leaves, in extreme cases, flowering significantly decreases or preliminary maturation may occur which leads to a reduction in the yield of the crop [12]. Just like nitrogen, phosphorous also play a significant role in various metabolic activity. Phosphorous is the main component of nucleic acid, phospholipids and some other factors that effectively promote activation of the production of amino acid. These amino acids are further processed in protein synthesis, nucleic acids and ATP. When soil is deficient in phosphorous, it may lead to a decrease in chloroplast carbon fixation [13]. Lavender growth is also affected greatly by the pH values of soil. *Lavandula angustifolia* and *lavandula astoechasc* can grow on soil with pH ranges between 5.5 and 8.5 but optimum germination of seeds and photosynthesis are recorded at 7 pH. Although these two species can grow on acidic, neutral as well as basic pH, optimal growth is observed on soil with neutral pH. Minimal values were recorded for the germination of seed when lavender was grown on basic soil. The results of germination of seeds obtained from *L. astoechasc* were highly affected by the change in pH. Photosynthesis taking place in these two lavender species was also affected by the change in the pH of the soil. Minimal values of photosynthesis were observed by leaves of *L. stoechas* grown on soil with basic pH. Whereas leaves of L. angustifolia present the lowest values of photosynthesis when they are grown in acidic soil. Chlorophyll contents in the leaves also vary depending on soil pH. Their chlorophyll content decreases on basic pH. This happens so because nutritional imbalance occurs at different pH. The transpiration rate of a plant also differs in different pH conditions of the soil. Minimal values of the extent of transpiration are recorded on acidic and minimum on basic soil for both lavender species [14]. Salinity is another factor that directly interferes with the growth of a lavender plant many disorders in respiration, mineral intake, oxidative stress and photosynthesis are due to salinity of the soil, which leads to a reduction in yield as well as the quality of plants. When salinity is not compatible with the soil, its osmotic pressure is increased that resulting in lowering water potential, toxicity, and imbalance of certain nutrient ions. Increasing salinity leads to the increase of $Na+$ and $Cl-$ that respond to reduced uptake of some important ions e.g. potassium, calcium or phosphorous [15].

37.3.2 Climate

Lavender plants require a minimum of 6–8 h of daily full sun exposure, these plants prefer a dry and moderately warm climate, gentle winters, and cloudless summers. Its origin was the Mediterranean regions, where temperature averagely remains between 20 and 30 °C [16]. Moreover, soil temperature of more than 18 °C favors good growth and also regeneration after the harvest. Although lavender loves to grow in above climatic conditions it can bear lower temperatures. If lavender is grown in high humidity it will be more suspected to development of fungal diseases, while the quality of essential oil is negatively affected due to extensive summer heat effects. A lavender plant can only prosper at high altitudes up to 1500 m as well as in inclined fields. U.S. Department of plant harshness zones from 5 to 8 are the zones that are best suited for English lavender. The minimum temperature range of these zones varies from −20 to 10 °F [16]. Lavender plants will not thrive if coldness increases. Climates with low humidity are also preferable for lavender blossoms. Regions within these plant harshness zones include cool coastal or hilly areas, inland valleys and deserts. The effects of humidity can be reduced by topping the plant with soil, which will not carry moisture. There are also some other conditions to grow lavender in borderline climates. If lavender is suspected to root rot, it can be protected by placing lavender in full sun and using a bright-colored shell as mulch. In this way, soil can be kept warm and dry. Similarly, if we want to grow lavender in a colder climate, lavender must be planted in a container and should be moved indoors in front of a sunny window in winter. If there is high precipitation in the climate, lavender must be planted in rocky soil, through which water will be drained quickly if rocky soil is not present then a place must be chosen where there is good air circulation, in this way plants will be dried out after rain quickly. During the winter season, lavender likes it cool but dislikes a drafty environment. The temperature must not be warmer than 18 °C and should not be colder than 5 °C Lavender plants must be kept away from the heater as its heat will dry them out as well as away from drafty windows so a cool breeze may not chill them. If a plant is to be planted inside the home in winter, it must be kept in a place there is a cool but non-freezing environment, with proper sunlight and air cross ventilation. Dry soil is preferable for a lavender plant, and the quantity of rain that normally falls will be significant for the plant. English lavender is known suited to milder weather, like those in the south, especially in the upper and middle South. Lavenders do not flourish in high humidity but it likes to thrive in a hot, dry climate with plenty of suns. English lavenders are hardy in the zone from 5 to 10 and contain a very sunny disposition. These plants significantly prefer properly drained soil, especially in the winter season. Poor draining soils in winter ultimately kill this English lavender. Among all the species of lavender, only English lavender varieties can regularly survive the winter season in colder regions, despite frosts and snowfall. Hidcote which is a popular English lavender has been reported to survive in extreme conditions up to −20 °F in winter. French (*Lavandula dentata*) and Spanish lavenders (*Lavandula stoechas*) are usually not cold hardy and cannot live in winter in pots in cold weather. English lavender can bear cold weather with freezing temperatures and can grow again in the spring season. But Spanish and French lavenders may not be able to come back after a chilling winter with freezing temperatures. Most of the lavender species can be grown in warm or moderately cool climates. But some of the modern hybrids are resistant to humidity, they can be grown in the tropics if preferable conditions like good drainage, excessive air circulation and full sun are provided to them. Lavenders do not thrive in high humidity. In the case of lavender planting, the level of watering is of great concern. Overwatering can lead to the yellowing of leaves, firstly on the lower leaves, providing them with a rotting smell. While watered plants will droop and soil will feel entirely dried out. Watering is not required in winter for outdoor lavenders. But potted lavenders which are brought inside because of winter protection need watering once every 4–6 weeks. Permanent indoor lavenders will require watering every 2 weeks at moderate temperatures and as often as after every 10 days at high temperatures. The most common causes of the yellowing of lavender leaves are too much nitrogen in the soil, excessive use of fertilizers, overwatering, poor soil draining, or too much humidity. Overwatering and poor soil drainage can be resolved by mixing the soil with sand to make a proportion of a minimum of 70% soil to 30% sand. French lavender is more aggressive in fragrance and appearance than other lavenders. The color lasts the longest out of any type of lavender, throughout the whole summer and significantly through fall [17, 18].

37.3.3 Planting

English lavender is *Lavandula angustifolia* (formerly named *L. officinalis*) which is a most common form in planting. *L. stoechas*, *L. dentata*, and *L. multifida* (Egyptian lavender) are other commonly grown decorative species. English lavender can only be planted in gardens worldwide but these ornamental species are found growing wild as the garden escapes well beyond their natural range. This wild and uncontrolled growth of some species of Lavandula made them invasive. In Australia, *Lavandula stoechas* became a reason for observation when it was found growing widely in the whole continent, since 1920 it has been considered a deadly weed in Victoria. In some parts of Spain, it is also declared as a weed [19]. Lavenders proliferate best in dry, well-drained and sandy soil in full sun [20]. All lavandula species require little or no fertilizer and proper cross-ventilation. Root rot can be a problem due to fungus infection in the area of high humidity. Moisture can be trapped by organic mulches around the plant's bases, which encourages root rot. Crushed rocks give better look to their growth [21].

37.3.4 Pests and Diseases

Lavender is included in the list of plants that are well known for their hardiness. But despite its hardness, lavender still falls victim to pests. These pests cause different diseases in them. Some of the diseases cure on their own but very few become deadly for plants. Some strains of *Candidatus Phytoplasma solani* are considered to be hazardous to lavender growth. In France, the lavender decline has some relation with chronic infection caused by certain strains of *Candidatus Phytoplasma solani*. Lavender decline compromised French lavender production and earlier studies revealed the involvement of *Candidatus Phytoplasma solani* in the cause of disease. To determine the quantitative analysis of "Ca. *P. solani*" epidemiological role, a survey was carried out in France. "Ca. *P. solani*" was detected in 19–56% of the declining plants that depend on seasons and cultivars and its spread was intercepted based on the severity of symptoms. The Autumn season was preferably more important than spring for detecting the phytoplasma, and "*Ca. P. solani*" presence was found higher in *Lavandula angustifolia* than in *Lavandula intermedia* hybrids. Detection of the phytoplasma that fluctuates over time suggests the chronicity of infection and it was also proved that this chronicity varied according to the season. Three 'Ca. P. solani' secY genotypes S17, S16 and S14 were most common in lavender plants and were also found in gardens which suggested that lavender was the main pathogen reservoir of the epidemic. Adults of the planthopper vector *Hyalesthes obsoletus* were commonly found in lavender plants and were shown to mainly harbor the prevalent phytoplasma genotypes found in lavenders. An important survey of lavender fields concludes that *Candidatus Phytoplasma solani* was infecting declining lavenders chronically and also has a relation with infectious populations of *Hyalesthes obsoletus* planthoppers present on the crops. It has been reported that Lavender is the main host for lavender-specific phytoplasma strains which is an important feature for this phytoplasma that propagates from weeds to different economically important plants. These findings pointed out the ways to save young lavender plants from the initial phytoplasma inoculum coming from other lavender fields and from infected nurseries and to promote the agricultural practice to inhibit the development of *H. obsoletus* vector populations [22]. Root rot is also one of the most commonly occurring problems with lavender. Some pathogens like *Fusarium* spp., *Phytophthora* spp., *Pythium* spp. and *Rhizoctonia* spp. can be a cause of this aggressive root rotting. These pathogens most commonly exist in wet, thrilled and waterlogged soils of the Pacific Northwest. Lavender grows in a watery environment when root rot occurs it causes the yellowing or browning of leaves and the whole plant seems to be droopy. Halted growth can also be noticed in root rot. Discolored and mushy roots that are blackened or smelly may be observed upon removing soil around the roots. Due to extreme infection roots become unable to absorb nutrients from the soil, unfortunately causing deaths.

Root rot specifically infects lavender plants during rainy winters, especially in the regions where the soil lacks drainage. This pathogen damages the herb, especially in cold and wet soils. The biggest cause of root rot is overwatering from excessive rainfall and too much irrigation. To minimize the risk factor of root rot, lavender should be grown in well-drained soil. Before plantation softly amends the soil with sand, gravel, or peat moss and double dig or use a broad fork to slacken the soil. Avoid the use of heavy clay soils as it holds extra moisture content. Root rot may be controlled effectively by adopting cultural preventative factors. If plants have already been diagnosed to root rot, the disease can be stopped from invading by pruning away infectious tissues and by applying organic fungicides like baking soda, horsetail or sulfur to the soil but if *Phytophtora* or *Pythium*, are the causes of root rot then fungicides will not work because these species are oomycetes. Removal of the entire shrub will be the best option, to prevent invading If the root rot has initiated to extremely affect the plants. Crown rot is also a common disease in lavender plants that is analogous to root rot which spreads by species of *Phytophthora*. In crown root target of the pathogen is above the ground core of the plant [23].

Alfalfa mosaic virus is also a deadly pathogen of the lavender plant. The main source of transmission of this virus is contaminated garden tools, insects, and bodily touching between plants. Lavender plants that are contaminated with this virus may be seen with a smear of yellow leaves that look deformed and transfigure. Alfalfa mosaic virus often attacks plants on which aphids live. The most usual vectors of this virus are human hands and aphids. To prevent plants from being attacked by mosaic viruses, methods should be adopted to obstruct aphid invasion, also removing and destroy contaminated plants so the virus may not spread to other plants [24].

The occurrence of another disease named xylella is also very common in lavender plants. It is caused by the bacteria *Xylella fastidiosa*. Xylella disease is extensively deadly and usually spread by sap-sucking insects. Sensible signs of xylella are undersized growth and foliage that seems singe or droop. Once the plant is affected by *Xylella* disease, it becomes hard to take control of or eliminate the bacteria [25].

Another very common pathogen that raids the lavender plants is the fungus *Phomopsis lavandulae*. This pathogen mostly damages the stems and branches of plants. It is notorious for its well-known disease called shab. This deadly fungus may take months or years to show its signs and symptoms. The shoot starts drooping and dying. The main source of its transmission is wind and nectar-sucking insects. In spring, shoots that suddenly start drooping, getting brown color or developing blackened marks are at higher risk to have shab. As shab invades down, the branches became brown and wilt. Black spots have also been observed on barks as well. Studies have proved that some species of lavender plant (*Lavandula dentanta*) are immune to the fungus. Shab-infected plants must dig up and burn care to prevent other plants from this disease. There are no effective fungicides or sprays for this pathogen so the only key is the prevention of disease [15, 26].

37.3.5 Important Chemical Constituents

Lavender is extensively used in landscaping and essential oil is a significant raw material used on the commercial level. Eighty-two constituents were isolated from fresh flowers of thirteen Ukrainian cultivars of *L. angustifolia*. These identified components were explored by GC-MS. All of the thirteen cultivars were having two principle constituents common in them which were named Linalool and Linalyl acetate. Although their concentrations were different in all the samples but on average, their ranges were from respectively. 11.4–46.7% and 7.4–44.2%, other major constituents were terpinen-4-ol, lavandulol, lavandulyl acetate, 1,8-cineole, limonene, *cis*- and *trans*-β-ocimene [27]. The essential oil of lavender is identified by its high content of two mostly occurring components linalool and linalyl acetate, and a minor quantity of other constituents, terpiene-4-ol, lavandulyl acetate, lavandulol and varying composition of eucalyptol and camphor. Out of more than 100 components, these are the few constituents that play role in the physical, chemical and biological properties of the essential oil of lavender. The chemical composition of the essential oil of *L. angustifolia* is usually calculated by plant genotype. But studies have shown that there are many factors including environmental, ontogenetic, and regional factors as well as the condition of cultivation and harvesting procedures also affect the chemical composition of essential oil [28]. Linalool is a natural plant by-product with well-known antifungal, antimicrobial and insecticidal abilities. It has a very well-reported history of repulsive properties, especially insects.

Linalool is a suitable pesticide to kill ticks and fleas. It is also a major constituent in basil oil active against tephritid flies. Linalool acts as insecticidal by inhibiting feeding, oviposition and egg hatching in the *Musca domestica* L. which is a common housefly [29]. Linalool is a volatile compound with a pleasant scent associated with the fragrance of laurel. Because of this property, it is useful in industrial products like flavoring agents, scents, and cosmetics, while linalool is generally considered safe for these purposes. In vitamin E manufacturing linalool is an important intermediate. Genotoxicity is effectively suppressed by linalool which is mediated by radical scavenging activity. Some researchers have revealed that linalool has effects on the central nervous system. Linalool also performs reversible inhibition of acetylcholinesterase. Plants that exhibit linalool have been used in the era of folk medicines for their anticonvulsant, sedative and antinociception effects [30]. Linalool is a monoterpene composite found in different plants. The enantiomer naturally present in all essential oils is linalool.

Many reports have described the effect of linalool on the brain neurotransmitters like glutamic acid, −y-aminobutyric acid (GASA), acetylcholine and dopamine. Moreover, linalool has been revealed to exhibit a general anesthetic activity, analogous to its effects on the nicotinic receptor-ion channel, and to be operative against several bacteria and fungi. Spasmolytic effects and antioxidant abilities have also been explained. Studies recently carried out have shown that linalool exhibits anti-inflammatory, antinociceptive and antihyperalgesic effects in several animal models. These characteristics have been ascribed to the capability of linalool to activate the opioidergic, cholinergic M2 and dopaminergic D2 systems, as well as to its interaction with potassium ion (K+) channels. In addition, the effects of linalool on pain reflexes are mediated, at least in portion, by the functioning of adenosine Al and A2A receptors and by nitric oxide (NO) production or release, which might be through mechanisms involving opioidergic, cholinergic and glutamatergic systems. Linalool is a drug that is capable to alleviate, in some degree pain reflexes mediated by a broad range of neurotransmitter systems. Receptors modulated by these constituents are dominated in both the peripheral and the central nervous system (CNS). Pain research is now a day targeting a whole complex system of receptors, ion channels, as well as their modulators to establish newer and safer methods for pain relief of pain [29, 30].

Linalyl acetate is an organic compound that is the acetate ester of linalool. This phytochemical is mostly found in different flowers and spice plants. It is one of the major and principal constituents of the essential oils of bergamot and lavender plants. It mostly occurs along with linalool. Its IUPAC name is 3,7-Dimethylocta-1,6-dien-3-yl acetate.

Also called Bergamiol, Bergamol and Linalool acetate. Its chemical formula is $C_{12}H_{20}O_2$. Having a molar mass of 196.290 g·mol^{-1}. The density of linalyl acetate is 0.895 g/cm^3. Physically linalyl acetate is a colorless and clear fluid. Its specific gravity at 20 c ranges from 0.897–0.910. It is a perfumed constituent that is used in several fragrance substances. It is found in fragrances that are used in decorative cosmetics, shampoo, soaps and many different toiletries as well as found in noncosmetic articles like household cleaners and detergents. On average worldwide usage of linalyl acetate per annum is more than 1000 metric tons. Linalyl acetate is low acute toxicity to mammals, acute oral LD50 is >13,360 mg/kg and acute inhalation LC50 is >2740 mg/m^3. It has zero or very low potential to irritate the human skin and no information is available on possible eye irritation. Based on the mild use of this compound in cosmetics in the form of constituents of perfumes, it is suspected that linalyl acetate has effective sensitizing abilities. In the gastrointestinal tract, it is assumed that linalyl acetate, hydrolyses linalool and acetic acid [31].

In a 28 days oral rate study, it was observed that this hydrolysis product, linalool (72.9%) increased the weight of the liver and kidney, also thickened lobes of the liver and pastel regions on both kidneys but in women just hepatocellular cytoplasmic vacuolization observed. Other signs had about the localized annoyance of the GIT tract. A NOAEL of 148 mg/kg bw/d was derived for linalyl acetate based on the significant effects on the liver and kidney. No effects on gonads were found in both males and females in this 28-day oral rat study. Linalyl acetate was also examined in the reproduction screening test. Based on clinical indications and effects on the weight of the body. The NOAEL for maternal toxicity was calculated to be 464 mg/kg bw/d for linalyl acetate. This compound is readily biodegradable. Linalyl acetate possesses an extreme ability for sorption to soil that is determined to be log Koc 2.9, however, it is unlikely that effective uptake may occur given the rapid hydrolysis of linalyl acetate. The vapor pressure of linalyl acetate is 0.61 Pa at 25 °C [32, 33]. The half-life measured for photo-oxidation of linalyl acetate is 1.1 h. This compound is observed to vanish from the check medium within 180 min at pH values of 4 and 7 at a temperature of 50 °C [17, 32].

Linalool and acetic acid are the two products of linalyl acetate produced upon its hydrolysis. Linalyl acetate is highly toxic to aquatic animals and daphnia. In fish 4 days LC50 is 11 mg/L. In daphnia 2 days EC50 is 6.2 mg/L. A decrease of biomass was observed at 1.2 mg/L, in a test carried out with algae and the exposure of linalyl acetate to algae was 3 days. EbC50 for 3 days was 4.2 mg/L and the ErC50 was 16 mg/L. When ethanol was used as a dispersant, EC50 measured was 415 mg/L for microorganism inhibition. When swiss mice were exposed to inhalation of 2.74 mg linalyl acetate/L for one and half hours, it led to reduced motor activity. This reduction in motor activity was more severe in aged swiss mice of 6–8 weeks about 100% reduction was observed in them and about 81% reduction was reported in mice of 6 weeks. 1000–5000 tonnes of linalyl acetate are being produced per annum. Within the united states of America, 49 tonnes are being produced, exported and imported yearly. In Canada since 1986, 3–39 tonnes of linalyl acetate are imported [32].

It is assumed that in Norway total amount used or manufactured of this compound is less than 1 tonne. Linalyl acetate is usually considered safe for human consumption. Its daily accepted intake is 0.25 mg/day. During production, workers may be exposed to linalyl acetate, but the consumers are more expected to its exposure, and chances of dermal and oral exposures occur more repeatedly. In the American states on average 155,491 workers at 11,043 industrial points are potentially exposed to it. This exposure was observed in the fields, where linalyl acetate was being manufactured and applied in the manufacturing of perfumes. It is reported that linalyl acetate hydrolysis more fastly in gastric fluids at low pH. Its half-life is 5.5 and 52.6 min in gastric and pancreatic fluid respectively. Its hydrolyzed products are linalool and acetic acid, thus it is assumed that linalool is the compound that will enter the blood circulation when taken linalyl acetate orally which is then changed into geraniol, 1,5-dimethyl-hexadiene-1,6-dicarboxylic acid and 7-carboxy-5-methylocto-6-enoic acid [32, 33].

37.4 Medicinal Uses

Lavender is highly recommended for beauty, skin products, shampoos, and fragrances. Some varieties are also added to foods and baked goods to enhance flavors and can be bought over the counter from drug stores. However, many medicinal properties are associated with it (Table 37.1).

37.5 Myths, Legends, Tales, Folklore, and Interesting Facts

Lavender was first farmed by ancient Arabia, and they distilled its oil. But the lavender plant was brought to England by Romans. Its name lavender was also suggested by Romans. Romans steeped stock of the buds and herbs in water that was used for bathing. Through their bathhouses, they spread lavender's use throughout Europe. Lavender has been used during the mummification process by ancient Egyptians, in embalming the corpse with scent. While lavender was used for insomnia treatment and get relief from back pain in ancient Greeks. Lavender plants were also being used as a source to ward off hazardous diseases in the seventeenth century during the bubonic plague [7, 14].

Shakers were the first lavenders to be grown and harvested successfully, shakers developed farms of lavender and produced their herbs to sell to different markets worldwide. Different colors of lavender originated from *Lavandula angustifolia.* Lavender plants also act as bug repellent; they can protect other plants from different pests as well as certain creepy crawlers. Lavender is a scented plant that can survive pretty well on neglect, specifically in any poorly conditioned soil. The lavender plant uses as a sign of purity, loyalty, calmness and grace when it comes to planting symbolism. Some so many people showed their love for lavender species

Table 37.1 Medicinal properties of different parts and varieties of Lavander

Botanical name	Part	Solvent/dose	activity	Model	Biomarkers affected	Reference
Lavandula angustifolia	Flower	Diazepam (1 mg/kg)	Anti-anxiety agents	Anxiolytic effects of Lavandula angustifolia odour on the Mongolian gerbil elevated plus maze	Long term disclosure to lavende smell causes reduce in anxiety in females, increases anxiolytic effect	[33]
	Leaf	Ethanol/100–400 mg/kg	Brain diseases	Contusive model of spinal cord injury in Wistar rats	200 and 400 doses enhance fuctional and structural recovery after injury, ecrease number of motor neurons, decrease lesion post injury, cavity areas decreased	[34]
	Aerial parts	Anhydrous Na_2SO_4 (100 mg/kg)	Cardiomyopathies	Isoproterenol-induced acute myocardial infarction in rat	Increase: Strength of myocardial membrane, protection of myocardium, maintainance of architecture of heart cells, efficiency of antioxidant system Decrease: lipid peroxidation, responses of pro-inflammation, tissue damage of heart.	[35]
Lavandula bipinnata	Leaves	Petroleum ether (10 µl oil solution)	Anti microbial agents	11 different microbial strains used for checking antimicrobial properties of lavandula bipinnata oil	At concentration of 0.125–8 µg/µl of oil there was maximum decline in growth rate, gram positive bacteria show maximum inhibition of growth than gram positive bacteria.	[36]
Lavandula gibsoni	Aerial parts	Acetone/2 mg/cm^2	Insecticides	Larvae of Aedes aegypti, Anopheles stephensi and Culex quinquefasciatus for larvicidal activity of Lavandula gibsoni	Oil obtained from plant shows maximum deaths at 150 mg/L against concerned species. Repellant activity of insects was also checked as at 2 mg/L oil shows maximum protection against mosquitoes.	[37]

Botanical name	Part	Solvent/dose	activity	Model	Biomarkers affected	Reference
Lavandula pedunculata		Ethanol/5 mg/ml	Neurological disorders	The in vitro screening for acetylcholinesterase inhibition and antioxidant activity using different plant extracts	Increase: scavenging activity, antioxidant activity in polar solvents, inhibitory activity of ACE. Decrease: oxygen reactive species, inflammation.	[38]
Lavandula stoechas	Flower	Water/0.6 ml/kg/day	Antidepressive agents	Sprague Dawley Rats *for checking antidepressive and anxiolytic effect*	Increase: tryptophan in blood and brain, serotonine level, antidepressant and anxiolytic effect, anticonvulsant activity Decrease: rate of serotonin turnover, depression, insomnia, convulsions, immobility rate, level of 5HIAA in brain.	[39]
Lavandula stoechas	Aerial parts	Methanol/325 mg/kg/p.o	Memory enhancemnet	Scopolamine-induced hyper-amnesic albino mice	Increase: inflexion ratio, retention behavior, memory, level of choline acetyltransferase, catalase, superoxide dismutase, cholinergic transmission, capacity of learning, Decrease: level of acetylcholinesterase and malondialdehyde in brain, reactive oxygen species, memory loss	[40]
	Aerial parts	Na_2SO_4 (200 mg/kg)	Hepatoprotectivity and renoprotectivity	Malathion-induced oxidative stress in young male mice	Increase: hepatoprotective and nephroprotective effect,weight of kidney and liver, scavenging capacity, hemodynamic parameter regulation Decrease: melathion induced deregulation, H_2O_2 levels, Kupffer cell activation, and formation of fibrogenic mediators, lipoperoxidation, and oxidative stress.	[41]

being their beauty, it is a very unsurprisingly fact that queen Victoria was also a big fan of lavender. She ordered that all of the furniture should be completely cleaned with solutions that must be lavender based, and to help ease her stomach she used to take lavender-infused tea [8, 41].

The aroma of lavender plants can be explained as a mixture of honey, balsam fir, and thyme. Lavender is counted as a perennial flowering plant that will flower and grow year after year. A delightful cup of herbal tea can be prepared by using 4 teaspoons of lavender buds. The earliest famous use of the lavender plant as a constituent in perfumes and body scents was popularized during the early 1900s. Interestingly this graceful and precious plant has also been used as currency during medieval times in France. It is said that the fragrance of lavender has a relaxing effect that acts to calm the mental state and relax the body and spirit. Ancient Romans considered gardening lavender in and around their homes as a good luck charm. They used it for its healing and therapeutic abilities; they also used to burn lavender buds as incense in their worship places. Because of its beauty ancient Greeks used lavender plants to decorate their temples. Lavender is included in the list of very few plants that are considered to be versatile because of their numerous uses including culinary, medical and ornamental welfare. For skin treatments in conditions like acne and acute dermatitis, lavender has been used by ancient Greeks because of its nonsteroidal and anti-inflammatory capabilities that act by blocking the release of histamine from cells present in the skin. It was also thought in the Middle Ages that lavender is associated with death and evil spirits. Terpene present in the essential oil gives lavender a unique smell [12, 42].

Ancient from the middle ages used to bathe with it. Lavender was considered to be a holy plant by Egyptians because of its charming look, fragrance and medicinal capabilities. At the start of the 1900s, women used to carry a packet of dried lavender flowers to protect themselves from different diseases like diphtheria and malaria. Essential oils obtained from lavender plants are still known for their antioxidant abilities and their potential for anti-oxidation is increasing over time. Ancient Greeks named lavender buds nardus, after the Syrian city of Naarda. In the 1800s lavender was also commonly known as nard. Lavender was included in the holy herbs that were used in temples of Jerusalem to pray for the essence of God. The Latin name of lavender was lavandarius derived from Lavanda obtained from the verb lavare. Early on it was discovered by the Greeks that lavender would release a calming fume on burning if flowers are crushed and treated correctly. Powdered lavender has also been used as a food condiment during medieval times. During World War 1, lavender was used as an antiseptic for washing during surgery and to clean deep wounds. Lavender was carried into North America by colonists. There it was grown commercialized by a different sect of Quakers for more use as it had been in the ancient world. When tariffs on Asian tea was too high, lavender was added to herbal tea blends as a tea alternative [42, 43]. One of the most beloved herbs, Lavender has been in use for more than 2000 years. Romans are usually credited with naming this significantly aromatic herb some believe that it derived from bluish called Lavandula. This beautiful and glorious plant has been *spikenard,* in ancient Greece, India and also in the bible. Today lavender is mostly related to

England, but in truth, it is not native to Europe. It was used under domesticate production in south Arabia. It spread into Europe from Greece in 600 BC. Romans later used lavender in their water baths. By the late middle ages, washerwomen were known as lavenders for invading outfits to dry upon lavender bushes and using lavender for perfuming cleaned outfits in storage. Well, know pharmacist of the Middle Ages Hildegard von Bingen used lavender water to make a mixture of lavender and brandy, as a therapy for acute migraine. Most of the myths regarding lavender are prehistoric. It is said that Cleopatra who ruled Egypt, used to wear its fragrance to induce roman dictator Julius Caesar. Another very well-known tale about this graceful plant is associated with the expulsion of Adam and Eve from Heaven. Both of them are credited with taking the lavender plant with them at the time of their expulsion from the garden of Eden [43]. Many Christians use crosses made of lavender bushes to get rid of bad evils. Some Christians believe that lavender got its fragrance from the outfits of Jesus when his sister hung them on a bush to dry. Today there are more than 115 species of lavender, being cultivated all over the world. Lavender-made products are economical and readily available. Lavender brings color and fragrance into our lives. We must bring the charm of this glorious plant into our lives [44].

37.6 Conclusion

The revitalized concern in lavender and its essential oils as a potential natural remedy is due to the recent raise in the attractiveness of natural products and alternative medicine. The present chapter will be helpful to improve the knowledge of lavender, and its pharmacological properties ll improve clinical research plans. Very limited data regarding the diverse administration of lavender such as aromatherapy, massage or oral etc. is available. The oral administration of lavender is promising; although independent long-term data is inadequate. However, adcpting rigorous standard methodology and well-reported along with experimental pharmacological research in a future clinical trial could be helpful to clarify the therapeutic value of lavender.

References

1. Biswas, K.-K., Foster, A.-J., Aung, T., & Mahmoud, S.-S. (2009). Essential oil production: Relationship with abundance of glandular trichomes in aerial surface of plants. *Acta Physiologiae Plantarum, 31*, 13–19.
2. Tzortzakis, N.-G., & Economakis, C.-D. (2007). Antifungal activity of lemongrass (*Cympopogon citratus* L.) essential oil against key postharvest pathogens *Innovative Food Science & Emerging Technologies, 8*(2), 253–258.
3. Kotsiris, G., Nektarios, P.-A., & Paraskevopoulou, A.-T. (2012). Lavandula angustifolia growth and physiology is affected by substrate type and depth when grown under Mediterranean semi-intensive green roof conditions. *HortScience, 47*(2), 311–317.

4. Miliauskas, G., Venskutonis, P.-R., & Van Beek, T.-A. (2004). Screening of radical scavenging activity of some medicinal and aromatic plant extracts. *Food Chemistry, 85*(2), 231–237.
5. Hassanpouraghdam, M.-B., Hassani, A., Vojodi, L., Asl, B.-H., & Rostami, A. (2011). Essential oil constituents of Lavandula officinalis Chaix. from Northwest Iran. *Chemija, 22*(3), 167–171.
6. Cassella, S., Cassella, J. P., & Smith, I. (2002). Synergistic antifungal activity of tea tree (Melaleuca alternifolia) and lavender (Lavandula angustifolia) essential oils against dermatophyte infection. *International Journal of Aromatherapy, 12*, 2–15.
7. Vakilian, K., Atarha, M., Bekhradi, R., & Chaman, R. (2011). Healing advantages of lavender essential oil during episiotomy recovery: A clinical trial. *Complementary Therapies in Clinical Practice, 17*(1), 50–53.
8. Wu, P. A., & James, W. D. (2011). Lavender. *DERM, 22*(6), 344–347.
9. Lehrner, J., Marwinski, G., Lehr, S., Johren, P., & Deecke, L. (2005). Ambient odors of orange and lavender reduce anxiety and improve mood in a dental office. *Physiology & Behavior, 86*(1-2), 92–95.
10. Talaei, G.-H., Gholami, S., Pishva, Z.-K., & Dehaghi, M.-A. (2014). Effects of biological and chemical fertilizers nitrogen on yield quality and quantity in cumin. *Journal of Chemical Health Risks, 4*(2), 55–64.
11. Chrysargyris, A., Panayiotou, C., & Tzortzakis, N. (2016). Nitrogen and phosphorus levels affected plant growth, essential oil composition and antioxidant status of lavender plant (*Lavandula angustifolia* Mill.). *Industrial Crops and Products, 83*, 577–586.
12. Uchida, R. (2000). Essential nutrients for plant growth: Nutrient functions and deficiency symptoms. *Plant Nutrient Management in Hawaii's Soils, 4*, 31–55.
13. Rouached, H., Arpat, A.-B., & Poirier, Y. (2010). Regulation of phosphate starvation responses in plants: Signaling players and cross-talks. *Molecular Plant, 3*(2), 288–299.
14. Dragomir, L.-B., & Nicolae, I. (2015). Researches regarding the physiology of lavender plants grown on soils with different ph values. *Annals of the University of Craiova-Agriculture, Montanology, Cadastre Series, 45*(1), 45–50.
15. Khatami, S.-A., Kasraie, P., Oveysi, M., Moghadam, H.-R.-T., & Ghooshchi, F. (2022). Mitigating the adverse effects of salinity stress on lavender using biodynamic preparations and bio-fertilizers. *Industrial Crops and Products, 183*, 114985.
16. Crummitt, K., & Drost, D. (2012). *English Lavender in the garden*. USA, Utah State, published of May, 8.
17. Adam, K. L. (2006). *Lavender production, products, markets, and entertainment farms*. Retrieved November, 5, 2006.
18. Smetan, Ș., Nita, S., David, G., Imbrea, I.-M., & Pop, G. (2017). Partial results regarding the influence of climatic conditions on the quality of Lavender and Sage production. *Research Journal of Agricultural Science, 49*, 3.
19. Guimont, C., Brisson, C., Dagenais, G. R., Milot, A., Vézina, M., Mâsse, B., et al. (2006). Effects of job strain on blood pressure: A prospective study of male and female white-collar workers. *American Journal of Public Health, 96*(8), 1436–1443.
20. Ganesh, K.-S., & Deivanai, S.-N. (2014). Prevalence and risk factors of hypertension among bank employees in urban Puducherry, India.
21. Seidler-Łożykowska, K., Mordalski, R., Kucharski, W., Kędzia, B., & Bocianowski, J. (2014). Yielding and quality of lavender flowers (*Lavandula angustifolia* Mill.) from organic cultivation. *Acta Scientiarum Polonorum Hortorum Cultus, 13*(6), 173–183.
22. Sémétey, O., Gaudin, J., Danet, J. L., Salar, P., Theil, S., Fontaine, M., et al. (2018). Lavender decline in France is associated with chronic infection by lavender-specific strains of "*Candidatus Phytoplasma solani*". *Applied and Environmental Microbiology, 84*(24), e01507–e01518.
23. Davino, S., Cacciola, S. O., Pennisi, A.-M., & Nicosia, M.-L.-D. (2002). Phytophthora palmivora a new pathogen of lavender in Italy. *Plant Disease, 86*(5), 561–561.

24. Stanković, I. M., Vrandečić, K., Ćosić, J., Milojević, K. N., Bulajić, A. R., & Krstić, B. B. (2014). The spreading of alfalfa mosaic virus in lavandin in Croatia. *Pesticides and Phytomedicine/Pesticidi i fitomedicina, 29*(2).
25. Loconsole, G., Zicca, S., Manco, L., El Hatib, O., Altamura, G., Potere, O., et al. (2021). Diagnostic procedures to detect *Xylella fastidiosa* in nursery stocks and consignments of plants for planting. *Agriculture, 11*(10), 922.
26. Metcalfe, C.-R. (1931). The "shab" disease of lavender. *Transactions of the British Mycological Society, 16*(2–3), 149-IN5.
27. Pokajewicz, K., Białoń, M., Svydenko, L., & Fedin., R., & Hudz, N. (2021). Chemical composition of the essential oil of the new cultivars of *Lavandula angustifolia* Mill. Bred in Ukraine. *Molecules, 26*(18), 5681.
28. Détár, E., Németh, É.-Z., Gosztola, B., Demján, I., & Pluhár, Z. (2020). Effects of variety and growth year on the essential oil properties of lavender (*Lavandula angustifolia* Mill.) and lavandin (*Lavandula* x *intermedia* Emeric ex Loisel.). *Biochemical Systematics and Ecology, 90*, 104020.
29. Mitić-Ćulafić, D., Žegura, B., Nikolić, B., Vuković-Gačić, B., Knežević-Vukčević, J., & Filipič, M. (2009). Protective effect of linalool, myrcene and eucalyptol against t-butyl hydroperoxide induced genotoxicity in bacteria and cultured human cells. *Food and Chemical Toxicology, 47*(1), 260–266.
30. Bickers, D., Calow, P., Greim, H., Hanifin, J.-M., Rogers, A.-E., Saurat, J.-H., et al. (2003). A toxicologic and dermatologic assessment of linalool and related esters when used as fragrance ingredients. *Food and Chemical Toxicology, 41*(7), 919–942.
31. Letizia, C.-S., Cocchiara, J., Lalko, J., & Api, A.-M. (2003). Fragrance material review on linalyl acetate. *Food and Chemical Toxicology, 41*(7), 965–976.
32. Rai, V.-K., Sinha, P., Yadav, K.-S., Shukla, A., Saxena, A., Bawankule, D.-U., et al. (2020). Anti-psoriatic effect of Lavandula angustifolia essential oil and its major components linalool and linalyl acetate. *Journal of Ethnopharmacology, 261*, 113127.
33. Bradley, B.-F., Starkey, N.-J., Brown, S.-L., & Lea, R.-W. (2007). Anxiolytic effects of *Lavandula angustifolia* odour on the Mongolian gerbil elevated plus maze. *Journal of Ethnopharmacology, 111*(3), 517–525.
34. Kaka, G., Yaghoobi, K., Davoodi, S., Hosseini, S.-R., Sadraie, S.-H., & Mansouri, K. (2016). Assessment of the neuroprotective effects of *Lavandula angustifolia* extract on the contusive model of spinal cord injury in Wistar rats. *Frontiers in Neuroscience, 10*, 25.
35. Ziaee, M., Khorrami, A., Ebrahimi, M., Nourafcan, H., Amiraslanzadeh, M., Rameshrad, M., et al. (2015). Cardioprotective effects of essential oil of *Lavandula angustifolia* on isoproterenol-induced acute myocardial infarction in rat. *Iranian Journal of Pharmaceutical Research: IJPR, 14*(1), 279.
36. Hanamanthagouda, M.-S., Kakkalameli, S.-B., Naik, P.-M., Nagella, P., Seetharamareddy, H.-R., & Murthy, H.-N. (2010). Essential oils of *Lavandula bipinnata* and their antimicrobial activities. *Food Chemistry, 118*(3), 836–839.
37. Kulkarni, R.-R., Pawar, P.-V., Joseph, M.-P., Akulwad, A.-K., Sen, A., & Joshi, S.-P. (2013). *Lavandula gibsoni* and *Plectranthus mollis* essential oils: Chemical analysis and insect control activities against *Aedes aegypti, Anopheles sfttephensi* and *Culex quinquefasciatus*. *Journal of Pest Science, 86*, 713–718.
38. Ferreira, A., Proença, C., Serralheiro, M.-L.-M., & Araujo, M.-E.-M. (2006). The in vitro screening for acetylcholinesterase inhibition and antioxidant activity of medicinal plants from Portugal. *Journal of Ethnopharmacology, 108*(1), 31–37.
39. Ahmed, M., & Azmat, A. (2017). Decreased brain serotonin turnover rate following administration of Sharbat-e-Ahmed Shah produces antidepressant and anxiolytic effect in rats. *Metabolic Brain Disease, 32*, 1785–1790.
40. Mushtaq, A., Anwar, R., Gohar, U.-F., Ahmad, M., Marc, R.-A., Mureşan, C.-C., et al. (2021). Biomolecular evaluation of *Lavandula stoechas* L. for nootropic activity. *Plants, 10*(6), 1259.

41. Selmi, S., Jallouli, M., Gharbi, N., & Marzouki, L. (2015). Hepatoprotective and renoprotective effects of lavender (*Lavandula stoechas* L.) essential oils against malathion-induced oxidative stress in young male mice. *Journal of Medicinal Food, 18*(10), 1103–1111.
42. Putnam, J. K. (1977). Lavender, David, "California: A bicentennial history" (book review). *The Historian, 39*(4), 815.
43. Platt, E. S. (2009). *Lavender: How to grow and use the fragrant herb*. Stackpole Books.
44. Tisserand, R. (2021). Lavender oil and pregnancy. *Robert Tisserand.* Retrieved, 20.

Chapter 38
Tulsi

Abida Parveen, Shagufta Perveen, Mobeen Ahmad, Farah Naz, and Muhammad Riaz

38.1 Introduction

A. Parveen (✉) · S. Perveen · M. Ahmad · F. Naz
Department of Botany, Government College University, Faisalabad, Pakistan

M. Riaz
Department of Pharmacy, Shaheed Benazir Bhutto University, Sheringal, Pakistan

M. Zia-Ul-Haq et al. (eds.), *Essentials of Medicinal and Aromatic Crops*,
https://doi.org/10.1007/978-3-031-35403-8_38

Plant profile

Family	Lamiaceae
English name	Holy Basil
Indian name	Tulsi, Tulasi, Krishna tulsi, Vana tulsi, Kapoor tulsi, Rama tulsi
Species and Varieties	Ocimum tenuiflorum
	O. basilicum
	O. canum
	O. gratissimum
	O. camphora
	O. kilimandschricum
	O. americanum
	O. micranthum
Distribution	India, Southeast Asia, Africa, Australia
Uses	Medicine (Ayurveda, Siddha)

The aromatic perennial plant (Ocimum sanctum L.) belongs to the Lamiaceae family (Labiatae). Ocimum sanctum [1] is known as "Holy Basil" in English and "Tulsi" or "Tulasi" in Hindi and is regarded as a sacred plant in Hindu culture. Tulsi is a popular plant in India, especially in religious gardens and residential locations Tulsi is naturally grown. Basil is a member of the genus Ocimum L. which contains about 160 species, including O. basilicum, O. canum, O. gratissimum, O. camphora, and O. americanum, O.micranthum [2].

Tulsi (Ocimum sanctum L) is a common herb in Indian traditional medicine. It is also known as "Queen of Herbs." [3, 4]. Tulsi is a Sanskrit word that means "incomparable one" [5]. It is known as "Vishnupriya" in Indian culture and is worshiped very religiously. O. sanctum, O. canum, O. basilicum, O. ameicanum, O. camphora, are among the 160 species of this plant that are used for medicinal purposes [6, 7]. In Hindi culture, O. tenuiflorum as Krishna tulsi, O. gratissimum as Vana Tulsi and O. sanctum as Rama Tulsi are the three most often used types of Tulsi [8]. This plant has 2n = 16 chromosomes, according to an analysis [9]. The genome of this medicinal plant is the first to have been fully sequenced and examined. Combining the sequence reads from 4 libraries across 3 NGS platforms produced the generated data [10].

The O. sanctum is likely the domestic herb that Indians utilize the most and revere the most. The plant is frequently planted in gardens and is also found close to temples. Not just in Ayurveda, but also in the Greek and Unani systems of medicine, the plant is said to have medicinal powers [1]. In India, the traditional medical systems of Unani, Ayurveda, and Siddha use the dried leaf, seed, and complete plant individually [11]. This plant contains numerous phytochemical constituents that are responsible for its medicinal use in both modern and traditional medicine systems, such as Ayurveda, Unani, Siddha, Greek, and Roman. These plant-derived phytochemicals are taken from the plant and utilized to treat various diseases [12].

O. sanctum has a complicated chemical make-up. Alkaloids, flavonoids, tannins, polysaccharides, urosolic acid, and others are some generally known biologically active phytochemical components. They engage in pharmaceutical operations against various illness states. Different plant extracts from this species have diverse anti-infective capabilities, as evidenced by the plant's antibacterial activities against a wide range of bacteria, fungi, and viruses [13]. Because of the plant's growing and harvesting methods, the presence of phytochemicals may vary. It is referred to as "the elixir of life" in Ayurveda and is thought to enhance life span. Ocimum sanctum has a pleasant perfume and is well-known for its therapeutic qualities. Its essential oils have a variety of medical characteristics, which is why it is utilized as a medicinal plant and for herbal tea throughout South Asia [14, 15]. Tulsi is a common herb used in ayurveda medicine to treat cardiovascular disease, congestion, headaches, stomach problems, malaria, different poisonings, and the common cold. Tulsi is typically consumed in a variety of ways, such as herbal tea, dried powder, fresh leaf, or combined with honey or ghee. This common shrub can be identified by its square stem and distinct perfume [16]. Although this herb is revered by many religions and is used to treat a variety of illnesses and problems, it still presents a difficulty to identify phytomolecules with specific therapeutic properties [10].

Many Ocimum species tissues are used to make purified essential oil, dried, frozen, or fresh forms of the tissues. Due to "their distinctive scent, fresh leaves of Ocimum species are frequently used in French, Greek, Italian, and Mexican cuisines. Such fresh fragrant leaves are suitable as seasonings or flavours in salad dressings, stews, and sauces. It may be used to prepare different types of food, including beef, fishes, yogurt, cream, and drinks [17].

38.2 Origin and Distribution

Ocimum sanctum has been recognized as one of the most versatile medicinal plants for over 2000 years, with a wide range of biological activity [18]. Plants of the genus Ocimum belong to the old world and are indigenous to central and west Africa and India [19]. It's also difficult to define Ocimum itself. An estimated 30–160 annual and perennial plants and shrubs native to the tropics as well as subtropical regions of South America, Central America, Asia and Africa were found [20]. In various parts of Asia and Africa, it is widely grown. This Plant is grown in India from the Himalayas to the Andaman and Nicobar Islands. It is extensively spread in tropical areas and can be easily found in the Himalayan region at a height of 1800 m. The majority of Ocimum species are found in tropical and subtropical areas of the world [21–23]. O. sanctum is widely grown for religious and therapeutic uses across the ancient world, but mainly in the tropics. It is extensively known as a medicinal plant and herbal tea throughout the vast region of southeast Asia, where it is frequently utilized in Ayurveda [13].

There are numerous cultivars and chemotypes of this plant, which is traditionally cultivated primarily in India. These cultivars and chemotypes vary in their habitats, heights, and composition of active substances, especially essential oil, where eugenol is most frequently the major compound and other essential oil constituents include -caryophyllene, citrol, geraniol, camphor, linalool, and methyl cinnamate [24]. Ocimum sanctum has a wide range of biological activities and has been regarded as one of the most adaptable medicinal herbs for further than 2000 years. It is a well-known sacred plant on the Indian subcontinent that is also referred to as tulsi or holy basi [18]. Basils are native to tropical Asia and most likely came from India. It is an annual plant that is often spread by seeds. It could be easily located at an altitude of 1800 metres in the Himalayan region and is extensively dispersed in tropical places [22].

The plant has been widely distributed across the world due to its numerous uses and benefits. According to a study published in the Journal of Medicinal Plants Research, "Ocimum sanctum is one of the most widely cultivated medicinal plants in the world" [25]. The plant has been mentioned in ancient Indian scriptures, including the Vedas and the Puranas. It is considered a sacred plant in Hinduism and is worshipped in many households. The leaves of the plant are used in religious rituals and ceremonies. In India, Ocimum sanctum is commonly grown in home gardens, and its leaves are used for medicinal purposes, including treating respiratory illnesses, fever, and digestive disorders. The plant is also used in traditional Ayurvedic medicine for its anti-inflammatory, antioxidant, and immunomodulatory properties [26]. The plant has also been introduced and cultivated in other parts of the world, including Africa, Australia, Europe, and the Americas. It is commonly used in traditional medicine practices in these regions as well [26].

38.3 Description of the Plant

Ocimum sanctum is a much branched, fragrant, and erect plant that grows 30–60 cm tall and has bushy stems and leaves are green in a simple opposing direction with a strong perfumed scent. The leaves are elliptical, up to 5 cm long, and slightly serrated, with petioles. The leaf color varies from pale green (Rama tulsi) to dark purple (Krishna tulsi). Tulsi flowers are small in size and are dark purple to reddish in color, and grow in small clusters on cylinder-shaped spikes. At the base of each cluster, there are fewer stalkless heart-shaped bracts. The inside of the sepal cup is not hairy. Flowers are rarely more than 5 mm long, with a bearded calyx tube on the outside at the base. The tube of the flower is hairy. The seeds are yellow to reddish in color and the fruits are small. After the rainy season, this plant is sown and harvested after a few months [1, 27].

Ocimum sanctum is a 75 cm tall, broadly branched, upright, sturdy, and scented plant. This little herb is grown close to Hindu homes and temples and may be found all across India. This plant's roots, seeds, and leaves have all been employed in Ayurveda medicine. Tulsi has a very intricate chemical makeup that includes several nutrients and other biologically active substances. Tulsi standardisation has so far escaped contemporary science because of the intrinsic botanical and biochemical complexity of the herb. Eugenol (an essential oil) and ursolic acid are two of the numerous active chemicals that have been discovered and extracted that are maybe the best known [28]. Ocimum sanctum, often known as sacred basil or holy basil, is a sacred herb that is frequently employed in the pharmaceutical and cosmetic sectors and is prized for its many therapeutic benefits. During the growing seasons of 2015 and 2016 at the ICAR-Directorate of Medicinal and Aromatic Plants Research, Boriavi, Anand, Gujarat, a field experiment was carried out to examine the effects of organic sources of nutrients on the growth, yield, and quality of sacred basil (Purple kind). [29]. Tulsi is a preventative shrub with a height of around 75 cm. Hindus frequently find it close to temples and homes, and they worship it every day [30].

38.4 Species and Varieties

Tulsi has two varieties: Black (Krishna Tulsi) and Green (Krishna Tulsi) (Ram Tulsi). They have chemical and medicinal qualities that are similar. Ocimum is a genus of annual and perennial herbs and shrubs with over 200 species. The majority of the species are found in tropical and warm temperate climates, including India. Ocimum sanctum L (Tulsi), O. gratissimum (Ram Tulsi), O. canum (Dulal Tulsi), O. bascilicum (Ban Tulsi), O. kilimandschricum, O. americanum, O. camphora, and O. micranthum are some of the species of the genus Ocimum. They are extensively known for their medicinal properties and are grown in various parts of the world [27, 31].

Ocimum species rank highly in a number of pharmacological activities, including antimicrobial, immunomodulatory, stress-relieving, anti-inflammatory, antiulcer, antidiabetic, hepatoprotective, chemoprotective, antihyperlipidemic, cardioprotective, antioxidant, antitussive, radioprotective, memory-enhancing, anti-arthritic, antifertility, antihypertensive, anticoagulant, anticataract, and as well as antinociceptive actions[32]. The different Ocimum species, sometimes referred to as the "king of the herbs," are the main subject of the current review. A full report on the botany of the more than 50 herb and shrub species that make up this genus is included, as well as information on traditional usage and cultivation methods [33]. The Ocimum genus has over 160 species, which are widely scattered throughout the world's warm climates [6].

38.5 Ecology

Ocimum sanctum is found throughout Asia's tropical and subtropical regions. The plant is native to India and also grown in Himalayas. Its other native regions includes Malaysia and Pacific. The Caribbean, and various regions of Africa are also native to this plant. It is grown in almost every state in India. It is found near temples and places of worship due to various religious beliefs. There is currently no information on specific habitats where the species can be found naturally [34].

38.6 Types

There are three main types of O. sanctum namely Rama Tulsi, Krishna Tulsi and Vana Tulsi. Rama Tulsi is another name for Sri Tulsi. It is the most prevalent kind of all, with light purple blooms and green-colored foliage. It is often known as Ocimum tenuiflorum and is mostly utilized for religious purposes. The chemical Eugenol gives it a clove-like odor. The violet-leaved Krishna Tulsi is favored for medicinal usage and designated as O. uniform L because it matches the hue of the Hindu god Krishna. It tastes spicy and smells like cloves. The best tulsi for medicine is Krishna Tulsi because of its increased potency. It's also been employed as a conventional kind of contraception. The word "forest" is Vana Tulsi. It can be seen from tulsi that it frequently grows in the wild. Its natural leaves are a brilliant or light green colour. Native to several regions of Asia and northeast Africa, they grow in the wild.

A fourth kind of Tulsi, known as Kapoor Tulsi, is available in addition to the three primary varieties. Yet, the genus Ocimum tenuiflorum has approximately 60 distinct species that are found all over the world. Amrita Tulsi, which means "immortality," is one of the varieties of Tulsi that is most commonly found in the Indian Subcontinent. Other varieties include sweet basil, lime basil, Greek basil,

Thai basil, cinnamon basil or Mexican basil, lettuce leaf basil, Christmas basil, Napoletano basil, Summerlong basil, Genovese basil, Ararat basil, etc. [35].

38.7 Soil

Ocimum species grow well in a wide range of soils. Rich loam to poor laterite, as well as saline and alkaline to moderately acidic soils, are appropriate for cultivation. Tulsi yields better when the soil pH is between 5.5 and 7.0. Improved vegetative growth can be facilitated by well-drained soil. Waterlogged conditions and poor soils, on the other hand, are not favorable to its cultivation [19].

38.8 Climate

Ocimum sanctum is usually cultivated between April to June. Ocimum species may thrive in a variety of climates. Long days and high temperatures have been observed to stimulate plant growth and oil production, while basil thrives best in places with moderate to high rainfall and humidity. O. sanctum may be cultivated in partially shaded conditions, although the oil output is minimal. Because of the factors mentioned, tropical and sub-tropical climates are ideal for basil cultivation [19].

38.9 Cultivation

Tulsi grows well in warm, tropical climates. The herb was invented in India and is now grown and marketed all throughout the nation. It is grown up to 1800 metres above sea level [36]. Plants are grown all throughout India, from the Himalayas to the Andaman & Nicobar Islands. In various regions of Asia and Africa, it is commonly cultivated. In naturally damp soil, it flourishes. The kind of soil and fluctuations in rainfall affect the plant's size, shape, and medicinal qualities. In Asia's tropical areas, the Ocimum genus has around 150 species [4].

38.10 Propagation

Seeds or cuttings can be used to propagate the Tulsi plant. Seeds are the traditional way of propagation for these species. Their multiplication is limited, however, due to their low germination capacity. Unfortunately, neither of the plant species can be propagated through vegetative reproduction. The technology of micropropagation provides an alternate option for cloning these plants [37, 38]. The study provide techniques for axillary buds-based rapid propagation of O. americanum and O. sanctum. This might be helpful for both in vitro germplasm preservation and large-scale multiplication of these two potent medicinal plants [39].

The best time to sow seeds is during the monsoon season, which is from June to September in India. Another way to propagate Ocimum Sanctum is through cuttings. A stem cutting of 5–6 inches in length can be taken from the mother plant, and the leaves at the bottom should be removed. The cutting can then be dipped in a rooting hormone and planted in a pot filled with well-draining soil. The pot should be kept in a warm and moist environment until the cutting develops roots and starts to grow. Layering is another method of propagation that can be used for Ocimum Sanctum. A low branch of the mother plant can be selected, and a small section of the stem can be bent down to the ground. The stem should be covered with soil, leaving the tip exposed. The covered stem should be kept moist, and after a few weeks, the stem will develop roots. The stem can then be cut from the mother plant and transplanted into a pot. In conclusion, propagation of Ocimum Sanctum is a fairly easy process that can be done by various methods such as seeds, cuttings, and layering. Proper care and attention should be given to the plant during the propagation process to ensure a healthy and successful growth [14, 15],

38.11 Manuring

Manure is important for plant growth because it provides all of the required macro and micronutrients for mineralization and improves the physical and chemical qualities of the soil [40]. Farmyard manures are a decomposed mixture of farm animal dung and urine, as well as litter and leftover material from roughages or other greens provided to the farm animals. It has 0.5% nitrogen, 0.2% P_2O_5, and 0.5% potassium [41]. Biofertilizers in combination with organic manures could be a technique for achieving sustainable production in Tulsi in the near future [42]. Indian basil yield and parameters were impacted by the use of biofertilizers which includes Azotobacter and mycorrhiza as well as organic fertilizers (Vermicompost and Neem cake). Vemicompost is thought to provide a plentiful supply of readily available plant nutrients, growth regulators, enzymes, and antifungal and antibacterial compounds [43]. The benefits of organic manure and biofertilizer treatment on plant fresh and dry herbage output are significant. It was discovered that the use of organic and biofertilizers improved the amount of fresh and dried herbage produced by tulsi. In order to boost output while maintaining sustainability, using biofertilizers in conjunction with organic manures may be a technique in Tulsi in the near future [42].

The application of organic manure significantly increased the growth and yield of Ocimum sanctum. The study found that the use of cow dung, poultry manure, and vermicompost resulted in a significant increase in plant height, leaf area, and fresh and dry weight of the leaves. In addition, the application of organic manure also improved the quality of the leaves by increasing the content of secondary metabolites such as phenols, flavonoids, and alkaloids. This is important as these compounds are responsible for the medicinal properties of Ocimum sanctum. The study also noted that the application of chemical fertilizers alone did not result in significant growth or yield improvement. However, the combined application of chemical fertilizers and organic manure resulted in a synergistic effect, further improving the growth and yield of the plant [44].

38.12 Classification

Tulsi has a very complex chemical nature that includes a wide range of nutrients and other physiologically active substances, the amounts of which can differ significantly across strains and even between plants in the same area. The entire herb's natural, traditionally-used pharmacological and nutritional characteristics are the consequence of a synergistic various diverse active phytochemical interactions. As a result, single chemicals or extracts cannot effectively mimic the entire benefits of Tulsi. Tulsi standardisation has so far escaped contemporary science due to its inborn botanical and biochemical complexity.

The volatile oil from the leaf includes:

1. Eugenol (1-hydroxy-2-methoxy-4-allylbenzene)
2. Euginal (also called eugenic acid)
3. Urosolic acid (2,3,4,5,6,6a,7,8,8a,10,11,12,13,14b-tetradecahydro-1 h-picene-4a-carboxylic acid),
4. Carvacrol (5-isopropyl-2-methylphenol)
5. Linalool (3,7-dimethylocta-1,6-dien-3-ol)
6. Caryophyllene and limatrol (4,11,11-trimethyl-8-methylene-bicyclo[7.2.0]undec-4-ene)
7. Methyl Carvicol (also called estragol: 1-allyl-4-methoxybenzene) [45].

38.13 Properties

The following characteristics of ayurvedic herbs are clearly recognisable:

1. The Ayurveda herb can be divided into categories based on how simple or difficult they are to digest: Guna
2. Based on preference? sweet, salty, savoury, bitter, pungent, and astringent: Rasa
3. The assimilation of herbs through digestion, metabolism, and assimilation: Vipaka
4. Virya, based on the herb's usefulness for the stomach
5. What type of bodily activity does the meal stimulate: Karma [46].

38.14 Medicinal Uses of Ocimum sanctum

herbal plant species tend to be vital for mankind, since they provide variety of and modern medicines for the healthcare industry [47]. Herbal components, which have therapeutic potential, are increasingly being used to create cutting-edge medical treatments [48]. Researchers have expressed an interest in the wide variety of phytochemicals due to their substantial multi-target medicinal activities. The therapeutic benefits of medicinal plants include their affordability, safety, efficacy, and access [49]. As medicinal plants are rich in a range of secondary metabolites, including tannins, phenolics, alkaloids, and flavonoids, which enhance human development, innate immune response, and disease resistance, they have recently become a viable choice. Ocimum sanctum (holy basil), a member of the Lamiaceae family, is regarded as a major source of pharmaceuticals, medications with bioactive molecules, and essential oils advised for treating various illnesses [50].

Ocimum sanctum L. has a number of therapeutic qualities. Different parts of the Tulsi plant, such as the leaves, flowers, stem, root, and seeds, have been used as expectorant, analgesic, anticancer, antiasthmatic, antiemetic, diaphoretic, antidiabetic, antifertility, hepatoprotective, hypotensive and antistress agents by traditional medical practitioners. Fever, bronchitis, arthritis, convulsions, and other illnesses have also been treated with tulsi [51–53]. The leaves help to calm the nerves and improve memory. Tulasi leaves can be chewed to treat mouth infections and ulcers. A few leaves sprinkled in food or drink can kill bacteria and purify the substance. Holy basil is excellent for strengthening the immune system. It defends against almost all bacterial, fungal, viral, and protozoal illnesses. According to recent studies, it also aids in preventing the spread of HIV as well as cancerous cells [13]. Ocimum sanctum has a long history of use as a respiratory and common cold treatment. Tulsi is a well-known, traditional home treatment for fever and the common cold. In general, tulsi may help to prevent infections in the respiratory tract. Studies conducted throughout time show that Tulsi 's phytochemicals have anti-inflammatory, antibacterial, antiviral, and antipyretic properties. [54].

The dried powdered leaves, when combined with mustard oil and made into a paste, are an effective treatment for pyorrhea, bad odours, and other dental issues. In problems of the genito-urinary system, the seeds are utilizeds [31]. Tulsi leaf extract is used to lessen scars, acne, and pimples. Due to its effects on vata, tulsi has anti-inflammatory qualities. Therefore, applying it externally to a sore area helps to lessen discomfort and swelling. Tulsi treats a variety of skin conditions. It works well for itching, skin irritation, and animal bites. This plant's trees can treat ring worm infections and lucoderma successfully [16].

Tulsi is considered as a sacred herb, and it has long been thought that eating Tulsi leaves on an empty stomach boosts immunity. Alcoholic Tulsi extract influences immunity, according to experimental investigations [55, 56]. According to reports, Tulsi essential oil has 100% larvicidal action against Culex mosquitoes. Its extracts exhibit pronounced mosquito-targeting insecticidal action. Its repellent effect lasts for around 2 h [57].

Patients with gastric and hepatic problems are given an aqueous decoction of Tulsi leaves. Herbal formulations containing Tulsi has been in use to treat the duration of sickness in patients with viral hepatitis. Juice of Ocimum sanctum L. with a combination of Triphala is used in treatment of various eye disorders. Three drops of tulsi oil mixed with honey might be added to one's daily routine to help with eye sight. Patients are also given fresh leaf juice to treat persistent fever, diarrhoea, bleeding, and dyspepsia [58]. A infusion of Tulsi leaves is a common cold treatment [59]. In traditional medicine, tulsi has been prescribed as an antidote for dog bites, scorpion bites, and insect bites [60].

Although this therapeutic plant has long been used by Indian traditional healers for the treatment of numerous diseases since ancient times, nothing is known about its mechanism of action, and there is no rational way to combine this traditional medical practice with modern medicine. Indian scientists have done a lot of work to demonstrate the role of essential oils in Tulsi plant which has a great use in the treatment of many diseases [58, 59]. Ursolic acid has anticancer, antibacterial, anti-viral and many other therapeutic properties [61, 62]. Due to the presence of beta caryophyllene, a sesquiterpenoid, and eugenol, a phenylpropanoid, in the hydrodistilled leaf essential oil, holy basil has a strong, somewhat more pungent flavour than other basils. Applying the oil helps to alleviate body rashes, irritation, and aches and pains [31]. Numerous studies on the essential oils of several Ocimum species have revealed the biological properties of the oils. Antibacterial, antimicrobial, and antifungal activities are crucial among them [63].

The primary element of the oil was eugenol, a high-quality essential oil isolate that is extremely valuable for flavoring various food products during commercial food manufacturing. Additionally, eugenol is used to create vanillin, which is the flavouring used the most in food goods worldwide. The medicinal industry places a high value on eugenol, which is often derived from clove buds and cinnamomum leaves and bark [64].

Eugenol is a phenolic chemical that is found in abundance in essential oils derived from various parts of the Tulsi plant. Eugenol and essential oils also have properties to lower blood sugar and cholesterol levels in people with diabetes. The content of essential oil derived from Tulsi leaves has also shown seasonal variation. Ocimum sanctum L. has been found to produce the most eugenol in the month of November [58, 65]. Another medicinal use of eugenol isolated from the tulsi plant includes antiulcerogenic properties to treat gastric ulcers. The lowering of uric acid levels by eugenol and essential oils suggests that Basil has therapeutic properties to treat rheumatoid arthritis as well [51, 59, 66]. It has been proven that Eugenol and essential oils have qualities that stimulate the immune system, implying that Ocimum sanctum L. has therapeutic potential in immunological illnesses related with immunosuppression.

Antifungal and antiviral properties are also found in the essential oils derived from Tulsi leaves. Eugenol and essential oils have been discovered to have membrane stabilizing effects on synaptosomes, erythrocytes, and mast cells, which explains Tulsi's medicinal potential in the therapy of neurological and allergy disorders [58, 59]. Recent research has demonstrated that O. sanctum possesses strong adaptogenic qualities. Tulsi leaf tea can help prevent and lessen both physical and mental stress [67]. Tulsi functions as a rejuvenator, lowering stress levels, calming the body, and enhancing memory [13]. It has been suggested to use Ocimum sanctum Linn. for the treatment of bronchitis, malaria, diarrhoea, dysentery, skin conditions, arthritis, eye conditions, insect bites, and other ailments. Moreover, the O. sanctum L. may have adaptogenic, analgesic, antispasmodic, cardioprotective, anticancer, anti-fungal, antibacterial, and anti-cancer effects [68].

For instance, a review articlehighlighted the plant's anti-inflammatory, antioxidant, and immunomodulatory properties. The authors also reported that Ocimum sanctum can be used to manage various health conditions, including diabetes, cardiovascular diseases, and cancer [26]. Another study investigated the radioprotective properties of Ocimum sanctum. The authors reported that the plant can protect against the harmful effects of radiation, such as oxidative stress and DNA damage. They attributed this effect to the plant's antioxidant and free radical scavenging activity [69]. Furthermore, a randomized, double-blind, placebo-controlled trial investigated the effects of Ocimum sanctum on stress and anxiety in humans. The results showed that the plant can reduce stress and anxiety levels in healthy individuals, suggesting its potential use as a natural remedy for stress-related disorders [55, 56].

Tulsi affects the kidneys to improve their capacity to function.Taking tulsi juice and honey consistently for 6 months can help the kidneys eliminate any renal stones that may be presen urinary system [70]. Ocimum sanctum used as skin care in such a way a paste made from tulsi leaves is applied to the afflicted region to treat ringworm and other associated conditions including leucoderma. Tulasi leaves are used topically to the affected area in order to treat chicken pox [71].

The herb helps with dental issues. Its leaves can be used to wash teeth after being sun-dried and powdered.Moreover, it may be used as toothpaste by making a paste with mustard oil. This is excellent for preventing bad breath, preserving oral health, and massaging the gums. Moreover, it helps with pyorrhea and other dental issues [72].

38.15 Nutritional Content

Seems to include chlorophyll and several other phytonutrients, as well as vitamins C and A, minerals including calcium, zinc, and iron. Also improves how well food and other herbs are digested, absorbed, and used for their benefits. 30 Kcal, 4.2 g of protein; 0.5 g of fat; 2.3 g of carbohydrates; 25 mg of calcium; 287 mg of phosphorus; 15.1 mg of iron; and 25 mg of vitamin C in the edible component per 100 g [72].

38.16 Chemical Constituents

Ocimum sanctum plays a significant function in medicine and is a rich source of secondary metabolites. Phytochemicals present in this plant includes fixed oil, essential oil, mineral contents and alcoholic extract [16]. It has been discovered that leaves of basil crop is a great source of volatile oil. Eugenol around 71% and methyl eugenol around 20% have been found in this crop. Carvacrol and sesquiterpine hydrocarbon caryophyllene are also found in the volatile oil. Phenolics, flavonoids, terpenoids, and fatty acids are among the other chemical compounds present. The plant's seeds are high in fixed oil which is around 18–22%. Polysaccharides mucilage and βsitosterol are also present in plant's seeds in good amount. Linoleic acid has said to be the most abundant fatty acid in seed oil [73].

The Tulsi plant is a multi-nutrient complex with certain well-known biologically active components that are involved in pharmacological activity against a variety of diseases [74]. Many components from the leaves of basil plants have been discovered. These components were detected from different preperations, which were prepared from leaves of basil plant. The preperations include, acetonic, petroleum ether extracts, methanolic and dry leaf powder. Around 49 components were discovered from the dried leaf powder in which 1-Methyl eugenol and 2-Eugenol being the most prominent were obtained. In methanolic extract, the predominent components were 1-Methyl eugenol and 2-Eugenol. 1-Stigmast-5-en-3-ol, 2-Stigmast-5, 22-dien-3-ol and 3-Methyl eugenol were identified in majority from acetonic extract while 1- Methyl eugenol and 2-Neophytadiene were found in majority in Petroleum ether extract. 1-Methyl eugenol, 2-Octadecane, and 3-caryophylene were detected in most of the samples [75].

The volatile oils of the leaves contain many chemical constituents which includes eugenol, caryophyllene and euginal (also known as eugenic acid). Urosolic acid, limatrol, linalool, carvacrol and methyl carvicol (also called Estragol) are also present in volatile oils of leaves. While fatty acids and sitosterol are found in the seed volatile oil. Aside from these, Tulsi seed mucilage contains some carbohydrates, and the green leaves contain anthocyans. Xylose and polysaccharides are two primary sugars found in plants [76, 77]. Three coumarin compounds isolated from the basil plant are aeculetin, aesculin, and ocimarin [78]. Alkaloids, flavonoids, tannins, and carbohydrates were discovered in an aqueous extract of O. sanctum leaves [7, 79]. Moreover, thi plant is a source of vitamin A, vitamin C, calcium, phosphorus, iron and zinc [16].

Ocimum sanctum's essential oil, which is a concentrated heterogeneous composition of phenylpropanoids and terpenoid derivatives, has been credited with supplying most of the plant's medicinal and therapeutic potential [80]. O. sanctum contains apigenin ($C_{15}H_{10}O_5$) having the antimicrobial, anti-viral, and antioxidant properties. It also contains Linalool ($C_{10}H_8O$) which having the Anti-cancer, anti-viral, anti-bacterial, and anti-fungal properties [46]. Ocimum species have phenolic-rich essential oils in addition to phenolic-rich natural products like flavonoids. Ocimum sanctum is a worldwide annual plant that may reach heights of 1800 m in India. Due to its multiple health advantages, it is regarded as a sacred plant. Many recent research studies have found that ocimum sanctum possesses antipyretic, analgesic, antiesthetic, and chemo preventive properties [81]. The Ayurveda medical system has employed ocimum sanctum L. (syn. O. tenuiflorum L.) extensively for a variety of diseases, including its capacity to reduce plasma glucose [71]. Stem of O. sanctum contains Phytochemicals Flavonoids, phenols, tannins, saponins, triterpenoids and Seeds contains Sitosterol, fatty acids [82].

Holy basil seeds are said to be extremely important because 10–20 gm of the seeds, which have been crushed and then immersed in about six times of water, stop piles from leaking when combined with cumin, sugar, or both [83].

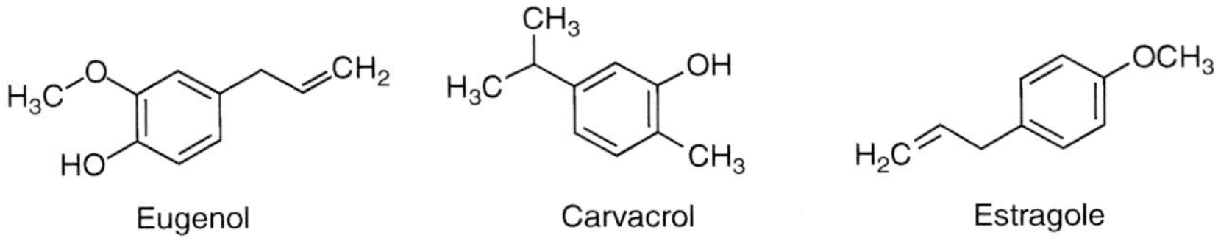

Almatroodi et al. [84]

38.17 Ocimum sanctum (basil) Negative Effects

According to several clinical studies, holy basil doesn't appear to have any adverse reactions. According to a comprehensive study, the only possible negative effects of holy basil are brief, moderate nausea. [85]. Holy basil has a very limited number of adverse effects and has been utilized by Asians for a very long time. In the event that someone is taking an antidepressant, this plant has the capacity to thin the blood.

Holy basil can cause the blood to become even more diluted, thus he must avoid using it as a clotting agent. Similar to this, anyone consuming holy basil leaves while also taking antidiabetic medications should speak with a doctor beforehand. When they significantly lower blood sugar levels, they might cause fainting or dizziness. Holy basil ingestion reduces fertility, according to animal studies [83].

38.18 Safety Measures

If you are extremely sensitive to holy basil or any of its ingredients, stay away from this herb. Pregnant or lactating women may not participate in studies with holy basil [86].

38.19 Supplements

There are several dose forms for holy basil, such as pills and capsules. There are alternative dose forms of holy basil if you have trouble swallowing tablets like in packets, in liquid- form and powder tea. Vegan as well as vegetarian goods are also offered using holy basil [87].

38.20 Crop Description

Ocimum sanctum L. is found in India and other Asian countries in the tropical and semitropical regions [88, 89]. It is a valuable therapeutic plant that many Indians regard as sacred and worship. Several names have been given to this plant which includes "Holy Basil" and "Sacred Basil". Ocimum sanctum is also known as "Queen of Herbs". This plant has been cultivated in India for ages for a variety of purposes. It produces nearly 2% essential oil from its flowers as well as leaves the majority of which is eugenol and for a long time, this plant has been known as a great source of fragrant essential oil which is primarily employed in the pharmaceutical, cosmetic as well as diet and flavour sectors [90, 91].

The natural oils of O. sanctum grown in various regions of India have been shown to contain varying amounts of eugenol. The range of eugenol content in O. sanctum essential oil is around 40% to 70%. The amount of eugenol in leaves dramatically decreases during the rainy season in India's climatic condition, notably for genotypes of O. sanctum. Despite the fact that these plant sources are abundant in eugenol, the commercial extraction of eugenol from them is difficult due to their higher costs [64]. O. sanctum, in contrast to these sources, is a less expensive source for industrial eugenol extraction.

Basil's stomach-friendly, expectorant and stimulating properties have been used in folk medicine. It is also known to be used as a pesticide to ward against snake and insect bites as well as to repel fleas and moths. Fresh and dried plant leaves have become increasingly popular in the food and spice industries in recent years. Basil's leaves and flowers have historically been used as medicines for a variety of illnesses, including headaches, coughing, vomiting, indigestion, tumors, fever, and renal dysfunction, as well as for their carminative, antispasmodic, and anti-malarial febrifuge effects. Its essential oils are produced synthetically and kept in glandular hairs where they are employed as flavourings in foods and drinks, scents in personal care goods like mouthwash and dental lotions, and fungicides or insecticides in industrial and pharmaceutical items [92].

About 80% of the patents on Ocimum sp. are for chemical extracts or the use of plant parts, with an emphasis on the constituents of essential oils. Thus, the entire nuclear and chloroplast genomes were sequenced in an effort to recognise the complete metabolic potential of this revered herb [9]. The plant has been shown to be beneficial in the treatment of respiratory syndrome, colds and fevers. In less than a decade, commercial production of basil plant began on a huge scale.This production began for mass consumption of leaves from Madhya Pradesh and other southern Indian states. Ocimum sanctum is cultivated around 25,000 hactares in India, with an annual production of around 250 to 300 tonnes of essential oil. Sacred basil has a lot of potential for large-scale production because it takes short duration to grow.

Moreover this is an aromatic and economically practicable medicinal crop that fits in well with the present cropping pattern [19, 93]. Tulsi plants are becoming more and more recognized as a key source of livelihood chances for the rural poor as well as a source of revenue for the government. Many rural groups, especially women, ancient forest-dependent tribes, impoverished and oppressed farmers, rely on the collecting and production of tulsi as their primary source of immediate income. The current situation suggests that traditional medicine is becoming more generally regarded across all societal segments in various nations, including India, claims trade site of India. The main causes of the rising demand for plant-based medications are the rapid rise in population and growing urbanisation [10].

According to a study published in the journal *Phytotherapy Research*, Ocimum sanctum has a variety of therapeutic properties, including anti-inflammatory, antioxidant, antimicrobial, and analgesic effects. In addition, Ocimum sanctum has been shown to have antioxidant properties, which can help protect the body from oxidative stress and free radical damage. This is due to the presence of flavonoids and phenolic compounds in the plant (Singh et al. 1999). Because of its widespread use in a variety of biochemical industrial sectors, including cosmetics, agrochemicals, food additives, pharmaceuticals, and nutraceuticals, reports have been made in recent years suggesting that nanotechnology may be a technology to enhance the action of phytoactive compounds in medicinal aromatic plants [94].

In another study, chitosan nanoparticles were added to eugenol, the primary component of Ocimum species, to increase its bioavailability and stability for medicinal application [95]. Foliar nutrition (NPK) in conjunction with salicylic acid had an impact on thyme plants in addition to impacting EO content and production [96]. Furthermore, it has recently been found that using nanohormones like kinetin rather than bulk kinetin administration significantly increases the output of essential oils in Mentha arvensis [97].

The effectiveness of Ocimum sanctum in treating wounds in rats was also demonstrated using two different concentrations (200 and 400 mg/kg). The excise, incise, and dead space wound models were the types of wounds employed in this study. It has been discovered that Ascorbic acid, Hexose amine, L-Hydroxyproline, and Malondialdehyde extracted from Tulsi have wound healing activity by histological investigation of granuloma tissue utilising Van Gieson and Masson Trichome strains. There are several studies that support the use of tulsi as a complementary therapy for the care of burn wounds [98].

38.21 Pharmaceutical Actions

38.21.1 Radioprotectant

Rat salivary glands were exposed to radioiodine to test Tulsi's radiation protection ability, and its results were compared to those of Amifostine, a well-known radioprotectant. Because they have an antioxidant effect on erythrocytes, flavonoids extracted from Ocimum sanctum leaves have radioprotective properties in cases of oral cancer. The ability of tulsi to mediate repair processes and offer protection against radiation poisoning has shown promise.

Vicenin and orientin, two flavonoids found in the leaves of O. sanctum, provide protection against the clastogenic effects of radiation on human lymphocytes at low and non-toxic levels. The sanctum leaf extract provides bone marrow protection when given in combination with WR-2721 and reduced its toxicity in higher concentrations [5].

38.21.2 Healing of a Wound

The effectiveness of Ocimum sanctum in treating wounds in rats was also demonstrated using two different concentrations (200 and 400 mg/kg). The excise, incise, and dead space wound models were the types of wounds employed in this study. It has been discovered that Ascorbic acid, Hexose amine, L-Hydroxyproline, and Malondialdehyde extracted from Tulsi have wound healing activity by histological investigation of granuloma tissue utilising Van Gieson and Masson Trichome strains. There are several studies that support the use of tulsi as a complementary therapy for the care of burn wounds [98].

38.21.3 Bacterial Resistance

Ocimum sanctum leaf extracts and oils were tested for their antibacterial properties against E. coli, P. aeruginosa, S. typhimurium, and S. aureus. Extract gotten both harmful gram-positive and gram-negative bacteria from O. sanctum were revealed to be equally efficient against them [99]. Tulsi‘s fresh leaf essential oil has demonstrated stronger antibacterial qualities than its dried leaf counterpart, while its antifungal characteristics are the exact opposite [100].

38.21.4 Adaptogenic

Tulsi is a good resource of adaptogenic characteristics that can regulate chaotic moods and promote mental calmness and clarity. The two most important adaptogen components in tulsi‘s chemical components are eugenol and caryophyllene. They work wonders in reducing corticosterone levels, which are the major source of stress. Also, it boosts memory and reduces the likelihood of mental health issues brought on by ageing. Both ursolic acid and oleanolic acid serve as adaptogens and are highly effective in reducing stress levels [72].

38.22 Lipid-Lowering Action

Atherosclerosis, hyperlipidemia, and other associated illnesses are increasingly considered to be serious health issues. Rats with acute hyperlipidemia brought on by triton WR1339 had their levels of total cholesterol, triglycerides, and LDL cholesterol reduced by an aqueous extract of O. basilicum. In a research on rabbits, adding 1–2% of fresh Tulsi leaves to the diet for 28 days reduced total cholesterol levels [101].

38.23 Anti-diabetic Action

Ocimum sanctum is said to have excellent anti-diabetic effects. At dosage levels of 250 and 500 mg/kg body weight, the hydroalcoholic extract of O. tenuiflorum was reported to have considerable anti-diabetic action against streptozotocin and nicotimanide induced diabetes in rats. This efficacy was equivalent to glibenclamide [102]. In both short-term and long-term feeding tests, ethanol extract of O. sanctum was proven to lower hyperglycemia in alloxan diabetic rats [103]. In a different research, the ethanol extract and three partitions (ethylacetate, butanol, and aqueous) of O. sanctum were administered to rats' pancreas and were found to have significant effects on insulin secretion. Acute insulin-release experiments with isolated rat islets revealed similar results [104].

38.24 Low-Lipoprotein Activity

Ocimum basilicum is extremely helpful in preventing hyperlipidemia and the cardiovascular diseases that are associated with it since it has been discovered to exhibit hypolipidemic effects through the use of certain polar phytoconstituents. When applied to animals with induced hyperlipidemia, aqueous basil extract significantly reduced triglyceride and total cholesterol levels in the liver and plasma [105].

38.25 Impact of Anti-aging

Tulsi's main chemical components, oleanolic acid (OA) and UA, are crucial for many of the plant's therapeutic properties. In order to slow down cell proliferation and expansion, methodologies for quick detection of UA, OA, and their oxidation products from Tulsi leaves have been developed [106].

Sr. No	Species	Collection	Activity	Oil/ Extract	Plant Parts	Major Compound	Refrences
1	O.sanctum	India Manipal	Anticarcinogenic and antioidant	Etract	–	–	Devi [107]
2	O.sanctum	India, tamilnadu	Antioidant	Etract	Leaves, Stem and inflorescence	Rosmarinic acid	Hakkim et al. [108]
3	O.sanctum	India Delhi	Antifungal	Aqueous extracts and oils	–	Eugenol, methyl eugenol, linalool and 1, 8-cineole	Khan et al. [109]
4	O.sanctum	India north kamataka	Antimicrobial and antioidant	Essential oil	Flowering arial parts	Eugenol (75.1%) and methyl eugenol (92.4%)	Joshi and Karna [80]
5	O.sanctum	Central European Romania	Antimicrobial	Extract oil	Arial	Linallol (38.60%), eugenol (10.20%) tau-cadinol (10.20%)	Stefan e al. [110]
6	O.sanctum	Australia	Antimicrobial	Extract oil	Fressh leaves and inflorescence		Yamani et al. [111]

38.26 Myths, Legends, Tales, Folklore and Interesting Facts

The plant is associated with religious beliefs all across the world, especially in India. The plant is said to have flourished near the crucifixion site of Jesus Christ [112], despite the fact that there is no literature about basil in the Bible [113]. Hindu mythology believes that Tulsi was one of the 14 “Ratnas” (gems or treasures) that emerged from the ocean as the supreme spiritual plant to promote health and ward off disease. Due of its exceptional qualities, tulsi is referred regarded as the Queen of herbs [16].

It is mainly respected in Hindu folklore. Tulsi is said to be the goddess’s manifestation, having been raised from her embers. There are several legends about Tulsi, but the most well-known is from the Shiva Purana. Indra and Brihaspati were on their way to see Lord Shiva at the Kailash peak when they were stopped by a sage. Shiva himself was the sage. To put Indra and Bri haspati to the test, Lord Shiva changes his appearance to include tangled hair and a radiant face. Indra didn’t recognize Lord Shiva and became enraged when he didn’t move out of the path. Indra threatened him with his thunderbolt to get him out of his way. Lord Shiva was enraged by this act and became angry. His third eye was opened to kill Indra as a result of his rage, but the Bri haspati recognized Lord Shiva at the same time. He begged Shiva to forgive Indra. Bri Haspati convinced Lord Shiva, and he focused the fire towards the sea from his eye. The result of the meeting of seawater and fire was a boy named Jalandhara. Jalandhara grew up to be a great demon king by Guru Sukra. Vrinda, the daughter of the demon Kalanemi, was the woman he married. Lord Vishnu, as well as other gods, were pitted against him by Bhrugu. Jalandhara and Vishnu engaged in a battle that ended in a stalemate. Jalandhara's bravery influenced Lord Vishnu, who asked him in kshira sagara, and Jalandhara conquered in the lack of all gods. Devas were determined about avoiding being governed by him. The Devas sought advice from sage Narada in order to meet Jalandhara. However, throughout his visit, he demonstrated Kailasa's magnificence. Narada went on to discuss the charm and beauty of Shiva's residence and the goddess Parvati [114] which impressed Jalandhara and he transformed himself to deceive goddess Parvati But he was identified by the goddess and was about to be attacked by her. Jalandhara somehow managed to escape. After this, the goddess addressed Vishnu and requested him to decieve Vrinda the same way Jalandhara tried to decieve her. Vrinda was highly virtuous to Lord Vishnu for restoring Jalandhara after Shiva had destroyed it. Vrinda embraced Jalandhara, who is Vishnu in actuality. When Vrinda learned what had happened, she cursed Lord Vishnu, saying that someone would kidnap his wife and turn her into a stone (as was eventually done in the Ramayana). He accepted the curse in order to respect his true devotee. And it's now known as the stone "Shaligram shila" in Gandaki River in Nepal. Vrinda eventually immolated herself by stepping into the fire. Jalandhara grew enraged after learning of his wife's death and turned into a battleground. As a result, Lord Shiva killed Jalandhara. Vrinda's soul joined with Parvati, and His soul fused with Shiva. Because of her devotion and faithfulness for Lord Vishnu, Vrinda has been given the name Tulsi. Tulsi died as a result of her suffering. She was promised by Vishnu to be "worshipped by women for her faithfulness" [115, 116]. It is thus seen as a sign of love, eternal life, purification, and protection in Hinduism. It is also utilized for purification in funeral ceremonies [10].

References

1. Joseph, B., & Nair, V. M. (2013). Ethanopharmacological and phytochemical aspects of Ocimum sanctum Linn-the elixir of life. *British Journal of Pharmaceutical Research, 3*(2), 273–292.
2. Javanmardi, J., et al. (2002). Chemical characterization of basil (Ocimum basilicum L.) found in local accessions and used in traditional medicines in Iran. *Journal of Agricultural and Food Chemistry, 50*(21), 5878–5883.
3. Rahal, A. & Kumar, A. (2009). *Tulsi: A miracle herb in the hands of traditional house lady.*
4. Kayastha, B. L. (2014). Queen of herbs tulsi (Ocimum sanctum) removes impurities from water and plays disinfectant role. *Journal of Medicinal Plants Studies, 2*, 2.

5. Kadian, R., & Parle, M. (2012a). Therapeutic potential and phytopharmacology of tulsi. *International Journal of Pharmacy & Life Sciences, **3***, 7.
6. Kirtikar, K. R., & Basu, B. D. (1965a). *Ocimum sanctum in Indian Medicinal Plants*. Published by LB Basu.
7. Gupta, S., Prakash, J, & Srivastava, S., (2002). *Validation of traditional claim of Tulsi, Ocimum sanctum Linn. as a medicinal plant.*
8. Bhamra, S., et al. (2015). DNA authentication of tulsi (Ocimum tenuiflorum) using the nuclear ribosomal internal transcribed spacer (ITS) and the chloroplast intergenic spacer trnH-psbA. *Planta Medica, 81*(16), PW_20.
9. Rastogi, V., et al. (2014). *Anti-Bacterial effect of extracts of ocimum on imipenem resistant gram negative bacteria isolated from burn wound infections.*
10. Shasany, A. K. (2016). The Holy basil (Ocimum sanctum L.) and its genome. *Indian Journal of History of Science, 51*(2), 343–350.
11. Ved, D., & Goraya, G. (2008). Demand and supply of medicinal plants. *Medplant-ENVIS Newsletter on Medicinal Plants, 1*(1), 2–4.
12. Garodia, P., et al. (2007). From ancient medicine to modern medicine: Ayurvedic concepts of health and their role in inflammation and cancer. *Journal of the Society for Integrative Oncology, 5*(1), 25–37.
13. Kumar, A., et al. (2013). Ocimum sanctum (Tulsi): A miracle herb and boon to medical science – A review. *International Journal of Agronomy and Plant Production, 4*(7), 1580–1589.
14. Pandey, P., Swami, A., & Bhatt, A. (2014a). Combined Effect of Pseudomonas fluorescens and Glomus fasciculatum with Zinc on Growth of Indian Basil (Ocimum sanctum Linn.). *Global Journal of Current Research Pandey, **2***, 2.
15. Pandey, B., Agrawal, M., & Singh, S. (2014b). Coal mining activities change plant community structure due to air pollution and soil degradation. *Ecotoxicology, 23*, 1474–1483.
16. Kulkarni, K. V., & Adavirao, B. V. (2018a). A review on: Indian traditional shrub Tulsi (Ocimum sanctum): The unique medicinal plant. *Journal of Medicinal Plants Studies, 6*(2), 106–110.
17. Gurav, T. P., Dholakia, B. B., & Giri, A. P. (2022). A glance at the chemodiversity of Ocimum species: Trends, implications, and strategies for the quality and yield improvement of essential oil. *Phytochemistry Reviews, 21*(3), 879–913.
18. Devi, P. U., Bisht, K., & Vinitha, M. (1998a). A comparative study of radioprotection by Ocimum flavonoids and synthetic aminothiol protectors in the mouse. *The British Journal of Radiology, 71*(847), 782–784.
19. Pushpangadan, P., & Bradu, B. (1995). Medicinal and aromatic plants. *Advances in Horticulture, 11*, 627–657.
20. Paton, A., Harley, R., & Harley, M. (1999). Ocimum: An overview of classification and relationships. *Basil*, 11–46.
21. Jain, S., Mohan, R., & Rai, R. (2015). Ocimum sanctum as an herbal medicine: A Review. *International Journal of Maxillofacial Research, 1*(1), 1–12.
22. Prakash, P., & Gupta, N. (2005a). Therapeutic uses of Ocimum sanctum Linn (Tulsi) with a note on eugenol and its pharmacological actions: A short review. *Indian Journal of Physiology and Pharmacology, 49*(2), 125.
23. Balyan, S., & Pushpangadan, P. (1988). A study on the taxonomical status and geographic distribution of the genus Ocimum. *Pafai of Journal, 10*(2), 13–19.
24. Pavela, R., Kaffková, K., Smékalová, K., Vrchotová, N., Bednář, J., & Tříska, J. (2023). Biomass yield potential of Tulsi (Ocimum sanctum L.) in European conditions. *Industrial Crops and Products, 194*, 116365.
25. Singh, S., & Majumdar, D. K. (1999). Evaluation of the gastric antiulcer activity of fixed oil of Ocimum sanctum (Holy Basil). *Journal of Ethnopharmacology, 65*(1), 13–19.
26. Cohen, M. M. (2014). Tulsi-Ocimum sanctum: A herb for all reasons. *Journal of Ayurveda and integrative medicine, 5*(4), 251.

27. Buddhadev, S. G., Buddhadev, S. S., & Mehta, N. D. (2014). A review article on Ocimum Sanctum Linn. *International Peer Review of Ayurveda of Journal, 2*(2), 1–6.
28. Mohan, L., Amberkar, M. V., & Kumari, M. (2011). Ocimum sanctum linn.(TULSI)-an overview. *International Journal of Pharmaceutical Sciences Review and Research, 7*(1), 51–53.
29. Smitha, G. R., Basak, B. B., Thondaiman, V., & Saha, A. (2019). Nutrient management through organics, bio-fertilizers and crop residues improves growth, yield and quality of sacred basil (Ocimum sanctum Linn). *Industrial Crops and Products, 128*, 599–606.
30. Pole, S. (2006). *Ayurvedic medicine: The principles of traditional practice*. Elsevier Health Sciences.
31. Bhasin, M. (2012). Ocimum-Taxonomy, medicinal potentialities and economic value of essential oil. *Journal of Biosphere, 1*, 48–50.
32. Mahajan, N., Rawal, S., Verma, M., Poddar, M., & Alok, S. (2013). A phytopharmacological overview on Ocimum species with special emphasis on Ocimum sanctum. *Biomedicine & Preventive Nutrition, 3*(2), 185–192.
33. Makri, O., & Kintzios, S. (2008). Ocimum sp.(basil): Botany, cultivation, pharmaceutical properties, and biotechnology. *Journal of Herbs, Spices & Medicinal Plants, 13*(3), 123–150.
34. Govaerts, R. (2014). *World Checklist of Lamiaceae*. Royal Botanic Gardens.
35. Mishra, A., & Agarwal, R. (2021). *Extraction of Active Components from Crude Ocimum Sanctum and Maintenance of C.* Elegans.
36. Kumar, P. K., Kumar, M. R., Kavitha, K., Singh, J., & Khan, R. (2012). Pharmacological actions of Ocimum sanctum–review article. *International Journal of Advances in Pharmacy, Biology and Chemistry, 1*(3), 2277–4688.
37. Kukreja, A., Mathur, A., & Zaim, M. (1990). Mass production of virus-free patchouli plants (Pogostemon cablin (Blanco) Benth.) by in vitro culture. *Tropical Agriculture, Trinidad and Tobago, 67*(2), 101–104.
38. Sen, J., & Sharma, A. (1991). Micropropagation of Withania somnifera from germinating seeds and shoot tips. *Plant Cell, Tissue and Organ Culture, 26*, 71–73.
39. Pattnaik, S. (1996). and P.K. Chand, *In vitro propagation of the medicinal herbs Ocimum americanum L. syn. O. canum Sims.(hoary basil) and Ocimum sanctum L.(holy basil). Plant Cell Reports, 15*, 846–850.
40. Chaterjee, B., et al. (2005). Effect of organic nutrition in sprouting broccoli (Brassica oleracea L. var. italica Plenck). *Vegetable Science, 33*(1), 51–54.
41. Gaur, A. (1991). Bulky organic manure and crop residues. In *Fertilizers organic matter recyclable wastes and biofertilziers*. H. LS, Tondon, Fertilizer Development and Consulation Organization.
42. Ram, R., et al. (2019). Influence of organic manures and biofertilizers on yield and yield attributes of Indian Basil (Ocimum sanctum L.) cvs-Cim-Ayu and Cim-Angana. *Journal of Pharmacognosy and Phytochemistry, 8*(5), 1809–1813.
43. Arancon, N., et al. (2004). Influences of vermicomposts on field strawberries: 1. Effects on growth and yields. *Bioresource Technology, 93*(2), 145–153.
44. Du, S., Bhattacharya, C. B., & Sen, S. (2011). Corporate social responsibility and competitive advantage: Overcoming the trust barrier. *Management Science, 57*(9), 1528–1545.
45. Deshmukh, R. M. (2022). Ocimum sanctum–A Medicine from my courtyard. *Journal Homepage: www.ijrpr.comISSN*, 2582, 7421.
46. Tyagi, M., Tyagi, N., & Tyagi, J. (2021). *A Review on: The Major Constituents of Tulsi (Ocimum Sanctum) and their Antimicrobial and Antioxidant Activities against Different Microbes*.
47. Alves, R. R., & Rosa, I. M. (2007). Biodiversity, traditional medicine and public health: where do they meet? *Journal of Ethnobiology and Ethnomedicine, 3*, 1–9.
48. Muzammil, A., Waqas, M., Umar, A., Sufyan, M., Rehman, A., Haider, A., et al. (2021). Anti-aging natural compounds and their role in the regulation of metabolic pathways leading to longevity. *Mini Reviews in Medicinal Chemistry, 21*(18), 2630–2656.
49. Tomlinson, T. R., & Akerele, O. (Eds.). (2015). *Medicinal plants: Their role in health and biodiversity*. University of Pennsylvania press.

50. Tabassum, S., Khalid, H. R., Haq, W. U., Aslam, S., Alshammari, A., Alharbi, M., et al. (2022). Implementation of System Pharmacology and Molecular Docking Approaches to Explore Active Compounds and Mechanism of Ocimum Sanctum against Tuberculosis. *Processes, 10*(2), 298.
51. Sarkar, A., Pandey, D., & Pant, M. (1990). A report on the effects of Ocimum sanctum (Tulsi) leaves and seeds on blood and urinary uric acid, urea and urine volume in normal albino rabbits. *Indian Journal of Physiology and Pharmacology, 34*(1), 61–62.
52. Reghunandanan, P., et al. (1995). Effect of Ocinium Sanctum Linn (Tulsi) extract on testicular function. *Indian Journal of Medical Sciences, 49*(4), 83–87.
53. Sethi, J., et al. (2003). Protective effect of Tulsi (Ocimum Sanctum) on lipid peroxidation in stress induced by anemic hypoxia in rabbits. *Indian Journal of Physiology and Pharmacology, 47*(1), 115–119.
54. Mohapatra, P. K., Chopdar, K. S., Dash, G. C., Mohanty, A. K., & Raval, M. K. (2023). In silico screening and covalent binding of phytochemicals of Ocimum sanctum against SARS-CoV-2 (COVID 19) main protease. *Journal of Biomolecular Structure and Dynamics, 41*(2), 435–444.
55. Mondal, S., et al. (2011a). Double-blinded randomized controlled trial for immunomodulatory effects of Tulsi (Ocimum sanctum Linn.) leaf extract on healthy volunteers. *Journal of Ethnopharmacology, 136*(3), 452–456.
56. Mondal, S., Varma, S., Bamola, V. D., Naik, S. N., Mirdha, B. R., Padhi, M. M., et al. (2011b). Double-blinded randomized controlled trial for immunomodulatory effects of Tulsi (Ocimum sanctum Linn.) leaf extract on healthy volunteers. *Journal of Ethnopharmacology, 136*(3), 452–456.
57. Singh, S., Mahour, K., & Prakash, S. (2009). Evaluation of mosquito repellent efficacy of Ocimum sanctum plant extract. *Journal of Herbal Medicine and Toxicology, 3*(1), 87–90.
58. Rajeshwari, S. (1992). *Ocimum sanctum. The Indian home remedy* (Current Medical Scene). Cipla Ltd.
59. Sen, P. (1993). Therapeutic potentials of Tulsi: From experience to facts. *Drugs News & Views, 1*(2), 15–21.
60. Khanna, N., & Bhatia, J. (2003). Antinociceptive action of Ocimum sanctum (Tulsi) in mice: Possible mechanisms involved. *Journal of Ethnopharmacology, 88*(2–3), 293–296.
61. Rajeshwar, S. (1992). *Ocimum sanctum: The Indian home remedy in current medical scene.*
62. Liu, J. (1995). Pharmacology of oleanolic acid and ursolic acid. *Journal of Ethnopharmacology, 49*(2), 57–68.
63. Kalita, J., & Latif Khan, M. (2013). Commercial potentialities of essential oil of Ocimum members growing in North East India. *International Journal of Pharmacy & Life Sciences, **4***, 4.
64. Mukherji, S. (1987). Ocimum–a cheap source of eugenol. *Science Reporter 31*, 599.
65. Mukherji, S. (1995). Ocimum-a cheap source of eugenol. In *Science reporter, 31, 599. Eugenol-rich Ocimum variety released* (Vol. 45, p. 256). CSIR NEWS Published by PID CSIR.
66. Mandal, S., et al. (1993). Ocimum sanctum Linn-A study on gastric ulceration and gastric secretion in rats. *Indian Journal of Physiology and Pharmacology, 37*, 91–91.
67. Singh, N., et al. (2012). Therapeutic potential of Ocimum sanctum in prevention and treatment of cancer and exposure to radiation: An overview. *International Journal of Pharmaceutical Sciences and Drug Research, 4*(2), 97–104.
68. Pattanayak, P., Behera, P., Das, D., & Panda, S. K. (2010). Ocimum sanctum Linn. A reservoir plant for therapeutic applications: An overview. *Pharmacognosy Reviews, 4*(7), 95.
69. Jagetia, G. C. (2007). Radioprotective potential of plants and herbs against the effects of ionizing radiation. *Journal of Clinical Biochemistry and Nutrition, 40*(2), 74–81.
70. Chandra, S., Dwivedi, P., Arti, K. M., & Shinde, L. P. (2016). An industrial cultivation of Tulsi (Ocimum sanctum) for medicinal use and rural empowerment. *Journal of Medicinal Plants Studies, 4*(6), 213–218.

71. Joshi, R. K., Setzer, W. N., & Da Silva, J. K. (2017). Phytoconstituents, traditional medicinal uses and bioactivities of Tulsi (Ocimum sanctum Linn.): A review. *American Journal of Essential Oils and Natural Products, 5*(1), 18–21.
72. Bhadra, P., & Sethi, L. (2020). A review paper on the Tulsi plant (Ocimum sanctum). *Indian Journal of Natural Sciences, 10*(60), 20854–20860.
73. Naji-Tabasi, S., & Razavi, S. M. A. (2017). Functional properties and applications of basil seed gum: An overview. *Food Hydrocolloids, 73*, 313–325.
74. Rahal, A. (2006). Herbal Therapy: An untapped treasure in livestock production. In *Seminar on "Improving socioeconomic status of livestock farmers through animal productivity" held at CVA Sc., Pantnagar. UKD. India* (pp. 24–25).
75. Wagner, H., Nörr, H., & Winterhoff, H. (1994). Plant adaptogens. *Phytomedicine, 1*(1), 63–76.
76. Kelm, M., et al. (2000). Antioxidant and cyclooxygenase inhibitory phenolic compounds from Ocimum sanctum Linn. *Phytomedicine, 7*(1), 7–13.
77. Shishodia, S., et al. (2003). Ursolic acid inhibits nuclear factor-κB activation induced by carcinogenic agents through suppression of IκBα kinase and p65 phosphorylation: correlation with down-regulation of cyclooxygenase 2, matrix metalloproteinase 9, and cyclin D1. *Cancer Research, 63*(15), 4375–4383.
78. Anandjiwala, S., Kalola, J., & Rajani, M. (2006). Quantification of eugenol, luteolin, ursolic acid, and oleanolic acid in black (Krishna Tulasi) and green (Sri Tulasi) varieties of Ocimum sanctum Linn. using high-performance thin-layer chromatography. *Journal of AOAC International, 89*(6), 1467–1474.
79. Aswar, M. K., & Joshi, R. H. (2010). Anti-cataleptic activity of various extracts of Ocimum sanctum. *International Journal of Pharmaceutical Research and Development, 2*(6), 7.
80. Joshi, S., & Karna, A. K. (2013). Analysis of phytoconstiuents and cytotoxic activities of different parts of Ocimum sanctum. *International Journal of Applied Sciences and Biotechnology, 1*(3), 137–144.
81. Ganasoundari, A., Devi, P. U., & Rao, M. N. A. (1997). Protection against radiation-induced chromosome damage in mouse bone marrow by Ocimum sanctum. *Mutation Research/ Fundamental and Molecular Mechanisms of Mutagenesis, 373*(2), 271–276.
82. Siva, M., Shanmugam, K. R., Shanmugam, B., Venkata, S. G., Ravi, S., Sathyavelu, R. K., & Mallikarjuna, K. (2016). Ocimum sanctum: A review on the pharmacological properties. *International Journal of Basic and Clinical Pharmacology, 5*, 558–565.
83. Kumar, V., Andola, H. C., Lohani, H., & Chauhan, N. (2011). Pharmacological review on Ocimum sanctum Linnaeus: A queen of herbs. *Journal of Pharmaceutical Research, 4*, 366–368.
84. Almatroodi, S. A., Alsahli, M. A., Almatroudi, A., & Rahmani, A. H. (2020). Ocimum sanctum: Role in diseases management through modulating various biological activity. *Pharmacognosy Journal, 12*(5), 1198–1205.
85. Jamshidi, N., & Cohen, M. M. (2017). The clinical efficacy and safety of Tulsi in humans: A systematic review of the literature. *Evidence-Based Complementary and Alternative Medicine, 2017*, 13.
86. Lopresti, A. L., Smith, S. J., Metse, A. P., & Drummond, P. D. (2022). A randomized, double-blind, placebo-controlled trial investigating the effects of an Ocimum tenuiflorum (Holy Basil) extract (HolixerTM) on stress, mood, and sleep in adults experiencing stress. *Frontiers in Nutrition, 9*, 965130.
87. Saldanha, L. G., Dwyer, J. T., & Bailen, R. A. (2021). Modernization of the national institutes of health dietary supplement label database. *Journal of Food Composition and Analysis, 102*, 104058.
88. Godhwani, S., Godhwani, J., & Was, D. (1988). Ocimum sanctum – A preliminary study evaluating its immunoregulatory profile in albino rats. *Journal of Ethnopharmacology, 24*(2–3), 193–198.
89. Pattanayak, P., et al. (2010). Ocimum sanctum Linn. A reservoir plant for therapeutic applications: An overview. *Pharmacognosy Reviews, **4***(7), 95.

90. Charles, D. J., & Simon, J. E. (1990). Comparison of extraction methods for the rapid determination of essential oil content and composition of basil. *Journal of the American Society for Horticultural Science, 115*(3), 458–462.
91. Yaldiz, G., Gul, F., & Kulak, M. (2015). Herb yield and chemical composition of basil (Ocimum basilicum L) essential oil in relation to the different harvest period and cultivation conditions. *African Journal of Traditional, Complementary and Alternative Medicines, 12*(6), 71–76.
92. Mondello, L., et al. (2002). Studies on the essential oil-bearing plants of Bangladesh. Part VIII. Composition of some Ocimum oils O. basilicum L. var. purpurascens; O. sanctum L. green; O. sanctum L. purple; O. americanum L., citral type; O. americanum L., camphor type. *Flavour and Fragrance Journal, 17*(5), 335–340.
93. Ajjan, N., et al. (2009). Economics of cultivation and marketing of Tulsi (Ocimum sanctum) in Tamil Nadu. *Indian Journal of Arecanut, Spices and Medicinal Plants, 11*(2), 52–59.
94. Tahir, M., Pan, L., Idrees, F., Zhang, X., Wang, L., Zou, J. J., & Wang, Z. L. (2017). Electrocatalytic oxygen evolution reaction for energy conversion and storage: A comprehensive review. *Nano Energy, 37*, 136–157.
95. Woranuch, S., & Yoksan, R. (2013). Eugenol-loaded chitosan nanoparticles: I. Thermal stability improvement of eugenol through encapsulation. *Carbohydrate Polymers, 96*(2), 578–585.
96. Pavela, R., Žabka, M., Vrchotová, N., & Tříska, J. (2018). Effect of foliar nutrition on the essential oil yield of Thyme (Thymus vulgaris L.). *Industrial Crops and Products, 112*, 762–765.
97. Khan, M. M. A., Quasar, N., & Afreen, R. (2023). Nanotized form of indole acetic acid improve biochemical activities, the ultrastructure of glandular trichomes and essential oil production in Ocimum tenuiflorum L. *Industrial Crops and Products, 193*, 116117.
98. Bano, N., Ahmed, A., Tanveer, M., Khan, G. M., & Ansari, M. T. (2017). Pharmacological evaluation of Ocimum sanctum. *Journal of Bioequivalence & Bioavailability, 9*(3), 387–392.
99. Mishra, P., & Mishra, S. (2011). *Study of antibacterial activity of Ocimum sanctum extract against gram positive and gram negative bacteria* (No. RESEARCH).
100. Mirdha, B. R., Naik, S. N., & Mahapatra, S. C. (2007, March). Antimicrobial activities of essential oils obtained from fresh and dried leaves of Ocimum sanctum (L.) against enteric bacteria and yeast. In *International Symposium on Medicinal and Nutraceutical Plants* (Vol. *756*, pp. 267–270). ISHS.
101. Verma, S. (2016). Chemical constituents and pharmacological action of Ocimum sanctum (Indian holy basil-Tulsi). *The Journal of Phytopharmacology, 5*(5), 205–207.
102. Parasuraman, S., Balamurugan, S., Christapher, P. V., Petchi, R. R., Yeng, W. Y., Sujithra, J., & Vijaya, C. (2015). Evaluation of antidiabetic and antihyperlipidemic effects of hydroalcoholic extract of leaves of Ocimum tenuiflorum (Lamiaceae) and prediction of biological activity of its phytoconstituents. *Pharmacognosy Research, 7*(2), 156.
103. Vats, V., Yadav, S. P., & Grover, J. K. (2003). Effect of T. foenumgraecum on glycogen content of tissues and the key enzymes of carbohydrate metabolism. *Journal of Ethnopharmacology, 85*(2–3), 237–242.
104. Hannan, J. M. A., Marenah, L., Ali, L., Rokeya, B., Flatt, P. R., & Abdel-Wahab, Y. H. A. (2006). Ocimum sanctum leaf extracts stimulate insulin secretion from perfused pancreas, isolated islets and clonal pancreatic beta-cells. *Journal of Endocrinology, 189*(1), 127–136.
105. Harnafi, H., Caid, H. S., el Houda Bouanani, N., Aziz, M., & Amrani, S. (2008). Hypolipemic activity of polyphenol-rich extracts from Ocimum basilicum in Triton WR-1339-induced hyperlipidemic mice. *Food Chemistry, 108*(1), 205–212.
106. Srivastava, A. K., & Singh, V. K. (2021). Tulsi (Ocimum sanctum): A potent adaptogen. *Clinical Research Notes, 2*(2), 01–05.
107. Devi, P. U. (2001). Radioprotective, anticarcinogenic and antioxidant properties of the Indian holy basil, Ocimum sanctum (Tulasi). *Indian Journal of Experimental Biology, 39*(3), 185–190.

108. Hakkim, F. L., Shankar, C. G., & Girija, S. (2007). Chemical composition and antioxidant property of holy basil (Ocimum sanctum L.) leaves, stems, and inflorescence and their in vitro callus cultures. *Journal of Agricultural and Food Chemistry, 55*(22), 9109–9117.
109. Khan, A., Ahmad, A., Manzoor, N., & Khan, L. A. (2010). Antifungal activities of Ocimum sanctum essential oil and its lead molecules. *Natural Product Communications, 5*(2), 1934578X1000500235.
110. Stefan, M., Zamfirache, M. M., Padurariu, C., Trută, E., & Gostin, I. (2013). The composition and antibacterial activity of essential oils in three Ocimum species growing in Romania. *Central European Journal of Biology, 8*, 600–608.
111. Yamani, H. A., Pang, E. C., Mantri, N., & Deighton, M. A. (2016). Antimicrobial activity of Tulsi (Ocimum tenuiflorum) essential oil and their major constituents against three species of bacteria. *Frontiers in Microbiology, 7*, 681.
112. Meyers, M. (2003). *Basil: An herb society of America guide* (pp. 6–7). The Herb Society of America.
113. Simon, J. E., Quinn, J., & Murray, R. G. (1990). *Basil: A source of essential oils* (pp. 484–489). Timber Press Portland.
114. Stella, K. (1992). *The presence of Siva*. Princeton University Press.
115. Gupta, S. M. (1971). *Plant myths and traditions in India* (p. Brill Archive).
116. Dyer, T. (1889). *Thiselton. The folk-lore of plants*. Appleton and Co.

Chapter 39
Chamomile

Abida Parveen, Shagufta Perveen, Farah Naz, Mobeen Ahmad, and Mohammad Khalid

39.1 Introduction

Plant Profile

Family	Asteraceae
English name	German chamomile, Crown mayweed
Common name	Hungarian, wild and blue chamomile,scented mayweed.
Species and varieties	Matricaria chamomilla L. (2n = 18) Matricaria suaveolens L. Chamomilla recutita L.
Distribution	Europe. Germany, Hungary, France, Russia,
Yugoslavia, and Brazil	
Uses	Cosmetics, food industry, medicine

Chamomilla recutita normally branded as German, Hungarian (kamilla) blue and wild chamomile as well as scented mayweed [1]. It can adapt to different kind of soil texture for growing, temperatures, and soil alkalinity. It is known as medicinal herb due to wide range of uses in medicines. It is s God gifted medicinal plant since Egyptians times.

It is known as essential medicinal herb, member of family Asteraceae. That is used in gastrointestinal problems. It is used in irritation of the skin [2]. Chamomile has been used in herbal remedies [3]. The drug chamomile is listed in the pharmacopoeias of 26 different countries. The word "chamomile"derived from the Greek

A. Parveen (✉) · S. Perveen · F. Naz · M. Ahmad
Department of Botany, Government College University, Faisalabad, Pakistan

M. Khalid
Department of Pharmacognosy, College of Pharmacy, Prince Sattam Bin Abdulaziz University, Al-Kharj, Saudi Arabia

M. Zia-Ul-Haq et al. (eds.), *Essentials of Medicinal and Aromatic Crops*,
https://doi.org/10.1007/978-3-031-35403-8_39

word "chamaimlon," which means "earth-apple." This word comes from the words "chamai," which means "on the ground," and "mlon," which means "apple." its apple-like scent gives it its name [4]. M. chamomilla has been used in traditional recipes for millennia. Indeed, chamomile, known as the "star among medicinal species" [5]. Germany named it "the medicinal plant of the year" [6]. It is called "Star" herb because of its constructive medical properties [7]. Its flower is used as a herbal tea and extracts used as a herbal remedies in various regions [5]. German chamomile oil has several medical properties like flavones, polysaccharides and lipophilic. Its basic use found in food industry due to their capabilities of being anti-inflammatory, antiulcerogenic, antibacterial, antiseptic, antispasmodic, sedative, immunomodulatory [8].its oil play an important role in medicine industry. It is found in a traditional, herbal, and homoeopathic medicines [9]. It's used to treat gripes and fever [10]. The extracted essential oil of M. *chamomilla* starts from 0.2% to 1.9% [11]. Chamomile is used as an antibacterial, antispasmodic and mildly medicaments. Flower of chamomile consist of flavonoids terpenes, and essential element such as protein, lipids carbohydrates, lipids and amino acid and mineral element [12].

Persian writers, along with Arab doctors, reported that the scent of chamomile could help people sleep and feel less anxious. They even mentioned that bathing with chamomile tea had aphrodisiac and labor-inducing effects [13]. Additionally, chamomile tea was used orally as a diuretic and menstrual inducer by Unani doctors practicing Persian and Arab traditional medicine [14] tea of plant chamomile was also used topically as an antiseptic and anti-inflammatory agent. Chamomile was used by Hindu doctors to treat dyspepsia, chlorosis, and flatulent colic further east. It was said that the tea extracts it contained had significant pain relief properties for headaches, menstrual pain, and digestive issues. chamomile oil has been shown to alleviate asthma, cough, colic, and diarrhea in infants [15].

39.2 Origin and Distribution

The chamomile flower is also known as medicated herb whose native from different countries of world like Europe, Germany Russia France, Hungry and Brazil. It's also grown in different region of India, Africa, Asia, America, Australia, and New Zealand. Hungary produce the maximum biomass. It grows more on degraded lands in Hungary. Flowers are shipped in quantity to Germany for oil extraction. In world oil market Poland, Hungary, Germany, Argentina, and Czechoslovakia, are major producer of chamomile plant, they have recently started extensive plant development initiatives to create plants with high concentrations of essential oils that have a specific chemical make-up [16].

Chamomile is native to Europe and has been produced by many manufacturers in the south of Brazil [17]. In India, it was cultivated in Lucknow and then to Punjab during the Mughal dynasty about 300 years ago. Chamomile flowers are in high demand [18] .

39.3 Description of the Plant

It has a branched, smooth and erect stem. The bipinnate or tripinnate leaves are present that are long and narrow. The scales are present in hollow receptacle. In early to midsummer, the flowers bloom. In the flowers blue essential oil is present. The plant is known as a blue chamomile due to its color of the oil because it contains chamazulene. Its achene has a yellowish-brown fruit [5]. True chamomile has roots that are spindle-shaped. Yellowish tubular with teeth finish in a tube. White brooms plant 6–11 mm, 3–5 mm long and wide respectively. The receptacle is broad, flat at first and then cone-shaped, hollow and paleae-free [19]. According to a 1927 annual report on medical plant in Poland, chamomile production was 44 tons, the cultivation area was 0.01 ha, showing that the majority of the chamomile was grown in the wild. It is used in phytotherapy and have many properties like anti allergic, anti-inflammatory etc. Chamomile has a domestic and international market that is slowly growing. It is grown in Europe, North and South America, Hungary and Croatia. chamomile is produced in Argentina, Egypt, Bulgaria, and Hungary on the worldwide market for the commercial uses.

Spain, Czech Republic, Germany, Brazil, Chile, and Peru manufacture chamomile for the worldwide market to a smaller extent. Because of its properties, its blossoms are used to make tea. Flower extract mostly used in.

cosmetics industry to flavor personal hygiene items. Chamomile oil has been rising continuously in popularity around the world [20].

39.4 Species and Varieties

It is diploid species (2n = 18). Their nature is allogamous. They have broad dispersion. Chamomile is an old medicinal plant that contain names including Baboonig, Babuna, Babuna camornile, Babunj, Hungarian, Roman, English, Single, Sweet false pinheads and Scented mayweed are the names of chamomile plant [21].

39.5 Soil

They grow best on soil with good topsoil [22]. Chamomile is both tolerant and easy to grow. The chamomile species grow in sandy, loamy and acidic soil. It has been cultivated at the National Botanical Research Institute's Banthra farm in Lucknow on soil with a pH of 9. Chamomile has a strong resistance for alkalinity in the soil. The plants collect a sodium, which that cause reduction of salt levels in the soil.

39.6 Climate

The seed germination temperature is 10 and 20 degrees Celsius. The time to start seedlings is after the monsoon season which is in September.

Within 4–5 days of seeding, seed germination begins, and seedlings are ready for transplantation in 4–5 weeks [23]. The highest yields generated by chamomile if it is seeded in autumn.

39.7 Land Preparation

The thermal model was used to determine the best time and manner for sowing. Crop transplanting yielded better yields and it is more useful than sowing directly, and ideal period that relocate crop was October 10 and 18. Transplantation should not be postponed until October end. Different time zone for the sowing directly affected the period of harvesting but oil and chamazulene was not affected [23].

39.8 Cultivation

39.8.1 Propagation

The crop can be grown in two ways: directly and indirectly. Sowing and transplanting the seed moisture levels in the environment. A good field for direct seeding is required [23]. The maximum yields of flowers were obtained by crop transplantation at 15, 20, and 30 cm spacing, according to crop design research. The best yields of oil content were found under 30 cm2 spacing, according to Dutta and Singh. A greater spacing of 40 cm^2 is recommended for cultivars [24].

39.8.2 Planting

Chamomile seeds are so minute, seeding is done with a special sowing machine. Seeding is usually done in rows. The seedbed must be level and weed-free before sowing and recompressed when seeding is done. The image compression is necessary and soil rolled after finishing the seeding. The chamomile plant requires a lot of moisture during its germination. Germination begins after 2 weeks. If seeding finish in the autumn, September is the best month. The biggest yields come from chamomile sown in the autumn. The flower begins in the May and June in Central Europe. In March and May, planting takes place. Around the second half of July, the crop can be harvested. The plants are chopped and after the harvesting, it remained on the field [25].

39.8.3 Manuring

The addition of organic matter to soil raises its humus, which increases crops yield. Nitrogen application at a higher rate resulted in a reduction in the chamazulene percentage. Phosphorous and potassium applied before sowing in the autumn and Nitrogen in the spring was necessary for good crop. Singh discovered plants responded well to N and P fertilisers in saline alkaline soils. Content of oil have less influence among the herbicides but remarkable differences were found in quantitative composition [26].

39.8.4 Irrigation

Watering during the bloom phase helps in increasing flower production. Irrigation improved the production [27]. The crop is irrigated on alkaline soils with 5–7 irrigations necessary during the cycle of crop. If the soil is kept wet, good crop yield can be obtained.

39.8.5 Weedling

For a good crop, 3, 4 weeding's are typically needed. Weeds were controlled for four weeks. Herbicides such as atrazine, propyzamide, chloropropham, mecoprop, trifluralin, and linurones, provide weed control [28].

39.8.6 Diseases and Pests

The crop was found to have an outbreak of Alternaria spp. leaf damage. The sickness was managed with a 0.1% Benlate spray. M. chamomilla was being eaten by black bean aphids (Aphis fabae). M. chamomilla flowers were lost by the insect Nysius minor and the plant was defoliated by Autographa chryson. Aphid infestation on chamomile was mitigated with a single application of fosfothion 0.2%. In the dehydrated herb of chamomile, methyl bromide in the Ephestia elutella Hb was effective as an antiseptic against the pest Ephestia elutella Hb [29].

A well-known plant hormone signal molecule called salicylic acid (SA) is involved in a variety of growth responses and disease resistance [30]. Some plant species, including wheat, soy, and maize, have shown evidence of stimulation of growth following SA exposure [31] by promoting highly branched metabolic responses, it can also improve stress tolerance [32].

Exogenous SA's effects are influenced by a number of variables, including the variety and stage of development, the application method, and the SA concentration [33]. There are many known physiologic responses of plants to SA treatment [34]. demonstrated that exogenous SA administration at low concentration influenced the synthesized procedure of coumarin-related chemicals in chamomile. At doses of 50 and 250 M, respectively, SA was found to have both growth-promoting and development effects. The latter is associated with a drop in water content, soluble proteins, and chlorophylls. Based on the fast rise in PAL activity, which is following by a rise in total soluble phenolic content and lignin buildup, it appears that the larger SA dose has a detrimental effect on phenolic metabolism. At a dosage of 250 M SA, GPX activity increased. With continued exposure to SA, however, PAL activity reduced, demonstrating its inhibition. SA had no effect on the accumulation of coumarin-related substances (umbelliferone and herniarin), but at 250 M SA, the rosettes produced more (Z)- and (E)-GCMAs [34].

39.9 Chemistry

39.9.1 Bio-chemical Active Constituents

German chamomile's flower heads are its most precious component economically because they contain essential oil, which is made up of numerous significant biochemical substances. However, aroma-therapeutic proposals also use the leaf extract in this plant. German chamomile's chemical makeup changes from location to location according on the texture of soil environmental factors and various genotypic backgrounds [35] More than 120 substances of secondary metabolites were found in some early investigations on the German chamomile [36].

Twenty-eight types of Terpenoids and 36 types of flavonoids are regarded as the two most significant component classes among these active chemicals. A thorough investigation of the main chemical components, their makeup in blue oil, and the demographic origin of chamomile.

39.9.2 Essential oil Extraction and Constituents

German chamomile flower heads that are still fresh have premium oil quality and then flower heads that have been dried, which may be because drying lowers the amounts of chamazulene and bisabolol [37]. Maximum essential oil content was obtained under lower and shady drying temperature but minimum essential oil content was obtained with high fame and temperature for drying higher. Microwave drying and oven drying, respectively, were used to get the highest and lowest chamazulene content.

After the flower petals are almost fully flowered, the essential oils of chamomile can be harvested by steam distillation or hydro-distillation [38]. Frabboni et al., Using a solar thermal method assisted by microwaves and a subcritical CO2 fluid extraction technique [39]. The simplest and least expensive distillation technique uses a Clevenger-style distillation device and is called hydro-distillation. In a round-bottom flask, flower heads are completely submerged in water and cooked for 4 h. Fresh flower heads are hydro-distilled to reveal essential oil constituents such (Z)—farnesene, D-limonene, and -bisabolol oxide A [40]. The hydro-distillation of dry flowers of chamomile and analysis of the resulting essential oil revealed the presence of the essential oil constituents azulene-7-ethyl-1,4-dimethyl, bisabolol oxides A and B, and isobornyl isobutyrate [41]. which were identifies as essential oil compounds, in chamollia α-bisabolol oxide A and B, 1,8-cineole, α-terpineol and germacrene D are compounds [42]. Flower of chamomollia that were grown in Egypt used for study the essential oil and analyzed the remarkable presences of α-bisabolol oxide-B, chamazulene, bisabolol oxide A and bisabolol oxide as main compounds [43]. German chamomilla flower used for hydro-distillation extraction where α-bisabolol oxide A and chamazulene were identified [39].

Several distillation techniques, including batch steam distillation, continuous steam distillation, and quasi-continuous steam type distillation, have been recorded. However, in underdeveloped nations, batch steam distillation is the most widely utilized method. Moreover, the flower head of chamomile organically dried for the extract the essential oil from the plant. The batch steam method's longer heating times, however, led to higher azulene and lower bisabolol levels in the extracted oils. In order to prevent prolonged water contact, the standardized distillation duration of water. The substance needs to be thoroughly crushed before distillation can be done. When using fresh German chamomile, the average moisture level not more than the 80% and when utilizing the extraction leftovers, it should be, on average, 60% (water-alcohol). According to reports, chamazulene is not a true component of German chamomile; rather, it is created during the steam distillation process as a result of dehydration, and decarboxylation [44]. High-quality chamomile essential oil contains chamazulene, which gives it its dark blue colour [45]. There are two types of techniques used for extraction the essential oil from the flower heads of German chamomile. The solvent extraction and traditional Clevenger distillation procedures were used to extract oils. For solvent extraction, solvents including hexane, acetone, dichloromethane (DCM), ethyl acetate, and methanol were used. DCM produced the highest yield of essential oils of all of them [46].

Other developing method for chamomile extraction is CO_2 fluid extraction and extract oil shows the properties of matricine, α-bisabolol and farnesene [47]. In water extraction method for extraction of oil in plant chamomile was studied and noted that a temperature at 150 °C and flow rate at 4 mL/min for 120 min shows pure oil extract with a vital range of oxygenated compounds [48].

Subcritical CO_2 fluid extraction is a different quick technique created for chamomile extraction, and presences oil.

The benefit of distillation, however, is that it enables a nearly full extraction of the other parts of the plants. In addition, extracting the German chamomile essential oil is quite difficult. Hence, only about half of the oils in the product that include bisabolol and chamazulene are recovered in the extract with the sensitive spiroethers when essential oil is extracted using 45% ethanol. The essential oil contains a high concentration of thermos labile spiroethers as a result of steam distillation. Flower head and stem of chamomile plant are the preferable plant pieces for steam and aqueous distillation. The ideal German chamomile plant components for extraction are the flower heads. As stated above, an aqueous organic solvent such as ethanol, isopropanol, methanol, etc. is preferred when extracting German chamomile flower heads, whether they are fresh, dried, or deep frozen. Preferably, the water content should be at least 20% (v/v). With a high concentration of Spiroethers, German chamomile oil can be produced in this manner. The extraction technique should be economically feasible and make the greatest use of the German chamomile leftovers.

Together with increasing the yield resulted in significant cost savings on the raw materials used to make extracts, lowering the entire cost of operation. German chamomile extract formulations can be made using the acquired essential oil without further purification. Traditional mobile or fixed distillation vessels can be used for distillation.

39.9.3 Ethan Pharmacological Properties and Uses

The above-mentioned active components of German chamomile's blue essential oil have used in more than 20 medicine formation that make it star herb. There are many Anti-difference effects on cancer, microbial, allergic, inflammatory a hepatoprotective. In a dose-dependent manner, chamomile extracts can control tumors angiogenesis by down-regulating it [49]. Traditional home remedies for a number of gastrointestinal problems, bodily aches, and stress reduction also include the use of chamomile tea. Moreover, shampoo, other hair care products, and bath soaps all contain the essential oil as a medicinal and aromatic ingredient. In the food sector, chamomile extract, essential oil, and dry flower heads are also used to give food items flavor, color, and taste. It is interesting to note that a recent study on chicken feed revealed that adding 1 g of chamomile flower powder per kg of feed during the laying season increases productivity, net income, and relative economic returns while decreasing feed consumption [50].

39.9.4 Diseases and Pests

The crop was found to have an outbreak of Alternaria spp. leaf damage. The sickness was managed with a 0.1% Benlate spray. M. chamomilla was being eaten by black bean aphids (Aphis fabae). M. chamomilla flowers were lost by the insect

Nysius minor and the plant was defoliated by Autographa chryson. Aphid infestation on chamomile was mitigated with a single application of fosfothion 0.2%. In the dehydrated herb of chamomile, methyl bromide in the Ephestia elutella Hb was effective as an antiseptic against the pest Ephestia elutella Hb [29].

A well-known plant hormone signal molecule called salicylic acid (SA) is involved in a variety of growth responses and disease resistance [30]. Some plant species, including wheat, soy, and maize, have shown evidence of stimulation of growth following SA exposure [31] by promoting highly branched metabolic responses, it can also improve stress tolerance [32].

Exogenous SA's effects are influenced by a number of variables, including the variety and stage of development, the application method, and the SA concentration [33]. There are many known physiologic responses of plants to SA treatment [34]. demonstrated that SA administration at a concentration of 2 mM influenced the accumulation of coumarin-related chemicals in chamomile. At doses of 50 and 250 M, respectively, SA was found to have both growth-promoting and development effects. The latter is associated with a drop in water content, soluble proteins, and chlorophylls. Based on the fast rise in PAL activity, which is following by a rise in total soluble phenolic content and lignin buildup, it appears that the larger SA dose has a detrimental effect on phenolic metabolism. At a dosage of 250 M SA, GPX activity increased. With continued exposure to SA, however, PAL activity reduced, demonstrating its inhibition. SA had no effect on the accumulation of coumarin-related substances (umbelliferone and herniarin), but at 250 M SA, the rosettes produced more (Z)- and (E)-GCMAs [34].

39.9.5 Harvesting and Yield

Flowers can be collected during the prime flowering season, which lasts 3–6 weeks in March–April. Flowering is so abundant that at least 30–40 units will be necessary to pluck the blooms from 0.25–0.3 ha area about every other day. Flower plucking is a process that is selective because the plants produce flowers at all phases, including buds, moderately buds and flowers in all phases of blossom. February onwards, flowers will be harvested. In North India, the peak plucking season is between the second and third weeks of March. Singh reported a production of 7637 kg flowers in regular soils, that contained 3500–4000 kg/ha yield. It is challenging to obtain regular updates of statistics regarding the demand and supply, collecting, and economics of chamomile in India due to the disorganised nature of the country's medicinal plant industry. Additionally, cultivation at small scale and typically do not mention chamomile separate from herbs make it impossible to isolate global production estimates. The annual yield of dried flowers from industrial farming was estimated in 1995 to reach over 500 t. This amount increased to 1000 t of dried flowers annually in 1998 due to industrial farming.

39.9.6 Chemical Constituents

More than 120 secondary metabolites identified in flower of chamomile [51]. Blue essential oil extract from the flower. Essential chemicals such as terpenes, bisabolol, farnesene, chamazulene, flavonoids, apigenin, quercetin, patuletin, luteolin and coumarin are the part of essential oil [2]. Flower of chamomile plant contain 0.3–1.5% oil [22].

Chamomilla contain not only herniarin, umbelliferone but also some other minor coumarins [52]. Herniarin is source of glucoside where GMCA used as natural chemical in chamomile. Eleven types of phenolic extract from the chamomile such as hernarin, quercetin, rutin (flavonol) and finally naringenin (flavanone). Chamomile have 52 types of chemicals other than 28 types of terpenoids and 36 flavonoids [53].

Blue oil is found naturally in German chamomile (essential oil). Flower and head of flower mostly used for the production of oil. Chamomile flower oil is made up of sesquiterpene, with low amount of monoterpenes. Polyynes made 20% of the oil. Chamazulene is a derivative of the hydro jndistillation which is present in flowers. The quality of the oil is obtained by its shade. Sesquiterpene is responsible for oil's blue tint. The amount of chamazulene varies depending on the native area and chamomile plant age. Quality of the essential oil indicated by the blue colour of the oil. The presence of flavonoid, terponoid, primarily chamazulene and -bisabolol chemically indicated the colour and better quality of oil. During the storage of the flowers, it reduces [54]. Bisabolol is used in cosmetic preparations [55].

For chamomile extracts that used for antiphlogistic actions, plants high in bisabolol and chamazulene will be used. Bactericidal components include bisabolol, cyclic ethers [3]. fungistatic umbelliferone, antiseptic chamazulene, and antiseptic–bisabolol [56]. The presence of bisabolol and its oxides in the flowers has emerged as a key determinant of analyze the quality of drug and value as effective technologies for identifying the drug's ingredients and effectiveness have been developed. Qualitative and quantitative makeup of the essential oil, four fundamental varieties of chamomile—A, B, C, and D—are thus distinguished [57].

39.10 Genetic Improvement

39.10.1 Breeding Challenges

German chamomile currently comes in a number of diploid and tetraploid forms, however the most of them are collection of old local cultivars or may be mixed diverse population, that's why theses cultivars have less advance for mechanical harvesting. Furthermore, little is known about the the yield of dry flower also produced by mechanic harvesting. Additionally, standard plant-breeding methods for improving the genetics of German chamomile are hampered by their size of flower, seed cracking and irregular anthesis flushes. Nonetheless, in chamomile intra- specific self-incompatibility and pollen sterility were marked in diploid cultivars [58].

39.10.2 Reproductive Biology

The flower head of the German chamomile comprises of pistilate ray florets and bisexual disc florets. The slightly winged ray florets have a white corolla. Ray florets have deflexed laminas that are 6–15 mm long and 2.5–3.5 mm broad ([59]. The number of disc florets is large, the number of ray florets ranges from 16 to 24, and tubular shape with greenish yellow, 1.5 to 2.5 mm while 0.5 mm long and wide respectively, and has five teeth at the tube's end [60]. At the time of anthesis, anthers are longitudinally dehiscing and linearly bi-lobed. German chamomile's anthesis might continue for up to 10–15 days. At the first flush of anthesis, the anthers dehisce, which is followed by stigma receptivity flushes in the ray and disc florets. The stigma of newly flush mostly moves upper side that's are the flower head tip, but it has occasionally been seen to begin close to the center and proceed in both directions. German chamomile has tiny (2–3 mm) florets that prevent hand emasculation. Yet, by repeatedly removing the female bud's disc in a crossing pattern, polygamy may aid in controlled pollination.

39.10.3 Genetic Diversity

German chamomile genetic diversity evaluation studies are extremely rare. In order to test 24 different characteristics of German chamomile, Solouti used a Duncan test. According to the test no of flower's head per plant, yield and oil concentration had the highest coefficients of variance, whereas the diameter and height of the flower heads had the lowest diversity variance coefficient. Round about 369 band were marked and studied by using the 29 RAPD primers, also study the diversity of chamomile among 25 Iranian populations. Of these, 314 (85.09%) were determined to have high polymorphism [61], Additionally, 93.18% polymorphism (205 fragments out of 220) was observed in a study of 25 chamomile population employing 18 RAPD markers [62]. Moreover, 18 RAPDs and 3 AFLP molecular markers to examine the genetic diversity in di- and tetraploid German chamomile and classified all of the germplasm into two main groups according to the provenance of the germplasm [63]. Genetic diversity in tetraploids is less than as compare to diploids utilizing. Using five ISSR primers, an ISSR analysis, which was determined to be the most cost-effective among PCR-based markers, was used to detect 85.4% polymorphism in 15 genotypes of chamomile [64]. Nei's genetic diversity index (h) has also been used to determine the average genetic diversity in Iranian chamomile [65]. Range of this index 0.16–0.36 and 0.23–0.5 respectively [62]. Differences between five commercial types and eleven native Greek populations. The researcher revealed that the instinctive populations in the area had a high variance in a-bisabolol and chamazulene levels and were more stable for phenotypic features; hence, these populations might be exploited in the future for breeding [66] (Ahmadi et al., 2014) Using 10 ISSR primers to analyse the genetic diversity of 23 chamomile populations, four main groupings were identified according to the geographic distribution of the germplasm. In terms of phenotypic features and interests, the survey found that Europe and Iran native plant populations are the most diversified [67]. The Esfahan population had a greater percentage of essential oils and oil output [68]. Plant height and bloom productivity varied greatly between Italian populations [69]. By a D2 study of dry flower production, blossom diameter, disc floret periphery, disc floret height, and the number of flower heads per plant, according to Singh et al. a comprehensive range of genetic variability by D^2 analysis of yiels and diameter of flower, number of flower heads, and disc floret height [70]. Chamomile polygenic traits is a unique variation generated by gamma radiation [71].

Germplasm organization at genomic level on the base of genetic diversity classification, selection at an early stage, and selection for harmful traits, among other tasks, plant breeders found that molecular markers were the most reliable way. More precisely, molecular markers from chamomile could aid in determining the genetic diversity of the plant and early selections for features related to its flowering. German chamomile's RAPD, AFLP, and ISSR molecular markers have received the most attention to date [67]. Nonetheless, a few studies also discussed the utility of SNPs in German chamomile population structure and gene mapping [72]. Ahmadi et al. [67] demonstrated that the morphological trait-based categorization of the germplasm was not consistent with an ISSR molecular marker-based classification and that there was a weak association between the examined primers and phenotypic attributes. Using bulk segregating analysis, Wagner et al. [63] discovered three AFLP-markers connected to the—bisabolol locus (BSA). They also discovered one RAPD marker and 17 AFLP markers connected to the presence of chamazulene in the blue essential oil. These molecular markers could be applied to German chamomile pre-flowering selections.

39.10.4 Breeding Approaches

Little flies are the main pollinators of German chamomile, an allogamous crop. It is crucial to have a higher dry flower production, synchronous blossoming, a high oil content, and better blue essential oil. Understanding the crop's breeding behaviour is necessary to comprehend the systemic implications of conventional breeding methods. According to earlier studies, selfing and open pollination conditions resulted in 82–100% seed production [58]. Reports on the evolution of advanced generations and trait inheritance, however, are rare. Most of the features in this plant's heredity have not yet been researched. A single recessive gene for ()-bisabolol and chamazulene was discovered through the study of a few features that govern chemical components and contribute to essential oil quality [63]. With the direct and indirect selection of component features like the quantity of flower heads, the length of the flowering period, the size of the flowers, etc., the flower yield and oil content may take quantitative traits into consideration and be enhanced. According to a study in this area by Golparvar and Pirbalouti [73] indirect selection for flower heads per plant and days to 50% blossoming could increase German chamomile's dry flower output and essential oil percentage. They showed that the relationship between oil content and dry matter is positive.

The number of flower heads and flower yield were both positively correlated with oil content, according to Pirkhezri et al. [62] demonstrating the possibility of simultaneously improving oil content and flower yield in a breeding programme. Conversely, oil content was negatively correlated with days to maturity, highlighting the significance of the early and ideal flower harvesting stage in chamomile [68]. The most popular techniques for varietal development and improvement for flower production and oil in this crop, in particular, a selection from local landraces and polyploidy breeding. Using the application of colchicine to German chamomile seeds, researchers created tetraploids [74]. Tetraploid cultivars have larger flowers in terms of both weight and diameter [69]. Height of plant and length of flower pick increase in tetraploids which causes difficulties in harvest by machine [75]. The availability of any instruments for hybridization, such as male sterility and self-incompatibility, etc., is necessary for the production of triploid chamomile hybrid seeds. On the basis of oil %, diploids and tetraploids were shown to be statistically comparable [69]. Albrecht et al. [76] combined 30 lines from progeny test-crossing based on single plant selection for up to four generations (L1L4 generations).

German chamomile has been used to build a gene transfer technique that makes use of cationic carbon nanotubes (CNTs), which have a strong affinity for DNA and effectively shield it from destruction caused by digestion and ultrasound enzymes. The simultaneous use of CNTs and ultrasonic dramatically enhanced the ability of nanoparticles to transfer ssDNA-FITC, as shown by the fluorescence microscopy.

The availability of triploid chamomile hybrid seeds, however, Higher terpene synthase expressions correlated with sesquiterpenoids in German chamomile, identifying a more sesquiterpenoid concentration, according to gene co-expression networks analyses of the terpenoid biosynthesis pathways in German chamomile (M. Chamomilla) and Roman chamomile (Chamaemelum nobile). Because German chamomile has 10 times more unigenes related to plant-pathogen interactions than Roman chamomile, its volatile chemicals are affected by unigenes that were considerably enriched in pathways connected to plant-pathogen interactions, increasing its resistance to pathogens [77]. In German chamomile (M. chamomilla L.), three aphid alarm pheromone genes were cloned and functionally analysed to examine the in vitro and in vivo activities of three sesquiterpene synthases. A technique for transforming hairy roots was also created, and it was shown that excessive sesquiterpene synthase gene expression caused -muurolene to accumulate in hairy roots. The role of aphid alarm pheromones in the production and control of volatile terpenes is the subject of molecular research. In early flowers, qPCR indicated high expression of these genes, which also shown a stronger association with the concentrations of the essential oil components (E)—farnesene, −elemene, and germacrene D [78]. Three main conventional breeding methods for improving German chamomile were mass selection, polyploidization breeding, and mutation breeding using chemical radiation mutagens. There are now proven biotechnological methods for generating genotype fingerprints and examining the connections among species

[72]. In order to prevent overharvesting from the wild and the depletion of genetic resources, it may also be possible to conserve germplasm at seed banks and field gene banks. Future research on German chamomile needs to pay close attention to a number of critical areas, including uniform blooming, equal pick length, genotypes with higher essential oil content, and increased production from the use of relevant agrotechnologies. To enhance secondary metabolites, several researchers are trying to genetically modify the plant's growth process [79].

39.10.5 Trade and Adulteration

M. chamomilla is related to Anthemis cotula L., a poisonous plant with an offensive odour that makes identification difficult. Three species named Matricatia recutita L., Anthemis nobilis L., and Corchorus depressus L.are mentioned in the Unani system of medicine under the same vernacular name, "Babuna," which caused confusion when the species were identified. By examining taxonomic and anatomical components, Ghauri et al. [170] seek to resolve the species ambiguity and discover that M. chamomilla is the rightful owner of the term "Babuna," whereas A. nobilis L. and C. depressus are utilised as adulterants [5]. Many techniques were created using near infrared spectroscopy and chemometrics to identify, measure, and authenticate some of the typical hazardous adulterants in German beer. Mahgoub et al. [80] The orthogonal projection of latent structures models and a soft independent modelling of class analogies were utilised to authenticate samples with 100% sensitivity, but partial least squares models demonstrated strong predictive power for each adulterant in adulteration mixes individually. A common component of herbal teas, chamomile tea is made from the dried flower heads of German chamomile, however adulteration is a significant production problem. In a comparison of chromatographic methods for evaluating adulteration in accordance with the European Pharmacopoeia, it was discovered that HPTLC was more effective than HPLC for the detection of adulteration [81].

39.11 Medicinal Properties of Chamomile Plant

Using databases such as a Espacenet patent search, Canada patents, Google patents, SciFinder, etc., a study on patents on German chamomile was undertaken in the European Patent Society and World Intellectual Property Organization from 1993 to 2015. The following is a list of patents relating to German chamomile's therapeutic uses and methods of extraction.

39.11.1 Extraction of Sub-Critical Water

The process described in the invention was effective for obtaining German chamomile and other medicinal plants'extracts with good pharmacological properties without the need for an organic solvent. Subcritical water was used as part of an extraction technique to create an extract with a comparable structure to that seen in methanol or alcohol combinations. In order to increase the bioavailability and effectiveness of the medicinal components of the extract, the invention additionally comprised oral and topical preparations [82].

39.11.2 Production of High Natural Spiro Ethers from Oil Content

Steam and water distillation method making chamomile oil that contains at least 1% of natural spiro ethers by weight was created. Natural cis and trans spiro ether were discovered to contain at least 1 and 0.5% of the total weight, respectively [82].

39.11.3 Extraction for the Treatment of Hypertension

A composition for treating hypertension, its symptoms, and disorders linked to it was created through the method. The only active principle ingredient in the extract is chamomile plants [82].

39.11.4 Endotoxin Reduced by the Oil Extraction of Chamomile and Black Cumin

It has been discovered that a steam-distilled chamomile extract can lower DNA synthesis in human cancer cells and suppress the generation of leukotrienes and IL-6 (interleukin 6). Whether used separately or in conjunction with black cumin seed oil, chamomile's volatile oil possesses anti-inflammatory and anticancer qualities [83].

39.11.5 Flower of Chamomile Have Properties of Acaricidal

Chamomile flower have the fraction of ethanoic, hexane, and chloroform with acaricidal characteristics, a secure and reliable extraction technique was created. Using 8–10% of the extract, oviposition of Rhipicephalus (Boophilus) microplus was shown to be 80–90% lethal and 63–100% inhibited. By utilising 4–5% of hexane, chloroform, and ethanolic extract, respectively, against R. (Boophilus) microplus, an efficacy of above 90%, 95%, and 95% was demonstrated. A death of deltamethrin-resistant ticks was also discovered in the invention using an ethanolic extract (60–70%) of chamomile flowers.

39.11.6 Treatment of Skin by Oil Blend

The procedure produced a blend of oils that was largely anhydrous including German chamomile oil, which is beneficial for treating skin conditions. German chamomile oil was effective for the skin diseases treatments and prevention of skin ailments, like diaper rash, eczema, and in soothing the skin. It was combined with other oils, such as virgin coconut olive, jojoba tea tree vitamin E and calendula oils [84].

39.11.7 Extract of Chamomile and Nettle Used for Hemostatic Dressing

The procedure required making a hemostatic mixture from flower of chamomile and dioecious nettle leaves. A biocompatible polymeric base (alignate) was included in the formulation to create a broad spectrum hemostatic extract with coordinated antiseptic and anti-inflammatory effects. The innovation also provided techniques and tools for application in various hemostatic situations, as well as in inventive compositions to halt bleeding. Hemostatic dressings are made up of formulation components applied as a polymeric layer to a fabric [85].

39.11.8 Medicinal Uses

Chamomile is a medicinal plant that has been known as "the star of therapeutic species".It is most abundant and widely used therapeutic herb in number of medicines. Many past years and scientific us confirmed its multitherapeutic, appreciative and nutritious values. The powdered medicine can be used to treat slow-healing injuries, skin ulcers, illnesses including herpes and boils, haemorrhoids and irritation of the skin. It is efficient for swelling of the urinary system. It is used to cure irregular menstruation [86]. German chamomile can be used for a sore stomach as a herbal medicine, skin care and bowel syndrome. Pills of chamomile are used. Herbal tea of chamomile is used. Oil extract from the flowers that are (farnesol) E-farnesene terpene alcohol, chamazulene, bisabolol oxides A and B. Bisabolol reduce the activity of pepsin proteolytic enzyme in stomach but not affect the stomach acid, that is useful for gastric and intestine illnesses. It induces epithelialization and granulation. The drug chamomile is classified in the pharmacopoeias of 26 different countries. It is found in many traditional, unani, and homoeopathic medicines [87]. It's used to treat cramps, gripes, frenzyand fever as a medication [10]. It has anti-inflammatory and antibacterial properties. For moderate sedative and digestion, it is very useful. It is antispasmodic and sudorific [86]. Chamomilla is consider as the member of medicinal herbs family. Tisane was used for stomach problems, digestion, and vomiting.

The powdered form of medicine used to treat wounds, infections and enanthem. In Europe, chamomile flower products are collected and used as medicine for many diseases [88]. The medication industry today offers products with a variety of origins and therapeutic benefits. The content of chamazulene and oil concertation of the plant material, among other factors, were used to assess its therapeutic efficacy [3]. Only those chamomile varieties with wide range of bisabolol and the synthetic racemic bisabolol should be used to produce chamomile extracts with antiphlogistic effectiveness. As a result, chamomile with a certain chemical make-up is utilised as medicine because it has a particular pharmacological activity [89].

39.11.9 Antibacterial Properties

Many investigations have been conducted to explore the antibacterial effectiveness of chamomile EOs and extracts. In general, technique of agar diffusion, which employs discs or wells, is the most commonly employed to test the bactericidal activity of EOs and extracts. Stanojevic et al. [35] used this approach to report the activity of antibacterial in chamomile extract. Staphylococcus aureus was the most susceptible strain, while Pseudomonas aeruginosa was the most resistant. Similarly, employing the disc diffusion approach, Owlia et al. [90] found no efficacy against P. aeruginosa. The EO, on the other hand, was able to diminish biofilm formation and alginate synthesis, demonstrating its efficacy in reducing biofilm-producing bacteria. Bacillus subtilis, on the other hand, was shown to be the most susceptible bacterium to both diffusion and dilution procedures. In addition, the EO had the smallest MIC and MBC values against S. aureus and the largest inhibition zone against B. cereus [91]. According to Silva et al.'s findings, less MIC values in G. positive bacteria [92]. While G. negative bacteria have more rigid and complex membrane with lipopolysaccharide that inhibit the access of antimicrobial molecules, these results could be explained by differences in the structure of the cell wall [93]. M. chamomilla EO has also been shown to be effective against a number of Streptococcus species [94]. Additionally, the commercial EO demonstrated activity against a number of Gram-positive bacteria [95]. The findings demonstrated that, in comparison to reference strains, the EO was more effective against S. aureus MRSA. P. aeruginosa, on the other hand, was the most sensitive bacteria in both micro-dilution and diffusion assays [96]. This suggests that Gram-negative strains of bacteria may be more sensitive to EO than Gram-positive strains. Furthermore, Shakya et al. [97] discovered that M. chamomilla flower extract significantly inhibited the growth of Enterococcus faecalis. Surprisingly, Satyal et al. M. chamomilla EO had the strongest activity against both S. aureus and P. aeruginosa, according to [98]. In addition, no activity was found against P. aeruginosa, but Hartmann and Onofre [99] reported the lowest MIC values against Escherichia coli. When the EO was used without dilution (100%), the largest inhibition zone was observed against S. aureus. Höferl et al. found that plant origin and other factors can affect EO antibacterial activity [100]. In fact, south African and Hungary produced the highest ativity of EO against the bacterial activity. Roby et al. compared the antibacterial properties of chamomilla EO and other four extracts [101], found that EO was the most effective against all of the bacteria tested. As the concentrations of EO and extracts increased, so did the antibacterial activity. Lowest antibacterial activity reported in diethyl ether extract, indicating that extraction solvent influences activity. In a similar vein, according to Ismail et al. [102] different concentration of methanol and ethanol that was tested were only effective against S. aureus at the highest concentration. At all concentrations tested, the aqueous extract was also ineffective against all strains. Contrarily, the aqueous extract was the most potent against the majority of the strains examined by El Mihyaoui et al., [103] with the exception of P. spp., which responded to methanol extract more strongly. Abdalla and Abdelgadir [104] examine the different extract and then they found that petroleum ether extract had the most activity (diameters between 22 and 26 mm), while ethyl acetate extract had no effect on any of the bacteria they tested. On the other hand, Mariod et al.'s [105] study The antibacterial activity of Sudanese and Egyptian M. chamomilla was confirmed by experiments on methanol, n-hexane, chloroform, ethyl acetate, and aqueous extracts. In fact, at different concentrations of Sudanese methanolic extract had the greatest activity against S. aureus (W17). Antibacterial activity of ethyl acetate extract had a great ability as compare to ethanol extract [106]. MBC also varied depending on the percentage of ethanol used (MBC = 125 g/mL with 99.8% ethanol). However, cyclohexane extract lacked antibacterial activity against any of the bacteria tested, while ethanolic extract was only effective against P. aeruginosa [107].

39.11.10 Antifungal Activity

Chamomile plants part and its extract have been documented for the antifungal properties, indicating that they are highly effective against a wide range of fungal strains. The majority of studies looked into how chamomilla EO affected Candida sp. Höferl et al. reported that, EO from Indian cultivated chamomile plant that was rich in bisabolol oxide A 25%, C. albicans growth reduced by the origins of EO [100].This demonstrates the connection between antifungal activity and the different concentration of essential oil. M. chamomilla EO also inhibited fluconazole-resistant and susceptible C. albicans strains derived from HIV-positive patients with oropharyngeal candidiasis, with a stronger effect on susceptible strains [108]. Toward the EO C. albicans is more sensitive as compare to Aspergillus sp. [98]. Additionally, EL-Hefny et al. [40] discovered that EO's antifungal activity was dependent on the dose, with A. niger showing the best results. Mekonnen et al.'s findings on antibacterial activity are comparable to two Aspergillus sp., were not affected by M. chamomilla EO from Ethiopia. two species of Trichophyton [109] Strains AbdoulLatif et al. compared the antifungal properties of M. chamomilla EO and extracts [96]. discovered that EO outperformed methanolic extract in terms of activity. Nevertheless, Roby et al. found that the activity of essential oils (EOs) and various extracts was dependent on dose concentration [101]. Only chloroform had antifungal activity towards C. albicans and Fusarium sp., out of aqueous, methanol, and chloroform extracts [110]. This demonstrates how extracts affect antifungal activity. Hameed et al. conducted another study in found that A. terreus was affected by methanolic plant extract [111], whereas according to Lavanya et al. no effect on four Candida species (C. krusei, C. parapsilosis, C. albicans, and C. tropicalis) [112]. However, ethanolic flower extract also demonstrated antifungal properties [[113]. In addition, Saccharomyces cerevisiae cell survival and growth were significantly reduced by the hydroalcoholic extract of M. chamomilla [114]. Aqueous extracts also demonstrated activity against fungi [115]. However, the antifungal properties of seed aqueous extracts made at varying pHs against A. niger and P. citrinum were identical [116]. Despite having a higher phenolic content, their sulfated derivatives also displayed close antifungal activity. Seyedjavadi et al. in another study [117] isolated AMP1, a novel peptide with antifungal activity against C. albicans and Aspergillus sp., from M. chamomilla. This demonstrates that M. chamomilla has the potential to produce intriguing antifungal molecules.

39.11.11 Antiparasitic and Insecticidal Properties

The ability of chamomilla and its extracts to prevent the insects and parasitic unwanted growth. Tunisian M. chamomilla essential oils'leishmanicidal activity was examined in vitro [118]. The findings demonstrated that EOs had a favorable effect on the promastigotes and that -bisabolol was able to trigger the promastigote's effects of programmed cell death. Hajaji et al. conducted another study, the effect that -bisabolol has on Acanthamoebacastellani has been studied. With an IC50 of 20.83 g/mL and an IC90 of 46.60 g/mL, the findings demonstrated that -bisabolol was capable of increase the permeability of plasma membrane while level of ATP decrease [119]. Nepal origin chamomile were tested for larvicidal, insecticidal, and nematicidal activity against Caenorhabditis elegans and Artemiasalina, as well as larvicidal activity against glassworm (Chaoborusplumicornis) [98]. These organisms were not significantly harmed by the findings. M. chamomilla EOs, on the other hand, were found to be insecticidal against the desert locust Schistocercagregaria third nymphal instar [120]. M. chamomilla EOs were effective at killing the red spider mite Tetranychusurticae (LC50 = 0.65%) [121], but they were less effective at killing the cattle fever tick Rhipicephalusannulatus (LC50 > 8%) [122].M. chamomilla essential oils high in -bisabolol oxide A were found to have effective activity against the parasite [123]. Likewise, Höferl et al. tested the effectiveness of six EOs from M. chamomilla against Aedesaegypti mosquito larvae and adults. The activity of EOs from South Africa was the highest (LD50 = 2.9 ppm), and the results varied depending on plant origin. These results were related to the high concentration of essential oils (EOs) in steroidal spiroethers (12%) [100]. On Acanthamoebacastellanii, flower extracts of Tunisian M. chamomilla were tested for their anti-Acanthamoeba activity [124]. A mixture of coumarins is responsible for the methanolic extract's potent anti-acanthamoeba activity (IC50 = 66.23 g/mL). The egg and adult stages of the gastrointestinal parasite Haemonchuscontortus, which is found in small ruminants, were used to test the extracts from M. chamomilla flowers for their anti-helminthic activity [125]. With an IC50 of 1.52 and 2.55 mg/mL, respectively, methanolic extracts inhibited egg hatching more effectively than aqueous extracts. At the highest concentration tested, 8 mg/mL, the methanolic extract caused 91.77% mortality after 8 hours, while the aqueous extract only caused 75.05% mortality at the same concentration.

Methanol extract, measured at 1024 g/mL, was found to be the most effective anti-helminthic against Haemonchuscontortus in another study [125]. In the egg hatch test, 37.5% of ovicidal activity was observed, while in the larval development test, 84% of larvicidal activity was observed. Al-Mekhlafi et al. tested the combination of hexane extracts from M. chamomilla and Foeniculum vulgare against Culexpipiens for their ability to kill mosquitoes. After 72 hours of exposure, the obtained mixture displayed larvicidal activity with an LC50 of 100.3 mg/mL [126]. Ethyl acetate extract with the concentration of 90% from the chamomile shows more effeteness' toward the culexpipiens after 48 h of exposure [127].

39.11.12 Antidiabetic Activity

M. chamomilla extract, isolated apigenin, apigenin-7-O-glucoside, and cis and trans-2-hydroxy-4-methoxycinnamic acid glucosides were tested for their ability to inhibit beta-amylase and maltase [128]. The findings demonstrated that the compounds and the extract inhibited both enzyme activities in a concentration-dependent manner. The apigenin and apigenin-7-O-glucosides had the greatest inhibition of -amylase and maltase, respectively. Additionally, these two flavonoids were able to regulate sugar absorption and limit the transport of sucrose and glucose. Additionally, chamomile extracts such as hydro methanolic as well as some isolated compounds were found to inhibit rat lens aldose reductase activity [129]. Under high-glucose conditions, O-di-caffeoylquinic acid and luteolin glucuronide inhibited sorbitol accumulation in rat lenses, and luteolin inhibited the synthesis of end product glycation. Additionally, M. chamomillaethanolic extract exhibited anti-glycation properties with an IC50 of 264.2 g/mL for lipa.

39.11.13 Anti-tumor Activity

The anti-tumor effects of M. chamomilla extracts and essential oils on a variety of cancer cell lines have also been investigated. Ali [130] conducted an evaluation which proves that chamomile EO have an anticancer activity on the human breast carcinoma cell line. Cell proliferation inhibit after 48 hours of high dose concentration of EO [131]. It is also reported as, EOs were tested for their anticancer activity against two human promyelocytic leukemia cell lines with high dead concentration towards the NB4 cells as compare to HL-60 cells. The Eos have ability to inhibit both cell growth. In addition, human breast cancer cells were found to be inhibited from proliferating by M. chamomilla hydro alcoholic extracts from aerial parts or roots [132]. Fraihat et al. tested the chamomile methanolic extract on two lines of melanocyte tumor cells of human [133], WM1361A and A375.S2. at the end only the melanotic WM1361A cell line's proliferation was inhibited (IC50 = 25.2 g/mL). another study of Cvetanovic et al., [113] found that the way M. chamomilla extracts were extracted had an effect on how effective they were against cancer. Indeed, the most effective cytotoxic activity was found in subcritical water extracts against the murine fibroblast cell line (IC50 = 19.65 g/mL), Hep2c, a human cervical carcinoma cell line, and rhabdomyosarcoma cell line (IC50 = 30.54 g/mL) [134]. Three different level of concentration the Ehrlich ascites carcinoma cell line's growth was slightly slowed by each extract. M. chamomilla's anticancer properties appear to be linked to oncogenic cells' decreased capacity for migration and invasion, apoptosis, and necrosis [134].

39.11.14 Anti-inflammatory Activity

M. chamomilla extracts have been shown to reduce inflammation [135]. The extraction of ethanolic from the chamomile that inhibit the production of nitric oxides and visibility of cell, whereas on lymphocytes, that increased an anti-inflammatory cytokine, and a decrease in cell viability. In contrast, M. chamomilla aqueous extract was an effective T helper Th2 suppressor by disrupting the balance between Th1 and Th2, resulting in a decrease in nitric oxide production and an increase in cell viability in macrophages. The presence of distinct active compounds could account for the distinction between these two extracts. Singh et al. conducted another study in investigated M. chamomilla tea extract's anti-inflammatory properties. The findings indicated that the extract had anti-inflammatory properties because it prevented protein denaturation and stabilized the membrane of human red blood cells [136]. Animal models were also used to investigate M. chamomilla's anti-inflammatory properties. Wu et al. state that M. chamomilla's extracted oil and water of flower, as well as volatile and non-volatile components, and aqueous extract can all significantly reduce xylene-induced swelling of mouse ears, carrageenan-induced pedal swelling in rats, and celiac capillary vessel permeability in mice. Additionally, they significantly inhibited the carrageenan-induced rise in prostaglandin E2 and nitric oxide in rat pedal edema [94]. In addition, chamomile hydro-alcoholic extraction used to check its effectiveness at inflammatory level of blood indicator in rats [137]. Serum concentrations of tumor necrosis factor (TNF), C-reactive protein (CRP), interleukin 6 (IL-6) and fibrinogen were not significantly elevated when treated with hydro-alcoholic extract. However, ethanolic extract and diclofenac or indomethacin, two non-steroidal anti-inflammatory medications, were found to have interesting synergistic anti-inflammatory effects on rats' paw inflammation and stomach damage caused by carrageenan [103].

39.11.15 Essential oil Compositon Effects by Biological and Abiotic Stress

Quality of essential oil strongly affected by the environmental factors such as light, temperature, moisture, salinity of soil and nutrients [138]. Both normal and high temperature increase the concentration of chamazulene and α-bisabolol in All three types of M. chamomilla on the other hand heat stress did not affect the bisabolol oxide A quality [139]. Medicianal value of chamomilla measured by the concentration of α-bisabolol. Content of α- bisabolol increased by the foliar spray of 5-aminolevulinic acidzzz on chamomilla leaves [140]. Changing in chamomilla morphology and phytochemical properties under salt water irrigation. Nitrogen not only regulates the plant growth but also maintain the plant metabolism [141]. Under low limited supply of nitrogen plant maintains the metabolism by carbon rich metabolites [142] Phenylalanine ammonia-lyase (PAL) is play a vital role in synthesis of flavonoids, benzoic acid and it's also inhibit the phenolic accumulation. Plant growth, chlorophyll concentration and phenolic metabolites with antioxidant activity (umbelliferone) in chamomilla increased with increased the range of nitrogen concentration in field [143]. Heavy metal stress not only reduces the oxidation state of chamomilla but also affects the water balance and capacity antioxidant [144]. Herniasin known as coumarin- related element that is unaffected by heavy metal stress but presences of this compound slightly increase the zinc oxides and GMCAs [145]. Low molecular weight antioxidant (LMAOs) present in high concentration in chamomilla plant that selected from high and medium levels of heavy metal areas [146]. Under biological stress secondary metabolites play a vital role to maintain the defense system of plant and release the terpenoids. Primary and secondary plant metabolites reflect the remarkable changes when they affected from the attacked of insects. Increase in toxic substances and also increase the production of anti-digestion and anti-nutritional compound [147]. Chamomilla directly affected by viruses, bacteria and fungi. It is the host plant for lettuce vein and cabbage black ring virus, these viruses causes injury in the root, shoot, flower, leaves and decrease the quality of essential oil [148].

39.11.16 Compositon of Chamomilla Effects by Geographical Origin

Geographical origin plays a vital role in chemical composition of chamomilla plant. The concentration of α-bisabolol oxide B was more than as compare the oxide A in chamomilla that originated from the Lative and Poland. Artemisinin ketone and (E)-β-farnesene high in concentration of United States originated chamomilla. Chamomilla oil that from Estonian and Polish have high chamazulene content as compare to American samples [149]. Chamomilla with α-bisabolol, α-bisabolol oxide A and B, (E)-β-farnesene, are the main component of Estonian originated plant [150].

39.11.17 Chamomile Composition Effected by Seedling Aging

The volatile oil components of M. chamomilla from various seedling ages and geographical locations showed substantial variations, While other terpenes including-ocimene, citronella, epoxybutane, and germacrene D fluctuated greatly during seedling growth [151]sesquiterpene, −caryophyllene, and -farnesene were often dominant in the first 30 days of seedling development in M. chamomilla. This growth procedure did not produce any -bisabolol or its oxides, in contrast to Mohammad et al. [151]

The formation of a significant number of sesquiterpenes at the start of chamomile seedling development may reduce the generation of ROS due to changing in physiological conditions. After 30 days, plant of M. chamomilla in Manzana and Pnos started producing spiroethers, whereas those in the other five locations continued to build up significant quantities for the first 40 days. A young plant can respond to many types of stress better than an older plant since it has a stronger defense mechanism [152]. Total metabolites level significantly lower in stem leaves bud and flower at seedling stage of plant chamomilla as compare to mature plant. The synthesis of secondary metabolites is gradually decrease in stage of germination, and as seedlings grow, primary metabolite production may become more focused on growth and development. The complex terpene synthesized in vegetative part of chamomilla while it did not contain any type of –bisabolol and there were notable variations in the production of terpene metabolites [148]. All interactions within the ecosystem were impacted by the unpredictability of their metabolite production. No matter what type of M. chamomilla you have, the stems and leaves lack -bisabolol and its oxides, and the immature organs have more secondary metabolites than the mature leaves or stems do [153].

39.12 Regions Differences Shows the Metabolic Differences in Mature Varieties

Generally speaking, the quantity of metabolites varies greatly amongst germplasms. Sesquiterpene was discovered in stem and leaves of plant as the primary metabolite, while the content of the sole monoterpene, −ocimene, varied significantly among different germplasm resources. The leaves have 13 types of sources for germplasm, with the exception of Bodegold and Germania, were higher [148]. Aristolochia, longifolene, and -terpineol are typically found in high quantities in the stem whereas sesquiterpene dicycloguimarene, −farnesene, and -farnesene are typically found in high concentrations in the leaves. The control of sesquiterpene is separated in both leaves and stem. Sesquiterpene concentrations were high in the Bona stems as compare to the leaves of Goral and Lazur. The concentrations of -caryophyllene, −farnesene, and germacrene D were higher in the Bode gold, while the amount of bicyclic germ acrene in the stems and leaves varied significantly in chamomilla species. The amount of -farnesene in the leaves was higher than it was in the stems in Bona, Germania, Lutea, and Pnos, but ZlotyLan had a greater differential. In general, the high concentration of Spiro-ether in leaves of Bodegold, Bohemia, and Goral. The stem concentration of of Spiro-ether much lesser as compare to leaves in Lazur, Lutea, and Margaritar. Bohemia, Goral, Lazur, and Lutea had considerably varied amounts of -gurjunene, aristolene, and longipinene in the stems and leaves. Leaves have higher concentration of bi-cyclic ether as compare to other parts of the plant, indicating that there are considerable differences between species in in the stems [148]. Sesquiterpene makes up the majority of the essential and coloured oil elements in floral buds and blooms. Monoterpenes including -ocimene and terpinene were discovered in the examined samples, but only a small number of samples included sabinene, −geranene, and -pinene in even lower concentrations. The concentration of monoterpenes varies from person to person and is often highest in flowers. Spiro-ethers and di-cyclic ethers are more prevalent in flower as well as in buds than other monoterpenes and sesquiterpenes, and they are more prevalent in flowers than in buds. Of the two ethers, Margaritar contains the most [148] With the exception of Bohemia and Bona, −bisabolol and its oxides have quite different analytical properties. The concentration of -bisabolol oxide A in flowers is heigher as compare to buds.The buds of chamomilla did not contain any -bisabolol oxide A, although Argenmilla, Goral, and Pohorelicky velkovety did have a minor amount of bisabolol oxide B. With the exception of Bohemia, flowers contained more bisabolol than flower buds did [148].

Research Perspectives: M. chamomilla, commonly referred to "star of herbal medicine," it is one of the top most popular medicianal herb, it also a significant ingredient in food, medicine, and cosmetics. It has significant developmental and commercial worth as a medicinal and aromatic plant that can withstand stress. In contrast to Western nations, M. chamomilla study and interest in China began much later, and the majority of both domestic and international studies on the plant concentrate on its pharmaceutical properties. Only a few research has revealed the flavonoids and other components of M. chamomilla, despite the fact that many terpenes have been separated and identified from the plant. Future research should advance understanding of M. chamomilla's synthetic regulatory mechanism and provide a theoretical foundation for enhancing both its therapeutic and commercial potential.

References

1. Blamey, M., Fitter, R., & Fitter, A. (1978). *The Wild Flowers of Britain and Northern Europe*. Book Club Associates.
2. McKay, D. L., & Blumberg, J. B. (2006). A review of the bioactivity and potential health benefits of chamomile tea (Matricaria recutita L.). *Phytotherapy Research: An International Journal Devoted to Pharmacological and Toxicological Evaluation of Natural Product Derivatives, 20*(7), 519–530.
3. Issac, O. J. P. (1989). Recent progress in chamomile research-medicines of plant origin in modern therapy. *Prague, Czeco-Slovakia, 7*, PMC3210003.
4. Sharafzadeh, S., & Alizadeh, O. (2011). German and Roman chamomile. *Journal of Applied Pharmaceutical Science, 2*, 1–5.
5. Singh, O., et al. (2011). Chamomile (Matricaria chamomilla L.): an overview. *Pharmacognosy Reviews, 5*(9), 82.
6. El Joumaa, M. M., & Borjac, J. M. (2022). Matricaria chamomilla: A valuable insight into recent advances in medicinal uses and pharmacological activities. *Phytochemistry Reviews, 21*, 1–28.
7. Upadhyay, R., et al. (2016). New agro-technology to increase productivity of chamomile (Matricaria chamomilla L.). *Industrial Crops and Products, 89*, 10–13.
8. Gupta, V., et al. (2010). Pharmacological potential of Matricaria recutita-A review. *International Journal of Pharmaceutical Sciences and Drug Research, 2*(1), 12–16.
9. Kumar, S., Das, M., Singh, A., Ram, G., Mallavarapu, G. R., & Ramesh, S. (2001). *Journal of Medicinal and Aromatic Plant Studies, 23*, 617–623.
10. Tyihak, E., Sarkany-Kiss, I., & Verzar-Petri, G. J. P. (1962). Pflanzenchemische Untersuchung der Apigeninglykoside der echten Kamille (Matricaria chamomilla L.). *Pharmazie, 17*, 301–304.
11. Mann, C., et al. Department of Medicinal Chemistry and Pharmacognosy, College of Pharmacy, University of Minnesota, Minneapolis, MN 55455. *Herbs, Spices, and Medicinal Plants: Recent Advances in Botany, Horticulture, and Pharmacology, 1992*(1), 235.
12. Bradley, P. R. (1992). *British herbal compendium. Volume 1. A handbook of scientific information on widely used plant drugs*. Companion to Volume 1 of the British Herbal Pharmacopoeia. British Herbal Medicine Association.
13. Zargaran, A., et al. (2016). A review on the management of migraine in the Avicenna's Canon of Medicine. *Neurological Sciences, 37*, 471–478.
14. Das, M. (2014). *Chamomile: medicinal, biochemical, and agricultural aspects*. CRC Press.
15. Chauhan, E. S., & Jaya, A. (2017). Chamomile an ancient aromatic plant-A review. *Journal of Ayurveda Medical Sciences, 2*(4), 251–255.
16. Svab, J. J. H. P. (1979). New aspects of cultivating camomile. *Herba Polonica, 25*(1), 35–39.
17. Mills, S. Y. (1991). *The essential book of herbal medicine*. Arkana Penguin.
18. Chandra, V. (1973). Cultivation of plants for perfumery industry at Lucknow. *Indian Perfumer, 16*(1), 40–44.
19. Franz, C., et al. (2007). Study on the assessment of plants/herbs, plant/herb extracts and their naturally or synthetically produced components as 'additives' for use in animal production. *EFSA Supporting Publications, 4*(4), 1–297.

20. Lawrence, B. M. (1996). Progress in essential oils-mandarin oil, fennel oil and artemisia annua oil. *Perfumer and Flavorist, 21*(2), 25–32.
21. Franke, R., & Schilcher, H. (2005). *Chamomile: industrial profiles*. CRC press.
22. Dettmer, B., & Sauer, T. (2014). Regionalökonomische Auswirkungen eines geplanten Pumpspeicherkraftwerks: Eine Input-Output-Analyse für den Freistaat Thüringen. *Zeitschrift für Energiewirtschaft, 4*(38), 255–268.
23. Zalecki, R. (1971). Cultivation and fertilizing of the tetraploid Matricaria chamomilla LI The sowing time. *Herba Polonica, 17*, 367–375.
24. Dutta, P., & Singh, A. (1964). Effect of different spacings on fresh flower and oil yield of Matricaria chamomilla. *Indian Journal of Agronomy, 9*, 1–12.
25. Franke, R., & Hannig, H. J. (2012). Kamille (Matricaria recutita L.). *Handbuch des Arznei- und Gewürzpflanzenbaus, 4*, 618–648.
26. Singh, L. B. (1970). Utilisation of saline-alkali soils for agro-industry without prior reclamation. *Economic Botany, 24*(4), 439–442.
27. Kerches, J. (1966). Experiments with the cultivation of chamomile (Matricaria chamomilla). *Herba Hungarica, 5*, 141–147.
28. Reichling, J., Beiderbeck, R., & Becker, H. (1979). Vergleichende Untersuchungen über sekundäre Inhaltsstoffe bei Pflanzentumoren, Blüte, Kraut und Wurzel von Matricaria chamomilla L. *Planta Medica, 36*(8), 322–332.
29. Handa, H., & Kaul, M. K. (1997). *Supplement to cultivation and utilization of aromatic plants*. Regional Research Laboratory/Council of Scientific & Industrial Research.
30. Park, Y., et al. (2003). Characterization of salicylic acid-induced genes in Chinese cabbage. *Plant Cell Reports, 21*, 1027–1034.
31. Gunes, A., et al. (2007). Salicylic acid induced changes on some physiological parameters symptomatic for oxidative stress and mineral nutrition in maize (Zea mays L.) grown under salinity. *Journal of Plant Physiology, 164*(6), 728–736.
32. Horváth, E., Szalai, G., & Janda, T. (2007). Induction of abiotic stress tolerance by salicylic acid signaling. *Journal of Plant Growth Regulation, 26*, 290–300.
33. Vanacker, H., et al. (2001). A role for salicylic acid and NPR1 in regulating cell growth in Arabidopsis. *The Plant Journal, 28*(2), 209–216.
34. Pastírová, A., Repčák, M., & Eliašová, A. (2004). Salicylic acid induces changes of coumarin metabolites in Matricaria chamomilla L. *Plant Science, 167*(4), 819–824.
35. Stanojevic, L. P., et al. (2016). Chemical composition, antioxidant and antimicrobial activity of chamomile flowers essential oil (Matricaria chamomilla L.). *Journal of Essential Oil Bearing Plants, 19*(8), 2017–2028.
36. Salehi, A., & Hazrati, S. (2017). How essential oil content and composition fluctuate in German chamomile flowers during the day? *Journal of Essential Oil Bearing Plants, 20*(3), 622–631.
37. Bucko, D., & Salamon, I. (2006). The essential oil quality of chamomile, Matricaria recutita L., after its large-scale distillation. in I International Symposium on Chamomile Research. *Development and Production, 749*, 269–273.
38. Frabboni, L., et al. (2019). Bio-herbicidal effects of oregano and rosemary essential oils on chamomile (Matricaria chamomilla L.) crop in organic farming system. *Agronomy, 9*(9), 475.
39. Homami, S. S., et al. (2016). Comparative studies of different extraction methods of essential oil from Matricaria recutita L. in Iran. *Journal of the Chilean Chemical Society, 61*(2), 2982–2984.
40. EL-Hefny, M., et al. (2019). Essential and recovery oils from Matricaria chamomilla flowers as environmentally friendly fungicides against four fungi isolated from cultural heritage objects. *Processes, 7*(11), 809.
41. Pirzad, A., et al. (2006). Essential oil content and composition of German chamomile (Matricaria chamomilla L.) at different irrigation regimes. *Journal of Agronomy, 5*(3), 451–455.

42. Amiri, S., & Sharafzadeh, S. (2014). Essential oil components of German chamomile cultivated in Firoozabad, Iran. *Oriental Journal of Chemistry, 30*(1), 365–367.
43. Kawthar, A., Ashour, H., & Fatma, S. J. (2017). Growth characters and some chemical constituents of Matricaria chamomilla L. plants in relation to green manure and compost fertilizer in sandy soil. *Middle East Journal, 6*(1), 76–86.
44. Stahl, E. (1954). Über das Cham-Azulen und dessen Vorstufen, III. *Mitteil.: Zur Konstitution der Cham-Azulencarbonsäure., 87*(11), 1626–1628.
45. Council, N. R. (1930). *Food chemicals codex: First supplement to the* (2nd ed.). The Convention.
46. Mwaniki, J. M. (2015). Analysis of blue chamomile essential oil produced by multi-solvent solvent extraction clevenger distillation method. *Africa Journal of Physical Sciences ISSN, 2*(1), 1–10.
47. Taraj, K., et al. (2017). Eco-extraction of albanian chamomile essential oils by liquid Co2 at different temperatures and characterisation by FTIR spectroscopy. *Journal of Environmental Protection and Ecology, 18*(1), 117–124.
48. Noori, H., & Khajenoori, M. (2013). Subcritical Water Extraction of Essential Oils from Matricaria Chamomilla L. *International Journal of Engineering, 26*(5), 489–494.
49. Al-Dabbagh, B., et al. (2019). Antioxidant and anticancer activities of chamomile (Matricaria recutita L.). *BMC Research Notes, 12*(1), 1–8.
50. El-Shhat, A., et al. (2021). Effect of dietary addition of chamomile flower (Matericaria chamomilla L.) powder on productive performance. *Hatching Traits and Economic Efficiency of Sudani Duck Breeders., 12*(3), 119–124.
51. Pino, J. A., et al. (2002). Essential oil of chamomile Chamomilla recutita (L.) Rausch. from Iran. *Journal of Essential Oil Research, 14*(6), 407–408.
52. Redaelli, C., Formentini, L., & Santaniello, E. (1981). HPLC determination of coumarins in Matricaria chamomilla. *Planta Medica, 43*(12), 412–413.
53. Achterrath-Tuckermann, U., et al. (1980). Pharmacological investigations with compounds of chamomile. V. Investigations on the spasmolytic effect of compounds of chamomile and Kamillosan on the isolated guinea pig ileum, *39*(1), 38–50.
54. Guenther, E. (1952). The essential oils. In E. Robert (Ed.), *Florida* (Vol. 5). Krieger Publishing Company.
55. Jellinek, J. J. P. C. A. (1984). α-Bisabolol–an agent, anti-inflammatory pour products cosmetique. *Perfume Cosmet Aromes, 57*, 55–57.
56. Duke, J. A. (1990). CRC Handbook of medicinal herbs. *International Clinical Psychopharmacology, 5*(1), 74.
57. Schilcher, H. (1973). Recent knowledge in quality evaluation of camomile blossoms respectively camomile oil. 2. Quality evaluation of the volatile oil in Flores Chamomillae. Grading of commercial camomiles into 4 respectively 5 chemical types. *Planta Medica, 23*(2), 132–144.
58. Faehnrich, B., et al. (2013). Self-incompatibility and male sterility in six Matricaria recutita varieties. *Journal of Applied Botany and Food Quality, 86*(1).
59. Buswell, J. A., & Chen, M. J. (2005). Cultivation, biochemical, molecular biological and medical aspects of the culinary-medicinal straw mushroom Volvariella volvacea (Bull.: Fr.) Singer (Agaricomycetideae). *International Journal of Medicinal Mushrooms, 7*(1&2), 157–166.
60. Srivastava, S. (2015). *Flora of cold deserts of Western Himalaya*. Botanical Survey of India.
61. Solouki, M., et al. (2008). Study of genetic diversity in Chamomile (Matricaria chamomilla) based on morphological traits and molecular markers. *Scientia Horticulturae, 117*(3), 281–287.
62. Pirkhezri, M., Hassani, M., & Hadian, J. (2010). Genetic diversity in different populations of Matricaria chamomilla l. Growing in southwest of Iran, based on morphological. *Research Journal of Medicinal Plant, 4*(1), 1–13.

63. Wagner, C., et al. (2005). Molecular analyses on the genetic diversity and inheritance of (−)-α-bisabolol and chamazulene content in tetraploid chamomile (Chamomilla recutita (L.) Rausch.). *Plant Science, 169*(5), 917–927.
64. Otto, L.-G., et al. (2017). Use of genotyping-by-sequencing to determine the genetic structure in the medicinal plant chamomile, and to identify flowering time and alpha-bisabolol associated SNP-loci by genome-wide association mapping. *Acta Scientiarum Polonorum. Hortorum Cultus, 18*, 1–18.
65. Okoń, S., Surmacz-Magdziak, A., & Paczos-Grzęda, E. (2013). Genetic diversity among cultivated and wild chamomile germplasm based on ISSR analysis. *Acta Scientiarum Polonorum. Hortorum Cultus, 12*(2), 43–50.
66. Tsivelika, N., et al. (2018). Phenotypic variation of wild Chamomile (Matricaria chamomilla L.) populations and their evaluation for medicinally important essential oil. *Biochemical Systematics and Ecology, 80*, 21–28.
67. Ahmadi, H., et al. (2014). Assessment of the genetic variation of chamomile (Matricaria chamomilla L.) populations using phytochemical, morphological and ISSR markers. *Biochemical Systematics and Ecology, 54*, 190–197.
68. Adeli, N., et al. (2013). Evaluation of essential oil yield, morphological and phenological traits in some populations of two Chamomile species (Matricaria recutita and M. aurea). *Journal of Medicinal Plants and By-product, 2*(2), 153–158.
69. D'Andrea, L. (2002). Variation of morphology, yield and essential oil components in common chamomile (Chamomilla recutita (L.) Rauschert) cultivars Grown in Southern Italy. *Journal of Herbs, Spices & Medicinal Plants, 9*(4), 359–365.
70. Singh, S., et al. (2007). Genetic divergence in German chamomile [Chamomilla recutita (L.) Rauch.]. *Journal of Spices and Aromatic Crops, 16*(2), 125–128.
71. Lal, R. K., et al. (2019). Induced polygenic variations through γ-rays irradiation and selection of novel genotype in chamomile (Chamomilla recutita [L.] Rauschert). *International Journal of Radiation Biology, 95*(9), 1242–1250.
72. Irmisch, S., et al. (2012). The organ-specific expression of terpene synthase genes contributes to the terpene hydrocarbon composition of chamomile essential oils. *BMC Plant Biology, 12*(1), 1–13.
73. Golparvar, A., & Pirbalouti, A. G. (2011). Genetic improvement of essence percent and dry flower yield using indirect selection in German chamomile (Matricaria chamomilla L.). *Journal of Medicinal Herbs, 1*(4), 33–40.
74. Oravec, V., Sr., Oravec, V., Jr., & Oravec, V. (2006). Breeding of bisabolol diploid and tetraploid varieties of chamomile in slovakia. In *I international symposium on chamomile research, development and production* (Vol. 749). ISHS.
75. Otto, L. G., et al. (2015). Towards breeding of triploid chamomile (Matricaria recutita L.)–Ploidy variation within German chamomile of various origins. *Plant Breeding, 134*(4), 485–493.
76. Albrecht, S., Sonnenschein, M., & Plescher, A. J. (2016). Breeding of a high yielding chamomile variety (Matricaria recutita L.) with improved traits for machine harvesting. *Julius-Kühn-Archiv, 453*, 135.
77. Tai, Y., et al. (2020). Analysis of terpenoid biosynthesis pathways in German chamomile (Matricaria recutita) and Roman chamomile (Chamaemelum nobile) based on co-expression networks. *Genomics, 112*(2), 1055–1064.
78. Ling, C., et al. (2020). Cloning and functional analysis of three aphid alarm pheromone genes from German chamomile (Matricaria chamomilla L.). *Plant Science, 294*, 110463.
79. Wang, H.-M., et al. (2017). In vitro regeneration, Agrobacterium-mediated transformation, and genetic assay of chalcone synthase in the medicinal plant Echinacea pallida. *Microchemical Journal, 130*, 117–130.
80. Mahgoub, Y. A., et al. (2020). Near-infrared spectroscopy combined with chemometrics for quality control of German chamomile (Matricaria recutita L.) and detection of its adulteration by related toxic plants. *Microchemical Journal, 158*, 105153.

81. Guzelmeric, E., et al. (2017). Quality assessment of marketed chamomile tea products by a validated HPTLC method combined with multivariate analysis. *Journal of Pharmaceutical and Biomedical Analysis, 132*, 35–45.
82. Chauhan, R., et al. (2022). A comprehensive review on biology, genetic improvement, agro and process technology of German chamomile (Matricaria chamomilla L.). *Plants, 11*(1), 29.
83. Kreuter, M. H. (2013). *Composition containing oils of chamomile flower and black cumin with reduced endotoxins*. Google Patents.
84. Henderson, A. (2015). *Oil blend for skin treatment*. Google Patents.
85. Oltarshevskaya, N. D., Savilova, L. B., & Krichevsky, G. E. (2013). *Hemostatic dressing comprising extract of chamomile and nettle*. Google Patents.
86. Mericli, A. H. (1990). The lipophilic compounds of a Turkish Matricaria chamomilla variety with no chamazulene in the volatile oil. *International Journal of Crude Drug Research, 28*(2), 145–147.
87. Kumar, S., et al. (2001). Composition of the essential oils of the flowers, shoots and roots of two cultivars of Chamomilla recutita. *Journal of Medicinal and Aromatic Plant Sciences, 23*(4), 617–623.
88. Subiza, J., et al. (1990). Allergic conjunctivitis to chamomile tea. *Annals of Allergy, 65*(2), 127–132.
89. Jakovlev, V., et al. (1979). Pharmacological investigations with compounds of chamomile ii. new investigations on the antiphlogistic effects of (-)-alpha-bisabolol and bisabolol oxides (author's transl). *Planta Medica, 35*(2), 125–140.
90. Owlia, P., et al. (2007). Retardation of biofilm formation with reduced productivity of alginate as a result of Pseudomonas aeruginosa exposure to Matricaria chamomilla essential oil. *Pharmacognosy Magazine, 3*(10), 83–89.
91. Kazemi, M. (2015). Chemical composition and antimicrobial activity of essential oil of Matricaria recutita. *Journal of Food Properties, 18*(8), 1784–1792.
92. Silva, N., et al. (2012). Antimicrobial activity and phytochemical analysis of crude extracts and essential oils from medicinal plants. *Natural Product Research, 26*(16), 1510–1514.
93. Chouhan, S., Sharma, K., & Guleria, S. J. M. (2017). Antimicrobial activity of some essential oils—present status and future perspectives. *Medicines, 4*(3), 58.
94. Owlia, P., Saderi, H., & Rasooli, I. (2007). Antistreptococcal and antioxidant activity of essential oil from Matricaria chamomilla L. *Research Journal of Biological Sciences, 2*, 155–160.
95. Sakkas, H., et al. (2018). Antibacterial efficacy of commercially available essential oils tested against drug-resistant gram-positive pathogens. *Applied Sciences, 8*(11), 2201.
96. Abdoul-Latif, F. M., et al. (2011). Antimicrobial and antioxidant activities of essential oil and methanol extract of Matricaria chamomilla L. from Djibouti. *Journal of Medicinal Plants Research, 5*(9), 1512–1517.
97. Shakya, V. K., et al. (2019). A relative assessment of essential oil of Chrysopogon zizanioides and Matricaria chamomilla along with calcium hydroxide and chlorhexidine gel against Enterococcus faecalis in ex vivo root canal models. *Journal of Conservative Dentistry: JCD, 22*(1), 34.
98. Satyal, P., Shrestha, S., & Setzer, W. N. (2015). Composition and bioactivities of an (E)-β-farnesene chemotype of chamomile (Matricaria chamomilla) essential oil from Nepal. *Natural Product Communications, 10*(8), 1934578X1501000835.
99. Hartmann, K. C., & Onofre, S. B. (2010). Atividade antimicrobiana de óleos essenciais da camomila (Matricaria chamomilla L.). *Saúde E Pesquisa, 3*, 3.
100. Höferl, M., et al. (2020). Biological activity of Matricaria chamomilla essential oils of various chemotypes. *Planta Medica International Open, 7*(03), e114–e121.
101. Roby, M. H. H., et al. (2013). Antioxidant and antimicrobial activities of essential oil and extracts of fennel (Foeniculum vulgare L.) and chamomile (Matricaria chamomilla L.). *Industrial Crops and Products, 44*, 437–445.

102. Ismail, M. C., et al. (2013). Synergistic interaction between chamomile flower (Matricaria chamomilla L.) extracts and tetracycline against wound infection bacteria. *Al-Nahrain Journal of Science, 16*(3), 191–195.
103. El Mihyaoui, A., et al. (2022). Chamomile (Matricaria chamomilla L.): A review of ethnomedicinal use, phytochemistry and pharmacological uses. *Life, 12*(4), 479.
104. Abdalla, R. M., & Abdelgadir, A. E. (2016). Antibacterial activity and phytochemical constituents of Cinnamomum verum and Matricaria chamomilla from Sudan. *Bio Bulletin, 2*(2), 01-0.
105. Mariod, A. A., Kakil, E. S., & Elneel, Y. F. H. (2019). Antibacterial activity of Zingiber officinale, Matricaria chamomilla and Nigella sativa extractions on the growth of pathogenic bacteria isolated from different clinical specimens. *Acta Scientific Nutritional Health, 3*, 26–32.
106. Malm, A., et al. (2015). Anti-Helicobacter pylori activity in vitro of chamomile flowers, coneflower herbs, peppermint leaves and thyme herbs–a preliminary report. *Current Issues in Pharmacy and Medical Sciences, 28*(1), 30–32.
107. Carvalho, A., et al. (2014). Avaliação da atividade antibacteriana de extratos etanólico e de ciclohexano a partir das flores de camomila (Matricaria chamomilla L.). *Revista Brasileira de Plantas Medicinais, 16*, 521–526.
108. Sharifzadeh, A., & Shokri, H. J. (2016). Antifungal activity of essential oils from Iranian plants against fluconazole-resistant and fluconazole-susceptible Candida albicans. *Avicenna Journal of Phytomedicine, 6*(2), 215.
109. Mekonnen, A., et al. (2016). In vitro antimicrobial activity of essential oil of Thymus schimperi, Matricaria chamomilla, Eucalyptus globulus, and Rosmarinus officinalis. *International Journal of Microbiology Research, 2016*, 9545693.
110. Boudıeb, K., et al. (2018). Phytochemical characterization and antimicrobial potentialities of two medicinal plants, Chamaemelum nobile (L.) All and Matricaria chamomilla (L.). *International Journal of Innovative Science and Research Technology, 2*, 126–139.
111. Hameed, R. H., Mohammed, G. J., & Hameed, I. H. (2018). Matricaria chamonbmilla: Bioactive compounds of methanolic fruit extract using GC-MS and FTIR techniques and determination of its antimicrobial properties. *Indian Journal of Public Health Research & Development, 9*(3), 223–228.
112. Lavanya, J., et al. (2016). Antioxidant and antimicrobial activity of selected medicinal plants against human oral pathogens. *International Journal of Pharmacy and Pharmaceutical Sciences, 8*(9), 71–78.
113. Cvetanović, A., et al. (2015). Antioxidant and biological activity of chamomile extracts obtained by different techniques: perspective of using superheated water for isolation of biologically active compounds. *Industrial Crops and Products, 65*, 582–591.
114. Hosseinpour, M., et al. (2013). Antiproliferative effects of Matricaria chamomilla on Saccharomyces cerevisiae. *Journal of HerbMed Pharmacology, 2*(2), 49–51.
115. Rahman, H., & Chandra, A. (2015). Microbiologic evaluation of matricaria and chlorhexidine against E. faecalis and C. albicans. *Indian Journal of Dentistry, 6*(2), 60.
116. Osman, M. Y., et al. (2016). Screening for antioxidant, antifungal, and antitumor activities of aqueous extracts of chamomile (Matricaria chamomilla). *Egyptian Pharmaceutical Journal, 15*(2), 55.
117. Seyedjavadi, S. S., et al. (2019). Isolation, functional characterization, and biological properties of MCh-AMP1, a novel antifungal peptide from Matricaria chamomilla L. *Chemical Biology & Drug Design, 93*(5), 949–959.
118. Hajaji, S., et al. (2018). Leishmanicidal activity of α-bisabolol from Tunisian chamomile essential oil. *Parasitology Research, 117*, 2855–2867.
119. Hajaji, S., et al. (2018). Anthelmintic activity of Tunisian chamomile (Matricaria recutita L.) against Haemonchus contortus. *Journal of Helminthology, 92*(2), 168–177.
120. Mansour, S. A., et al. (2015). Toxicity of essential plant oils, in comparison with conventional insecticides, against the desert locust, Schistocerca gregaria (Forskål). *Industrial Crops and Products, 63*, 92–99.

121. Afify, A. E.-M. M., Ali, F. S., & Turky, A. F. (2012). Control of Tetranychus urticae Koch by extracts of three essential oils of chamomile, marjoram and Eucalyptus. *Asian Pacific Journal of Tropical Biomedicine, 2*(1), 24–30.
122. Pirali-Kheirabadi, K., & Razzaghi-Abyaneh, M. (2007). Biological activities of chamomile (Matricaria chamomile) flowers' extract against the survival and egg laying of the cattle fever tick (Acari Ixodidae). *Journal of Zhejiang University Science B, 8*(9), 693–696.
123. del Carmen Romero, M., et al. (2012). Activity of Matricaria chamomilla essential oil against anisakiasis. *Phytomedicine, 19*(6), 520–523.
124. Hajaji, S., et al. (2017). Correlation of radical-scavenging capacity and amoebicidal activity of Matricaria recutita L.(Asteraceae). *Experimental Parasitology, 183*, 212–217.
125. Váradyová, Z., et al. (2018). Ovicidal and larvicidal activity of extracts from medicinal-plants against Haemonchus contortus. *Experimental Parasitology, 195*, 71–77.
126. Al-Mekhlafi, F. A., et al. (2021). Target and non-target effects of Foeniculum vulgare and Matricaria chamomilla combined extract on Culex pipiens mosquitoes. *Saudi Journal of Biological Sciences, 28*(10), 5773–5780.
127. Al-Mekhlafi, F. A., et al. (2020). Inhibition of the growth and development of mosquito larvae of Culex pipiens L.(Diptera: Culicidae) treated with extract from flower of Matricaria chamomilla (Asteraceae). *Entomological Research, 50*(3), 138–145.
128. Villa-Rodriguez, J., et al. (2015). German Chamomile (Matricaria chamomilla) Extract and Its Major Polyphenols Inhibit Intestinal α-glycosidases in vitro. *The FASEB Journal, 29*, LB323.
129. Hwang, S. H., et al. (2018). Evaluation of aldose reductase, protein glycation, and antioxidant inhibitory activities of bioactive flavonoids in Matricaria recutita L. and their structure-activity relationship. *Journal of Diabetes Research, 2018*, 3276162.
130. Ali, E. M. (2013). Phytochemical composition, antifungal, antiaflatoxigenic, antioxidant, and anticancer activities of Glycyrrhiza glabra L. and Matricaria chamomilla L. essential oils. *Journal of Medicinal Plant Research, 7*(29), 2197–2207.
131. Romeilah, R. M. (2009). Anticancer and antioxidant activities of Matricaria chamomilla L. and Marjorana hortensis essential oils. *Research Journal of Medicine and Medical Sciences, 4*(2), 332–339.
132. Kamali, A. M., et al. (2014). In vitro cytotoxic activity of Matricaria chamomilla root extract in Human breast cancer cell line MCF-7. *Life Science Journal, 11*, 403–406.
133. Fraihat, A., et al. (2018). Inhibitory effects of methanol extracts of selected plants on the proliferation of two human melanoma cell lines. *Tropical Journal of Pharmaceutical Research, 17*(6), 1081–1086.
134. Wu, Y. N., Xu, Y., & Yao, L. (2012). Anti-inflammatory and anti-allergic effects of German chamomile (Matricaria chamomilla L.). *Journal of Essential Oil Bearing Plants, 15*(1), 75–83.
135. Asadi, Z., et al. (2020). Anti-inflammatory effects of Matricaria chamomilla extracts on BALB/c mice macrophages and lymphocytes. *Iranian Journal of Allergy, Asthma and Immunology, 19*(S1), 63–73.
136. Singh, K. G., Sonia, S., & Konsoor, N. (2018). In-vitro and ex-vivo studies on the antioxidant, anti-inflammatory and antiarthritic properties of Camellia sinensis, Hibiscus rosa sinensis, Matricaria chamomilla, Rosa SP., Zingiber officinale tea extracts. *Inflammation, 49*, 50.
137. Nargesi, S., et al. (2018). The effects of Matricaria chamomilla L. hydroalcoholic extract on atherosclerotic plaques, antioxidant activity, lipid profile and inflammatory indicators in rats. *Biomedical Research and Therapy, 5*(10), 2752–2761.
138. Banchio, E., et al. (2005). Quantitative variations in the essential oil of Minthostachys mollis (Kunth.) Griseb. in response to insects with different feeding habits. *Journal of Agricultural and Food Chemistry, 53*(17), 6903–6906.
139. Ghasemi, M., et al. (2016). Increase of Chamazulene and α-Bisabolol contents of the essential oil of german chamomile (Matricaria chamomilla L.) using salicylic acid treatments under normal and heat stress conditions. *Food, 5*(3), 56.

140. Xiaomeng, L., et al. (2018). Effects of 5-aminolevulinic acid on the photosynthesis, antioxidant system, and α-bisabolol content of Matricaria recutita. *Notulae Botanicae Horti Agrobotanici Cluj-Napoca, 46*(2), 418–425.
141. Kusano, M., et al. (2011). Metabolomic approaches toward understanding nitrogen metabolism in plants. *Journal of Experimental Botany, 62*(4), 1439–1453.
142. Matt, P., et al. (2002). Decreased Rubisco activity leads to dramatic changes of nitrate metabolism, amino acid metabolism and the levels of phenylpropanoids and nicotine in tobacco antisense RBCS transformants. *The Plant Journal, 30*(6), 663–677.
143. Repčák, M., Imrich, J., & Franeková, M. (2001). Umbelliferone, a stress metabolite of Chamomilla recutita (L.) Rauschert. *Journal of Plant Physiology, 158*(8), 1085–1087.
144. Maksymiec, W. J. A. P. P. (2007). Signaling responses in plants to heavy metal stress. *Acta Physiologiae Plantarum, 29*, 177–187.
145. Kováčik, J., et al. (2006). Matricaria chamomilla is not a hyperaccumulator, but tolerant to cadmium stress. *Plant Growth Regulation, 50*, 239–247.
146. Prokop'ev, I., et al. (2014). Impact of urban anthropogenic pollution on seed production, morphological and biochemical characteristics of chamomile. *Matricaria chamomila L., 45*, 18–23.
147. Kessler, A., & Baldwin, I. T. (2002). Plant responses to insect herbivory: the emerging molecular analysis. *Annual Review of Plant Biology, 53*(1), 299–328.
148. Mondal, P. Biosynthesis and Regulation of Terpene Production in Accessions of Chamomile (Matricaria recutita L.). Ph.D. Thesis, Martin-Luther-Universität Halle-Wittenberg, Halle, Germany, 2020
149. Raal, A., et al. (2011). Content and composition of essential oils in some Asteraceae species. *Proceedings of the Estonian Academy of Sciences, 60*(1), 55.
150. Orav, A., Kailas, T., & Ivask, K. (2001). Volatile constituents of Matricaria recutita L. from Estonia. In *Proceedings-Estonian academy of sciences chemistry*. TRUEKITUD OU.
151. Mohammad, R., et al. (2010). Effects of planting date and seedling age on agro-morphological characteristics, essential oil content and composition of German chamomile (Matricaria chamomilla L.) grown in Belgium. *Industrial Crops and Products, 31*(1), 145–152.
152. Mertens, D., et al. (2021). Predictability of biotic stress structures plant defence evolution. *Trends in Ecology & Evolution, 36*(5), 444–456.
153. Sarrou, E., et al. (2016). Metabolite profiling and antioxidative activity of Sage (Salvia fruticosa Mill.) under the influence of genotype and harvesting period. *Industrial Crops and Products, 94*, 240–250.

Chapter 40
Bhumyamalaki

Misbah Hameed, Hafiza Nasreen Aslam, Romina Alina Marc, and Marius Irimie

40.1 Introduction

"*Bhumyamalaki*," is the herbal plant drug belongs to the family Euphorbiaceae which is used in Indian system of medicine for various liver disorders. *Phyllanthus niruri Linn* is endemic to West Indies and is not found in India. "Bhumyamalaki" helps to treat, gonorrhea, frequent menstruation, jaundice, dysentery, and diabetes skin ulcers, sores, swelling, etc. and also it emphases on its budding for fighting viruses specially with hepatitis B virus. Phyllanthus also inhibit the growth and replication of the virus and decrease the amount of hepatitis B virus found in the blood stream. It is also effective in overwhelming the symptoms combat with hepatitis B virus. Phyllanthus also support the general well-being of the liver [1]. Ethanol and aqueous extract of the plant were studied for their antimicrobial activity by agar well diffusion method isolates from human samples [2]. It grows as a weed, in gardens, in farms and waste-lands. P. niruri is an annual plant, originate in Konkan region.It contains phyllanthin and other constituents, exhibit hepato-protective activity. Its roots, leaves, fruits, milky juice and powder of dried plant are used in ayurvedic preparations. Its milky latex is used for external application, young leaves for fevers, fresh roots against hepatitis,as leaf and seed decoction, and seeds effective against leucorrhoea, irregular menses, malaria, against burning micturition and boils [3]. It contains Phenolic compounds, Saponins, Flavonoids, Terpenoids,

M. Hameed (✉) · H. N. Aslam
Department of Pharmaceutics, Lahore College for Women University, Lahore, Pakistan

R. A. Marc
Food Engineering Department, Faculty of Food Science and Technology, University of Agricultural Sciences and Veterinary Medicine Cluj-Napoca, Cluj-Napoca, Romania

M. Irimie
Faculty of Medicine, Transilvania University of Brasov, Brasov, Romania

M. Zia-Ul-Haq et al. (eds.), *Essentials of Medicinal and Aromatic Crops*,
https://doi.org/10.1007/978-3-031-35403-8_40

Alkaloids, Tannins, Cardioglycosides, Steroids, Reducing Sugars, Anthraquinones and Resins [4].

40.1.1 Scientific Classification

Kingdom:	Plantae
Order:	Malpighiales
Family:	Phyllanthaceae
Genus:	Phyllanthus
Species:	P. niruri
Clade:	Tracheophytes
Clade:	Angiosperms
Clade:	Eudicots
Clade:	Rosids
Botanical name:	Phyllanthus niruri [5]

The Spanish name of the plant *P. nuriri* is chanca piedra, means "stone breaker" or "shatter stone or gale of the wind."

In Brazil, the plant is also known as quebra-pedra or arranca-pedras (which also decodes to "break-stone"). In adding to kidney stones, the plant is engaged in numerous other conditions by the ethnic peoples, including colic, diabetes, malaria, dysentery, fever, flu, tumors, jaundice, vaginitis, gonorrhea, and dyspepsia. Based on its long documented history of use in the region, the plant is generally engaged to reduce pain, expel intestinal gas, to stimulate and promote digestion, to expel worms, as a mild laxative. Chanca piedra is used in herbal medicine systems in every tropical country where it grows. Its core uses are for many types of biliary and urinary conditions including kidney and gallbladder stones; for hepatitis, colds, flu, tuberculosis, and other viral infections; liver diseases and disorders including anemia, jaundice and liver cancer; and for bacterial infections such as cystitis, prostatitis, venereal diseases and urinary tract infections [6].

40.1.2 Nutritive Value

Plant chemical characteristics have been found to be responsible for the nutritional and therapeutic qualities of plants [7]. For the mineral and vitamin contents, the plant leaf is rich in potassium (K), phosphorus (Ph) and adequate amount of calcium (Ca), magnesium (Mg) and sodium (Na); but low in copper (Cu), Iron (Fe), zinc (Zn) and manganese (Mn) [8]. The leaves of plant had greater dry matter content (91.06%) and are high in glucose,energy, protein, crude fibre, moderate fat, and a little amount of fat and ash, according to proximate analysis. Feed fibres have been proven to help in the absorption of dietary minerals while also lowering cholesterol absorption. Proteins are required for the synthesis and repairing of bodily tissues and as enzymes, and the crude protein content of the leaf (14.74%) exceeded that of most therapeutic plants [9] (Tables 40.1 and 40.2).

Manifestation of minerals such as iron,zinc, and magnesium in the leaf maybe contribute to immune modulatory action, since these substances have been involved in modulating immune system [10]. And it may also increase the effect of antioxidant enzymes [11]. Meanwhile these elements help as cofactors for such enzymes and finally modify the immune system activity [12].

Table 40.1 Proximate composition (g/100 g DM) of Phyllanthus niruri leaf (PNL)

Nutrients	% Dry matter
Dry matter	91.06
Crude fibre	16.90
Crude protein	14.74
Ether extract	7.55
Ash	7.33
Nitrogen free extract	44.54
Neutral detergent fibre	43.81
Acid detergent fibre	25.49
Metabolizable energy (MJ/kg)	7.74
In vitro organic matter digestibility	57.40

Table 40.2 Mineral and vitamin composition of Phyllanthus niruri leaf

Parameters	Concentration
Sodium (ppm)	1947
Calcium (ppm)	2911
Phosphorus (ppm)	6978
Potassium (ppm)	18,252
Magnesium (ppm)	2914
Manganese (ppm)	94
Ferric (ppm)	665
Copper (ppm)	12
Zinc (ppm)	37
Vitamin A (μg/100 g)	812.78
Vitamin B6 (mg/100 g)	0.234
Vitamin D (μg/100 g)	34.79
Vitamin E (mg/100 g)	57.18
Vitamin K (mg/100 g)	19.83

40.2 Chemical Constituents

Different parts of plants such as fruits, leaves, and roots of *P. nuriri* have rich phytochemistry. Many of the "active" constituents are accredited to biologically active lignands, glycosides, flavonoids, alkaloids, ellagitannins, and phenylpropanoids found in the leaf, stem, and root of the plant. Common lipids, sterols, and flavonols also present in the plant. The important plant chemicals include alkaloids, astragalin, brevifolin, carboxylic acids, corilagin, cymene, ellagic acid, ellagitannins, gallocatechins, geraniin, hypophyllanthin, lignans, lintetralins, lupeols, methyl salicylate, niranthin, nirtetralin, niruretin, nirurin, nirurine, niruriside, norsecurinines, phyllanthin, phyllanthine, phyllanthenol, phyllochrysine, phyltetralin, repandusinic acids, quercetin, quercetol, quercitrin, rutin, saponins, triacontanal, and tricontanol [13]. The phytochemical content of Phyllanthus niruri leaf is presented in Table 40.3. The plant is rich in alkaloids (3.67 g/100 g), oxalate (22.69 mg/g), total phenolics (11.85 mgGAE/100 g) but modest in phytate and flavonoid. Presence of alkaloids, flavonoids, phenols and triterpenes in the plant had a great importance [14]. The presence of high levels of alkaloids and total phenolics in the leaves of *P. niruri* recommends that the plant may have worthy blood glucose lowering properties by reducing dietary glucose absorption in the gastrointestinal tract, as well as noticeable effects on modulating immune system and prevention of oxidative stress [15]. Saponins are expectorants, cough suppressants and administered for hemolytic activities [16]. Presence of Saponin in the leaf (0.2 g/100 g) might result in the permeabilization of plasma membranes due to the bipolar structure of saponins, in which the lipophilic components assimilate easily and complex with the lipid fraction of plasma membranes, while the hydrophilic glycosidic portion forms complexes with transmembrane proteins, thus causing irreversible disorder and interruption of the plasma membrane [17]. The leaf containing phenols and flavonoids may help in eluding oxidative stress by sifting free radicals and bioactivation of carcinogens for excretion in the liver, between other things [18].

Proximate analysis indicated that the leaves contained high percentage of dry matter (91.06%), crude fibre (16.90%), crude protein (14.74%), ether extract (7.55%), ash (7.33%), neutral detergent fibre (43.81%), acid detergent fibre (25.49%), metabolizable energy (7.74 MJ/Kg) and in vitro organic matter

Table 40.3 Phytochemical contents of Phyllanthus niruri leaf

Parameter	Concentration
Total Phenolics (mgGAE/100 g)	11.85
Tannin (g/100 g)	0.001
Alkaloid (g/100 g)	3.67
Saponin (g/100 g)	0.20
Phytate (g/100 g)	5.40
Oxalate (mg/g)	22.69
Flavonoid (g/100 g)	1.40

digestibility (57.40%). Mineral and vitamin analysis showed that Phyllanthus niruri leaves contained macro minerals (ppm) such as Na (1947), Ca (2911), P (6978), K (18252), Mg (2914), micro minerals (ppm) such as Mn (94), Fe (665), Cu (12), Zn (37) and high content of vitamins A (812.78 μg/100 g), D (34.79 μg/100 g) and E (57.18 mg/100 g) respectively. Phytochemical evaluation revealed that Phyllanthus niruri leaf have high contents of total phenolic (11.85mgGAE/100 g), oxalate (22.69 mg/g), phytate (5.40 g/100 g), moderate in alkaloid (3.67 g/100 g), but low in tannin (0.001 g/100 g), saponin (0.20 g/100 g), and flavonoid (1.40 g/100 g). The results presented that *Phyllanthus niruri* leaves contains high nutritional quality due to high crude protein, vitamin A, mineral contents especially potassium with the resulting phyto-chemicals features that could aid as feed additives in poultry production. Thus the leaves o f Phyllanthus niruri was evaluated for the proximate, minerals, vitamins and phytochemical composition [19] (Table 40.4).

40.3 Cultivation

Phyllanthus niruri annual herbs are stiff, 10–60 cm elevated having simple or branched main stem and younger parts are turret smooth or rough. Cataphylls and stipules are 1.5–1.9 mm long having 1–1.5 mm long deltoid acuminate blade and subulate acuminate. The Leaves are 3–11 × 1.5–6 mm elliptic, rhombus,obviate, oblong, or even obtuse, or microscopically having a short sharply pointed tip at apex, base is obtuse or slightly inequilateral, 0.3–0.5 mm long petioles and stipules are 0.8–1.1 mm long triangular accuminate. Flowers transcripts, proximal 2–3 axis with unisexual cymules, both consisting of 1 male and 1 female or 2–3 males and female or 1 male and 2 females flower or combination thereof; male flowers pedicals are 1 mm long. 5 Calyx lobe, subequal each ca 0.7 × 0.3 mm elliptic or oblong elliptic and sharply critical at apex hyaline with unbranched mid ribs. 5 Disc segments, 3 (rarely 2) roundish stems. Filaments connect into a column which are 0.2–0.3 mm high longitudinally. Female flowers having pedicles 0.3–1 mm long, obtusely 4-gonous, dialated above, ca 1.5 mm in fruits, five calyx lobes, subequal. Lobes sometimes toothed at apex. 3 Stylesthat are free, more or less spreading, and superficially bifid at apex; arms opposite [21]. The seed capsules on stalks are 1–2 mm long and are round, smooth, 2 mm wide, with six seeds. When the fruits are ruptured, the open seeds are heaved away. Seeds are triangular (like an orange segment); light brown, 1 mm long, with 5–6 ribs on the back [22]. Flowers are Yellowish in appearance, very numerous, axillary. The male flowers are 1–3 in number while the female flowers are lonely in nature. Capsules are 2.5 mm in diameter, depressed globose, smooth narrowly lobed [23].

Table 40.4 Phytoconstituents reported in P. niruri [20]

Class	Compound
Alkaloid	4-methoxy-nor-securinine Nirurine Ent-norsecurinine
Benzenoid	Gallic acid
Coumarins	Ellagic acid Ethyl brevifolin carboxylate Methyl brevifolin carboxylate
Flavonoid	Quercetin Rutin Astragalin Quercitrin Isoquercitrin Eridictyol-7-rhamnopyranoside Fisetin-4-O-glucoside Nirurin (prenylated flavanone) Gallocatechin Niruriflavone Querceto
Lignan	Phyllanthin Hypophyllanthin Niranthin Nirtetralin Phyltetralin Hinokinin Lintetralin Isolintetralin Linnanthin Nirphyllin Phyllnirurin Demethylenedioxyniranthin
Tannin	Geraniin Repandusinic acid Corilagin
Triterpene	Limonene p-Cymene Lupeol acetate Lupeol Phyllanthenol Phyllanthenone Phyllantheol
Sterol	B-sitosterol Estradiol Isopropyl-24-cholesterol
Phytallate	Phyllester
Lipid	Ricinoleic acid
Saponins	Diosgenin
Miscellaneous	Niruriside Triacontanal

40.4 Cultivars

It is a ground weed which is found blooming throughout tropical and subtropical regions of Asia, America, and China. The genus Phyllanthus (L.) Murr. Comprises of 600–700 species with minor unique geographies among them. Phyllanthus niruri is an yearly herb that grows in the wild after first cascades of monsoon in different states of India like Jharkhand, Bihar, Chhattisgarh, etc. Though, it has also been reported to grow generally in coastal areas. it usually grows during second week of June and starts behavior as fruits up to mid July or August in Indian states. It remains in the barren up to the end of the rainy season. Though, it can grow and survive upto mid-winter under safe conditions [24]. Local people of Chhattisgarh and Jharkhand has widely usage it for the treatment of skin diseases, indigestion, cough, ulcers etc. [25]. Its Vernacular names are Assamese: Holpholi; Poram-lokhi, Bengali: Bhui amla, Hindi: Chalmeri, Harfarauri, Bhuiaonla. Kannada: Kirunelli, Nela Nelli, Konkani: Bhuin-avalae,Telugu: Ratsavusirike, Nela Usiri, Tamil: Arunelli, Keela Nelli, Malayalam: Arinelli, Kizhanelli, Nellipuli Marathi: Rayavali, Bhuiavli, Oriya: Narakoli,Sanskrit: Amala, Bhumyamlaki, Sukshmadala, Vitunika, Bhoodati [21]. A few significant types that can be grown in various regions of the nation are described as follows:

- Admiralty Islands
- Argentina
- Bimini
- Brazil
- Dominican Republic
- East Africa
- East Indies
- Fiji
- French Guiana
- Haiti
- Puerto Rico
- Philippines
- Peru
- Papau-New Guinea
- Mexico
- Malaysia
- India [26]

40.4.1 Seed Propagation

For seed propagation the soil should be plow up, prepared, and tugged into a fine tilth. Twenty tones of farm yard manure (FYM) are used during land preparation. Farm yard manure is assorted in the soil at a 10 t/ha rate of for nursery beds, organized with 100 g each of Phosphobacter,Azospirillum, and Trichoderma as a basal medium [27]. After the shower of first monsoon, 15–20 day hoary seedlings of 8–10 cm height are uprooted in rows in the field at 30 cm arrangement. When it is not raining, the field should be dampened as soon as possible after transplanting. For the area of one-hectare, round about 4.0 lac seedlings at a positioning of 25 × 25 cm are essential. Phyllanthus niruri, similar to other plants undergoes through a generational cycle. The sporophytic (diploid) phase of Stonebreaker's lifecycle is ruling, whereas the gametophyte (haploid) phase is moderated. The flowers of *P. niruri* are monoecious, that means they need both male and female parts on the same plant. Gametophytes microspores, are mature into pollen grains (sperm) during meiosis stage, these are also seen in *P. niruri*. Through meiosis Megaspores will grow into female parts. The microsporangium forms microspores on the anther of each flower, that are shifted to the ovule in the megasporangium through wind pollination or cross-pollination with insects. The two gametophytes unite together after pollination to generate a diploid zygote. The zygote matures into an embryo, which is enclosed in a seed with a food supply. After the seed germinates the embryo will undergo mitosis until *P. niruri* reaches adulthood [28].

40.4.2 Nursery Preparation

The Seeds of *Phylanthus niruri* are seeded in elevated nursery beds by resulting rains in June, and seedlings are lately transplanted on ridges in well-planned plots. On 1 ha of land for producing seedlings for planting, nearby 4 kg seed might be required. The Seed should not be pretreated in any way [29].

40.4.3 Orchard Establishment

The *P. niruri* tree starts to yield fruit 3–4 years after planting, and they reach a specific level after 10–12 years, where they are mature enough to continue making fruit for commercial purposes for another 60–70 years. This layout is passed out on land that has been extensively ploughed, leveled, and cleared of any vegetation. Planting typically begins round the middle of the August in subtropical environment, and is completed by the completion of the month. For Early seeding it is important that the plants collect a lengthy period of rain that is crucial for the initial growth and maturation of plantlets. However, the seedlings planted about mid-July that had the most success in blooming in both March and July [30]. Mature *P. niruri* trees have a loftiness of between 8 and 18 m. Consequently, square planting is done with 8–10 m gaps among and within the rows to allow for enough amount of light, easy culturing procedures (like pruning), or sufficient fruiting. In this era, inaugurating hedgerows with spacing of 8 m among lines and 4–5 m between plants is also being considered. *P. niruri* is normally spread directly by saplings grown in appropriate containers in unfavorable soil conditions, which are then uprooted to a permanent location where in situ budding is performed. Two cultivars must be planted in alternating rows to overcome self-incompatibility. Before planting, the fields are set out and designated. After that, soil is left in its current state during the first few raindrops to fall and settled properly. Before providing regular irrigation after planting, immediate watering is done to aid in the appropriate establishment of the plants [31].

40.4.4 Orchard Management

Taking proper care of the fertilizer and water supply, canopy architectural maintenance, field cleaning, and timely implementation of plant protection measures are all part of managing an *P. niruri* orchard. Young plants often emerge with a particular level of yearly vegetative growth to provide an initial canopy. The plants develop a proper canopy to yield fruits after 2 years. Flowers and fruits need to be cut back in the initial 2 years, though, to maintain healthier development, and regular spraying, hoeing, tidying, plant fortification, etc. should be done as needed. Trees in young orchards of fruits (2–7 years) require additional nutrients to maintain healthy development and ripening. Early trees need adequate and judicious trimming, ideally between March and April, to allow the chief branches to develop to an altitude of 0.75–1 m from the soil since their excessive vegetative growth inhibits regular fruiting. Eventually, only 4–6 carefully chosen branches are permitted to continue growing [32].

40.4.5 Water Management

In established orchards with typical soils, *phylanthus niruri* is grown as a moisture tree, thus irrigation is not necessary, especially during the winter and rainy months. The first watering should be applied to the fruit-bearing plant only after fertilizer has been applied between January and February. But, it is advised to avoid applying water throughout the flowering phase (mid-March to mid-April). For *p. niruri*, a basin irrigation system works well. Pitcher irrigation is typically advised for the growth of orchards in places with water shortages, while drip irrigation is also a viable approach. Substantial stock width and plant height can be observed during a normal dry season with the use total of 9 irrigations together with the mulch, which finally can save up to 20 cm of irrigation water (four irrigations) on the basis of net area. The water saved in this case might be used to plant an extra area of the orchard [33].

40.5 Pests and Diseases

40.5.1 Pests

40.5.1.1 Root-Knot Nematodes

It belongs to *Meloidogyne* spp. which are the most recurrently detected and damaging plant-parasitic nematodes in vegetable production [34]. Galls are formed as a consequence of physiological disturbances in the root tissues caused by the trophic interactions of female nematodes. Root-knot nematodes effect on economic environment and cause significant economic losses. An average 10% of loss in yield is frequently cited for vegetables [35]. Nematicides are highly toxic to both human health and the environment [36].

40.5.2 Treatment

Several another techniques are deliberated, including sanitation, soil management, organic amendments, fertilization, biological control and heat-based methods for prevention of fruit from pest diseases [37]. Various other techniques for management of pests control are *Pseudomonas fluorescent*, *Trichoderma viride*, neem cake and farm yard manure (FYM) as soil application and humic acid and salicylic acid as foliar application on growth of bhumyamalaki. All the treatments resulted in increased plant growth parameters and less galls production and improve the biomass of these medicinal crops [38]. Insecticides are the chemical substances that are used with the overt objective to stop or abolish the insect pests; various insecticides are used including pyrethroids, organophosphates, acephates, cyfluthrin, and neonicotinoids. Imidacloprid is a member of new chemical class of neonicotinoids, that is developed with the aim of high toxicity against insects but low to the mammals. It is used most commonly to control soil and sap-sucking insects [39]. Some plant extracts also used such as Silybum marianum and Ginkgo biloba to treat pesticide toxicity [40]. Imidacloprid is also an insect neurotoxin which is commonly used in agriculture to control the insect pests [41].

40.6 Medicinal Uses

40.6.1 Acute Venereal Disease

Hot water extract of dried bark and leaves is taken orally for treating acute venereal disease. The 500 ml extract is taken twice daily for up to 6 months.

40.6.2 Fevers and as a Laxative

Hot water extract of the entire plant is taken orally to reduce fevers and as a laxative.

40.6.3 Emmenagogue

Hot water extract of leaves is taken orally after a miscarriage and used as an emmenagogue.

40.6.4 Galactagogue

Water extract of roots is taken orally as a galactagogue [42].

40.6.5 Chronic Dysentery

The young shoots of the plant are administered in the form of an infusion for the treatment of chronic dysentery.

40.6.6 Gastro Intestinal Disorders

It has greater efficacy in the arena of gastro intestinal disorders like dyspepsia, colic, diarrhea and constipation. It is also used in leucorrhoea, menorrhagia and mammary abscess.

40.6.7 Skin Conditions

It is used to treat Scabbies or crusty lesions, bruises, wounds, scabies, offensive ulcers and sores, oedematous swellings, tubercular ulcers and ringworm, with good effect since many years.

The plant is bitter, astringent, have cooling effect, diuretic, stomachic, febrifuge and antiseptic.

40.6.8 Hepatitis B Virus Infection

The powdered leaves of Phyllanthus niruri (bhumyamalaki) is used in clinical studies weighing its usefulness in patients suffering from chronic damage to the liver due to the prolonged hepatitis B virus infection. This type of infection upshots in inability of the body's immune system to eradicate the virus from the liver cells. This condition is labelled as a carrier state. Some of the components of the virus detectable in the carrier state in the blood are as following listed as HBsAg or the surface antigen of the virus and HBeAg or the envelope antigen of the virus. Moreover, the carrier state may be long-established by the presence of antibodies directed against the staple of the virus or the antiHBc antibodies. It is used in a dose of 200 mg three times a day for 30 days [43].

40.6.9 Pharmacological Activity

It is used for complications of the stomach, genitourinary system, liver, kidney, and spleen etc. The medicinal qualities and pharmacological activities validated to various parts of Phyllanthus niruri which are detailed as follows:

40.6.10 Hepatoprotective Effect

Hepatitis B is one of the most important diseases perpetrating in human population. The Conventional treatment of hepatitis B with interferon –alpha is very costly and has many severe side effects. Substitute herbal medicine using for treatment are the extracts of Phyllanthus niruri and Phyllanthus urinaria which have been described to be active against Hepatitis B and other viral infections. The quantitative determination of the anti-viral effect of these herbs is well-defined in vitro systems [44]. Phyllanthus niruri exhibits noticeable anti-hepatitis B virus surface antigen activity in in-vivo and in-vitro studies. Infectious hepatitis is caused due to the incapability of the immune system of body to eradicate the virus from the liver cells: therefore called the "carrier state". An infection with the virus is recognized by measurable levels of various viral antigens in the blood, which includes, HbaAg (the surface antigen of the virus) as well as the antibodies to the core of virus (HBc antibodies) [45]. Research in Japan and India in the 1980s has revealed that the Phyllanthus niruri has liver -healing properties. The primary compounds accountable having healing properties are phyllanthin, hypophyllanthin and triacontanal. Glycosides found in Phyllanthus niruri proved that Aldose reductase (AR) has inhibitory activity [46].

40.6.11 Anti Cancerous & Cellular Protective Actions

Phyllanthus niruri has great potential to prevent the growth and intiation of cancerous cells [47].

40.6.12 Action of Kidney Stones & Uric Acid

Kidney stone is a very common problem that amasses the calcium oxalate crystals, which leads to urinary calculi formation, nucleation, growth, and aggregation of crystals in kidney. Phyllanthus niruri's extract impedes in the growth and accumulation of calcium oxalate [CaOx] crystals in the renal calculi and in the early stages of stone formation [48]. The CaOx metastable boundary was reduced by the treatment of [5% [v/v]] extractof *Phyllanthus niruri* and it can also dispossess the CaOx crystals and formation of nucleation [49]. The extract has the great ability to inhibit the growth of calculi and it can also change the shape and texture of the calculi. When it is treated on the preformed calculi it can form a matrix like material on its surface so it can transform the appearance and texture of the calculus [50]. *P. niruri* extract is also directed in hyper calciuric patients so it can lessening the urinary calcium levels [51]. It can also lessens the excess uric acid in hyperuricemic people by the lignans with uricosuric action in the extract [52].

40.6.13 Anti-inflammatory Activity

When The Hexane Extract (HE), the lignans phyltetralin, the Lignan-Rich Fraction (LRF), nirtetralin, niranthin of Phyllanthus niruri is given orally so repressed the carrageenan (Cg)-encouraged paw oedema and neutrophil influx. It is also inhibited the increase of IL1-β tissue levels induced by Cg. Platelet Activating Factor (PAF), Bradykinin (BK), and endothelin-1 (ET-1)-induced paw oedema were pointedly repressed by the HE or LRF. Lastly, nirtetralin or phyltetralin caused inhibition of paw oedema induced by PAF or ET-1. These results showed that the HE, the LRF and the lignans niranthin, phyltetralin and nirtetralin showed noticeable anti-inflammatory properties [53].

40.6.14 Antioxidant Activity

The Total Phenolic Content (TPC) and antioxidant activity of fresh and dried Phyllanthus niruri were assessed by 2,2-diphenyl-1-picrylhydrazyl (DPPH) free radical scavenging activity, FolinCiocalteau method and Ferric Reducing Antioxidant Power (FRAP) assays. Different drying treatments led to significant reduction in antioxidant properties of Phyllanthus niruri methanolic extracts, and microwave drying method causing the highest decrease in TPC and antioxidant activity exhibited by the decrease in both FRAP and radical scavenging activity. This demonstrates that it has strong free radical scavenging activity [54].

40.6.15 Anti-spasmodic Activity

An alkaloid named as phyllanthoside is present in the *Phylanthus niruri* leaves and stem exhibiting strong antispasmodic activity. It attended as a relaxing agent for smooth muscles and it is also concluded that its spasmolytic accomplishment perhaps accounted for the efficacy of Phyllanthus niruri in dismissing stones [55].

40.6.16 Analgesic Activity

Methanol extract of dried callus tissue of this fruit exhibits strong analgesic activity [43].

40.6.17 Antibacterial Activity

Extracts of the root and leaf of *P. niruri* exhibits the antibacterial activity and it was evaluated against extend spectrum lactamase (ESBL) making Escherichia coli and isolated from the stool samples of HIV seropositive patients by using Bauer disc diffusion method. The strains are isolated from both the HIV sero-positive patients. This demonstrates the antibacterial activity of the extract [56].

40.6.18 Immune Modulatory Actions

An arabinogalactan (AG) enzyme which was obtained from Phyllanthus niruri tea preparations was found to have immunological properties. The glycoside showed the same activity when it is subjected to acidic and neutral gastric conditions using human gastric fluids and aq. HCL solution [57].

40.6.19 Aphrodisiac Activity

The methanolic extract of the leaves of *Phyllanthus niruri exhibits* Aphrodisiac activity [58].

40.6.20 Anti-ulcer Activity

The acidic heteroxylan and another polysaccharide exhibited anti-ulcer activity. These compounds lessened the gastric lesions encouraged by 65% and 78% ethanol. Phyllanthus niruri has proven to be an efficient activity against peptic ulcers [59].

40.6.21 Contraceptive Effect

An alcoholic extract of the whole plant of Phyllanthus niruri is used as antifertility effect [60].

40.6.22 Lipid Lowering Activity

Phyllanthus niruri alcoholic extracts exhibits lipid lowering activity in triton induced hyperlipidaemia. It was observed that administration of triton in rat caused increase in serum cholesterol by 3.5-fold, phospholipid 2-fold and triglyceride 1.2-fold. Administration of Phyllanthus niruri at the dose of 200 mg/kg simultaneously with triton lowered the level of total cholesterol, phospholipid and triglyceride by 27%, 25% and 24% respectively. Phyllanthus niruri at a dose of 100 mg/kg decreased the elevated level of low-density lipoprotein lipids in hyperlipidemic and drug fed animals [61].

40.6.23 Anticonvulsant Activity

Epilepsy is a foremost neurological disorder that is characterized by the manifestation of recurrent seizures. The two widely anticipated mechanisms involved are alterations in the voltage dependent ion channels such as reduction in inhibitory GABA and mediated drive or increase in excitatory glutamate mediated inputs. This chronic progressive CNS disorder affects a large population of the world. In search of herbal treatment, aqueous and ethanolic extract of *Phyllanthus niruri* were assessed for anticonvulsant effect using pentylenetetrazole (PTZ) and maximal electroshock - induced seizures (MES). The result showed that ethanolic and aqueous extract of leaves and stem of *Phyllanthus niruri* considerably effective in abolishing hind limb extension induced by MES as well as PTZ induced seizures [62].

40.6.24 Chemoprotective Activity

Seventy-five percent methanolic extract of the Phyllanthus niruri plant was studied for determining the chemoprotective activity effect of against cyclophosphamide (CTX) induced toxicity. Administration of CTX created significant myelosuppression which is seen from the decreased WBC count and bone marrow cellularity. Administration of Phyllanthus niruri extract at doses 250 and 750 mg/kg body weight meaningfully abridged the myelosuppression and enhanced the WBC count, bone marrow cellularity as well as the number of maturing monocytes that accounted for its chemoprotective activity. Furthermore, extract from the plant of Phyllanthus niruri was found to lessening the activity of phase I enzyme that exhibited powerful chemoprotective potential of the plant [63].

40.6.25 Dosage

- For Adults (18 years and older) generally, an infusion or tea of *Phyllanthus niruri* has been taken by mouth. Conventionally, individuals drink 1–3 cups daily or weekly. Certain pharmacies in South America sell concentrated extracts with a daily dose of 26 ml, taken twice or thrice daily.
- For treat acute viral hepatitis, 900 mg of powdered *Phyllanthus niruri* capsules has been taken by mouth three times daily for 7 days.
- To treat diabetes, 100 ml of *Phyllanthus niruri of extract has been taken* by mouth twice daily for 1 week. Moreover, two pellets of *Phyllanthus niruri*, composed of 0.8 g which have been taken three times daily by mouth for 10 days.
- For treating hepatitis B, 100 mg of dried *Phyllanthus niruri* has been taken by mouth three times daily for up to 3 months.
- For treating high blood pressure, two 0.8-g pellets have been taken by mouth three times daily for 10 days.
- For treating liver disease, 3 g of *Phyllanthus niruri* powder has been taken by mouth three times daily for 3045 days.
- For treating urinary stones, 450 mg of *Phyllanthus niruri* has been taken by mouth three times daily for 3 months. There is no demonstrated safe or effective dose for *Phyllanthus niruri in* children [64].

40.6.26 Side Effects and Warnings

- In overall, side effects were missing in human studies. *Phyllanthus niruri* is probably safe when in use up to 400 mg three times daily for 1 month, underneath the attention of a healthcare practitioner.

- *Phyllanthus niruri* might lessens the blood sugar levels. Attention is recommended in people with diabetes or hypoglycemic patients, and in those who are taking drugs, herbs, or supplements that affect blood sugar. Blood glucose levels may required to be examined by a qualified healthcare professional, including a pharmacist and so medication modifications may be compulsory.
- *Phyllanthus niruri* may upsurge the risk of bleeding. Attention is recommended in people with bleeding disorders and in those taking drugs that may increase the risk of bleeding. Dosing modifications may be essential.
- *Phyllanthus niruri* may lowers blood pressure and Carefulness is recommended in people taking agents that lesser the blood pressure.
- Drowsiness or sedation may occur. Use caution if driving or operating heavy machinery.
- It is recommended for use of *P. niruri* carefully in persons with liver disease.
- It is recommended for use of *P. niruri* carefully in individuals taking diuretics, and agents that lower cholesterol or agents that distress the immune system.
- It should be avoided in pregnant or breastfeeding women, and in those women trying to become pregnant.
- It is avoided in people with known allergy or sensitivity to *Phyllanthus niruri*; their parts, or participants of the Euphorbiaceae family [65].
- Phyllanthus niruri may also cause abdominal pain or discomfort; altered immune system function; altered levels of certain white blood cells; anorexia; chills; diarrhea; disturbed sleep; dizziness; enlarged lymph nodes; fatigue; fever; headache; hives; increased sodium, potassium, and chloride in the urine; increased urine volume; joint pain; lung disease; malaise (discomfort); muscle pain; nausea; rash; skin prickling; and sore mouth [66].

40.6.27 Pregnancy and Breastfeeding

There is a deficiency of scientific confirmation on the use of *Phyllanthus niruri* during pregnancy or breastfeeding.

40.6.28 Interactions with Drugs

- *Phyllanthus niruri* may lesser the blood sugar levels. Attentiveness is recommended when using medications that may also lower blood sugar. People who are taking drugs for diabetes by mouth or insulin should be watched closely by a qualified healthcare professional, including a pharmacist. Medication alterations may be obligatory.
- *Phyllanthus niruri* may increase the risk of bleeding when taken with drugs that upturn the risk of bleeding. Several examples include aspirin and anticoagulants (blood thinners) for example warfarin (Coumadin®) or heparin, antiplatelet drugs

such as clopidogrel (Plavix®), and nonsteroidal anti-inflammatory drugs such as ibuprofen (Motrin®, Advil®) or naproxen (Naprosyn®, Aleve®).

- *Phyllanthus niruri* may upturn the amount of drowsiness instigated by some drugs. It includes benzodiazepines such as lorazepam (Ativan®) or diazepam (Valium®), barbiturates such as phenobarbital, narcotics such as codeine, some antidepressants, and alcohol. Attention is directed while driving or operative machinery [67].
- *Phyllanthus niruri* may intermingle with agents for inflammation, pain relief, or wound healing; with agents that amend immune functions; agents that damage the liver; agents that proliferates the urine output; agents that impedes angiotensin-converting enzyme (ACE), angiotensinconverting enzyme receptor (ARB), or endothelin; agents that satirist the acetylcholine (a neurotransmitter); agents that defend against radiation; agents that guard against lethal effects of chemotherapy; agents that treat illnesses of the digestive tract,blood,, eyes, kidneys,heart, or agents that treat gout or inhibit xanthine oxidase (enzyme that breaks down purine); agents that expand blood vessels; agents that are used for cancer, diarrhea, fever, HIV/AIDS, malaria, obesity, or urinary stones; antibiotics, antivirals, cholesterol lowering agents, fertility agents, interferons (man-made version of an immune system protein), and nonsteroidal anti-inflammatories (NSAIDs, medications that reduce swelling and pain).

40.6.29 Toxicology

Phyllanthus niruri is less toxic, and it exhibited toxicity to batrachians and fishes when extract is water and alcohol based. It is also very lesser amount of toxic to mammals [68].

40.7 Conclusion

Phyllanthus niruri has been used ever since ages by the folk because of its rich medicinal standards. The comprehensive range of its medicinal use includes interpretations for the chemical investigation of the herb. Hereafter, the chemical standardization of the raw material and the formulations containing *Phyllanthus niruri* is under massive innovation and thus additional work is mandatory to inaugurate *Phyllanthus niruri* as a advantageous herb for management of numerous technical hitches. Combined with developments in methodologies for natural-product segregation, synthesis and characterization, this might be foundational door to a new epoch in the exploration of natural products in academic world and in productiveness. There are still many geographies of research on this herb that is requisite and to be considered for larger sample sizes, toxicological studies, molecular analyses, and mechanism studies.Modern confirmation is largely restricted to correlation between identified phytochemicals and their biological activities. There is a deficiency of mechanisms of action studies to understand the interaction of bioactive phytochemicals from *P. niruri* and their corresponding molecular targets. Fundamentally, more vigorous scientific approaches are compulsory before confirmatory decisions can be made on the prospective of *P. niruri*. A key step before clinical trials may be well-thought-out.

References

1. Sharma, S. K., Sheela, M. A., & Pharmacognostic. (2011). Evaluation of leaves of certain Phyllanthus species used as a botanical source of Bhumyamalaki in Ayurveda. *Ayu, 32*(2), 250–253.
2. Bolape Adeola Oyekanmi, Innocent Bamidele Osho (2016) Antimicrobial, phytochemical and pharmacological properties of Phyllanthus niruri Linn.
3. Kurian, J. C. (2004). *Plants that heal* (p. 112). Oriental Watchman Publishing House.
4. Sirajudeen, J., Ahamath, J. M., & Vahith, R. A. (2017). Phytochemical screening and antimicrobial activity of Phyllanthus niruri. *Journal of Advanced Applied Scientific Research (JOAASR)-ICCER-2017, 1*(7).
5. Patel, J. R., Tripathi, P., Sharma, V., Chauhan, N. S., & Dixit, V. K. (2011). Phyllanthus amarus: Ethnomedicinal uses, phytochemistry and pharmacology: A review. *Journal of Ethnopharmacology, 138*(2), 286–313. https://doi.org/10.1016/j.jep.2011.09.040
6. Bello, I., et al. (2020, March). Blood pressure lowering effect and vascular activity of Phyllanthus niruri extract: The role of NO/cGMP signaling pathway and ß-adrenoceptor mediated relaxation of isolated aortic rings. *Journal of Ethnopharmacology, 250*, 112461.
7. Hasan, H. H., Habib, H. I., Gonaid, M. H., & Islam, M. (2011). Comparative phytochemical and antimicrobial investigation of some plants growing in Al Jabal AlAkhdar. *Journal of Natural Products Plant Resources, 1*(1), 15–23.
8. Oyewo, E. B., Akanji, M. A., & Adekunle, A. S. (2012). Immunomodulation capabilities of aqueous leaf extract of Phyllanthus amarus in male Wistar Rats. *Report and Opinion, 4*(1).
9. Vanghan, J. G., & Judd, P. A. (2003). *The oxford book of health foods: A comprehensive guide to natural remedies*. Oxford University Press.
10. Ravalglia, G., Forti, P., & Maioli, F. (2000). Effects of micronutrient status on natural killer cell immune function in healthy free-living subjects age >90 years. *American Journal of Clinical Oncology, 3*(2), 871.
11. Prasaid, A. S. (2000). Effects of zinc deficiency on Th1 and Th2 cytokinesshift. *Journal of Infectious Diseases., 182*, 62–68.
12. De la Fuente, M., Hernanz, A., & Vallejo, M. C. (2005). The immune system in the oxidative stress conditions of aging and hypertension: Favourable effects of antioxidants and physical exercise. *Antioxidant Redox Signal, 7*, 1356–1366.
13. Pucci, N., et al. (2018). Effect of *Phyllanthus niruri* on metabolic parameters of patients with kidney stone: A perspective for disease prevention. *International Brazilian Journal of Urology, 44*(4), 758–764.
14. Saranraj, P., & Sivasakthivelan, P. (2012). Screening of antibacterial activity of the medicinal plant Phyllanthus amarus against urinary tract infection causing bacterial pathogens. *Applied Journal of Hygiene, 1*(3), 19–24. https://doi.org/10.5829/idosi.ajh.2012.1.3.71111
15. Adelowotan, O., Aibinu, I., Adenipekun, E., & Odugbemi, T. (2008). The invitro antimicrobial activity of Abrus precatorius (L) fabaceae extract on some clinical isolates. *Nigerian Postgraduate Medical Journal, 15*, 32–37.

16. Okwu, D. E. (2005). Phytochemicals, vitamins and mineral contents of two Nigerian medicinal plants. *International Journal of Molecular Medicine and Advance Sciences, 1*, 372–381.
17. Choi, Y. E., Yang, D. C., Kusano, T., & Sano, H. (2001). Rapid and efficient Agrobacterium mediated genetic transformation by plasmolyzing pretreatment of cotyledon in Panax ginseng. *Plant Cell Reports, 20*, 616–621.
18. Khanna, A. K., Rizvi, F., & Chander, R. (2002). Lipid lowering activity of Phyllanthus niruri in hyperlipideamia rats. *Journal of Ethanopharmacology, 82*(1), 19–22.
19. Olufayo, O. O., Tayo, G. O., Olumide, M. D. and Akintunde, A. O (2021). Assessment of the nutritive value of Phyllanthus niruri Linn. (stonebreaker) leaves. Nigerian Journal of Animal Science 23 (3): 108–115.
20. Nathanael, Y., Lee, S., et al. (2016). Pharmacology of P. nirur. *Royal Pharmaceutical Society, Journal of Pharmacy and Pharmacology, 68*, 953–969.
21. Mitra, R. L., & Jain, S. K. (1985). Concept of Phyllanthus niruri (Euphorbiaceae) in Indian floras. *Bulletin of Botanical Survey of India, 27*, 161–176.
22. Morton, J. F. (1981). *Atlas of Medicinal Plants of Middle America. Library of Congress Cataloging in Publication Data* (p. 1420). Thomas Books.
23. Prajapati, N. D., Purohit, S. S., & Sharma, A. K.. *A handbook of medicinal plants – A complete source book*. Agrobios; 1st edition, 392.
24. Maryadle, O. J., Annsmith, Patricia, H. E., John Jr., O. R., Jo Ann, G. R., & Mary, D. A.. *The Merck index*. Merck Research Laboratories, 139, 312, 483, 599, 624, 631, 737, 1428, 1456, 5515, 9142.
25. Oakes, A. J., & Morris, M. P. (1958). The West Indian Weedwoman of the United States Virgin Islands. *Bulletin of the History of Medicine, 32*, 164.
26. Quisumbing, E. (1951). *Medicinal plants of the Philippines*. Tech Bull 16, Rep Philippines, Dept Agr Nat Resources, Manilla 1951; 1.
27. Meselhy, M. R., Abdel-Sattar, O. E., El-Mekkawy, S., El-Desoky, A. M., Mohamed, S. O., Mohsen, S. M., & El-Halawany, A. (2020). Preparation of Lignan-rich extract from the aerial parts of Phyllanthus niruri using nonconventional methods. *Molecules, 25*, 1179.
28. Chandana, G., Manasa, R., Vishwanath, S., Shekhara Naik, R., & Ms, M. (2020). Antimicrobial activity of Phyllanthus niruri (Chanka piedra). *Journal of Nutrition, Metabolism and Health Science, 3*, 103–108.
29. Colpo, E., Vilanova, C. D. D. A., Pereiraa, R. P., et al. (2014). Antioxidant effects of Phyllanthus niruri tea on healthy subjects. *Asian Pacific Journal of Tropical Medicine, 7*, 13–118.
30. Banyal, S. K., & Banyal, A. K. (2019). Refinement of propagation techniques of aonla (Emblica officinalis Gaertn.) in north western himalayan region. *International Journal of Bio-resource and Stress Management, 10*, 87–91.
31. Gantait, S., Mahanta, M., Bera, S., & Verma, S. K. (2021). Advances in biotechnology of Emblica officinalis Gaertn. syn. Phyllanthus emblica L.: A nutraceuticals-rich fruit tree with multifaceted ethnomedicinal uses. 3. *Biotech, 11*(2), 1–25.
32. Wali, V., Bakshi, P., Jasrotia, A., Bhushan, B., & Bakshi, M. (2015). Aonla. Directorate of Extension, SKUAST-Jammu, 1–30.
33. Vashisht, B., Singh, C., & Biwalkar, N. (2018). Establishment and growth of Aonla (Emblica officinalis) as affected by irrigation and mulching in the Shivaliks of Punjab. *Journal of Soil and Water Conservation, 17*, 98–101.
34. Oka, Y., et al. (2007). Control of root-knot nematodes in organic farming systems by organic amendments and soil solarization. *Crop Protection, 26*, 1556–1565.
35. Gallaher, R. N., et al. (1988). Tillage and multiple cropping systems and population-dynamics of phytoparasitic nematodes. *The Journal of Nematology*.
36. Akhtar, M., et al. (2000). Roles of organic soil amendments and soil organisms in the biological control of plant-parasitic nematodes: A review. *Bioresource Technology*.
37. Collange, B., et al. (2011, October). Root-knot nematode (Meloidogyne) management in vegetable crop production. *The challenge of an agronomic system analysis., 30*(10), 1251–1262.

38. Stalin, C., Ramakrishnan, S., & Jonathan, E. I. (2007). Management of root knot nematode *Meloidogyne incognita* in bhumyamalaki (*Phyllanthus amarus*) and makoy (*Solanum nigrum*). 2(2), 119–122, ref.8.
39. El-Naggar, J. B., & Zidan, N. E. H. A. (2013). Field evaluation of imidacloprid and thiamethoxam against sucking insects and their side effects on soil fauna. *Journal of Plant Protection Research, 53*, 375–387.
40. Jindal, R., Sinha, R., & Brar, P. (2019). Evaluating the protective efficacy of Silybum marianum against deltamethrin induced hepatotoxicity in piscine model. *Environmental Toxicology and Pharmacology, 66*, 6268. https://doi.org/10.1016/j.etap.2018.12.014
41. Khandia, R., et al. (2020, October–December). Evaluation of the ameliorative effects of Phyllanthus niruri on the deleterious insecticide imidacloprid in the vital organs of chicken embryos. *Journal of Ayurveda and Integrative Medicine, 11*(4), 495–501.
42. Ross, V. A. *Medicinal plants of the World, Vol. 1: Chemical constituents, traditional and modern medicinal uses* (2nd ed.). © Humana Press Inc..
43. Akinjogunla, O. J., Eghafona, N. O., Enabulele, I. O., Mboto, C. I., & Ogbemudia, F. O. (2010). Antibacterial activity of ethanolic extracts of Phyllanthus niruri against extended spectrum β-lactamase producing Escherichia coli isolated from stool samples of HIV sero-positive patients with or without diarrhoea. *African Journal of Pharmacy and Pharmacology, 4*, 402407.
44. Thyagarajan, S. P., & Subramanian, S. (1988). Effect of Phyllanthus niruri on chronic carriers of hepatitis B virus. *Lancet, 2*, 7646.
45. Latha, U., & Rajesh, M. G. (1999). Hepatoprotective effect of an Ayurvedic medicine. *Indian Drugs, 36*(7), 470–473.
46. Mehrotra, R., Rawat, S., Kulshreshtha, D. K., Goyal, P., Patnaik, G. K., & Dhawan, B. N. (1991). In vitro effect of Phyllanthus niruri on Hepatitis B virus. *The Indian Journal of Medical Research, 93*, 71–73.
47. Sharma, P., Parmar, J., Verma, P., Sharma, P., & Goyal, P. K. (2010). Chemopreventive effect of Phyllanthus niruri on DMBA induced skin papillomagenesis in swiss albino mice. *International Journal of Biological and Medical Research, 1*(4), 158–164.
48. Ramsout, R., Rodgers, A., & Webber, D. (2011). Investigation of the effects of Phyllanthus niruri on in vitro calcium oxalate crystallization. *European Urology Supplements, 10*, 461–474.
49. Barros, M. E. (2006). Effect of extract of Phyllanthus niruri on crystal deposition in experimental urolithiasis. *Urological Research, 34*(6), 351–357.
50. Nishiura, J. L., Campos, A. H., Boim, M. A., Heilberg, I. P., & Schor, N. (2004). Phyllanthus niruri normalizes elevated urinary calcium levels in calcium stone forming (CSF) patients. *Urological Research, 32*(5), 362–366.
51. Murugaiyah, V. (2009). Mechanisms of antihyperuricemic effect of Phyllanthus niruri and its lignan constituents. *Journal of Ethnopharmacology, 124*(2), 233–239.
52. Kassuya, C. A. L., Daniela, F. P. L., Lucilia, V. M., Vera-Lucia, G. R., & Joao, B. C. (2005). Anti-inflammatory properties of extract, fractions and lignans isolated from Phyllanthus niruri. *Planta Medica, 71*, 721–726.
53. Lim, Y. Y., & Murtijaya, J. (2007). Antioxidant properties of Phyllanthus niruri extracts as affected by different drying methods. *LWT-Food Science and Technology, 40*, 1664–1669.
54. Grewal, R. C. (1984). *Medicinal plants* (1st ed., pp. 298–304). Campus Book International.
55. Santos, A. R. (1994). Analgesic effects of callus culture extracts from selected species of Phyllanthus in mice. *The Journal of Pharmacy and Pharmacology, 46*(9), 755–759.
56. Mellinger, C. G., Cipriani, T. R., Noleto, G. R., Carbonero, E. R., Maria Benigna, M., et al. (2008). Chemical and immunological modifications of an arabinogalactan present in tea preparations of Phyllanthus niruri after treatment with gastric fluid. *International Journal of Biological Macromolecules, 43*, 115–120.
57. Obianime, A. W., & Uche, F. I. (2009). The phytochemical constituents and the effects of methanol extracts of Phyllanthus niruri leaves (kidney stone plant) on the hormonal parameters of male Guinea pigs. *Journal of Applied Sciences and Environmental Management, 13*, 5–9.

58. Cipriani, T. R., Mellinger, C. G., de Souza, L. M., & Iacomini, M. (2008). Acidic heteroxylans from medicinal plants and their anti-ulcer activity. *Carbohydrate Polymers, 74*, 274–278.
59. Rao, M. V., & Alice, K. M. (2001). Contraceptive effects of Phyllanthus niruri in female mice. *Phytotherapy Research, 15*, 265–267.
60. Chandra, R. (2000). Lipid lowering activity of Phyllanthus niruri. *Journal of Medicinal & Aromatic Plant Sciences., 22*(1), 29–30.
61. Manikkoth, S., Deepa, B., Joy, A. E., & Rao, S. (2011). Anticonvulsant activity of Phyllanthus niruri in experimental animal models. *International Journal of Applied Biology and Pharmaceutical Technology, 4*, 144–149.
62. Kumar, K. B., & Kuttan, R. (2005). Chemoprotective activity of an extract of Phyllanthus niruri against cyclophosphamide induced toxicity in mice. *Phytomedicine, 12*, 494500.
63. Wright, C. I., Van-Buren, L., Kroner, C. I., & Koning, M. M. (2007). Antiallodynic and anti-oedematogenic properties of the extract and lignans from Phyllanthus niruri in models of persistent inflammatory and neuropathic pain. *J Ethnopharmacol, 114*, 1–31.
64. Calixto, J. B., et al. (1998). A review of the plants of the genus Phyllanthus: Their chemistry, pharmacology, and therapeutic potential. *Medicinal Research Reviews, 18*, 225–258.
65. Dhar, M. L., et al. (1968). Screening of Indian plants for biological activity: I. *Indian Journal of Experimental Biology, 6*, 232–247.
66. Venkateswaran, P. S., et al. (1987). Effects of an extract from Phyllanthus niruri on hepatitis B and woodchuck hepatitis viruses: In vitro and in vivo studies. *Proceedings of the National Academy of Sciences of the United States of America, 84*, 274–278.
67. Syamasundar, K. V., et al. (1985). Antihepatotoxic principles of Phyllanthus niruri herbs. *Journal of Ethnopharmacology, 14*, 41–44.
68. Kamruzzaman, H. M., & Hoq, M. O. (2016). A review on ethnomedicinal, phytochemical and pharmacological properties of Phyllanthus niruri. *Journal of Medicinal Plants Studies, 4*(6), 173–180.

Chapter 41
Moringa

Shahzeena Arshad, Bazghah Sajjad, Arusa Aftab, Zubaida Yousaf, and Modhi O. Alotaibi

41.1 Introduction

Moringa oleifera Lam. also known as "Wonder tree". It can grow in subtropical and tropical regions of the world, but it is thought to be indigenous to Pakistan, India, Bangladesh and Afghanistan [1]. *Moringa oleifera Lam. belongs to the family Moringaceae, the most broadly harvested species of the tropical blossoming flora, consisting of 13 various species* [1]. Out of 13 species (*M. oleifera, M. arborea, M. rivae, M. ruspoliana, M. drouhardii, M. hildebrandtii, M. concanensis, M. borziana, M. longituba, M. pygmaea, M. ovalifolia, M. peregrina,* and *M. stenopetala*) *M. oleifera* is well recognized for its usage in fertilizer, biogas generation, food, medicine and other industries [2].

In the Dravidian tongue of India, *M. oleifera* Lam. is referred to as Morunga, meaning "generic root." Mulangay, Nebeday, Kelor, Marango, Sauna, Saijhan, Mlonge, Benzolive as well as Moonga, are additional localized names. Radish tree, West Indian Ben tree, never dies tree, Horseradish tree and Drumstick tree are some of its common name [3]. *M. oleifera* has been cultivated in India for hundreds of thousands of years, even though the name "Shigon" for this tree first appears in the "Shushruta Sanhita" at the beginning of the first century A.D [2]. The Moringa plant has been around since 150 B.C. According to historical evidence, former monarchs and queens consumed Moringa fruit and leaves to keep their skin and minds sharp. Indian soldiers of the Maurian era were fed on extracts of Moringa leaves on the battlefield [4].

S. Arshad · B. Sajjad · A. Aftab · Z. Yousaf (✉)
Department of Botany, Lahore college for women university, Lahore, Pakistan
e-mail: zubaida.yousaf@lcwu.edu.pk

M. O. Alotaibi
Department of Biology, College of Science, Princess Nourah bint Abdulrahman University, Riyadh, Saudi Arabia

M. Zia-Ul-Haq et al. (eds.), *Essentials of Medicinal and Aromatic Crops*,
https://doi.org/10.1007/978-3-031-35403-8_41

South Asia is the region where M. oleifera originates, and it can be found growing in the Himalayan slopes from North-Eastern Pakistan to North-Western Bengal, India [5]. In other regions of the world, such as East and West Africa, the species were recognized and accepted [6]. It is a quickly expanding, drought-resistant plant that is perfectly suited to drier tropical regions [7]. and embraces temperatures varying from −1 to 48 °C, an average rainfall that varies from 250 to 3000 mm, and a pH level of 5.5–7.5 [8]. It develops best in desiccated sandy loam soil and flourished in equatorial meteorological conditions [8]. The plant is cultivated by seedlings or 1–2 m tall branch pieces are planted [9]. Since all of the Moringa plant's parts are a rich source of vitamins, protein, carotenoids, and minerals. It has been widely used for nutritional, economic, and therapeutic purposes around the world [10]. Being enriched with nutrients it has a potential to be used as a highly valuable edible agricultural product, medication, and animal feed, especially in developing nations [1].

41.2 Classification

Kingdom:	Plantae
Division:	Magnoliophyta
Class:	Magnoliopsida
Order:	Brassicales
Family:	Moringaceae
Genus:	*Moringa*
Specie:	*Oleifera*

M. oleifera is an annual or deciduous tree that typically reaches a height of 9 m. Its wood is soft and white, and its bark is corky and sticky. Even though the stalk is typically straight, it is occasionally ineffectively framed. The plant grows a straight, short stalk that reaches a height of 1.5–2 m and can go as high as 3.0 m before branching out. Its broadened limbs are scrambled, and its shelter resembles an umbrella [11, 12]. The leaves are vertically split into 30–75 centimeters lengthy, with a branch-jointed main stem and glandular joints. The segments of the leaves are entire and smooth and its roots have the taste of horseradish [13]. The leaflets have entire (no toothed) margins, are rounded or blunt-pointed at the apex, and are short-pointed at the base. The top surface of the leaflets is finely hairy, green, and nearly hairless; the underside is paler and hairless. The midveins have a crimson tint. White, fragrant flowers are arranged in broad, axillary panicles. Pendulous, ribbed capsules hold the seeds, which are triangular (Fig. 41.1) [14].

Fig. 41.1 Different parts of the Moringa oleifera plant (**a**) Leaves (**b**) Bark (**c**) Rhizome and (**d**) seeds

Table 41.1 Growth parameters of *Moringa oleifera*

Parameters	Requirement or range
Climate	Best growth in tropical and sub-tropical regions [15]
Rainfall	25–1500 mm is advantageous for its growth [16]
Altitude	Below 600 m and up to 1200 m [17]
Soil type	Sandy loamy, loamy, heavier clay loam soil [18]
Soil pH	5–9 [19]
Temperature	25–35° C under sunlight [20]

41.3 Cultivation

41.3.1 Climate and Soil Requirements

M. oleifera is classified as a tropical tree, adapted to a broad range of climatic conditions. It is mainly grown in tropical, subtropical and semi-arid regions (Table 41.1). Temperature, rainfall, latitude and elevation are the key factors that influence the growth and cultivation of Moringa [15]. Although it grows best at a temperature range of 25–35 °C but, 48 °C in shade and even light frost is also favorable for its growth. Temperature below 10 °C hinder down its growth. Annual rainfall of 25–1500 mm is advantageous for its growth, but some species can tolerate rainfall up to 3000 mm. Some cultivars are sensitive to flooding and waterlogged conditions [11, 21]. Altitudes below 600 m and up to 1200 m are best for its growth [17].

Moringa growth is favored by a variety of soil types such as well drained sandy loam soil, loamy soil and heavier clay loam soil, whereas poor drainage and continuous flooding did not favor its growth [18]. Moringa can grow successfully in soils with a pH range of 5–9, tolerating alkaline conditions up to a pH of 9 [19, 22]. Since Moringa has a deep root system that allows efficient absorption of soil nutrients, it can grow in moist soils without fertilizers addition [23]. For the formation of proteins application of nitrogen source is recommended [24].

41.4 Land Preparation

Moringa requires proper soil and seed bed preparation. MoringaTall trees, bushes, herbs and grasses should be removed from the field before it is ploughed and tilled to a maximum depth of 30 cm. Growing Moringa on ridges and furrows can reduce the risk of waterlogging [25]. In Taiwan at the world vegetable center Moringa is grown on elevated beds that are 30 cm high with maximum widths between 60 and 200 cm to improve drainage [26]. To avoid weed growth beds are covered with plastic mulch and holes are dug at a depth of 15–20 cm for replanting Moringa [18].

41.5 Propagation

41.5.1 Planting Materials and Methods

In order to make plantation, maintenance and harvesting easier, plant spacing should be 15 × 15 cm (or) 20 × 10 cm, with conveniently spaced alleys. They could also be cultivated by spacing the seed lines 45 cm apart and sowing every 5 cm on those lines [27]. It is difficult to prevent viruses and weeds from infecting Moringa because of its high susceptibility to them. The distance between plants in semi-intensive approach ranges from 50 cm to 1 m apart and this gives beneficial results [28].

Direct seeding, transplanting, or stem cutting are the three methods used to plant Moringa [29]. When there is plenty of seed available, but labor is limited, direct seedling (2 or 3 seeds per hill at a depth of 2 cm) is preferred. Transplanting (seeds are germinated in a nursery or green house) allows for greater flexibility in the field planting but requires more time and money to raise seedlings. When labor is plentiful and seedling is limited then, stem cuttings (branches of tree) are used [30].

41.5.2 Plant Spacing and Density

The spacing and pant density are determined by what Moringa plant parts are produced and for what purpose. Closer spacing and high plant density is required if the purpose is to produce a high amount of leaf biomass. Wider spacing is required if seed production is the primary objective. Plant spacing should be intermediate if the objective is to produce both leaf and seed [31].

41.5.3 Propagation by Tissue Culture

41.5.3.1 *In Vitro* Culture

Due to increasing demand for planting material and shortage of seed alternative propagation methods are being explored to enhance rapid multiplication. The use of clonal micropropagation (use of young nodal seedlings) and tissue culture (use of immature seeds) has been reported [32–34].

41.6 Irrigation

The Moringa plant can withstand drought and doesn't require frequent irrigation. When seeds are sown in moist conditions, which is common during the rainy season, they will germinate and grow well [18, 35]. It is advised to irrigate often for the first 2–3 months following sowing or planting in order to ensure optimal growth [36]. Irrigation is required to keep the trees producing leaves throughout the dry season. The trees grow slowly and shed their leaves without irrigation throughout the dry season. Irrigation is not required once the trees are completely established in the humid tropics where rainfall is evenly distributed throughout the year [37]. Watering cans, sprinklers, hoses, and low cost, gravity flow drip systems can all be utilized to irrigate small to large scale plantation [38]. During the dry season, drip irrigation can be used for commercial cultivation, with daily application rates of 12–16 l of water per tree, and half this rate during the wet seasons [19]. Moring species can also be grown using wastewater if pH is below 8.5 [23].

41.7 Pest and Disease Management

Most pests and diseases don't affect Moringa, although under some circumstances outbreaks might happen. For example, severe wilting and death of plants occur in waterlogged soils due to presence of diplodia root rot disease. During the wet season stem rot is also very common in Moringa [24]. During dry and cool weather mite populations also increase. Yellowing of leaves occurs due to pests but plants usually recover during warm weather. Some of the pests are termites, aphids, leaf miners, whiteflies, grasshoppers, crickets and caterpillars [39]. These insects cause destruction of leaves, buds, shoots, flowers, seeds and pods by chewing parts of plants as well as interrupt sap flow. Ants damage the seeds by attacking mature and dry pods. These outbreaks are common in dry areas where Moringa leaves attract pests and insects, especially during the dry season when they are unable to locate other green foliage to feed on [16].

Brown spots and Alternaria are the most serious fungal diseases in Moringa cultivation. Brown spots (*Cercospora* spp. and *Septoria lycopersici*) damage the leaves forming small spots turning the leaves yellow and eventually leading to dry and dead tissues [16]. *Alternaria solani* infects leaves as well as branches forming angular dark brown and black spots with concentric circles. Effective fungicides are either mancozeb or maneb [40]. Demand for organic Moringa is increasing day by day which can be achieved by using beneficial organisms, natural enemies and organic biopesticides. *Bacillus thuringiensis* and neem seed extracts are effective biopesticides against insects and pests of Moringa [41]. When there is a severe infestation, chemical pest treatment should only be employed. Only those pesticides should be used which are effective for only a few days and are degradable. Animals like cattle, pigs, sheep, pigs and goats feed on Moringa plants therefore, it is necessary to install a fence or grow a hedge around the plot to keep cattle away from the Moringa seedlings [42].

41.8 Weed Management

Weeds and Moringa compete for soil nutrients, water, and light, and if they are not handled in a timely manner, the competition can be quite detrimental, particularly during the seedling and plant establishment stages. Soil should thoroughly prepare to eliminate all weeds before planting [18]. Weeds can be managed by using hand hoes (loosen soil for aeration), mulching (organic or synthetic) [43].

41.9 Manuring

Moringa is efficient in mining nutrients from the soil due to its deep root system and can grow well even without fertilizers [18]. It can produce large amounts of green biomass if it is provided with organic soil amendments and manure [44]. For the best results fertilizers are applied at the time of planting. Commercial NPK fertilizer is applied at 300 g per tree and around the base of the plant holes and trenches are dug (10–20 cm). farmyard manure or compost (1–2 kg per tree) can be applied in case fertilizers are not available [19, 45]. For commercial fertilizers 100–100-100 kg /ha of K_2O, P_2O and N is applied per year. If trees become productive for 5 consecutive years, it is recommended to fertilize once or twice each year. Application of nitrogen fertilizer increases quality, leaf tissue N, chlorophyll content, protein content and leaf biomass yield [46, 47]. Foliar application of K and N also enhance chemical content (vitamin C, N, Ca, K and total chlorophyll content) of leaves, yield and pod quality as compared to control [48].

41.10 Ecology

Moringa colonizes easily in areas of savannah and stream banks where the soil is well-drained, and the water table is consistently high. Although it tolerates drought well, it produces significantly less foliage when it is consistently under water stressed. Frost cannot affect it, but a freeze can kill it back to the ground. When cut, it instantly sprouts new growth from the root; when frozen, it sprouts new growth from the ground [49].

41.11 Bioactive Constituents

Different parts of *Moringa oleifera* plant have different chemical constituents. *Moringa oleifera* leaves comprise mainly vitamin C, saponins, anthraquinones, phenols, esters, fatty acids, hydrocarbons and alcohols [50, 51]. 1, 2, 3-propanetriyl ester-9 octadecenoic acid, 1, 2-benzenedicarboxylic acid, 4, 8, 12, 16-tetramethyl heptadecan-4-olide, 3–5-bis (1, 1-dimethylethyl)-phenol, 1-hexadecanol, 3, 7, 11 and 15-tetramethyl-2 hexadecene-1-ol, N-(−1-methylethyllidene)- benzene ethanamine, 3, 4-epoxyethanone, 9-Octadecenoic acid, Ascorbic acid- 2, 6- dihexadecanoate, 14 –methyl −8- Hexadecenal, 4- hydroxyl-4-methyl 2-pentanone, 3-ethyl-2, 4-dimethylpentane, octadea methylcyclononasiloxane and phytol are the major chemical constituents [52, 53].

Seeds of *Moringa oleifera* comprise mainly hydrocarbons, sterols, alkaloids, alcohols, esters, tannins, amino acids and fatty acids [54]. 4-(4'-O-acetyl-α-Lrhamnosyloxy) benzyl isothiocynate, 3–4-(α-L-rhamnosyloxy) benzyl isothiocyanate and niazimicin, triolein, β-sitosteryl oleate, 1 and 4-(3'-O-acetyl-α-L-rhamnosyloxy) benzyl isothiocyanate, β-sitosterol and stigmasterol, oleic acid, 1-octadecane, moringyne, 4-(á-L-rhamnosyloxy) benzyl isothiocyanate and several amino acids [55, 56].

Isolation of water soluble polysacahride which contains D-galactose, 6-O-Me-D-galactose, D-galacturonic acid, L-arabinose, and L-rhamnose, glucosinolates has been reported form pods of *Moringa oleifera* [57, 58].

The gum of *M. oleifera* contains aldotriouronic acid, O-(β-D-glucopyranosyluronic acid) (1 → 6)-β-D-galactopyranosyl (1 → 6)-D-galactose [13].

The flowers of *M. oleifera* are rich in antioxidants (α- and γ-tocopherol), calcium, potassium, amino acids, D-glucose, sucrose, alkaloids, wax, quercetin, kaempherol, rhamnetin, isoquercitrin and kaempferitrin [59, 60].

Glucosinolates are also found in the bark of *M. oleifera* and 4-(alpha-l-rhamnopyranosyloxy)-benzylglucosinolate are also detected in the bark tissue [61].

The roots of *M. oleifera* contain high concentrations of both 4-(alpha-l-rhamnopyranosyloxy)-benzylglucosinolate and benzyl glucosinolate and aurantiamide acetate 4 and 1, 3-dibenzyl urea 5 [61, 62].

41.12 Description of Parts of *Moringa oleifera* Plant

The leaves, seeds, roots, and flowers of the Moringa tree are all safe for ingestion by both human beings and animals. Being highly nutritive leaves are used in traditional medicinal systems. Instead, the seeds have piqued scientists' interest since *M. oleifera* seed kernels feature a substantial amount of oil (up to 40%) with a high-quality fatty acid composition (oleic acid >70%) and, after being refined, a notable resistance to oxidative degradation [3]. *M. oleifera* inflorescences are axillary loose panicles that range in length from 10 to 25 cm. The up to 12 mm long, fragrant, white or cream colored, bisexual zygomorphic blooms have five pale green sepals, five white petals, five stamens with anthers, and five stamens without anthers (staminoid) [11]. Due to the delayed stigma receptivity, the flowers are heavily cross-pollinated, and several insects are needed for effective pollination [14].

M. oleifera seeds are three angle globular shaped, with a diameter of 1 cm. Their average weight is 0.3 g. Seeds are winged that grows from the base of the seed to the tip and are 2–2.5 cm long and 0.4–0.7 cm broad. 70–75% weight of seeds is made up by the kernels (Fig. 41.2). In the second year of development, lobes and pods are formed. Approximately after 3 months of flowering, 20–60 cm long trilobite capsule-shaped fruit (pod), matures. On maturation, pods turn brown, dry and longitudinally divide into three sections. Typically, each pod contains 12–35 spherical seeds with a diameter of 1 cm. The seeds have three papery, pale "wings" arranged around the brownish, semi-permeable shell at 120° intervals. A single tree can produce almost 15,000–25,000 seeds of an average weight of 0.3 g. However, some kinds take longer than a year to yield pods, early blooming variants do it in 6 months [63].

41.12.1 Ethnobotical Uses

The seed oil of *M. oleifera* can be used for culinary, making makeup, or even as medicine [64]. Due to its ability to absorb and keep scents, perfume made from seed oil is highly prized by perfumers, especially to produce deodorant products [65, 66].

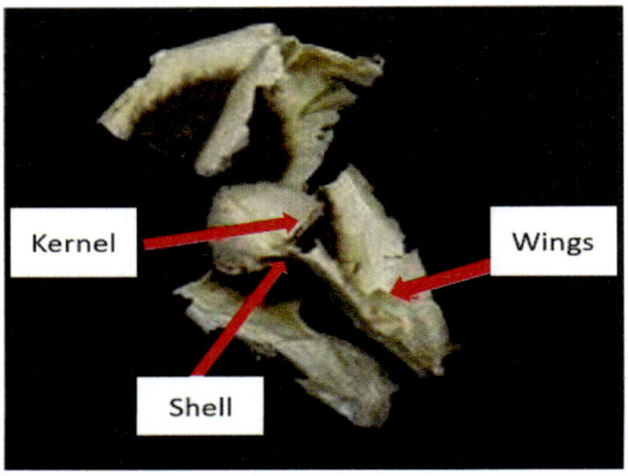

Fig. 41.2 Different parts of *M. oleifera* seeds

Specialized protein fractions for skin and hair maintenance are also present in the seeds. Anti-pollution proteins from the *M. oleifera* seed safeguard the skin from pollutants and fight early skin aging [67–69]. The seed extract is a widely recognized invention and an effective hair care product [70]. In addition, seed flour cake can be utilized to disinfect water, lowering the incidence of waterborne illnesses that are a major cause of mortality in developing nations [71]. Additionally, seed oil is of exceptional quality [72] and can be utilized as a raw material for making biodiesel [73]. The juvenile pods are the most valuable and nutrient dense as they comprise all the necessary amino acids as well as a variety of supplements and other calories [63].

Pods may be consumed by both humans and animals [74–76]. Mature pods are typically cooked and have a peanut flavor, whereas juvenile pods can be consumed raw or cooked as legumes or bean sprouts and are said to taste like asparagus. They also generate 38–40% of the Ben Oil, a nutritious oil used for cleaning watches and other sensitive machinery as well as in the art [77]. Its flavor cannot change because this oil is transparent, bitter, and odorless [78]. Overall, its nutritional content is like that of olive oil, and it is noteworthy that it contains anti-inflammatory properties that reduce inflammation and pain brought on by gout, arthritis, and rheumatism [79, 80].

The roots can also be used in cuisine as a garnish or a sauce because they have a taste that resembles horseradish [75, 81]. The bark can be used to make fibers, dyes, and tannins for skin tanning [63]. The wood of *M. oleifera* is valuable as a building material because of its capacity to retain heat, and it can be utilized as a windscreen to stop soil erosion [4, 82, 83].

The fresh leaves are used as spices, in salads, or veggie curries. They may also possess therapeutic qualities or be prepared as soups and stews [84]. Moringa leaves have milk and eggs in terms of protein quality and contain more potassium than bananas, vitamin C than oranges, iron than spinach, calcium than milk, and vitamin A than carrots [85]. They should be collected early in the morning and sold that same day if they are to be consumed fresh. The leaves are ground up and used for cleaning surfaces and dishes. It's fascinating to use dried leaves to make foods that are more nutrient-dense by combining them with grains and legumes to create a full protein source. Older leaves should have removed their tough, woody stalks because they are better for drier leaf granules [86]. This powder is used to enhance cuisine and can be kept at ambient temperature for several months without lacking its nutritional content [87].

Fresh *M. oleifera* leaves have a beneficial impact on animal feed because they encourage higher levels of metabolizable energy efficiency by boosting microbial growth as well as higher levels of pasture energy efficiency [88–90].

Stem and flowers can also be eaten fresh or cooked, the dried flowers are used for tea and are high in potassium and calcium [91, 92]. The larvicidal, ovicidal, and insecticidal action is present in the seeds, leaves, and particularly the flowers [84]. The production of newspapers and the cloth industry both use stem pulp [93].

41.13 Medicinal Uses

It is well known that *M. oleifera* leaves are an efficient source of various antioxidants [94], which may explain how *M. oleifera* leaves affect various cancers (Table 41.2). To shield important biomolecules from the oxidative harm caused by free radicals, both ancient and juvenile leaves have oxidation resistance [95]. The cancerous potential of *M. oleifera* leaves are directly attributed to the significantly high antioxidant contents [96]. The extracts from the leaves have been shown to have anti-proliferative effects, which can stop the development of cancer cells [80].

Table 41.2 Medicinal uses and nutritional value of various parts of plant

Part of plant	Medicinal uses of plant	Nutritional value
Leaves	Asthma, hyperglycemia, dyslipidemia, influenza, heartburn, syphilis, malaria, pneumonia, diarrhoea, headaches, scurvy, skin conditions, bronchitis, eye and ear infections, and migraines can all be treated with moringa leaves. Also lowers blood pressure and cholesterol while acting as a neuroprotectant, anticancer agent, antibacterial, antioxidant, antidiabetic, and anti-atherosclerotic agent	Fiber, fat, protein, and minerals including Ca, Mg, P, K, Cu, Fe, and S are all present in moringa leaves. There are several vitamins present, including vitamin A (beta-carotene), vitamin B-choline, vitamin B1-thiamine, riboflavin, nicotinic acid, and ascorbic acid. Many amino acids are found, including Arg, His, Lys, Trp, Phe, Thr, Leu, Met, Ile, and Val. There are phytochemicals such flavonoids like quercitin, isoquercitin, kaemfericitin, isothiocyanates, and glycoside compounds as well as tannins, sterols, saponins, trepenoids, phenolics, and saponins
Seeds	Moringa seeds have antibacterial and anti-inflammatory properties, making them useful in the treatment of a variety of conditions including hyperthyroidism, Chrohn's disease, anti-herpes-simplex virus joint problems, gout, rheumatism, epilepsy, and sexually transmitted illnesses	Leaves contains fatty acids like linoleic acid, linolenic acid, and Behenic acid, as well as phytochemicals including tannins, saponin, phenolics, phytate, flavanoids, terpenoids, and lectins. It also contains an antibiotic called pterygospermin. Besides them, foods also include lipids, fibre, proteins, minerals, vitamins A, B, and C, and amino acids
Root bark	Root bark has anti-ulcer, anti-inflammatory, and cardiac-stimulating properties	Minerals include calcium, magnesium, and sodium, as well as alkaloids like morphine and moriginine
Flower	Moringa flowers have hypocholesterolemic, anti-arthritic, and anti-viral properties	It includes amino acids, calcium, and potassium. They also have nectar in them
Pods	Moringa pods are used to alleviate joint discomfort, liver and spleen issues, and diarrhoea	Moringa pods are used to alleviate joint discomfort, liver and spleen issues, and diarrhoea

According to their antioxidant properties, *M. oleifera* leaf has the potential to treat both Type 1 and Type 2 diabetes [97]. The anti-inflammatory, neuroprotective, cardioprotective, hepatoprotective, and anti-ulcer action of *M. oleifera* leaves are also attributed to the plant's antioxidant properties [98].

The wealth of vitamins also gives *M. oleifera* leaves the ability to treat a variety of symptoms. Vitamin C can scavenge free radicals and contribute to the production of vitamin E. The latter can both avoid and treat symptoms of scurvy like rashes, gum conditions, and ozostomia. *M. oleifera* leaves can aid in the treatment of night blindness because vitamin A serves as a kind of essential nutrient for regular vision, particularly in low-light conditions [99].

Additionally, studies on rodents fed leaf extracts showed that they reduced urea and creatinine levels and increased blood protein levels, protecting the kidneys from nickel-induced nephrotoxicity [100]. *M. oleifera* leaf preparations can lower acetylcholine esterase activity, which enhances cholinergic function and memory [101]. *M. oleifera* leaves are a useful agent for boosting defenses and aiding in the prevention of illnesses like HIV and AIDS [102].

M. oleifera blossoms have pharmacological benefits for hepatoprotection, diuretic action, and the prevention of spleen, muscle, and tetanus diseases [103]. Rats' rheumatoid symptoms can be reduced by *M. oleifera* blossom hydro-alcoholic extract, which can also treat arthritic [104]. Flower infusion has been identified as a type of herbal stimulant with a variety of therapeutic benefits for treating infertility and dysfunctional testosterone in both men and women [105].

Due to their high protein and fiber composition, *M. oleifera* pods are useful for treating undernutrition, diarrhea, and digestive issues, and preventing colon cancer [106]. Pods can help breastfeeding women in lactation produce more milk, combat malnutrition [107] and treat endocrine disorders [108]. Strong anti-inflammatory benefits of the ethanolic extract of pods have been documented [109].

The seeds of *M. oleifera* have phenolics and flavonoids, which have antioxidant properties, similar to the foliage and pods of the plant [110]. Seeds have the potential to be a source of hypoglycemic hypotensive, anti-inflammatory, antidiabetic, antidyslipidemic, and anticancer agents [111] for the treatment of liver fibrosis, skin ulcer, spasm, rheumatism, and arthritis [112] as well as boost immune response [113]. The seed extract has anti-rhinitis and anti-asthmatic properties as well as the ability to relax bronchioles and heal the respiratory tract. Moringine and moringinine are present in seeds [114]. The extracts of M. oleifera seeds have antimicrobial effects that inhibit the development of bacteria like Hortaea werneckii, Candida spp., and some other food-borne microbes [115].

The extract of *M. oleifera* barks can also be used as a cardio activator due to its high level of alkaloids, and *M. oleifera* barks can greatly lower the level of acid in stomach ulcers and thus aid in curing ulceration [116]. *M. oleifera* roots can be used to treat a variety of conditions, including cardiac complaints, arthritis, flatulence, dyspepsia, otalgia, edema, toothache dental caries eye disease, tumors, anabrosis, common cold, and thyroid nodules [117]. The root has therapeutic effects for the treatment of diarrhea. The gum developed from the bark has antimicrobial, anti-inflammatory, and antifungal effects [118]. The gum is additionally used in dyeing, textile printing, and grocery stores [119]. The gum is a very excellent option for a natural medicinal excipient that can improve the stability and disintegration of drugs to increase therapeutic efficacy [120].

41.13.1 Antihypertensive Activity

Two glycosides (thiocarbamite and isothiocyanate) identified from ethanolic extract of *Moringa oleifera* pods showed antihypertensive activity. Infusion of Moringa roots, leaves, flowers, gum and aqueous seed also reported in lowering blood pressure [121].

41.13.2 Antimicrobial Activity

Root and leaf extract exhibits potent antimicrobial profile; it is due to presence of pterygospermine. Bark possess antifungal potential [122] and extract of stem bark showed antibacterial potential against *Staphylococcus aureus*. Seeds documented to have the ability to stop the reproduction of bacteriophages [122].

41.13.3 Anticancer Activity

Niaziminin, a thiocarbamate is present in the leaves, showed to inhibit the activation of the Epstein-Barr virus caused by tumour promoters [95].

41.13.4 Cholesterol Lowering Activity

A study observed *in vivo* blood cholesterol lowering potential of *M. oleifera* leaves. Significant low blood cholesterol level was found in blood of rats on a high fat diet. It was observed that ß-sitosterol present in leaves is responsible for lowering cholesterol level. Fruit was reported to lower the lipid profile and enhances faecal chlosterol excretion in hypercholesteremic rabbits [123].

41.13.5 Antispasmodic Activity

The traditional usage to treat diarrhoea is based on the antispasmodic effect of the compound 4-[α-[L-rhamnosyloxy] benzyl]-o-methylthiocarbamate (trans) in the ethanol extract of leaves [124].

41.13.6 Hepatoprotective Activity

Aqueous leaf extract showed antiulcer effect. A well-known flavonoid i.e. quercetin, identified in flowers has been shown to have hepatoprotective properties, in both their aqueous and alcohol extracts [125].

41.13.7 Anthelmintic Action

M. oleifera gum is also known as anti-filarial substance. *in vitro* anti protozoan efficacy of *M. oleifera* had been evalualted. Aqueous extract showed larvicidal, pupicidal, and adult mosquito-killing capabilities against *Culex quinquefasciatus* [126].

41.13.8 Uterotonic Activity

Cold and hot aqueous extracts of *M. oleifera* leaves (MOL), which are often employed in Nigerian traditional medicine, have both in vitro and in vivo uteroactive and contraceptive potentials [108, 127].

41.13.9 Central Nervous System Activity

To evaluate the protection efficacy of *M. oleifera* against the monoaminergic deficiency related to Alzheimer disease, a rat model was prepared. For this Purpose, Alzhemier disease was induced in rats by injecting colchicines intracerebrally [128]. Methanolic root extract exhibits dose-dependent central nervous system (CNS) depressive effect [129] but it also offered protection from convulsions brought on by strychnine and leptazol.

41.13.10 Wound-Healing Potential

Significant wound-healing potenial was observed by the aqueous and ethyl acetate extract of *M. oleifera* leaves leaves [130].

41.13.11 Anti-hyperglycemic Activity

The aqueous extract exhibits hypoglycemic and anti-diabetic effects in rats [131].

41.13.12 Anti-pyretic Activity

In albino rats, ethanol seed extract reduced the normal body temperature in a dose-dependent manner [132].

41.13.13 Anti-asthmatic Activity

In vivo experiment showed that ethanolic seed extract inhibits the immune-mediated inflammatory reactions that lead to asthma in Wistar rats. Dried seed kernels can treat bronchial asthma [133].

41.13.14 Anti-inflammatory, Antiarthritic, and Analgesic Activity

From the roots, 1,3-dibenzyl urea and aurantiamide acetate were isolated. These two bioactive compounds showed anti-inflammatory and analgesic effects by inhibiting tumour necrosis factor-alpha, inter-leukin, and other cytokines. Rheumatoid factor levels in the serum were also noted to be reduce [134].

41.13.15 Anti-thyroid Activity

The leaf extract inhibited the peripheral conversion of tetraiodothyronine to triiodothyronine in adult Swiss rats [135].

41.13.16 Anti-allergic Behavior

An anti-anaphylactic effect of *M. oleifera* seed ethanolic extract has been observed [136–138].

41.13.17 Radio-Protective Activity

M. oleifera leaf extract in methanol had radioprotective effects that were seen in pretreatment, radiotreated Swiss albino mice [139].

41.13.18 Anti-fertility Activity

A study reported that the anti-implantational properties of *M. oleifera* root may be due to its distinct antiprogestational activity [140].

41.13.19 Anti-oxidant Activity

The ethanolic extract reported to have significant metal chelation characteristics with the potential to prevent DNA nicking, however, the phenolic component in the leaves imparts free-radical scavenging capacity [141].

41.14 Traditional Uses

The herb has conventionally utilized as an expectorant, diuretic, stimulant, and antispasmodic. Its gum is used as an adhesive. The acrid kernel acts as a stimulant. The wood has antibacterial and antifungal properties. Its flowers are thought to be diuretics, tonics, stimulants, and helpful for boosting bile flow. The herb is also used as a disinfectant and heart circulatory tonic. The beans are used to treat diabetes and are thought to be antipyretic and anthelmintic. Moringa leaves can lower thyroid hormone levels [108, 142]. To maintain healthy domestic animals, farmers added Moringa leaves to their livestock feed [85]. The root liquid is used for the treatment of asthma, enlarged liver, spleen, nervous debility, antiepileptic, and cardiac tonic purposes. Almost all plant parts can be used to make nourishment. Blue pigment is produced by *M. oleifera* wood. People cook tiny leaflets and consume them like spinach in many different cultures. In Africa, the Moringa tree is frequently established as a living fence (Hausa) tree [143].

Malnutrition, particularly in young children and nursing mothers, has been combated by using Moringa plants. Approximately 23% of the iron, 40% of the calcium, 14% of the protein, and nearly all of the vitamin A needs for a kid aged 1–3 are met by one level tablespoon (8 g) of leaf powder. During maternity and lactating, six circular spoonfuls of leaf powder provided nearly all of a woman's daily iron and calcium requirements [144]. To increase breast milk output, Philippine women use Moringa leaves in shellfish and chicken soup [59].

41.15 Myth

The majority of Indonesians think that if someone has been ill for a long time but hasn't yet passed away, there are supernatural abilities in that person that should be eliminated from his body. The person is typically bathed in *Moringa oleifera* leaves to dispel the mysterious supernatural forces until the sick person can pass away quietly. The individual bathed his body and then washed it once more with *M. oleifera* leaves to get rid of any remaining magical items and creatures. It's also thought that *M. oleifera* rejects the existence of ghosts. As a result, some homes in Indonesia have a bundle of *M. oleifera* in the main entryway as some protective supports. Because of those magical myths, Indonesians do not prefer to eat *M. oleifera* as their primary source of nutrition [145].

41.16 Summary

Moringa oleifera is a rapidly growing tree that belongs to the family Moringnacaeae. –Globally it has been exploited for many purposes. Up till now its ethnopharmacological, pharmacological activities, phytochemistry, phytopharmaceutical formulations, clinical investigations, toxicity analysis had been evaluated. The therapeutic properties of *M. oleifera* are due to the presence of alkaloids, phenolic acid, glycosides, sterols, glucosinolates, flavonoids, terpenes, and fatty acids. It is also rich in substances like vitamins, minerals, and carotenoids, which raises its therapeutic usefulness and acceptance as a superfood. Pharmacological research demonstrates that the plant's active ingredients have successfully treated a number of disorders, including cancer, hypertension, diabetes, obesity, and neuropathic pain. Yet, it is still unknown if some phytochemicals have any medicinal potential (Table 41.3).

Table 41.3 list of phytoactive compounds present in various parts of plant

Plant part	Constituents	Structure
Leaves	Isoquercetin	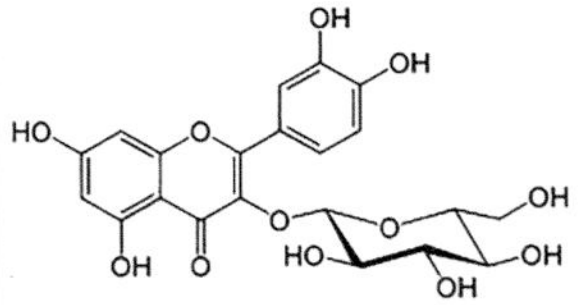
	Astragalin	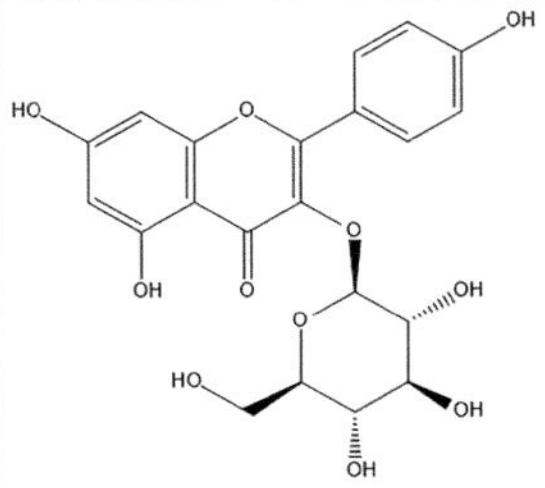
	Isorhamnetin	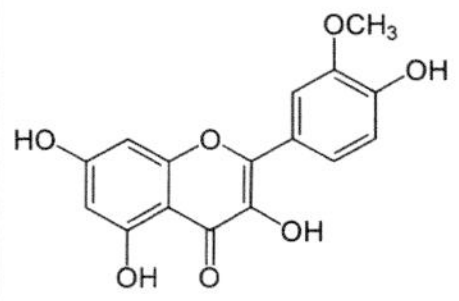

(continued)

Table 41.3 (continued)

Plant part	Constituents	Structure
	Daidzein	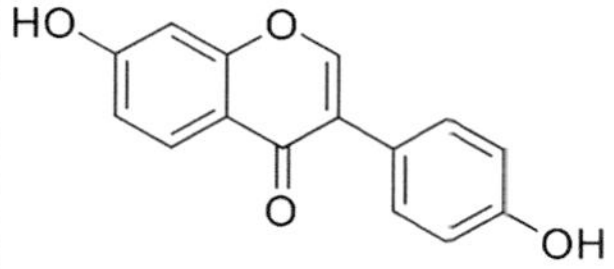
	Apigenin	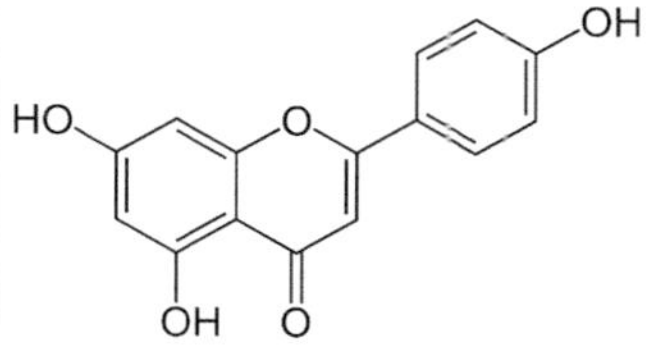
	Luteolin	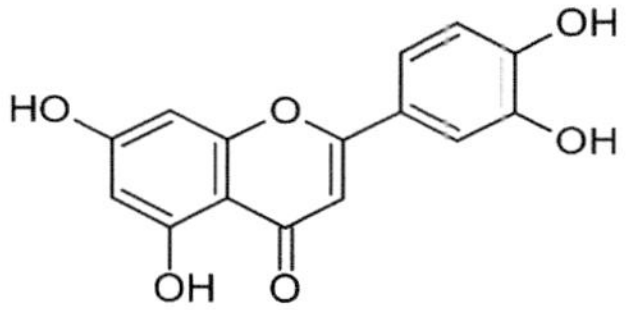
	Genistein	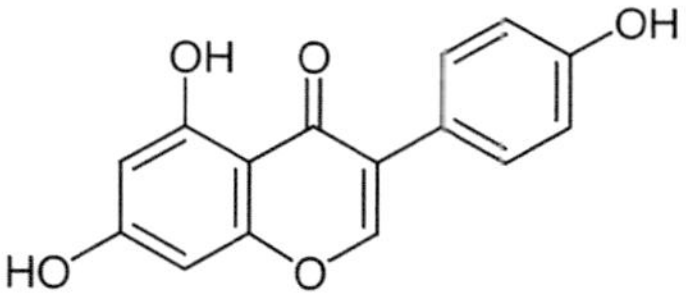
	4-(α-L-rhamnopyranosyloxy) benzyl glucosinolate	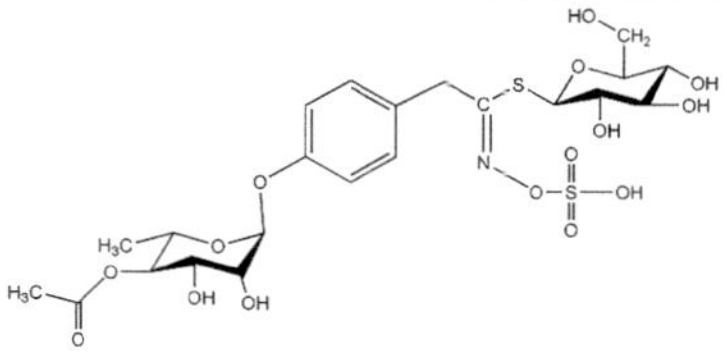
	4-[(2′ -O-acetyl-α-L-rhamnosyloxy) benzyl] Glucosinolate	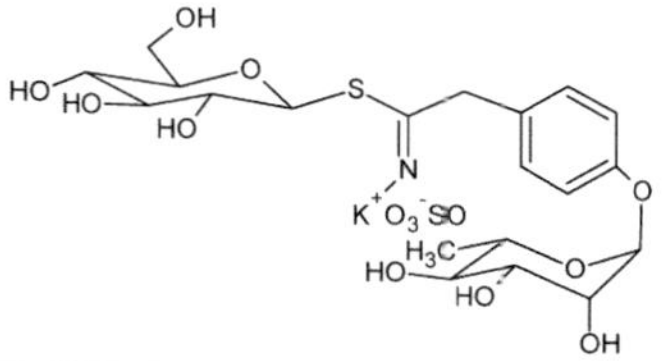
	Epicatechin	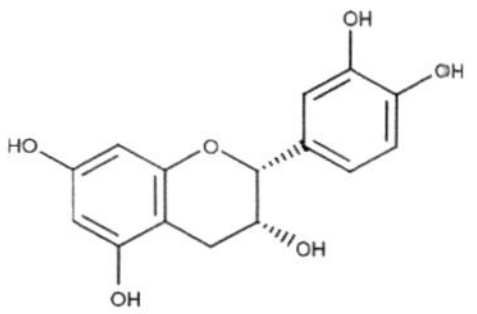
	Ferulic acid	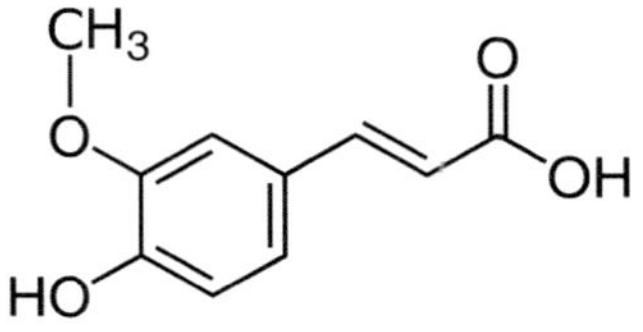

(continued)

Table 41.3 (continued)

Plant part	Constituents	Structure
	Caffeic acid	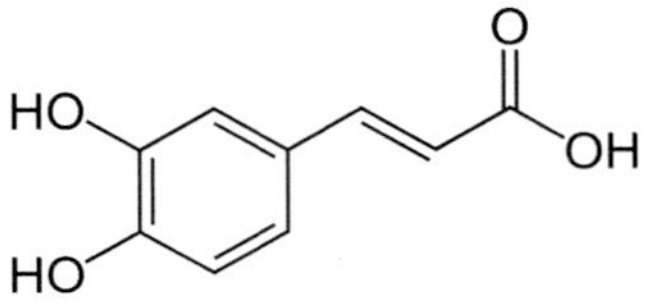
	Ellagic acid	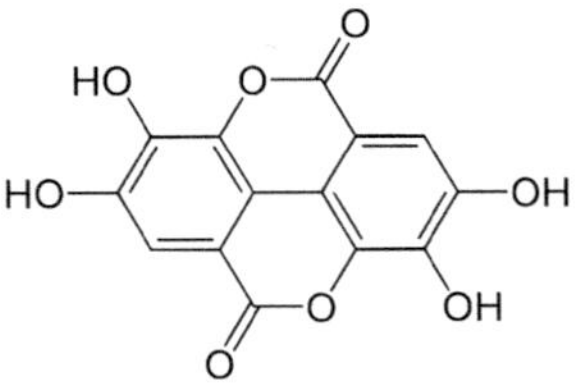
	Sinalbin	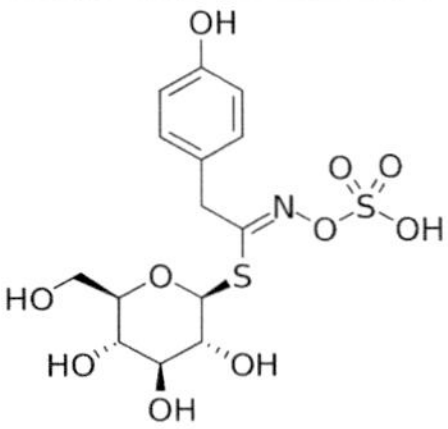
	Sinapic acid	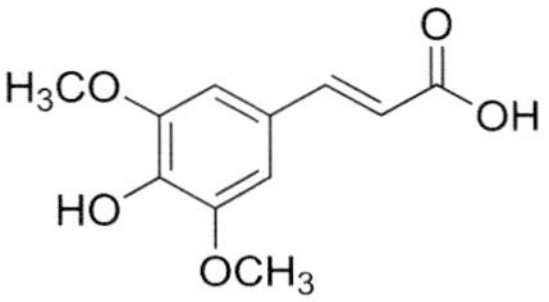
	Chlorogenic acid	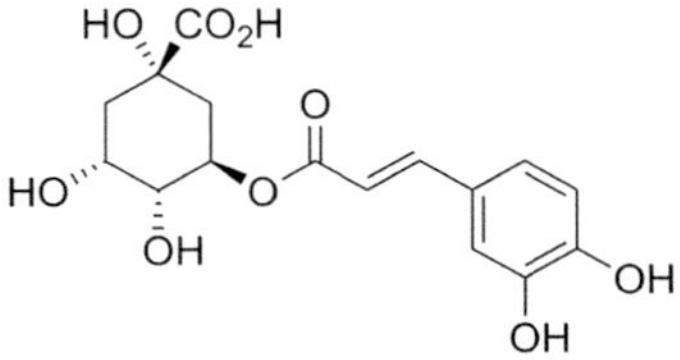
	Gallic acid	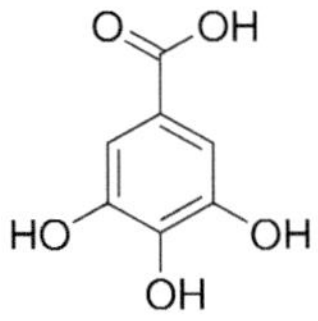
	Salicylic acid	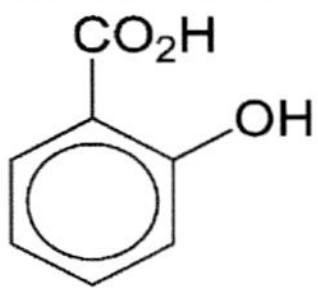

(continued)

Table 41.3 (continued)

Plant part	Constituents	Structure
	Vicenin-2	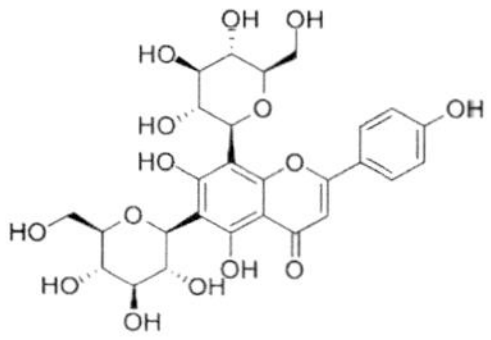
	Quercetin-3-O-(6″-malonyl) glucoside	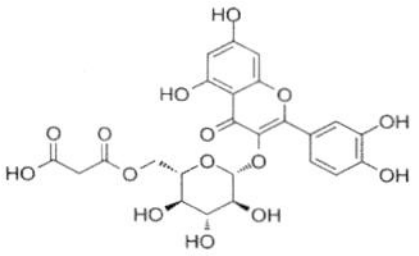
	L Pyrrolemarumine-4″ -O-α-Lrhamnopyranoside	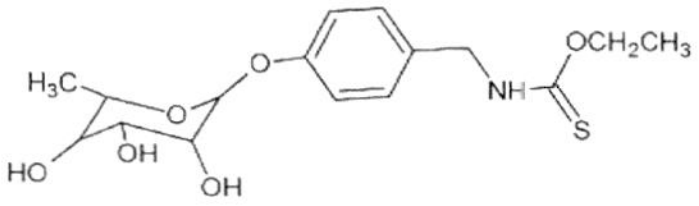
Seeds	Niazimicin	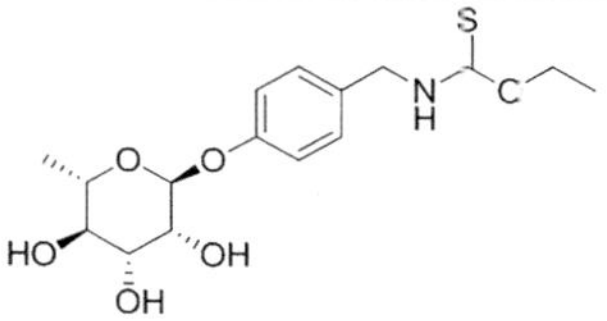
	Niazirin	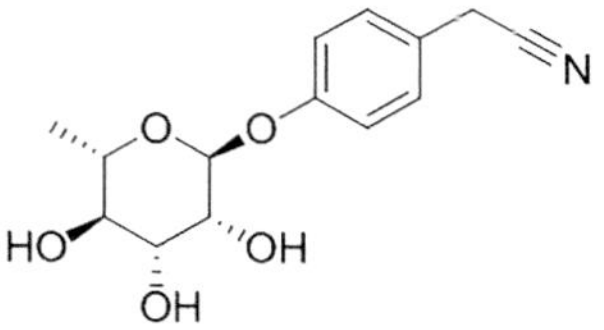
Roots	Arachidic acid	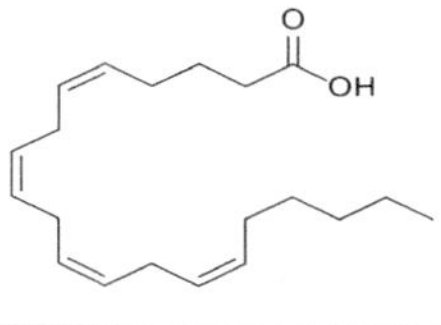
Bark	β-Sitosterol-3-O-β-D-galactopyranoside	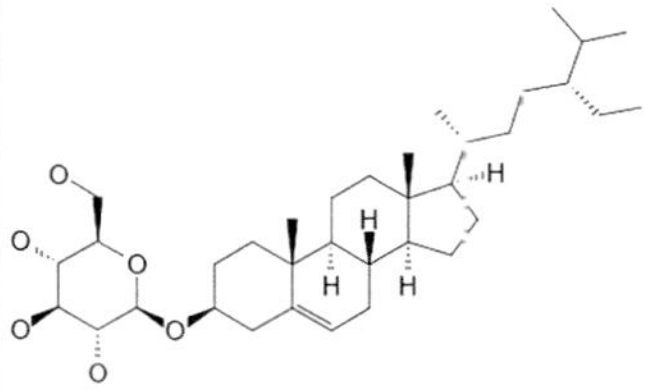

References

1. Shahzad, U., Khan, M. A., Jaskani, M. J., Khan, I. A., & Korban, S. S. (2013). Genetic diversity and population structure of *Moringa oleifera*. *Conservation Genetics, 14*, 1161–1172.
2. Sujatha, B. K., & Patel, P. (2017). Moringa oleifera–nature's gold. *Imperial Journal of Interdisciplinary Research, 3*(5), 1175–1179.
3. Rebecca, H. S. U., Sharon, M., Arbainsyah, A., Rebecca, H. S. U., Midcap, S., Arbainsyah, A., & Witte, L. D. (2006). Moringa oleifera: medicinal and socio-economic uses. In *International course on economic botany* (pp. 2–13). National Herbarium Leiden.
4. Dhakar, R. C., Maurya, S. D., Pooniya, B. K., Bairwa, N., & Gupta, M. (2011). *Moringa*: The herbal gold to combat malnutrition. *Journal of Sciences, 2*, 119.
5. Sharma, V., Paliwal, R., Sharma, P., & Sharma, S. (2011). Phytochemical analysis and evaluation of antioxidant activities of hydroethanolic extract of *Moringa oleifera* Lam. Pods. *Journal of Pharmacy Research, 4*, 554–557.
6. Paliwal, R., Sharma, V., Pracheta, S. S., Yadav, S., & Sharma, S. (2011). Antinephrotoxic effect of administration of *Moringa oleifera* Lam in amelioration of DMBA-induced renal carcinogenesis in Swiss albino mice. *Biology and Medicine, 3*(2), 27–35.
7. Keatinge, J. D. H., Ebert, A. W., Hughes, J. A., Yang, R. Y., & Curaba, J. (2015). Seeking to attain the UN's sustainable development goal 2 worldwide: The important role of *Moringa oleifera*. *l International Symposium on Moringa, 11*(58), 1–10.
8. Patricio, H. G., & Palada, M. C. (2015). Adaptability and horticultural characterization of different moringa accessions in Central Philippines. *l International Symposium on Moringa, 11*(58), 45–54.
9. Ebert, A. W., & Palada, M. C. (2015). Moringa-a vegetable tree for improved nutrition, health, and income of smallholder farmers. *I International Symposium on Moringa, 11*(58), 309–316.
10. Olayemiv, A. T., Olanrewaju, M. J., & Oloruntoba, A. C. (2016). Toxicological evaluation of *Moringa oleifera* Lam. seeds and leaves in Wistar rats. *Pharmacognosy Communications, 6*(2), 100–111.
11. Chukwuebuka, E. (2015). *Moringa oleifera* "the mother's best friend". *International Journal of Nutrition and Food Sciences, 4*(6), 624–630.
12. Swatią, A. K., Kumari, J., Garg, P., Thakur, P. A., Attri, A., & Kulshrestha, S. A. (2018). Moringa oleifera-A never die tree: An overview. *Asian Journal of Pharmaceutical and Clinical Research, 11*(12), 57–65.
13. Mishra, G., Singh, P., Verma, R., Kumar, S., Srivastav, S., Jha, K. K., & Khosa, R. L. (2011). Traditional uses, phytochemistry and pharmacological properties of Moringa oleifera plant: An overview. *Der Pharmacia Lettre, 3*(2), 141–164.
14. Roloff, A., Weisgerber, H., Lang, U., & Stimm, B. (2009). *Enzyclopädie der Holzgewächse-Handbuch. Atlas der Dendrologie* (pp. 1–8). Wiley.
15. Olson, M. E. (2002). Combining data from DNA sequences and morphology for a phylogeny of *Moringaceae* (Brassicales). *Systematic Botany, 27*(1), 55–73.
16. Palada, M., & Foidl, N. (2019). Harvesting, postharvest technology and processing. In *The miracle tree*. Moringa Oleifera.
17. Price, M. L. (2007). The moringa tree. *ECHO Technical Note, 17391*, 1–19.
18. Palada, M. C., & Chang, L. C. (2003). Suggested cultural practices for Moringa. *International Cooperators, 3*, –545.
19. Prabhakar, M., Shankara, H. S., & Devarja, M. (2003). Effect of integrated nutrient management on yield, yield components and economics of drumstick (*Moringa oleifera* Lam.) grown under rainfed condition. *Vegetable Science, 30*, 187–189.
20. Saini, R. K., Sivanesan, I., & Keum, Y. S. (2016). Phytochemicals of *Moringa oleifera*: A review of their nutritional, therapeutic and industrial significance. *Biotechnology, 3*(6), 1–14.
21. Rastogi, R. P., & Mehrotra, B. N. (1990). *Compendium of Indian medicinal plants*. Central Drug Research Institute.

22. Mridha, M. A. U. (2015). Prospects of moringa cultivation in Saudi Arabia. *Journal of Applicable Environmental Biology, 5*(3), 39–46.
23. Padilla, C., Valenciaga, N., Crespo, G., Toral, O., González, D., Reino, J., & Santana, H. (2017). Agronomy of *Moringa oleifera* (Lam.) in agricultural systems in Latin America and the Caribbean region. In *Mulberry, moringa and tithonia in animal feed, and other uses. Results in Latin America and the Caribbean* (p. 95). FAO.
24. Pérez-Rivera, E. P., Montes-Ávila, J., Castro-Tamayo, C. B., Portillo-Loera, J. J., & Castillo-López, R. I. (2021). Agronomical aspects of *Moringa oleifera* (Moringa). In *Biological and pharmacological properties of the genus Moringa* (pp. 39–63). CRC Press.
25. El-Hack, A., Mohamed, E., Alagawany, M., Elrys, A. S., Desoky, E. S. M., Tolba, H., & Swelum, A. A. (2018). Effect of forage *Moringa oleifera* L.(moringa) on animal health and nutrition and its beneficial applications in soil, plants and water purification. *Agriculture, 8*(9), 145.
26. Vijayakumar, R. M. (2001). Studies on influence of months of sowing and growth regulation on annual Moringa. *Moringa pterygosperma Gaertn, 5*(2), 16.
27. Raja, R. R., Sreenivasulu, M., Vaishnavi, S., Navyasri, D. M., Samatha, G., & Geethalakshmi, S. (2016). *Moringa oleifera*-An overview. *Journal of Applicable Resources, 2*(9), 620–624.
28. Parihar, S., Chattarpal, S., & Hooda, S. (2022). *Moringa oleifera* extract- A miracle tree. *Journal of Pharmacology, 11*(1), 1–5.
29. Olorukooba, M. M., Mohammed, R., Sodimu, A. I., & Abdullahi, M. U. (2013). Influence of planting methods and pinching on growth and vegetative yield of drumstick (*Moringa oleifera*. Lam). *Agrosearch, 13*(1), 143–148.
30. Leone, A., Spada, A., Battezzati, A., Schiraldi, A., Aristil, J., & Bertoli, S. (2015). Cultivation, genetic, ethnopharmacology, phytochemistry and pharmacology of *Moringa oleifera* leaves: An overview. *International Journal of Molecular Sciences, 16*(6), 12791–12835.
31. Patricio, H. G., Palada, M. C., Deloso, H. E., & Garcia, D. E. (2015). Biomass yield of *Moringa oleifera* as influenced by plant density and harvest frequency. *I International Symposium on Moringa, 1158*, 97–104.
32. Islam, S., Jahan, M. A. A., & Khatun, R. (2005). In vitro regeneration and multiplication of year-round fruit-bearing Moringa oleifera L. *Journal of Biological Sciences, 5*(2), 145–148.
33. Marfori, E. C. (2010). Clonal micropropagation of *Moringa oleifera* L. *Philippine Agricultural Sciences, 93*, 4.
34. Steinitz, B., Tabib, Y., Gaba, V., Gefen, T., & Vaknin, Y. (2009). Vegetative micro-cloning to sustain biodiversity of threatened Moringa species. *In Vitro Cellular & Developmental Biology-Plant, 45*, 65–71.
35. Moringa, C. O. *Growing Moringa for personal or community use.*
36. Saint Sauveur, A. D., & Broin, M. (2010). *Growing and processing moringa leaves*. Published by Moringa association of Ghana. Pp 36.
37. Mendieta-Araica, B., Spörndly, E., Reyes-Sánchez, N., Salmerón-Miranda, F., & Halling, M. (2013). Biomass production and chemical composition of *Moringa oleifera* under different planting densities and levels of nitrogen fertilization. *Agroforestry Systems, 87*, 81–92.
38. De Bon, H., Holmer, R. J., & Aubry, C. (2015). Urban horticulture. In *Cities and agriculture* (pp. 236–272). Routledge.
39. Mani, M. (2022). Organic pest management in horticultural crops. In *Trends in horticultural entomology* (pp. 211–241). Springer.
40. El-Mohamedy, R. S., & Abdalla, A. M. (2014). Evaluation of antifungal activity of Moringa oleifera extracts as natural fungicide against some plant pathogenic fungi *in vitro*. *Journal of Agricultural Technology, 10*(4), 963–982.
41. Parajuli, S., Shrestha, J., Subedi, S., & Pandey, M. (2022). Biopesticides: a sustainable approach for pest management: Biopesticides in sustainable pest management. *Journal of Agriculture, 20*(1), 1–13.

42. Gandji, K., Chadare, F. J., Idohou, R., Salako, V. K., Assogbadjo, A. E., & Kakaï, R. G. (2018). Status and utilisation of Moringa oleifera Lam: A review. *African Crop Science Journal, 26*(1), 137–156.
43. Shulner, I., Asaf, E., Ben-Simhon, Z., Cohen-Zinder, M., Shabtay, A., Peleg, Z., & Lati, R. N. (2021). Optimizing weed management for the new super-forage *Moringa oleifera*. *Agronomy, 11*(6), 1055.
44. Aslam, M. F., Basra, S. M., Hafeez, M. B., Khan, S., Irshad, S., Iqbal, S., & Akram, M. Z. (2020). Inorganic fertilization improves quality and biomass of *Moringa oleifera* L. *Agroforestry Systems, 94*, 975–983.
45. Durai, J., & Long, T. T. (2019). *Manual for sustainable management of clumping bamboo forest*. INBAR Technical Report.
46. Dash, S., & Gupta, N. (2009). Effect of inorganic, organic and bio fertilizer on growth of hybrid *Moringa oleifera*. *Science, 4*, 630–635.
47. Ratshilivha, N., Du Toit, E. S., Vahrmeijer, J. T., & Eloff, J. N. (2015). Yield and quality responses of *Moringa oleifera*. Lam. to nitrogen fertilization. *In I International Symposium on Moringa, 1158*, 201–208.
48. Abdelwanise, F. M., Saleh, S. A., Ezzo, M. I., Helmy, S. S., & Abodahab, M. A. (2015). Response of Moringa plants to foliar application of nitrogen and potassium fertilizers. *In I International Symposium on Moringa, 11*(58), 187–194.
49. Orwa, C., Mutua, A., Kindt, R., Jamnadass, R., & Simons, A. (2009). *Agroforestree database: A tree reference and selection guide. Version 4*. World Agroforestry Centre.
50. Shanmugavel, G., Prabakaran, K., & George, B. (2018). Evaluation of phytochemical constituents of *Moringa oleifera* (Lam.) leaves collected from Puducherry region, South India. *International Journal of Biosciences, 3*(1), 1–8.
51. Kasolo, J. N., Bimenya, G. S., Ojok, L., Ochieng, J., & Ogwal-Okeng, J. W. (2010). Phytochemicals and uses of *Moringa oleifera* leaves in Ugandan rural communities. *Journal of Medicinal Plant Research, 4*, 9.
52. Aja, P. M., Nwachukwu, N., Ibiam, U. A., Igwenyi, I. O., Offor, C. E., & Orji, U. O. (2014). Chemical constituents of *Moringa oleifera* leaves and seeds from Abakaliki, Nigeria. *American Journal of Phytomedicine and Clinical Therapeutics, 2*(3), 310–321.
53. Nepolean, P., Anitha, J., & Emilin, R. R. (2009). Isolation, analysis and identification of phytochemicals of antimicrobial activity of *Moringa oleifera* Lam. *Current biotica, 3*(1), 33–37.
54. Rageeb, M., Usman, M., & Barhate, S. D. (2012). Chemical investigation and studies of analgesic and antipyretic activity of Moringa oleifera Lam. seeds extract. *International Journal of Pharmacological Chemistry and Sciences, 1*, 221–230.
55. Ragasa, C. Y., Ng, V. A. S., & Shen, C. C. (2016). Chemical constituents of *Moringa oleifera* Lam. Seeds. *International Journal of Pharmacognosy and Phytochemical Research, 8*(3), 495–498.
56. Gupta, A. K. (2003). *Quality standards of Indian medicinal plants* (Vol. 1). Indian Council of Medicinal Research.
57. Roy, S. K., Chandra, K., Ghosh, K., Mondal, S., Maiti, D., Ojha, A. K., & Islam, S. S. (2007). Structural investigation of a heteropolysaccharide isolated from the pods (fruits) of Moringa oleifera (Sajina). *Carbohydrate Research, 342*(16), 2380–2389.
58. Amaglo, N. K., Bennett, R. N., & Curto, R. B. (2010). Profiling selected phytochemicals and nutrients in different tissues of the multipurpose tree *Moringa oleifera* L., grown in Ghana. *Journal of Food Chemistry, 122*, 1047–1054.
59. Siddhuraju, P., & Becker, K. (2003). Antioxidant properties of various solvent extracts of total phenolic constituents from three different agroclimatic origins of drumstick tree (*Moringa oleifera* lam.) leaves. *Journal of Agricultural Food Chemistry, 51*, 2144–2155.
60. Sánchez-Machado, D. I., Lopez-Cervantes, J., & Vázquez, N. R. (2006). High-performance liquid chromatography method to measure α-and γ-tocopherol in leaves, flowers and fresh beans from *Moringa oleifera*. *Journal of Chromatography A, 1105*, 111–114.

61. Bennett, R. N., Mellon, F. A., Foidl, N., Pratt, J. H., Dupont, M. S., Perkins, L., & Kroon, P. A. (2003). Profiling glucosinolates and phenolics in vegetative and reproductive tissues of the multi-purpose trees *Moringa oleifera* L. (horseradish tree) and Moringa stenopetala L. *Journal of Agricultural and Food Chemistry, 51*(12), 3546–3553.
62. Sashidhara, K. V., Rosaiah, J. N., Tyagi, E., Shukla, R., Raghubir, R., & Rajendran, S. M. (2009). Rare dipeptide and urea derivatives from roots of Moringa oleifera as potential anti-inflammatory and antinociceptive agents. *European Journal of Medicinal Chemistry, 44*(1), 432–436.
63. Foidl, N., Makkar, H. P. S., & Becker, K. (2001). The potential of *Moringa oleifera* for agricultural and industrial uses. In *The Miracle Tree: The Multiple Attributes of Moringa* (pp. 45–76). Academia.
64. Maurya, S. K., & Singh, A. K. (2014). Clinical efficacy of Moringa oleifera Lam. stems bark in urinary tract infections. *International Research, 2014*, 43–68.
65. Balemie, K., & Kebebew, F. (2006). Ethnobotanical study of wild edible plants in Derashe and Kucha districts, South Ethiopia. *Journal of Ethnobiology and Ethnomedicine, 2*, 1–9.
66. Sandeep, G., Anitha, T., Vijayalatha, K. R., & Sadasakthi, A. (2019). *Moringa* for nutritional security (*Moringa oleifera* Lam.). *International Journal of Botanical Studies, 4*, 21–24.
67. Kuete, V. (2017). Moringa oleifera. In *Medicinal spices and vegetables from Africa* (pp. 485–496). Academic.
68. Korsor, M., Ntahonshikira, C., Bello, H. M., & Kwaambwa, H. M. (2017). Comparative proximate and mineral composition of *Moringa oleifera* and *Moringa ovalifolia* grown in Central Namibia. *Sustainable Agriculture Research, 6*, 526–691.
69. Makita, C., Chimuka, L., Steenkamp, P., Cukrowska, E., & Madala, E. (2016). Comparative analyses of flavonoid content in *Moringa oleifera* and *Moringa ovalifolia* with the aid of UHPLC-qTOF-MS fingerprinting. *South African Journal of Botany, 105*, 116–122.
70. Toma, A., & Deyno, S. (2014). Phytochemistry and pharmacological activities of *Moringa oleifera*. *International Journal of Pharmacolgy, l*, 222–231.
71. Zaku, S. G., Emmanuel, S., Tukur, A. A., & Kabir, A. (2015). *Moringa oleifera*: An underutilized tree in Nigeria with amazing versatility. *African Journal of Food Sciences, 9*, 456–461.
72. Raman, J. K., Alves, C. M., & Gnansounou, E. (2018). A review on *Moringa* tree and vetiver grass–potential biorefinery feedstocks. *Journal of Bioresource and Technology, 249*, 1044–1051.
73. Fernandes, D. M., Sousa, R. M., de Oliveira, A., Morais, S. A., Richter, E. M., & Munoz, R. A. (2015). *Moringa oleifera*: A potential source for production of biodiesel and antioxidant additives. *Fuel, 146*, 75–80.
74. Santhi, K., & Sengottuvel, R. (2016). Qualitative and quantitative phytochemical analysis of Moringa concanensis Nimmo. *International Journal of Current Microbiology and Applied Sciences, 5*(1), 633–640.
75. Padayachee, B., & Baijnath, H. (2012). An overview of the medicinal importance of Moringaceae. *Journal of Medicinal Plants Research, 6*(48), 5831–5839.
76. Boopathi, N. M., & Abubakar, B. Y. (2021). Botanical descriptions of Moringa spp. In *The Moringa Genome* (pp. 11–20). Springer.
77. Asensi, G. D. (2017). *Moringa oleifera*: Revisión sobre aplicaciones y usos en alimentos. *Archivos Latinoamericanos de Nutricion, 67*, 86–97.
78. Liu, Y., Wang, X. Y., Wei, X. M., Gao, Z. T., & Han, J. P. (2018). Values, properties, and utility of different parts of *Moringa oleifera*. *Journal of Chinese Herbal Medicines, 10*, 371–378.
79. Kemble, B. (2021). Moringas of Madagascar. *Journal of Cactus and Succulent, 93*(2), 90–97.
80. Gopalakrishnan, L., Doriya, K., & Kumar, D. S. (2016). Moringa oleifera: A review on nutritive importance and its medicinal application. *Food Science and Human Wellness, 5*(2), 49–56.
81. Said-Al Ahl, H. A., Hikal, W. M., & Mahmoud, A. A. (2017). Biological activity of *Moringa peregrina*: A review. *American Journal of Food Science and Health, 3*(4), 83–87.
82. Olson, M. E., & Razafimandimbison, S. G. (2000). *Moringa hildebrandtii* (Moringaceae): A tree extinct in the wild but preserved by indigenous horticultural practices in Madagascar. *Adansonia, 22*(2), 217–221.

83. Stadtlander, T., & Becker, K. (2017). Proximate composition, amino and fatty acid profiles and element compositions of four different *Moringa* species. *Journal of Agricultural Science, 9*(7), 46–45.
84. Brilhante, R. S. N., Sales, J. A., Pereira, V. S., Castelo, D. D. S. C. M., de Aguiar Cordeiro, R., de Souza Sampaio, C. M., & Rocha, M. F. G. (2017). Research advances on the multiple uses of *Moringa oleifera*: A sustainable alternative for socially neglected population. *Asian Pacific Journal of Tropical Medicine, 10*(7), 621–630.
85. Fahey, J. W. (2005). *Moringa oleifera*: A review of the medical evidence for its nutritional, therapeutic, and prophylactic properties. *Trees for life Journal, 1*(5), 1–15.
86. Quintanilla-Medina, J. J., Garay-Martínez, J. R., Alvarado-Ramirez, E. R., Hernández-Meléndez, J., Mendoza-Pedroza, S. I., Rojas-Garcia, A. R., & Hernández-Garay, A. (2018). Time and temperature on the loss of moisture and protein content in *Moringa oleifera* Lam. leaves. *Journal of Agroproductividad, 11*(5), 88–92.
87. Srinivasamurthy, S., Yadav, U., Sahay, S., & Singh, A. (2017). Development of muffin by incorporation of dried *Moringa oleifera* (Drumstick) leaf powder with enhanced micronutrient content. *Development, 2*(4), 65–71.
88. Makanjuola, B. A., Obi, O. O., Olorungbohunmi, T. O., Morakinyo, O. A., Oladele-Bukola, M. O., & Boladuro, B. A. (2014). Effect of *Moringa oleifera* leaf meal as a substitute for antibiotics on the performance and blood parameters of broiler chickens. *Livestock Research for Rural Development, 26*(8), 144.
89. Gupta, S., Kachhwaha, S., Kothari, S. L., Bohra, M. K., & Jain, R. (2020). Surface morphology and physicochemical characterization of thermostable moringa gum: A potential pharmaceutical excipient. *ACS Omega, 5*(45), 29189–29198.
90. Anwar, F., Latif, S., Ashraf, M., & Gilani, A. H. (2007). *Moringa oleifera*: A food plant with multiple medicinal uses. *Phytotherapy Research: An International Journal Devoted to Pharmacological and Toxicological Evaluation of Natural Product Derivatives, 21*(1), 17–25.
91. Trigo, C., Castello, M. L., Ortola, M. D., Garcia-Mares, F. J., & Desamparados Soriano, M. (2020). Moringa oleifera: An unknown crop in developed countries with great potential for industry and adapted to climate change. *Food, 10*(1), 31.
92. Maurya, S. K., & Singh, A. K. (2014). Clinical efficacy of *Moringa oleifera* Lam. stems bark in urinary tract infections. *International scholarly research notices, 14*, 20.
93. Bhargave, A., Pandey, I., Nama, K. S., & Pandey, M. (2015). *Moringa oleifera* Lam. Sanjana (horseradish tree) A miracle food plant with multipurpose uses in Rajasthan-India an overview. *International Journal of Pure Application of Biosciences, 3*(6), 237–248.
94. He, T. B., Huang, Y. P., Huang, Y., Wang, X. J., Hu, J. M., & Sheng, J. (2018). Structural elucidation and antioxidant activity of an arabinogalactan from the leaves of *Moringa oleifera*. *International Journal of Biological Macromolecules, 112*, 126–133.
95. Sreelatha, S., & Padma, P. R. (2009). Antioxidant activity and total phenolic content of *Moringa oleifera* leaves in two stages of maturity. *Plant Foods for Human Nutrition, 64*, 303–311.
96. Hermawan, A., Nur, K. A., Dewi, D., Putri, P., & Meiyanto, E. (2012). Ethanolic extract of Moringa oleifera increased cytotoxic effect of doxorubicin on HeLa cancer cells. *Journal of Natural remedies, 12*, 108–114.
97. Tang, Y., Choi, E. J., Han, W. C., Oh, M., Kim, J., Hwang, J. Y., & Kim, E. K. (2017). *Moringa oleifera* from Cambodia ameliorates oxidative stress, hyperglycemia, and kidney dysfunction in type 2 diabetic mice. *Journal of Medicinal Food, 20*(5), 502–510.
98. Okumu, M. O., Ochola, F. O., Mbaria, J. M., Kanja, L. W., Gakuya, D. W., Kinyua, A. W., & Kiama, S. G. (2017). Mitigative effects of *Moringa oleifera* against liver injury induced by artesunate-amodiaquine antimalarial combination in wistar rats. *Clinical Phytoscience, 3*(1), 1–8.
99. Lopez-Teros, V., Ford, J. L., Green, M. H., Tang, G., Grusak, M. A., Quihui-Cota, L., & Astiazaran-Garcia, H. (2017). Use of a "super-child" approach to assess the vitamin A equivalence of *Moringa oleifera* leaves, develop a compartmental model for vitamin A kinet-

ics, and estimate vitamin A total body stores in young Mexican children. *The Journal of Nutrition, 147*(12), 2356–2363.
100. Adeyemi, O. S., & Elebiyo, T. C. (2014). *Moringa oleifera* supplemented diets prevented nickel-induced nephrotoxicity in wistar rats. *Journal of Nutrition and Metabolism, 14*, 958621.
101. Sutalangka, C., Wattanathorn, J., Muchimapura, S., & Thukham-mee, W. (2013). *Moringa oleifera* mitigates memory impairment and neurodegeneration in animal model of age-related dementia. *Oxidative Medicine and Cellular Longevity, 13*(20), 695936.
102. Tété-Bénissan, A., Quashie, M. A., Lawson-Evi, K., Gnandi, K., Kokou, K., & Gbeassor, M. (2013). Influence of *Moringa oleifera* leaves on atherogenic lipids and glycaemia evolution in HIV-infected and uninfected malnourished patients. *Journal of Applied Biosciences, 62*, 4610–4619.
103. Toppo, R., Roy, B. K., Gora, R. H., Baxla, S. L., & Kumar, P. (2015). Hepatoprotective activity of *Moringa oleifera* against cadmium toxicity in rats. *Veterinary World, 8*(4), 537.
104. Mahajan, S. G., & Mehta, A. A. (2009). Anti-arthritic activity of hydroalcoholic extract of flowers of *Moringa oleifera* lam. in wistar rats. *Journal of Herbs, spices and Medicinal Plants, 15*(2), 149–163.
105. Mehta, J., Shukla, A., Bukhariya, V., & Charde, R. (2011). The magic remedy of *Moringa oleifera*: An overview. *International Journal of Biomedical and Advance Research, 2*(6), 215–227.
106. Mehta, K., Balaraman, R., Amin, A. H., Bafna, P. A., & Gulati, O. (2003). Effect of fruits of *Moringa oleifera* on the lipid profile of normal and hypercholesterolaemic rabbits. *Journal of Ethnopharmacology, 86*(2), 191–195.
107. Dillard, C. J., & German, J. B. (2000). Phytochemicals: Nutraceuticals and human health. *Journal of the Science of Food and Agriculture, 80*(12), 1744–1756.
108. Tahiliani, P., & Kar, A. (2000). Role of *Moringa oleifera* leaf extract in the regulation of thyroid hormone status in adult male and female rats. *Pharmacological Research, 41*(3), 319–323.
109. Luqman, S., Srivastava, S., Kumar, R., Maurya, A. K., & Chanda, D. (2012). Experimental assessment of *Moringa oleifera* leaf and fruit for its antistress, antioxidant, and scavenging potential using in vitro and *in vivo* assays. *Evidence-based Complementary and Alternative Medicine, 12*(20), 519084.
110. Ramabulana, T., Mavunda, R. D., Steenkamp, P. A., Piater, L. A., Dubery, I. A., & Madala, N. E. (2016). Perturbation of pharmacologically relevant polyphenolic compounds in *Moringa oleifera* against photo-oxidative damages imposed by gamma radiation. *Journal of Photochemistry and Photobiology B: Biology, 156*, 79–86.
111. Jaja-Chimedza, A., Graf, B. L., Simmler, C., Kim, Y., Kuhn, P., Pauli, G. F., & Raskin, I. (2017). Biochemical characterization and anti-inflammatory properties of an isothiocyanate-enriched moringa *(Moringa oleifera)* seed extract. *PLoS One, 12*(8), 182–658.
112. Agbogidi, O. M., & Ilondu, E. M. (2012). *Moringa oleifera* Lam: its potentials as a food security and rural medicinal item. *International Journal of Biology, 1*(6), 156–167.
113. Anudeep, S., Prasanna, V. K., Adya, S. M., & Radha, C. (2016). Characterization of soluble dietary fiber from *Moringa oleifera* seeds and its immunomodulatory effects. *International Journal of Biological Macromolecules, 91*, 656–662.
114. Gupta, S., Jain, R., Kachhwaha, S., & Kothari, S. L. (2018). Nutritional and medicinal applications of *Moringa oleifera* Lam. Review of current status and future possibilities. *Journal of Herbal Medicine, 11*, 1–11.
115. Rocha, M. F. G., Alencar, L. P. D., Brilhante, R. S. N., Sales, J. D. A., Ponte, Y. B. D., Rodrigues, P. H. D. A., & Sidrim, J. J. C. (2014). *Moringa oleifera* inhibits growth of Candida spp. and Hortaea werneckii isolated from Macrobrachium amazonicum prawn farming with a wide margin of safety. *Ciência Rural, 44*(12), 2197–2203.
116. Choudhary, M. K., Bodakhe, S. H., & Gupta, S. K. (2013). Assessment of the antiulcer potential of *Moringa oleifera* root-bark extract in rats. *Journal of Acupuncture and Meridian Studies, 6*(4), 214–220.

117. Adejumo, O. E., Kolapo, A. L., & Folarin, A. O. (2012). *Moringa oleifera* Lam. (Moringaceae) grown in Nigeria: *In vitro* antisickling activity on deoxygenated erythrocyte cells. *Journal of Pharmacy & Bioallied Sciences, 4*(2), 118.
118. Shank, L. P., Riyathong, T., Lee, V. S., & Dheeranupattana, S. (2013). Peroxidase activity in native and callus culture of *Moringa oleifera* Lam. *Journal of Medicinal Biology and Enginering, 2*, 3.
119. Popoola, J. O., & Obembe, O. O. (2013). Local knowledge, use pattern and geographical distribution of *Moringa oleifera* Lam. (Moringaceae) in Nigeria. *Journal of Ethnopharmacology, 150*(2), 682–691.
120. Choudhary, P. D., & Pawar, H. A. (2014). Recently investigated natural gums and mucilages as pharmaceutical excipients: An overview. *Journal of Pharmaceutics, 14*, 204849.
121. Guevara, A. P., Vargas, C., Sakurai, H., Fujiwara, Y., Hashimoto, K., Maoka, T., & Nishino, H. (1999). An antitumor promoter from *Moringa oleifera* Lam. *Mutation Research/Genetic Toxicology and Environmental Mutagenesis, 440*(2), 181–188.
122. Caceres, A. (1991). Pharmacological properties of *Moringa oleifera*. 3. Effect of seed extracts in the treatment of experimental pyodermia. *Fitoterapia, 62*, 449–450.
123. Gassenschmidt, U., Jany, K. D., Bernhard, T., & Niebergall, H. (1995). Isolation and characterization of a flocculating protein from *Moringa oleifera* Lam. *Biochimica et Biophysica Acta (BBA)-General Subjects, 1243*(3), 477–481.
124. Madsen, M., Schlundt, J., & El Fadil, E. O. (1987). Effect of water coagulation by seeds of *Moringa oleifera* on. *Journal of Tropical Medicine and Hygiene, 90*, 101–109.
125. Pal, S. K., Mukherjee, P. K., Saha, K., & Pal., & Saha, B. P. (1995). Antimicrobial action of the leaf extract of *moringa oleifera* lam. *Ancient Science of Life, 14*(3), 197.
126. Sharma, P., Kumari, P., Srivastava, M. M., & Srivastava, S. (2006). Removal of cadmium from aqueous system by shelled *Moringa oleifera* Lam. seed powder. *Bioresource Technology, 97*(2), 299–305.
127. Kalogo, Y., Rosillon, F., Hammes, F., & Verstraete, W. (2000). Effect of a water extract of Moringa oleifera seeds on the hydrolytic microbial species diversity of a UASB reactor treating domestic wastewater. *Letters in Applied Microbiology, 31*(3), 259–264.
128. Martyn, C. N., Osmond, C., Edwardson, J. A., Barker, D. J. P., Harris, E. C., & Lacey, R. F. (1989). Geographical relation between Alzheimer's disease and aluminium in drinking water. *The Lancet, 333*(8629), 59–62.
129. Fatima, T., Sajid, M. S., Jawad-ul-Hassan, M., Siddique, R. M., & Iqbal, Z. (2014). Phytomedicinal value of *Moringa oleifera* with special reference to antiparasitics. *Pakistan Journal of Agricultural Sciences, 51*(1), 251–262.
130. Singh, D., Choudhury, S., Singh, T. U., & Garg, S. K. (2008). Pharmacodynamics of uterotonic effect of *Moringa oleifera* flowers extract. *Journal of Veterinary Pharmacology and Toxicology, 7*(1–2), 12–15.
131. Ganguly, R., & Guha, D. (2008). Alteration of brain monoamines & EEG wave pattern in rat model of Alzheimer's disease & protection by *Moringa oleifera*. *Indian Journal of Medical Research, 128*(6), 744–751.
132. Gupta, M., & kanti Mazumder, U., & Chakrabarti, S. (1999). CNS activities of methanolic extract of *Moringa oleifera* root in mice. *Fitoterapia, 70*(3), 244–250.
133. Hukkeri, V. I., Nagathan, C. V., Karadi, R. V., & Patil, B. S. (2006). Antipyretic and wound healing activities of *Moringa oleifera* Lam. in rats. *Indian Journal of Pharmaceutical Sciences, 68*(1), 124–126.
134. Ndong, M., Uehara, M., Katsumata, S. I., & Suzuki, K. (2007). Effects of oral administration of *Moringa oleifera* Lam on glucose tolerance in Goto-Kakizaki and Wistar rats. *Journal of Clinical Biochemistry and Nutrition, 40*(3), 229–233.
135. Sutar, N. G., Patil, V. V., Deshmukh, T. A., Jawle, N. M., Sr., Patil, V. R., Sr., & Bhangale, S. C. (2009). Evaluation of anti-pyretic potential of seeds of *Moringa oleifera* Lam. *International Journal of Green Pharmacy (IJGP), 3*, 2.
136. Agrawal, B., & Mehta, A. (2008). Antiasthmatic activity of Moringa oleifera Lam: A clinical study. *Indian Journal of pharmacology, 40*(1), 28–31.

137. Shukla, S., Mathur, R., & Prakash, A. O. (1988). Antifertility profile of the aqueous extract of Moringa oleifera roots. *Journal of Ethnopharmacology, 22*(1), 51–62.
138. Nath, D., Sethi, N., Singh, R. K., & Jain, A. K. (1992). Commonly used Indian abortifacient plants with special reference to their teratologic effects in rats. *Journal of Ethnopharmacology, 36*(2), 147–154.
139. Mahajan, S. G., Mali, R. G., & Mehta, A. A. (2007). Protective effect of ethanolic extract of seeds of Moringa oleifera Lam. against inflammation associated with development of arthritis in rats. *Journal of Immunotoxicology, 4*(1), 39–47.
140. Mahajan, S. G., & Mehta, A. A. (2007). Inhibitory action of ethanolic extract of seeds of Moringa oleifera Lam. on systemic and local anaphylaxis. *Journal of Immunotoxicology, 4*(4), 287–294.
141. Rao, A. V., Devi, P. U., & Kamath, R. (2001). In vivo radioprotective effect of Moringa oleifera leaves. *Indian Journal of ExPermiental Biology, 39*, 9.
142. Chumark, P., Khunawat, P., Sanvarinda, Y., Phornchirasilp, S., Morales, N. P., Phivthong-Ngam, L., & Klai-upsorn, S. P. (2008). The *in vitro* and *ex vivo* antioxidant properties, hypolipidaemic and antiatherosclerotic activities of water extract of Moringa oleifera Lam. leaves. *Journal of Ethnopharmacology, 116*(3), 439–446.
143. Dixit, S., Tripathi, A., & Kumar, P. (2016). Medicinal properties of *Moringa oleifera:* A review. *International Journal of education and Science research review, 3*(2), 173–185.
144. Mishra, S. P., Singh, P., & Singh, S. (2012). Processing of *Moringa oleifera* leaves for human consumption. *Bulletin of Environment, Pharmacology and life sciences, 2*(1), 28–31.
145. Irawaty, D. K. (2021). Conquering the myth of *Moringa oleifera* tress of Indonesian people. *In Al Insyirah International Scientific Conference on Health, 2*, 165–172.

Chapter 42
Saffron

Sana Javed, Samina Hanif, Arusa Aftab, Zubaida Yousaf, and Marius Moga

42.1 Introduction

Crocus sativus L. (saffron) is a representative member of the family Iridaceae with over 80 species and is a significant medicinal plant [1]. Its name is derived from the Persian word "Zaafaran," which denotes yellow flowers [2]. Saffron is mainly cultivated in Pakistan, (Kashmir), Iran, Japan, China, Spain, and India [3]. Saffron can be grown in dry and semi-arid areas with elevated altitudes and the least yearly rainfall [4]. Saffron is highly valuable, referred to as "red gold or gold condiment." [5, 6].

The classification of Saffron is complicated because of its sterility, triploidy ($2n = 3x = 24$), and variety of both its morphological characteristics and cytogenetic records. [7].

Only 7% of a saffron flower is made up of stigma; the other 93% is made up of petals, stamens, and styles, with sepals making up 78.4% [8]. The only component of the flower that is used to make commercial saffron is the stigma (female organ) [9].

Dried three branched stigmas present on the upper aerial portion of the pistil make up the spice. Dried stigmas are prepared (filaments or powdered) for marketing in different commercial categories viz., bulk or packed following the need of buyer and seller [10]. With a total area under cultivation of 5.707 thousand hectares and a yield of 2.8 kg/ha in 1997, saffron produced 15.95 tonnes annually. Contraily, it has since reduced to 3.674 thousand hectares with 9.6 tonnes of annual production and output of 2.61 kg/ha [11].

S. Javed · S. Hanif · A. Aftab (✉) · Z. Yousaf (✉)
Department of Botany, Lahore college for women university, Lahore, Pakistan
e-mail: arusa.aftab@lcwu.edu.pk; zubaida.yousaf@lcwu.edu.pk

M. Moga
Faculty of Medicine, Transilvania University of Brasov, Brasov, Romania

M. Zia-Ul-Haq et al. (eds.), *Essentials of Medicinal and Aromatic Crops*,
https://doi.org/10.1007/978-3-031-35403-8_42

42.2 Biological Classification

Kingdom	**Plantae**
Phylum	Magnoliophyta
Class	Liliopsida
Order	Liliales
Family	Iridaceae
Genus	*Crocus*
Species	*S.sativus*
Botanical name	*Crocus sativus* L.
Common name	Saffron or zaffaran or Kesar

Crocin, picrocrocin, and safranal are the three fundamental saffron constituents, that are involved in the spice‘s color, flavor, and odor, respectively. [13]. Terpenes and carotenoids are abundant in saffron [14]. It is commonly used as a seasoning, perfuming, flavoring, and dying agent (Fig. 42.1) [15].

Saffron has been used as a medicine since ancient times, it is also included in the Ayruvedic medicinal system. Saffron is known as an aphrodisiac, stimulant, anti-poison, lactogogue, nervine tonic, heart tonic, digestive, immunological stimulator, diaphoretic, diuretic, sedative, relaxant, febrifuge, anti-stress, and antianxiety medication. Saffron has furthermore shown beneficial effects against respiratory problems, gastrointestinal disorders, measles, smallpox, scarlet fever, and skin illnesses [16]. Saffron is used in conventional/traditional medicinal systems for different purposes like it has been used as a relaxant, expectorant, energizing agent, digestive stimulant, and spasm calmative. Moreover, it has also been reported to be used to treat cholera, diabetes, measles, liver and spleen problems, blood diarrhea, fever, and measles. Moreover, Saffron has also been used in allopathic medications (saffron tincture, saffron glycerin, and codex saffron syrup) [17].

Fig. 42.1 *Crocus sativus*

42.3 Agronomy

42.3.1 Geographical Representation

C. sativus is grown in a variety of geographical areas and altitudes around the world. Saffron is found between the latitudes of 30°–50° N and 10° W-80° E [18]. The optimal altitude for saffron growth is between 200 and 2000 AMSL (meters above mean sea level). It is better suited to hills, plateaus, and mountains between 600 and 1700 m AMSL. Saffron is planted at an altitude of 650–1100 m AMSL in Italy while in Morocco at elevations ranging from 1200 to 1400 meters above sea level. Saffron is grown in moderate, semi-arid, and desert climates at heights fluctuating from 1500 to 2800 meters above sea level. Favorable climatic conditions to high saffron yields include gentle heat, pleasant summers, and rainy autumns [19].

42.3.2 Climate

Saffron prefers a controlled, dry climate with plenty of sunshine. Flower production is greater in October and November when average temperatures range from 15–20 °C during the day and 6–8 °C at night. Autumn rains increase flower productivity, while spring rains encourage corm duplication [11]. It needs warm summers with little or no precipitation and cool to cold winters with spring showers. The temperature during flower appearance has a significant impact on flower size. Flowering requires a much lower temperature (about 17 °C or slightly more, but less than 20 °C) than sprouting [12]. Temperature of 27 °C resulted in the best vegetative development and a temperature of 17 °C resulted in the best flowering. The plant requires 1100 cooling hours during the vegetative phase, which is sufficient for vernalization [20]. A temperature series of 23 °C to 25 °C is good for vegetative development in the early stages of growth, while a temperature less than 16 °C is suitable for reproducing new corms [21].

42.3.3 Soil Conditions

Calcareous soils with more organic content that is friable, well-watered, loose, and well-drained are preferred for saffron. To increase the soil's organic content, 20–30 Mg ha^{-1} of farmyard manure (FYM) is used. Sandy or loamy textured soils are the best soil for the production of saffron [11]. Saffron growth favors soil pH values between 6.3–8.3 with average value of 7.5 and electrical conductivities (EC) between 0.09–0.30 dS m^{-1} with a average value of 0.17 dS m^{-1} [22].

42.3.4 Land Preparation and Planting Time

For the cultivation of saffron, proper land preparation is necessary. Loose and friable texture of a land depth is 30 cm, the field must be plowed 3 to 5 times in May, June, and July. For corm multiplication, about 10 Mg ha^{-1} of degraded FYM is required [11]. At 12–15 day interval, top dressings with an extreme amounts of FYM (10 Mg ha^{-1}), P_2O_5 (60 kg ha^{-1}), K_2O (60 kg ha^{-1}), and 1/fourth amount of nitrogen (22.5 kg ha^{-1}) should be applied. Using cross-cultural tasks like a plow, hoeing, leveling the ground, manure, and fertilizer applications should be managed in August [12].

From the second week of August to the first fortnight of September, the saffron corms should be planted. After the bed formation, corms are manually planted behind the plow by hand. Saffron prefers cold winters, rainy autumns, springtime precipitation, and hot, dry summers [11]. In the Mashhad region of Iran, June through July are the optimum months for propagating saffron corms [23]. Delay in plantation can reduce the growth of saffron When the corms were propagated in June and October, the highest and lowest floral yields were obtained, respectively. Owing to corm dormancy during this time, spring sowing results in increased growth and saffron production [24].

42.3.5 Manuring

C. sativus is a perennial crop that has been adapted to fertilizers, particularly nitrogen fertilizers, which have a substantial impression on production. The use of chemical fertilizers in unproductive soil can enhance the production as well as the generation of daughter corms [25]. The soil fecundity (C/N ratio, accessible phosphorus, mineral nitrogen, and exchangeable potassium) accounts for 20 to 80% of saffron yield [20].

Currently, worldwide agricultural management is shifting towards an organic farming system, which can be drawn from conventional agriculture forms or used after the creation of fresh plantations [26]. As a primary source of nutrients for the saffron crop, a large amount of farmyard manure (animal manures, straw, and compost) has traditionally been applied. Research revealed that applying 40–60 t ha^{-1} animal dung not only provided plant nutrient supplies but also increased soil fertility. This led to a decrease in the usage of chemical fertilizers (organic systems), which had a consistent impact on the amount and quality of saffron yield [20]. Cow dung (20–30 t ha^{-1}) applied to the soil surface and mixed to a depth of 30 cm enhanced soil organic matter (OM), electrical conductivity (EC), pH, and cation exchange capacity considerably ($P = 0.05$). There was a favorable association between the amount of manure applied and the OM content of the soil [26].

42.3.6 Irrigation

The drought resistance of saffron and its quiescent season, which does not need irrigation; make it a crop with a lower water requirement than other crops [27]. There are three primary methods of irrigation are applied in saffron filed i;e surface, aerosol, and drip irrigation, and they are selected according to climatic conditions and the kind of soil [28]. Saffron requires six applications of irrigation management: the first in the middle of summer to encourage flowering, the second in early October to expedite harvest, the third in November, which requires the collection of flowers and the emergence of leaves, the fourth in late December to early January after weeding and fertilizer application, the fifth in early March, and the sixth in early April to complete the development of new corms [29]. Early flowering was caused by irrigation applied in September, but irrigation applied in late August increased stigma output by 17–40% [30].

42.3.7 Plant Propagation

C. sativus is vegetatively reproduced by development of daughter corms from the parent corm. *C. sativus* blooms are infertile and consequently unable to produce viable seeds. Under natural conditions, a saffron corm can generate up to 5 new corms, which may result in the formation of new plants [31]. Saffron planting material is limited because of a stumpy rate of corm duplication, pathogen infestation, living and non living stressors, and inadequate crop management [11]. As a consequence, natural saffron propagation is relatively complex. Because of the restrictions of traditional growing methods, as well as the problem of corm availability, it is required to grow effective and viable ways for the quick and large-scale production of selected saffron consents [1].

42.3.8 Pest and Disease Management

Corm rots disease in saffron are caused by *Rhizoctonia crocorum, Phoma crocophila, Fusarium moniliforme, Macrophomina phaseolina, Fusarium oxysporum, F. solani, F. pallidoroseum, F. equiseti, Mucor* sp. and *Penicillium* sp. [32].

Infected corms have dark-brown gaunt and uneven spots beneath the corm scales, particularly in the bud and root areas. Infected corms also have dark-brown sunken and uneven patches beneath the corm scales, particularly in the root and bud regions. In adverse conditions, the whole corm transforms into a dark powdered substance. The leaves of diseased corms displays 'die-back' symptoms [33] Every year, a significant amount of the product is lost due to these diseases.

To control saffron corm rot, soak the corms in a fungicidal solution comprising Mancozeb 75WP (0.3%) and Carbendazim 50WP (0.1%) for 5–10 min before spreading them on a cloth and allowing them to dry in the shade for another 10–15 min [34]. In separate research, Blitox (copper-oxychloride), Indofil M-45 (mancozeb), Difolatoan (captafol), Folpat (captafol), Bavistin (carbendazim), and Tecto (thiobendazole) (0.2% each) as a dint or drench provided 100% disease control. In separate research, Blitox (copper-oxychloride), Indofil M-45 (mancozeb), Difolatoan (captafol), Folpat (captafol), Bavistin (carbendazim), and Tecto (thiobendazole) (0.2% each) as a dip or drench provided 100% disease control. With exception of corm rot, plant parasitic nematodes of several species invading saffron-growing soil induces corm harm by sucking the sap. Sap imbibing promotes necrosis in the roots and predisposes saffron corms to corm rot, resulting in significant yield losses. In spite of this, farmers are not currently employing any systematic control techniques [33, 34]. As a soil treatment, Chlorpyriphos 10G (at 1000 g a.i. ha^{-1}) or Fenvalerate 0.4% (at 120 g a.i. ha^{-1}) substantially lowers insect population. Identifying effective biological control agents would be an environmentally sound solution [34].

42.3.9 Weed Management

A crop needs to be managed for weeds to develop healthily. Annual crops are hand weeded, while perpetual crops are handled using the weedkiller Simazine (Gesatop 50) or Atrazine (Gesaprim 50) at 1.0 kg ha^{-1} [35]. Torsion and finger weeders are difficult to utilize in saffron due to the presence of fresh corms. Saffron vegetative development follows from October/November to April, while weeds grow in bare arenas during the dormant phase of saffron. The following weed species have been described in the saffron fields of J&K (Union territory of Jammu and Kashmir), *Chenopodium album, Tulipa stellata, Papaver rhoae, Erodium cicutarium Euphorbia helioscopia, Filago arvense, Galium tricorne, Lepidium virginicum Polygonum aviculare, Salvia moorcroftiana* [33].

42.3.10 Crop Description

Various ecological and genetic components affect the size and proportion of the various floral components. In Khorasan, 78.5 kg of fresh flowers (or over 170,000 flowers) are needed to produce 1 kg of dry stigmata and style [33]. The quantity of flowers per corm is significantly influenced by the size of the corm [35]. In the saffron farm, hand harvesting is necessary every day because a maximum of three flowers are produced in one plant. Approximately 15,000 to 16,000 flowers are required to produced 1 kg of saffron spice Stigma is its final commercially valuable product [12].

42.3.11 Harvesting

Saffron harvesting includes removing flowers and separating stigmas. The flower collection begins when they bloom in the field [27]. Flowers are collected from October to November, however, the timing differs from country to country due to temperature and initial irrigation time [36]. Flowers are picked and then take placed under the covering to separate the filaments from the flower. Separated portions of the saffron's flower are then dried for 2–3 days in the shade (on a sunny day). Saffron plants typically flower in the autumn, approximately 40 days afterward planting [37].

42.3.12 Chemical Constituents in Crocus sativus

Secondary metabolites are responsible for their distinctive flavor, aroma, and color [38]. Crocin is responsible for a color while, safranal for aroma and picrocrocin for bitterness are the important bioactive element of *C. sativus* [39]. Safranal, also known as 2,6,6-trimethyl-1,3-cyclohexadiene-1-carboxaldehyde, is the primary volatile compound that is responsible for aroma 4-(b-D-glucopyranosyloxy)-2,6,6-trimethyl-1-cyclohexene-1-carboxaldehyde, also known as picrocrocin [40]. Picrocrocin, which makes up between 1% and 13% of safron's dry matter and is the second common compound afterward crocin, gives taste to *C. sativus* [41]. Crocin, (8,8-diapocarotene-8,8-dioic acid) gives *C. sativus* its color. The melting point of α-crocin is 186 °C, which is a trans-crocetin di-(b-D-gentiobiosyl) ester. The bulk of dried saffron may also contain more than 10% -crocin [41].

Above 150 volatile and aroma-producing chemical constituents are present in *C. sativus*. Moreover, there are numerous non-volatile active components in it, many of which are carotenoids including zeaxanthin, carotenes, and other lycopene [42]. More than 34 substances, including terpenes, terpene alcohols, and their esters, constitute the volatiles that possesses a potent odor [43]. Crocins responsible for the red or reddish brown color of stigma, as well as carotenes, crocetin, picrocrocin (a glycosidic precursor of safranal), the vicious substance, and safranal, the primary organoleptic element of stigmas, are non-volatile substances of saffron [44]. Chemical analysis has identified over 150 components in the stigma of saffron, including hydrophilic and lipophilic, minerals, proteins, mucilage, carbohydrates, gums, vitamins, starch, many pigments including crocin, and alkaloids, saponins, carotenoid, carotenes mangicrocin, xanthone, safranal, and picrocrocin [45].

Isoorientin, tectoridin, rutin, caffeic acid, ferulic acid, rutin, and caffeic acid are the antioxidant compound that is present in saffron. Hydroxycinnamic acids, xanthones, flavonoids, isoflavonoids, tectoridin, iristectorigenin B, nigericin, and irigenin. Various essential amino acids ILE, LEU, LYS, MET, PHE, THR, TRP, VAL in large quantity are present in *C. sativus* while isoleucine is the dominant compound [46].

42.3.13 Crocetin

Crocetin ($C_{20}H_{24}O_4$) is a peculiar lipophilic carotenoid made up of conjugates of several unsaturated olefin acids [45]. Moreover, crocetin has been shown to have positive effects on the cardiovascular system [47], as well as anti-cancer and anti-depressant action [48].

42.3.14 Crocin

Crocin ($C_{44}H_{64}O_{24}$), which is typically a bright red color, and readily mix in water to give color, making it a helpful natural cooking coloring agent. The antioxidant capabilities of crocin are also well recognized for trapping free radicals and preventing tissue and cell damage. Depending on the cultivar used, the environment, and the method of processing, the total dry matter of saffron, the water-soluble crocins account for between 6 and 16% of the carotenoids [12, 49].

42.3.15 Safranal

Safranal ($C_{10}H_{14}O$) is the primary element of saffron‘s crucial oil. It is a source of saffron‘s odor [40]. Some samples' total volatile fraction may include safranal at a percentage of up to 70% [49]. Saffron is recognized to have a strong antioxidant capacity along with being cytotoxic to cancer cells in vitro (Tables 42.1 and 42.2) [50].

42.3.16 Ethnobotanical Uses of C. sativus

Saffron (*Crocus sativus*), is the most expensive therapeutic food item. The dried stigmas of the plant are used to make a well-known spice i.e. saffron. It is also used in the pharmaceutical, textile dye industries, and as well as in cosmetic products. The ethnobotanical and medicinal use of saffron are described in Tables 42.3 and 42.4.

Table 42.1 Chemical constituents in *Crocus sativus*

Chemical constituents	chemical structure	Effects	References
Crocin ($C_{44}H_{64}O_{24}$)	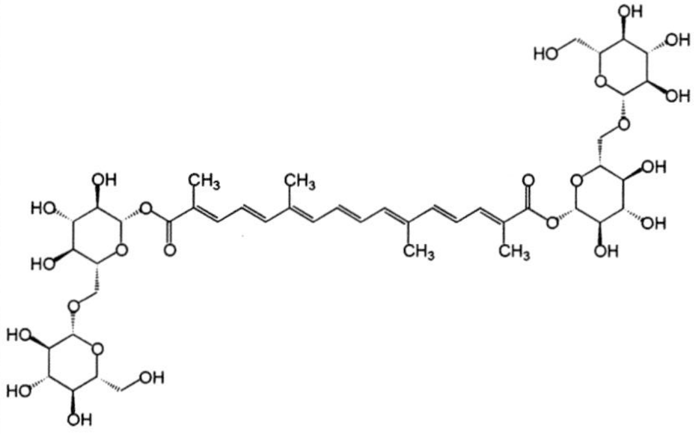	A possible cause of the inhibitory effect on the production of cancer cells is a decline the production of Deoxyribonucleic acid (DNA) and ribonucleic acid (RNA) and proteins in tumor cells.	[12]
Crocetin ($C_{20}H_{24}O_4$)	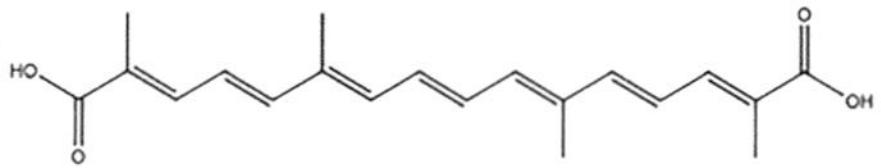	Anti-cancerous,antic-diabetic, anti-inflammatory, hepatoprotective	[49]
Safranal ($C_{10}H_{14}O$)	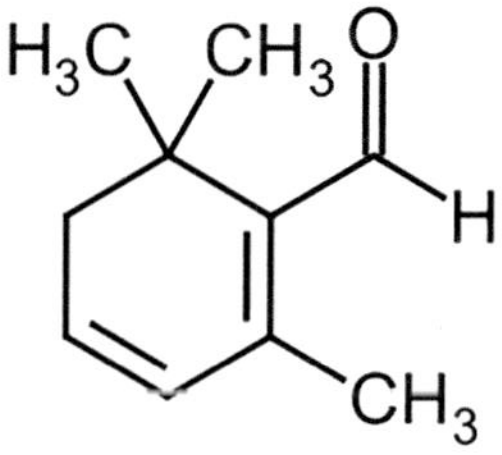	Cardioprotective, effective against neurological disorders	[49]
Picrocrocin ($C_{16}H_{26}O_7$)	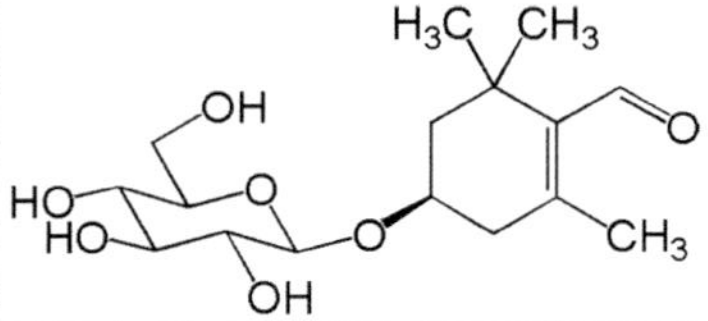	Effective against Alzheimer's disease	[51]

Table 42.2 Anticancer properties of saffron extract and its primary components in cancer cell lines and animal models

Active constituents	Animal model and cancer Cell Lines	References
Saffron ethanol extracts		
Ethanol extract	MCF-7 breast cancer cells	[104, 118]
Safranal and crocin	Myelocytic leukaemia cells, K-562.	[116]
Crocin, crocetin and Safranal	MCF-7 and MDA-MB 231 breast cancer cells	[100, 101]
Crocetin	MCF-7 and MDA-MB 231 breast cancer cells	[100, 101]
	SW480 colorectal cancer	[108]
	P388, K-562, L1210, NB4, HL60 human leukemia cells	[114]
Crocetin and crocin	SPC-A1 AND A549 lung cancer cells	[113]
Crocin	MCF-7 breast cancer cells	[96–98]
	HCC70 and HCC1806 breast cancer cells	[99]
	HO-8910-ovarian cancer cells	[104]
	AGS gastric carcinogen	
	HCT-116, HT-29 and SW-480, Colon cancer cells	[59]
	HCT116 colorectal cancer	[105]
	HGC-27 and AGS stomach cancer cells	[106]
	AGS gastric cancer cells	[107]
	Prostate (adenocarcinomas) cancer cells LnCaP, CWR22, 22rv1, LAPC-4, PC3, C4–2B, BPH-1, and DU145	[95]
Aqueous extract of saffron		
Aqueous extract	Mice received xenografts of 4 T1 cells.	[110]
Crocin	N-Nitroso-N-Methyluria administration to female Wistar albino rats	[102]
	In mice, xenografted 4 T1 breast cancer cells were used	[103]
	Xenografted T24 cells in BALB/c nude mice	[117]
	To cause chemical colitis linked with colorectal cancer in mice, azoxymethane and dextran sodium sulphate were used	[109]
	Polyposis with adenomas. Models for familial adenomatous polyposis in humans: ApcMin/+ mice	[111]
Crocin and crocetin	Male nude rats with xenografted PC3 and 22rv1 prostate cancer cells	[112]

Table 42.3 Traditional and pharmaceutical uses of *Crocus sativus*

Medicinal properties	Effects	References
Anti cancerous	Appropriate to the presence of crocin and crocetin, it is effective against ductal carcinoma, colon cancer, lymphoma, and tumours, and gastrointestinal tumour	[16, 60]
Infection disease	Antimicrobial, disinfectant, measles	[79]
Pulmonary disease	Cold, sinus infection, sneezing, and bronchitis	[79]
Dermatitis	Used to treat wounds, skin issues, and acne It can also make the body appear brighter	[80]
Neuroprotective	Used for treatment of ischemia, Parkinson's, and Alzheimer's and to improve cognitive dysfunction	[39, 75]
Anti-atherosclerotic	Due to the presence of crocetin, it has an anti-atherosclerotic effect	[68]
Anti-diabetic	Crocin, crocetin, and safranal are the active anti-diabetic component for diabetes mellitus	[63]
Cardioprotective	Safranal reduces oxidative stress and myocardial infarction	[38]
Antidepressant	Effective for depression, anxiety, headache, and nausea	[78]

Table 42.4 Other uses of *Crocus sativus*

Common uses	Part used	Chemical constituent	References
Culinary purposes	Whole stigma and petals	Crocin,safranl,and picrocrocin	[15]
Coloring agent	Dried stigma and whole flower	Crocin, dicarboxylic acids, esters, and nitrogen compounds	[81, 88]
Cosmetics	Whole flower	Crocin and Safranal	[11, 87]
Staining dye for histopathology	Whole plant	Crocin	[89]
Fragrance industry	Dried stigma and dried flower	Crocin and safranal	[85]

42.3.17 Medicinal Properties of Crocus sativus

Saffron is a significant medicinal herb that has great aphrodisiac, anti-spasmodic, expectorant, anti-diabetic, anti-inflammatory, anti-depressant, antioxidant, anti-tumor, and anti-cancerous properties [52]. Saffron is utilized as a medication to cure a variety of human health issues, including tussis (coughs), Gastrointestinal diseases, sleeplessness, chronic uterine bleeding, scarlet fever, smallpox, colds, breathe suffocation, colic, and cardiovascular issues [53]. Saffron can also be used to treat wounds and soothe infants who are teething [54]. The neurological, circulatory, and pulmonary systems have all been treated with crocin extracts from saffron [43]. There is evidence that certain saffron extract constituents are effective antidepressants and can help control mental problems [55]. Saffron extract has recently undergone successful tests against both mental illnesses and cancer [53]. Several researchers have reported the chemopreventive and tumoricidal effects of saffron extracts on cancer after successful *in-vitro* and *in-vivo* tests [56]. During chemopreventative therapy, saffron can be utilized as a drug. Saffron exhibits antitumor and anticancer characteristics, which were proved by its potential to prevent the production of DNA and RNA but not protein, that eliminate free radicals, involve the processes of transforming carotenoids into retinoids, and stimulate the interaction mediated by lectins [57].

42.4 Anti-cancerous Effect of *Crocus sativus*

Saffron has been shown to have various beneficial properties for medical purposes, including anti-mutagenic and immune-modulating properties as well as radical scavenging properties. Saffron has proved to be effective against human cancer cell lines in culture and animal models and its primary compounds have anticancer and cancer-preventive properties (Table 42.2). its components, including crocin, safranal, and crocetin, have cytotoxic effects on several cancer cell lines, including those from the breast, lung, and liver [58]. Saffron has exhibited potential anticancer properties by inducing apoptosis, reducing cell proliferation, and blocking angiogenesis against a sort of cancer, including breast, lung, ovarian, and tomour [59].

The research also showed that saffron extract reduced the activity of NF-kB, a transcription factor that causes the initiation and development of cancer [60]. Saffron and its bioactive ingredients have the potential to be used as chemotherapeutic medicines to treat various cancers [58].

42.4.1 Cardioprotective Activity

C. sativus has great anti-inflammatory, anti-hypertensive, antioxidant, and hypolipidemic activities that help to cure and prevent various cardiovascular diseases [61]. Saffron's chemical compounds (safranal, crocin crocetin and picrocrocin) exhibit cardioprotective efficiency in isoproterenol-induced myocardial infarction because they lessen oxidative stress [38]. According to research conducted on animals, crocetin protects by inhibiting the production of malondialdehyde, can prevent against cardiac ischemia-reperfusion injury (MDA) obstructing the activity of TNF (tumor necrosis factor-alpha), and lower the effect of myocardium apoptosis and infarct size [47]. In ischemia-reperfusion rat models, saffron was also observed significantly lessen the susceptibility and occurrence of severe ventricular arrhythmia [62]. Saffron has been linked to several health advantages for atherosclerosis, hypertension, dyslipidemia, and type 2 diabetes, all of which are significant menace factors for the expansion of heart diseases.

42.4.2 Anti Diabetic Properties

Saffron has a great therapeutic approach for Diabetes mellitus. The primary active components that provide an anti-diabetic response are crocin, crocetin, and safranal, which have insulin-sensitizing effects but no discernible impact on blood serum concentrations of creatinine, glutamic-pyruvic transaminase (SGPT), or glutamic-oxaloacetic transaminase (SGOT) [63]. Patients with type 2 diabetes who took saffron supplements for 12 weeks had significant drops in their fasting blood glucose and HbA1c values.

According to the study, saffron may be used as an alternative therapy in the treatment of diabetes [64]. Saffron extract can increase the capability of adipocytes to absorb glucose and become more sensitive to insulin, which may help patients control their blood sugar levels [65]. Saffron has additionally been discovered to have a preventive impact against diabetic problems such as retinopathy, neuropathy, and nephropathy [66].

42.4.3 Anti–atherosclerotic Effect

Crocetin is present in the *C. sativus* that helps to lower the level of LDL (low density lipoprotein) and TBARs (thiobarbituric reactive acid substances), expression, foam cell formation, NF-B activation, vascular cell adhesion molecule-1 (VCAM-1) and the progression of aortic atherosclerosis tumour [67]. The reticence of NF-B activation and VCAM-1 expression, a member of the cytokine-induced immunoglobulin gene superfamily implicated in atherogenesis by promoting monocyte adhesion to vascular endothelium, was part of the explanation for crocetin's anti-atherosclerotic effect [68]. The decrease in IL-6, tumour necrosis factor (TNF), and monocyte chemoattractant protein-1 (MCP-1) expressions, which most likely resulted in less foam cell generation, was one explanation for the anti-atherosclerotic effects [69].

42.5 Neuroprotective

Saffron is enriched with essential phytochemicals due to which it has great potential for aplastic anemia and another neurological disease [70]. The antioxidant potential of saffron may contribute to a decrease in brain ischemia [71]. Active constituents of saffron (crocin) decrease oxidative stress by decreasing the malondialdehyde (MDA) and increasing the activity of antioxidative enzymes that decreased oxidation and thus improve antioxidant capability. Safranal, crocin, and kaemferol are the active phytoconstituents responsible for neuroprotective activity [71].

According to reports, crocin in the saffron has amphiphilic characteristics that help to successfully stop the formation of harmful amyloid. Crocin is the primary phytochemical improving cognitive dysfunction and has a neuroprotective effect [72].The destruction and degeneration of neurons in the black nucleus commonly called 'substantia nigra' are attributed to Parkinson's disease (SN) Because of its inhibitory action on stress-induced cytotoxicity in the endoplasmic reticulum, crocin reduced cellular apoptosis, and mitochondrial dysfunctionality. Moreover, crocin enhances the levels of GSH, and TSH, and reduced the activity of acetylcholinesterase [73]. It is also effective against neuroinflammatory conditions [74]. The evidence suggests that the interactions between saffron's major biochemical constituents and dopaminergic and cholinergic systems are what give it its neuroprotective qualities against conditions like Alzheimer's and Parkinson's disease [75].

42.5.1 Anti Depression

Crocus sativus is mostly utilized as a medicinal plant over the world to treat neurodegenerative illnesses, such as depression [76]. According to tests conducted on humans, *C. sativus* has effects on the level of the brains's neurotransmitters, is useful in treating mild to moderate depression, and has an impact on opioids [77]. Clinical research showed that saffron much more efficient than both placebos and other synthetic antidepressants, with few side effects such as edginess, pain, and sickness [78].

42.6 Learning and Memory-Enhancing Effect

Iranian herbalists frequently treat cognitive disorders with *Crocus sativus*. The plant is currently utilized to relax smooth muscles and treat specific neurological illnesses [79]. Saffron's potential role in memory improvement may be due to its capacity to alter neurotransmitter systems, raise cerebral blood flow, and reduce oxidative stress [78].

42.6.1 Culinary Uses

Due to the aroma of *C. sativus*, it has been used for cooking purposes. Chefs and saffron experts say it has a honey-like aroma with metallic undertones. Iran, Spain, India, and other nations utilize saffron as a rice flavoring. It is a common ingredient in many Spanish cuisines, including the rice-based Paella Valenciana and the fish-based zarzuela. Moreover, it is utilized in saffron cake, a spicy fish soup, Italian Milanese risotto, and French bouillabaisse. Also, it is utilized in various candies like Kulfi and Gulab Jamun [81]. Saffron is a spice that is used in Morocco to make a variety of traditional foods, such as koftas (meatballs with tomatoes) and Grazia. It is also substituted for mint in Moroccan tea. Moreover, some component of the mixture of chermoula plants which flavor many Moroccan meals is saffron [82].

42.6.2 Coloring Agent

Due to the solubility of crocin in water, saffron is used as an unconventional dye that is favorable in the agriculture and food industries [83]. Hence, for a very protracted period, cheeses, cooking oil, pasta, and oleomargarines have been colored using the potent coloring ability of saffron, which might also be utilized in cosmetics. Saffron is a citrine color that is used in clothing and paintings. Despite being in an acidic or alkaline environment, saffron solutions typically remain stable. Dicarboxylic acids, esters, nitrogen compounds, and Crocin's pKa (acid dissociation constant) are all responsible for this feature. Solutions containing saffron buffers slow down cellulose oxidation. Buddhist monks' clothing, as well as silk, wool, and oriental carpets, are still dyed with saffron today. Natural dyes are less harmful and allergic than synthetic dyes, and they are more environmentally friendly and biodegradable [81].

42.6.3 Perfume Industry

The spice releases a pleasing perfume when dried, which Aristophanes calls a "sensual odor" [84]. Safranal is the primary odoriferous compound of *C. sativus*. Saffron was a royal color and was practically applied as a perfume in salons, palaces, theatres, and toilets in ancient Greece [85]. The Parthian Kingdom use this spice in their royal scent and also use it as cooling face oil for monarchs and ceremonial leaders [86].

42.6.4 Cosmetics

Many commercially available cosmetics contain synthetic colorants, which, when used for an extended period, may have negative side effects. Yet, the current fashion is for these cosmetic products to have healthful natural elements. Due to its expensive price, saffron has only been used sparingly in cosmetics. When turmeric would fade from exposure to light, it has been used as a replacement. It is also utilized as a tartrazine replacement [87].

42.6.5 *Origin and History of* Crocus sativus

The word "Saffron" originated from the Persian term, sahafaran". As stated by some researchers, the Arabic term "Zafaran," which indicates yellow, gave rise to the name "saffron." The precise history of saffron has been obscured by many myths, and different experts have different theories on where it came from. While some experts, like Vavilov, claim that saffron hails from the Middle East, others place its origin in Asia Minor or the South-Western Greek islands [90]. The genetic makeup of saffron was examined recently using the genome-wide single-nucleotide polymorphisms approach, and it was confirmed that it originated from *Crocus cartwrightianus* in Attica, Greece [91]. In Castilla-Mancha, Spain, artifacts dating to around 2400 B.C. were found that show the use of saffron as a fabric dye. Then, saffron was utilized as a condiment under the rule of Hammurabi (1800–1700 BC) [62]. The Minoan culture, as recorded in the Papyrus Eber, existed around 1550 B.C., which is a far later period for the first evidence of saffron production. The Old Testament of the Bible (twelfth century B.C.), as well as the Hebrew Bible (also known as the Tanakh), which was the first Jewish classic, were the earliest religious literature that first explain this [92].

Saffron was woven into carpets and shrouds for the monarchs in Iran, then known as Persia. (tenth century B.C.).Later, the Persians brought saffron to the Kashmir region. The Persians first used saffron baths to ease exhaustion or cool off, and when Alexander the Great encouraged Greek soldiers to use them, the practise spread to Macedonia (356 B.C.–323 B.C.) [56]. A notable yearly saffron harvest festival was later established in Safranbolu, a city in northern Turkey, as a result of the introduction of saffron farming. Since the late twentieth century, the municipality of Consuegra, Spain, has also hosted [93] an annual "Saffron Rose Festival" that is comparable. It takes place on the final weekend in October [94].

42.7 Summary

Saffron (*Crocus sativus* L.), is a perennial stemless herb of the family Iridaceae. It is known as king of condiments. The dried red stigma with a little piece of the yellowish style make up commercial saffron. Saffron is the most expensive condiment. Based on the usage of healthy and natural components dried stigmas are also used in the pharmaceutical, textile dye, and cosmetics industries, saffron cultivation is becoming more popular. The key environmental factors influencing the production of saffron include altitude, soil properties, temperature, photoperiod, and topographical locations. There is a need to use climatic data to examine the impact of saffron quality for each geographical area. Saffron significantly improved depression symptoms, cognitive function, and sexual dysfunction symptoms when compared to controls, and significantly decreased fasting blood glucose, waist circumference, diastolic blood pressure, total cholesterol, and low-density lipoprotein cholesterol concentrations (mainly placebos). Additionally, big sample size, high-quality randomised trials in several nations are required to demonstrate the therapeutic effects of saffron.

References

1. Tahiri, A., Mazri, M. A., Karra, Y., Ait Aabd, N., Bouharroud, R., & Mimouni, A. (2023). Propagation of saffron (*Crocus sativus* L.) through tissue culture: A review. *The Journal of Horticultural Science and Biotechnology, 98*(1), 10–30.
2. Ramadan, A., Soliman, G., Mahmoud, S. S., Nofal, S. M., & Abdel-Rahman, R. F. (2012). Evaluation of the safety and antioxidant activities of Crocus sativus and Propolis ethanolic extracts. *Journal of Saudi Chemical Society, 16*(1), 13–21.
3. Emam, V., Eghbal, M. K., Lar, M. M. S., Khalaj, K. N., Peknejad, F., & Rohami, B. (2012). The effect of planting density and different nitrogen and phosphorus application rates on saffron yield. *Journal of Basic and Applied Scientific Research, 2*(3), 2400–2404.
4. Khilare, V., Tiknaik, A., Prakash, B., Ughade, B., Korhale, G., Nalage, D., & Khedkar, G. (2019). Multiple tests on saffron find new adulterant materials and reveal that Ist grade saffron is rare in the market. *Food Chemistry, 272*, 635–642.
5. Mohtashami, L., Amiri, M. S., Ramezani, M., Emami, S. A., & Simal-Gandara, J. (2021). The genus crocus l.: A review of ethnobotanical uses, phytochemistry and pharmacology. *Indian Crops Production, 171*, 113923.
6. Mykhailenko, O., Desenko, V., Ivanauskas, L., & Georgiyants, V. (2020). Standard operating procedure of Ukrainian saffron cultivation according to good agriculture and collection practices to assure quality and traceability. *India Crops Production, 151*, 112376.
7. Hajyzadeh, M., Olmez, F., & Khawar, K. M. (2020). Molecular approaches to determine phylogeny in saffron. In *Saffron* (pp. 57–68). Elsevier.
8. Serrano-Díaz, J., Sánchez, A. M., Martínez-Tomé, M., Winterhalter, P., & Alonso, G. L. (2013). A contribution to nutritional studies on Crocus sativus flowers and their value as food. *Journal of Food Composition and Analysis, 31*(1), 101–108.
9. Bakshi, R. A., Sodhi, N. S., Wani, I. A., Khan, Z. S., Dhillon, B., & Gani, A. (2022). Bioactive constituents of saffron plant: Extraction, encapsulation and their food and pharmaceutical applications. *Applied Food Research, 100076*, 100076.
10. Tsimidou, M. Z. (2023). On the importance of the starting material choice and analytical procedures adopted when developing a strategy for the Nanoencapsulation of saffron (Crocus sativus L.). *Bioactive Antioxidants Antioxidants, 12*(2), 496.
11. Menia, M., Iqbal, S., Zahida, R., Tahir, S., Kanth, R. H., Saad, A. A., & Hussian, A. (2018). Production technology of saffron for enhancing productivity. *Journal of Pharmacognosy and Phytochemistry, 7*(1), 1033–1039.
12. Kothari, D., Thakur, R., & Kumar, R. (2021). Saffron (*Crocus sativus* L.): Gold of the spices—a comprehensive review. *Horticulture, Environment, and Biotechnology, 62*(5), 661–677.
13. Muzaffar, S., Rather, S. A., Khan, K. Z., & Akhter, R. (2016). Nutritional composition and in-vitro antioxidant properties of two cultivars of Indian saffron. *Journal of Food Measurement and Characterization, 10*, 185–192.
14. El-Midaoui, A., Ghzaiel, I., Vervandier-Fasseur, D., Ksila, M., Zarrouk, A., Nury, T., & Lizard, G. (2022). Saffron (*Crocus sativus* L.): A source of nutrients for health and for the treatment of neuropsychiatric and age-related diseases. *Nutrients, 14*(3), 597.
15. Zeka, K., Ruparelia, K. C., Continenza, M. A., Stagos, D., Vegliò, F., & Arroo, R. (2015). Petals of *Crocus sativus* L. as a potential source of the antioxidants crocin and kaempferol. *Fitoterapia, 107*, 128–134.

16. Mousavi, S. Z., & Bathaie, S. Z. (2011). Historical uses of saffron identifying potential new avenues for modern research. *Avicenna Journal Phytomed, 1*, 57–66.
17. Moghaddasi, M. S. (2010). Saffron chemicals and medicine usage. *Journal of Medicinal Plants Research, 4*(6), 427–430.
18. Yildirim, M. U., Asil, H., Hajyzadeh, M., Sarihan, E. O., & Khawar, K. M. (2017). Effect of changes in planting depths of saffron (*Crocus sativus* L.) corms and determining their agronomic characteristics under warm and temperate (Csa) climatic conditions of Turkish province of Hatay. *Acta Horticulturae, 1184*, 47–53.
19. Rahimi, H., Shokrpour, M., Tabrizi-Raeini, L., & Esfandiari, E. (2017). A study on the efects of environmental factors on vegetative characteristics and corm yield of saffron (*Crocus sativus*). *Iran Journal of Horticulture Science, 48*, 45–52.
20. Koocheki, A., Nassiri, M., & Behdani, M. A. (2006). Agronomic attributes of saffron yield at agroecosystems. *Acta Hortulturae, 739*, 24–33.
21. Zahmati, R., Shekari, H. A., & Fotokian, M. H. (2018). Growth and development of saffron (*Crocus sativus* L.) in response to temperature pre-treatment and environmental conditions. *Journal of Bioscience and Biotechnology, 7*, 47–50.
22. Nehvi, F. A. (2010). Forthcoming challenges for improving safron farming systems in Kashmir. *Acta Horticulturae, 850*, 281–286.
23. Bayat, M., Rahimi, M., & Ramezani, M. (2016). Determining the most effective traits to improve saffron (*Crocus sativus* L.) yield. *Physiological Molecular Biology Plants, 22*, 153–161.
24. Koocheki, A., Rezvani, M. P., & Fallahi, H. R. (2016). Effects of planting dates, irrigation management and cover crops on growth and yield of saffron (*Crocus sativus* L.). *Agroecology, 8*, 435–451.
25. Mohammad, M., Amiri, M. E., & Sharghi, Y. (2012). Respond of saffron (*Crocus sativus* L.) to animal manure application. *Journal of Medicinal Plants Research, 6*(7), 1323–1326.
26. Eyhorn, F., Muller, A., Reganold, J. P., Frison, E., Herren, H. R., Luttikholt, L., Mueller, A., Sanders, J., Scialabba, N. E.-H., Seufert, V., & Smith, P. (2019). Sustainability in global agriculture driven by organic farming. *Natural Sustainability, 2*, 253–255.
27. Kafi, M., Koocheki, A., & Rashed, M. H. (2006). *Saffron (Crocus sativus): Production and processing*. Science Publishers.
28. Gresta, F., Lombardo, G. M., Siracusa, L., & Ruberto, G. (2008). Effect of mother corm dimension and sowing time on stigma yield, daughter corms and qualitative aspects of saffron (*Crocus sativus* L.) in a mediterranean environment. *Journal of the Science of Food and Agriculture, 88*(7), 1144–1150.
29. Yarami, N., Kamgar-Haghighi, A. A., Sepaskhah, A. R., & Zand-Parsa, S. (2011). Determination of the potential evapotranspiration and crop coefficient for saffron using a water-balance lysimeter. *Archives of Agronomy and Soil Science, 57*(7), 727–740.
30. Cardone, L., Castronuovo, D., Perniola, M., Cicco, N., & Candido, V. (2020). Saffron (*Crocus sativus* L.), the king of spices: An overview. *Scientia Horticulturae, 272*, 109560.
31. Devi, K., Sharma, M., & Ahuja, P. S. (2014). Direct somatic embryogenesis with high frequency plantlet regeneration and successive cormlet production in saffron (*Crocus sativus* L.). *South African Journal of Botany, 93*, 207–216.
32. Ahmad, M., & Sagar, V. (2007). Integrated management of corm/tuber rot of saffron and Kalazeera. In *Horticulture Mini Mission-1, Indian Council for Agricultural Research (ICAR)* (p. 22).
33. Husaini, A. M., Hassan, B., Ghani, M. Y., Teixeira da Silva, J. A., & Kirmani, N. A. (2010). Saffron (*Crocus sativus* Kashmirianus) cultivation in Kashmir: Practices and problems. *Functional Plant Science and Biotechnology, 4*(2), 108–115.
34. Ghani, M. Y. (2002). Corm rot disease of saffron and its management. In *Proceedings of seminar-cum-workshop on saffron (Crocus sativus)* (pp. 107–112). SKUAST-K.
35. Dar, M. H., Groach, R., Razvi, S. M., & Singh, N. (2017). Saffron crop (golden crop) in modern sustainable agricultural systems. *International Journal for Research in Applied Science and Engineering Technology, 5*, 247–259.

36. Kafi, M., & Showket, T. (2007). A comparative study of saffron agronomy and production systems of Khorasan (Iran) and Kashmir (India). *Acta Horticulturae, 739*, 123–132.
37. Dar, M. H., Groach, R., Razvi, S. M., & Singh, N. (2017). Saffron crop (golden crop) in a modern sustainable agricultural system. *International Journal for Research in Applied Science and Engineering Technology, 5*, 247–259.
38. Mehdizadeh, R., Parizadeh, M. R., Khooei, A. R., Mehri, S., & Hosseinzadeh, H. (2013). Cardioprotective effect of saffron extract and safranal in isoproterenol-induced myocardial infarction in wistar rats. *Iranian Journal of Basic Medical Sciences, 16*(1), 56–63.
39. Zhang, A., Shen, Y., Cen, M., Hong, X., Shao, Q., Chen, Y., & Zheng, B. (2019). Polysaccharide and crocin contents, and antioxidant activity of safron from diferent origins. *Indian Crops Production, 133*, 111–117.
40. Maggi, L., Carmona, M., Zalacain, A., Kanakis, C. D., Anastasaki, E., Tarantilis, P. A., Polissiou, M. G., & Alonso, G. L. (2010). Changes in saffron volatile profle according to its storage time. *Food Research International Journal, 43*, 1329–1334.
41. Shahi, T., Assadpour, E., & Jafari, S. M. (2016). Main chemical compounds and pharmacological activities of stigmas and tepals of 'red gold'; safron. *Trends Food Science Technology, 58*, 69–78.
42. Liakopoulou-Kyriakides, M., & Kyriakidis, D. A. (2002). Croscus sativus-biological active constitutents. *Studies in Natural Products Chemistry, 26*, 293–312.
43. Abe, K., & Saito, H. (2000). Effects of saffron extract and its constituent crocin on learning behavior and long-term potentiation. *Phytotherapy Research, 14*(3), 149–152.
44. Wallis, T. E. (2005). *Textbook of pharmacognosy*. CBS.
45. Samarghandian, S., Borji, A., Farahmand, S. K., Afshari, R., & Davoodi, S. (2013). *Crocus sativus L. (safron) stigma aqueous extract induces apoptosis in alveolar human lung cancer cells through caspase-dependent pathways activation* (pp. 1–12). Biomed Research International.
46. Mykhailenko, O., Ivanauskas, L., Bezruk, I., Sidorenko, L., Lesyk, R., & Georgiyants, V. (2021). Characterization of phytochemical components of Crocus sativus leaves: A new attractive by-product. *Scientia Pharmaceutica, 89*(2), 28.
47. Wang, Y., Sun, J., Liu, C., & Fang, C. (2014). Protective efects of crocetin pretreatment on myocardial injury in an ischemia/reperfusion rat model. *European Jornal of Pharmacology, 741*, 290–296.
48. Ohba, T., Ishisaka, M., Tsujii, S., Tsuruma, K., Shimazawa, M., Kubo, K., Umigai, N., Iwawaki, T., & Hara, H. (2016). Crocetin protects ultraviolet A-induced oxidative stress and cell death in skin in vitro and in vivo. *European Journal Pharmacology, 789*, 244–253.
49. Kabiri, M., Rezadoost, H., & Ghassempour, A. (2017). A comparative quality study of saffron constituents through HPLC and HPTLC methods followed by isolation of crocins and picrocrocin. *LWT-Food Science and Technology, 84*, 1–9.
50. Assimopoulou, A. N., Sinakos, Z., & Papageorgiou, V. P. (2005). Radical scavenging activity of *Crocus sativus* L. extract and its bioactive constituents. *Phytotherapy Research, 19*, 997–1000.
51. Samarghandian, S., & Borji, A. (2014). Anticarcinogenic effect of saffron (*Crocus sativus* L.) and its ingredients. *Pharmacognosy Research, 6*, 99–107.
52. Jan, S., Wani, A. A., Kamili, A. N., & Kashtwari, M. (2014). Distribution, chemical composition and medicinal importance of saffron (*Crocus sativus* L.). *African Journal of Plant Science, 8*(12), 537–545.
53. Abdullaev, F. (2003). *Crocus sativus* against cancer. *Archives of Medical Research, 4*(34), 354.
54. Abdullaev, F. I., & Espinosa-Aguirre, J. J. (2004). Biomedical properties of saffron and its potential use in cancer therapy and chemoprevention trials. *Cancer Detection and Prevention, 28*(6), 426–432.
55. Lechtenberg, M., Schepmann, D., Niehues, M., Hellenbrand, N., Wünsch, B., & Hensel, A. (2008). Quality and functionality of saffron: Quality control, species assortment and affin-

ity of extract and isolated saffron compounds to NMDA and σ1 (sigma-1) receptors. *Planta Medica, 74*(07), 764–772.
56. Magesh, V., Singh, J. P. V., Selvendiran, K., Ekambaram, G., & Sakthisekaran, D. (2006). Antitumour activity of crocetin in accordance to tumor incidence, antioxidant status, drug metabolizing enzymes and histopathological studies. *Molecular and cellular biochemistry, 287*, 127–135.
57. Rangarajan, P., Subramaniam, D., Paul, S., Kwatra, D., Palaniyandi, K., Islam, S., & Dhar, A. (2015). Crocetinic acid inhibits hedgehog signaling to inhibit pancreatic cancer stem cells. *Oncotarget, 6*(29), 27661–27673.
58. Bhandari, P. R. (2015). *Crocus sativus* L.(saffron) for cancer chemoprevention: A mini review. *Journal of traditional and complementary medicine, 5*(2), 81–87.
59. Aung, H. H., Wang, C. Z., Ni, M., Fishbein, A., Mehendale, S. R., Xie, J. T., & Yuan, C. S. (2007). Crocin from *Crocus sativus* possesses significant anti-proliferation effects on human colorectal cancer cells. *Experimental Oncology, 29*(3), 175–180.
60. Zeinali, M., Zirak, M. R., Rezaee, S. A., Karimi, G., & Hosseinzadeh, H. (2019). Immunoregulatory and anti-inflammatory properties of *Crocus sativus* (saffron) and its main active constituents: A review. *Iranian Journal of Basic Medical Sciences, 22*(4), 334–344.
61. Xing, B., Li, S., Yang, J., Lin, D., Feng, Y., Lu, J., & Shao, Q. (2021). Phytochemistry, pharmacology, and potential clinical applications of saffron: A review. *Journal of Ethnopharmacology, 281*, 114555.
62. Joukar, S., Ghasemipour-Afshar, E., Sheibani, M., Naghsh, N., & Bashiri, A. (2013). Protective effects of saffron (*Crocus sativus*) against lethal ventricular arrhythmias induced by heart reperfusion in the rat: A potential anti-arrhythmic agent. *Pharmaceutical biology, 51*(7), 836–843.
63. Kianbakht, S., & Hajiaghaee, R. (2011). Anti-hyperglycemic effects of saffron and its active constituents, crocin and safranal, in alloxan-induced diabetic rats. *Journal of Medicinal Plants, 10*(39), 82–89.
64. Abu-Izneid, T., Rauf, A., Khalil, A. A., Olatunde, A., Khalid, A., Alhumaydhi, F. A., & Rengasamy, K. R. (2022). Nutritional and health beneficial properties of saffron (*Crocus sativus* L): A comprehensive review. *Critical Reviews in Food Science and Nutrition, 62*(10), 2683–2706.
65. Sepahi, S., Mohajeri, S. A., Hosseini, S. M., Khodaverdi, E., Shoeibi, N., Namdari, M., & Tabassi, S. A. S. (2018). Effects of crocin on diabetic maculopathy: A placebo-controlled randomized clinical trial. *American Journal of Ophthalmology, 190*, 89–98.
66. Sanaie, S., Nikanfar, S., Kalekhane, Z. Y., Azizi-Zeinalhajlou, A., Sadigh-Eteghad, S., Araj-Khodaei, M., & Andalib, S. (2023). Saffron as a promising therapy for diabetes and Alzheimer's disease: Mechanistic insights. *Metabolic Brain Disease, 38*(1), 137–162.
67. Zheng, S., Qian, Z., Tang, F., & Sheng, L. (2005). Suppression of vascular cell adhesion molecule-1 expression by crocetin contributes to attenuation of atherosclerosis in hypercholesterolemic rabbits. *Biochemical pharmacology, 70*(8), 1192–1199.
68. Zheng, S., Qian, Z., Sheng, L., & Wen, N. (2006). Crocetin attenuates atherosclerosis in hyperlipidemic rabbits through inhibition of LDL oxidation. *Journal of Cardiovascular Pharmacology, 47*(1), 70–76.
69. Christodoulou, E., Kadoglou, N. P. E., Stasinopoulou, M., Konstandi, O. A., Kenoutis, C., Kakazanis, Z. I., & Valsami, G. (2018). *Crocus sativus* L. aqueous extract reduces atherogenesis, increases atherosclerotic plaque stability, and improves glucose control in diabetic atherosclerotic animals. *Atherosclerosis, 268*, 207–214.
70. Saleem, S., Ahmad, M., Ahmad, A. S., Yousuf, S., Ansari, M. A., Khan, M. B., Ishrat, T., & F., & Islam. (2006). Effect of saffron (*Crocus sativus*) on neurobehavioral and neurochemical changes in cerebral ischemia in rats. *Journal of Medicinal Food, 9*(2), 246–253.
71. Zheng, Y. Q., Liu, J. X., Wang, J. N., & Xu, L. (2007). Effects of crocin on reperfusion-induced oxidative/nitrative injury to cerebral microvessels after global cerebral ischemia. *Brain Research, 1138*, 86–94.

72. Khalili, M., & Hamzeh, F. (2010). Effects of active constituents of *Crocus sativus* L., crocin on streptozocin-induced model of sporadic Alzheimer's disease in male rats. *Iranian Biomedical Journal, 14*(1–2), 59–65.
73. Zhang, G. F., Zhang, Y., & Zhao, G. (2015). Crocin protects PC12 cells against MPP(+)-induced injury through inhibition of mitochondrial dysfunction and ER stress. *Neurochemistry International, 89*, 101–110.
74. Deslauriers, A., Afkhami-Goli, A. M., Paul, R. K., Bhat, S., Acharjee, K. K., Ellested, F., Noorbaksh, M., & Michalak-Power, C. (2011). Neuroinflammation and endoplasmic reticulum stress are coregulated by crocin to prevent demyelination and neurodegeneration. *The Journal of Immunology, 187*(9), 4788–4799.
75. Purushothuman, S., Nandasena, C., Peoples, C. L., El Massri, N., Johnstone, D. M., Mitrofanis, J., & Stone, J. (2013). Saffron pre-treatment offers neuroprotection to Nigral and retinal dopaminergic cells of MPTP-treated mice. *Journal of Parkinson's Disease, 3*(1), 77–83.
76. Jalali-Heravi, M., Parastar, H., & Ebrahimi-Najafabadi, H. (2009). Characterization of volatile components of Iranian saffron using factorial-based response surface modeling of ultrasonic extraction combined with gas chromatography-mass spectrometry analysis. *Journal of Chromatography, 1216*(33), 6088–6097.
77. Khazdair, M. R., Boskabady, M. H., Hosseini, M., Rezaee, R., & Tsatsakis, A. M. (2015). The effects of *Crocus sativus* (saffron) and its constituents on nervous system: A review. *Avicenna Journal of Phytomedicine, 5*(5), 376–391.
78. Kell, G., Rao, A., Beccaria, G., Clayton, P., Inarejos-García, A. M., & Prodanov, M. (2017). Saffron a novel saffron extract (*Crocus sativus* L.) improves mood in healthy adults over 4 weeks in a double-blind, parallel, randomized, placebo-controlled clinical trial. *Complementary Therapies in Medicine, 33*, 58–64.
79. Hosseinzadeh, H., Sadeghnia, H. R., Ghaeni, F. A., Motamedshariaty, V. S., & Mohajeri, S. A. (2012). Effects of saffron (*Crocus sativus* L.) and its active constituent, crocin, on recognition and spatial memory after chronic cerebral hypoperfusion in rats. *Phytotherapy Research, 26*, 381–386.
80. Mousavi, S. Z., & Bathaie, S. Z. (2011). Historical uses of saffron: Identifying potential new avenues for modern research. *Avicenna Journal of Phytomedicine, 1*(2), 57–66.
81. Mzabri, I., Addi, M., & Berrichi, A. (2019). Traditional and modern uses of saffron (*Crocus sativus*). *Cosmetics, 6*(4), 63.
82. Modaghegh, M. H., Shahabian, M., Esmaeili, H. A., Rajbai, O., & Hosseinzadeh, H. (2008). Safety evaluation of saffron (*Crocus sativus*) tablets in healthy volunteers. *Phytomedicine, 15*(12), 1032–1037.
83. Ramadan, A., Soliman, G., Mahmoud, S. S., Nofal, S. M., & Abdel-Rahman, R. F. (2010). Evaluation of the safety and antioxidant activities of *Crocus sativus* and propolis ethanolic extracts. *Journal of Saudi Chemistry Society, 16*, 13–21.
84. Li, C. Y., & Wu, T. S. (2002). Constituents of the pollen of *Crocus sativus* L. and their tyrosinase inhibitory activity. *Chemical and pharmaceutical bulletin, 50*(10), 1305–1309.
85. Giaccio, M. (2004). Crocetin from saffron: An active component of an ancient spice. *Critical Reviews in Food Science and Nutrition, 44*(3), 155–172.
86. Dadkhah, M. R., Ehtesham, M., & Fekrat, H. (2003). *Iranian saffron an unknown jewel* (pp. 1–20). Shahr Ashoob Publication.
87. Colledge, M. A. R. (2009). *The Parthians, translated to Persian by Masoud Rajabnia.* Hirmand Publication.
88. Raja, A. S. M., Pareek, P. K., Shakyawar, D. B., Wani, S. A., Nehvi, F. A., & Sof, A. H. (2012). Extraction of natural dye from saffron flower waste and its application on pashmina fabric. *Advances in Applied Science Research, 3*, 156–161.
89. Bathaie, S. Z., Farajzade, A., & Hoshyar, R. (2014). A review of the chemistry and uses of crocins and crocetin, the carotenoid natural dyes in safron, with particular emphasis on

applications as colorants including their use as biological stains. *Biotech Histochemistry, 89*, 401–411.
90. Winterhalter, P., & Straubinger, M. (2000). Saffron—renewed interest in an ancient spice. *Food Reviews International, 16*(1), 39–59.
91. Nemati, Z., Harpke, D., Gemicioglu, A., Kerndorff, H., & Blattner, F. R. (2019). Saffron (Crocus sativus) is an autotriploid that evolved in Attica (Greece) from wild Crocus cartwrightianus. *Molecular Phylogenetics and Evolution, 136*, 14–20.
92. Halvorson, S. (2008). Saffron cultivation and culture in Central Spain. *FOCUS on Geography, 51*(1), 17–24.
93. Melnyk, J. P., Wang, S., & Marcone, M. F. (2010). Chemical and biological properties of the world's most expensive spice: Saffron. *Food Research International Journal, 43*, 1981–1989.
94. Dai, R. C., Nabil, W. N. N., & Xu, H. X. (2021). The history of saffron in China: From its origin to applications. *Chinese Medicine and Culture, 4*(4), 228–234.
95. D'Alessandro, A. M., Mancini, A., Lizzi, A. R., De-Simone, A., Marroccella, C. E., Gravina, G. L., Tatone, C., & Festuccia, C. (2013). Crocus sativus stigma extract and its major constituent crocin possess significant antiproliferative properties against human prostate cancer. *Nutrition and Cancer, 65*, 930–942.
96. Lu, P., Lin, H., Gu, Y., Li, L., Guo, H., Wang, F., & Qiu, X. (2015). Antitumor effects of crocin on human breast cancer cells. *International Journal of Clinical and Experimental Medicine, 8*(11), 20316–20322.
97. Bakshi, H. A., Hakkim, F. L., & Sam, S. (2016). Molecular mechanism of crocin induced caspase mediated MCF-7 cell death: In vivo toxicity profiling and ex vivo macrophage activation. *Asian Pacific Journal of Cancer Prevention, 17*(3), 1499–1506.
98. Mostafavinia, S. E., Khorashadizadeh, M., & Hoshyar, R. (2016). Antiproliferative and pro-apoptotic effects of crocin combined with hyperthermia on human breast cancer cells. *DNA and Cell Biology, 35*(7), 340–347.
99. Hire, R. R., Srivastava, S., Davis, M. B., Kumar Konreddy, A., & Panda, D. (2017). Antiproliferative activity of crocin involves targeting of microtubules in breast cancer cells. *Scientific Reports, 7*(1), 1–11.
100. Chryssanthi, D. G., Lamari, F. N., Iatrou, G., Pylara, A., Karamanos, N. K., & Cordopatis, P. (2007). Inhibition of breast cancer cell proliferation by style constituents of different crocus species. *Anticancer Research, 27*(1A), 357–362.
101. Ashrafi, M., Bathaie, S. Z., Abroun, S., & Azizian, M. (2015). Effect of crocin on cell cycle regulators in N-nitroso-N-methylurea-induced breast cancer in rats. *DNA and Cell Biology, 34*(11), 684–691.
102. Arzi, L., Farahi, A., Jafarzadeh, N., Riazi, G., Sadeghizadeh, M., & Hoshyar, R. (2018). Inhibitory effect of crocin on metastasis of triple-negative breast cancer by interfering with Wnt/β-catenin pathway in murine model. *DNA and Cell Biology, 37*(12), 1068–1075.
103. Xia, D. (2015). Ovarian cancer HO-8910 cell apoptosis induced by crocin in vitro. *Natural Product Communications, 10*(2), 1934578X1501000208.
104. Hoshyar, R., & Mollaei, H. (2017). A comprehensive review on anticancer mechanisms of the main carotenoid of saffron, crocin. *Journal of Pharmacy and Pharmacology, 69*(11), 1419–1427.
105. Amin, A., Bajbouj, K., Koch, A., Gandesiri, M., & Schneider-Stock, R. (2015). Defective autophagosome formation in p53-null colorectal cancer reinforces crocin-induced apoptosis. *International Journal of Molecular Sciences, 16*(1), 1544–1561. [
106. Zhou, Y., Xu, Q., Shang, J., Lu, L., & Chen, G. (2019). Crocin inhibits the migration, invasion, and epithelial-mesenchymal transition of gastric cancer cells via miR-320/KLF5/HIF-1α signaling. *Journal of Cellular Physiology, 234*(10), 17876–17885.
107. Akbarpoor, V., Karimabad, M. N., Mahmoodi, M., & Mirzaei, M. R. (2020). The saffron effects on expression pattern of critical self-renewal genes in adenocarcinoma tumor cell line (AGS). *Gene Reports, 19*, 100629.

108. Li, C. Y., Huang, W. F., Wang, Q. L., Wang, F., Cai, E., Hu, B., & Li, H. H (2012). Crocetin induces cytotoxicity in colon cancer cells via p53-independent mechanisms. *Asian Pacific Journal of Cancer Prevention, 13*(8), 3757–3761.
109. Amerizadeh, F., Rezaei, N., Rahmani, F., Hassanian, S. M., Moradi-Marjaneh, R., Fiuji, H., Boroumand, N., Nosrati-Tirkani, A., Ghayour-Mobarhan, M., Ferns, G. A., et al. Crocin synergistically enhances the antiproliferative activity of 5-flurouracil through Wnt/PI3K pathway in a mouse model of colitis-associated colorectal cancer. *Journal and Cell Biochemistry, 119*, 10250–10261.
110. Fujimoto, K., Ohta, T., Yamaguchi, H., Tung, N. H., Fujii, G., Mutoh, M., & Shoyama, Y. (2019). Suppression of polyps formation by saffron extract in adenomatous polyposis coli-Min/+ mice. *Pharmacognosy Research, 11*(1), 98.
111. Festuccia, C., Mancini, A., Gravina, G. L., Scarsella, L., Llorens, S., Alonso, G. L., et al. (2014). Antitumor effects of saffron-derived carotenoids in prostate cancer cell models. *BioMed Research International, 2014*, 135048.
112. Chen, S., Zhao, S., Wang, X., Zhang, L., Jiang, E., Gu, Y., et al. (2015). Crocin inhibits cell proliferation and enhances cisplatin and pemetrexed chemosensitivity in lung cancer cells. *Translational Lung Cancer Research, 4*(6), 775–783.
113. Moradzadeh, M., Kalani, M. R., & Avan, A. (2019). The antileukemic effects of saffron (Crocus sativus L.) and its related molecular targets: A mini review. *Journal of Cellular Biochemistry, 120*(4), 4732–4738.
114. Moradzadeh, M. A. L. I. H. E. H., Tabarraei, A. L. I. J. A. N., Ghorbani, A. H. M. A. D., Hosseini, A. Z. A. R., & Sadeghnia, H. R. (2018). Short-term in vitro exposure to crocetin promotes apoptosis in human leukemic HL-60 cells via intrinsic pathway. *Acta Poloniae Pharmaceutica Drug Research, 75*(2), 445–451.
115. Geromichalos, G. D., Papadopoulos, T., Sahpazidou, D., & Sinakos, Z. (2014). Safranal, a Crocus sativus L constituent suppresses the growth of K-562 cells of chronic myelogenous leukemia. In silico and in vitro study. *Food and Chemical Toxicology, 74*, 45–50.
116. Shakeri, M., Tayer, A. H., Shakeri, H., Jahromi, A. S., Moradzadeh, M., & Hojjat-Farsangi, M. (2020). Toxicity of saffron extracts on cancer and normal cells: A review article. *Asian Pacific Journal of Cancer Prevention: APJCP, 21*(7), 1867–1875.
117. Mousavi, S. H., Tavakkol-Afshari, J., Brook, A., & Jafari-Anarkooli, I. (2009). Role of caspases and Bax protein in saffron-induced apoptosis in MCF-7 cells. *Food and Chemical Toxicology, 47*(8), 1909–1913.
118. Mousavi, M., & Baharara, J. (2014). Effect of Crocus sativus L. on expression of VEGF-A and VEGFR-2 genes (angiogenic biomarkers) in MCF-7 cell line. *Zahedan. Journal of Research in Medical Sciences, 16*(12), 9–15.

Chapter 43
Barbados

Adeeba Mushtaq, Nayyab Naeem, Zubaida Yousaf, Arusa Aftab, and Modhi O. Alotaibi

43.1 Introduction

Aloe vera, (synonym *Aloe barbadensis* Mill) is a perennial, pea-green, xerophytic, shrubby or arborescent plant of the Asphodelaceae family.It is a succulent plant that can tolerate drought. It primarily grows in arid regions of the world namely Asia, Africa, America and Europe America. Since ages, *Aloe vera* has been recognized and used by many people because of its health benefits. The word *Aloe vera* is arised from the Arabic word "Alloeh," which means "shining bitter material, while the Latin word "vera" indicates "truth." Around 2000 years ago, Greek scientists began to acknowledge *Aloe vera* as a universal, all-purpose treatment. The Egyptians called *Aloe vera* as "the herb of immortality." Now a day, the plant is utilized in dermatology to cure a variety of diseases.

Aloe vera has a significant historical role in traditional indigenous medical systems, including unani, ayurveda, siddah and homoeopathy [1]. *Aloe vera*It is commonly known as Sanskrit [8], Bengali (Ghritakalmi), Assamese (Musabhar, Machambar), Nepali (Gheeukumari), Gujrati (Eliyo, Eariyo), English (Indian Aloe), Hindi (Musabhar, Elva) [26]. Leaves of the plant are lance-shaped, pointy, jagged, and edged [2].

A. Mushtaq · N. Naeem · Z. Yousaf (✉) · A. Aftab
Department of Botany, Lahore College for Women University, Lahore, Pakistan
e-mail: zubaida.yousaf@lcwu.edu.pk

M. O. Alotaibi
Department of Biology, College of Science, Princess Nourah Bint Abdulrahman University, Riyadh, Saudi Arabia

M. Zia-Ul-Haq et al. (eds.), *Essentials of Medicinal and Aromatic Crops*,
https://doi.org/10.1007/978-3-031-35403-8_43

43.2 Classification

Kingdom	Plantae
Order	Asparagales
Division	Spermatophyte
Subdivision	Angiospermae
Class	Monocotyledoneae
Genus	*Aloe*
Specie	*Barbadensis mill*

Carl Linnaeus first described the plant in 1753 as *Aloe perfoliata* var. Later on, in April 6, 1768, Nicolas Laurens Burman redescribed the plant as as *Aloe vera* is listed in Greenery Indicia and is described as Aloe barbadensis by Philip Mill operator 10 days later in the Gardener's Dictionary [9]. This succulent perennial herb has a pea-green colour and rosette-shaped, fleshy, serrated leaves that are 30 to 50 cm long and 10 cm wide at the base (Fig. 43.1). It also has a triangular, sessile stem and a shallow root system. The bright yellow tubular flowers range in length from 25 to 35 cm, have an axillary spike, and the fruits frequently have many seeds [3]. The leaf's margin contains tiny white teeth and a serrated texture. Each flower is pendulous and has a yellow tubular corolla that is 2–3 cm (0.8–1.2 in) long. The blooms are produced in the summer on a spike that can reach a height of 90 cm (35 in) [4, 5]. *Aloe* crude fabric's market value is estimated to be at $125 million. The market for wrapped *Aloe vera* products are estimated to be worth over $110 billion [5]. The plant can endure temperatures as high as 104 degrees Fahrenheit and can withstand below-freezing conditions without suffering root damage [6].

Fig. 43.1 *Aloe vera*

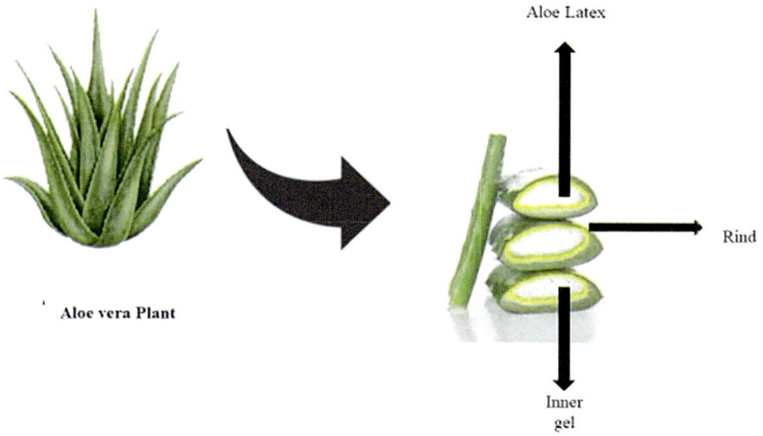

Fig. 43.2 different layers of *Aloe vera* leaf

Aloe vera leaf consists of the three layers (Fig. 43.2):

1. **Inner gel:** Vitamins, sterols, lipids, glucomannans, and amino acids make up the remaining 1% of the gel that is made up of 99% water.
2. **Latex:** In the middle layer of latex, which is the bitter yellow sap, glycosides and anthraquinones are found.
3. **Rind:** The rind is substantial, 15–20 cell outer layer that serves as protection and produces carbohydrates and proteins. Vascular bundles in the rind are in charge of transporting water and starch (phloem) (xylem) [3].

In the history of cosmetics, aloe plants have been used extensively, as an emollient and moisturizer, as well as a topical healing agent for burns and abrasions. Similar uses for *Aloe vera* exist across cultures. The latex, which flows from the plant when it is cut, is applied to the affected region of skin. Alternatively, the leaf may be split lengthwise and either deposited directly on the skin or its inner gel may be removed and used as an ointment. *Aloe vera* has historically been farmed in Western nations, particularly in the U.S., primarily to provide the pharmaceutical sector with the latex component of the leaf [7]. The most often used application of *Aloe vera* as a remedy is to speed up the healing process of wounds and other restorative processes. Wound healing potential of *Aloe vera* has not undergone a thorough analysis. Since *Aloe vera* yields lucrative components, it is the most significant and valuable plant for producing homemade drugs and other items. *Aloe vera* [5].

43.3 Agronomy

43.3.1 Soil Requirements

The cultivation of *Aloe vera* is possible in a range of soil types, from marginal to sub-marginal soils, but it prefers well-drained sandy or loamy soils with high levels of organic matter. *Aloe vera* may withstand poor, salty, and/or sodic soil as well [10, 11]. The plants prefer to bear high pH conditions with lots of K and Na salts. However it has been noted that its growth was more rapid in fertile heavier soils, medium, such the black cotton soils of central India. While moderate fertility soils and a pH of up to 8.5 and well-drained loam to coarse sandy loam are preferable for its commercial cultivation [12].

43.3.2 Climatic Condition

Aloe vera can flourish in hot, humid climates with lots of rainfall. The plant grows best in alkaline, nitrogen-rich soil that receives more than 50 cm of rain annually. The optimal range for soil nitrogen should be between 0.40% and 0.50%. But the majority of aloe species often grow in sandy soils [2].

43.3.3 Varieties

Many genus species, including *Aloe vera* Linn, have been known by the common name *Aloe*. Among the members of the family Liliaceae family, *A. perryi* Baker, *A. ferox* Miller, *A. barbadensis* Miller, *A. indica* Royle, *A. chinensis* Baker, etc. are the widely used species. The consensus among these is that *A. vera* Linn. syn. *A. barbadensis* Miller is the genuine *Aloe*'s botanical source. Although *A. vera* Linn is the approved scientific name for this species, *A. barbadensis* Miller is still commonly used as the term in reference books [12].

43.3.4 Land Preparation

True *Aloe vera* does not need to be planted in very deep soil because its roots do not extend below 20–30 cm. Depending on the agroclimatic conditions and the type of soil, 1–2 ploughing and laddering operations are required to make the soil weed-free and friable. Field can be divided into plots that are the right size (10–15 m by 3 m), taking into account the slope and irrigation supply [2, 11].

It can be propagated vegetatively from cuttings of rhizome or root suckers. 15–18 cm long root suckers were planted by burying two-thirds portion of them in soil. For one hectare land preparation, approximately 15,000 root suckers of 60 × 60 cm spacing is required.

43.3.5 Plant Nutrient

Prior to soil preparation, 8–10 tonnes of FYM/ha (Farm Yard Manure) is applied. After the last ploughing, 35 kg of nitrogen, 70 kg of phosphorous, and 70 kg of potassium are added per hectare. 350–400 kg of neem cake per hectare may be used to control termite problems. A top dressing of 35–40 kg of Nitrogen (N) may be used in September through October. N dose can be decreased if the soil is organically rich [2].

43.3.6 Planting

For better plant development and field survival, suckers are commonly sown in July and August during the monsoon season. With the exception of the winter, planting is feasible year-round (November–February) [12].

43.3.7 Manuring

Compost and farmyard manure can be applied, as the crop responds well to both. During the first year of plantation, when the land is being prepared, FYM @ 20 t/ha is applied, and the same is done in succeeding years. Another choice is vermicompost at 2.5 tonnes per hectare [12].

43.3.8 Irrigation

Aloe vera doesn't need excessive water, whereas, irrigated land is needed soon after plantation. For a healthy leaf (approx. 1 kg), almost 150 cc of water is the required of the plant in a month. For a high crop production, irrigation is crucial at specific times during plant cultivation [13].

Both irrigated and rainfed areas are suitable for growing Aloe vera. Irrigation during the summer and just after planting will assure a fruitful yield. The plants, however, are sensitive to moist environments [12].

43.3.9 Plant Protection

In any region of the world, diseases and insect pests have not been a significant problem for this crop. Nonetheless, leaf spots, mealy bug, and anthracnose have been reported in various areas of the globe. Providing a light irrigation can easily reduce the sporadic appearance of a termite issue. To disseminate it, rhizome cuttings can also be employed. In this case, when the crop is harvested, the subterranean rhizome is also dug out and cut into 5–6 cm long cuttings with at least 2–3 nodes. It is transplanted once it starts to sprout after being rooted in specially prepared pots or sand beds. A nursery of 1 hectare in size typically requires about 25,000 suckers (10,000 for 1 acre) [12].

43.3.10 Harvesting

Aloe vera leaves are typically harvested 7–8 months after plantation. The outermost leaves are taken by pushing them away from the stalk of plant and cutting them at the white base. Avoiding microbiological contamination requires, the outer rind should be handled with the greatest care, and the seal at the leaf's base should be kept intact. The months of October and November are said to be the optimum for harvesting. The second year is best for maximizing yield [2].

43.3.11 Pests and Diseases

Aloe vera pests and illnesses are rarely discussed in scientific literature, but numerous frequent infections of the plant have been documented. *Phakopsora pachyrhizi*, the fungus that causes aloe rust, causes orange spore masses to occasionally be discovered on the underside of leaves. Aloe rust is characterized by the emergence of pale yellow spots on the leaves that spread and turn brown. Eventually, these could make the leaves fall off. It is believed that the fungus, which is frequently found in places with cool temperatures and high humidity, does not need to be treated. Another fungus illness brought on by *Colletotrichum gloeosporiodes* is anthracnose disease. Warm, moist weather, especially water splash, favors this disease to spread more readily. It begins as small, round to oval, dark green water-soaked spots that subsequently turn into reddish brown to brown spherical lesions that cluster together to form a large necrotic region. For treatment, appropriate fungicides are frequently needed. Another fungus, *Fusarium spp.*, attacks the base of *Aloe vera,* turning it reddish brown and black before it eventually rots. Cuttings taken above the rotting stem may be used to save parts of the plant as cold, moist circumstances are favourable for the formation of this disease. *Pectobacterium chrysanthemi*, a dangerous disease of *A. vera*, causes bacterial soft rot, which is characterized by watery, rotting, wilting, and collapsing young leaves. This bacterium survives in plant waste on the field, and hot, wet weather favors the disease's appearance. Overwatering the plant may make it worse. Aphids that feed on *Aloe vera* (*Aloephagus myersi*) secrete honeydew, which encourages the growth of mould. When there is a severe infestation, the plant grows slowly and becomes stunted. By using insecticidal soap, this pest may be organically controlled. It is advised that scientific investigations be conducted with the goal of accurately recognizing the many diseases and histories of *Aloe vera* [13–15].

43.4 Description of Crops

Aloe vera has the ability to store significant amounts of water in its tissue, is able to exploit Crassulacean acid metabolism (CAM), an adaptation to synthesis malic acid in the photosynthetic pathway, and can adapt to settings with limited or unpredictable water supplies [16]. The leaf pulp's vascular bundles transport latex along the edges of the leaves for storage, as well as water and minerals from the roots to the leaves, synthesized materials, and latex. Depending on the size of leaves and the plant's age, there are different numbers of vascular bundles [17]. Regarding products made from *Aloe vera*, there is a lot of inconsistency in both the scientific and popular literature. The first issue is that the three different liquids extracted from leaves of *Aloe vera* are all mistakenly referred to as "aloe juice" in the literature, which has led to misunderstanding. In order to avoid confusion, the phrase "Aloe juice" should only ever refer to the latex component of the pericycle. in accordance with pharmacopoeial definitions [18, 19]. The main characteristic of the plant *Aloe vera* is its high content of water, which ranges from 99% to 99.5%. However, the 0.5–1.0% solid portion is said to contain over 200 different potentially active substances, including simple and complex polysaccharides, organic acids, enzymes, minerals, phenolic compounds, and vitamins [16, 20]. The pulp and rind together make up 70–80% of the weight of the entire leaf, according to compositional research on the operational elements of *Aloe vera* plant leaf parts. The rind and pulp, measured in dry weight, have fat contents of 2.7% and 4.2%, and protein contents of 6.3% and 7.3%, respectively [21]. The percentages of ash (13.5% and 15.4%), in particular calcium, and soluble sugars (11.2% and 16.5%), largely as glucose, were both comparatively high in the rind and pulp, respectively. The majority of each leaf fraction was discovered to be non-starch polysaccharides and lignin, which made up 57.6% and 62.3% of the dry weight of the pulp and rind, respectively. The physical and chemical components of products manufactured from *Aloe vera* plants vary based on the source (such as a part of the plant), the species of the plant, the climatic conditions, seasonal impacts, and grower influences [16].

43.5 Production Process

A sandy loam is ideal for *Aloe vera* growth. While the plant requires warm, semi-tropical environments, excessive sun exposure leads to stunted plants with minimal gel output. As a result, fruit trees and *Aloe vera* are frequently planted together. Because to variations in growing, harvesting, processing, and storing practices, the quality of products made from *Aloe vera* plants varies greatly [16]. It could also be influenced by the regulatory framework that governs the sale of the product. The USA, China, Thailand, and Mexico were listed as the top producing nations, followed by the remainder of Latin America [20]. In Arizona and the southern Texas Rio Grande Valley, *Aloe vera* has grown to be a significant plant crop [16]. A number of steps are involved in the preparation of products from *Aloe vera* plant, including pressing, crushing, or grinding the plant, filtration, decolorization, stabilization, heat processing, and sometimes the addition of preservatives and stabilizers [22]. Because it is challenging to maintain the stability of processed extracts, which could lead to variations in product potency, the gel or whole leaf extracts can go through a stabilization step before being packaged. Pasteurization, ultraviolet stabilization, hydrogen peroxide chemical oxidation, chemical preservatives and additives addition, concentration, and/or drying are some of the possible steps in this procedure [16].

43.6 Trade Status

Aloe vera ranked 20th among the top-selling dietary supplements in the USA, according to the Nutrition Industry Journal Annual Report for 2012. From US$ 31 million in 2000 to US$ 72 million in 2011, sales have generally increased [23, 24].

The global market for raw aloe species was projected to be worth roughly US$ 125 million in 2006, while the market for finished products incorporating *Aloe vera* was worth about US$ 110 billion [22].

IMS Health MIDAS data showed that sales of products derived from Aloe species worldwide reached $351 million in 2012. *Aloe vera* has been listed as (90%) source of the majority of products. The International *Aloe vera* Science Council (IASC) estimates that the global market for *Aloe vera* products has reached 13 billion, with a projected 35% rise in raw material sales over the next 5 years [25].

43.7 Chemical Constituents

Chromone and anthraquinone and its glycoside derivatives are the two primary classes of active components of the *Aloe vera* plant extract, along with phenylpyrone derivatives, flavonoids, phenylpropanoids, coumarins, phytosterols, naphthalene analogues, lipids, and vitamins Table 43.1 [47, 58].

Table 43.1 Properties of the Bio-active ingredients in *Aloe vera*

Bioactive component	Properties
Enzymes	Cellulase, lipase, cylooxygenase, peroxidase, alkaline phosphatase, carboxypeptidase, oxygenase and bradykinase. When applied topically to the skin, bradykinase helps to minimise excessive inflammation while the other enzymes plays a role in oxidative defense mechanism and aid in the digestion of carbohydrates, proteins, and fats.
Micro/macro nutrients	Chromium, copper, selenium, manganese, potassium, sodium, selenium, magnesium, and zinc. A few minerals function as antioxidants, while others are necessary for the normal functioning of numerous enzyme systems in diverse metabolic pathways.
Vitamins	By scavenging free radicals, vitamins defend the body against oxidative stress. Antioxidants include vitamins E, C, and A (beta-carotene). Moreover, it also includes folic acid, choline, and vitamins B1, B12, B2, and B6.
Carbohydrates	Glucose and fructose (monosaccharides); glucomannans and polymannose (polysaccharides). Mannose-6-phosphate is the most abundant monosaccharide, whereas glucomannans [beta-(1,4)-acetylated mannan] are the most prevalent polysaccharides. Also discovered is Acemannan, a well-known glucomannan. Recently two new anti-inflammatory compounds: C-glucosyl chromone and alprogen, and a glycoprotein with anti-allergic properties have been isoloated from *Aloe vera* gel.
Organic acids	Sorbate, salicylic acid, and uric acid. Salicylic acid exhibited antibacterial and anti-inflammatory potential.
Anthraquinones	Aloin, barbaloin, isobarbaloin, anthranol, *Aloe vera*tic acid, *Aloe vera*-emodin, ester of cinnamic acid, resistannol, chrysophannic acid, and emodin have laxative properties. Aloin and emodin have analgesic, anti-bacterial and antiviral effects.
Hormones	Auxins and gibberellins, which have anti-inflammatory properties and helps in wound healing.
Non-essential aminoacids	Proline, glycine, tyrosine, alanine, arginine, histidine, glutamic acid, aspartic acid, and hydroxyl proline are included on non-essential amino acids list.
Essential aminoacids	Phenylalanine, isoleucine, leucine, valine, threonine, and lysine are among essential the amino acids.
Fatty acids and steroids	Campesterol, lupeol, β-sisosterol, and cholesterol. Fatty acids, such as arachidonic and linolenic. They all have anti-inflammatory effects, and lupeol has antibacterial and analgesic qualities as well.
Others	**Lignin**, an inert chemical, increases the other compounds' ability to penetrate the skin when administered as topical creams. Around 3% of the gel is made up of soap-like **saponins**, which have antibacterial and cleaning characteristics.

43.7.1 Chromone and Its Glycoside Derivatives

A total of *Aloe vera* 29 chromone compounds were extracted and identified from *Aloe vera* (Table 43.2). The four main active components of *Aloe vera* are *Aloe*sin,

Table 43.2 Chromone and its glycoside derivatives isolated and identified from *Aloe vera*

No.	Constituents	Molecular formula
1	Aloesin	$C_{19}H_{22}O_9$
2	Neoaloesin A	$C_{19}H_{22}O_9$
3	8-C-glucosyl-(R)-aloesol	$C_{19}H_{24}O_9$
4	8-C-glucosyl-7-methoxy-(R)-aloesol	$C_{20}H_{26}O_9$
5	8-C-glucosyl-(S)-aloesol	$C_{19}H_{24}O_9$
6	8-C-glucosyl-7-methoxy-(S)-aloesol	$C_{20}H_{26}O_9$
7	8-C-glucosyl-7-O-methylaloediol	$C_{20}H_{26}O_{10}$
8	8-glucosyl-(2′-O-cinnamoyl)-7-O-methylaloediol A	$C_{29}H_{32}O_{12}$
9	8-glucosyl-(2′-O-cinnamoyl)-7-O-methylaloediol B	$C_{29}H_{32}O_{12}$
10	C-20 -decoumaroyl-aloeresin G	$C_{20}H_{24}O_8$
11	Aloeresin E	$C_{29}H_{32}O_{10}$
12	Isoaloeresin D	$C_{29}H_{32}O_{11}$
13	Iso-rabaichromone	$C_{29}H_{32}O_{12}$
14	8-[C-β-D-[2-O-(E)-cinnamoyl] glucopyranosyl]-2-[(R)-2-hydroxypropyl]-7-methoxy-5-methylchromone	$C_{29}H_{32}O_{10}$
15	Aloeresin D	$C_{29}H_{32}O_{11}$
16	Rabaichromone	$C_{29}H_{32}O_{12}$
17	Allo-aloeresin D	$C_{29}H_{32}O_{11}$
18	Aloeresin K	$C_{31}H_{34}O_{12}$
19	Aloeresin J	$C_{30}H_{34}O_{11}$
20	8-C-glucosyl-noreugenin	$C_{16}H_{18}O_9$
21	4′-O-glucosyl-isoaloeresin DI	$C_{35}H_{42}O_{16}$
22	4′-O-glucosyl-isoaloeresin DII	$C_{35}H_{42}O_{16}$
23	Aloeresin A	$C_{28}H_{28}O_{11}$
24	7-O-methyl-aloeresin A	$C_{29}H_{30}O_{11}$
25	9-dihydroxyl-2′-O-(Z)-cinnamoyl-7-methoxy-aloesin	$C_{29}H_{30}O_{12}$
26	60-O-coumaroyl-aloesin	$C_{28}H_{28}O_{12}$
27	7-methoxy-60-O-coumaroyl-aloesin	$C_{29}H_{30}O_{12}$
28	Aloeveraside B	$C_{28}H_{28}O_{12}$
29	Aloeveraside A	$C_{29}H_{30}O_{12}$

formerly known as, aloeresin A, aloeresin B, aloeresin E, and isoaloeresin D. Three aloediols, were also extracted from it, but their exact configuration has not yet been determined [58].

43.7.2 Anthraquinone and Its Glycoside Derivatives

Aloe vera about 32 anthraquinones and their glycoside derivatives have been identified, extracted and characterized from the *Aloe vera* (Table 43.3). The most prevalent active components of *Aloe vera* are the isomers of aloin A, aloin B and two

Table 43.3 Anthraquinone and its glycoside derivatives isolated and identified from *Aloe vera*

No.	Constituents	Molecular formula
1	Aloin A	$C_{21}H_{22}O_9$
2	Aloin B	$C_{21}H_{22}O_9$
3	60 -O-acetyl-aloin A	$C_{23}H_{24}O_{10}$
4	60 -O-acetyl-aloin B	$C_{23}H_{24}O_{10}$
5	10-hydroxyaloins A	$C_{21}H_{22}O_{10}$
6	10-hydroxyaloins B	$C_{21}H_{22}O_{10}$
7	Aloinoside A	$C_{27}H_{32}O_{13}$
8	Aloinoside B	$C_{27}H_{32}O_{13}$
9	7-hydroxyaloin A	$C_{21}H_{22}O_{10}$
10	7-hydroxyaloin B	$C_{21}H_{22}O_{10}$
11	7-hydroxy-8-O-methylaloin A	$C_{22}H_{24}O_{10}$
12	7-hydroxy-8-O-methylaloin B	$C_{22}H_{24}O_{10}$
13	60 -malonylnataloin A	$C_{24}H_{24}O_{12}$
14	60 -malonylnataloin B	$C_{24}H_{24}O_{12}$
15	Homonataloside B	$C_{28}H_{34}O_{14}$
16	Elgonica dimer A	$C_{36}H_{30}O_{14}$
17	Elgonica dimer B	$C_{36}H_{30}O_{14}$
18	Aloindimer A	$C_{42}H_{42}O_{18}$
19	Aloindimer B	$C_{42}H_{42}O_{18}$
20	Aloindimer C	$C_{42}H_{42}O_{18}$
21	Aloindimer D	$C_{42}H_{42}O_{18}$
22	-emodin-11-O-rhamnoside	$C_{21}H_{20}O_9$
23	Chrysophanol	$C_{15}H_{10}O_4$
24	Emodin	$C_{15}H_{10}O_5$
25	Physcione	$C_{16}H_{12}O_5$
26	Aloe-emodin	$C_{15}H_{10}O_5$
27	Natemodin	$C_{15}H_{10}O_5$
28	Aloesaponarin I	$C_{17}H_{12}O_6$
29	Aloesaponarin II	$C_{15}H_{10}O_4$
30	Madagascine	$C_{20}H_{18}O_5$
31	3-Geranyloxyemodin	$C_{25}H_{24}O_5$
32	Rhein	$C_{15}H_8O_6$

anthraquinone glucosides. The four main anthraquinone aglycones are chrysophanol, emodin, physcione, and aloe-emodin. Six anthraquinone dimmers *Aloe vera* also identified from the plant [58].

43.7.3 Flavonoids

Many flavonoids and their derivatives (glycoside derivatives), including three different types—flavone, flavonol, and flavan-3-ol—had been identified and extracted (Table 43.4) [58].

Table 43.4 Flavonoids identified and isolated from *Aloe vera*

No.	Constituents	Molecular formula
1	Apigenin	$C_{15}H_{10}O_5$
2	Luteolin	$C_{15}H_{10}O_6$
3	Isovitexin	$C_{21}H_{20}O_{10}$
4	Isoorientin	$C_{21}H_{20}O_{11}$
5	Saponarin	$C_{27}H_{30}O_{15}$
6	Lutonarin	$C_{27}H_{30}O_{16}$
7	Kaempferol	$C_{15}H_{10}O_6$
8	Quercetin	$C_{15}H_{10}O_7$
9	Myricetin	$C_{15}H_{10}O_8$
10	Quercitrin	$C_{21}H_{20}O_{11}$
11	Rutin	$C_{27}H_{30}O_{16}$
12	Catechin	$C_{15}H_{14}O_6$
13	Epicatechin	$C_{15}H_{14}O_6$

Table 43.5 Phenylpropanoids and coumarins isolated and identified from *Aloe vera*

No.	Constituents	Molecular formula
1	Cinnamic acid	$C_9H_8O_2$
2	p-coumaric	$C_9H_8O_3$
3	Caffeic acid	$C_9H_8O_4$
4	Ferulic acid	$C_{10}H_{10}O_4$
5	Sinapic acid	$C_{11}H_{12}O_5$
6	5-p-coumaroylquinic	$C_{16}H_{18}O_8$
7	Chlorogenic	$C_{16}H_{18}O_9$
8	5-feruloylquinic	$C_{17}H_{20}O_9$
9	Caffeoylshikimic	$C_{16}H_{16}O_8$
10	5-p-cis-coumaroylquinic	$C_{16}H_{18}O_8$
11	3-(4-hydroxyphenyl) propanoic acid	$C_9H_{10}O_3$
12	Methyl 3-(4-hydroxyphenyl) propionate	$C_{10}H_{12}O_3$
13	7-demethylsiderin	$C_{11}H_{10}O_4$
14	Feralolide	$C_{18}H_{16}O_7$
15	Dihydrocoumarin	$C_{22}H_{18}O_7$
16	Dihydrocoumarin ethyl ester	$C_{25}H_{26}O_7$

43.7.4 Phenylpropanoids and Coumarins

Phenylpropanoids and Coumarins present in *Aloe vera* are presented in Table 43.5. Approximately 12 phenylpropanoid acids and their ester derivatives, and four coumarins, were isolated and identified from *Aloe vera* [58].

Table 43.6 Phenylpyrone and phenol derivatives isolated and identified from *Aloe vera*

No.	Constituents	Molecular formula
1	Aloenin A	$C_{19}H_{22}O_{10}$
2	Aloenin B	$C_{34}H_{38}O_{17}$
3	p-coumaroyl aloenin	$C_{28}H_{28}O_{12}$
4	Aloveroside A	$C_{30}H_{40}O_{17}$
5	Feroxidin	$C_{11}H_{14}O_3$
6	1-(2,4-dihydroxy-6-methylphenyl) ethanone	$C_9H_{10}O_3$
7	p-anisaldehyde	$C_8H_8O_2$
8	Salicylaldehyde	$C_7H_6O_2$
9	p-cresol	C_7H_8O
10	Pyrocatechol	$C_6H_6O_2$
11	Gentisic acid	$C_7H_6O_4$
12	Gallic acid	$C_7H_6O_5$
13	Vanillic acid	$C_8H_8O_4$
14	Syringic acid	$C_9H_{10}O_5$
15	Ascorbic acid	$C_6H_8O_6$

43.7.5 Phenylpyrone and Phenol Derivatives

. Nine phenolic compounds, vitamin C and three phenylpyrone derivatives (triglucosylated naphthalene derivative called aloveroside A, and one 1-methyltetralin derivative called feroxidin) had been identified and extracted from the plant (Table 43.6) [58].

43.7.6 Phytosterols and Others

Five phytosterols, including cycloartanol, 24-ethyl-lophenol, lophenol, 24-methylene-cycloartanol and 24-methyl-lophenol were extracted from *Aloe vera* leaf gel (Table 43.7). Furthermore, prostanoids and a few polar and nonpolar lipids were also isolated from *Aloe vera* leaves. Chemical study of the principal substances found in *Aloe vera* leaves—lipids, minerals, organic acids, free sugars, polysaccharides, fibre, protein, and ash—shows that moisture is present. The three main free sugars (sucrose, glucose and fructose) were also identified. The principal organic acids were isocitric, oxalic, fumaric acid, lactic, *L*-malic, acetic, citric, and lactone [58].

Table 43.7 Phytosterols isolated and identified from *Aloe vera*

No.	Constituents	Molecular formula
1	Cycloartanol	$C_{30}H_{52}O$
2	24-methylene-cycloartanol	$C_{31}H_{52}O$
3	Lophenol	$C_{28}H_{48}O$
4	24-methyl-lophenol	$C_{29}H_{50}O$
5	24-ethyl-lophenol	$C_{30}H_{52}O$

43.8 Health Benefits

- **Helps in digestion**
  Natural digestive system cleansing is made possible by drinking *Aloe vera* juice. If someone is constipated, it aids with elimination and promotes the intestines to move. It will also aid in curing diarrhea [12].
- **Increase energy levels**
  Many ingredients in our diets have been shown to contribute to fatigue and exhaustion. Regular use of *Aloe vera* juice ensures a stronger sense of well-being, increases aids in maintaining a healthy body weight and energy levels [12].
- **Builds immunity**
  In *Aloe vera* juice the polysaccharides boost up the white blood cells that fight viruses, it is especially beneficial for people with chronic immunological illnesses like polysaccharides or fibromyalgia [12].
- **Detoxification**
  Aloe vera juice is a wonderful all-natural detoxifier. We all need to occasionally cleanse our systems because of how hectic our lives are, how polluted the environment is, and how much junk food we consume. Drinking *Aloe vera* juice offers a delightfully rich cocktail of minerals, vitamins, and trace elements to assist our bodies cope with these daily pressures and strains, [12].
- **Reduces inflammation**
  Joint flexibility is increased, and it aids in body cell regeneration. It increases the strength of the muscles that support joints, which lessens discomfort and inflammation in weak or arthritic joints [12].

43.9 Uses of *Aloe vera*

Deobstuent, carminative, depurative, cathartic, diuretic, Antineoplastic, stomachic, and emmenagoge are all properties of *Aloe vera*. To treat splenopathy, amenorrhea, hindrance, dyspepsia, colic, hyperdenosis, smoulders, hepatopathy, run, stomach, menorrhea, tumours, dropsy carbunles, sciatica, lumbago and tooting, juice is often used. *Aloe vera* gel is especially helpful in cases of mouth ulcers and ulcerative colitis [27].

- Ulcerative colitis
- Mild to moderate burns
- Psoriasis vulgaris
- Oral lichen planus infections
- Erythema
- Genital herpe
- Type 2 diabetes
- Seborrheic dermatitis
- Skin moisturizer
- Angina pectoris

43.10 Therapeutic Use

There are three stages to the dynamic process of wound healing. Inflammation, hyperemia, and leukocyte infiltration characterize the first phase. The elimination of dead tissue makes up the second stage. The development of fibrous tissue and epithelial regeneration comprise the third stage of proliferation [46]. According to a more recent evaluation, the body of evidence favors the use of *Aloe vera* for treating first- to second-degree burns [41]. Mannose-6-phosphate has been proposed as the source of gel's ability of *Aloe vera* plant to heal wounds [47]. When aloe is applied topically or taken orally, glucomannan and plant growth hormone gibberellins interact with fibroblast growth factor receptors and boost their activity and proliferation to improve collagen formation. For wound contraction and improved breaking strength, aloe administration influences collagen composition and increased collagen cross-linking. Moreover, it boosts the production of dermatan sulphate and hyaluronic acid in the granulation tissue of a wound that is healing [6]. Acemannan, which is a long chain of acetylated mannose, is thought to be the primary functional element of *Aloe vera* [44, 45]. This complex carbohydrate hastens the healing of wounds and lessens skin responses brought on by radiation. Radiation burns and radiation ulcers have been treated with aloe gel. The fresh gel worked better than the cream [3]. Several *in vivo* and *in vitro* research have demonstrated the bradykinase activity of *Aloe vera* gel to be the anti-inflammatory activity. It has been demonstrated that the bradykinin, an inflammatory compound that causes pain, is broken down by the bradykinase peptidase, which was isolated from *Aloe vera*. C-glucosyl chromone, a new anti-inflammatory molecule, was discovered in gel extracts. The aloe sterol contains cholesterol, campesterol, sitosterol, lupeol, and other compounds that have anti-inflammatory properties and function as natural analgesics. Aloe's additional aspirin-like component is what gives it its anti-inflammatory and antibacterial benefits [6].

Alprogen blocks the entry of calcium into labrocytes, avoiding the mast cell release of leukotriene and histamine that is arbitrated by interactions of antigen-antibody. Muco-polysaccharides assist the ability of skin to hold wetness. Moreover, Amino acids soften hard skin cells, and zinc acts as an astringent to tighten pores

[48]. Both a cooling and moisturizing impact are provided by aloe gel. Moreover, it helps regenerate ageing skin and has applications in gerontology. The biogenic nature of aloe explains this characteristic. The cosmetics industry uses aloe vera as a skin tonic [6]. *Aloe vera* gel contains a number of glycoproteins that have been shown to have anticancer and antiulcer properties as well as to promote the proliferation of healthy human skin cells [27, 49]. Unfortunately, there are extremely less frequently conflicting statistically meaningful clinical research on the effectiveness of *Aloe vera* gel on human health. A powerful laxative, anthraquinones found in latex stimulate mucus secretion, raise intestinal water content, and promote intestinal peristalsis [6].

43.11 Medicinal Uses

An *Aloe vera* also has these properties like emmenagogue, aperient, deobstruent, depurative, emmenagogue, diuretic, carminative, stomachic, and anthelmintic. Juice is used in the treatment of abdominal tumours, dropsy carbuncles, sciatica, lumbago, burns, colic, hyperadenosis, hepatopathy, splenopathy, constipation, span menorrhagia, and verbosity. A substance derived from the juice of this plant, is a purgative, anthelmintic, and emmenagogue known as elio, used to treat childhood helminthiasis. *Aloe vera* gel contains a number of glycoproteins that have been shown to have anticancer and antiulcer properties as well as to promote the proliferation of healthy human skin cells [27, 49]. Both ulcerative colitis and pressure ulcers can benefit from gel [50]. Traditional uses for *Aloe vera* gel include treating wounds, small burns, and skin irritations topically, as well as orally for the treatment of constipation, coughs, ulcers, diabetes, migraines, arthritis, and immune system inadequacies [51]. Greece, Egypt, India, Mexico, Japan, and China are just a few of the societies that have employed *Aloe vera* for thousands of years as a medicine. *Aloe vera* was utilized by the Egyptians to treat TB and to create scrolls that resembled papyrus. Despite this, advertisements for the cosmetic and alternative medicine sectors frequently tout the calming, hydrating, and healing benefits of *Aloe vera* [52]. The bioactive compounds are effective in treating stomach disorders, gastrointestinal issues, skin diseases, constipation, radiation injury, wound healing, burns, dysentery, diarrhea, and in the treatment of skin diseases. They are also used as astringents, hemostatics, antidiabetics, antiulcer, antiseptic, antibacterial, anti-inflammatory, antioxidant, and anticancer agents. It is used in ayurvedic formulations to treat cough, colds, piles, debility, dyspnea, asthma, and jaundice as well as an appetite stimulant, purgative, emmenogogue, and antihelminthic [53]. Application *Aloe vera* and its gel are applied topically to treat acne. In order to prevent flaky skin and scalp in arid weather, *Aloe vera* is also used to soothe the skin and keep it moist [56]. Aloe sugars are additionally utilized in moisturizing formulations. It creates wonderful skin smoothing moisturisers, sun block lotions, and a variety of other beauty products when combined with particular essential oils [53].

Nowadays, *Aloe vera* is broadly used in cosmetics, nutraceuticals, and skin care [57]. *Aloe vera* has six antibacterial substances, including lupeol, salicylic acid, urea nitrogen, cinnamonic acid, phenols, and sulphur, which give it its antiseptic properties [72]. These substances have an inhibiting effect on bacteria, fungi, and viruses. Controlled trials are necessary to establish its efficacy in treating all ailments, even though the majority of these uses appear intriguing [6]. Aloe juice aids in the efficient operation of the body's systems [54]. It lessens physiological and biochemical changes in the body and lessens cell damage caused by stress [55]. Chemical reactions that modify a compound's oxidative state are referred to as oxidative stress. Some antioxidants are produced by the body as part of its normal regulatory system, while others are obtained through diet. *Aloe vera* is a prime example of a functional food that is essential for oxidative stress defense [53].

For the formulation of anti-cancer medicines like paclitaxel, docetaxel, etoposide, topotecan, and irinotecan, which have high toxicity for non-cancerous cells and tissues, phytochemicals or bioactive substances from natural sources have been studied [54]. *Aloe vera* active components had showed Apoptosis induction [55, 56], anti-inflammatory [57], of tumour suppressor genes [58], downregulation of oncogenes, regulation of hormonal levels, growth factor regulation, and suppression of invasion and metastasis. Several biological components with anti-oxidant, anti-bacterial/viral, anti-inflammatory, and immunomodulatory effects have been found in *Aloe vera*, according to reports [59]. Figure 43.3 demonstrated that acemannan has an inhibitory impact and that it possesses chemo-preventive properties by preventing B[a]P absorption [60–62].

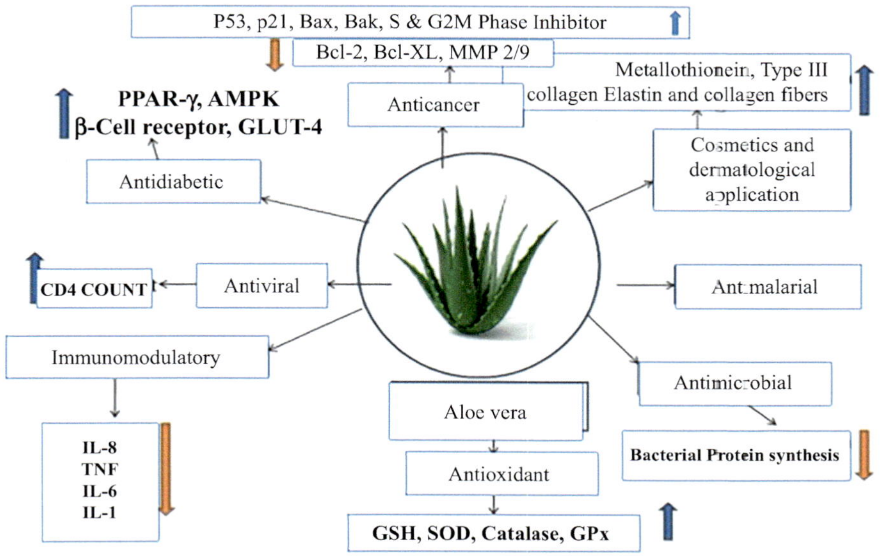

Fig. 43.3 Biological properties of *Aloe vera*

43.12 Traditional Uses

43.12.1 Indian System of Medicine

Aloe is referred to as kumari, or a young lady, in Ayurveda because it is said to restore femininity and youthful vigor. For the female reproductive system, aloe is utilized as a tonic. In Ayurveda, aloe is considered to provide alliterative, tonic, rejuvenating, purgative, and vulnerary effects. The three Ayurvedic constitutions of Pitta, Kapha, and Vata are all thought to be toned by *Aloe*. It is used to treat infections, worm infestations, colic, skin disorders, and constipation in conventional Indian medicine. Internal use of *Aloe* includes laxative, helminthic, hemorrhoid therapy, and uterine stimulating effects (menstrual regulator). It is applied topically to treat eczema or psoriasis, frequently in conjunction with liquorice root. In Tamil Nadu, a state in India, people frequently make a curry with *A. vera* that is eaten with rice or nan bread, a type of Indian bread [28, 29].

43.12.2 Chinese System of Medicine

To treat constipation brought on by an excess of heat, the Chinese utilized the skin and inner lining of *Aloe* leaves as a chilly and bitter cure. It is advised to use the gel to cure fungi illnesses because it is thought to be chilly and moist. Chinese herbalists utilized *Aloe* to get rid of worms, ease constipation, and normalize bowel motions since they understood the plant's potential as a purgative, just like their Ayurveda counterparts [29, 30].

43.12.3 Egyptian System of Medicine

Aloe was mentioned in Mesopotamian and Egyptian Papyrus as being beneficial for treating skin conditions, laxative usage, and curing infections. *Aloe* cream was reportedly a part of Cleopatra's beauty routine [31, 32].

43.12.4 Arabian System of Medicine

In Arabian medicine, the fresh gel is used topically to the body to reduce fever and to the forehead to relieve headaches. It is also used directly to treat conjunctivitis, mend wounds, act as a laxative and disinfectant [28].

43.12.5 Western System of Medicine

A. vera is one of the few herbal remedies that are widely used in Western culture, and it has applications in the food, drug, and cosmetic industries. It is applied topically and orally in therapeutic settings [33].

43.12.6 Greek System of Medicine

The Greek Herbal of Dioscorides (about 70 AD) provides a detailed description of the *A. vera* plant and recommends its usage for the treatment of hair loss, wounds, hemorrhoids and genital ulcers [34].

43.12.7 Spanish Medicine System

Hippocrates and Arab physicians both utilized *Aloe*, which Spanish explorers brought to the Western Hemisphere to cure soldiers who had been injured in battle [35].

43.12.8 United States System of Medicine

A. vera was first used clinically in the 1930s to treat radiation burns, skin and mucous membrane burns, and as a purgative and skin protectant by the U.S. pharmacopoeia in 1820. It is unclear how *Aloe's* was used by Native Americans. They learned the information from the Spanish explorers who brought *Aloe* along. The same way as their European predecessors, contemporary native healers in America utilize *Aloe* [36–38].

43.12.9 Mexican System of Medicine

Diabetes mellitus type 2 is treated with *Aloe vera* [39].

43.12.10 Trinidad and Tobago

Used for the treatment of hypertension (high blood pressure) [40].

43.12.11 Roman System of Medicine

Since the Roman era or even earlier, *A. vera* gel has been utilized for a variety of uses. One of the main uses of *A. vera* gel is to treat burns, which is practiced in many nations [41].

43.12.12 Japan System of Medicine

Used frequently as a component in yogurt that is sold commercially. Also, there are numerous companies that create *A. vera* beverages [42].

43.12.13 Philippines

Used in combination with milk to treat kidney infections [42].

43.12.14 Russia System of Medicine

Used to treat a variety of skin inflammations, including cold sores, sunburns, mild burns, and cuts and scrapes [43].

43.13 Proposed Mechanism of Action

Activation of fibroblasts and macrophages, as well as an increase in the production of collagen and proteoglycan [63].

Fibroblasts' growth factor receptors are bound by mannose-6-phosphate, which makes them more active [32].

Acemannan increases nitric oxide synthase activity, which activates macrophages and causes the production of fibrogenic cytokines [64–66].

Acemannan upregulates macrophage phagocytosis and fungicidal activity [67].

Acemannan and other cell wall biomaterials may extend stimulation of granulation tissue and enhance growth factor stability [68, 69].

Thromboxan A236,53 inhibition.

The phytosterols lophenol, cycloartenol, and their alkylated derivatives may boost hypoglycemic impact by regulating membrane-bound enzyme activity of phosphatases and hydrolases and increasing glucose metabolism [70, 71].

Endothelial cells may exhibit angiogenic activity in response to activator [73].

Lupeol, campesterol, and β-sitosterol are plant sterols that have an anti-inflammatory action by activating bradikinase, producing prostaglandin F2 and E2, inhibiting thromboxane A2, and inhibiting the release of IL-10.

Reduced intracellular free calcium levels have an inhibitory influence on human neutrophils' production of reactive oxygen species [74].

Increased metalloproteinase and plasminogen activator mRNA expression may cause endothelial cells to become angiogenic [75].

43.13.1 Summary

Aloe vera, commonly known as the Barbados or Curaçao Aloe, is a conventional medicinal plant with a long history of use among many different civilizations. The succulent plant thrives in arid and subtropical climates and is best known for two distinct preparations: the thick sap of the leaves that turns yellow-brown has potent laxative effects that warn against its use, and the clear mucilaginous gel that is widely used for the treatment of minor burns, especially sunburns. The transparent mucilaginous gel has a variety of traditional uses, including local treatments to decrease sweating and oral dosage for diabetes and a number of gastrointestinal conditions. Aloe vera gel has been shown to be useful in healing burn wounds, genital herpes, and seborrheic dermatitis in clinical trials. Little clinical research with frequently lax methods are the greatest constraint on existing clinical information regarding *Aloe vera* gel. In order to better assess the effectiveness of *Aloe vera* gel in treating a range of illnesses and to support existing traditional applications of the plant extract, several clinical experiments should be carried out. More clinical evidence must be provided through well planned trials using specific aloe extracts and placebo controls in order to support the use of aloe vera gel or its components for the treatment of a number of illnesses and disorders.

References

1. Baby, J., & Justin, S. R. (2010). Pharmacognostic and phytochemical properties of *Aloe vera* linn –an overview. *International Journal of Pharmaceutical Sciences Review and Research, 4*, 106.
2. Manvitha, K., & Bidya, B. (2014). Aloe vera: A wonder plant its history, cultivation and medicinal uses. *Journal of Pharmacognosy and Phytochemistry, 2*(5), 85–88.
3. Yeh, G. Y., Eisenberg, D. M., Kaptchuk, T. J., & Phillips, R. S. (2003). Systematic review of herbs and dietary supplements for glycemic control in diabetes. *Diabetes Care, 26*(4), 1277–1294.
4. Surjushe, A., Vasani, R., & Saple, D. G. (2008). *Aloe vera*: A short review. *Indian Journal of Dermatology, 53*(4), 163.
5. Pegu, A. J., & Sharma, M. A. (2019). Review on *Aloe vera. International Journal of Trend in Scientific Research and Development, 3*(4), 35–40.

6. Sahu, P. K., Giri, D. D., Singh, R., Pandey, P., Gupta, S., Shrivastava, A. K., & Pandey, K. D. (2013). Therapeutic and medicinal uses of *Aloe vera*: A review. *Pharmacology and Pharmacy, 4*(08), 599.
7. Lee, K. Y., Weintraub, S. T., & Yu, B. P. (2000). Isolation and identification of a phenolic antioxidant from *Aloe barbadensis*. *Free Radical Biology and Medicine, 28*(2), 261–265.
8. Nadkarni, K. M. (2004). *Indian plants & drugs* (pp. 28–29). Srishti Book Distributors.
9. Newton, L. E. (1979). In defence of the name *Aloe vera*. *Cactus and Succulent Journal of Great Britain, 41*, 29–30.
10. Rahi, T. S., Singh, K., & Singh, B. (2013). Screening of sodicity tolerance in *Aloe vera*: An industrial crop for utilization of sodic lands. *Industrial Crops and Products, 44*, 528–533.
11. Chowdhury, T., Rahman, M. A., Nahar, K., Chowdhury, M. A. H., & Khan, M. S. I. (2018). Growth and yield performance of *Aloe vera* grown in different soil types of Bangladesh: Yield performance of *Aloe vera* in different soils. *Journal of the Bangladesh Agricultural University, 16*(3), 448–456.
12. Rajeswari, R., Umadevi, M., Rahale, C. S., Pushpa, R., Selvavenkadesh, S., Kumar, K. S., & Bhowmik, D. (2012). *Aloe vera*: The miracle plant its medicinal and traditional uses in India. *Journal of Pharmacognosy and Phytochemistry, 1*(4), 118–124.
13. Das, N., & Chattopadhay, R. N. (2004). *Commercial cultivation of Aloe, 3*(2), 85–87.
14. Lawal, O. I., I Oyediran, R., B Olaniyi, M., F Okanlawon, T., O Oyeleye, A., O Akanni, F., & L Ifeanyichukwu, S. (2021). A review on the diverse uses, conservation measures and agronomic aspect of *Aloe vera* (L.). *Burm. f. European Journal of Medicinal Plants, 32*(9), 39–51.
15. UC Davis Botanical Conservatory. The Genus Aloe. Botanical Notes. 2009; 1(1). Available: http://greenhouse.ucdavis.edu/files/botnot_01-01.00.pdf on September 14, 2021.
16. Boudreau, M. D., Beland, F. A., Nichols, J. A., & Pogribna, M. (2013). Toxicology and carcinogenesis studies of a nondecolorized [corrected] whole leaf extract of Aloe barbadensis Miller (*Aloe vera*) in F344/N rats and B6C3F1 mice (drinking water study). *National Toxicology Program Technical Report Series, 577*, 1–266.
17. Ni, Y., & Tizard, I. R. (2004). Analytical methodology: The gel-analysis of aloe pulp and its derivatives. *Aloes, 1*(4), 129–144.
18. Organizaci6n Mundial de la Salud, World Health Organization, & WHO. (1999). *WHO monographs on selected medicinal plants* (Vol. 1, pp. 33–49). World Health Organization.
19. Hosokawa, R. (2011). The ministry of health, labour and welfare ministerial notification no. 65. *The Japanese Pharmacopoeia, Sixteenth Edition, [online]*, 1–2319.
20. Rodríguez, E. R., Martín, J. D., & Romero, C. D. (2010). Aloe vera as a functional ingredient in foods. *Critical Reviews in Food Science and Nutrition, 50*(4), 305–326.
21. Femenia, A., Sanchez, E., Simal, S., & Rossello, C. (1999). Compositional features of polysacchardies from Aloe vera (*Aloe barbadensis Miller*) plant tissues. *Carbohydrate Polymers, 39*(2), 109–117.
22. Ahlawat, K. S., & Khatkar, B. S. (2011). Processing, food applications and safety of aloe vera products: A review. *Journal of Food and Science Technology, 48*(5), 525–533.
23. Nutrition Business Journal. (2010). *NBJ's Supplement Business Report. An analysis of markets, trends, competition and strategy in the U.S. dietary supplement industry*. Penton Media, Inc.
24. Nutrition Business Journal. (2012). *NBJ's Supplement Business Report. An analysis of markets, trends, competition and strategy in the U.S. dietary supplement industry*. Penton Media, Inc.
25. IMS Health. (2012). *Multinational Integrated Data Analysis (MIDAS). Worldwide sales of selected products*. Plymouth Meeting, Pennsylvania: IMS Health.
26. The Ayurvedic Pharmacopoeia of India. In: Ayush Do, editor: Government of India Ministry of Health And Family Welfare. p. 104.
27. Yagi, A., Kabash, A., Mizuno, K., Moustafa, S. M., Khalifa, T. I., & Tsuji, H. (2003). Radical scavenging glycoprotein inhibiting Cyclooxygenase-2 and thromboxane A2 synthase from *Aloe Vera* gel. *Planta Medica, 69*(3), 269–271.
28. Ghazanfar, S. A. (1994). *Handbook of Arabian medicinal plants*. CRC Press.

29. Heber, D. (2007). Physicians' desk reference for herbal medicines. *Thomson Heath Care*, 515–518.
30. Babaei, A., Manafi, M., & Tavafi, H. (2013). Study on effect of *Aloe vera* leaf extracts on growth of Aspergillus flavus. *Annual Research and Review in Biology, 3*, 1091–1097.
31. Haller, J. S., Jr. (1990). A drug for all seasons. Medical and pharmacological history of Aloe. *Bulletin of the New York Academy of Medicine, 66*(6), 647.
32. Shelton, R. M. (1991). *Aloe vera*: Its chemical and therapeutic properties. *International Journal of Dermatology, 30*(10), 679–683.
33. Foster, M., Hunter, D., & Samman, S. (2011). Evaluation of the nutritional and metabolic effects of Aloe vera. *Herbal Medicine: Biomolecular and Clinical Aspects*, 37–54.
34. Davis, R. H. (1997). *Aloe vera: A scientific approach* (pp. 109–111).
35. Atherton, P. (1998). *Aloe vera*: Magic or medicine? *Nursing Standard (Through 2013), 12*(41), 49.
36. Collins, C. E., & Collins, C. (1935). Roentgen dermatitis treated with fresh whole leaf of *Aloe vera*. *American Journal of Roentgenology, 33*(3), 396–397.
37. Mandeville, F. B. (1939). *Aloe vera* in the treatment of radiation ulcers of mucous membranes. *Radiology, 32*(5), 598–599.
38. Park, Y. I., & Jo, T. H. (2006). Perspective of industrial application of Aloe vera. In *New perspectives on Aloe* (pp. 191–200). Springer.
39. Coronado, G. D., Thompson, B., Tejeda, S., & Godina, R. (2004). Attitudes and beliefs among Mexican Americans about type 2 diabetes. *Journal of Health Care for the Poor and Underserved, 15*(4), 576–588.
40. Lans, C. A. (2006). Ethnomedicines used in Trinidad and Tobago for urinary problems and diabetes mellitus. *Journal of Ethnobiology and Ethnomedicine, 2*, 1–11.
41. Maenthaisong, R., Chaiyakunapruk, N., Niruntraporn, S., & Kongkaew, C. (2007). The efficacy of aloe vera used for burn wound healing: A systematic review. *Burns, 33*(6), 713–718.
42. Calvin, J. (2008). *Aloe vera*: Plant history uses and benefits. P 356. Chinnusamy K, Nandagopal T, Nagaraj K, Sridharan S (2009). Aloe vera induced oral mucositis: A case report. *The Internet Journal of Pediatrics and Neonatology, 9*, 2–10.
43. Kumar, S., & Yadav, J. P. (2014). Ethnobotanical and pharmacological properties of Aloe vera: A review. *Journal of Medicinal Plants Research, 48*(8), 1387–1398.
44. Lee, J. K., Lee, M. K., Yun, Y. P., Kim, Y., Kim, J. S., Kim, Y. S., Kim, K., Han, S. S., & Lee, C. K. (2001). Acemannan purified from *Aloe vera* induces phenotypic and functional maturation of immature dendritic cells. *International Immunopharmacology, 1*(7), 1275–1284.
45. Djeraba, A., & Quere, P. (2000). In vivo macrophage activation in chickens with Acemannan, a complex carbohydrate extracted from Aloe vera. *International Journal of Immunopharmacology, 22*(5), 365–372.
46. Reddy Uma, C. H., Reddy, S. K., & Reddy, J. (2011). *Aloe vera*—A wound healer. *Asian Journal of Oral Health and Allied Sciences, 1*, 91–92.
47. Maenthaisong, R., Chaiyakunapruk, N., & Niruntraporn, S. (2007). The efficacy of Aloe vera for burn wound healing: A systematic review. *Burns, 33*(6), 713–718.
48. West, D. P., & Zhu, Y. F. (2003). Evaluation of Aloe vera gel gloves in the treatment of dry skin associated with occupational exposure. *American Journal of Infection Control, 31*(1), 40–42.
49. Choi, S. W., Son, B. W., Son, Y. S., Park, Y. I., Lee, S. K., & Chung, M H. (2001). The wound healing effect of a glycoprotein fraction isolated from Aloe vera. *British Journal of Dermatology, 145*(4), 535–545.
50. Langmead, L., Feakins, R. M., & Goldthorpe, S. (2004). Randomized, Doubleblind, placebo-controlled trial of oral Aloe vera gel for active ulcerative colitis. *Alimentary Pharmacology & Therapeutics, 19*(7), 739–747.
51. Eshun, K., & He, Q. (2004). *Aloe vera*: A valuable ingredient for the food, pharmaceutical and cosmetic industries—A review. *Critical Reviews in Food Science and Nutrition, 44*(2), 91–96.
52. Boudreau, M. D., & Beland, F. A. (2006). An evaluation of the biological and toxicological properties of Aloe barbadensis (miller), *Aloe vera*. *Journal of Environmental Science and Health, 24*, 103–154.

53. Joseph, B., & Raj, S. J. (2010). Pharmacognostic and phytochemical properties of aloe vera Linn—An overview. *International Journal of Pharmaceutical Sciences Review & Research, 4*(2), 106–110.
54. Saroj, P. L., Dhandar, D. G., & Singh, R. S. (2004). *Indian Aloe*. Central Institute for Arid Horticulture.
55. El-Shemy, H. A., Aboul-Soud, M. A., Nassr-Allah, A. A., Aboul-Enein, K. M., Kabash, A., & Yagi, A. Antitumor properties and modulation of antioxidant enzymes' activity by Aloe vera leaf active principles isolated via supercritical carbon dioxide extraction. *Current Medicinal Chemistry, 17*(2), 129–138.
56. Barcroft & Myskja. (2003). *Aloe vera: Nature's Silent Healer*. BAAM.
57. Gordon, M. C., & David, J. N. (2001). Natural product drug discovery in the next millennium. *Pharmaceutical Biology, 39*, 8–17.
58. Kahramanoğlu, İ., Chen, C., Chen, J., & Wan, C. (2019). Chemical constituents, antimicrobial activity, and food preservative characteristics of Aloe vera gel. *Agronomy, 9*(12), 831.
59. Majumder, R., Das, C. K., & Mandal, M. (2019). Lead bioactive compounds of *Aloe vera* as potential anticancer agent. *Pharmacological Research, 148*, 104416.
60. Sampedro, M. C., Artola, R. L., Murature, M., Murature, D., Ditamo, Y., Roth, G. A., & Kivatinitz, S. (2004). Mannan from Aloe saponaria inhibits tumoral cell activation and proliferation. *International Immunopharmacology, 4*(3), 411–418.
61. Acevedo-Duncan, M., Russell, C., Patel, S., & Patel, R. (2004). Aloe–emodin modulates PKC isozymes, inhibits proliferation, and induces apoptosis in U-373MG glioma cells. *International Immunopharmacology, 4*(14), 1775–1784.
62. Lee, H. Z., Hsu, S. L., Liu, M. C., & Wu, C. H. (2001). Effects and mechanisms of aloe-emodin on cell death in human lung squamous cell carcinoma. *European Journal of Pharmacology, 431*(3), 287–295.
63. Davis, R. H., Donato, J. J., Hartman, G. M., & Haas, R. C. (1994). Anti-inflammatory and wound healing activity of a growth substance in Aloe vera. *Journal of the American Podiatric Medical Association, 84*(2), 77–81.
64. Tizard, I. R., Busbee, D., Maxwell, B., & Kemp, M. C. (1995). Effects of acemannan, a complex carbohydrate, on wound healing in young and aged rats. *Wounds, 6*, 201–209.
65. Roberts, D. B., & Travis, E. L. (1995). Acemannan-containing wound dressing gels reduce radiation-induced skin reactions in C3H mice. *International Journal of Radiation Oncology, Biology, Physics, 15*, 1047–1052.
66. McCauly, R. (1990). Frostbite-methods to minimize tissue loss. *Postgraduate Medical Journal, 88*, 67–70.
67. Rajasekaran, S., Sivagnanam, K., Ravi, K., & Subramanian, S. (2004). Hypoglycemic effect of Aloe vera gel on streptozotocin-induced diabetes in experimental rats. *Journal of Medical Food, 7*(1), 61–66.
68. Rajasekaran, S., Sriram, N., Arulselvan, P., & Subramanian, S. (2007). Effect of aloe vera leaf gel extract on membrane bound phosphatases and lysosomal hydrolases in rats with streptozotocin diabetes. *Pharmazie, 62*(3), 221–225.
69. Fujita, K., Teradaira, R., & Nagatsu, T. (1976). Bradykinase activity of aloe extract. *Biochemical Pharmacology, 25*(2), 205.
70. Davis, R. H., & Maro, N. P. (1989). Aloe vera and gibberellin. Anti-inflammatory activity in diabetes. *Journal of the American Podiatric Medical Association, 79*(1), 24–26.
71. Barrantes, E., & Guinea, M. (2003). Inhibition of collagenase and metalloproteinases by aloins and aloe gel. *Life Sciences, 72*(7), 843–850.
72. t Hart, L. A, Nibbering, P. H., van den Barselaar, M. T., van Dijk, H., van den Berg, A. J., & Labadie, R. P. (1990). Effects of low molecular constituents from Aloe vera gel on oxidative metabolism and cytotoxic and bactericidal activities of human neutrophils. *International Journal of Immunopharmacology, 12*(4), 427–434.
73. Vazquez, B., Avila, G., Segura, D., & Escalante, B. (1996). Antiinflammatory activity of extracts from Aloe vera gel. *Journal of Ethnopharmacology, 55*(1), 69–75.

74. Lee, M. J., Lee, O. H., Yoon, S. H., et al. (1998). In vitro angiogenic activity of Aloe vera gel on calf pulmonary artery endothelial (CPAE) cells. *Archives of Pharmacal Research, 21*(3), 260–265.
75. Karaca, K., Sharma, J. M., & Nordgren, R. (1995). Nitric oxide production by chicken macrophages activated by Acemannan, a complex carbohydrate extracted from Aloe vera. *International Journal of Immunopharmacology, 17*(3), 183–188.

Chapter 44
Tea

Rabia Sabri, Mahwish Hussain, Shadma Wahab, and Muhammad Zia-Ul-Haq

44.1 Introduction

The tea infusions that a significant portion of the global population enjoys are made from the green leaves of the *Camellia sinensis* plant. The most widely produced tea leaf varieties are green tea, black tea, or oolong tea, with black tea making up majorly of all consumed tea-based products. Increased antioxidant levels from regular tea consumption may reduce the risk of heart disease and mutagenicity. Tea has a therapeutic function in addition to antimicrobial, antibacterial, anticarcinogenic, antihypertensive, neuroprotective, and thermogenic characteristics. The large beverage industry is believed to include tea. A vast range of teas are produced in Asia, the Near East, South America, Africa, and South America. Due to this and its reputation for good quality on the international market, Asia now accounts for a significant share of all tea imports. Huge populations in Asia, Africa, the Middle East, and the UK (Fig. 44.1) [1].

R. Sabri (✉)
Department of Biochemistry, University of Agriculture, Faisalabad, Pakistan

M. Hussain
Department of Pharmacology, Lahore College for Women University, Lahore, Pakistan

S. Wahab
Department of Pharmacognosy, College of Pharmacy, King Khalid University, Abha, Saudi Arabia

Complementary and Alternative Medicine Unit, College of Pharmacy, King Khalid University, Abha, Saudi Arabia

M. Zia-Ul-Haq
Office of Research, Innovation and Commercialization, University of Engineering and Technology, Lahore, Pakistan

M. Zia-Ul-Haq et al. (eds.), *Essentials of Medicinal and Aromatic Crops*,
https://doi.org/10.1007/978-3-031-35403-8_44

Fig. 44.1 Camellia sinensis plant leaves and buds

44.2 Scientific Classification

Kingdom	Plantae
Subkingdom	Viridiplantae
Infrakingdom	Streptophyta
Superdivision	Embryophyta
Division	Tracheophyta
Subdivision	Spermatophytin
Class	Magnoliopsida
Superorder	Asteranae
Order	Ericales
Family	Theaceae
Genus	Camellia L
Species	*Camellia sinensis*
Common name	Tea

Tea leaves also include the stimulant caffeine, essential oils that provide flavor, and tannin, which gives tea its flavour and colour. The United Kingdom Tea Council claims that the most frequent associations between tea and benefits are political and health-related. Politically, the most illustrious episode took place in 1773 when a tax on tea sparked the Boston Tea Party, which eventually turned into the American Revolution. The Tea Party movement was founded in 2008 as a result of American political discontent that was reminiscent of the Boston Tea Party. The aforementioned explanations explain why tea is so well-liked around the world, especially given that it offers many countries that produce it with a very important source of money [2].

Tea plants are multiplied from seeds and cuttings; it takes a plant 4–12 years to yield seeds and 3 years to mature into a harvestable plant. According to the degree of fermentation, tea is divided into three separate variations or production types:

I. nonfermented tea (green tea)
II. semifermented tea (oolong tea)
III. fully fermented tea (black tea)

In addition, China has three other distinct types of tea: yellow tea, white tea (tea that has undergone minor fermentation), and dark tea. (postfermented tea). Tea fermentation is described as the hydrolysis and oxidation of plucked leaf components by specially endogenous enzymes, which is completely distinct from microbial alcoholic fermentations. Green tea's flavour is primarily influenced by how the leaves are processed, i.e. the selection of clone, time of plucking, shoot maturity, and cultivation method. Green tea is made by steaming or pan-firing of *Camellia sinensis* leaves right after plucking, which inhibits enzyme action and preserves its endogenous constituents in the leaves largely unchanged. Oolong tea is made by gently rolling fresh leaves that have been withered indoors for several hours after a brief period of time in the sun. The most crucial chemical processes in black tea are withering, fermenting, and drying, whereas the most important physical processes are water removal/dehydration and determining particle size of leave. Therefore, it is during these procedures that the flavour of black tea begins to develop. A well-liked substitute for the brewed goods is "instant tea," which comes in both hot and cold forms. A different concept of Instant tea was introduced in the 1930s and is similar to freeze-dried instant coffee but does not require the process of boiling water. The steps involved in making instant tea are: choosing the raw ingredients, extraction, scent stripping, processing the cream, concentration, and drying. Every procedure will be taken into account separately [1].

44.3 Nutritive Values

Fermented tea leaves had significantly more energy when refers to per 100 g dry weight (DW) leaves than steamed tea and fresh tea leaves (389.89 kcal), according to nutritional values. (386.49 kcal and 372.38 kcal, respectively). The high fat and carbohydrate contents of fermented tea leaves responsible for high energy level. Fresh and steamed tea leaves had low levels of fat (1.76 and 1.75 g, respectively), whereas fermented tea leaves had a noticeably greater level. (3.20 g). Steamed and fermented tea leaves had a lot of carbohydrates (67.99 and 67.69 g, respectively), whereas fresh tea leaves contained a lot less. (38.74 g). The highest amount of dietary fibre overall, as well as soluble and insoluble fibres, was found in fresh tea leaves. (88.06, 15.23, and 72.83 g, respectively). Fresh samples had around 4.8 times as much insoluble dietary fibre of leaves as soluble dietary fibre. Insoluble dietary fibre was 2.6–2.9 times more abundant in steamed and fermented tea leaves than soluble dietary fibre of leaves. Additionally, fresh tea leaves possesses much high protein (50.40 g) when compared to steamed and fermented tea leaves. (24.70 g and 22.58 g, respectively). All samples had a tiny amount of ash. (5.57 g – 9.10 g).

All tea samples had significant amounts of the minerals calcium, sodium, potassium, and magnesium. The maximum concentrations of calcium (587.70 mg), potassium (2726 mg), and magnesium (244 mg) were found in the fresh tea leaves, but these concentrations were dramatically reduced throughout the steaming process to 511 mg of calcium, 1432 mg of potassium, and 146 mg of magnesium. Calcium and magnesium levels were further reduced by fermentation to 453.65 mg and 127.82 mg, respectively. However, potassium rose following fermentation in contrast to the steaming procedure. (1745.96 mg). Interestingly, fermented tea leaves had the highest salt concentration (706.04 mg), followed by fresh (227.15 mg) and steaming (106.98 mg) tea leaves. All samples contained traces of iron and zinc. 7.11 mg of iron and 8.65 mg of zinc were found in fresh tea leaves. Both minerals were lessened in quantity by the steaming and fermentation procedures. Fermented *Camellia sinensis* leaves had 6.72 mg of iron (Fe) and 2.01 mg of zinc, compared to steamed tea leaves' 0.96 mg and 1.75 mg.

Steamed tea leaves had much more vitamin B1 (thiamine) (4.17 mg) than fresh tea leaves (2.52 mg), which had vitamin B1 (thiamine) content equivalent to fermented tea leaves. (2.56 mg). The highest concentrations of vitamin B2 (Riboflavin), vitamin B3 (Niacin), and vitamins C (Ascorbic acid) were discovered to reside in freshly ground tea leaves. (1.45 mg, 10.78 mg, and 58.69 mg, respectively). While no vitamin C (ascorbic acid) was found in fermented tea leaves, the concentrations of these vitamins were further diminished by the steaming and fermentation processes [3].

44.4 Types of *Camellia sinensis*

The most popular traditional teas are black tea (also known as "cay"), green tea (also known as "unoxidized tea"), yellow tea (also known as "huángchá in Chinese"), red tea (heicha in Chinese), dark tea, and white tea ("Scholars tea"). Green, Black, and oolong teas are produced and consumed in huge quantities in many regions of the world. Although all three kinds are made from the same plant (C. sinensis), they differ due to different processing methods that result in varying the extend of fermentation process in the final product [4].

44.4.1 Green Tea/Unoxidized Tea

Green tea is unfermented tea. The leaves are wilted in hot air and then steamed at 95–100 °C for 30–45 s to inactivate the enzymes (such as polyphenol oxidase) and stop the biological processes that would otherwise maintain the original polyphenols. The chemical composition resembles that of freshly picked tea leaves as a result [5].

44.4.2 Oolong Tea

Also known as qingcha (Chinese: pinyin: qīngchá) or "dark green teas". The phrase "blue tea" (French: thé bleu) is equivalent to the word "oolong" in English. Additionally popular in the East is oolong tea in its slightly fermented state. After being sun-wilted, crushed, and left to oxidise until the margins turn dark in color, the tea leaves are heated and dried.

44.4.3 Black Tea

Black tea has finished the fermenting process. Withered leaves undergo metabolic reactions and fermentation stages during manufacturing process before being dried. This type of tea is different from fresh tea leaves in terms of content due to oxidation, polymerization, and other processing changes that take place [5].

44.5 Cultivars

Many cultivars of tea are known. Some cultivars are Japanese cultivars which include:

1. **Benifuuki**
   The word "Benifuuki" is a *Camellia sinensis* cultivar which has the technique, with strong resistance to disease (it can be grown by using low quantity of pesticide), strong vigour, high production yield, and very unique distinct, and splendid fragrance/aroma.
   Mechanism of action: By blocking cellular protein tyrosine phosphorylation, myosin II regulatory light chain (MRLC) phosphorylation, and the production of FcRI, EGCG3"Me significantly suppresses mast cell activation. The inhibitory impact of EGCG3"Me (Epigallocatechin gallate) on histamine release is partly due to the suppression of MRLC phosphorylation caused by cell-surface attachment to the 67 kDa Laminin Receptor (67-LR). Therefore, it is hypothesised that these protective actions prevent mast cell degranulation (histamine release and interleukin secretion after FcRI cross-linking via IgE antigen complexes).
2. **Fushun**
   This cultivar type means maturity at a late stage with high cold resistance. Cutting survival after planting and initial growth are both great, and cutting germination is also good. Strong cold resistance and tree vigour are combined with favourable cultivating traits. For anthracnose and ring spot disease, this cultivar has greater disease resistance than "Yabukita," however caution must be exercised because red blight has been seen on cuttings and young trees in Kanaya, where it is grown. Young trees should be tailored in the same way as "Yabukita" because the tree stands upright.

3. **Kanayamidori**
   A cultivar called Kanayamidori is renowned for producing sencha with substance and a milky odor. It's probable that this cultivar will be used more frequently in the future because many tea research centers have supported its adoption.
4. **Meiryoku**
   A tea cultivar recognised for its robustness in the sense that it's simple to grow and maintain is Meiryoku.
   In Chinese, the word "mei" is frequently used to denote tea, whereas "ryoku" denotes green. It is a typical budding cultivar that has an earlier harvest window than Yabukita.
5. **Saemidori**
   An organic sencha grown on a small family farm at the base of Mount Fuji.
   For a fukamushi, this Saemidori produces a highly clear and limpid liquor. It has a subtle oiliness, sweet flavours of cooked veggies, umami, and just enough freshness and bite to keep everything in harmony. A really well-made daily, clean and resourceful.
6. **Okumidori**
   A late-maturity cultivar called "Okumidori" has sprouting periods that start 11 days and 8 days later than those of "Yabukita" and "Yabukita," respectively.
   Its sprouts have a high output, grow nicely, and harden gradually. The tea that is produced is of outstanding quality; it has an easy translucent look, a dark green colour and sheen, and a clean, crisp flavor. It doesn't have a particularly special quality, but it is a straightforward cultivar without any flaws or distinguishing traits.
7. **Yabukita**
   No other cultivar comes close to the popularity of the Yabukita cultivar in Japan.
   This cultivar can grow in a variety of soil types and climates, generates a decent yield, and is resistant to cold weather. The best part is that when converted into sencha, its leaves have a potent scent and tasty flavor [6].

44.6 Cultivation

A long-cultivated crop, tea (*Camellia sinensis*) dates back thousands of years. Although some triploid cultivar have been documented, it is a woody perennial tree crop with a chromosomal number of 30 (2n = 30) [7].

44.6.1 Habitat and Geography

The most productive tea-growing area in the world is southeastern Asia, specifically China, Tibet, and northern India, where *Camellia sinensis* is indigenous.

Air temperatures between 18 and 25 °C are ideal for tea plant growth. Shoot growth has been discovered to be reduced by temperatures below 13 °C and above 30 °C.Additionally, it is raised at a height of 2200 m.a.s.l. Tea plants require 1200 mm of annual rainfall as a minimum, but 2500–3000 mm is thought to be the ideal amount.

The best soil for growing *Camellia sinensis* is slightly rocky; additionally, graveled soil is the second-best soil for growing tea. Tea can thrive best in tropical and subtropical climates. Tea requires both warmth and shade to grow successfully. Rain must be plentiful, averaging 80–100 in. annually, for tea to flourish. Additionally, monsoons promote quicker and more plentiful leaf growth, which results in a loss of tea flavour. Another important factor in the growth of tea plants is elevation; the optimum elevation for growing tea is between 3000 and 6000 feet above sea level [8].

The **pH of the soil** must be between 4.5 and 5.6 for the tea plant to thrive. The optimum soil for growing tea plants is one with more than 2% organic content, is thick, well-drained, and well-aerated [9].

Leaf type, growth circumstances, plucking protocol, interval, and season, manufacturing procedures, volume of ground tea leaves, and infusion preparation are the primary determinants of tea quality [10].

44.6.2 Propagation

Tea is commercially propagated by vegetative material of the *Camellia sinensis* plant. Tea plants can be grown through seed germination, cuttings, and tissue culture (also known as micropropagation). Nowadays, less seed is used for propagation due to the advent of quick, affordable, and practical methods for Vegetative propagation (VP), which makes cultivar production simple. However, tea-breeding seed barriers can provide open-pollinated seeds upon request. Tissue culturing is quick and space-efficient. However, using it for micropropagation is expensive and more suitable for breeding purposes [11].

Nursery For raising vegetatively propagated materials, sleeve nurseries are advised. The nursery's proximity to a reliable water source and the availability of cover from the wind are two aspects to be taken into account when choosing a location. Availability of soil that is deep, well-draining, friable, and has a pH range of 5.0–5.6—all characteristics that are suitable for nursery development. Avoid low-lying areas that are extremely wet during rain or that freeze during dry months. Sleeves made of 250 gauge that are 25 cm long and 10 cm wide, or 5.25 cm in diameter when laid flat. For 1200 sleeves, combine 8 wheelbarrows of the subsoil/topsoil mixture with 1/4 kg of DAP. With a mixture of subsoil, topsoil, and fertilizer, the sleeves are filled to a depth of 17 cm – 17.5 cm (bottom 1/4 of sleeve). The remaining 1/4 of the sleeve is just covered with subsoil. The soil should be packed reasonably firmly; it shouldn't be too loose or too tightly packed, and it should always be dap. For nursery propagation, strong and robust cuttings from mother bushes that have been allowed to develop naturally for 5–6 months following trimming should be carefully chosen. For a period of 8–12 months, the propagated materials are ready for transplanting. When the roots have extended through the sleeves and have at least 20 cm (8 in) of space between them, sleeved plants are prepared for transplantation for further plant growth. Stems to stay on the parent bush for longer than 7 months because the material gets hard and the cuttings that originate from it don't grow well. After pruning, don't cover mother bushes. Apply 300 kg N/Ha/yr. of fertilizer to mother bushes, which is twice as much as you would to bushes that had been harvested at the time. Then every year, add the fertilizer in at least two doses. Followed this 2–3 months after each trimming if they are pruned every 5–7 months. After the cuttings have been done, return any branches of the plant and shoot material to the mother plants as mulch.

44.6.3 Nursery Maintenance

Nursery upkeep should done by Check all beds at least once in a week, to check plants for its growth and any signs of disease from insects, pests, or both, and treat as necessary. Always pick weeds by hand. Keep the soil surrounding the polythene moist during dry times. When the sheet is recorded as dry or 21 days after planting, water the beds. Adapt shade according to the weather at the time. 50% of the incoming sunlight should be permitted by the shade. As an alternative, UV nets are advised for their effectiveness and long lifespan, making them very cost-effective.

44.6.4 Hardening Off

The procedure of hardening off is to release the polythene sheeting at both ends of the bed about 3–4 months after propagation (when the new shoots are about 20 cmcc). The polythene should be rolled up at both ends and left in that position for a week to allow air to freely circulate. The polythene covering should be rolled up 120 cm at each end after a further week, and 30 cm at each end after that. The weekly opening should be increased by 1.2 m each week until the entire bed is exposed. While the earth is being hardened off, avoid letting it dry out. Start applying fertilizer once the polythene sheet has been completely removed by applying NPK(S) fertilizer on a weekly basis as a solution of 1 g/m2 of nitrogen in 1.3 l of water (10 g NPK(S) in 10 L, or 1tablespoon of NPK(S) in a 10 L watering can. To prevent burning, immediately spray young plants with water to remove the fertilizer solution from their leaves. When they are 8–12 months old, plants are prepared for transplanting to the main field [11].

44.6.5 *Withering*

The main goal of withering freshly plucked tea shoots was to prepare them for the rest of the black tea production process. After plucking, numerous biochemical and physiological processes take place. It is a process that causes physical and chemical changes in the young shoots to yield quality [12].

44.7 Crop Management

Site Selection and Other Fundamental Factors

No other tropical crop requires the kind of precision that tea does in order to produce its maximum yield. Tea needs a climate with certain constraints on certain characteristics, unique soil, and proper clearing and site preparation before planting. Therefore, it is crucial that careful consideration be given to the soil and climate needs of the tea plant while choosing a location before deciding if the place is suited for tea. Keep in mind that once established, tea plants can live for a century or longer. Tea requires temperatures between 130C and 300C, with 300C being the ideal temperature, and rainfall between 1200MM and 2200MM that is evenly spread throughout the year. The ideal altitude for tea planting is between 1500 and 2250 metres above sea level. Wind breaks lessen wind speed, which reduces moisture loss from the soil through evaporation and from the plants through evapotranspiration [11]. Tree belts should be 85 metres apart and 10 metres tall to be effective. Hakea saligna and Grevillea robusta are two trees that are suggested for use as windbreaks. Tea hedges may also be used to protect tea bushes. Tea grows well on deep, well-drained red volcanic soil that is at least 2 metres (6 feet) deep and has a pH range of 4–5.6. Tea is a soil-specific crop that prefers acidic soil, a moist environment, and does not withstand prolonged droughts. The best tea is grown at high elevations where frost is avoided. In order to prevent the illness armillaria root rot, crop care should be made when preparing the field to remove all forest trees and plants by ring barking [11].

Field Planting

When roots of *Camellia sinensis* plant have reached the bottom of the sleeves and there is at least 20 cm (8 in) of top growth, sleeved plants are ready to be transplanted. The soil cylinder in the sleeves shouldn't be dry when transplanting. In order to prevent shattering of the soil cylinder and maybe the roots, the plants must be handled cautiously. They should also be placed properly and snugly on any vehicle transporting them to the field. A wheelbarrow may transport several containers. By doing this, all pointless handling of the sleeves is avoided. Until planting is finished, the sleeves should always be shielded from the sun to avoid damaging the roots. Although the minimum should be 25, the holes should be at least 25 cm in diameter and 15–20 cm deeper than the sleeves. For sleeves that are 6.25 cm in diameter and 25 cm long, the holes will be 40 cm × 25 cm. For each planting hole, use 30 g of single super phosphate or 15 g of diammonium phosphate (DAP)/triple super phosphate. Combine the appropriate fertilizers thoroughly in soil with planting hole dirt.

For both Young Plants and Mature Plants Fertilizers
The shrubs in question are tea bushes, which are frequently exploited as cutting supplies. In comparison to plucked tea, mother bushes lose nutrients at a significantly faster rate. Bushes that are compromised by nutritional deficiency (or because of pests/insects, plants diseases, hail, extreme climate changes like drought, or cold) give birth to fewer cuttings, which are more difficult to develop in nurseries and strike more slowly than those from bushes that have vigorous branch growth following pruning. The amount of fertilizer that should be applied to mother bushes each year should be double that of plucked bushes of the same age. At least twice a year, apply the fertilizers. Two or three months after each pruning, these might be made. Each shrub can receive fertilizer right away after being pruned if a few are done each day. If it's expected or forecasted that there won't be any rain for 2 or 3 months following pruning, fertilizer should be sprayed right away.

Placing Fertilizer in Planting Holes
Provision super phosphate (fertilizer) is added to the soil in the planting holes, transplants will establish and flourish more quickly comparatively. Because it contains sulfur, single super phosphate (SSP) is preferred to double super phosphate (DSP), and it should be incorporated into the soil at rates that vary depending on the size of the holes (Table 44.1).

On all soils of planting site, with the exception of intensely wealthy and hutch site soils, carefully combine fertilizers by dirt starting implanting cells. (pH 5.7 and beyond). Diammonium phosphate must be applied instead of single, double, or triple super phosphate because soils that were previously utilized for super grass or unfertilised arable yields need nitrogen and phosphate. (TSP). Do not fill the planting holes with just NPKS 25:5:5:5.

44.7.1 *For Infills, Fertilizer*

Fertilizers including nitrogen (N), phosphate (Ph), and potash must be added to the planting hole in percentage to the size of the hole for rapid infill establishment. Use 115g DAP and 115g sulphate of potash to fill a hole that is 60 centimeters in diameter and 60 centimeters deep [3]. Apply the nitrogen, phosphorus, potassium and sulfur

Table 44.1 Fertilizers ratio according to area

Sr no	Planting hole (Depth × Width)	Single super phosphate in grams	Triple super phosphate in grams	Double super phosphate in grams
1	0.45 m × 0.22 m	30	15	15
2	0.50 m × 0.25 m	40	20	20
3	0.60 m × 30 m	54	27	27

in 25:5:5 ratio at a rate of 50g per plant 3 months after planting, and then treat remaining of the field as usual.

44.7.2 Young Tea Fertilizers

Young tea is tea that has been harvested for less than 3 years and is any age linking the time of transplanting and the time of pruning at the conclusion of its pioneer cycle. (total of 5 years). The plants require nutrients to sustain their health during these 5 years as well as additional fertilizer to promote the growth of robust rootle and stem systems that will support vigorous cropping at maturity The fertilizer needs to be a substance or combination that supplies nitrogen, phosphorus, potassium and Sulphur in a ratio of 5: 1: 1: 1, or one that is higher in phosphorus and potasium. Young tea must be kept free of weeds, and more fertilizer is applied to the other crops produced there in addition to the tea. Prior to the initial application of mulch to a ground, any convenient nitrogenous N fertilizer should be spread into the ground surface to deliver nitrogen at a rate of 12 kg/ha. This is see to to make up for the short-term loss of nitrogen from the top soil as the mulch decomposes [11].

44.7.3 The First-Year Application

Plants with sleeves have active roots and leaf shoots, and they can act to fertilizers as almost immediately as 6 weeks afterward transferring. However, the progress of plants this period can also be regulated by "over applications" of even as minor as 36 g of NPK fertilizer practical in a single dose. Delay beyond this time can diminish the plants' potential for growth. Therefore, starting 6 weeks after planting, the plants should receive tiny but regular doses of 1.5 g N (6 g N.P.K.(S) – 1 soda flask top) per plant. Repeat every eight work week (2 months) or so.

Precaution Applying during dry spells is not advised. Never make the fertilizer less than 10 cm wide when you spread it around each plant in the outside ring. The ring should stretch from 5 cm from the shrub stem to just outside the stretched of the shoots in order to avoid fertilizer touching the plant's branch. To a distance downwards of 5 cm, dribble the fertilizer into the soil. Mulch should be repositioned if necessary so that the fertilizer can be administered, and then it should be replaced.

44.7.4 Application for the Following Year

Sleeved plants will benefit more from multiple little treatments in the second year following transplanting than from a single large application. According to the rates listed below, this would be put on in four dose up (every three calendar month).

Spacing in the field	Kilograms per ha	Kilograms per acre	Grams per plant
4 × 2	160 (3bags~)	65 (1.3bags~)	12 gm
4 × 2.5	160 (3bags~)	65 (1.3bags~)	15 gm
5 × 2.5	160 (3bags~)	65 (1.3bags~)	19 gm

44.7.5 Application for Third Year

Apply as a single dosage by broadcasting at the following rates in the interrow spaces:

Spacing in the field	Kilograms per ha	Kilograms per acre	Grams per plant
4 × 2	720 (approximately 14 bags)	292 (approximately 6bags)	54 gm
4 × 2.5	720 (approximately 14bags)	292 (approximately 6bags)	67 gm
5 × 2.5	720 (approximately 14bags)	292 (approximately 6bags)	84 gm

44.7.6 Four-year Application of Fertilizer

Put in a one dose by evenly distribution across the plane in the intervals between rows at the values listed underneath.

Spacing in the field	Kilograms per ha	Kilograms per acre	Grams per plant
4 × 2	920 (approximately 18 bags)	373 (approximately 8 bags)	68 gm
4 × 2.5	920 (approximately 18 bags)	373 (approximately 8 bags)	86 gm
5 × 2.5	920 (approximately 18 bags)	373 (approximately 8 bags)	107 gm

Available places with a single rainy season in both the fourth and fifth years, fertilizer can be applied once, ideally towards the beginning of the rains. It is beneficial to apply two semi-applications, one at the beginning of each rain season, in regions with two separate raining seasons. Applying fertilizer during times of very heavy rain is not recommended since some main nutrients will be lost to plane runoff. Spreading the compost evenly across the top soil surface is recommended; the immediate vicinity of the plant's stems should be avoided.

44.7.6.1 Time of Application of Fertilizers

Tea that is severely nutritionally stressed should as soon as practically possible get a therapeutic fertilizer application. If nitrogen is the nutrient that has to be added, the farmer should wait to fertilise until that rain will fall soon after. If fertilizers like phosphate and potassium are left on the soil surface under dry conditions, there is minimal chance that they will be lost chemically or biologically. Regular fertilizer applications should be avoided during extended cold or wet seasons, and if necessary, they had better be postponed awaiting it looks like rainwater will fall within a limited days if they must be made during dry weather [11].

Split Applications In mature tea, split applications barely increase yield. However, it would be simple to create a strategy creäte on a high-analysis complex fertilizer and a straight fertilizer so that the fertilizers were assigned to distinct seasons. If this is the case, it is advised that the multi-nutrient nourishment be applied preceding to the core cropping period. The final application of a cycle should get the same fertilizer, if it is practical to do so.

To reduce the possibility of increasing already abundant crop in some seasons, the yearly fertilizer program might be divided. In terms of the quality of yield produced, this could lower the fertilizer's overall effectiveness.

44.7.7 Plant Spacing in the Field and Population Guidelines for Better Growth

Spacing in feet	plants per acre (in numbers)	plants per Ha (in numbers)
4 × 2 ft	5379~	13,448~
3 × 3 ft	4788~	11,970~
31/4 × 31/4 ft	4762~	10,000~
4 × 21/2 ft	4000~	10,776~
31/2 × 31/2 ft	3510~	8784~
4 × 3 ft	3590~	8975~

Systems of Bringing Into Bearing
These systems were created in order to produce effective permanent frames.

1. **Formative pruning**
2. ***Pegging***
3. **Continuous Tipping**
4. **Free growth**

44.7.8 Cutting Selection

Well and vigour cuttings from main mother bushes that have been allowed to develop naturally for 5–6 months following trimming should be carefully considered for nursery propagation. The mother bush chosen must be of a certain cultivar designated for mother bushes. Even if cuttings are only required once, mother bushes should still be pruned twice a year. If the bushes haven't been pruned before, cut 40 cm above the prior level or 2.5 cm above it using a straight regular cut across procedure. Only once a year, during one of the prunes, should you clean out (remove weak and crossed branches). Five to seven months after pruning, new shoots are prepared for cutting.

44.7.9 Plucking Standards for Tea

Plucking standards are crucial in determining the grade of black tea. The standard of plucking has a significant impact on tea flavor and yield. The plucking specification can commonly be classified by way of fine, medium or coarse. Selected leaves plucking only removes the first two shrubberies and the shoot, whereas coarse plucking typically removes three or four shrubberies/leaves and the bloom [13].

However, a coarse plucking standard lowers the frequency of plucking because it takes longer for new shoots to develop to that standard. The benefits of cumulative biomass production over a long length of time may not be significant. Fine plucking standards of young shoots also increases yields. The tea's caffeine, tea-flavins, total ash along with total water soluble solids contents decreases while thearubigins, crude fiber of leaves and florid increased with course plucking standards [14].

The plucking conditions of a bud, one leaf, and a sprout yielded very superior-quality infusions as evidenced by the concentrations of Group II Volatile Flavor Compounds, the Flavor Index, caffeine, and total water-soluble solids. The study recommended against practicing plucking dual leaves and a bud, which degrades equally the production and quality traits [10].

44.7.10 Cutting Preparation

Cut branches or cuttings (whips) for cuttings should be brought immediately to a shelter close to the nursery water after being wrapped in moist sacking. All cutting should be done in the shade and under cover. Use only strong, young branches that are 5–7 months old while making the cuttings. Throw away the branches' extremely fragile tips and their extremely hard lower portions. Bark is growing in these areas. Each cutting should have one leaf and three to four centimetres of stem below the leaf. Utilize extremely sharp blades to prepare cuttings. Before planting, soak cuttings for about 30 min, just after they are prepared, in a frame full of fungicide mancozeb. It is not advisable to apply cuttings with broken leaves.

When **planting cuttings**, never let the leaf or bud contact the ground. The branches should be placed into the soil at an direction or angle if cuttings' plants are naturally deflexed, which means they bend backward instead of upward. Fingertips shouldn't contact the top or else bottom cuttings of the stems of plants when planting because the moisture after the fingertips could harm the plant's survival. Throughout planting, water the cutting frequently to keep it moist. Gentle watering is recommended because powerful jets could knock clippings around. To prevent any air exchange, bury the clear polythene sheeting (250,500 gauge) 1 foot deep in the ground over the hoops.

44.7.11 Plucking Intervals

The distribution of shoots, as well as the quality and yield of the crop, may be influenced by the choice of a plucking interval. Regular plucking periods lead to better yields since it accelerates the growth of auxiliary buds. Since the beheading of apical shoots is a more common method of overcoming apical dominance. Concerning the influence of plucking intervals on yields, contrasting results have been found. Appropriate to the lower level of residue and fiber, teas manufactured from 5-day plucking rounds are regarded as being particularly good organoleptically. Alternatively, black teas from the 7-day plucking round are regarded as good quality tea by tea analysts because they have a balance of VFC, ash, soluble solids present, and caffeine compounds in it. The most significant distinctive composites in black tea are theaflavine (TF) and thearubigines (TR), simply as the plucking interval lengthened, so did their concentration.

44.7.12 Plucking Seasons

The fundamental determinants of the quality of *Camellia sinensis* are the phenolic conponents present in immature plant shoots. Black teas with low total polyphenol concentration are of lower grade. As a result, the total amount of polyphenols and how much of them are present in the young tea shoots have an effect on the quality of produced black tea. [5, 15]

During cold season, plant had the highest concentration of flavanols in the newly formed apical shoots. In comparison to catechin gallates, tea shoots harvested during winter season had a greater concentration of simple catechins compounds, with epigallocatechin (EGC) being the most severely impacted. The overall flavanol concentration, on the other hand, is highest in the northern hemisphere throughout the apex of the budding season (summer), and it is lowest at the conclusion of the season. (late autumn). Black teas produced during the slow-growing chilly seasons are often of high quality and low yield. Black teas produced during warm, wet seasons are known for their rapid development, high production, and poor quality [16].

44.7.13 Withering

There are two goals for the withering stage. The moment leaves of tree plants were removed, chemical withering starts from that moment. Complex larger chemical molecules of leave are disintegrating into smaller substances throughout this process. The other method used for this is called physical withering, which involves drying off the tea leaf. The turgid leaf deforms into a flaccid leaf throughout this process. The sap is also concentrated in the cells of the tea leaf as a result of this process. Aerating the heaped leaves will remove the required amount of moisture [17]. Physical withering is mostly influenced by 3 main factors which are time of withering, treated temperature, and relative humidity during process. A certain amount of the leaf's solid content and moisture are lost when it withers. This can mainly consist of 3–4% of the entering leaf's dry weight(DW) and is caused by the respiration-related loss of carbon dioxide. The connection between the withering process and the elements that give black tea its distinctive aroma. The beginning of the anaerobic or catabolic phase, chemical withering or senescence, determines the quality of black tea. Protein levels drop, amino acids rise, soluble proteins become more permeable, and the cell membrane becomes more permeable during withering. Tea shoots' polyphenol oxidase (PPO) activity decreases when they wither, which has an impact on the oxidative condensation of flavanols. The amount of lipids that are degraded during withering is 20% [18].

44.7.14 Rolling and Chopping

The main goal of rolling is reducing the size of selected leaves along with cell damage that allows the helps fermentation stage to expose new surfaces to air. Additionally, it pushes out the juice and coats each leaf particle with a thin film of juice to accelerate chemical reactions. Yellow to red-brown colors are produced during the process of rolling as cytoplasmic content of leaves especially flavonoids gradually oxidized into quinones as a result of chloroplast polyphenol oxidase and cell wall peroxidase [19]. Cytoplasmic flavonoids gradually oxidise into quinones throughout rolling, due to cell wall peroxidase and chloroplast polyphenol oxidase, producing yellow to red-brown colours [20]. As the leaf splits during rolling, the enzymes are released, exposing the fluids to oxidation in a natural way. The raw materials rolled at various lengths of time. The dholes (crush particles. As the leaf splits during rolling, the enzymes are released, exposing the fluids to oxidation in a natural way. The raw materials were rolled for various lengths of time. The dholes (crush particles) then put through a "Rotor vane machine" for additional crushing after the rolling operation were finished. This processed material was then sent through a C.T.C. which is short form of Crush, Tear, and Curl machine, for producing finer-grained material. Next step was to break twisted balls and slow the fermentation process, the same materials put through a roll breaker. The after fermentation process will last for much long period and soften the liquor when exhaust temperature was less than 49 °C. The rate with which moisture is removed is excessively rapid when the exhaust temperature is higher than 57.2 ° C, which leads to case hardened tea granules, which has particles that are hard on the exterior but not fully dry from inside of the granules. These tea produce bitter liquor and don't store in good condition.

In comparison to rolling stays of less than and more than such minutes, a length of 25 minutes produced greater benefits.

44.7.15 Fermentation Step

Fermentation is the main step for the production of black tea since considerable chemical changes take place at this stage. The creation of TFs and TRs, which are factors in the unique properties of tea liquors, with the enzymatic oxidatiOn of polyphenols of tea leaves, particularly tea catechins, during the fermentation stage.

When opposed to fermented tea, nonfermented tea has less volatile flavourings. The essential oil made from fermented leaves contains linalool oxides, but fresh leaf homogenates do not. The synthesis of volatile flavour compounds in tea leaves appears to be hindered by polyphenols that oxidise quickly. In order to ensure the best possible generation of flavour components, fermentation should be managed for both time and temperature. Despite the fact that temperature range should be in between 24 and 28 °C is regarded to be optimal/necessary, the enzymatic oxidations in fermentation proceed at their maximum at 28 °C [21].

At varying fermentation temperatures and for various fermentation times, brightness and astringency levels were reached. Maximum sensory assessment ratings were attained at low fermentation temperatures. Thus, maintaining a low fermentation temperature, even though it demands a much longer fermentation period, demands that the resulting teas were of higher quality. High grade black teas can thus only be produced at lower fermentation temperatures for longer periods of time [21].

44.7.16 Drying

Three main goals of drying fermented tea include: ceasing biochemical processes by heating enzymes to denaturation; lowering moisture to boost shelf stability and shelf life of tea; and, finally, enhancing chemically this processes that give tea its distinctive flavour and character.

Tea of good grade is produced by drying black tea at a temperature of 110 °C and a dryer speed of 1.5 rpm. To remove 95–97% of the moisture, all of the lots dried a second time at a low temperature that is nearly 80 °C, which produced better keeping and storing condition. Black tea that has a great moisture content (greater than 6%) degrades owing to post-drying fermentation. High moisture contents, however, have a short shelf life.

44.7.17 Pests Attacking Tea Plant and Their Management

I. **Pest Crevice/Scarlet Mite**

Symptom leaves corkiness, brown in color especially from on the underside of leaves, leaf fall occurs during early stage before maturity of plant, but during severe drought small leaves on the plant are attacked.

Management optimize tea nutrition, Use of resistant clones e.g. clone 31/8 is classified as sensitive to red crevice mites-Pruning is one of the vital cultural practice Pruning can be done if due when severe, spray with Omite (Propargite) at 3 ml/litre of water Note Allow 14 days interval plucking

II. **Pest, disease** Red spider mite (*Olygonnychus coffeae*)

Symptom upper side/surface of adult/mature leaves**,** darkens, and brown in colour

Management -Mainly adequate nutrition -Use of resistant gene clones e.g.- BBLK 152, TRFK 6/8 nd TRFK 7/9 are. classified as resistant or tolerant to Pruning is among one of the major practices. Pruning can be done if due-when severe, spray with pesticide Omite (Propargite) at 3 ml/l of water.

III. **Pest** Yellow Tea Mite (*Polyphagotarsonemus latus*)

Symptom Young leaves grow deformed, corky, and curl inward.

Management Pruning is one of the key cultural practises. Generally appropriate nourishment (Not beyond suggested rates below/none at all). Pruning can be done when necessary; for severe cases, use Omite (Propargite) at a rate of 3 ml/l of water.

Name of Pest/Disease Tobako crikets (Gryltalpaafricana)

Symptom When young plants are severed from the soil, the plant perishes.

Management Chemical bait formulation

44.8 Medicinal Values and Therapeutic Uses of Tea (*Camellia sinensis*)

Due to their phytochemical components' potential therapeutic benefits, traditionally uses medicinal plants are imperative components of medicine. This plant contains tens of thousands of different biological active substances, including lipid, amino acids, protein, volatile substances, carbohydrate, especially carotene, fluoride, the flavonoids, with its essential properties and thearubigins (TRs) and theaflavins (TFs) and catechins, amino acids like (L. theanine), ascorbic acid, retinoid acid and vitamin K, phenolic acids (caffeic acid (CA), gal1ic acid (GA), ch1orogenic acid (CGA), and cauramic acid), Although the molecular evidence of black tea's ability to decrease cholesterol and act as an antioxidant in humans has grown exponentially, little is known about its pharmacological effects. This review study emphasises widening the deep insight on *camellia sinensis* that might be employed as a nontoxic food additive in order to close this information opening. This essay also discusses the intriguing function black tea plays as a herbal remedy and the need to eliminate synthetic health promoters from human health care in the future. Additionally, this knowledge would be beneficial for the affordable usage of biological medications with no remarkable side effects and for natural human defense. In addition, more research at the molecular level is required to determine the MOA, particularly for

Table 44.2 key components of black tea (Sources and phytochemistry)

Active biological components	Active pharmaceutical ingredients	Percentage by dry weight
Catechin	-EGCG	10–12%
Thearubigins		12–18%
Methylxanthine	Caffeine present	8–11a%
Theaflavins	Resultant from the oxidation of catechins during making processing of tea	3–6%
Phenolic acid (PA)	CA, Quinic acid, GA	Not available
Amino acids	Theanine	5–10%

the hypocholesterolemia especially of black tea, which can prevent cardiac illnesses, have fewer adverse effects (Table 44.2) [22].

Flavonoids with comprises of catechins, TFs, along with TRs, phenolic acids like CGA, CA, GA, and cauramic acid, methylxanthines includes caffeine, amino acids (theanine), carbohydrates, lipids present, proteins, carotene, volatile oils compounds fluoride, as well as traces of ascorbic acid, retinoids, retinols and vitamin k, and folate are just a few of the active components in tea. Six types of polyphenol chemicals, including

(a) phenolic acids (PA),
(b) flavones,
(c) flavonols,
(d) anthocyanins,
(e) flavanols, and
(f) hydroxyl-4 flavanols.

Black tea contains more than 50% of all amino acids, and the main unique amino acid is ltheanine (−glutamylethylamide). Black tea's flavour and aroma are produced as a result of theanine breakdown. Tea gains bitterness from other AA like alanine and arginine. According to certain reports, (+)-catechin is more readily absorbed than (−)-catechin, therefore the gut bacteria converts (+)-EC to (+)-catechins. Black tea and its polyphenolic components are unabsorbed by human metabolism. The researchers finds that tea have mostly addressed flavan-3-ol absorption. The metabolites that are 3-Omethylgallic acid, 3,4-di-O-methylgallic acid, GA, and 4-O-methylgallic acid in urine after excretion are tested to assess black tea use. However, various fruits, such grapes, and tea wines were said to create the same compounds in urine [22].

44.9 Mechanism of Action of Tea Components

It is widely known that Free Radicals (FRs) and Reactive Oxygen Species (ROS) is responsible for directly harm cells through DNA degradation and lipid peroxidation (LPO), which in turn causes apoptotic cell death. Numerous research suggested that the polyphenols in black tea have anti-apoptotic and cytoprotective properties. By preventing the nitration of tyrosine, black tea catechins have been shown to significantly hunt FRs like hydroxyl, peroxyl, 1,1-diphenyl-3-picrylhydrazyl, and superoxide radicals as well as singlet oxygen, lipid, nitric oxide, and peroxynitrite radicals. Catechins are known to have an anti-inflammatory effect by scavenging NO and decreasing NO synthase activity. The neuroprotective activity of black tea poly-phenols have been interconnected to their capacity to reduce FRs. According to reports, catechins increased the brain's capacity to scavenge superoxide, which decreased cell death after ischemia. By catalysing the breakdown of Arachidonic Acid (AA) by the enzyme cyclo-oxygenase (COX), another important way that results in inflammation and worsens post-cerebral ischemia, catechins are also known to stimulate the synthesis of eicosanoids. The creation of stable semiquinone FRs, which lowers the FRs deaminating ability, is another potential way by which catechins exercise their antioxidative and anti-inflammatory effects. By altering the expression of pro-apoptotic and anti-apoptotic (B-cells lymphoma-extra large genes of chromosome), black tea polyphenols may affect apoptosis. Black tea polyphenols have been shown in a rat study to protect against immunological suppression, neutrophil and lymphocyte infiltration, and the reduction of antigen-presenting cells (APCs), such as dendritic cells and macrophages in immune system. The activation of "vascular cell adhesion molecule-1 (VCAM-1)" in the cell by tumour necrosis factor (TNF) along with interleukin-1 (IL-1) has also been found to be inhibited by (-)-EGCG, resulting in reduced monocyte adherence in the blood. The reduction of aortic atheromatous regions, lipid peroxides, and aortic cholesterol by (-)-EGCG was reported to lower plasma lipids in mice with apoprotein [3] deficiency, which is a mouse model of atherosclerosis. Additionally, (-)-EGCG has been shown to successfully cross the BBB of brain circulation, which results in neuroprotective health benefits by inhibiting LPO, hydrogen peroxide (H_2O_2), and an influx of calcium Ca^{+2} ions into the neuron cell of the central nervous system. LPO is shown by lower malondialdehyde (MDA) levels. TFs present in black tea are strong inducers of apoptosis in human leukaemia cells, according to recent literature.

Responsible mechanism reported is, the synthesis of cytochrome-c and subsequent apoptosis were caused by the down-regulation of Bcl-2 genes in the chromatin material, which in turn uplift or increases the translocation of Bax from the cytosol to the mitochondrial membrane of the tissues. Through the two important phases of cell division called G0/G1 phase cell cycle inhibited along with inhibition of nuclear factor kappa-B (NF-) and TNF- gene production, mitochondrial depolarization, besides commencement of caspase-3 and caspase-8 causing, tea poly-phenol (-)-EGCG is also known to trigger programmed cell death in tumour cells. *Camellia sinensis* also have potential against the risk of fractures in the osteoporosis population, as the tea polyphenols retard the inflammatory cytokine in immune system, IL-17, and arthritis-inducing proteins in the blood, Bhsp65, along with upregulation of anti-inflammatory substances (e.g., IL-10). Many researcher support this idea that the catechins possess anti-inflammatory effects via retardation of inducible nitric oxide synthase [23], as (-)-EGCG obstructed the induction of iNOS mRNA after treatment with TNF-α, lipopolysaccharides (LPS), IL-1 and interferon-alpha (IFN-α). It is also known that *Camellia Sinensis* polyphenols also possesses the capability to impact on many stages of the inflammatory cascade in the blood stream. One of the main pathways that causes swelling and increased cerebral ischemia CI is the creation of eicosanoids by AA degradation by mostly COX1 and COX2 enzyme. Additionally, it has been suggested that (-)-EGCG could protect the membrane integrity of mitochondria in the tissue (as indicated by the activity of citrate syntheses) and stop the leakage of essential mitochondrial matrix components, preventing cell apoptosis [22].

44.10 Pharmacological Activity and Therapeutic Properties of Tea

Some reported medicinal value and pharmacological activity of tea is listed below.

44.10.1 Cardiovascular Disease

Studies shows that Black tea has a stout antioxidant activity [24]. Polyphenolsis mainly responsible for this activity. In terms of preventing atherosclerosis, flavonoids may be helpful, and these findings offer a mechanism for the epidemiology research demonstrating the heart disease advantages of flavonoids [24].

Black /green tea flavonoids prevent LDL Oxidation (atherosclercsis). It was found that catecholaimes in tea inhibits Cu + 2 induced oxidation of lipoprotein in blood. it was also evident that antifibrinolytic and hypolipaemic effects also helps reduced CVS complications. Thromboxane and 8-epi-PGF2α (isoprostaglandins) levels was also reduced this improve platelet functions [25].

44.10.2 Antimutagenic Activity of Black Tea

Caffeine and polyphenols found in tea extract have been shown to have antimutagenic properties. The antimutagenic properties of tea seem to be regulated by both intracellular and extracellular pathways. These include altering metabolism, blocking or suppressing, altering the impacts of DNA replication and repair, restraining invasion and metastasis, and inducing unique processes.

44.10.3 Anticancer Activity

In the animal modeling for study purposes, hamster buccal pouch carcinogenesis (HBP) model, black tea extract was found to inhibits 7, 12, and DMBA-induced carcinogenesis. Active ingredients of tea Theaflavins and thearubigins may be the reason, tea poly-phenols were discovered to be added effective than green tea polyphenols. Some researches on the impact of black tea have produced contradictory findings. While some statistical studies show that tea has anticancer properties, others do not. In colonic epithelium, tea was found to rise glutathione S transferase activity and decrease quinone reductase activity.

44.10.4 Diabetes Mellitus and Effect on Obesity

Tea consumption has been broadly researched for its influence on various diseases that include DM, which has confirmed tea's substantial ability to prevent diabetes mellitus.According to reports, the poly-phenols in black tea may reduce the hazard of getting non-insulin-dependent diabetes. Black tea polyphenols have insulin mimetic activity, which lowers blood glucose levels, and black tea theaflavins increases the insulin/glucose ratio, which leads to better insulin action [26].

Obesity is considered to be an main etiological factor for the increase of cardiovascular CVS disease and diabetes. Black tea significantly lowers blood glucose levels fasting and random, and body weight, and body mass index (BMI) in both old and younger people. Tea intake is seen with reduced levels of fasting blood glucose level in diabetic patients [27].

44.10.5 GIT and Antimicrobial

Black tea has a strong ability to treat oral and stomach ulcers, according to reports. 80% of stomach ulcers are thought to be brought on by Helicobacter pylori. Epidemiological research have shown that *Camellia sinensis* has powerful anti-Helicobacter pylori effects [28]. Black tea gargling may prevent Haemophilus influenzae infection, according to some data. Because the bacteria in the stomach are inhibited, tea polyphenols inhibits the urease enzyme, which actually converts urea into ammonia. In addition, it is known to reduce hunger by increasing gastric emptying following intake of cheese fondue and work as appetite suppressants [29, 30].

44.10.6 Antihistaminic Properties/Asthma and Allergy

Research indicates that black tea may have antihistaminic properties. Histamine is a chemical that is biologically active and is produced by mast cells that have been sensitized during allergic reactions such as dermatitis and urticaria and inflammation. Black tea has been proposed as a potential treatment for asthma because it contains 54–71% of quercetin type flavanol glycosides, which is identified as strong antiallergy properties; antihistaminic and anti-inflammatory [31].

44.10.7 Toxicology of Camellia sinensis

Several research work proved that green tea [32], *Camellia sinensis*, has a great therapeutic activity to manage central nervous system CNS, cardiovascular CVS, and metabolic diseases like diabetes and thyroid disorder and treat tumor and inflammatory disorders but it's crucial to point out that "natural" is not always "safe," though. A few pertinent publications discussed the negative health impacts and side effects of GT. The purpose of this study is to present a classified report on the acute, subacute, sub-chronic, and chronic toxicity of GT and its primary ingredients. The cytotoxicity of tea, genotoxicity, mutagenicity, carcinogenicity, and developmental toxicity of gastrointestinal track and its primary ingredients are also reported. Liver toxicity and gastrointestinal problems, particularly when ingested on an empty stomach, have been identified as the most serious side effects. Neither GT nor its primary constituents are significant teratogens, mutagens, or carcinogens. There isn't much information on using them during pregnancy, therefore pregnant women, nursing mothers, and other susceptible individuals should use them with caution Due to the wide range of pharmacological interactions that GT and its primary components have, caution should be used when co-administering them with medications that have a narrow therapeutic index. Additionally, they cause malignant cells to exhibit selective cytotoxicity, which means they may be used as an adjuvant drug in cancer therapy [33].

References

1. Hicks, A. J. A. J. (2009). Current status and future development of global tea production and tea products. *Assumption University Journal of Technology, 12*(4), 251–264.
2. Li, X., & Zhu, X. (2016). Tea: Types, production, and trade. In *Encyclopedia of food and health*. Elsevier.
3. Chupeerach, C., et al. (2021). The effect of steaming and fermentation on nutritive values, antioxidant activities, and inhibitory properties of tea leaves. *Food, 10*(1), 117.
4. Kuroda, Y., & Hara, Y. (1999). Antimutagenic and anticarcinogenic activity of tea polyphenols. *Mutation Research, 436*(1), 69–97.
5. Yamanishi, T. (1995). Special issue on tea: Flavor of tea. *Food Reviews International, 11*, 477–525.
6. Kondo, T. (2008). *Varietal differences in the adaptability of tea [Camellia sinensis] cultivars to light nitrogen application*. Food and Agriculture Organization. Retrieved 1 March 2023.
7. Devarumath, R., et al. (2002). RAPD, ISSR and RFLP fingerprints as useful markers to evaluate genetic integrity of micropropagated plants of three diploid and triploid elite tea clones representing Camellia sinensis (China type) and C. assamica ssp. assamica (Assam-India type). *Plant Cell Reports, 21*, 166–173.
8. Hajiboland, R. (2017). Environmental and nutritional requirements for tea cultivation. *Folia Horticulturae, 29*(2), 199–220.
9. De Silva, M. S. D. L. (2007). The effects of soil amendments on selected properties of tea soils and tea plants (Camellia sinensis L.). In *Australia and Sri Lanka*. James Cook University.
10. Teshome, K. (2019). Effect of tea processing methods on biochemical composition and sensory quality of black tea (Camellia sinensis (L.) O. Kuntze): A review. *Journal of Horticulture and Forestry, 11*(6), 84–95.

11. Kamunya, S., et al. (2019). *Tea growers guide*. Kenya Agricultural & Livestock Research Organization.
12. Mobegi, V., et al. (2012). Influence of new technology on financial performance; a case of small scale tea industry in Kebirigo, Kenya. *Global Advanced Research Journal of Economics, Accounting and Finance, 2*(4), 073–085.
13. Wright, L. P. (2006). *Biochemical analysis for identification of quality in black tea (Camellia sinensis)*. University of Pretoria.
14. Owuor, P. O., & Kwach, B. O. (2012). Quality and yields of black tea Camellia sinensis LO Kuntze in responses to harvesting in Kenya: A review. *Asian Journal of Biological and Life Sciences, 1*(1), 1–7.
15. Harbowy, M. E., et al. (1997). Tea chemistry. *Critical Reviews in Plant Sciences, 16*(5), 415–480.
16. Owuor, P. O., & Obanda, M. J. F. C. (2001). Comparative responses in plain black tea quality parameters of different tea clones to fermentation temperature and duration. *Food Chemistry, 72*(3), 319–327.
17. Deb, S., & Pou, K. R. J. (2016). A review of withering in the processing of black tea. *Journal of Biosystems Engineering, 41*(4), 365–372.
18. Hatanaka, A., et al. (1987). Enzymic oxygenative-cleavage reaction of linolenic acid in leaves—Chloroplastic lipoxygenase and fatty acid hydroperoxide lyase in tea leaves. In *The metabolism, structure, and function of plant lipids* (pp. 391–398). Springer.
19. Javed, A.J.C.o.T., GB Pant University of Agriculture and P.I.h.w.r.n.p. Technology, *Role of processing conditions in determining tea quality*. 2015.
20. Harris, N., & Ellis, R. T. (1981). Black tea manufacture. *Effects on leaf structure of different processing systems. Annals of Applied Biology, 99*(3), 359–366.
21. Mike, J. (1998). Development corporation project No: UQ-61A, In *The evaluation of volatile quality factors in black tea*. (98/46).
22. Naveed, M., et al. (2018). Pharmacological values and therapeutic properties of black tea (Camellia sinensis): A comprehensive overview. *Biomedicine & Pharmacotherapy, 100*, 521–531.
23. Escandón, P., et al. (2018). Notes from the field: Surveillance for Candida auris—Colombia, September 2016–may 2017. *Morbidity and Mortality Weekly Report, 67*(15), 459.
24. Vinson, J. A., et al. (1995). Plant flavonoids, especially tea flavonols, are powerful antioxidants using an in vitro oxidation model for heart disease. *Journal of Agricultural and Food Chemistry, 43*(11), 2800–2802.
25. Sharma, V., & Rao, L. J. M. (2009). A thought on the biological activities of black tea. *Critical Reviews in Food Science and Nutrition, 49*(5), 379–404.
26. Abeywickrama, K., Ratnasooriya, W., & Amarakoon, A. M. T. (2011). Oral hypoglycaemic, anti-hyperglycaemic and antidiabetic activities of Sri Lankan broken Orange pekoe Fannings (BOPF) grade black tea (Camellia sinensis L.) in rats. *Journal of Ethnopharmacology, 135*(2), 278–286.
27. Polychronopoulos, E., et al. (2008). Effects of black and green tea consumption on blood glucose levels in non-obese elderly men and women from Mediterranean Islands (MEDIS epidemiological study). *European Journal of Nutrition, 47*, 10–16.
28. Boyanova, L., et al. (2015). Honey and green/black tea consumption may reduce the risk of helicobacter pylori infection. *Diagnostic Microbiology and Infectious Disease, 82*(1), 85–86.
29. Heinrich, H., et al. (2010). *Effect on gastric function and symptoms of drinking wine, black tea, or schnapps with a Swiss cheese fondue: Randomised controlled crossover trial* (p. 341). BWJ.
30. Iwata, M., et al. (1997). Prophylactic effect of black tea extract as gargle against influenza. *The Journal of the Japanese Association for Infectious Diseases, 71*(6), 487–494.
31. Blanc, P. D., et al. (2001). Alternative therapies among adults with a reported diagnosis of asthma or rhinosinusitis: Data from a population-based survey. *Chest, 120*(5), 1461–1467.
32. Colombo, A. L., et al. (2006). Epidemiology of candidemia in Brazil: A nationwide sentinel surveillance of candidemia in eleven medical centers. *Journal of Clinical Microbiology, 44*(8), 2816–2823.
33. Bedrood, Z., Rameshrad, M., & Hosseinzadeh, H. J. P. R. (2018). Toxicological effects of Camellia sinensis (green tea): A review. *Phytotherapy Research, 32*(7), 1163–1180.

Chapter 45
Celery

Mahwish Hussain, Rabia Sabri, Muhammad Zia-Ul-Haq, and Muhammad Riaz

45.1 Introduction

Apium graveolens, (2n = 2× = 22) the celeriac and celery are wide-reaching aromatic, fertilized plants and vegetables and are significant for pharmaceutical properties. It belongs to the parsley family, Apiaceae or Umbelliferae, and includes three botanical varieties: *var. secalinum*, for its low production and mostly available in Asia, *var. rapaceum*, which is popular in Europe and known as celeriac, been cultivated for its eatable hypocotyl, *var. dulce*, populous in America and western Europe and cultivated for its crisp leafstalk. Celery is also cultivated for its edible seed-like fruit, famous for its aromatic flavoring. *Apium graveolens* have been used worldwide for centuries for cultivation, while the other species have been domesticated only in the last 200–300 years [1]. It provide number of values for the ecosystem and to human beings, as a source of nutritious food for humans and animals such as rabbits and Lepidoptera. Years of human creativity and hard work have made them to learn to cultivate this vegetable and to produce many varieties and lastly prepare them in many different ways.

Celery is made up of leaf-topped stalks grouped in a tapering shape and connected at a mutual base. It grows to a height of 12–16 in. Small celery seeds have an

M. Hussain (✉)
Department of Biochemistry, University of Agriculture, Faisalabad, Pakistan

R. Sabri
Department of Pharmacology, Lahore College for Women University, Lahore, Pakistan

M. Zia-Ul-Haq
Office of Research, Innovation and Commercialization, University of Engineering and Technology, Lahore, Pakistan

M. Riaz
Department of Pharmacy, Shaheed Benazir Bhutto University, Sheringal, Pakistan

M. Zia-Ul-Haq et al. (eds.), *Essentials of Medicinal and Aromatic Crops*,
https://doi.org/10.1007/978-3-031-35403-8_45

approximate length of 1.3 mm, are tan in color, oval, and ridged. Each of the two connected carpels that make up the fruit has a seed within. The spice has a little spicy flavor and a lovely distinctive aroma [2]. Celery is used to flavor meals in a variety of ways, including as a new herb, stalk, leaves, seeds, seed oil, and oleoresin. In the eighteenth century, celery gained popularity in Europe, and in the nineteenth century, it was familiarized to the United States.

45.2 Scientific Classification

Kingdom: Plantae
Division: Spermatophytes
Sub-division: Angiospermae
Phylum: Magnoliophyta
Class: Magnoliopsida
Subclass: Rosidae
Order: Apiales
Family: Apiaceae
Genus: *Apium*
Species: *graveolens*
Common name: Celery

45.3 Origin/Geographical Distribution

Apium graveolens is a biennial plant and commonly known as celery, smallage and wild celery. Previously the herb is known as *apium dulce*, *apium rapaceum* and *celery graveolens*. The generic name, *Apium*, is Latin word which means "bees". The name is given to it because bees were fascinated to this herb. Wild celery is usually erect biennial or perennial herb. Celery seeds originates from the Wild celery plant but the remaining plant is unpleasant to taste and inedible. It is grown in temperate climates. The celery plant is innate to temperate and Mediterranean areas of North Africa, Europe and Asia. The herb characteristically found in moist regions with low elevations. The herb prefers full sun, moist, rich and well-drained soil. Celery can be reproduce by its seeds. The plant grown areas may become adopted from self-seeding. The celery plant grows up to 1.5–3 feet tall and grows erect with ascending stems. The leaves are green, pinnatifide, simple and lobed. The flowers are greenish-white and seems to be in round clusters. The fruits are black, tiny, ovoid-shaped schizocarp. The shape and chemical make-up of this plant's stem, leaves, and flowers are all variable. Celery is incompatible with self-pollination but readily cross-pollinates [3]. The common name "celery"only applies to one type of plant. Different names for *Apium graveolens* are used in various parts of the world. It is frequently referred to as "celery"in English, "Karafs" in Persian, and "Apio" in

Spanish. In German, it is frequently referred to as "Sellerie". It is referred to as "Alkarafs" in Arabic. In Thailand, celery is referred to as "Khuen chaai" [4]. In Urdu, it is referred to as "Tukhme karafs". The fruit of Apium graveolens is usually referred to as "celery seeds" and the plant is also known as "Ajmod" in Hindi.

45.4 Nutritive Value

Celery is a decent source of volatile oil, flavonoids, and antioxidants as well as vitamins, protein, carotene, cellulose, and other minerals. Celery is also used in the pharmaceutical and chemical sectors. The use of celery and its potential health benefits have been continuously explored and improved as living standards and health consciousness have increased. Studies on celery have chiefly concentrated on breeding, cultivation, and biochemical makeup. Vegetable crops have been studied using a massive amount of statistics resources and effective research techniques thanks to the rapid development of molecular biology and the appearance of new technologies, such as sequencing of transcriptome, small RNA sequencing, sequencing of protein, genome sequencing, and digital expression profile. Celery has remained extensively studied around the world and many pertinent parts have been clarified because it is a significant vegetable in the Apiaceae family. In order to give a valuable probable reference for celery breeding and usage, research achievements in genetic breeding, genome detections, function genes, and chemical configuration are summarized in this review (Fig. 45.1) [3].

Celery and other species' essential oil concentration and chemical makeup are greatly influenced by a variety of factors, including genetic, ontogenetic, environmental, and agronomic factors like fertilization, irrigation, cultivation technique, and harvesting technique. Celery seed oil is produced in California, India, and France and is mostly used to flavour soups, juices, vegetables, and other meals including pickles and meat. In small amounts, it is also utilised in perfumes.

Fig. 45.1 Celery seed

Celery, raw (g/100 g edible portion)
Dietary fiber total: 1.5
Dietary fiber insoluble: 1.0
Dietary fiber soluble: 0.5

Dietary fibre is the word used to describe plant material in food that is resistant to enzymatic digestion. It contains lignin, cellulose, hemicellulose, pectin materials, gums, and mucilage, among other things. Cereals, fruits, vegetables, nuts, and mueslis all naturally contain dietary fiber. The constancy, surface, rheological behavior, and sensory qualities of the finished goods can all be altered by the addition of fiber.

45.5 Functions of Dietary Fiber

- Increase your diet's bulk to feel full faster.
- During digestion, it draws water and transforms into a gel, trapping carbohydrates and delaying the absorption of glucose.
- reduces both LDL and total cholesterol
- controls blood pressure
- increases the rate at which food is absorbed by the digestive system
- bulks up the stool
- Stimulates the formation of short chain fatty acids through intestinal fermentation and balances intestinal pH.

45.6 Varieties

Vilmorin claims that in the seventeenth century, France had access to two varieties of solid stem Italian celery: "striped rose" and "solid celery." Chemin chose the "Solid Golden White Celery"line in 1875, which later gave rise to the variety "Paris Golden Yellow Self-blanching." From this variety, Pascal chose a green celery line in 1884 that became known as Pascal celery in the seed trade. Both types were presented to North America in 1887; the self-blanching variety was called "White Plume," and the green variety was called "Giant Pascal". The common varieties of celery are:

- *Apium graveolens var. dulce*

Its common name is Celery. These plants grow thick, succulent petioles that are primarily eaten uncooked in salads or as celery "sticks" and are frequently cooked in soups and stews. In both the Americas and Europe, celery is widely farmed. Approximately 16,000 ha are used to grow celery in the USA. With a usual production of 64 tonnes/ha, California harvests 67% of the celery cultivated in the United States. For other nations, there are no production statistics available.

Fig. 45.2 *Apium graveolens* botanical varieties: (**a**) *dulce*, (**b**) *rapeceum*, and (**c**) *secalinum*

- *Apium graveolens var. rapaceum*

Common name is celeriac or root celery. The growth of larger hypocotyl and tissue of roots is its defining feature. The resulting flavorful globe-like structure is mainly eaten broiled in stews and soups or shredded for salads. Northern and Eastern Europe are the main growing regions for this crop.

- *Apium graveolens var. secalinum*

Its common name is smallage or leafy celery. These plants are largely used for their leaves, which are also used for medicinal purposes. They have thin, leafy, and frequently hollow petioles. Smallage is primarily consumed in Asian and Mediterranean nations. This type is also grown annually for the seed, which is utilized as a condiment and a beverage ingredient (Fig. 45.2).

A. graveolens was probably grown for therapeutic purposes for the first time as a crop in Egypt and the Roman Empire around 400 BC. It was used by the Greeks to create a celery wine known as selinites. At the Isthmus of Corinth, celery stalks and leaves were fashioned into crowns and given to the champions of sporting competitions.

45.7 Chemical Constituents

Celery has a high calorie content and contains fat. It is also renowned for being a decent source of several minerals and vitamin C. Its seed is made of starch, ash, carbohydrates, volatile oil, proteins, crude fibers, moisture, and stable oil. Oleic acid, palmitic acid, linolenic acid, stearic acid, linoleic acid, and petroselenic acid are the fatty acids found in fixed oil [5]. This plant is a upright source of minerals like calcium, magnesium, and potassium. It also has a lot of salt. Chopped celery leaves contain about 100 mg of salt per cup. Salience, sesquiterpenes, limonene, and distinctive scent make up the essential oil. Folic acid, potassium, sodium, fibers, alpha-carotene, magnesium, silica, and chlorophyll are all abundant in it.

In addition to alkaloids and steroids, plants also contain essential oils or fixed oils. Steroids, glycosides, flavonoids, and carbohydrates can be found in the seed extract. The plant extract also contains additional furocoumarins, including apigravrin, isopimpinellin, apiumoside, celerin, apiumetin, isoimperatorin, bergapten, and celereoside. There are other phenols such apigenin, tannins, isoquercitrin, phytic acid, and graveobioside. Sesquiterpene, alcohol (1–3%), and fatty acids are the main components of the essential oil found in celery leaves, stems, and seeds (Fig. 45.3) [6].

Camphene, limonene, sedanenolide, stearic acids, linoleic acid, santalol, oleic acid, terpinene, p-cymene, myristic acid, myristoleic acid, sabinene, and terpinolene are among the constituents found in this plant. Selinene (10%), glycosides, frocoumarin flavonoids, and glycosides make up the majority of the compounds found in celery seed. Celery seeds are consumed as diuretics and used to treat kidney, rheumatic, and arthritic illnesses. Celery kernels are also used to make brew that promotes relaxation and better sleep.

45.8 Celery Seed Composition

Around 8.0% of celery seed is made up of moisture [7]. High amounts of salt are also present in celery. Approximately 100 mg of sodium can be found in a full cup of chopped celery leaves from two stalks.

45.9 Cultivation

Soil: Except for saline, alkaline, and waterlogged soils, celery can be successfully grown on many types of soil; loamy soils work best. Celery can tolerate high soil response conditions. A soil's PH ranges between 5 and 7.

Climate: Cold and dry climates are ideal since they allow for 80% seed development over a two-week period when day and night temperatures are between 12 and 25 °C. The crop is available for harvest in November after the seeds are sowed in March–April, transplanted in May.

Fertilizer: About 30–50 ha of FYM are added before the seeding. On medium soil, the crop receives roughly 80–200 kg of nitrogen, 30–40 kg of phosphorus, and 20 kg of potassium per hectare. The full doses of nitrogen, phosphorus, and potassium are administered at planting, and the remaining nitrogen is applied as a topdressing one month later. During the crop time, the crop required 10–12 irrigations [8].

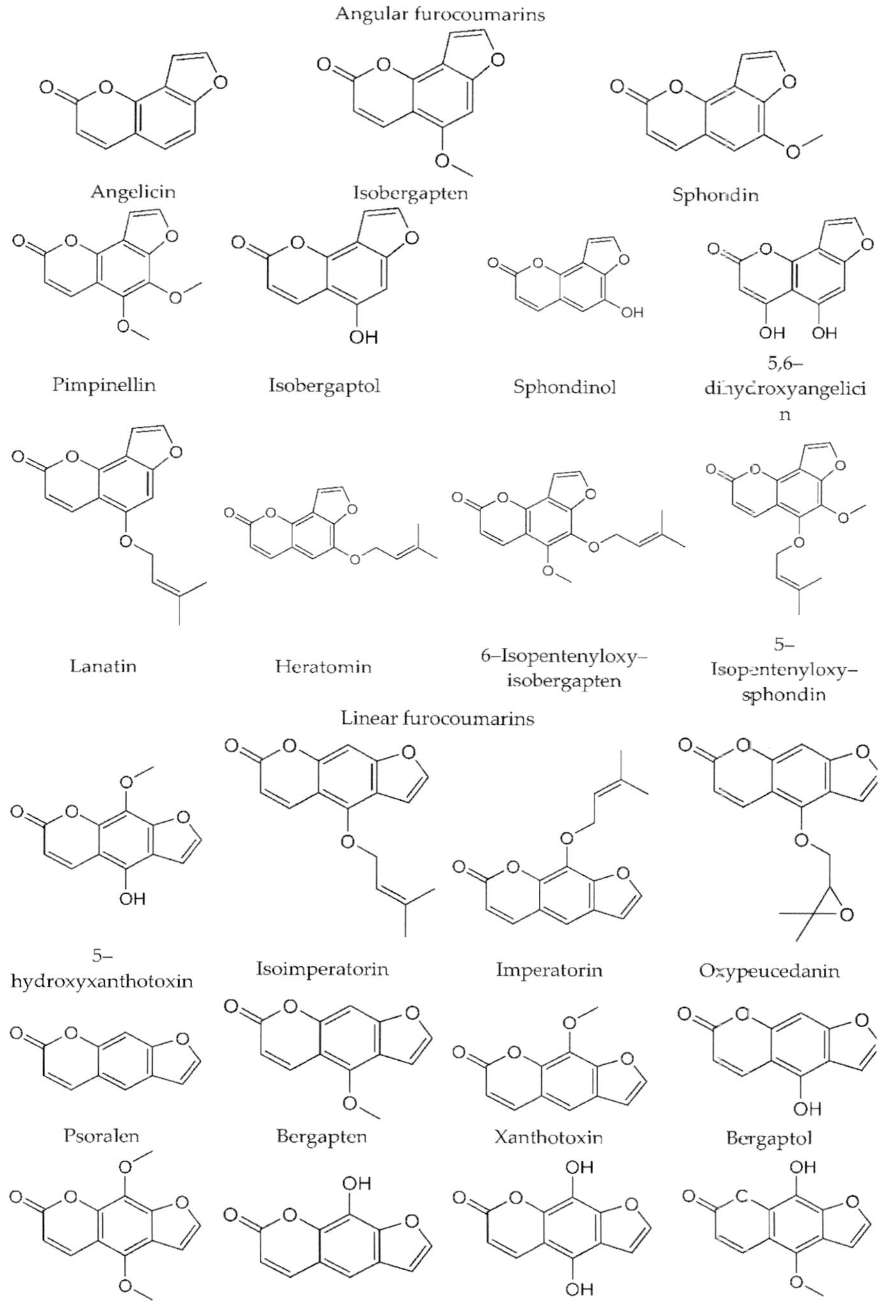

Fig. 45.3 Chemical constituents of celery

45.10 Harvesting

Fields are seeded with grown seedlings. The following year is when the harvest is done. Celery greeneries are used in salads, and to keep celery, it should be placed in a refrigerator-safe container that is well closed, or it can be wrapped in a plastic bag or moistened cloth [9]. If peeled celery must be preserved, it must be kept dry and free of water, as this can cause some of its nutrients to be lost. Celery must not be frozen if it is going to be utilized in a dish that calls for cooking because freezing will cause it to wilt. The plants are divided, threshed and then dried in the field. After being fully dried, cleaned, and divided into large and small seeds, the seeds are bagged.

45.11 Post-harvest Technique

Celery is typically best harvested during the mild winter months. The plant needs to be shielded from hoarfrost and harvested before the chilly season in areas with extremely cold winters. By building up the soil surrounding the plant's locations, the celery plant can likewise be sheltered from the cold [10]. The plant should be harvested when the plant stalks measure 6 in. or more from the soil to the first stroke of leaves. Individual stalks or the entire plant might be cut below the soil line. More nutrients are present in the stem if it is green. Uncooked, the inner portion of the stalk is flavorful and more delicate than the outside portion.

For brief times, celery‘s fresh herbs are kept in storage to boost their marketability. The best temperatures for storing its renewed herb are 0 °C and high (95% or more) relative humidity. Celery herbs can be kept in marketable condition for a very long time by adopting controlled atmospheric storage. This storage requires a constant removal of ethylene from the atmosphere coupled with a 0 °C temperature, 4% CO_2, 1–2% O_2, and high comparative humidity. The celery seed crop is harvested, collected, indorsed to cure and arid for 1 or 2 days in thin layer, then transported to the threshing floor where the seed is separated by mild threshing. Compared to seeds dried out in the sun, seeds dried in the shade contain more oil. Seeds are scrubbed using a screening mill and gravity separator, categorized by sieving, and then kept in gunny bags in a cool, dry environment.

45.11.1 Celery Essential Oil

The essential oil found in 2% of the seeds is used in the flavor and fragrance industries. Celery essential oil adds warm, clinging tones to oriental perfumes and gives them a floral scent [11]. Celery volatile oil, and gas chromatography-mass spectrum analysis has revealed that d-limonene and selinene make up roughly 60% and 20% of the oil, respectively. Although they are existing in very small amounts, 3-n-butyl-4,5-dihydrophthalide (sedanenolide), 3-n-butyl phthalide, sedanolide, and sedanonic anhydride are the key flavor constituents of the oil that give it its distinctive aroma.

The produced by steam distillation celery seed oil comprises of seven main compounds: piperitone, eugenol, ß-pinene, terpinolene 3-carene, myrcene, and menthone [12]. Fresh celery juice has been distillated, and flavoring substances including phthalides and hydrophthalides have been identified. There have been reports of myrcene, limonene, butyl phthalide, pentyl benzene, and ß-caryophyllene in the oil of a particular Indian type of celery seeds. There has been an evaluation on celery that includes the oil composition.

45.11.2 Flaking Examination

Celery seed has a complex concentration of fixed oil, which can lead to issues during grinding processes such overheating, mill blockage, and volatile loss. Flaking or peeling of the seeds for the extraction of volatile oil and its impact on the harvest and physicochemical properties of the volatile oil have remained examined as an alternative to standard grinding [13]. Prior to steam distillation, celery seeds could be flaked to produce a larger harvest of volatile oil (1.76%) than with traditional powdering (1.42%), and the yielding of oil could be further increased (2.2%) by precooling and flaking. Flaking has no impact on the oil's flavor character. Studies using scanning electron microscopy have shown that the microstructure of flakes and powder differ noticeably. In the occurrence of flakes, the cells had ruptured and a complete destruction had been seen, which permitted the release of more oil in a petite length of time. In the situation of powder, lump-like structures are seen, and the particles uphold their spherical shapes with minimum cell breakage [13]. As a result, the output of oil is reduced and the distillation process takes longer. Selective assortment of volatile oil at various steam distillation time intervals results in fractions with various flavor profiles. When the condensate as of steam distillation is exposed to an additional hydro distillation, oil by means of a different flavor profile from the steam-distilled seed oil is recovered (approximately 40–45 mL per 10 kg of celery powder).

45.11.3 Enzyme Pretreatment

There have been reports of the usage of enzymes to extract flavor commencing a few spices, including fenugreek, pepper, mustard, chilli, and citrus peels. It has remained investigated how several enzymes, including cellulase, pectinase, protease, and viscozyme, affect the extraction of celery's volatile oil. In the case of pretreatment with cellulase, pectinase, protease, and viscozyme, the oil manufacture ranged from 1.9% to 2.3% as opposed to 1.8% in the steam-distilled control sample [13]. The production of oil was higher with a particular enzyme, cellulase, at 0.5% than with a mixture of enzymes. With minimal change in the oil's flavor profile or physicochemical qualities, enzymatic pretreatment of celery seeds increased the output of volatile oil (17–22%) in celery.

45.11.4 Celery Resin

There are two ways to get celery oleoresin. (1) Steam distillation is used to separate the volatile oil from celery powder or flakes that have been finely pulverised. The leftover powder or flakes after the extraction of volatile oils are crushed to 30 mesh once more and used in percolators to remove solvent. Five to six times of the extraction are done, each time allowing an hour of powder contact time. Miscella is created when extract is combined and subjected to distillation to remove solvent. At the conclusion of the distillation, high vacuum is used to remove any remaining solvent. To create oleoresin, the obtained resin is blended in various ratios with volatile oil. Oleoresins that have been extracted from stale seeds may taste oxidized. (2) Oleoresin can alternatively be made by extracting celery powder using a solvent (such as hexane or acetone) rather than steam distilling it first and then removing the solvent. By applying a high vacuity at the supposition of the distillation to eliminate solvent remnants, the finer scent of low boiling mixtures may be lost in the latter phases of the distillation of oleoresin produced using this approach.

Equally the fixed oil and the essential oil are extracted from seeds using hexane as a solvent. These seeds can then be more fractionated utilizing 90% alcohol, producing (1) an oleoresin that contains 25% volatile oil and (2) a fatty oil that contains roughly 3% volatile oil. Depending on whether a product with a strong flavor or one with a moderate flavor is needed, either product can be used to flavor meals.

45.11.5 Celery Oil and Oleoresin Chemistry

A valuable commodity in the flavor and fragrance sectors is celery seed oil. According to research on celery seed oil, its main flavoring agents include 3-n-butyl phthalide, 3-n-butyl-4,5-dihydrophthalide (sedanenolide), and sedanolide.

Researchers have also looked into the volatile chemicals in celery foliage juice from harvest waste [14]. It is possible to create a phtalide-enriched fraction from celery seed oil using a straightforward procedure. The manufacture of phthalide-enriched fraction from celery seed oil has been covered by a patent. Celery is also used to make leaf oil, root and seed extracts, and seed oleoresin. Compared to the seed oil, celery leaf oil has more monoterpenoids and fewer sesquiterpenoids.

Celery oleoresin is a viscous, thick substance that is insoluble in water and contains 20–30% volatile oil [13]. For convenience of use in foods, oleoresin is either dissolved in liquid form by accumulating diluents like propylene glycol or in solid form on dextrose to produce powder. Oleoresin is made up of wax, resin, volatile and fixed oils, as well as artefacts that serve as fixatives for the volatiles. Since there is no significant flavor ingredient existing in the resin ratio of the celery oleoresin, unlike other significant flavor oleoresins, the volatile oil content of the oleoresin serves as the standard for quality [15].

45.11.6 Celery Byproducts

Ruperez and Toledano [16] report on a studies on celery byproducts as a foundation of mannitol. For the purpose of determining the presence of soluble sugars and mannitol, two types of celery remains from the industry—stalks and stalks with leaves—were examined. To solubilize soluble sugars and mannitol, celery residues, including stalks alone and stalks including leaves, were extracted with hot 85% ethanol. High-performance liquid chromatography (HPLC) was used to identify and measure the low-molecular-weight carbohydrates in the extracts. Similar levels of sucrose (5.7, 5.9%) were identified in both celery residues, but the ratios of hexose (glucose and fructose) to mannitol varied. The amounts of mannitol and total sugar in the stalks were higher (45.5% and 15.2%, respectively) than in the residues made up of the stalk and leaf (33.9% and 13.3%, respectively). In celery wastes, mannitol made up 33.5–39.3% of all the carbohydrates. Alcohol removes from celery byproducts are suggested as a natural source of mannitol and soluble sugars, and the foodi industry may also use alcohol-insoluble celery byproducts to make dietary fiber-rich food supplements. Similar research has revealed celery stalks to be a substantial source of mannitol in their ethanol-insoluble residue. Myoinositol, glucose, and sucrose were all present in significant levels in the ethanolsoluble residue. According to reports, mannitol absorbs from the small intestine gradually and insufficiently; it can also be thought of as a possible prebiotic component of functional foods. Mannitol is abundant in celery, which makes up to 50% of the plant's total carbohydrate content.

45.12 Breeding

45.12.1 Male Sterility and Hybrid Reproduction

One of the main objectives of cultivating celery and celeriac across Europe continues to be hybrid breeding. Heterosis appeals to breeders and producers alike due to its excellent uniformity, enhanced yield, vigor, and disease resistance [17]. Breeding for celery and celeriac is primarily done through propagation, resulting in considerable plant-to-plant diversity and has a negative financial impact. In comparison to open pollinated (OP) types, there are still few hybrid cultivars on the marketplace in Europe and the USA. The inability to emasculate umbelliferous blooms has made it difficult to obtain celery and celeriac F1 hybrid seeds. *Apium* blooms, which are small and arranged in umbel-like inflorescences, are very susceptible to self-pollination [18].

The use of completely male sterile lines is the most effective emasculation technique. Male sterility can be genetically or cytoplasmically driven and is characterized by the plant's capability to produce viable pollen. These lines are vegetative propagated in breeding programmes because genetic male sterility segregates [19]. Contrarily, cytoplasmic male sterility (CMS) and cytoplasmic genic male sterility (CGMS), which are caused by mitochondrial genes and are therefore transmitted from the mother, are advantageous from an economic standpoint. There are few reports on CMS sources in *Apium*. Dawson claimed that he had unstablely introduced CMS into celery from an anonymous wild celery plant in the "Grower" magazine in 1993. Some of the hybrid celery and celeriac cultivars available today were created via CMS, but the breeding corporations have kept the techniques and origin a secret.

45.12.2 Breeding for Disease Control

The main celery and celeriac leaf sickness caused by *Septoria apiicola*, late blight, is extremely harmful when crops are managed poorly. Celery acquired innate resistance from *Petroselinum hortense* (parsley) [20]. In contrast, a different study found that the offspring of celery and parsley hybrids were sensitive to *Septoria* [21]. It was hypothesised that *Septoria* resistance was bred out with each subsequent backcrossing. The natural species *Apium nodiflorum, Apium chilense*, and *Apium panul*—the later two of which were crossed with celery—were found to be resistant to *septoria*. These crosses produced F1 hybrids that were somewhat less resilient than their resistant parent, demonstrating that the resistance trait was not completely dominant [22].

Fusarium oxysporum, There is no cure for the soil-borne fungus *Fusarium oxysporum*, which causes *Fusarium* yellow disease happening in celery and celeriac [23]. In the middle of the 1980s, the sensitive cultivar "Tall Utah 52-70R" of celery had been crossed to a landrace of celeriac that was *Fusarium* resistant. One dominant gene and other autonomous quantitative genes together determined the resistance attribute. The resistant line “UC1,” which was back-crossed in elite cultivars in the early 1990s, was created using this resistance. UC8-1, UC10-1, and UC26-1, the resultant resistant lines, were used to create commercial cultivars. Through semi-clonal variation and selection, more *Fusarium*-resistant celery plants were created during regeneration from cell suspensions. A stable immune line, MSU-SHK5, was produced after five compeers of self-pollination paired with *Fusarium* resistance selection, although there is no information on its utilization in commercial cultivars.

F. oxysporum f. sp. apii race 4 was a brand-new race that was discovered in California (USA) in 2013; no commercial cultivars are resistant to it. Searching for sources of resistance to this new race was the subject of investigation at the University of California, Davis [24]. The following *Apium* species were examined: 120 celery, 66 celeriac, 25 smallage, 5 celery x celeriac, and 15 other *Apium* species. Three possibly resistant celeriac accessions from Turkey and a celery from China were found.

Early blight, a highly contagious *Apium* disease brought on by the mold *Cercospora apii,* calls for a comprehensive and pricey approach to disease treatment. A number of cultivars for the Florida region were produced as a consequence of early studies by Wolf and Scully in the 1990s. When compared to other resistant cultivars, “Floribelle M9” demonstrated improved early blight resistance at the time of its development. Its resistance to Cercospora can be connected to the use of celeriac from Turkey [25]. Further use of “Floribelle M9” led to the creation of the initial blight-resistant cultivar “FBL 5-2M.”

The furthermost prevalent viral ailment of celery is called Celery Mosaic Virus (CeMV), and aphids are the vectors of this disease. In 2001, investigation found a solitary recessive locus for CeMV resistance [26]. In 2001, it is discovered that markers are connected to the CMV gene. These indicators can aid in the selection of virus resistance and the introgression into different types of celery [27]. Using posttranscriptional gene silencing technologies, McCormick bade to create celery and carrot plants that were resistant to CeMV and carrot virus Y (CarVY). Due to the employed celery cultivars' resistance to Agrobacterium-mediated transformation, they were unable to produce celery resistant plants.

45.12.3 Breeding for Insect Pest Control

Celery and celeriac pest *Spodoptera exigua* is challenging to chemically control. *Apium prostratum* was shown to contain insect resistance. Additionally, *Fusarium* yellow-resistant celery somaclonal lines exhibit a substantial increase in beetroot armyworm resistance. An evaluation of 13 cultivars of the *var. rapaceum, dulce*, and *secalinum* for resistance to the beetroot armyworm shows that the ‘Kockanski’ cultivar had a stronger resistance to *S. exigua* when its level of a sedanolide molecule increased [28] (3-n-butyl-4,5-dihydro-isobenzofuranone).

Thirteen different types of parasitic nematodes can infest *apium*. *Meloidogyne javanica, Meloidogyne incognita*, and *Meloidogyne hapla* are frequent species in temperate and cooler climates, respectively [29]. These nematodes are among the most contagious nematodes. The abundance of plant species that *Meloidogyne* nematodes use as hosts contributes to the difficulties in managing these parasitic infestations [30].

M. hapla, a northern root-knot nematode, can reduce celery yields by up to 5%. Although it also affects other vegetable crops [30], this insect is difficult to treat with broad-spectrum pesticides. Some examination explained that five different celery cultivars for *M. hapla* resistance and found that two of them, "Green Boy" and "Dutchess," produced noticeably less adult nematodes.

The related Apiaceae crops may provide as a probable source of nematode resistance genes. Research on carrot confrontation to *M. hapla* has been thorough, and resistant lines have been found [31]. Two distinct homozygous recessive genes control the resistance.

Celery can occasionally become infected with *M. incognita* and *M. javanica,* though this is uncommon. However, no sources or defences against opposition were mentioned. One or two connected genes that have been replicated and linked together, known as locus Mj-1, control carrot's resistance to parasitic nematodes [32]. In the carrot breeded material produced from Asian germplasm, another resistance to *M. javanica*-region, Mj-2, was discovered. These resistance sources, which are present in carrot germplasm, could be used to breed *Apium* with resistance, coupled with the markers that have been found to be related with resistance [33].

An aggressive nematode called Meloidogyne chitwoodi is currently under quarantine in Europe. By entering the enlarged hypocotyl, it can impair celeriac production by causing skin irregularities and browning beneath the skin. The Flanders Research Institute for Agriculture, Fisheries, and Food is actively evaluating the resistance status of a number of cultivars. (ILVO).

45.12.4 Late Bolting Breeding

For celery and celeriac to flower, a prolonged cold spell is necessary. Bolting happens through the first year of growth as the foremost stem starts to create a flowering shoot. Premature bolting might occur as a result of stress-related reduction of vegetative type of growth, which has a significant impression on celery harvesting and quality since the inflorescence shoot uses up energy that would otherwise be used by the leaves and petioles.

In *Apium*, the reliance on a cold stage is highly varied and hereditarily predetermined. In general, biennial celery cultivars show poor to strong resistance to bolting, but annual cultivars bolt easily. Early bolting resistance breeding in celery activated with the crossovers carried out [34].

There are significant areas where growers must deal with early bolting even though the majority of European celery and the plant cultivars are now biannual and sufficiently unaffected to bolting in the first growing season. The plants sown in the fall in Southern Europe (Spain) have a good overwintering duration and are encouraged to flower the following year.

45.12.5 Breeding for Petiole Trait

The *secalinum* and *rapaceum* types' petioles are tinier and muffled, with bitter flavors and fibrous in textures, whereas the *dulce* kinds' petioles are lengthy with solid stems. Numerous competing petiole cultivars with a wide range of color, shape, crispness, pithiness, stringiness, and flavor have been created throughout time [21].

The consumer's preference will determine the color of the celery petiole. It can range from white to light yellow to light green to light green [18]. Celery can be divided into two main categories: green cultivars through bottomless green petioles and leaves like Utah kinds and summer Pascal types, and yellow/golden reputed "self-blanching" cultivars with nearly colorless petioles and foliage, such "Golden Self Blanching" and "Celebrity" [35].

45.12.6 Enlarged Hypocotyl Traits Breeding

Unlike celery, celeriac is only found in a small area in northern Europe. The top nations for celeriac breeding are the Netherlands, France, Germany, Belgium, and the UK. The development of a larger hypocotyl and root, which creates a large, rounded structure with a delicious flavor, is the primary attribute of celeriac. The enlarged hypocotyls can be processed for the frozen food business or sold on the fresh market. The breeding objectives for this crop are likewise influenced by these applications [36].

To avoid injury during the digging for harvesting, the roots ought to be thin, few in number, and centered on the basal region of the expanded hypocotyl. Harvesting is made easier by the narrower implantation of the roots in the soil found in cultivars like "Claire," "Diamant," "Camus" and "Rex" [37].

45.12.7 Intergeneric Breeding

Intergeneric crossings, like as reciprocal breeding with parsley that produced intermediate of phenotypes, were included in attempts at hybridising celery [38]. The goal of the crosses between celery and *A. nodiflorum*, which produced fertile hybrids, and between celery and *A. chilense,* which produced low fertility hybrids as a consequence of chromosomal rearrangements, was late blight resistant introgression. To introduce leaf miner resistance, interspecific crossovers between the celery cultivar "Tall Utah 52-75" and A. prostratum were used [39]. Celeriac 'Cupidon' and celery 'Samurai' were crossed intraspecifically in 2009. The F1 hybrids had longer leaf blades and taller plants, while most of the other features had intermediate phenotypes [40].

As many wild accessions have distinctive features and disease resistances, crossings with wild kinds are extremely fascinating for increasing variety. Even after a few generations of backcrossing, these crossovers can occasionally lead to decreased petiole quality [41]. In other instances, a desired characteristic transferred from a wild variety is correlated with an undesirable trait. Genome excision techniques can be used to remove the undesired features.

45.12.8 Breeding Meant for Nutraceutical and Food Safety

Low in calories and high in fiber, minerals like calcium, phosphorus, iron, potassium, and magnesium, vitamins like B1, B2, and C, and flavonoids, celery and celeriac are well-known therapeutic plants. The creation and characterization of cultivars with higher nutritional contents advance gradually [6].

The cultivars "Liuhe Huangxinqin" and "Ventura" have different levels of vitamin C and linked metabolic pathway, with the latter having higher intensities of ascorbic acid in both the leaves and the petioles [42]. It is abundantly obvious from the research that different celery cultivars differ in their vitamin C content, and gene expression screening can be easily applied to breeding programs.

There is an association between the effectiveness of antioxidants and phenol composition in their study of 11 celery varieties [43]. Three celery cultivars from Pakistan and a wild concurrence were compared in terms of their nutraceutical composition [44]. As in the case of the wild *Apium,* which showed higher concentration of tannins and phenols, variation in the content of different elements was seen. An examination shows the selenines, limonene, tannins, phenolic acids, anthocyanins, chlorophyll, carotenoids, and other essential oils present in a variety of celeries. Between cultivars, the amount of carotenoids was consistent, but there were considerable differences in the other components [45]. By adding more health-promoting substances to the breeding programs, local cultivars and wild accessions may eventually make them better.

45.12.9 Breeding Due to Abiotic Stress and Phytoremediation

Abiotic stress comprises of salt, high levels of substantially heavy metals, high light concentration, and UV radiations in addition to heat, drought, and freezing which may get worse in some places as a result of climate change. Reactive oxygen species

(ROS) buildup is brought on by each of these abiotic stressors in plants. Many plants can flourish in such harsh environments and can withstand some types of toxicity.

Owing to the rise in mannitol production, which is regulated by mannose-6-phosphate reductase, celery has great salt tolerance. (M6PR) [46]. Two transporter genes of sucrose, SUC uptake transport 1 and 2 (AgSUT1 and 2), as well as MANNITOL TRANSPORTER 1 (AgMaT1), a mannitol transporter protein producing gene, have been identified in celery.

There are a number of studies available that discuss the potential for various cultivars of celery and celeriac on the way to accumulate heavy metals and contaminants. An assessment shows the accumulation of heavy metals in the petioles and leaves of white and celeriac cultivated on soils contaminated with heavy metals within the allowable limits [18]. Between the examined cultivars, different amounts of cadmium, manganese, and zinc were accumulated. Twenty-seven Chinese *Apium* accessions were cultivated in soil contaminated with cadmium and lead, and the amount of heavy metals in the stalks was assessed [47]. Differences between the cultivars were noted, and 'Shuanggangkangbing' accumulated the least amount of cadmium and lead.

Future breeding of celery and celeriac will be nearly entirely devoted to the creation of hybrid cultivars. F1 hybrids are becoming more and more intriguing when it comes to improving crop productivity and lowering agriculture's total environmental effect. A small amount of OP variants will be produced, mostly for markets where these cultivars have a long history (the popularity of the "Tango" cultivar on US markets is one such example). The majority of markets, however, do not prefer OP types due to their unpredictability in harsh environments.

45.13 Applications of Celery

Celery has been utilized as an emmenagogue, aphrodisiac, anthelmintic, antispasmodic, carminative, diuretic, laxative, stimulant, sedative, and poisonous substance. As a moderate diuretic and urinary antibacterial, celery is also used to treat flatulence and gnawing pains. Celery oil, celery oil extract, ground celery seed, and ground celery root are promoted as herbal supplements and dietary supplements that "promote and regulate" good blood pressure, joint health, and uric acid levels. Root tinctures have been used to treat hypertension and urinary disorders by acting as diuretics.

Fish, salads, and eggs can all benefit from the flavor of "celery salt," which is made by combining the crushed seed with salt. There is also celery with pepper powder available. To flavor canned soups, sauces, pickles, tomato products, and meats, celery herb, seeds, volatile oil, and oleoresins are utilized. Celery essential oil adds warm, clinging tones to oriental perfumes and gives them a floral scent. Celery's oleoresin is frequently used to flavor dishes. Both fresh celery and infusions of seeds, powder, and extracts are ingested. In western countries, the leaves are eaten as a salad. Celery seed is frequently used in preserving and is delicious in bread, rolls, pastries, omelets, stewed tomatoes, tomato juice, clam juice, tomato sauce, potato and green salads, salad dressings, tuna and salmon salad, vegetables, stuffings, sauerkraut, and other egg dishes.

Celery seeds are also utilized for their medicinal properties, such as in the treatment of liver illnesses and as an asthma booster and tonic. Seeds are consumed as a sedative and a nerve stimulant in domestic medicine. Rheumatoid arthritis has been treated effectively with celery seed oil.

Anti-fungal activity: *Apium* exhibits anti-fungal effect against Bactria, including *Salmonella typhi, Staphylococcus aureus, Staphylococcus album,* and *Shigella dysentery*.

Lowers blood pressure and cholesterol: Over the course of 4 weeks, a daily dosage of a chemical extracted from celery seed experimentally resulted in a 12% decrease in their blood pressure.

Healthy joints: Celery has been utilized and recommended as an alternative treatment for gout, arthritis, and rheumatism. Celery is been discovered to possess anti-inflammatory qualities that lessen inflammation and soreness near joints.

Anti-septic property: Celery seed possesses a diuretic propriety to aid with fluid retention as well as an antibacterial activity that makes it beneficial for the healthiness of the urinary tract. It helps with uric acid removal.

Acts as diuretics: In addition, celery helps gout and arthritis sufferers lose water weight by flushing out uric acid crystals that hoard around the joints.

Menstrual discomfort: Celery might accelerate the initiation of the menstrual cycle.

Cancer agents: Celery contain eight different families of anti-cancer compounds that degrade carcinogens in cigarette smoke, such as phthalide and polyacetylens.

Medicinal virtues: Where there is stop period or for removing stone and gravel, the roots are impacted and cause the urine. Additionally, they eliminate vaginal blockage and relieve dropsy and jaundice by opening up liver and spleen obstructions.

Reproduction condition: Damiana and kola nut aphrodisiacs restore diminished sexual arousal brought on by illness.

Genitourinary condition: Celery seed is a beneficial herb for genitourinary conditions because of its diuretic activity and antibacterial component.

45.14 Applications of Celery Seed Extract

Since the Middle Ages, celery has been used extensively as a food and a medicine. The first medicinal preparations start to appear in the late nineteenth century, and they typically contained celery seed juice, which helps with protein digestion when consumed with food. Celery is frequently used as primary and secondary. Arthritis, back pain, anxiety, and rheumatism are the primary main uses. Bright's disease, post nasal edema, migraines, and insomnia are among its secondary uses.

Nutrition beverage a nutritious cleaning drink made with organic carrot juice and celery is beneficial for many chronic illnesses. Celery seeds from glue ears are helpful for respiratory conditions including bronchitis and asthma. It is often used in conjunction with other medications that lower blood pressure. As an emmenagogue, or substance that encourages menstruation flow, celery seed extract is utilised. The kidneys are stimulated by celery seed, which increases urine production. They function to lower activity throughout the body and assist the kidney in eliminating ureates and other undesired waste products. By creating a combing effect, celery seed has been discovered to assist with nervous system regulation. Marsh water parsley is another name for celery. It produces drowsy effects while stimulating the sees drives.

45.14.1 Bioactivity

Celery is being employed as a tonic, aphrodisiac, antihelmintic, antispasmodic, carminative, diuretic, laxative, sedative, and sedative-hypnotic plant. The plant's roots, leaves, and seeds are all therapeutic [48]. Celery preparations are also used to treat kidney and gallstone stones, regulate bowel movements for evacuation, stimulate the glands, and purify the blood. Celery seeds are consumed to treat gout, rheumatoid arthritis, and other rheumatic disorders [49]. Celery seed oil contains nitrogenous chemicals that might shown to have impacts on the central nervous system. Celery is excellent at lowering blood pressure despite having a high sodium content because 3-n-butyl phthalide has been shown to lessen the smooth muscles that line blood vessels. Vitamin C, a vitamin that supports the immune system, is abundant in celery. Vitamin C is known to be a cold-fighter, and more than 20 scientific studies have found that it also lessens the severity of inflammatory diseases such as rheumatoid arthritis, osteoarthritis, and asthma by preventing the free radical damage that sets off the inflammatory cascade. Vitamin C is helpful for promoting cardiovascular health since free radicals could oxidize cholesterol and result in plaques which may rupture and cause heart attacks or strokes.

45.14.2 Antioxidant Activity

Using the flourescence spectrophotometric technique, antioxidant bioassays were conducted on liposome oxidation using crude celery extracts [50]. By using in vitro and in vivo testing, the possible protective effects of extracts of parts of celery leaves and roots in four different solvents with varying polarity, including ether, chloroform, ethyl acetate, n-butanol, and water, were investigated. Root and leaf extracts were discovered to be potential OH and DPPH radical scavengers in addition to liposomal peroxidation inhibitors [51]. The authors claim that all of the extracts had some level of protective activity, with n-butanol extract demonstrating the most activity. Researchers looked at how various extracts affected the antioxidant systems of mouse liver and blood when combined with and without CCL_4. The combined action of extracts and CCL_4 demonstrated both positive and negative synergism, promoting and inhibiting CCL_4 effects.

45.14.3 Anti-inflammatory Effect

Apium graveolens Linn. seed fractions have been extracted, and their antioxidant, cyclooxygenase, and topoisomerase inhibitory properties have been examined [52]. It was investigated how feeding rats with genetically elevated cholesterol levels aqueous and butanol extracts of celery affected their lipid levels. According to reports, celery extracts have an anti-inflammatory impact on rats as shown by the inhibition of carageenan-induced paw edoema [53]. However, the precise ingredient that causes the impact has not been located. For the management and anticipation of inflammation and gastrointestinal discomfort, celery seed extracts have been studied [54].

For prophylaxis and therapy of joint and connective tissue problems in vertebrates, specific herbal arrangements containing phytochemicals from ginger, cayenne pepper, turmeric, yucca, devil's claw, nettle leaf, alfalfa and celery seeds have been employed. The above-mentioned health advantages are said to be caused by n-butyl phthalide, the main flavor-impacting component of celery's volatile oil [55]. The phthalide-enriched fraction of celery oil would be in great demand and helpful for the management of hypertension and cardiac conditions due to the health benefits of celery's phthalides as well as the desire for additional sources of nutraceuticals [56]. Another claim about 3-n-butyl phthalide is that it relaxes muscles.

45.14.4 Anti-cancer Effects

The most important bioactive substances found in celery are its phthalides, which have been shown to protect against cholesterol, high blood pressure, and cancer. According to reports, sedanolide is the phthalide that is most effective at reducing tumours in lab animals [57]. Female mice's target tissues were highly responsive to glutathione S transferase (GST) induction by celery seed oil-derived sesdanolide and 3-nbutyl phthalide. The incidence of tumours was decreased after treatment with 3-n-butyl phthalide and sedanolide, from 68% to 30% and 11%, respectively. Three-n-butyl phthalide and sedanolide both showed activity in inhibiting tumour growth, with GST tests showing a link among the inhibitory activity and the capacity to induce GST, respectively, by about 67% and 83% reduction in tumour multiplicity. These findings imply that the bioactive natural product class known as phthalides, which is present in edible umbelliferae plants, may be an efficient chemopreventive.

Celery seed methanolic extracts are said to have hepatoprotective properties. Rats with liver damage caused by di-(2-ethylxyl) phthalate (DEHP) were used to test the extract. Celery seeds produces 300 mg/kg body weight/day administered orally for 6 weeks protected DEHP-induced mice from developing elevated levels of the blood marker enzymes glutamate oxaloacetate transaminase (SGOT), serum glutamate pyruvate transaminase (SGPT), alkaline phosphatases (ALP), and bilirubin. Similarly, after treatment with A. greaveolens extract, a decrease in oxidative stress indicators was seen in rat testes [58].

Celery includes substances known as coumarins that assist in preventing free radicals in harming cells, so reducing the mutations that raise the likelihood that cells would develop into malignant cells [59]. Additionally, coumarins increase the activity of some white blood cells, which work as immune defenders and hunt down and destroy potentially hazardous cells, such as cancer cells. Additionally, coumarin compounds lower blood pressure, tone the vascular system, and may be useful for treating migraines. Celery-based juices are excellent electrolyte replacement drinks when drunk after exercise because of the high levels of potassium and salt they contain. Celery may decrease cholesterol and inhibit cancer by enhancing detoxification, according to studies. It has been demonstrated that a celery extract with 85% 3-n-butyl phthalide is beneficial in treating "rheumatism," the umbrella word for arthritic and muscle aches and pains. Additionally, it has been demonstrated that acetylenic chemicals found in celery can inhibit the development of tumor cells.

45.14.5 Anti-microbial Activity

Celery volatile oil has been demonstrated to have anti-fungal action and is effective against a wide range of bacteria, including *Pseudomonas solanacearum, Shigella dysenteriae, Salmonella typhi, Streptococcus faecalis*, and *Staphylococcus aureus*. Both *Pseudomonas aeruginosa* and *Escherichia coli* showed no signs of action [60].

45.14.6 Blood Platelets Aggregation and Anti-hyperlipidemia Activity

It has been documented that celery extract inhibits blood platelet aggregation in vitro. According to reports, apigenin, not phthalides, was what caused the inhibition [61]. Celery aqueous extracts were found to lower low-density lipoprotein and total cholesterol. For 8 weeks, one group of rats received a high-protein diet together with aqueous celery extract, while the other group received simply a high-protein diet. After 8 weeks, it was shown that the group of rats given supplements containing celery extract had significantly lower levels of blood cholesterol, low-density of lipoprotein cholesterol, and triglycerides. When compared to control rats, the celery-treated rats had considerably lower hepatic triglycerol lipase activity. Examination of the celery aqueous extract revealed that this was devoid of the active ingredient 3-nphthalide, which was beforehand thought to be the catalyst for the lipid-lowering effect [62]. It might be caused by additional active celery extract ingredients that have an impact on lipid and serum levels.

45.14.7 Impacts on the Kidney

Celery seeds shows a direct effect on the kidneys, enhancing water excretion and accelerating the removal of built-up toxins from the joints, which is beneficial for the condition of arthritis [63]. In order to improve the effectiveness of liver and renal elimination, celery is given in combination with Taraxacum radix (Dandelion).

45.14.8 Celery-Spent as a Source of Dietary Fiber

Since the solvents engaged for oleoresin extraction might not extract the aforementioned components, celery-spent attained after the volatile oil and oleoresin withdrawal contains protein, fiber, carbs, and minerals. Obtainable from the spice oleoresins businesses each year are 1234 tonns of celery-spent residue that are not used for commercial purposes. It is combined in modest amounts with veterinarian and poultry diets. The majority of it is consumed as a boiler feed to produce heat [64]. Since ancient times, dietary fiber (DF) has been consumed and is known to provide numerous health advantages. Consuming foods high in DF is supplemented with improved laxation, decreased cholesterol, increased stool bulk, attenuated glycemic and insulin response, and attenuated glycemic and insulin response [65, 66]. Diets with enough fiber have a lower energy density, a higher volume, and a quicker onset of satiety. Celery spent residue was investigated as a potential new dietary fiber source [13]. Total dietary fibre (TDF) was found to be 60% in the wasted residue, with 52% of it being insoluble dietary fibre (IDF), 7% of it being soluble, 7% of it being ash, 5% of it being crude fat, 19% of it being protein, and 9% of it being starch. SEM analysis has been used to examine the microstructure of celery powder, celery-spent residue (residue left over afterward oil and oleoresin extraction), and the insoluble and soluble fiber extracted from wasted byproducts [13]. Before volatile oil and oleoresin extraction, scanning electron micrographs of celery powder revealed a few spherical-shaped starch granules within a fiber matrix. After extraction, the starch granules became gelatinized. Unlike soluble fiber, which has a smooth, porous surface, insoluble fiber has a coiled helix structure that resembles a rod. Hexane or ethanol could be used to wash out any remaining fat from the material. Celery wasted residue can be used as a source of protein, minerals, dietary fiber, and other nutrients in a variety of cuisine preparations.

45.14.9 Antihypotensive Agent

In spontaneously hypertensive rats, the role of 3-n-butylphthalide in lowering blood pressure and acting as a vasorelaxant was investigated. A brief hypotensive effect was achieved by a 13-day intraperitoneal fermentation of butyl phthalide at doses of 2.0 and 4.0 mg/day. The blocking of calcium entrance to receptor calcium channels, which lowers the rats' systolic blood pressure, is thought to be the cause of butyl phthalide's vasorelaxant effects [62].

References

1. QUIROS, C. F. (1993). Celery: Apium graveolens L. In Genetic improvement of vegetable crops (pp. 523–534). Pergamon.
2. Lewis, D. A., Tharib, S. M., & Veitch, G. B. A. (1985). The anti-inflammatory activity of celery Apium graveolens L.(Fam. Umbelliferae). *International Journal of Crude Drug Research, 23*(1), 27–32.
3. Li, M. Y., Hou, X. L., Wang, F., Tan, G. F., Xu, Z. S., & Xiong, A. S. (2017). https://doi.org/10.1080/07388551.2017.131227538, 172–183.
4. 'Apium graveolens (celery)' (2022) PlantwisePlus Knowledge Bank. CABI International. https://doi.org/10.1079/pwkb.species.6599
5. Keller, F., & Phytologist, P. M.-N. (1989). undefined 1989, Wiley Online. *Library., 113*, 291–299.
6. Fazal, S., R.S.-I.G. J. of Pharmaceutical, undefined 2012, Academia. Edu 2 (2012) 36–42.
7. Destaillats, F., & Angers, P. (2002). Base-catalyzed derivatization methodology for FA analysis. Application to milk fat and celery seed lipid TAG. *Lipids, 37*, 527–532.
8. Chandel, K. P. S., Shukla, G., & Sharma, N. (1996). Biodiversity in medicinal and aromatic plants in India.
9. Dyduch, J. (1994). *Acta Horticulturae*, 275–282.
10. Ashworth, S. (2002). Seed to seed: Seed saving and growing techniques for vegetable gardeners.
11. Wilson, C. W. (1970). Relative recovery and identification of carbonyl compounds from celery essential oil. *Journal of Food Science, 35*, 766–768.
12. Uhlig, J. W., Chang, A., & Jen, J. J. (1987). Effect of Phthalides on Celery Flavor. *Journal of Food Science, 52*, 658–660.
13. Sowbhagya, H. B., Sampathu, S. R., & Krishnamurthy, N. (2007). Evaluation of size reduction on the yield and quality of celery seed oil. *Journal of food engineering, 80*(4), 1255–1260.
14. Verghese, J. (1990). In the kaleidoscope–celery. *Perfumer and flavorist, 15*, 55–59.
15. Buttkus, H. A. (1978). Celery leaf juice: Evaluation and utilization of a product from harvest debris. *Journal of Agricultural and Food Chemistry, 26*, 827–830.
16. Rupérez, P., & Toledano, G. (2003). Celery by-products as a source of mannitol. *European Food Research and Technology, 216*, 224–226.
17. Alessandro, M. S., Galmarini, C. R., Iorizzo, M., & Simon, P. W. (2013). Molecular mapping of vernalization requirement and fertility restoration genes in carrot. *Theoretical and Applied Genetics, 126*, 415–423.
18. Bruznican, S., De Clercq, H., Eeckhaut, T., Van Huylenbroeck, J., & Geelen, D. (2020). *Frontiers in Plant Science, 10*.
19. Colombo, N., & Galmarini, C. R. (2017). The use of genetic, manual and chemical methods to control pollination in vegetable hybrid seed production: a review. *Plant Breeding, 136*, 287–299.

20. Naqvi, S. A. M. H. (2004). Diseases of Fruits and Vegetables.
21. Honma, S., & Lacy, M. L. (1980). Hybridization between pascal celery and parsley. *Euphytica, 29*, 801–805.
22. Irwanto, R., Ginting, W. M., & Novia, R. (2023). Ethanol extract liquid soap formulation leaves of celery (Apium graveolens L.) AGAINST Escherichia coli BACTERIA. *Jurnal Farmasimed (JFM), 5*(2), 157–165.
23. Toth, K. F. (1991). Increasing resistance in celery to *Fusarium oxysporum* f. sp. apii Race 2 with somaclonal variation. *Plant Disease, 75*, 1034.
24. Wolf, E. A., & Scully, B. (2019). 'Floribelle M9': An autumn celery cultivar for Florida. *HortScience, 27*, 1235–1236.
25. Scully, B., Nuessly, G. S., Raid, R. N., & Stubblefield, R. E. (2019). 'Wolf-249' and FBL 5-2M: An autumn celery cultivar and breeding line for Florida. *HortScience, 30*, 1104–1105.
26. D'Antonio, V., Falk, B., & Quiros, C. F. (2001). Inheritance of resistance tocelery mosaic virusin celery. *Plant Disease, 85*, 1276–1277.
27. Ochoa, O., & Quiros, C. F. (1989). Apium wild species: Novel sources for resistance to late blight in celery. *Plant Breeding, 102*, 317–321.
28. Meade, T., & Daniel Hare, J. (1991). Differential performance of beet armyworm and cabbage looper (Lepidoptera: Noctuidae) larvae on selected Apium graveolens cultivars. *Environmental Entomology, 20*, 1636–1644.
29. Trumble, J. T., Diawara, M. M., & Quiros, C. F. (1998, August). Breeding resistance in Apium graveolens to Liriomyza trifolii: Antibiosis and linear furanocoumarin content. In XXV International Horticultural Congress, Part 3: Culture Techniques with Special Emphasis on Environmental Implications, 513 (pp. 29–38).
30. Sikora, R.A., & Fernández, E.. (2005). *Plant parasitic nematodes in subtropical and tropical agriculture* (2nd Ed., pp. 319–392).
31. Wang, M., & Goldman, I. L. (1996). Resistance to root knot nematode (Meloidogyne hapla Chitwood) in carrot is controlled by two recessive genes. *The Journal of Heredity, 87*, 119–123.
32. Vovlas, N., Lucarelli, G., Sasanelli, N., Troccoli, A., Papajova, I., Palomares-Rius, J. E., & Castillo, P. (2008). *Wiley Online Library, 57*, 981–987.
33. Ali, A., Matthews, W. C., Cavagnaro, P. F., Iorizzo, M., Roberts, P. A., & Simon, P. W. (2014). Inheritance and mapping of Mj-2, a new source of root-knot nematode (Meloidogyne javanica) resistance in carrot. *Journal of heredity, 105*(2), 288–291.
34. Li, M., Zhou, J., Du, J., Li, X., Sun, Y., Wang, Z., ... & Tang, H. (2022). Comparative physiological and transcriptomic analyses of improved heat stress tolerance in celery (Apium graveolens l.) caused by exogenous melatonin. *International Journal of Molecular Sciences, 23*(19), 11382.
35. Malhotra, S. K. (2006). Celery. In Handbook of herbs and spices (pp. 317–336). Woodhead Publishing.
36. Kaiser, A., Hartmann, K. I., Kammerer, D. R., & Carle, R. (2013). Evaluation of the effects of thermal treatments on color, polyphenol stability, enzyme activities and antioxidant capacities of innovative pasty celeriac (Apium graveolens L. var. rapaceum (Mill.) DC.) products. *European Food Research and Technology, 237*, 353–365.
37. Aydemir, T., & Akkanli, G. (2006). Partial purification and characterisation of polyphenol oxidase from celery root (Apium graveolens L.) and the investigation of the effects on the enzyme activity of some inhibitors. *International Journal of Food Science and Technology, 41*, 1090–1098.
38. D.J. Madjarova, M.G. Bubarova, Acta Hortic. (1978) 65–72.
39. Trumble, J. T., Diawara, M. M., & Quiros, C. F. (1998). Breeding resistance in Apium graveolens to Liriomyza trifolii: Antibiosis and linear furanocoumarin content. *Acta Horticulturae, 513*, 29–37.
40. Ivanova, M. I., Bukharov, A. F. (2009). Analysis of F1 and F2 populations resulting from cross of curled celery (Apium graveolens convar. secalinum Alef. var crispum Alef.) and turnip-rooted celery (Apium graveolens L. var rapaceum (Mill.) Gaud.). AGRIS Sci. 259–261.

41. Bolon, Y. T., Stec, A. O., Michno, J. M., Roessler, J., Bhaskar, P. B., Ries, L., ... & Stupar, R. M. (2014). Genome resilience and prevalence of segmental duplications following fast neutron irradiation of soybean. *Genetics, 198*(3), 967–981.
42. Huang, W., Wang, G. L., Li, H., Wang, F., Xu, Z. S., & Xiong, A. S. (2016). Transcriptional profiling of genes involved in ascorbic acid biosynthesis, recycling, and degradation during three leaf developmental stages in celery. *Molecular Genetics and Genomics, 291*, 2131–2143.
43. Yao, Y., Sang, W., Zhou, M., & Ren, G. (2010). *Journal of Food Science, 75*.
44. Ali Shad, A., Ullah Shah, H., Bakht, J., Iqbal Choudhary, M., & Ullah, J. (2011). *Academicjournals.Org, 5*, 5160–5166.
45. Al-Din, A., Pill Baek, J., Mady, E., Eldekashy, M., & Craker, L. (2015). *Scholarworks.Umass. Edu, 4*, 1–7.
46. Gao, Z., & Loescher, W. H. (2000). NADPH supply and mannitol biosynthesis. Characterization, cloning, and regulation of the non-reversible glyceraldehyde-3-phosphate dehydrogenase in celery leaves. *Plant Physiology, 124*(1), 321–330.
47. Zhang, K., Wang, J., Yang, Z., Xin, G., Yuan, J., Xin, J., & Huang, C. (2013). Genotype variations in accumulation of cadmium and lead in celery (Apium graveolens L.) and screening for low Cd and Pb accumulative cultivars. *Frontiers of Environmental Science & Engineering, 7*, 85–96.
48. Bradley, P. (2006). British herbal compendium. Volume 2: a handbook of scientific information of widely used plant drugs.
49. Bjeldanes, L. F., & Kim, I. S. (1977). Phthalide components of celery essential oil. *The Journal of Organic Chemistry, 42*, 2333–2335.
50. Arora, A., Nair, M. G., & Strasburg, G. M. (1998). Structure–Activity relationships for antioxidant activities of a series of flavonoids in a liposomal system. *Free Radical Biology & Medicine, 24*, 1355–1363.
51. Popović, M., Kaurinović, B., Trivić, S., Mimica-Dukić, N., & Bursać, M. (2006). Effect of celery (Apium graveolens) extracts on some biochemical parameters of oxidative stress in mice treated with carbon tetrachloride. *Phytherapy Research, 20*, 531–537.
52. Momin, R. A., & Nair, M. G. (2002). Antioxidant, cyclooxygenase and topoisomerase inhibitory compounds from Apium graveolens Linn. seeds. *Phytomedicine, 9*, 312–318.
53. Al-Hindawi, M. K., Al-Deen, I. H. S., Nabi, M. H. A., & Ismail, M. A. (1989). Anti-inflammatory activity of some Iraqi plants using intact rats. *Journal of Ethnopharmacology, 26*, 163–168.
54. Powanda, M. C., & Rainsford, K. D. (2011). A toxicological investigation of a celery seed extract having anti-inflammatory activity. *Inflammopharmacology, 19*, 227–233.
55. Gold, H. J., Wilson, C. W., & Food, J. (1963). The volatile flavor substances of celery. *Journal of Food Science, 28*, 484–488.
56. Tsi, D., & Tan, B. K. H. (1997). Cardiovascular pharmacology of 3-n-butylphthalide in spontaneously hypertensive rats. Phytotherapy Research: An International Journal Devoted to Medical and Scientific Research on Plants and Plant Products, 11(8), 576–582.
57. Zheng, G. Q., Kenne, P. M., Zhang, J., & Lam, L. K. T. (1993). Chemoprevention of benzo[a]pyrene-induced forestomach cancer in mice by natural phthalides from celery seed oil. *Nutrition and Cancer, 19*, 77–86.
58. Hamza, A. A., & Amin, A. (2007). Apium graveolens modulates sodium valproate-induced reproductive toxicity in rats. *Journal of Experimental Zoology Part A: Ecological Genetics and Physiology, 307*, 199–206.
59. Sy, K. V., Murray, M. B., Harrison, M. D., & Beuchat, L. R. (2005). Evaluation of gaseous chlorine dioxide as a sanitizer for killing Salmonella, Escherichia coli O157: H7, Listeria monocytogenes, and yeasts and molds on fresh and fresh-cut produce. *Journal of food protection, 68*(6), 1176–1187.
60. Atta, A. H., & Alkofahi, A. (1998). Anti-nociceptive and anti-inflammatory effects of some Jordanian medicinal plant extracts. *Journal of ethnopharmacology, 60*(2), 117–124.

61. Teng, N. N. H., Kaplan, H. S., Hebert, J. M., Moore, C., Douglas, H., Wunderlich, A., & Braude, A. I. (1985). Protection against gram-negative bacteremia and endotoxemia with human monoclonal IgM antibodies. *Proceedings of the National Academy of Sciences of the United States of America, 82*, 1790–1794.
62. Tsi, D., Das, N. P., & Tan, B. K. H. (1995). Effects of aqueous celery (Apium graveolens) extract on lipid parameters of rats fed a high fat diet. *Planta Medica, 61*(01), 18–21.
63. Sandek, A., Bauditz, J., Swidsinski, A., Buhner, S., Weber-Eibel, J., von Haehling, S., Schroedl, W., Karhausen, T., Doehner, W., Rauchhaus, M., Poole-Wilson, P., Volk, H. D., Lochs, H., & Anker, S. D. (2007). Altered intestinal function in patients with chronic heart failure. *Journal of the American College of Cardiology, 50*, 1561–1569.
64. Mathew, A., Peters, U., Chatterjee, N., Kulldorff, M., & Sinha, R. (2004). Fat, fiber, fruits, vegetables, and risk of colorectal adenomas. *International Journal of Cancer, 108*, 287–292.
65. Bourdon, I., Yokoyama, W., Davis, P., Hudson, C., Backus, R., Richter, D., Knuckles, B., & Schneeman, B. O. (1999). Postprandial lipid, glucose, insulin, and cholecystokinin responses in men fed barley pasta enriched with β-glucan. *The American Journal of Clinical Nutrition, 69*, 55–63.
66. Humbert, M., Beasley, R., Ayres, J., Slavin, R., J.H.- Allergy, undefined (2005), Wiley Online Library. 60, 309–316.

Chapter 46
Dioscorea

Muhammad Zulqurnain Haider, Asia Shaheen, Saqib Mahmood, Aisha Tariq, Hira Rafique, and Umar Farooq Gohar

46.1 Introduction

Dioscorea belongs to the family *Dioscoreaceae*; and generally known as yam in most areas of the world, as an invasive, deciduous, and herbaceous vine can grow upto 18 m in length. It is a diverse genus with a number of species names such as water yam, winged yam, greater yam, air yam, air potato, etc. It has a wide distribution from Africa, the Pacific, Asia, the Caribbean, and Latin America [1]. Its nutritive material is stored in tubers (single or multiple) of variable shapes and colours. This genus is from division monocotyledon of flowering plants with more than 350 species, as a herbaceous nature that may be annual or perennial and survive in the form of climbing or trailing vines. From variety to variety, its length may reach up to 10 m. The texture of yams are usually smooth or prickly. Simple leaves with oval or heart-like shapes and long petioles. Spikes may be present or absent. Male flowers lie in the leaf axil and may be simple or branched; while the female flowers are small and clustered. *Dioscorea* yams are an important food source for vast numbers of people around the world, especially in West Africa and Asia. Yams have cultural significance and are used in traditional medicine, in addition to their nutritional value [2].

M. Z. Haider · A. Shaheen · S. Mahmood (✉) · H. Rafique
Department of Botany, Government College University Faisalabad, Faisalabad, Pakistan
e-mail: saqibmahmood@gcuf.edu.pk

A. Tariq
Department of Nutritional Sciences, Government College University Faisalabad, Faisalabad, Pakistan

U. F. Gohar
Institute of Industrial Biotechnology, Government College University Lahore, Lahore, Pakistan

M. Zia-Ul-Haq et al. (eds.), *Essentials of Medicinal and Aromatic Crops*,
https://doi.org/10.1007/978-3-031-35403-8_46

Dioscorea plants have a rich history of use in traditional medicine, and studies have revealed that they have a variety of pharmacological effects, including anti-inflammatory, antiseptic, and antioxidant properties [3]. Some species have also been discovered to have therapeutic potential for diabetes and cancer [4]. In terms of cultivation, *Dioscorea* plants can be challenging to grow, requiring specific environmental conditions and careful management. However, they have significant potential as a viable crop for small-scale farmers in developing countries, and efforts are underway to improve their cultivation techniques and maximize yields.

46.2 Agronomic Requirements

46.2.1 Soil and Climate

The most suitable environment for yams is a tropical or subtropical climate. Temperatures below 22 °C (71.6 °F) conducive to its growth; they may even cause frost damage. Optimum growth is observed from 25 to 30 °C (77 to 86 °F) with a pH of 5.5–6.5 in the presence of full sun or part shade. It's important to note that excessive irrigation can promote tuber rot, so careful water management is crucial for successful cultivation.

46.2.2 Season

Planting season is from May to June.

46.2.3 Seed Rate

Tubers, whole or in part, at a rate of 1875–2500 Kg/ha.

46.2.4 Field Preparation

The vegetative part of tubers is used to propagate yam. It necessitates a well-ploughed and harrowed field. To plant tubers, the trenches (15 cm deep), at least 15 cm apart from neighbouring plants (at least 30 cm between individual plants and 1.5 m between rows) are prepared. Soil is frequently mounded or ridged around plants to aid drainage. As yam plants are climbers, they need support, typically in the form of stakes or trellises, with heights ranging from 2 to 4 meters. Ploughing is an essential step in field preparation, particularly the formation of ridges and furrows separated by at least 75 cm.

46.2.5 Planting (Sowing)

Mini setts (about 25 g) that can be planted directly or raised by nurseries and transplanted are recommended for plantations (60-day-old seedlings). Planting distances of 75 cm are maintained, whether in beds, ridges, or mounds.

46.2.6 Irrigation Requirement

Watering is required on a weekly basis.

46.2.7 Following Cultivation

The vines will require some poles to be trained on. Weeding could be scheduled if necessary can be completed as needed. It can be grown alongside other crops such as coffee, banana, rubber, and coconut.

46.2.8 Trailing

Plants can expose maximum their leaves to capture sunlight by trailing. It can also boost yam yield.

46.2.9 Crop Protection

Yam's scale protects tubers both in the field and after harvest. Before plantation scale can be removed with chemical treatment such as methyl demeton (0.25%).

46.2.10 Harvest

Mature yams take 8–10 months to mature from plantation. Time varies according to species and variety. Damage can be minimized by careful digging during harvest.

46.2.11 Yield

In the first 240 days (around 8 months) of tuber life, the production rate of yam ranges from 20 to 25 tons per acre.

46.2.12 Economic Significance

The *Dioscorea* flowering plant genus has significant economic importance on global scale. Yams, which are the edible tubers produced by these plants, are a staple food in many tropical and subtropical regions, providing a critical source of nutrition for millions of people. In West Africa, yams are a primary crop, with Nigeria stands as leading global producer of yams. Yams are also hold agricultural significance in other regions of Africa, Asia, and the Americas. In addition to their use as food, yams have cultural significance and are used in traditional medicine [4]. The global yam trade is worth billions of dollars, with yams being exported to markets worldwide. The demand for yams continues to grow, particularly in Asia, where they are becoming increasingly popular. This demand has led to efforts to improve yam cultivation and increase yields, as well as to develop new yam varieties that are more resistant to pests and diseases.

Moreover, the pharmacological properties of *Dioscorea* plants have led to increased interest in their potential application in the pharmaceutical industry [5]. Extensive research has shown that certain species of *Dioscorea* hold potential as treatments for diabetes, cancer, and other diseases. This has led to the development of yam-based pharmaceutical products, which have the potential to generate significant revenue for the industry.

In conclusions, the *Dioscorea* plant genus has significant economic importance, both as a food crop and as a source of potential pharmaceutical products. Its cultivation and trade have significant implications for global food security and human health [6]. It is a diverse genus with numerous species rich in nutritious constituents and medicinally recognized metabolites.

The following medicinal and nutritive properties of various *Dioscorea* spp. are prominent in the literature.

46.3 *Dioscorea persimilis*

46.3.1 Nutritive Value

Dioscorea persimilis, also known as the cinnamon vine yam, is a root vegetable that is a good source of several important nutrients. It is particularly high in dietary fiber, which supports digestive health and helps regulate blood sugar levels. It is also a

Table 46.1 Morphographic overview and dry mass based nutritive profile of *D. persimillis*

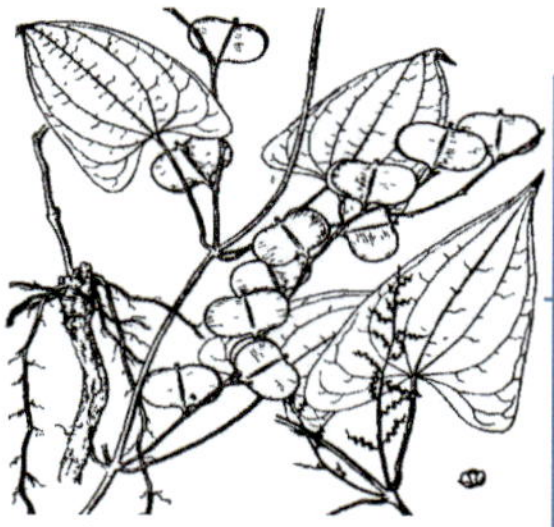

Dioscorea persimilis

Carbohydrates	70% to 80% of dry weight
Protein	5% to 10%
Fat	Less than 1%

good source of vitamin C, which supports immune function and wound healing, as well as potassium, which is important for regulating blood pressure and heart function [7]. Additionally, it contains significant amounts of iron, which is essential for oxygen transport in the body, and copper, which is important for maintaining healthy connective tissues and red blood cells (Table 46.1).

46.3.2 Therapeutic Potential

Dioscorea persimilis tubers have been used to treat renal failure, metritis, spermatorrhea, diarrhea, long term dysentery, back discomfort, wooziness, and night sweating, in addition to being used as a food source [7].

46.4 *Dioscorea polystachya*

46.4.1 Nutritive Value

Dioscorea polystachya, formaly known as *D. batatas* also known as the Chinese wild yam, is a root vegetable that is a good source of several important nutrients. It is particularly high in dietary fiber, which supports digestive health and helps regulate blood sugar levels. It is also a good source of vitamin C, which supports immune function and wound healing, as well as potassium, which is important for regulating blood pressure and heart function. Additionally, it contains significant amounts of manganese, which is essential for bone health and metabolism, and copper, which is important for maintaining healthy connective tissues and red blood cells. See the Table 46.2 for more details.

Table 46.2 Morphographic overview and dry mass based nutritive profile of *D. polystachya*

Dioscorea polystachya

Calories	118 Kcal
Carbohydrates	27.9 grams
Protein	1.5 grams
Fat	0.2 grams
Dietary Fiber	4.1 grams
Vitamin C	17.1 mg
Phosphorus	50 mg
Potassium	816 mg
Iron	0.5 mg
Folate	23 μg
Calcium	17 mg
Thiamine (vitamin B1)	0.11 mg
Riboflavin (vitamin B2)	0.04 mg
Niacin (vitamin B3)	0.6 mg
Vitamin B6	0.38 mg
Magnesium	21 mg
Zinc	0.4 mg

46.4.2 Therapeutic Potential

Histamine is the compound that induces allergic reactions in the body including skin itching [7]. Investigations have shown that the dioscorin is produced by *D. batatas* (currently named as *D. polystachya*) It has the potential to produce allergic reactions, even though histamine is the main allergen in yam [8]. The rhizome of *D. polystachya* treats dysentery and consumptive coughing, and also helps gastric motility and digestion [9].

The antioxidant properties as anti-lipid test, OH^- scavenging activity and reducing power of *D. polystachya* was investigated in a in vitro comparitive study [10]. The results showed the inhibitory effect of lectin from *D. polystachya* on cancers cell lines (including nasopharyngeal carcinoma CNE2 cells, hepatoma HepG2 cells and breast cancer MCF7 cells). It also found that lectin produced by *Dioscorea* spp. may participate in induction of apoptosis in MCF7 cells [11]. Moreover, it has been shown that diosgenin dramatically reduces the development of sarcoma-180 tumour cells in vivo whereas, increasing the phagocytic capacity of macrophages in vitro, indicating the possible role of diosgenin in enhancement of cellular immune reactions [12]. When in study rats suffering from diabetes were subjected to the treatment of phytoextracts of *D. polystachya* it showed hypoglycemic effect and bone fragility [13].

46.5 *Dioscorea praehensilis*

46.5.1 Nutritive Value

Dioscorea praehensilis, also known as the forest yam, is a root vegetable that is a good source of several important nutrients. It is particularly high in dietary fiber, which helps regulate digestion and prevent constipation. It is also a good source of potassium, which is important for regulating blood pressure and heart function, and vitamin C, which supports immune function and wound healing [14]. Additionally, it contains significant amounts of iron, which is important for oxygen transport in the body, and calcium, which is essential for strong bones and teeth. See the Table 46.3 for more details.

46.5.2 Therapeutic Potential

D. prachensilis, along with some other *Dioscorea* species (*D. alata*, *D. bulbifera*, *D. dumetorum*, and *D. burkilliana*), has been used to treat diabetes. *D. praehensilis* has anti-diabetic activity similar to *D. bulbifera* [15]. *D. praehensilis* has

Table 46.3 Morphographic overview and dry mass based nutritive profile of *D. praehensilis*

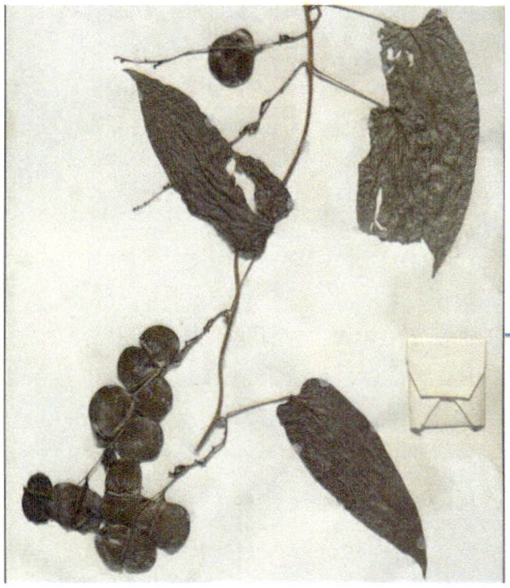

Dioscorea praehensilis

Calories	137.68 Kcal
Carbohydrates	28.56 grams
Protein	3.17 grams
Fat	0.67 grams
Dietary fiber	6.84 grams
Vitamin C	28.31 mg
Phosphorus	53.83 mg
Potassium	496.17 mg
Iron	2.35 mg
Ash	1.77 grams
Sodium	27.24 mg

hypoglycemic activity, according to a mouse study [16]. The bioactivity of Dioscorea as determined by the ABTS and DPPH assays was positive, but it had less antioxidant capacity than *D. bulbifera*. It also demonstrated that antioxidant activity is not dependent on the phenolic content of the species, as *D. dumetorum* had a lower antioxidant response (but a higher phenolic content) than *D. praehensilis*, which had a higher antioxidant response but a lower phenolic content [15].

46.6 *Dioscorea pyrifolia*

46.6.1 *Nutritive Value*

Dioscorea pyrifolia, also known as Chinese yam, is a nutritious root vegetable that is widely consumed in Asia. It is a good source of several important nutrients, including dietary fiber, vitamin C, potassium, and manganese (Table 46.4).

Table 46.4 Morphographic overview and dry mass based nutritive profile of *D. pyrifolia*

Dioscorea pyrifolia

Calories	92 Kcal
Carbohydrate	21.3 grams
Protein	1.5grams
Fat	0.1 grams
Dietary Fiber	3.0 grams
Vitamin C	7 mg
Potassium	370 mg
Calcium	8 mg
Iron	0.3 mg
Vitamin B1	0.02 mg
Vitamin B2	0.01 mg
Niacin	0.6 mg
Phosphorus	35 mg
Zinc	0.4 mg

46.6.2 Therapeutic Potential

D. pyrifolia powder and its extract produced favorable antibacterial photochemistry screening findings against the chemical classes of secondary metabolites like alkaloids, flavonoids, glycosides, saponins, triterpenoids, and tannins. Moreover, anthraquinone glycosides and cyanogenic glycosides Results from various concentrations display a range of outcomes [17].

46.7 *Dioscorea rotundata* Poir.

46.7.1 Nutritive Value

Dioscorea rotundata Poir., also known as white yam or sweet yam, is a root vegetable that contains several important nutrients. It is high in carbohydrates, fiber, potassium, and vitamin C [18]. See the Table 46.5 for more details.

Table 46.5 Morphographic overview and dry mass based nutritive profile of *D. rotundata* Poir.

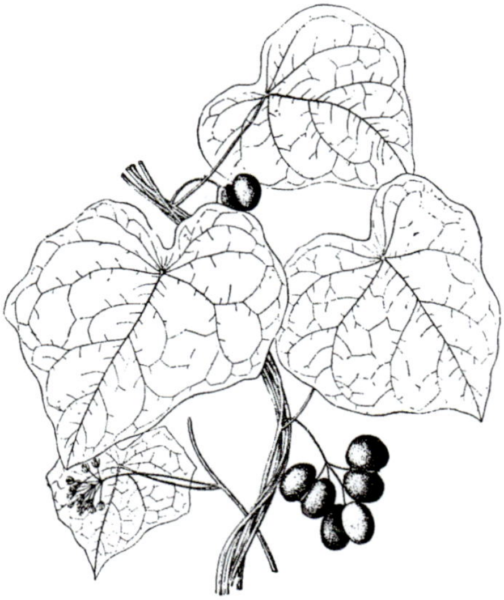

***Dioscorea rotundata* Poir.**

Calories	118 Kcal
Carbohydrates	27.9 grams
Protein	1.5 grams
Fat	0.2 grams
Dietary Fiber	4.1 grams
Vitamin C	17.1 mg
Phosphorus	50 mg
Potassium	816 mg
Iron	0.5 mg
Folate	23 µg
Calcium	17 mg
Thiamine (vitamin B1)	0.11 mg
Riboflavin (vitamin B2)	0.04 mg
Niacin (vitamin B3)	0.6 mg
Vitamin B6	0.38 mg
Magnesium	21 mg
Zinc	0.4 mg

46.7.2 Therapeutic Potential

The antioxidant activity of *Dioscorea rotundata* Poir. (white yam) was investigated. In addition, the study looked into the effect of paclobutrazol (PBZ) application on its metabolic system, which is linked to antioxidant activity. Where low concentrations of PBZ increased non-enzymatic and enzymatic antioxidant potential [19]. It was discovered that certain plant growth regulators can improve the natural antioxidant activities of this species.

46.8 *Dioscorea spicata*

46.8.1 Nutritive Value

Dioscorea spicata, also known as the air potato, is a root vegetable that contains several important nutrients. It is high in carbohydrates and fiber, and also contains significant amounts of potassium, vitamin C, and manganese. Overall, it is a nutritious source of energy and may support digestive and immune health. However, it is important to note that the air potato is considered an invasive species in some regions and should not be consumed without proper preparation to remove any potential toxins. See the Table 46.6 for more details.

46.8.2 Therapeutic Potential

Dioscorea spicata, also known as cinnamon vine yam, is a species of yam that is commonly consumed in Asia. However, there is limited information available on the nutritional values of this specific yam species. Based on the available information, it is likely that *D. spicata* is a good source of carbohydrates and dietary fiber, like other yam species. Additionally, some studies have suggested that certain compounds found in yams, such as diosgenin and dioscorin, may have possible vigor-related benefits, including anti-inflammation and antioxidant effects. A powerful source of natural antioxidants is the tuber of the *D. spicata* plant. In a research study, in vitro models including DPPH, hydroxyl radical, superoxide radical, ABTS radical cation, and reducing power were used to assess the in vitro antioxidant and free radical scavenging activity of methanol extract from tubers. The results showed that the effect depended on the concentration of the extract; the higher the extract concentration, the greater the antioxidant response [20].

Table 46.6 Morphographic overview and dry mass based nutritive profile of *D. spicata*

Dioscorea spicata

Calories	118 Kcal
Carbohydrates	27.9 grams
Protein	1.5grams
Fat	0.2 grams
Dietary Fiber	4.1 grams
Vitamin C	17.1 mg
Potassium	816 mg
Calcium	17 mg
Iron	0.5 mg
Thiamine (vitamin B1)	0.11 mg
Riboflavin (vitamin B2)	0.04 mg
Niacin (vitamin B3)	0.6 mg
Vitamin B6	0.38 mg
Folate	23 µg
Magnesium	21 mg
Phosphorus	50 mg
Potassium	816 mg
Zinc	0.4 mg

46.9 *Dioscorea steriscus*

46.9.1 Nutritive Value

There is limited information available on the nutrient value of *Dioscorea steriscus*, however, it is known to be a source of dietary fiber and contains some amounts of potassium, vitamin C, and calcium. See the Table 46.7 for more details.

46.9.2 Therapeutic Potential

It has been reported that *D. steriscus* tuber extract obtained via solvent cold percolation has anti-obesity properties. The extract from *D. steriscus* tubers have shown a much stronger anti-obesity action when compared to anti-obesity drug (herbex) that is available in market. The bioactive components of *D. steriscus* tubers, known for their role in anti-obesity therapeutics includes some lipase and amylase inhibitors [21].

Table 46.7 Morphographic overview and dry mass based nutritive profile of *D. steriscus*

Dioscorea steriscus

Calories	118 Kcal
Carbohydrates	27.9 grams
Protein	1.5 grams
Fat	0.2 grams
Dietary Fiber	4.1 grams
Vitamin C	17.1 mg
Potassium	816 mg
Calcium	17 mg
Iron	0.5 mg
Thiamine (vitamin B1)	0.11 mg
Riboflavin (vitamin B2)	0.04 mg
Niacin (vitamin B3)	0.6 mg
Vitamin B6	0.38 mg
Folate	23 µg
Magnesium	21 mg
Phosphorus	50 mg
Potassium	816 mg
Zinc	0.4 mg

46.10 *Dioscorea subhastata*

46.10.1 Nutritive Value

D. subhastata plant contains carbohydrates, protein, vitamins and fat contents. See the Table 46.8 for more details.

46.10.2 Therapeutic Potential

The main medicinal uses of *Dioscorea subhastata* are for the treatment of digestive disorders. It is believed to be effective in relieving symptoms such as bloating, constipation, and diarrhea. Additionally, it has been used for anti-inflammation and

Table 46.8 Morphographic overview and dry mass based nutritive profile of *D subhastata*

Dioscorea subhastata

Calories	118 kcal
Carbohydrates	27.9 grams
Protein	1.5grams
Fat	0.2 grams
Dietary Fiber	4.1 grams
Vitamin C	17.1 mg
Potassium	816 mg
Calcium	17 mg
Iron	0.5 mg
Thiamine (vitamin B1)	0.11 mg
Riboflavin (vitamin B2)	0.04 mg
Niacin (vitamin B3)	0.6 mg
Vitamin B6	0.38 mg
Folate	23 µg
Magnesium	21 mg
Phosphorus	50 mg
Potassium	816 mg
Zinc	0.4 mg

radical scavenging, which may possibly relieve stress in the body. *D subhastata* has also been studied for its potential for reducing the level of blood sugar in patients with diabetes. The plant contains compounds that may help to regulate insulin sensitivity and glucose metabolism, making it a promising natural treatment for this condition [22]. Other potential medicinal uses of *D. subhastata* include its ability to improve cognitive function, enhance immune system function, and reduce the risk of certain tumor-related activities. However, further research is required for comprehensive information regarding its benefits. It is important to note that while *D. subhastata* has shown promising medicinal potential, it should not be used as a replacement for traditional medical treatment without the advice of a healthcare professional.

46.11 *Dioscorea tomentosa*

46.11.1 Nutritive Value

D. tomentosa plant are the known source of dietary fiber and contains some amounts of potassium, iron, and calcium. See the Table 46.9 for more details.

Table 46.9 Morphographic overview and dry mass based nutritive profile of *D. tomentosa*

Dioscorea tomentosa

Calories	118 Kcal
Carbohydrates	27.9 grams
Protein	1.5 grams
Fat	0.2 grams
Dietary fiber	4.1 grams
Vitamin C	17.1 mg
Potassium	816 mg
Calcium	17 mg
Iron	0.5 mg
Thiamine (vitamin B1)	0.11 mg
Riboflavin (vitamin B2)	0.04 mg
Niacin (vitamin B3)	0.6 mg
Vitamin B6	0.38 mg
Folate	23 µg
Magnesium	21 mg
Phosphorus	50 mg
Potassium	816 mg
Zinc	0.4 mg

46.11.2 Therapeutic Potential

Dioscorea tomentosa tuber's antioxidant capacity in methanol extracts has been evaluated in a study using a variety of models, including hydroxyl, superoxide, DPPH, hydroxyl, ABTS, and reducing power, where it was compared with previously reported antioxidants like vitamin C and troloxin. The results showed that the effect depended on the concentration of the extract; the higher the extract concentration, the greater the antioxidant response [23]. In another study, children were provided with 10 g of the cooked, peeled tuber of this species. The observation was about the relief of gastrointestinal symptoms [24].

46.12 *Dioscorea trifida*

46.12.1 Nutritive Value

Dioscorea trifida, also known as cush-cush yam or sweet yam, is a root vegetable that is high in carbohydrates and a good source of dietary fiber. It also contains some amounts of potassium, vitamin C, and manganese. Overall, it is a nutritious source of energy and may support digestive and immune health. However, it is important to note that some varieties of *D. trifida* may contain potentially harmful levels of toxic substances and should be properly prepared before consumption. See the Table 46.10 for more details.

46.12.2 Therapeutic Potential

In a research study, ovalbumin-induced food allergy in mice was treated with a *D. trifida* extract that had anti-inflammatory properties. Moreover, its leaf extracts, rhizomes, and bulbils showed anti-inflammation [25].

46.13 *Dioscorea triphylla*

46.13.1 Nutritive Value

D. triphylla plants are known to be a source of dietary fiber and contains some amounts of potassium and calcium. See the Table 46.11 for more details.

Table 46.10 Morphographic overview and dry mass based nutritive profile of *D. trifida*

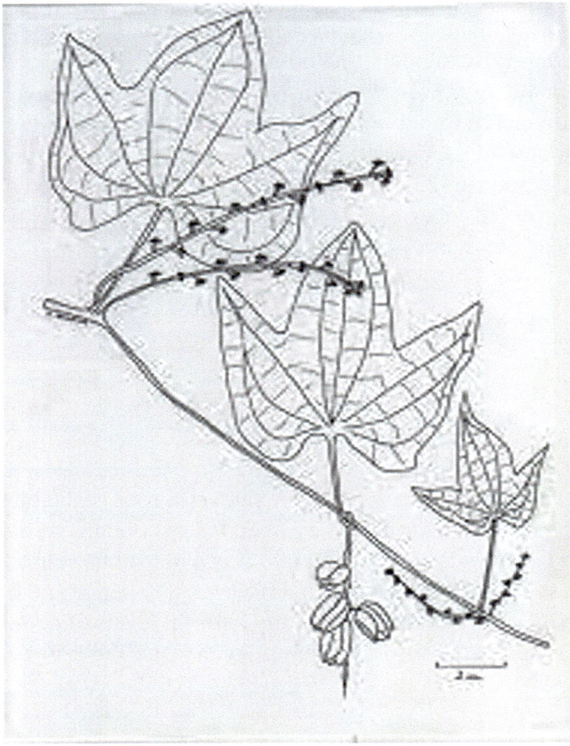

Dioscorea trifida

Calories	118 Kcal
Carbohydrates	27.9 grams
Protein	1.5 grams
Fat	0.2 grams
Dietary Fiber	4.1 grams
Vitamin C	17.1 mg
Potassium	816 mg
Calcium	17 mg
Iron	0.5 mg
Thiamine (vitamin B1)	0.11 mg
Riboflavin (vitamin B2)	0.04 mg
Niacin (vitamin B3	0.6 mg
Vitamin B6	0.38 mg
Folate	23 µg
Magnesium	21 mg
Phosphorus	50 mg
Potassium	816 mg
Zinc	0.4 mg

46.13.2 Therapeutic Potential

Dioscorea triphylla, also known as five leaf yam, has been traditionally used in herbal medicine for various therapeutic purposes. It is believed to have anti-inflammatory and anti-oxidative properties, and may help to regulate blood sugar levels and support digestive health. Additionally, some studies have suggested that it may have potential in the treatment of cancer and cardiovascular diseases, although further research is needed to fully understand its therapeutic potential.

Table 46.11 Morphographic overview and dry mass based nutritive profile of *D. triphylla*

Dioscorea triphylla

Calories	118 Kcal
Carbohydrates	28 grams
Protein	1.5 grams
Fat	0.2 grams
Fiber	3.9 grams
Vitamin C	17.1 mg
Potassium	670 mg
Vitamin B6	0.3 mg
Iron	1.2 mg
Thiamin (Vitamin B1)	0.2 mg
Riboflavin (Vitamin B2)	0.1 mg
Magnesium	40 mg

46.14 *Dioscorea versicolor*

46.14.1 Nutritive Value

D. versicolor plant is known to be a source of dietary fiber and contains some amounts of potassium and calcium. See the Table 46.12 for more details.

46.14.2 Therapeutic Potential

The findings of a study revealed that there were various levels of α -amylase inhibitors present. Among the yam tubers examined, *D. versicolor* was noted for its maximum (147 IU/g DM) for α -amylase inhibition [26]. A natural purpose for α -amylase

Table 46.12 Morphographic overview and dry mass based nutritive profile of *D. versicolor*

Dioscorea versicolor

Calories	99 Kcal
Carbohydrates	23 grams
Protein	1.4 grams
Fat	0.2 grams
Dietary fiber	3.4 grams
Vitamin C	2.4 mg
Potassium	511 mg
Calcium	17 mg
Iron	0.5 mg

inhibitors is the regulation of endogenous amylase activity or defense against infections and pests. Certain inhibitors are reportedly antinutritional agents [27]. An increasing number of studies in recent years have assessed the potential effectiveness of α -amylase inhibitors in the treatment of diabetes and obesity [28].

46.15 *Dioscorea villosa*

46.15.1 Nutritive Value

Dioscorea villosa, also known as wild yam, is a root vegetable that is low in calories and fat, but high in carbohydrates and dietary fiber. It also contains some amounts of potassium and manganese. While it is not a significant source of other vitamins or minerals, it may be a good addition to a balanced diet as a source of complex carbohydrates and dietary fiber. See the Table 46.13 for more details.

Table 46.13 Morphographic overview and dry mass based nutritive profile of *D. villosa*

Dioscorea villosa

Calories	118 Kcal
Carbohydrates	27.9 grams
Protein	1.5 grams
Fat	0.2 grams
Dietary fiber	4.3 grams
Vitamin C	17.3 mg
Potassium	670 mg
Calcium	17 mg
Iron	0.4 mg

46.15.2 Therapeutic Potential

Diosgenin from *D. villosa* has been used in the synthesis of cortisone and hormonal medications such as sex hormone, progesterone, and other steroids [10, 29]. The use of diosgenin is associated with its pharmacology and medicinal activities. Properties, which include reduction of oxidative stress, induction of apoptosis, suppression of malignant transformation, prevention of inflammatory events, promotion of cellular differentiation and proliferation, and control of the T-cell immune response that leads towards anti-diabetic, anti-cancer, neuro- and cardiovascular protection, immunomodulation, estrogenic activity, and skin protection [18, 30–33]. A study also investigated the skin aging effect of diosgenin extracted from *D. composita* or *D. villosa* and showed that the diosgenin has the potential to enhance DNA synthesis in the skin by restorating keratinocyte proliferation in aged skin [29]. The rhizome of *D. villosa* is also known to have effects against menstrual complaints and perimenopausal symptoms [10, 35].

46.16 *Dioscorea wallichii*

46.16.1 Nutritive Value

D. wallichii plant is a known to be a source of carbohydrates and dietary fiber. See the Table 46.14 for more details.

Table 46.14 Morphographic overview and dry mass based nutritive profile of *D. wallichii*

Dioscorea wallichii

Calories	118 Kcal
Carbohydrates	27.9 grams
Protein	1.5 grams
Fat	0.2 grams
Dietary fiber	4.1 grams
Vitamin C	17.1 mg
Potassium	816 mg
Calcium	17 mg
Iron	0.5 mg
Thiamine (vitamin B1)	0.11 mg
Riboflavin (vitamin B2)	0.04 mg
Niacin (vitamin B3)	0.6 mg
Vitamin B6	0.38 mg
Folate	23 µg
Magnesium	21 mg
Phosphorus	50 mg
Potassium	816 mg
Zinc	0.4 mg

46.16.2 Therapeutic Potential

A study documented that the radical-scavenging activity was demonstrated by the methanol extract of *D. wallichii*'s leaf. The results showed that *D. wallichii*'s methanolic leaf extracts had substantial free radical scavenging or antioxidant activity. It also showed antibacterial activity against some bacterial pathogens. It also exhibited anti-fungal activity [31]. *D. wallichii* is also helpful in conditions like flatulence and stomach pain [18, 31].

46.17 *Dioscorea zingiberensis*

46.17.1 Nutritive Value

D. zingiberensi plant is known to be a source of carbohydrates and dietary fiber. See the Table 46.15 for more details.

Table 46.15 Morphographic overview and dry mass based nutritive profile of *D. zingiberensis*

Dioscorea zingiberensis

Calories	118 Kcal
Carbohydrates	27.9 grams
Protein	1.5 grams
Fat	0.2 grams
Dietary Fiber	4.1 grams
Vitamin C	17.1 mg
Potassium	816 mg
Calcium	17 mg
Iron	0.5 mg
Thiamine (vitamin B1)	0.11 mg
Riboflavin (vitamin B2)	0.04 mg
Niacin (vitamin B3)	0.6 mg
Vitamin B6	0.38 mg
Folate	23 µg
Magnesium	21 mg
Phosphorus	50 mg
Potassium	816 mg
Zinc	0.4 mg

46.17.2 Therapeutic Potential

The rhizome of *D. zingiberensis* is useful for treatments against cough, anthrax, rheumatic heart disease, rheum, arthritis, and tumefaction. The major bioactive components were steroids of this species, which have been utilized for many years in China to treat critical cardiac problems [34]. Moreover, dioscin is a steroid belonging to the family of saponins [38]. It is also derived from *D. zingiberensis*. Dioscin showed its potential to fight against the cancer cell line K562 [39].

46.18 *Dioscorea persimilis*

46.18.1 Nutritive Value

D. praehensilis is a root vegetable that is low in fat and calories, but high in carbohydrates and dietary fiber. See the Table 46.16 for more details.

Table 46.16 Morphographic overview and dry mass based nutritive profile of *D. persimilis*

Dioscorea persimilis

Carbohydrates	70% to 80% of dry weight
Protein	5% to 10%
Fat	Less than 1%.

46.18.2 Therapeutic Potential

Dioscorea persimilis, also recognized as the wild yam, is a South American yam species. It, like other *Dioscorea* family members, have been utilized in conventional healers for a variety of purposes. Menstrual spasms and other menstrual-related issues are one of the main medical applications of *D. persimilis*. The plant contains compounds known as diosgenin and saponins, which are thought to relax the muscles of the uterus, resulting in menstrual cramp relief. Furthermore, *D. persimilis* has been studied for its anti-inflammation and antioxidant defense properties. These properties may aid in the reduction of inflammation and oxidative activities in a stressed body, potentially leading to health benefits like better overall health and weight loss [18].

46.19 *Dioscorea lijangensis*

46.19.1 Nutritive Value

D. lijangensis plant is a good source of carbohydrates, mineral, vitamins and dietary fiber. See the Table 46.17 for more details.

46.19.2 Therapeutic Potential

Dioscorea lijiangensis is a species of yam that is native to China. It has its place in traditional Chinese medicine for various medicinal purposes. Some of the medicinal benefits of *D. lijiangensis* are as follows: Recent studies found that the roots of *D. lijiangensis* contain a variety of compounds that have been shown to boost

Table 46.17 Morphographic overview and dry mass based nutritive profile of *D. lijiangensis*

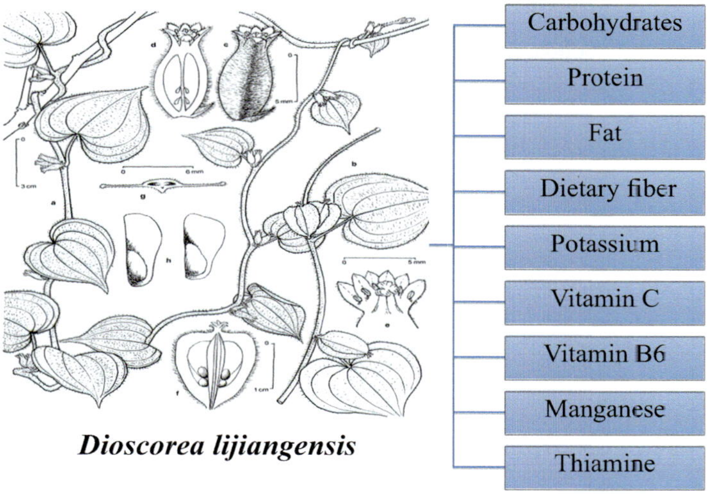

immunity and combat infectious ailments. The root extract of *D. lijiangensis* has been reported as an anti-inflammatory source that can help in the reduction of inflammation in the body, providing relief from pain and swelling. Research has also shown that the root extract of *D. lijiangensis* contains compounds that have anti-cancer properties, which may help in preventing the growth and spread of cancer cells [18]. The root extract of *D. lijiangensis* has been found to possess properties that can help in regulating blood sugar levels, which may be beneficial for people with diabetes. It contains antioxidants that can help protect the body from damage caused by free radicals, which are known to contribute to the development of chronic diseases.

46.20 *Dioscorea monadelpha*

46.20.1 Nutritive Value

D. monadelpha plant contains carbohydrates, protein, various mineral, vitamins and fat contents. See the Table 46.18 for more details.

46.20.2 Therapeutic Potential

The Chinese yam is a plant that has been traditionally used in Chinese medicine for its medicinal properties. Here are some of the medicinal uses and benefits of *Discorea monadelpha*: it contains compounds that play a role in anti-inflammation

Table 46.18 Morphographic overview and dry mass based nutritive profile of *D. monadelpha*

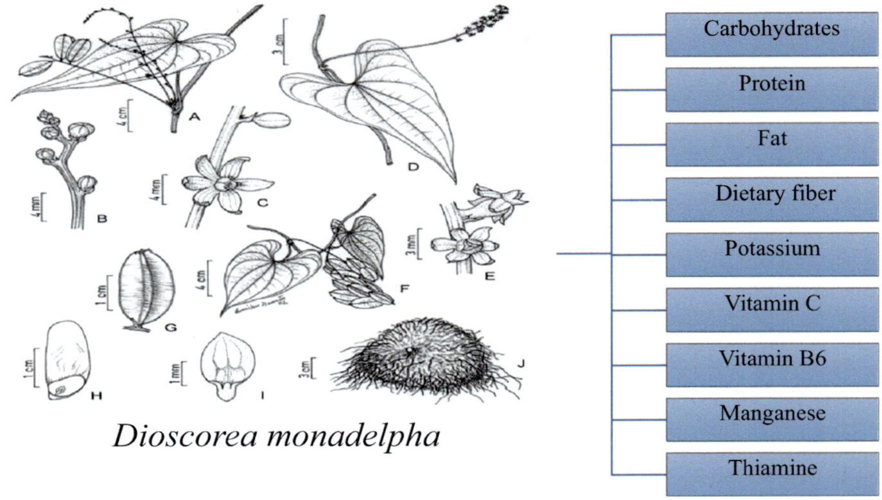

and pain relief in conditions such as arthritis. The plant contains antioxidants that can help protect the body against oxidative stress and prevent cell damage. The root of *D. monadelpha* is rich in fiber, which can help improve digestive health and prevent constipation. The plant contains compounds that can help boost the immune system and improve overall health. *D. monadelpha* has been shown to have anti-diabetic properties, regulating blood sugar levels. The plant has been used traditionally to treat respiratory conditions such as coughs, asthma, and bronchitis. It has been used to treat skin conditions such as eczema, acne, and dry skin.

46.21 *Dioscorea alata*

46.21.1 Nutritive Value

D. alata also known as purple yam or ube, is a highly nutritious root vegetable. It is a rich source of carbohydrates, dietary fiber, vitamins, and minerals. Purple yam is also known to contain antioxidants, which help to protect the body from damage caused by free radicals. Additionally, it has a low glycemic index, making it a great option for people with diabetes or those trying to manage their blood sugar levels [40]. The high nutrient content of purple yam makes it a valuable addition to a healthy and balanced diet. See the Table 46.19 for more details.

Table 46.19 Morphographic overview and dry mass based nutritive profile of *D. alata*

Dioscorea alata

Calories	114 Kcal
Carbohydrates	27.88 grams
Protein	1.53 grams
Fat	0.17 grams
Dietary fiber	4.1 grams
Vitamin C	17.1 mg
Potassium	816 mg
Calcium	17 mg
Iron	0.54 mg

46.21.2 Therapeutic Potential

Dioscorea alata, also known as purple yam, has several medicinal properties. It contains compounds that have potent anti-inflammatory properties, which can help reduce inflammation and pain in the body. *D. alata* is rich in antioxidants, which can help prevent damage to cells caused by free radicals [40]. This can help reduce the risk of chronic diseases such as cancer, heart disease, and Alzheimer's disease. Some studies have suggested that *D. alata* may have anti-diabetic properties, which can help in reducing blood sugar levels and improving insulin sensitivity. It contains compounds that have been shown to have anti-cancer properties, which can help in preventing the growth and spread of cancer cells. *D. alata* is rich in dietary fiber, which can help promote the digestive system and reduce the risk of digestive disorders such as constipation and diarrhea. *D. alata* is rich in compounds that have been shown to have anti-aging properties, which can help in reducing the appearance of wrinkles and fine lines on the skin [40, 41].

46.22 *Dioscorea sansibarensis*

46.22.1 Nutritive Value

D. sansibarensis also known as the Zanzibar yam, is a type of yam found in East Africa. It is a good source of carbohydrates, dietary fiber, and essential minerals such as potassium and magnesium. It also contains small amounts of vitamins C and B6 [42]. The tubers of this plant are traditionally used in East African cuisine. See the Table 46.20 for more details.

Table 46.20 Morphographic overview and dry mass based nutritive profile of *D. sansibarensis*

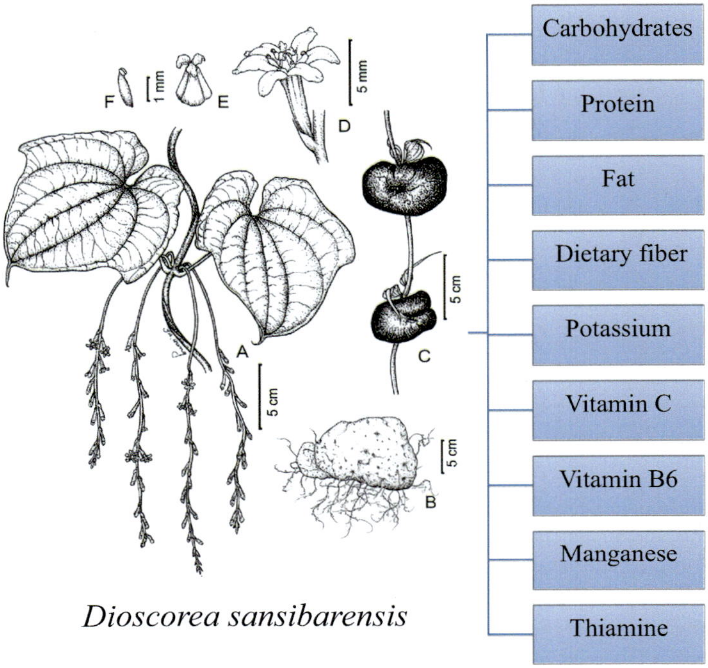

46.22.2 Therapeutic Potential

Plant species that belongs to the family *Dioscoreaceae,* are commonly found in tropical regions of Africa, particularly in Tanzania and Kenya. While there is limited scientific research on the medicinal properties of this plant, it has been traditionally used for its therapeutic benefits. Some of the potential medicinal properties of *D. sansibarensis* include: Some studies have suggested that extracts of *D. sansibarensis* have antimicrobial activity against certain strains of bacteria and fungi. The plant also contains compounds that may help to reduce inflammation, which could make it useful for treating conditions such as arthritis and other inflammatory diseases. The root of *D. sansibarensis* has traditional use as a cure for digestive problems such as diarrhea and stomach pain. Research has suggested that the plant contains compounds that have antioxidant properties, which could help to protect against cellular damage and potentially reduce the risk of certain diseases. After some preliminary research, it was found that it may have anti-cancer properties, From the water and methanolic extracts of *D. sansibarensis* (Pax yam) gathered in Tanzania, many phenanthrenes (1–5), phenolics (6–8), and steroidal sapogenins (9–11) were initially identified. Using 1D and 2D nuclear magnetic resonance spectral techniques, the chemical structures of all the isolates were identified. Utilizing in vitro inhibitory assays for the cyclooxygenase enzymes (COX-1 and -2), the anti-inflammatory activity of all pure isolates was assessed. Although more research is needed to confirm these findings [37, 39, 42]. Overall, while *D. sansibarensis* has some potential medicinal benefits, more research is needed to fully understand its therapeutic properties and determine its efficacy and safety for human use.

46.23 *Dioscorea bulbifera*

46.23.1 Nutritive Value

D. bulbifera commonly known as air potato, bitter yam or cheeky yam, is a type of yam with edible bulbils or aerial tubers [43]. It is a good source of carbohydrates, dietary fiber, and vitamins, particularly vitamin C and vitamin B6. It also contains minerals such as potassium, calcium, iron, and manganese. See the Table 46.21 for more details.

46.23.2 Therapeutic Potential

Dioscorea bulbifera, also known as the "air potato," has a long history of use in traditional medicine. Various parts of the plant, including the tubers and leaves, have been used for their medicinal properties. In traditional Chinese medicine, the tubers

Table 46.21 Morphographic overview and dry mass based nutritive profile of *D. bulbifera*

Dioscorea bulbifera

Calories	11 Kcal
Carbohydrates	28.12 grams
Protein	1.68 grams
Fat	0.18 grams
Dietary fiber	3.9 grams
Vitamin C	17.3 mg
Potassium	670 mg
Calcium	17 mg
Iron	0.39 mg

of *D. bulbifera* have been used to treat a variety of diseases, like abnormalities in digestion and arthritis. The tubers are also believed to have anti-inflammatory and analgesic properties. In Ayurvedic medicine, *D. bulbifera* is used to treat asthma, bronchitis, and other respiratory problems. The plant is also believed to have anti-inflammatory and anti-rheumatic properties. Research has suggested that *D. bulbifera* may have potential as a treatment for diabetes, as it has been found to have hypoglycemic effects in animal studies [37]. Additionally, the plant has been found to have antioxidant and antimicrobial properties. Bioactive components from this species have exhibited antioxidant, anti-inflammatory, antibacterial, plasmid-curing, anti-diabetic, and anti-cancer activities [44].

46.24 *Dioscorea colletti*

46.24.1 Nutritive Value

D. colletti plant contains carbohydrates, protein, some vitamins, dietary fibers and fat contents [36]. See the Table 46.22 for more details.

Table 46.22 Morphographic overview and dry mass based nutritive profile of *D. colletti*

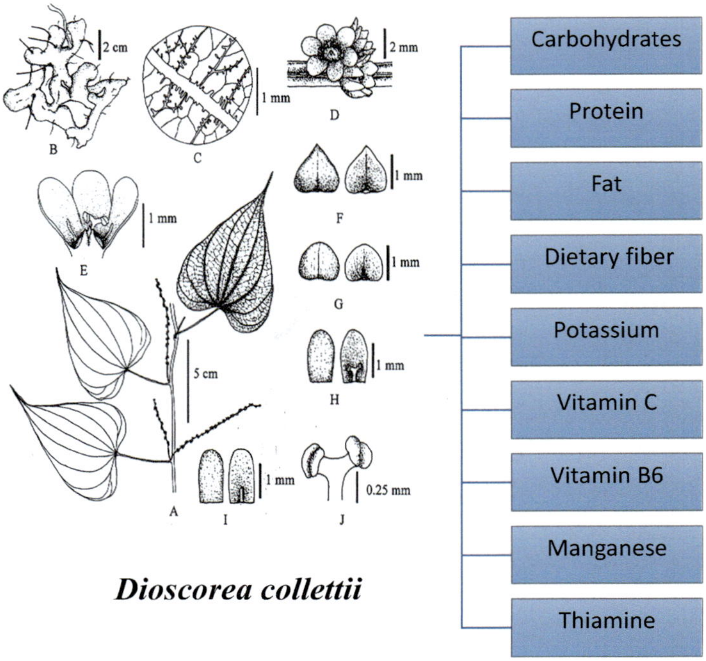

46.24.2 *Therapeutic Potential*

Dioscorea colletti is a species of yam found in southern Africa. As for the medicinal value of *D. colletti*, it has traditionally been used by various African tribes to treat a range of ailments, including diarrhea, stomachaches, and respiratory infections. It is also believed to have anti-inflammatory properties and has been used to alleviate joint pain and inflammation. Traditional medicine makes considerable use of the rhizomes of *D. colletti* to treat arthritic conditions, particularly gouty arthritis (GA). It is described by the meridian points of Gan (liver), Wei (stomach), and Pang Guang (bladder) as being neutral in nature and bitter in flavor. In clinical settings, it was frequently used to treat chyloid stranguria, gonorrhea, leucorrhagia, rheumatism, arthralgia discomfort, disabled joints, and lumbar knee pain syndrome [37, 45].

46.25 Conclusions

Because of their great nutritional content, *Dioscorea* species, commonly referred as as yams, are extensively consumed as a valuable food source all over the world. The diverse array of macronutrients, micronutrients, and phytochemicals in various yam species contribute to their potential health benefits. Many of these species have traditional medicinal applications, and recent scientific investigations have shown their promising therapeutic potential in treating diseases such as diabetes, cancer, and inflammation. More study, however, is required to fully comprehend the processes underlying these benefits, as well as to determine optimum dosages and safety profiles. The final conclusion is that including a range of yams, especially those with established therapeutic potential, into a varied and balanced diet can supply key nutrients while also potentially providing additional health advantages. Before making any significant modifications to one's diet or treatment plan, as with any dietary supplement or complementary therapy, it is critical to check with a healthcare expert. The *Dioscorea* genus represents a captivating convergence of agriculture, nutrition, and medicine. Its economic importance as a food crop and potential source of pharmaceutical products cannot be overlooked. The cultivation, trade, and utilization of yams have significant implications for global food security, human health, and economic growth. Continued research, innovation, and collaboration between the agricultural, pharmaceutical, and healthcare sectors are vital to fully harness the potential of *Dioscorea* plants and ensure their sustainable and beneficial integration into human lives.

References

1. Padhan, B., & Panda, D. (2020). Potential of neglected and underutilized yams (Dioscorea spp.) for improving nutritional security and health benefits. *Frontiers in Pharmacology, 11*, 496.
2. Asiedu, R., & Sartie, A. (2010). Crops that feed the world 1. Yams: Yams for income and food security. *Food Security, 2*, 305–315.

3. Li, H., Zhang, X., & Wang, J. (1999). Progress of basic research on Dioscorea spp. in China. *Economic for Research, 17*, 44–48.
4. Shan, N., Wang, P., Zhu, Q., Sun, J., Zhang, H., Liu, X., Cao, T., Chen, X., Huang, Y., & Zhou, Q. (2020). Comprehensive characterization of yam tuber nutrition and medicinal quality of Dioscorea opposita and D. alata from Different Geographic Groups in China. *Journal of Integrative Agriculture, 19*, 2839–2848.
5. Kim, H., Cao, T. Q., Yeo, C. E., Shin, S. H., Kim, H., Hong, D. H., & Hahn, D. (2022). Development and validation of quantitative analysis method for phenanthrenes in peels of the Dioscorea genus. *Journal of Microbiology and Biotechnology, 32*, 976–981.
6. Adomėnienė, A., & Venskutonis, P. R. (2022). Dioscorea spp.: Comprehensive review of antioxidant properties and their relation to phytochemicals and health benefits. *Molecules, 27*(8), 2530.
7. Shim, W.-S., & Oh, U. (2008). Histamine-induced itch and its relationship with pain. *Molecular Pain, 4*, 1744-8069-4-29.
8. Hur, G.-Y., et al. (2008). Identification of Dioscorea batatas (sanyak) allergen as an inhalant and oral allergen. *Journal of Korean Medical Science, 23*(1), 72–76.
9. Obidiegwu, J. E., Lyons, J. B., & Chilaka, C. A. (2020). The Dioscorea genus (Yam)—An appraisal of nutritional and therapeutic potentials. *Food, 9*(9), 1304.
10. Liu, Y.-H., et al. (2006). Comparisons of in vitro antioxidant activities of storage proteins in tuber of two Dioscorea species. *Botanical Studies, 47*, 231–237.
11. Chan, Y. S., & Ng, T. B. (2013). A lectin with highly potent inhibitory activity toward breast cancer cells from edible tubers of Dioscorea opposita cv. nagaimo. *PLoS One, 8*(1), e54212.
12. He, Z., et al. (2012). Anti-tumour and immunomodulating activities of diosgenin, a naturally occurring steroidal saponin. *Natural Product Research, 26*(23), 2243–2246.
13. Zhao, Z., Wang, L., Ruan, Y., Wen, C., Ge, M., Qian, Y., & Ma, B. (2023). Physicochemical properties and biological activities of polysaccharides from the peel of Dioscorea opposita Thunb. extracted by four different methods. *Food Science and Human Wellness, 12*(1), 130–139.
14. Mierza, V., Haro, G., & Suryanto, D. (2019). Influence of variation extraction methods (classical procedure) for antibacterial activity of Rarugadong (Dioscorea pyrifolia Kunth.) tuber. *Journal of Innovations in Applied Pharmaceutical Science (JIAPS)*, 01–06.
15. Jaleel, C. A., et al. (2007). Responses of antioxidant potentials in Dioscorea rotundata Poir. following paclobutrazol drenching. *Comptes Rendus Biologies, 330*(11), 798–805.
16. Mohan, V. (2013). Antioxidant activity of Dioscorea spicata roth of using various in vitro assay models. *Pharma Science Monitor, 4*(2).
17. Dzomba, P., & Musekiwa, C. (2014). Antiobesity and antioxidant activity of dietary flavonoids from dioscorea steriscus tubers. *Journal of Coastal Life Medicine, 2*(6), 465–470.
18. Rajalakshmi, K., & Mohan, V. (2013). In vitro antioxidant activity of Dioscorea tomentosa Koen ex. Spreng. *Pharma Science Monitor, 4*(2).
19. Muthukumarasamy, S., et al., Traditional medicinal practices of Palliyar tribe of Srivilliputhur in antenatal and post-natal care of mother and child. 2004.
20. Mollica, J. Q., et al. (2013). Anti-inflammatory activity of American yam Dioscorea trifida Lf in food allergy induced by ovalbumin in mice. *Journal of Functional Foods, 5*(4), 1975–1984.
21. Bhandari, M. R., & Kawabata, J. (2004). Assessment of antinutritional factors and bioavailability of calcium and zinc in wild yam (Dioscorea spp.) tubers of Nepal. *Food Chemistry, 85*(2), 281–287.
22. Svensson, B., et al. (2004). Proteinaceous α-amylase inhibitors. *Biochimica et Biophysica Acta (BBA)-Proteins and Proteomics, 1696*(2), 145–156.
23. Mahmood, N. (2016). A review of α-amylase inhibitors on weight loss and glycemic control in pathological state such as obesity and diabetes. *Comparative Clinical Pathology, 25*(6), 1253–1264.
24. Wang, Z., Zhao, S., Tao, S., Hou, G., Zhao, F., Tan, S., & Meng, Q. (2023). Dioscorea spp.: Bioactive compounds and potential for the treatment of inflammatory and metabolic diseases. *Molecules, 28*(6), 2878.

25. Jesus, M., et al. (2016). Diosgenin: Recent highlights on pharmacology and analytical methodology. *Journal of Analytical Methods in Chemistry, 2016*, 1–16.
26. Yan, C., et al. (2015). Advances in the pharmacological activities and mechanisms of diosgenin. *Chinese Journal of Natural Medicines, 13*(8), 578–587.
27. Raju, J., & Rao, C. V. (2012). Diosgenin, a steroid saponin constituent of yams and fenugreek: Emerging evidence for applications in medicine. *Bioactive compounds in phytomedicine, 125*, 143.
28. Escobar-Sánchez, M. L., Sánchez-Sánchez, L., & Sandoval-Ramírez, J. (2015). Steroidal saponins and cell death in cancer. In *Cell death-autophagy, apoptosis and necrosis*.
29. Tada, Y., et al. (2009). Novel effects of diosgenin on skin aging. *Steroids, 74*(6), 504–511.
30. Soffa, V. M. (1996). Alternatives to hormone replacement for menopause. *Alternative Therapies in Health and Medicine, 2*(2), 34–39.
31. Irulandi, K., Geetha, S., & Mehalingam, P. (2016). Antioxidant, antimicrobial activities and phytochemical analysis of leaves extracts of Dioscorea wallichii Hook. f. *Journal of Applied Pharmaceutical Science, 6*(11), 070–074.
32. Rout, S., & Panda, S. (2010). Ethnomedicinal plant resources of Mayurbhanj district, Orissa. *Indian Journal of Traditional Knowledge, 9*(1), 68–72.
33. Dutta, B. (2015). Food and medicinal values of certain species of Dioscorea with special reference to Assam. *Journal of Pharmacognosy and Phytochemistry, 3*(5), 15–18.
34. Zhu, J., et al. (2010). Characterization of steroidal saponins in crude extracts from Dioscorea zingiberensis CH Wright by ultra-performance liquid chromatography/electrospray ionization quadrupole time-of-flight tandem mass spectrometry. *Journal of Pharmaceutical and Biomedical Analysis, 53*(3), 462–474.
35. Siddiqui, M. A., Ali, Z., Chittiboyina, A. G., & Khan, I. A. (2018). Hepatoprotective effect of steroidal glycosides from Dioscorea villosa on hydrogen peroxide-induced hepatotoxicity in HepG2 cells. *Frontiers in Pharmacology, 9*, 797.
36. Hu, K., et al. (1996). Antineoplastic agents; I. Three spirostanol glycosides from rhizomes of Dioscorea collettii var. hypoglauca. *Planta Medica, 62*(06), 573–575.
37. Cheng, Y., et al. (2015). Enhanced production of diosgenin from Dioscorea zingiberensis in mixed culture solid state fermentation with Trichoderma reesei and Aspergillus fumigatus. *Biotechnology & Biotechnological Equipment, 29*(4), 773–778.
38. Yoon, K. D., & Kim, J. (2008). Preparative separation of dioscin derivatives from Dioscorea villosa by centrifugal partition chromatography coupled with evaporative light scattering detection. *Journal of Separation Science, 31*(13), 2486–2491.
39. Tao, X., Yin, L., Xu, L., & Peng, J. (2008). Dioscin: A diverse acting natural compound with therapeutic potential in metabolic diseases, cancer, inflammation and infections. *Pharmacological Research, 137*, 259–269.
40. Nan, S. H. A. N., Wang, P. T., Zhu, Q. L., Sun, J. Y., Zhang, H. Y., Liu, X. Y., et al. (2020). Comprehensive characterization of yam tuber nutrition and medicinal quality of Dioscorea opposita and D. alata from different geographic groups in China. *Journal of Integrative Agriculture, 19*(11), 2839–2848.
41. Lebot, V., Lawac, F., & Legendre, L. (2022). The greater yam (Dioscorea alata L.): A review of its phytochemical content and potential for processed products and biofortification. *Journal of Food Composition and Analysis, 115*, 104987.
42. Walsh, M. (2009). The use of wild and cultivated plants as famine foods on Pemba Island Zanzibar. *Études Océan Indien, 42-43*, 217–241.
43. Kundu, B. B., Vanni, K., Farheen, A., Jha, P., Pandey, D. K., & Kumar, V. (2021). Dioscorea bulbifera L.(Dioscoreaceae): A review of its ethnobotany, pharmacology and conservation needs. *South African Journal of Botany, 140*, 365–374.
44. Chaniad, P., Tewtrakul, S., Sudsai, T., Langyanai, S., & Kaewdana, K. (2020). Anti-inflammatory, wound healing and antioxidant potential of compounds from Dioscorea bulbifera L. bulbils. *PLoS One, 15*(12), e0243632.
45. Jing, S. S., Wang, Y., Li, X. J., Li, X., Zhao, W. S., Zhou, B., Zhao, C.-C., Huang, L.-Q., & Gao, W. Y. (2017). Phytochemical and chemotaxonomic studies on Dioscorea collettii. *Biochemical Systematics and Ecology, 71*, 10–15.

Index

M. Zia-Ul-Haq et al. (eds.), *Essentials of Medicinal and Aromatic Crops*, https://doi.org/10.1007/978-3-031-35403-8

Printed by Printforce, the Netherlands